ENCYCLOPAEDIA OF MATHEMATICS

Volume 6

ENCYCLOPAEDIA OF MATHEMATICS

Managing Editor

M. Hazewinkel

Scientific Board

ENCYCLOPAEDIA OF MATHEMATICS

Volume 6

Lobachevskiĭ Criterion (for Convergence) – Optional Sigma-Algebra

An updated and annotated translation of the Soviet
'Mathematical Encyclopaedia'

KLUWER ACADEMIC PUBLISHERS

Dordrecht / Boston / London

Library of Congress Cataloging-in-Publication Data CIP

Matematicheskaia entsiklopediia. English.
Encyclopaedia of mathematics.

1. Mathematics––Dictionaries. I. Hazewinkel, Michiel. II. Title.
QA5.M3713 1987 510′.3′21   87–26437
ISBN 1–55608–010–7 (set)
ISBN 1–55608–005–0 (v. 6)

Published by Kluwer Academic Publishers,
P.O. Box 17, 3300 AA Dordrecht, The Netherlands.

Kluwer Academic Publishers incorporates
the publishing programmes of
D. Reidel, Martinus Nijhoff, Dr W. Junk and MTP Press.

Sold and distributed in the U.S.A. and Canada
by Kluwer Academic Publishers,
101 Philip Drive, Norwell, MA 02061, U.S.A.

In all other countries, sold and distributed
by Kluwer Academic Publishers Group,
P.O. Box 322, 3300 AH Dordrecht, The Netherlands.

Printed in The Netherlands

SOVIET MATHEMATICAL ENCYCLOPAEDIA

Editor-in-Chief

I. M. Vinogradov

Editorial Board

S. I. Adyan, P. S. Aleksandrov, N. S. Bakhvalov, A. V. Bitsadze, V. I. Bityutskov (Deputy Editor-in-Chief), L. N. Bol'shev, A. A. Gonchar, N. V. Efimov, V. A. Il'in, A. A. Karatsuba, L. D. Kudryavtsev, B. M. Levitan, K. K. Mardzhanishvili, E. F. Mishchenko, S. P. Novikov, E. G. Poznyak, Yu. V. Prokhorov (Deputy Editor-in-Chief), A. I. Shirshov, A. G. Sveshnikov, A. N. Tikhonov, P. L. Ul'yanov, S. V. Yablonskiĭ

Translation Arrangements Committee

V. I. Bityutskov, R. V. Gamkrelidze, Yu. V. Prokhorov

'Soviet Encyclopaedia' Publishing House

PREFACE

This ENCYCLOPAEDIA OF MATHEMATICS aims to be a reference work for all parts of mathematics. It is a translation with updates and editorial comments of the Soviet *Mathematical Encyclopaedia* published by 'Soviet Encyclopaedia Publishing House' in five volumes in 1977–1985. The annotated translation consists of ten volumes including a special index volume.

There are three kinds of articles in this ENCYCLOPAEDIA. First of all there are survey-type articles dealing with the various main directions in mathematics (where a rather fine subdivision has been used). The main requirement for these articles has been that they should give a reasonably complete up-to-date account of the current state of affairs in these areas and that they should be maximally accessible. On the whole, these articles should be understandable to mathematics students in their first specialization years, to graduates from other mathematical areas and, depending on the specific subject, to specialists in other domains of science, engineers and teachers of mathematics. These articles treat their material at a fairly general level and aim to give an idea of the kind of problems, techniques and concepts involved in the area in question. They also contain background and motivation rather than precise statements of precise theorems with detailed definitions and technical details on how to carry out proofs and constructions.

The second kind of article, of medium length, contains more detailed concrete problems, results and techniques. These are aimed at a smaller group of readers and require more background expertise. Often these articles contain more precise and refined accounts of topics and results touched upon in a general way in the first kind of article.

Finally, there is a third kind of article: short (reference) definitions.

Practically all articles (all except a few of the third kind) contain a list of references by means of which more details and more material on the topic can be found. Most articles were specially written for the encyclopaedia and in such cases the names of the original Soviet authors are mentioned. Some articles have another origin such as the *Great Soviet Encyclopaedia* (*Bol'shaya Sovetskaya Entsiklopediya* or BSE).

Communication between mathematicians in various parts of the world has certainly greatly improved in the last decennia. However, this does not mean that there are so-to-speak 'one-to-one onto' translations of the terminology, concepts and tools used by one mathematical school to those of another. There also are varying traditions of which questions are important and which not, and what is considered a central problem in one tradition may well be besides the point from the point of view of another. Even for well-established areas of mathematical inquiry, terminology varies across languages and even within a given language domain. Further, a concept, theorem, algorithm, . . . , which is associated with one proper name within one tradition may well have another one in another, especially if the result or idea in question was indeed discovered independently and more-or-less simultaneously. Finally, mathematics is a very dynamic science and much has happened since the original articles were finalized (mostly around 1977). This made updates desirable (when needed). All this, as well as providing additional references to Western literature when needed, meant an enormous amount of work for the board of experts as a whole; some indeed have done a truly impressive amount of work. I must stress though that I am totally responsible for what is finally included and what is not of all the material provided by the members of the board of experts.

Many articles are thus provided with an editorial comment section in a different and somewhat smaller typeface. In particular, these annotations contain additional material, amplifications, alternative names, additional references, ... Modifications, updates and other extra material provided by the original Soviet authors (not a rare occurrence) have been incorporated in the articles themselves.

The final (10-th) volume of the ENCYCLOPAEDIA OF MATHEMATICS will be an index volume. This index will contain all the titles of the articles (some 6600) and in addition the names of all the definitions, named theorems, algorithms, lemmas, scholia, constructions, . . . , which occur in the various articles. This includes, but is by no means limited to, all items which are printed in bold or italic. Bold words or phrases, by the way, always refer to another article with (precisely) that title.

All articles have been provided with one or more AMS classification numbers according to the 1980 classification scheme (not, for various reasons, the 1985 revision), as have all items occurring in the index. A phrase or word from an article which is included in the index always inherits all the classification numbers of the article in question. In addition, it may have been provided with its own classification numbers. In the index volume these numbers will be listed with the phrase in question. Thus e.g. the Quillen – Suslin theorem of algebraic K-theory will have its own main classification numbers (these are printed in bold; in this case that number is 18F25) as well as a number of others, often from totally different fields, pointing e.g. to parts of mathematics where the theorem is applied, or where there occurs a problem related to it (in this case e.g. 93D15). The index volume will also contain the inversion of this list which will, for each number, provide a list of words and phrases which may serve as an initial description of the 'content' of that classification number (as far as this ENCYCLOPAEDIA is concerned). For more details on the index volume, its structure and organisation, and what kind of things can be done with it, cf. the (future) special preface to that volume.

Classifying articles is a subjective matter. Opinions vary greatly as to what belongs where and thus this attempt will certainly reflect the tastes and opinions of those who did the classification work. One feature of the present classification attempt is that the general basic concepts and definitions of an area like e.g. 55N (Homology and Cohomology theories) or 60J (Markov processes) have been assigned classification numbers like 55NXX and 60JXX if there was no finer classification number different from ...99 to which it clearly completely belongs.

Different parts of mathematics tend to have differences in notation. As a rule, in this ENCYCLOPAEDIA in a given article a notation is used which is traditional in the corresponding field. Thus for example the (repeated index) summation convention is used in articles about topics in fields where that is traditional (such as in certain parts of differential geometry (tensor geometry)) and it is not used in other articles (e.g. on summation of series). This pertains especially to the more technical articles.

For proper names in Cyrillic the British Standards Institute transcription system has been used (cf. Mathematical Reviews). This makes well known names like S. N. Bernstein come out as Bernshteĭn.

In such cases, especially in names of theorems and article titles, the traditional spelling has been retained and the standard transcription version is given between brackets.

Ideally an encyclopaedia should be complete up to a certain more-or-less well defined level of detail. In the present case I would like to aim at the completeness level whereby every theorem, concept, definition, lemma, construction which has a more-or-less constant and accepted name by which it is referred to by a recognizable group of mathematicians occurs somewhere, and can be found via the index. It is unlikely that this completeness ideal will be reached with this present ENCYCLOPAEDIA OF MATHEMATICS, but it certainly takes substantial steps in this direction. Everyone who uses this ENCYCLOPAEDIA and finds items which are not covered, which, he feels, should have been included, is invited to inform me about it. When enough material has come in this way supplementary volumes will be put together.

The ENCYCLOPAEDIA is alphabetical. Many titles consist of several words. Thus the problem arises how to order them. There are several systematic ways of doing this of course, for instance using the first noun. All are unsatisfactory in one way or another. Here an attempt has been made to order things according to words or natural groups of words as they are daily used in practice. Some sample titles may serve to illustrate this: **Statistical mechanics, mathematical problems in; Lie algebra; Free algebra; Associative algebra; Absolute continuity; Abstract algebraic geometry; Boolean functions, normal forms of.** Here again taste plays a role (and usages vary). The index will contain all permutations. Meanwhile it will be advisable for the reader to try out an occasional transposition himself. Titles like K-theory are to be found under K, more precisely its lexicographic place is identical with 'K theory', i.e. '-' = 'space' and comes before all other symbols. Greek letters come before the corresponding Latin ones, using the standard transcriptions. Thus χ^2-distribution (chi-squared distribution) is at the beginning of the letter C. A $*$ as in C^*-algebra and $*$-regular ring is ignored lexicographically. Some titles involve Greek letters spelled out in Latin. These are of course ordered just like any other 'ordinary' title.

This volume has been computer typeset using the (Unix-based) system of the CWI, Amsterdam. The technical (mark-up-language) keyboarding was done by Rosemary Daniëls, Chahrzade van 't Hoff and Joke Pesch. To meet the data-base and typesetting requirements of this ENCYCLOPAEDIA substantial amounts of additional programming had to be done. This was done by Johan Wolleswinkel. Checking the translations against the original texts, and a lot of desk editing and daily coordination was in the hands of Rob Hoksbergen. All these persons, the members of the board of experts, and numerous others who provided information, remarks and material for the editorial comments, I thank most cordially for their past and continuing efforts.

The original Soviet version had a printrun of 150,000 and is completely sold out. I hope that this annotated and updated translation will turn out to be comparably useful.

Bussum, August 1987   MICHIEL HAZEWINKEL

LOBACHEVSKIĬ CRITERION (FOR CONVERGENCE) - A series $\sum_{n=1}^{\infty} a_n$ with positive terms a_n tending monotonically to zero converges or diverges according as the series

$$\sum_{m=0}^{\infty} p_m 2^{-m}$$

converges or diverges, where p_m is the largest of the indices of the terms a_n that satisfy the inequality $a_n \geqslant 2^{-m}$, $n = 1, \ldots, p_m$.

It was proposed by N.I. Lobachevskiĭ in 1834 - 1836.

V.I. Bityutskov

AMS 1980 Subject Classification: 40A05

LOBACHEVSKIĬ FUNCTION - 1) The *angle of parallelism* in **Lobachevskiĭ geometry** is a function that expresses the angle α between the line u_1 (or u_2) (see Fig.) and the segment OA perpendicular to a line a parallel to u_1 (or u_2) in terms of the length l of the segment OA:

$$\alpha = \Pi(l) = 2 \arctan e^{-l/R},$$

where R is a positive constant that corresponds to the scale of measurement of distances.

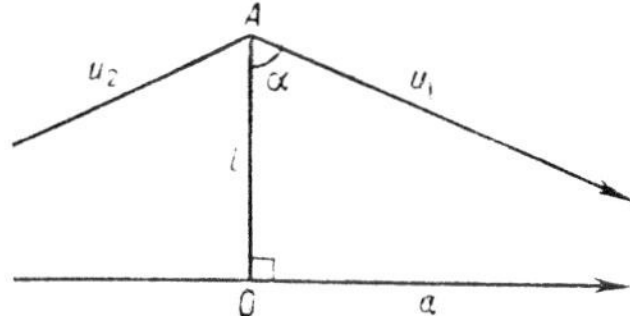

The Lobachevskiĭ function is a continuous monotone decreasing function with values between $\pi/2$ and 0:

$$\lim_{l \to 0} \Pi(l) = \frac{\pi}{2}, \quad \lim_{l \to \infty} \Pi(l) = 0.$$

It was introduced by N.I. Lobachevskiĭ in 1826.

References

[1] KAGAN, V.F.: *The foundations of geometry*, 1, Moscow-Leningrad, 1949 (in Russian).

[2] EFIMOV, N.V.: *Höhere Geometrie*, Deutsch. Verlag Wissenschaft., 1960 (translated from the Russian).

2) The special function (cf. **Special functions**) defined for real x by

$$L(x) = -\int_0^x \ln \cos t \, dt.$$

The Lobachevskiĭ function can be represented as a series

$$L(x) = x \ln 2 - \frac{1}{2} \sum_{k=1}^{\infty} (-1)^{k-1} \frac{\sin 2kx}{k^2}.$$

The main relations are:

$$L(-x) = -L(x), \quad -\frac{\pi}{2} \leqslant x \leqslant \frac{\pi}{2},$$

$$L(\pi - x) = \pi \ln 2 - L(x),$$

$$L(\pi + x) = \pi \ln 2 + L(x).$$

It was introduced by N.I. Lobachevskiĭ in 1829.

References

[1] RYZHIK, I.M. and GRADSHTEĬN, I.S.: *Tables of integrals, series, and products*, Acad. Press, 1980 (translated from the Russian).

A.B. Ivanov

Editorial comments. For the Lobachevskiĭ function in the sense of 1) (i.e. the angle of parallelism) see also [A1] - [A4].

For Lobachevskiĭ's function as defined in 2) see also [A5].

References

[A1] GREENBERG, M.: *Euclidean and non-Euclidean geometries*, Freeman, 1974.

[A2] COXETER, H.S.M.: 'Parallel lines', *Canad. Math. Bull.* 21 (1978), 385-397.

[A3] COXETER, H.S.M.: *Non-Euclidean geometry*, Toronto Univ. Press, 1957.

[A4] BONOLA, R.: *Non-Euclidean geometry*, Dover, reprint, 1955.

[A5] COXETER, H.S.M.: *Twelve geometric esays*, Carbondale, 1968, Chapt. 1.

AMS 1980 Subject Classification: 51M10, 33A70

LOBACHEVSKIĬ GEOMETRY - A geometry based on the same fundamental premises as **Euclidean geometry**, except for the axiom of parallelism (see **Fifth postulate**). In Euclidean geometry, according to this axiom, in a plane through a point P not lying on a straight line $A'A$ there passes precisely one line $B'B$ that does not intersect $A'A$. The line $B'B$ is called a *parallel* to $A'A$. It is sufficient to require that there is at most one straight line, since the existence of a non-intersecting line can be proved by successively drawing lines $PQ \perp A'A$ and $PB \perp PQ$.

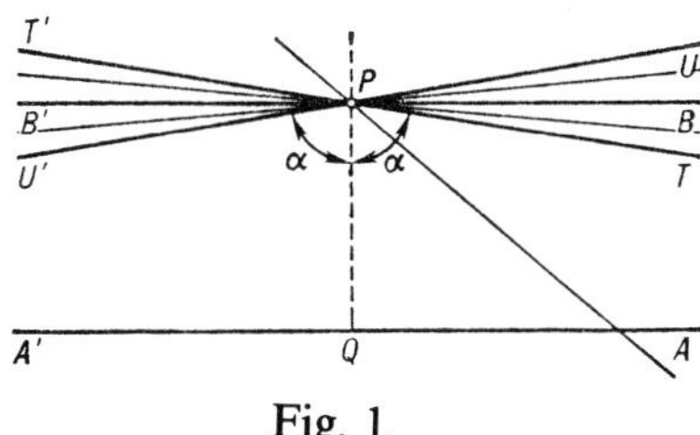

Fig. 1.

In Lobachevskiĭ geometry the axiom of parallelism requires that through a point P (Fig. 1) there passes more than one line not intersecting $A'A$. The non-intersecting lines fill the part of the pencil with vertex P lying inside a pair of vertically opposite angles TPU and $U'PT'$ situated symmetrically with respect to the perpendicular PQ. The lines that form the sides of the vertically opposite angles separate the intersecting lines from the non-intersecting lines, and they themselves are also non-intersecting. These limiting lines are called the *parallels* at P to the line $A'A$ in its two directions: $T'T$

is parallel to $A'A$ in the direction $A'A$, and UU' is parallel to $A'A$ in the direction AA'. The remaining non-intersecting lines are called *ultraparallels* to $A'A$ (for details see below).

The angle α, $0<\alpha<\pi/2$, that a parallel at P makes with the perpendicular PQ, $\angle QPT = \angle QPU' = \alpha$, is called the *angle of parallelism* of the interval $PQ = a$ and is denoted by $\alpha \equiv \Pi(a)$. For $a = 0$, $\alpha = \pi/2$; as a increases the angle α decreases, so that for every given α, $0<\alpha<\pi/2$, there is a definite value of a. This dependence is called the *Lobachevskiĭ function*:

$$\Pi(a) = 2\arctan(e^{-a/k}),$$

where k is a constant that determines the fixed scale of measurement. It is called the *radius of curvature of the Lobachevskiĭ space*.

Euclidean geometry can be obtained as a limiting case of Lobachevskiĭ geometry when the two parallels passing through P merge into one, that is, when the set of all lines passing through P and not intersecting the given line $A'A$ reduce to a unique line. Then the angle $\alpha = \pi/2$ for any a. This condition is equivalent to the requirement that $k = \infty$. In small regions of space, that is, when the linear dimensions of figures are infinitesimal with respect to k, all relations of Lobachevskiĭ geometry are approximated by relations of the Euclidean geometry obtained in the limit.

Two distinct lines of the plane form a pair of one of three types.

Intersecting lines. The distance from points of one line to the other line increases without limit as the distance from the intersection of the lines increases. If the lines are not perpendicular, then each is projected orthogonally on the other into an open interval of finite size.

Parallel lines. Coplanar non-intersecting lines that have no common perpendicular. In Fig. 1, $PT\|QA$ and $PU'\|QA'$, in P, with angle of parallelism α. Parallelism is transitive (if $a\|b$ and $b\|c$ in one direction, then $a\|c$ in the corresponding direction). In the direction of parallelism parallels approach each other without limit (in the sense of the distance from a moving point of one line to the other line). The orthogonal projection of one line on the other is an open half-line.

Ultraparallels. They have one common perpendicular, the length of which gives the shortest distance. On both sides of the perpendicular the lines diverge without limit. Each line projects onto the other in an open interval of finite length.

To the three types of lines there correspond in the plane three types of pencils of lines, each of which covers the whole plane: a *pencil of the first kind* — the set of all lines passing through one point (the *centre of the pencil*); a *pencil of the second kind* — the set of all lines perpendicular to one line (the *base of the pencil*); and a

pencil of the third kind — the set of all lines parallel to one line in a given direction, including this line.

The orthogonal trajectories of the lines of these pencils form analogues of circles of the Euclidean plane: a *circle* in the proper sense; an *equi-distant*, or *curve of equal distances* (if one does not consider the base), which is concave on the side of the base; or a *limiting curve*, or **horocycle**, which can be regarded as a circle with its centre at infinity. Horocycles are congruent. They are not compact and are concave on the side of parallelism. Two horocycles generated by the same pencil are concentric (they cut off equal intervals on the lines of the pencil). The ratio of the lengths of concentric arcs included between two lines of the pencil decreases on the side of parallelism like an exponential function of the distance x between the arcs:

$$\frac{s'}{s} = e^{-x/k}.$$

Each of the analogues of a circle can slide on itself, generating three types of one-parameter motions of the plane: rotation about a proper centre; translation (one trajectory is the base, and the others are equi-distants); and parallel displacement (all the trajectories are horocycles).

Rotation of the analogues of circles about a line of the generating pencil leads to analogues of the sphere: a *proper sphere*, a *surface of equal distances* and a *horosphere*, or *limiting surface*.

On a sphere the geometry of great circles is ordinary spherical geometry; on a surface of equal distances it is the geometry of equi-distants, which is Lobachevskiĭ planimetry, but with a larger value of k; and on a horosphere it is the Euclidean geometry of horocycles.

The connection between lengths of arcs and chords of horocycles and Euclidean trigonometric relations on horospheres make it possible to derive trigonometric relations in the plane, that is, trigonometric formulas for rectilinear triangles. For example, the formula for the area σ of a triangle is:

$$\sigma = k^2(\pi - A - B - C);$$

for the perimeter of a circle the formula is:

$$l = 2\pi k \sinh\frac{r}{k}.$$

The trigonometric formulas of Lobachevskiĭ geometry can be obtained from the formulas of **spherical geometry** by replacing the radius R by the imaginary number ki.

The proof of the consistency of Lobachevskiĭ geometry is carried out by constructing an interpretation (a model). The first such interpretation was the **Beltrami interpretation**, which establishes that in Euclidean space the intrinsic geometry of a surface of constant negative Gaussian curvature coincides locally with Lobachevskiĭ geometry (the role of straight lines is

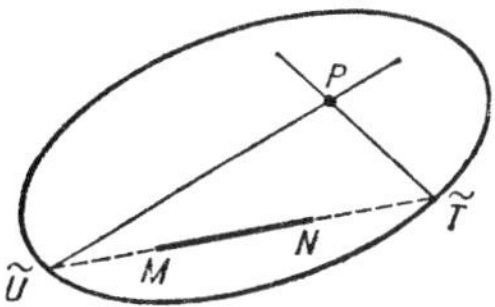

Fig. 2.

played by the geodesic curves of the surface). A surface of this type is called a *pseudo-sphere*.

Another interpretation of Beltrami consists of a geodesic mapping of a surface of constant negative curvature into the interior of a disc.

However, the Beltrami interpretations model only part of the Lobachevskiĭ plane. The first interpretation of the whole Lobachevskiĭ plane was the **Klein interpretation**, in which Cayley's **projective metric** was used. In this interpretation (Fig. 2) the straight lines of Lobachevskiĭ space are realized by chords of the absolute (without the end-points), and perpendicular lines are realized by conjugate chords. Distances and angles are expressed by means of the cross ratios (cf. **Cross ratio**) of quadruples of points (the end-points M, N of an interval and the end-points $\tilde{U}$, $\tilde{T}$ of the chord on which the interval lies) and the corresponding quadruples of lines (the sides of the angle and the (imaginary) tangents to the absolute that pass through the vertex). The parallels through a point P to the line MN are realized by the lines $P\tilde{T}$ and $P\tilde{U}$ that intersect MN

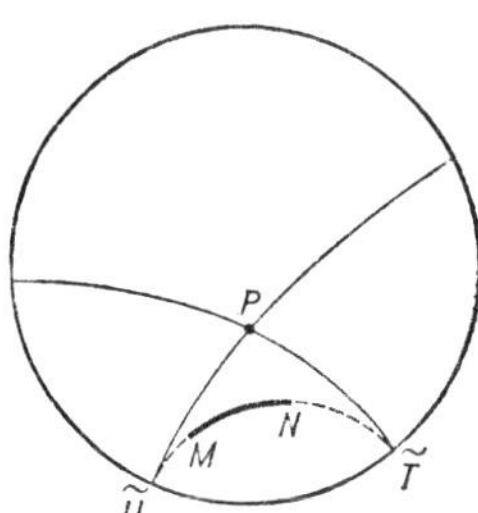

Fig. 3.

at points on the absolute. The points of the absolute model the 'points at infinity', which are not points of the Lobachevskiĭ plane.

In 1882, H. Poincaré, in constructing a theory of automorphic functions, arrived at two other models, in a disc and on a half-plane (cf. **Poincaré model**). In the first model (Fig. 3) the Lobachevskiĭ plane is realized by the interior of a disc, and lines by the inner parts of arcs of circles that intersect the main disc orthogonally. The metric is introduced by means of cross ratios, and the values of angles on the model are the same as those on the Lobachevskiĭ plane (a conformal model).

The introduction of coordinates makes it possible to obtain different analytical models of the Lobachevskiĭ plane. Poincaré, in 1887, proposed a model of Lobachevskiĭ geometry as the geometry of plane diametral sections of one of the sheets of a two-sheet

hyperboloid, which can also be treated as the geometry of a sphere of purely imaginary radius in a **pseudo-Euclidean space**. These models can be generalized to the case of an n-dimensional space.

Like **elliptic geometry**, Lobachevskiĭ geometry is the geometry of a Riemannian space of constant curvature.

The origin of the creation of Lobachevskiĭ geometry was the problem of parallels, that is, attempts to prove Euclid's fifth postulate concerning parallels. N.I. Lobachevskiĭ (1826, published in 1829 - 1830) showed that the assumption of a postulate different from Euclid's postulate makes it possible to construct Lobachevskiĭ geometry, which is more general than Euclidean geometry. Independently of Lobachevskiĭ, J. Bolyai arrived at the same discovery in 1832. Not having obtained the open support of C.F. Gauss, Bolyai did not continue his research.

Gauss worked out the beginnings of the new geometry much earlier, but he did not publish this research, and never spoke openly about these ideas. However, in private correspondence he regarded the work of Bolyai and Lobachevskiĭ highly, but he did not say so openly in print.

Applications of Lobachevskiĭ geometry. In the first work on his geometry, Lobachevskiĭ, relying on the yearly parallax of stars first measured by the astronomers of the time, showed that if his geometry is realized in physical space, then within the limits of the Solar System the deviations from Euclidean geometry would be several orders smaller than possible measuring errors. Thus, the first application of Lobachevskiĭ geometry was a justification of the practical accuracy of Euclidean geometry.

Lobachevskiĭ applied his geometry to mathematical analysis. Going over from one coordinate system to another in his space, he found the values of about 200 distinct definite integrals. Other mathematical applications were found by Poincaré (1882), who successfully applied Lobachevskiĭ geometry to the development of the theory of automorphic functions.

The significance of Lobachevskiĭ geometry for cosmology was first explained by A.A. Friedman. In 1922 he found a solution of the **Einstein equations** which implied that the universe expands in the course of time. This conclusion was subsequently confirmed by observations of E. Hubble (1929), who discovered the scattering of distant nebuli. The metric found by Friedman gives, for a fixed time, the Lobachevskiĭ space. The velocity space of the special theory of relativity is a Lobachevskiĭ space.

Lobachevskiĭ geometry has been used successfully in the study of collisions of elementary particles and in the development of other questions of nuclear research.

Visual perception of nearby regions of space by man provides the effect of inverse perspective, which is explained by the fact that the geometry of these regions of perspective space are close to Lobachevskiĭ geometry with radius of curvature of about 15 metres.

The creation of Lobachevskiĭ geometry was an important step in the development of studies on the possible properties of space. It also had special significance for the foundations of mathematics, since the principles of the modern axiomatic method were worked out to a significant extent thanks to the appearance of Lobachevskiĭ geometry.

References

[1] LOBACHEVSKIĬ, N.I.: *Zwei geometrische Abhandlungen*, Teubner, 1898 (translated from the Russian).

[2] BOLYAI, J.: 'Appendix. Scientiam spatii absolute veram exhibens', in *Tentamen*, Vol. 1, M. Vásárhelyini Die, 1833.

[3] ALEKSANDROV, A.D.: 'Abstract spaces', in *Mathematics, its Content, Methods and Significance*, Vol. 3, Amer. Math. Soc., 1962 (translated from the Russian).

[4] EGOROV, I.P.: *Introduction to non-Euclidean geometries*, Penze, 1972 (in Russian).

[5] EFIMOV, N.V.: *Höhere Geometrie*, Deutsch. Verlag Wissenschaft., 1960 (translated from the Russian).

[6] KAGAN, V.F.: *The foundations of geometry*, 1, Moscow-Leningrad, 1949 (in Russian).

[7] LAPTEV, B.L.: *Nikolaĭ Ivanovich Lobachevskiĭ*, Kazan', 1976 (in Russian).

[8] NORDEN, A.P.: *Elementare Einführung in die Lobatschewskische Geometrie*, Deutsch. Verlag Wissenschaft., 1958 (translated from the Russian).

[9] NUT, YU.YU.: *Lobachevskiĭ geometry in an analytic setting*, Moscow, 1961 (in Russian).

[10] RAUSHENBAKH, B.V.: *Spatial constructions in old Russian paintings*, Moscow, 1975 (in Russian).

[11] ROZENFEL'D, B.A.: *Non-Euclidean spaces*, Moscow, 1969 (in Russian).

[12] SHIROKOV, P.A.: *A sketch of the fundamentals of Lobachevskian geometry*, Noordhoff, 1964 (translated from the Russian).

B.L. Laptev

Editorial comments. Lobachevskiĭ geometry is also called *hyperbolic geometry*. Just as in spherical geometry it is natural to use a sphere of radius $R=1$, in Lobachevskiĭ geometry one usually assumes $k=1$, thereby simplifying somewhat the formulas. (E.g. $\Pi(a)=2\arctan e^{-a}$, $\sigma=\pi-A-B-C$, $l=2\pi\sinh r$.)

Beltrami could have anticipated F. Klein, by combining his two interpretations.

For the introduction in Poincaré's first model (Fig. 3) of a metric avoiding cross ratios see [A2].

Concerning some history see also [A9]. This reference also includes translations of a paper by Lobachevskiĭ and of the *Appendix* [2] by Bolyai.

References

[A1] COXETER, H.S.M.: *Non-Euclidean geometry*, Univ. Toronto Press, 1965.

[A2] COXETER, H.S.M.: 'Parallel lines', *Canad. Math. Bull.* 21 (1978), 385-397.

[A3] COXETER, H.S.M.: 'The non-Euclidean symmetry of Escher's picture 'Circle Limit III'', *Leonardo* 12 (1979), 19-25; 32.

[A4] COXETER, H.S.M.: 'Angles and arcs in the hyperbolic plane', *Math. Chronicle (New Zealand)* 9 (1980), 7-33.

[A5] BERGER, M.: *Geometry*, 1-2, Springer, 1987 (translated from the French).

[A6] GREENBERG, M.: *Euclidean and non-Euclidean geometries*, Freeman, 1974.

[A7] SOMMERVILLE, D.M.Y.: *Bibliography of non-Euclidean geometry*, Chelsea, reprint, 1970.

[A8] COXETER, H.S.M.: *Introduction to geometry*, Wiley, 1961.

[A9] BONOLA, R.: *Non-Euclidean geometry*, Dover, reprint, 1955.

AMS 1980 Subject Classification: 51M10

LOBACHEVSKIĬ METHOD, *Graeffe method, Dandelin method* - A method for simultaneously calculating all roots of a polynomial. Suppose that the roots $r_1, \ldots, r_n$ of the polynomial

$$f(z) = a_0 z^n + \cdots + a_{n-1}z + a_n = \tag{1}$$
$$= a_0(z-r_1) \cdots (z-r_n), \quad a_0 \neq 0,$$

satisfy the inequalities

$$|r_1| \gg \cdots \gg |r_n|. \tag{2}$$

As approximations to the roots one can take the ratios a_i/a_{i-1}, $i=1, \ldots, n$.

Now suppose that the roots of $f(z)$, although not satisfying (2), are all distinct in absolute value. The Lobachevskiĭ method consists in applying to the equation $f(z)=0$ the process of squaring, which after sufficiently many repetitions leads to an equation with roots satisfying (2). *Squaring* consists in going over from a polynomial $f_k(z)$ to a polynomial $f_{k+1}(z)$ of the same degree but with as roots the squares of the roots of $f_k(z)$. The transition is carried out by recurrence formulas.

The Lobachevskiĭ method can be applied if there are groups of roots equal in absolute value, though this leads to complications in the logic and formulas of the method. The advantage of the method is that it does not require initial approximations to the roots of the polynomial. In the case of roots distinct in absolute value the rate of convergence of the process is asymptotically quadratic.

However, the Lobachevskiĭ method is numerically unstable, since the process of squaring leads to very rapid accumulation of the computational error. In this connection, attempts have been made to give the Lobachevskiĭ method a self-correcting form. For example, for the calculation of the roots of (1) a sequence of polynomials $g_k(z)$ of degree $\leq n-1$ is constructed, connected by the relations

$$g_{k+1}(z) = zg_k(z) - \phi_k f(z), \quad k=0,1,\ldots; \quad g_0(z) = 1;$$

hence

$$g_k(z) = z^k \pmod{f(z)}.$$

For every fixed k one looks for polynomials $g_{k,p}(z)$ defined as follows:

$$g_{k,1}(z) = g_k(z);$$

for $p \geq 2$, $g_{k,p}(z)$ is a polynomial of the form

$$\psi_{p-2}(z)f(z) + \phi_{p-1}(z)g_k(z),$$

having degree $\leqslant n - p$. If the roots of $f(z)$ satisfy the inequalities

$$|r_1| \geqslant \cdots \geqslant |r_p| > |r_{p+1}| \geqslant \cdots \geqslant |r_n|,$$

then

$$\lim_{k \to \infty} g_{k,p} = \frac{f^*(z)}{(z - r_1) \cdots (z - r_p)},$$

where $f^*(z)$ is the polynomial $f(z)$ normalized by dividing by the coefficient at the leading term. Thus, from the original polynomial one picks out the factors corresponding to groups of roots equal in absolute value (see [3]). The method was proposed by N.I. Lobachevskiĭ in 1834 (see [1]).

References

[1] LOBACHEVSKIĬ, N.I.: *Collected works*, 4, Moscow-Leningrad, 1948 (in Russian).
[2] BEREZIN, I.S. and ZHIDKOV, N.P.: *Computing methods*, Pergamon, 1973 (translated from the Russian).
[3] SEBASTIÃO E SILVA, J.: 'Sur une méthode de approximation semblable à celle de Gräffe', *Portug. Math.* 2 (1941), 271-279.
[4] HOUSEHOLDER, A.S. and STEWART, G.W.: 'The numerical factorization of a polynomial', *SIAM Rev.* 13 (1971), 38-46.

Kh.D. Ikramov

Editorial comments. The basic idea of squaring the roots rests on the simple observation that

$$(-1)^n f(-z) f(z) = a_0^2 \prod_{i=1}^{n} (z^2 - r_i^2)$$

is a polynomial in z^2 with roots $r_1^2, \ldots, r_n^2$. For a systematic treatment in the case of complex roots cf. also [A1]. The method is most often referred to as *Graeffe's root squaring method*.

References

[A1] BRODETSKY, S. and SMEAL, G.: 'On Graeffe's method for complex roots', *Proc. Cambridge Philos. Soc.* 22 (1924), 83-87.
[A2] HILDEBRAND, F.B.: *Introduction to numerical analysis*, McGraw-Hill, 1974.

AMS 1980 Subject Classification: 65H05, 26C10, 12D10

LOBACHEVSKIĬ SPACE - A space whose geometry is defined by the axioms of **Lobachevskiĭ geometry**. In a wider sense a Lobachevskiĭ space is a non-Euclidean hyperbolic space whose definition is connected with concepts of the geometry of a **pseudo-Euclidean space**. Let $^1R_{n+1}$ be the Lorentz$-$Minkowskian $(n+1)$-space with one time-like direction. A sphere of time-like radius is analogous to a hyperboloid of two sheets. One sheet (say the 'future' sheet) is isometric to a Lobachevskiĭ n-space ^1S_n. This definition of a Lobachevskiĭ space makes it possible to include this space in the projective classification of non-Euclidean spaces. The space ^1S_n in the projective space P_n is represented by the interior of an oval $(n-1)$-quadric that is the intersection of an n-sphere of time-like radius with the hyperplane at infinity of the space

$^1R_{n+1}$ that completes this space to the projective space P_{n+1}. The points of the oval $(n-1)$-quadric are the points at infinity of ^1S_n, that is, the quadric is the **absolute** of this space. The outside of this quadric, which completes ^1S_n to the complete space P_n, is called the *ideal domain* of ^1S_n. This interpretation is called the *Cayley$-$Klein projective interpretation*. It can also be obtained by projecting an n-sphere of time-like radius in $^1R_{n+1}$ from its centre to a tangent n-plane, which is a Euclidean n-space; the space ^1S_n is represented by the inside of an n-ball in this n-plane, and the boundary of the n-ball is the absolute of ^1S_n (the latter interpretation of ^1S_n in the Euclidean space R_n is sometimes called the *Beltrami$-$Klein interpretation*).

The projective interpretation of Lobachevskiĭ 3-space makes it possible to verify the axioms of Lobachevskiĭ geometry, to give a representation of all figures of this geometry, and to establish their properties; in particular, in this interpretation it is easy to establish the geometrical properties of the Lobachevskiĭ 2-plane that follow from the axioms of Lobachevskiĭ geometry.

When the hyperbolic space ^1S_n is imbedded in the projective space P_n, an m-flat ($m<n$) is said to be *proper* if it intersects the absolute in an $(m-1)$-quadric; an m-flat that touches the absolute is *isotropic*; and an m-flat that does not intersect the absolute is *ideal*. The poles of proper hyperplanes are *ideal points*, and the *proper points* are the poles of ideal hyperplanes. More generally, the polar $(n-m-1)$-flats of proper m-flats of the Lobachevskiĭ space ^1S_n are ideal $(n-m-1)$-flats, and the polar $(n-m-1)$-flats of ideal m-flats are proper $(n-m-1)$-flats.

In the space ^1S_n, as coordinates of a point X one takes the components of the corresponding vector $\mathbf{x}$ of this point in $^1R_{n+1}$. These **Weierstrass coordinates** must satisfy the condition

$$(x^0)^2 - \sum_t (x^t)^2 = 1, \quad t > 0.$$

In ^1S_n one may introduce instead coordinates $u^1, \ldots, u^n$, analogous to spherical polar coordinates, which are connected with the coordinates x^t by the relations

$$x^0 = \cosh u^1 \cosh u^2 \cdots \cosh u^n,$$

$$x^1 = \sinh u^1 \cosh u^2 \cdots \cosh u^n,$$

$$\cdots \cdots \cdots$$

$$x^t = \sinh u^t \cosh u^{t+1} \cdots \cosh u^n,$$

$$\cdots \cdots \cdots \cdots$$

$$x^n = \sinh u^n,$$

The distance δ between two points of ^1S_n is then defined, in terms of their Weierstrass coordinates, by

$$\cosh \delta = x^0 y^0 - \sum x^t y^t.$$

The angle ϕ ($\leqslant \pi/2$) between two intersecting hyperplanes $X_0 x^0 + \sum X_t x_t = 0$ and $Y_0 x^0 + \sum Y_t x^t = 0$ (where $-X_0^2 + \sum X_t^2 = 1 = -Y_0^2 + \sum Y_t^2$) can be identified with the space-like distance between their poles $(-X_0, X_1, \ldots, X_n)$ and $(-Y_0, Y_1, \ldots, Y_n)$, and is thus given by

$$\cos\phi = |-X_0 Y_0 + \sum X_t Y_t|.$$

Similarly, the distance δ between two ultraparallel hyperplanes is given by

$$\cosh\delta = |-X_0 Y_0 + \sum X_t Y_t|.$$

The distance between points and the values of the angles between planes admit expressions in terms of the cross ratios (cf. **Cross ratio**) of points, using points of the absolute.

In the Lobachevskiĭ space $^1 S_n$ one can define spheres (balls), equi-distant surfaces, horospheres (horocycles for $n=2$, cf. **Horocycle**), m-simplexes, etc.

The classification of motions of the Lobachevskiĭ space $^1 S_n$ as collineations that take points of the absolute (oval quadric) into itself reduces to the classification of motions fixing one point of the Lorentz$-$Minkowskian space $^1 R_{n+1}$ without interchanging 'future' and 'past'. (This is a **Lie group**.) In order to specify a motion of $^1 S_n$ it is sufficient to give the images of $n+1$ points that do not lie in one hyperplane.

There are several conformal interpretations of a Lobachevskiĭ space, one of which is the **Poincaré model**. It is also possible to have a conformal interpretation of the space on one of its hyperplanes. Apart from these there are interpretations in complex spaces. In particular, for the space $^1 S_n$ one can construct the **Kotel'nikov interpretation** of manifolds of lines.

By means of projective interpretations, quadrics in $^1 S_n$, and particularly in the 2-plane $^1 S_2$, can be classified more completely.

The space $^1 S_n$ is a Riemannian n-space of constant negative curvature $-1/\sigma^2$, where σi is the radius of curvature of the space. The geometry of a Lobachevskiĭ space in sufficiently small neighbourhoods of points is close to the geometry of the Euclidean space of the same dimension.

In the large, the space $^1 S_n$ is homeomorphic to the space R_n; it extends indefinitely in all directions. Any m-flat of $^1 S_n$, $m<n$, is a space $^1 S_m$. Also, straight lines of $^1 S_n$ are geodesics, and m-flats are totally geodesic m-surfaces of this space.

In the projective classification of metrics of non-Euclidean spaces a Lobachevskiĭ space is also classified with respect to the metrics on lines, pencils of planes and m-flats. In particular, on a 2-flat of a Lobachevskiĭ space the projective metric on a line is hyperbolic, and the metric in pencils of lines is elliptic.

References

[1] EFIMOV, N.V.: *Höhere Geometrie*, Deutsch. Verlag Wissenschaft., 1960 (translated from the Russian).
[2] KLEIN, F.: *Vorlesungen über Nicht-Euklidische Geometrie*, Springer, 1928.
[3] KAGAN, V.F.: *The foundations of geometry*, 1-2, Moscow-Leningrad, 1949-1956 (in Russian).
[4] ROZENFEL'D, B.A.: *Non-Euclidean spaces*, Moscow, 1969 (in Russian).

L.A. Sidorov

Editorial comments.

References

[A1] ROSENFELD, B.A. [B.A. ROZENFEL'D]: *A history of non-Euclidean geometry*, Springer, 1988, Chapt. 6 (translated from the Russian).
[A2] ROBB, A.A.: *Geometry of time and space*, Cambridge Univ. Press, 1936, p. 406.
[A3] COXETER, H.S.M.: *Non-Euclidean geometry*, Univ. Toronto Press, 1965, p. 209.
[A4] BERGER, M.: *Geometry*, 1-2, Springer, 1987 (translated from the French).
[A5] BUSEMANN, H. and KELLY, P.: *Projective geometry and projective metrics*, Acad. Press, 1953.

AMS 1980 Subject Classification: 51M10

LOBATTO QUADRATURE FORMULA - A quadrature formula of highest algebraic degree of accuracy for the interval $[a, b] = [-1, 1]$ and weight $p(x) = 1$ with two fixed nodes: the end-points of $[-1, 1]$. The Lobatto quadrature formula has the form

$$\int_{-1}^{1} f(x)\,dx \cong A[f(-1) + f(1)] + \sum_{j=1}^{n} C_j f(x_j).$$

The points x_j are the roots of the polynomial $P_n^{(1,1)}(x)$ (a Jacobi polynomial), orthogonal on $[-1, 1]$ with respect to the weight $1 - x^2$, $A = 2/(n+1)(n+2)$ and $C_j > 0$. The algebraic degree of accuracy is $2n+1$. A table of nodes and coefficients of the Lobatto quadrature formula for $n = 1\,(1)\,15$ (n varies from 1 to 15 with step 1) was given in [2] (see also [3]).

The formula was established by R. Lobatto (see [1]).

References

[1] LOBATTO, R.: *Lessen over de differentiaal- en integraalrekening*, 1-2, 's Gravenhage, 1851-1852.
[2] KRYLOV, V.I.: *Approximate calculation of integrals*, Macmillan, 1962 (translated from the Russian).
[3] MICHELS, H.H.: 'Abscissas and weight coefficients for Lobatto quadrature', *Math. Comp.* **17** (1963), 237-244.

I.P. Mysovskikh

Editorial comments. For the notion of algebraic degree of accuracy of a quadrature formula see **Quadrature formula**.

References

[A1] STROUD, A.H. and SECREST, D.: *Gaussian quadrature formulas*, Prentice-Hall, 1966.

AMS 1980 Subject Classification: 65D32

LOCAL AND RESIDUAL PROPERTIES - Certain abstract properties (that is, properties preserved under isomorphism) of algebraic systems or universal algebras. If P is an abstract property of algebras, one says that an algebra A *locally has the property* P if there is a local system of subalgebras of A each of which has the property P. A *local system of subalgebras of A* is a system of non-empty subalgebras, directed by inclusion, whose union coincides with A. If every algebra of some class that locally has the property P actually has the property P itself, then P is called a *local property* of the algebras of this class. For example, the property of being an Abelian group is a local property in the class of all groups, but the property of being a finite group is not local. For more details about the local nature of properties, see **Mal'tsev local theorems**.

One says that an algebra A *residually has the property* P if there is a separating family of congruences $(q_\lambda)_{\lambda \in \Lambda}$ on A such that every quotient algebra A/q_λ has the property P. A family $(q_\lambda)_{\lambda \in \Lambda}$ is called a *separating family of congruences* if the intersection of all the q_λ is the diagonal congruence (the equality relation) on the given algebra. An algebra residually has the property P if and only if it can be represented as a subdirect product of algebras of the appropriate type having the property P. A property P is said to be residual in a class of algebras if every algebra of this class that residually has the property P actually has the property P itself. In the class of all groups the property of being Abelian is residual, but finiteness is not residual. Every residual property of algebras that is preserved under transition to homomorphic images is local.

References

[1] COHN, P.M.: *Universal algebra*, Reidel, 1981.

O.A. Ivanova

Editorial comments. Of course, the precise meaning of the phrase 'local property' depends on the local system of algebras that is used to define it.

A most important example of a local system of subalgebras of A is the system of all finitely-generated subalgebras of A; see also e.g. **Locally free group**; **Locally solvable group**; **Locally solvable algebra**; **Local property**; **Locally nilpotent group**; **Locally nilpotent algebra**; **Locally finite group**; **Locally finite algebra**.

The term 'local property' is used in topology as well as algebra. If P is an abstract property of topological spaces, one says that a space X *locally has the property* P if every point of X possesses a base of neighbourhoods having property P. (For example, the properties 'locally compact' and 'locally connected' may be defined in this way.) One says that P is a *local property* if the spaces which locally have the property P are exactly the P-spaces: for example, the T_1 separation axiom is a local property, but the Hausdorff axiom is not.

AMS 1980 Subject Classification: 08C99, 20E25, 20E26, 54D99

 - A measure of approximation (in particular, **best approximation**) of a function f on a set $E \subset \mathbf{R}^m$, regarded as a function of this set. The main interest is in the behaviour of a local approximation of a function as $\mathrm{mes}\, E \to 0$. In certain cases it is possible to characterize the degree of smoothness of the function to be approximated in terms of a local approximation of the function. Let $E_n(f; (\alpha, \beta))$ be the best approximation of a function $f \in C[a, b]$ by algebraic polynomials of degree n on an interval (α, β), $a \leqslant \alpha < \beta \leqslant b$. The following assertion holds: A necessary and sufficient condition for a function f to have a continuous derivative of order $n + 1$ at all points of $[a, b]$ is that

$$\frac{E_n(f; (\alpha, \beta))}{(\beta - \alpha)^{n+1}} \to \lambda(x), \quad a \leqslant x \leqslant b,$$

uniformly for $\beta \to x$, $\alpha \to x$, $\alpha < x < \beta$, where the continuous function λ is defined by

$$(n+1)! 2^{2n+1} \lambda(x) = |f^{(n+1)}(x)|.$$

References

[1] RAĬKOV, D.A.: 'On the local approximation of differentiable functions', *Dokl. Akad. Nauk SSSR* **24**, no. 7 (1939), 653-656 (in Russian).
[2] BERNSHTEĬN, S.N.: *Collected works*, 2, Moscow, 1954 (in Russian).
[3] BRUDNYĬ, YU.A.: 'Spaces defined by means of local approximations', *Trans. Moscow Math. Soc.* **24** (1974), 73-139. (*Trudy Moskov. Mat. Obshch.* **24** (1971), 69-132)

N.P. Koreneĭchuk
V.P. Motornyĭ

Editorial comments. According to [3], which is a valuable survey paper with a rather extensive bibliography, the first result characterizing a space of smooth functions in terms of local approximations was obtained by D.A. Raĭkov [1].

References

[A1] PEETRE, J.: 'On the theory of $L_{p,\lambda}$ spaces', *J. Funct. Anal.* **4** (1969), 71-87.

AMS 1980 Subject Classification: 41A50, 41A25

LOCAL COHOMOLOGY *with values in a sheaf of Abelian groups* - A **cohomology** theory with values in a **sheaf** and with supports contained in a given subset. Let X be a **topological space**, $\mathscr{F}$ a sheaf of Abelian groups on X and Z a locally closed subset of X, that is, a closed subset of some subset V open in X. Then $\Gamma_Z(X, \mathscr{F})$ denotes the subgroup of $\Gamma(V, \mathscr{F}|_V)$ consisting of the sections of the sheaf $\mathscr{F}|_V$ with supports in Z. If Z is fixed, then the correspondence $\mathscr{F} \to \Gamma_Z(X, \mathscr{F})$ defines a left-exact functor from the category of sheaves of Abelian groups on X into the category of Abelian groups. The value of the corresponding i-th right **derived functor** on $\mathscr{F}$ is denoted by $H^i_Z(X, \mathscr{F})$ and is called the i-th local cohomology group of X with values in $\mathscr{F}$, with respect to Z. One has

$$H^0_Z(X, \mathscr{F}) = \Gamma_Z(X, \mathscr{F}).$$

Let $\mathscr{H}^0_Z(\mathscr{F})$ be the sheaf on X corresponding to the pre-sheaf that associates with any open subset $U \subset X$ the group $\Gamma_{Z \cap U}(U, \mathscr{F}|_U)$. The correspondence $\mathscr{F} \to \mathscr{H}^0_Z(\mathscr{F})$ is a left-exact functor from the category of sheaves of Abelian groups on X into itself. The value of its i-th right derived functor on $\mathscr{F}$ is denoted by $\mathscr{H}^i_Z(\mathscr{F})$ and is called the i-th local cohomology sheaf of $\mathscr{F}$ with respect to Z. The sheaf $\mathscr{H}^i_Z(\mathscr{F})$ is associated with the pre-sheaf that associates with an open subset $U \subset X$ the group $H^i_{Z \cap U}(U, \mathscr{F}|_U)$.

There is a spectral sequence $E^{p,q}_r$, converging to $H^{p+q}_Z(X, \mathscr{F})$, for which $E^{p,q}_2 = H^p(X, \mathscr{H}^q_Z(\mathscr{F}))$ (see [2], [3]).

Let Z be a locally closed subset of X, Z' a closed subset of Z and $Z'' = Z \setminus Z'$; then there are the following exact sequences:

$$0 \to H^0_{Z'}(X, \mathscr{F}) \to \cdots \to H^i_{Z'}(X, \mathscr{F}) \to \qquad (1)$$

$$\to H^i_Z(X, \mathscr{F}) \to H^i_{Z''}(X, \mathscr{F}) \to H^{i+1}_{Z'}(X, \mathscr{F}) \to \cdots;$$

$$0 \to \mathscr{H}^0_{Z'}(\mathscr{F}) \to \cdots \to \mathscr{H}^i_{Z'}(\mathscr{F}) \to \qquad (2)$$

$$\to \mathscr{H}^i_Z(\mathscr{F}) \to \mathscr{H}^i_{Z''}(\mathscr{F}) \to \mathscr{H}^{i+1}_{Z'}(\mathscr{F}) \to \cdots.$$

If Z is the whole of X and Z' is a closed subset of X, then the sequence (2) gives the exact sequence

$$0 \to \mathscr{H}^0_{Z'}(\mathscr{F}) \to \mathscr{F} \to \mathscr{H}^0_{X \setminus Z'}(\mathscr{F}) \to \mathscr{H}^1_{Z'}(\mathscr{F}) \to 0$$

and the system of isomorphisms

$$\mathscr{H}^i_{X \setminus Z'}(\mathscr{F}) \cong \mathscr{H}^{i+1}_{Z'}(\mathscr{F}), \quad i \geq 1.$$

The sheaves $\mathscr{H}^i_{X \setminus Z'}(\mathscr{F})$ are called the i-th *gap sheaves* of $\mathscr{F}$ and have important applications in questions concerning the extension of sections and cohomology classes of $\mathscr{F}$, defined on $X \setminus Z'$, to the whole of X (see [4]).

If X is a locally **Noetherian scheme**, $\mathscr{F}$ is a **quasi-coherent sheaf** on X and Z is a closed subscheme of X, then $\mathscr{H}^i_Z(\mathscr{F})$ are quasi-coherent sheaves on X. If $\mathscr{Y}$ is a **coherent sheaf** of ideals on X that specifies the subscheme Z, then one has the isomorphisms

$$\varinjlim_n \mathrm{Ext}^i_{\mathscr{O}_X}(\mathscr{O}_X / \mathscr{Y}^n, \mathscr{F}) \cong \mathscr{H}^i_Z(\mathscr{F}).$$

The following criteria for triviality and coherence of local cohomology sheaves are important for applications (see [3], [4]).

Let X be a locally Noetherian scheme or a complex-analytic space, Z a locally closed subscheme or analytic subspace of X, $\mathscr{F}$ a coherent sheaf of $\mathscr{O}_X$-modules, and $\mathscr{Y}$ a coherent sheaf of ideals that specifies Z. Let

$$\mathrm{prof}_Z \mathscr{F} = \min_{x \in Z} \mathrm{prof}_{\mathscr{Y}_{X,x}} \mathscr{F}_x,$$

where $\mathrm{prof}_{\mathscr{Y}_{X,x}} \mathscr{F}_x$ is the maximal length of a sequence of elements of $\mathscr{Y}_{X,x}$ that is regular for $\mathscr{F}_x$, or ∞ if $\mathscr{F}_x = 0$. Then the equality $\mathscr{H}^i_Z(\mathscr{F}) = 0$ for $i < n$ is

equivalent to the condition $\mathrm{prof}_Z \mathscr{F} \geq n$. Let $\mathrm{codh}_x \mathscr{F}_x = \mathrm{prof}_{\mathfrak{m}_x} \mathscr{F}_x$ (where $\mathfrak{m}$ is the maximal ideal of the ring $\mathscr{O}_{X,x}$) and let $S_m(\mathscr{F}) = \{x \in X : \mathrm{codh}_x \mathscr{F}_x \geq m\}$. If X is a complex-analytic space or an algebraic variety, then all sets $S_m(\mathscr{F})$ are analytic or algebraic, respectively. If $\mathscr{F}$ is a coherent sheaf on X and Z is an analytic subspace or subvariety, respectively, then coherence of the sheaves $\mathscr{H}^i_Z(\mathscr{F})$ for $0 \leq i \leq q$ is equivalent to the condition

$$\dim Z \cap \overline{S_{k+q+1}(\mathscr{F}|_{X \setminus Z})} \leq k$$

for any integer k.

In terms of local cohomology one can define *hyperfunctions*, which have important applications in the theory of partial differential equations [5]. Let Ω be an open subset of $\mathbf{R}^n$, which is naturally imbedded in $\mathbf{C}^n$. Then $\mathscr{H}^p_\Omega(\mathbf{C}^n, \mathscr{O}_{\mathbf{C}}) = 0$ for $p \neq n$. The pre-sheaf $\Omega \to \mathscr{H}^n_\Omega(\mathbf{C}^n, \mathscr{O}_{\mathbf{C}})$ on $\mathbf{R}^n$ defines a **flabby sheaf**, called the sheaf of hyperfunctions.

An analogue of local cohomology also exists in **étale cohomology** theory [3].

References

[1] DOLGACHEV, I.V.: 'Abstract algebraic geometry', *Russian Math. Surveys* **2**, no. 3 (1974), 264-303. (*Itogi Nauk. i Tekhn. Algebra. Topol. Geom.* **10** (1972), 47-112)
[2] GROTHENDIECK, A.: *Local cohomology*, Springer, 1967.
[3] GROTHENDIECK, A.: *Cohomologie locale des faisceaux cohérents et théorèmes de Lefschetz locaux et globaux*, North-Holland, 1968.
[4] SIU, Y.-T. and TRAUTMANN, G.: *Gap-sheaves and extension of coherent analytic subsheaves*, Springer, 1971.
[5] SCHAPIRA, P.: *Théorie des hyperfonctions*, Springer, 1970.
[6] BANICA, C. and STANASILA, O.: *Algebraic methods in the global theory of complex spaces*, Wiley, 1976 (translated from the Rumanian).

D.A. Ponomarev

Editorial comments. See also **Hyperfunction** for the sheaf of hyperfunctions.

For an ideal $\mathfrak{a}$ in a commutative ring R with unit element the local cohomology can be described as follows. Let A be the set of prime ideals in R containing $\mathfrak{a}$. For an R-module M the submodule $\Gamma_A(M)$ is defined as $\{m: \mathrm{support}(m) \subset A\}$. Thus,

$$\Gamma_A(M) = \{m: \mathrm{rad}(\mathrm{Ann}(m)) \supset \mathfrak{a}\} =$$
$$= \{m: \mathfrak{a}^n m = 0 \text{ for } n \text{ large enough}\} \simeq$$
$$\simeq \varinjlim_n \mathrm{Hom}_R(R/\mathfrak{a}^n, M).$$

$M \mapsto \Gamma_A(M)$ is a covariant, left-exact, R-linear functor from the category of R-modules into itself. Its derived functors are the local cohomology functors $\mathscr{H}^i_A(M)$ (of M with respect to A (or $\mathfrak{a}$)). These cohomology functors can be explicitly calculated using Koszul complexes, cf. **Koszul complex**.

References

[A1] SIU, Y.-T.: *Techniques of extension of analytic objects*, M. Dekker, 1974.

AMS 1980 Subject Classification: 55N25, 14F05, 32C36

LOCAL DECOMPOSITION, *local cut* - A closed set Φ in the space $\mathbf{R}^n$ is a local cut if there are a point a (a point at which the set Φ cuts the space) and a positive number ϵ such that for any number $\delta > 0$ there is in the open set $O(a, \delta) \setminus \Phi$, where $O(a, \delta)$ is the (open) ball of radius δ with centre at a, a pair of points with the following property: Any **continuum** lying in $O(a, \epsilon)$ and containing this pair of points has a non-empty intersection with Φ. K. Menger and P.S. Urysohn proved that a closed set Φ lying in a plane has dimension 1 if and only if it does not contain interior (with respect to the plane) points and locally cuts the plane (at one point a at least).

A similar characterization of closed $(n-1)$-dimensional sets in the n-dimensional space $\mathbf{R}^n$ was given by P.S. Aleksandrov (see **Local linking**).

A.A. Mal'tsev

AMS 1980 Subject Classification: 54F45, 46AXX, 28AXX, 54CXX, 52A07

LOCAL DIFFERENTIAL GEOMETRY - The part of differential geometry that studies properties of geometrical forms, in particular curves and surfaces, 'in the small'. In other words, the structure of a geometrical form is studied in a small neighbourhood of an arbitrary point of it.

Suppose that in three-dimensional Euclidean space E^3 a curve γ is specified by its equation

$$\mathbf{r} = \mathbf{r}(t).$$

The study of this curve reduces to the discovery of quantities that are invariant with respect to the group of motions of E^3. The position vector $\mathbf{r}$ of a point M on the curve is not invariant, but its derivatives

$$\frac{d\mathbf{r}}{dt}, \frac{d^2\mathbf{r}}{dt^2}, \ldots, \tag{*}$$

are invariant. The *differential neighbourhood of order n* of a point M of a curve γ is the totality of all concepts and properties connected with the curve that can be expressed in terms of the first n vectors of the sequence (*). Thus, to the differential neighbourhood of order one belong the concepts of the tangent to the curve and its normal plane. To the differential neighbourhood of order two belong the concepts of curvature, the osculating plane, the Frénet trihedron, and the osculating circle of the curve. The concept of the torsion of the curve belongs to the differential neighbourhood of order three. The curvature and torsion of a curve form a complete system of invariants of it in the sense that any invariant of the curve is a function of curvature, torsion and their derivatives of some orders. The local theory of surfaces of E^3 is constructed similarly. The local theory of curves and surfaces of E^3 is the oldest part of local differential geometry, mainly created in the 18-th

and 19-th centuries. Already in the 19-th century various generalizations of this theory had begun to appear. One of these generalizations is connected with the concept of a **homogeneous space**. In an arbitrary differential-geometric homogeneous space G/H one can construct a local theory of curves and surfaces of various dimensions, similar as above for E^3; namely as the theory of invariants of the fundamental group G. In this direction the greatest development occurred in **affine differential geometry** and **projective differential geometry**.

A generalization of the concept of the first fundamental form of a surface in E^3 led to the theory of Riemannian spaces. The local theory of Riemannian spaces had already appeared in the middle of the 19-th century and continues to be developed, finding numerous applications.

The concept of parallel displacement of a vector along a curve on a surface in space led to the theory of spaces of **affine connection**. In turn, this was the start of the development of the general theory of connections (cf. **Connection**).

References

[1] FAVARD, J.: *Course de géométrie differentiélle locale*, Gauthier-Villars, 1957.

A.S. Fedenko

Editorial comments.

References

[A1] BERGER, M. and GOSTIAUX, B.: *Differential geometry: manifolds, curves and surfaces*, Springer, 1988 (translated from the French).
[A2] KLINGENBERG, W.: *A course in differential geometry*, Springer, 1978 (translated from the German).
[A3] SPIVAK, M.: *A comprehensive introduction to differential geometry*, 1-5, Publish or Perish, 1979.

AMS 1980 Subject Classification: 53BXX, 53A04, 53A05

LOCAL DIMENSION *of a normal topological space X* - The topological invariant $\operatorname{locdim} X$, defined as follows. One says that $\operatorname{locdim} X \leqslant n$, $n = -1, 0, 1, \ldots$, if for any point $x \in X$ there is a neighbourhood O_x for which the **Lebesgue dimension** of its closure satisfies the relation $\dim \overline{O}_x \leqslant n$. If $\operatorname{locdim} X \leqslant n$ for some n, then the local dimension of X is finite, so one writes $\operatorname{locdim} X < +\infty$ and puts

$$\operatorname{locdim} X = \min\{n : \operatorname{locdim} X \leqslant n\}.$$

Always $\operatorname{locdim} X \leqslant \dim X$; there are normal spaces X with $\operatorname{locdim} X < \dim X$; in the class of paracompact spaces always $\operatorname{locdim} X = \dim X$. If in the definition of local dimension the Lebesgue dimension $\dim \overline{O}_x$ is replaced by the large **inductive dimension** $\operatorname{Ind} \overline{O}_x$, then one obtains the definition of the *local large inductive dimension* $\operatorname{locInd} X$.

Editorial comments. See [A1] for a construction of a space with locdim $X<\dim X$ and — as an application — a hereditarily normal space Y with $\dim Y=0$ yet Y contains subspaces of arbitrary high dimension.

For the notions of the *local dimension at a point* of an analytic space, algebraic variety or scheme cf. **Analytic space; Dimension** of an associative ring; **Analytic set,** and **Spectrum of a ring.**

References

[A1] POL, E. and POL, R.: 'A hereditarily normal strongly zero-dimensional space containing subspaces of arbitrarily large dimension', *Fund. Math.* **102** (1979), 137-142.
[A2] ENGELKING, R.: *Dimension theory*, PWN & North-Holland, 1978.

AMS 1980 Subject Classification: 54F45

LOCAL FIELD - A **field** that is complete with respect to a discrete valuation and has finite residue field. The structure of a local field K is well known: 1) if the characteristic of K is 0, then K is a finite extension of the field $\mathbf{Q}_p$ of p-adic numbers (cf. *p*-**adic number**); 2) if the characteristic of K is greater than 0, then K is isomorphic to the field $k((T))$ of formal power series over a finite field k. Such fields are called local, in contrast to global fields (finite extensions of the fields $\mathbf{Q}$ or $k(T)$), and are means for studying the latter. For cohomological properties of Galois extensions of local fields see [1], and also **Adèle; Idèle;** and **Class field theory.**

To construct a class field theory of multi-dimensional schemes one uses a generalization of the concept of a local field. Namely, an *n-dimensional local field* is a sequence $O_0, \ldots, O_n$ of complete discrete valuation rings together with isomorphisms

$$k(O_i) \xrightarrow{\sim} K(O_{i+1}),$$

where k is the residue field and K is the field of fractions of a ring O. Moreover, $k(O_n)$ must be finite. There exists a structure theory for n-dimensional local fields (see [3]).

References

[1] SERRE, J.-P.: *Local fields*, Springer, 1979 (translated from the French).
[2] CASSELS, J.W.S. and FRÖHLICH, A. (EDS.): *Algebraic number theory*, Acad. Press, 1986.
[3] PARSHIN, A.N.: 'Abelian coverings of arithmetic schemes', *Soviet Math. Dokl.* **19**, no. 6 (1978), 1438-1442. (*Dokl. Akad. Nauk SSSR* **243** (1978), 855-858)

V.I. Danilov

Editorial comments. The concept of a local field is sometimes extended to include that of discretely valued fields with arbitrary residue fields. There is a class field theory for local fields with perfect residue fields in terms of a certain *fundamental group* [A1], [A2]. For an account of the class field theory of n-dimensional local fields (in terms of algebraic K-theory) see also [A3] - [A5].

References

[A1] SERRE, J.-P.: 'Sur les corps locaux à corps résiduel algébriquement clos', *Bull. Soc. Math. France* **89** (1961), 105-154.
[A2] DEMAZURE, M. and GABRIEL, P.: *Groupes algébriques*, North-Holland, 1971, pp. 648-674. (Appendix: M. Hazewinkel, Classes de corps local)
[A3] KATO, K.: 'Class field theory and algebraic K-theory', in M. Raynaud and T. Shioda (eds.): *Algebraic geometry*, Lecture notes in math., Vol. 1016, Springer, 1983, pp. 109-126.
[A4] KATO, K.: 'Vanishing cycles, ramification of valuations and class field theory', *Duke Math. J.* **55** (1987), 629-661.
[A5] PARŠIN, A.N. [A.N. PARSHIN]: 'Local class field theory', *Proc. Steklov Inst. Math.* **165** (1985), 157-185. (*Trudy Mat. Inst. Steklov.* **165** (1984), 143-170)

AMS 1980 Subject Classification: 12J10

LOCAL HOMEOMORPHISM - A mapping $f: X \to Y$ between topological spaces such that for every point $x \in X$ there is a neighbourhood O_x that maps homeomorphically into Y under f (cf. **Homeomorphism**). Sometimes in the definition of a local homeomorphism the requirement $fX = Y$ is included and f is also assumed to be open (cf. **Open mapping**). Examples of local homeomorphisms are: a continuously-differentiable mapping with non-zero Jacobian on an open subset of an n-dimensional Euclidean space into the n-dimensional Euclidean space; a covering mapping, in particular the natural mapping of a topological group onto its quotient space with respect to a discrete subgroup. If the mapping $f: X \to Y$ of a Čech-complete space, in particular a locally compact Hausdorff space, onto a Tikhonov space Y is open and countable-to-one, that is, $|f^{-1}y| \leqslant \aleph_0$, $y \in Y$, then on some open everywhere-dense set in X the mapping f is a local homeomorphism.

B.A. Pasynkov

Editorial comments.

References

[A1] ARKHANGEL'SKIĬ, A.V. and PONOMAREV, V.I.: *Fundamentals of general topology: problems and exercises*, Reidel, 1984 (translated from the Russian).

AMS 1980 Subject Classification: 54CXX

LOCAL HOMOLOGY - The homology groups (cf. **Homology group**)

$$H_p^x = H_p^c(X, X \setminus x; G),$$

defined at points $x \in X$, where H_p^c is **homology with compact support**. These groups coincide with the direct limits

$$\lim_{\to} H_p^c(X, X \setminus U; G)$$

over open neighbourhoods U of x, and for homologically locally connected X they also coincide with the inverse limits

$$\lim_{\leftarrow} H_{p-1}^c(U \setminus x; G).$$

The homological dimension of a finite-dimensional metrizable locally compact space X over G (cf. **Homological dimension of a space**) coincides with the largest value of n for which $H_n^x \neq 0$, and the set of such points $x \in X$ has dimension n.

Let $\mathscr{C}_*$ be the differential sheaf over X defined by associating with each open set $U \subset X$ the chain complex $C_*(X, X \setminus U; G)$. The groups H_p^x are the fibres of the derived sheaves $\mathscr{H}_p = H_p(\mathscr{C}_*)$. For generalized manifolds, $H_p^x = 0$ for $p \neq n = \dim X$. In this case the homology sequence of the pair (X, A) with coefficients in G coincides with the cohomology of the pair $(X, X \setminus A)$ with coefficients in the sheaf $\mathscr{H}_n$ (*Poincaré$-$Lefschetz duality*). The similar facts for the *local cohomology* of locally compact spaces do not hold.

References

[1] SKLYARENKO, E.G.: 'On the theory of generalized manifolds', *Math. USSR Izv.* **5**, no. 4 (1971), 845-858. (*Izv. Akad. Nauk SSSR Ser. Mat.* **35** (1971), 831-843)

[2] HARLAP, A.E. [A.E. KHARLAP]: 'Local homology and cohomology, homology dimension and generalized manifolds', *Math. USSR Sb.* **25**, no. 3 (1975), 323-349. (*Mat. Sb.* **96** (1975), 347-373)

E.G. Sklyarenko

Editorial comments.

References

[A1] DOLD, A.: *Lectures on algebraic topology*, Springer, 1972, Sect. IV.3.

AMS 1980 Subject Classification: 55N25, 55Q07, 54F43, 55M10

LOCAL LIMIT THEOREMS *in probability theory* - Limit theorems for densities, that is, theorems that establish the convergence of the densities of a sequence of distributions to the density of the limit distribution (if the given densities exist), or a classical version of local limit theorems, namely local theorems for lattice distributions, the simplest of which is the local **Laplace theorem**.

Let $X_1, X_2, \ldots$, be a sequence of independent random variables that have a common distribution function $F(x)$ with mean a and finite positive variance σ^2. Let $F_n(x)$ be the distribution function of the normalized sum

$$Z_n = \frac{1}{\sigma \sqrt{n}} \sum_{j=1}^{n} (X_j - a)$$

and let $\Phi(x)$ be the normal $(0, 1)$-distribution function. The assumptions ensure that $F_n(x) \to \Phi(x)$ as $n \to \infty$ for any x. It can be shown that this relation does not imply the convergence of the density $p_n(x)$ of the distribution of the random variable Z_n to the normal density

$$\frac{1}{\sqrt{2\pi}} e^{-x^2/2},$$

even if the distribution F has a density. If Z_n, for some $n = n_0$, has a bounded density $p_{n_0}(x)$, then

$$p_n(x) \to \frac{1}{\sqrt{2\pi}} e^{-x^2/2} \qquad (*)$$

uniformly with respect to x. The condition that $p_{n_0}(x)$ is bounded for some n_0 is necessary for $(*)$ to hold uniformly with respect to x.

Let $X_1, X_2, \ldots$, be a sequence of independent random variables that have the same non-degenerate distribution, and suppose that X_1 takes values of the form $b + Nh$, $N = 0, \pm 1, \pm 2, \ldots$, with probability 1, where $h > 0$ and b are constants (that is, X_1 has a **lattice distribution** with step h).

Suppose that X_1 has finite variance σ^2, let $a = EX_1$ and let

$$P_n(N) = P\left\{ \sum_{j=1}^{n} X_j = nb + Nh \right\}.$$

In order that

$$\sup_N \left| \frac{\sigma \sqrt{n}}{h} P_n(N) - \frac{1}{\sqrt{2\pi}} \exp\left\{ -\frac{1}{2} \left[\frac{nb + Nh - na}{\sigma \sqrt{n}} \right]^2 \right\} \right| \to 0$$

as $n \to \infty$ it is necessary and sufficient that the step h should be maximal. This theorem of B.V. Gnedenko is a generalization of the local Laplace theorem.

Local limit theorems for sums of independent non-identically distributed random variables serve as a basic mathematical tool in classical statistical mechanics and quantum statistics (see [7], [8]).

Local limit theorems have been intensively studied for sums of independent random variables and vectors, together with estimates of the rate of convergence in these theorems. The case of a limiting normal distribution has been most fully investigated (see [3], Chapt. 7); a number of papers have been devoted to local limit theorems for the case of an arbitrary **stable distribution** (see [2]). Similar investigations have been carried out for sums of dependent random variables, in particular for sums of random variables that form a **Markov chain** (see [5], [6]).

References

[1] GNEDENKO, B.V. and KOLMOGOROV, A.N.: *Limit distributions for sums of independent random variables*, Addison-Wesley, 1954 (translated from the Russian).

[2] IBRAGIMOV, I.A. and LINNIK, YU.V.: *Independent and stationary sequences of random variables*, Wolters-Noordhoff, 1971 (translated from the Russian).

[3] PETROV, V.V.: *Sums of independent random variables*, Springer, 1975 (translated from the Russian).

[4] PROKHOROV, YU.V. and ROZANOV, YU.A.: *Probability theory*, Springer, 1969 (translated from the Russian).

[5] SIRAZHDINOV, S.KH.: *Limit theorems for homogeneous Markov chains*, Tashkent, 1955 (in Russian).

[6A] STATULYAVICHUS, V.A.: 'Limit theorems and asymptotic expansions for non-stationary Markov chains', *Litovsk. Mat. Sb.* **1** (1961), 231-314 (in Russian). English abstract.

[6B] STATULYAVICHUS, V.A.: 'Limit theorems for sums of random variables that are connected in a Markov chain I', *Litovsk. Mat. Sb.* **9** (1969), 345-362 (in Russian). English abstract.

[7] KHINCHIN, A.YA.: *Mathematical foundations of statistical mechanics*, Dover, reprint, 1949 (translated from the Russian).
[8] KHINCHIN, A.YA.: *Mathematical foundations of quantum statistics*, Moscow-Leningrad, 1951 (in Russian).

V.V. Petrov

Editorial comments.

References
[A1] BHATTACHARYA, R.N. and RANGA RAO, R.: *Normal approximations and asymptotic expansions*, Wiley, 1976.
[A2] PAULAUSKAS, V. and RAČKAUSKAS, A.: *Approximation theory in the central limit theorem*, Kluwer, 1989 (translated from the Russian).

AMS 1980 Subject Classification: 60F05

LOCAL LINKING - A property of the disposition of a closed set Φ close to a point a of it in a Euclidean space $\mathbf{R}^n$. It consists of the existence of a number $\epsilon > 0$ such that, for any positive number δ, in the open set $O(a, \delta) \setminus \Phi$ there lies a q-dimensional cycle Z^q, $q < n$, with integer coefficients, having the following property: Any compact set P lying in $O(a, \epsilon)$ in which Z^q is homologous to zero has non-empty intersection with Φ. Here $O(a, \delta)$ and $O(a, \epsilon)$ are spheres with centre a and radii δ and ϵ. Without changing the content of this definition one can restrict oneself to compact sets P that are polyhedra. For $q = 0$ the concept of a local linking goes over to the concept of a local cut (cf. **Local decomposition**). *Aleksandrov's obstruction theorem*: In order that $\dim \Phi = p$ it is necessary and sufficient that the number $n - p - 1$ should be the smallest integer q for which there is a q-dimensional linking of Φ in $\mathbf{R}^n$ close to some point $a \in \Phi$. An analogous theorem has been proved concerning obstructions 'modulo m', which characterizes sets Φ that have **homological dimension** p 'modulo m'.

Far-reaching generalizations of obstruction theorems are theorems on the **homological containment** of compact sets.

References
[1] ALEKSANDROV, P.S.: *An introduction to homological dimension theory and general combinatorial topology*, Moscow, 1975 (in Russian).
[2] SITNIKOV, K.: 'On homological girdling of compacta in Euclidean space', *Dokl. Akad. Nauk SSSR* **81** (1951), 153-156 (in Russian).

A.A. Mal'tsev

AMS 1980 Subject Classification: 55M10, 55Q07, 54F43, 54F45, 57N25

LOCAL PROPERTY *in commutative algebra* - A property P of a **commutative ring** A or an A-module M that is true for A (or M) if and only if a similar property holds for the localizations (cf. **Local ring**) of A (or M) with respect to all prime ideals of A, that is, a property that holds globally if and only if it holds locally everywhere. Often, instead of the set of all prime ideals one can restrict oneself to the set of maximal ideals of A. This terminology becomes clear if one associates to the ring A the topological space $\operatorname{Spec} A$ (the spectrum of A) consisting of all prime ideals of A. Then the assertion 'P is true for A' is equivalent to the assertion 'P holds on the whole space $\operatorname{Spec} A$', and the assertion 'P is true for all $A_{\mathfrak{P}}$' is equivalent to the assertion 'every point $\mathfrak{P}$ of $\operatorname{Spec} A$ has a neighbourhood in which P holds'.

Examples of local properties. An integral domain A is integrally closed in its field of fractions if and only if the localizations $A_{\mathfrak{P}}$ are integrally closed for all maximal ideals $\mathfrak{P}$ of A. A homomorphism of A-modules $f: M \to N$ is an isomorphism (monomorphism, epimorphism, null morphism) if and only if the mapping of localized modules $f_{\mathfrak{P}}: M_{\mathfrak{P}} \to N_{\mathfrak{P}}$ is an isomorphism (monomorphism, epimorphism, null morphism) for all maximal ideals of A.

However, the property of an A-module M of being free is not local.

References
[1] BOURBAKI, N.: *Elements of mathematics. Commutative algebra*, Addison-Wesley, 1972 (translated from the French).

L.V. Kuz'min

Editorial comments. For the term 'local property' in algebraic systems (such as groups) as well as in topology see Local and residual properties.

AMS 1980 Subject Classification: 13-XX

LOCAL RING - A **commutative ring** with a unit that has a unique maximal ideal. If A is a local ring with maximal ideal $\mathfrak{m}$, then the quotient ring $A / \mathfrak{m}$ is a field, called the *residue field* of A.

Examples of local rings. Any field or valuation ring is local. The ring of formal power series $k[[X_1, \ldots, X_n]]$ over a field k or over any local ring is local. On the other hand, the polynomial ring $k[X_1, \ldots, X_n]$ with $n \geqslant 1$ is not local. Let X be a topological space (or a differentiable manifold, an analytic space or an algebraic variety) and let x be a point of X. Let A be the ring of germs at x of continuous functions (respectively, differentiable, analytic or regular functions); then A is a local ring whose maximal ideal consists of the germs of functions that vanish at x.

Some general ring-theoretical constructions lead to local rings, the most important of which is localization (cf. **Localization in a commutative algebra**). Let A be a commutative ring and let $\mathfrak{p}$ be a prime ideal of A. The ring $A_{\mathfrak{p}}$, which consists of fractions of the form a / s, where $a \in A$, $s \in A \setminus \mathfrak{p}$, is local and is called the *localization of the ring A at $\mathfrak{p}$*. The maximal ideal of $A_{\mathfrak{p}}$ is $\mathfrak{p} A_{\mathfrak{p}}$, and the residue field of $A_{\mathfrak{p}}$ is identified with the field of fractions of the integral quotient ring $A / \mathfrak{p}$. Other constructions that lead to local rings are Henselization (cf. **Hensel ring**) or **completion** of a ring with respect to a maximal ideal. Any quotient ring of a local ring is also local.

A property of a ring A (or an A-module M, or an A-algebra B) is called a *local property* if its validity for A (or M, or B) is equivalent to its validity for the rings $A_\mathfrak{p}$ (respectively, modules $M \otimes_A A_\mathfrak{p}$ or algebras $B \otimes_A A_\mathfrak{p}$) for all prime ideals $\mathfrak{p}$ of A (see **Local property**).

The powers $\mathfrak{m}^n$ of the maximal ideal $\mathfrak{m}$ of a local ring A determine a basis of neighbourhoods of zero of the so-called *local-ring topology* (or $\mathfrak{m}$-*adic topology*). For a Noetherian local ring this topology is separated (*Krull's theorem*), and any ideal of it is closed.

From now on only Noetherian local rings are considered (cf. also **Noetherian ring**). A local ring is called a *complete local ring* if it is complete with respect to the $\mathfrak{m}$-adic topology; in this case $A = \lim_{\leftarrow n} A / \mathfrak{m}^n$. In a complete local ring the $\mathfrak{m}$-adic topology is weaker than any other separated topology (*Chevalley's theorem*). Any complete local ring can be represented as the quotient ring of the ring $S[[X_1, \ldots, X_n]]$ of formal power series, where S is a field (in the case of equal characteristic) or of a complete discrete valuation ring (in the case of different characteristic). This theorem makes it possible to prove that complete local rings have a number of specific properties that are absent in arbitrary Noetherian local rings (see [5]); for example, a complete local ring is an **excellent ring**.

A finer quantitative investigation of a local ring A is connected with the application of the concept of the *adjoint graded ring* $\mathrm{Gr}(A) = \oplus_{n \geqslant 0}(\mathfrak{m}^n / \mathfrak{m}^{n+1})$. Let $H_A(n)$ be the dimension of the vector space $\mathfrak{m}^n / \mathfrak{m}^{n+1}$ over the residue field $A / \mathfrak{m}$; as a function of the integer argument n it is called the *Hilbert $-$ Samuel function* (or *characteristic function*) *of the local ring A*. For large n this function coincides with a certain polynomial $\overline{H}_A(n)$ in n, which is called the *Hilbert $-$ Samuel polynomial of the local ring A* (see also **Hilbert polynomial**). This fact can be expressed in terms of a Poincaré series: The formal series

$$P_A(t) = \sum_{n \geqslant 0} H_A(n) \cdot t^n$$

is a rational function of the form $f(t)(1-t)^{-d(A)}$, where $f(t) \in \mathbf{Z}[t]$ is a polynomial and $d(A) - 1$ is the degree of $\overline{H}_A$. The integer $d(A)$ is the (*Krull*) *dimension* $\dim A$ of the ring A and is one of the most important invariants of a ring. Moreover, $d(A)$ is equal to the least number of elements $a_1, \ldots, a_d \in A$ for which the quotient ring $A / (a_1, \ldots, a_d)$ is Artinian (cf. **Artinian ring**). If these elements can be chosen in such a way that they generate the maximal ideal $\mathfrak{m}$, then A is called a *regular local ring*. The regularity of A is equivalent to the fact that $\dim(\mathfrak{m} / \mathfrak{m}^2) = \dim A$. For a d-dimensional regular ring A,

$$H_A(n) = \begin{bmatrix} n+d-1 \\ d-1 \end{bmatrix}$$

and $P_A(t) = (1-t)^{-d}$. Geometrically, regularity means that the corresponding point of the (analytic or algebraic) variety is non-singular.

Besides the characteristic function H_A and the dimension and multiplicity connected with it, a local ring has various invariants of a homological kind. The main one of these is the depth $\operatorname{depth} A$ (see **Depth of a module**); the condition $\operatorname{depth} A = \dim A$ distinguishes among local rings the so-called Cohen $-$ Macaulay rings (cf. **Cohen $-$ Macaulay ring**). It is not known (1989) whether there is a module M with $\operatorname{depth} M = \dim A$ for an arbitrary or a complete local ring A. Other homological invariants are the so-called *Betti numbers $b_i(A)$ of a local ring A*, that is, the dimensions of the k-spaces $\mathrm{Tor}_i^A(k, k)$, where k is the residue field of A. The question of the rationality of the Poincaré series $\sum_{n \geqslant 0} b_n(A) t^n$ is open, although for many classes of rings an affirmative answer is known. There are also invariants of an algebraic-geometrical nature; for their definition one uses resolution of the singularity corresponding to the local ring.

A similar theory has been constructed for *semi-local rings*; that is, rings that have finitely many maximal ideals. The role of a maximal ideal for them is played by the **Jacobson radical**.

References

[1] KRULL, W.: 'Dimensionstheorie in Stellenringen', *J. Reine Angew. Math.* **179** (1939), 204-226.
[2] CHEVALLEY, C.: 'On the theory of local rings', *Ann. of Math.* (2) **44** (1943), 690-708.
[3] COHEN, I.S.: 'On the structure and ideal theory of complete local rings', *Trans. Amer. Math. Soc.* **59** (1946), 54-106.
[4] SAMUEL, P.: *Algèbre locale*, Gauthier-Villars, 1953.
[5] NAGATA, M.: *Local rings*, Interscience, 1962.
[6] ZARISKI, O. and SAMUEL, P.: *Commutative algebra*, 2, Springer, 1975.
[7] SERRE, J.-P.: *Algèbre locale. Multiplicités*, Springer, 1965.
[8] BOURBAKI, N.: *Elements of mathematics. Commutative algebra*, Addison-Wesley, 1972 (translated from the French).
[9] ATIYAH, M.F. and MACDONALD, I.G.: *Introduction to commutative algebra*, Addison-Wesley, 1969.

V.I. Danilov

Editorial comments. For the notion of Krull dimension see **Dimension** of an associative ring.

A counter-example to the question of the rationality of the Poincaré series was given by D. Anick [A1].

References

[A1] ANICK, D.: 'Construction d'espaces de lacets et d'anneaux locaux à séries de Poincaré $-$ Betti non rationelles', *C.R. Acad. Soc. Paris* **290** (1980), 1729-1732. English abstract.

AMS 1980 Subject Classification: 13HXX, 14-XX

LOCAL STRUCTURAL STABILITY of a compact invariant set F of a smooth dynamical system - The preservation of all topological properties of the system in some neighbourhood of F under any sufficiently small (in the C^1 sense) perturbation of the system. More precisely, local structural stability consists in the following: There are neighbourhoods $U \supset V \supset F$ and for any $\epsilon > 0$ there is a $\delta > 0$ such that under a perturbation of the original system in U at a distance at most δ from it in the C^1-metric there is a homeomorphic imbedding $V \to U$ that shifts points by at most ϵ and takes segments of trajectories of the original system lying in V to segments of trajectories of the perturbed system. (Thus, strictly speaking, local structural stability is a property not of the set F itself but of the system considered in a neighbourhood of F.)

If F is an equilibrium position of a **flow (continuous-time dynamical system)** (or a fixed point of a *cascade*, that is, a dynamical system with discrete time), then the local structural stability of F implies the preservation of the topological properties of the system under linearization at the point F. It is almost obvious that in this case a necessary condition for local structural stability is that the eigen values of the linearized system are outside the imaginary axis (respectively, do not lie on the unit circle). The condition is also sufficient (the *Grobman—Hartman theorem*, see [1], Chapt. IX). From this it is easy to derive a necessary and sufficient condition for the local structural stability of a periodic trajectory of a flow: Only one multiplier (cf. **Multipliers**) of the variational equation lies on the unit circle. There are also results about the local structural stability of certain hyperbolic sets (see **Hyperbolic set**, and [2], [3]).

References

[1] HARTMAN, P.: *Ordinary differential equations*, Birkhäuser, 1982.
[2] ANOSOV, D.V.: 'On a class of invariant sets in smooth dynamical systems', in *Proc. Fifth Internat. Conf. Non-Linear Oscillations*, Vol. 2, Kiev, 1970, pp. 39-45 (in Russian).
[3A] ROBINSON, C.: 'Structural stability of vector fields', *Ann. of Math. (2)* **99** (1974), 154-175.
[3B] ROBINSON, C.: 'Correction to 'Structural stability of vector fields'', *Ann. of Math. (2)* **101** (1975), 368.

D.V. Anosov

Editorial comments. Instead of 'multiplier of the variational equation' one uses also '*Floquet multiplier*'.

AMS 1980 Subject Classification: 58F10

LOCAL STRUCTURE OF TRAJECTORIES *of a quadratic differential* - A description of the behaviour of the trajectories of a **quadratic differential** on an oriented **Riemann surface** in a neighbourhood of any point of this surface. Let R be an oriented Riemann surface and let $Q(z)\,dz^2$ be a quadratic differential on R; let C be the set of all zeros and simple poles of $Q(z)\,dz^2$ and let H be the set of all poles of $Q(z)\,dz^2$ of order ≥ 2. The trajectories of $Q(z)\,dz^2$ form a regular family of curves on $R \setminus (C \bigcup H)$. Under an extension of the concept of a regular family of curves this remains true on $R \setminus H$ also. The behaviour of the trajectories in neighbourhoods of points of H is significantly more complicated. A complete description of the local structure of trajectories is given below.

a) For any point $P \in R \setminus (C \bigcup H)$ there is a neighbourhood N of P on R and a homeomorphic mapping of N onto the disc $|w| < 1$ $(w = u + iv)$ such that a maximal open arc of each trajectory in N goes to a segment on which v is constant. Consequently, through each point of $R \setminus (C \bigcup H)$ there passes a trajectory of $Q(z)\,dz^2$ that is either an open arc or a Jordan curve on R.

b) For any point $P \in C$ of order μ ($\mu > 0$ if P is a zero and $\mu = -1$ if P is a simple pole) there is a neighbourhood N of P on R and a homeomorphic mapping of N onto the disc $|w| < 1$ such that a maximal arc of each trajectory in N goes to an open arc on which $\operatorname{Im} w^{(\mu+2)/2}$ is constant. There are $\mu + 2$ trajectories with ends at P and with limiting tangential directions that make equal angles $2\pi/(\mu+2)$ with each other.

c) Let $P \in H$ be a pole of order $\mu > 2$. If a certain trajectory has an end at P, then it tends to P along one of $\mu - 2$ directions making equal angles $2\pi/(\mu-2)$. There is a neighbourhood N of P on R with the following properties: 1) every trajectory that passes through some point of N in each of the directions either tends to P or leaves N; 2) there is a neighbourhood N^* of P contained in N and such that every trajectory that passes through some point of N^* tends to P in at least one direction, remaining in N^*; 3) if some trajectory lies entirely in N and therefore tends to P in both directions, then the tangent to this trajectory as P is approached in the corresponding direction tends to one of two adjacent limiting positions. The Jordan curve obtained by adjoining P to this trajectory bounds a domain D containing points of the angle formed by the two adjacent limiting tangents. The tangent to any trajectory that has points in common with D tends to these adjacent limiting positions as P is approached in the two directions. By means of a suitable branch of the function $\zeta = \int [Q(z)]^{1/2}\,dz$ the domain D is mapped onto the half-plane $\operatorname{Im} \zeta > c$ (where c is a real number); and 4) for every pair of adjacent limiting positions there is a trajectory having the properties described in 3).

d) Let $P \in H$ be a pole of order two and let z be the local parameter in terms of which P is the point $z = 0$. Suppose that $[Q(z)]^{-1/2}$ has (for some choice of the branch of the root) the following expansion in a neighbourhood of $z = 0$:

$$[Q(z)]^{-1/2} = (a + ib)z\{1 + b_1 z + b_2 z^2 + \cdots\},$$

where a and b are real constants and $b_1, b_2, \ldots$, are complex constants. The structure of the images of the trajectories of the differential $Q(z)\,dz^2$ in the z-plane is determined by which of the following three cases holds.

Case I: $a \neq 0$, $b \neq 0$. For sufficiently small $\alpha > 0$ the image of each trajectory that intersects the disc

$|z|<\alpha$ tends to $z=0$ in one direction, and leaves $|z|<\alpha$ in the other direction. Both the modulus and the argument of z vary monotonically on the image of the trajectory in $|z|<\alpha$. Each image of a trajectory twists around the point $z=0$ and behaves asymptotically like a **logarithmic spiral**.

Case II: $a\neq0$, $b=0$. For sufficiently small $\alpha>0$ the image of every trajectory that intersects the disc $|z|<\alpha$ tends to $z=0$ in one direction and leaves $|z|<\alpha$ in the other direction. The modulus of z varies monotonically on the image of the trajectory in $|z|<\alpha$. Different images of trajectories have different limiting directions at the point $z=0$.

Case III: $a=0$, $b\neq0$. For each $\epsilon>0$ there is a number $\alpha(\epsilon)>0$ such that for $0<\alpha\leqslant\alpha(\epsilon)$ the image of a trajectory that intersects the circle $|z|=\alpha$ is a Jordan curve lying in the circular annulus $\alpha(1+\epsilon)^{-1}<|z|<\alpha(1+\epsilon)$.

References

[1] JENKINS, J.A.: *Univalent functions and conformal mapping*, Springer, 1958.

G.V. Kuz'mina

Editorial comments. Cf. Quadratic differential for the notion of the trajectory of a quadratic differential.

This description is taken essentially from section 3.2 of [1]. For a detailed treatment of quadratic differentials see also [A1].

For the global structure see Global structure of trajectories.

References

[A1] STREBEL, K.: *Quadratic differentials*, Springer, 1984 (translated from the German).
[A2] GARDINER, F.P.: *Teichmüller theory and quadratic differentials*, Wiley, 1987.

AMS 1980 Subject Classification: 30CXX, 30F30

LOCAL TOPOLOGICAL GROUP - A **topological group** in which the group operations are defined only for elements sufficiently close to the identity. The introduction of local topological groups was inspired by the study of the local structure of topological groups (that is, their structure in an arbitrary small neighbourhood of the identity, see [1]). The precise definition of a local topological group is as follows.

Let G be a topological space, e an element of it, Θ and Ω open subsets of G and $G\times G$, respectively, where $e\in\Theta$, and let $i:\Theta\to G$ and $m:\Omega\to G$ be continuous mappings. Then the system (G,e,Θ,Ω,i,m) is called a local topological group if the following conditions are satisfied:

1) (e,g) and $(g,e)\in\Omega$ for any $g\in G$ and $m((e,g))=m((g,e))=g$;

2) if $g,h,t\in G$ and (g,h), (h,t), $((g,h),t)$, $(g,(h,t))\in\Omega$, then $m((m((g,h)),t))=m((g,m((h,t))))$;

3) $(g,i(g))$ and $(i(g),g)\in\Omega$ for any $g\in\Theta$ and $m((g,i(g)))=m((i(g),g))=e$.

The local topological group (G,e,Θ,Ω,i,m) is usually denoted simply by G; the element $m((g,h))$ is denoted by gh and called the *product* of g and h; the element $i(g)$ is denoted by g^{-1} and called the *inverse* of g; the element e is called the *identity* of G. If $(g,h)\in\Omega$, one says that the product of g and h is defined; if $g\in\Theta$, one says that an inverse element is defined for g.

These operations on G (which are not defined for all elements) induce the structure of a local topological group in an arbitrary neighbourhood of the identity e of G. Let G_1 and G_2 be two local topological groups. A *local homomorphism* of G_1 into G_2 is a continuous mapping f of a neighbourhood U_1 of the identity e_1 of G_1 into a neighbourhood U_2 of the identity e_2 of G_2 such that $f(e_1)=e_2$ and for any elements $g,h\in U_1$ whose product is defined in G_1 the product of the elements $f(g)$ and $f(h)$ is also defined in G_2 and $f(gh)=f(g)f(h)$. Two local homomorphisms of G_1 into G_2 are said to be *equivalent* if they coincide in a neighbourhood of the identity of G_1. Suppose that the local homomorphism f is a **homeomorphism** of the neighbourhoods U_1 and U_2 and that the inverse mapping $f^{-1}:U_2\to U_1$ is a local homomorphism of G_2 to G_1. Then f is called a *local isomorphism* of G_1 and G_2. Two local topological groups between which there is a local isomorphism are said to be *locally isomorphic*. For example, any local topological group is locally isomorphic to an arbitrary neighbourhood of the identity of it.

As an example of a local topological group one can take any topological group (and hence any neighbourhood of the identity of it). In the theory of local topological groups the main question is to what extent this example has a general character; that is, whether any local topological group is locally isomorphic to some topological group. In the general case the answer is negative (see [4]), but in the important special case of finite-dimensional local Lie groups (cf. **Lie group, local**) it is affirmative.

Just as in the theory of topological groups, in the theory of local topological groups one can define the concepts of (local) subgroups, normal subgroups, cosets, and quotient groups. For example, let (G,e,Θ,Ω,i,m) be a local topological group and let H be a subset of G containing e such that in a neighbourhood U of e in G the set $U\cap H$ is closed. Suppose also that for any $g\in H\cap\Theta$ the element $i(g)$ belongs to H and that the set

$$\Omega_H=\{(g,h)\in\Omega\cap(H\times H):m((g,h))\in H\}$$

is open in $H\times H$ (under the assumption that H is endowed with the topology induced from G). Then the system

$$(H,e,\Theta\cap H,\Omega_H,i|_H,m|_{\Omega_H})$$

is a local topological group, called a *local subgroup* of G. For the definitions of a normal subgroup, cosets with respect to a subgroup and a quotient group, see [1].

References

[1] PONTRYAGIN, L.S.: *Topological groups*, Princeton Univ. Press, 1958 (translated from the Russian).
[2] BOURBAKI, N.: *Elements of mathematics. Lie groups and Lie algebras*, Addison-Wesley, 1975 (translated from the French).
[3] SERRE, J.-P.: *Lie algebras and Lie groups*, Benjamin, 1965.
[4] LIE, S. and ENGEL, F.: *Theorie der Transformationsgruppen*, 1-3, Teubner, 1930.

V.L. Popov

AMS 1980 Subject Classification: 22-XX

LOCAL UNIFORMIZATION - For a **local ring**, this is the determination of a regular local ring birationally equivalent to it. For an irreducible algebraic variety (cf. **Irreducible variety**) V over a field k a *resolving system* is a family of irreducible projective varieties $\{V_\alpha\}$ birationally equivalent to V (that is, such that the rational function fields $k(V_\alpha)$ and $k(V)$ are isomorphic) and satisfying the following condition: For any valuation (place) v of $k(V)$ there is a variety $V' \in \{V_\alpha\}$ such that the centre P' of v on V' is a non-singular point. The existence of a resolving system (the *local uniformization theorem*) was proved for arbitrary varieties over a field of characteristic zero (see [1]), and also for two-dimensional varieties over any field and three-dimensional varieties over an algebraically closed field of characteristic other than 2, 3 or 5 (see [2]). The existence of a resolving system for V consisting of a single variety implies resolution of the singularities of V and can be obtained from the local uniformization theorem in dimension $\leqslant 3$. In the general case the local uniformization theorem implies the existence of a finite resolving system (see [3]).

References

[1] ZARISKI, O.: 'Local uniformization on algebraic varieties', *Ann. of Math.* (2) **41** (1940), 852-896.
[2] ABHANKAYAR, S.: *Resolution of singularities of embedded algebraic surfaces*, Acad. Press, 1966.
[3] HODGE, W.V.D. and PEDOE, D.: *Methods of algebraic geometry*, 3, Cambridge Univ. Press, 1954.
[4] ZARISKI, O. and SAMUEL, P.: *Commutative algebra*, 2, Springer, 1975.

V.I. Danilov

Editorial comments. The resolution of singularities for algebraic varieties of arbitrary dimension over an algebraically closed field of characteristic zero has been achieved by H. Hironaka in 1964 [A1]. Over algebraically closed fields of characteristic $p > 0$ resolution of singularities for varieties of dimension 2, and for varieties of dimension 3 provided $p > 5$, has been proved by S.S. Abhyankar [A2].

For (local) uniformization in analytic geometry and in the theory of functions of a complex variable (Riemann surfaces) cf. **Uniformization**.

References

[A1] HIRONAKA, H.: 'Resolution of singularities of an algebraic variety over a field of characteristic zero', *Ann. of Math.* 79 (1964), 109-326.
[A2] ABHYANKAR, S.S.: *Resolution of singularities of arithmetic surfaces*, Harper & Row, 1965.

AMS 1980 Subject Classification: 14B07

LOCAL UNIFORMIZING PARAMETER, *local uniformizer, local parameter* - A complex variable t defined as a continuous function $t_{p_0} = \phi_{p_0}(p)$ of a point p on a **Riemann surface** R, defined everywhere in some neighbourhood $V(p_0)$ of a point $p_0 \in R$ and realizing a homeomorphic mapping of $V(p_0)$ onto the disc $D(p_0) = \{t \in \mathbf{C}: \ |t| < r(p_0)\}$, where $\phi_{p_0}(p_0) = 0$. Here $V(p_0)$ is said to be a distinguished or *parametric neighbourhood*, $\phi_{p_0}: V(p_0) \to D(p_0)$ a distinguished or *parametric mapping*, and $D(p_0)$ a distinguished or *parametric disc*. Under a parametric mapping any point function $g(p)$, defined in a parametric neighbourhood $V(p_0)$, goes into a function of the local uniformizing parameter t, that is, $g(p) = g[\phi_{p_0}^{-1}(t)] = G(t)$. If $V(p_0)$ and $V(p_1)$ are two parametric neighbourhoods such that $V(p_0) \cap V(p_1) \neq \varnothing$, and t_{p_0} and t_{p_1} are the two corresponding local uniformizing parameters, then $t_{p_1} = \phi_{p_1}[\phi_{p_0}^{-1}(t_{p_0})]$ is a univalent holomorphic function on some subdomain of $D(p_0)$ realizing a biholomorphic mapping of this subdomain into $D(p_1)$.

If $R = R_F$ is the Riemann surface of an analytic function $w = F(z)$ and p_0 is a regular element of $F(z)$ with projection $z_0 \neq \infty$, then $t_{p_0} = z - z_0$; $t_{p_0} = 1/z$ for $z_0 = \infty$. If p_0 is a singular, or algebraic, element of $F(z)$, corresponding to the **branch point** z_0 of order $k - 1 > 0$, then $t_{p_0} = (z - z_0)^{1/k}$ for $z_0 \neq \infty$ and $t_{p_0} = 1/z^{1/k}$ for $z_0 = \infty$. In a parametric neighbourhood of an element p_0 the local uniformizing parameter t actually realizes a local **uniformization**, generally speaking, of the many-valued relation $w = F(z)$, according to the formulas (for example, for $z_0 \neq \infty$):

$$z = z_0 + t^k, \quad w = F(z_0 + t^k) = w(t), \quad k \geqslant 1.$$

In the case when R is a Riemann surface with boundary, for points p_0 belonging to the boundary of R the local uniformizing parameter $t_{p_0} = \phi_{p_0}(p)$ maps the parametric neighbourhood $V(p_0)$ onto the half-disc

$$D(p_0) = \{t \in \mathbf{C}: |t| < r(p_0), \ \mathrm{Im}\, t \geqslant 0\}.$$

If R is a a **Riemannian domain** over a complex space $\mathbf{C}^n$, $n > 1$, then the local uniformizing parameter

$$t_{p_0} = \phi_{p_0}(p) = (t_1, \ldots, t_n)_{p_0} = (\phi_1(p), \ldots, \phi_n(p))_{p_0}$$

realizes a homeomorphic mapping of the parametric neighbourhood $V(p_0)$ onto the polydisc

$$D(p_0) =$$

$$= \{t = (t_1, \ldots, t_n) \in \mathbf{C}^n : |t_1| < r_1(p_0), \ldots, |t_n| < r_n(p_0)\}.$$

If $V(p_0) \cap V(p_1)$ is not empty, then the mapping $t_{p_1} = \phi_{p_1}[\phi_{p_0}^{-1}(t_{p_0})]$ biholomorphically maps a certain subdomain of $D(p_0)$ into $D(p_1)$.

References

[1] MARKUSHEVICH, A.I.: *Theory of functions of a complex variable*, 2, Chelsea, 1977 (translated from the Russian).
[2] SPRINGER, G.: *Introduction to Riemann surfaces*, Addison-Wesley, 1957.
[3] SHABAT, B.V.: *Introduction to complex analysis*, 1-2, Moscow, 1976 (in Russian).

E.D. Solomentsev

Editorial comments.

References

[A1] FARKAS, H.M. and KRA, I.: *Riemann surfaces*, Springer, 1980.

AMS 1980 Subject Classification: 30F10, 32G15, 14H15

LOCAL VARIATIONS, METHOD OF - A **direct method** for numerically solving problems of optimal control with constraints on the phase coordinates and control functions, based on variation in the state space (see [1] - [3]).

In the method of local variations the original problem of optimal control, given as a **Lagrange problem**, is made discrete with respect to the argument t and the phase vector x. The original problem is thus replaced by the problem of minimizing the additive functional

$$I = \sum_{k=0}^{N-1} F^0(x_k, x_{k+1}, u_k, t_k, t_{k+1}) \tag{1}$$

under the constraints

$$x_{k+1} - x_k = \tau F(x_k, x_{k+1}, u_k, t_k, t_{k+1}), \tag{2}$$
$$k = 0, \ldots, N-1,$$
$$(x_k, u_k) \in G_k, \quad k = 0, \ldots, N, \tag{3}$$

where x_k and u_k are the vectors of the phase coordinates and controls at the node t_k (having dimensions n and m, respectively), G_k are given domains in the $(n+m)$-dimensional space (G_0 and G_N describe the boundary conditions), and $\tau = (T - t_0)/N$ is the step of the partition of the original interval $[t_0, T]$ for the independent variable. An essential condition for the method of local variations is that the dimensions n and m of x and u are equal, in which case the construction of an elementary operation turns out to be quite simple. An elementary operation is the determination of a control u_k that takes the system from the point (t_k, x_k) to a neighbouring point (t_{k+1}, x_{k+1}). If the dimensions of x and u are equal, and under certain additional constraints, the control u_k is defined on every interval (t_k, t_{k+1}) by the solution of the system of N equations (2), obtained as a result of a finite-difference approxi-

mation of the system of differential equations of the original variational problem.

Suppose that for the initial approximation there is specified a polygonal curve $\Gamma_0 : (x_0^0, \ldots, x_N^0)$ for which (2) and (3) are satisfied. The algorithm of the method of local variations consists in successively improving the position of the nodes through which Γ_0 passes, realized as a result of successive local variation of each j-th component of the vector x_k. On each part from t_k to t_{k+2}, $k = 0, 1, \ldots$, for fixed x_k and x_{k+2} each j-th component of x_k in turn is varied with step $h_j > 0$. If as a result of this variation the value of the functional (1) decreases (with (2) and (3) satisfied), then a similar variation is carried out with the next $(j+1)$-st component of x_k, otherwise the j-th component is varied with step $(-h_j)$. This local variation is carried out successively for all the nodes of Γ_0. As a result, at the instant of ending the iteration one obtains a new polygonal curve Γ_1 on which the functional (1) takes a value not greater than on the initial approximation Γ_0. Successive iterations are carried out similarly. If necessary one decreases the steps h_j and τ. The approximate solution of the original variational problem is determined by interpolation with respect to the values of the control found at each step.

For $m = n = 1$ the solution obtained by the method of local variations for given values of τ and h satisfies the finite-difference approximation of the **Euler equation** up to terms of order $\max(h, h/\tau, h/\tau^2)$ (see [3]).

If the variation of the original polygonal curve Γ_0 is not restricted to successive variation of its nodes, and is carried out on a more complete graph corresponding to the chosen steps τ and h_j and including Γ_0, one talks of the **travelling-tube method**.

For $m < n$, performing an elementary operation presents certain difficulties. In this case instead of the method of local variations one can use the **travelling-wave method**, which is close to it.

The method of local variations can be directly generalized to variational problems with non-additive functionals, in which the constraints have the character of isoperimetric conditions (see [3]). The method of local variations can be extended to variational problems in which the unknown functions depend on several independent variables and the corresponding functionals are given in the form of integrals over domains of different dimensions (see [2]).

References

[1] MOISEEV, N.N.: *Computational methods in the theory of optimal systems*, Moscow, 1971 (in Russian).
[2] KRYLOV, I.A. and CHERNOUS'KO, F.L.: 'Solution of problems of optimal control by the method of local variations', *USSR Comp. Math. Math. Phys.* **6**, no. 2 (1966), 12-31. (*Zh. Vychisl. Mat. i Mat. Fiz.* **6** (1966), 203-217)

[3] BANICHUK, N.V., PETROV, V.M. and CHERNOUS'KO, F.L.: 'The method of local variations for variational problems involving non-additive functionals', *USSR Comp. Math. Math. Phys.* **9**, no. 3 (1969), 66-76. (*Zh. Vychisl. Mat. i Mat. Fiz.* **9** (1969), 548-557)

I.B. Vapnyarskiĭ

AMS 1980 Subject Classification: 49AXX

LOCALE

Editorial comments. A complete Heyting algebra (see **Brouwer lattice**) regarded as a 'generalized topological space'. The name 'locale' is due to J.R. Isbell [A1], although the concept had been studied by a number of earlier writers: the basic idea is that, for any topological space X, the lattice $\mathcal{O}(X)$ of open subsets of X is complete and satisfies the *infinite distributive law*

$$U \cap \bigcup \{V_i: i \in I\} = \bigcup \{U \cap V_i: i \in I\}$$

(equivalently, it is a *Heyting algebra*), and many important topological properties of spaces (compactness, connectedness, etc.) are in fact properties of their open-set lattices. Thus one may regard any complete lattice satisfying the infinite distributive law (such a lattice is commonly called a *frame*) as if it were the open-set lattice of a space, irrespective of whether it possesses enough 'points' to be representable as an actual lattice of open subsets. A *frame homomorphism* is a mapping preserving finite meets (intersections) and arbitrary joins (unions). A locale is extensionally the same thing as a frame, but intensionally different: the difference resides in the fact that a morphism (or *continuous mapping*) of locales from X to Y is defined to be a frame homomorphism from Y to X. (To emphasize the intensional difference, some authors write $\mathcal{O}(X)$ for the frame corresponding to a locale X. Other authors — e.g. those of [A2] — use a different terminology: they redefine 'space' to mean what is called a locale above, and use 'locale' to mean a frame in the terminology above. The sense in which 'locale' is used in this article is the original one, as used by Isbell.)

A frame is representable as the open-set lattice of a space if and only if every element is expressible as a meet of prime elements; locales with this property are called *spatial*. The space corresponding to a spatial locale is not uniquely determined, but it becomes so if one requires that it be *sober*, i.e. that every prime open set should be the complement of the closure of a unique point. (Every Hausdorff space is sober, and every sober space satisfies the T_0 separation axiom.) The passage from sober spaces to locales is a full imbedding of categories; it does not preserve products in general, but it does so if one of the factors is locally compact (for a more general form of this result, see [A3]). Many familiar topological properties can be extended from the category of (sober) spaces to the category of locales. For example, one defines a locale to be *locally compact* if the corresponding frame is a **continuous lattice**; it can be shown [A4] that every locally compact locale is spatial, and in fact the category of locally compact locales is equivalent to that of locally compact sober spaces.

There is a notion of *sublocale* (or *quotient frame*) which corresponds to that of a topological subspace. Every locale can be represented as a sublocale of a spatial locale, but in general sublocales behave rather differently from subspaces: this is most evident in the fact that the intersection of any number of dense sublocales of a given locale is dense. It is this property (often in conjunction with the discrepancy between the two notions of product) which gives rise to most of the significant differences between the category of locales and that of spaces; as was first pointed out by Isbell [A1], these differences often have the effect of making the former a pleasanter category to work in than the latter. Three examples of this: the property of paracompactness for locales is inherited by arbitrary locale products [A1], the Lindelöf covering property for locales is equivalent to realcompactness [A5], and every localic subgroup of a localic group is closed [A6], [A7] (localic groups are to topological groups as locales are to spaces).

As well as localic groups, uniform locales have received some attention [A1], [A8]: a uniformity on a locale is usually defined as a family of coverings (i.e. subsets of the frame whose join is the top element) satisfying appropriate closure properties. Locales also play an important part in the constructive approach to general topology [A9], [A10]. For a general account of locale theory, see [A11].

References

[A1] ISBELL, J.R.: 'Atomless parts of spaces', *Math. Scand.* **31** (1972), 5-32.

[A2] JOYAL, A. and TIERNEY, M.: 'An extension of the Galois theory of Grothendieck', *Mem. Amer. Math. Soc.* **309** (1984).

[A3] ISBELL, J.R.: 'Product spaces in locales', *Proc. Amer. Math. Soc.* **81** (1981), 116-118.

[A4] BANASCHEWSKI, B.: 'The duality of distributive continuous lattices', *Canad. J. Math.* **32** (1980), 385-394.

[A5] MADDEN, J. and VERMEER, J.: 'Lindelöf locales and realcompactness', *Math. Proc. Cambridge Philos. Soc.* **99** (1986), 473-480.

[A6] ISBELL, J.R., KŘÍŽ, I., PULTR, A. and ROSICKÝ, J.: 'Remarks on localic groups', in F. Borceux (ed.): *Categorical Algebra and its Applications*, Lecture notes in math., Vol. 1348, Springer, 1988, pp. 154-172.

[A7] JOHNSTONE, P.T.: 'A simple proof that localic subgroups are closed', *Cahiers Top. et Géom. Diff. Catégoriques* **29** (1988), 157-161.

[A8] PULTR, A.: 'Pointless uniformities', *Comment. Math. Univ. Carolinae* **25** (1984), 91-120.

[A9] FOURMAN, M.P. and GRAYSON, R.J.: 'Formal spaces', in A.S. Troelstra and D. van Dalen (eds.): *The L.E.J. Brouwer Centenary Symposium*, Studies in Logic, Vol. 110, North-Holland, 1982, pp. 107-122.

[A10] JOHNSTONE, P.T.: 'Open locales and exponentiation', in J.W. Gray (ed.): *Mathematical Applications of Category Theory*, Contemporary Math., Vol. 30, Amer. Math. Soc., 1984, pp. 84-116.

[A11] JOHNSTONE, P.T.: *Stone spaces*, Cambridge Univ. Press, 1982.

P.T. Johnstone

AMS 1980 Subject Classification: 06D20, 54A05

LOCALITY PRINCIPLE

LOCALITY PRINCIPLE - A collective concept that combines a number of assertions related mainly to elliptic (in some cases to hypo-elliptic) equations (operators) and that follows from the pointwise character of the singularity of a fundamental solution for this class of equations. For example, an elliptic operator $L(D, x)$ with variable coefficients, written in the form

$$L(D, x) \equiv \sum_{|\alpha| \leq m} a_\alpha(x) D^\alpha, \quad x \in \mathbf{R}^n,$$

can be represented, in an appropriate sense, in a neighbourhood of a point x_0 as a sum

$$L(D, x) = \sum_{|\alpha| \leq m} a_\alpha(x_0) D^\alpha + L'(x),$$

where the first term is an operator with constant coefficients, and $L'(x)$ is 'sufficiently small' in the given neighbourhood.

References
[1] DUNFORD, N. and SCHWARTZ, J.T.: *Linear operators. Spectral theory*, 2, Interscience, 1963 .

A.A. Dezin

AMS 1980 Subject Classification: 35A25

LOCALIZATION IN A COMMUTATIVE ALGEBRA - A transition from a commutative ring A to the ring of fractions (cf. **Fractions, ring of**) $A[S^{-1}]$, where S is a subset of A. The ring $A[S^{-1}]$ can be defined as the solution of the problem of a universal mapping from A into a ring under which all elements of S become invertible. However, there are explicit constructions for $A[S^{-1}]$:

1) as the set of fractions of the form a/s, where $a \in A$ and s is a product of elements of S (two fractions a/s and a'/s' are regarded as equivalent if and only if there is an s'' that is a product of elements of S and is such that $s''(sa' - s'a) = 0$; fractions are added and multiplied by the usual rules);

2) as the quotient ring of the ring of polynomials $A[X_s]$, $s \in S$, with respect to the ideal generated by the polynomials $sX_s - 1$, $s \in S$;

3) as the **inductive limit** of an inductive system of A-modules (A_i, ϕ_{ij}), where i runs through a naturally-ordered free commutative monoid $N^{(S)}$. All the A_i are isomorphic to A, and the homomorphisms $\phi_{ij} : A_i \to A_j$ with $j = i + n_1 s_1 + \cdots + n_k s_k$ coincide with multiplication by $s_1^{n_1} \cdots s_k^{n_k} \in A$.

The ring A is canonically mapped into $A[S^{-1}]$ and converts the latter into an A-algebra. This mapping $A \to A[S^{-1}]$ is injective if and only if S does not contain any divisor of zero in A. On the other hand, if S contains a nilpotent element, then $A[S^{-1}] = 0$.

Without loss of generality the set S can be assumed to be closed with respect to products (such a set is known as *multiplicative*, or as a *multiplicative system*). In this case the ring $A[S^{-1}]$ is also denoted by $S^{-1}A$ or A_S. The most important examples of multiplicative systems are the following:

a) the set $\{s^n\}$ of all powers of an element of A;

b) the set $A \setminus \mathfrak{P}$, that is, the complement of a prime ideal $\mathfrak{P}$. The corresponding ring of fractions is local and is denoted by $A_{\mathfrak{P}}$;

c) the set R of all non-divisors of zero in A.

The ring $R^{-1}A$ is called the *complete ring of fractions* of A. If A is integral, then $R^{-1}A = A_{(0)}$ is a field of fractions.

The operation of localization carries over with no difficulty to arbitrary A-modules M if one sets

$$M[S^{-1}] = M \otimes_A A[S^{-1}].$$

The transition from M to $M[S^{-1}]$ is an exact functor. In other words, the A-module $A[S^{-1}]$ is flat. Localization commutes with direct sums and inductive limits.

From the geometrical point of view localization means transition to an open subset. More precisely, for $s \in A$ the spectrum $\operatorname{Spec} A[s^{-1}]$ is canonically identified with the open (in the **Zariski topology**) subset $D(s) \subset \operatorname{Spec} A$ consisting of the prime ideals $\mathfrak{P}$ not containing s. Moreover, this operation makes it possible to associate with each A-module M a quasi-coherent sheaf $\tilde{M}$ on the affine scheme $\operatorname{Spec} A$ for which

$$\Gamma(D(s), \tilde{M}) = M[S^{-1}].$$

Localization can be regarded as an operation that makes it possible to invert morphisms of multiplication by an $s \in S$ in the category of A-modules. In this approach the operation of localization admits a wide generalization to arbitrary categories (see **Localization in categories**).

References
[1] BOURBAKI, N.: *Elements of mathematics. Commutative algebra*, Addison-Wesley, 1972 (translated from the French).

V.I. Danilov

AMS 1980 Subject Classification: 13-XX

LOCALIZATION IN CATEGORIES - A construction associated with special radical subcategories; it first appeared in Abelian categories in the description of the so-called Grothendieck categories in terms of categories of modules over rings (cf. **Grothendieck category**). Let $\mathfrak{A}$ be an **Abelian category**. A full subcategory $\mathfrak{A}'$ of $\mathfrak{A}$ is said to be *thick* if it contains all subobjects and quotient objects of its objects and is closed with respect to extension, that is, in an exact sequence

$$0 \to A \to B \to C \to 0,$$

$B \in \operatorname{Ob} \mathfrak{A}'$ if and only if $A, C \in \operatorname{Ob} \mathfrak{A}'$. The quotient category $\mathfrak{A}/\mathfrak{A}'$ is constructed in the following way. Let $(R, \mu]$ be a subobject of the direct sum $A \oplus B(\pi_1, \pi_2)$, where π_1 and π_2 are projections, and suppose that the square

$$\begin{array}{ccc} & \pi_2\mu & \\ R & \to & B \\ \pi_1\mu \downarrow & & \downarrow \beta \\ A & \to & C \\ & \alpha & \end{array}$$

is a pushout. The subobject $(R, \mu]$ is called an $\mathfrak{A}'$-subobject if $\operatorname{Coker} \pi_1 \mu, \operatorname{Ker} \beta \in \operatorname{Ob} \mathfrak{A}'$. Two $\mathfrak{A}'$-subobjects are equivalent if they contain an $\mathfrak{A}'$-subobject. By definition, the set $H_{\mathfrak{A}/\mathfrak{A}'}(A, B)$ consists of equivalence classes of $\mathfrak{A}'$-subobjects of the direct sum $A \oplus B$. Ordinary composition of binary relations in an Abelian category is compatible with the equivalence introduced, which makes it possible to define the quotient category $\mathfrak{A}/\mathfrak{A}'$. This quotient category turns out to be an Abelian category. An exact functor $T : \mathfrak{A} \to \mathfrak{A}/\mathfrak{A}'$ can be defined by associating with each morphism $\alpha : A \to B$ its graph in $A \oplus B$. A thick subcategory $\mathfrak{A}'$ is called a *localizing subcategory* if the functor T has a full and faithful right adjoint $S : \mathfrak{A}/\mathfrak{A}' \to \mathfrak{A}$. A localizing subcategory is always the subcategory of all radical objects for some hereditary radical.

In the category of Abelian groups the subcategory of all torsion groups is a localizing subcategory. The quotient category of any category of modules with respect to a localizing subcategory is a Grothendieck category. Conversely, any Grothendieck category is equivalent to a quotient category of a suitable category of modules.

The concept of a localizing subcategory can also be defined for non-Abelian categories [3]. However, in the non-Abelian case there usually are few such subcategories. For example, in the category of associative rings there are only the two trivial localizing subcategories, namely the whole category and the full subcategory of it that contains only trivial rings.

References

[1] BUCUR, I. and DELEANU, A.: *Introduction to the theory of categories and functors*, Wiley, 1968.

[2] GABRIEL, P.: 'Des catégories abéliennes', *Bull. Soc. Math. France* **90** (1962), 323-448.

[3] SHUL'GEĬFER, E.G.: 'Localizations and strongly hereditary strict radicals in categories', *Trans. Moscow Math. Soc.* **19** (1969), 299-331. (*Trudy Moskov. Mat. Obshch.* **19** (1968), 271-301)

M.Sh. Tsalenko

Editorial comments. The term 'dense subcategory' is sometimes used in place of 'thick subcategory'; but '*dense subcategory*' has another, conflicting, meaning. The term '*Serre class*' is also used for this concept, particularly by algebraic topologists (cf. [A1]). A thick subcategory $\mathfrak{A}'$ is a localizing subcategory if and only if: 1) every object of $\mathfrak{A}$ has a largest subobject in $\mathfrak{A}'$; and 2) given an object A for which this greatest subobject is 0, there exists a monomorphism $A \to B$, where B has the property that each morphism $C \to B$ in the quotient category $\mathfrak{A}/\mathfrak{A}'$ derives from a unique morphism $C \to B$ in $\mathfrak{A}$ (see [A2]). The quotient category $\mathfrak{A}/\mathfrak{A}'$ may also be defined as a *category of fractions* (cf. [A3]), in which one formally adjoins inverses for these morphisms in $\mathfrak{A}$ which are 'isomorphisms modulo $\mathfrak{A}'$' in the sense that their kernels and cokernels both belong to $\mathfrak{A}'$. The class of all isomorphisms modulo $\mathfrak{A}'$ admits both a calculus of left fractions and a calculus of right fractions; this corresponds to the fact that the canonical functor $T : \mathfrak{A} \to \mathfrak{A}/\mathfrak{A}'$ is exact.

Localizations of module categories have been extensively used in non-commutative ring theory and in the attempts to develop a 'non-commutative algebraic geometry'; see [A4], [A5].

In the context of non-Abelian categories, a localization of a category C is generally taken to mean a functor $T : C \to D$ which is exact (i.e. preserves finite limits and colimits) and has a full and faithful right adjoint S; equivalently, the localization of C may be identified with the (full, reflective) subcategories of C which are the images of these right adjoints. Such localizations cannot be classified by localizing subcategories, as in the Abelian case, but various techniques have been developed for handling them in many particular cases of interest. For example, the '*little Giraud theorem*' classifies localizations of a functor category $[C^{op}, \operatorname{Set}]$ in terms of Grothendieck topologies on C [A6]; more generally, the localizations of an arbitrary (elementary) **topos** E are classified by Lawvere—Tierney topologies in E [A7]. (See also [A8] for a topos-theoretic analogue of the notion of Serre class.) For localizations of algebraic categories (and more generally of locally presentable categories), see [A9] and [A10]. [A11] studies the ordered set of localizations of a given category; it turns out that under reasonable hypotheses this set is a **complete lattice** satisfying an infinite distributive law.

References

[A1] SERRE, J.P.: 'Groupes d'homotopie et classes de groupes abéliens', *Ann. of Math.* **58** (1953), 258-294.

[A2] POPESCU, N.: *Abelian categories with applications to rings and modules*, Acad. Press, 1973.

[A3] GABRIEL, P. and ZISMAN, M.: *Categories of fractions and homotopy theory*, Springer, 1967.

[A4] GOLAN, J.S.: *Localization of noncommutative rings*, M. Dekker, 1975.

[A5] GOLAN, J.S.: *Torsion theories*, Longman, 1986.

[A6] ARTIN, M., GROTHENDIECK, A. and VERDIER, J.L.: *Théorie des topos (SGA 4, vol. I)*, Lecture notes in math., 269, Springer, 1972.

[A7] JOHNSTONE, P.T.: *Topos theory*, Acad. Press, 1977.

[A8] ADELMAN, M. and JOHNSTONE, P.T.: 'Serre classes for toposes', *Bull. Austral. Math. Soc.* **25** (1982), 103-115.

[A9] BORCEUX, F. and VAN DEN BOSSCHE, G.: *Algebra in a localic topos with applications to ring theory*, Lecture notes in math., 1038, Springer, 1983.

[A10] BORCEUX, F. and VEIT, B.: 'On the left exactness of orthogonal reflections', *J. Pure Appl. Alg.* **49** (1987), 33-42.

[A11] BORCEUX, F. and KELLY, G.M.: 'On locales of localizations', *J. Pure Appl. Alg.* **46** (1987), 1-34.

AMS 1980 Subject Classification: 18E35

LOCALIZATION PRINCIPLE - For any **trigonometric series** with coefficients tending to zero, the convergence or divergence of the series at some point depends on the behaviour of the so-called *Riemann function* in a neighbourhood of this point.

The Riemann function F of a given trigonometric series

$$\frac{a_0}{2} + \sum_{n=1}^{\infty} a_n \cos nx + b_n \sin nx$$

is the result of integrating it twice, that is,

$$F(x) = \frac{a_0}{4} x^2 + Cx + D - \sum_{n=1}^{\infty} \frac{a_n \cos nx + b_n \sin nx}{n^2}.$$

There is a generalization of the localization principle for series with coefficients that do not tend to zero (see [2]).

References

[1] BARY, N.K. [N.K. BARI]: *A treatise on trigonometric series*, Pergamon, 1964 (translated from the Russian).
[2] ZYGMUND, A.: *Trigonometric series*, 1, Cambridge Univ. Press, 1988.

M.I. Voĭtsekhovskiĭ

AMS 1980 Subject Classification: 42A20, 42A63

LOCALLY COMPACT SKEW-FIELD - A set K endowed with both the algebraic structure of a **skew-field** and a locally compact topology (cf. **Locally compact space**). It is required that the algebraic operations, that is, addition, multiplication and transitions to negative and inverse elements (the latter is defined only on the set of non-zero elements $K^* = K \setminus 0$) are continuous in the given topology. Since any skew-field is locally compact with respect to the discrete topology, it is assumed that the topology of K is not discrete.

The study of locally compact skew-fields is based on the existence of a **Haar measure** on the locally compact group K_+ (the additive group of the skew-field). Let μ be a Haar measure on K_+ and let $S \subset K$ be a compact set in K of positive measure. Then the formula

$$\mathrm{mod}_K(a) = \frac{\mu(aS)}{\mu(S)}$$

defines a homomorphism (the modulus) of the multiplicative group K^* into the multiplicative group $\mathbf{R}^*_+$ of positive real numbers. By definition one puts $\mathrm{mod}_K(0) = 0$.

The 'modulus' function satisfies the inequality

$$\mathrm{mod}_K(a+b) \leqslant A \sup(\mathrm{mod}_K(a), \mathrm{mod}_K(b))$$

with some constant $A > 0$. If this inequality holds for $A = 1$, then K is said to be *non-Archimedean*, or *ultrametric*. Otherwise K is called an *Archimedean skew-field*. A skew-field K is Archimedean if and only if it is connected. Any Archimedean skew-field is isomorphic to either the field of real numbers, the field of complex numbers or the skew-field of quaternions.

An ultrametric skew-field K is totally disconnected (cf. **Totally-disconnected space**). The 'modulus' function determines a non-Archimedean metric on K. Any such skew-field is an extension of finite degree of either the field $\mathbf{Q}_p$ of rational p-adic numbers for some prime number p (in the case when K has characteristic 0) or the field $\mathbf{F}_p((X))$ of **formal power series** over the field $\mathbf{F}_p$ of p elements (in the case when K has characteristic p). The field $\mathbf{Q}_p$ (respectively, the field $\mathbf{F}_p((X))$) lies in the centre of K. In each of these cases K is called a *p-skew-field*, or a *p-field*.

An ultrametric skew-field K contains a unique maximal subring R, defined by the condition

$$R = \{a \in K : \mathrm{mod}_K(a) \leqslant 1\}.$$

This ring is local (cf. **Local ring**). Its maximal ideal P is defined by the condition

$$P = \{a \in R : \mathrm{mod}_K(a) < 1\},$$

and all elements with modulus 1 are invertible in R. P is a **principal ideal**, and the residue field R/P is a finite field of characteristic p.

In the case when the p-skew-field is not commutative, it has dimension n^2 over its centre K_0 and ramification index n over K_0. Also, there is an intermediate field K_1 such that $K \supset K_1 \supset K_0$, where K_1 is an unramified extension of K_0 of degree n, and all automorphisms of K_1 over K_0 are induced by inner automorphisms of K.

References

[1] BOURBAKI, N.: *Elements of mathematics. Commutative algebra*, Addison-Wesley, 1972 (translated from the French).
[2] WEIL, A.: *Basic number theory*, Springer, 1974.
[3] WAERDEN, B.L. VAN DER: *Algebra*, 1-2, Springer, 1967-1971 (translated from the German).
[4] CASSELS, J.W.S. and FRÖLICH, A. (EDS.): *Algebraic number theory*, Acad. Press, 1986.
[5] PONTRYAGIN, L.S.: *Topological groups*, Princeton Univ. Press, 1958 (translated from the Russian).

L.V. Kuz'min

AMS 1980 Subject Classification: 12JXX, 16A80

LOCALLY COMPACT SPACE - A topological space at every point of which there is a neighbourhood with compact closure. A locally compact Hausdorff space X is a **completely-regular space**. The partially ordered set of all its Hausdorff compactifications (cf. **Compactification**) is a complete lattice. Its minimal element is the **Aleksandrov compactification** αX. The class of locally compact Hausdorff spaces coincides with the class of open subsets of Hausdorff compacta. For a locally compact Hausdorff space X its remainder $bX \setminus X$ in any Hausdorff compactification bX is a Hausdorff compactum. Every connected paracompact locally compact space is the sum of countably many compact subsets.

The most important example of a locally compact space is n-dimensional Euclidean space. A topological Hausdorff **vector space** E (not reducing to the zero element) over a complete non-discretely normed division ring k is locally compact if and only if k is locally compact and E is finite-dimensional over k.

V.V. Fedorchuk

Editorial comments. A product $\prod_\alpha X_\alpha$ of topological spaces is locally compact if and only if each separate coordinate space X_α is locally compact and all but finitely many are compact.

References

[A1] KELLEY, J.L.: *General topology*, v. Nostrand, 1955, pp. 146-147.

AMS 1980 Subject Classification: 54D45

LOCALLY CONNECTED CONTINUUM - A continuum that is a **locally connected space**. Examples of locally connected continua are the n-dimensional cube, $n = 0, 1, \ldots$, the **Hilbert cube**, and all Tikhonov cubes (cf. **Tikhonov cube**). The union of the graph of the function

$$y = \sin\frac{1}{x}, \quad 0 < x \leqslant 1,$$

and the interval $I = \{(0, y): -1 \leqslant y \leqslant 1\}$ gives an example of a continuum that is not locally connected (at the points of I). A metrizable continuum is locally connected if and only if it is a curve in the sense of Jordan (cf. **Line (curve)**). Any metrizable locally connected continuum is path-connected (cf. **Path-connected space**). Moreover, any two distinct points of such a continuum K are contained in a simple arc lying in K.

B.A. Pasynkov

AMS 1980 Subject Classification: 54F15

LOCALLY CONNECTED SPACE - A topological space X such that for any point x and any neighbourhood O_x of it there is a smaller connected neighbourhood U_x of x. Any open subset of a locally connected space is locally connected. Any connected component of a locally connected space is open-and-closed. A space X is locally connected if and only if for any family $\{A_\iota\}$ of subsets of X,

$$\partial \bigcup_\iota A_\iota \subset \overline{\bigcup_\iota \partial A_\iota}$$

(here ∂B is the boundary of B and $\overline{B}$ is the closure of B). Any **locally path-connected space** is locally connected. A partial converse of this assertion is the following: Any complete metric locally connected space is locally path-connected (the *Mazurkiewicz — Moore — Menger theorem*).

S.A. Bogatyĭ

Editorial comments.

References

[A1] KELLEY, G.L.: *General topology*, v. Nostrand, 1955, p. 61.
[A2] ČECH, E.: *Topological spaces*, Interscience, 1966, § 21B.

AMS 1980 Subject Classification: 54D05

LOCALLY CONVEX LATTICE - A real **topological vector space** E that is simultaneously a **vector lattice** and whose topology is a **locally convex topology**, while the mappings of $E \times E$ into E defined by

$$(x, y) \to \sup(x, y), \quad (x, y) \to \inf(x, y), \quad x, y \in E,$$

are continuous. General questions in the theory of locally convex lattices are the following: The study of the connections between topological properties and order properties; in particular, the topological properties of bands and positive cones in a locally convex lattice and connections between lattice properties and topological properties of completeness in a locally convex lattice. The study of properties of the strong dual of a locally convex lattice and properties of the imbedding of a locally convex lattice E into its second dual. The construction of a theory of extension of positive functionals and linear mappings between locally convex lattices.

The most important example of a locally convex lattice is a **Banach lattice**.

References

[1] KANTOROVICH, L.V. and AKILOV, G.P.: *Functional analysis*, Pergamon, 1982 (translated from the Russian).
[2] DAY, M.M.: *Normed linear spaces*, Springer, 1958.
[3] SCHAEFER, H.H.: *Topological vector spaces*, Macmillan, 1966.

A.I. Shtern

Editorial comments.

References

[A1] LUXEMBURG, W.A.J. and ZAANEN, A.C.: *Riesz spaces*, I, North-Holland, 1971.
[A2] ZAANEN, A.C.: *Riesz space*, II, North-Holland, 1983.
[A3] SCHAEFER, H.H.: *Banach lattices and positive operators*, Springer, 1974.

AMS 1980 Subject Classification: 46A40, 46B30

LOCALLY CONVEX SPACE - A Hausdorff **topological vector space** over the field of real or complex numbers in which any neighbourhood of the zero element contains a convex neighbourhood of the zero element; in other words, a topological vector space E is a locally convex space if and only if the topology of E is a Hausdorff **locally convex topology**. Examples of locally convex spaces (and at the same time classes of locally convex spaces that are important in the theory and applications) are normed spaces, countably-normed spaces and Fréchet spaces (cf. **Normed space; Countably-normed space; Fréchet space**).

A number of general properties of locally convex spaces follows immediately from the corresponding properties of locally convex topologies; in particular, subspaces and Hausdorff quotient spaces of a locally convex space, and also products of families of locally convex spaces, are themselves locally convex spaces. Let A be an upward directed set of indices and $\{E_\alpha: \alpha \in A\}$ a family of locally convex spaces (over the same field) with topologies $\{\tau_\alpha: \alpha \in A\}$; suppose that for any pair (α, β), $\alpha \leqslant \beta$, $\alpha, \beta \in A$, there is defined a continuous linear mapping $g_{\alpha\beta}: E_\beta \to E_\alpha$; let E be the subspace of the product $\prod_{\alpha \in A} E_\alpha$ whose elements $x = (x_\alpha)$ satisfy the relations $x_\alpha = g_{\alpha\beta}(x_\beta)$ for all $\alpha \leqslant \beta$; the space E is called the *projective limit* of the family $\{E_\alpha\}$ with respect to $\{g_{\alpha\beta}\}$ and is denoted by $\lim g_{\alpha\beta} E_\beta$ or $\lim_{\leftarrow} E_\alpha$; the topology of E is the projective topology with respect to the family $\{E_\alpha, \tau_\alpha, f_\alpha\}$, where f_α is the restriction to the subspace E of the projection $(\prod_{\beta \in A} E_\beta) \to E_\alpha$. On the other hand, suppose that for any pair (α, β), $\alpha \leqslant \beta$, $\alpha, \beta \in A$, there is defined a continuous linear mapping $h_{\alpha\beta}: E_\alpha \to E_\beta$; let g_α, $\alpha \in A$, be the canonical imbedding of E_α in the direct sum $\oplus_{\alpha \in A} E_\alpha$ and let H be the subspace of $\oplus_{\alpha \in A} E_\alpha$ generated by the images of all spaces E_α under the mappings $g_\alpha - g_\beta \circ h_{\alpha\beta}$, where (α, β) runs through all pairs in $A \times A$ for which $\alpha \leqslant \beta$. If H is closed in $\oplus_{\alpha \in A} E_\alpha$, then the locally convex space $(\oplus_{\alpha \in A} E_\alpha)/H$ is called the *inductive limit* of the family $\{E_\alpha\}$ with respect to $\{h_{\alpha\beta}\}$, and is denoted by $\lim h_{\alpha\beta} E_\alpha$ or $\lim_\rightarrow E_\alpha$. If $\{E_\alpha\}$ is a family of subspaces of a vector space E, ordered by inclusion, and the topology τ_β induces τ_α on E_α for $\alpha \leqslant \beta$, then the inductive limit of the family $\{E_\alpha\}$ is said to be *strict*. A locally convex space is metrizable if and only if its topology is induced by a sequence of semi-norms (cf. **Semi-norm**); a locally convex space is normable if and only if it contains a bounded open set (*Kolmogorov's theorem*). Any finite-dimensional subspace of a locally convex space has a complemented closed subspace. The completion of a locally convex space is a locally convex space, and any complete locally convex space is isomorphic to the projective limit of a family of Banach spaces. The space $L(F, E)$ of continuous linear mappings from a topological vector space F into a locally convex space E is naturally endowed with the structure of a locally convex space (see also **Operator topology**) with respect to a given family γ of bounded subsets of F for which the linear hull of its union is dense in F. A basis of neighbourhoods of zero of the corresponding topology is the family of sets $\{f: f \in L(F, E), f(S) \subset V\}$, where S runs through γ and V runs through a basis of neighbourhoods of zero in E.

A central topic in the theory of locally convex spaces (and also in the theory of topological vector spaces) is the study of the relation of the space with its dual or **adjoint space**. The foundation of this theory of **duality** for locally convex spaces is the **Hahn $-$ Banach theorem**, which implies, in particular, that if E is a locally convex space, then its dual space E' separates the points of E.

An essential part of the theory of locally convex spaces is the theory of compact convex sets in a locally convex space. The convex hull $\operatorname{co} K$ and the convex balanced hull (cf. also **Balanced set**) of a pre-compact set K in a locally convex space E are pre-compact; if E is also quasi-complete, then the closed convex hull $\overline{\operatorname{co}} K$ of K and its closed convex balanced hull are compact. If A and K are disjoint non-empty convex subsets of a locally convex space E, where A is closed and K is compact, then there is a continuous real **linear functional** f on E such that for some real number α the inequalities $f(x) > \alpha$, $f(y) < \alpha$ hold for all $x \in A$, $y \in K$, respectively. In particular, a non-empty closed convex set A in a locally convex space is the intersection of all closed half-spaces containing it. A non-empty closed convex subset B of a closed convex set A is called a *face* (or *extremal subset*) of A if any closed segment in A with an interior point in B lies entirely in B; a point $x \in A$ is called an *extreme point* of A if the set $\{x\}$ is a face of A. If K is a compact convex set in a locally convex space E and $\partial_e K$ is the set of its extreme points, then the following conditions are equivalent for a set $X \subset K$: 1) $\overline{\operatorname{co}} X = K$; 2) $\overline{X} \supset \partial_e K$; and 3) $\sup f(X) = \sup f(K)$ for any continuous real linear functional f on E. In particular, $\overline{\operatorname{co}} \partial_e K = K$ (the *Kreĭn $-$ Mil'man theorem*). The set $\partial_e K$ is a **Baire space** in the induced topology (that is, the intersection of any sequence of open subsets of $\partial_e K$ that are dense in $\partial_e K$ is dense in ∂K), and for any $x \in K$ there is a probability measure μ on K such that $x = \int_K d\mu(y)$ and the measure μ vanishes on all Baire subsets $X \subset K$ that do not intersect $\partial_e K$ (if K is metrizable, then $\mu(\partial_e K) = 1$) (*Choquet's theorem*). Any continuous mapping of a compact convex set K in a locally convex space into itself has a fixed point (the *Schauder $-$ Tikhonov theorem*); a commuting family of continuous affine transformations of K onto itself (and an equicontinuous group of continuous affine transformations of K onto itself) has a fixed point (the *Markov $-$ Kakutani theorem*).

One quite important branch of the theory of locally convex spaces is the theory of linear operators on a locally convex space; in particular, the theory of compact (also called completely-continuous), nuclear and Fredholm operators (cf. **Compact operator**; **Fredholm operator**; **Nuclear operator**). Closed-graph and open-mapping theorems have far-reaching generalizations in the theory of locally convex spaces. A locally convex space E is said to have the *approximation property* if the identity mapping of E into itself can be uniformly approximated on pre-compact sets of E by finite-rank continuous linear mappings of E into itself. If a locally convex space has the approximation property, then it has a number of other remarkable properties. In particular, in such a space any nuclear operator has a uniquely defined **trace**. There are separable Banach spaces that do not have the approximation property, but Banach spaces with a Schauder basis and subspaces of projective limits of Hilbert spaces do have the approximation property. Some versions of this property are of interest in the theory of completely-continuous and Fredholm operators.

A notable role in the theory of locally convex spaces is played by methods of homological algebra connected with the study of the category of locally convex spaces and their continuous mappings, and also some subcategories of this category. In particular, homological methods have made it possible to solve a number of problems connected with the extension of linear mappings and with the existence of a linear mapping into a given space that lifts a mapping into a quotient space of this space, and also to study properties of completions of quotient spaces E/F in relation to the completions of the spaces E and F.

Other important questions in the theory of locally convex spaces are: the theory of integration of vector-valued functions with values in a locally convex space (as a rule, a barrelled space); the theory of differentiation of non-linear mappings between locally convex spaces; the theory of topological tensor products of locally convex spaces and the theory of Fredholm operators and nuclear operators. There is a detailed theory of a number of special classes of locally convex spaces, such as a barrelled spaces (cf. **Barrelled space**), bornological spaces (on which any semi-norm that is bounded on bounded sets is continuous), reflexive and semi-reflexive spaces (the canonical mapping of which into the strong second dual is a topological or linear isomorphism, respectively), nuclear spaces (cf. **Nuclear space**), etc.

References

[1] KOLMOGOROV, A.N.: 'Zur Normierbarkeit eines allgemeinen topologischen linearen Raumes', *Studia Math.* **5** (1935), 29-33.
[2] BOURBAKI, N.: *Elements of mathematics. Topological vector spaces*, Addison-Wesley, 1977 (translated from the French).
[3] BOURBAKI, N.: *Elements of mathematics. Integration*, Addison-Wesley, 1975, Chapt. 6; 7; 8 (translated from the French).
[4] SCHAEFER, H.H.: *Topological vector spaces*, Macmillan, 1966.
[5] DAY, M.M.: *Normed linear spaces*, Springer, 1973.
[6] DUNFORD, N. and SCHWARTZ, J.T.: *Linear operators. General theory*, 1, Interscience, 1958.
[7] PIETSCH, A.: *Nuclear locally convex spaces*, Springer, 1972 (translated from the German).
[8] ROBERTSON, A.P. and ROBERTSON, W.J.: *Topological vector spaces*, Cambridge Univ. Press, 1973.
[9] PHELPS, R.R.: *Lectures on Choquet's theorem*, v. Nostrand, 1966.
[10] ENFLO, P.: 'A counterexample to the approximation problem in Banach spaces', *Acta. Math.* **130** (1973), 309-317.
[11] FRÖLICHER, A. and BUCHER, W.: *Calculus in vector spaces without norm*, Lecture notes in math., 30, Springer, 1966.
[12] PALMODOV, V.P.: 'Homological methods in the theory of locally convex spaces', *Russian Math. Surveys* **26**, no. 1 (1971), 1-64. (*Uspekhi Mat. Nauk* **26**, no. 1 (1971), 3-65)

A.I. Shtern

Editorial comments. Locally convex spaces arise in great profusion throughout such fields of analysis as measure and integration theory, complex analysis in one, several or an infinite number of variables, partial differential equations, integral equations, approximation theory, operator and spectral theory, as well as probability theory. Many sequence spaces, spaces of holomorphic, continuous or measurable functions, spaces of measures, test functions and distributions have a natural locally convex topology.

The powerful duality theory of locally convex spaces provides an important tool to translate a problem on the space (or on linear operators between locally convex spaces) into one concerning the linear forms. The fundamental results of duality theory include the *bipolar theorem* (a form of the Hahn−Banach theorem), the *Alaoğlu−Bourbaki theorem* (on equicontinuous sets in the dual) and the *Mackey−Arens theorem* (characterizing the topologies which are compatible with a given dual pair). By means of duality theory, the surjectivity of linear operators and the existence of continuous linear right inverses (leading to solution operators for partial differential equations) can be studied; V.P. Palamodov developed the homological methods with such applications in mind. There is an abstract duality between topology and bornology, and equicontinuous sets provide an important example of compactology.

Part of the classical structure theory of locally convex spaces can be viewed as a generalization of the (basic) theory of Banach spaces (cf. **Banach space**) and of its main theorems (which are often consequences of the Hahn−Banach theorem and of the Baire category theorem, cf. **Baire theorem**). This development led to the introduction of special classes of locally convex spaces, of which the most important ones are: Fréchet and (DF)-spaces, barrelled and bornological spaces, reflexive spaces, (LF)-spaces (i.e., countable inductive limits of Fréchet spaces), nuclear, Schwartz and Montel spaces.

Topological tensor products were introduced as a tool to study spaces of operators and spaces of vector-valued functions and distributions. A. Grothendieck [A4] investigated the nuclear spaces in this context and raised the approximation problem which was solved by P. Enflo [10] by giving the first example of a Banach space without the approximation property. Later on, A. Szankowski proved that the space of all bounded linear operators on a Hilbert space does not have the approximation property.

Besides compact convex sets (Choquet's theory has important applications in abstract potential theory), also weakly compact sets are studied (cf. [A3]).

References [A5] - [A8] are general monographs on locally convex spaces and duality theory. [A1], [A9] and [A10] are devoted to more specialized topics, while [A2] is a monograph on infinite-dimensional holomorphy and its connections to the theory of locally convex spaces.

References

[A1] DEWILDE, M.: *Closed graph theorems and webbed spaces*, Pitman, 1978.
[A2] DINEEN, S.: *Complex analysis in locally convex spaces*, North-Holland, 1981.
[A3] FLORET, K.: *Weakly compact sets*, Lecture notes in math., 801, Springer.
[A4] GROTHENDIECK, A.: *Produits tensoriels topologiques et espaces nucléaires*, Amer. Math. Soc., 1955.
[A5] GROTHENDIECK, A.: *Topological vector spaces*, Gordon & Breach, 1973 (translated from the French).
[A6] HORVÁTH, J.: *Topological vector spaces and distributions*, I, Addison-Wesley, 1966.
[A7] JARCHOW, H.: *Locally convex spaces*, Teubner, 1981 (translated from the German).
[A8] KÖTHE, G.: *Topological vector spaces*, 1-2, Springer, 1969-1979.
[A9] PÉREZ CARRERAS, P. and BONET, J.: *Barrelled locally convex spaces*, North-Holland, 1987.
[A10] SCHMETS, J.: *Espaces de fonctions continues*, Lecture notes in math., 519, Springer, 1976.
[A11] CHOQUET, G.: *Lectures on analysis*, 1-3, Benjamin, 1969.

AMS 1980 Subject Classification: 46AXX, 46E10, 46E40, 46F05, 46G20, 46M05, 46M20, 46M40

LOCALLY CONVEX TOPOLOGY - A (not necessarily Hausdorff) topology τ on a real or complex **topological vector space** E that has a basis consisting of convex sets and is such that the linear operations in E are continuous with respect to τ. A locally convex topology τ on a vector space E is defined analytically by a family of semi-norms (cf. **Semi-norm**) $\{p_\alpha: \alpha \in A\}$ as the topology with basis of neighbourhoods of zero consisting of the sets of the form $\{n^{-1}U\}$, where n runs through the natural numbers and U is the family of all finite intersections of the sets of the form $\{x \in E: p_\alpha(x) \leqslant 1\}$, $\alpha \in A$; such a family of semi-norms is said to *be a generator for* τ or to *generate* τ. The topology induced by a given locally convex topology on a vector subspace, the quotient topology on a quotient space and the topology of a product of locally convex topologies, are also locally convex topologies. A topology τ on a topological vector space E is a locally convex topology if and only if τ is the topology of **uniform convergence** on equicontinuous subsets of the **adjoint space** E^*.

Let E and E_α, $\alpha \in A$, be vector spaces over **R** or **C**, let f_α (respectively g_α) be a linear mapping of E into E_α (respectively, of E_α into E) and let τ_α be a locally convex topology on E_α, $\alpha \in A$. The weakest topology on E for which all f_α are continuous mappings of E into (E_α, τ_α) is called the *projective topology* on E with respect to the family $\{(E_\alpha, \tau_\alpha, f_\alpha): \alpha \in A\}$. The projective topology is a locally convex topology. In particular, the least upper bound of a family of locally convex topologies on a given vector space, the induced topology on a subspace and the topology of a product of locally convex topologies are projective topologies (and therefore locally convex topologies). The strongest locally convex topology on E with respect to which all g_α, $\alpha \in A$, are continuous mappings of (E_α, τ_α) into E is called the *inductive topology* on E with respect to the family $\{(E_\alpha, \tau_\alpha, g_\alpha): \alpha \in A\}$. In particular, the quotient topology of a given locally convex topology and the topology of a direct sum of locally convex topologies are inductive topologies (and therefore locally convex topologies). The concepts of projective and inductive locally convex topologies make it possible to define the operations of projective and inductive limits in the category of locally convex spaces and their linear mappings.

References

[1] BOURBAKI, N.: *Elements of mathematics. Topological vector spaces*, Addison-Wesley, 1977 (translated from the French).
[2] SCHAEFER, H.H.: *Topological vector spaces*, Macmillan, 1966.
[3] KANTOROVICH, L.V. and AKILOV, G.P.: *Functional analysis*, Pergamon, 1982 (translated from the Russian).

A.I. Shtern

Editorial comments.

References

[A1] KÖTHE, G.: *Topological vector spaces*, I, Springer, 1969.
[A2] JARCHOW, H.: *Locally convex spaces*, Teubner, 1981.

AMS 1980 Subject Classification: 46A05

LOCALLY FINITE ALGEBRA - An **algebra** in which every subalgebra with finitely many generators has finite dimension over the ground field.

It is convenient to represent a locally finite algebra as the union of an increasing chain of finite-dimensional subalgebras. The class of locally finite algebras is closed with respect to taking homomorphic images and transition to subalgebras. If one restricts oneself to the consideration of associative algebras (cf. **Associative rings and algebras**), then the extension of a locally finite algebra by a locally finite algebra is again a locally finite algebra. Therefore, in any algebra the sum of the locally finite ideals is the largest locally finite ideal containing all locally finite ideals, and is called the *locally finite radical*.

In the associative case any locally finite algebra is algebraic (cf. **Algebraic algebra**). The converse is false (see [6]). Nevertheless, an algebraic algebra that satisfies a polynomial identity is locally finite. It is not known (1989) whether an algebraic division algebra is locally finite. There is a conjecture that a finitely-defined algebraic algebra is finite-dimensional. The Jacobson radical of a locally finite associative algebra coincides with the upper nil radical. The Jacobson radical of a locally finite Jordan algebra is also a nil ideal. Every alternative or special Jordan algebraic algebra (cf. **Jordan algebra**; **Algebraic algebra**) of bounded index (the degrees of minimal annihilating polynomials of all elements are uniformly bounded) over a field of characteristic $\neq 2$ is locally finite. A solvable algebraic Lie algebra (the inner derivations of all elements are algebraic) is locally finite. An algebraic Lie algebra (cf. **Lie algebra, algebraic**) of bounded index is locally finite.

References

[1] JACOBSON, N.: *Structure of rings*, Amer. Math. Soc., 1956.
[2] HERSTEIN, I.: *Noncommutative rings*, Math. Assoc. Amer., 1968.
[3] SHIRSHOV, A.I.: 'On certain non-associative nil rings and algebraic algebras', *Mat. Sb.* **41** (1957), 381-394 (in Russian).
[4] McCRIMMON, K.: 'The radical of a Jordan algebra', *Proc. Nat. Acad. Sci. USA* **62** (1969), 671-678.
[5] LYU, S.S.: 'On the splitting of locally finite algebras', *Mat. Sb.* **39** (1956), 385-396 (in Russian).
[6] GOLOD, E.S.: 'On nil algebras and residually finite p-groups', *Izv. Akad. Nauk SSSR Ser. Mat.* **28** (1964), 273-276 (in Russian).

V.N. Latyshev

AMS 1980 Subject Classification: 16A46, 16A38, 17A99, 17B30, 17B65, 17C65

LOCALLY FINITE COVERING - A covering (cf. **Covering (of a set)**) of a topological space by subsets of it such that every point has a neighbourhood that intersects only finitely many elements of this covering. One cannot select a locally finite covering from every open covering of a straight line: it is sufficient to consider a monotone sequence of intervals that increase in length without limit. It turns out that the possibility of selecting a locally finite covering from any open covering of a space is equivalent to **compactness** of the space. The idea of local finiteness in conjunction with the concept of refinement carries an essentially new meaning. *A.H. Stone's theorem* asserts that any open covering of an arbitrary **metric space** can be refined to a locally finite covering. Hausdorff spaces that have the latter property are said to be *paracompact* (cf. **Paracompact space**). Locally finite coverings are important not only because of their participation in the definition of paracompactness. The requirement of local finiteness plays an essential role in constructions belonging to dimension theory and in the statements and proofs of addition theorems of various kinds. The existence in a **regular space** of a base that splits into a union of a countable family of locally finite open coverings is equivalent to the metrizability of this space. Open locally finite coverings of a **normal space** serve as a construction of a partition of unity on this space, subordinate to this covering. By means of partitions of unity it has been possible to construct, in particular, standard mappings of manifolds into Euclidean spaces. The requirement of local finiteness of a covering is not necessarily connected with the assumption that it is open. Local finiteness of a covering of a space automatically implies that in this covering there are 'sufficiently many' sets that are close in their properties to open sets. If any open covering of a regular space can be refined to a locally finite covering, that space is paracompact. Locally finite families of sets in a space, defined similarly but not necessarily covering the space, have also been considered. A special case of them are *discrete families of sets*: families of sets such that each point in the whole space has a neighbourhood that intersects at most one element of this family. Discrete families are important in connection with the study of separation in a space. Thus, collectively-normal spaces are distinguished by the requirement that any discrete family of sets is separated by a discrete family of neighbourhoods. This condition is directly connected with the problem of the combinatorial extension of locally finite families of sets to locally finite families of open sets.

References

[1] ENGELKING, R.: *General topology*, PWN, 1977.
[2] ARKHANGEL'SKIĬ, A.V. and PONOMAREV, V.I.: *Fundamentals of general topology: problems and exercises*, Reidel, 1984 (translated from the Russian).
[3] ALEXANDROFF, P. [P.S. ALEKSANDROV]: 'Sur les ensembles de la première classe et les espaces abstraits', *C.R. Acad. Sci. Paris* **178** (1924), 185-187.
[4] STONE, A.H.: 'Paracompactness and product spaces', *Bull. Amer. Math. Soc.* **54** (1948), 977-982.
[5] MICHAEL, E.A.: 'A note on paracompact spaces', *Proc. Amer. Math. Soc.* **4** (1953), 831-838.

A.V. Arkhangel'skiĭ

Editorial comments. A *partition of unity* on a space X is a family of continuous functions $\{f_i\}_i$ from X to $[0, 1]$ such that $\sum_i f_i(x) = 1$ for all $x \in X$. It is said to be *subordinate* to a covering $\mathcal{U}$ if the open covering $\{f_i^{-1}[(0, 1]]\}_i$ refines $\mathcal{U}$.

A (locally finite) family $\{U_i\}_i$ of (open) sets is a *combinatorial extension* of a (locally finite) family $\{F_i\}_i$ of sets if $F_i \subseteq U_i$ for all i and for every set I of indices $\bigcap_{i \in I} F_i = \varnothing$ implies $\bigcap_{i \in I} U_i = \varnothing$.

References

[A1] BURKE, D.K.: 'Covering properties', in K. Kunen and J.E. Vaughan (eds.): *Handbook of Set-Theoretic Topology*, North-Holland, 1984, Chapt. 9; pp. 347-422.

AMS 1980 Subject Classification: 54D18

LOCALLY FINITE FAMILY *of sets in a topological space* - A family F of sets such that every point of the space has a neighbourhood that intersects only finitely many elements of F. Locally finite families of open sets and locally finite open coverings are important. Thus, a **regular space** is metrizable if and only if has a base that splits into countably many locally finite families. Any open covering of a **metric space** can be refined to a locally finite open covering. Spaces that have this property are called *paracompact* (cf. **Paracompact space**).

A.V. Arkhangel'skiĭ

Editorial comments. See also **Locally finite covering**.

AMS 1980 Subject Classification: 54D18

LOCALLY FINITE GROUP - A group in which every finitely-generated subgroup is finite. Any locally finite group is a torsion group (cf. **Periodic group**), but not conversely (see **Burnside problem**). An extension of a locally finite group by a locally finite group is again a locally finite group. Every locally finite group with the minimum condition for subgroups (and even for Abelian subgroups) has an Abelian subgroup of finite index [3] (see **Group with a finiteness condition**). A locally finite group whose Abelian subgroups have finite rank (cf. **Rank of a group**) has itself finite rank and contains a locally solvable subgroup (cf. **Locally solvable group**) of finite index.

References

[1] KUROSH, A.G.: *The theory of groups*, 1-2, Chelsea, 1955-1956 (translated from the Russian).
[2] CHERNIKOV, S.N.: 'Finiteness conditions in general group theory', *Uspekhi Mat. Nauk* **14**, no. 5 (1959), 45-96 (in Russian).
[3] SHUNKOV, V.P.: 'On locally finite groups with a minimality condition for Abelian subgroups', *Algebra and Logic* **9**, no. 5 (1970), 350-370. (*Algebra i Logika* **9**, no. 5 (1970), 579-615)
[4] SHUNKOV, V.P.: 'On locally finite groups of finite rank ', *Alge-

bra and Logic **10**, no. 2 (1971), 127-142. (*Algebra i Logika* **10**, no. 2 (1971), 199-225)

A.L. Shmel'kin

AMS 1980 Subject Classification: 20E25

LOCALLY FINITE SEMI-GROUP - A semi-group in which every finitely-generated sub-semi-group is finite. A locally finite semi-group is a **periodic semi-group** (torsion semi-group). The converse is false: There are even torsion groups that are not locally finite (see **Burnside problem**). Long before the solution of the Burnside problem for groups, examples had been constructed of semi-groups that are torsion but not locally finite in classes of semi-groups remote from groups; above all, in the class of nil semi-groups (cf. **Nil semi-group**). These are, for example, a free semi-group with two generators in the variety given by $x^3 = 0$, and a free semi-group with three generators in the variety given by $x^2 = 0$. Moreover, for a number of classes of semi-groups the conditions of periodicity and local finiteness are equivalent. A trivial example is given by commutative semi-groups. A band of locally finite semi-groups (see **Band of semi-groups**) is itself a locally finite semi-group [1]; moreover, a semi-group that has a decomposition into locally finite groups is a locally finite semi-group; in particular, a semi-group of idempotents (cf. **Idempotents, semi-group of**) is a locally finite semi-group [7]. If n is such that any group satisfying the law $x^n = 1$ is locally finite, then any semi-group with the law $x^{n+1} = x$ is locally finite [6]. A semi-group that has a decomposition into locally finite semi-groups need not be a locally finite semi-group [3], but if ρ is a congruence on a semi-group S such that the quotient semi-group S / ρ is locally finite and every ρ-class that is a sub-semi-group is locally finite, then S is a locally finite semi-group (see [4], [5]); in particular, an ideal extension of a locally finite semi-group by a locally finite semi-group is itself a locally finite semi-group. If S is a periodic semi-group of matrices over a skew-field and all subgroups of S are locally finite, then S is locally finite [8], which implies that any periodic semi-group of matrices over an arbitrary field is locally finite.

If S is a periodic inverse semi-group of matrices over a field and, moreover, the periods of all its elements (see **Monogenic semi-group**) are uniformly bounded and are not divided by the characteristic of the field, then S is finite [2].

References

[1] SHEVRIN, L.N.: 'On locally finite semigroups', *Soviet Math. Dokl.* **6** (1965), 769-772. (*Dokl. Akad. Nauk SSSR* **162** (1965), 770-773)

[2] SHNEPERMAN, L.B.: 'Periodic inverse linear semigroups', *Vesci Akad. Nauk BSSR Ser. Fiz. Mat. Nauk.* **4** (1976), 22-28 (in Russian).

[3] BROWN, T.C.: 'On locally finite semigroups', *Ukr. Math. J.* **20** (1968), 631-636. (*Ukr. Mat. Zh.* **20**, no. 6 (1968), 732-738)

[4] BROWN, T.C.: 'A semigroup union of disjoint locally finite subsemigroups which is not locally finite', *Pacific J. Math.* **22**, no. 1 (1967), 11-14.

[5] BROWN, T.C.: 'An interesting combinatorial method in the theory of locally finite semigroups', *Pacific J. Math.* **36**, no. 2 (1971), 285-289.

[6] GREEN, J.A. and REES, D.: 'On semi-groups in which $x^r = x$', *Proc. Cambridge Philos. Soc.* **48**, no. 1 (1952), 35-40.

[7] MCLEAN, D.: 'Idempotent semigroups', *Amer. Math. Monthly* **61**, no. 2 (1954), 110-113.

[8] MCNAUGHTON, R. and ZALCSTEIN, Y.: 'The Burnside problem for semigroups', *J. of Algebra* **34**, no. 2 (1975), 292-299.

L.N. Shevrin

AMS 1980 Subject Classification: 20M10

LOCALLY FLAT IMBEDDING - An imbedding (cf. **Immersion**) q of one topological manifold $M = M^m$ into another $N = N^n$ such that for any point $x \in M$ there are charts in a neighbourhood U of x and in a neighbourhood V of the point qx in N in which the restriction of q to U linearly maps U to V. In other words, q is locally linear in suitable coordinate systems. Equivalently: There are neighbourhoods U of a point $x \in M$ and V of the point $qx \in N$ such that the pair (V, qU) can be mapped homeomorphically onto a standard pair (D^n, D^m) or (D^n, D^m_+), where D^k is the unit ball of the space $\mathbf{R}^k$ with centre at the origin and D^k_+ is the intersection of this ball with the half-space $x_k \geq 0$.

Any imbedding of a circle and an arc into a plane is locally flat; however, a circle or an arc can be imbedded in $\mathbf{R}^k$ with $k \geq 3$ in a manner that is not locally flat (see **Wild imbedding**; **Wild sphere**). Any smooth imbedding is locally flat in the smooth sense (that is, in the definition the coordinates can be chosen to be smooth). A piecewise-linear imbedding need not be locally flat, not only in the piecewise-linear sense, but even not in the topological sense; for example, a cone with vertex in $\mathbf{R}^4_+$ over a closed polygon knotted in the bounding plane $\mathbf{R}^3$. For $n \neq 4$ and $m \neq n - 2$ there is a homotopy criterion for an imbedding to be locally flat: For every point $x \in M$ and neighbourhood U of the point qx there is a neighbourhood $V \subset U$ such that any loop in $V \setminus qM$ is homotopic to zero in $U \setminus qM$ (local simple connectedness). If $m = n - 2$, then such a criterion holds for $n \neq 4$, but is essentially more complicated. For $m = 4$ the question remains unsettled (1989). For $m = n - 1$ and $m = n - 2$ a locally flat imbedding has a topological **normal bundle**.

A.V. Chernavskiĭ

Editorial comments.

References

[A1] CANTRELL, J.C. and EDWARDS, C.H., JR. (EDS.): *Topology of manifolds*, Markham, 1970.

AMS 1980 Subject Classification: 57N35

LOCALLY FREE GROUP - A group in which every finitely-generated subgroup is free (see **Finitely-generated group; Free group**). Thus, a countable locally free group is the union of an ascending sequence of free subgroups.

One says that a locally free group has finite rank n if any finite subset of it is contained in a free subgroup of rank n, n being the smallest number with this property. The class of locally free groups is closed with respect to taking free products, and the rank of a free product of locally free groups of finite ranks equals the sum of the ranks of the factors.

References

[1] KUROSH, A.G.: *The theory of groups*, 1-2, Chelsea, 1955-1956 (translated from the Russian).

A.L. Shmel'kin

AMS 1980 Subject Classification: 20E05

LOCALLY FREE SHEAF - A **sheaf** of modules that is locally isomorphic to the direct sum of several copies of the structure sheaf. More precisely, let $(X, \mathcal{O}_X)$ be a **ringed space**. A sheaf of modules $\mathcal{F}$ over $\mathcal{O}_X$ is said to be *locally free* if for every point $x \in X$ there is an open neighbourhood $U \subset X$, $x \in U$, such that the restriction $\mathcal{F}|_U$ of $\mathcal{F}$ to U is a free sheaf of modules over $\mathcal{O}_X|_U$, that is, it is isomorphic to the direct sum of a set $I(x)$ of copies of the structure sheaf $\mathcal{O}_X|_U$. If X is connected and $I(x)$ is finite, for example consisting of n elements, then n does not depend on the point x and is called the rank of the locally free sheaf $\mathcal{F}$. Let V be a vector bundle of rank n on X and let $\mathcal{F}$ be the sheaf of germs of its sections. Then $\mathcal{F}$ is a locally free sheaf of rank n. Conversely, for every locally free sheaf $\mathcal{F}$ of rank n there is a vector bundle V of rank n on X such that $\mathcal{F}$ is the sheaf of germs of its sections (see [1], [2]); hence there is a natural one-to-one correspondence between the isomorphy classes of locally free sheaves of rank n and the isomorphy classes of vector bundles of rank n on X.

Example. Let X be a smooth connected algebraic variety of dimension n. Then the sheaf of regular differential forms Ω^1_X is a locally free sheaf of rank n.

Let $X = \operatorname{Spec} A$, a connected **affine scheme**, be the spectrum of the commutative ring A (cf. **Spectrum of a ring**), let $\mathcal{F}$ be a locally free sheaf of rank n and let $M = \Gamma(X, \mathcal{F})$ be the A-module of its global sections. Then the A-module M is projective and the mapping $\mathcal{F} \mapsto \Gamma(X, \mathcal{F})$ establishes a one-to-one correspondence between the set of classes (up to isomorphisms) of locally free sheaves of rank n and the set of classes (up to isomorphisms) of projective A-modules of rank n (see [2]).

References

[1] GODEMENT, R.: *Topologie algébrique et théorie des faisceaux*, Hermann, 1958.

[2] HARTSHORNE, R.: *Algebraic geometry*, Springer, 1977.

V.A. Iskovskikh

AMS 1980 Subject Classification: 14F05

LOCALLY INTEGRABLE FUNCTION *at a point M* - A function that is integrable in some sense or other in a neighbourhood of M. If a real-valued function f, defined on the interval $[a, b]$, is the pointwise finite derivative of a function F, real-valued and defined on this interval, then f is locally Lebesgue integrable at the points of an open everywhere-dense set on $[a, b]$. In the two-dimensional case (see [2]) there is a real-valued function f, defined on the square $[0, 1] \times [0, 1]$, that is the pointwise finite mixed derivative in either order $\partial^2 F / \partial x \partial y = \partial^2 F / \partial y \partial x = f(x, y)$ and that is not locally Lebesgue integrable at any point of the square.

References

[1] SAKS, S.: *Theory of the integral*, Hafner, 1952 (translated from the Polish).

[2] TOLSTOV, G.P.: 'On the curvilinear and iterated integral', *Trudy Mat. Inst. Steklov.* **35** (1950), 1-101 (in Russian).

I.A. Vinogradova

AMS 1980 Subject Classification: 26BXX, 26AXX

LOCALLY NILPOTENT ALGEBRA - An algebra in which any finitely-generated subalgebra is nilpotent (cf. **Nilpotent algebra; Finitely-generated group**). It is convenient to represent a locally nilpotent algebra as the union of an increasing chain of nilpotent subalgebras. A locally nilpotent **algebra with associative powers** is a **nil algebra**. A locally nilpotent Lie algebra is an **Engel algebra**. The class of locally nilpotent algebras is closed with respect to homeomorphic images and transition to subalgebras.

In the case of associative algebras an extension of a locally nilpotent algebra by a locally nilpotent algebra is again a locally nilpotent algebra. Therefore, the sum of all locally nilpotent ideals of an associative algebra is the largest locally nilpotent ideal that contains all locally nilpotent ideals; it is called the *Levitskiĭ radical*. An analogue of the Levitskiĭ radical can be defined in an Engel Lie algebra of bounded index. A locally nilpotent algebra cannot be simple (cf. **Simple algebra**).

References

[1] JACOBSON, N.: *Structure of rings*, Amer. Math. Soc., 1956.

[2] HERSTEIN, I.: *Noncommutative rings*, Math. Assoc. Amer., 1968.

[3] KOSTRIKIN, A.I.: 'Lie rings satisfying the Engel condition', *Izv. Akad. Nauk SSSR Ser. Mat.* **21** (1957), 515-540 (in Russian).

V.N. Latyshev

AMS 1980 Subject Classification: 16A22, 17A99, 17B30, 17B65

LOCALLY NILPOTENT GROUP - A group in which every finitely-generated subgroup is nilpotent (see **Nil-**

potent group; Finitely-generated group). In a locally nilpotent group the elements of finite order form a **normal subgroup**, the torsion part of this group (cf. **Periodic group**). This subgroup is the direct product of its Sylow subgroups, and the quotient group with respect to it is torsion-free. A locally nilpotent torsion-free group (cf. **Group without torsion**) has the uniqueness-of-roots property: If, for elements a and b and any integer $n \neq 0$, one has $a^n = b^n$, then $a = b$. Every locally nilpotent torsion-free group G has a *Mal'tsev completion*, that is, it can be imbedded in a unique locally nilpotent torsion-free group G^* such that all equations of the form $x^n = g$ are solvable in G^*, where $n \neq 0$ and g is any element of G. This completion is functorial, that is, any homomorphism $f: G_1 \to G_2$ of locally nilpotent torsion-free groups G_1 and G_2 can be uniquely extended to a homomorphism $f^*: G_1^* \to G_2^*$.

References

[1] KUROSH, A.G.: *The theory of groups*, 1-2, Chelsea, 1955-1956 (translated from the Russian).
[2] KARGAPOLOV, M.I. and MERZLYAKOV, YU.I.: *Fundamentals of the theory of groups*, Springer, 1979 (translated from the Russian).

A.L. Shmel'kin

AMS 1980 Subject Classification: 20D15, 20F18

LOCALLY NORMAL GROUP - A group G in which any finite subset is contained in a finite **normal subgroup** of G.

AMS 1980 Subject Classification: 20FXX

LOCALLY PATH-CONNECTED SPACE - A topological space X in which for any point $x \in X$ and any neighbourhood O_x of it there is a smaller neighbourhood $U_x \subset O_x$ such that for any two points $x_0, x_1 \in U_x$ there is a continuous mapping $F: I \to O_x$ of the unit interval $I = [0, 1]$ into O_x with $f(0) = x_0$ and $f(1) = x_1$. Any locally path-connected space is locally connected. Any open subset of a locally path-connected space is locally path-connected. A connected locally path-connected space is a **path-connected space**.

Locally path-connected spaces play an important role in the theory of covering spaces. Let $p: (\tilde{X}, \tilde{x}_0) \to (X, x_0)$ be a **covering** and let Y be a locally path-connected space. Then a necessary and sufficient condition for a mapping $f: (Y, y_0) \to (X, x_0)$ to admit a lifting, that is, a mapping $g: (Y, y_0) \to (\tilde{X}, \tilde{x}_0)$ such that $f = p \circ g$, is that

$$f_\#(\pi_1(Y, y_0)) \subset p_\#(\pi_1(\tilde{X}, \tilde{x}_0)),$$

where π_1 is the **fundamental group**. If X is a locally simply-connected (locally 1-connected, see below) space and $x_0 \in X$, then for any subgroup H of $\pi_1(X, x_0)$ there is a covering $p: (\tilde{X}, \tilde{x}_0) \to (X, x_0)$ for which $p_\#((\tilde{X}, \tilde{x}_0)) = H$.

The higher-dimensional generalization of local path-connectedness is *local k-connectedness* (local connectedness in dimension k). A space X is said to be locally k-connected if for any point $x \in X$ and any neighbourhood O_x of it there is a smaller neighbourhood $U_x \subset O_x$ such that any mapping of an r-dimensional sphere S^r into U_x is homotopic in O_x to a constant mapping. A metric space X is locally k-connected if and only if any mapping $f: A \to X$ from an arbitrary closed subset A in a **metric space** Y with $\dim Y \leq k + 1$ can be extended to a neighbourhood of A in Y (the *Kuratowski − Dugundji theorem*).

S.A. Bogatyĭ

Editorial comments.

References

[A1] MILL, J. VAN: *Infinite-dimensional topology*, North-Holland, 1988.

AMS 1980 Subject Classification: 54D05, 54F35

LOCALLY SOLVABLE ALGEBRA - An **algebra** for which any finitely-generated subalgebra is solvable (cf. **Solvable group**). It is convenient to represent a locally solvable algebra as the union of an increasing chain of solvable subalgebras. The class of locally solvable algebras is closed with respect to transition to subalgebras and taking homomorphic images.

References

[1] JACOBSON, N.: *Lie algebras*, Interscience, 1962 (also: Dover, reprint, 1979).

V.N. Latyshev

AMS 1980 Subject Classification: 17B05

LOCALLY SOLVABLE GROUP - A group in which every finitely-generated subgroup is solvable (see **Finitely-generated group; Solvable group**). The class of locally solvable groups is closed with respect to taking subgroups and homomorphic images, but it is not closed under extension.

A locally solvable torsion group is locally finite.

References

[1] KUROSH, A.G.: *The theory of groups*, 1-2, Chelsea, 1955-1956 (translated from the Russian).

A.L. Shmel'kin

AMS 1980 Subject Classification: 20D10, 20F16

LOCALLY TRIVIAL FIBRE BUNDLE - A fibre bundle (cf. **Fibre space**) $\pi: X \to B$ with fibre F such that for any point of the base $b \in B$ there is a neighbourhood $U \ni b$ and a homeomorphism $\phi_U: U \times F \to \pi^{-1}(U)$ such that $\pi \phi_U(u, f) = u$, where $u \in U$, $f \in F$. The mapping $h_U = \phi_U^{-1}$ is called a *chart* of the locally trivial bundle. The totality of charts $\{h_U\}$ associated with a covering of the base $\{U\}$ forms the atlas of the locally trivial bundle. For example, a **principal fibre bundle** with a locally compact space and a Lie group G is a locally trivial fibre bundle, and any chart h_U satisfies the relation

$$h_U(gx) = gh_U(x), \quad x \in \pi^{-1}(U),$$

where G acts on $G \times U$ according to the formula $g(g', u) = (gg', u)$. For any locally trivial fibre bundle $\pi: X \to B$ and continuous mapping $f: B_1 \to B$ the **induced fibre bundle** is locally trivial.

References

[1] SPANIER, E.H.: *Algebraic topology*, McGraw-Hill, 1966.
[2] STEENROD, N.E.: *The topology of fibre bundles*, Princeton Univ. Press, 1951.
[3] HU, S.-T.: *Homotopy theory*, Acad. Press, 1959.
[4] HUSEMOLLER, D.: *Fibre bundles*, McGraw-Hill, 1966.

M.I. Voĭtsekhovskiĭ

AMS 1980 Subject Classification: 55R10

LOG-NORMAL DISTRIBUTION - See Logarithmic normal distribution.

AMS 1980 Subject Classification: 60E10, 62E05

LOGARITHM OF A NUMBER N *to a base* a - The degree m to which the number a (the *base of the logarithm*) must be raised in order to obtain N; it is denoted by $\log_a N$; that is, $m = \log_a N$ means $a^m = N$. To every positive number N and given base $a > 0$, $a \neq 1$, there corresponds a unique real logarithm (logarithms of negative numbers are complex numbers). The main properties of the logarithm are:

$$\log_a(MN) = \log_a M + \log_a N,$$

$$\log_a \frac{M}{N} = \log_a M - \log_a N,$$

$$\log_a N^k = k \log_a N,$$

$$\log_a N^{1/k} = \frac{1}{k} \log_a N.$$

These make it possible to reduce multiplication and division of numbers to the addition and subtraction of their logarithms, and the raising to powers and extraction of roots to the multiplication and division of the logarithm by the index of the power or root.

In accordance with the decimal system, the most commonly used are decimal logarithms ($a = 10$), denoted by $\lg N$. For rational numbers other than 10^k with k an integer, the decimal logarithms are transcendental numbers (cf. **Transcendental number**), which can be approximately expressed by finite decimal fractions. The integer part of a decimal logarithm is called the *characteristic*, and the fractional part is called the *mantissa*. Since $\lg(10^k N) = k + \lg N$, the decimal logarithms of numbers that differ by a multiple of 10^k have the same mantissa and differ only in the characteristics. This property is the basis for the construction of logarithm tables, which contain only the mantissa of the logarithms of integers.

The *natural logarithms* also have great significance; the base of these is the transcendental number $e = 2.71828 \cdots$; they are denoted by $\ln N$. The transi-

tion from one base to another is carried out by the formula $\log_b N = \log_a N / \log_a b$; the factor $1 / \log_a b$ is called the *modulus of transition* from the base a to the base b. The formulas for the transition from natural logarithms to decimal logarithms and vice versa are:

$$\ln N = \frac{\lg N}{\lg e}, \quad \lg N = \frac{\ln N}{\ln 10},$$

$$\frac{1}{\lg e} = 2.30258 \cdots, \quad \frac{1}{\ln 10} = 0.43429 \cdots.$$

See also **Logarithmic function**.

Material of the article of the same name in BSE-3.

Editorial comments. Western writers of post-calculus mathematics almost always use $\log N$ instead of $\ln N$ to denote the *natural logarithm* (base e) of N. The number e is given by $e = \lim_{n \to \infty}(1 + 1/n)^n$ and by $e = \sum_{n=0}^{\infty} 1/n!$ (cf. e (number)).

On the other hand, writers of calculus and pre-calculus books as well as manufacturers of calculators usually use $\log N$ for the decimal (common) logarithm and $\ln N$ for the natural logarithm. Cf. also **Logarithmic function**.

AMS 1980 Subject Classification: 26A09

LOGARITHMIC BRANCH POINT, *branch point of infinite order* - A special form of a **branch point** a of an analytic function $f(z)$ of one complex variable z, when for no finite number of successive circuits in the same direction about a the **analytic continuation** of some element of $f(z)$ returns to the original element. More precisely, an isolated singular point a is called a logarithmic branch point for $f(z)$ if there exist: 1) an annulus $V = \{z: 0 < |z - a| < \rho\}$ in which $f(z)$ can be analytically continued along any path; and 2) a point $z_1 \in V$ and an element of $f(z)$ in the form of a power series $\Pi(z_1; r) = \sum_{\nu=0}^{\infty} c_\nu (z - z_1)^\nu$ with centre z_1 and radius of convergence $r > 0$, the analytic continuation of which along the circle $|z - a| = |z_1 - a|$, taken arbitrarily many times in the same direction, never returns to the original element $\Pi(z_1; r)$. In the case of a logarithmic branch point at infinity, $a = \infty$, instead of V one must consider a neighbourhood $V' = \{z: |z| > \rho\}$. Logarithmic branch points belong to the class of transcendental branch points (cf. **Transcendental branch point**). The behaviour of the Riemann surface R of a function $f(z)$ in the presence of a logarithmic branch point a is characterized by the fact that infinitely many sheets of the same branch of R are joined over a; this branch is defined in V or V' by the elements $\Pi(z_1; r)$.

See also **Singular point** of an analytic function.

References

[1] MARKUSHEVICH, A.I.: *Theory of functions of a complex variable*, 2, Chelsea, 1977, Chapt. 8 (translated from the Russian).

E.D. Solomentsev

Editorial comments. The function $\mathrm{Ln}(z - z_0)$ has a logarithmic branch point at z_0, where Ln is the (multiple-valued) logarithmic function of a complex variable.

References

[A1] AHLFORS, L.V.: *Complex analysis*, McGraw-Hill, 1979, Chapt. 8.

AMS 1980 Subject Classification: 30A05

LOGARITHMIC CAPACITY - See **Capacity**.

AMS 1980 Subject Classification: 31A15

LOGARITHMIC CONVERGENCE CRITERION - A

series of numbers $\sum_{n=1}^{\infty} a_n$, where $a_n > 0$, converges if there is a number $\alpha > 0$ such that

$$\frac{\ln 1/a_n}{\ln n} \geqslant 1 + \alpha \quad \text{for } n \geqslant n_0,$$

and diverges if

$$\frac{\ln 1/a_n}{\ln n} \leqslant 1 \quad \text{for } n \geqslant n_0.$$

V.I. Bityutskov

AMS 1980 Subject Classification: 40A05

LOGARITHMIC DERIVATIVE - The derivative of

the logarithm of a given function.

Editorial comments. I.e., let $f : [a, b] \to \mathbb{R}$ be a positive function. Then its logarithmic derivative equals

$$(\ln f)' = \frac{f'}{f}.$$

AMS 1980 Subject Classification: 26A24

LOGARITHMIC FUNCTION, *logarithm* - The func-

tion inverse to the **exponential function**. The logarithmic function is denoted by

$$y = \ln x; \tag{1}$$

its value y, corresponding to the value of the argument x, is called the *natural logarithm* of x. From the definition, relation (1) is equivalent to

$$x = e^y. \tag{2}$$

Since $e^y > 0$ for any real y, the logarithmic function is defined only for $x > 0$. In a more general sense a logarithmic function is a function

$$y = \log_a x,$$

where $a > 0$ ($a \neq 1$) is an arbitrary base of the logarithm; this function can be expressed in terms of $\ln x$ by the formula

$$\log_a x = \frac{\ln x}{\ln a}.$$

The logarithmic function is one of the main elementary functions; its graph (see Fig.) is called a *logarithmic curve*.

The main properties of the logarithmic function follow from the corresponding properties of the exponential function and logarithms; for example, the loga-

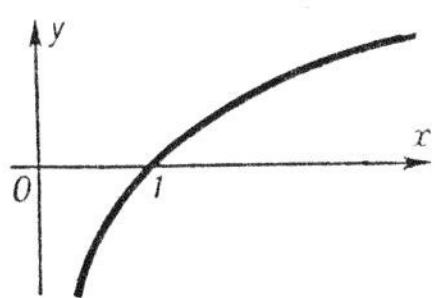

rithmic function satisfies the functional equation

$$\ln x + \ln y = \ln xy.$$

The logarithmic function $y = \ln x$ is a strictly-increasing function, and $\lim_{x \downarrow 0} \ln x = -\infty$, $\lim_{x \to \infty} \ln x = +\infty$. At every point $x > 0$ the logarithmic function has derivatives of all orders and in a sufficiently small neighbourhood it can be expanded in a power series, that is, it is an **analytic function**. For $-1 < x \leqslant 1$ the following expansion of the (natural) logarithmic function is valid:

$$\ln(1+x) = x - \frac{x^2}{2} + \frac{x^3}{3} - \frac{x^4}{4} + \cdots.$$

The derivative of the logarithmic function is

$$(\ln x)' = \frac{1}{x}, \quad (\log_a x)' = \frac{\log_a e}{x} = \frac{1}{x \ln a}.$$

Many integrals can be expressed in terms of the logarithmic function; for example:

$$\int \frac{dx}{x} = \ln |x| + C,$$

$$\int \frac{dx}{\sqrt{x^2 + a}} = \ln(x + \sqrt{x^2 + a}) + C.$$

The dependence between variable quantities expressed by the logarithmic function was first considered by J. Napier in 1614.

The logarithmic function on the complex plane is an infinitely-valued function, defined for all values of the argument $z \neq 0$, and is denoted by $\operatorname{Ln} z$ (or $\ln z$ if no confusion arises). The single-valued branch of this function defined by

$$\ln z = \ln |z| + i \arg z,$$

where $\arg z$ is the principal value of the argument of the complex number z, $\pi < \arg z \leqslant \pi$, is called the *principal value* of the logarithmic function. One has

$$\operatorname{Ln} z = \ln z + 2k\pi i, \quad k = 0, \pm 1, \ldots .$$

All values of the logarithmic function for negative real z are purely imaginary complex numbers. The first satisfactory theory of the logarithmic function for complex arguments was given by L. Euler in 1749; he started from the definition

$$\operatorname{Ln} z = \lim_{n \to \infty} n(z^{1/n} - 1).$$

References

[1] NIKOL'SKIĬ, S.M.: *A course of mathematical analysis*, 1, Mir, 1977 (translated from the Russian).
[2] MARKUSHEVICH, A.I.: *Theory of functions of a complex variable*, 1, Chelsea, 1977 (translated from the Russian).

BSE-3

Editorial comments. The principal value of the logarithm maps the punctured complex z-plane ($z \neq 0$) onto the strip $-\pi < \mathrm{Ln}\, z \leqslant \pi$ in the complex w-plane. To fill the w-plane one has to map infinitely many copies of the z-plane, where for the n-th copy one has $-\pi + 2n\pi < \arg z \leqslant \pi + 2n\pi$, $n = 0, \pm 1, \ldots$. In this case 0 is a **branch point**. The copies make up the so-called **Riemann surface** of the logarithmic function. Clearly, $\ln z$ is a one-to-one mapping of this surface ($z \neq 0$) onto the w-plane. The derivative of the principal value is $1/z$ (as in the real case) for $-\pi < \arg z < \pi$.

Instead of ln and Ln, many Western writers of post-calculus mathematics use log and Log (see also (the editorial comments to) **Logarithm of a number**).

References

[A1] CONWAY, J.B.: *Functions of one complex variable*, Springer, 1973.
[A2] MARSDEN, E.: *Basic complex analysis*, Freeman, 1973.
[A3] TITCHMARSH, E.C.: *The theory of functions*, Oxford Univ. Press, 1979.
[A4] SAKS, S. and ZYGMUND, A.: *Analytic functions*, PWN, 1952 (translated from the Polish).
[A5] STROMBERG, K.: *An introduction to classical real analysis*, Wadsworth, 1981.

AMS 1980 Subject Classification: 26A09

LOGARITHMIC NORMAL DISTRIBUTION - The continuous probability distribution, concentrated on $(0, \infty)$, with density

$$p(x) = \begin{cases} \dfrac{\log e}{\sigma x \sqrt{2\pi}} e^{-(\log x - a)^2 / 2\sigma^2}, & x > 0, \\[2mm] 0, & x \leqslant 0, \end{cases} \qquad (*)$$

where $-\infty < a < \infty$, $\sigma^2 > 0$. A random variable X is subject to the logarithmic normal distribution with density (*) if its logarithm $\log X$ has the **normal distribution** with parameters a and σ^2. Thus, $a = \mathsf{E}\log X$ and $\sigma^2 = \mathsf{D}\log X$. The logarithmic normal distribution is a unimodal distribution and has positive asymmetry. The moments of a random variable X with logarithmic normal distribution with parameters a and σ^2 are given by the formula

$$\mathsf{E}X^k = e^{ka + k^2 \sigma^2 / 2};$$

hence the mean and variance are, respectively, equal to

$$\mathsf{E}X = e^{a + \sigma^2/2} \quad \text{and} \quad \mathsf{D}X = e^{2a + \sigma^2}(e^{\sigma^2} - 1).$$

The logarithmic normal distribution is one of the simplest examples of a distribution that is not defined uniquely by its moments. The properties of the logarithmic normal distribution are determined by the properties of the corresponding normal distribution. An important property of the logarithmic normal distribution is the following: The product of independent random variables with a logarithmic normal distribution is again subject to a logarithmic normal distribution. There is an analogue of the **central limit theorem**: The distribution of the product of n independent positive random variables tends, under certain general conditions, to a logarithmic normal distribution as $n \to \infty$. The logarithmic normal distribution arises as a limit distribution and in certain other schemes (for example, in models of branching of particles, models of growth, etc.).

References

[1] KOLMOGOROV, A.N.: 'Ueber das logarithmische normale Verteilungsgesetz der Dimensionen der Teilchen bei Zerstückelung', *Dokl. Akad. Nauk SSSR* **31**, no. 2 (1941), 99-101.
[2] CRAMÉR, H.: *Mathematical methods of statistics*, Princeton Univ. Press, 1946.
[3] AITCHISON, J. and BROWN, J.A.C.: *The lognormal distribution*, Cambridge Univ. Press, 1957.

A.V. Prokhorov

Editorial comments. The logarithmic normal distribution, or *lognormal distribution*, is infinitely divisible (cf. **Infinitely-divisible distribution**), cf. [A2].

References

[A1] JOHNSON, N.L. and KOTZ, S.: *Distributions in statistics: continuous univariate distributions*, 1, Wiley, 1970.
[A2] THORIN, O.: 'On the infinite divisibility of the lognormal distribution', *Skand. Aktuartidskr.*, no. 3 (1977), 121-148.

AMS 1980 Subject Classification: 60E10, 62E05

LOGARITHMIC PAPER, *double logarithmic paper* - A special form of ruled paper; it is usually made typographically (Fig. 1) as follows: On each of the axes of a rectangular (u, v)-coordinate system one marks the decimal logarithms of numbers u (on the horizontal axis) and v (on the vertical axis) (cf. also **Logarithm of a number**); then lines are drawn through the resulting points (u, v) parallel to the axes.

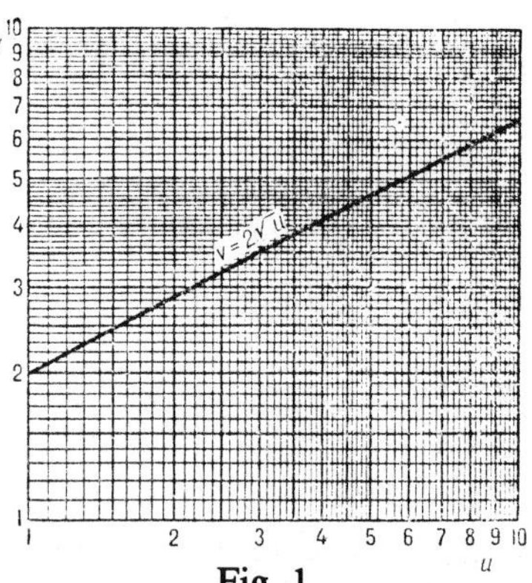

Fig. 1.

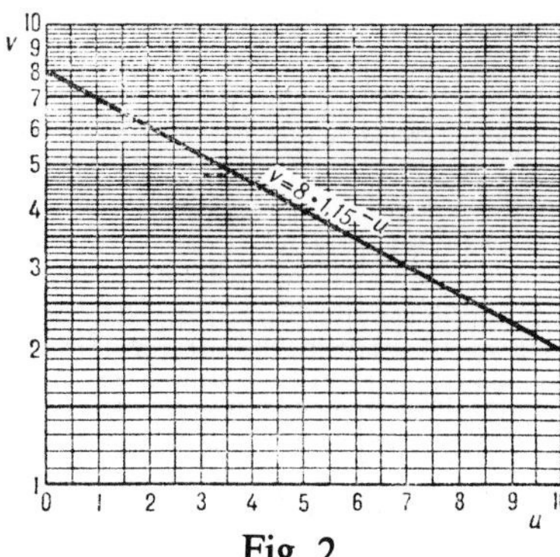

Fig. 2.

There is also *semi-logarithmic paper* (*single logarithmic paper*) (Fig. 2): On one of the axes of a rectangular (u, v)-coordinate system the values of numbers u are marked, and on the other the decimal logarithms of numbers v.

Logarithmic paper and semi-logarithmic paper are used for drawing the graphs of functions which (in those coordinates) may take a simpler and more obvious form, and in many cases are straight lines. On logarithmic paper straight lines represent functions given by equations of the form $v = au^b$, $u > 0$, where $a > 0$ and b are constant coefficients; on semi-logarithmic paper straight lines represent functions given by equations of the form $v = ab^u$.

BSE-3

AMS 1980 Subject Classification: 26A09

LOGARITHMIC POTENTIAL - A potential with the *logarithmic kernel* $\ln 1 / |x - y|$, where $|x - y|$ is the distance between the points x and y of the Euclidean plane $\mathbf{R}^2$, that is, a potential of the form

$$u(x) = \int \ln \frac{1}{|x-y|} \, d\mu(y), \qquad (1)$$

where, generally, speaking, the integration is carried out with respect to an arbitrary **Borel measure** μ on $\mathbf{R}^2$ with compact support $S = S(\mu)$. Physically one can assume that the logarithmic potential arises from the **Newton potential** of the forces of gravitation when the distribution of the attracting masses in the Euclidean space $\mathbf{R}^3$ of points $y = (y_1, y_2, y_3)$ does not depend, for example, on the coordinate y_3. Of course the total mass is infinite, but if one performs a regularization of the resulting attracting force F, which can be regarded as acting in the plane $(x_1, x_2, 0)$, consisting in discarding the infinite term, then the potential of the finite part of F will invariably have the form (1) (see [2]). In contrast to the Newton kernel, the logarithmic kernel has a singularity not only as $|x - y| > 0$, but also as $|x - y| \to \infty$, which causes some differences in the behaviour of the logarithmic potential as compared with the Newton potential. They occur mainly in the solution of exterior boundary value problems (see **Exterior and interior boundary value problems**). The main applications of the logarithmic potential occur in the solution of planar **boundary value problems in potential theory** (see also **Boundary value problem, elliptic equations**).

The main properties of the logarithmic potential are: 1) outside the support S of the measure μ the logarithmic potential is a regular solution of the **Laplace equation** $\Delta u = 0$, that is, u is a **harmonic function** on the open set $\mathbf{R}^2 \setminus S$, but is not regular at infinity, however; 2) if the measure μ is absolutely continuous, that is, the integral (1) takes the form

$$u(x) = \int\limits_D f(y) \ln \frac{1}{|x-y|} \, d\sigma(y), \qquad (2)$$

where D is a finite domain, $d\sigma$ is the area element of D and the density f belongs to the class $C^1(D \bigcup \partial D)$, then the second derivatives of u are continuous in D and satisfy the **Poisson equation** $\Delta u = -2\pi f$.

If the integral in (2) extends along a closed Lyapunov curve L (see **Lyapunov surfaces and curves**), that is,

$$u(x) = \int\limits_L f(y) \ln \frac{1}{|x-y|} \, ds(y), \qquad (3)$$

one talks of the *logarithmic potential of a single* (or *simple*) *layer*, distributed on L. If $f \in C^1(L)$, then the logarithmic potential of the single layer (3) is continuous everywhere in $\mathbf{R}^2$. Its normal derivative has limits from the inside and the outside of L, respectively:

$$\lim_{x \to y_0} \frac{du}{dn_0}\bigg|_i = \frac{du(y_0)}{dn_0} + \pi f(y_0),$$

$$\lim_{x \to y_0} \frac{du}{dn_0}\bigg|_o = \frac{du(y_0)}{dn_0} - \pi f(y_0),$$

where

$$\frac{du(y_0)}{dn_0} = \int\limits_L f(y) \frac{\cos(y - y_0, n_0)}{|y - y_0|} \, ds(y), \quad y_0 \in L, \qquad (4)$$

is the so-called direct value of the normal derivative of the logarithmic potential of a single layer and $(y - y_0, n_0)$ is the angle between the vector $y - y_0$ and the outward normal n_0 to L at the point $y_0 \in L$. The integral (4) is continuous on L.

The *logarithmic potential of a double layer* has the form

$$v(x) = \int\limits_L g(y) \frac{\cos(y - x, n)}{|y - x|} \, ds(y), \qquad (5)$$

where n is the outward normal to L at $y \in L$. If $g \in C^1(L)$, then the logarithmic potential of the double layer (5) is a regular harmonic function inside and outside L and has normal (non-angular) limits from the inside and the outside of L, respectively:

$$\lim_{x \to y_0} v(x)\big|_i = v(y_0) + \pi f(y_0),$$

$$\lim_{x \to y_0} v(x)\big|_o = v(y_0) - \pi f(y_0),$$

where

$$v(y_0) = \int\limits_L g(y) \frac{\cos(y - y_0, n)}{|y - y_0|} \, ds(y), \quad y_0 \in L,$$

is the direct value of the logarithmic potential of the double layer at the point $y_0 \in L$. The normal derivative of the logarithmic potential of a double layer is continuous under transition through L.

The listed boundary properties of the logarithmic potential of a simple and a double layer are completely analogous to the corresponding properties of the Newton potential (see also **Potential theory**). From (5) it is obvious that the logarithmic potential of a double layer is a harmonic function that is regular at infinity.

The logarithmic potential is also directly connected with **boundary value problems of analytic function theory**, since an integral of Cauchy type can be expressed in terms of the logarithmic potential of a single and a double layer (see [3]).

References

[1] SOBOLEV, S.L.: *Partial differential equations of mathematical physics*, Pergamon, 1964 (translated from the Russian).
[2] WEBSTER, A.G.: *Partial differential equations of mathematical physics*, Hafner, 1955.
[3] MUSKHELISHVILI, N.I.: *Singular integral equations*, Wolters-Noordhoff, 1972 (translated from the Russian).
[4] VALLÉE-POUSSIN, CH.J. DE LA: *Le potentiel logarithmique, balayage et réprésentation conforme*, Librarie Univ., 1949.
[5] EVANS, G.C.: *The logarithmic potential, discontinuous Dirichlet and Neumann problems*, New York, 1927.

E.D. Solomentsev

Editorial comments. See also Lyapunov theorem.

References

[A1] HAYMAN, W.K. and KENNEDY, P.B.: *Subharmonic functions*, 1, Acad. Press, 1976.
[A2] KRÁL, J.: *Integral operators in potential theory*, Lecture notes in math., 823, Springer, 1980.

AMS 1980 Subject Classification: 31A15

LOGARITHMIC RESIDUE of a meromorphic function $w = f(z)$ at a point a of the extended complex z-plane - The residue

$$\operatorname{res}_a \frac{f'(z)}{f(z)}$$

of the logarithmic derivative $f'(z)/f(z)$ at the point a. Representing the function $\ln f(z)$ in a neighbourhood $V(a)$ of a point $a \neq \infty$ in the form $\ln f(z) = A \ln(z-a) + \phi(z)$, where $\phi(z)$ is a regular function in $V(a)$, one obtains

$$\operatorname{res}_a \frac{f'(z)}{f(z)} = A.$$

The corresponding formulas for the case $a = \infty$ have the form

$$\ln f(z) = A \ln\left[\frac{1}{z}\right] + \phi(z),$$

$$\operatorname{res}_\infty \frac{f'(z)}{f(z)} = A.$$

If a is a zero or a pole of $f(z)$ of multiplicity m, then the logarithmic residue of $f(z)$ at a is equal to m or $-m$, respectively; at all other points the logarithmic residue is zero.

If $f(z)$ is a meromorphic function in a domain D and Γ is a rectifiable Jordan curve situated in D and not passing through the zeros or poles of $f(z)$, then the *logarithmic residue of $f(z)$ with respect to the contour Γ is the integral*

$$\frac{1}{2\pi i} \int_\Gamma \frac{f'(z)}{f(z)} \, dz = N - P, \tag{1}$$

where N is the number of zeros and P is the number of poles of $f(z)$ inside Γ (taking account of multiplicity). The geometrical meaning of (1) is that as Γ is traversed in the positive sense, the vector $w = f(z)$ performs $N - P$ rotations about the origin $w = 0$ of the w-plane (see **Argument, principle of the**). In particular, if $f(z)$ is regular in D, that is, $P = 0$, then from (1) one obtains a formula for the calculation of the index of the point $w = 0$ with respect to the image $\Gamma^* = f(\Gamma)$ of Γ by means of the logarithmic residue:

$$\operatorname{ind}_0 \Gamma^* = \frac{1}{2\pi i} \int_\Gamma \frac{f'(z)}{f(z)} \, dz. \tag{2}$$

Formula (2) leads to a generalization of the concept of a logarithmic residue to regular functions of several complex variables in a domain D of the complex space $\mathbf{C}^n$, $n \geq 1$. Let $w = f(z) = (f_1, \ldots, f_n): D \to \mathbf{C}^n$ be a **holomorphic mapping** such that the **Jacobian** $J_f(z) \neq 0$ and the set of zeros $E = f^{-1}(0)$ is isolated in D. Then for any domain $G \subset \overline{G} \subset D$ bounded by a simple closed surface Γ not passing through the zeros of f one has a formula for the index of the point $w = 0$ with respect to the image $\Gamma^* = f(\Gamma)$:

$$\operatorname{ind}_0 \Gamma^* = \frac{1}{(2\pi i)^n} \int_{\Gamma_\epsilon} \frac{df_1 \wedge \cdots \wedge df_n}{f_1 \cdots f_n} = N, \tag{3}$$

where the integration is carried out with respect to the n-dimensional frame $\Gamma_\epsilon = \{z \in G: \ |f_\nu(z)| = \epsilon, \ \nu = 1, \ldots, n\}$ with sufficiently small $\epsilon > 0$. The integral in (3) also expresses the sum of the multiplicities of the zeros of f in G (see [2]).

References

[1] MARKUSHEVICH, A.I.: *Theory of functions of a complex variable*, 1, Chelsea, 1977 (translated from the Russian).
[2] SHABAT, B.V.: *Introduction to complex analysis*, 1-2, Moscow, 1976 (in Russian).

E.D. Solomentsev

Editorial comments. The *index* of the origin with respect to a curve in the complex plane (also called the *winding number* of the curve, cf. **Winding of a curve**) is the number of times that the curve encircles the origin. More precisely, it is the change in the argument of $\ln w$ as w traverses the curve (cf. [A1], [A3]). In higher dimensions, the index of a point with respect to a closed surface may be defined as the number N such that the surface is homologous to N times the boundary of a ball centred at the point (cf. [A2], [A4]).

References

[A1] BURCKEL, R.B.: *An introduction to classical complex analysis*, 1, Acad. Press, 1979.
[A2] AÏZENBERG, L.A. and YUZHAKOV, A.P.: *Integral representations and residues in multidimensional complex analysis*, Amer. Math. Soc., 1983 (translated from the Russian).
[A3] AHLFORS, L.V.: *Complex analysis*, McGraw-Hill, 1979.
[A4] MILNOR, J.: *Topology from the differentiable viewpoint*, Univ. Press Virginia, 1969.

AMS 1980 Subject Classification: 30DXX, 32A27

LOGARITHMIC SPIRAL - A plane **transcendental**

curve whose equation in polar coordinates has the form

$$\rho = a^{\phi}, \quad a>0.$$

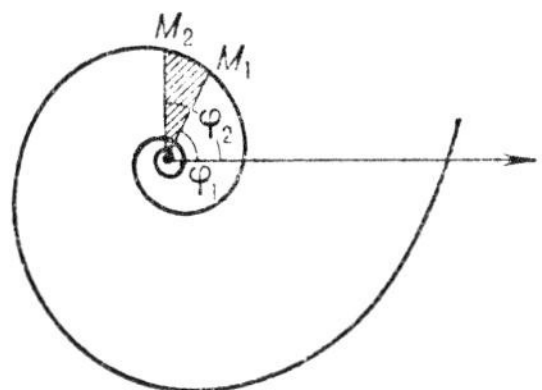

If $a>1$, as $\phi \to +\infty$ the logarithmic spiral evolves anti-clockwise, and as $\phi \to -\infty$ the spiral twists clockwise, tending to its asymptotic point 0 (see Fig.). If $a<1$, the twisting behaviour is opposite. The angle formed by the tangent at an arbitrary point of the logarithmic spiral and the position vector of that point depends only on the parameter a. The length of the arc between two points $M_1(\rho_1, \phi_1)$ and $M_2(\rho_2, \phi_2)$ is:

$$l = \rho_2 \frac{\sqrt{1+\ln^2 a}}{\ln a} - \rho_1 \frac{\sqrt{1+\ln^2 a}}{\ln a}.$$

The radius of curvature is $r=\sqrt{1+\ln^2 a}$. The **natural equation** is $s=kr$, where $k=1/\ln a$. Logarithmic spirals go into logarithmic spirals under linear isometries, similarities and inversions of the plane. The logarithmic spiral is related to the so-called pseudo-spirals (see **Spirals**).

References

[1] SAVELOV, A.A.: *Plane curves*, Moscow, 1960 (in Russian).

D.D. Sokolov

Editorial comments.

References

[A1] BERGER, M.: *Geometry*, 1-2, Springer, 1987 (translated from the French).
[A2] COXETER, H.S.M.: *Introduction to geometry*, Wiley, 1963.
[A3] FLADT, K.: *Analytische Geometrie spezieller ebener Kurven*, Akad. Verlagsgesellschaft, 1962.

AMS 1980 Subject Classification: 53A04, 51M15

LOGARITHMIC SUMMATION METHOD - One of the methods for summing series of numbers. A series $\sum_{k=0}^{\infty} a_k$ with partial sums s_n is *summable by the logarithmic method* to the sum s if the *logarithmic mean*

$$\sigma_m = \frac{1}{\sum_{k=0}^{m} \dfrac{1}{k+1}} \left[s_0 + \frac{s_1}{2} + \cdots + \frac{s_m}{m+1} \right]$$

converges to s as $m \to \infty$. The logarithmic summation method is the **Riesz summation method** (R, p_n) with $p_n = 1/(n+1)$. It is equivalent to and compatible (cf. **Compatibility of summation methods**) with the Riesz summation method $(R, \lambda_n, 1)$ with $\lambda_n = \ln(n+1)$ and is more powerful than the summation method of arithmetical averages (cf. **Arithmetical averages, summation method of**).

References

[1] RIESZ, F.: 'Sur la sommation des séries de Dirichlet', *C.R. Acad. Sci. Paris* **149** (1909), 18-21.
[2] HARDY, G.H.: *Divergent series*, Clarendon Press, 1949.

I.I. Volkov

AMS 1980 Subject Classification: 40GXX

LOGARITHMICALLY-SUBHARMONIC FUNCTION - A positive function $u(x)$ in a domain of the Euclidean space $\mathbf{R}^n$, $n \geq 2$, whose logarithm $\log u(x)$ is a **subharmonic function**. For example, the modulus $|f(z)|$ of an analytic function $f(z)$ of a complex variable is a logarithmically-subharmonic function, but there are continuous logarithmically-subharmonic functions in planar domains that cannot be represented as the modulus of any analytic function. The logarithmically-subharmonic functions constitute a subclass of the strongly-subharmonic functions (cf. **Subharmonic function**). For $n=1$ they correspond to logarithmically-convex functions.

The main property of logarithmically-subharmonic functions is that not only the product, but also a positive linear combination, of several logarithmically-subharmonic functions is a logarithmically-subharmonic function.

References

[1] PRIVALOV, I.I.: *Subharmonic functions*, Moscow-Leningrad, 1937, Chapt. 3 (in Russian).

E.D. Solomentsev

Editorial comments.

References

[A1] RONKIN, L.I.: *Introduction to the theory of entire functions of several variables*, Amer. Math. Soc., 1974, p. 36 (translated from the Russian).
[A2] HAYMAN, W.K. and KENNEDY, P.B.: *Subharmonic functions*, 1, Acad. Press, 1976.

AMS 1980 Subject Classification: 31A05, 31B05, 31C05

LOGICAL AXIOM - A logical system S generally consists of a language L and a set T of sentences of L, called provable in S. T is defined inductively, as being the smallest set of sentences of L which contains a given set A of L-sentences and closed under certain specified operations. The elements of A are called the logical axioms of S.

References

[1] MENDELSON, E.: *Introduction to mathematical logic*, v. Nostrand, 1964.
[2] SHOENFIELD, J.R.: *Mathematical logic*, Addison-Wesley, 1967.

V.E. Plisko

Editorial comments. The phrase 'logical axiom' is often more specifically used to distinguish those axioms, in a formal theory, which are concerned with securing the meaning of the logical connectives and quantifiers (cf. **Logical calculus**), as opposed to the 'non-logical axioms' which are the standing hypotheses about the interpretation of the particular function and predicate symbols in the language in which the theory is formulated (cf. **Logico-mathematical calculus**).

AMS 1980 Subject Classification: 03BXX

LOGICAL CALCULUS - A formalization of a meaningful logical theory. The derivable objects of a logical calculus are interpreted as statements, formed from the simplest ones (generally speaking, having subject-predicate structure) by means of propositional connectives and quantifiers. The most frequently used connectives are 'not', 'and', 'or', 'if $\cdots$, then $\cdots$', and the existential and universal quantifiers. Logical calculi are distinguished from arbitrary calculi (cf. **Calculus**) by the purely logical character of interpretations and derivation rules, and from logico-mathematical calculi (cf. **Logico-mathematical calculus**) by the absence in the language of symbols for specific mathematical predicates and functions (except for the symbol '='', the addition of which is interpreted as the introduction of equality and is usually supposed not to violate the logical character of the calculus). These differences have a relative character, since logical calculi remain pure formal systems, and the semantics of any possible interpretation of them must be regarded as something external, having heuristic but not conclusive value in the study of properties of the calculus.

One of the most important logical calculi is the *classical predicate calculus with function symbols*. The language of this calculus, apart from parentheses and the logical symbols $\neg$, $\&$, $\vee$, $\supset$, $\exists$, $\forall$, contains three potentially infinite lists: lists of object variables, predicate variables and function variables. (Each of the predicate and function variables is endowed with information about its dimension, where for predicate variables the least dimension is 1 and for function variables the least dimension is 0.) *Terms* are defined as follows: 1) any object variable and any function variable of dimension 0 is a term; 2) if $T_1, \ldots, T_l$ are terms and f is a function variable of dimension l, then $f(T_1, \ldots, T_l)$ is also a term. If $T_1, \ldots, T_k$ are terms and P is a predicate variable of dimension k, then $P(T_1, \ldots, T_k)$ is called an *atomic formula. Formulas* are defined as follows: 1) any atomic formula is a formula; 2) if F and G are formulas and x is an object variable, then the expressions

$$\neg F, \quad (F \& G), \quad (F \vee G), \quad (F \supset G), \quad \exists x\, F, \quad \forall x\, F$$

are also formulas. In the last two formulas all occurrences of the variable x are said to be *bound*; occurrences of variables that are not associated with quantifiers in the process of constructing a formula are called *free*. A term T is said to be *free for x in F* if no free occurrence of x in F is in a subformula of the form $\exists y\, G$ or $\forall y\, G$, where y is one of the variables that occurs in T; $[F]^x_T$ denotes the result of substituting T for all free occurrences of x in F.

Let x be an arbitrary object variable, let A, B, C, D be arbitrary formulas, where D does not contain x freely, and let T be an arbitrary term, free for x in A. The axioms of the calculus in question are all formulas of the following 10 kinds (each of which is called an *axiom scheme*):

1. $(A \supset (B \supset A))$,
2. $((A \supset B) \supset ((A \supset (B \supset C)) \supset (A \supset C)))$,
3. $(A \supset (B \supset (A \& B)))$,
4a. $((A \& B) \supset A)$, 4b. $((A \& B) \supset B)$,
5a. $(A \supset (A \vee B))$, 5b. $(B \supset (A \vee B))$,
6. $((A \supset C) \supset ((B \supset C) \supset ((A \vee B) \supset C)))$,
7. $((A \supset B) \supset ((A \supset \neg B) \supset \neg A))$,
8. $(\neg \neg A \supset A)$,
9. $(\forall x\, A \supset [A]^x_T)$,
10. $([A]^x_T \supset \exists x\, A)$.

In addition, this calculus has three derivation rules: 'from A and $(A \supset B)$ one can obtain B'; 'from $(D \supset A)$ one can obtain $(D \supset \forall x\, A)$'; and 'from $(A \supset D)$ one can obtain $(\exists x\, A \supset D)$'. *Provable formulas* (or *theorems*) of the calculus in question are any formulas that can be obtained from the axioms of the calculus as a result of applying (possibly repeatedly) the given rules (see **Derivation, logical**).

A basic interpretation of the predicate calculus. The domain of values of object variables is a non-empty set of objects M, that of function variables consists of functions from M^l into M, and that of predicate variables consists of functions from M^k into $\{0, 1\}$ (one of the values is interpreted as 'truth', the other as 'falsehood'), where the number k corresponds to the dimension of the predicate variable. Now for any atomic formula, fixing the value of the predicate variables in it and the values of the object and function variables that occur in it, one can talk of the truth or falsehood of this formula. Similarly, using truth tables for propositional connectives and the usual interpretation of quantifiers (as infinite conjunctions and disjunctions), one can judge the truth of an arbitrary formula of the language in question for the chosen M and the chosen values of the predicate, function and free object variables that occur in it. A formula is said to be *universally valid* (*generally valid*) if it is true for any such choice. Thus, whatever the values of a two-place predicate variable P and one-place function variable f, from the fact that for some x and any y the formula $P(f(x), f(y))$ is true it follows that there is a z for which $P(z, f(z))$ is true. Consequently, the formula

$$(\exists x\, \forall y\, P(f(x), f(y)) \supset \exists z\, P(z, f(z)))$$

is universally valid. One can prove that a formula is derivable in the calculus thus constructed if and only if it is universally valid (the so-called **Gödel completeness theorem**). This interpretation relies on rather complicated set-theoretic abstractions and is therefore inadmissible from the point of view of certain philosophies of mathematics and meta-mathematical theories (for example, **intuitionism**; **finitism**; **constructive mathematics**). In the framework of these theories one can obtain a completeness theorem by changing the semantics of the logical calculus.

Numerous logical calculi are obtained by modification of the logical calculus constructed above. Thus, the addition to the language of the symbol '$=$' together with the schemes

11. $(T = T)$,

12. $((T_1 = T_2) \supset ([A]_{T_1}^x = [A]_{T_2}^x))$

(here A and T are arbitrary, and T_1 and T_2 are free for x in A) leads to the classical *predicate calculus with equality*. Exclusion from the language of function variables leads to the *pure* (or *narrow* or *restricted*) *predicate calculus*. The axiom schemes 1) - 8) in conjunction with the first derivation rule give the *classical propositional calculus*; since the subject-predicate structure of inferences cannot be analyzed by the tools of predicate calculi, instead of various types of variables in the language of these calculi one usually needs only one type — propositional variables, each of which acts as an atomic formula. The rejection of scheme 8 from all the calculi mentioned above leads to *minimal logical calculi*, and the rejection of schemes 7 and 8 leads to *positive logical calculi*. Other *partial logical calculi* are possible, for example those obtained by fixing part of the logical symbols or part of the variables of the language (in combination with a possible reconstruction of the system of axioms) while preserving all classically-derivable formulas consisting of these symbols and variables; these are the **implicative propositional calculus** (the only symbol is $\supset$), the pure *one-place (monadic) predicate calculus* (in the language there are only object variables and one-place predicate variables), etc. More meaningful examples of partial logical calculi are the *intuitionistic (constructive) calculi*, which are obtained from the classical calculi mentioned above by replacing scheme 8 by the scheme

8'. $(A \supset (\neg A \supset B))$.

The names of logical calculi are naturally formed from the terms mentioned; thus, the schemes 1 - 7, 8', 11, 12 define *intuitionistic propositional calculus with equality*.

One also considers *many-sorted logical calculi* (and terms), where the substitution of terms of one kind for those of another is not allowed. In simple cases the domains of values of terms of different kinds are interpreted as different sets of objects (thus, convenient formalizations of plane geometry can be based on logical calculi with object variables of two kinds — 'points' and 'straight lines'). But one can successively consider first a calculus with a unique domain of objects, then a calculus with an additional domain of objects, namely predicates over the first domain (that is, in the second calculus one admits quantifiers with respect to the predicate variables of the first), etc. Thus there arise *higher-order logical calculi* (the logical calculus mentioned earlier is of the first order). The tendency to formalize logical theories with a more powerful supply of concepts leads to a number of generalizations of logical calculi. The consideration, together with 'truth' and 'falsehood', of various degrees of indeterminacy leads to various formalizations of many-valued logics (cf. **Many-valued logic**) and *calculi of partial predicates*. The latter are closely related to calculi of logical consequences and the **strict implication calculus**, which arose as a result of attempts to formalize the common use of the expression 'A implies B' by removing the paradoxes of material implication and rejecting its definition in the form of a table. *Modal logical calculi* serve to formalize the distinction, studied in **modal logic**, between 'necessary', 'possible' and 'contingent' assertions.

Together with the specification of a logical calculus in terms of axiom schemes, one often meets formulations with finitely many specific axioms, but with the addition of various rules of substitution for variables. Reformulations of a logical calculus in the form of a **Gentzen formal system** are convenient in many questions of proof theory.

A calculus is a completely adequately formalized theory if derivability of a formula in it is equivalent to its identical truth in the basic interpretation. The truth of derivable formulas is connected with the consistency (soundness, cf. **Sound rule**) of the calculus, and the derivability of all true formulas is connected with its completeness. All logical calculi mentioned above are sound, and many of them are complete in one sense or another (see **Gödel completeness theorem**). An important property of logical calculi is decidability (see **Solvable problem**): almost-all propositional calculi that have been constructed are decidable; on the other hand, all the predicate calculi mentioned above (except the monadic one) are undecidable. Nevertheless, there are algorithms for undecidable logical calculi that for each derivable formula establish its derivability, but for certain underivable formulas they need not terminate.

References

[1] HILBERT, D. and BERNAYS, P.: *Grundlagen der Mathematik*, 1, Springer, 1968.
[2] KLEENE, S.C.: *Introduction to metamathematics*, North-Holland, 1951.

[3] NOVIKOV, P.S.: *Elements of mathematical logic*, Oliver & Boyd and Acad. Press, 1964 (translated from the Russian).
[4] CHURCH, A.: *Introduction to mathematical logic*, 1, Princeton Univ. Press, 1956.
[5] *Mathematical theory of logical derivation*, Moscow, 1967 (in Russian). Collection of translations.

S.Yu. Maslov

Editorial comments.

References
[A1] KLEENE, S.C.: *Mathematical logic*, Wiley, 1967.
[A2] KREISEL, G. and KRIVINE, J.L.: *Elements of mathematical logic*, North-Holland, 1967 (translated from the French).
[A3] WÓJCICKI, R.: *Theory of logical calculi*, Kluwer, 1988.

AMS 1980 Subject Classification: 03BXX

LOGICAL CONSEQUENCE *of a given set of premises* - A proposition that is true for any interpretation of the non-logical symbols (that is, the names (cf. **Name**) of objects, functions, predicates) for which the premises are true. If a proposition A is a logical consequence of a set of propositions Γ, one says that Γ logically implies A, or that A follows logically from Γ.

If Γ is a set of statements of a formalized first-order **logico-mathematical language** and A is a proposition of this language, then the relation 'A is a logical consequence of Γ' means that any model for Γ is a model for A. This relation is denoted by $\Gamma \vDash A$. The **Gödel completeness theorem** of classical predicate calculus implies that the relation $\Gamma \vDash A$ coincides with the relation $\Gamma \vdash A$, that is, $\Gamma \vDash A$ if and only if A is deducible from Γ by the methods of classical predicate calculus.

References
[1] RASIOWA, H. and SIKORSKI, R.: *The mathematics of metamathematics*, PWN, 1963.
[2] GÖDEL, K.: 'Die Vollständigkeit der Axiome des logischen Funktionenkalküls', *Monatsh. Math. Phys.* **37** (1930), 349-360.

V.E. Plisko

Editorial comments.
The phrase '*semantic entailment*' is sometimes used instead of 'logical consequence'; thus, the expression $\Gamma \vDash A$ is read as 'Γ semantically entails A'. The expression $\Gamma \vdash A$ is similarly read as 'Γ syntactically entails A'.

References
[A1] JOHNSTONE, P.T.: *Notes on logic and set theory*, Cambridge Univ. Press, 1987.
[A2] GRZEGORCZYK, A.: *An outline of mathematical logic*, Reidel, 1974.

AMS 1980 Subject Classification: 03BXX

LOGICAL FORMULA - An expression in the language of formal logic. An exact definition of a logical formula is given for each specific logical language. As a rule, the definition of a formula has an inductive character: one distinguishes a class of statements, called *atomic formulas*, and states rules that make it possible to construct new formulas from formulas already constructed, using the symbols for logical operations (cf.

Logical operation). For example, the formulas of propositional logic are defined as follows. Any propositional variable is an (atomic) formula. If A and B are formulas, then $(A\&B)$, $(A \vee B)$, $(A \supset B)$, $(\neg A)$ are formulas. The formulas of predicate logic are constructed from propositional, predicate and object variables by using logical connectives, quantifiers and auxiliary symbols (brackets and commas). Atomic formulas are propositional variables and expressions of the form $P(y_1, \ldots, y_n)$, where P is an n-place predicate variable and $y_1, \ldots, y_n$ are object variables. The *formulas of predicate calculus* are defined as follows: a) any atomic formula is a formula; and b) if A and B are formulas and y is an object variable, then $(\neg A)$, $(A\&B)$, $(A \vee B)$, $(A \supset B)$, $(\forall y\, A)$, $(\exists y\, A)$ are formulas.

V.E. Plisko

Editorial comments.
The term '*well-formed formula*' (sometimes abbreviated to '*wff*' or '*wf*') is in fairly common use.

References
[A1] GRZEGORCZYK, A.: *An outline of mathematical logic*, Reidel, 1974.

AMS 1980 Subject Classification: 03BXX

LOGICAL FUNCTION - An n-place function defined on the set of truth values (cf. **Truth value**) $\{T, F\}$ and taking values in this set. With every **logical operation** $\mathfrak{A}$ is associated a logical function $f_{\mathfrak{A}}$: If $V_1, \ldots, V_n$ are truth values, then $f_{\mathfrak{A}}(V_1, \ldots, V_n)$ is the truth value of the proposition $\mathfrak{A}(P_1, \ldots, P_n)$, where $P_1, \ldots, P_n$ are propositions such that the truth value of P_i is equal to V_i, $i=1, \ldots, n$.

A logical function is sometimes defined as an n-place function defined on a set M and taking values in the set $\{T, F\}$. Such functions are used in mathematical logic as an analogue of the concept of a **predicate**.

V.E. Plisko

AMS 1980 Subject Classification: 03BXX

LOGICAL LAW in mathematical logic - A **logical formula** that becomes a true proposition under any interpretation of the variables for propositions and predicates that occur in it. Such formulas are called *generally valid*, *universally valid* or *tautologies*. For example, the tautology $A \vee \neg A$ expresses the **law of the excluded middle**.

V.E. Plisko

Editorial comments.

References
[A1] GRZEGORCZYK, A.: *An outline of mathematical logic*, Reidel, 1974.

AMS 1980 Subject Classification: 03AXX

LOGICAL MATRIX - A system

$$\mathfrak{M} = <M; D, \&, \vee, \supset, \neg>,$$

where M is a non-empty set; $D \subseteq M$; $\&, \vee, \supset$ are binary operations; and $\neg$ is a unary operation on M. Any formula of propositional logic, constructed from propositional variables $p_1, \ldots, p_n$ by means of the logical connectives $\&, \vee, \supset, \neg$, can be regarded as an n-place function on M if $p_1, \ldots, p_n$ are assumed to be variables with range of values M and the logical connectives are interpreted as the corresponding operations of the logical matrix $\mathfrak{M}$. A formula $\mathfrak{A}$ is said to be *generally valid* in $\mathfrak{M}$ if for any values of the variables in M the value of $\mathfrak{A}$ belongs to D. A logical matrix $\mathfrak{M}$ is said to be *characteristic* for a propositional calculus K if the formulas that are generally valid in $\mathfrak{M}$ are exactly those that are deducible in K. An example of a logical matrix is the system

$$<\{0, 1\}; \{1\}, \&, \vee, \supset, \neg>,$$

where
$$x \& y = \min\{x, y\},$$
$$x \vee y = \max\{x, y\},$$
$$x \supset y = \max\{1-x, y\},$$
$$\neg x = 1 - x.$$

This logical matrix is characteristic for the classical propositional calculus. K. Gödel proved that it is impossible to construct a logical matrix with a finite set M that is characteristic for the intuitionistic propositional calculus.

V.E. Plisko

Editorial comments.

References

[A1] WÓJCICKI, R.: *Theory of logical calculi*, Kluwer, 1988.

AMS 1980 Subject Classification: 03BXX

LOGICAL OPERATION - A method for constructing a compound formula from given formulas, such that the **truth value** of the compound formula is completely determined by the truth values of the subformulas. Examples of logical operations are **conjunction; disjunction; implication;** and **negation.** Quantifiers (cf. **Quantifier**) are also related to logical operations.

V.E. Plisko

AMS 1980 Subject Classification: 03AXX

LOGICISM - One of the trends in the foundations of mathematics whose object is to justify mathematics by reducing its initial concepts to those of logic. The idea of reducing mathematics to logic was expressed by G. Leibniz (end of the 17-th century). The practical realization of the thesis of logicism was undertaken at the end of the 19-th century and the beginning of the 20-th century in papers of G. Frege and B. Russell (see [1], [2]). The view of mathematics as part of logic depends

on the fact that in an axiomatic system any mathematical theorem can be regarded as an assertion about logical consequences. It just remains to define all the constants that occur in such assertions in terms of logical terms. At the end of the 19-th century different forms of numbers, including complex numbers, were defined in mathematics in terms of the natural numbers and operations on them. An attempt to reduce the natural numbers to logical concepts was undertaken by Frege. In Frege's interpretation the natural numbers were the cardinal numbers of certain concepts. However, Frege's system was not free from contradictions. This was clarified when Russell discovered a contradiction in Cantor's set theory (Russell's **antinomy**) in trying to reduce it to logic. This contradiction stimulated Russell to a reconsideration of the views on logic, which he stated in the form of a ramified theory of types (cf. **Types, theory of**). However, the construction of mathematics on the basis of the theory of types required the assumption of axioms that would be unnatural to regard as purely logical. One is, for example, the *axiom of infinity*, which asserts that there are infinitely many individuals, that is, objects of the lowest type. On the whole the attempt to reduce mathematics to logic was not successful. As K. Gödel showed [3], no formalized system of logic can be an adequate basis for mathematics (cf. **Gödel incompleteness theorem**).

References

[1] FREGE, G.: *Grundgesetze der Arithmetik, begriffsschriftlich abgeleitet*, 1-2, Olms, reprint, 1962.
[2] WHITEHEAD, A.N. and RUSSEL, B.: *Principia Mathematica*, 1-2, Cambridge Univ. Press, 1910-1913.
[3] GÖDEL, K.: 'Ueber formal unentscheidbare Sätze der Principia Mathematica und verwandter Systeme', *Monatsh. Math. Phys.* **38** (1931), 173-198.
[4] CURRY, H.B.: *Foundations of mathematical logic*, McGraw-Hill, 1963.
[5] FRAENKEL, A.A. and BAR-HILLEL, Y.: *Foundations of set theory*, North-Holland, 1958.

V.E. Plisko

AMS 1980 Subject Classification: 03A05

LOGICO-MATHEMATICAL CALCULUS, *applied calculus* - A formalization of a mathematical theory. A logico-mathematical calculus is specified by its language and list of postulates (these elements form the **syntax**) and in most cases it is endowed with a **semantics.** The essential features that distinguish logico-mathematical calculi from axiomatic theories of traditional mathematics are: 1) the study of the logical tools used in the formulation of axioms and derivation rules (cf. **Derivation rule**) that make it possible to deduce one statement from another; and 2) the transition from informal language to an exact **formal language.** A logico-mathematical calculus is usually constructed on the basis of a **logical calculus** (the *basic logical calculus*). The language of the logico-mathematical calculus is obtained from the language of this logical calculus by adding symbols for particular functions and predicates (and possibly removing the predicate variables and variables for functions). The list of postulates of the logico-mathematical calculus is obtained by adding to the list of postulates of the basic logical calculus (understood in relation to the new language) some postulates that describe the properties of the added functions and predicates. For example, the language of elementary group theory is obtained according to this scheme from the language of classical predicate calculus with equality: one adds the symbols $\cdot$ (multiplication), inv (inversion) and e (the identity), and discards all predicate symbols except equality. The additional postulate

$$\forall x\, \forall y\, \forall z\, (e{\cdot}x = x\ \&\ \mathrm{inv}(x){\cdot}x = e\ \&\ (x{\cdot}y){\cdot}z = x{\cdot}(y{\cdot}z))$$

asserts that e is the identity of the group, $\mathrm{inv}(x)$ is the element inverse to x and multiplication is associative.

Logico-mathematical calculi constructed on the basis of predicate calculus with equality serve to describe the most frequently encountered mathematical structures. Among them are formal arithmetic (cf. **Arithmetic, formal**), formalized analysis (with quantifiers of orders one and two), **axiomatic set theory**, etc. The semantics of a logico-mathematical calculus gives interpretations of the variables, mathematical symbols (symbols of predicates and functions) and logical operations. These interpretations determine *models of the logico-mathematical calculus*. The truth of a formula in every model in which the axioms of some logico-mathematical calculus based on classical predicate calculus with equality are true implies, by the **Gödel completeness theorem**, that this formula can be deduced in the logico-mathematical calculus in question.

The simplest in structure are *quantifier-free logico-mathematical calculi*. They are used most frequently to describe properties of different classes of computable functions. The language of a quantifier-free logico-mathematical calculus is constructed on the basis of the language of the propositional calculus with equality, or even consists of single equalities. In the first case the logical apparatus of the logico-mathematical calculus is classical propositional calculus, and in the second case it is designed as a calculus of equalities. In both cases the following are regarded as postulates: 1) defining equalities of the functions in question (for example, $x{\cdot}0 = 0$, $x{\cdot}y' = x{\cdot}y + x$); 2) the basic properties of equality; 3) arguments by the method of mathematical induction (most frequently on the pattern 'from $f(0) = g(0)$, $f(x') = h(x, f(x))$, $g(x') = h(x, g(x))$ one can derive $f(x) = g(x)$'); and 4) an argument corresponding to $\forall$-removal: 'given $A(x)$, to derive $A(t)$' (the rule of substituting an object variable for a free variable). An

example is the *system PRA* (*primitive recursive arithmetic*).

The object variables of PRA are $(a), (aa), (aaa), \ldots$; the function variables are $(f), (ff), (fff), \ldots$; the natural numbers are $0, 0', 0'', \ldots$. The function symbols (functors) are constructed from the primitive symbols, $'$ ('successor'), Z (the constant function with value zero), $[J, n, m]$ (the m-place function whose value is equal to the n-th argument), where n, m are natural numbers, $n \leqslant m$, by means of substitution S and primitive recursion R: If ϕ is an n-place functor and $\psi_1, \ldots, \psi_n$ are m-place functors, then $S[\phi, \psi_1, \ldots, \psi_n]$ (the result of substituting $\psi_1, \ldots, \psi_n$ in ϕ) is an m-place functor; if ψ is an n-place functor and ψ is an $(n+2)$-place functor, then $R[\phi, \psi]$ is an $(n+1)$-place functor (primitive recursion:

$$R[\phi, \psi](0, X) = \phi(X);$$
$$R[\phi, \psi](y', X) = \psi(y, X, R[\phi, \psi](y, X))).$$

The terms of PRA are 0, object variables and expressions of the form s', $\phi(s_1, \ldots, s_n)$, where $s, s_1, \ldots, s_n$ are terms and ϕ is a functor.

The formulas of PRA are $r = s$, where r and s are terms. Admissible values of object variables are natural numbers, and admissible values of function variables are primitive recursive functions (sometimes wider classes of computable functions, cf. **Primitive recursive function**).

In the description of partial (not everywhere-defined) functions, apart from the equality predicate there appears the predicate $\uparrow$ or $!$ ('defined'); in this case $r = s$ is interpreted as meaning that r is defined if and only if s is, and that they take the same value whenever they are defined. One also adds means of representing a function that is universal for the class in question: Either a symbol for this function, or the rule: if t is a term, then $<t>$ is a functor (whose number in some previously fixed enumeration of the class in question is equal to t). The postulates of $\forall$-removal and $\exists$-introduction are modified:

$$A(t)\,\&\,!t \rightarrow \exists x\, A(x);\quad \forall x\, A(x)\,\&\,!t \rightarrow A(t).$$

One adds the axiom $!t$, where t is an object variable or a constant, and also axioms of the form

$$!\phi(t) \rightarrow !t.$$

Logico-mathematical calculi are also used to describe computable functions of different types: 0 is a type (the objects of type 0 are the natural numbers); if σ and τ are types, then $(\sigma \rightarrow \tau)$ is a type (of operations that transform objects of type σ into objects of type τ). These are the finite types (see **Types, theory of**). One also considers transfinite types. For every type one introduces variables and constants of this type, usually including a symbol for the operation all values of which are equal to 0, and also the object $'$ of type $(0 \to 0)$. For every σ, among the constants of the type $(\sigma \to ((0 \to (\sigma \to \sigma)) \to (0 \to \sigma)))$ one often includes the operator of primitive recursion. Terms of type σ are variables and constants of type σ, expressions of the form $r(s)$, where r is a term of type $(\tau \to \sigma)$ and s is a term of type τ, and also expressions of the form $\lambda x r$, which is interpreted as the notation of the functional that transforms x into $r(x)$, where r is of type β, x is of type α and $\sigma = (\alpha \to \beta)$. Quantifier-free logico-mathematical calculi for functionals of finite types are used for the mathematical investigation of quantifier logico-mathematical calculi. In particular, by means of a quantifier-free system of primitive recursive functionals it has been possible to prove the relative consistency of formal arithmetic; the addition of the so-called bar-recursion operator makes it possible to prove the relative consistency of formal analysis.

An important property of some logico-mathematical calculi is *completeness*. This means that every formula not containing free variables is deducible or refutable. The completeness of a logico-mathematical calculus implies the solvability of the deducibility problem — the existence of an algorithm that determines for every formula whether it is deducible or not. An example of a complete logico-mathematical calculus is the theory of algebraically closed fields (the *Tarski system*). According to the **Gödel incompleteness theorem**, complete theories are rare: Any consistent, recursively-presented logico-mathematical calculus that contains a very restricted fragment of arithmetic is incomplete. For a wider class of logico-mathematical calculi the deducibility problem is algorithmically unsolvable.

An important characteristic of a logico-mathematical calculus is its expressive capacity. It is often possible to introduce expressive tools that do not occur explicitly in the language of the calculus in question. Thus, in some quantifier-free languages it is possible to introduce logical connectives and restricted quantifiers:

$$x = y \ \& \ u = v \text{ means } |x - y| + |u - v| = 0,$$

$$\forall x \leqslant a (f(x) = g(x)) \text{ means } \sum_{x \leqslant a} |f(x) - g(x)| = 0.$$

In the language of formal arithmetic one can talk of finite sets, partial recursive functions, etc. Some logical connectives are expressible in terms of others; thus, in second-order calculi (including those based on intuitionistic logic) all connectives are expressible in terms of $\forall$ and $\to$; for example, $\exists x A(x)$ is equivalent to $\forall P(\forall x(Ax \to P) \to P)$. The principal restrictions on the expressive capacity of the language are given by *Tarski's theorem*: Under the natural enumeration of the formulas of a language containing a minimum of arithmetic, it is impossible to give a formula $T(x)$ of this language such that $T(n)$ is true if and only if n is the number of a true formula. For some logico-mathematical calculi based on constructive (intuitionistic) logic the disjunction theorem holds: The deducibility of a closed formula $A \vee B$ implies the deducibility of one of the formulas A and B. In the investigation of the structure of such logico-mathematical calculi various analogues of the concept of **realizability** are used. Questions of the consistency of logico-mathematical calculi, the independence of individual postulates, the existence of *individual axiomatics* (that is, such that every deducible formula A has a derivation that uses only the postulates for implication and the symbols that occur in A), and the existence of interpretations of some logico-mathematical calculi in others, are investigated in **proof theory**.

References

[1] Hilbert, D. and Bernays, P.: *Grundlagen der Mathematik*, 1, Springer, 1968.
[2] Novikov, P.S.: *Elements of mathematical logic*, Oliver & Boyd and Acad. Press, 1964 (translated from the Russian).
[3] Kleene, S.C.: *Introduction to metamathematics*, North-Holland, 1951.
[4] Church, A.: *Introduction to mathematical logic*, 1, Princeton Univ. Press, 1956.
[5] Lyndon, R.C.: *Notes on logic*, v. Nostrand, 1966.

G.E. Mints

Editorial comments. The term 'logico-mathematical calculus' is not in common use in English: a more usual name for this concept is '(*first-order*) *theory*'.

References

[A1] Kleene, S.C.: *Mathematical logic*, Wiley, 1967.
[A2] Chang, C.C. and Keisler, H.J.: *Model theory*, North-Holland, 1973.

AMS 1980 Subject Classification: 03B10

LOGISTIC DISTRIBUTION - A probability distribution with distribution function $\psi(ax + b)$, where a is scale parameter, b is a shift and

$$\psi(x) = \frac{1}{1 + e^{-x}}.$$

The function $\psi(x)$ satisfies the differential equation

$$\frac{d\psi}{dx} = \psi(1 - \psi).$$

The logistic distribution is close to the **normal distribution**:

$$\sup_x |\psi(1.7x) - \Phi(x)| < 0.01,$$

where $\Phi(x)$ is the normal distribution function with mean 0 and variance 1. To test the hypothesis of coincidence of the distribution functions of two samples of a logistic distribution with possibly different shifts the **Wilcoxon test** (the **Mann−Whitney test**) is asymptotically optimal. The logistic distribution is sometimes more convenient than the normal distribution in data processing and the interpretation of inferences. In applications the multi-dimensional logistic distribution is also used.

References

[1] KENDALL, M.G. and STUART, A.: *The advanced theory of statistics*, 2. Inference and relationship, Griffin, 1979.
[2] COX, D.R. and HINKLEY, D.V.: *Theoretical statistics*, Chapman & Hall, 1974.

A.I. Orlov

Editorial comments.

References

[A1] JOHNSON, N.L. and KOTZ, S.: *Distributions in statistics*, 1. Continuous univariate distributions, Wiley, 1970.

AMS 1980 Subject Classification: 62E10

LOGISTICS - A term used to denote systems of logic characterized by an attempt to reduce logical arguments to formal calculations. In Antiquity and in the Middle Ages the term 'logistics' meant practical operations of arithmetical calculations. G. Leibniz (end of the 17-th century) used the term 'logistics' to denote the calculus of inferences. At the beginning of the 20-th century 'logistics' meant **mathematical logic**.

V.E. Plisko

AMS 1980 Subject Classification: 03-03, 01A35

LOMMEL FUNCTION - A solution of the non-homogeneous **Bessel equation**

$$x^2 y'' + xy' + (x^2 - \nu^2)y = x^\rho.$$

If $\rho = \nu + 2n$, where n is a natural number, then

$$y = (-1)^{n-1}(n-1)! 2^{\nu+2n-2} \times$$

$$\times \sum_{k=0}^{n-1} (-1)^k \left(\frac{x}{2}\right)^{\nu+2k} \frac{\Gamma(\nu+n)}{k!\Gamma(\nu+k+1)}.$$

If the numbers $\rho + \nu \geqslant 0$ and $\rho - \nu \geqslant 0$ are not integers, then

$$y = 2^{\rho-2}\Gamma\left(\frac{\rho+\nu}{2}\right)\Gamma\left(\frac{\rho-\nu}{2}\right) \times$$

$$\times \sum_{k=0}^{\infty} \frac{(-1)^k(x/2)^{\rho+2k}}{\Gamma(k+1+(\rho+\nu)/2)\Gamma(k+1+(\rho-\nu)/2)}.$$

If $\rho = \nu - 2n$, where $n \geqslant 0$ is an integer and ν is not an integer $\leqslant n$, then

$$y = \frac{\Gamma(\nu-n)}{n! 2^{-\nu+2n+2}}\left[2J_\nu(x)\ln\frac{x}{2} +\right.$$

$$-\sum_{k=0}^{n-1} \frac{(n-k-1)!}{\Gamma(\nu-n+k+1)}\left(\frac{x}{2}\right)^{\nu-2n+2k} +$$

$$\left. -\sum_{k=0}^{\infty} \frac{(-1)^k(x/2)^{\nu+2k}}{k!\Gamma(\nu+k+1)}\left[\frac{\Gamma'(k+1)}{\Gamma(k+1)} + \frac{\Gamma'(\nu+k+1)}{\Gamma(\nu+k+1)}\right]\right].$$

Here, for $n = 0$ the first sum is taken to be zero, and $J_\nu(x)$ is a Bessel function (cf. **Bessel functions**). Lommel functions in two variables are also known.

See also **Anger function; Weber function; Struve function**.

Lommel functions were studied by E. Lommel [1].

References

[1] LOMMEL, E.: 'Zur Theorie der Bessel'schen Funktionen IV', *Math. Ann.* **16** (1880), 183-208.
[2] WATSON, G.N.: *A treatise on the theory of Bessel functions*, 1, Cambridge Univ. Press, 1952.
[3] KAMKE, E.: *Differentialgleichungen. Lösungsmethoden und Lösungen*, 1. Gewöhnliche Differentialgleichungen, Chelsea, reprint, 1947.

E.D. Solomentsev

AMS 1980 Subject Classification: 33A65

LOMMEL POLYNOMIAL - The polynomial $R_{m,\nu}(z)$ of degree m in z^{-1} which for $m = 0, 1, \ldots$ and any ν is defined by

$$R_{m,\nu}(z) =$$

$$= \frac{\pi z}{2\sin\nu\pi}[J_{\nu+m}(z)J_{-\nu+1}(z) + (-1)^m J_{-\nu-m}(z)J_{\nu-1}(z)]$$

or

$$R_{m,\nu}(z) = \frac{\Gamma(\nu+m)}{\Gamma(\nu)}\left(\frac{2}{z}\right)^m \times$$

$$\times {}_2F_3\left[\frac{1-m}{2}, -\frac{m}{2}; \nu, -m, 1-\nu-m; -z^2\right].$$

Here $J_\mu(z)$ is the Bessel function (cf. **Bessel functions**) and ${}_2F_3$ is the **hypergeometric series**. The Lommel polynomials satisfy the relations

$$J_{\nu+m}(z) = J_\nu(z)R_{m,\nu}(z) - J_{\nu-1}(z)R_{m-1,\nu+1}(z),$$

$$R_{0,\nu}(z) = 1, \quad m = 1, 2, \ldots..$$

References

[1] MAGNUS, W., OBERHETTINGER, F. and SONI, R.P.: *Formulas and theorems for the special functions of mathematical physics*, Springer, 1966.

A.B. Ivanov

AMS 1980 Subject Classification: 33A65

LONGMAN METHOD - A method for the approximate calculation of a definite integral

$$I = \int_a^b f(x)\,dx,$$

where f has exactly n roots x_i inside the interval $[a, b]$,

$$x_0 = a < x_1 < \cdots < x_n < b = x_{n+1},$$

and satisfies the conditions stated below. Let

$$v_i = (-1)^i \int_{x_i}^{x_{i+1}} f(x)\,dx, \quad i = 0, \ldots, n,$$

then $I = S$, where

$$S = \sum_{j=0}^{n}(-1)^j v_j.$$

It is assumed that f preserves its sign on the interval $[x_i, x_{i+1}]$, has different signs on adjacent intervals, and $v_i \neq 0$, $i = 0, \ldots, n$. Such a function f is said to be *oscillatory*. The calculation of I by means of a quadrature formula for large n is difficult, since a good approximation of an oscillatory function on the whole interval $[a, b]$ is impossible in practice. The use of the equality $I = S$ leads to the need to calculate all integrals v_j, which is also inadvisable in the case of large n.

The approximate calculation of I in Longman's method is based on the equality ($n \geqslant p$)

$$S = \sum_{k=0}^{p-1} (-1)^k 2^{-k-1} \Delta^k v_0 + \qquad (1)$$
$$+ (-1)^n \sum_{k=0}^{p-1} 2^{-k-1} \Delta^k v_{n-k} + 2^{-p} (-1)^p \sum_{k=0}^{n-p} (-1)^k \Delta^p v_k.$$

In (1) the finite differences of v_j as functions of the discrete argument j occur:

$$\Delta v_j = v_{j+1} - v_j, \quad j = 0, \ldots, n-1,$$
$$\Delta^{r+1} v_j = \Delta^r v_{j+1} - \Delta^r v_j,$$
$$r = 1, \ldots, p-1; \quad j = 0, \ldots, n-r-1.$$

If v_j is such that on the right-hand side of (1) one can neglect terms containing finite differences of order p, then the approximate equality

$$S \cong \sum_{k=0}^{p-1} (-1)^k 2^{-k-1} \Delta^k v_0 + (-1)^n \sum_{k=0}^{p-1} 2^{-k-1} \Delta^k v_{n-k} \qquad (2)$$

can be used to calculate S. To calculate the right-hand side of (2) it is sufficient to know the first p values v_j, that is, the values $v_0, \ldots, v_{p-1}$, and the last p values $v_n, \ldots, v_{n-p+1}$. Longman's method consists in the use of (2) for an approximate calculation of the sum S.

If in the integral I the upper limit of integration $b = +\infty$ and

$$I = \sum_{i=0}^{\infty} (-1)^i v_i,$$

then instead of (1) one must use the equality

$$\sum_{i=0}^{\infty} (-1)^i v_i = \sum_{i=0}^{\infty} (-1)^i 2^{-i-1} \Delta^i v_0$$

(the *Euler transform*) and replace the series on the right-hand side by a partial sum.

The method was proposed by I.M. Longman [1].

References

[1] LONGMAN, I.M.: 'A method for the numerical evaluation of finite integrals of oscillatory functions', *Math. Comput.* **14**, no. 69 (1960), 53–59.
[2] DAVIS, P.J. and RABINOWITZ, P.: *Methods of numerical integration*, Acad. Press, 1984.

I.P. Mysovskikh

AMS 1980 Subject Classification: 65D30

LOOP - A **quasi-group** with an identity, that is, with an element e such that $xe = ex = x$ for every element x of the quasi-group. The significance of loops in the theory of quasi-groups is determined by the following theorem: Any quasi-group is isotopic (see **Isotopy**) to a loop. Therefore, one of the main problems in the theory of quasi-groups is to describe the loops to which the quasi-groups of a given class are isotopic.

With every loop there are associated three kernels. The set

$$N_l = \{a: ax \cdot y = a \cdot xy \text{ for all } x, y \in Q\}$$

of elements of a loop $Q(\cdot)$ is called the *left kernel*. Similarly one defines the *middle* and *right* kernels. They always exist in a loop. Their intersection is called the *kernel of the loop*. Every kernel is an associative subloop, that is, a subgroup of $Q(\cdot)$. Corresponding kernels of isotopic loops are isomorphic. There are loops with any preassigned kernels. A loop Q isotopic to a group $Q(\cdot)$ is itself a group and is isomorphic to the group $Q(\cdot)$ (*Albert's theorem*). In particular, isotopic groups are isomorphic. Some other classes of loops also have this property, for example free loops. A loop $Q(\cdot)$ is called a *G-loop* if any loop isotopic to $Q(\cdot)$ is isomorphic to it.

Many concepts and results of group theory can be extended to loops. However, some of the usual properties of a group need not hold for loops. Thus, in finite loops Lagrange's theorem (that the order of a subgroup divides the order of the group) does not hold, generally speaking. If, nevertheless, Lagrange's theorem does hold for a loop, then the loop is said to be *Lagrangian*. If every subloop of a loop $Q(\cdot)$ is Lagrangian, one says that $Q(\cdot)$ has the *property L'*. A necessary and sufficient condition for a loop $Q(\cdot)$ to have the property L' is the following: $Q(\cdot)$ must have a normal chain

$$Q = Q_0 \supset Q_1 \supset \cdots \supset Q_n = e,$$

where Q_i is a normal subloop in Q_{i-1}, such that Q_{i-1} / Q_i has the property L' for all $i = 1, \ldots, n$.

The loops that have been most studied and are closest to groups are Moufang loops (cf. **Moufang loop**). The main theorem about them (*Moufang's theorem*) is: If three elements a, b, c of such a loop are connected by the associative law, that is, if

$$ab \cdot c = a \cdot bc,$$

then they generate an associative subloop, that is, a group. In particular, any Moufang loop is *di-associative*, that is, any two elements of it generate an associative subloop. The property of being a Moufang loop is *universal*, that is, it is invariant under isotopy: Any loop isotopic to a Moufang loop is itself a Moufang loop.

One of the most general classes of loops is the class of *IP-loops*, or *loops with the invertibility property*. They are defined by the identities

$$^{-1}x \cdot xy = y \quad \text{and} \quad yx \cdot x^{-1} = y.$$

Here ^{-1}x and x^{-1} are, respectively, the left and right inverse elements of x. Any Moufang loop is an IP-loop. In an IP-loop all kernels coincide. The kernel of a Moufang loop is a normal characteristic subloop. The property of being an IP-loop is not universal. Moreover, if any isotope of an IP-loop $Q(\cdot)$ is an IP-loop, then $Q(\cdot)$ is a Moufang loop.

More general than the class of IP-loops is the class of *WIP-loops*, or *loops with the weak invertibility property*. They are defined by the identity $x(yx)^{-1}=y^{-1}$. This identity is universal if

$$(yx)(\theta_y z \cdot y) = (y \cdot xz)y$$

for all $x, y, z \in Q$, where θ_y is an automorphism. In this case the kernel of the WIP-loop is normal and the quotient loop with respect to the kernel is a Moufang loop. A special case of WIP-loops are the *CI-loops*, or *cross-invertible loops*, defined by the identity $(xy)x^{-1}=y$.

A generalization of Moufang loops are the (left) *Bol loops*, in which the identity

$$x(y \cdot xz) = (x \cdot yx)z$$

holds. They are invariant under isotopy and are *mono-associative*, that is, every element of such a loop generates an associative subloop.

An important concept in the theory of quasi-groups and loops is the concept of a *pseudo-automorphism*. A permutation ϕ of a loop $Q(\cdot)$ is called a *left pseudo-automorphism* if there is an element $a \in Q$ such that

$$\phi(xy) \cdot a = \phi(x)(\phi y \cdot a) \quad \text{for all } x, y \in Q,$$

and is called a *right pseudo-automorphism* if there is an element $b \in Q$ such that

$$b \cdot \phi(xy) = (b \cdot \phi x)\phi y \quad \text{for all } x, y \in Q.$$

If ϕ is both a left and a right pseudo-automorphism, then ϕ is called a *pseudo-automorphism*, and the elements a and b are called the *left* and *right companions*, respectively. For loops an automorphism is a special case of a pseudo-automorphism. Every pseudo-automorphism of an IP-loop induces an automorphism in its kernel, and in a commutative Moufang loop any pseudo-automorphism is an automorphism.

In the theory of loops a significant role is played by inner permutations. A permutation α of the associated group G of a loop $Q(\cdot)$ with an identity e is said to be *inner* if $\alpha e = e$. The set I of all inner permutations is a subgroup of G and is called the *group of inner permutations*. The group I is generated by permutations of three types:

$$L_{x,y} = L_{xy}^{-1}L_x L_y; \quad R_{x,y} = R_{xy}^{-1}R_y R_x; \quad T_x = R_x^{-1}L_x.$$

By means of inner permutations one can define *A-loops* as loops for which all inner permutations are automorphisms. If an A-loop is also an IP-loop, then it is di-associative. Commutative di-associative A-loops are Moufang loops. For commutative Moufang loops inner permutations are automorphisms.

Some definitions of group theory carry over to loops. Thus, a loop is said to be *Hamiltonian* if any subgroup of it is normal. Abelian groups are also regarded as Hamiltonian loops. Mono-associative Hamiltonian loops with elements of finite order are direct products of Hamiltonian p-loops (a p-loop is defined in the same way as a p-group). Di-associative Hamiltonian loops are either Abelian groups or direct products: $A \times T \times H$, where A is an Abelian group whose elements have odd order, T is an Abelian group of exponent 2 and H is a non-commutative loop satisfying additional conditions.

A loop $Q(\cdot)$ is said to be *totally (partially) ordered* if Q is a totally (partially) ordered set (with respect to $\leqslant$) and if $a \leqslant b$ implies

$$ac \leqslant bc, \quad ca \leqslant cb,$$

and conversely. If the centre of a totally ordered loop $Q(\cdot)$ has finite index, then $Q(\cdot)$ is centrally nilpotent. Lattice-ordered loops with the minimum condition for elements are free Abelian groups.

Loops have also been studied by means of associated groups. It has been proved, for example, that there is a one-to-one correspondence between the normal subloops of a loop and the normal subgroups of the corresponding associated group.

For references see **Quasi-group**.

V.D. Belousov

AMS 1980 Subject Classification: 20N05

LOOP, ANALYTIC - An **analytic manifold** M endowed with the structure of a **loop** whose basic operations (multiplication, left and right division) are analytic mappings of $M \times M$ into M. If e is the identity of the loop M, and $g(t)$ and $h(t)$ are analytic paths starting from e and having tangent vectors a and b at e, then the tangent vector $c = ab$ at e to the path $k(t)$, where

$$k(t^2) = (g(t)h(t))/(h(t)g(t)),$$

where $/$ stands for right division, is a bilinear function of the vectors a and b. The tangent space $T(M)$ at e with the operation of multiplication $c = ab$ is called the *tangent algebra of the loop M*. In some neighbourhood U of the element $e = (0, \ldots, 0)$ the coordinates $(x^1, \ldots, x^n)$ are said to be *canonical of the first kind* if for any vector $a = (a^1, \ldots, a^n)$ the curve $x(t) = (a^1 t, \ldots, a^n t)$ is a local one-parameter subgroup $(\,|\,t\,| \leqslant \epsilon)$ with tangent vector a at e (see [1]). A power-associative analytic loop (cf. **Algebra with associative power**) has canonical coordinates of the first kind [2]. In this case the mapping $a \to x(1)$, defined for sufficiently small a, makes it possible to identify U with a neighbourhood of the origin in $T(M)$ and to endow $T(M)$ with the structure of a local analytic loop M_0. If an analytic loop M is alternative, that is, if any two elements of it generate a subgroup, then the tangent algebra $T(M)$ is a **binary Lie algebra**, and the multiplication $(x, y) \rightarrow x \circ y$ in M_0 can be expressed by the **Campbell–Hausdorff formula**. Any finite-dimensional binary Lie algebra over the field $\mathbf{R}$ is the tangent algebra of one and only one (up to local isomorphisms) local alternative analytic loop [1].

The most fully studied are analytic Moufang loops (cf. **Moufang loop**). The tangent algebra of an analytic Moufang loop satisfies the identities

$$x^2 = 0, \quad J(x, y, xz) = J(x, y, z)x,$$

where

$$J(x, y, z) = (xy)z + (yz)x + (zx)y;$$

such algebras are called *Mal'tsev algebras*. Conversely, any finite-dimensional Mal'tsev algebra over $\mathbf{R}$ is the tangent algebra of a simply-connected analytic Moufang loop M, defined uniquely up to an isomorphism (see [2], [3]). If M' is a connected analytic Moufang loop with the same tangent algebra, and hence is locally isomorphic to M, then there is an epimorphism $M \rightarrow M'$ whose kernel H is a discrete normal subgroup of M; the fundamental group $\pi(M')$ of the space M' is isomorphic to H. If ϕ is a local homomorphism of a simply-connected analytic Moufang loop M into a connected analytic Moufang loop M', then ϕ can be uniquely extended to a homomorphism of M into M'. The space of a simply-connected analytic Moufang loop with solvable Mal'tsev tangent algebra is analytically isomorphic to the Euclidean space $\mathbf{R}^n$ (see [3]).

References

[1] MAL'TSEV, A.I.: 'Analytic loops', *Mat. Sb.* **36**, no. 3 (1955), 569-578 (in Russian).
[2] KUZ'MIN, E.N.: 'On the relation between Mal'tsev algebras and analytic Moufang loops', *Algebra and Logic* **10**, no. 1 (1971), 1-14. (*Algebra i Logika* **10**, no. 1 (1971), 3-22)
[3] KERDMAN, F.S.: 'On global analytic Moufang loops', *Soviet Math. Dokl.* **20** (1979), 1297-1300. (*Dokl. Akad. Nauk SSSR* **249**, no. 3 (1979), 533-536)

E.N. Kuz'min

Editorial comments.

References

[A1] CHEIN, O., PFLUGFELDER, H. and SMITH, J.D.H. (EDS.): *Theory and application of quasigroups and loops*, Heldermann, 1989.

AMS 1980 Subject Classification: 20N05, 57-XX

LOOP (IN TOPOLOGY) - A closed **path**. In detail, a loop f is a **continuous mapping** of the interval $[0, 1]$ into a topological space X such that $f(0) = f(1)$. The set of all loops in a space X with a distinguished point $*$ for which $f(0) = f(1) = *$ forms the **loop space**.

M.I. Voĭtsekhovskiĭ

AMS 1980 Subject Classification: 55P35

LOOP SPACE - The space ΩX of all loops (cf. **Loop**

(in topology)) at the point $*$ in a topological space X with distinguished point $*$, endowed with the compact-open topology. A loop space is a fibre in the **Serre fibration** (E, p, X) over the space X (here E is the **path space**).

A.F. Kharshiladze

Editorial comments.

References

[A1] ADAMS, J.F.: *Infinite loop spaces*, Princeton Univ. Press, 1978.

AMS 1980 Subject Classification: 55P35, 55P47

LORENTZ FORCE - The force that a given electromagnetic field exerts on a moving electrically-charged particle. An expression of the Lorentz force $\mathbf{F}$ was first given by H.A. Lorentz (see [1]):

$$\mathbf{F} = e\mathbf{E} + \frac{e}{c}[\mathbf{V}, \mathbf{B}], \tag{1}$$

where $\mathbf{E}$ is the electric field strength, $\mathbf{B}$ is the magnetic induction, $\mathbf{V}$ is the velocity of the charged particle with respect to the coordinate system in which the quantities $\mathbf{E}, \mathbf{B}, \mathbf{F}$ are calculated, e is the charge of the particle, and c is the velocity of light in vacuum. The expression for the Lorentz force is relativistically invariant (that is, it holds in any inertial reference system); it makes it possible to connect the equations for an electromagnetic field with the equations of motion of charged particles.

In a constant and uniform magnetic field the motion of a particle with mass m and charge e in a non-relativistic approximation ($\mathbf{V} \ll c$) is described by the equation

$$m\frac{d\mathbf{V}}{dt} = \frac{e}{c}[\mathbf{V}, \mathbf{B}]. \tag{2}$$

In a rectangular coordinate system with z-axis directed along the outward magnetic field B, the solution of (2) has the form

$$x = x_0 + r\sin(\omega_L t + \alpha), \quad y = y_0 + r\cos(\omega_L t + \alpha),$$

$$z = z_0 + \mathbf{V}_{0z}t,$$

where $\omega_L = e\,|\,\mathbf{B}\,|\,/mc$ is the Larmor frequency of rotation of the particle, $r = |\,\mathbf{V}_{0t}\,|\,/\omega_L$ is the radius of rotation of the particle (the **Larmor radius**), α is the initial phase of the rotation, and $\mathbf{V}_0$ is the initial velocity of the particle. Thus, in a uniform magnetic field the charge moves along a helix with axis along the magnetic field.

If the electric field $\mathbf{E}$ is not equal to zero, the motion has a more complicated character. There occurs a displacement of the centre of rotation of the particle across the field $\mathbf{B}$ (so-called *drift*). The mean value of drift in vector form is

$$\mathbf{V} = c\frac{[\mathbf{E}, \mathbf{B}]}{|\,\mathbf{B}\,|^2}.$$

The unaveraged motion of the particle in the xy-plane in this case takes place along a trochoid.

References

[1] LORENTZ, H.A.: *The theory of electrons and its applications to the phenomena of light and radiant heat*, Teubner, 1909.
[2] LANDAU, L.D. and LIFSCHITZ, E.M.: *The classical theory of fields* , Pergamon, 1975 (translated from the Russian).

V.V. Parail

Editorial comments.

References

[A1] LEVICH, B.G.: *Theoretical physics*, 1, North-Holland, 1970.
[A2] HYLLERAAS, E.A.: *Mathematical and theoretical physics*, 2, Wiley, 1970.
[A3] CLEMMOW, P.C. and DOUGHERTY, J.P.: *Electrodynamics of particles and plasmas*, Addison-Wesley, 1969.

AMS 1980 Subject Classification: 78A25

LORENTZ TRANSFORMATION - A coordinate transformation that connects two Galilean coordinate systems (cf. **Galilean coordinate system**) in a **pseudo-Euclidean space**; in other words, a Lorentz transformation preserves the square of the so-called interval between events. A Lorentz transformation is an analogue of an orthogonal transformation (or a generalization of the concept of a motion) in Euclidean space. The Lorentz transformations form a group, called the *Lorentz group* (or the *general Lorentz group*), which is denoted by L. Lorentz transformations find applications in the four-dimensional space-time of the special theory of relativity, for which in Galilean coordinates x, y, z, t the interval has the form

$$s^2 = c^2(\Delta t)^2 - (\Delta x)^2 - (\Delta y)^2 - (\Delta z)^2,$$

where c is the velocity of light in vacuum.

One often considers more restricted classes of Lorentz transformations. Thus, the Lorentz transformations that preserve the sign of the coordinate t form the so-called *orthochronous Lorentz group* $L_\uparrow$. The Lorentz transformations whose matrices have positive determinants are called *proper Lorentz transformations* and form the *proper Lorentz group* L_+. The intersection of $L_\uparrow$ and L_+ is often simply called the *Lorentz group*.

The general Lorentz group consists of combinations of spatial reflections, reflections in time, spatial rotations, and transformations that from the physical point of view are the transformations of transition from one inertial reference system to another, moving with respect to the first with velocity V, and from the mathematical point of view are hyperbolic rotations through an angle ψ in a plane with a pseudo-Euclidean metric. The existence of the latter type of transformations is a specific feature of the group of Lorentz transformations.

For the transition from the Galilean coordinate system x', y', z', t' to the Galilean coordinate system x, y, z, t, moving with respect to the first with velocity V parallel to the x'-axis, these transformations have the form

$$x = \frac{x' - Vt'}{\sqrt{1 - V^2/c^2}}; \ y = y'; \ z = z'; \ t = \frac{t' - Vx'/c^2}{\sqrt{1 - V^2/c^2}}.$$

If one introduces the angle of hyperbolic rotation ψ by the formulas

$$\sinh\psi = -\frac{V/c}{\sqrt{1 - V^2/c^2}}, \ \cosh\psi = \frac{1}{\sqrt{1 - V^2/c^2}},$$

then the Lorentz transformations take the form

$$x = x'\cosh\psi + ct'\sinh\psi;$$
$$ct = x'\sinh\psi + ct'\cosh\psi;$$
$$y = y';$$
$$z = z'.$$

These transformations are often simply called Lorentz transformations. They do not form a group: the action of three hyperbolic rotations with non-parallel velocity vectors may give an ordinary spatial rotation, the so-called *Thomas precession*.

One often supplements the general Lorentz transformations with displacements of the origin, thus obtaining the so-called *Poincaré transformations*, which form the *Poincaré group*.

The properties of the group of Lorentz transformations are similar to the properties of the orthogonal groups (cf. **Orthogonal group**). The differences are connected with the existence of two types of reflections (space and time) and with the fact that the group of Lorentz transformations is not compact (since the unit sphere in a pseudo-Euclidean space, that is, the set of points for which the modulus of the interval up to the origin is equal to one, is not compact).

The physical applications of Lorentz transformations are connected with Einstein's **relativity principle**, according to which all physical laws, except the law of gravitation, are invariant under Lorentz transformations. In a number of cases, for example in axiomatic quantum field theory, the use of this and other equally general postulates makes it possible to make far-reaching deductions about the forms of functional dependencies between different physical quantities.

In various branches of physics (particularly in the theory of elementary particles) representations of both the homogeneous Lorentz group and the Poincaré group find wide application. In accordance with Einstein's relativity principle physical quantities with different transformation properties — vectors, spinors, tensors — are transformed according to various representations of the Lorentz group. It turns out that relevant unitary representations of the Poincaré group can be characterized by two invariants, to be identified with mass and spin of the particles, the quantum-mechanical states of which transform according to the unitary representation in question.

Infinitesimal Lorentz transformations, that is, rotations through an infinitesimal angle, are often used to obtain various conservation laws.

There are also applications of Lorentz transformations in the tangent space of a pseudo-Riemannian space; these transformations are related to the so-called *local space-time symmetries*.

Lorentz transformations take their name from the works of H. Lorentz in electron theory, which have played an important role in the formulation of this theory.

References

[1] LANDAU, L.D. and LIFSCHITS, E.M.: *The classical theory of fields*, Pergamon, 1975 (translated from the Russian).
[2] NAĬMARK, M.A.: *Les répresentations linéaires du groupe de Lorentz*, Dunod, 1962 (translated from the Russian).
[3] *Encyclopeadic dictionary of physics*, 3, Moscow, 1963 (in Russian).

D.D. Sokolov

Editorial comments.

References

[A1] WIGNER, E.P.: 'Unitary representations of the inhomogeneous Lorentz group including reflections', in *Istambul Summer School of Theoretical Physics, 1962*, Gordon & Breach, 1964, pp. 37-80.
[A2] WIGHTMAN, A.S.: 'L'invariance dans la mécanique quantique relativiste', in *Ecole d'Eté de Physique Théorique: LesHouches, 1960*, Hermann & Wiley, 1960, pp. 159-226.
[A3] RUHL, W.: *The Lorentz group and harmonic analysis*, Benjamin, 1970.
[A4] SAXL, R.U. and URBANTKE, H.K.: *Relativität, Gruppen, Teilchen*, Springer, 1976.

AMS 1980 Subject Classification: 83A05

LORENZ ATTRACTOR - A compact invariant set L in the three-dimensional phase space of a smooth flow $\{S_t\}$ which has the complicated topological structure mentioned below and is asymptotically stable (that is, it is Lyapunov stable and all trajectories in some neighbourhood of L tend to L as $t \to \infty$). The concept of an attractor, that is, an attracting set, often includes only the latter of these two properties; however, both the Lorenz attractor and other practically important attractors have both these properties.

The Lorenz attractor first appeared in numerical experiments of E. Lorenz [1], who investigated the behaviour of the trajectories of the system

$$\dot{x} = -\sigma x + \sigma y, \quad \dot{y} = rx - y - xz, \quad \dot{z} = -bz + xy \quad (*)$$

for certain specific values of the parameters σ, r, b. (This system was initially introduced as the first non-trivial Galerkin approximation for certain hydrodynamical problems; this also motivated the choice of values $\sigma = 10$, $r = 28$, $b = 8/3$, but it also arises in other physi-

cal problems, see [2], [3].) In [4] the results of [1] and newer data of numerical experiments were compared in a general way with theoretical ideas in the theory of smooth dynamical systems, and in [5] the results of [1] were interpreted as an indication of the existence in the system (*) of an attractor (called the *Lorenz attractor*), which in many ways is analogous to hyperbolic sets (cf. **Hyperbolic set**), but is not such a set (the main difference lies in the fact that a Lorenz attractor contains an equilibrium position of saddle type with one positive eigen value; for the system (*) this equilibrium position is the origin). The existence of a Lorenz attractor and a number of its properties follow from specific properties of the **Poincaré return map** on some surface Π (for the system (*) one uses the plane $z = 27$), the exact formulation of which, however, is very unwieldy (see [6], [9], [12], [13]) and in the verification of which for specific systems, including (*), one must recourse to numerical integration. Correspondingly, investigations of two types are devoted to Lorenz attractors.

1) In investigations of theoretical character it is assumed from the very beginning that the flows in question have on some surface Π a suitable Poincaré return map, and from this consequences about the properties of the Lorenz attractor are derived. Its structure is described as follows [8]. Consider a 'branched manifold' L_0, on which for $t \geqslant 0$ there is defined a flow ϕ_t, as shown in the figure. Suppose that the Poincaré return map on the branch line ab is (uniformly) expanding, that is, at all points (except c, where it is discontinuous) its derivative is greater than some $\lambda > 1$ (any such λ is suitable, but the investigation is simplified if $\lambda > \sqrt{2}$, which is compatible with the computational data concerning the system (*) for the given σ, r, b). The pairs $\{(L_0, \phi_t)\}$ naturally form an inverse spectrum of topological spaces and mappings (cf. **Spectrum of spaces**); its limit is $(L, S_t|_L)$. (Further study of the structure of the Lorenz attractor, [8], [11], is based on this description, which is therefore naturally included in the definition of the Lorenz attractor, particularly if one does not relate the definition with special properties of the Poincaré map.) Lorenz attractors have **topological transitivity** and the set of their periodic trajectories is dense in L. Under a small (in the sense of C^1) perturbation of such a flow having a suitable Poincaré map, the perturbed flow has a Lorenz attractor close to the Lorenz attractor of the original flow, but generally speaking not homeomorphic to it. In this sense a Lorenz attractor is preserved under small perturbations (in the theory of smooth dynamical systems only two classes of compact invariant sets are known (1982) with this property and whose structure is more-or-less well-studied: Lorenz attractors and locally maximal hyperbolic sets, cf. **Hyperbolic set**), but Lorenz

attractors (in contrast to the latter) do not have the property of **local structural stability**. Ergodic properties of the Lorenz attractor with respect to some 'natural' invariant measures are studied in [7] and [14].

2) In order to discover a Lorenz attractor in a specific system of the type (*) and to determine its properties more exactly, one must use numerical integration together with various theoretical arguments (see [7], [9]). In this way bifurcations arising in the system (*) as r or σ vary and leading to the appearance of Lorenz attractors have been investigated [9]. Naturally, numerical integration by itself gives some information about the attractor (since the trajectories approximate it with time, in a figure, showing successive points of intersection of Π with the trajectories being calculated, these points are situated significantly 'more densely' along the attractor than far from it, and this immediately strikes the eye).

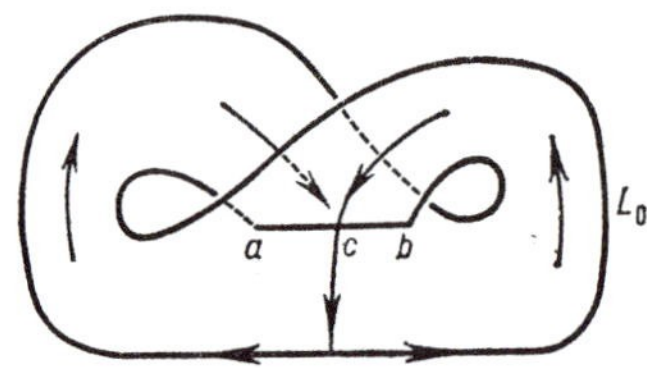

Material on this topic is very extensive; thus, new ranges of values of parameters for which the system (*) has a Lorenz attractor have been found [10]. Here, however, the details of the topological structure, which can differ from the structure of 'standard' Lorenz attractors described above, remain unclarified.

A theoretical interpretation of the data of numerical experiments for systems of order greater than three has so far (1982) not been given.

References

[1]   LORENZ, E.N.: 'Deterministic non-periodic flow', *J. Atmos. Sci.* **20**, no. 2 (1963), 130-141.
[2]   MONIN, A.S.: 'On the nature of turbulence', *Soviet Phys. Uspekhi* **21** (1978), 429-442. (*Uspekhi Fiz. Nauk* **125**, no. 1 (1978), 97-122)
[3]   RABINOVICH, M.I.: 'Stochastic self-oscillations and turbulence', *Soviet Phys. Uspekhi* **21** (1978), 443-469. (*Uspekhi Fiz. Nauk* **125**, no. 1 (1978), 123-168)
[4A] McLUGHLIN, J.B. and MARTIN, P.C.: 'Transition to turbulence of a statically stressed fluid system', *Phys. Rev. Letters* **33** (1974), 1189-1192.
[4B] McLUGHLIN, J.B. and MARTIN, P.C.: 'Transition to turbulence of a statically stressed fluid system', *Phys. Rev. Ser. A* **12** (1975), 186-203.
[5]   RUELLE, D.: 'The Lorentz attractor and the problem of turbulence ', in R. Temam (ed.): *Turbulence and Navier—Stokes equations*, Lecture notes in math., Vol. 565, Springer, 1976, pp. 146-158.
[6]   MARSDEN, J.E., MacCRACKEN, M. and OSTER, G.F.: *The Hopf bifurcation and its applications*, Springer, 1976. (The Russian translation, Mir (1980), contains an appendix on the Lorenz attractor.)
[7A] SINAÏ, YA.G.: 'Stochasticity of dynamical systems', in *Non-linear waves*, Moscow, 1979, pp. 192-212 (in Russian).
[7B] BUNIMOVICH, Z.A. and SINAÏ, YA.G.: 'Stochasticity of the attractor in the Lorenz model', in *Non-linear waves*, Moscow, 1979, pp. 212-226 (in Russian).
[8A] WILLIAMS, R.F.: 'The structure of Lorenz attractors', in R. Temam (ed.): *Turbulence and Navier—Stokes equations*, Lecture notes in math., Vol. 565, Springer, 1976, pp. 146-158.
[8B] WILLIAMS, R.F.: 'The structure of Lorenz attractors', *Publ. Math. IHES* **50** (1979), 73-100.
[9]   AFRAÏMOVICH, V.S., BYKOV, V.V. and SHIL'NIKOV, L.P.: 'On the origin and structure of the Lorenz attractor', *Dokl. Akad. Nauk SSSR* **234**, no. 2 (1977), 336-339 (in Russian).
[10] MORIOKA, N. and SHIMIZU, T.: 'Transition between turbulent and periodic states in the Lorenz model', *Phys. Letters* **66A**, no. 6 (1978), 447-449.
[11] RAND, D.: 'The topological classification of Lorenz attractors', *Math. Proc. Cambridge Phil. Soc.* **83**, no. 3 (1978), 451-460.
[12] AFRAÏMOVICH, V.S., BYKOV, V.V. and SHIL'NIKOV, Z.P.: 'On structurally unstable attracting limit sets of Lorenz-attractor type', *Trans. Moscow Math. Soc.* **44** (1982), 153-216. (*Trudy Moscov. Mat. Obshch.* **44** (1982), 150-122)
[13] GUCKENHEIMER, J. and WILLIAMS, R.F.: 'Structural stability of Lorenz attractors', *Publ. Math. IHES* **50** (1979), 59-72.
[14] PESIN, YA.: 'Ergodic properties and dimensionlike characteristics of strange attractors that are close to hyperbolic', in *Proc. Internat. Congress Mathematics, Berkely 1986*, Amer. Math. Soc., 1987, pp. 1195-1209.

D.V. Anosov

AMS 1980 Subject Classification: 58F12, 54H20

LOSS FUNCTION - In a problem of statistical decision making, a non-negative function indicating the loss (cost) to an experimenter given a particular state of the world and a particular decision. Let X be a random variable taking values in a sample space $(\mathcal{X}, \mathcal{B}, P_\theta)$, $\theta \in \Theta$, and let $D = \{d\}$ be the space of all possible decisions that can be taken on the basis of an observed X. In the theory of statistical decision functions, any non-negative function L defined on $\Theta \times D$ is called a *loss function*. The value of a loss function L at an arbitrary point $(\theta, d) \in \Theta \times D$ is interpreted as the cost incurred by taking a decision $d \in D$, when the true parameter is θ, $\theta \in \Theta$.

References

[1]   WALD, A.: *Statistical decision functions*, Wiley, 1950.
[2]   LEHMANN, E.L.: *Testing statistical hypotheses*, Wiley, 1986.

M.S. Nikulin

Editorial comments.

References

[A1] BERGER, J.: *Statistical decision theory and Bayesian analysis*, Springer, 1985.

AMS 1980 Subject Classification: 62C05

LÖWENHEIM—SKOLEM THEOREM - See **Gödel completeness theorem**.

AMS 1980 Subject Classification: 03C35, 03-XX

LOWER BOUND *of a set* - See **Upper and lower bounds**.

AMS 1980 Subject Classification: 54-XX, 26-XX, 04-XX, 40-XX

LOWER BOUND OF A FAMILY OF TOPOLOGIES *F* (*given on a single set X*) - The set-theoretical intersection of this family, that is, $\cap F$. It is usually denoted by $\wedge F$ and is always a topology on X. If $\mathcal{T}_1$ and $\mathcal{T}_2$ are two topologies on X and if $\mathcal{T}_1$ is contained (as a set) in $\mathcal{T}_2$, then one writes $\mathcal{T}_1 \leqslant \mathcal{T}_2$.

The topology $\wedge F$ has the following property: If $\mathcal{T}_1$ is a topology on X and if $\mathcal{T}_1 \leqslant \mathcal{T}$ for all $\mathcal{T} \in F$, then $\mathcal{T}_1 \leqslant \wedge F$. The free sum of the spaces that are obtained when all the individual topologies in F are put on X can be mapped canonically onto the space $(X, \wedge F)$. An important property of this mapping is that it is a **quotient mapping**. On this basis one proves general theorems on the preservation of a number of properties under the operation of intersecting topologies.

References

[1] ARKHANGEL'SKIĬ, A.V. and PONOMAREV, V.I.: *Fundamentals of general topology: problems and exercises*, Reidel, 1984 (translated from the Russian).

A.V. Arkhangel'skiĭ

Editorial comments. The article above actually defines the *infimum of the family of topologies*, which is a particular (the largest) lower bound for the family; a lower bound being any topology $\leqslant$ this infimum.

AMS 1980 Subject Classification: 54AXX

LOWER LIMIT *of a sequence* $A_1, \ldots, A_n, \ldots$, *of sets of a topological space X* - The collection of points $p \in X$ every neighbourhood of which intersects all elements of the sequence from some N onwards.

For the *lower limit of a sequence of numbers* see **Upper and lower limits**.

A.A. Mal'tsev

Editorial comments. The lower limit of $\{A_i\}_i$ is denoted by Li A_n.

References

[A1] KURATOWSKI, C.: *Topology*, 1, PWN & Acad. Press, 1966 (translated from the French).

AMS 1980 Subject Classification: 54AXX

LÖWNER EQUATION - A differential equation of the form

$$\frac{dw}{dt} = -w\frac{1+e^{i\alpha(t)}w}{1-e^{i\alpha(t)}w},$$

where $\alpha(t)$ is a real-valued continuous function on the interval $-\infty < t < \infty$. A generalization of the Löwner equation is the *Kufarev−Löwner equation*:

$$\frac{dw}{dt} = -wP(w, t),$$

where $P(w, t)$, $|w| < 1$, $-\infty < t < \infty$, is a function measurable in t for fixed w and regular in w, with positive real part, normalized by the condition $P(0, t) = 1$. The Löwner equation and the Kufarev−Löwner equation, which arise in the theory of univalent functions, are the basis of the **variation-parametric method** of

investigating extremal problems on conformal mapping.

The solution $w(t, z, \tau)$, $w(\tau, z, \tau) = z$, of the Kufarev−Löwner equation, regarded as a function of the initial value z, for any $t > \tau$ maps the disc $|z| < 1$ conformally onto a one-sheeted simply-connected domain belonging to the disc $|w| < 1$. From the formula

$$f(z) = a + b \lim_{t \to \infty} e^t w(t, z, 0),$$

by a suitable choice of $P(w, t)$ in the Kufarev−Löwner equation and complex constants a, b one can obtain an arbitrary regular univalent function in the disc $|z| < 1$. In this way the Löwner equation generates, in particular, the conformal mappings of the disc onto domains obtained from the whole plane by making a slit along some Jordan arc (see [1] - [4]).

The partial differential equation

$$\frac{\partial f(z, \tau)}{\partial \tau} = z\frac{\partial f(z, \tau)}{\partial z}P(z, \tau),$$

which is satisfied by the function

$$f(z, \tau) = \lim_{t \to \infty} e^t w(t, z, \tau),$$

is also called the Kufarev−Löwner equation.

The Löwner equation was set up by K. Löwner [1]; the Kufarev−Löwner equation was obtained by P.P. Kufarev (see [5]).

References

[1] LÖWNER, K.: 'Untersuchungen über schlichte konforme Abbildungen des Einheitskreises, I', *Math. Ann.* **89** (1923), 103-121.
[2] KUFAREV, P.P.: 'A theorem on solutions of a differential equation', *Uchen. Zap. Tomsk. Gos. Univ.* **5** (1947), 20-21 (in Russian).
[3] POMMERENKE, C.: 'Ueber die Subordination analytischer Funktionen', *J. Reine Angew. Math.* **218** (1965), 159-173.
[4] GUTLYANSKIĬ, V.YA.: 'Parametric representation of univalent functions', *Soviet Math. Dokl.* **11** (1970), 1273-1276. (*Dokl. Akad. Nauk SSSR* **194**, no. 4 (1970), 750-753)
[5] KUFAREV, P.P.: 'On one-parameter families of analytic functions', *Mat. Sb.* **13** (1943), 87-118 (in Russian).
[6] GOLUZIN, G.M.: *Geometric theory of functions of a complex variable*, Amer. Math. Soc., 1969 (translated from the Russian).

V.Ya. Gutlyanskiĭ

Editorial comments. For more information see also Löwner method.

AMS 1980 Subject Classification: 30C75

LÖWNER METHOD, *Löwner's method of parametric representation of univalent functions, Löwner's parametric method* - A method in the theory of univalent functions that consists in using the **Löwner equation** to solve extremal problems. The method was proposed by K. Löwner [1]. It is based on the fact that the set of functions $f(z)$, $f(0) = 0$, that are regular and univalent in the disc $E = \{z: |z| < 1\}$ and that map E onto domains of type (s) (cf. **Smirnov domain**), which are obtained from the disc $|w| < 1$ by making a slit along a part of a Jordan arc starting from a point on the circle $|w| = 1$

and not passing through the point $w=0$, is complete (in the topology of uniform convergence of functions inside E) in the whole family of functions $f(z)$, $f(0)=0$, that are regular and univalent in E and are such that $|f(z)|<1$ in E. Associating the length of the arc that has been removed with a parameter t, it has been established that a function $w=f(z)$, $f(0)=0$, that maps E univalently onto a domain D of type (s) is a solution of the differential equation (see **Löwner equation**)

$$\frac{\partial f(z,t)}{\partial t} = -f(z,t)\frac{1+k(t)f(z,t)}{1-k(t)f(z,t)}, \qquad (*)$$

$f(z,t_0)=f(z)$, satisfying the initial condition $f(z,0)=z$. Here $t\in[0,t_0]$ and $k(t)$ is a continuous complex-valued function on the interval $[0,t_0]$ corresponding to D with $|k(t)|=1$. Löwner used this method to obtain sharp estimates of the coefficients c_3 and b_n, $n=2,3,\dots$, in the expansions

$$w = f(z) = z + \sum_{n=2}^{\infty} c_n z^n$$

and

$$z = f^{-1}(w) = w + \sum_{n=2}^{\infty} b_n w^n$$

in the class S of functions $w=f(z)$, $f(0)=0$, $f'(0)=1$, that are regular and univalent in E.

The Löwner method has been used (see [3]) to obtain fundamental results in the theory of univalent functions (distortion theorems, reciprocal growth theorems, rotation theorems). Let S' be the subclass of functions $f(z)$ in S that have in E the representation

$$f(z) = \lim_{t\to\infty} e^t f(z,t),$$

where $f(z,t)$, as a function of z, is regular and univalent in E, $|f(z,t)|<1$ in E, $f(0,t)=0$, $f'_z(0,t)>0$, and as a function of t, $0<t<\infty$, is a solution of the differential equation (*) satisfying the initial condition $f(z,0)=z$; $k(t)$ in (*) is any complex-valued function that is piecewise continuous and has modulus 1 on the interval $[0,\infty)$. To estimate any quantity on the class S it is sufficient to estimate it on the subclass S', since any function $f(z)$ of class S can be approximated by functions $f_n(z)$, $f_n(0)=0$, $f'_n(0)>0$, each of which maps E univalently onto the w-plane with a slit along a Jordan arc starting at ∞ and not passing through $w=0$, and hence by functions $f_n(z)/f'_n(0)\in S'$. Under this approximation the quantities to be estimated for the approximating functions converge to the same quantity as for the function $f(z)$.

Löwner's method has been used in work on the theory of univalent functions (see [3]); it often leads to success in obtaining explicit estimates, but as a rule it does not ensure the classification of all extremal functions and does not give complete information about their uniqueness. For a complete solution of extremal problems Löwner's method is usually combined with a variational method (see [3] and **Variation-parametric method**). Löwner's method has been extended to doubly-connected domains. A generalized equation of the type of Löwner's equation has been obtained for multiply-connected domains and for automorphic functions (see [4]).

References

[1] LÖWNER, K.: 'Untersuchungen über schlichte konforme Abbildungen des Einheitskreises, I', *Math. Ann.* **89** (1923), 103-121.
[2] PESCHL, E.: 'Zur Theorie der schlichten Funktionen', *J. Reine Angew. Math.* **176** (1936), 61-94.
[3] GOLUZIN, G.M.: *Geometric theory of functions of a complex variable*, Amer. Math. Soc., 1969 (translated from the Russian).
[4] ALEKSANDROV, I.A.: *Parametric extensions in the theory of univalent functions*, Moscow, 1976 (in Russian).

E.G. Goluzina

Editorial comments. The Löwner equation has been used to solve the **Bieberbach conjecture**, [A1]; cf. [A2]. Further references on the method include [A3] - [A6].

References

[A1] DE BRANGES, L.: 'A proof of the Bieberbach conjecture', *Acta. Math.* **154** (1985), 137-152.
[A2] FITZGERALD, C.H. and POMMERENKE, C.: 'The de Branges theorem on univalent functions', *Trans. Amer. Math. Soc.* **290** (1985), 683-690.
[A3] HAYMAN, W.K.: *Multivalent functions*, Cambridge Univ. Press, 1958.
[A4] POMMERENKE, C.: *Univalent functions*, Vandenhoeck & Ruprecht, 1975.
[A5] DUREN, P.L.: *Univalent functions*, Springer, 1983.
[A6] BRANNAN, D.A.: 'The Löwner differential equation', in D.A. Brannan and J.G. Clunie (eds.): *Aspects of Contemporary Complex Analysis*, Acad. Press, 1980, pp. 79-95.

AMS 1980 Subject Classification: 30C75

LOXODROME - A curve on a **surface of revolution** that cuts all the meridians at a constant angle α. If α is an acute or obtuse angle, then a loxodrome performs infinitely many windings about the pole, getting closer and closer to it.

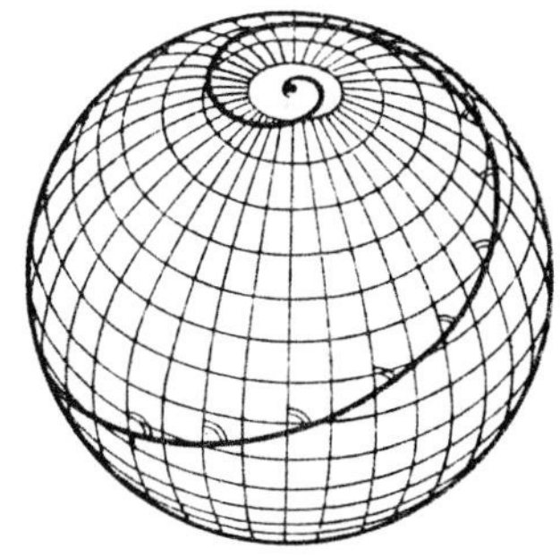

For surfaces of revolution whose **first fundamental form** can be written as

$$ds^2 = du^2 + G(u)\,dv^2,$$

the equation of a loxodrome is

$$v\cot\alpha = \pm\int_{u_0}^{u}\frac{du}{\sqrt{G(u)}}.$$

For a sphere with first fundamental form

$$ds^2 = R^2(du^2 + \cos^2 u\, dv^2)$$

the equation of a loxodrome is

$$v \cot\alpha = R \ln\tan\left(\frac{\pi}{4} + \frac{u}{2R}\right).$$

A.B. Ivanov

Editorial comments.

References

[A1] STRUBECKER, K.: *Differentialgeometrie*, 2. Theorie der Flächenmetrik, de Gruyter, 1958.

AMS 1980 Subject Classification: 53A04, 53A05

LÜROTH PROBLEM - The problem of characterizing subfields of a field of rational functions.

In 1876 J. Lüroth [1] (see also [2]) proved that any subfield of a field $k(x)$ of rational functions in one variable, containing k and distinct from k, is isomorphic to the field $k(x)$ (*Lüroth's theorem*). The question of whether a similar assertion is true for subfields R of the field $k(x_1, \ldots, x_n)$, $R \supset k$, $R \neq k$, $n \geq 2$, is known as the *Lüroth problem*.

Let X be an **algebraic variety** that is a model (see **Minimal model**) of the field R; then the imbedding $R = k(X) \subset k(x_1, \ldots, x_n)$ defines a rational mapping $f: P^n \to X$ whose image is dense in X. Varieties for which there is such a mapping of projective space onto them are said to be *unirational* (cf. **Unirational variety**). Varieties that are birationally isomorphic to P^m are said to be *rational* (cf. **Rational variety**). In geometrical language Lüroth's problem can be stated as follows: Is any unirational variety X rational? Without loss of generality one may assume that $\dim X = n$, that is, that R has transcendence degree n.

In the case $n = 1$ an affirmative solution of Lüroth's problem for any ground field k is given by Lüroth's theorem stated above. For $n = 2$ and an algebraically closed field k of characteristic 0 the problem was solved affirmatively by G. Castelnuovo in 1893. Castelnuovo's rationality criterion implies also an affirmative solution of Lüroth's problem for surfaces X over an algebraically closed field of arbitrary characteristic for which there is a **separable mapping** $f: P^2 \to X$ (see [7]). For non-separable mappings f there are examples that give a negative solution of Lüroth's problem for fields of prime characteristic. In the case of an algebraically non-closed field k such examples are the minimal cubic surfaces in P^3 that have k-points.

For three-dimensional varieties Lüroth's problem has also been solved negatively (see [4], [5], [6]). It has been proved [5] that a three-dimensional **cubic hypersurface**, which is known to be unirational, is not rational. For the proof a new method was found, based on the comparison of the **intermediate Jacobian** of the cubic with the Jacobians of curves. It has been proved [4] that smooth three-dimensional quadrics are not rational. In [6], for the construction of counter-examples the **Brauer group** of the variety (the torsion subgroup of the three-dimensional cohomology group) was used as an invariant. This birational invariant has also been used in the construction of counter-examples in all dimensions $n \geq 3$.

References

[1] LÜROTH, J.: *Math. Ann.* **9** (1876), 163-165.
[2] WAERDEN, B.L. VAN DER: *Algebra*, 1-2, Springer, 1967-1971 (translated from the German).
[3] MANIN, YU.N.: *Cubic forms. Algebra, geometry, arithmetic*, North-Holland, 1974 (translated from the Russian).
[4] ISKOVSKIKH, V.A. and MANIN, YU.I.: 'Three-dimensional quartics and counterexamples to the Lüroth problem', *Math. USSR Sb.* **15**, no. 1 (1971), 141-166. (*Mat. Sb.* **86**, no. 1 (1971), 140-166)
[5] CLEMENS, C.H. and GRIFFITHS, P.: 'The intermediate Jacobian of the cubic threefold', *Ann. of Math.* **95** (1972), 281-356.
[6] ARTIN, M. and MUMFORD, D.: 'Some elementary examples of unirational varieties which are not rational', *Proc. London Math. Soc.* **25**, no. 1 (1972), 75-95.
[7] ZARISKI, O.: 'The problem of minimal models in the theory of algebraic surfaces', *Amer. J. Math.* **80** (1958), 146-184.

V.A. Iskovskikh

AMS 1980 Subject Classification: 14G05

LUXEMBURG NORM - A function

$$\| x \|_{(M)} = \inf\left\{\lambda: \lambda > 0,\ \int_G M(\lambda^{-1} x(t))\, dt \leq 1\right\},$$

where $M(u)$ is an even convex function that increases for positive u,

$$\lim_{u \to 0} u^{-1} M(u) = \lim_{u \to \infty} u(M(u))^{-1} = 0,$$

$M(u) > 0$ for $u > 0$, and G is a bounded set in $\mathbf{R}^n$. The properties of this norm were studied by W.A.J. Luxemburg [1]. The Luxemburg norm is equivalent to the Orlicz norm (see **Orlicz space**), and

$$\| x \|_{(M)} \leq \| x \|_M \leq 2 \| x \|_{(M)}.$$

If the functions $M(u)$ and $N(u)$ are complementary (or dual) to each other (see **Orlicz class**), then

$$\| x \|_{(M)} = \sup\left\{\int_G x(t) y(t)\, dt: \| y \|_{(N)} \leq 1\right\}.$$

If $\chi_E(t)$ is the characteristic function of a measurable subset $E \subset G$, then

$$\| \chi_E \|_{(M)} = \frac{1}{M^{-1}(1/\operatorname{mes} E)}.$$

References

[1] LUXEMBURG, W.A.J.: *Banach function spaces*, T.U. Delft, 1955. Thesis.
[2] KRASNOSEL'SKIĬ, M.A. and RUTITSKIĬ, YA.B.: *Convex functions and Orlicz spaces*, Noordhoff, 1961 (translated from the Russian).

E.M. Semenov

AMS 1980 Subject Classification: 46EXX, 46E30

LUZIN C-PROPERTY - A characteristic property of a **measurable function** that is finite almost-everywhere on its domain of definition. A function f, finite almost-everywhere on $[0, 1]$, *has the C-property on* $[0, 1]$ if for every $\epsilon > 0$ there is a **perfect set** Q in $[0, 1]$ with measure $> 1 - \epsilon$ on which f is continuous if considered only on Q. The C-property was introduced by N.N. Luzin [1], who also proved that for a function to have the C-property it is necessary and sufficient that it be measurable and finite almost-everywhere on $[0, 1]$. This theorem of Luzin (the **Luzin criterion**) can be generalized to the case of functions of several variables (see [3], [4]) and is one of the main theorems in the metric theory of functions.

References
[1] LUZIN, N.N.: *Mat. Sb.* **28** (1912), 266-294.
[2] LUZIN, N.N.: *Collected works*, 1, Moscow, 1953 (in Russian).
[3] SAKS, S.: *Theory of the integral*, Hafner, 1952 (translated from the Polish).
[4] KAMKE, E.: *Das Lebesgue−Stieltjes Integral*, Teubner, 1960.

A.A. Konyushkov

Editorial comments. See Luzin criterion for additional references and comments.

AMS 1980 Subject Classification: 28A15, 28A20

LUZIN CRITERION *for measurability of a function of a real variable* - For a function f, defined on the interval $[a, b]$ and almost-everywhere finite, to be measurable it is necessary and sufficient that for any $\epsilon > 0$ there is a function ϕ, continuous on $[a, b]$, such that the measure of the set

$$\{x \in [a, b] : f(x) \neq \phi(x)\}$$

is less than ϵ. It was proved by N.N. Luzin [1]. In other words, an almost-everywhere finite function is measurable if and only if it becomes continuous if one neglects a set of arbitrary small measure.

References
[1] LUSIN, N.N. [N.N. LUZIN]: 'Sur les propriétés des fonctions mesurables', *C.R. Acad. Sci. Paris* **154** (1912), 1688-1690.
[2] NATANSON, I.P.: *Theory of functions of a real variable*, 1-2, F. Ungar, 1955-1961 (translated from the Russian).

B.A. Efimov

Editorial comments. In the West, Luzin's criterion is known as *Luzin's theorem* (in spite of an ambiguity — cf. Luzin theorem) and is generally stated a little bit differently, more like in **Luzin C-property** (but with a compact set instead of a perfect set). The tightness of the measure and the normality of the space makes all these formulations equivalent.

The Luzin criterion remains true if the interval $[a, b]$ is replaced by any **completely-regular space** and the (restriction of the) Lebesgue measure by any tight bounded measure on the Borel σ-field. In this general setting the Luzin property may be used in order to give an alternative definition of the notion of measurability (cf. [A1]) or, in recent works, a more adequate definition of this notion when f is no longer a real-valued function but, for example, a Banach-valued function.

The Luzin criterion is intimately related to the **Egorov theorem** and to the notion of measurability according to Carathéodory (cf. **Carathéodory measure**).

References
[A1] BOURBAKI, N.: *Elements of mathematics. Integration*, Addison-Wesley, 1975, Chapt. 6; 7; 8 (translated from the French).
[A2] HALMOS, P.: *Measure theory*, v. Nostrand, 1950.
[A3] RUDIN, W.: *Real and complex analysis*, McGraw-Hill, 1966.
[A4] HEWITT, E. and STROMBERG, K.: *Real and abstract analysis*, Springer, 1965.

AMS 1980 Subject Classification: 28A20, 28C15

LUZIN − DENJOY THEOREM - See **Denjoy − Luzin theorem**.

AMS 1980 Subject Classification: 42A28, 43A50

LUZIN EXAMPLES *in the theory of functions of a complex variable* - Examples that characterize boundary **uniqueness properties of analytic functions** (see [1], [2]).

1) For any set E of measure zero on the unit circle $\Gamma = \{z : |z| = 1\}$, N.N. Luzin constructed (1919, see [1]) a function $f(z)$ that is regular, analytic and bounded in the unit disc $D = \{z : |z| < 1\}$ and is such that $f(z)$ does not have radial boundary values along each of the radii that end at points of E.

A similar example of Luzin and I.I. Privalov (1925, see [2], [3]) differs only by insignificant changes.

2) Luzin also constructed (1925, see [2]) regular analytic functions $f_1(z)$ and $f_2(z) \neq 0$ in D that tend, respectively, to infinity and zero along all radii that end at points of some set of full measure 2π on Γ. This set E is of the first Baire category (cf. **Baire classes**) on Γ.

See also **Boundary properties of analytic functions; Luzin − Privalov theorems; Cluster set**.

References
[1] LUZIN, N.N.: *Collected works*, Vol. 1, Moscow, 1953, pp. 267-269 (in Russian).
[2] LUZIN, N.N.: *Collected works*, Vol. 1, Moscow, 1953, pp. 280-318 (in Russian).
[3] PRIWALOW, I.I. [I.I. PRIVALOV]: *Randeigenschaften analytischer Funktionen*, Deutsch. Verlag Wissenschaft., 1956 (translated from the Russian).
[4] LOHWATER, A.: 'The boundary behaviour of analytic functions', *Itogi Nauki i Tekhn. Mat. Anal.* **10** (1973), 99-259 (in Russian).

E.D. Solomentsev

Editorial comments.

References
[A1] COLLINGWOOD, E.F. and LOHWATER, A.J.: *The theory of cluster sets*, Cambridge Univ. Press, 1966.

AMS 1980 Subject Classification: 30D40

LUZIN HYPOTHESIS *in set theory* - The **cardinality**

of the continuum is the cardinality of the set of all subsets of the countable ordinals, that is, $2^{\aleph_0} = 2^{\aleph_1}$. Luzin's hypothesis is compatible with the Zermelo$-$Fraenkel system of axioms of set theory and the axiom of choice. N.N. Luzin [1] considered this hypothesis as an alternative to the **continuum hypothesis**, that is, $2^{\aleph_0} = \aleph_1 < 2^{\aleph_1}$. Martin's axiom (cf. **Suslin hypothesis**) and the negation of the continuum hypothesis together imply the Luzin hypothesis. The negation of the Luzin hypothesis, $2^{\aleph_0} < 2^{\aleph_1}$, is also sometimes called the Luzin hypothesis. The Luzin hypothesis, denoted by (HL), or its negation, which is denoted by (LH), are used in the proof of a number of theorems in general topology. For example, (LH) is equivalent to one of the following assertions: any **compact space** of cardinality not exceeding the cardinality of the continuum has an everywhere-dense subspace that satisfies the **first axiom of countability**; any dyadic compact Hausdorff space of cardinality not exceeding the cardinality of the continuum is metrizable. The following propositions follow from (LH): any **normal space** that satisfies the first axiom of countability and the **Suslin condition** is collection-wise normal; any separable normal **Moore space** is metrizable.

References

[1] LUSIN, N.N. [N.N. LUZIN]: 'Sur les ensembles analytiques nuls', *Fund. Math.* **25** (1935), 109-131.

[2] MOSTOWSKI, A.: *Constructible sets ands applications*, North-Holland, 1969.

B.A. Efimov

Editorial comments. For the consistency of Luzin's hypothesis see also [A1].

References

[A1] KUNEN, K.: *Set theory*, North-Holland, 1980.

AMS 1980 Subject Classification: 03E65, 03E50, 04A30

LUZIN N-PROPERTY, *'null-property'*, of a function f, continuous on an interval $[a, b]$ - For any set $E \subset [a, b]$ of measure mes $E = 0$, the image of this set, $f(E)$, also has measure zero. It was introduced by N.N. Luzin in 1915 (see [1]). The following assertions hold.

1) A function $f \not\equiv$ const on $[a, b]$ such that $f'(x) = 0$ almost-everywhere on $[a, b]$ does not have the Luzin N-property.

2) If f does not have the Luzin N-property, then on $[a, b]$ there is a **perfect set** P of measure zero such that mes $f(P) > 0$.

3) An absolutely continuous function has the Luzin N-property.

4) If f has the Luzin N-property and has bounded variation on $[a, b]$ (as well as being continuous on $[a, b]$), then f is absolutely continuous on $[a, b]$ (the *Banach$-$Zaretskiĭ theorem*).

5) If f does not decrease on $[a, b]$ and f' is finite on $[a, b]$, then f has the Luzin N-property.

6) In order that $f(E)$ be measurable for every measurable set $E \subset [a, b]$ it is necessary and sufficient that f have the Luzin N-property on $[a, b]$.

7) A function f that has the Luzin N-property has a derivative f' on the set for which any non-empty **portion** of it has positive measure.

8) For any perfect nowhere-dense set $P \subset [a, b]$ there is a function f having the Luzin N-property on $[a, b]$ and such that f' does not exist at any point of P.

The concept of Luzin's N-property can be generalized to functions of several variables and functions of a more general nature, defined on measure spaces.

References

[1] LUZIN, N.N.: *The integral and trigonometric series*, Moscow-Leningrad, 1915 (in Russian).

A.A. Konyushkov

Editorial comments. There is another property intimately related to the Luzin N-property. A function f continuous on an interval $[a, b]$ has the *Banach S-property* if for all Lebesgue-measurable sets $E \subset [a, b]$ and all $\epsilon > 0$ is a $\delta > 0$ such that

$$\text{mes}(E) < \delta \Rightarrow \text{mes}(f(E)) < \epsilon.$$

This is clearly stronger than the N-property. S. Banach proved that a function f has the S-property (respectively, the N-property) if and only if (respectively, only if — see below for the missing 'if') the inverse image $f^{-1}(\{x\})$ is finite (respectively, is at most countable) for almost-all x in $f([a, b])$. For classical results on the N- and S-properties, see [A3].

Recently a powerful extension of these results has been given by G. Mokobodzki (cf. [A1], [A2]), allowing one to prove deep results in potential theory. Let Ω and T be two compact metrizable spaces, Ω being equipped with a probability measure P. Let F be a Borel subset of $\Omega \times T$ and, for any Borel subset E of Ω, define the subset $F(E)$ of T by $F(E) = \{t \in T:$ there is an $\omega \in \Omega$ such that $(\omega, t) \in F\}$ (if F is the graph of a mapping $f : \Omega \to T$, then $F(E) = f(E)$). The set F is said to have the property (N) (respectively, the property (S)) if there exists a measure λ on T (here depending on F) such that for all $E \subset \mathscr{B}(\Omega)$,

$$P(E) = 0 \Rightarrow \lambda(F(E)) = 0$$

(respectively, for all $\epsilon > 0$ there is a $\delta > 0$ such that for all $E \in \mathscr{B}(\Omega)$ one has

$$P(E) < \delta \Rightarrow \lambda(F(E)) < \epsilon).$$

Now F has the property (N) (respectively, the property (S)) if and only if the section $F(\omega)$ of F is at most countable (respectively, is finite) for almost-all $\omega \in \Omega$.

References

[A1] DELLACHERIE, C., FEYEL, D. and MOKOBODZKI, G.: 'Intégrales de capacités fortement sous-additives', in *Sem. Probab. Strasbourg XVI*, Lecture notes in math., Vol. 920, Springer, 1982, pp. 8-28.

[A2] LOUVEAU, A.: *Minceur et continuité séquentielle des sous-mesures analytiques fortement sous-additives*, Sem. Initiation à l'Analyse, 66, Univ. P. et M. Curie, 1983-1984.

[A3] SAKS, S.: *Theory of the integral*, Hafner, 1952 (translated from the Polish).

[A4] HEWITT, E. and STROMBERG, K.: *Real and abstract analysis*, Springer, 1965.

AMS 1980 Subject Classification: 28A20, 26B30

LUZIN — PRIVALOV THEOREMS *in the theory of functions of a complex variable* - Classical results of N.N. Luzin and I.I. Privalov that clarify the character of a boundary uniqueness property of analytic functions (cf. **Uniqueness properties of analytic functions**) (see [1]).

1) Let $f(z)$ be a meromorphic function of the complex variable z in a simply-connected domain D with rectifiable boundary Γ. If $f(z)$ takes angular boundary values zero on a set $E \subset \Gamma$ of positive Lebesgue measure on Γ, then $f(z) \equiv 0$ in D. There is no function meromorphic in D that has infinite angular boundary values on a set $E \subset \Gamma$ of positive measure.

2) Let $w = f(z)$ be a meromorphic function in the unit disc $D = \{z: |z| < 1\}$ other than a constant and having radial boundary values (finite or infinite) on a set E situated on an arc σ of the unit circle $\Gamma = \{z: |z| = 1\}$ that is metrically dense and of the second Baire category (cf. **Baire classes**) on σ. Then the set W of its radial boundary values on E contains at least two distinct points. *Metric density* of E on σ means that every **portion** of E on σ has positive measure. This implies that if the radial boundary values of $f(z)$ on a set E of the given type are equal to zero, then $f(z) \equiv 0$ in D. Moreover, there is no meromorphic function in the unit disc that takes infinite radial boundary values on a set E of the given type.

Luzin and Privalov (see [1], [2]) constructed examples to show that neither metric density nor second Baire category are by themselves sufficient for the assertion in 2 to hold.

See also **Boundary properties of analytic functions; Luzin examples; Cluster set; Privalov theorem; Riesz theorem.**

References

[1] LUSIN, N.N. [N.N. LUZIN] and PRIWALOFF, I.I. [I.I. PRIVALOV]: 'Sur l'unicité et la multiplicité des fonctions analytiques', *Ann. Sci. Ecole Norm. Sup.* (3) **42** (1925), 143-191.
[2] PRIWALOW, I.I. [I.I. PRIVALOV]: *Randeigenschaften analytischer Funktionen*, Deutsch. Verlag Wissenschaft., 1956 (translated from the Russian).
[3] LOHWATER, A.: 'The boundary behaviour of analytic functions', *Itogi Nauk. i Tekhn. Mat. Anal.* **10** (1973), 99-259 (in Russian).

E.D. Solomentsev

Editorial comments.

References

[A1] COLLINGWOOD, E.F. and LOHWATER, A.J.: *The theory of cluster sets*, Cambridge Univ. Press, 1966.

AMS 1980 Subject Classification: 30D40

LUZIN PROBLEM - 1) A problem in the theory of trigonometric series. It consists in proving *Luzin's conjecture*, stating that the Fourier series

$$a_0(f) + \sum_{n=1}^{\infty} \{a_n(f)\cos nx + b_n(f)\sin nx\} \qquad (*)$$

of a Lebesgue-measurable function f, defined on the interval $[0, 2\pi]$, with finite integral

$$\int_0^{2\pi} |f(x)|^2 \, dx,$$

converges almost everywhere on $[0, 2\pi]$. The conjecture was made by N.N. Luzin in 1915 in his dissertation (see [1]). Luzin's problem was solved in 1966 in the affirmative sense by L. Carleson (see **Carleson theorem**). Until Carleson's paper [2] it was not even known whether the Fourier series of a continuous function on the interval $[0, 2\pi]$ converges at least at one point.

References

[1] LUZIN, N.N.: *The integral and trigonometric series*, 1, Moscow-Leningrad, 1953, p. 219 (in Russian).
[2] CARLESON, L.: 'Convergence and growth of partial sums of Fourier series', *Acta Math.* **116** (1966), 135-157.

B.S. Kashin

2) One of a number of fundamental problems in set theory posed by N.N. Luzin [1], for the solution of which he proposed the *method of resolvents*. Namely, a problem P of set theory is posed in a resolvent if one can indicate a set of points E such that P is solved affirmatively every time one can indicate a point of E, and is solved negatively if one can prove that E is empty. The set E itself is called the *resolvent of the problem P*.

Problem 1. Are all co-analytic sets (cf. $C\mathscr{A}$-set) countable or do they have the cardinality of the continuum? The resolvent E of this problem is a **Luzin set** of class at most 3; that is, if one can find a point of E, then there is an uncountable co-analytic set without perfect part, while if E is empty, then there are no such co-analytic sets.

Problem 2. Do there exists Lebesgue-unmeasurable Luzin sets?

Problem 3. Does there exist a Luzin set without the **Baire property**?

Luzin conjectured that the Problems 1, 2, 3 are undecidable. This conjecture has been confirmed (see [3], [4]). Connections between these problems have been established. For example, from the existence of an unmeasurable set of type A_2 follows the existence of an uncountable set of type $C\mathscr{A}$ not containing a perfect subset. I. Novak [5] obtained an affirmative solution of Luzin's problem about parts of the series of natural numbers, starting from the **continuum hypothesis** or the negation of the **Luzin hypothesis**.

References

[1] LUSIN, N.N. [N.N. LUZIN]: 'Sur le problème de M. Emile

Borel et la méthode des résolvants', *C.R. Acad. Sci. Paris* **181** (1925), 279-281.

[2] LUZIN, N.N.: *Collected works*, 2, Moscow, 1958 (in Russian).

[3] NOVIKOV, P.S.: 'On the non-contradictibility of certain propositions in descriptive set theory', *Trudy Mat. Inst. Steklov.* **38** (1951), 279-316 (in Russian).

[4] SOLOVAY, R.: 'A model of set theory in which every set of reals is Lebesgue measurable', *Ann. of Math. (2)* **92**, no. 1 (1970), 1-56.

[5] NOVAK, J.: 'On some problems of Lusin concerning the subsets of natural numbers', *Czechoslovak. Math. J.* **3** (1953), 385-395.

B.A. Efimov

Editorial comments. See **Luzin set** for usual terminology. For other problems of Luzin see **Luzin theorem**.

References

[A1] JECH, T.: *Set theory*, Acad. Press, 1978.

[A2] MOSCHOVAKIS, Y.N.: *Descriptive set theory*, North-Holland, 1980.

AMS 1980 Subject Classification: 42A20, 04A15, 03E15

LUZIN SEPARABILITY PRINCIPLES, *Luzin separation principles* - Two theorems in **descriptive set theory**, proved by N.N. Luzin in 1930 (see [1]). Two sets E and E_1 without common points, lying in a Euclidean space, are called *B-separable* or *Borel separable* if there are two Borel sets H and H_1 without common points containing E and E_1, respectively. The first Luzin separation principle states that two disjoint analytic sets (cf. $\mathscr{A}$-set; **Analytic set**) are always *B*-separable. Since there are two disjoint co-analytic sets (cf. $C\mathscr{A}$-set) that are *B*-inseparable, the following definition makes sense: Two sets E_1 and E_2 without common points are separable by means of co-analytic sets if there are two disjoint sets H_1 and H_2 containing E_1 and E_2, respectively, each of which is a co-analytic set. Luzin's second separation principle asserts that if from two analytic sets one removes their common part, then the remaining parts are always separable by means of co-analytic sets.

References

[1] LUZIN, N.N.: *Leçons sur les ensembles analytiques et leurs applications*, Gauthier-Villars, 1930.

B.A. Efimov

Editorial comments. Both principles are still valid when Euclidean space is replaced by a Polish space. The first separation principle, which was already implicitly proved by M.Ya. Suslin (1917) while proving that a set H is Borel if and only if H and its complement are analytic, has numerous applications in analysis. Following C. Kuratowski, the second one is generally stated now as a *reduction theorem* for co-analytic sets (i.e. complements of analytic sets): If C_1 and C_2 are two co-analytic sets, then there exist two disjoint co-analytic sets $D_1 \subset C_1$ and $D_2 \subset C_2$ such that $D_1 \cup D_2 = C_1 \cup C_2$. This is related to the use of countable ordinals in descriptive set theory and has some deep applications in analysis. For more details and references see **Descriptive set theory**.

Under extra set-theoretical hypotheses (Gödel's constructibility axiom, large-cardinal and, especially, determinacy hypotheses), much more is known nowadays on the separation principle at higher levels of the projective hierarchy, cf. [A3], [A4].

References

[A1] LUZIN, N.N.: 'Sur les ensembles analytiques', *Fund. Math.* **10** (1927), 1-92.

[A2] KURATOWSKI, C.: *Topology* , I, PWN & Acad. Press, 1966 (translated from the French).

[A3] JECH, T.: *Set theory*, Acad. Press, 1978.

[A4] MOSCHOVAKIS, Y.N.: *Descriptive set theory*, North-Holland, 1980.

AMS 1980 Subject Classification: 04A15, 28A05, 03E15, 54H05

LUZIN SET, *projective set* - A subset of a complete separable metric space, defined by induction as follows. The Luzin sets of class 0 are the Borel sets (cf. **Borel set**). The Luzin sets of class $2n+1$ are continuous images of Luzin sets of class $2n$. The Luzin sets of class $2n$ are complements of Luzin sets of class $2n-1$. In particular, Luzin sets of class 1, that is, continuous images of Borel sets, are called *analytic sets*, or $\mathscr{A}$-sets or *Suslin sets* (cf. **A-set; Analytic set**). The concept of a Luzin set is due to N.N. Luzin [1]. If the sets P_i are Luzin sets of class n, then $\bigcup_{i=1}^{k} P_i$ and $\bigcap_{i=1}^{k} P_i$ are also Luzin sets of class n. If the sets $P_i \subset X_i$ are Luzin sets of class n lying in complete separable metric spaces X_i, then the direct product (finite or countable) $\prod_i P_i$ is a Luzin set of class n in the space $\prod_i X_i$. A Luzin set of odd class n situated in a space X coincides with the projection of a set of class $n-1$ situated in $X \times X$. The space X of irrational numbers in the interval $[0, 1]$ contains, for any $n > 0$, a Luzin set of class n that is not a Luzin set of class $< n$; the space X also contains sets that are not Luzin sets.

References

[1] LUSIN, N.N. [N.N. LUZIN]. 'Sur un problème de M. Emile Borel et les ensemble projectifs de M. Henri Lebesgue', *C.R. Acad. Sci. Paris* **180** (1925), 1318-1320.

[2] LUSIN, N.N. [N.N. LUZIN]: *Leçons sur les ensembles analytiques et leurs applications*, Gauthier-Villars, 1930.

[3] KURATOWSKI, C.: *Topology*, 1, PWN & Acad. Press, 1966 (translated from the French).

B.A. Efimov

Editorial comments. The term 'Luzin set' is in the West almost exclusively used for a subset of the real line whose intersection with every nowhere-dense set is countable (see **Luzin space**). The sets discussed in the main article above are almost exclusively called *projective sets* (cf. **Projective set**). The sets of class $2n+1$ are generally called Σ_n^1-sets and those of class $2n$ are called Π_n^1-sets. See **Descriptive set theory**.

All important problems about projective sets have received satisfactory answers during the last three decades, see **Descriptive set theory** and Luzin problem.

References
[A1] MOSCHOVAKIS, Y.N.: *Descriptive set theory*, North-Holland, 1980.
[A2] JECH, T.: *Set theory*, Acad. Press, 1978.

AMS 1980 Subject Classification: 03E15, 04A15, 54H05, 28A05

LUZIN SIEVE - An arbitrary mapping $W: \mathbf{Q}_0 \to 2^X$ that puts each dyadic fraction $r \in \mathbf{Q}_0$ into correspondence with a subset $W_r \subset X$. As a rule, X is assumed to be a complete separable **metric space**. It was introduced by N.N. Luzin [1]. The set A of points $x \in X$ such that there is an infinite sequence $r_1 < r_2 < \cdots$ that satisfies the condition $x \in W_{r_1} \cap W_{r_2} \cap \cdots$ is said to be *sifted* through the Luzin sieve W. For every $\mathscr{A}$-**operation** there is a Luzin sieve W such that the result of this $\mathscr{A}$-operation is sifted through W. The main result concerning the Luzin sieve is that a **Luzin set** of the n-th class (or of the projective class L_n) is invariant under the operation of sifting through the Luzin sieve for $0 \neq n \neq 2$.

References
[1] LUZIN, N.N.: 'Sur les ensembles analytiques', *Fund. Math.* **10** (1927), 1-95.
[2] KURATOWSKI, C.: *Topology*, 1, PWN & Acad. Press, 1966 (translated from the French).

B.A. Efimov

Editorial comments. A Luzin set in the sense of the article is invariably called a **projective set** in the West. The Luzin sieve has been an extremely powerful tool in descriptive set theory; it gave rise, with other techniques, to the modern use of countable ordinals in this theory. For more details and references see **Descriptive set theory**.

This notion has nothing to do with the notion of sieve used by N. Bourbaki [A1] while proving one of the Luzin theorems. A *Bourbaki sieve* is just a way to write a disjoint Suslin scheme.

If A is an **analytic set** and $W = \{W_r: r \in \mathbf{Q}\}$ is a **Luzin sieve** for A consisting of closed sets, then, as one readily sees, $X \setminus A = \{x: M_x \text{ is well-ordered by } \geq\}$, where $M_x = \{r: x \in W_r\}$. The sets $A_\alpha = \{x \in X \setminus A: \text{the order type of } M_x \text{ is } \alpha\}$, where $\alpha < \omega_1$, are called the *constituents* of the set $X \setminus A$ determined by the sieve W.

References
[A1] BOURBAKI, N.: *Elements of mathematics. General topology*, Addison-Wesley, 1966, Chapt. 10 (translated from the French).

AMS 1980 Subject Classification: 03E15, 04A15, 54H05

LUZIN SPACE - An uncountable topological T_1-space without isolated points in which every nowhere-dense subset is countable. The existence of a Luzin space on the real line follows from the **continuum hypothesis**. From the negation of the continuum hypothesis and Martin's axiom (cf. **Suslin hypothesis**) together it follows that no Luzin space exists. In partic-ular, this is compatible with the Zermelo – Fraenkel system of axioms of set theory and the **axiom of choice**. The existence of metrizable Luzin spaces has been proved under very general assumptions about the place of the **cardinality** of the continuum in the scale of alephs. Any Luzin space X that lies in a separable metric space Y has the following property: For any sequence $\{\lambda_n\}$ of positive numbers there is a sequence of sets $\{A_n\}$ such that $X = \bigcup_{n=1}^{\infty} A_n$ and $\delta(A_n) < \lambda_n$, where $\delta(A)$ is the diameter of the set A. This property is invariant under continuous mappings. Any continuous image of a Luzin space lying in Y has Lebesgue measure zero and dimension zero. Moreover, it is totally imperfect, that is, it does not contain a Cantor set. The continuum hypothesis implies that there is a regular hereditarily-separable, hereditarily Lindelöf, extremally-disconnected Luzin space of countable π-weight and with the cardinality of the continuum.

References
[1] LUSIN, N.N. [N.N. LUZIN]: 'Sur un problème de M. Baire', *C.R. Acad. Sci. Paris* **158** (1914), 1258-1261.
[2] KURATOWSKI, C.: *Topology*, 1, PWN & Acad. Press, 1966 (translated from the French).

B.A. Efimov

Editorial comments. Three slightly different definitions of Luzin space are still in use (apart from whether they must be T_1, T_2 or T_3): An uncountable space all of whose nowhere-dense sets are countable, with 1) no isolated points; or 2) at most ω isolated points; or 3) any number of isolated points.

References
[A1] KUNEN, K.: 'Luzin spaces', *Topology Proceedings* 1 (1977), 191-199.
[A2] ROITMAN, J.: 'Basic S and L', in K. Kunen and J.E. Vaughan (eds.): *Handbook of Set-Theoretic Topology*, North-Holland, 1984, pp. 295-326.
[A3] WEISS, W.: 'Versions of Martin's axiom', in K. Kunen and J.E. Vaughan (eds.): *Handbook of Set-Theoretic Topology*, North-Holland, 1984, pp. 827-886.
[A4] MILLER, A.W.: 'Special subsets of the real line', in K. Kunen and J.E. Vaughan (eds.): *Handbook of Set-Theoretic Topology*, North-Holland, 1984, pp. 201-233.

AMS 1980 Subject Classification: 54G15, 54A25, 54A35

LUZIN THEOREM - 1) Luzin's theorem in the theory of functions of a complex variable (the *local principle of finite area*) is a result of N.N. Luzin that reveals a connection between the boundary properties of an analytic function in the unit disc and the metric of the Riemann surface onto which it maps the disc (see [1], [2]).

Let V be any domain inside the unit disc $D = \{z: |z| < 1\}$ of the complex z-plane adjoining an arc σ of the unit circle $\Gamma = \{z: |z| = 1\}$, and let

$$w = f(z) = \sum_{k=0}^{\infty} c_k z^k$$

be a regular analytic function in D. If the area of the Riemann surface that is the image of V under the mapping $w = f(z)$ is finite, then the series

$$\sum_{k=0}^{\infty} c_k z^k$$

converges almost-everywhere on σ.

In connection with this theorem Luzin made a conjecture, also known as *Luzin's problem*. A point $e^{i\theta_0} \in \Gamma$ is called a *Luzin point* of the function $w = f(z)$ if $w = f(z)$ maps every disc touching Γ from the inside at $e^{i\theta_0}$ onto a domain of infinite area on the Riemann surface of $w = f(z)$. The *Luzin conjecture* is that there are bounded analytic functions in D such that every point of Γ is a Luzin point for them. The Luzin conjecture was first confirmed completely in 1955 (see [3]).

References

[1] LUZIN, N.N.: 'On localization of the finite area principle', *Dokl. Akad. Nauk SSSR* **56** (1947), 447-450 (in Russian).
[2] LUZIN, N.N.: *Collected works*, 1, Moscow, 1953, pp. 318-330 (in Russian).
[3] LOHWATER, A.: 'The boundary behaviour of analytic functions', *Itogi Nauki i Tekhn. Mat. Anal.* **10** (1973), 99-259 (in Russian).

E.D. Solomentsev

Editorial comments. The reference for the solution of Luzin's problem is [A1].

References

[A1] LOHWATER, A.J. and PIRANIAN, G.: 'On a conjecture of Luzin', *Michigan Math. J.* 3 (1955), 63-68.

2) Luzin's theorems in descriptive set theory are, by convention, split into three parts. The first and main part is directed towards the study of effective sets (analytic, Borel, Luzin (projective) sets). Here one is concerned with the **Luzin separability principles** and the theorem on the existence of Luzin sets of arbitrary class (cf. **Luzin set**). The second part is the study of problems lying on the path to the solution of the **continuum hypothesis** and the problem of the cardinality of $C\mathscr{A}$-sets (cf. $C\mathscr{A}$-**set**). Here one distinguishes the Luzin$-$Sierpiński theorem on partitioning an interval into $\aleph_1$ Borel sets, determined by the corresponding **Luzin sieve**, and also Luzin's covering theorem: Let E and A be disjoint analytic sets (cf. $\mathscr{A}$-**set**; **Analytic set**) and let

$$E \subset X \setminus A = \bigcup_{\alpha < \omega_1} A_\alpha$$

be a decomposition of $X \setminus A$ into constituents; then there is an index $\alpha_0 < \omega_1$ such that

$$E \subset \bigcup_{\alpha < \alpha_0} A_\alpha.$$

The third part contains results obtained by the use of the **axiom of choice**. Here one borders on philosophical work in set theory. One distinguishes Luzin's theorem on the existence of an uncountable set of the first category (cf. **Category of a set**) in any **perfect set**, and

on partitioning an interval into an uncountable number of non-measurable sets. To complete this part there is Luzin's theorem on subsets of the set of natural numbers, which reflects some properties of the remainder $\beta N \setminus N$ of the **Stone$-$Čech compactification** βN of the natural number series N.

References

[1] LUZIN, N.N.: *Collected works*, 2, Moscow, 1958 (in Russian).

B.A. Efimov

Editorial comments. See **Luzin sieve** for the definition of constituents.

'Luzin's theorem on subsets of the set of natural numbers' states that there is a family $\{A_\alpha\}_{\alpha < \omega_1}$ of infinite subsets of N such that $A_\alpha \cap A_\beta$ is finite for $\alpha = \beta$ and such that for any two uncountable disjoint subsets E and F of ω_1 there is no subset C of N such that for all $\alpha \in E$: $A_\alpha \setminus C$ is finite and for all $\alpha \in F$: $A_\alpha \cap C$ is finite. A family like this usually called a *Luzin family*. See [A2].

In the West, the name 'Luzin theorem' refers almost always to a result in measure theory; see **Luzin criterion**. It may also refer to the following result of Luzin in **descriptive set theory**: If P is a Polish space, Q a separable metrizable space and f is an injective Borel mapping from P into Q, then the direct image $f(B)$ of any Borel subset B of P is a Borel subset of Q. Luzin's covering theorem in descriptive set theory is usually called the (classical) *boundness theorem* in the West; it gave rise, together with the Luzin$-$Sierpiński theorem, Luzin sieves, etc., to the modern use of countable ordinals in this theory. See also **Descriptive set theory**.

References

[A1] KURATOWSKI, C.: *Topology*, PWN & Acad. Press, 1966 (translated from the French).
[A2] DOUWEN, E.K. VAN: 'The integers and topology', in K. Kunen and J.E. Vaughan (eds.): *Handbook of Set-Theoretic Topology*, North-Holland, 1984, pp. 111-167.
[A3] ENGELKING, R.: 'Hausdorff's gaps and limits and compactifications', in *Theory of Sets and Topology (in honour of F. Hausdorff)*, Deutsch. Verlag Wissenschaft., 1972, pp. 89-94.

AMS 1980 Subject Classification: 04AXX, 30D40, 28-XX, 541 I05

LYAPUNOV CHARACTERISTIC EXPONENT of a solution of a linear system - The limes superior

$$\lambda_{x(t)} = \overline{\lim_{t \to +\infty}} \frac{1}{t} \ln |x(t)|,$$

where $x(t) \neq 0$ is a solution of the linear system of ordinary differential equations

$$\dot{x} = A(t)x; \tag{1}$$

here $x \in \mathbf{R}^n$ and $A(\cdot)$ is a mapping $\mathbf{R} \to \mathrm{Hom}(\mathbf{R}^n, \mathbf{R}^n)$, summable on every interval. In coordinates,

$$x(t) = (x^1(t), \ldots, x^n(t)),$$

$$\dot{x}^i = \sum_{j=1}^{n} a_j^i(t)x^j, \quad i = 1, \ldots, n,$$

where $a_j^i(t)$ are functions summable on every interval

and

$$| x | = \sqrt{\sum_{i=1}^{n} | x^i |^2}$$

(or any other equivalent norm; $\lambda_{x(t)}$ does not depend on the choice of the norm in $\mathbf{R}^n$ or in $\mathbf{C}^n$).

Lyapunov's theorem. Suppose that

$$\overline{\lim_{t \to +\infty}} \frac{1}{t} \int_0^t \| A(\tau) \| \, d\tau < +\infty;$$

equivalently:

$$\overline{\lim_{t \to +\infty}} \frac{1}{t} \int_0^t | a_j^i(\tau) | \, d\tau < +\infty, \quad i,j=1,\dots,n.$$

Then for any solution $x(t) \neq 0$ of the system (1) the Lyapunov characteristic exponent $\lambda_{x(t)}$ is a real number (that is, $\neq \pm \infty$). The following assertions hold for the Lyapunov characteristic exponents of non-zero solutions of (1):

1) $\lambda_{\alpha x(t)} = \lambda_{x(t)}$, $\alpha \neq 0$;

2) $\lambda_{x_1(t)+x_2(t)} \leqslant \max(\lambda_{x_1(t)}, \lambda_{x_2(t)})$;

3) there exists a set of linearly independent solutions of (1), denoted by $\{x_i(t)\}_{i=1}^n$, such that for any other n linearly independent solutions $\hat{x}_i(t)$, $i=1,\dots,n$, of (1), numbered in decreasing order of the Lyapunov characteristic exponents, that is, $\lambda_{\hat{x}_i(t)} \geqslant \lambda_{\hat{x}_j(t)}$ for $i \leqslant j$, the following inequalities hold:

$$\lambda_{\hat{x}_i(t)} \geqslant \lambda_{x_i(t)}, \quad i=1,\dots,n.$$

A **fundamental system of solutions** $\{x_i(t)\}_{i=1}^n$ with this property is called *normal*. Such a normal system has the properties:

a) the family of numbers $\lambda_i(A) = \lambda_{x_i(t)}$, $i=1,\dots,n$, does not depend on the choice of the normal fundamental system;

b) for any solution $x(t) \neq 0$ of (1) its Lyapunov characteristic exponent $\lambda_{x(t)}$ is equal to some $\lambda_i(A)$;

c) $\lambda_i(A) \geqslant \lambda_j(A)$, $i \leqslant j$.

The numbers $\lambda_1(A) \geqslant \cdots \geqslant \lambda_n(A)$ are called the Lyapunov characteristic exponents of the system (1); the number $\lambda_1(A)$ is often called the *leading Lyapunov characteristic exponent* of (1).

The set of all Lyapunov characteristic exponents of non-zero solutions of (1) is called the *spectrum*.

Special case. 1) A system with constant coefficients (that is, $A(t) \equiv A(0)$). In this case the $\lambda_i(A)$ are equal to the real parts of the eigen values of the operator $A(0)$ (the matrix $\| a_j^i \|$).

2) A system with periodic coefficients (that is, $A(t+T) \equiv A(t)$, $T > 0$). In this case

$$\lambda_i(A) = \frac{1}{T} \ln | \mu_i |,$$

where μ_i are the **multipliers** of the system (1), numbered in non-increasing order of their moduli (each is taken as many times as its multiplicity).

The role of the Lyapunov characteristic exponent in the theory of Lyapunov stability is based on the following assertion: If $\lambda_1(A) < 0$ (>0), then the solutions of (1) are asymptotically stable (respectively, unstable, cf. **Asymptotically-stable solution**). From $\lambda_1(A) < 0$ it does not follow that the null solution of the system

$$\dot{x} = A(t)x + O(| x |^2)$$

is Lyapunov stable; however, if it is also known that the system (1) is a **regular linear system**, then this conclusion is valid (*Lyapunov's theorem*).

Suppose that the system $\dot{x} = B(t)x$ is obtained by a small perturbation of a system (1) satisfying the condition

$$\sup_{t \in \mathbf{R}} \| A(t) \| < +\infty;$$

that is, the distance between them, defined by the formula

$$d(A, B) = \sup_{t \in \mathbf{R}} \| A(t) - B(t) \|, \tag{2}$$

is small. For $n > 1$ this does not imply that the quantity

$$| \lambda_1(A) - \lambda_1(B) |$$

is small (it is implied if the system (1) has constant or periodic coefficients, and also for certain other systems); in other words, the functionals $\lambda_i(A)$ are not everywhere continuous on the space of systems (1) $(\sup_{t \in \mathbf{R}} \| A(t) \| < +\infty)$, endowed with the given metric (2).

Lyapunov characteristic exponents were introduced by A.M. Lyapunov, not only for solutions of the system (1), but also for arbitrary functions on $\mathbf{R}^+$ (see [1]).

References

[1] LYAPUNOV, A.M.: *Stability of motion*, Acad. Press, 1966 (translated from the Russian).
[2] BYLOV, B.F., VINOGRAD, R.E., GROBMAN, D.M. and NEMYTSKIĬ, V.V.: *The theory of the Lyapunov exponent and its application to questions of stability*, Moscow, 1966 (in Russian).
[3] IZOBOV, N.A.: 'Linear systems of ordinary differential equations', *J. Soviet Math.* **5** (1976), 46-96. (*Itogi Nauk,i Tekhn. Mat. Anal.* **12** (1974), 71-146)

V.M. Millionshchikov

Editorial comments. Presently, Lyapunov (characteristic) exponents are used at a much wider scale. For a survey see [A6]. First, the matrix A may be a stochastic time-dependent function. Lyapunov exponents are also used in relation with a system of non-linear differential equations

$$x = f(x), \quad x = (x^1(t), \dots, x^n(t)),$$

having a **strange attractor** (or repellor) $s(t)$ as limit solution, see [A7]. The system linearized in $s(t)$ is of the form (1) with

$$A(t) = \left\{ \frac{\partial f_i}{\partial x_j}(s(t)) \right\}_{(n \times n)}.$$

One of the exponents is zero. The occurrence of one or more positive exponents indicates that $s(t)$ is a strange attractor. For a conservative system the sum of the Lyapunov exponents is zero, while for a dissipative system the sum is negative. The *capacity D of a strange attractor* is a fractal number related to the **Hausdorff dimension**. J.L. Kaplan and J.A. Yorke made the conjecture

$$D = j - \frac{1}{\lambda_{j+1}} \sum_{i=1}^{j} \lambda_i$$

with $0 < \sum_{i=1}^{j} \lambda_i < -\lambda_{j+1}$ $(\lambda_1 \geqslant \cdots \geqslant \lambda_n)$.

The concept of Lyapunov exponents extends to non-linear stochastic systems, as well as to iteration mappings

$$x(t+1) = A(t)x(t),$$

see [A6] and [A7].

In a yet more general deterministic setting, let X be a compact subset of a Hilbert space H, and let $f: X \to H$ be a mapping such that $f(X) = X$. The mapping f is supposed to satisfy the following uniform differentiability condition: For each $x \in X$ there is a linear compact operator $L(x): H \to H$ such that

$$\sup_{x,y} \frac{\| f(y) - f(x) - L(x)(y-x) \|}{\| y - x \|} \to 0$$

as $\epsilon \to 0$, where $\| \cdot \|$ denotes the norm in H, and where the supremum is taken over all x, y with $0 < \| x - y \| \leqslant \epsilon$.

For a compact linear operator L, let $\alpha_1(L) \geqslant \alpha_2(L) \geqslant \cdots$ be the eigen values of $(L^* L)^{1/2}$. For each positive integer d, let $\omega_d(L) = \alpha_1(L) \cdots \alpha_d(L)$, while for a non-integer positive real number $d = n + s$, $0 < s < 1$, define $\omega_d(L) = \omega_n(L)(\alpha_{n+1}(L))^s$.

Suppose that f and L are such that $\sup_{x \in X} \| L(x) \| < \infty$.

For each $x \in X$, let $L_p(x) = L(f^{p-1}(x)) \circ \cdots \circ L(f(x)) \circ L(x)$, where f^r denotes the r-th iterate of f. Define the (local) *Lyapunov numbers* and *Lyapunov exponents* by

$$\Lambda_j(x) = \limsup_{p \to \infty} \{\alpha_j(L_p(x))\}^{1/p},$$

$$\mu_j(x) = \ln \Lambda_j(x).$$

The *uniform Lyapunov exponents* μ_i and *uniform Lyapunov numbers* Λ_i in this setting are defined as follows:

$$\bar{\omega}_j(p) = \sup_{x \in X} \omega_j(L_p(x)),$$

$$\Pi_j = \lim_{p \to \infty} (\bar{\omega}_j(p))^{1/p},$$

$$\Lambda_1 \cdots \Lambda_j = \Pi_j, \quad j = 1, 2, \ldots,$$

$$\mu_j = \ln \Lambda_j, \quad j = 1, 2, \ldots.$$

Let n be the smallest integer such that $\mu_1 + \cdots + \mu_n \geqslant 0$ and $\mu_1 + \cdots + \mu_n + \mu_{n+1} < 0$. The number $d_L(X) = n + | \mu_{n+1} |^{-1}(\mu_1 + \cdots + \mu_n)$ is then called the *Lyapunov dimension* of X. One has (see [A2]) $d_H(X) \leqslant d_L(X)$, where $d_H(X)$ is the **Hausdorff dimension** of X.

References

[A1] KLIEMANN, W.: 'Analysis of nonlinear stochastic systems', in W. Schiehlen and W. Wedig (eds.): *Analysis and estimation of stochastic mechanical systems*, Springer (Wien), 1988, pp. 43-102.
[A2] CONSTANTIN, P., FOIAS, C. and TEMAM, R.: *Attractors representing turbulent flows*, Amer. Math. Soc., 1985.
[A3] YOUNG, L.S.: 'Capacity of attractors', *Ergod. Th. Dynam. Systems* 1 (1981), 381-383.
[A4] YOUNG, L.S.: 'Dimension, entropy, and Lyapunov exponents', *Ergod. Th. Dynam. Systems* 2 (1982), 109-124.
[A5] FREDRICKSON, P., KAPLAN, J.L., YORKE, E.D. and YORKE, J.A.: 'The Lyapunov dimension of strange attractors', *J. Diff. Eq.* 49 (1983), 185-207.
[A6] ARNOLD, L. and WIHSTUTZ, V. (EDS.): *Lyapunov exponents*, Lecture notes in math., 1186, Springer, 1986.
[A7] SCHUSTER, H.G.: *Deterministic chaos*, Physik-Verlag, 1988.
[A8] CODDINGTON, E.A. and LEVINSON, N.: *Theory of ordinary differential equations*, McGraw-Hill, 1955.
[A9] GUCKENHEIMER, J. and HOLMES, PH.: *Nonlinear oscillations, dynamical systems and bifurcations of vector fields*, Springer, 1983.

AMS 1980 Subject Classification: 34D05, 34D20, 34A30, 34C11, 58F13, 60H10

LYAPUNOV FUNCTION - A function defined as follows. Let x_0 be a *fixed point* of the system of differential equations

$$\dot{x} = f(x, t)$$

(that is, $f(x_0, t) \equiv 0$), where the mapping $f(x, t): U \times \mathbf{R}^+ \to \mathbf{R}^n$ is continuous and continuously differentiable with respect to x (here U is a neighbourhood of x_0 in $\mathbf{R}^n$). In coordinates this system is written in the form

$$\dot{x}^i = f^i(x^1, \ldots, x^n, t), \quad i = 1, \ldots, n.$$

A differentiable function $V(x): U \to \mathbf{R}$ is called a Lyapunov function if it has the following properties:

1) $V(x) > 0$ for $x \neq x_0$;

2) $V(x_0) = 0$;

3) $0 \geqslant \dfrac{dV(x)}{dx} f(x, t) = \sum_{i=1}^{n} \dfrac{\partial V(x^1, \ldots, x^n)}{\partial x^i} f^i(x^1, \ldots, x^n, t).$

The function $V(x)$ was introduced by A.M. Lyapunov (see [1]).

Lyapunov's lemma holds: If a Lyapunov function exists, then the fixed point is Lyapunov stable (cf. **Lyapunov stability**). This lemma is the basis for one of the methods for investigating stability (the so-called *second method of Lyapunov*).

References

[1] LYAPUNOV, A.M.: *Stability of motion*, Acad. Press, 1966 (translated from the Russian).
[2] BARBASHIN, E.A.: *Lyapunov functions*, Moscow, 1970 (in Russian).

V.M. Millionshchikov

Editorial comments. For additional references see **Lyapunov stability**.

AMS 1980 Subject Classification: 34D20

LYAPUNOV – SCHMIDT EQUATION - A nonlinear integral equation of the form

$$u(x) - \int_{\Omega} K(x, s)u(s)\, ds = \qquad (1)$$

$$= U_{01} \begin{bmatrix} x \\ v \end{bmatrix} + \sum_{m+n \geqslant 2} U_{mn} \begin{bmatrix} x \\ u, v \end{bmatrix}, \quad x \in \Omega,$$

where

$$U_{01} \begin{bmatrix} x \\ v \end{bmatrix} = K_0(x)v(x) + \int_{\Omega} K_1(x, s)v(s)\, ds,$$

$$U_{mn}\begin{bmatrix} x \\ u,v \end{bmatrix} = \sum_{v=1}^{n} \int_{\Omega} \cdots \int_{\Omega} K^{(v)}(x,s_1,\ldots,s_i)\times$$

$$\times u^{\alpha_0}(x)u^{\alpha_1}(s_1)\cdots u^{\alpha_i}(s_i)v^{\beta_0}(x)v^{\beta_1}(s_1)\cdots v^{\beta_i}(s_i)\,ds_1\cdots ds_i,$$

$\alpha_0,\ldots,\alpha_i,\beta_0,\ldots,\beta_i$ are non-negative integers,

$$\alpha_0 + \cdots + \alpha_i = m, \quad \beta_0 + \cdots + \beta_i = n,$$

Ω is a closed bounded set in a finite-dimensional Euclidean space, v and the function K are given continuous functions of their arguments, $s_1,\ldots,s_i\in\Omega$, and u is the unknown function. The sum on the right-hand side of (1) may be finite or it may represent an infinite series. In the latter case the series is called an *integro-power series* of two functional arguments. It is assumed that the series converges absolutely and uniformly.

If 1 is not a characteristic number of the kernel $K(x,s)$, then equation (1) has a unique small solution in the class of continuous functions for sufficiently small $|v(x)|$. This solution can be represented as an integro-power series. The case when 1 is a characteristic number of the kernel K is more complicated. In this case one constructs a system of equations — the *branching equations (bifurcation equations)*:

$$\omega_k(\xi_1,\ldots,\xi_n,v) = 0, \quad k=1,\ldots,n, \qquad (2)$$

where ω_k are known power series and n is the multiplicity of the characteristic number 1. In the general case the system (2) has a non-unique solution. Whatever the fixed sufficiently small function v, to every small continuous solution of (2) (a continuous solution of (2) is said to be small if $\xi_i(0)=0$) there corresponds a small solution of (1) that can be represented as an integro-power series.

An equation of the type (1) was first considered by A.M. Lyapunov in 1906, and later — in a more general form — by E. Schmidt in 1908.

References

[1] VAĬNBERG, M.M. and TRENOGIN, V.A.: *Theory of branching of solutions of non-linear equations*, Noordhoff, 1974 (translated from the Russian).
[2] SMIRNOV, N.S.: *Introduction to the theory of non-linear integral equations*, Leningrad-Moscow, 1936 (in Russian).

B.V. Khvedelidze

Editorial comments.

References

[A1] CHOW, S. and HALE, J.: *Methods of bifurcation theory*, Springer, 1982.

AMS 1980 Subject Classification: 45H05

LYAPUNOV STABILITY *of a point relative to a family of mappings*

$$\{f_t\}_{t\in G^+}: E \to E \qquad (1)$$

of a certain space E - *Equicontinuity* of this family of mappings at the point (here G^+ is the set of non-negative numbers in G; for example, the real numbers

$G=\mathbf{R}$ or the integers $G=\mathbf{Z}$). Lyapunov stability of a point relative to the family of mappings (1) is equivalent to the continuity at this point of the mapping $x\mapsto x(\cdot)$ of a neighbourhood of this point into the set of functions $x(\cdot)$ defined by the formula $x(t)=f_t(x)$, equipped with the topology of uniform convergence on G^+. Lyapunov stability of a point relative to a mapping is defined as Lyapunov stability relative to the family of non-negative powers of this mapping. Lyapunov stability of a point relative to a dynamical system f^t is Lyapunov stability of this point relative to the family $\{f^t\}_{t\in G^+}$. Lyapunov stability of the solution $x_0(\cdot)$ of an equation $x(t+1)=g_t(t)$ given on $t_0+\mathbf{Z}^+$ is Lyapunov stability of the point $x_0(t_0)$ relative to the family of mappings $\{f_t=g_{t_0+t}\cdots g_{t_0}\}_{t\in\mathbf{Z}^+}$.

Lyapunov stability of the solution $x_0(\cdot)$ of a differential equation $\dot{x}=f(x,t)$ given on $t_0+\mathbf{R}^+$ is Lyapunov stability of the point $x_0(t_0)$ relative to the family of mappings $\{X(t_0+t,t_0)\}_{t\in\mathbf{R}^+}$, where $X(\theta,\tau)$ is the **Cauchy operator** of this equation. Lyapunov stability of the solution $y(\cdot)$ of a differential equation

$$y^{(m)} = g(y,\dot{y},\ldots,y^{(m-1)},t)$$

of order m, given on $t_0+\mathbf{R}^+$, is Lyapunov stability of the solution $x(\cdot)=(y(\cdot),\dot{y}(\cdot),\ldots,y^{(m-1)}(\cdot))$ of the corresponding first-order differential equation $\dot{x}=f(x,t)$, given on $t_0+\mathbf{R}^+$, where

$$x = (x_1,\ldots,x_m),$$

$$f(x,t) = (x_2,\ldots,x_m,g(x_1,\ldots,x_m,t)).$$

The definitions 1 - 7 given below are some concrete instances of the above and related definitions.

1. Let a differential equation $\dot{x}=f(x,t)$ be given, where x lies in an n-dimensional normed space E. A solution $x_0(\cdot): t_0+\mathbf{R}^+\to E$ of this equation is called *Lyapunov stable* if for every $\epsilon>0$ there exists a $\delta>0$ such that for every $x\in E$ satisfying the inequality $|x-x_0(t_0)|<\delta$, the solution $x(\cdot)$ of the Cauchy problem

$$\dot{x} = f(x,t), \quad x(t_0) = x$$

is unique, defined on $t_0+\mathbf{R}^+$, and for each $t\in t_0+\mathbf{R}^+$ satisfies the inequality $|x(t)-x_0(t)|<\epsilon$. If, in addition, one can find a $\delta_0>0$ such that for every solution $x(\cdot)$ of the equation $\dot{x}=f(x,t)$ whose initial value satisfies the inequality

$$|x(t_0)-x_0(t_0)| < \delta_0,$$

the equation

$$\lim_{t\to+\infty} |x(t)-x_0(t)| = 0$$

holds (respectively, the inequality

$$\varlimsup_{t\to+\infty} \frac{1}{t}\ln|x(t)-x_0(t)| < 0$$

holds; here and elsewhere one puts $\ln 0=\infty$), then the solution $x_0(\cdot)$ is called *asymptotically* (respectively, *exponentially*) *stable*.

A solution of the equation

$$\dot{x} = f(x, t), \qquad (2)$$

where $x \in \mathbf{R}^n$ or $x \in \mathbf{C}^n$, is called *Lyapunov stable* (*asymptotically*, *exponentially stable*) if it becomes such after equipping the space $\mathbf{R}^n$ (or $\mathbf{C}^n$) with a norm. This property of the solution does not depend on the choice of the norm.

2. Let a mapping $f: S \to S$ be given, where (S, d) is a metric space. The point $x_0 \in S$ is called *Lyapunov stable* relative to the mapping f if for every $\epsilon > 0$ there exists a $\delta > 0$ such that for any $x \in S$ satisfying the inequality $d(x, x_0) < \delta$, the inequality

$$d(f^t x, f^t x_0) < \epsilon$$

holds for each $t \in \mathbf{N}$. If, moreover, one can find a $\delta_0 > 0$ such that for each $x \in S$ satisfying $d(x, x_0) < \delta_0$ one has the equation

$$\lim_{t \to +\infty} d(f^t x, f^t x_0) = 0$$

(the inequality

$$\overline{\lim_{t \to +\infty}} \frac{1}{t} \ln d(f^t x, f^t x_0) < 0,$$

respectively), then the point x_0 is called *asymptotically* (respectively, *exponentially*) *stable* relative to f.

Let f be a mapping from a compact topological space S into itself. A point $x_0 \in S$ is called *Lyapunov stable* (*asymptotically stable*) relative to f if it becomes such after equipping S with a metric. This property of the point does not depend on the choice of the metric.

If S is a compact differentiable manifold, then a point $x_0 \in S$ is called *exponentially stable* relative to a mapping $f: S \to S$ if it becomes such after equipping S with a certain Riemannian metric. This property of the point does not depend on the choice of the Riemannian metric.

3. Suppose that a differential equation (2) is given, where x lies in a topological vector space E. A solution $x_0(\cdot): t_0 + \mathbf{R}^+ \to E$ of this equation is called *Lyapunov stable* if for each neighbourhood of zero $U \subset E$ there is a neighbourhood V of $x_0(t_0)$ in E such that for every $x \in E$ the solution $x(\cdot)$ of the Cauchy problem (2), $x(t_0) = x$, is unique, defined on $t_0 + \mathbf{R}^+$ and satisfies the relation $x(t) - x_0(t) \in U$ for all $t \in t_0 + \mathbf{R}^+$. If, in addition, one can find a neighbourhood $V_0 \subset E$ of the point $x_0(t_0)$ such that for every solution $x(\cdot)$ of (2) satisfying $x(t_0) \in V_0$ one has the equation

$$\lim_{t \to +\infty} (x(t) - x_0(t)) = 0$$

(respectively,

$$\lim_{t \to +\infty} e^{\alpha t}(x(t) - x_0(t)) = 0$$

for a certain $\alpha > 0$), then the solution $x_0(\cdot)$ is called *asymptotically* (respectively, *exponentially*) *stable*. If E is a normed space, then this definition may be formulated as in 1 above, if as norm $|\cdot|$ one takes any norm compatible with the topology on E.

4. Let a differential equation (2) be given on a Riemannian manifold U (for which a Euclidean or a Hilbert space can serve as a model) or, in a more general situation, on a Finsler manifold U (for which a normed space can serve as a model); the distance function in U is denoted by $d(\cdot, \cdot)$. A solution $x_0(\cdot): t_0 + \mathbf{R}^+ \to U$ of this equation is called *Lyapunov stable* if for each $\epsilon > 0$ one can find a $\delta > 0$ such that for each $x \in U$ satisfying $d(x, x_0(t_0)) < \delta$, the solution $x(\cdot)$ of the Cauchy problem (2), $x(t_0) = x$, is unique, defined for $t_0 + \mathbf{R}^+$ and satisfies the inequality $d(x(t), x_0(t)) < \epsilon$ for all $t \in t_0 + \mathbf{R}^+$. If, in addition, one can find a $\delta_0 > 0$ such that for every solution $x_0(\cdot)$ of (2) whose initial value satisfies the inequality $d(x(t_0), x_0(t_0)) < \delta_0$ one has the equation

$$\lim_{t \to +\infty} d(x(t), x_0(t)) = 0$$

(the inequality

$$\overline{\lim_{t \to +\infty}} \frac{1}{t} \ln d(x(t), x_0(t)) < 0,$$

respectively), then the solution $x_0(\cdot)$ is called *asymptotically* (respectively, *exponentially*) *stable*.

Suppose that the differential equation (2) is given on a compact differentiable manifold V^n. A solution of this equation is called *Lyapunov stable* (*asymptotically*, *exponentially stable*) if it becomes such when the manifold V^n is equipped with some Riemannian metric. This property of the solution does not depend on the choice of the Riemannian metric.

5. Let E be a **uniform space**. Let

$$f_t: U \to E, \quad t \in G^+ \quad (G = \mathbf{R} \text{ or } = \mathbf{Z}),$$

be a mapping defined on an open set $U \subset E$. A point $x_0 \in U$ is called *Lyapunov stable* relative to the family of mappings $\{f_t\}_{t \in G^+}$ if for every entourage W there exists a neighbourhood V of x_0 such that the set of all $x \in U$ satisfying $(f_t x, f_t x_0) \in W$ for all $t \in G^+$, is a neighbourhood of x_0. If, in addition, there exists a neighbourhood V_0 of x_0 such that for every $x \in V_0$ and every entourage W one can find a $t(x, w) \in G^+$ such that $(f_t x, f_t x_0) \in W$ for all $t \in t(x, w) + \mathbf{R}^+$, then the point x_0 is called *asymptotically stable*.

If E is a compact topological space and $f_t: U \to E$, $t \in G^+$, is a mapping given on some open set $U \subset E$, then the point $x_0 \in U$ is called *Lyapunov stable* (*asymptotically stable*) relative to the family of mappings $\{f_t\}_{t \in G^+}$ if it becomes such after the space E is equipped with the unique uniform structure that is compatible with the topology on E.

6. Let E be a topological space and U an open subspace in it. Let $f_t: U \to E$, $t \in G^+$, where G is $\mathbf{R}$ or $\mathbf{Z}$, be

a mapping having x_0 as fixed point. The fixed point x_0 is called *Lyapunov stable* relative to the family of mappings $\{f_t\}_{t \in G^+}$ if for every neighbourhood V of x_0 there exists a neighbourhood W of the same point such that $f_t W \subset V$ for all $t \in G^+$. If, in addition, there exists a neighbourhood V_0 of x_0 such that $\lim_{t \to +\infty} f_t x = x_0$ for every $x \in V_0$, then the point x_0 is called *asymptotically stable* relative to the family of mappings $\{f_t\}_{t \in G^+}$.

7. *Lyapunov stability* (*asymptotic, exponential stability*) of a solution $y_0(\cdot)$ of an equation of arbitrary order, $y^{(m)} = g(y, \dot{y}, \ldots, y^{(m-1)}, t)$, is understood to mean Lyapunov stability (respectively asymptotic, exponential stability) of the solution $x_0(\cdot) = (y_0(\cdot), \dot{y}_0(\cdot), \ldots, y_0^{(m-1)}(\cdot))$ of the corresponding first-order equation (2), where $x = (x_1, \ldots, x_m)$, $f(x, t) = (x_2, \ldots, x_m, g(x_1, \ldots, x_m, t))$.

Definitions 1, 2, 4, 6, 7 include stable motions of systems with a finite number of degrees of freedom (where the equations on manifolds arise naturally when considering mechanical systems with a constraint). Definitions 2 - 7 include stable motions in the mechanics of continuous media and in other parts of physics, stable solutions of operator equations, functional-differential equations (in particular, equations with retarded arguments) and other equations.

Study of the stability of an equilibrium position of an autonomous system. Let $\dot{x} = f(x)$ be an autonomous differential equation defined in a neighbourhood of a point $x_0 \in \mathbf{R}^n$, where the function $f(\cdot)$ is continuously differentiable and vanishes at this point. If the real parts of all eigen values of the derivative df_{x_0} are negative, then the fixed point x_0 of $\dot{x} = f(x)$ is exponentially stable (*Lyapunov's theorem on stability in a first approximation*); to facilitate the verification of the condition in this theorem one applies criteria for stability. If under these conditions at least one of the eigen values of the derivative df_{x_0} has positive real part (this condition may be checked without finding the eigen values themselves, cf. **Stability criterion**), then the fixed point of the differential equation $\dot{x} = f(x)$ is unstable.

Example. The equation of the oscillation of a pendulum with friction is

$$\ddot{y} + a\dot{y} + b \sin y = 0, \quad a, b > 0.$$

The lower equilibrium position $y = \dot{y} = 0$ is exponentially stable, since the roots of the characteristic equation $\lambda^2 + a\lambda + b = 0$ of the variational equation (cf. **Variational equations**) have negative real parts. The upper equilibrium position $y = \pi, \dot{y} = 0$ is unstable, since the characteristic equation $\lambda^2 + a\lambda - b = 0$ of the variational equation $\ddot{y} + a\dot{y} - by = 0$ has a positive root. This instability takes place even in the absence of friction ($a = 0$). The lower equilibrium position of a pendulum without

friction is one of the so-called *critical cases*, when all eigen values of the derivative df_{x_0} are contained in the left complex half-plane, and at least one of them lies on the imaginary axis.

For the study of stability in critical cases, A.M. Lyapunov proposed the so-called *second method* for studying stability (cf. **Lyapunov function**). For a pendulum without friction,

$$\ddot{y} + b \sin y = 0, \quad b > 0,$$

the lower equilibrium position is Lyapunov stable, since there exists a **Lyapunov function**

$$V(y, \dot{y}) = \frac{1}{2}\dot{y}^2 + b(1 - \cos y)$$

— the total energy of the pendulum; the condition of non-positivity for the derivative of this function is a consequence of the law of conservation of energy.

A fixed point x_0 of a differentiable mapping $f: \mathbf{R}^n \to \mathbf{R}^n$ is exponentially stable relative to f if all eigen values of the derivative df_{x_0} are less than 1 in modulus, and it is unstable if at least one of them has modulus > 1.

The study of the stability of periodic points of differentiable mappings reduces to the study of stability of fixed points relative to the powers of these mappings. Periodic solutions of autonomous differential equations are not asymptotically stable (cf. **Orbit stability**; **Andronov − Witt theorem**).

It should not be believed that exponential stability of the null solution of the variational equation of the autonomous differential equation $\dot{x} = f(x)$ along a solution $x(\cdot)$ implies stability of the solution. This is shown by *Perron's example* (cf. [2], [3]):

$$\left. \begin{aligned} \dot{u} &= -au, \\ \dot{v} &= (\sin \ln t + \cos \ln t - 2a)v + u^2; \end{aligned} \right\} \quad (3)$$

for $a > 1/2$ the null solution of the system of variational equations,

$$\left. \begin{aligned} \dot{u} &= -au, \\ \dot{v} &= (\sin \ln t + \cos \ln t - 2a)v, \end{aligned} \right\} \quad (4)$$

of the system (3) (along the null solution) is exponentially stable (the Lyapunov characteristic exponents of the system (4) are $-a, 1 - 2a$, cf. **Lyapunov characteristic exponent**), but for $a \in (1/2, (2 + e^{-\pi})/4)$ the null solution of the system (3) is unstable. However, stability in the first approximation is typical, in a sense explained below.

Let S be the set of diffeomorphisms f of a Euclidean space E^n onto itself having uniformly continuous derivatives that satisfy the inequality

$$\sup_{x \in E^n} \max\{\| df_x \|, \| (df_x)^{-1} \|\} < +\infty.$$

For each diffeomorphism $j \in S$ denote by S_j the set of all $f \in S$ satisfying the inequality

$$\sup_{x \in E^n} | fx - jx | < +\infty;$$

one endows S_j with the distance function

$$d(f, g) = \sup_{x \in E^n} (| fx - gx | + \| df_x - dg_x \|).$$

For each $j \in S$ there is in $S_j \times E^n$ an everywhere-dense set D_j of type G_δ with the following property: If $(f, x) \in D_j$ is such that for every $g \in T_x E^n$ the inequality

$$\overline{\lim_{m \to +\infty}} \frac{1}{m} \ln | df^m g | < 0$$

holds, then there is a neighbourhood U of (f, x) in $S_j \times E^n$ such that for every $(g, y) \in U$ the point y is exponentially stable relative to the diffeomorphism g.

For a dynamical system given on a compact differentiable manifold, an analogous theorem can be formulated more simply and as a differential-topologically invariant statement. Let V^n be a closed differentiable manifold. The set S of all diffeomorphisms f of class C^1 mapping V^n to V^n can be equipped with the C^1-topology. In the space $S \times V^n$ there is an everywhere-dense set D of type G_δ with the following property: If for an $(f, x) \in D$ the inequality

$$\lim_{m \to +\infty} \frac{1}{m} \ln | df^m g | < 0$$

holds for all $g \in T_x V^n$, then there is a neighbourhood U of (f, x) in $S \times V^n$ such that for each $(g, y) \in U$ the point y is exponentially stable relative to the diffeomorphism g.

The concepts of Lyapunov stability, asymptotic stability and exponential stability were introduced by Lyapunov [1] in order to develop methods for studying stability in the sense of these definitions (cf. **Lyapunov stability theory**).

References

[1] LYAPUNOV, A.M.: *Stability of motion*, Acad. Press, 1966 (translated from the Russian).
[2] PERRON, O.: 'Ueber Stabilität und asymptotisches Verhalten der Integrale von Differentialgleichungssystemen', *Math. Z.* **29** (1928), 129-160.
[3] BELLMAN, R.: *Stability theory of differential equations*, Dover, reprint, 1969.

V.M. Millionshchikov

Editorial comments. For stability questions of differential equations with discontinuous right-hand sides cf. [A6].

References

[A1] LASALLE, J.P. and LEFSCHETZ, S.: *Stability by Lyapunov's direct method with applications*, Acad. Press, 1961.
[A2] HIRSCH, M.W. and SMALE, S.: *Differential equations, dynamical systems and linear algebra*, Acad. Press, 1974.
[A3] HAHN, W.: *Stability of motion*, Springer, 1967.
[A4] BHATIA, N.P. and SZEGÖ, G.P.: *Stability theory of dynamical systems*, Springer, 1970.
[A5] RABINOVICH, M.I. and TRUBETSKOV, D.I.: *Oscillations and waves*, Kluwer, 1989, Chapt. 6-7 (translated from the Russian).
[A6] FILIPPOV, A.F.: *Differential equations with discontinuous righthand sides*, Kluwer, 1989, § 15 (translated from the Russian).

AMS 1980 Subject Classification: 34D20, 58F10, 93D05

LYAPUNOV STABILITY THEORY - A theory of the stability of motion, constructed by A.M. Lyapunov at the turn into 20-th century (see [1]). At the basis of it lie the concepts of **Lyapunov stability** and asymptotic stability (see **Asymptotically-stable solution**), introduced by Lyapunov, Lyapunov's theorem on stability in the first approximation (on which Lyapunov's first method for investigating stability is based) and Lyapunov's second method (see **Lyapunov function**). The results and methods of Lyapunov, developed for the construction of stability theory, found important and numerous applications in mathematics, mechanics and engineering. See also **Stability theory**.

References

[1] LYAPUNOV, A.M.: *Stability of motion*, Acad. Press, 1966 (translated from the Russian).

V.M. Millionshchikov

AMS 1980 Subject Classification: 73HXX

LYAPUNOV STOCHASTIC FUNCTION - A nonnegative function $V(t, x)$ for which the pair $(V(t, X(t)), F_t)$ is a supermartingale for some random process X up to the instant t (cf. also **Martingale**). If $X(t)$ is a **Markov process**, then the Lyapunov stochastic function is a function for which the Lyapunov stochastic operator

$$LV(t,x) =$$
$$= \lim_{h \to 0} \frac{1}{h} E[V(t+h, X(t+h)) - V(t, X(t)) | X(t) = x]$$

is non-positive. The operator L is the **infinitesimal operator** of the process $(t, X(t))$, and so the verification of the condition $LV \leq 0$ is easily carried out in specific cases. The operator L goes into the usual Lyapunov operator $dV(t, X(t))/dt$ when the process X is determinate and is described by a system of differential equations. By means of the Lyapunov stochastic function it is possible to verify a number of qualitative properties of the trajectories of $X(t)$; their role in the theory of random processes is similar to the role of the classical **Lyapunov function** in the theory of systems of differential equations.

Functions $V(t, x)$ for which $(V(t, X(t)), F_t)$ is not a supermartingale, but from which one can readily form a supermartingale, are sometimes also called Lyapunov stochastic functions. Below typical results are presented on the qualitative behaviour of trajectories of Markov processes in terms of a Lyapunov stochastic function.

1) If $X(t)$ is a right-continuous strong Markov process in $\mathbf{R}^k$, defined up to the instant τ of first leaving an arbitrary compact set, and if there is a Lyapunov

stochastic function $V(t, x)$, $t>0$, $x \in \mathbf{R}^k$, and a constant c such that

$$\inf_{t, |x|>R} V(t, x) \to \infty \text{ as } R \to \infty, \quad LV \leqslant cV,$$

then

$$P\{\tau < \infty \mid X(0) = x\} = 1$$

for any $x \in \mathbf{R}^k$; that is, the process X is defined for all $t>0$ (is indefinitely extendable).

2) For the stationary Markov process in $\mathbf{R}^k$ corresponding to a **transition function** $P(t, x, A)$ to exist it is sufficient that there should be a function $V(x) \geqslant 0$ for which

$$\sup_{|x|>R} LV(x) \to -\infty$$

as $R \to \infty$.

By means of the Lyapunov stochastic function one can carry over to Markov processes the main theorems of the direct Lyapunov method; these functions have also found application in the investigation of processes in discrete time.

References
[1] KUSHNER, H.J.: *Stochastic stability and control*, Acad. Press, 1967.
[2] HAS'MINSKIĬ, R.Z. [R.Z. KHAS'MINSKIĬ]: *Stochastic stability of differential equations*, Sythoff & Noordhoff, 1980 (translated from the Russian).
[3] KALASHNIKOV, V.V.: *Qualitative analysis of the behaviour of complex systems by the method of test functions*, Moscow, 1978 (in Russian).

R.Z. Khas'minskiĭ

Editorial comments. The phrase *stochastic Lyapunov function* is more common than 'Lyapunov stochastic function'.

Recently, stochastic Lyapunov functions have been used to prove convergence of recursive algorithms driven by stochastic processes. Convergence problems of this type arise in *system identification* and *adaptive control*.

References
[A1] GOODWIN, G.C., RAMAGADGE, P.J. and CAINES, P.E.: 'Discrete time stochastic adaptive control', *SIAM J. Control Optim.* **19** (1981), 829-853.
[A2] METIVIER, M. and PRIOURET, P.: 'Applications of a Kushner and Clark lemma to general classes of stochastic algorithms', *IEEE Trans. Inform. Theory* **30** (1984), 140-151.
[A3] SOLO, V.: 'The convergence of AML', *IEEE Trans. Autom. Control* **24** (1979), 958-962.
[A4] SCHUPPEN, J.H. VAN: 'Convergence results for continuous-time stochastic filtering algorithms', *J. Math. Anal. Appl.* **96** (1983), 209-225.

AMS 1980 Subject Classification: 60GXX, 60J25, 60J60, 93E15, 93E12, 93C40

LYAPUNOV SURFACES AND CURVES - A class of surfaces and curves that have quite good smoothness properties; it was introduced in potential theory by A.M. Lyapunov at the turn into the 20-th century.

A surface S in the three-dimensional Euclidean space $\mathbf{R}^3$ is called a *Lyapunov surface* if it satisfies the following three conditions (*Lyapunov's conditions*): 1) at every point of S there is a well-defined tangent plane, and consequently a well-defined normal; 2) there is a number $r>0$, the same for all points of S, such that if one takes the part Σ of S lying inside the *Lyapunov sphere* $B(y_0, r)$ with centre at an arbitrary point $y_0 \in S$ and radius r, then the lines parallel to the normal to S at y_0 meet Σ at most once; and 3) there are two numbers $A>0$ and λ, $0<\lambda \leqslant 1$, the same for the whole of S, such that for any two points $y_1, y_2 \in S$,

$$|\theta| < A|y_1 - y_2|^\lambda, \qquad (*)$$

where θ is the angle between the normals to S at y_1 and y_2. Sometimes these three conditions are supplemented by the requirement that S is closed and that the solid angle under which any part σ of S is visible at an arbitrary point $x \in \mathbf{R}^3$ is uniformly bounded.

The Lyapunov conditions can be generalized to hypersurfaces in $\mathbf{R}^n$, $n \geqslant 3$.

Similarly, a simple continuous curve L in the plane $\mathbf{R}^2$ is called a *Lyapunov curve* if it satisfies the following conditions: $1'$) at every point of L there is a well-defined tangent, and consequently a well-defined normal; and $3'$) there are two numbers $A>0$ and λ, $0<\lambda \leqslant 1$, the same for the whole of L, such that for any two points $y_1, y_2 \in L$ (*) holds, where θ is the angle between the tangents or normals to L at y_1 and y_2. Here Lyapunov's condition 2) follows from $1'$) and $3'$). The Lyapunov curves are a subclass of the simple smooth curves.

References
[1] LYAPUNOV, A.M.: 'On certain questions connected with the Dirichlet problem', in *Collected works*, Vol. 1, Moscow, 1954, pp. 45-47; 48-100 (in Russian).
[2] SOBOLEV, S.L.: *Partial differential equations of mathematical physics*, Pergamon, 1964 (translated from the Russian).
[3] VLADIMIROV, V.S.: *Equations of mathematical physics*, Mir, 1984, Chapt. 5 (translated from the Russian).
[4] MUSKHELISHVILI, N.I.: *Singular integral equations*, Wolters-Noordhoff, 1972, Chapt. 1 (translated from the Russian).

E.D. Solomentsev

Editorial comments. A Lyapunov surface is necessarily C^1, and on the other hand a compact surface of class C^2 is a Lyapunov surface. Lyapunov surfaces are used in the study of simple- and double-layer potentials.

AMS 1980 Subject Classification: 31AXX, 31BXX

LYAPUNOV THEOREM - 1) *Lyapunov's theorem in probability theory* is a theorem that establishes very general sufficient conditions for the convergence of the distributions of sums of independent random variables to the **normal distribution**. The precise statement of Lyapunov's theorem is as follows: Suppose that the independent random variables $X_1, X_2, \ldots$, have finite means EX_k, variances DX_k and absolute moments $E|X_k - EX_k|^{2+\delta}$, $\delta > 0$, and suppose also that $B_n = \sum_{k=1}^n DX_k$ is the variance of the sum of

$X_1, \ldots, X_n$. Then if for some $\delta > 0$,

$$\lim_{n \to \infty} \frac{\sum_{k=1}^{n} \mathsf{E} |X_k - \mathsf{E}X_k|^{2+\delta}}{B_n^{1+\delta/2}} = 0, \qquad (1)$$

the probability of the inequality

$$x_1 < \frac{\sum_{k=1}^{n} (X_k - \mathsf{E}X_k)}{\sqrt{B_n}} < x_2 \qquad (2)$$

tends to the limit

$$\frac{1}{\sqrt{2\pi}} \int_{x_1}^{x_2} e^{-x^2/2} \, dx \qquad (3)$$

as $n \to \infty$, uniformly with respect to all values of x_1 and x_2. Condition (1) is called the *Lyapunov condition*. Lyapunov's theorem was stated and proved by A.M. Lyapunov in 1901 and was the final step in research of P.L. Chebyshev, A.A. Markov and Lyapunov on conditions for the applicability of the **central limit theorem** of probability theory. Later, conditions were established that extend Lyapunov's conditions and that are not only sufficient but also necessary. A final solution of the question in this direction was obtained by S.N. Bernstein [S.N. Bernshteĭn], J. Lindeberg and W. Feller. The power of the method of characteristic functions was demonstrated for the first time in Lyapunov's theorem.

Lyapunov also gave an upper bound (for $\delta \leqslant 1$) for the absolute value of the difference Δ between the probability of (2) and its approximate value (3). This bound can be expressed in the following form: For $\delta < 1$,

$$|\Delta| \leqslant C_1 L_{n,\delta},$$

and for $\delta = 1$,

$$|\Delta| \leqslant C_2 L_{n,1} \left| \log \frac{1}{L_{n,1}} \right|,$$

where C_1 and C_2 are absolute constants and $L_{n,\delta}$ is the fraction (the *Lyapunov fraction*) under the limit sign in (1). See also **Berry – Esseen inequality**.

References

[1] LYAPUNOV, A.M.: *Collected works*, 1, Moscow, 1954, pp. 157-176 (in Russian).
[2] BERNSHTEĬN, S.N.: *Probability theory*, Moscow-Leningrad, 1946 (in Russian).
[3] FELLER, W.: *An introduction to probability theory and its applications*, 2, Wiley, 1966.

A.V. Prokhorov

Editorial comments.

References

[A1] LAHA, R.G. and ROHATGI, V.K.: *Probability theory*, Wiley, 1979.

2) *Lyapunov's theorems in potential theory* are theorems on the behaviour of potentials and the solution of the **Dirichlet problem**, obtained by A.M. Lyapunov in 1886 - 1902 (see [1]).

The *theorem on the body of greatest potential*: If there is a homogeneous body T in the Euclidean space $\mathbf{R}^3$, the energy of whose **Newton potential**, that is, the integral

$$E(T) = \iint_{T\,T} \frac{dx\,dy}{|x-y|},$$

attains its greatest value for a given volume, then this body is a ball.

The integral (1) is the *energy* of a homogeneous mass distribution of density 1 on the body T. Later, T. Carleman (1919) proved that a body T for which the energy $E(T)$ attains its greatest value for a given volume actually exists.

First theorem on the normal derivatives of a double-layer potential: Let S be a closed Lyapunov surface in $\mathbf{R}^3$, let $f(y)$ be the density of the mass distributed on S, and suppose that one of the following two conditions is satisfied: a) $f(y)$ is continuous on S, and the exponent $\lambda = 1$ in the Lyapunov condition on the angle θ between the normals to S at two points $y_1, y_2 \in S$, that is, $|\theta| < A |y_1 - y_2|$ (see **Lyapunov surfaces and curves**); or b) $f(y)$ is Hölder continuous with exponent 1, that is, $|f(y_1) - f(y_2)| < A |y_1 - y_2|$; then if for the **double-layer potential**

$$W(x) = \int_S f(y) \frac{\cos(y - x, n_y)}{|x-y|^2} \, dy$$

one of the normal derivatives dW_i / dn_{y_0} inside S or dW_e / dn_{y_0} outside S at a point $y_0 \in S$ exists, then the other also exists, and these derivatives coincide.

Second theorem on the normal derivatives of a double-layer potential: Under the assumptions of the previous theorem, suppose also that the density $f(y)$ satisfies the *Lyapunov condition*

$$\int_0^{2\pi} |f(\rho, \phi) - f(y_0)| \, d\phi < a\rho^{1+\nu}, \quad a, \nu > 0,$$

where (ρ, ϕ, z) are cylindrical coordinates inside the Lyapunov sphere (see **Lyapunov surfaces and curves**) with origin at a point $y_0 \in S$ and z-axis directed along the normal n_{y_0}. Then the double-layer potential (2) has both normal derivatives at y_0.

Theorem on the first derivatives of a simple-layer potential: Let S be a closed Lyapunov surface and suppose that the density $f(y)$ is Hölder continuous, that is,

$$|f(y_1) - f(y_2)| < A |y_1 - y_2|^\lambda, \quad 0 < \lambda < 1.$$

Then the first-order partial derivatives $\partial V / \partial x_i$, $i = 1, 2, 3$, $x = (x_1, x_2, x_3)$, of the **simple-layer potential**

$$V(x) = \int_S f(y) \frac{dy}{|x-y|}$$

are Hölder continuous with the same exponent λ in the closed interior domain $\overline{D}_i$ and closed exterior domain $\overline{D}_e$.

In this theorem the Hölder continuity was only stated by Lyapunov; the proof was completed by N.M. Gunther (see [2]).

These theorems served Lyapunov as a basis for the construction of a strict theory of solvability of the Dirichlet problem by the method of integral equations. A monograph of Gunther was devoted to the development of the ideas of Lyapunov (see [2]); for generalizations to potentials of a more general form see [3].

References

[1A] LYAPUNOV, A.M.: *Collected works*, Vol. 1, Moscow, 1954, pp. 26-32 (in Russian).

[1B] LYAPUNOV, A.M.: *Collected works*, Vol. 1, Moscow, 1954, pp. 33-44 (in Russian).

[1C] LYAPUNOV, A.M.: 'On certain questions connected with the Dirichlet problem', in *Collected works*, Vol. 1, Moscow, 1954, pp. 45-47; 48-100 (in Russian).

[1D] LYAPUNOV, A.M.: *Collected works*, Vol. 1, Moscow, 1954, pp. 101-122 (in Russian).

[2] GUNTHER, N.M.: *Potential theory and its applications to basic problems of mathematical physics*, F. Ungar, 1967 (translated from the French).

[3] MIRANDA, C.: *Partial differential equations of elliptic type*, Springer, 1970 (translated from the Italian).

E.D. Solomentsev

Editorial comments.

References

[A1] KRÁL, J.: *Integral operators in potential theory*, Lecture notes in math., 823, Springer, 1980.

AMS 1980 Subject Classification: 31A15, 31B15, 60F05

LYAPUNOV TRANSFORMATION - A non-degenerate linear transformation $L(t): \mathbf{R}^n \to \mathbf{R}^n$ (or $L(t): \mathbf{C}^n \to \mathbf{C}^n$), smoothly depending on a parameter $t \in \mathbf{R}$, that satisfies the condition

$$\sup_{t \in \mathbf{R}}[\| L(t) \| + \| L^{-1}(t) \| + \| \dot{L}(t) \|] < +\infty.$$

It was introduced by A.M. Lyapunov in 1892 (see [1]). The Lyapunov transformation is widely used in the theory of linear systems of ordinary differential equations. In many cases the requirement

$$\sup_{t \in \mathbf{R}} \| \dot{L}(t) \| < +\infty$$

can be discarded.

References

[1] LYAPUNOV, A.M.: *Stability of motion*, Acad. Press, 1966 (translated from the Russian).

V.M. Millionshchikov

Editorial comments.

References

[A1] HAHN, W.: *Stability of motion*, Springer, 1967.

AMS 1980 Subject Classification: 34-XX

M

m-DEPENDENT PROCESS

Editorial comments. A discrete-time stochastic process $(X_n)_{n \in \mathbf{Z}}$ is *m-dependent* if for all k the joint stochastic variables $(X_n)_{n \leq k}$ are independent of the joint stochastic variables $(X_n)_{n \geq k+m+1}$.

Such processes arise naturally as limits of rescaling transformations (renormalizations) and (hence) as examples of processes with scaling symmetries [A1]. Examples of *m*-dependent processes are given by $(m+1)$-*block factors*. These are defined as follows. Let $(Z_n)_{n \in \mathbf{Z}}$ be an independent process and ϕ a function of $m+1$ variables; let $X_n = f(Z_n, \ldots, Z_{n+m})$; then the $(m+1)$-block factor X_n is an *m*-dependent process.

There are *one-dependent processes* which are not 2-block factors, [A2].

References

[A1] O'BRIEN, G.L.: 'Scaling transformations for $\{0, 1\}$-valued sequences', *Z. Wahrscheinlichkeitstheorie Verw. Gebiete* 53 (1980), 35-49.
[A2] AARONSON, J., GILAT, D., KEANE, M. and VALK, V. DE: 'An algebraic construction of a class of one-dependent processes', *Ann. Probab.* 17 (1988), 128-143.
[A3] JANSON, S.: 'Runs in *m*-dependent sequences', *Ann. Probab.* 12 (1984), 805-818.
[A4] HAIMAN, G.: 'Valeurs extrémales de suites stationaires de variable aléatoires *m*-dépendantes', *Ann. Inst. H. Poincaré Sect. B (N.S.)* 17 (1981), 309-330.

AMS 1980 Subject Classification: 60KXX

MACDONALD FUNCTION, *modified cylinder function, Bessel function of imaginary argument* - A function

$$K_\nu(z) = \frac{\pi}{2} \frac{I_{-\nu}(z) - I_\nu(z)}{\sin \nu\pi},$$

where ν is an arbitrary non-integral real number and

$$I_\nu(z) = \sum_{m=0}^{\infty} \frac{\left(\dfrac{z}{2}\right)^{\nu+2m}}{m! \Gamma(\nu+m+1)}$$

is a cylinder function with pure imaginary argument (cf. **Cylinder functions**). They have been discussed by H.M. Macdonald [1]. If n is an integer, then

$$K_n(z) = \lim_{\nu \to n} K_\nu(z).$$

The Macdonald function $K_\nu(z)$ is the solution of the differential equation

$$z^2 \frac{d^2 y}{dz^2} + z \frac{dy}{dz} - (z^2 + \nu^2)y = 0 \qquad (*)$$

that tends exponentially to zero as $z \to \infty$ and takes positive values. The functions $I_\nu(z)$ and $K_\nu(z)$ form a **fundamental system of solutions** of (*).

For $\nu \geq 0$, $K_\nu(z)$ has roots only when $\operatorname{Re} z < 0$. If $\pi/2 < |\arg z| < \pi$, then the number of roots in these two sectors is equal to the even number nearest to $\nu - 1/2$, provided that $\nu - 1/2$ is not an integer; in the latter case the number of roots is equal to $\nu - 1/2$. For $\arg z = \pm\pi$ there are no roots if $\nu - 1/2$ is not an integer.

Series and asymptotic representations are:

$$K_{n+1/2}(z) = \left[\frac{\pi}{2z} \right]^{1/2} e^{-z} \sum_{r=0}^{n} \frac{(n+r)!}{r!(n-r)!(2z)^r},$$

where n is a non-negative integer;

$$K_0(z) = -\ln \left[\frac{z}{2} \right] I_0(z) + \sum_{m=0}^{\infty} \left[\frac{z}{2} \right]^{2m} \frac{1}{(m!)^2} \psi(m+1),$$

$$\psi(1) = -C, \quad \psi(m+1) = 1 + \frac{1}{2} + \cdots + \frac{1}{m} - C,$$

where $C = 0.5772157 \cdots$ is the Euler constant;

$$K_n(z) = \frac{1}{2} \sum_{m=0}^{n-1} \frac{(-1)^m (n-m-1)!}{m!(z/2)^{n-2m}} +$$

$$+ (-1)^{n-1} \sum_{m=0}^{\infty} \frac{(z/2)^{n+2m}}{m!(n+m)!} \left\{ \ln \left[\frac{z}{2} \right] - \frac{\psi(m+1) - \psi(n+m+1)}{2} \right\},$$

where $n \geq 1$ is an integer;

$$K_\nu \sim$$

$$\sim \left[\frac{\pi}{2z} \right]^{1/2} e^{-z} \left[1 + \frac{4\nu^2 - 1^2}{1!8z} + \frac{(4\nu^2 - 1^2)(4\nu^2 - 3^2)}{2!(8z)^2} + \cdots \right],$$

for large z and $|\arg z| < \pi/2$.

Recurrence formulas:

$$K_{\nu-1}(z) - K_{\nu+1}(z) = -\frac{2\nu}{z} K_\nu(z),$$

$$K_{\nu-1}(z) + K_{\nu+1}(z) = -2 \frac{dK_\nu(z)}{dz}.$$

References
    [1] MACDONALD, H.M.: 'Zeroes of the Bessel functions', *Proc. London Math. Soc.* **30** (1899), 165-179.
    [2] WATSON, G.N.: *A treatise on the theory of Bessel functions*, 1-2, Cambridge Univ. Press, 1952.

V.I. Pagurova

AMS 1980 Subject Classification: 33A40

MACH NUMBER - One of the basic criteria of aerodynamic similarity when the viscosity of the gas cannot be neglected. Mach's number $M = v/a$ is the ratio of the velocity of the gas flow and the velocity of sound at the same point of the flow (or the ratio of the velocity of a body in the gas and the velocity of sound in this medium). The number was called thus to honour E. Mach.

BSE-3

Editorial comments.

References
    [A1] HOWARTH, L. (ED.): *Modern developments in fluid dynamics, high speed flow*, 1-2, Oxford Univ. Press, 1953.

AMS 1980 Subject Classification: 76-XX, 76NXX

MACH PRINCIPLE - The statement that the inertial properties of each physical body are determined by all the other physical bodies present in the Universe. In classical mechanics it is, on the contrary, considered that the inertial properties of a body, for instance, its mass, are not affected by the presence or absence of other bodies. The principle was formulated by E. Mach in 1896 (see [1]) as the outcome of a critical analysis of the foundations of classical mechanics; however, Mach did not provide a mathematical formulation of this principle. Several non-equivalent mathematical formulations of the principle are known.

References
    [1] MACH, E.: *The science of mechanics*, Open Court, 1893.
    [2] EINSTEIN, A.: 'Ernst Mach', *Phys. Zeitschr.* **17**, no. 7 (1916), 101-104.

D.D. Sokolov

Editorial comments.

References
    [A1] WHEELER, J.A.: 'Geometrodynamics and the issue of the final state', in C. DeWitt and B. DeWitt (eds.): *Relativité, Groupes and Topologie*, Gordon & Breach, 1964, pp. 315-520.

AMS 1980 Subject Classification: 70A05, 83CXX

MACHINE (in mathematics) - An abstract device realizing **information** processing. The terms '*abstract machine*' and '*automaton*' are also used. Abstract machines are particular cases of control systems (cf. **Control system**). They arose in connection with the analysis of the concept of an algorithm, starting in the middle of the nineteen-thirties, with the development of computers and with the construction of mathematical models of biological systems. Machines processing discrete information became most widespread. Typical representatives of these are finite automata (cf. **Automaton, finite**) and Turing machines (cf. **Turing machine**). Abstract machines are easily visualized, they can be easily combined in various ways, and they have simple operating steps. Machines are being studied within the frame of the theory of algorithms (cf. **Algorithms, theory of**) and mathematical **cybernetics** and their study aims at an analysis and formalization of the concept of an algorithm and at modelling real devices and processes in a mathematical way. There is a fruitful relation between abstract machines and real computers. The ideas of constructing computers and of programming them relies, to a large extent, on ideas from the theory of algorithms and mathematical cybernetics. In turn, practical work with computers poses new problems and suggests models of machines in order to solve these problems.

Common to all abstract machines is the presence of a finite control device that may be in one of the states $q_1, \ldots, q_k$. A machine has a potentially-unbounded exterior memory and has a means for reading and writing information into the exterior memory. Reading (writing) of information is, as a rule, carried out locally. The operation of a machine is determined by a program, consisting of commands of the form $q_i a \rightarrow q_j b$, where q_i, q_j are states of the control device, a stands for ordered information of local character from the exterior memory and b is the notation for both the change of the contents of the exterior memory and the position of the reading device in it. The machine operates in discrete time. At each step t the machine uses the reading device to move information a from the exterior memory to the control device. If at step t the control device of the machine is in state q_i and the program contains the command $q_i a \rightarrow q_j b$, then at the next step $t+1$ the control device is in state q_j, and the contents of the exterior memory and the position of the reading device are changed according to b. Usually one distinguishes one or several initial or final states among the states of the control device. Before starting to operate, the control device is in an initial state, and the end of the operation is determined by the final state(s).

In many cases an abstract machine is constructed to compute numerical (word) functions and predicates. The computation of a function $f(x_1, \ldots, x_n)$ on a machine M is understood as follows. For an arbitrary choice of the arguments $(x_1, \ldots, x_n)$ an initial configuration $K_1(x_1, \ldots, x_n)$ of M is indicated, this being a filling of the external memory and a position of the reading device with respect to it corresponding to $x_1, \ldots, x_n$. It is determined which configurations of M are final, i.e. when the computation of f on M is terminated. It is determined what value of the function to be computed is obtained for each final configuration. The computation of the value of $f(x_1, \ldots, x_n)$ consists of a series of transitions from the initial state $K_1(x_1, \ldots, x_n)$ to the next state $K_2(x_1, \ldots, x_n)$, etc., in correspondence with the program of the machine M, until a final configuration is reached. If the value of $f(x_1, \ldots, x_n)$ is not defined, M will never reach a final configuration when starting from $K_1(x_1, \ldots, x_n)$. In this case the computing process may continue indefinitely.

Often one uses an ordinary Turing machine in the definition of an abstract machine; the differences consist of adding new possibilities or restrictions on the Turing machine. The modification of a Turing machine is usually conducted along the following three lines.

1. The organization of the exterior memory. The introduction of several tapes is most common (multi-tape Turing machines). One also studies machines with one-sided or multi-dimensional tapes, with an exterior memory in the form of an infinite **graph** (e.g. in the form of an infinite binary **tree**). Another approach consists in replacing tapes by registers or calculating frames, suitable for containing natural (integral) numbers or words of arbitrary length. Such are *Shepherdson − Sturgis machines* [6], the *Minski machine* [3] and a *random-access machine*.

2. The storage, transformation and obtaining of information from the external memory. One considers machines for which a part of the information in the external memory to which the control device has had no access for some time (depending on the input) becomes totally inaccessible to the reading device for the rest of the time. The definition of a Turing machine with tapes of storage type is based on a closely related idea (*pushdown automata*).

For machines taking and transforming information from the external memory symbol-by-symbol, a typical restriction is the prohibition to print some symbols next to certain others. Such are, for example, the non-erasing Turing machines and two-sided automata equivalent to finite automata. The important class of finite automata also belongs here. However, restrictions on the size of the memory, the operating time, etc., are more usual. For example, a linearly bounded automaton is a Turing machine for which the length of the tape used in the computations is bounded by a linear function in the length of the input.

Often, the information from the external memory is obtained by using several reading heads. A more usual approach is where general recursive predicates have been defined on the external memory. The set of these predicates determines the information arriving at the control device of the machine. This idea is used in Shepherdson − Sturgis machines, where the addition of information into the external memory is realized by arbitrary general recursive functions.

3. The means of functioning. Here one distinguishes between deterministic, non-deterministic and probabilistic machines. For a *deterministic machine* every operating step is uniquely determined by the state of the control device and the information from the external memory obtained by the reading device. The program of a non-deterministic machine may contain different commands with the same left-hand side. Therefore, for a non-deterministic machine one considers not one but the set of all computations compatible with the program for a given input. A *probabilistic machine* is a machine that is provided with a random-number generator or one that has a program in which the transition from one command to another is realized with a given probability.

In the non-deterministic case one usually considers the computation of predicates. If a non-deterministic machine M computes a predicate $p(x_1, \ldots, x_n)$, then $p(x_1, \ldots, x_n)$ is true if and only if among all possible computations of M starting at the configuration $K_1(x_1, \ldots, x_n)$ there is a computation containing a final configuration.

In general, the computing possibilities of deterministic, non-deterministic and probabilistic machines are the same. However, in separate narrow classes of machines, non-deterministic and probabilistic machines may prove to be more powerful than deterministic machines. For many types of abstract machines with restrictions either on the volume of the external memory or on the operating time, the problem of 'determinization' of non-deterministic machines is an interesting open (1989) problem.

See also **Computer, abstract.**

References

[1] SHANNON, C.E. and McCARTHY, J. (EDS.): *Automata studies*, Princeton Univ. Press, 1956.
[2] MAL'TSEV, A.I.: *Algorithms and recursive functions*, Wolters-Noordhoff, 1970 (translated from the Russian).
[3] MINSKY, M.: *Computation: finite and infinite machines*, Prentice-Hall, 1967.
[4] *Problems in mathematical logic*, Moscow, 1970 (in Russian). Collection of translations.
[5] *Complexity of computations and algorithms*, Moscow, 1974 (in Russian). Collection of translations.
[6] SHEPHERDSON, J.C. and STURGIS, H.E.: 'Computability and recursive functions', *J. Assoc. Comp. Mach.* **10**, no. 2 (1963), 217-255.

S.S. Marchenko

Editorial comments. The basic distinction which has to be made among contemporary machine models is the distinction between sequential and parallel machines.

In a *sequential machine* model there is a single control unit which operates on some potentially infinite storage structure. For the case of the Turing machine, the storage structure consists of linearly ordered tapes which are accessed by read-write heads. The various Turing machine models differ with respect to the number of tapes, the number of heads on a tape and the dimensionality of the worktapes. Some special tapes may have restrictions on their use, like being a read-only input tape, a write-only output tape, etc.

In the *random access machine* (RAM) the memory consists of an infinite collection of computer words, each with an unbounded word-length, thus capable of storing arbitrary integers. The memory is accessed by means of direct and indirect load and store instructions. Arguments fetched from memory into the central processor can be subjected to arithmetic instructions like additions, subtractions and comparisons. More powerful models also include multiplicative instructions and parallel bitwise operations based on the binary representation of the numerical values of the arguments.

A third class of sequential models operates on directed or undirected graphs by creating nodes and redirecting pointers between nodes. The oldest of these models is the *Kolmogorov–Uspenskiĭ machine* (which operates on undirected graphs); machines operating on directed graphs are frequently called *pointer machines* or *storage modification machines*.

Parallel machine models have a potentially infinite number of processing units which interact either by means of communication or by a shared memory. The diversity of parallel models even exceeds that of the sequential models. The *cellular array model* consists of finite state processors connected in a network. Even though the components are finite states devices, the network obtains universal computational capabilities. Other models have for each processor a private storage structure; the processors communicate by exchanging messages. In a contrasting model a growing number of processors share a single (potentially infinite) storage structure; a crucial parameter in such a model is the formal treatment of write-conflicts (disallow, arbitrary choice of winner, priority, . . .). Other models combine local and global storage.

During a computation only a finite number of processors become active. The different models have various mechanisms for controlling process activation. Some *message passing models* create processes dynamically in a graph-shaped interconnection pattern. *Central memory based machines* specify for each instruction the number of processors which are activated during this instruction. A third control mechanism is based on recursive subroutines, where several recursive branches are executed in parallel on processors created at calling time; these processors again are deleted after having returned their answer to their calling processor.

Complexity considerations. Machine models play a crucial role in defining the basic notions like running-time or space consumption in computational complexity, since these running times and space consumption ultimately have to be expressed in terms of computations on some idealized machine. It is therefore crucial to investigate the extent to which the complexity notions thus obtained become model dependent. This issue is closely related to the question of the efficiency of the mutual simulations between the different models — since these models all have universal computational power such mutual simulations exist.

It can be shown that the basic sequential models all simulate each other with polynomial overhead in time and constant factor overhead in space (provided space measures are defined correctly). As a consequence, the notions of logspace and/or polynomial time computability become mathematically well-defined machine-independent notions, whereas the notion of computability within quadratic time is inherently model dependent. This invariance property should be considered as a criterion for determining the standard class of sequential machine models.

For a large variety of parallel models it has been established that there exists an equivalence between polynomial time on the parallel machine and polynomial space in the sequential world. This equivalence does not hold for all parallel models proposed in the literature. Some degenerate parallel models have the power of computing arbitrarily complex functions in constant time. On the other hand, there exist some sequential models which turn out to be equivalent to the parallel models, due to the fact that they can operate on exponentially large objects in unit time. Again, the above equivalence, which is traditionally referred to under the name of the *parallel computation thesis*, should be considered to be the characterization of some basic mode of parallelism.

A basic open question about sequential models is the power of non-determinism. The notorious $\mathscr{P} = \mathscr{NP}$? problem (which is machine independent as long as one considers the basic sequential models only) can be understood as asking for the overhead required for making a non-deterministic sequential device deterministic. Similarly, the unsolved $\mathscr{P} = \mathscr{PSPACE}$? problem is related to the relation between time and space as a computational resource, and, in the light of the parallel computation thesis, to the power of parallelism.

An alternative mode of using machine computations, called *alternation*, was discovered in 1976 by A.K. Chandra, D.C. Kozen and L.J. Stockmeyer. *Sequential alternating machines* turn out to be equivalent to the standard parallel machine models, but the alternating models have the additional feature that logarithmic space in the alternating model is equivalent to polynomial deterministic time in the sequential world; similarly, polynomial alternating space is equivalent to deterministic sequential exponential time.

References

[A1] AHO, A.V., HOPCROFT, J.E. and ULLMAN, J.D.: *The design and analysis of computer algorithms*, Addison Wesley, 1974.
[A2] CHANDRA, A.K., KOZEN, D.C. and STOCKMEYER, L.J.: 'Alternation', *J. Assoc. Comp. Mach.* 28 (1981), 114-133.
[A3] HOPCROFT, J.E. and ULLMAN, J.D.: *Introduction to automata theory, languages and computation*, Addison-Wesley, 1979.
[A4] KOLMOGOROV, A.N. and USPENKIĬ, V.A.: 'On the definition of an algorithm', *Transl. Amer. Math. Soc. (2)* 29 (1963), 217-245. (*Uspehi Mat. Nauk* 13 (1958), 3-28)
[A5] WAGNER, K. and WECHSUNG, G.: *Computational complexity*, Reidel, 1986.
[A6] VAN EMDE BOAS, P.: 'Machine models and simulations', in J.

van Leeuwen (ed.): *Handbook of Theoretical Computer Science*, North-Holland, 1990.

[A7] HARRISON, M.A.: *Introduction to switching and automata theory*, McGraw-Hill, 1965.

[A8] ARBIB, M.A.: *Theories of abstract automata*, Prentice-Hall, 1969.

[A9] KALMAN, R.E., FALB, P.L. and ARBIB, M.A.: *Topics in mathematical system theory*, McGraw-Hill, 1969.

[A10] HOPCROFT, J.E. and ULLMAN, J.D.: *Formal languages and their relation to automata*, Addison-Wesley, 1969.

AMS 1980 Subject Classification: 68DXX, 68F10, 68C40, 93-XX

MACHINE-ORIENTED LANGUAGE - A programming language that allows one, when compiling programs, to take into account the peculiarities of the systems of commands and representation of information in the object computer. Machine-oriented languages, in contrast to universal problem-oriented languages (cf. **Problem-oriented language**) that realize a mapping of the set P of input programs into the set M of machine programs, try to map P onto M.

The simplest machine-oriented languages are *assemblers*, which, while completely preserving the structure of the machine programs, allow one to use a symbolic notation for commands and memory addresses, as well as to collect programs from a couple of separately-described parts. For additional possibilities concerning substitutions in the text and other simple transformations when compiling the text of programs, one needs *macro-assemblers*. Machine-oriented languages of a higher level, like universal languages, have a phrase-structure allowing for compound objects and defining operations. This structure also contains additional means for describing elementary objects and basic operations in terms of the machine structure.

References

[1] STRUBLE, G.: *Assembler language programming: the IBM system/360*, Addison-Wesley, 1969.

[2] BROWN, P.: *Macro processors and techniques for portable software*, Wiley, 1974.

[3] KATKOV, V.L. and RAR, A.F.: *Programming in Epsilon*, Novosibirsk, 1972 (in Russian).

A.P. Ershov

Editorial comments. A typical contemporary example of a higher-level programming language which supports access to machine-level features is the language C, which is widely used for general programming in a UNIX environment.

References

[A1] KERNIGHAN, B.W. and RITCHIE, D.M.: *The C programming language*, Prentice-Hall, 1978.

AMS 1980 Subject Classification: 68B05

MACKEY–BOREL STRUCTURE - A Borel structure (i.e. a **Borel system of sets**) on the spectrum $\hat{A}$ of a separable C^***-algebra** A (cf. also **Spectrum of a C^*-algebra**), defined as follows. Let H_n, $n = 1, 2, \ldots$, be a Hilbert space of dimension n, and let $\mathrm{Irr}_n(A)$ be the set of non-zero irreducible representations (cf. **Irreducible representation**) of the C^*-algebra A on H_n equipped with the topology of pointwise convergence in the weak topology. Let on $\mathrm{Irr}_n(A)$ be given the Borel structure generated by its topology (that is, the smallest Borel structure relative to which all mappings $\pi \to (\pi(x)\xi, \eta)$, $x \in A$, $\xi, \eta \in H_n$, $\pi \in \mathrm{Irr}_n(A)$, are Borel functions) and let $\mathrm{Irr}(A)$ be the union of the subspaces $\mathrm{Irr}_n(A)$, $n = 1, 2, \ldots$, provided with the Borel structure such that a subset of $\mathrm{Irr}(A)$ is a Borel set if and only if its intersection with each $\mathrm{Irr}_n(A)$ belongs to the Borel structure on the latter. Let ϕ be the mapping of the Borel space $\mathrm{Irr}(A)$ into the spectrum $\hat{A}$ of A which maps a representation to its unitary equivalence class. The Borel structure on $\hat{A}$ generated by the sets whose inverse images under ϕ are Borel sets in $\mathrm{Irr}(A)$ is called the *Mackey–Borel structure* on $\hat{A}$. The Mackey–Borel structure contains all sets of the Borel structure generated by the topology of $\hat{A}$; each point of $\hat{A}$ is a Borel set in the Mackey–Borel structure. The following conditions are equivalent: 1) the Mackey–Borel structure is standard (i.e. it is isomorphic as a Borel structure to the Borel structure generated by the topology of some complete separable metric space); 2) the Mackey–Borel structure coincides with the Borel structure generated by the topology on $\hat{A}$; 3) the Mackey–Borel structure on $\hat{A}$ is countably separated; and 4) if A is a CGR-algebra, then a Mackey–Borel structure can also be introduced on the quasi-spectrum of a separable C^*-algebra.

References

[1] DIXMIER, J.: *C^* algebras*, North-Holland, 1977 (translated from the French).

[2] GARDNER, L.T.: 'On the Mackey Borel structure', *Canad. J. Math.* **23**, no. 4 (1971), 674-678.

[3] HALPERN, H.: 'Mackey Borel structure for the quasi-dual of a separable C^*-algebra', *Canad. J. Math.* **26**, no. 3 (1974), 621-628.

A.I. Shtern

Editorial comments.

References

[A1] ARVESON, W.: *An invitation to C^*-algebras*, Springer, 1976, Chapts. 3-4.

AMS 1980 Subject Classification: 46LXX, 54H05

MACKEY INTERTWINING NUMBER THEOREM, *intertwining number theorem*.

Editorial comments. Let G be a finite group. The *intertwining number* between two representations $\pi_i: G \to \mathrm{Aut}(E_i)$, $i = 1, 2$, is, by definition, the dimension of the space of G-homomorphisms $E_1 \to E_2$: $i(\pi_1, \pi_2) = \dim(\mathrm{Hom}_G(E_1, E_2))$.

Now let H_1, H_2 be subgroups of G, and D a (H_2, H_1)

double coset in G (i.e. D is a set of the form $H_2 x H_1$ for some $x \in G$). Let π_i be a unitary representation of H_i and let $\text{Ind}_{H_i}^G(\pi_i)$ be the corresponding **induced representation** of G, $i = 1, 2$. Consider the intertwining number between the unitary representations $g \mapsto \pi_1(g)$ and $g \mapsto \pi_2(xgx^{-1})$ of the subgroup $H_1 \cap x^{-1} H_2 x$ for some $x \in D$. Then this number only depends on D (and π_1, π_2). It is denoted by $i(\pi_1, \pi_2, D)$.

For the intertwining number between the induced representations $\text{Ind}_{H_i}^G(\pi_i)$ of G, $i = 1, 2$, one now has the *intertwining number formula*

$$i(\text{Ind}_{H_1}^G(\pi_1), \text{Ind}_{H_2}^G(\pi_2)) = \sum_D i(\pi_1, \pi_2, D),$$

where the sum is over the set of all (H_2, H_1) double cosets.

The *Frobenius reciprocity theorem* $i(\pi, \text{Ind}_H^G(\sigma)) = i(\text{Res}_H^G(\pi), \sigma)$ (cf. **Induced representation**) for representations π of G and σ of a subgroup H of G is an immediate consequence.

For a discussion of the intertwining number theorem for locally compact groups cf. [A2].

References

[A1] CURTIS, C.W. and REINER, I.: *Representation theory of finite groups and associative algebras*, Interscience, 1962, §44.
[A2] WARNER, G.: *Harmonic analysis on semi-simple Lie groups*, I, Springer, 1972, Chapt. V.

AMS 1980 Subject Classification: 20C05

MACKEY TOPOLOGY $\tau(F, G)$ on a space F, being in duality with a space G (over the same field) - The topology of uniform convergence on the convex balanced subsets of G that are compact in the weak topology $\sigma(F, G)$ (defined by the duality between F and G). It was introduced by G.W. Mackey [1]. The Mackey topology is the strongest of the separated locally convex topologies (cf. **Locally convex topology**) which are compatible with the duality between F and G (that is, separated locally convex topologies $\mathcal{T}$ on F such that the set of all continuous linear functionals on F endowed with the topology $\mathcal{T}$ coincides with G). The families of sets in F which are bounded relative to the Mackey topology and bounded relative to the weak topology coincide. A convex subsets of G is equicontinuous when F is endowed with the Mackey topology if and only if it is relatively compact in the weak topology. If a separated **locally convex space** E is barrelled or bornological (in particular, metrizable) and E' is its dual, then the Mackey topology on E (being dual with E') coincides with the initial topology on E. For pairs of spaces (F, G) in duality the Mackey topology $\mathcal{T}$ is not necessarily barrelled or metrizable. A weakly-continuous linear mapping of a separated locally convex space E into a separated locally convex space F is continuous relative to the Mackey topologies $\tau(E, E')$ and $\tau(F, F')$. A locally convex space E is called a *Mackey space* if the topology on E is $\tau(E, E')$. Completions, quotient spaces and metrizable subspaces, pro-

ducts, locally convex direct sums, and inductive limits of families of Mackey spaces are Mackey spaces. If E is a Mackey space and ϕ is a weakly-continuous mapping of E into a locally convex space F, then ϕ is a continuous linear mapping of E into F. If E is a quasi-complete Mackey space and the space dual to E equipped with the strong E-topology is semi-reflexive, then E is reflexive.

References

[1] MACKEY, G.W.: 'On convex topological linear spaces', *Trans. Amer. Math. Soc.* **60** (1946), 519-537.
[2] BOURBAKI, N.: *Elements of mathematics. Topological vector spaces*, Addison-Wesley, 1977 (translated from the French).
[3] SCHAEFFER, H.: *Topological vector spaces*, Springer, 1971.

A.I. Shtern

Editorial comments.

References

[A1] JARCHOW, H.: *Locally convex spaces*, Teubner, 1981 (translated from the German).

AMS 1980 Subject Classification: 46A20, 46AXX, 46A05

MACLAURIN FORMULA - A particular case of the **Taylor formula**. Let a function f have n derivatives at $x = 0$. Then in some neighbourhood U of this point f can be represented in the form

$$f(x) = \sum_{k=0}^{n} \frac{f^{(k)}(0)}{k!} x^k + r_n(x), \quad x \in U,$$

where $r_n(x)$, the n-th order remainder term, can be represented in some form or other.

The term 'MacLaurin formula' is also used for functions of m variables $x = (x_1, \ldots, x_m)$. In this case k in the MacLaurin formula is taken to be a multi-index, $k = (k_1, \ldots, k_m)$ (see **MacLaurin series**). The formula is named after C. MacLaurin.

L.D. Kudryavtsev

Editorial comments. For some expressions for the remainder $r_n(x)$ and for estimates of it see **Taylor formula**.

References

[A1] RUDIN, W.: *Principles of mathematical analysis*, McGraw-Hill, 1976.

AMS 1980 Subject Classification: 30B10

MACLAURIN SERIES *for a function $f(z)$* - The power series

$$f(z) = \sum_{k=0}^{\infty} \frac{f^{(k)}(0)}{k!} z^k.$$

It was studied by C. MacLaurin [1]. If a function $f(z)$ analytic at zero is expanded as a **power series** around zero, then this series coincides with the MacLaurin series. When a function depends on m variables, the MacLaurin series is a multiple power series:

$$\sum_{|k|=0}^{\infty} \frac{f^{(k_1)}(0) \cdots f^{(k_m)}(0)}{k_1! \cdots k_m!} z_1^{k_1} \cdots z_m^{k_m}$$

in which the summation is over the multi-indices $k = (k_1, \ldots, k_m)$, $|k| = |k_1| + \cdots + |k_m|$, and k_i are non-negative integers. A MacLaurin series is a special case of a **Taylor series**.

References

[1] MacLaurin, C.: *A treatise of fluxions*, 1-2, Edinburgh, 1742.

L.D. Kudryavtsev

AMS 1980 Subject Classification: 30B10

MAGIC SQUARE - A square $n \times n$ array $\| a_{ij} \|$ composed of the integers from 1 up to n^2 and satisfying the following conditions:

$$\sum_{i=1}^{n} a_{ij} = \sum_{j=1}^{n} a_{ij} = \sum_{i=1}^{n} a_{ii} = \sum_{i=1}^{n} a_{i,n+1-i} = s, \qquad (*)$$

where $s = n(n^2+1)/2$. There are also more general magic squares, in which $1 \leq a_{ij} \leq n^2$ is not required.

Any number a, $1 \leq a \leq n^2$, is uniquely characterized by a pair of residues $(\alpha, \beta) \bmod n$ (the digits to base n of $a-1$), that is, by the points of the two-dimensional space $(\mathbf{Z}/n)^2$ over the ring $\mathbf{Z}/n$ of residues modulo n. Since the coordinates (i, j) of the cells of the square may also be regarded as the elements of $(\mathbf{Z}/n)^2$, it follows that any distribution of the numbers from 1 up to n^2 in an array $\| a_{ij} \|$ is given by a mapping

$$(\mathbf{Z}/n)^2 \to (\mathbf{Z}/n)^2,$$

that is, by a pair of functions $\alpha = \alpha(i, j) \in \mathbf{Z}/n$, $\beta = \beta(i, j) \in \mathbf{Z}/n$, where $i, j \in \mathbf{Z}/n$. The problem is to investigate those pairs that give magic squares. In general this has been done (see [1]) only under the additional assumption of linearity of α and β. It turns out, in particular, that magic squares with linear α and β exist for odd n only.

Already in the Middle Ages a number of algorithms for constructing magic squares of odd order n had been found. Each such algorithm is characterized by six residues $i_0, j_0, p, q, \bar{p}, \bar{q}$, and is described by the following rules: 1) the number 1 is put into the cell (i_0, j_0); and 2) if a was put into (i, j), then $a+1$ is put into $(i+p, j+q)$ if that cell is still empty or into $(i+\bar{p}, j+\bar{q})$ if $(i+p, j+q)$ is occupied.

The residues $i_0, j_0, p, q, \bar{p}, \bar{q}$ are not arbitrary but must satisfy certain conditions to ensure not only that $(*)$ holds, but also that the algorithm is feasible, that is, that $(i+\bar{p}, j+\bar{q})$ is empty when $(i+p, j+q)$ is occupied. These conditions are easily found (see [1]). Moreover, it turns out that a magic square can be constructed by an algorithm of this type if and only if the functions α and β describing the square are linear.

Many algorithms for constructing magic squares are known (resulting in squares with non-linear α and β), but there is no general theory for them (1989). Even the number of magic squares of order n is unknown (for $n \geqslant 5$; for $n = 3$ there is, up to obvious symmetries, only one magic square, whereas for $n = 4$ there are 880 magic squares).

Magic squares having additional symmetry have also been investigated, again only in very special circumstances (for example, for $n \leqslant 5$; see [2]).

References

[1] Postnikov, M.M.: *Magic squares*, Moscow, 1964 (in Russian).
[2] Gurevich, E.Ya.: *The secret of the Ancient Talisman*, Moscow, 1969 (in Russian).

M.M. Postnikov

Editorial comments. Magic squares have been considered since ancient times. For instance, the magic square of order 3 was known in China around 2000 B.C. Dürer's famous 'Melancholy' shows a magic square of order 4.

There is a close connection between (pairs of orthogonal) Latin squares (cf. **Latin square**; **Orthogonal Latin squares**) and magic squares, which has been studied since L. Euler (see [A1] and [A2]). See also [A3] and the references given there.

References

[A1] Euler, L.: 'De quadratis magicis', in *Opera Omnia, Sér. 1*, Vol. 7, Teubner, 1923, pp. 441-457.
[A2] Euler, L.: 'Recherches sur une nouvelle espèce de quarrés magiques', in *Opera Omnia, Sér. 1*, Vol. 7, Teubner, 1923, pp. 291-392.
[A3] Dénes, J. and Keedwell, A.D.: *Latin squares and their applications*, English Univ. Press, 1974.

AMS 1980 Subject Classification: 05B15

MAGNETO-HYDRODYNAMICS, MATHEMATICAL PROBLEMS IN - The problems connected with the investigation of the motion of conducting fluids and gases in the presence of a magnetic field.

Editorial comments.

References

[A1] Cowling, T.G.: *Magnetohydrodynamics*, Interscience, 1957.
[A2] Chandrasekhar, S.: *Hydrodynamics and hydrodynamic stability*, Dover, reprint, 1981.
[A3] Alfvén, H. and Falthammer, C.G.: *Cosmical electrodynamics*, Clarendon Press, 1963.

AMS 1980 Subject Classification: 76W05

MAHALANOBIS DISTANCE - The quantity

$$\rho(X, Y \mid A) = \{(X - Y)^T A (X - Y)\}^{1/2},$$

where X, Y are vectors and A is a matrix (and T denotes transposition). The Mahalanobis distance is used in **multi-dimensional statistical analysis**; in particular, for testing hypotheses and the classification of observations. It was introduced by P. Mahalanobis [1], who used the quantity

$$\rho(\mu_1, \mu_2 \mid \Sigma^{-1})$$

as a distance between two normal distributions with expectations μ_1 and μ_2 and common covariance matrix Σ. The Mahalanobis distance between two samples (from distributions with identical covariance matrices), or between a sample and a distribution, is defined by replacing the corresponding theoretical moments by sampling moments. As an estimate of the Mahalanobis distance between two distributions one uses the Mahalanobis distance between the samples extracted from these distributions or, in the case [5] where a linear discriminant function is utilized — the statistic $\Phi^{-1}(\alpha)+\Phi^{-1}(\beta)$, where α and β are the frequencies of correct classification in the first and the second collection, respectively, and Φ is the normal distribution function with expectation 0 and variance 1.

References

[1] MAHALANOBIS, P.: 'On tests and measures of group divergence I. Theoretical formulae', *J. and Proc. Asiat. Soc. of Bengal* **26** (1930), 541-588.

[2] MAHALANOBIS, P.: 'On the generalized distance in statistics', *Proc. Nat. Inst. Sci. India (Calcutta)* **2** (1936), 49-55.

[3] ANDERSON, T.W.: *Introduction to multivariate statistical analysis*, Wiley, 1958.

[4] AÏVAZYAN, S.A., BEZHAEVA, Z.I. and STAROVEROV, O.V.: *Classifying multivariate observations*, Moscow, 1974 (in Russian).

[5] ORLOV, A.I.: 'On the comparison of algorithms for classifying by results observations of actual data', *Dokl. Moskov. Obshch. Isp. Prirod. 1985, Otdel. Biol.* (1987), 79-82 (in Russian).

A.I. Orlov

AMS 1980 Subject Classification: 62H30

MAHLER PROBLEM - A conjecture in the metric theory of Diophantine approximation (cf. **Diophantine approximation, metric theory of**) stated by K. Mahler [1]: For almost-all (in the sense of the Lebesgue measure) numbers $\omega \in \mathbf{R}$ the inequality

$$| P(\omega) | < | H(P) |^{-n-\epsilon}$$

has a finite number of solutions in polynomials $P \in \mathbf{Z}[x]$ of degree not exceeding n. Here $\epsilon > 0$, n is a natural number and $H(P)$ is the maximum modulus of the coefficients of P. An equivalent formulation is: For almost-all $\omega \in \mathbf{R}$ the inequality

$$\max(\| \omega q \|, \ldots, \| \omega^n q \|) < q^{-1/n-\epsilon}$$

has a finite number of solutions in integers q ($\| \alpha \|$ is the distance from α to the nearest integer).

Mahler's problem was solved affirmatively in 1964 by V.G. Sprindzhuk [2]. He also proved similar results for complex and p-adic numbers, and also for power series over finite fields.

References

[1] MAHLER, K.: 'Ueber das Mass der Menge aller *S*-Zahlen', *Math. Ann.* **106** (1932), 131-139.

[2] SPRINDZHUK, V.G.: *Mahler's problem in metric number theory*, Amer. Math. Soc., 1969 (translated from the Russian).

Yu. V. Nesterenko

Editorial comments. The original paper of Sprindzhuk is [A1].

References

[A1] SPRINDZHUK, V.G.: 'A proof of Mahler's conjecture on the measure of the set of *S* numbers', *Izv. Akad. Nauk SSSR Ser. Mat.* **29** (1965), 379-436.

AMS 1980 Subject Classification: 10FXX

MAJORANT AND MINORANT - 1) Two functions, the values of the first being not less and the values of the second being not greater than the corresponding values of a given function (for all admissible values of the independent variable).

2) For functions represented by a power series, a *majorant* is e.g. the sum of a power series with positive coefficients which are not less than the absolute values of the corresponding coefficients of the given series.

3) A majorant (minorant) of a subset X of an ordered set E is an element $y \in E$ such that $y \geq x$ ($x \geq y$) for every $x \in X$.

4) In the theory of integral and differential equations, a *majorant* (*minorant*) or *majorant function* (*minorant function*) for some function f is a continuous function whose **Dini derivative** at each point t is not less (not greater) than $f(t)$ and is different from $-\infty$ ($+\infty$). The difference between any majorant and any minorant is a non-decreasing function. Any summable function on an interval has absolutely-continuous majorants and minorants which are arbitrarily close to its indefinite **Lebesgue integral**. The notion of majorant and minorant can be generalized to the case of additive set functions and also to the case when the derivatives are taken in some generalized sense.

References

[1] BOURBAKI, N.: *Elements of mathematics. Theory of sets*, Addison-Wesley, 1968 (translated from the French).

[2] SAKS, S.: *Theory of the integral*, Hafner, 1952 (translated from the Polish).

V.A. Skvortsov

Editorial comments. The terms 'majorant' and 'minorant' are rarely used in the sense 3) in English; instead the terms '*upper bound*' and '*lower bound*' are in common use.

See also **Lower bound; Upper and lower bounds**.

AMS 1980 Subject Classification: 06AXX, 41-01

MAJORIZATION ORDERING

Editorial comments. Let $p=(p_1, \ldots, p_n)$ and $q=(q_1, \ldots, q_n)$ be n-tuples of non-negative real numbers of the same l_1-norm, i.e.

$$| p | = p_1 + \cdots + p_n =$$
$$= | q | = q_1 + \cdots + q_n.$$

Then p is said to be *majorized* by q if and only if $\bar{p}_1 \leq \bar{q}_1$, $\bar{p}_1 + \bar{p}_2 \leq \bar{q}_1 + \bar{q}_2, \ldots, \bar{p}_1 + \cdots + \bar{p}_n \leq \bar{q}_1 + \cdots + \bar{q}_n$, where $(\bar{p}_1, \ldots, \bar{p}_n)$ is a reordering of the n-tuple $(p_1, \ldots, p_n)$ such that $\bar{p}_1 \geq \cdots \geq \bar{p}_n$. This defines a **partial order** which occurs under various names in various parts of mathematics: *majority ordering, majorization ordering, specialization ordering, Snapper ordering, Ehresmann ordering, dominance ordering, mixing ordering, natural ordering, ...*

The symbol $p \prec q$ denotes that q majorizes p. A few results involving the majorization ordering are as follows.

If $p \prec q$, then for all continuous convex functions ϕ of one variable, $\sum \phi(p_i) \leqslant \sum \phi(q_i)$.

A matrix Q of non-negative real numbers is said to be *doubly stochastic* if all its rows and all its columns sum to 1: $\sum_i q_{ik} = 1$, $\sum_j q_{kj} = 1$ for all $k = 1, \ldots, n$. Then $p \prec q$ if and only if there is a doubly-stochastic matrix Q such that $p = Qq$.

Let A be a Hermitian $(n \times n)$-matrix, $\lambda = (\lambda_1, \ldots, \lambda_n)$ its n-tuple of eigen values, and $a = (a_{11}, \ldots, a_{nn})$ its n-tuple of diagonal elements. Then $a \prec \lambda$, [A1]. Conversely, if $a \prec \lambda$, then there exists a real symmetric $(n \times n)$-matrix with eigen values λ and diagonal elements a, [A2], [A3]. The Schur result, [A1], can be reformulated to say that a is in the convex hull of $S_n \lambda = \{\sigma \lambda : \sigma$ a permutation (matrix)$\}$. In this form the result generalizes as follows. Let G be a compact **Lie group** with **Lie algebra** $\mathfrak{g}$; let T be a maximal torus in G and W the corresponding **Weyl group**. Consider the adjoint action of G on $\mathfrak{g}$. Then W-orbits in $\mathrm{Lie}(T) \subset \mathfrak{g}$ correspond to G-orbits in $\mathfrak{g}$. Fix a G-invariant metric on $\mathfrak{g}$. Then the orthogonal projection of a G-orbit onto $\mathrm{Lie}(T)$ is the convex hull of the corresponding W-orbit, [A18]. For a more general result in the context of symplectic geometry cf. [A19].

A function $f : I^n \to \mathbf{R}$, where I is an open interval in $[0, \infty)$, is said to be *Schur convex* if $p \prec q$, $p, q \in I^n$, implies $f(p) \leqslant f(q)$. Then f is Schur convex if and only if it is symmetric, i.e. $f(\sigma p) = f(p)$ for all permutations σ, and

$$(p_i - p_j) \left[\frac{\partial f}{\partial p_i} - \frac{\partial f}{\partial p_j} \right] \geqslant 0$$

for all $i \neq j$. This condition is often called the *Schur condition*.

For each n-tuple of non-negative real numbers p, define a function $[p](x)$ by

$$[p](x) = \frac{1}{n!} \sum_\sigma x_1^{p_{\sigma(1)}} \cdots x_n^{p_{\sigma(n)}},$$

where the sum is over all permutations σ of $\{1, \ldots, n\}$. Then *Muirhead's inequality* says that $p \prec q$ if and only if $[p](x) \leqslant [q](x)$ for all n-tuples of non-negative real numbers x. The *arithmetic-mean geometric-mean inequality* corresponds to the special case $(n^{-1}, \ldots, n^{-1}) \prec (1, 0, \ldots, 0)$.

Let α be a partition of n. The corresponding Young subgroup of the symmetric group S_n is $S_\alpha = S_{\alpha_1} \times \cdots \times S_{\alpha_k} \subset S_n$, where S_{α_j} permutes the α_j elements $\alpha_1 + \cdots + \alpha_{j-1} + 1, \ldots, \alpha_1 + \cdots + \alpha_j + \alpha_{j+1}$. Let ρ_0 be the trivial representation of S_α and ρ_1 the *alternating* or *sign representation*, which assigns to a permutation $\sigma \in S_\alpha$ its sign $\mathrm{sgn}(\sigma) \in \{+1, -1\}$, i.e. if σ can be written as a product of t transpositions then $\mathrm{sgn}(\sigma) = (-1)^t = \rho_1(\sigma)$.

Let α and β be two partitions of n, extended with zeros if necessary to make up a vector of length n.

The *Snapper − Liebler − Vitale − Lam − Young theorem* says that the representation $\mathrm{Ind}_{S_\alpha}^{S_n}(\rho_0)$ is a subrepresentation of $\mathrm{Ind}_{S_\beta}^{S_n}(\rho_0)$ if and only if $\alpha \succ \beta$, [A13].

The *Rugh − Schönhofer theorem* says that the **intertwining number** $i(\mathrm{Ind}_{S_\alpha}^{S_n}(\rho_0), \mathrm{Ind}_{S_\beta}^{S_n}(\rho_1))$ is non-zero if and only if $\alpha \prec \beta^*$. Here β^* is the *dual partition (conjugate partition)* of β, i.e. β_i^* is the number of elements in $\{j : \beta_j \geqslant i\}$, [A10].

The *Gale − Ryser theorem* says that there exists a matrix of zeros and ones whose rows sum to the vector α and whose columns sum to the vector β if and only if $\beta \prec \alpha^*$ (or, equivalently, $\alpha \prec \beta^*$), [A12].

Consider the space $\mathcal{N}_n$ of all complex nilpotent $(n \times n)$-matrices. Let $\mathrm{GL}_n(\mathbf{C})$ act on $\mathcal{N}_n$ by similarity, and for each partition $\alpha = (\alpha_1, \ldots, \alpha_m)$ of n, let $Q(\alpha)$ be the orbit containing the Jordan matrix with Jordan blocks of size α_i and zero eigen value. Then the *Gerstenhaber − Hesselink theorem* says that the closure of an orbit $Q(\alpha)$ contains an orbit $Q(\beta)$, $\overline{Q(\alpha)} \supset Q(\beta)$, if and only if $\beta \prec \alpha$, [A6].

Let (A, B) be a pair consisting of an $(n \times n)$-matrix A and an $(n \times m)$-matrix B. The pair (A, B) is called *completely reachable* if the column vectors of the matrices $A^i B$, $i = 0, \ldots, n-1$, span all of $\mathbf{R}^n$. This is equivalent to the property that in the *linear control system* $\dot{x} = Ax + Bu$ the origin can be steered to any point in $\mathbf{R}^n$ by suitable controls $u(t) \in \mathbf{R}^m$. The *Kronecker indices* or *controllability indices* of a completely-reachable pair (A, B) are defined as follows. Let λ_j be the dimension of the space spanned by the column vectors of $A^i B$, $i = 0, \ldots, j$. Let $\mu_{j+1} = \lambda_j - \lambda_{j-1}$, $j = 0, \ldots, n-1$, $\lambda_{-1} = 0$, $\mu = (\mu_1, \ldots, \mu_n)$. Then the Kronecker indices $\kappa_i(A, B)$ of (A, B) are the elements of the partition $\kappa = \mu^*$ dual to μ. (This concept of a Kronecker index should not be confused with the Kronecker index of a function at a point, cf. **Lefschetz formula**). The partition $\kappa(A, B)$ is invariant under the following transformations of (A, B): $(A, B) \to (A + BK, B)$, where K is an arbitrary $(m \times n)$-matrix; $(A, B) \mapsto (SAS^{-1}, SB)$, where S is an invertible $(n \times n)$-matrix; and $(A, B) \mapsto (A, BT^{-1})$, where T is an invertible $(m \times m)$-matrix. Together these transformations make up the *feedback group* (of the control system $\dot{x} = Ax + Bu$) and the *Brunowski − Kalman − Morse − Wonham theorem* says that $\kappa(A, B)$ is a complete invariant of the action of the feedback group, i.e. the orbits of the action are labelled by partitions of n. Moreover, the closure $\overline{U(k)}$ of the orbit labelled by k contains the orbit $U(\lambda)$ if and only if $k \prec \lambda$, [A6].

Let E be a holomorphic **vector bundle** (cf. also **Vector bundle, analytic**) over the Riemann sphere $S^2 = \mathbf{P}_1(\mathbf{C})$. Then by *Grothendieck's theorem*, E splits as a direct sum of complex line bundles $E = L(\kappa_1) \oplus \cdots \oplus L(\kappa_m)$, where $L(i)$ is the unique (up to isomorphism) complex line bundle of first Chern number i. Now consider a holomorphic family of complex vector bundles E_t over S^2. Then by *Shatz' theorem*, $\kappa(E_t) \prec \kappa(E_0)$ for t small enough, and, conversely, if $\lambda \prec k$, then there is a holomorphic family such that $\kappa(E_t) = \lambda$ for t small and $\neq 0$ and $\kappa(E_\kappa) = \kappa$.

All these manifestations of the majorization order are far from unrelated, cf. [A4], [A6], [A9]. There are generalizations of certain of the theorems above to the case of other Weyl groups (than S_n) and relations with the *Bruhat ordering* on the Weyl group defined by the inclusions $\overline{BwB} \supset Bw'B$ of the closure of the parts of the **Bruhat decomposition** $G = \bigcup_w BwB$ of a simple Lie group, [A5], [A7].

Let Γ be a directed graph (cf. **Graph, oriented**). For a vertex x, let the *out-degree* (*demi-degree outwards*), $d_\Gamma^+(x)$, be defined as the number of arcs starting in x, and the *in-degree* (*demi-degree inwards*), $d_\Gamma^-(x)$, as the number of arcs terminating in x. The *degree* at x, $d_\Gamma(x)$, is $d_\Gamma(x) = d_\Gamma^+(x) + d_\Gamma^-(x)$. A *k-graph* is a graph with $d_\Gamma^+(x) \leqslant k$ for all vertices x. Let $(r_1, s_1), \ldots, (r_n, s_n)$ be pairs of elements of $\mathbf{N} \bigcup \{0\}$ and $k_j \in \mathbf{N}$. Define $t_j = \sum_{i=1}^n \min(k_j, r_i)$. Then there exists a graph Γ with $(d_\Gamma^+(x_i), d_\Gamma^-(x_i)) = (r_i, s_i)$ if and only if $t \succ s$ in the majorization ordering, [A17].

References

[A1] SCHUR, I.: 'Ueber ein Klasse von Mittelbildungen mit Anwendungen auf der Determinantentheorie', *Sitzungsber. Berliner Math. Ges.* **22** (1923), 9-20.

[A2] HORN, A.: 'Doubly stochastic matrices and the diagonal of a rotation matrix', *Amer. J. Math.* **76** (1954), 620-630.

[A3] MIRSKY, L.: 'Matrices with prescribed characteristic roots and diagonal elements', *J. London Math. Soc.* **33** (1958), 14-21.

[A4] MARSHALL, A.W. and OLKIN, J.: *Inequalities: majorization and its applications*, Acad. Press, 1979.

[A5] KRAFT, H.-P.: 'Conjugacy classes and Weyl group representations', *Astérisque* **87/88** (1981), 191-206.

[A6] HAZEWINKEL, M. and MARTIN, C.F.: 'Representations of the symmetric group, the specialization order, systems, and Grassmann manifolds', *Enseign. Math.* **29** (1983), 53-87.

[A7] COCINI, C. DE and PROCESI, C.: 'Symmetric functions, conjugacy classes and the flag variety', *Invent. Math.* **64** (1981), 203-220.

[A8] HARDY, G.H., LITTLEWOOD, J.E. and PÓLYA, G.: *Inequalities*, Cambridge Univ. Press, 1952.

[A9] KERBER, A.: 'The diagram lattice as structural principle in mathematics', in P. Kramer and A. Rieckers (eds.): *Group Theoretical Methods in Physics (Tübingen 1977)*, Springer, 1978, pp. 53-71.

[A10] MEAD, A., RUCH, E. and SCHÖNHOFER, A.: 'Theory of chirality functions, generalized for molecules with chiral ligands', *Theor. Chin. Acta* **29** (1973), 269-304.

[A11] MITRINOVIĆ, D.S., PEČARIĆ, J.E. and VOLENEC, V.: *Recent advances in geometric inequalities*, Kluwer, 1989.

[A12] RYSER, H.J.: *Combinatorial mathematics*, Math. Assoc. Amer., 1963.

[A13] LIEBLER, R.A. and VITALE, M.R.: 'Ordering the partition characters of the symmetric group', *J. of Algebra* **25** (1973), 487-489.

[A14] MACDONALD, J.G.: *Symmetric functions and Hall polynomials*, Oxford Univ. Press, 1979.

[A15] BULLEN, P.S., MITRINOVIĆ, D.S. and VASIĆ, P.M.: *Means and their inequalities*, Reidel, 1988.

[A16] BRYLAWSKI, T.: 'The lattice of integer partitions', *Discrete Math.* **6** (1973), 201-209.

[A17] BERGE, C.: *Graphs et hypergraphes*, Dunod, 1970, Chapt. 6.

[A18] KOSTANT, B.: 'On convexity, the Weyl group and the Iwasawa decomposition', *Ann. Sci. Ec. Norm. Sup.* **6** (1973), 413-455.

[A19] ATIYAH, M.F.: 'Convexity and commuting Hamiltonians', *Bull. London Math. Soc.* **14** (1982), 1-15.

AMS 1980 Subject Classification: 20C30, 26DXX, 15A51, 14F05, 05CXX, 05B20, 93B25, 20G20

MAL'TSEV ALGEBRA, *Moufang−Lie algebra* - An algebra over a field satisfying the identities

$$x^2 = 0, \quad J(x, y, xz) = J(x, y, z)x,$$

where $J(x, y, z) = (xy)z + (zx)y + (yz)x$ is the Jacobian of x, y, z. Mal'tsev algebras are a natural generalization of Lie algebras. Any Mal'tsev algebra is a **binary Lie algebra**.

Mal'tsev algebras were introduced by A.I. Mal'tsev [1], who called them Moufang−Lie algebras because of the connection with analytic Moufang loops (cf. **Moufang loop**). The tangent algebra of a locally analytic Moufang loop is a Mal'tsev algebra. The converse is also true: Any finite-dimensional Mal'tsev algebra over a complete normed field of characteristic zero is the tangent algebra of some locally analytic Moufang loop.

There is a close connection between Mal'tsev algebras and alternative algebras (see **Alternative rings and algebras**). The commutator algebra of an arbitrary alternative algebra, that is, the algebra obtained by replacing the original multiplication by the commutator operation

$$[x, y] = xy - yx,$$

is a Mal'tsev algebra.

Every simple Mal'tsev algebra (cf. **Simple algebra**) of characteristic $p \neq 2$ is either a Lie algebra or is a 7-dimensional algebra over its centroid. Every primary Mal'tsev algebra (for $p \neq 2$, cf. also **Primary ring**) is either a Lie algebra or can be imbedded as a subring in a suitable 7-dimensional simple algebra over some field. An arbitrary semi-primary Mal'tsev algebra (for $p \neq 2$) can be isomorphically imbedded as a subalgebra in the commutator algebra of an alternative algebra. The question of imbedding an arbitrary Mal'tsev algebra in the commutator algebra of an alternative algebra is open (1989).

Let $Z(A)$ be the Lie centre of a Mal'tsev algebra A:

$$Z(A) = \{n \in A : J(n, a, b) = 0 \text{ for all } a, b \in A\}.$$

For any ideal I of a semi-primary Mal'tsev algebra A (for $p \neq 2$), $Z(I) = Z(A) \bigcap I$.

The properties of an algebraic Mal'tsev algebra (cf. **Algebraic algebra**) are similar to the properties of an algebraic Lie algebra (cf. **Lie algebra, algebraic**). In any algebraic Mal'tsev algebra (for $p \neq 2$) there is a locally finite radical, that is, a maximal locally finite ideal such that the quotient algebra with respect to it does not contain locally finite ideals. A Mal'tsev algebra of characteristic $p \geqslant n$ or $p = 0$ satisfying the n-th Engel condition (see **Engel algebra**) is locally nilpotent (cf. **Locally nilpotent algebra**). The difference between Mal'tsev algebras and Lie algebras manifests itself in the passage from local nilpotency to global. There is, for example, a Mal'tsev algebra $(p = 0)$ satisfying the third Engel condition and which is solvable of index 2, but is not nilpotent (cf. **Nilpotent algebra**).

For a Mal'tsev algebra there is an analogue of Engel's theorem, which played a major role in the structure theory of Lie algebras: A Mal'tsev algebra satisfying the Engel condition and the maximum condition for subalgebras is nilpotent. This result also holds in the more general case of binary Lie algebras.

In every free Mal'tsev algebra (for $p \neq 2$) there is a non-zero Lie centre. A free Mal'tsev algebra (for $p \neq 2$) with three or more generators is not a primary algebra. A free Mal'tsev algebra (for $p \neq 2$) with nine or more generators contains trivial ideals.

If R_n is the variety of Mal'tsev algebras generated by the free Mal'tsev algebra on n generators and $p = 0$, then the chain of varieties

$$R_1 \subseteq R_2 \subseteq \cdots$$

does not stabilize at any finite stage.

The theory of finite-dimensional Mal'tsev algebras and their representations is well-developed. The fundamental results are similar to the results in the theory of Lie algebras. There are analogues of Lie's classical theorems: if ρ is a split representation of a solvable Mal'tsev algebra of characteristic 0, then all matrices $\rho(x)$ can be simultaneously reduced to triangular form; if ρ is a split representation of a nilpotent Mal'tsev algebra on a space V, then V decomposes into a direct sum of weight subspaces V_α, and the matrices of the bounded operators $\rho(x)$ in V_α can be simultaneously reduced to triangular form with the numbers $\alpha(x)$ on the main diagonal (cf. **Representation of a Lie algebra**).

The following results are similar to Cartan's solvability criterion and semi-simplicity of Lie algebras: if ρ is a **faithful representation** of a Mal'tsev algebra A ($p = 0$) and if the bilinear form on A associated with ρ is trivial, then A is solvable; if ρ is a representation of a semi-simple Mal'tsev algebra, then the trace form associated with ρ is non-degenerate. If the **Killing form** of A is non-degenerate, then A is semi-simple.

Any representation of a semi-simple Mal'tsev algebra with $p = 0$ is completely reducible (cf. **Reducible representation**). If S is the *radical* (maximal solvable ideal) of a Mal'tsev algebra A, N the *nil radical* (maximal nilpotent ideal), then for any derivation (cf. **Derivation in a ring**) D on A, $SD \subseteq N$.

An arbitrary finite-dimensional Mal'tsev algebra A of characteristic zero is the direct sum (as linear spaces) of its radical S and a semi-simple subalgebra B isomorphic to the quotient algebra of A by S. Two semi-simple factors are conjugate by inner automorphisms (an analogue of the Levi $-$ Mal'tsev $-$ Harish-Chandra theorem for Lie algebras).

References

[1] MAL'TSEV, A.I.: 'Analytic loops', *Mat. Sb.* **36**, no. 3 (1955), 569-576 (in Russian).
[2] SAGLE, A.: 'Malcev algebras', *Trans. Amer. Math. Soc.* **101**, no. 3 (1961), 426-458.
[3] KUZ'MIN, E.N.: 'Algebraic sets in Mal'tsev algebras', *Algebra and Logic* **7**, no. 2 (1968), 95-97. (*Algebra i Logika* **7**, no. 2 (1968), 42-47)
[4] KUZ'MIN, E.N.: 'Mal'tsev algebras and their representations', *Algebra and Logic* **7**, no. 4 (1968), 233-244. (*Algebra i Logika* **7**, no. 4 (1968), 48-69)
[5] KUZ'MIN, E.N.: 'On the relation between Mal'tsev algebras and analytic Mufang loops', *Algebra and Logic* **10**, no. 1 (1971), 1-14. (*Algebra i Logika* **10**, no. 1 (1971), 3-22)
[6] KUZ'MIN, E.N.: 'Levi's theorem for Mal'tsev algebras', *Algebra and Logic* **16**, no. 4 (1977), 286-291. (*Algebra i Logika* **16**, no. 4 (1977), 424-431)
[7] FILIPPOV, V.T.: 'On Engelian Mal'tsev algebras', *Algebra and Logic* **15**, no. 1 (1976), 57-71. (*Algebra i Logika* **15**, no. 1 (1976), 89-109)
[8] FILIPPOV, V.T.: 'Mal'tsev algebras', *Algebra and Logic* **16**, no. 1 (1977), 70-74. (*Algebra i Logika* **16**, no. 1 (1977), 101-108)
[9] GRISHKOV, A.N.: 'Analogues of Levi's theorem for Mal'tsev algebras', *Algebra and Logic* **16**, no. 4 (1977), 260-265. (*Algebra i Logika* **16**, no. 4 (1977), 389-396)
[10] SHESTAKOV, I.P.: 'A problem of Shirshov', *Algebra and Logic* **16**, no. 2 (1977), 153-166. (*Algebra i Logika* **16**, no. 2 (1977), 227-246)

V.T. Filippov

Editorial comments. For an arbitrary algebra A over a field F, the *centroid E* of A is the set of elements of the module endomorphisms $\text{Hom}_F(A, A)$ of A over F which commute for all $x, y \in A$ with the left and right multiplications $L_x, R_y \in \text{Hom}_F(A, A)$. If A is simple over F, E is an extension field of F, [A3].

References

[A1] SAGLE, A.A.: 'Simple Malčev algebras over a field of characteristic zero', *Pacific J. Math.* **12** (1962), 1057-1078.
[A2] JACOBSON, N.: *Lie algebras*, Interscience, 1962, p. 291.

AMS 1980 Subject Classification: 17D10

MAL'TSEV LOCAL THEOREMS - Theorems on transferring properties of local parts of a **model (in logic)** to the whole model, established by A.I. Mal'tsev. A system $\{M_i: i \in I\}$ of subsets of a set is called a local covering of it if each element of the set is contained in some M_i and any two M_i, M_j are contained in a third subset M_k. Examples of local coverings are: the system of all finite subsets of a set, and the system of all finitely-generated subgroups of a given group. A model M locally has a property σ if there is a local covering of M consisting of submodels with the property σ. A local theorem holds for a property σ (and a corresponding class of models) if every model locally having property σ has the property in the large.

A source of a great variety of local theorems is the following *fundamental local theorem of Mal'tsev* (or the compactness theorem of the restricted predicate calculus) [1]: If each finite subsystem of an infinite system of axioms of the restricted predicate calculus is consistent, then the whole system is consistent. Mal'tsev [2] gave a general method for obtaining concrete local theorems in group theory with the help of the fundamental local theorem, thus making a major contribution to **model theory**. Later, by improving the method itself, he proved [3] a local theorem for any property described by so-called quasi-universal axioms. The question of the validity of a local theorem for a property σ, which had previously been investigated separately for each σ, has thus been reduced to a common and quite 'grammatical' question: Is it possible to describe σ by universal axioms?

References

[1] MAL'TSEV, A.I.: 'Investigation in the realm of mathematical logic', *Mat. Sb.* **1**, no. 3 (1936), 323-336 (in Russian).
[2] MAL'TSEV, A.I.: 'A general method for obtaining local theorems in group theory', *Uchen. Zap. Ivanovsk. Gos. Ped. Inst.* **1**, no. 1 (1941), 3-9 (in Russian).
[3] MAL'TSEV, A.I.: 'Model correspondences', *Izv. Akad. Nauk. SSSR Ser. Mat.* **23**, no. 3 (1959), 313-336 (in Russian).
[4] KARGAPOLOV, M.I. and MERZLJAKOV, YU.I. [YU.I. MERZLYAKOV]: *Fundamentals of the theory of groups*, Springer, 1979 (translated from the Russian).

Yu.I. Merzlyakov

Editorial comments. English translations of references [1], [2] and [3] may be found as Chapts. 1, 2 and 11 in [A1].

References

[A1] MAL'CEV, A.I. [A.I. MAL'TSEV]: *The metamathematics of algebraic systems*, North-Holland, 1971. Collection of articles translated from the Russian.
[A2] CHANG, C. and KEISLER, H.J.: *Model theory*, North-Holland, 1973.

AMS 1980 Subject Classification: 03C07, 20E25

MAL'TSEV PRODUCT - An operation on the class of all groups (denoted by $\bigcirc$), hereditary on passing to subgroups of the factors; that is, if

$$G = \prod_{i \in I}^{\bigcirc} G_i$$

and if a subgroup H_i is chosen in each factor G_i, then the subgroups H_i, $i \in I$, generate a subgroup H of G which is the same as the product of the H_i:

$$H = \prod_{i \in I}^{\bigcirc} H_i.$$

Direct sums and free products of groups are Mal'tsev products. There exist other Mal'tsev products, but *Mal'tsev's problem* on the existence of Mal'tsev products (other than direct or free) satisfying the associative law and certain other natural conditions is still (1989) open. (The Mal'tsev product originated in connection with this problem.)

References

[1] KUROSH, A.G.: *The theory of groups*, 1, Chelsea, 1955-1956 (translated from the Russian).

A.L. Shmel'kin

Editorial comments.

References

[A1] GOLOVIN, O.N. and BRONSHTEĬN, M.A.: 'An axiomatic classification of exact operations', in *Selected Questions of Algebra and Logic*, Novosibirsk, 1973, pp. 40-96 (in Russian).

AMS 1980 Subject Classification: 08B25, 20E06, 20F99

MANGOLDT FUNCTION - The **arithmetic function** defined by

$$\Lambda(n) = \begin{cases} \ln p & \text{if } n = p^m, \ p \text{ prime}, \ m \geq 1, \\ 0 & \text{otherwise.} \end{cases}$$

The function $\Lambda(n)$ has the following properties:

$$\sum_{d \mid n} \Lambda(n) = \ln n,$$

$$\Lambda(n) = \sum_{d \mid n} \mu(d) \ln \frac{n}{d},$$

where the sums are taken over all divisors d of n. The Mangoldt function is closely connected with the Riemann **zeta-function** $\zeta(s)$. In fact, the generating series for $\Lambda(n)$ is the logarithmic derivative of $\zeta(s)$:

$$-\frac{\zeta'(s)}{\zeta(s)} = \sum \frac{\Lambda(n)}{n^s} \quad (\operatorname{Re} s > 1).$$

The Mangoldt function was proposed by H. Mangoldt in 1894.

S.A. Stepanov

Editorial comments. In the article above, μ denotes the Möbius function.

References

[A1] HARDY, G.H. and WRIGHT, E.M.: *An introduction to the theory of numbers*, Oxford Univ. Press, 1979, Sect. 17.7.

AMS 1980 Subject Classification: 10A20

MANIFOLD - A geometric object which locally has the structure (topological, smooth, homological, etc.) of $\mathbf{R}^n$ or some other vector space. This fundamental idea in mathematics refines and generalizes, to an arbitrary dimension, the notions of a line and a surface. The introduction of this idea was influenced by various requirements, both of mathematics itself and of other sciences. In mathematics, manifolds arose first of all as sets of solutions of non-degenerate systems of equations and also as various sets of geometric and other objects allowing local parametrization (see below); for example, the set of planes of dimension k in $\mathbf{R}^n$. They also appear as solutions of multi-dimensional variational problems (soap films), as integral manifolds of Pfaffian systems and of dynamical systems, as groups of geometric transformations and their homogeneous spaces, etc. In physics they play the role of models of space-time, in mechanics they act as phase spaces, energy levels, etc., in economics as surfaces of indifference, in psychology as spaces of sensations (for example, colours), etc.

Although the initial idea underlying the definition of a manifold is that of a local structure ('the very same as $\mathbf{R}^n$'), this idea admits a whole series of global features typical for manifolds: (non-) orientability, homological **Poincaré duality**, the possibility of defining the **degree of a mapping** of one manifold onto another of the same dimension, etc. Of particular significance is the introduction of the **tangent bundle** and its related invariants.

The local structure of a manifold also allows the use of geometric techniques: putting into **general position**, the construction of Morse functions (cf. **Morse function**), etc., which are used for the geometric study of the global structure of manifolds, and, roughly speaking, consist of constructing a possibly simpler image of a manifold as a union of simple pieces; for instance, simplices or handles.

To make use of the idea of a manifold a transition from the local to the global point of view is usually made. The first stage is the introduction of a *parametrization*, that is, a representation of the 'state space' of a given problem as a domain in a space of numbers R^n. This gives one the possibility to describe by a set of numbers the coordinates of the corresponding point (the coordinate method). In the large the state space need not admit a similar description, that is, it need not be homeomorphic to a region in R^n. If one does not resort to parametrizations with degeneracies (as in polar coordinates and their generalizations), then two paths are possible: either to introduce from the outset a larger than necessary number of parameters, and to find then the actual space implicitly using a system of equations ('state equations'), or to parametrize the space locally ('in the small'). For example, the set of lines in a plane is covered by two subsets : Π_1, consisting of lines with equations of the form $y = kx + b$, and Π_2, consisting of lines with equations of the form $x = \bar{k}y + \bar{b}$; both of these are homeomorphic to R^2 with parametrizations by the pairs (k, b) and $(\bar{k}, \bar{b})$, respectively. However, in the large, this set is homeomorphic to an open **Möbius strip**, and hence not like R^2.

When manifolds appear naturally in some domain or other, they are sure to carry some additional structure, which also is a topic of study in the domain. However, a major role is played by the topological structure, restricting the a priori possibilities. Conversely, in topology, the local and global properties of manifolds are studied taking into consideration the extra structures (for example, smooth) as tools.

The basis of the general idea of a manifold is the definition of a topological manifold as a topological space in which each point has a neighbourhood $\mathfrak{X}$ and a homeomorphism $\phi: \mathfrak{X} \rightarrow U$ onto a domain in R^n or into the half-space $R^n_+ = \{x \in R^n: \ x_n \geqslant 0\}$; the homeomorphism ϕ is called a *local parametrization*, or *chart*, in $\mathfrak{X}$. For a connected manifold M, the dimension $n = \dim M$ is an invariant. For a disconnected manifold the components are usually taken to be of the same dimension. A manifold decomposes into an *interior* Int M and a *boundary* ∂M (also called *edge*): points of the boundary correspond in the charts to the points of the boundary of R^n_+ in R^n. The boundary is an $(n-1)$-dimensional manifold without boundary and need not be empty. A connected manifold without

boundary is called *open* if it is non-compact, and *closed* if it is compact. The simplest examples of the four possible types of manifold are R^n, R^n_+, the ball B^n, and its boundary S^{n-1}. Although non-Hausdorff manifolds occur in certain situations (for example, the total space of a sheaf), it is usually assumed that a manifold is Hausdorff, paracompact, has a countable base, and, in particular, is metrizable.

The global specification of a manifold is accomplished by an atlas: A set of charts covering the manifold. To use manifolds in mathematical analysis it is necessary that the coordinate transitions from one chart to another are differentiable. Therefore differentiable manifolds (cf. **Differentiable manifold**) are most often considered. A more general form is introduced by the idea of a Γ-structure on a manifold, given by atlases $\{\phi_i: \mathfrak{X}_i \rightarrow U_i\}$, in which the coordinate transitions $h_{ij} = \phi_j \phi_i^{-1}$ are homeomorphisms in a system Γ of mappings of domains in R^n_+ that is closed relative to composition. If Γ consists of r times continuously-differentiable mappings, then one says that the smoothness class of the manifold is C^r. In a similar way one defines analytic manifolds (cf. **Analytic manifold**), and piecewise-linear, Lipschitz, etc., types of manifolds. Two Γ-atlases give a single Γ-*structure* if their union is a Γ-atlas. The classification of Γ-structures is the most important problem in the geometry of manifolds. A mapping $f: M \rightarrow N$ of one Γ-manifold into another is called a Γ-*mapping* if locally it has a 'coordinate representation' $f_{ij} = \phi_j f \phi_i^{-1}$, where ϕ_j, ϕ_i are charts in M and N and $f_{ij} \in \Gamma$. In particular, there is the notion of a Γ-*homeomorphism* (a C^r-*diffeomorphism* if $\Gamma = C^r$).

Since, in mathematical analysis, manifolds are important as carriers of differentiable mappings, they are sometimes defined (see [12]) by their set of smooth functions, defined in neighbourhoods of points (see **Sheaf**). The development of this idea has led to the idea of a pre-manifold or a **ringed space** (a sheaf of rings over a space), and further to the notion of a **scheme**. Replacing R^n by vector or other spaces, one arrives at various generalizations of a manifold such as, for example, complex-analytic manifolds. *Infinite-dimensional manifolds* arose in mathematical analysis and topology as spaces of mappings and sections of bundles, as spaces of homeomorphisms, spaces of closed subsets, etc. Their local models are vector spaces (Banach spaces, etc.) and spaces such as the **Hilbert cube**. The notions of smooth and other structures on infinite-dimensional spaces have not been sufficiently studied. A difficulty arises here because of the absence of technical theorems on approximations, existence of partitions of unity (a stock of small smooth functions), implicit functions, etc.

Manifolds being implicitly given as sets of solutions of systems of equations (inequalities in the case of a non-empty boundary) arise as subsets of $\mathbf{R}^n$. These manifolds are given all at once, and not in pieces, as with the assignment of an atlas. However, a necessary condition is non-degeneracy, otherwise every closed set can be given by one equation. The existence of a local parametrization is provided by the implicit-function theorem from the condition of maximal rank of the **Jacobi matrix** of the system. The equations are used as a language for expressing, by mathematical analysis, the properties of a manifold and are used to define the manifold. For example, the property of orthogonality of $(n \times n)$-matrices is given by a system of $n(n+1)/2$ equations relative to the entries of the matrix. The system turns out to be non-degenerate, and the group of orthogonal matrices is a smooth submanifold in $\mathbf{R}^m$.

In a mathematical system with given coordinates, smaller systems are chosen by equations or inequalities expressing restrictions or 'constraints'. If the non-degeneracy condition for a system $F_i = 0$ is satisfied at all points of a manifold, then the gradients of the functions F_i form a *framing* or *frame* (a k-frame orthogonal to the tangent plane at a point of the manifold and depending continuously on the point). Manifolds admitting a framing form the rather narrow class of *stably-parallelizable manifolds* (for example, they have orientations, cf. **Orientation**). But, locally, any differentiable manifold in $\mathbf{R}^n$ may be given by a non-degenerate system, and by using a partition of unity it is possible to construct a system of constant (but not maximal) rank which defines the manifold.

For a manifold given by an atlas, the problem arises of realizing it as a submanifold in $\mathbf{R}^n$ with regard to some Γ-structure. Any topological, smooth or piecewise-linear manifold M *is imbeddable*, that is, is Γ-homeomorphic to a submanifold in $\mathbf{R}^{2n}$, and in $\mathbf{R}^{2n+1}$ the set of imbeddings is dense in the space of all continuous mappings. For other classes the question is significantly more complicated. It has been studied intensively, for example, for Riemannian manifolds. Algebraic varieties (cf. **Algebraic variety**), realized in a complex projective space (replacing here $\mathbf{R}^n$), constitute a very special class (Hodge varieties, cf. **Hodge variety**). If degenerate systems of equations are allowed, then *manifolds with singularities* arises. In algebraic geometry both ideas combine: An **algebraic space** is defined by joining pieces which are locally given by polynomial systems with degeneracies. In **cobordism** theory, in topology, under the name of manifold with singularities one discusses spaces which are constructed locally as products of cones over manifolds. Subsequent generalizations are connected with the consideration of continuous families of manifolds (see **Fibre space**; **Fibration**; **Stratification**).

The study of manifolds by pure topological means is very complicated and, until recently, was restricted to local properties, that is, virtually properties of $\mathbf{R}^n$ (**continuous decomposition**; **wild imbedding**; etc.). The global study of manifolds necessarily requires the use of either homological or other means from algebraic topology, or smooth, piecewise-linear or other structures. The fundamental homological property of a manifold is Poincaré duality, which is also preserved for homology manifolds (cf. **Homology manifold**), and is used for generalizations of the notion of a manifold, such as Poincaré complexes. The degree of a mapping is homological in nature; this is a basic idea which is also defined for *pseudo-manifolds*, i.e. spaces (usually complexes) which have open connected dense subsets that are manifolds.

The geometric study of manifolds, based on the use of the various structures, comes first of all from the local equivalence of manifolds with $\mathbf{R}^n$ and from the possibility, by approximating a mapping by smooth or piecewise-linear mappings, of putting them into general position, etc. The first problem here is the representation of a manifold as a space made up of simple pieces. The initial idea was that of **triangulation**, developing into the general idea of a **complex**. The difficult problems of triangulability and equivalence of triangulations were cleared up in the 1960's and 1970's (see **Topology of manifolds**). A more flexible tool has turned out to be handle decomposition of a manifold, equivalent to considering Morse functions. The basic theorems (on *h*-**cobordism**, the generalized **Poincaré conjecture**, etc.) were proved using the technique of successive simplification of this decomposition. This technique also acts as a geometric foundation for the classification theorems on smooth and piecewise-linear structures in a given homotopy type of manifolds. These theorems, however, required the use of subtle invariants, connected with the tangent bundle of the manifold (see below). Of great significance was also the discovery of new homology theories: *K*-**theory** and the theory of **cobordism**. Two manifolds are cobordant if together they bound a third manifold. Here the various structures are taken into account. This equivalence relation is based on the initial definition of homology, when cycles were considered naively as piecewise-smooth manifolds. In cobordism theory this idea was restored, but submanifolds were replaced by mappings of manifolds. The fundamental groups play a special role in questions on the classification of structures on manifolds, due to which the connection of these questions with **algebraic *K*-theory** has been established.

The most important invariant of a smooth manifold M is its tangent bundle τM. The subject of analysis on manifolds is the study of sections of τM and the various bundles associated with it. From the topological point of view τM is characterized by the existence of the **exponential mapping** $\exp: \tau M \to M$, diffeomorphically mapping small balls in the fibres $\tau_x M$ onto neighbourhoods of the corresponding points $x \in M$. It can be described as a projection of a **tubular neighbourhood** of the diagonal in $M \times M$ onto a factor. In this form the definition of τM can be transferred to the topological, piecewise-linear, etc., cases (see **Micro-bundle**).

Different reductions of the structure group of the tangent bundle are called *structures on the manifold*. If an atlas can be chosen such that the Jacobi matrices of its coordinate transitions belong to a reduced structure group, and thus determine this reduction, then such a structure is called *integrable*. The question of classification of structures and their integrability is related to the fundamental questions of the **differential geometry of manifolds**. In the topology of manifolds the question on the classification of smooth, piecewise-linear and other structures reduces mainly to the classification of the corresponding structures on the tangent bundle and, by the same token, lead to homology problems.

The major role of the tangent bundle is connected with the fact that there are cohomology classes invariantly connected with it (cf. **Characteristic class**) in some homology theory, carrying the most essential information on the global structure of the manifold and connecting the properties of the different constructions and structures on M with its topological properties.

Research into the topology of manifolds in this direction, up to the 1970's, essentially used the smooth and piecewise-linear structures on manifolds (more precisely, the **homotopy type** of manifolds). The transition to pure topological results became possible only after the proof of difficult and deep theorems, beginning with the proof of the topological invariance of the (rational) characteristic classes (see **Topology of manifolds**). At the end of the 1970's this direction was merging with the above-mentioned pure topological study of manifolds. A vivid example is the proof of the conjecture that the dual **suspension** over a three-dimensional homology sphere is a manifold (a sphere). This allowed one to give a topological classification of manifolds (known up to that time only for one-dimensional and two-dimensional manifolds, cf. **Two-dimensional manifold**), to clarify the question of which polyhedra are manifolds (the only obstacle here is the, as yet unproved (1990), **Poincaré conjecture** in dimension 4) and others (see **Topology of manifolds** and [19]).

Historical sketch. The first period in the study of manifolds is related to the analysis of the idea of a multi-dimensional parametrization, with research into the geometry of the physical world (the surface of the Earth) and into geometric axiomatics. The two methods for specifying a manifold in $\mathbf{R}^n$ (local parametrization and equations) were first considered by C.F. Gauss (see

[1], p. 127) for surfaces in $\mathbf{R}^3$, and in the multi-dimensional case by H. Poincaré (see [3]). J. Plücker [5] studied local coordinates in manifolds formed from curves, surfaces, etc. H. Grassmann arrived in [6] at the general idea of a 'multi-dimensional extension', which, under the name 'manifold', was introduced into mathematics by B. Riemann in his famous lecture *Über die Hypothesen, welche der Geometrie zuGrunde liegen* (see [2]). The properties of various special coordinates were studied by C.G.J. Jacobi, G. Lamé and others (see [8]).

Gauss (see [1], p. 123), in connection with his own work on geodesy, began a systematic study of surfaces, introducing the notion of intrinsic geometry, and thus that of manifold, not depending on the ambient number space, and in fact the notion of a structure on a manifold. This idea was completely understood only in the theory of characteristic classes, constructed in the mid 20-th century. Riemann translated the ideas of Gauss to multi-dimensional manifolds. On the basis of Riemannian geometry, tensor analysis was created by G. Ricci, T. Levi-Civita, E.B. Christoffel, and others, the further development of which went in close connection with relativity theory. Another geometric line of development of the notion of a manifold was made initially in the discovery of the possibility of **non-Euclidean geometries**, and in the construction of geometry on the basis of the notion of a motion (H. Helmholtz, see [1], p. 366). This idea was transformed in a general program of group-theoretic constructions of geometry by F. Klein (see [1], p. 399 and [8]) and led to the sophisticated work of S. Lie on the theory of continuous groups. The line of Helmholtz$-$Klein$-$Lie remained for a long time apart from the line of Gauss$-$Riemann$-$Ricci. It borrowed the notion of curvature from the latter, but was interested only in Klein spaces (cf. **Klein space**). However, important questions arose in the global structure of Lie groups and their homogeneous spaces and this attracted the attention to the global structure of manifolds. An important fact, producing a deep impression, was the discovery by Klein of **elliptic geometry**, locally equivalent to spherical geometry, but globally having essentially different properties, and also the discovery by A.F. Möbius and Klein of the phenomenon of non-orientability.

The synthesis of both directions took place in the papers of E. Cartan (see [1], p. 483). Starting from the research of G. Darboux on the theory of surfaces, he considered the **moving-frame method** for arbitrary manifolds in $\mathbf{R}^n$ and came to the theory of structure equations, a distant relative of Darboux's theory including the work of Lie. In Cartan's method the notion of a G-structure unified the ideas of Riemannian geometry and the theory of Lie groups. Essentially, Cartan introduced the notion of a tangent bundle and its structure group, which finally took shape only in the 1940's (see [13]). This idea also allowed a unification of mathematical analysis on manifolds with the topological study of manifolds. The foundation was provided by the de Rham isomorphism (see **de Rham theorem**) — the final form of Poincaré's idea (see [3]) on the connection between real cohomology and differential forms. The most important subsequent stage was the introduction of characteristic classes and their expression as integrals of forms, expressed in terms of the curvature form (an example here is the expression of the Euler characteristic in the **Gauss—Bonnet theorem**, originally due to W. van Dyck, see [14]).

The topological study of manifolds was initiated with the discovery of Riemann surfaces (cf. **Riemann surface**) in connection with the representation of complex-analytic functions as integrals, as an attempt to free oneself from the multivalency of these functions. The 'periods' of integrals led to the notion of a connectivity number and, in the end, to homology. The notion of a multi-dimensional generalization of this idea and the idea of a global homological study of manifolds is due to Riemann (see [2]). This study was begun by Poincaré, who made a number of important discoveries and proved **Poincaré duality**. A stimulating role, following Riemann's discovery, was played by the study of two-dimensional manifolds (in the first place by Möbius and C. Jordan, see [14]), leading to their complete classification. Finally, it proved possible to progress only after clarifying the idea of a 'pure' homeomorphism (Poincaré, for example, essentially used piecewise-smooth homeomorphisms). This clarification was one of the results of the analysis of the numerical continuum, undertoken at the end of the 19-th century. Of great significance in this direction was the posing of Hilbert's fifth problem and the work of L.E.J. Brouwer [9], who proved theorems (on invariance of domain and invariance of dimension) which permitted H. Weyl [4] to formulate the notion of a topological manifold. However, in higher dimensions the topological study of manifolds for a long time was conducted within the limits of smooth and piecewise-linear structures. Smooth structures, introduced in [20], were mainly analyzed by H. Whitney [21], and also by G. Whitehead and others. Piecewise-linear structures were introduced by Brouwer and analyzed by J.W. Alexander [22] and also by M. Newman and Whitehead. For a long time they were considered only as an auxiliary means for the topological study of manifolds. Only in the 1950's the non-uniqueness of differentiable structures, even on spheres, was discovered, and at the end of the 1960's the possibility of non-uniqueness of piecewise-linear structures (for example, on tori) was discovered. Since the 1950's the study of manifolds has moved to the synthesis of ideas of topology and analysis, based in the first place on the idea of characteristic classes (see [17]).

References
[1]   *On the foundation of geometry*, Moscow, 1956 (in Russian).
[2A]   RIEMANN, B.: 'Ueber die Hypothesen, welche der Geometrie zuGrunde liegen', in *Das Kontinuum und andere Monographien*, Chelsea, reprint, 1973.
[2B]   RIEMANN, B.: *Gesammelte mathematische Werke - Nachträge*, Teubner, 1892-1902 (translated from the German).
[3A]   POINCARÉ, H.: *Oeuvres de Henri Poincaré*, Vol. 3, Gauthier-Villars, 1916-1965.
[3B]   POINCARÉ, H.: *Oeuvres de Henri Poincaré*, Vol. 2, Gauthier-Villars, 1916-1965.
[4]   WEYL, H.: *Die Idee der Riemannschen Fläche*, Teubner, 1955.
[5]   PLÜCKER, J.: *Neue Geometrie des Raumes gegründet auf die Betrachtung der geraden Linie als Raumelement*, 1-2, Teubner, 1868-1869.
[6]   GRASSMAN, H.: 'Die Ausdehnungslehre von 1844', in *Gesammelte Werke*, Vol. 1, Springer, pp. 1-319.
[7]   KRONECKER, L.: 'Ueber Systeme von Functionen mehrere Variabeln I', *Monatsber. Preuss. Akad. Wiss.* (1869), 159-193.
[8]   KLEIN, F.: *Vorlesungen über höhere Geometrie*, Springer, 1926.
[9A]   BROUWER, L.E.J.: 'Beweis der Invarianz der Dimensionszahl', *Math. Ann.* **70** (1911), 161-165.
[9B]   BROUWER, L.E.J.: 'Ueber Abbildungen von Mannigfaltigkeiten', *Math. Ann.* **71** (1911), 97-115.
[9C]   BROUWER, L.E.J.: 'Zur Invarianz des n-dimensionalen Gebiets', *Math. Ann.* **72** (1912), 55-56.
[10]   WEYL, H.: *Mathematische Analyse des Raumproblems*, Springer, 1923.
[11]   STEENROD, N.E.: *The topology of fibre bundles*, Princeton Univ. Press, 1951.
[12]   CHEVALLEY, C.: *Theory of Lie groups*, 1, Princeton Univ. Press, 1946.
[13]   LICHNEROWICZ, A.: *Global theory of connections and holonomy groups*, Noordhoff, 1976 (translated from the French).
[14]   HIRSCH, M.: *Differential topology*, Springer, 1976.
[15]   MILNOR, J. and STASHEF, J.: *Characteristic classes*, Princeton Univ. Press, 1974.
[16]   NIJENHUIS, A.: *Theory of the geometric object*, Univ. of Amsterdam, 1952. Thesis.
[17]   HIRZEBRUCH, F.: *Topological methods in algebraic geometry*, Springer, 1978 (translated from the German).
[18]   SULLIVAN, D.: *Geometric topology*, M.I.T., 1971.
[19]   CANNON, J.W.: 'The recognition problem: what is a topological manifold?', *Bull. Amer. Math. Soc.* **84**, no. 5 (1978), 832-866.
[20]   VEBLEN, O. and WHITEHEAD, G.: *The foundations of differential geometry*, Cambridge Univ. Press, 1953.
[21]   WHITNEY, H.: 'Differentiable manifolds', *Ann. of Math.* **37** (1936), 645-680.
[22]   ALEXANDER, J.W.: 'Combinatorial analysis situs', *Trans. Amer. Math. Soc.* **28** (1926), 301-329.

See also the references to the articles **Differential geometry of manifolds; Differential topology; Differentiable manifold.**

A.V. Chernavskiĭ

Editorial comments. The four-dimensional Poincaré conjecture was solved in 1984, see **Poincaré conjecture**.

For infinite-dimensional manifolds see also [A1].

References
[A1]   LANG, S.: *Introduction to differentiable manifolds*, Interscience, 1967.

AMS 1980 Subject Classification: 57JXX, 58BXX

MANIFOLD OF FIGURES (LINES, SURFACES, SPHERES, . . .) - A manifold whose constituent elements are various figures in a certain **homogeneous space**. From an analytic point of view the simplest figures are algebraic curves and surfaces. So the manifolds that have been investigated (as a rule, in Euclidean, affine and projective spaces) mainly have as constituent elements points, straight lines, planes, circles, spheres, conics, quadrics, and multi-dimensional analogues of these.

An m-dimensional manifold $\mathfrak{M}_m$ of figures F of rank N is defined by a closed system of differential equations

$$\Omega^a = \lambda_i^a \Omega^i, \quad \Delta\lambda_i^a \wedge \Omega^i = 0, \quad i, j = 1, \ldots, m; \qquad (1)$$
$$a, b = m+1, \ldots, N,$$

where Ω^J ($J = 1, \ldots, N$) are the left-hand sides of the isotropy equations of a figure F and $\Delta\lambda_i^a$ are the Pfaffian forms arising upon closing the Pfaffian equations of the system (1), with $\Omega^1 \wedge \cdots \wedge \Omega^m \neq 0$ (cf. **Pfaffian form; Pfaffian equation**). After carrying out a sequence of prolongations of the system (1), one obtains a sequence of fundamental objects of $\mathfrak{M}_m$, selects from it a basic object of the manifold, and carries out an invariant construction of the differential geometry of the manifold.

The differential geometry of ruled manifolds has been developed in depth. The simplest manifolds with non-linear constituent elements are manifolds of conics. With every one-dimensional manifold C_1^2 of conics in a three-dimensional space (Euclidean, affine or projective) there is associated a *torse T*, which is the envelope of the planes of the conics. A manifold C_1^2 is called *focal* or *non-focal*, depending on whether a generator of the torse is tangent to the conics or not. A two-dimensional manifold (a *congruence*) of conics in a three-dimensional space has, in general, six focal surfaces and six focal families. All the conics in the congruence are tangent to these surfaces. Congruences of conics with indefinite focal families (every two adjacent conics of which intersect up to second order) are characterized by all the conics in the congruence belonging to one quadric. Congruences of conics whose planes form a one-parameter family have one quadruplicate focal family to which correspond four focal points of intersection of two adjacent conics belonging to one plane. The two other focal points are points of intersection of a conic with a characteristic of its plane.

A congruence K_2 of quadrics in P_3 has, in general, eight focal surfaces, to which all the quadrics in the congruence are tangent. A point of a quadric $F = 0$ of the congruence of K_2 defined along any direction by the system of equations $F = 0, dF = 0, \ldots, d^m F = 0$, is called a focal point of order m of this quadric. A focal point of the second order is a fourfold point of the first

order; a focal point of third order is a focal point of arbitrary order $m > 3$. On each conic C of a three-dimensional manifold (a *complex*) of conics there are, in general, six invariant points (*t-focal points of the conic*). For each conic C of a complex whose planes form a two-parameter family, there is a unique conic C^* passing through the characteristic point of the plane of C and the four points of intersection of the conic with an adjacent conic in the same plane. The geometric properties of multi-parameter families of conics depend essentially on the number of parameters that characterize the planes of the conics of these families.

An immediate generalization of a conic in P_3 is a quadratic element — an $(n-2)$-dimensional non-degenerate quadric in P_n ($n > 3$). An m-parameter family of quadratic elements whose hyperplanes form an h-parameter family is called an $(h, m, n)^2$-manifold in P_n. $(h, h, 3)^2$-manifolds, $h = 1, 2, 3$, are the most general one-parameter families, congruences and complexes of conics in P_3. With each local quadratic elements of an $(h, h, n)^2$-manifold for $h < n$ is associated an $(n-h-1)$-dimensional characteristic subspace and an $(h-1)$-dimensional polar subspace. The *rank of an $(h, h, n)^2$-manifold* is the number R equal to $n - h - 1 - \nu$, where ν is the dimension of the subspace in which the characteristic subspace intersects its polar subspace.

The differential geometry of an $(n, n, n)^2$-manifold can be regarded as the geometry of some regular hypersurface of an $(n+1)$-dimensional centro-projective space P_0^{n+1} in which the original n-dimensional space plays the role of the fixed point.

A quadric of dimension $p \leqslant n - 2$ in P_n always lies in a $(p+1)$-dimensional subspace. Algebraic surfaces of order $k > 2$ do not have this property, in general. In investigating manifolds of algebraic surfaces it turns out to be advisable to select families of surfaces that lie in planes of dimension higher by one. A non-degenerate $(n-2)$-dimensional algebraic surface of order k belonging to a hyperplane P_{n-1} of P_n is called a *plane algebraic element* of order k. An m-dimensional manifold of plane algebraic elements of order k whose hyperplanes form an n-parameter family is called an $(h, m, n)^k$-manifold. A fundamental object of the first order is the basic object of an $(h, m, n)^k$-manifold.

In applications of geometry to hydromechanics, field theory and differential equations one uses manifolds of lines and surfaces that are not algebraic, in general. The study of such manifolds is of interest for geometry in its own right. A *curvilinear congruence* in a three-dimensional space is a two-dimensional family Γ_t of curves $x = x(t, u, v)$ such that through each point of the space passes, in general, a unique curve of the family. For fixed u and v a curve C_t of the family Γ_t is distinguished; for fixed t and variable u and v a surface

S_{uv} is distinguished which is called a *transversal surface* of the congruence Γ_t. Points of C_t at which $(xx_t x_u x_v) = 0$ are called *focal points*. The set of focal points of the curves C_t of Γ_t is called the *focal surface* of the congruence. A surface $v = v(u)$ of Γ_t on which the curves C_t have an envelope is called a *principal surface* of Γ_t.

References

[1] MALAKHOVASKiĭ, V.S.: 'Differential geometry of manifolds of figures', *J. Soviet Math.* **21**, no. 2 (1983), 127-150. (*Itogi Nauk. i Tekhn. Probl. Geom.* **12** (1981), 31-60)

V.S. Malakhovskiĭ

Editorial comments.

References

[A1] GRIFFITHS, P. and HARRIS, S.: *Principles of algebraic geometry*, Wiley, 1978.

AMS 1980 Subject Classification: 53C30, 53A35, 53A20, 53A07

MANN THEOREM - A theorem giving an estimate of the density of the sum of two sequences (cf. **Density of a sequence**). Let $A = \{0, a_1, \ldots, a_j, \ldots\}$ be an increasing sequence of integers and let

$$A(n) = \sum_{\substack{0 < a_j \leq n \\ a_j \in A}} 1.$$

The *density* of the sequence A is the quantity

$$d(A) = \inf_n \frac{A(n)}{n}.$$

The *arithmetic sum* of two sequences A and B is the sequence $C = A + B$ consisting of all possible sums $c = a + b$, where $a \in A$ and $b \in B$. Mann's theorem asserts that

$$d(A + B) \geqslant \min\{d(A) + d(B), 1\}.$$

Mann's theorem implies that if A is a sequence of positive density less than 1 and B is another sequence of positive density, then on addition of A and B the density is increased. Another important consequence of Mann's theorem is: Each sequence of positive density is a **basis** for the sequence of natural numbers. Mann's theorem essentially strengthens a similar theorem of Shnirel'man (cf. **Shnirel'man method**). It was proved by H.B. Mann [1].

References

[1] MANN, H.B.: 'A proof of the fundamental theorem on the density of sums of sets of positive integers', *Ann. of Math.* **43** (1942), 523-527.
[2] OSTMANN, H.H.: *Additive Zahlentheorie*, Springer, 1956.
[3] GEL'FOND, A.O. and LINNIK, YU.V.: *Elementary methods in the analytic theory of numbers*, M.I.T., 1966 (translated from the Russian).

B.M. Bredikhin

AMS 1980 Subject Classification: 10L02

MANN—WHITNEY TEST - A statistical test for testing the hypothesis H_0 of homogeneity of two samples $X_1, \ldots, X_n$ and $Y_1, \ldots, Y_m$, all $m + n$ elements of which are mutually independent and have continuous distributions. This test, suggested by H.B. Mann and D.R. Whitney [1], is based on the statistic

$$U = W - \frac{1}{2}m(m+1) = \sum_{i=1}^{n} \sum_{j=1}^{m} \delta_{ij},$$

where W is the statistic of the **Wilcoxon test** intended for testing the same hypothesis, equal to the sum of the ranks of the elements of the second sample among the pooled order statistics (cf. **Order statistic**), and

$$\delta_{ij} = \begin{cases} 1 & \text{if } X_i < Y_j, \\ 0 & \text{otherwise.} \end{cases}$$

Thus, U counts the number of cases when the elements of the second sample exceed elements of the first sample. It follows from the definition of U that if H_0 is true, then

$$\mathrm{E}U = \frac{nm}{2}, \quad \mathrm{D}U = \frac{nm(n+m+1)}{12}, \qquad (*)$$

and, in addition, this statistic has all the properties of the Wilcoxon statistic W, including asymptotic normality with parameters $(*)$.

References

[1] MANN, H.B. and WHITNEY, D.R.: 'On a test whether one of two random variables is statistically larger than the other', *Ann. Math. Stat.* **18** (1947), 50-60.

M.S. Nikulin

Editorial comments. Instead of Mann—Whitney test, the phrase *U-test* is also used.

AMS 1980 Subject Classification: 62G10

MANY-VALUED LOGIC - A branch of **mathematical logic** studying mathematical models of **propositional logic**. These models reflect the two fundamental features of the latter: the plurality of truth values of propositions, and the possibility of constructing new, more complicated, propositions from given ones using logical operations, which also allow the truth values of the compound propositions to be established with respect to the truth values of initial propositions. Examples of many-valued propositions are statements with a modal outcome ('yes', 'no', 'may be') and statements of a probabilistic nature; examples of logical operations are logical connectives of the type 'and', 'or', 'if $\cdots$ then'. In general, models of many-valued logic are generalizations of the **algebra of logic**. It is important to note that in the algebra of logic, propositions take only two truth values ('yes', 'no'); thus, in general, they may not reflect the full variety of logical structures encountered in practice. In a fairly broad interpretation of many-valued logic, logical calculi (cf. **Logical calculus**) are also sometimes included.

Historically, the first models of many-valued logics were the two-valued logic of G. Boole (mid 19-th century), also called the algebra of logic, the three-valued logic of J. Łucasiewicz (1920) and the m-valued logic of E. Post (1921). The investigation of these models was an important stage in the development of the theory of many-valued logics.

Basic models of many-valued logics. Many-valued logics have a specific distinguishing feature, which is the discussion of the problems and methods arising in the investigation of many-valued logics from the point of view of mathematical logic, mathematical cybernetics and algebra. Thus, from the point of view of mathematical cybernetics, models of many-valued logics are considered as languages describing the functioning of complex control systems the components of which may be in a certain number of different states, and from the point of view of algebra, models of many-valued logics are algebraic systems (cf. **Algebraic system**) having both applied and theoretical interest.

The construction of models of many-valued logics is done by analogy with the construction of a two-valued logic. Thus the individual propositions of the logic, separated into classes with the same truth value, lead to the idea of a set E, a constant of the model, which in fact identifies all individual propositions, replacing them by their corresponding truth values; variable propositions become variable quantities $x_1, x_2, \ldots$, which take values in E; the logical connectives become elements of a set M of elementary functions (operations) the arguments of which take values from E. Compound propositions, constructed from individual and variable propositions and also the logical connectives, become elements of the set $<M>$ of formulas over M. The truth value from E of a compound proposition is a function of the corresponding truth values of the propositions figuring in the given compound proposition. In the model this function is associated with a formula corresponding to the given compound proposition; the formula is also said to *realise* the function. Functions mapping a **tuple** of elements of E into E are called *functions of m-valued logic*, where m denotes the cardinality of the set E. The set of all functions of m-valued logic is denoted by P_m. The set of formulas $<M>$ leads to a set $[M]$ of functions realizing the formulas from $<M>$ called *superpositions* (or *compositions*) over M. The set $[M]$ is called the *closure of the set M*. The specification of the concrete model of a many-valued logic M_F is to be regarded as equivalent to indicating the sets E, M, $<M>$, and $[M]$; the model is then said to be generated by M; if M is finite, the model is called *finitely generated*. This model is called a *formula model*, and also an *m-valued logic*.

The distinctive approach of mathematical cybernetics to many-valued logics is to regard models of many-valued logics as control systems. The elementary func-

tions here are the elements producing the operations, and formulas are interpreted as schemes constructed from the elements which also process input information into output. Control systems of this kind, known in cybernetics as *diagrams of functional elements* (without branching), are widely utilized in theoretical and practical questions of cybernetics.

At the same time there are a number of problems of logic and cybernetics connected with the study of relationships between the sets M and $[M]$ in which the role of $<M>$ is somewhat in the background, regarded just as a means of defining the second set from the first. This situation leads to another model of a many-valued logic as an algebra whose elements are functions taking values and arguments from E. It is common to use as operations in this algebra a special set of operations equivalent in meaning to the relationship of M and $[M]$ to the set of formulas constructed from functions on M; that is, compound functions are obtained from single functions whose arguments are other functions. These algebras are called the *algebras of m-valued logic*. They may be arrived at concretely, for example, by the introduction of the following one-place operations: $\zeta, \tau, \Delta, \nabla$, and a binary operation $\star$. If one assumes that, taking into account fictitious variables, each function f from P_E depends on $x_1, \ldots, x_n$, where n is determined by f, then these operations can be given as:

$$(\zeta f)(x_1, \ldots, x_n) = f(x_2, \ldots, x_n, x_1),$$

$$(\tau f)(x_1, \ldots, x_n) = f(x_2, x_1, x_3, \ldots, x_n),$$

$$(\Delta f)(x_1, \ldots, x_{n-1}) = f(x_1, x_1, x_2, \ldots, x_{n-1})$$

if $n > 1$, and $\zeta f = \tau f = \Delta f = f$ if $n = 1$;

$$(\nabla f)(x_1, \ldots, x_n, x_{n+1}) = f(x_2, x_3, \ldots, x_{n+1})$$

and for $f(x_1, \ldots, x_n)$ and $g(x_1, \ldots, x_m)$,

$$(f \star g)(x_1, \ldots, x_{n+m-1}) =$$
$$= f(g(x_1, \ldots, x_m), x_{m+1}, \ldots, x_{m+n-1}).$$

The algebras $M_E = <M, \zeta, \tau, \Delta, \nabla, \star>$ are sometimes called *Post algebras*.

Problems in many-valued logic. Among the problems characteristic of formula models of many-valued logics is the problem of 'description', that is, the question of giving all formulas of $<M_1>$ which realise functions from M_2, for a given set $M_2 \subseteq [M_1]$. A particular case of this problem is the important question in mathematical logic of giving all formulas which realise a given constant; for example, in propositional calculus, this is equivalent to the construction of all identically-true propositions.

A question at the interface between mathematical logic and algebra, related to the problem of description, is the *problem of identical transformations*. Here for a given M it is required to select the simplest, in some sense, subset of pairs of equal (i.e. realizing the same function) formulas from $<M>$ (identities) which by substitution of one formula for the other allows one to obtain from any formula all formulas equal to it (a complete system of identities).

A similar place is occupied by the so-called *completeness problem*, which is to give all subsets M_1 of a given closed (i.e. equal to its closure) set M for which $[M_1] = M$; that is, the *completeness property* (or the *functional completeness property*) of M_1 in M holds. Close to this problem is the *basis problem*, which is to give each complete subset M_1 in M no proper subset of which is complete in M; *complete independent systems of functions* are also called bases. There are two approaches to the solution of the completeness problem — the algorithmic and the algebraic. In the first case there is the question of the existence of an algorithm which establishes the completeness or incompleteness of a system of functions described in some language; in the second, one passes to the study of the properties of the lattice of subalgebras of a given algebra of m-valued logic and solves the completeness question using these properties. An important idea in the algebraic approach is that of a criterial system of subalgebras. A system N of subalgebras of an algebra $M = <M, \zeta, \tau, \Delta, \nabla, \star>$ is called *criterial* if the following property holds: A set $M' \subseteq M$ is complete if and only if it is not a subset of any subalgebra in N. For example, the set of all proper subalgebras of the algebra M is criterial. In general, the latter criterial system is too large. Each criterial system contains all maximal subalgebras of the algebra M (all *pre-complete classes*), and this enables one to move to the consideration of more economical criterial systems which do not contain subalgebras of maximal subalgebras. Thus, in completeness problems one may confine oneself to the use of criterial systems of the form $N = \Sigma_1 \bigcup \Sigma_2$, where Σ_1 is the set of all maximal subalgebras and Σ_2 consists of some of those subalgebras that are not subalgebras of any maximal subalgebra. If Σ_2 is empty, then the completeness problem reduces to the description of all maximal subalgebras of the algebra N.

A global problem in many-valued logics is the description of the lattice of closed classes of a given model of a many-valued logic.

The characteristic question in the theory of complexity of control systems arises naturally in relation to the formulas and functions from a many-valued logic. Typical for this approach is the following *problem on the complexity of realization*. By some means a numerical measure (the *complexity of a formula*) is introduced on the set of all elementary formulas; it is then extended to the set of all formulas, for example, by summing the measures of all elementary formulas entering in a given formula. For a given function it is required to give the (simplest) formula which realizes the function with least complexity, and also to clarify how this complexity depends on the properties of the function considered. Various generalizations of the problem have been studied.

There is a wide circle of questions connected with the realization of functions by formulas with preassigned properties. Here one has the problem on the realization of functions of the algebra of logic by disjunctive normal forms (cf. **Disjunctive normal form**) and the related minimization problem; also there is the problem of the realization of functions by formulas which have *bounded depth* (that is, by formulas in which the chain of formulas substituted in each other has bounded length, this bound being connected with realizability and speed of calculation), the *decomposition problem*, that is, the realization of functions of variables using formulas constructed from elementary formulas which realize functions of fewer variables, and a number of other problems.

The solution of all the problems above depends essentially on the cardinalities of the sets E and M generating the given model of the many-valued logic.

The most important examples of many-valued logics. The most important examples of many-valued logics are the finite-valued logics (that is, m-valued logics for which m is finite). For these it is common to assume that $E = E_m = \{0, \ldots, m-1\}$, and the m-valued logic M_E is denoted by M_m. The case $m = 2$ has been investigated most deeply; here the functions are also called Boolean functions (cf. **Boolean function**). The most important result is the complete description, by Post (see [1]), of the lattice of closed classes. Their set turns out to be countable, and each class and the lattice of classes, ordered by inclusion, can be effectively constructed. These classes are called *Post classes*. Each closed class has a finite basis of cardinality not exceeding 4. These results lead to solutions of the problems on description, completeness and bases, and also of the problem on identical transformations. With respect to complete finite systems, for 'almost-all' functions the behaviour of the measure of complexity of the simplest formulas realizing these functions has been given and corresponding algorithms for the synthesis of the formulas have been constructed (see [2]). The problem of the construction of optimal formulas with respect to complexity, realizing functions reliably and sufficiently quickly, and also the question of the complexity of realization for a large number of special classes of functions and individual functions have been studied.

One of the intensively researched problems for two-valued logics is the *minimization problem*. Here a special language for the specification of the functions of a two-valued logic — the language of disjunctive normal forms — has been studied. The idea of the *complexity of a disjunctive normal form*, the number of letters in it, has been introduced, and the problems connected with the finding, construction and metric properties of the disjunctive normal form that is 'simplest' in the sense of this measure and that realizes a given function have been investigated (see [3]). The decomposition problem of Boolean functions is the clarification of conditions under which a given function of n variables may be realized by a formula constructed from functions of fewer variables, where all functions substituted into each other during the construction of the formula do not depend on common variables. Such formulas are called *non-iterative*. It has been shown that there is no finite system of functions such that each function can be realized by a non-iterative formula over this system, and even that 'almost-all' functions do not have a realization by non-iterative formulas. On the whole, two-valued logics, because of special properties, are the main objects in which the general formulation of problems is discussed.

For arbitrary finite-valued logics, which for $m>2$ are essentially different from two-valued logics, there are effective solutions of the problems on description for finite systems. It has been shown that for $m\geqslant 3$ there are m-valued logics with a finite basis and which, at the same time, do not have a finite complete system of identities (see [4]), whereas in two-valued logics each closed class has a finite complete system of identities (see [5]). For finite-valued logics with a finite basis there is an effective solution of the completeness problem. It is obtained in the following way. One says that a function $f(x_1, \ldots, x_n)$ *preserves* a set $K = \{g_1(x_1, \ldots, x_n), \ldots, g_s(x_1, \ldots, x_n)\}$ if for any set $g_{i_1}, \ldots, g_{i_k}$, where $g_{i_j} \in K$, one has $f(g_{i_1}, \ldots, g_{i_k}) \in K$. A set K is called *regular* if the choice function $g_j(x_1, \ldots, x_n) = x_j$ is contained in K for all j, $1 \leqslant j \leqslant n$, and if each function from K preserves K. If $M = [K]$, then in M one selects all systems K' with the following properties: the functions in them depend only on $x_1, \ldots, x_n$; the addition to them of all choice functions makes these systems regular; they do not contain K as a subset. This procedure of selection of all regular sets $K_1, \ldots, K_t$, $t \leqslant 2^{m^*}$, is effective. The system $\{U(K_1), \ldots, U(K_t)\}$, where $U(K_i)$ (the preservation class of K_i) consists of all functions from M which preserve K_i, $1 \leqslant i \leqslant t$, is criterial. Inclusion of an arbitrary finite set $K \subseteq M$ in each of the classes $U(K_i)$ can also be verified effectively. Each pre-complete class is a preservation class and the set of all pre-complete classes in this case forms a criterial system. It has been shown that for $m \geqslant 3$ there is a continuum of closed

classes in P_m, and that there exist closed classes with bases of given finite or infinite cardinality, and also classes which do not have bases; here the families of classes without a basis, or with a countable basis, have the cardinality of the continuum. Examples of classes without a basis or having a countable basis are $[\{f_0, \ldots, f_n, \ldots\}]$ or $[\{g_1, \ldots, g_n, \ldots\}]$, where $f_n \equiv 0$ for $n = 0$ and for $n > 0$, $f_n(x_1, \ldots, x_n)$ is equal to 0 on all collections except $(2, \ldots, 2)$, at which it is equal to 1; $g_n(x_1, \ldots, x_n)$ is equal to 0 on all collections except the collections of the form $(2, \ldots, 2, 1, 2, \ldots, 2)$ at which it is equal to 1 (see [6]).

The completeness problem in the m-valued logic P_m is of particular interest. In P_m there are finite complete systems; this allows one to select a finite complete subsystem for each complete system and to reduce the problem to the study of finite complete systems. There also exist complete systems consisting of one function; such functions are called *Sheffer functions*. An example is the *Wenn function* $\max(x_1, x_2) + 1 \pmod{m}$. Because of finite generation, the completeness problem in P_m can be solved effectively.

In P_m all pre-complete classes, which are preservation classes for special predicates, have been effectively constructed. Giving these predicates forms the content of the completeness theorem in finite-valued logics (see [7]). One says that a function $f(x_1, \ldots, x_n)$ *preserves a predicate* $R: (E_m)^h \to \{0, 1\}$ if the formula

$$R(z_{11}, \ldots, z_{1h}) \& \cdots \& R(z_{n1}, \ldots, z_{nh}) \to$$
$$\to R(f(z_{11}, \ldots, z_{n1}), \ldots, f(z_{1h}, \ldots, z_{nh}))$$

is identically equal to 1 for all values of the variables z_{ij}, $1 \leqslant i \leqslant n$, $1 \leqslant j \leqslant h$. The class $U(R)$ of all functions preserving a predicate R is called the *preservation class of the predicate R*. It has been shown that for each pre-complete class N it is possible to choose a predicate R, depending on at most $m + 2$ variables, and for $m \geqslant 3$ on at most m variables, such that $N = U(R)$. These predicates may be partitioned into six families: H, S, E, L, Z, U. The families H, S and E consist of all two-place predicates or order relations on E_m having one maximal and one minimal element in the first case; giving permutations on E_m which decompose into cycles of identical prime number length (without invariant elements) in the second case; and giving non-identity and non-universal equivalence relations on E_m in the third case. The family L is not empty for $m = p^k$, p a prime, and consists of all four-place predicates $R_G(y_1, y_2, y_3, y_4)$ for which $R_G(y_1, y_2, y_3, y_4) = 1$ is equivalent to $y_1 + y_2 = y_3 + y_4$, where G is an Abelian group in which each non-zero element has order p. The predicate $R(y_1, \ldots, y_h)$, $1 \leqslant h \leqslant m$, is reflexive if the fact that not all the numbers $a_1, \ldots, a_h$ are distinct implies that $R(a_1, \ldots, a_h) = 1$. It is symmetric if for

mapping ϕ defined on a subset A of X by the equality $\phi(x)=f(x)$, $x\in A$, is called the *restriction of the mapping* $f:X\to Y$ *to* A; this restriction is often denoted by f_A. A mapping F defined on a set $E\supset X$ and satisfying the equality $F(x)=f(x)$ for all $x\in X$ is called an *extension* (or *continuation*) *of the mapping* to E. If three sets X, Y, Z are given, if a mapping f with values in Y is defined on X, and a mapping g with values in Z is defined on Y, then there exists a mapping h with domain of definition X, taking values in Z, and defined by the equality $h(x)=g[f(x)]$. This mapping is called the *composite of the mappings* f *and* g, while f and g are called *component (factor) mappings*. The mapping h is also called the *compound mapping* (*composite mapping, composed mapping*), consisting of the interior mapping f and the exterior mapping g. The composed mapping is denoted by $g\circ f$, where the order of the notation is vital (for functions of a real variable, the term *superposition* is also used). The concept of a compound mapping can be generalized to any finite number of components of the mapping.

A mapping f, defined on X and taking values in Y, gives rise to a new mapping defined on the subsets of X and taking subsets of Y as values. In fact, if $A\subset X$, then

$$f(A) = \{f(x): x\in A\}.$$

The set $f(A)$ is called the *image* of A. If $A=\{x\}$, the initial mapping $f(x)$ is obtained; thus, $f(A)$ is an extension of $f(x)$ from the set X to the set $\mathfrak{P}(X)$ of all subsets of X if a one-element set is identified with the element comprising it. When $Y=X$, a set A is called an *invariant subset* for f if $f(A)\subset A$, while a point x is called a *fixed point* for f if $f(x)=x$. Invariant sets and fixed points are important in solving functional equations of the form $f(x)=a$ or $x-f(x)=a$.

Every mapping $f:X\to Y$ gives rise to a mapping defined on the subsets of the set $f(X)$ or Y and taking subsets of the set X as values. In fact, for every $B\subset f(X)$ (or $\subset Y$), the set $\{x: f(x)\in B\}$ is denoted by $f^{-1}(B)$, and is called the *complete inverse image* (*complete pre-image*) of B. If $f^{-1}(y)$ for each $y\in f(X)$ consists of a single element, then f^{-1} is a mapping of elements, is defined on $f(X)$, and takes values in X. It is also called the *inverse mapping* for f. The existence of an inverse mapping is equivalent to the solvability of the equation $f(x)=y$, $y\in f(X)$, for a unique x when y is given.

If the sets X and Y have certain properties, then interesting classes can be distinguished in the set $F(X, Y)$ of all mappings from X into Y. Thus, for partially ordered sets X and Y, the mapping f is *isotone* if $x<y$ implies $f(x)\leqslant f(y)$ (cf. **Isotone mapping**). For complex planes X and Y, the class of holomorphic mappings is naturally selected. For topological spaces X and Y, the class of continuous mappings between these spaces is distinguished naturally; an extended theory of differentiation of mappings (cf. **Differentiation of a mapping**) has been constructed. For mappings of a scalar argument and, in the most general case, for mappings defined on a **measure space**, the concept of (weak or strong) measurability can be introduced, and various Lebesgue-type integrals can be constructed (for example, the **Bochner integral** and the **Daniell integral**).

A mapping is called a *multi-valued mapping* if subsets $Y_x\subset Y$ consisting of more than one element are assigned to certain values of x. Examples of this type of mappings include multi-sheeted functions of a complex variable, multi-valued mappings of topological spaces, and others.

References

[1] BOURBAKI, N.: *Elements of mathematics. Theory of sets*, Addison-Wesley, 1968 (translated from the French).
[2] BOURBAKI, N.: *Elements of mathematics. General topology*, Addison-Wesley, 1966 (translated from the French).
[3] KELLEY, J.L.: *General topology*, Springer, 1975.

V.I. Sobolev

Editorial comments. For a mapping $f:X\to Y$, the set X is also called the *source of f*, while Y is also called the *target of f*, [A3].

References

[A1] HALMOS, P.R.: *Measure theory*, v. Nostrand, 1950.
[A2] HALMOS, P.R.: *Naive set theory*, v. Nostrand, 1961.
[A3] COHN, P.M.: *Universal algebra*, Reidel, 1981.

AMS 1980 Subject Classification: 04A05, 26-01

MAPPING-CONE CONSTRUCTION - The construction that associates with every continuous mapping $f:X\to Y$ of topological spaces the topological space $C_f\supset Y$ obtained from the topological sum (disjoint union) $CX\oplus Y$ (here $CX=(X\times[0, 1])/(X\times\{0\})$ is the **cone** over X) by identifying $x\times\{1\}=f(x)$, $x\in X$. The space C_f is called the *mapping cone* of f. If X and Y are pointed spaces with distinguished points $x\in X$, $y\in Y$, then the generator $x\times[0, 1]$ of CX is contracted to a point, and C_f is said to be the *collapsed mapping cone* of f. For an arbitrary pointed topological space K, the sequence $X\to^f Y\subset C_f$ induces an exact sequence

$$[X, K] \leftarrow [Y, K] \leftarrow [C_f, K]$$

of pointed sets. The mapping f is homotopic to the constant mapping to the distinguished point if and only if Y is a retract of C_f (cf. **Retract of a topological space**).

References

[1] SPANIER, E.H.: *Algebraic topology*, McGraw-Hill, 1966.
[2] MOSHER, R.E. and TANGORA, M.K.: *Cohomology operations and their applications in homotopy theory*, Harper & Row, 1968.

A.F. Kharshiladze

Editorial comments. The *algebraic analogue of the mapping-cone construction* is as follows.

Let $u: K. \to L.$ be a morphism of complexes, i.e. $u = (u_n)_{n \in \mathbb{Z}}$ and $u_{n-1} \partial_n = \partial_n u_n$, where $\partial_n^K: K_n \to K_{n-1}$. The mapping cone of u is the complex $C(u).$ defined by

$$C(u)_n = K_{n-1} \oplus L_n, \quad \partial(k, l) = (-\partial k, \partial l + uk).$$

The injections $L_n \to C(u)_n$ define a morphism of complexes and if $K[-1]$ denotes the complex with $K[-1]_n = K_{n-1}$ and $\partial_n^{K[-1]} = -\partial_{n-1}^K$, then the projections $C(u)_n \to K_{n-1}$ yield

$$0 \to L_n \to C(u)_n \to K[-1]_n \to 0,$$

which fit together to define a short exact sequence of complexes

$$0 \to L. \xrightarrow{i} C(u). \xrightarrow{p} K[-1]. \to 0,$$

and there results a long exact homology sequence

$$\cdots \to H_n(L.) \xrightarrow{i_*} H_n(C(u).) \xrightarrow{p_*} H_{n-1}(K.) \xrightarrow{u_*} H_{n-1}(L.) \to \cdots.$$

By turning a complex $K.$ into a 'co-complex' $K^{\cdot}$, $K^n = K_{-n}$, the analogous constructions and results in a cohomological setting are obtained.

The complex $K[-1].$ is called the *suspension of the complex* $K..$

References

[A1] MACLANE, S.: *Homology*, Springer, 1963, Sect. II.4.

AMS 1980 Subject Classification: 54B17, 55P99

MAPPING CYLINDER - A construction associating with every continuous mapping of topological spaces $f: X \to Y$ the topological space $I_f \supset Y$ that is obtained from the topological sum (disjoint union) $X \times [0, 1] \amalg Y$ by the identification $x \times \{1\} = f(x)$, $x \in X$. The space I_f is called the *mapping cylinder* of f, the subspace Y is a **deformation retract** of I_f. The imbedding $i: X = X \times \{0\} \subset I_f$ has the property that the composite $\pi \circ i: X \to Y$ coincides with f (here π is the natural retraction of I_f onto Y). The mapping $\pi: I_f \to Y$ is a homotopy equivalence, and from the homotopy point of view every continuous mapping can be regarded as an imbedding and even as a **cofibration**. A similar assertion holds for a **Serre fibration**. For any continuous mapping $f: X \to Y$ the fibre and cofibre are defined up to a homotopy equivalence.

References

[1] SPANIER, E.H.: *Algebraic topology*, McGraw-Hill, 1966.
[2] MOSHER, R.E. and TANGORA, M.C.: *Cohomology operations and their application in homotopy theory*, Harper & Row, 1968.

A.F. Kharshiladze

Editorial comments. The literal translation from the Russian yields the phrase '*cylindrical construction*' for the mapping cylinder. This phrase sometimes turns up in translations.

References

[A1] WHITEHEAD, G.W.: *Elements of homotopy theory*, Springer, 1978, p. 22, 23.

AMS 1980 Subject Classification: 55P05

MAPPING METHOD, *method of images* - A method in potential theory for solving certain boundary value problems for partial differential equations in a domain D, where the fulfillment of the boundary conditions on the boundary $\partial D = \Gamma$ is achieved by the choice of additional field sources situated outside D and called *image sources*.

The greatest development of the mapping method occurred in electrostatics. Suppose, for example, that it is required to solve the **Dirichlet problem** for the **Poisson equation** $\Delta u = -2\pi \rho(x, y)$ in the half-plane $D = \{(x, y): y > 0, -\infty < x < +\infty\}$ with a given function $\psi(x)$ on the boundary $\Gamma = \{(x, y): y = 0, -\infty < x < +\infty\}$, that is, it is required to find the potential of electrostatic charges of density $\rho(x, y)$ situated in D, with the condition that the potential $\psi(x)$ is fixed on Γ. It is known that in order to solve this problem it suffices to know the **Green function** $G(x, y; x_0, y_0)$ representing the potential of a unit point charge at the point $(x_0, y_0) \in D$ when the boundary Γ is earthed, that is, $G(x, 0; x_0, y_0) = 0$. The solution $u(x, y)$ of the original problem is expressed as follows in terms of $G(x, y; x_0, y_0)$:

$$u(x, y) = \iint_D \rho(x_0, y_0) G(x, y; x_0, y_0) \, dx_0 \, dy_0 + \tag{1}$$

$$+ \frac{1}{2\pi} \int_\Gamma \psi(x_0) \frac{\partial G(x, y; x_0, 0)}{\partial y_0} \, dx_0.$$

When there is no boundary, the potential of a point charge can be written as a fundamental solution $\ln(1/\sqrt{(x - x_0)^2 + (y - y_0)^2})$ of the **Laplace equation**. Adding a negative unit charge image at the point $(x_0, -y_0)$ and forming the sum of the potentials of these two charges, one obtains the desired Green function:

$$G(x, y; x_0, y_0) = \tag{2}$$

$$= \ln \frac{1}{\sqrt{(x - x_0)^2 + (y - y_0)^2}} - \ln \frac{1}{\sqrt{(x - x_0)^2 + (y + y_0)^2}}.$$

In the case of a strip $D = \{(x, y): 0 < y < b, -\infty < x < +\infty\}$, on reflecting a unit charge at $(x_0, y_0) \in D$ with respect to the lines $y = 0$ and $y = b$ one gets an infinite sequence of charges $-1, -1, +1, +1, \ldots$, at the points $(x_0, -y_0)$, $(x_0, 2b - y_0)$, $(x_0, -2b + y_0)$, $(x_0, 2b + y_0), \ldots$, respectively. The Green function in this case can be expressed in the form of an infinite series of potentials of point charges.

For a domain D in the form of a disc, $D = \{(\rho, \phi): 0 \leq \rho < a, 0 \leq \phi < 2\pi\}$, the image of a unit charge at (ρ_0, ϕ_0) is a negative unit charge at $(a^2/\rho_0, \phi_0)$, which is the image of (ρ_0, ϕ_0) under **inversion** relative to the circle $\rho = a$.

Other configurations of boundaries are also possible, consisting of lines and circles, and the solution is

If, in addition, $\lim_{t\to 0}\psi(t)/t=\infty$, then M_ψ^0 is the closure in M_ψ of the set of all bounded functions of compact support. In this case the dual of M_ψ^0 is isomorphic to the Lorentz space and, consequently, M_ψ is isomorphic to the second dual space of M_ψ^0.

If $\mathfrak{M}$ is a space with a σ-finite measure μ defined on its σ-algebra of measurable sets, then for each measurable function $x(m)$ its rearrangement $x^*(s)$, $0<s<\infty$, is defined, which makes it possible to introduce the Marcinkiewicz space $M_\psi(\mathfrak{M},\mu)$ with the norm (1).

References

[1] MARCINKIEWICZ, J.: 'Sur l'interpolation d'opérations', *C.R. Acad. Sci. Paris* **208** (1939), 1272-1273.
[2] KREĬN, S.G., PETUNIN, YU.I. and SEMENOV, E.M.: *Interpolation of linear operators*, Amer. Math. Soc., 1982 (translated from the Russian).
[3] STEIN, E.M. and WEISS, G.: *Introduction to Fourier analysis on Euclidean spaces*, Princeton Univ. Press, 1975.

S.G. Kreĭn

Editorial comments. Let f be a continuous function on $[0, 1]$. The left-continuous decreasing *rearrangement* f^* of f is defined by the properties:

i) f^* is decreasing (i.e. non-increasing);

ii) f^* is left-continuous;

iii) $\{x: f(x)>s\}$ and $\{x: f^*(x)>s\}$ have the same measure for all s.

Alternatively one considers left-continuous or right-continuous decreasing (or increasing) rearrangements. The right-continuous decreasing rearrangement can be described as follows. Let $m(y)$ be the measure of the set $\{u: f(u)>y\}$. Then

$$f^*(x) = \sup\{y: m(y) > x\}.$$

The notion is a continuous analogue of putting a finite sequence of real numbers in decreasing (or increasing) order. This last construction is of importance in the context of the **majorization ordering** and there are in fact various continuous analogues of results connected with that ordering, such as the *Muirhead inequalities* and the result linking the majorization ordering and *doubly-stochastic matrices*, cf. [A1] - [A3].

The *Lorentz space* defined by means of a function ψ as above is the space of all measurable functions f such that

$$\| f \|_\psi = \int_0^\infty f^*(t)\, d\tilde\psi(t) < \infty,$$

where f^* is the decreasing rearrangement of $|f|$ and $\tilde\psi$ is the least concave majorant of ψ. More generally one also considers Lorentz spaces based on L_p norms (instead of L_1 like above).

The analogous *Lorentz sequence spaces* are defined as follows. For every non-increasing sequence of positive numbers $w=(w_n)_{n=1}^\infty$ and every $p\geq 1$, let $d(w, p)$ be the space of all sequences $x=(a_i)$ of scalars for which

$$\| x \|_{(w,p)} = \left[\sum_{n=1}^\infty |a_{\pi(n)}|^p w_n\right]^{1/p} < \infty,$$

where π is a permutation of $\{1, 2, \ldots\}$ such that $(|a_{\pi(n)}|)_{n=1}^\infty$ is a non-increasing sequence. If $\inf w_n>0$, then $d(w, p)$ is isomorphic to l_p, and if $\sum_{n=1}^\infty w_n<\infty$, then $d(w, p)\simeq l_\infty$. These two 'trivial' cases are sometimes excluded. For a great deal of material on Lorentz sequence spaces see [A4].

References

[A1] MARSHALL, A.W. and OLKIN, J.: *Inequalities: theory of majorization and its applications*, Acad. Press, 1979.
[A2] BULLEN, P.S., MITRINOVIĆ, D.S. and VASIĆ, P.M.: *Means and their inequalities*, Reidel, 1988.
[A3] HARDY, G.H., LITTLEWOOD, J.E. and PÓLYA, G.: *Inequalities*, Cambridge Univ. Press, 1952.
[A4] LINDENSTRAUSS, J. and TZAFRIRI, L.: *Classical Banach spaces*, I: Sequence spaces, Springer, 1977.

AMS 1980 Subject Classification: 46E30

MARGINAL DISTRIBUTION - The distribution of a random variable, or set of random variables, obtained by considering a component, or subset of components, of a larger random vector (see **Multi-dimensional distribution**) with a given distribution. Thus the marginal distribution is the projection of the distribution of the random vector $X=(X_1,\ldots,X_n)$ onto an axis x_1 or subspace defined by variables $x_{i_1},\ldots,x_{i_k}$, and is completely determined by the distribution of the original vector. For example, if $F(x_1,x_2)$ is the distribution function of $X=(X_1,X_2)$ in $\mathbf{R}^2$, then the distribution function of X_1 is equal to $F_1(x_1)=F(x_1,+\infty)$; if the two-dimensional distribution is absolutely continuous and if $p(x_1,x_2)$ is its density, then the density of the marginal distribution of X_1 is

$$p_1(x_1) = \int_{-\infty}^{+\infty} p(x_1,x_2)\,dx_2.$$

The marginal distribution is calculated similarly for any component or set of components of the vector $X=(X_1,\ldots,X_n)$ for any n. If the distribution of X is normal, then all marginal distributions are also normal. When $X_1,\ldots,X_n$ are mutually independent, then the distribution of X is uniquely determined by the marginal distributions of the components $X_1,\ldots,X_n$ of X:

$$F(x_1,\ldots,x_n) = \prod_{i=1}^n F_i(x_i)$$

and

$$p(x_1,\ldots,x_n) = \prod_{i=1}^n p_i(x_i).$$

The marginal distribution with respect to a probability distribution given on a product of spaces more general than real lines is defined similarly.

References

[1] LOÈVE, M.: *Probability theory*, Springer, 1977.
[2] CRAMÉR, H.: *Mathematical methods of statistics*, Princeton Univ. Press, 1946.

A.V. Prokhorov

AMS 1980 Subject Classification: 62EXX, 60EXX

MARKOV CHAIN - A **Markov process** with finite or

countable state space. The theory of Markov chains was created by A.A. Markov who, in 1907, initiated the study of sequences of dependent trials and related sums of random variables [1].

Let the state space be the set of natural numbers $\mathbf{N}$ or a finite subset thereof. Let $\xi(t)$ be the state of a Markov chain at time t. The fundamental property of a Markov chain is the **Markov property**, which for a discrete-time Markov chain (that is, when t takes only non-negative integer values) is defined as follows: For any $i, j \in \mathbf{N}$, any non-negative integers $t_1 < \cdots < t_k < t$ and any natural numbers $i_1, \ldots, i_k$, the equality

$$P\{\xi(t)=j \mid \xi(t_1)=i_1, \ldots, \xi(t_k)=i_k\} = \tag{1}$$

$$= P\{\xi(t)=j \mid \xi(t_k)=i_k\}$$

holds.

The Markov property (1) can be formulated as follows. The time t and the related events of the form $\{\xi(t)=j\}$ are called the 'present' of the process; events determined by the values $\xi(u)$, $u<t$, are called the 'past' of the process; events determined by $\xi(u)$, $u>t$, are called the 'future' of the process. Then (1) is equivalent to the following: For any $t \in \mathbf{N}$ and fixed 'present' $\xi(t)=j$, any 'past' event A and 'future' event B are conditionally independent, given the present. That is,

$$P\{A \cap B \mid \xi(t)=j\} =$$

$$= P\{A \mid \xi(t)=j\}P\{B \mid \xi(t)=j\}.$$

In the probabilistic description of a Markov chain $\xi(t)$ the **transition probabilities**

$$P\{\xi(t+1)=j \mid \xi(t)=i\} \tag{2}$$

play a major role. When (2) does not depend on t, the Markov chain is called *homogeneous* (in time); otherwise it is called *non-homogeneous*. Only homogeneous Markov chains are considered below. Let

$$p_{ij} = P\{\xi(t+1)=j \mid \xi(t)=i\}.$$

The matrix $P = \| p_{ij} \|$ with entries p_{ij} is called the *matrix of transition probabilities*. The probability of an arbitrary trajectory $\xi(k)=i_k$, $k=0, \ldots, t$, is given by the transition probabilities p_{ij} and the initial distribution $P\{\xi(0)=i\}$ as follows:

$$P\{\xi(k)=i_k, \ k=0, \ldots, t\} =$$

$$= P\{\xi(0)=i_0\} \prod_{k=1}^{t} p_{i_{k-1}i_k}.$$

Along with the transition probabilities p_{ij}, in a Markov chain the t-step transition probabilities $p_{ij}(t)$ are also discussed:

$$p_{ij}(t) = P\{\xi(t_0+t)=j \mid \xi(t_0)=i\}.$$

These transition probabilities satisfy the **Kolmogorov − Chapman equation**

$$p_{ij}(t_1+t_2) = \sum_{k} p_{ik}(t_1)p_{kj}(t_2).$$

Using transition probabilities it is possible to give the following classification of states. Two states i and j are called *communicating* if there are $t_1, t_2 > 0$ such that $p_{ij}(t_1)>0$ and $p_{ji}(t_2)>0$. A state k is called *inessential* if there is a state l such that $p_{kl}(t_1)>0$ for some $t_1 \geq 1$ and $p_{lk}(t)\equiv 0$ for all $t \in \mathbf{N}$. All remaining states are called *essential*. Thus, the complete state space of a Markov chain decomposes into inessential and essential states. The set of all essential states decomposes into disjoint classes of communicating states, such that any two states from one class communicate with each other, and for any two states i, j from distinct classes, $p_{ij}(t)\equiv 0$, $p_{ji}(t)\equiv 0$. A Markov chain for which all states belong to one class of communicating states is called *non-decomposable* (cf. **Markov chain, non-decomposable**); otherwise the Markov chain is called *decomposable* (see **Markov chain, decomposable**). If the state space is finite, then its decomposition into these classes determines, to a great extent, the asymptotic properties of the Markov chain. For example, for a finite non-decomposable Markov chain the limit

$$\lim_{T \to \infty} \frac{1}{T+1} \sum_{t=0}^{T} p_{ij}(t) = p_j, \tag{3}$$

where $\sum_{j} p_j = 1$, always exists. If, in addition, the Markov chain is *aperiodic*, that is, if for some t_0, for all $t \geq t_0$ and all states i and j, $p_{ij}(t)>0$ (see also **Markov chain, periodic**), then there is the stronger result

$$\lim_{t \to \infty} p_{ij}(t) = p_j \tag{4}$$

(see also **Markov chain, ergodic**).

If the state space of a Markov chain is countable, then its asymptotic properties depend on more subtle properties of the classes of communicating states. The series

$$\sum_{t} p_{ii}(t) \tag{5}$$

simultaneously diverge or converge for all states of a given class. A class of states is called *recurrent* if for any state i of the class the series (5) diverges and non-recurrent if (5) converges. In a recurrent class, with probability 1 a Markov chain returns to any of its states, in a non-recurrent class the probability of recurrence is less than 1. If the mean return time in a recurrent class is finite, then the class is called *positive*; otherwise it is called *zero* (see **Markov chain, class of positive states of a; Markov chain, class of zero states of a**). If i and j belong to the same class of positive states, then the limit (3) exists, and in the aperiodic case the limit (4) exists. If j belongs to a class of zero states or is inessential, then $p_{ij}(t) \to 0$ as $t \to \infty$.

Let $f(\cdot)$ be a real-valued function defined on the states of a Markov chain $\xi(t)$. If the Markov chain is non-decomposable and its states form a positive class, then for the sum

enables one to find the $\{p_j\}$ without calculating the limits in (1).

Let
$$\tau_{jj} = \min\{t \geqslant 1 : \xi(t) = j \mid \xi(0) = j\}$$
be the moment of first return to the state j (for a discrete-time Markov chain), then
$$\mathsf{E}\tau_{jj} = p_j^{-1}.$$
A similar (more complicated) relation holds for a continuous-time Markov chain.

The trajectories of an ergodic Markov chain satisfy the ergodic theorem: If $f(\cdot)$ is a function on the state space of the chain $\xi(t)$, then, in the discrete-time case,
$$\mathsf{P}\left\{ \lim_{n \to \infty} \frac{1}{n} \sum_{t=0}^{n} f(\xi(t)) = \sum_i p_i f(j) \right\} = 1,$$
while in the continuous-time case the sum on the left is replaced by an integral. A Markov chain for which there are $\rho < 1$ and $C_{ij} < \infty$ such that for all i, j, t,
$$|p_{ij}(t) - p_j| \leqslant C_{ij} \rho^t, \tag{2}$$
is called *geometrically ergodic*. A sufficient condition for geometric ergodicity of an ergodic Markov chain is the *Doeblin condition* (see, for example, [1]), which for a discrete (finite or countable) Markov chain may be stated as follows: There are an $n < \infty$ and a state j such that $\inf_i p_{ij}(n) = \delta > 0$. If the Doeblin condition is satisfied, then for the constants in (2) the relation $\sup_{i,j} C_{ij} = C < \infty$ holds.

A necessary and sufficient condition for geometric ergodicity of a countable discrete-time Markov chain is the following (see [3]): There are numbers $f(j)$, $q < 1$ and a finite set B of states such that:
$$\mathsf{E}\{f(\xi(1)) \mid \xi(0) = i\} \leqslant qf(i), \quad i \notin B,$$
$$\max_{i \in B} \mathsf{E}\{f(\xi(1)) \mid \xi(0) = i\} < \infty.$$

References
[1] DOOB, J.L.: *Stochastic processes*, Wiley, 1953.
[2] CHUNG, K.L.: *Markov chains with stationary transition probabilities*, Springer, 1967.
[3] POPOV, N.N.: 'Conditions for geometric ergodicity of countable Markov chains', *Soviet Math. Dokl.* **18**, no. 3 (1977), 676-679. (*Dokl. Akad. Nauk SSSR* **234**, no. 2 (1977), 316-319)

A.M. Zubkov

Editorial comments.

References
[A1] FREEDMAN, D.: *Markov chains*, Holden-Day, 1975.
[A2] IOSIFESCU, M.: *Finite Markov processes and their applications*, Wiley, 1980.
[A3] KEMENY, J.G. and SNELL, J.L.: *Finite Markov chains*, v. Nostrand, 1960.
[A4] KEMENY, J.G., SNELL, J.L. and KNAPP, A.W.: *Denumerable Markov chains*, Springer, 1976.
[A5] REVUZ, D.: *Markov chains*, North-Holland, 1975.
[A6] ROMANOVSKY, V.I. [V.I. ROMANOVSKIĬ]: *Discrete Markov chains*, Wolters-Noordhoff, 1970 (translated from the Russian).
[A7] SENETA, E.: *Non-negative matrices and Markov chains*, Springer, 1981.

AMS 1980 Subject Classification: 60J27, 60J10

MARKOV CHAIN, GENERALIZED - A sequence of random variables ξ_n with the properties:

1) the set of values of each ξ_n is finite or countable;

2) for any n and any $i_0, \ldots, i_n$,
$$\mathsf{P}\{\xi_n = i_n \mid \xi_0 = i_0, \ldots, \xi_{n-s} = i_{n-s}, \ldots, \xi_{n-1} = i_{n-1}\} = (*)$$
$$= \mathsf{P}\{\xi_n = i_n \mid \xi_{n-s} = i_{n-s}, \ldots, \xi_{n-1} = i_{n-1}\}.$$

A generalized Markov chain satisfying (*) is called *s-generalized*. For $s = 1$, (*) is the usual **Markov property**. The study of s-generalized Markov chains can be reduced to the study of ordinary Markov chains. Consider the sequence of random variables η_n whose values are in one-to-one correspondence with the values of the vector
$$(\xi_{n-s+1}, \xi_{n-s+2}, \ldots, \xi_n).$$
The sequence η_n forms an ordinary **Markov chain**.

References
[1] DOOB, J.L.: *Stochastic processes*, Wiley, 1953.

V.P. Chistyakov

Editorial comments.

References
[A1] FREEDMAN, D.: *Markov chains*, Holden-Day, 1975.
[A2] KEMENY, J.G. and SNELL, J.L.: *Finite Markov chains*, v. Nostrand, 1960.
[A3] REVUZ, D.: *Markov chains*, North-Holland, 1975.
[A4] ROMANOVSKY, V.I. [V.I. ROMANOVSKIĬ]: *Discrete Markov chains*, Wolters-Noordhoff, 1970 (translated from the Russian).
[A5] SENETA, E.: *Non-negative matrices and Markov chains*, Springer, 1981.
[A6] BLANC-LAPIERRE, A. and FORTET, R.: *Theory of random functions*, 1-2, Gordon & Breach, 1965-1968 (translated from the French).

AMS 1980 Subject Classification: 60J27, 60J10

MARKOV CHAIN, NON-DECOMPOSABLE - A **Markov chain** whose transition probabilities $p_{ij}(t)$ have the following property: For any states i and j there is a time t_{ij} such that $p_{ij}(t_{ij}) > 0$. The non-decomposability of a Markov chain is equivalent to non-decomposability of its matrix of transition probabilities $P = \|p_{ij}\|$ for a discrete-time Markov chain, and of its matrix of transition probability densities $Q = \|q_{ij}\|$, $q_{ij} = dp_{ij}(t)/dt \big|_{t=0}$ for a continuous-time Markov chain. The state space of a non-decomposable Markov chain consists of one class of communicating states (cf. **Markov chain**).

V.A. Sevast'yanov

Editorial comments.
Cf. also **Markov chain** and **Markov chain, decomposable** for references.

AMS 1980 Subject Classification: 60J27, 60J10

MARKOV CHAIN, PERIODIC - A non-decomposable homogeneous **Markov chain** $\xi(n)$, $n = 1, 2, \ldots$, in which each state i has period larger than 1, that is,
$$d_i = \text{g.c.d.}\{n : \mathsf{P}\{\xi(n) = i \mid \xi(0) = i\} > 0\} > 1.$$

In a non-decomposable Markov chain (cf. **Markov chain, non-decomposable**) all states have the same period. If $d=1$, then the Markov chain is called *aperiodic*.

V.P. Chistyakov

Editorial comments. Cf. also **Markov chain** and **Markov chain, decomposable** for references.

AMS 1980 Subject Classification: 60J27, 60J10

MARKOV CHAIN, RECURRENT - A **Markov chain** in which a random trajectory $\xi(t)$, starting at any state $\xi(0)=i$, returns to that state with probability 1. In terms of the **transition probabilities** $p_{ij}(t)$, recurrence of a discrete-time Markov chain is equivalent to the divergence for any i of the series

$$\sum_{t=0}^{\infty} p_{ii}(t).$$

In a recurrent Markov chain a trajectory $\xi(t)$, $0 \leqslant t < \infty$, $\xi(0)=i$, returns infinitely often to the state i with probability 1. In a recurrent Markov chain there are no inessential states and the essential states decompose into recurrent classes. An example of a recurrent Markov chain is the symmetric random walk on the integer lattice on the line or plane. In the symmetric walk on the line a particle moves from position x to $x \pm 1$ with probabilities $1/2$; in the symmetric walk on the plane a particle moves from (x,y) to one of the four points $(x \pm 1, y)$, $(x, y \pm 1)$ with probabilities $1/4$. In these examples a particle, starting the walk at an arbitrary point, returns to that point with probability 1. The symmetric walk on the integer lattice in the three-dimensional space, when the probability of transition from (x,y,z) to a neighbouring point $(x \pm 1, y, z)$, $(x, y \pm 1, z)$, $(x, y, z \pm 1)$ is equal to $1/6$, is not recurrent. In this case the probability of return of the particle to its initial point is approximately 0.35.

References

[1] FELLER, W.: *An introduction to probability theory and its applications*, 1, Wiley, 1966.

V.A. Sevast'yanov

Editorial comments.

References

[A1] FREEMAN, D.: *Markov chains*, Holden-Day, 1975.

[A2] IOSIFESCU, M.: *Finite Markov processes and their applications*, Wiley, 1980.

[A3] KEMENY, J.G. and SNELL, J.L.: *Finite Markov chains*, v. Nostrand, 1960.

[A4] KEMENY, J.G., SNELL, J.L. and KNAPP, A.W.: *Denumerable Markov chains*, Springer, 1976.

[A5] REVUZ, D.: *Markov chains*, North-Holland, 1975.

[A6] ROMANOVSKY, V.I. [V.I. ROMANOVSKIĬ]: *Discrete Markov chains*, Wolters-Noordhoff, 1970 (translated from the Russian).

[A7] SENETA, E.: *Non-negative matrices and Markov chains*, Springer, 1981.

[A8] SPITZER, V.: *Principles of random walk*, v. Nostrand, 1964.

AMS 1980 Subject Classification: 60J27, 60J10

MARKOV CRITERION *for best integral approximation* - A theorem which in some cases enables one to give effectively the polynomial and the error of best integral approximation of a function f. It was established by A.A. Markov in 1898 (see [1]). Let $\{\phi_k(x)\}$, $k=1,\ldots,n$, be a system of linearly independent functions continuous on the interval $[a,b]$, and let the continuous function ψ change sign at the points $x_1 < \cdots < x_r$ in (a,b) and be such that

$$\int_a^b \phi_k(x)\,\mathrm{sgn}\,\psi(x)\,dx = 0, \quad k=1,\ldots,n.$$

If the polynomial

$$P_n^*(x) = \sum_{k=1}^n c_k^* \phi_k(x)$$

has the property that the difference $f - P_n^*$ changes sign at the points $x_1,\ldots,x_r$, and only at those points, then P_n^* is the polynomial of best integral approximation to f and

$$\inf_{\{c_k\}} \int_a^b \left| f(x) - \sum_{k=1}^n c_k \phi_k(x) \right| dx =$$

$$= \int_a^b \left| f(x) - P_n^*(x) \right| dx = \left| \int_a^b f(x)\,\mathrm{sgn}\,\psi(x)\,dx \right|.$$

For the system $\{1, \cos x, \ldots, \cos nx\}$ on $[0, \pi]$, ψ can be taken to be $\cos(n+1)x$; for the system $\{\sin x, \ldots, \sin nx\}$, $0 \leqslant x \leqslant \pi$, ψ can be taken to be $\sin(n+1)x$; and for the system $\{1, x, \ldots, x^n\}$, $-1 \leqslant x \leqslant 1$, one can take $\psi(x) = \sin((n+2)\arccos x)$.

References

[1] MARKOV, A.A.: *Selected works*, Moscow-Leningrad, 1948 (in Russian).

[2] ACHIEZER, N.I. [N.I. AKHIEZER]: *Theory of approximation*, F. Ungar, 1956 (translated from the Russian).

[3] DAUGAVET, I.K.: *Introduction to the theory of approximation of functions*, Leningrad, 1977 (in Russian).

N.P. Korneĭchuk
V.P. Motornyĭ

Editorial comments.

References

[A1] CHENEY, E.W.: *Introduction to approximation theory*, Chelsea, reprint, 1982.

[A2] MÜLLER, M.W.: *Approximationstheorie*, Akad. Verlagsgesellschaft, 1978.

[A3] RICE, J.R.: *The approximation of functions*, I, Addison-Wesley, 1964.

AMS 1980 Subject Classification: 41A50, 42A10

MARKOV FORM - An indefinite **binary quadratic form** $f = f(x,y)$ with Markov constant $\mu(f) < 3$ (see **Markov spectrum problem**).

A.V. Malyshev

AMS 1980 Subject Classification: 10CXX

MARKOV FUNCTION SYSTEM - A system $\{\phi_\nu(x)\}_{\nu=1}^n$ ($n \leqslant \infty$) of linearly independent real-valued continuous functions defined on a finite interval $[a,b]$ and satisfying the condition: For any finite $k \leqslant n$ the functions $\phi_1(x), \ldots, \phi_k(x)$ form a **Chebyshev system**

is, its trajectories can be chosen so). The existence of a continuous Markov process is guaranteed by the condition $\alpha_\epsilon(h) = o(h)$ as $h \downarrow 0$ (see [9], [11]).

In the theory of Markov processes most attention is given to homogeneous (in time) processes. The corresponding definition assumes one is given a system of objects a) - d) with the difference that the parameters s and u may now only take the value 0. Even the notation can be simplified:

$$P_x = P_{0x}, \quad F_t = F_t^0, \quad P(t, x, B) = P(0, x; t, B),$$
$$x \in E, \quad t \geqslant 0, \quad B \in \mathcal{B}.$$

Subsequently, homogeneity of Ω is assumed. That is, it is required that for any $\omega \in \Omega$ and $s \geqslant 0$ there is an $\omega' \in \Omega$ such that $x_t(\omega') = x_{t+s}(\omega)$ for $t \geqslant 0$. Because of this, on the σ-algebra N, the smallest σ-algebra in Ω containing the events $\{\omega: x_s \in B\}$, the time shift operators θ_t are defined, which preserve the operations of union, intersection and difference of sets, and for which

$$\theta_t\{\omega: x_s \in B\} = \{\omega: x_{t+s} \in B\},$$

where $s, t \geqslant 0$, $B \in \mathcal{B}$.

The collection $X(t) = (x_t, F_t, P_x)$ is called a (*non-terminating*) *homogeneous Markov process* given on $(E, \mathcal{B})$ if P_x-almost certainly

$$P_x\{\theta_t \Lambda \,|\, F_t\} = P_{x_t}\{\Lambda\} \tag{4}$$

for $x \in E$, $t \geqslant 0$ and $\Lambda \in N$. The transition function of $X(t)$ is taken to be $P(t, x, B)$, where, unless otherwise indicated, it is required that $P(0, x, \{x\}) \equiv 1$. It is useful to bear in mind that in the verification of (4) it is only necessary to consider sets of the form $\Lambda = \{\omega: x_s \in B\}$, where $s \geqslant 0$, $B \in \mathcal{B}$, and in (4), F_t may always replaced by the σ-algebra $\bar{F}_t$ equal to the intersection of the completions of F_t relative to all possible measures $P_x\{x \in B\}$. Often, one fixes on $\mathcal{B}$ a probability measure μ (the 'initial distribution') and considers a random Markov function (x_t, F_t, P_μ), where P_μ is the measure on F_∞ given by

$$P_\mu\{\cdot\} = \int P_x\{\cdot\} \mu(dx).$$

A Markov process $X(t) = (x_t, F_t, P_x)$ is called *progressively measurable* if for each $t > 0$ the function $x(s, \omega) = x_s(\omega)$ induces a measurable mapping from $([0, t] \times \Omega, \mathcal{B}_t \times F_t)$ to $(E, \mathcal{B})$, where $\mathcal{B}_t$ is the σ-algebra of Borel subsets of $[0, t]$. A right-continuous Markov process is progressively measurable. There is a method for reducing the non-homogeneous case to the homogeneous case (see [11]), and in what follows homogeneous Markov processes will be discussed.

The strong Markov property. Suppose that, on a measurable space $(E, \mathcal{B})$, a Markov process $X(t) = (x_t, F_t, P_x)$ is given. A function $\tau: \Omega \to [0, \infty]$ is called a **Markov moment** (*stopping time*) if $\{\omega:$

$\tau \leqslant t\} \in F_t$ for $t \geqslant 0$. Here a set $\Lambda \subset \Omega_\tau = \{\omega: \tau < \infty\}$ is considered in the family F_τ if $\Lambda \cap \{\omega: \tau < t\} \in F_t$ for $t \geqslant 0$ (most often F_τ is interpreted as the family of events connected with the evolution of $X(t)$ up to time τ). For $\Lambda \in N$, set

$$\theta_\tau \Lambda = \bigcup_{t \geqslant 0} [\theta_t \Lambda \cap \{\omega: \tau = t\}].$$

A progressively-measurable Markov process X is called a strong Markov process if for any Markov moment τ and all $t \geqslant 0$, $x \in E$ and $\Lambda \in N$, the relation

$$P_x\{\theta_\tau \Lambda \,|\, F_\tau\} = P_{x_\tau}\{\Lambda\} \tag{5}$$

(the *strong Markov property*) is satisfied P_x-almost certainly in Ω_τ. In the verification of (5) it suffices to consider only sets of the form $\Lambda = \{\omega: x_s \in B\}$ where $s \geqslant 0$, $B \in \mathcal{B}$; in this case $\theta_\tau \Lambda = \{\omega: x_{s+\tau} \in B\}$. For example, any right-continuous Feller–Markov process on a topological space E is a strong Markov process. A Markov process is called a *Feller–Markov process* if the function

$$P^t f(\cdot) = \int f(y) P(t, \cdot, dy)$$

is continuous whenever f is continuous and bounded.

In the case of strong Markov processes various subclasses have been distinguished. Let the Markov transition function $P(t, x, B)$, given on a locally compact metric space E, be stochastically continuous:

$$\lim_{t \downarrow 0} P(t, x, U) = 1$$

for any neighbourhood U of each point $x \in B$. If P^t maps the class of continuous functions that vanish at infinity into itself, then $P(t, x, B)$ corresponds to a *standard Markov process* X. That is, a right-continuous strong Markov process for which: 1) $F_t = \bar{F}_t$ for $t \in [0, \infty)$ and $F_t = \bigcap_{s > t} F_s$ for $t \in [0, \infty)$; 2) $\lim_{n \to \infty} x_{\tau_n} = x_\tau$, P_x-almost certainly on the set $\{\omega: \tau < \infty\}$, where $\tau = \lim_{n \to \infty} \tau_n$ and τ_n $(n \geqslant 1)$ are Markov moments that are non-decreasing as n increases.

Terminating Markov processes. Frequently, a physical system can be best described using a non-terminating Markov process, but only in a time interval of random length. In addition, even simple transformations of a Markov process may lead to processes with trajectories given on random intervals (see **Functional of a Markov process**). Guided by these considerations one introduces the notion of a terminating Markov process.

Let $\tilde{X}(t) = (\tilde{X}_t, \tilde{F}_t, \tilde{P}_x)$ be a homogeneous Markov process in a phase space $(\tilde{E}, \tilde{\mathcal{B}})$, having a transition function $\tilde{P}(t, x, N)$, and let there be a point $e \in \tilde{E}$ and a function $\zeta: \Omega \to [0, \infty)$ such that $\tilde{x}_t(\omega) = e$ for $\zeta(\omega) \leqslant t$ and $\tilde{x}_t(\omega) \neq e$ otherwise (unless stated otherwise, take $\zeta > 0$). A new trajectory $x_t(\omega)$ is given for $t \in [0, \zeta(\omega))$ by the equality $x_t(\omega) = \tilde{x}_t(\omega)$, and F_t is defined as the trace of $\tilde{F}_t$ on the set $\{\omega: \zeta > t\}$.

The collection $X(t)=(x_t, \zeta, F_t, \tilde{P}_x)$, where $x \in E = \tilde{E} \setminus \{e\}$, is called the *terminating Markov process* obtained from $\tilde{X}(t)$ by censoring (or killing) at the time ζ. The variable ζ is called the *censoring time* or *lifetime* of the terminating Markov process. The phase space of the new process is $(E, \mathscr{B})$, where $\mathscr{B}$ is the trace of the σ-algebra $\tilde{\mathscr{B}}$ in E. The transition function of a terminating Markov process is the restriction of $\tilde{P}(t, x, B)$ to the set $t \geq 0$, $x \in E$, $B \subset \mathscr{B}$. The process $X(t)$ is called a *strong Markov process* or a *standard Markov process* if $\tilde{X}(t)$ has the corresponding property. A non-terminating Markov process can be considered as a terminating Markov process with censoring time $\zeta \equiv \infty$. A non-homogeneous terminating Markov process is defined similarly.

M.G. Shur

Markov processes and differential equations. A Markov process of Brownian-motion type is closely connected with partial differential equations of parabolic type. The transition density $p(s, x, t, y)$ of a diffusion process satisfies, under certain additional assumptions, the backward and forward Kolmogorov equations (cf. **Kolmogorov equation**):

$$\frac{\partial p}{\partial s} + \sum_{k=1}^{n} a_k(s, x) \frac{\partial p}{\partial x_k} + \frac{1}{2} \sum_{k,j=1}^{n} b_{kj}(s, x) \frac{\partial^2 p}{\partial x_k \partial x_j} = \tag{6}$$

$$= \frac{\partial p}{\partial s} + L(s, x) p = 0,$$

$$\frac{\partial p}{\partial t} = -\sum_{k=1}^{n} \frac{\partial}{\partial y_k}(a_k(t, y)p) + \tag{7}$$

$$+ \frac{1}{2} \sum_{k,j=1}^{n} \frac{\partial^2}{\partial y_k \partial y_j}(b_{kj}(t, y)p) = L^*(t, y)p.$$

The function $p(s, x, t, y)$ is the Green's function of the equations (6) - (7), and the first known methods for constructing diffusion processes were based on existence theorems for this function for the partial differential equations (6) - (7). For a time-homogeneous process the operator $L(s, x) = L(x)$ coincides on smooth functions with the infinitesimal operator of the Markov process (see **Transition-operator semi-group**).

The expectations of various functionals of diffusion processes are solutions of boundary value problems for the differential equation (1). Let $E_{s,x}(\cdot)$ be the expectation with respect to the measure $P_{s,x}$. Then the function $E_{s,x}\phi(X(T)) = u_1(s, x)$ satisfies (6) for $s < T$ and $u_1(T, x) = \phi(x)$.

Similarly, the function

$$u_2(s, x) = E_{s,x} \int_s^T f(t, X(t)) \, dt$$

satisfies, for $s < T$,

$$\frac{\partial u_2}{\partial s} + L(s, x) u_2 = -f(s, x),$$

and $u_2(T, x) = 0$.

Let τ be the time at which the trajectories of $X(t)$ first hit the boundary ∂D of a domain $D \subset \mathbf{R}^n$, and let $\tau \wedge T = \min(\tau, T)$. Then, under certain conditions, the function

$$u_3(s, x) = E_{s,x} \int_s^{\tau \wedge T} f(t, X(t)) \, dt + E_{s,x}\phi(\tau \wedge T, X(\tau \wedge T))$$

satisfies

$$\frac{\partial u}{\partial s} + L(s, x) u = -f$$

and takes the value ϕ on the set

$$\Gamma = \{s < T, \ x \in \partial D\} \bigcup \{s = T, \ x \in D\}.$$

The solution of the first boundary value problem for a general second-order linear parabolic equation

$$\left. \begin{array}{r} \dfrac{\partial u}{\partial s} + L(s, x)u + c(s, x)u = -f(s, x), \\[2mm] u\big|_\Gamma = \phi, \end{array} \right\} \tag{8}$$

can, under fairly general assumptions, be described in the form

$$u(s, x) = E_{s,x} \int_s^{\tau \wedge T} \exp\left\{ \int_s^v c(t, X(t)) \, dt \right\} f(v, X(v)) \, dv + \tag{9}$$

$$+ E_{s,x}\left\{ \exp\left\{ \int_s^{\tau \wedge T} c(t, X(t)) \, dt \right\} \phi(\tau \wedge T, X(\tau \wedge T)) \right\}.$$

When the operator L and the functions c and f do not depend on s, a representation similar to (9) is possible also for the solution of a linear elliptic equation. More precisely, the function

$$u(x) = E_x \int_0^\tau \exp\left\{ \int_0^v c(X(t)) \, dt \right\} f(X(v)) \, dv + \tag{10}$$

$$+ E_x\left\{ \exp\left\{ \int_0^\tau c(X(t)) \, dt \right\} \phi(X(\tau)) \right\}$$

is, under certain assumptions, the solution of

$$L(x)u + c(x)u = -f(x), \quad u\big|_{\partial D} = \phi. \tag{11}$$

When L is degenerate ($\det b(s, x) = 0$) or ∂D is not sufficiently 'smooth', the boundary values need not be taken by the functions (9), (10) at individual points or on whole sets. The notion of a **regular boundary point** for L has a probabilistic interpretation. At regular points the boundary values are attained by (9), (10). The solution of (8) and (11) allows one to study the properties of the corresponding diffusion processes and functionals of them.

There are methods for constructing Markov processes which do not rely on the construction of solutions of (6) and (7). For example, the method of stochastic differential equations (cf. **Stochastic differential equation**), of absolutely-continuous change of measure, etc. This situation, together with the formulas (9) and (10), gives a probabilistic route to the construction and study of the properties of boundary value problems for (8) and

103

also to the study of properties of the solutions of the corresponding elliptic equation.

Since the solution of a stochastic differential equation is insensitive to degeneracy of $b(s, x)$, probabilistic methods can be applied to construct solutions of degenerate elliptic and parabolic differential equations. The extension of the averaging principle of N.M. Krylov and N.N. Bogolyubov to stochastic differential equations allows one, with the help of (9), to obtain corresponding results for elliptic and parabolic differential equations. It turns out that certain difficult problems in the investigation of properties of solutions of equations of this type with small parameters in front of the highest derivatives can be solved by probabilistic arguments. Even the solution of the second boundary value problem for (6) has a probabilistic meaning. The formulation of boundary value problems for unbounded domains is closely connected with recurrence in the corresponding diffusion process.

In the case of a time-homogeneous process (L is independent of s), a positive solution of $L^*q = 0$ coincides, under certain assumptions and up to a multiplicative constant, with the stationary density of the distribution of a Markov chain. Probabilistic arguments turn out to be useful even for boundary value problems for non-linear parabolic equations.

R.Z. Khas'minskii

References

[1]   MARKOV, A.A.: *Izv. Fiz.-Mat.Obshch. Kazan. Univ.* **15**, no. 4 (1906), 135-156.

[2]   BACHELIER, L.: *Ann. Sci. Ecole Norm. Sup.* **17** (1900), 21-86.

[3]   KOLMOGOROV, A.N.: 'Ueber die analytischen Methoden in der Wahrscheinlichkeitsrechnung', *Math. Ann.* **104** (1931), 415-458.

[4]   CHUNG, K.L.: *Markov chains with stationary transition probabilities*, Springer, 1967.

[5]   FELLER, W.: 'The general diffusion operator and positivity-preserving semi-groups in one dimension', *Ann. of Math.* **60** (1954), 417-436.

[6]   DYNKIN, E.B. and YUSHKEVICH, A.A.: 'Strong Markov processs', *Theor. Veroyatnost. i Primenen.* **1**, no. 1 (1956), 149-155 (in Russian). English abstract.

[7A]   HUNT, G.A.: 'Markov processes and potentials I', *Illinois J. Math.* **1** (1957), 44-93.

[7B]   HUNT, G.A.: 'Markov processes and potentials II', *Illinois J. Math.* **1** (1957), 316-369.

[7C]   HUNT, G.A.: 'Markov processes and potentials III', *Illinois J. Math.* **2** (1958), 151-213.

[8]   DELLACHERIE, C.: *Capacités et processus stochastiques*, Springer, 1972.

[9]   DYNKIN, E.B.: *Theory of Markov processes*, Pergamon, 1960 (translated from the Russian).

[10]   DYNKIN, E.B.: *Markov processes*, 1, Springer, 1965 (translated from the Russian).

[11]   GIHMAN, I.I. [I.I. GIKHMAN] and SKOROHOD, A.V. [A.V. SKOROKHOD]: *The theory of stochastic processes*, 2, Springer, 1979 (translated from the Russian).

[12]   FREĬDLIN, M.I.: 'Markov processes and differential equations', *Itogi Nauk. Teor. Veroyatnost. Mat. Statist. Teoret. Kibernet. 1966* (1967), 7-58 (in Russian).

[13]   KHAS'MINSKIĬ, R.Z.: 'Principle of averaging for parabolic and elliptic partial differential equations and for Markov processes

with small diffusion', *Theor. Probab. Appl.* **8** (1963), 1-21. (*Teor. Veroyatnost. i Primenen.* **8**, no. 1 (1963), 3-25)

[14]   VENTTSEL', A.D. and FREĬDLIN, M.I.: *Random perturbations of dynamical systems*, Springer, 1984 (translated from the Russian).

[15]   BLUMENTHAL, R.M. and GETOOR, R.K.: *Markov processes and potential theory*, Acad. Press, 1968.

[16]   GETOOR, R.K.: *Markov processes: Ray processes and right processes*, Lecture notes in math., 440, Springer, 1975.

[17]   KUZNETSOV, S.E.: 'Any Markov process in a Borel space has a transition function', *Theor. Prob. Appl.* **25** (1980), 384-388. (*Teor. Veroyatnost. i Primenen.* **25**, no. 2 (1980), 389-393)

Editorial comments.

References

[A1]   STROOCK, D.W. and VARADHAN, S.R.S.: *Multidimensional diffusion processes*, Springer, 1979.

[A2]   CHUNG, K.L.: *Lectures from Markov processes to Brownian motion*, Springer, 1982.

[A3]   DOOB, J.L.: *Stochastic processes*, Wiley, 1953.

[A4]   WENTZELL, A.D.: *A course in the theory of stochastic processes*, McGraw-Hill, 1981.

[A5]   ETHIER, S.N. and KURTZ, T.G.: *Markov processes*, Wiley, 1986.

[A6]   FELLER, W.: *An introduction to probability theory and its applications*, 1-2, Wiley, 1966.

[A7]   WAX, N. (ED.): *Selected papers on noise and stochastic processes*, Dover, 1954.

[A8]   KAC, M.: *Probability and related topics in physical sciences*, Wiley, 1959, Chapt. 4.

[A9]   LÉVY, P.: *Processus stochastiques et mouvement Brownien*, Gauthier-Villars, 1965.

[A10]   LOÈVE, M.: *Probability theory*, II, Springer, 1978.

AMS 1980 Subject Classification: 60J25

MARKOV PROCESS, STATIONARY - A Markov process which is a **stationary stochastic process**. There is a stationary Markov process associated with a homogeneous Markov **transition function** if and only if there is a stationary initial distribution $\mu(A)$ corresponding to this function, that is, $\mu(A)$ satisfies

$$\mu(A) = \int_X P(x, t, A)\mu(dx).$$

If the phase space X is finite, then a stationary initial distribution always exists, independent of whether the process has discrete ($t = 0, 1, \ldots$) or continuous time. For a process in discrete time and for a countable set X, a condition for existence of a stationary distribution has been found by A.N. Kolmogorov [1]: It is necessary and sufficient that there is class of communicating states $Y \subset X$ such that the mathematical expectation of the time for reaching $y_2 \in Y$ from $y_1 \in Y$ is finite for any $y_1 \in Y$. This criterion has been generalized to strong Markov processes with an arbitrary phase space X: For the existence of a stationary process it is sufficient that there is a compact set $K \subset X$ such that the expectation of the time of reaching K from x is finite for all $x \in X$. There is the following sufficient condition for the existence of a stationary Markov process in terms of Lyapunov stochastic functions (cf. **Lyapunov stochastic function**): If there is a function $V(x) \leqslant 0$ for which

$LV(x) \leqslant -1$ for $x \notin K$, then there is a stationary Markov process associated with the Markov transition function $P(x, t, A)$. Here L is the infinitesimal generator of the process.

When the stationary initial distribution μ is unique, the corresponding stationary process is ergodic. In this case the Cesàro mean of the transition probabilities converges weakly to μ. Under certain additional conditions,

$$\lim_{t \to \infty} P(x, t, A) = \mu(A) \quad \text{(weakly)}.$$

A stationary initial distribution satisfies the Fokker$-$Planck($-$Kolmogorov) equation $L^*\mu = 0$, where L^* is the adjoint operator to the infinitesimal operator of the process. For example, L^* is the adjoint operator to the generating differential operator of the process for diffusion processes. In this case μ has a density p with respect to the Lebesgue measure which satisfies $L^*p = 0$. In the one-dimensional case this equation can be solved by quadrature.

References

[1] KOLMOGOROV, A.N.: *Markov chains with a countable number of states*, Moscow, 1937 (in Russian).
[2] DOOB, J.L.: *Stochastic processes*, Wiley, 1953.
[3] SEVAST'YANOV, A.B.: 'An ergodic theorem for Markov processes and its application to telephone systems with refusals', *Theor. Probab. Appl.* **2** (1957), 104-112. (*Teor. Veroyatnost. i Primenen.* **2**, no. 1 (1957), 106-116)

R.Z. Khas'minskiĭ

Editorial comments.

References

[A1] CHUNG, K.L.: *Markov chains with stationary transition probabilities*, Springer, 1960.
[A2] FELLER, W.: *An introduction to probability theory and its applications*, 1-2, Wiley, 1966.
[A3] LÉVY, P.: *Processus stochastiques et mouvement Brownien*, Gauthier-Villars, 1965.
[A4] PARZEN, E.: *Stochastic processes*, Holden-Day, 1962.
[A5] ROZANOV, YU.A.: *Stationary random processes*, Holden-Day, 1967 (translated from the Russian).

AMS 1980 Subject Classification: 60JXX, 60G10

MARKOV PROPERTY for a real-valued **stochastic process** $X(t)$, $t \in T \subset \mathbf{R}$ - The property that for any set $t_1 < \cdots < t_{n+1}$ of times from T and any Borel set B,

$$P\{X(t_{n+1}) \in B \mid X(t_n), \ldots, X(t_1)\} = \qquad (*)$$
$$= P\{X(t_{n+1}) \in B \mid X(t_n)\}$$

with probability 1, that is, the conditional probability distribution of $X(t_{n+1})$ given $X(t_n), \ldots, X(t_1)$ coincides (almost certainly) with the conditional distribution of $X(t_{n+1})$ given $X(t_n)$. This can be interpreted as independence of the 'future' $X(t_{n+1})$ and the 'past' $(X(t_{n-1}), \ldots, X(t_1))$ given the fixed 'present' $X(t_n)$. Stochastic processes satisfying the property $(*)$ are called Markov processes (cf. **Markov process**). The Markov property has (under certain additional assump-

tions) a stronger version, known as the 'strong Markov property'. In discrete time $T = \{1, 2, \ldots\}$ the *strong Markov property*, which is always true for (Markov) sequences satisfying $(*)$, means that for each stopping time τ (relative to the family of σ-algebras $(F_n, n \geqslant 1)$, $F_n = \sigma\{\omega: X(1), \ldots, X(n)\}$), with probability one

$$P\{X(\tau+1) \in B \mid X(\tau), \ldots, X(1)\} =$$
$$= P\{X(\tau+1) \in B \mid X(\tau)\}.$$

References

[1] GIHMAN, I.I. [I.I. GIKHMAN] and SKOROHOD, A.V. [A.V. SKOROKHOD]: *Theory of stochastic processes*, 2, Springer, 1975 (translated from the Russian).

A.N. Shiryaev

Editorial comments.

References

[A1] CHUNG, K.L.: *Markov chains with stationary transition probabilities*, Springer, 1960.
[A2] DOOB, J.L.: *Stochastic processes*, Wiley, 1953.
[A3] DYNKIN, E.B.: *Markov processes*, 1, Springer, 1965.
[A4] ETHIER, S.N. and KURTZ, T.G.: *Markov processes*, Wiley, 1986.
[A5] FELLER, W.: *An introduction to probability theory and its applications*, 1-2, Wiley, 1966.
[A6] LÉVY, P.: *Processus stochastiques et mouvement Brownien*, Gauthier-Villars, 1965.
[A7] LOÈVE, M.: *Probability theory*, II, Springer, 1978.

AMS 1980 Subject Classification: 60JXX

MARKOV QUADRATURE FORMULA - See **Quadrature formula of highest algebraic accuracy**.

AMS 1980 Subject Classification: 65D32

MARKOV SPECTRUM - See **Markov spectrum problem**.

AMS 1980 Subject Classification: 10CXX

MARKOV SPECTRUM PROBLEM - A problem in number theory which arises in connection with the distribution of the normalized values of arithmetic minima of indefinite binary quadratic forms (cf. **Binary quadratic form**). Let

$$f = f(x, y) = \alpha x^2 + \beta xy + \gamma y^2, \quad \alpha, \beta, \gamma \in \mathbf{R},$$
$$\delta(f) = \beta^2 - 4\alpha\gamma > 0,$$

and let

$$m(f) = \inf |f(x, y)|, \quad x, y \in \mathbf{Z}^2, \quad (x, y) \neq (0, 0),$$

be the uniform arithmetic minimum of the form f. The number

$$\mu = \mu(f) = \frac{\sqrt{\delta(f)}}{m(f)}, \quad \mu \leqslant +\infty,$$

is called the *Markov constant* of f. The set $M = \{\mu(f)\}$, where f runs through all real indefinite quadratic forms, is called the *Markov spectrum*. The Markov constant and the Markov spectrum have been defined in various ways; in particular, A.A. Markov in [1] considered the set $\{2/\mu(f)\}$. It is known that $\mu(f)$ is an invariant of a ray F of classes of forms, that is, of a set

$$F = \{f': f' \simeq \tau f\,(\mathbf{Z}),\ \tau \in \mathbf{R},\ \tau > 0\}, \qquad (1)$$

since $\mu(f')=\mu(f)=\mu(F)$. Each ray of classes F is in one-to-one correspondence with a doubly-infinite (infinite in both directions) sequence

$$I_F = \{\ldots, a_{-1}, a_0, a_1, \ldots : a_k \in \mathbf{Z}\},$$

such that if one puts

$$\mu_k(I_F) = [a_k; a_{k+1}, a_{k+2}, \ldots] + [0; a_{k-1}, a_{k-2}, \ldots]$$

$([\,;\ldots\,]$ is the notation for a **continued fraction**), then

$$\mu(F) = \sup_{k \in \mathbf{Z}} \mu_k(I_F).$$

The Markov problem can be stated as follows: 1) describe the Markov spectrum M; and 2) for each $\mu \in M$, describe the set of forms $f=f(x,y)$ (or the rays F) for which $\mu(f)=\mu(F)=\mu$. The problem was solved by Markov for the initial part of the spectrum M defined by the condition $\mu(f)<3$. This part of the spectrum is a discrete set:

$$M \cap [0,3) =$$

$$= \left\{ \sqrt{9 - \frac{4}{m^2}} : m^2 + n^2 + p^2 = 3mnp,\ m,n,p \in \mathbf{N} \right\} =$$

$$= \left\{ \sqrt{5},\ \sqrt{8},\ \frac{\sqrt{221}}{5}, \ldots \right\}$$

with the unique limit point 3 (a condensation point of M); m, n and p run through all positive integer solutions of *Markov's Diophantine equation*

$$m^2 + n^2 + p^2 = 3mnp,\ m \geqslant n \geqslant p > 0. \qquad (2)$$

In this case there corresponds to each point of this part of the spectrum precisely one ray F_m, given by a Markov form $f_m = f_m(x,y)$, with

$$\mu(f_m) = \sqrt{9 - \frac{4}{m^2}}.$$

A solution (m,n,p) of (2) is called a *Markov triple*; the number m is called a *Markov number*. The Markov form f_m is associated to the Markov number $m = \max(m,n,p)$ as follows. Let $r,s \in \mathbf{Z}$ be defined by the conditions

$$nr \equiv p \ (\mathrm{mod}\ m),\ 0 \leqslant r < m,$$

$$r^2 + 1 = ms;$$

then, by definition,

$$f_m = f_m(x,y) = x^2 + \left[3 - \frac{2r}{m}\right]xy + \frac{s-3r}{m}y^2.$$

The set M is closed and there is a smallest number $\mu_0 = 4.5278 \cdots$ such that $[\mu_0, +\infty] \subset M$ and μ_0 borders the interval of contiguity of M.

The Markov problem is closely related to the *Lagrange−Hurwitz problem* on rational approximation

of a real number θ. The quantity

$$\lambda = \lambda(\theta) = \sup \tau,\ \lambda \leqslant +\infty,$$

where the least upper bound is taken over all $\tau \in \mathbf{R}$, $\tau > 0$, for which

$$\left| \theta - \frac{p}{n} \right| \leqslant \frac{1}{\tau q^2}$$

has an infinite set of solutions $p, q \in \mathbf{Z}$, $q > 0$, is called a *Lagrange constant*. The set $L = \{\lambda(\theta): \theta \in \mathbf{R}\}$ is called the *Lagrange spectrum*. It is natural to regard *Lagrange's theorem* as the first result in the theory of the Lagrange spectrum: All convergents of the continued fraction expansion of θ satisfy

$$\left| \theta - \frac{p}{q} \right| < \frac{1}{q^2}.$$

If $\theta' \sim \theta$, that is, if

$$\theta' = \frac{a\theta+b}{c\theta+d},\ a,b,c,d \in \mathbf{Z},\ |\,ad-bc\,| = 1,$$

then $\lambda(\theta')=\lambda(\theta)=\lambda(\Theta)$, where $\Theta = \{\theta': \theta' \sim \theta\}$ is an equivalence class of numbers. If θ is expanded as a continued fraction $\theta=[a_0; a_1, a_2, \ldots]$, then

$$\lambda(\theta) = \limsup_{k \to \infty} \lambda_k(\theta),$$

$$\lambda_k(\theta) = [0; a_{k+1}, a_{k+2}, \ldots] + [a_k; a_{k-1}, \ldots, a_1],$$

$$k = 1, 2, \ldots.$$

Thus, the Lagrange−Hurwitz problem can be stated as: a) describe the Lagrange spectrum L; and b) for each $\lambda \in L$, describe the set of numbers θ (or classes Θ) for which $\lambda(\theta)=\lambda(\Theta)=\lambda$.

For $\lambda(\theta)<3$ this problem reduces to the Markov problem; moreover,

$$L \cap [0,3) = M \cap [0,3),$$

and to each $\lambda \in L$, $\lambda < 3$, corresponds precisely one class Θ, described by the Markov form f_m. It has been proved that L, like M, is a closed set; that $L \subset M$ but $L \neq M$; that

$$L \cap [\mu_0, +\infty] = M \cap [\mu_0, +\infty] = [\mu_0, +\infty],$$

where μ_0 borders the interval of contiguity of L. Research into the structure of L and the connection between L and M is described in [6]. For generalizations and analogues of the Markov spectrum problem and 'isolation phenomena' see [2], [3], [7].

References

[1A] MARKOFF, A. [A.A. MARKOV]: 'Sur les formes quadratiques binaires indéfinies', *Math. Ann.* **15** (1879), 381-406.
[1B] MARKOFF, A. [A.A. MARKOV]: 'Sur les formes quadratiques binaires indéfinies', *Math. Ann.* **17** (1880), 379-400.
[2] CASSELS, J.W.S.: *An introduction to diophantine approximation*, Cambridge Univ. Press, 1957.
[3] DELONE, B.N.: *The Peterburg school of number theory*, Moscow-Leningrad, 1947 (in Russian).
[4] GORSHKOV, D.S.: 'Lobachevskiĭ geometry in connection with some problems of arithmetic', *Zap. Nauchn. Sem. Leningrad Otdel. Mat. Inst. Steklov.* **67** (1977), 39-85 (in Russian).

[5] Freĭman, G.A.: *Diophantine approximation and the geometry of numbers. (The Markov problem)*, Kalinin, 1975 (in Russian).

[6] Malyshev, A.V.: 'Markov and Lagrange spectra (a survey of the literature)', *Zap. Nauchn. Sem. Leningrad. Otdel. Mat. Inst. Steklov.* **67** (1977), 5-38 (in Russian).

[7] Venkov, B.A.: 'On an extremum problem of Markov for indefinite ternaire quadratic forms', *Izv. Akad. Nauk SSSR Ser. Mat.* **9** (1945), 429-494 (in Russian). French summary.

A.V. Malyshev

Editorial comments. In equation (1) in the article above, the notation $f \simeq f'$ (**Z**) refers to *equivalence of binary forms* over **Z**. More precisely, $f \simeq f'$ (**Z**) if and only if there are integers $a, b, c, d \in \mathbf{Z}$, $\det \left(\begin{smallmatrix} a & b \\ c & d \end{smallmatrix} \right) = \pm 1$ such that $f'(x, y) = f(ax + by, cx + dy)$.

The '*interval of contiguity*' of M is simply the maximal interval $[\mu_0, \infty]$ completely belonging to M. The intersections $M \cap [0, 3)$ and $(\mu_0, \infty] \cap M$ have been well-described. The structure of the portion between, i.e. $M \cap [3, \mu_0]$, is still (1989) unclear.

References

[A1] Zagier, D.: 'On the number of Markoff numbers below a given bound', *Math. Comp.* **39** (1982), 709-723.

[A2] Cusick, T.W. and Flahive, M.E.: *The Markoff and Lagrange spectra*, Amer. Math. Soc., 1989.

AMS 1980 Subject Classification: 10CXX

MARTIN BOUNDARY IN POTENTIAL THEORY -

The ideal boundary of a **Green space** Ω (see also **Boundary (in the theory of uniform algebras)**), which allows one to construct the characteristic representation of positive harmonic functions in Ω. Let Ω be a locally compact, non-compact, topological space, and let Φ be a family of continuous functions $f: \Omega \rightarrow [-\infty, +\infty]$. The *Constantinescu−Cornea theorem* [2] asserts that, up to a homeomorphism, there is a unique compact space $\hat{\Omega}$ with the following properties: 1) Ω is an everywhere-dense subspace of $\hat{\Omega}$; 2) each $f \in \Phi$ extends continuously to a function $\hat{f}$ on $\hat{\Omega}$, separating points on the ideal boundary $\Delta = \hat{\Omega} \setminus \Omega$ of Ω relative to Φ; and 3) Ω is an open set in $\hat{\Omega}$.

Now, let Ω be a bounded domain in a Euclidean space $\mathbf{R}^n$, $n \geq 2$, or, more generally, a Green space; let $G = G(x, y)$ be the **Green function** on Ω with pole $y \in \Omega$ and let $y_0 \in \Omega$ be fixed. The *Martin space* or *Martin compactification* $\hat{\Omega}$ of Ω is obtained via the Constantinescu−Cornea theorem by taking for Φ the family

$$\Phi = \left\{ x \in \Omega \rightarrow K(x, y) = \frac{G(x, y)}{G(x, y_0)} : y \in \Omega \right\},$$

where, by definition, $K(x_0, y_0) = 1$. The Martin boundary is the corresponding ideal boundary $\Delta = \hat{\Omega} \setminus \Omega$. The *Martin topology* T is the topology on the Martin space $\hat{\Omega}$. Two Martin spaces $\hat{\Omega}'$, $\hat{\Omega}''$ corresponding to different points $y_0', y_0'' \in \Omega$ are homeomorphic. The function $\hat{K}(\xi, y): \Delta \times \Omega \rightarrow [0, +\infty]$, the extension of $K(x, y)$,

is harmonic in y and jointly continuous in the variables (ξ, y); $\hat{\Omega}$ is a metrizable space. *Martin's fundamental theorem* [1] asserts: The class of all positive harmonic functions $u(y) \geq 0$ on Ω is characterized by the *Martin representation*:

$$u(y) = \int K(\xi, y) d\mu(\xi), \tag{*}$$

where μ is a positive **Radon measure** on Δ. The measure μ in (*) is not uniquely determined by the function u. A harmonic function $v \geq 0$ is called *minimal* in Ω if each harmonic function w such that $0 \leq w \leq v$ in Ω is proportional to v. Minimal harmonic functions $v \neq 0$ are proportional to $\hat{K}(\xi, y)$, the corresponding points $\xi \in \Delta$ are called minimal, and the set of minimal points $\Delta_1 \subset \Delta$ is called the *minimal Martin boundary*. If one poses the additional condition that μ in (*) be concentrated on Δ_1, one obtains the *canonical Martin representation*:

$$u(y) = \int \hat{K}(\xi, y) d\mu_1(\xi),$$

in which the measure $\mu_1 \geq 0$ is uniquely determined by u.

Examples. a) If $\Omega = \{ x \in \mathbf{R}^n : |x| < R \}$ is a ball of radius R in $\mathbf{R}^n$, $n \geq 2$, then

$$\hat{K}(\xi, y) = \frac{R^{n-2}(R^2 - |y|^2)}{|\xi - y|^n}$$

is the Poisson kernel, $\hat{\Omega}$ is the Euclidean closure $\hat{\Omega} = \bar{\Omega}$, the Martin boundary Δ is the sphere $\{ \xi \in \mathbf{R}^n : |\xi| = R \}$, all points of which are minimal. The representation (*) in this case reduces to the *Poisson−Herglotz formula* (see **Integral representation of an analytic function; Poisson integral**).

b) The Martin boundary Δ coincides with the Euclidean boundary $\Gamma = \bar{\Omega} \setminus \Omega$ whenever Γ is a sufficiently smooth hypersurface in $\mathbf{R}^n$, $n \geq 2$.

c) If Ω is a simply-connected domain in the plane, then the Martin boundary Δ coincides with the set of **limit elements**, or Carathéodory prime ends. Thus, an element of the Martin boundary $\xi \in \Delta$ can be considered as a generalization of the notion of a prime end to dimension $n \geq 2$.

References

[1] Martin, R.S.: 'Minimal positive harmonic functions', *Trans. Amer. Math. Soc.* **49** (1941), 137-172.

[2] Constantinescu, C. and Cornea, A.: *Ideale Ränder Riemannscher Flächen*, Springer, p. 1963.

[3] Brelot, M.: *On topologies and boundaries in potential theory*, Springer, 1971.

E.D. Solomentsev

Editorial comments. See also [A1], Chapt. 12, for a concise treatment. For Martin boundaries for the **heat equation** or in probabilistic potential theory, see [A3].

References

[A1] Brelot, M.: *Elements de la théorie classique du potentiel*, Sorbonne Univ., Centre Doc., 1959.

[A2] Brelot, M.: *Axiomatique des fonctions harmoniques*, Presses Univ. de Montréal, 1966.

[A3] DOOB, J.L.: *Classical potential theory and its probabilistic counterpart*, Springer, 1984.

AMS 1980 Subject Classification: 31B25, 31A20, 31D05

MARTIN BOUNDARY IN THE THEORY OF MARKOV PROCESSES

- The boundary of the state space of a **Markov process** or of its image in some compact space, constructed by a scheme similar to the Martin scheme (see [1]).

A probabilistic interpretation of Martin's construction was first proposed by J.L. Doob (see [4]), who discussed the case of discrete Markov chains.

Let $P(t, x, B)$ be the **transition function** of a homogeneous Markov process $X = (x_t, \zeta, F_t, P_x)$, given on a separable, locally compact space E, where $t \geqslant 0$, $x \in E$, $B \in \mathscr{B}$, and $\mathscr{B}$ is the family of Borel sets in E. A function $g_\alpha(x, y) \geqslant 0$ defined for $\alpha \geqslant 0$, $x \in E$, $y \in E$, which is $(\mathscr{B} \times \mathscr{B})$-measurable for fixed α is called a *Green's function* if for each $B \in \mathscr{B}$,

$$\int_B g_\alpha(x, y) m(dy) \equiv \int_0^\infty e^{-\alpha t} P(t, x, B)\, dt,$$

where m is a measure on $\mathscr{B}$. To avoid ambiguity in the definition of a Green's function, it can be required in addition that for any continuous function $f(x)$ with compact support, the function

$$g_\alpha(\cdot) = \int_E f(x) g_\alpha(x, \cdot) m(dx)$$

is Λ-continuous (meaning that there exists a function $F(t, \omega)$ which is left continuous in t and such that

$$P_x\{F(t, \omega) \neq g_\alpha(x_t(\omega))\} \equiv 0, \quad x \in E, \quad t > 0).$$

Fixing a measure γ in $\mathscr{B}$ and postulating the existence of a Green's function, one defines the *Martin kernel*

$$K_y^\alpha(x) = \frac{g_\alpha(x, y)}{q(y)},$$

where

$$q(y) = \int_E g_\alpha(x, y) \gamma(dx)$$

(here some restrictions must be introduced to ensure, in particular, the positivity and Λ-continuity of $q(y)$). If γ is the unit measure concentrated at some point and X is a **Wiener process** terminating at the first exit time for some domain, then the definition of $K_y^0(x)$ reduces to an analogous form [1]. Under broad conditions one can establish the existence of a compact set $\mathscr{E}$ (the 'Martin compactum'), a measure $K_y^\alpha(dx)$ on $\mathscr{B}$ ($\alpha \geqslant 0$, $y \in \mathscr{E}$) and a mapping $i: E \to \mathscr{E}$ for which: a) $i(E)$ is dense in $\mathscr{E}$; b) the function

$$K_y^\alpha(f) = \int_\mathscr{B} f(x) K_y^\alpha(dx)$$

separates points and is continuous on $\mathscr{E}$ as f runs through all continuous function in E with compact sup-

port; and c) the measure $K_{i(y)}^\alpha(dx)$ coincides with $K_y^\alpha(x) m(dx)$ if $y \in E$. The boundary of the set $i(E)$ in $\mathscr{E}$ is called the *Martin boundary* or *exit-boundary* (in the study of decompositions of excessive measures the dual object, the entrance-boundary, arises; see [3], [4]).

In order to describe the properties of $\mathscr{E}$ it is convenient to invoke h-processes in the sense of Doob: to each **excessive function** h is associated the transition function

$$P^h(t, x, B) = h^{-1}(x) \int_E h(y) P(t, x, dy)$$

on $(E^h, \mathscr{B}^h)$, where $E^h = \{x \in E: \ 0 < h(x) < \infty\}$ and $\mathscr{B}^h = \{A \in B: \ A \subset E^h\}$; the corresponding Markov process is an h-process. All h-processes can be realized, together with X, on the space of elementary events, so that they are distinguished only by the families of measures $\{P_x^h\}$. One constructs in $\mathscr{E}$ a modification of x_t, a left-continuous process z_t ($0 < t \leqslant \zeta$) for which $P_x^h\{z_t \neq i(x_t)\} \equiv 0$ if $h \in L_1(\gamma)$. In the topology of $\mathscr{E}$ the limit $z_\zeta = \lim_{t \uparrow \zeta} z_t$ exists almost certainly.

There is a set $U \subset \mathscr{E}$ (the 'exit space') such that: first, $P_x^h\{z_\zeta \in U\} \equiv 1$ for all $h(x)$ of the above form; secondly, the measure K_y^α for $y \in U$ has a density $k_y^\alpha(\cdot)$ with respect to m, where one can take for $k_y^0(\cdot)$ an excessive function whose spectral measure is the unit measure concentrated at y; and thirdly, $h(x)$ admits a unique integral decomposition of the form

$$h(x) = \int_U k_y^0(x) \mu(dx).$$

The measure μ in the decomposition is called the *spectral measure of the function* h; it is given by the formula

$$\mu(B) = \int_E h(x) P_x^h\{z_\zeta \in B\} \gamma(dx),$$

where B is a Borel set in $\mathscr{E}$.

In the theory of Markov processes other types of compactifications are also used, particularly those in which any function of the form

$$\int_0^\infty e^{-\alpha t}\, dt \int_E f(y) P(t, x, dy), \quad \alpha > 0, \quad x \in E,$$

has a continuous extension for a sufficiently general set of functions f.

References
[1] MARTIN, R.S.: 'Minimal positive harmonic functions', *Trans. Amer. Math. Soc.* **49** (1941), 137-172.

[2] MOTOO, M.: 'Application of additive functionals to the boundary problem of Markov processes. Lévy's system of U-processes', in *Proc. fifth Berkeley Symp. Math. Stat. Probab.*, Vol. 2, 1967, pp. 75-110.

[3] KUNITA, H. and WATANABE, T.: 'Some theorems concerning resolvents over locally compact spaces', in *Proc. fifth Berkeley Symp. Math. Stat. Probab.*, Vol. 2, 1967, pp. 131-164.

[4] DOOB, J.L.: 'Discrete potential theory and boundaries', *J. Math. and Mech.* **8**, no. 3 (1959), 433-458; 993.

[5] WATANABE, T.: 'On the theory of Martin boundaries induced by countable Markov processes', *Mem. Coll. Sci. Kyoto Univ. Ser. A* **33**, no. 1 (1960), 39-108.

[6] HUNT, G.A.: 'Markov chains and Markov boundaries', *Illinois J. Math.* **4** (1960), 313-340.

[7] HENNEQUIN, P.L. and TORTRAT, A.: *Théorie des probabilites et quelques applications*, Masson, 1965.

[8] KUNITA, H. and WATANABE, T.: 'Markov processes and Martin boundaries I', *Illinois J. Math.* **9**, no. 3 (1965), 485-526.

[9] SHUR, M.G.: *Trudy Moskov. Inst. Elektron. Mashinostr.* **5** (1970), 192-251.

[10] JEULIN, T.: 'Compactification de Martin d'un processus droit', *Z. Wahrsch. Verw. Gebiete* **42**, no. 3 (1978), 229-260.

[11] DYNKIN, E.B.: 'Boundary theory of Markov processes (the discrete case)', *Russian Math. Surveys* **24**, no. 2 (1969), 1-42. (*Uspekhi Mat. Nauk* **24**, no. 4 (1969), 89-152)

M.G. Shur

Editorial comments. One of the other types of compactifications used in the theory of Markov processes is the **Ray−Knight compactification**.

References

[A1] DOOB, J.L.: *Classical potential theory and its probabilistic counterpart*, Springer, 1984.

AMS 1980 Subject Classification: 60J50

MARTINGALE - A stochastic process $X=(X_t, \mathscr{F}_t)$, $t \in T \subseteq [0, \infty)$, defined on a probability space $(\Omega, \mathscr{F}, \mathrm{P})$ with a non-decreasing family of σ-algebras $(\mathscr{F}_t)_{t \in T}$, $\mathscr{F}_s \subseteq \mathscr{F}_t \subseteq \mathscr{F}$, $s \leqslant t$, such that $\mathrm{E}|X_t| < \infty$, X_t is $\mathscr{F}_t$-measurable and

$$\mathrm{E}(X_t | \mathscr{F}_s) = X_s \qquad (1)$$

(with probability 1). In the case of discrete time $T = \{1, 2, \ldots\}$; in the case of continuous time $T = [0, \infty)$. Related notions are stochastic processes which form a *submartingale*, if

$$\mathrm{E}(X_t | \mathscr{F}_s) \geqslant X_s,$$

or a *supermartingale*, if

$$\mathrm{E}(X_t | \mathscr{F}_s) \leqslant X_s.$$

Example 1. If $\xi_1, \xi_2, \ldots$, is a sequence of independent random variables with $\mathrm{E}\xi_j = 0$, then $X = (X_n, \mathscr{F}_n)$, $n \geqslant 1$, with $X_n = \xi_1 + \cdots + \xi_n$ and $\mathscr{F}_n = \sigma\{\xi_1, \ldots, \xi_n\}$ the σ-algebra generated by $\xi_1, \ldots, \xi_n$, is a martingale.

Example 2. Let $Y = (Y_n, \mathscr{F}_n)$ be a martingale (submartingale), $V = (V_n, \mathscr{F}_n)$ a predictable sequence (that is, V_n is not only $\mathscr{F}_n$-measurable but also $\mathscr{F}_{n-1}$-measurable, $n \geqslant 1$), $\mathscr{F}_0 = \{\varnothing, \Omega\}$, and let

$$(V \cdot Y)_n = V_1 Y_1 + \sum_{k=2}^{n} V_k \Delta Y_k, \quad \Delta Y_k = Y_k - Y_{k-1}.$$

Then, if the variables $(V \cdot Y)_n$ are integrable, the stochastic process $((V \cdot Y)_n, \mathscr{F}_n)$ forms a martingale (submartingale). In particular, if $\xi_1, \xi_2, \ldots$, is a sequence of independent random variables corresponding to a Bernoulli scheme

$$\mathrm{P}\{\xi_i = \pm 1\} = \frac{1}{2}, \quad Y_k = \xi_1 + \cdots + \xi_k,$$

$$\mathscr{F}_k = \sigma\{\xi_1, \ldots, \xi_k\},$$

and

$$V_k = \begin{cases} 2 & \text{if } \xi_1 = \cdots = \xi_{k-1} = 1, \\ 0 & \text{otherwise,} \end{cases} \qquad (2)$$

then $((V \cdot Y)_n, \mathscr{F}_n)$ is a martingale. This stochastic process is a mathematical model of a game in which a player wins one unit of capital if $\xi_k = +1$ and loses one unit of capital if $\xi_k = -1$, and V_k is the stake at the k-th game. The game-theoretic sense of the function V_k defined by (2) is that the player doubles his stake when he loses and stops the game on his first win. In the gambling world such a system is called a martingale, which explains the origin of the mathematical term 'martingale'.

One of the basic facts of the theory of martingales is that the structure of a martingale (submartingale) $X = (X_t, \mathscr{F}_t)$ is preserved under a random change of time. A precise statement of this (called the *optimal sampling theorem*) is the following: If τ_1 and τ_2 are two finite stopping times (cf. **Markov moment**), if $\mathrm{P}\{\tau_1 \leqslant \tau_2\} = 1$ and if

$$\mathrm{E}|X_{\tau_i}| < \infty, \quad \liminf_t \int_{\{\tau_i > t\}} |X_t| \, d\mathrm{P} = 0, \qquad (3)$$

then $\mathrm{E}(X_{\tau_2} | \mathscr{F}_{\tau_1})(\geqslant) = X_{\tau_1}$ (with probability 1), where

$$\mathscr{F}_{\tau_1} = \{A \in \mathscr{F}: A \cap \{\tau_1 \leqslant t\} \in \mathscr{F}_t \text{ for all } t \in T\}.$$

As a particular case of this the **Wald identity** follows:

$$\mathrm{E}(\xi_1 + \cdots + \xi_\tau) = \mathrm{E}\xi_1 \mathrm{E}\tau.$$

Among the basic results of the theory of martingales is *Doob's inequality*: If $X = (X_n, \mathscr{F}_n)$ is a non-negative submartingale,

$$X_n^* = \max_{1 \leqslant j \leqslant n} X_j,$$

$$\|X_n\|_p = (\mathrm{E}|X_n|^p)^{1/p}, \quad p \geqslant 1, \quad n \geqslant 1,$$

then

$$\mathrm{P}\{X_n^* \geqslant \epsilon\} \leqslant \frac{\mathrm{E}X_n}{\epsilon}, \qquad (4)$$

$$\|X_n\|_p \leqslant \|X_n^*\|_p \leqslant \frac{p}{p-1}\|X_n\|_p, \quad p > 1, \qquad (5)$$

$$\|X_n^*\|_p \leqslant \frac{e}{e-1}[1 + \|X_n \ln^+ X_n\|_p], \quad p = 1. \qquad (6)$$

If $X = (X_n, \mathscr{F}_n)$ is a martingale, then for $p > 1$ the *Burkholder inequalities* hold (generalizations of the inequalities of Khinchin and Marcinkiewicz−Zygmund for sums of independent random variables):

$$A_p \|\sqrt{[X]_n}\|_p \leqslant \|X_n\|_p \leqslant B_p \|\sqrt{[X]_n}\|_p, \qquad (7)$$

where A_p and B_p are certain universal constants (not depending on X or n), for which one can take

$$A_p = \left[\frac{18p^{3/2}}{p-1}\right]^{-1}, \quad B_p = \frac{18p^{3/2}}{(p-1)^{1/2}},$$

and

$$[X]_n = \sum_{i=1}^{n}(\Delta X_i)^2, \quad X_0 = 0.$$

Taking (5) and (7) into account, it follows that

$$A_p \|\sqrt{[X]_n}\|_p \leqslant \|X_n^*\|_p \leqslant \tilde{B}_p \|\sqrt{[X]_n}\|_p, \qquad (8)$$

where

$$\tilde{B}_p = \frac{18p^{5/2}}{(p-1)^{3/2}}.$$

When $p=1$ inequality (8) can be generalized. Namely, *Davis' inequality* holds: There are universal constants A and B such that

$$A \, \| \sqrt{[X]_n} \, \|_1 \leq \| X_n^* \|_1 \leq B \, \| \sqrt{[X]_n} \, \|_1.$$

In the proof of a different kind of theorem on the convergence of submartingales with probability 1, a key role is played by *Doob's inequality for the mathematical expectation* $\mathsf{E}\beta_n(a, b)$ of the number of upcrossings, $\beta_n(a, b)$, of the interval $[a, b]$ by the submartingale $X=(X_n, \mathscr{F}_n)$ in n steps; namely

$$\mathsf{E}\beta_n(a, b) \leq \frac{\mathsf{E}|X_n|+|a|}{b-a}. \qquad (9)$$

The basic result on the convergence of submartingales is *Doob's theorem*: If $X=(X_n, \mathscr{F}_n)$ is a submartingale and $\sup \mathsf{E}|X_n| < \infty$, then with probability 1, $\lim_{n\to\infty} X_n (=X_\infty)$ exists and $\mathsf{E}|X_\infty| < \infty$. If the submartingale X is uniformly integrable, then, in addition to convergence with probability 1, convergence in L_1 holds, that is,

$$\mathsf{E}|X_n - X_\infty| \to 0, \quad n\to\infty.$$

A corollary of this result is *Lévy's theorem on the continuity of conditional mathematical expectations*: If $\mathsf{E}|\xi| < \infty$, then

$$\mathsf{E}(\xi | \mathscr{F}_n) \to \mathsf{E}(\xi | \mathscr{F}_\infty),$$

where $\mathscr{F}_1 \subseteq \mathscr{F}_2 \subseteq \cdots$ and $\mathscr{F}_\infty = \sigma(\bigcup_n \mathscr{F}_n)$.

A natural generalization of a martingale is the concept of a *local martingale*, that is, a stochastic process $X=(X_t, \mathscr{F}_t)$ for which there is a sequence $(\tau_m)_{m \geq 1}$ of finite stopping times $\tau_m \uparrow \infty$ (with probability 1), $m \geq 1$, such that for each $m \geq 1$ the 'stopped' processes

$$X^{\tau_m} = (X_{t \wedge \tau_m} I(\tau_m > 0), \mathscr{F}_t)$$

are martingales. In the case of discrete time each local martingale $X=(X_n, \mathscr{F}_n)$ is a *martingale transform*, that is, can be represented in the form $X_n=(V \cdot Y)_n$, where V is a predictable process and Y is a martingale.

Each submartingale $X=(X_t, \mathscr{F}_t)$ has, moreover, a unique Doob$-$Meyer decomposition $X_t=M_t+A_t$, where $M=(M_t, \mathscr{F}_t)$ is a local martingale and $A=(A_t, \mathscr{F}_t)$ is a predictable non-decreasing process. In particular, if $m=(m_t, \mathscr{F}_t)$ is a square-integrable martingale, then its square $m^2=(m_t^2, \mathscr{F}_t)$ is a submartingale in whose Doob$-$Meyer decomposition $m_t^2=M_t+<m>_t$ the process $<m>=(<m>_t, \mathscr{F}_t)$ is called the *quadratic characteristic of the martingale m*. For each square-integrable martingale m and predictable process $V=(V_t, \mathscr{F}_t)$ such that $\int_0^t V_s^2 \, d<m>_s < \infty$ (with probability 1), $t>0$, it is possible to define the **stochastic integral**

$$(V \cdot m)_t = \int_0^t V_s \, dm_s,$$

which is a local martingale. In the case of a Wiener process $W=(W_t, \mathscr{F}_t)$, which is a square-integrable martingale, $<m>_t=t$ and the stochastic integral $(V \cdot W)_t$ is none other than the Itô stochastic integral with respect to the Wiener process.

In the case of continuous time the Doob, Burkholder and Davis inequalities are still true (for right-continuous processes having left limits).

References

[1] DOOB, J.L.: *Stochastic processes*, Chapman and Hall, 1953.
[2] GIKHMAN, I.I. and SKOROKHOD, A.V.: *The theory of stochastic processes*, 1, Springer, 1974 (translated from the Russian).

A.N. Shiryaev

Editorial comments. Stopping times are also called *optimal times*, or, in the older literature, *Markov times* or *Markov moments*, cf. **Markov moment**. The optimal sampling theorem is also called the *stopping theorem* or *Doob's stopping theorem*.

The notion of a martingale is one of the most important concepts in modern probability theory. It is basic in the theories of Markov processes and stochastic integrals, and is useful in many parts of analysis (convergence theorems in ergodic theory, derivatives and lifting in measure theory, inequalities in the theory of singular integrals, etc.). More generally one can define martingales with values in C, R^n, a Hilbert or a Banach space; Banach-valued martingales are used in the study of Banach spaces (Radon$-$Nikodým property, etc.).

References

[A1] DELLACHERIE, C. and MEYER, P.A.: *Probabilities and potential*, 1-3, North-Holland, 1978-1988, Chapts. V-VIII. Theory of martingales (translated from the French).
[A2] DOOB, J.L.: *Classical potential theory and its probabilistic counterpart*, Springer, 1984.
[A3] NEVEU, J.: *Discrete-parameter martingales*, North-Holland, 1975 (translated from the French).
[A4] VILLE, J.: *Etude critique de la notion de collectif*, Gauthier-Villars, 1939.
[A5] WALL, P. and HEYDE, C.C.: *Martingale limit theory and its application*, Acad. Press, 1980.

AMS 1980 Subject Classification: 60G42, 60G44, 60G46, 60G48

MASS - A physical quantity determining the inertial and gravitational properties of matter. In classical mechanics *inertial mass* is the coefficient of proportionality between force and acceleration in Newton's second law, which is a constant for a given body (cf. **Newton laws of mechanics**). The *gravitational mass* is defined as the coefficient of proportionality in the law of universal gravitation. According to the principle of equivalence, inertial and gravitational mass are proportional to each other and, in the usual system of units, are equal. In classical physics mass is additive: the mass of a system is equal to the sum of the masses of its parts. In the theory of special relativity mass (the so-called *rest mass*) can be defined as the coefficient of proportionality in the formula connecting the momentum p of a body and its velocity v,

$$p = \frac{mv}{\sqrt{1-v^2/c^2}},$$

where c is the velocity of light in vacuum. Sometimes one introduces the quantity

$$m_{\text{mot}} = \frac{m}{\sqrt{1-v^2/c^2}},$$

called the *motion mass* of the body. By this definition, in relativity theory the momentum and velocity are related by the classical formula $p=m_{\text{mot}} \cdot v$, where the mass depends on the velocity. The mass of a body is related to its energy E by the relation $E=m_{\text{mot}}c^2$. In relativity theory mass is not additive: the mass of a stable system is less than the sum of the masses of its parts by the quantity $\Delta m = \Delta E / c^2$, where ΔE is the binding energy of the system, equal to the energy produced on the formation of the system. The quantity Δm is called the *mass defect*. The rest masses of all known physical bodies are non-negative (the rest mass of a photon, for example, is equal to zero).

D.D. Sokolov

Editorial comments.

References

[A1] JAMMER, M.: *Concept of mass in classical and modern physics*, Harper & Row, 1964.

[A2] WEYL, H.: *Philosophy of mathematics and natural science*, Princeton Univ. Press, 1949.

AMS 1980 Subject Classification: 70D05, 83A05, 70H40

MASS AND CO-MASS - Adjoint norms (cf. **Norm**) in certain vector spaces dual to each other.

1) The *mass of an r-vector* α, i.e. an element of the r-fold exterior product of a vector space, is the number

$$|\alpha|_0 = \inf \left\{ \sum_i |\alpha_i| : \alpha = \sum \alpha_i,\ \alpha_i \text{ simple } r-\text{vectors} \right\}.$$

The *co-mass of an r-covector* ω is the number

$$|\omega|_0 = \sup_\alpha \{ |\omega \cdot \alpha| : \alpha \text{ a simple } r-\text{vector},\ |\alpha| = 1 \}.$$

Here $|\cdot|$ is the standard norm of an r-vector and $\omega \cdot \alpha$ is the scalar product of a vector and a covector.

The mass $|\alpha|_0$ and the co-mass $|\omega|_0$ are adjoint norms in the spaces of r-vectors $V_{[r]}$ and r-covectors $V^{[r]}$, respectively. In this connection:

a) $|\omega|_0 = \sup_\alpha \{ |\omega \cdot \alpha| : |\alpha|_0 = 1 \}$, $|\alpha|_0 = \sup_\alpha \{ |\omega \cdot \alpha| : |\omega|_0 = 1 \}$;

b) $|\alpha|_0 \geq |\alpha|$, $|\omega|_0 \geq |\omega|$, and equalities hold if and only if α (ω) is a simple r-(co)vector;

c) $|\alpha \vee \beta|_0 \leq |\alpha|_0 |\beta|_0$, $|\omega \vee \zeta|_0 \leq B |\omega|_0 |\zeta|_0$ for exterior products $\vee$, where for a simple multicovector ω (or ζ) $B=1$, and, in general, $B=\binom{r+s}{r}$ if $\omega \in V^{[r]}$ and $\zeta \in V^{[s]}$;

d) $|\omega \wedge \alpha|_0 \leq \tilde{B} |\omega|_0 |\alpha|_0$ for inner products $\wedge$, where $\tilde{B}=1$ for $r \geq s$ and $\tilde{B}=\binom{s}{r}$ for $r \leq s$, $\omega \in V^{[r]}$ and $\alpha \in V_{[s]}$.

These definitions enable one to define the mass and co-mass for sections of fibre bundles whose standard fibres are $V^{[r]}$ and $V_{[r]}$. For example, the *co-mass of a form* ω on a domain $G \subset E^n$ is

$$|\omega|_0 = \sup \{ |\omega(p)|_0 : p \in G \}.$$

2) The *mass of a polyhedral chain* $A = \sum a_i \sigma_i^r$ is

$$|A| = \sum |a_i| \, |\sigma_i^r|,$$

where $|\sigma_i^r|$ is the volume of the cell σ_i^r. For arbitrary chains the mass (finite or infinite) can be defined in various ways; for flat chains (see **Flat norm**) and sharp chains (see **Sharp norm**) these give the same value to the mass.

3) The *co-mass of a* (flat, in particular, sharp) *cochain* X is defined in the standard way:

$$|X| = \sup_{A \neq 0} \frac{|X \cdot A|}{|A|},$$

where A is a polyhedral chain and $X \cdot A$ is the value of the cochain X on the chain A.

For references see **Flat norm**.

M.I. Voĭtsekhovskiĭ

Editorial comments. A *simple r-vector* α is an element of the form $\alpha = \beta_1 \vee \cdots \vee \beta_r$ in the r-fold **exterior product** $V_{[r]}$ of a **vector space** V. Here '$\vee$' denotes exterior product and $\beta_1, \ldots, \beta_r \in V$.

References

[A1] FEDERER, H.: *Geometric measure theory*, Springer, 1969, Sect. 1.8.

AMS 1980 Subject Classification: 46A20, 55U15, 28A75

MASS OPERATOR, *operator of mass* - The operator taking account of the interaction of a particle with its own field and other fields. Let the state of a system be described by the quantity

$$\Psi(x) = \Psi_0(x)\psi(x),$$

where $\psi(x)$ is the field operator acting on the wave function Ψ_0 (the state vector) and x is a four-dimensional coordinate vector. If $\Psi(x)$ satisfies the equation

$$[L(x)+M(x)]\Psi(x) = 0, \qquad (*)$$

where the operator $L(x)$ corresponds to a free particle and $M(x)$ accounts for its interaction with the particle's own field and other fields, then $M(x)$ is called the *mass operator*. The mass operator is an integral operator with kernel $M(x, x')$:

$$M(x) = \Psi(x) = \int M(x, x')\Psi(x')dx'.$$

The mass operator is closely related to the one-particle **Green function** $G(x, x')$, which is a solution of an equation similar to (*) but with a δ-function source on the right-hand side:

$$[L(x)+M(x)]G(x, x') = \delta(x-x'),$$

where $\delta(x-x')$ is the four-dimensional **delta-function**.

References

[1] BOGOLYUBOV, N.N. and SHIRKOV, D.V.: *Introduction to the theory of quantized fields*, Interscience, 1959 (translated from the Russian).
[2] ABRIKOSOV, A.A., GOR'KOV, L.P. and DZYALOSHINSKIĬ, I.E.: *Methods of quantum field theory in statistical physics*, Moscow, 1962 (in Russian).

A.B. Ivanov

Editorial comments. The concept of a 'mass operator' can only be given some sense in the context of quantum field perturbation theory, and plays a minor role in that context.

AMS 1980 Subject Classification: 70-XX, 81EXX

MATHEMATICAL ANALYSIS - The part of mathematics in which functions (cf. **Function**) and their generalizations are studied by the method of limits (cf. **Limit**). The concept of limit is closely connected with that of an infinitesimal quantity, therefore it could be said that mathematical analysis studies functions and their generalizations by infinitesimal methods.

The name 'mathematical analysis' is a short version of the old name of this part of mathematics, 'infinitesimal analysis'; the latter more fully describes the content, but even it is an abbreviation (the name 'analysis by means of infinitesimals' would characterize the subject more precisely). In classical mathematical analysis the objects of study (analysis) were first and foremost functions. 'First and foremost' because the development of mathematical analysis has led to the possibility of studying, by its methods, forms more complicated than functions: functionals, operators, etc.

Everywhere in nature and technology one meets motions and processes which are characterized by functions; the laws of natural phenomena also are usually described by functions. Hence the objective importance of mathematical analysis as a means of studying functions.

Mathematical analysis, in the broad sense of the term, includes a very large part of mathematics. It includes **differential calculus**; **integral calculus**; the theory of functions of a real variable (cf. **Functions of a real variable, theory of**); the theory of functions of a complex variable (cf. **Functions of a complex variable, theory of**); **approximation theory**; the theory of ordinary differential equations (cf. **Differential equation, ordinary**); the theory of partial differential equations (cf. **Differential equation, partial**); the theory of integral equations (cf. **Integral equation**); **differential geometry**; **variational calculus**; **functional analysis**; **harmonic analysis**; and certain other mathematical disciplines. Modern **number theory** and **probability theory** use and develop methods of mathematical analysis.

Nevertheless, the term 'mathematical analysis' is often used as a name for the foundations of mathematical analysis, which unifies the theory of real numbers (cf. **Real number**), the theory of limits, the theory of **series**, differential and integral calculus, and their immediate applications such as the theory of maxima and minima, the theory of implicit functions (cf. **Implicit function**), **Fourier series**, and Fourier integrals (cf. **Fourier integral**).

Functions. Mathematical analysis began with the definition of a function by N.I. Lobachevskiĭ and P.G.L. Dirichlet. If to each number x, from some set F of numbers, is associated by some rule a number y, then this defines a function

$$y = f(x)$$

of one variable x. A function of n variables,

$$f(x) = f(x_1, \ldots, x_n),$$

is defined similarly, where $x=(x_1, \ldots, x_n)$ is a point of an n-dimensional space; one also considers functions

$$f(x) = (x_1, x_2, \ldots)$$

of points $x=(x_1, x_2, \ldots)$ of some infinite-dimensional space. These, however, are usually called functionals.

Elementary functions. In mathematical analysis the **elementary functions** are of fundamental importance. Basically, in practice, one operates with the elementary functions and more complicated functions are approximated by them. The elementary functions can be considered not only for real but also for complex x; then the conception of these functions becomes in some sense, complete. In this connection an important branch of mathematics has arisen, called the theory of functions of a complex variable, or the theory of analytic functions (cf. **Analytic function**).

Real numbers. The concept of a function is essentially founded on the concept of a real (rational or irrational) number. The latter was finally formulated only at the end of the 19-th century. In particular, it established a logically irreproachable connection between numbers and points of a geometrical line, which gave a formal foundation for the ideas of R. Descartes (mid 17-th century), who introduced into mathematics rectangular coordinate systems and the representation of functions by graphs.

Limits. In mathematical analysis a means of studying functions is the limit. One distinguishes between the limit of a sequence and the limit of a function. These concepts were finally formulated only in the 19-th century; however, the idea of a limit had been studied by the ancient Greeks. It suffices to say that Archimedes (3-rd century B.C.) was able to calculate the area of a segment of a parabola by a process which one would call a limit transition (see **Exhaustion, method of**).

Continuous functions. An important class of functions studied in mathematical analysis is formed by the continuous functions (cf. **Continuous function**). One of the possible definitions of this notion is: A function $y=f(x)$, of a variable x from an open interval (a, b), is called *continuous at the point x* if

$$\lim_{\Delta x \to 0} \Delta y = \lim_{\Delta x \to 0} [f(x+\Delta x)-f(x)] = 0.$$

A function is *continuous on the open interval* (a, b) if it is continuous at each of its points; its graph is then a curve which is continuous in the everyday sense of the word.

Derivative and differential. Among the continuous functions those having a **derivative** must be distinguished. The derivative of a function

$$y = f(x), \quad a<x<b,$$

at a point x is its rate of change at that point, that is, the limit

$$\lim_{\Delta x \to 0} \frac{\Delta y}{\Delta x} = \lim_{\Delta x \to 0} \frac{f(x+\Delta x)-f(x)}{\Delta x} = f'(x). \tag{1}$$

If y is the coordinate at the time x of a point moving along the coordinate axis, then $f'(x)$ is its instantaneous velocity at the time x.

From the sign of f' one can judge the nature of variation of f: If $f'>0$ ($f'<0$) in an interval (c, d), then f is increasing (decreasing) on this interval. If a function attains a local extremum (a maximum or a minimum) at x and has a derivative at this point, then the latter is equal to zero, $f'(x)=0$.

The equality (1) can be replaced by the equivalent equality

$$\frac{\Delta y}{\Delta x} = f'(x)+\epsilon(\Delta x), \quad \epsilon(\Delta x) \to 0 \text{ as } \Delta x \to 0,$$

or

$$\Delta y = f'(x)\Delta x + \Delta x \epsilon(\Delta x),$$

where $\epsilon(\Delta x)$ is an infinitesimal as $\Delta x \to 0$; that is, if f has a derivative at x, then its increment at this point decomposes into two terms. The first

$$dy = f'(x)\Delta x \tag{2}$$

is a linear function of Δx (is proportional to Δx), the second term tends to zero more rapidly than Δx.

The quantity (2) is called the **differential** of the function corresponding to the increment Δx. For small Δx it is possible to regard Δy as approximately equal to dy:

$$\Delta y \approx dy.$$

These arguments about differentials are characteristic of mathematical analysis. They have been extended to functions of several variables and to functionals.

For example, if a function

$$z = f(x_1, \ldots, x_n) = f(x)$$

of n variables has continuous partial derivatives (cf. **Partial derivative**) at a point $x=(x_1, \ldots, x_n)$, then its

increment Δz corresponding to increments $\Delta x_1, \ldots, \Delta x_n$ of the independent variables can be written in the form

$$\Delta z = \sum_{k=1}^{n} \frac{\partial f}{\partial x_k}\Delta x_k + \sqrt{\sum_{k=1}^{n} \Delta x_k^2}\; \epsilon(\Delta x), \tag{3}$$

where $\epsilon(\Delta x) \to 0$ as $\Delta x=(\Delta x_1, \ldots, \Delta x_n) \to 0$, that is, if all $\Delta x_k \to 0$. Here the first term on the right-hand side in (3) is the differential dz of f. It depends linearly on Δx and the second term tends to zero more rapidly than Δx as $\Delta x \to 0$.

Suppose one is given a functional (see **Variational calculus**)

$$J(x) = \int_{t_0}^{t_1} L(t, x, x')\,dt,$$

extended over the class $\mathfrak{M}$ of functions x having continuous derivatives on the closed interval $[t_0, t_1]$ and satisfying the boundary conditions $x(t_0)=x_0$, $x(t_1)=x_1$, where x_0 and x_1 are given numbers. Let, further, $\mathfrak{M}_0$ be the class of functions h having continuous derivatives on $[t_0, t_1]$ and such that $h(t_0)=h(t_1)=0$. Obviously, if $x \in \mathfrak{M}$ and $h \in \mathfrak{M}_0$, then $x+h \in \mathfrak{M}$.

In variational calculus it has been proved that under certain conditions on L the increment of $J(x)$ can be written in the form

$$J(x+h)-J(x) = \int_{t_0}^{t_1} \left[\frac{\partial L}{\partial x} - \frac{d}{dt}\left[\frac{\partial L}{\partial x'}\right] \right] h(t)\,dt + o(\|h\|) \tag{4}$$

as $\|h\| \to 0$, where

$$\|h\| = \max_{t_0 \leq t \leq t_1} |h(t)| + \max_{t_0 \leq t \leq t_1} |h'(t)|,$$

and, thus, the second term on the right-hand side of (4) tends to zero more rapidly than $\|h\|$, whereas the first term depends linearly on $h \in \mathfrak{M}_0$. The first term in (4) is called the variation of the functional $J(x, h)$ and is denoted by $\delta J(x, h)$.

Integrals. Side by side with the derivative, the integral has a fundamental significance in mathematical analysis. One distinguishes indefinite and definite integrals.

The indefinite integral is closely connected with primitive functions. A function F is called a *primitive function* of a function f on the interval (a, b) if, on this interval, $F'=f$.

The definite (Riemann) integral of a function f on an interval $[a, b]$ is the limit

$$\lim \sum_{j=0}^{N-1} f(\xi_j)(x_{j+1}-x_j) = \int_{a}^{b} f(x)\,dx$$

as $\max(x_{j+1}-x_j) \to 0$; here $a=x_0<x_1< \cdots <x_N=b$ and $x_j \leq \xi_j \leq x_{j+1}$ are arbitrary.

If f is positive and continuous on $[a, b]$, its integral on this segment is equal to the area of the figure bounded by the curve $y=f(x)$, the x-axis and the lines

$x = a$ and $x = b$.

The class of Riemann-integrable functions contains all continuous functions on $[a, b]$ and some discontinuous ones. But they are all necessarily bounded. For a slowly-growing unbounded function, and also for certain functions on unbounded intervals, the so-called **improper integral** has been introduced, requiring a double limit transition in its definition.

The concept of a Riemann integral of a function of one variable can be extended to functions of several variables (see **Multiple integral**).

On the other hand, the needs of mathematical analysis have led to a generalization of the integral in quite another direction, in the form of the **Lebesgue integral** or, more generally, the **Lebesgue−Stieltjes integral**. Essential in the definition of these integrals is the introduction for certain sets, called measurable, of their measure and, on this foundation, the notion of a measurable function. For measurable functions the Lebesgue−Stieltjes integral has been introduced. In this connection a broad range of different measures has been considered, together with the associated classes of measurable sets and functions. This provides an opportunity to adapt this or that integral to a definite concrete problem.

Newton−Leibniz formula. There is a connection between derivatives and integrals, expressed by the *Newton−Leibniz formula* (theorem):

$$\int_a^b f(x)\,dx = F(b) - F(a).$$

Here f is a continuous function on $[a, b]$ and F is its primitive function.

Taylor's formulas and series. Along with derivatives and integrals, the most important ideas (research tools) in mathematical analysis are the **Taylor formula** and **Taylor series**. If a function $f(x)$, $a < x < b$, has continuous derivatives up to and including order n in a neighbourhood of a point x_0, then it can be approximated in this neighbourhood by the polynomial

$$P_n(x) = f(x_0) + \frac{f'(x_0)}{1!}(x - x_0) + \cdots + \frac{f^{(n)}(x_0)}{n!}(x - x_0)^n,$$

called its *Taylor polynomial* (of degree n), in powers of $x - x_0$:

$$f(x) \approx P_n(x)$$

(*Taylor's formula*); here the error of approximation,

$$R_n(x) = f(x) - P_n(x),$$

tends to zero faster than $(x - x_0)^n$ as $x \to x_0$:

$$R_n(x) = o((x - x_0)^n) \text{ as } x \to x_0.$$

Thus, in a neighbourhood of x_0, f can be approximated to any degree of accuracy by very simple functions (polynomials), which for their calculation require only the arithmetic operations of addition, subtraction and multiplication.

Of special importance are the so-called analytic functions in a fixed neighbourhood of x_0; they have an infinite number of derivatives, $R_n(x) \to 0$ as $n \to \infty$ in the neighbourhood, and they may be represented by the infinite *Taylor series*

$$f(x) = f(x_0) + \frac{f'(x_0)}{1!}(x - x_0) + \cdots.$$

Taylor expansions are also possible, under certain conditions, for functions of several variables, functionals and operators.

Historical information. Up to the 17-th century mathematical analysis was a collection of solutions to disconnected particular problems; for example, in the integral calculus, the problems of the calculation of the areas of figures, the volumes of bodies with curved boundaries, the work done by a variable force, etc. Each problem, or special group of problems, was solved by its own method, sometimes complicated and tedious and sometimes even brilliant (regarding the prehistory of mathematical analysis see **Infinitesimal calculus**). Mathematical analysis as a unified and systematic whole was put together in the works of I. Newton, G. Leibniz, L. Euler, J.L. Lagrange, and other scholars in the 17-th and 18-th centuries, and its foundations, the theory of limits, was laid by A.L. Cauchy at the beginning of the 19-th century. A deep analysis of the original ideas of mathematical analysis was connected with the development in the 19-th and 20-th centuries of set theory, measure theory and the theory of functions of a real variable, and has led to a variety of generalizations.

References

[1] VALLEÉ-POUSSIN, CH.J. DE LA: *Cours d' analyse infinitésimales*, 1-2, Librairie Univ. Louvain, 1923-1925.
[2] IL'IN, V.A. and POZNYAK, E.G.: *Fundamentals of mathematical analysis*, 2, Mir, 1982 (translated from the Russian).
[3] IL'IN, V.A., SADOVNICHIĬ, V.A. and SENDOV, V.KH.: *Mathematical analysis*, Moscow, 1979 (in Russian).
[4] KUDRYAVTSEV, L.D.: *A course of mathematical analysis*, 1-3, Moscow, 1988-1989 (in Russian).
[5] NIKOL'SKIĬ, S.M.: *A course of mathematical analysis*, 1-2, Mir, 1977 (translated from the Russian).
[6] WHITTAKER, E.T. and WATSON, G.N.: *A course of modern analysis*, Cambridge Univ. Press, 1952.
[7] FICHTENHOLZ, G.M.: *Differential und Integralrechnung*, 1-3, Deutsch. Verlag Wissenschaft., 1964.

S.M. Nikol'skiĭ

Editorial comments. In 1961, A. Robinson provided truly infinitesimal methods in analysis with a logical foundation, and so vindicated the founders of the calculus, Leibniz in particular, against the now usual 'ϵ-δ' analysis. This new old way to look at analysis is spreading since twenty years and might become of first importance in a few years. See [A4] and **Non-standard analysis**.

References

[A1] BISHOP, E.: *Foundations of constructive analysis*, McGraw-Hill, 1967.

[A2] SHILOV, G.E.: *Mathematical analysis*, 1-2, M.I.T., 1974 (translated from the Russian).

[A3] COURANT, R. and ROBINS, H.: *What is mathematics?*, Oxford Univ. Press, 1980.

[A4] CUTLAND, N.: *Nonstandard analysis and its applications*, Cambridge Univ. Press, 1988.

[A5] HARDY, G.H.: *A course of pure mathematics*, Cambridge Univ. Press, 1975.

[A6] TITCHMARSH, E.C.: *The theory of functions*, Oxford Univ. Press, 1979.

[A7] RUDIN, W.: *Principles of mathematical analysis*, McGraw-Hill, 1976.

[A8] STROMBERG, K.R.: *Introduction to classical real analysis*, Wadsworth, 1981.

AMS 1980 Subject Classification: 26-XX

MATHEMATICAL ECONOMICS - The mathematical discipline whose subject concerns models of economic objects and processes, and methods for investigating them. However, the concepts, results and methods of mathematical economics are conveniently and commonly expounded in close connection with their economic derivations, interpretations and practical applications. Of particular significance is the connection with the science and practice of economics.

Mathematical economics, as a part of mathematics, began only in the 1920's. Earlier there was only sporadic research which cannot be strictly attributed to mathematics.

Peculiarities of economic-mathematical modelling. A peculiarity of economic modelling is the exceptional variety and diversity of the objects being modelled. In economics there are elements of controllability and spontaneity, rigid determinacy and essential ambiguity and freedom of choice, processes of technical nature, and social processes where human behaviour comes to the forefront. Different levels of economics (for example, shop economics and households) require essentially different descriptions. All this leads to a great diversity, in the models, of the mathematical apparatus. A delicate question is how to express the type of socio-economic systems which are modelled, taking account of the social structure. It often happens that an abstract mathematical model of some economic process or object can be successfully applied to both capitalist and socialist economies. This is accommodated in the method of utilization and interpretation of the results of analysis.

Production. Efficient production. Economics deals with wealth, or products, which are understood in an extremely broad sense in mathematical economics. For this one applies the general terminology of *ingredients* (*goods* or *commodities*). Ingredients are services, natural resources, the unfavourable influence on man of environmental factors, characteristics of the comfort of a present security system, etc. It is usually assumed that the number of ingredients is finite and the space of products is $\mathbf{R}^l$, Euclidean space, where l is the number of ingredients. A point z from $\mathbf{R}^l$, under appropriate conditions, can be considered as a 'production' method; positive components indicate the volume of output of the corresponding ingredients, and negative components the inputs. The word 'production' is put between quotes because it is to be understood in a very broad sense. The set of available (given, existing) production possibilities is $Z \subset \mathbf{R}^l$. A method of production $\bar{z} \in Z$ is *efficient* if there is no $z \in Z$ such that $z \geqslant \bar{z}$, with strict inequality in at least one component. The problem of discovering efficient methods is one of the most important in economics. Usually it is assumed, and in many cases this agrees well with reality, that Z is a compact convex set. By expanding the space of products the problem of the analysis of the efficient methods here may be reduced to the case where Z is a closed convex cone.

A typical problem is the *fundamental problem of production planning*. Given a set of production methods $Z \subset \mathbf{R}^l$ and a vector of requirements and resource limitations $b \in \mathbf{R}^{l-1}$, it is required to find a method $\bar{z} = (b, \bar{\mu}) \in Z$ such that $\bar{\mu} \geqslant \mu$ for all $(b, \mu) \in Z$. If Z is a closed convex cone, then this is the general problem of **convex programming**. If Z is given by a finite number of generators (the so-called basic methods), then this is the general problem of **linear programming**. A solution $\bar{z}$ lies on the boundary of Z. Let π be the coefficients for the supporting hyperplane for Z at the point $\bar{z}$, that is, $\pi z \leqslant 0$ for all $z \in Z$ and $\pi \bar{z} = 0$. The fundamental theorem of convex programming gives conditions under which $\pi_l > 0$. For example, a sufficient condition is: There is a vector $(b, \mu) \in \text{int } Z$ (the so-called *Slater condition*). The coefficients of π, which characterize the efficient method $\bar{z}$, have an important economic meaning. They can be interpreted as prices commensurate with the efficiency of the inputs and outputs of the different ingredients. A method is efficient if and only if the cost of the outputs is equal to the cost of the inputs. The given theory of efficient production methods and their characterization using π has exerted a revolutionary influence on the theory and practical planning of socialist economics. It has underpinned objective quantitative methods for the determination of prices and the social evaluation of resources, giving the possibility of choosing more efficient economic solutions under the conditions of a socialist economy. The theory generalizes naturally to an infinite number of ingredients. Then the space of ingredients is a suitably chosen function space.

Efficient growth. Ingredients relating to different times or time intervals can be formally regarded as distinct.

Therefore the description of production in dynamic form, in principle, is contained in the above scheme, which consists of objects $\{X, Z, b\}$, where X is the space of ingredients, Z is the space of production capacities and b is the specification of requirements and restrictions on the economy. However, the study of the truly dynamical aspects of production requires a more special form of the description of the production capacities.

The production capacities of a sufficiently general model of economic dynamics are given via a point-to-set mapping (many-valued function) $a: \mathbf{R}^l_+ \to 2^{\mathbf{R}^l_+}$. Here $\mathbf{R}^l_+$ is the (phase) space of the economy, $x \in \mathbf{R}^l_+$ is interpreted as the state of the economy at some time and x_k is the available quantity of the product k at that time. The set $a(x)$ consists of all states of the economy into which it may pass in unit time from x. One calls

$$Z = \{(x, y) \in \mathbf{R}^{2l}_+ : y \in a(x)\}$$

the *graph of the mapping a*. A point (x, y) is an admissible production process.

Different versions of the specification of possible trajectories of development of the economy have been considered. In particular, consumption by the population is allowed for either in the mapping a itself, or explicitly. For example, in the second case an admissible trajectory is a sequence $(X, C) = (x(t), c(t+1))^\infty_{t=0}$ such that $x(t+1) + c(t+1) \in a(x(t))$, $c(t) \geq 0$ for all t. Different concepts of efficiency of trajectories have been studied. A trajectory $(\overline{X}, \overline{C})$ is *efficient relative to consumption* if there does not exist an admissible trajectory (X, C) beginning from the same initial state for which $C \geq \overline{C}$. A trajectory $(\overline{X}, \overline{C})$ is *intrinsically efficient* if there is no other admissible trajectory (X, C), beginning from the same initial state, a time t_0 and a number $\lambda > 1$, such that

$$\lambda \overline{x}(t_0) = x(t_0).$$

Optimality of a trajectory is usually defined depending on a utility function $u: \mathbf{R}^l_+ \to \mathbf{R}^l_+$ and a coefficient for discounting utility in time $\mu \geq 1$ (see below, and also **Utility theory**, for something on utility functions). A trajectory $(\overline{X}, \overline{C})$ is called (u, μ)-*optimal* if

$$\varliminf_{t \to \infty} \left[\sum_{\tau=0}^t u(\overline{c}(\tau))\mu^{-\tau} - \sum_{\tau=0}^t u(c(\tau))\mu^{-\tau} \right] \geq 0$$

for any admissible trajectory (X, C) starting from the same initial state. There is a number of quite general existence theorems for the corresponding trajectories.

Trajectories which are efficient in different senses are characterized by a sequence of prices in exactly the same way as an efficient method $\overline{z}$ is characterized by the prices (the coefficients of the supporting hyperplane) π. That is, if for an efficient method the cost of input is equal to the cost of output at optimal prices,

then on an efficient trajectory the cost of a state is constant and maximal, and on all other admissible trajectories it cannot increase.

All of these definitions are easily generalized to the case when the production mapping a, the function u and the coefficient μ depend on time. Time itself may be continuous or, more generally, the parameter t may run through a set of quite arbitrary form.

From the economic point of view the interest is in trajectories which attain the maximum possible rate of economic growth and which can be sustained for an arbitrarily long time. It turns out that for a and u which do not vary in time such trajectories are *stationary*, that is, have the form

$$x(t) = x(0)\alpha^t, \quad c(t) = c(0)\alpha^t,$$

where α is the rate of growth (the expansion) of the economy. Stationary efficient, in some sense, and also stationary optimal, trajectories are called *turnpike trajectories*.

Under very broad assumptions, a theorem on turnpike trajectories asserts that every efficient trajectory, independent of the initial state, as time goes on approximates a turnpike trajectory. There is a large number of different theorems on turnpikes, which differ in their definitions of efficiency and optimality, the means of measuring the distance from a turnpike, the type of convergence, and, finally, on whether finite or infinite time intervals are involved.

The model of economic dynamics, in which production capacities are given by a polyhedral convex cone, is called the *von Neumann model*. A particular case of the von Neumann model is the *closed Leont'ev model*, or (in other terminology) the *closed dynamical interdepartmental balance model* (the term 'closed' is used here as a characteristic property of economics without non-reproducible products), which is given in terms of matrices Φ, A and B of order $l \times l$ with non-negative entries. A process $(x, y) \in Z$ if and only if vectors $v, w \in \mathbf{R}^l_+$ can be found such that

$$x \leq v\Phi, \quad v \geq vA + wB, \quad y \leq v\Phi + w.$$

The model of interdepartmental balance is more widespread because of the convenience in obtaining the initial information for its construction.

Models of economic dynamics are also discussed in continuous time. In fact, the first models to be studied were precisely models with continuous time. In particular, several works were devoted to the simplest one-product models, given by an equation

$$\dot{x} = f(x) - c,$$

where x is the volume of stock per unit of labour resource, c is the requirement per head of population and f is the production function (increasing and con-

cave). Non-negative functions $(x(t), c(t))_{t=0}^{\infty}$ satisfying this equation characterize an admissible trajectory. For a given utility function u and discount coefficient μ an optimal trajectory can be determined. Optimal trajectories (and only they) satisfy an analogue of the Euler equation:
$$u(x)\dot{x} = u(\bar{c}) - u(c),$$
where $\bar{c}$ is the largest number satisfying the condition $f(x) - c = x$.

The Leont'ev model was also initially formulated in continuous time as a system of differential equations
$$X = AX + B\dot{X} + C,$$
where X is the stream of products, A and B are the matrices of current and capital expenses, respectively, and C is the stream of finite requirements.

Efficient and optimal trajectories in models with continuous time are studied with the help of the methods of variational calculus, optimal control and mathematical programming in infinite-dimensional spaces. Models whose admissible trajectories are given by differential inclusions of the form $\dot{x} \in a(x)$, where a is the production mapping, are also discussed.

Rational behaviour of consumers. The tastes and goals of consumers, which determine their rational behaviour, are given in the form of some system of preferences in the space of products. Namely, for each consumer i there is defined a point-to-set mapping $P_i : Z \to 2^X$. Here Z is some subset of situations in which the consumer may find himself by the process of selection and X is the set of vectors accessible to the consumer, $X \subset \mathbf{R}^l$. In particular, Z may contain a subspace of $\mathbf{R}^l$. The set $P_i(z \mid x)$ consists of all vectors $\tilde{x} \in X$ which are (strictly) preferred to the vector x in the situation z. For example, the mapping P_i may be given by means of a utility function u, where $u(x)$ shows the utility to the consumer of the set of products x. Then
$$Z = \mathbf{R}_+^l, \quad P_i(z) = \{\tilde{x} \in \mathbf{R}_+^l : u(\tilde{x}) > u(x)\}.$$
In the description of the situation z the prices π of all products and the monetary income d of the consumer enter. Then $B_i(z) = \{x \in X : x\pi \leqslant d\}$ is the collection of sets from which the consumer may choose in the situation z. This set is called the *budget set*. Rationality of consumer behaviour is to choose a set x from $B_i(z)$ for which $P_i(z \mid x) \cap B_i(z) = \varnothing$. Let $D_i(z)$ be the collection of sets of products chosen by consumer i in situation z; D_i is called the *demand mapping* (or *demand function*, if $D_i(z)$ consists of one point). There are many investigations devoted to clarifying the properties of the mappings P_i, B_i and D_i. In particular, the case when P_i is a function has been studied at length. Conditions under which the mappings B_i and D_i are continuous have been determined. Of special interest is the study of the properties of the demand function D_i. The fact is that

sometimes it is more convenient to regard as primary only the demand functions D_i, and not the preferences P_i, since it is easy to construct them from the available information on consumer behaviour. For example, in an economy (a trade statistic) it is possible to observe quantities that approximate the partial derivatives
$$\frac{\partial D_{ik}(x)}{\partial \pi_\rho}, \quad \frac{\partial D_{ik}(x)}{\partial d},$$
where π_ρ is the price of the product ρ and d is the income.

Bordering on the theory of rational behaviour of consumers is the theory of group choice (*social choice*), concerning, as a rule, discrete variants. It is usually assumed that there is a finite number of participants in a group and a finite number of, for example, alternatives. The problem lies in the choice of a group solution of the selection of one variant, given the preference between the alternatives of each participant. Group choice provides various voting schemes; here axiomatic and game-theoretic approaches are also used.

Agreement of interests. The holders of interests are the individual parties of economic systems, and also society as a whole. As such parties one puts forward consumers (groups of consumers): enterprises, ministries, territorial organizations of administration, planning and financial organizations, etc. One distinguishes two mutually intertwined approaches to the problem of agreement of interests: the analytic, or constructive, and the synthetic, or descriptive. According to the first approach, initially there is a global criterion of optimality (a formalization of the interests of society at large). The problem is to derive the local (personal) criteria from the general one, taking account of personal interests. In the second approach there are just the personal interests, and the problem is to unify them into a single consistent system, the functioning of which leads to results which are satisfactory from the point of view of society as a whole.

Directly related to the first approach are the decomposition methods of mathematical programming. For example, in an economy let there be m producers, and let each producer j be given a set of production capacities Y_j, where $Y_j \subset \mathbf{R}^l$ is a compact convex set. Let there be given an objective function V for the society at large, where $V : \mathbf{R}_+^l \to \mathbf{R}_+^l$ is a concave function. The economy must be organized in such a way that the following problem of convex programming results: Find $\bar{y}$ from the conditions $y \geqslant 0$, $y \in \sum_j Y_j$, $V(y) \to \max$. From theorems on the characteristic behaviour of efficient production methods one may conclude that there are prices $p \in \mathbf{R}_+^l$ $(p \neq 0)$ such that
$$\bar{y}^{(j)} p = \max_{y^{(j)} \in Y_j} y^{(j)} p \text{ for all } j, \quad \bar{y} = \sum_j \bar{y}^{(j)}.$$

The quantity $y^{(j)}p$ is interpreted as the profit of the j-th producer with prices p. Hence it follows that the criterion of maximizing the profits for each producer does not conflict with the common goal if the operating prices are defined in a corresponding form. The second approach has been strongly developed within the area of models of economic equilibrium.

Economic equilibrium. Assume that the economy consists of individual parties having personal interests: producers listed by indices $j=1,\ldots,m$, and consumers listed by indices $i=1,\ldots,n$. Producer j is described by a set of production capacities $Y_j \subset \mathbf{R}^l$ and a mapping $F_j: Z \to 2^{Y_j}$, giving his or her system of preferences. Here Z is the set of possible states of the economy, made concrete below. Consumer i is described by a collection of possible sets of products available for consumption, $X_i \subset \mathbf{R}^l$, an initial supply of products $w^{(i)} \in \mathbf{R}^l_+$, preferences $P_i: Z \to 2^{X_i}$ and, finally, functions $\alpha_i: Z \to \mathbf{R}$ of income distribution, where $\alpha_i(z)$ shows the amount of money available to consumer i in the state z. The set of possible prices in the economy is Q. The set of possible states is $Z = \prod_i X_i \times \prod_j X_j \times Q$. The budget mapping B_i is defined here as:

$$B_i(z) = \{\tilde{x}^{(i)} \in X_i: \tilde{x}^{(i)}p \leqslant \alpha_i(z) + w^{(i)}p\}.$$

An equilibrium state of the described economy is a $\bar{z} \in Z$ satisfying the conditions

$$\sum_i \bar{x}^{(i)} = \sum_j \bar{y}^{(j)} + \sum_i w^{(i)},$$

$$\bar{x}^{(i)} \in B_i(\bar{z}), \quad B_i(\bar{z}) \cap P_i(\bar{z}) = \varnothing, \quad Y_j \cap F_j(\bar{z}) = \varnothing.$$

In essence, an equilibrium state of the economy is defined as a solution of a **non-cooperative game** with several players, in the sense of von Neumann–Nash, with the additional condition that there is a balance with respect to all products.

The existence of an equilibrium state has been proved under very general conditions on the initial economy. It is necessary to impose much stricter conditions to ensure that the equilibrium state be optimal, that is, is a solution to some global optimization problem with an objective function depending on the interests of the consumers. For example, let P_i be given by a concave continuous function $u_i: \mathbf{R}^l \to \mathbf{R}_+$, and let F_j be given by a function $py^{(j)}$; let

$$\alpha_i(z) = \sum_i \theta_{ij} y^{(j)} p, \quad \theta_{ij} \geqslant 0, \quad \sum_i \theta_{ij} = 1,$$

$$Q = \left\{ p \in \mathbf{R}^l_+: \sum_{k=1}^l p_k = 1 \right\},$$

where Y_j, X_i are convex compact sets, $0 \in Y_j$, $w^{(i)} \in \operatorname{int} X_i$. Any subset $S = \{i_1, \ldots, i_r\}$ of indices of consumers forms a subeconomy of the initial economy, in which to each consumer i_s from S corresponds one (and only one) producer, the set of production capaci-

ties of which is

$$\hat{Y}_{i_s} = \sum_{j=1}^m \theta_{i_s,j} Y_j.$$

The function of income distribution here has the form

$$\alpha_{i_s}(z) = y^{(i_s)}p.$$

A state $z \in Z$ is called *balanced* if

$$\sum_i x^{(i)} \leqslant \sum_i y^{(j)} + \sum_i w^{(i)}.$$

One says that a balanced state z of the initial economy blocks a coalition of consumers (cf. also **Coalition**) S if in the subeconomy determined by the coalition S there is a balanced state $\bar{z}^{(s)}$ such that $u_{i_s}(\bar{x}^{(i_s)}) \geqslant u_{i_s}(x^{(i_s)})$, for $s = 1, \ldots, r$, and for at least one index the inequality is strict. The *core of the economy* is the set of all balanced states that do not block any coalition of consumers. For economies with these properties there is the following theorem: Every equilibrium state belongs to the core. The converse is false; however, there is a number of sufficient conditions under which the set of equilibrium states and the core are close to each other or coincide completely. In particular, if the number of consumers tends to infinity and the influence of each consumer on the states of the economy becomes ever smaller, then the set of equilibrium states tends to the core. The coincidence of the core and the set of equilibrium states holds in economies with an infinite (continuum) number of consumers (*Aumann's theorem*).

Let the economy be a market model (that is, there are no producers), the set of participants (consumers) of which is the closed unit interval $[0, 1]$, denoted in the sequel by T. A state of the economy is $z = (x, p)$, where $p \in \{p \in \mathbf{R}^l_+: \sum_k p_k = 1\}$ and x is a function from T into $\mathbf{R}^l_+$ each component of which is Lebesgue integrable over the interval T. The initial distribution of products among the participants is given by a function w, $\int_T w > 0$, so that a balanced state z is such that $\int x = \int w$. A coalition of participants is a Lebesgue-measurable subset of T. If the subset has measure zero, then the corresponding coalition is called null. The core is the set of all balanced states which do not block any non-null coalition. A state $\bar{z} = (\bar{x}, \bar{p})$ is an equilibrium if for almost-all participants t,

$$u_t(\bar{x}(t)) = \max u_t(x(t)),$$

$$x(t) \in \{x: xp \leqslant pw(t)\}.$$

Aumann's theorem asserts that in this economy the core and the set of equilibrium states coincide.

The question of the structure of the set of equilibrium states is particularly interesting when the set is finite or consists of one point. Here one has the theorem of Debreu. Let the set of market models be $W = \{(w^{(i)}, D_i)_{i=1}^n\}$, where $w^{(i)} \in \mathbf{R}^l_+$ is the initial supply of products for the participants i and

$w = (w^{(1)}, \ldots, w^{(n)})$ is a parameter defining a concrete model from the set W, $e \in \mathbf{R}_+^{n \cdot l}$. The mapping $D_i: Q \times M \to \mathbf{R}_+^l$ represents the demand function for the participant i. The functions $D_1, \ldots, D_n$ are given (are fixed) for the whole set of economies W. Let W_0, $W_0 \subset W$, be the collection of economies for which the set of equilibrium states is infinite. *Debreu's theorem* asserts that if the functions $D_1, \ldots, D_n$ are continuously differentiable and if there are no points of saturation for at least one of the participants, then the closure of the set W_0 has (Lebesgue) measure zero in the space W.

On numerical methods. Mathematical economics has a close connection with computational mathematics. Linear programming and linear economic models have exerted a great influence on the computational methods of linear algebra. Essentially because of linear programming, inequalities in computational mathematics have become as much used as equations.

The calculation of economic equilibria is a difficult problem, having many aspects. For example, much work has been devoted to conditions for convergence to an equilibrium for systems of differential equations

$$\dot{p} = F(p), \qquad (*)$$

where p is the price vector and F is the excess demand function, that is, the difference between the demand and the supply functions. Equilibrium costs $\bar{p}$ provide, by definition, equality of demand and supply: $F(\bar{p}) = 0$. The surplus demand function F is given either directly, or by more primary concepts of the corresponding model of equilibrium. S. Smale [8] has studied a significantly more general dynamical system than (*), applied to a market model; alongside a variation in time of the prices p, a variation of the states x is also considered; here an admissible trajectory $(p(t), x(t))_{t=0}^{\infty}$ satisfies some differential inclusion of the form $\dot{p} \in K(p)$, $\dot{x} \subset C(p)$, where $K(p)$ and $C(p)$ are the sets of possible directions of variation of p and x, determined by the market model.

The economic equilibrium, the solution of a game, the solution of an extremal problem, all may be defined as a fixed point of an appropriate point-to-set mapping. Within the limits of research in mathematical economics, numerical methods for the computation of fixed points of various classes of mappings have been developed. The best known is *Scarf's method*, [6], which is a combination of the ideas of the **Sperner lemma** and the **simplex method** of solution of linear programming problems.

Related questions. Mathematical economics is closely connected with many mathematical disciplines. Sometimes it is difficult to determine the boundary between mathematical economics and mathematical statistics or convex analysis, functional analysis, topology, etc. One only has to mention, for example, the development of the theory of positive matrices, positive linear (and homogeneous) operators and the spectral properties of superlinear point-to-set mappings, under the influence of the requirements of mathematical economics.

References

[1] NEUMANN, J. VON and MORGENSTERN, O.: *Theory of games and economic behavior*, Princeton Univ. Press, 1947.
[2] KANTOROVICH, L.V.: *Economic calculation of the best use of resources*, Moscow, 1959 (in Russian).
[3] NIKAIDO, H.: *Convex structures and economic theory*, Acad. Press, 1968.
[4] MARKOV, V.L. and RUBINOV, A.M.: *The mathematical theory of economic dynamics and equilibrium*, Moscow, 1973 (in Russian).
[5] MIRKIN, B.G.: *Group choice*, Winston, 1979 (translated from the Russian).
[6] SCARF, H.: *The computation of economic equilibrium*, Yale Univ. Press, 1973.
[7] DANTZIG, G.B.: *Linear programming and extensions*, Princeton Univ. Press, 1963.
[8] SMALE, S.: 'A convergent process of price adjustment and global Newton methods', *J. Math. Economics* **2** (1976), 107-120.

L.D. Kantorovich
V.L. Makarov

Editorial comments. A classic on mathematical economics is [A8], and a useful general book from the optimization point of view is [A9]. Selected seminal papers on mathematical economics can be found in [A10] - [A12]. For more on dynamical systems as applied to economics, including (optimal) control and the calculus of variations, see also [A13] - [A15]. The books [A5] - [A7], [A16] - [A20] deal with prices, utility functions and general equilibrium theory. In its more advanced versions [A20] the latter makes sophisticated use of measure theory.

Besides the fields already mentioned, many other branches of mathematics can be fruitfully applied in economics, including bifurcation theory, Hamiltonian dynamical systems and the theory of Lie groups [A25].

The mathematical analysis of *social choice* and *voting systems* turned up a surprise in that a very reasonable sounding set of axioms for social choice (transitivity, independence of irrelevant alternatives, unanimity, no dictator) turns out to be contradictory. Such results (and there are many of them) are called *Arrow impossibility theorems* [A21] - [A24]. They have much to do with the *Kondortsev paradox*, which under simple majority voting gives a circular order in the case of three alternatives and three voters whose respective orderings of the three alternatives are $x > y > z$, $y > z > x$, $z > x > y$.

Scarf's method for the numerical calculation of Brouwer fixed points via Sperner's lemma developed into the *homotopy methods for solving equations*. In crude terms this amounts to deforming a given problem until a trivial problem is found, and then deforming back together with the solution, [A26]. This idea further developed into *continuation methods* for solving systems of equations (cf. Continuation method (to a parametrized family); Continuation method (to a parametrized family, for non-linear operators)).

Many previously known solving methods turned out to be special cases of these ideas. A selection of references is [A26] - [A29].

The relation between decomposition methods in mathematical programming and centrally-guided economic systems is extensively discussed in [A30].

References

[A1] *Handbook of mathematical economics*, North-Holland.

[A2] AUMANN, R.J.: 'Markets with a continuum of traders', *Econometrica* **32** (1964), 39-50.

[A3] DEBREU, G.: 'Economies with a finite set of equilibria', *Econometrica* **38** (1970), 387-392.

[A4] HILDENBRAND, W. and MAS-COLLELL, A. (EDS.): *Contributions to mathematical economics, in honour of Gérard Debreu*, North-Holland, 1986.

[A5] DEBREU, G.: *Theory of value*, Wiley, 1959.

[A6] ARROW, K.J. and HAHN, F.H.: *General competitive analysis*, Oliver & Boyd, 1971.

[A7] HILDENBRAND, W. and KIRMAN, A.P.: *Introduction to equilibrium analysis*, North-Holland, 1976.

[A8] SAMUELSON, P.A.: *Foundations of economic analysis*, Harvard Univ. Press, 1947.

[A9] INTRILLIGATOR, M.D.: *Mathematical optimization and economic theory*, Prentice Hall, 1971.

[A10] NEWMAN, P. (ED.): *Readings in mathematical economics, 1: Value theory*, Johns Hopkins Press, 1968.

[A11] NEWMAN, P. (ED.): *Readings in mathematical economics, 2: Capital and growth*, Johns Hopkins Press, 1968.

[A12] REITER, S. (ED.): *Studies in mathematical economics*, Math. Assoc. Amer., 1986.

[A13] GANDOLFO, G.: *Mathematical models and models in economic dynamics*, North-Holland, 1971.

[A14] CHOW, G.C.: *Analysis and control of dynamic economic systems*, Wiley, 1975.

[A15] HARDLEY, G. and KEMP, M.C.: *Variational methods in economics*, North-Holland, 1971.

[A16] LEVENSON, A.M. and SOLON, B.S.: *Outline of price theory*, Holt, Rinehart & Winston, 1964.

[A17] QUIRK, J. and SAPOSNIK, R.: *Introduction to general equilibrium theory and welfare economics*, McGraw-Hill, 1968.

[A18] SHEPHARD, R.W.: *Theory of cost and production functions*, Princeton Univ. Press, 1970.

[A19] NEGISHI, T.: *General equilibrium theory and international trade*, North-Holland, 1972.

[A20] HILDENBRAND, W.: *Core and equilibria of a large economy*, Princeton Univ. Press, 1974.

[A21] ARROW, K.J.: *Social choice and individual values*, Wiley, 1951.

[A22] SEN, A.K.: *Collective choice and social welfare*, Oliver & Boyd, 1970.

[A23] MURAKAMI, Y.: *Logic and social choice*, Routledge, Kegan, Paul, 1968.

[A24] KELLY, J.S.: *Arrow impossibility theorems*, Acad. Press, 1978.

[A25] SATO, R.: *Theory of technical change and economic invariance*, Acad. Press, 1981.

[A26] EAVES, B.C.: 'Homotopies for computation of fixed points', *Math. Progr.* **3** (1972), 1-22.

[A27] ALLGOWER, E.L. and GEORG, K.: 'Simplicial and continuation methods for approximating fixed points and solutions to systems of equations', *SIAM Rev.* **22** (1980), 28-85.

[A28] ALLGOWER, E.L. and GEORG, K.: *Continuation methods for numerically solving nonlinear systems of equations*, Springer, 1987.

[A29] EAVES, B.C., GOULD, F.J., PEITGEN, H.-O. and TODD, M.J. (EDS.): *Homotopy methods and global convergence*, Plenum, 1983.

[A30] RAZUMIKHIN, B.S.: *Physical models and equilibrium methods in programming and economics*, Reidel, 1984 (translated from the Russian).

AMS 1980 Subject Classification: 90AXX

MATHEMATICAL EXPECTATION, *mean value, of a random variable* - A numerical characteristic of the probability distribution of a random variable. In the most general setting, the mathematical expectation of a random variable $X(\omega)$, $\omega \in \Omega$, is defined as the **Lebesgue integral** with respect to a **probability measure** P on a given **probability space** $(\Omega, \mathscr{A}, \mathrm{P})$:

$$\mathrm{E}X = \int_{\Omega} X(\omega)\mathrm{P}(d\omega), \qquad (*)$$

provided the integral exists. The mathematical expectation of a real-valued random variable may be calculated also as the Lebesgue integral of x with respect to the probability distribution P_X of X:

$$\mathrm{E}X = \int_{\mathbf{R}} x\mathrm{P}_X(dx).$$

The mathematical expectation of a function in X is expressible in terms of the distribution P_X; for example, if X is a random variable with values in $\mathbf{R}$ and $f(x)$ is a single-valued **Borel function** of x, then

$$\mathrm{E}f(X) = \int_{\Omega} f(X(\omega))\mathrm{P}(d\omega) = \int_{\mathbf{R}^1} f(x)\mathrm{P}_X(dx).$$

If $F(x)$ is the distribution function of X, then the mathematical expectation of X can be represented as the Lebesgue$-$Stieltjes (or Riemann$-$Stieltjes) integral

$$\mathrm{E}X = \int_{-\infty}^{\infty} x\, dF(x);$$

here integrability of X in the sense of (*) is equivalent to the finiteness of the integral

$$\int_{-\infty}^{\infty} x\, dF(x).$$

In particular cases, if X has a discrete distribution with possible values x_k, $k=1, 2, \ldots,$ and corresponding probabilities $p_k = \mathrm{P}\{\omega: X(\omega) = x_k\}$, then

$$\mathrm{E}X = \sum_k x_k p_k;$$

if X has an absolutely continuous distribution with probability density $p(x)$, then

$$\mathrm{E}X = \int_{-\infty}^{\infty} x p(x)\, dx;$$

moreover, the existence of the mathematical expectation is equivalent to the absolute convergence of the corresponding series or integral.

Main properties of the mathematical expectation:

a) $\mathrm{E}X_1 \leqslant \mathrm{E}X_2$ whenever $X_1(\omega) \leqslant X_2(\omega)$ for all $\omega \in \Omega$;

b) $\mathrm{E}C = C$ for every real constant C;

c) $\mathrm{E}(\alpha X_1 + \beta X_2) = \alpha \mathrm{E}X_1 + \beta \mathrm{E}X_2$ for all real α and β;

d) $\mathrm{E}(\sum_{n=1}^{\infty} X_n) = \sum_{n=1}^{\infty} \mathrm{E}X_n$ if the series

$\sum_{n=1}^{\infty} \mathsf{E}|X_n|$ converges;

e) $g(\mathsf{E}X) \leqslant \mathsf{E}g(X)$ for convex functions g;

f) every bounded random variable has a finite mathematical expectation;

g) $\mathsf{E}(\prod_{k=1}^{n} X_k) = \prod_{k=1}^{n} \mathsf{E}X_k$ if the random variables $X_1, \ldots, X_k$ are mutually independent.

One can naturally define the notion of a random variable with an infinite mathematical expectation. A typical example is provided by the return times in certain random walks (see, e.g., **Bernoulli random walk**).

The mathematical expectation is used to define many numerical functional characteristics of probability distributions (as the mathematical expectations of appropriate functions in the given random variables), for example, the **generating function**, the **characteristic function** and the moments (cf. **Moment**) of all orders, in particular, the variance (cf. **Dispersion**) and the **covariance**.

The mathematical expectation is a characteristic of the location of the values of a random variable (the mean value of its distribution). Here, the mathematical expectation serves as a 'typical' value from the distribution and its role is analogous to the role played in mechanics by the statical momentum — the coordinates of the barycentre of a mass distribution. The mathematical expectation differs from other characteristics of location which describe the distribution in general terms — like the median (cf. **Median (in statistics)**) and the **mode**, by the higher importance that it and its corresponding scatter characteristic, the variance, have in limit theorems of probability theory. The meaning of the mathematical expectation is most completely revealed by the **law of large numbers** (see also **Chebyshev inequality in probability theory**) and the **strong law of large numbers**. In particular, if $X_1, \ldots, X_n$ is a sequence of mutually-independent identically-distributed random variables with finite mathematical expectation $a = \mathsf{E}X_k$, then, as $n \to \infty$ and for every $\epsilon > 0$,

$$\mathsf{P}\left(\left|\frac{X_1 + \cdots + X_n}{n} - a\right| > \epsilon\right) \to 0,$$

and, in addition,

$$\frac{X_1 + \cdots + X_n}{n} \to a$$

with probability one.

The notion of the mathematical expectation as the expected value of a random variable was first noticed in the eighteenth century in connection with the theory of games of chance. Initially the term 'mathematical expectation' was introduced as the expected pay-off of a player, equal to $\sum x_k p_k$ for possible pay-offs $x_1, \ldots, x_n$ with respective probabilities $p_1, \ldots, p_n$. Primary contributions in the generalization and utilization of the notion of the mathematical expectation in its contemporary meaning are due to P.L. Chebyshev.

References

[1] KOLMOGOROV, A.N.: *Foundations of the theory of probability*, Chelsea, reprint, 1950 (translated from the Russian).
[2] FELLER, W.: *An introduction to probability theory and its applications*, 1-2, Wiley, 1957-1971.
[3] LOÈVE, M.: *Probability theory*, Springer, 1978.
[4] CRAMÉR, H.: *Mathematical methods of statistics*, Princeton Univ. Press, 1946.

A.V. Prokhorov

AMS 1980 Subject Classification: 60-XX, 62-XX

MATHEMATICAL INDUCTION - A method of proving mathematical results based on the *principle of mathematical induction*: An assertion $A(x)$, depending on a natural number x, is regarded as proved if $A(1)$ has been proved and if for any natural number n the assumption that $A(n)$ is true implies that $A(n+1)$ is also true.

The proof of $A(1)$ is the *first step* (or *base*) *of the induction* and the proof of $A(n+1)$ from the assumed truth of $A(n)$ is called the *induction step*. Here n is called the *induction parameter* and the assumption of $A(n)$ for the proof of $A(n+1)$ is called the *induction assumption* or *induction hypothesis*. The principle of mathematical induction is also the basis for **inductive definition**. The simplest example of such a definition is the definition of the property: 'to be a word of length n over a given alphabet $\{a_1, \ldots, a_k\}$'.

The base of the induction is: Each symbol of the alphabet is a word of length 1. The induction step is: If E is a word of length n, then each word Ea_i, where $1 \leqslant i \leqslant k$, is a word of length $(n+1)$. Induction can also start at the zero-th step.

It often happens that $A(1)$ and $A(n+1)$ can be proved by similar arguments. In these cases it is convenient to use the following equivalent form of the principle of mathematical induction. If $A(1)$ is true and if for each natural number n from the assumption: $A(x)$ is true for any natural number $x < n$, it follows that $A(x)$ is also true for $x = n$, then $A(x)$ is true for any natural number x. In this form the principle of mathematical induction can be applied for proving assertions $A(x)$ in which the parameter x runs through any well-ordered set of a transfinite type (**transfinite induction**). As simple examples of transfinite induction one has induction over a parameter running through the set of all words over a given alphabet with the lexicographic ordering, and induction over the construction formulas in a given logico-mathematical calculus.

Sometimes, for the inductive proof of an assertion $A(n)$ one has to inductively prove, simultaneously with $A(n)$, a whole series of other results without which the induction for $A(n)$ cannot be carried out. In formal arithmetic one can also give results $A(n)$ for which, within the limits of the calculus considered, an induction can not be carried out without the addition of new auxiliary results depending on n (see [3]). In these cases one has to deal with the proof of a number of assertions by *compound mathematical induction*. All these statements could formally be united into one conjunction; however, in practice this would only complicate the discussion and the possibility of informal sensible references to specific induction assumptions would disappear.

In some concrete mathematical investigations the number of notions and results defined and proved by compound induction has reached three figures (see [4]). In this case, because of the presence in induction of a large number of cross references to the induction assumptions, for a concise (informal) understanding of any (even very simple) definition or results for a large value of the induction parameter, the reader must be familiar with the content of all induction ideas and properties of these ideas for small values of the induction parameter. Apparently, the only logically correct outcome of this circle of problems is the axiomatic presentation of all these systems of ideas. Thus, a great number of ideas defined by compound mathematical induction lead to the need for an application of the **axiomatic method** in inductive definitions and proofs. This is a visual example of the necessity of the axiomatic method for the solution of concrete mathematical problems, and not just for questions relating to the foundations of mathematics.

References

[1] HILBERT, D. and BERNAYS, P.: *Grundlagen der Mathematik*, 1-2, Springer, 1968-1970.
[2] KLEENE, S.C.: *Introduction to metamathematics*, North-Holland, 1951.
[3] CINMAN, L.L. [L.L. TSINMAN]: 'On the role of the principle of induction in a formal arithmetic system', *Math. USSR Sb.* **6**, no. 1 (1968), 65-95. (*Mat. Sb.* **77**, no. 1 (1968), 71-104)
[4] ADYAN, S.I.: *The Burnside problem and identities in groups*, Springer, 1979 (translated from the Russian).

S.I. Adyan

Editorial comments. In the article above, the natural numbers are 1, 2, . . . (i.e. excluding 0).

References

[A1] MACLANE, S. and BIRKHOFF, G.: *Algebra*, Macmillan, 1967.

AMS 1980 Subject Classification: 03-XX, 00A25

MATHEMATICAL LINGUISTICS - The mathematical discipline whose objective is the development and study of ideas forming the basis of a formal apparatus for the description of the structure of natural languages (that is, the metalanguage of linguistics). The origin of mathematical linguistics can be roughly placed in the 1950's; it was brought to life first of all by the internal needs of theoretical linguistics, which at that time was ripe for an elaboration of its basic ideas, and also by problems in the automatic processing of linguistic information (see **Automatic translation**). In mathematical linguistics methods of the theories of algorithms, automata and algebra are widely used. Keeping its applied sense, mathematical linguistics is constantly evolving along the path of transforming into a pure mathematical discipline, being essentially a branch of mathematical logic. At the same time the circle of applications of mathematical linguistics is expanding; its methods have found application in the theory of programming.

The linguistic concepts underlying the formal description of the structure of a language belong to **structural linguistics**. The most important of these concepts, the representation of a language as a 'system of pure relations', brings language near to the abstract systems studied in mathematics. This representation makes concrete the concept of the function of language as a transformation of certain abstract objects — 'meanings' — into objects of another type — 'texts' — and conversely. This suggests the idea of studying this transformation (after elaboration of the notions of 'meaning' and 'text') by mathematical means. The application of this approach is difficult if one attempts to consider the transformation 'in the large', because of its extraordinary complexity and also because of the difficulty of formalizing the notion of 'meaning'. However, a clear understanding suggests breaking down the transformation into steps. For example, as one of the roughest articulations of a certain stage can be included the passage from 'meaning' of statements to 'syntactic structures without linear order' — a set of statements joined by 'syntactic relations' but not yet arranged in linear sequence; in the next stage one obtains linear sequences of words, which are then transformed into a chain of sounds. For a more subtle articulation syntactic structures at several levels are introduced, each further removed from the 'meaning' and closer to the 'text'; the 'post-syntactic' stages also were subject to further breaking down. Such a stage already is easier to describe mathematically if one makes more precise the representation of objects at the intermediate levels and simulates the transition from one level to another by effective mappings. It is true, this transformation is ambiguous, as also are all, or almost all (depending on the method of articulation), intermediate steps. This is related to one of the most important peculiarities of language, the presence of *synonyms*, that is, the possibility of expressing one and the same content by different words. Therefore it is suitable not to construct deterministic effective systems (algorithms) but to construct non-deterministic systems (calculi), which allow either for a given object at some level to enumerate the corresponding objects in the next level or the objects (in the same level) synonymous with it, or to enumerate the set of 'correct' objects of the given level (that is, those which by some known correct method can be associated with objects of the preceding level), or to enumerate the set of associated pairs of objects of two given neighbourhood levels (for example, 'a sentence + its syntactic structure'), etc. Calculi of this kind are known as formal grammars (cf. **Grammar, formal**). Simultaneously with formal grammars which simulate the transformations of linguistic objects there arise constructions meant for the formal description of these same objects. In addition, in a set of objects of one level there arise classifications and relations, in many ways similar to the categories of traditional grammar (such as parts of speech, gender, case, etc.) and in a number of cases coinciding with them. Without the introduction of these classifications and relations the actual construction of formal grammars for natural languages is practically impossible.

Thus it is possible to separate three aspects of the formal description of a language: the description of the structure of linguistic objects of different levels, the description of certain special relations and classifications into sets of these objects and the description of the transformations of individual objects to others, and also the structure of the sets of 'correct' objects. Associated with these aspects are the three fundamental divisions of mathematical linguistics: 1) the development and study of ways of describing the structure of parts of speech; 2) the study of linguistically significant relations and classifications into sets of linguistic objects (a formal system constructed for this purpose is usually called an **analytic model of a language**); and 3) the theory of formal grammars.

To describe the structure of parts of speech syntactic structures are used which are represented by a graph or di-graph of a special form, usually with labelled vertices and/or edges. It is the best developed theory of description for the 'surface' levels (that is, those most remote from the 'meaning'). At these levels the structures are usually trees. The methods of description of the 'deeper' levels have been intensively studied. For this, in particular, the apparatus of so-called *lexical functions* was proposed, playing a role in the description of the meaning of combinations of words similar to that played by the traditional categories of gender, case, number, etc. in the description of syntactic combinations. There is not yet any means of strictly describing the 'meaning' level, but in many investigations it appears likely that to this end a 'successive approximation' may produce an approach to the formal description of meaning. This does not exclude other approaches; in particular, much research is devoted to

methods for expressing in natural language predicates, propositional links, quantifiers, and to the 'translation' of formal-logical languages into natural languages and conversely. Here one borders on work on the design of so-called *semantic languages*, in which meanings are associated with texts by simple and strictly formal methods.

Analytic models of languages are important, in particular, because they allow the logical nature of many ideas and categories of traditional linguistics to be made more precise. These models do not always have the nature of effective procedures, since they may contain such ideas as an (infinite) set of grammatically correct sentences of some language being regarded as given. However, in a number of models the initial data are represented as finite sets and finitary relations; in these cases the procedures in the model are effective. Bordering on the theory of analytic models is the theory of *linguistic deciphering*: its objective is the construction of procedures, applied in suitable analytic models to 'unordered' empirical data of the language, which are always effective and allow one to obtain not only abstract definitions but also concrete information on the structure of a concrete language (for example, algorithms realizing the automatic subdivision of the set of phonemes of a language into the classes of vowels and constants without using any information on the language apart from some sufficiently long text).

The theory of formal grammars occupies a central position in mathematical linguistic because it allows one to simulate the most essential aspects of the function of language — the processing of meaning into text and conversely — and because of this it serves as a connecting link between the remaining divisions of mathematical linguistics. The nature of the apparatus of the theory of formal grammars is in many ways close to the theory of algorithms and automata. The most developed of the other types of formal grammars are those which characterize the set of grammatically correct sentences of a language and attribute a syntactic structure to those sentences. Here, sentences are modelled by chains (words) over a finite alphabet whose elements are to be interpreted as the words of a natural language (therefore, in mathematical linguistics the term 'chain' is preferred to the term 'word' and also the alphabet is often called a *dictionary*) and the model of the set of grammatically correct sentences is some formal language. In particular, generative grammars are of this type (cf. **Grammar, generative**). A generative grammar is essentially a special case of a Post calculus: it consists of a finite alphabet divided into two parts, the basic and the auxiliary alphabet, a finite set of *deduction rules*, representing the substitution rule in the form $\phi \to \psi$ (ϕ and ψ are chains) and one axiom (usually) consisting of one auxiliary symbol, known as the *initial symbol*. The (formal) language generated by such a grammar is the set of chains over the basic alphabet derived from the axioms. The most important class of generative grammars for linguistic applications are context-sensitive grammars (cf. **Grammar, context-sensitive**), in which each rule has the form $\xi_1 A \xi_2 \rightarrow \xi_1 \theta \xi_2$, where ξ_1, ξ_2, θ are chains in the union of the basic and auxiliary alphabets, A is an auxiliary symbol and θ is non-empty. A context-sensitive grammar allows one to concatenate in a natural way the chains of the labelled systems of constituents generated by its language. This class of grammars is also most important in purely mathematical terms, since the languages generated by context-sensitive grammars are a simple and very important subclass of the primitive recursive sets. Among the context-sensitive grammars, in turn, of special importance, both from the theoretical as from the applied point of view, are context-free grammars (cf. **Grammar, context-free**), in which the rules have the form $A \rightarrow \theta$, where A is an auxiliary symbol. Close to context-free grammars are dominating grammars (cf. **Grammar, dominating**), which also generate formal languages but the constituent chains in these languages are subordination trees, and categorial grammars (cf. **Grammar, categorial**), which are characterized by a special method of assignment of information on the syntactical properties of words. The other principal types of formal grammars are transformational grammars (cf. **Grammar, transformational**); they realise the transformation of a syntactic structure not 'attached', in general, to chains; these grammars present greater perspectives for the description of the structure of natural language since they allow one to discuss syntactic and linear relations between words separately, which better expresses linguistic practice.

The theory of formal grammars, alongside its 'traditional' linguistic applications, has found application in the theory of programming for the description of programming languages and translators. Particularly widespread used for these aims are the context-free grammars, but grammars of a more general form are also used.

References

[1] Сномѕкіі, N.: *New in Linguistics* **2** (1962), 412-527.
[2] GLADKIĬ, A.V. and MEL'CHUK, I.A.: *Elements of mathematical linguistics*, Mouton, 1973 (translated from the Russian).

A.V. Gladkiĭ

Editorial comments. The theory of generative grammars has developed further, a standard work is [A4]. This theory is applied to several grammar formalisms that are developed in linguistics in [A3]. The complexity theory of linguistic grammar formalisms is investigated in [A1].

The relation between the meaning of natural language and formal grammars has been developed by the logician R.

Montague [A5], [A6]. An induction (especially for linguists) is [A2].

References

[A1] BARTON, G.E., BERWICK, R.C. and RISTAD, E.S.: *Computational complexity and natural language*, M.I.T., 1987.
[A2] DOWTY, D.R., WALL, R.E. and PETERS, S.: *Introduction to Montague semantics*, Reidel, 1981.
[A3] SAVITCH, J.W., BACH, E., MARSH, W. and SAFRAN-NAVEH, G.: *The formal complexity of natural language*, Reidel, 1977.
[A4] HOPCROFT, J.E. and ULLMAN, J.D.: *Introduction to automata theory, languages and computation*, Addison-Wesley, 1979.
[A5A] MONTAGUE, R.: 'Universal grammar', *Theoria* **36** (1970), 373-398.
[A5B] MONTAGUE, R.: 'Universal grammar', in R.H. Thomason (ed.): *Formal Philosophy. Selected Papers of Richard Montague*, Yale Univ. Press, 1974, pp. 222-246. Reprint of [A5A].
[A6A] MONTAGUE, R.: 'The proper treatment of quantification in ordinary English', in K.J.J. Hintikka, J.M.E. Moravcsik and P. Suppes (eds.): *Approaches to Natural Language*, Reidel, 1973, pp. 221-242.
[A6B] MONTAGUE, R.: 'The proper treatment of quantification in ordinary English', in R.H. Thomason (ed.): *Formal Philosophy. Selected Papers of Richard Montague*, Yale Univ. Press, 1974, pp. 222-246. Reprint of [A6A].

AMS 1980 Subject Classification: 03B65, 68FXX

MATHEMATICAL LOGIC, *symbolic logic* - The branch of mathematics concerned with the study of mathematical proofs and questions in the foundation of mathematics.

Historical sketch. The idea of constructing a universal language for the whole of mathematics, and of the formalization of proofs on the basis of such a language, was suggested in the 17-th century by G. Leibniz. But not until the middle of the 19-th century did there appear the first scientific work on the algebraization of Aristotelean logic (G. Boole, 1847, A. de Morgan 1858). After G. Frege (1879) and C. Peirce (1885) put the logic of predicates, variables and quantifiers into the language of algebra, it became possible to apply this language to questions in the foundations of mathematics.

On the other hand, the creation of non-Euclidean geometry in the 19-th century strongly shook the confidence of mathematicians in the absolute reliability of the geometrical intuition on which Euclidean geometry was founded. Doubts about the reliability of geometrical intuition were also promoted as a result of the fact that in the development of infinitesimal calculus, mathematicians came across unexpected examples of everywhere-continuous functions without derivatives. The need appeared to separate the notion of a real number from the vague notion of a 'variable', which was based on geometric intuition. This problem was solved in various ways by K. Weierstrass, R. Dedekind and G. Cantor. They showed the possibility of 'arithmetizing' analysis and function theory, as a result of which the arithmetic of integers came to be considered as the foundation of the whole of classical mathematics. Subsequently, the axiomatization of arithmetic was undertaken (Dedekind, 1888, and G. Peano, 1891). In this connection, Peano created a more suitable symbolic representation for the language of logic. Afterwards, this language was perfected in the joint work of B. Russell and A. Whitehead, *Principia Mathematica* (1910), in which they attempted to reduce the whole of mathematics to logic. But this attempt was not crowned with success, since it turned out to be impossible to deduce the existence of infinite sets from purely logical axioms. Although the logistic program of Frege—Russell on the foundations of mathematics never achieved its major aim, the reduction of mathematics to logic, in their papers they created a rich logical apparatus without which the appearance of mathematical logic as a valuable mathematical discipline would have been impossible.

At the turn of the 19-th into the 20-th century, antinomies (cf. **Antinomy**) related to the fundamental ideas of set theory were found. The strongest impression at that time was made by the *Russell paradox*. Let M be the set containing exactly those sets which are not an element of itself. It is easy to convince oneself that M is an element of itself if and only if M is not an element of itself. Certainly one may circumvent this contradiction by stating that such a set M cannot occur. However, if a set consisting precisely of all the elements satisfying some clearly defined condition, such as that given above in the definition of M, need not exist, then where is the guarantee that in everyday work one will not meet with sets which also need not exist? And, in general, what conditions must the definition of a set satisfy so that the set does exist? One thing was clear: Cantor's theory of sets must somehow be restricted.

L.E.J. Brouwer (1908) opposed the application of rules of classical logic to infinite sets. In his intuitionistic program it was suggested that the **abstraction of actual infinity** be removed from the discussion, that is, remove infinite sets as complete collections. While admitting the existence of arbitrarily large natural numbers, the intuitionist comes out against the consideration of the natural numbers as a complete set. They believe that in mathematics every proof concerning the existence of an object must be constructive, that is, must be accompanied by a construction of that object. If the premise that the object sought for does not exist results in a contradiction, then this, in the opinion of intuitionists, cannot be considered as a proof of its existence. Particular criticism on the part of intuitionists is aimed at the **law of the excluded middle**.

Since the law was initially considered in association with finite sets, and taking into account the fact that many properties of finite sets are not satisfied by infinite sets (for example, that each proper part is less than the whole), intuitionists regard the application of this law to infinite sets as inadmissible. For example, in order to assert that Fermat's problem has a positive or a negative solution, the intuitionist must give the corresponding solution. So long as Fermat's problem is unsolved, this disjunction is regarded as illegitimate. The same requirement is imposed on the meaning of each disjunction. This requirement of intuitionists may create difficulties even in the consideration of problems connected with finite sets. Imagine that, with eyes closed, a ball is taken from an urn in which there are three black and three white balls, and that the ball is then put back. If no one sees the ball, then one cannot possibly know what colour it was. However, it is doubtful whether one can seriously dispute the certainty of the assertion that the ball was either black or white in colour.

Intuitionists have constructed their own mathematics, with interesting distinctive peculiarities, but it has turned out to be more complicated and cumbersome than classical mathematics. The positive contribution of intuitionists to the investigation of questions in the foundations of mathematics is that they have once more decisively stressed the distinction between the constructive and the non-constructive in mathematics; they have made a careful analysis of the many difficulties which have been encountered in the development of mathematics, and, by the same token, they have contributed to overcoming them.

D. Hilbert (see the Appendices VII - X in [9]) planned another way to overcome the difficulties arising in the foundations of mathematics at the turn of the 19-th into the 20-th century. This route, based on the application of the axiomatic method in the discussion of formal models of interesting mathematics, and in the investigation of questions of consistency of such models by reliable finitary means, was given the name *Hilbert's finitism* in mathematics. Recognizing the unreliability of geometrical intuition, Hilbert first of all undertook a careful review of Euclidean geometry, liberating it from the appeal to intuition. As a result of this revision his *Grundlagen der Geometrie* (1910) appeared, [9].

Questions of consistency of various theories were essentially considered even before Hilbert. Thus the projective model of the non-Euclidean Lobachevskiĭ geometry constructed by F. Klein (1871) reduced the question of consistency of Lobachevskiĭ's geometry to the consistency of Euclidean geometry. The consistency of Euclidean geometry can similarly be reduced to the consistency of analysis, that is, the theory of real numbers. However, it was not clear how it would be possible to construct models of analysis and arithmetic for these consistency proofs. The merit of Hilbert is that he gave a direct way for the investigation of this question. Consistency of a given theory means that one cannot obtain a *contradiction* in it, that is, it is not possible to prove both an assertion A and its negation $\neg A$. Hilbert suggested representing the theory under discussion as a formal axiomatic system, in which those and only those assertions are derivable which are theorems of that theory. Then for the proof of consistency it suffices to establish the non-derivability in the theory of certain assertions. Thus a mathematical theory whose consistency one wishes to prove becomes an object of study in a mathematical science which Hilbert called *metamathematics*, or **proof theory**.

Hilbert wrote that the paradoxes of set theory have arisen not from the law of the excluded middle but 'rather that mathematicians have used inadmissible and meaningless formations of ideas which in my proof theory are excluded $\cdots$. To remove from mathematicians the law of the excluded middle is the same as taking the telescope away from astronomers or forbidding a boxer to use his fists' (see [9]). Hilbert suggested distinguishing between 'real' and 'ideal' assumptions of classical mathematics. The first have a genuine meaning but the second need not. Assumptions corresponding to the use of the actual infinite are ideal. Ideal assumptions can be added to the real ones in order that the simple results of logic be applicable even to arguments about infinite sets. This essentially simplifies the structure of the whole theory in much the same way as the addition of the line at infinity intersecting any two parallel lines in the projective geometry of the plane.

The program suggested by Hilbert for the foundation of mathematics, and his enthusiasm for it, inspired his contemporaries into an intensive development of the **axiomatic method**. Thus the formation of mathematical logic as an independent mathematical discipline is linked with Hilbert's initiative, at the start of 20-th century, and the subsequent development of proof theory based on the logical language developed by Frege, Peano and Russell.

The objective and fundamental branches of mathematical logic; relation to other areas of mathematics. The objective of modern mathematical logic is diverse. First of all one must mention the investigation of logical and logico-mathematical calculi founded on classical predicate calculus. In 1930 K. Gödel proved the completeness theorem for predicate calculus, according to which the set of all valid purely logical assertions of mathematics coincides with the set of all derivable formulas in predicate calculus (see **Gödel completeness**

theorem). This theorem showed that predicate calculus is a logical system on the basis of which mathematics can be formulated. Based on predicate calculus various logico-mathematical theories have been constructed (see **Logico-mathematical calculus**), representing the formalization of interesting mathematical theories: arithmetic, analysis, set theory, group theory, etc. Side-by-side with elementary theories (cf. **Elementary theory**), higher-order theories were also considered. In these one also admits quantifiers over predicates, predicates over predicates, etc. The traditional questions studied in these formal logical systems were the investigation of the structure of the deductions in the system, derivability of various formulas, and questions of consistency and completeness.

The **Gödel incompleteness theorem** on arithmetic, proved in 1931, destroyed the optimistic hopes of Hilbert for a complete solution of questions in the foundations of mathematics by the means mentioned above. According to this theorem, if a formal system containing arithmetic is consistent, then the assertion of its consistency expressed in the system cannot be proved by formalization within it. This means that with questions on foundations of mathematics the matter is not as simple as first desired or believed by Hilbert. But Gödel had already noted that the consistency of arithmetic could be proved by using fairly reliable constructive means, although still far from the means that are formalized in arithmetic. Similar proofs of the consistency of arithmetic were obtained by G. Gentzen (1936) and P.S. Novikov (1943) (cf. also **Gentzen formal system**).

As a result of the analysis of Cantor's set theory and the related paradoxes, various systems of **axiomatic set theory** were constructed with various restrictions on the formation of sets, to exclude the known inconsistencies. Within these axiomatic systems suitably extensive parts of mathematics could be developed. The consistency question for fairly rich axiomatic systems of set theory remains open. Of the most important results obtained in axiomatic set theory one must note Gödel's result on the consistency of the **continuum hypothesis** and the **axiom of choice** in the Bernays$-$Gödel system Σ (1939), and the results of P. Cohen (1963) on the independence of these axioms from the Zermelo$-$Fraenkel axioms ZF. One should note that these two axiom systems, Σ and ZF, are equiconsistent. For the proof of his result Gödel introduced the important idea of a constructible set (see **Gödel constructive set**) and proved the existence of a model of Σ consisting of those sets. Gödel's method was used by Novikov for the proof of the consistency of certain other results in **descriptive set theory** (1951). For the construction of models of the set theory ZF in which the negation of the continuum hypothesis or the axiom of choice holds, Cohen introduced the so-called **forcing method**, which subsequently became a fundamental method for constructing models of set theory with various properties (cf. also **Model theory**).

One of the most remarkable achievements of mathematical logic was the development of the notion of a **general recursive function** and the formulation of the **Church thesis**, asserting that the notion of a general recursive function makes precise the intuitive notion of an **algorithm**. Of the equivalent elaborations of the notion of an algorithm the most widely used are the idea of a **Turing machine** and a Markov **normal algorithm**. In essence all mathematics is connected with some algorithm or other. But the possibility of identifying undecidable algorithmic problems (cf. **Algorithmic problem**) in mathematics appeared only with the refinement of the notion of an algorithm. Undecidable algorithmic problems were discovered in many areas of mathematics (algebra, number theory, topology, probability theory, etc.). Moreover, it turned out that they could be connected with very widespread and fundamental ideas in mathematics. Research into algorithmic problems in various areas of mathematics, as a rule, is accompanied by the penetration of the ideas and methods of mathematical logic into the area, which then leads to the solution of other problems no longer of an algorithmic nature.

The development of a precise notion of an algorithm made it possible to refine the notion of effectiveness and to develop on that basis a refinement of the constructive directions in mathematics (see **Constructive mathematics**), which embodies certain features of intuitionism, but is essentially different from the latter. The foundations of constructive analysis, constructive topology, constructive probability theory, etc., were laid.

In the theory of algorithms itself it is possible to pick out research in the domain of recursive arithmetic, containing various classifications of recursive and recursively-enumerable sets, degrees of undecidability of recursively-enumerable sets, research into the complexity of description of algorithms and the complexity of algorithmic calculations (in time and extent, see **Algorithm, computational complexity of an**; **Algorithm, complexity of description of an**). An extensively developing area in the theory of algorithms is the theory of **enumeration**.

As noted above, the axiomatic method exerted a major influence on the development of many areas of mathematics. Of special significance was the penetration of this method into algebra. Thus at the junction of mathematical logic and algebra, the general theory of algebraic systems (cf. **Algebraic system**), or **model theory**, arose. The foundations of this theory were laid

by A.I. Mal'tsev, A. Tarski and their followers. Here one should note research into the elementary theory of classes of models, in particular, decidability questions in these theories, axiomatizability of classes of models, isomorphism of models, and questions of categoricity and completeness of classes of models.

An important place in model theory is occupied by studies on non-standard models of arithmetic and analysis. Even at the dawn of differential calculus, in the work of Leibniz and I. Newton, infinitely-small and infinitely-large quantities were regarded as numbers. Later the notion of a variable quantity appeared, and mathematicians turned away from the use of infinitely-small numbers, the modulus of which was different from zero and less than any positive real number, since their use required the loss of the Archimedean axiom. Only after three centuries, as a result of the development of the methods of mathematical logic, was it possible to establish that (non-standard) analysis with infinitely-small and infinitely-large numbers is consistent relative to the usual (standard) analysis of real numbers (cf. **Non-standard analysis**).

One could not conclude without mentioning the influence of the axiomatic method on intuitionistic mathematics. Thus, as long ago as 1930, A. Heyting introduced formal systems of **intuitionistic logic** of propositions and predicates (constructive propositional and predicate calculi). Later, formal systems of intuitionistic analysis were introduced (see, for example, [8]). Much of the research in intuitionistic logic and mathematics is concerned with formal systems. Special study was made of so-called intermediate logics (cf. **Intermediate logic**; also called super-intuitionistic logics), that is, logics lying between classical and intuitionistic logics. The notion of Kleene realizability of formulas is an attempt to interpret the idea of intuitionistic truth from the point of view of classical mathematics. However, it turned out that not every realizable formula of propositional calculus was derivable in intuitionistic (constructive) propositional calculus.

Modal logic has also been formalized. However, in spite of the large number of papers on formal systems of modal logic and its semantics (**Kripke models**), this can still be said to be an accumulation of uncoordinated facts.

Mathematical logic has a more applied value too; with each year there is a deeper penetration of the ideas and methods of mathematical logic into cybernetics, computational mathematics and structural linguistics.

References
[1] Hilbert, D. and Bernays, P.: *Grundlagen der Mathematik*, 1-2, Springer, 1968-1970.
[2] Kleene, S.C.: *Introduction to metamathematics*, North-Holland, 1951.

[3] MENDELSON, E.: *Introduction to mathematical logic*, v. Nostrand, 1964.

[4] NOVIKOV, P.S.: *Elements of mathematical logic*, Oliver & Boyd, 1964 (translated from the Russian).

[5] ERSHOV, YU.L. and PALYUTIN, E.A.: *Mathematical logic*, Moscow, 1979 (in Russian).

[6] SHOENFIELD, J.R.: *Mathematical logic*, Addison-Wesley, 1967.

[7] NOVIKOV, P.S.: *Constructive mathematical logic from the classical point of view*, Moscow, 1977 (in Russian).

[8] KLEENE, S.C. and VESLEY, R.E.: *The foundations of intuitionistic mathematics: especially in relation to recursive functions*, North-Holland, 1965.

[9] HILBERT, D.: *Grundlagen der Geometrie*, Springer, 1913.

[10] FRAENKEL, A.A., BAR-HILLEL, Y. and LEVY, A.: *Foundations of set theory*, North-Holland, 1973.

[11] *Mathematics of the 19-th century. Mathematical logic. Algebra. Number theory. Probability theory*, Moscow, 1978 (in Russian).

[12] MOSTOWSKI, A.: 'Thirty years of foundational studies', in *Foundational Studies: Selected Works*, Vol. 1, PWN & North-Holland, 1979, pp. 1-176.

See also the references to the articles on the various branches of mathematical logic.

S.I. Adyan

Editorial comments. A basic reference for mathematical logic as a whole is [A8]. The pioneering seminal work in non-standard analysis is due to A. Robinson, [A10], cf. also [A2].

References

[A1] HEYTING, A.: *Intuitionism, an introduction*, North-Holland, 1956.

[A2] ROBINSON, A.: *Non-standard analysis*, North-Holland, 1971.

[A3] MANIN, YU.I.: *A course in mathematical logic*, Springer, 1977.

[A4] BROUWER, L.E.J.: *Collected works*, North-Holland, 1975.

[A5] ŠANIN, N.A. [N.A. SHANIN]: 'On the constructive interpretation of mathematical judgements', *Transl. Amer. Math. Soc. (2)* 23 (1963), 109-190. (*Trudy Mat. Inst. Steklov.* 52 (1958), 226-311)

[A6] TROELSTRA, A.S. and DALEN, D. VAN: *Constructivism in mathematics, an introduction*, 1-2, North-Holland, 1989.

[A7] BISHOP, E.A.: *Foundations of constructive analysis*, McGraw-Hill, 1967.

[A8] BARWISE, J. (ED.): *Handbook of mathematical logic*, North-Holland, 1977.

[A9] MÜLLER, G. (ED.): *Bibliography of mathematical logic*, 1-4, Springer, 1988.

[A10] ROBINSON, A.: *Introduction to model theory and the metamathematics of algebra*, North-Holland, 1963.

AMS 1980 Subject Classification: 03-XX

MATHEMATICAL MODEL - A (rough) description of some class of events of the outside world, expressed using mathematical symbolism. A mathematical model is a powerful tool for understanding the outside world and for prediction and control. The analysis of a mathematical model allows the essence of a phenomenon to be penetrated. The process of *mathematical modelling*, that is, the study of phenomena with the aid of mathematical models, can be divided into four stages.

The first stage is the statement of the laws relating the basic objects of the model. This stage requires a broad knowledge of the facts relating to the phenomenon and a high penetration into their interconnections. This stage is completed with the description, in mathematical terms, of the formulated qualitative representations of the connections between the objects of the model.

The second stage is the investigation of the mathematical problems to which the mathematical model leads. The fundamental question here is the solution of the *direct problem*, that is, as a result of analysis of the model to obtain output data (theoretical consequences), which are then compared with the results of observation of the phenomenon. In this stage the mathematical apparatus necessary for the analysis of the mathematical model and the computational techniques — the powerful means for obtaining quantitative output of information as a result of the solution of complex mathematical problems — assume a major role. Frequently mathematical problems arising on the basis of various mathematical models turn out to be identical (for example, the basic problem of **linear programming** reflects various situations in nature). This provides a basis for considering these typical mathematical problems as independent objects abstracted from the phenomena.

The third stage is the clarification of whether the adopted (hypothetical) model satisfies the practical criterion, that is, whether the observed results agree with theoretical consequences of the model within the limits of accuracy of the observations. If the model was fully determined, all its parameters being given, then determination of the derivation of the theoretical consequences from the observations gives the solution to the direct problem with a posteriori estimates of the deviation. If the deviation lies outside the limits of precision of the observations, then the model cannot be accepted. Frequently, in the construction of models certain characteristics are left undetermined. Problems in which the characteristics of the model (parametric, functional) are determined so that the output information is comparable, within the limits of precision of the observations, with results of observations of the phenomenon, are called *inverse problems*. If a mathematical model is such that there is no choice of characteristic which satisfies these conditions, then the model is useless for the investigation of the phenomenon. The application of a practical criterion to appraise a mathematical model allows one to draw conclusions on the validity of the assumptions underlying the (hypothetical) model being studied. This is the only method of studying phenomena of the macro- and micro-world which are not directly accessible.

The fourth stage is the a posteriori analysis of the model in conjunction with the observation data of the phenomenon, and the updating of the model. In the process of development of science and technology, phenomenological data become more and more precise and the time comes when the output obtained on the basis of an accepted mathematical model does not correspond to knowledge of the phenomenon. Therefore the need arises to construct a new, more precise, mathematical model.

A typical example illustrating the characteristic stages in the construction of a mathematical model is the model of the solar system. Observation of the stars in the sky began in early Antiquity. A primary analysis of these observations allowed one to pick out the planets from the whole variety of celestial bodies. Thus, the first stage was the selection of the objects of study. The second stage was the determination of regularity in their motions (generally, the definition of the objects and their interconnections are the starting point, 'axioms', of the model). The model of the solar system during this development passed through a number of successive refinements. The first was the Ptolemeus model (second century B.C.), starting from the position that the planets and the Sun moved around the Earth (the geocentric model), and when this motion had been described by rules (formulas), these became increasingly complicated with the increase of observations.

The development of navigation raised, for astronomy, new requirements for precision in observation. N. Copernicus in 1543 suggested a fundamentally new basis for the laws of motion of planets by assuming that the planets rotate around the Sun in circles (the heliocentric system). This was a qualitatively (but not mathematically) new model of the solar system. However, the parameters of the system (the radii of the circles and the angular velocities of the motions) could not be measured, which resulted in quantitative conclusions of the theory not in proper correspondence with observations, so Copernicus was forced to introduce amendments to the motions of the planets by circles (epicycles).

The next stage in the development of the model of the solar system was the research by J. Kepler (1571 - 1630), in which the laws of planetary motion were formulated. The aims of Copernicus and Kepler were to give a kinematic description of the motion of each individual planet without touching upon the reasons for these motions.

A principally new stage was the work of I. Newton, who proposed, in the second half of the 17-th century, a dynamic model of the solar systems based on the law of universal gravitation (cf. **Newton laws of mechanics**). The dynamic model was consistent with the kinematic model suggested by Kepler, since Kepler's laws followed from the dynamical system of two bodies, 'Sun − planet'.

In the 1840's, conclusions of the dynamic models, the objects of which were the visible planets, were found in contradiction with the observations collected at that time. Namely, the observed motion of the planet Uranus deviated from the theoretically calculated motion. U. le Verrier in 1846 expanded the system of observed planets with a new hypothetical planet, named by him Neptune, and, by using the new model for the solar system, determined the mass and the law of motion of the new planet in order that in the new system the contradiction in the motion of Uranus was removed. The planet Neptune was discovered at the place stated by le Verrier. By a similar means, using the divergence of Neptune, the planet Pluto was discovered in 1930.

The method of mathematical modelling, reducing the investigation of the phenomena of the outside world to mathematical problems, occupies a leading position among methods of research, particularly in connection with the appearance of electronic computers. It allows one to design new technical means of working in optimal regimes for the solution of complex scientific and technical problems and to predict new phenomena. Mathematical models have been shown to be an important means of control. They are applied in very diverse domains of knowledge and have become a necessary tool in economic planning and an important element in automatic control systems (cf. **Automatic control theory**).

A.N. Tikhonov

Editorial comments. For models in logic (i.e. models of axiomatic systems), cf. **Model (in logic)** and **Model theory**.

References

[A1] BOCHNER, S.: *The role of mathematics in the rise of science*, Princeton Univ. Press, 1981.
[A2] KUHN, T.S.: *The Copernican revolution*, Cambridge, Mass., 1981.
[A3] NAGEL, E.: *The structure of science*, London, 1974.
[A4] ROSENBLUETH, A. and WIENER, N.: 'The role of models in science', *Philosophy of Science* 12 (1949).
[A5] BENDER, E.A.: *An introduction to mathematical modelling*, Wiley, 1978.

AMS 1980 Subject Classification: 00A69

MATHEMATICAL PHYSICS - The theory of mathematical models of physical events; it holds a special position, both in mathematics and physics, being found at the junction of the two sciences.

Mathematical physics is closely connected with the part of physics concerned with the construction of mathematical models and, at the same time, is a branch of mathematics, since the methods of investigation of these models are mathematical. Included in the notion of methods of mathematical physics are those mathematical methods which are used for the construction and study of mathematical models describing large classes of physical phenomena.

The methods of mathematical physics, as also the theory of mathematical models in physics, were first intensively developed by I. Newton in the creation of the foundations of classical mechanics, universal gravitation and the theory of light (cf. **Newton laws of mechanics**). Subsequent development of the methods of mathematical physics and their application to the study of mathematical models in an extensive domain of physical phenomena is connected with the names of J.L. Lagrange, L. Euler, J. Fourier, C.F. Gauss, B. Riemann, M.V. Ostrogradski, and many other scholars. A major contribution to the development of the methods of mathematical physics was made by A.M. Lyapunov and V.A. Steklov.

Beginning in the second half of the 19-th century, the methods of mathematical physics were successfully applied in the study of mathematical models of physical phenomena connected with the various physical field and wave processes in electrodynamics, acoustics, elasticity theory (cf. **Elasticity, mathematical theory of**), hydro- and aerodynamics (cf. **Hydrodynamics, mathematical problems in**), and a number of other research directions in the physics of continuous media. The mathematical models of this class of phenomena are most frequently described by partial differential equations, which have been given the name 'the equations of mathematical physics' (cf. **Mathematical physics, equations of**).

In addition to the differential equations of mathematical physics, in describing the mathematical models of physics one finds applications of integral equations and integro-differential equations, variational and probability-theoretical methods, potential theory, methods from complex function theory and from a number of other branches of mathematics. In connection with the vigorous development of computational mathematics, direct numerical methods, using computers (finite-difference methods and other computational algorithms for boundary value problems), acquire a special significance for the investigation of mathematical models of physics, and, by the methods of mathematical physics, have allowed the effective solution of new problems in gas dynamics (cf. **Gas dynamics, numerical methods of**), transport theory (cf. **Transport equations, numerical methods**), plasma physics, including the inverse problems for these most important directions of physical research.

Theoretical research in quantum physics (cf. **Quantum field theory**) and **relativity theory**, the widespread use of computers in various areas of mathematical physics, including inverse (ill-posed) problems, have required a significant expansion of the arsenal of mathematical methods used in mathematical physics.

Side by side with the traditional branches of mathematics there has been widespread application of operator theory, distribution theory, the theory of functions of several complex variables, and topological and algebraic methods. This intensive interaction of mathematical physics, mathematics and the use of computers in scientific research has led to a significant expansion of the subject, the creation of new classes of models and has raised modern mathematical physics to a new level. It has made a great contribution to the acceleration of scientific and technological progress.

The formulation of problems in mathematical physics results in the construction of mathematical models which describe the fundamental laws of the class of physical phenomena being studied. The formulation consists of deriving the equations (differential, integral, integro-differential, or algebraic) satisfied by the quantities characterizing the physical process. Here one proceeds from fundamental laws of physics, taking into account only the more essential features of the phenomenon and ignoring a number of secondary characteristics. Such laws are the usual laws of conservation of, for example, momentum, energy, the number of particles, etc. They are connected with symmetry groups, spatio-temporal ones and others. This leads to the fact that for the description of processes of different physical natures — but having common characteristic features — the same mathematical model can be used. For example, the mathematical problems for the simplest equation of hyperbolic type,

$$\frac{\partial^2 u}{\partial t^2} = a^2 \frac{\partial^2 u}{\partial x^2},$$

obtained originally by J. d'Alembert (1747) for the description of the free oscillations of a homogeneous string, turned out to be applicable also for the description of a broad range of wave processes in acoustics, hydrodynamics and other areas of physics. Similarly, the equation

$$\frac{\partial^2 u}{\partial x^2} + \frac{\partial^2 u}{\partial y^2} + \frac{\partial^2 u}{\partial z^2} = 0,$$

the boundary value problems of which were originally studied by P. Laplace (at the end of the 18-th century) in connection with the construction of a gravitation theory (cf. **Laplace equation**), has subsequently found application in the solution of many problems in electrostatics, elasticity theory, the problem of steady motion of an ideal fluid, etc. Each mathematical model of physics corresponds to a whole class of physics processes.

It is also characteristic of mathematical physics that many general methods used for the solution of the problems of mathematical physics were developed from particular methods of solution of concrete physical problems and, in their original form, did not have a rigorous mathematical foundation and sufficient perfection. This applies to such well-known methods of solution of problems in mathematical physics as the methods of Ritz and Galerkin (cf. **Galerkin method**; **Ritz method**), the methods of **perturbation theory**, the **Fourier transform**, and many others, including the method of separation of variables (cf. **Separation of variables, method of**). The effectiveness of the application of all of these methods for the solution of concrete problems is one of the reasons for giving them a mathematical proof and generalizations, leading in a number of cases to the appearance of new directions in mathematics.

The influence of mathematical physics on the various branches of mathematics is shown by the fact that the development of mathematical physics, reflecting the needs of the natural sciences and practical needs, has, on occasions, implied a re-orientation of the directions of research in certain already-established domains of mathematics. The posing of problems in mathematical physics connected with the use of mathematical models of real physical phenomena has led to a change in the fundamental problems considered in the theory of partial differential equations. The theory of boundary value problems (cf. **Boundary value problems, partial differential equations**) has arisen, subsequently allowing a connection between differential equations, integral equations and variational methods.

The study of mathematical models of physics by mathematical methods not only allows one to obtain quantitative characteristics of physical phenomena and to compute, with a given degree of accuracy, the course of real processes, but also provides the possibility of deep penetration into the very essence of the physical phenomena, an explanation of hidden laws and the prediction of new effects. The trend towards a more detailed study of physical phenomena leads to more complicated mathematical models describing these phenomena, which in turn make it impossible to apply analytic methods of research for these models. This is explained, in particular, by the fact that the mathematical models of real physical process are, as a rule, non-linear, that is, are described by non-linear equations of mathematical physics. For a detailed investigation of these models direct numerical methods using computers are successful. For typical problems in mathematical physics, the application of numerical methods leads, e.g., to replacing the equations of mathematical physics, for functions of continuous arguments, by algebraic equations for grid functions, given on a discrete set of points (a grid). In other words, instead of a continuous model of a medium one introduces a discrete analogue. The application of numerical methods in a number of cases allows one to replace complex, time-consuming

and expensive physical experiments by a significantly more economical mathematical (numerical) experiment. A mathematical experiment, properly carried out, is a basis for a choice of optimal conditions for a real physical experiment, the choice of parameters of complicated physical equipment, the determination of the conditions for the appearance of new physical effects, etc. Thus, numerical experiments usually expand the domain of effective utilization of mathematical models of physics phenomena.

Mathematical models of physics phenomena, as with all models, need not convey all the aspects of a phenomenon. Establishing the adequacy of an accepted model for investigating a phenomenon is possible only with the help of a practical criterion: comparing the results of theoretical research with experimental data.

In many cases it is possible to judge the adequacy of a model on the basis of the solution of an inverse problem of mathematical physics, when inferences on those properties of the phenomena which are inaccessible to direct observation are made on the basis of their indirect physical and observable effects.

A characteristic feature of mathematical physics is to construct mathematical models which not only describe and explain the already established physical phenomena in the domain of events being studied, but also allow one to predict phenomena not already discovered. The classical example of such a model is Newton's theory of universal gravitation, which not only unified the motion of the then known bodies of the solar system, but also permitted the prediction of the existence of new planets. On the other hand, new experimental data need not always fit into the accepted model. To take account of this a refinement of the model may be required.

References

[1] TICHONOFF, A.N. [A.N. TIKHONOV] and SAMARSKIĬ, A.A.: *Differentialgleichungen der mathematischen Physik*, Deutsch. Verlag Wissenschaft., 1959 (translated from the Russian).
[2] VLADIMIROV, V.S.: *Equations of mathematical physics*, Mir, 1984 (translated from the Russian).
[3] SOBOLEV, S.L.: *Partial differential equations of mathematical physics*, Pergamon, 1964 (translated from the Russian).
[4] COURANT, R. and HILBERT, D.: *Methods of mathematical physics*, 1-2, Interscience, 1953-1962 (translated from the German).
[5] MORSE, P.M. and FESHBACH, H.: *Methods of theoretical physics*, 1-2, McGraw-Hill, 1953.

A.N. Tikhonov
A.A. Samarskiĭ
A.G. Sveshnikov

Editorial comments. A recent development in field theories of elementary particles is the connection between the (non-linear) couplings and invariance properties under so-called gauge groups of transformations (cf. **Gauge transformation**).

References

[A1] MATHEWS, J. and WALKER, R.L.: *Mathematical methods of physics*, Benjamin, 1965.

[A2] CHOQUET-BRUHAT, Y., DEWITT-MORETTE, C. and DILLARD-BLEICH, M.: *Analysis, manifolds, physics*, North-Holland, 1977 (translated from the French).

[A3] REED, M. and SIMON, B.: *Methods of modern mathematical physics*, 1-4, Acad. Press, 1972-1978.

AMS 1980 Subject Classification: 70-XX, 73-XX, 81-XX, 82-XX, 83-XX, 35Q20

MATHEMATICAL PHYSICS, EQUATIONS OF - Equations which describe mathematical models of physical phenomena. The equations of mathematical physics are part of the subject of **mathematical physics.** Numerous phenomena of physics and mechanics (hydro- and gas-dynamics, elasticity, electro-dynamics, optics, transport theory, plasma physics, quantum mechanics, gravitation theory, etc.) can be described by boundary value problems for differential equations. A very wide class of models is reducible to such boundary value problems.

A complete description of the evolution of physical processes requires, first, the specification of the state of the process at some fixed moment of time (the *initial conditions*) and, secondly, the specification of the state on the boundary of the medium in which the process considered occurs (the *boundary conditions*). The initial and boundary conditions form the *boundary value conditions*, and the differential equations together with corresponding boundary value conditions define a boundary value problem of mathematical physics.

Below some examples of equations and corresponding boundary value problems are given.

The equation of oscillations

$$\rho\frac{\partial^2 u}{\partial t^2} = \text{div}(p\,\text{grad}\,u) - qu + f(x, t) \tag{1}$$

describes the small vibrations of strings, membranes, and acoustic and electromagnetic oscillations. In (1) the space variables $x = (x_1, \ldots, x_n)$ vary in a region $G \in \mathbf{R}^n$, $n = 1, 2, 3$, in which the physical process under consideration evolves; also, by their physical meaning the quantities appearing in (1) are such that $\rho > 0$, $p > 0$ and $q \geq 0$. Moreover, it is assumed that $\rho, q \in C(\overline{G})$ and $p \in C^1(\overline{G})$. Under these conditions (1) is a **hyperbolic partial differential equation.**

For $\rho = 1$, $p = a^2 = $const and $q = 0$, (1) becomes the **wave equation**

$$\frac{\partial^2 u}{\partial t^2} = a^2 \Delta u + f(x, t), \tag{2}$$

where Δ is the Laplace operator.

The diffusion equation

$$\rho\frac{\partial u}{\partial t} = \text{div}(p\,\text{grad}\,u) - qu + f(x, t) \tag{3}$$

describes processes of particle diffusion and heat transport in media. Equation (3) is a **parabolic partial dif-**ferential equation. For $\rho = 1$, $p = a^2 = $const and $q = 0$ it becomes the **thermal conductance equation** (*heat equation*):

$$\frac{\partial u}{\partial t} = a^2 \Delta u + f(x, t). \tag{4}$$

For stationary processes, in which there is no dependence on the time t, equation (1) and the diffusion equation (3) both take the form

$$-\text{div}(p\,\text{grad}\,u) + qu = f(x). \tag{5}$$

This is an **elliptic partial differential equation**. For $p = 1$ and $q = 0$ (5) is called the **Poisson equation**:

$$\Delta u = -f(x), \tag{6}$$

and for $f = 0$ — the **Laplace equation**:

$$\Delta u = 0. \tag{7}$$

Equations (6) and (7) are satisfied by various kinds of potentials: The Coulomb (Newton) potential, the potentials of the flows of incompressible fluids, etc.

If in the wave equation (2) the external perturbation f is periodic with frequency ω:

$$f(x, t) = a^2 f(x)e^{i\omega t},$$

then the amplitude $u(x)$ of a periodic solution with the same frequency ω,

$$u(x, t) = u(x)e^{i\omega t},$$

satisfies the **Helmholtz equation**

$$\Delta u + k^2 u = -f(x), \quad k^2 = \frac{\omega^2}{a^2}. \tag{8}$$

One is led to the Helmholtz equation by considering a scattering (diffraction) problem.

For a complete description of the oscillatory process it is necessary to give the initial perturbation and the initial velocity:

$$u\big|_{t=0} = u_0(x), \quad \frac{\partial u}{\partial t}\bigg|_{t=0} = u_1(x), \quad x \in \overline{G}. \tag{9}$$

In the case of a diffusion process it suffices to give the initial perturbation

$$u\big|_{t=0} = u_0(x), \quad x \in \overline{G}. \tag{10}$$

Moreover, on the boundary S of G the solution must take the prescribed values. In the simplest cases, physically-meaningful boundary conditions for equations (1), (3), (5) are described by the relations

$$k\frac{\partial u}{\partial \mathbf{n}} + hu\bigg|_S = v(x, t), \quad t > 0, \tag{11}$$

where k and h are given non-negative functions that do not vanish simultaneously, $\mathbf{n}$ is the outward normal to S, and v is a given function.

Thus, for a string the condition

$$u\big|_{x=x_0} = 0$$

means that the end x_0 of the string is fixed, whereas

the condition

$$\left.\frac{\partial u}{\partial x}\right|_{x=x_0} = 0$$

means that the end x_0 is free. For the heat equation the condition

$$u\,|_S = v_0(x,\,t) \qquad (12)$$

means that on the boundary S of G a prescribed temperature distribution is kept, whereas the condition

$$\left.\frac{\partial u}{\partial \mathbf{n}}\right|_S = v_1(x,\,t) \qquad (13)$$

prescribes the heat flow across S. In the case of unbounded regions, for example in the exterior of a bounded domain, the boundary conditions must be supplemented by a condition at infinity. Thus, for the Poisson equation (6) in space ($n = 3$), such a condition is

$$u(x) = o(1), \quad |x| \to \infty, \qquad (14)$$

whereas in the plane ($n = 2$) it is

$$u(x) = O(1), \quad |x| \to \infty. \qquad (15)$$

For the Helmholtz equation (8) one imposes at infinity the Sommerfeld radiation condition (cf. **Radiation conditions**)

$$u(x) = O(|x|^{-1}),$$

$$\frac{\partial u(x)}{\partial |x|} \mp iku(x) = o(|x|^{-1}), \quad |x| \to \infty, \qquad (16)$$

the sign '$-$' (respectively, '$+$') corresponds to outgoing (respectively, incident) waves.

A boundary value problem that involves only initial conditions (and hence does not contain boundary conditions, so that G is the whole space $\mathbf{R}^n$) is called a **Cauchy problem**. For the equation of oscillation (1) the Cauchy problem (1), (9) is posed as follows: To find a function $u(x,\,t)$ of class $C^2(t>0) \cap C^1(t \geq 0)$ which satisfies (1) for $t>0$ and the initial conditions (9) on the plane $t=0$. The Cauchy problem (3), (10) for the diffusion equation is posed in an analogous manner.

If a boundary value problem involves both initial and boundary conditions, then it is called a *mixed problem*. For equation (1) the mixed problem (1), (9), (11) is posed as follows: To find a function $u(x,\,t)$ of class

$$C^2(G \times (0,\,\infty)) \cap C^1(\overline{G} \times [0,\,\infty))$$

which satisfies equation (1) in the cylinder $G \times (0,\,\infty)$, the initial conditions (9) on its bottom base, $\overline{G} \times \{0\}$, and the boundary condition (11) on its lateral surface $S \times [0,\,\infty)$. The mixed problem (3), (10), (11) for the diffusion equation (3) is posed in an analogous manner. There exist also other formulations of boundary value problems, for example the **Goursat problem** and the **Tricomi problem**.

For the stationary equation (5) there are no initial conditions and the corresponding boundary value problem is posed as follows: To find a function $u(x)$ of

class $C^2(G) \cap C^1(\overline{G})$ that satisfies equation (5) in a region G and the boundary condition

$$\left.k\frac{\partial u}{\partial \mathbf{n}} + hu\right|_S = v(x) \qquad (11')$$

on the boundary S of G. For equation (5) the boundary value problem with boundary condition

$$u\,|_S = v_0(x) \qquad (12')$$

is called the **Dirichlet problem**, and with boundary condition

$$\left.\frac{\partial u}{\partial \mathbf{n}}\right|_S = v_1(x) \qquad (13')$$

— the **Neumann problem**. One distinguishes the exterior and the interior Dirichlet and Neumann problems. For the exterior problems the boundary conditions must be supplemented by conditions at infinity of the type (14), (15) or (16).

The following eigen value problems are also regarded as boundary value problems for equation (5): To find the values of the parameter λ (the eigen values) for which the homogeneous equation

$$Lu \equiv -\mathrm{div}(p\,\mathrm{grad}\,u) + qu = \lambda \rho u \qquad (17)$$

has non-trivial solutions (eigen functions) that satisfy the homogeneous boundary condition

$$\left.k\frac{\partial u}{\partial \mathbf{n}} + hu\right|_S = 0. \qquad (18)$$

If G is a bounded region with sufficiently smooth boundary S, then there exists a countable set of non-negative eigen values $\lambda_1, \lambda_2, \ldots,$ of problem (17), (18) ($0 \leq \lambda_1 \leq \lambda_2 \leq \cdots, \lambda_k \to \infty$), each λ_k of finite multiplicity, and the corresponding eigen functions $u_k(x)$, $Lu_k(x) = \lambda_k \rho u_k$, $k = 1, 2, \ldots,$ form a complete orthonormal system in $L_2(G; \rho(x)\,dx)$; moreover, every function of class $C^2(\overline{G})$ that satisfies the boundary condition (18) admits a regularly-convergent Fourier series expansion with respect to the system of eigen functions $\{u_k\}$.

The formulation of the boundary value problems discussed above assumes that the solutions are sufficiently regular in the interior of the region as well as up to the boundary. Such formulations of boundary value problems are termed *classical*. However, in many problems of physical interest one must relinquish such regularity requirements. Inside the region the solution may be a **generalized function** and satisfy the equation in the sense of generalized functions, while the boundary value conditions may be fulfilled in some generalized sense (almost everywhere, in L_p, in the weak sense, etc.). Such formulations are called *generalized*, and the corresponding solutions are called *generalized solutions*. For example, the generalized Cauchy problem for the wave equation is posed as follows. Let u be a classical solution of the Cauchy problem (2), (9). The functions u and f are extended by zero for $t<0$ and are denoted by $\tilde{u}$ and $\tilde{f}$, respectively. Then $\tilde{u}$ satisfies, as a generalized function in the entire space $\mathbf{R}^{n+1}$, the wave equation

$$\frac{\partial^2 \tilde{u}}{\partial t^2} = a^2 \Delta \tilde{u} + u_0(x) \times \delta'(t) + u_1(x) \times \delta(t) + \tilde{f}(x, t). \quad (19)$$

Here the initial perturbations u_0 and u_1 serve as external sources of the type of a double layer $u_0(x) \times \delta'(t)$ and a simple layer $u_1(x) \times \delta(t)$ acting instantaneously. This permits one to give the following definition. The *generalized Cauchy problem* for the wave equation with source $F \in D'(\mathbf{R}^{n+1})$, $F=0$ for $t<0$, is the problem of finding the generalized solutions $u(t, x)$ in $\mathbf{R}^{n+1}$ of the wave equation

$$\frac{\partial^2 u}{\partial t^2} = a^2 \Delta u + F(x, t) \quad (19')$$

that vanishes for $t<0$. The generalized Cauchy problem for the heat equation (4) is posed analogously.

Since the boundary value problems of mathematical physics describe real physical processes, they must meet the following natural requirements, formulated by J. Hadamard:

1) a solution must *exist* in some class of functions M_1;

2) the solution must be *unique* in, possibly, another class of functions M_2;

3) the solution must depend *continuously* on the data of the problem (the initial and boundary conditions, the free terms, the coefficients of the equation, etc.). This requirement is imposed in connection with the fact that, as a rule, the data of physical problems are determined experimentally only approximately, and hence it is necessary to be sure that the solution of the problem does not depend essentially on the measurement errors of these data.

A problem that meets the requirements 1) - 3) is called *well-posed*, and the set of functions $M_1 \cap M_2$ is the *well-posedness class*. Although requirements 1) - 3) seem natural at a first glance, they must nevertheless be proved in the framework of the mathematical model adopted. The proof of the well-posedness is the first validation of a mathematical model — the model is non-contradictory, does not contain parasitic solutions, and is weakly sensitive to measurement errors.

Finding well-posed boundary value problems of mathematical physics and methods for constructing their (exact or approximate) solutions is one of the main objectives of a branch of mathematical physics. It is known that all boundary value problems listed above are well-posed.

Example. The Cauchy problem $y'=f(x, y)$, $y(x_0)=y_0$, is well-posed if $f \in C^1$.

A problem that does not satisfy at least one of the conditions 1) - 3) is called an ill-posed problem (cf. **Ill-posed problems**). The importance of ill-posed problems in contemporary mathematical physics is increasing: in this class fall, in the first place, inverse problems, and also problems connected with the treatment and interpretation of results of observations.

An example of an ill-posed problem is the following Cauchy problem for the Laplace equation (*Hadamard's example*):

$$\Delta u(x, y) = 0, \quad u\big|_{y=0} = 0, \quad \frac{\partial u}{\partial y}\bigg|_{y=0} = \frac{\sin kx}{k}.$$

For $y>0$ the solution satisfies:

$$u(x, y) = \frac{1}{k^2} \sin kx \sinh ky \overset{x}{\nrightarrow} 0, \quad k \to \infty,$$

whereas

$$\frac{\sin kx}{k} \overset{x}{\Rightarrow} 0, \quad k \to \infty.$$

In order to solve approximately ill-posed problems one can resort to a **regularization method**, which utilizes supplementary information on the solution and which amounts to solving a sequence of well-posed problems.

An important role in the equations of mathematical physics is played by the notion of a **Green function**. The Green function of a linear differential operator

$$L(x, t; D) = \sum_{|a| \leq m} a_\alpha(x, t) D^\alpha,$$

$$D = \left[\frac{\partial}{\partial x_1}, \ldots, \frac{\partial}{\partial x_n}, \frac{\partial}{\partial t} \right],$$

with given (homogeneous) boundary value conditions on the boundary of the domain of variation of the variables (x, t) is, by definition, the function $G(x, t; \xi, \tau)$ which satisfies for each (ξ, τ) in this domain the equation

$$L(x, t; D) G(x, t; \xi, \tau) = \delta(x - \xi, t - \tau). \quad (20)$$

In physical situations the Green function $G(x, t; \xi, \tau)$ describes the disturbance produced by an instantaneous (at time τ) point source (placed at the point ξ) of intensity one (with the inhomogeneity of the medium and the effect of the boundary accounted for). In the case of operators with constant coefficients and in the absence of a boundary, the Green function for $\xi=0$ and $\tau=0$ is called a *fundamental solution* and is denoted by $E(x, t)$:

$$L(D)E(x, t) = \delta(x, t). \quad (20')$$

The existence of a fundamental solution in the spaces D' and S' has been established for any operator $L(D) \neq 0$.

Examples of fundamental solutions. For the wave equation:

$$E_1(x, t) = \frac{\theta(at - |x|)}{2a}, \quad E_2(x, t) = \frac{\theta(at - |x|)}{2\pi a \sqrt{a^2 t^2 - |x|^2}},$$

$$E_3(x, t) = \frac{1}{2\pi a}\delta_+(a^2 t^2 - |x|^2),$$

where $\theta(t)$ is the Heaviside function: $\theta(t)=0$ for $t<0$; $\theta(t)=1$ for $t\geqslant 0$.

For the heat equation:

$$E_n(x, t) = \frac{\theta(t)}{(2a\sqrt{\pi t})^n}e^{-|x|^2/4a^2 t}.$$

For the Laplace equation:

$$E_1(x) = \frac{|x|}{2}, \quad E_2(x) = \frac{\ln|x|}{2\pi}, \quad E_3(x) = -\frac{1}{4\pi|x|}.$$

Using the fundamental solution $E(x, t)$, the solution $u(x, t)$ of the equation

$$L(D)u = F(x, t) \tag{21}$$

with arbitrary right-hand side $F\in D'$, if it exists in D', is expressible in the whole space $\mathbf{R}^{n+1}$ as the convolution

$$u = F\star E. \tag{22}$$

The meaning of formula (22) in physical situations is as follows: The solution u is the result of superposition of the elementary disturbances $F(\xi, \tau)E(x-\xi, t-\tau)$ produced by the point sources $F(\xi, \tau)\delta(x-\xi, t-\tau)$ into which the source F is decomposed in view of the identity $F=F\star\delta$. The convolution $F\star E$ plays the role of the potential with source (density) F. This is the essence of the *method of point sources*, or *mapping method*, for solving linear problems of mathematical physics.

In particular, the solution of the generalized Cauchy problem for the wave equation (or heat equation) is given by the *wave (heat) potential*

$$u = F\star E_n. \tag{22'}$$

From this formula one can derive, under suitable assumptions on the smoothness of the source

$$F(x, t) = u_0(x)\delta'(t)+u_1(x)\delta(t)+f(x, t),$$

the classical formulas for the solution of the Cauchy problem. For the wave equation in three-dimensional space one has the **Kirchhoff formula**

$$u(x, t) = \frac{1}{4\pi a^2}\int_{|x-\xi|<at} f\left[\xi, t-\frac{|x-\xi|}{a}\right]\frac{d\xi}{|x-\xi|}+ \tag{23}$$

$$+\frac{1}{4\pi a^2 t}\int_{|x-\xi|=at}u_1(\xi)\,dS+\frac{1}{4\pi a^2}\frac{\partial}{\partial t}\left[\frac{1}{t}\int_{|x-\xi|=at}u_0(\xi)\,dS\right].$$

For the heat equation one has the **Poisson formula**

$$u(x, t) = \int_0^t\int\frac{f(\xi, \tau)}{[2a\sqrt{\pi(t-\tau)}]^n}e^{-|x-\xi|^2/4a^2(t-\tau)}\,d\xi\,d\tau+ \tag{24}$$

$$+\frac{1}{(2a\sqrt{\pi t})^n}\int u_0(\xi)e^{-|x-\xi|^2/4a^2 t}\,d\xi.$$

In the same manner, constructing the Green function for the Laplace equation for the sphere, one obtains the solution of the interior Dirichlet problem for the (three-dimensional) ball $|x|<R$ in the form of a **Poisson integral**:

$$u(x) = \frac{1}{4\pi R}\int_{|\xi|=R}\frac{R^2-|x|^2}{|x-\xi|^2}u_0(\xi)\,dS_\xi. \tag{25}$$

For the investigation and approximate solution of mixed problems one uses, under the assumption that the coefficients in the equation and in the boundary conditions do not depend on the time t, the **Fourier method** (separation of variables). The idea of the method applied, say, to the problem (3), (10), (18) is as follows. First, one expands the unknown solution $u(x, t)$ and the right-hand side $f(x, t)$ in Fourier series with respect to the eigen functions $\{u_k\}$ of the boundary value problem (17), (18):

$$u(x, t) = \sum_{k=1}^\infty b_k(t)u_k(x), \quad f(x, t) = \sum_{k=1}^\infty c_k(t)u_k(x). \tag{26}$$

Then, upon substituting formally these series in equation (3) one obtains for the unknown functions $b_k(t)$ the equations

$$b_k'(t)+\lambda_k b_k(t) = c_k(t), \quad k=1, 2, \ldots. \tag{27}$$

To ensure that the series (26) for u will satisfy the initial condition (10) it is necessary to set

$$b_k(0) = \int_G \rho(x)u_0(x)u_k(x)\,dx = a_k. \tag{28}$$

Solving the Cauchy problem (27), (28) one obtains a formal solution of the problem (3), (10), (18) in the form of a series:

$$u(x, t) = \sum_{k=1}^\infty\left[a_k e^{-\lambda_k t}+\int_0^t e^{-\lambda_k(t-\tau)}c_k(\tau)\,d\tau\right]u_k(x). \tag{29}$$

There arises the problem of substantiating the Fourier method, i.e. of determining when the formal series (29) yields a classical or generalized solution of the problem (3), (10), (18).

To substantiate the Fourier method, and, generally, for establishing the well posedness of the mixed problem for the diffusion equation (3), one resorts to the **maximum principle**. An analogue of the Fourier method is also used for the mixed problem (1), (9), (18) for the oscillation equation. In this case the method of the **energy integral** is found useful.

The method of separation of variables has also found use in solving boundary value problems for elliptic-type equations (5), in particular, for calculating the eigen functions and eigen values under the assumption that the domain G has enough symmetry.

For the investigation and approximate solution of boundary value problems for equation (5) one widely uses variational methods. For example, in the eigen value problems (17), (18) (for $\rho=1$) the eigen values λ_k satisfy the variational principle

$$\lambda_k = \inf_{\substack{(u,u_i)=0, \\ i=1,\ldots,k-1}}\frac{(Lu, u)}{\|u\|^2}, \tag{30}$$

where it is assumed that comparison functions $u(x)$

belong to the class $C^2(\overline{G})$ and satisfy the boundary condition (18); the infimum in (30) is attained on any of the eigen functions corresponding to the eigen value λ_k, and only on these.

When investigating boundary value problems for equation (5) (in particular, for harmonic functions) one applies the maximum principle.

The boundary value problems listed above do not exhaust the whole variety of boundary value problems of mathematical physics; they merely provide the simplest classical examples. The boundary value problems describing real physical processes may be very complicated: systems of equations, equations of higher order, or non-linear equations. Here the main examples are the **Schrödinger equation**, the equations of hydrodynamics, transport, and magneto-hydrodynamics, Maxwell's equation (cf. **Maxwell equations**), the equations of elasticity theory, the Dirac, Hilbert, Einstein, and Yang−Mills equations, etc. (cf. also **Dirac equation**; **Einstein equations**; **Yang−Mills field**).

In connection with the search for non-trivial models describing the interaction of quantum fields, there is an interest in classical non-linear equations, among them the **Korteweg−de Vries equation**

$$u_t - 6uu_x + u_{xxx} = 0, \tag{31}$$

the non-linear wave equation

$$u_{tt} - u_{xx} = gf(u), \; g > 0$$

(known as the Liouville equation for $f = e^u$ and as the sine-Gordon equation for $f = -\sin u$), and the non-linear Schrödinger equation:

$$iu_t + u_{xx} + \nu |u|^2 u = 0, \; \nu > 0.$$

A characteristic feature of such equations is that they admit solutions of 'solitary-wave' type (solitons, cf. **Soliton**). Thus, for equation (31) such a solution is

$$u(x, t) = \frac{a}{2\cosh^2[\sqrt{a}(x - at - x_0)/2]}, \; a > 0, \; x_0 \text{ arbitrary.}$$

This solution has finite energy.

References

[1] TIKHONOV, A.N. and SAMARSKIĬ, A.A.: *Differentialgleichungen der mathematischen Physik*, Deutsch. Verlag Wissenschaft., 1959 (translated from the Russian).
[2] VLADIMIROV, V.S.: *Equations of mathematical physics*, Mir, 1984 (translated from the Russian).
[3] TIKHONOV, A.N. and ARSENIN, V.YA.: *Solutions of ill-posed problems*, Winston, 1977 (translated from the Russian).
[4] HÖRMANDER, L.: *The analysis of linear partial differential operators*, 1-2, Springer, 1983.
[5] HADAMARD, .J.: *Lectures on Cauchy's problem in linear partial differential equations*, Dover, reprint, 1952.
[6] WHITHAM, G.B.: *Linear and non-linear waves*, Wiley, 1974.
[7] MIKHAĬLOV, V.P.: *Partial differential equations*, Moscow, 1983 (in Russian).
[8] LADYZHENSKAYA, O.A.: *The boundary value problems of mathematical physics*, Springer, 1985 (translated from the Russian).
[9] COURANT, R. and HILBERT, D.: *Methods of mathematical physics. Partial differential equations*, 2, Interscience, 1965 (translated from the German).
[10] VLADIMIROV, V.S.: *Generalized functions in mathematical physics*, Mir, 1979 (translated from the Russian).

V.S. Vladimirov

Editorial comments.

References

[A1] SOBOLEV, S.L.: *Partial differential equations of mathematical physics*, Pergamon, 1964 (translated from the Russian).
[A2] BUDAL, B.M., SAMARSKIĬ, A.A. and TIKHONOV, A.N.: *A collection of problems on mathematical physics*, Pergamon, 1964 (translated from the Russian).
[A3] MICHLIN, S.G. [S.G. MIKHLIN]: *Lehrgang der mathematischen Physik*, Akad. Verlag, 1972 (translated from the Russian).
[A4] MORSE, P.M. and FESHBACH, H.: *Methods of theoretical physics*, 1-2, McGraw-Hill, 1953.
[A5] ZAUDERER, E.: *Partial differential equations of applied mathematics*, Wiley, 1983.

AMS 1980 Subject Classification: 35Q20, 70-XX, 73-XX

MATHEMATICAL PROGRAMMING - The branch of mathematics concerned with the theory and methods for solving problems on finding the extrema of functions on sets defined by linear and non-linear constraints (equalities and inequalities) in a finite-dimensional vector space. Mathematical programming is a branch of **operations research**, which comprises a wide class of control problems the mathematical models of which are finite-dimensional extremum problems. The problems of mathematical programming find applications in various areas of human activity where it is necessary to choose one of the possible ways of action, e.g. in solving numerous problems of control and planning of production processes as well as in problems of design and long-term planning. The term 'mathematical programming' is connected with the fact that the goal of solving various problems is choosing *programs* of action.

The mathematical formulation of the problem of mathematical programming is: To minimize a scalar function $\phi(x)$ of a vector argument on the set

$$X = \{x : q_i(x) \geqslant 0, \; i = 1, \ldots, k;$$
$$h_j(x) = 0, \; j = 1, \ldots, m\},$$

where $q_i(x)$ and $h_j(x)$ are scalar functions. The function $\phi(x)$ is called the *objective function*, and also the *quality criterion*, the set X is called the *feasible set*, or the *set of plans*, a solution x^* of the mathematical programming problem is an *optimum point* (or vector), a *point of global minimum* and also an *optimal plan*.

In mathematical programming one customarily distinguishes the following branches. **Linear programming**: The objective function $\phi(x)$ and the constraints $q_i(x)$ and $h_j(x)$ are linear. **Quadratic programming**: The objective function is quadratic and convex, while the feasible set is defined by linear equalities and inequalities. **Convex programming**: The objective function and the constraints are convex. **Discrete programming**: The solution is sought only at discrete, say integer, points of X. **Stochastic programming**: In contrast to deterministic problems, here the data contain an element of indeterminacy. For example, in stochastic problems of minimization of a linear function

$$\sum_{j=1}^{n} c_j x_j$$

under linear constraints

$$\sum_{j=1}^{n} a_{ij} x_j \geqslant b_i, \quad i=1,2,\ldots,$$

the parameters c_j, a_{ij}, b_i, or only some of them, are random. The problems of linear, quadratic and convex programming have a common property: Local optimality implies global optimality. The so-called multiextremum problems, for which the indicated property does not hold, are both considerably more difficult and less investigated.

At the basis of the theory of convex, and, in particular, linear and quadratic, programming lies the *Kuhn−Tucker theorem*, which gives necessary and sufficient conditions for the existence of an optimum point x^*: In order that x^* be an optimum point, i.e.

$$\phi(x^*) = \min_{x \in X} \phi(x),$$

$$X = \{x: f_i(x) \geqslant 0, \ i=1,\ldots,k\},$$

it is necessary and sufficient that there exist a point $y^* = (y_1^*, \ldots, y_k^*)$ such that the pair x^*, y^* be a **saddle point** of the **Lagrange function**

$$L(x,y) = \phi(x) + \sum_{i=1}^{k} y_i f_i(x).$$

The latter means that

$$L(x^*, y) \leqslant L(x^*, y^*) \leqslant L(x, y^*)$$

for all x and all $y \geqslant 0$. If the constraints $f_i(x)$ are nonlinear, the theorem is valid under some additional assumptions on the feasible set, for example, the existence of a point $x \in X$ such that $f_i(x) > 0$, $i = 1,\ldots,k$ (*Slater's regularity condition*).

If $\phi(x)$ and $f_i(x)$ are differentiable functions, then the following relations characterize a saddle point:

$$\frac{\partial L}{\partial x_j} \geqslant 0, \quad j=1,\ldots,n;$$

$$\sum_{j=1}^{n} \frac{\partial L}{\partial x_j} x_j = 0; \quad \frac{\partial L}{\partial y_i} \leqslant 0, \quad i=1,\ldots,k;$$

$$\sum_{i=1}^{k} \frac{\partial L}{\partial y_i} y_i = 0; \quad y_i \geqslant 0, \quad i=1,\ldots,k.$$

Thus, the problem of convex programming reduces to the solution of a system of equations and inequalities.

On the basis of the Kuhn−Tucker theorem various iterative minimization methods were developed, which amount to searching for a saddle point of Lagrange's function.

In mathematical programming one of the main directions concerns computational methods for solving extremum problems. One of the widest used among these methods is the *method of feasible directions*. In this method a sequence $\{x_m\}$ of points of the set X is constructed by the formula $x_{p+1} = x_p + \alpha_p s_p$. At each iteration, to calculate the point x_{p+1} one has to choose a direction (a vector) s_p and a step length (a number) α_p. For s_p one selects from among the feasible directions (i.e. directions at the point x_p, a small displacement along which does not lead out of the set X) the one that makes an acute angle with the direction $g(x_p)$ of steepest descent of the objective function (with the vector $g(x_p) = -d\phi(x)/dx \,|_{x=x_p}$ if ϕ is differentiable). Therefore, along the direction s_p the function $\psi_p(\alpha) = \phi(x_p + \alpha s_p)$ is decreasing. The number α_p is determined from the conditions that $x_{p+1} \in X$ and $\phi(x_{p+1}) < \phi(x_p)$. To calculate s_p at x_p one defines the cone of feasible directions, given by a system of inequalities, and one formulates the problem (of linear programming) of finding the possible direction along which the objective function decreases most quickly. This problem is readily solved using, say, the standard **simplex method**. The step length α_p is determined by solving the minimization problem for the function $\psi_p(\alpha)$. Under rather general assumptions it has been shown that the distance from x_p to the set of stationary points of the original problem (i.e. the set of points at which the conditions necessary for a local minimum of $\phi(x)$ on X are satisfied) tends to zero as $p \to \infty$. In the case when the original problem is a problem in convex programming, then, as $p \to \infty$, x_p approaches the set of solutions (optimum points) of the original problem. The method of feasible directions admits an approximate solution of the indicated problems of determination of s_p and α_p, and in this sense it is stable with respect to computational errors.

For feasible sets of special structure (from the point of view of the simplicity of choosing s_p) one can apply minimization methods different from the method of feasible directions. Thus, in a *gradient-projection method*, as the descent direction at the point x_p one takes $s_p = y_p - x_p$, where y_p is the projection of $v_p = x_p + g(x_p)$, $g(x_p) = -d\phi(x)/dx \,|_{x=x_p}$, on the set X.

If X is given by linear constraints, one has the rather promising *method of the conditional gradient*: One linearizes the objective function at x_p by constructing the function $l(x) = \phi(x_p) + \langle g(x_p), x_p - x \rangle$, and then, minimizing $l(x)$ on the set X, one finds a minimum point y_p of it and one finally sets $s_p = y_p - x_p$.

Minimization methods for non-smooth functions were successfully elaborated. One of the representatives of this class is the *method of the generalized gradient*. By definition, a generalized gradient at x_p is any vector $\tilde{g}(x_p)$ such that the inequality $\phi(y) - \phi(x_p) \geqslant \langle \tilde{g}(x_p), y - x_p \rangle$ holds for all $y \in X$. This method differs from the projection-gradient method in that for the point that is projected one takes $v_p = x_p - \tilde{g}(x_p)$.

Stochastic methods of minimization are also widely used. In the simplest variant of the *method of random descent* one takes for s_p a random point on the n-dimensional unit sphere with centre at the coordinate origin, and if s_p is a feasible direction and if, in addition, $\phi(x)$ decreases along this direction, then a step is effected, i.e. one shifts to the point x_{p+1}. In the opposite case, the procedure for selecting s_p is carried out anew.

A characteristic feature of the computational aspect of the methods for solving the problems in mathematical programming is that the application of these methods is invariably connected with the utilization of electronic computers. The main reason for this is that the problems in mathematical programming that formalize situations of control of real systems involve a large amount of work which cannot be performed by manual computation.

One of the widespread methods for investigating problems in mathematical programming is the method of penalty functions (cf. **Penalty functions, method of**). This method essentially replaces the given mathematical programming problem by a sequence of parametric problems on unconditional minimization, such that when the parameter tends to infinity (in other cases, to zero) the solutions of these auxiliary problems converge to the solution of the original problem. Note that in an unconditional minimization problem every direction is feasible, and consequently the search for s_p requires less effort in comparison with the solution of the same problem by the method of feasible directions. (For example, in the method of steepest descent one takes $s_p = -d\phi(x)/dx \mid_{x = x_p}$.) The same is true concerning the search for α_p.

An important direction of investigation in mathematical programming is the problem of stability. Here essential importance is attached to the study of the class of stable problems, that is, problems for which small perturbations (errors) in the data result in perturbations of the solutions that are also small. In the case of unstable problems an important role is reserved for a procedure of approximating the unstable problem by a sequence of stable problems — this is known as the *regularization process*.

Along with finite-dimensional problems, problems of mathematical programming in infinite-dimensional spaces are considered as well. Among the latter there are various extremum problems in mathematical economics, technology, problems of optimization of physical characteristics of nuclear reactors, and others. In the terminology of mathematical programming one formulates the problems in the corresponding function space of **variational calculus** and **optimal control**.

Mathematical programming has crystallized as a science in the 1950-s until 1970-s. This is first of all connected with the development of electronic computers and, hence, the mathematical processing of large amounts of data. On this basis one solves problems of control and planning, in which the application of mathematical methods is connected mainly with the construction of mathematical models and related extremal problems, among them problems of mathematical programming.

References

[1] Moiseev, N.N., Ivanov, Yu.P. and Stolyarova, E.M.: *Optimization methods*, Moscow, 1978 (in Russian).
[2] Karmanov, V.G.: *Mathematical programming*, Moscow, 1975 (in Russian).
[3] Polak, E.: *Computational methods in optimization: a unified approach*, Acad. Press., 1971.
[4] Ermol'ev, Yu.M.: *Methods of stochastic programming*, Moscow, 1976 (in Russian).

V.G. Karmanov

Editorial comments.

References

[A1] Minoux, M.: *Mathematical programming: theory and algorithms*, Wiley, 1986.
[A2] Zoutendijk, G.: *Mathematical programming methods*, North-Holland, 1976.
[A3] Heesterman, A.R.G.: *Matrices and simplex algorithms*, Reidel, 1983.
[A4] Hadley, G.: *Nonlinear and dynamic programming*, Addison-Wesley, 1964.
[A5] Zangwill, W.I.: *Nonlinear programming*, Prentice-Hall, 1969.
[A6] Zionts, S.: *Linear and integer programming*, Prentice-Hall, 1973.

AMS 1980 Subject Classification: 90CXX

MATHEMATICAL STATISTICS - The branch of mathematics devoted to the study of mathematical methods for the organization, processing and utilization of statistical data for scientific and practical conclusions. Here, by statistical data is meant information on a number of objects in some, more or less extensive, collection, which have some specific properties.

The object and method of mathematical statistics. The statistical description of a collection of objects occupies an intermediate position between the individual description of each object in the collection, on the one hand, and the description of the collection by their common properties, with no individual breakdown into objects, on the other. By comparison with the first method, statistical data are always, to a greater or lesser extent, collective, and have only limited value in cases where the essence is the individual data (for example, a teacher getting to know a class obtains only a very preliminary orientation on the situation from the statistics on the number of excellent, good, adequate, and inadequate appraisals made by his or her predecessor). On the other hand, in comparison with data on a collection which is observed from the outside, and summarized by common properties, statistical data give a deeper penetration into the heart of the matter. For example, data on granulometric analysis of a rock (that is, data on the distribution of rock particles by size) gives valuable additional information when compared to measurements on the unfragmented form of the rock, which allows one, to some extent, to explain the properties of the rock, the conditions of its formation, etc.

The method of research, characterized as the discussion of statistical data on various collections of objects, is called *statistical*. The statistical method can be applied in very diverse areas of knowledge. However, the features of the statistical method in its applications to various kinds of objects are so specific that it would be meaningless to unify, for example, socio-economic statistics, physical statistics, stellar statistics, etc., in one science.

The common features of the statistical method in various areas of knowledge come down to the calculation of the number of objects in some group or other, the discussion of the distribution of quantitative attributes, the application of the sampling method (in cases where a detailed investigation of an extensive collection is difficult), the use of probability theory to estimate the adequacy of a number of observations for this or that conclusion, etc. This formal mathematical side of statistical research methods is indifferent to the specific nature of the objects being studied and comprises the topic of mathematical statistics.

The connection between mathematical statistics and probability theory. This connection is different in different cases. **Probability theory** studies not just any mass phenomenon, but phenomena which are random, to wit, 'probabilistically random'. That is, those for which it makes sense to talk of associated probability distributions. Nevertheless, probability theory plays a definite role in the statistical study of mass phenomena of any kind, even those unrelated to the category of probabilistically random phenomena. This comes about through the theories of the sampling method and errors (cf. **Errors, theory of; Sample method**), which are based on

probability theory. In these cases the phenomenon itself is not subject to probabilistic laws, but the means of investigation is.

A more important role is played by probability theory in the statistical investigation of probabilistically random phenomena. Here one finds in full measure the application of such probabilistically based parts of mathematical statistics as statistical hypotheses testing (cf. **Statistical hypotheses, verification of**), **statistical estimation** of probability distributions and their parameters, etc. The field of application of these deeper statistical methods is considerably narrower, since it is required that the phenomena themselves are subject to fairly definite probability laws. For example, the statistical study of turbulent regimes of water flow, or fluctuations in radio reception, is carried out on the basis of the theory of stationary stochastic processes. However, the application of this same theory to the analysis of economic time series may lead to gross errors, since the assumption of a time-invariant probability distribution in the definition of a stationary process is, as a rule, totally unacceptable in this case.

Probability laws gain a statistical expression on the strength of the law of large numbers (probabilities are realized approximately in the form of frequencies, and expectations in the form of averages).

The simplest modes of statistical description. A collection of n objects being studied may, relative to some qualitative property A, be divided into classes $A_1, \ldots, A_r$. The statistical distribution corresponding to this partition is given by the numbers (frequencies) $n_1, \ldots, n_r$ (where $\sum_{i=1}^{r} n_i = n$) of objects in the different classes. Instead of the number n_i one often gives the corresponding relative frequency $h_i = n_i / n$ (satisfying, obviously, $\sum_{i=1}^{r} h_i = 1$). If the investigation concerns some quantitative attribute, then its distribution in the collection of n objects may be given by directly listing the observed values of the attribute: $x_1, \ldots, x_n$; for example, in increasing order. However, for large n such a method is cumbersome and, at the same time, does not clearly reveal the essential properties of the distribution. For arbitrarily large n, in practice it is very unusual to compile complete tables of the observed values x_i, but rather to proceed in all subsequent work from tables which contain only the numbers in the classes obtained by grouping the observations into appropriate intervals.

Usually a grouping into 10 - 20 intervals, each containing no more than 15 - 20% of the values x_i, turns out to be sufficient for a fairly complete classification of the essential properties of the distribution and for an appropriate computation, relative to the numbers in the groups, of the basic characteristics of the distribution (see below). Forming a **histogram** with respect to the grouped data graphically portrays the distribution. A histogram formed on the basis of groups with small intervals obviously has many peaks and does not graphically reflect the essential properties of the distribution.

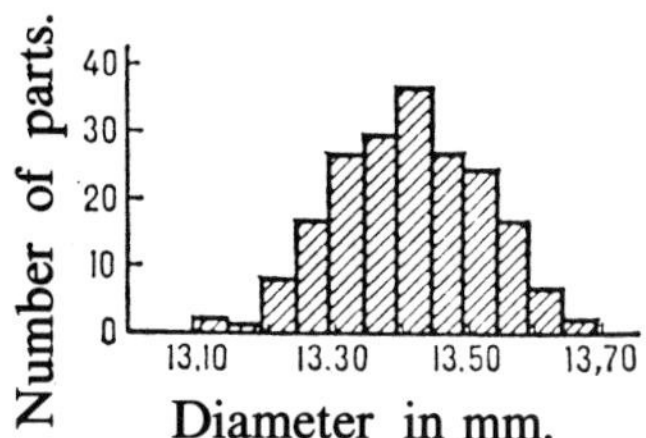

Fig. 1.

As an example, Fig. 1 is a histogram for the distribution of 200 diameters of certain parts (in mm), with group intervals of 0.05 mm, and Fig. 2 is the histogram of the same distribution with intervals of lengths 0.01 mm.

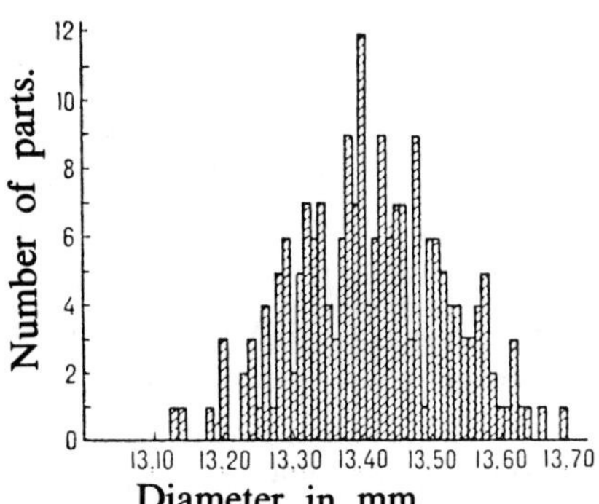

Fig. 2.

On the other hand, grouping into intervals which are too large may lead to a loss of clarity in the representation of the nature of the distribution, and to gross errors in the calculation of the mean and other characteristics of the distribution (see the corresponding histogram in Fig. 3).

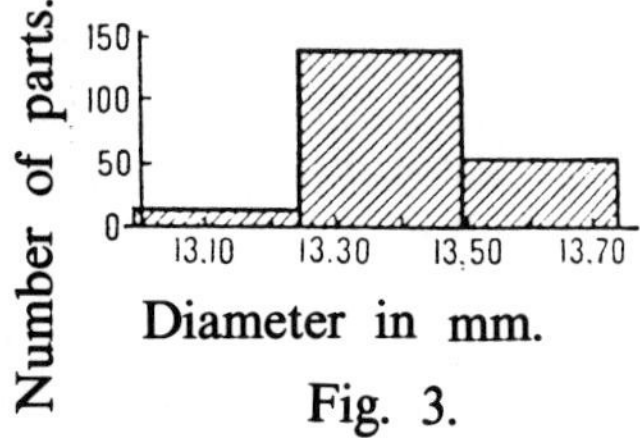

Fig. 3.

Within the limits of mathematical statistics, questions of grouping into intervals can only be considered from the formal point of view: the completeness of the mathematical description of a distribution, the precision of a calculation of means with respect to grouped data, etc.

The simplest summaries of the characteristics of the distribution of a single quantitative attribute are the mean

$$\bar{x} = \frac{1}{n}\sum_{i=1}^{n} x_i$$

and the mean-square deviation

$$D = \frac{S}{\sqrt{n}},$$

where

$$S^2 = \sum_{i=1}^{n}(x_i - \bar{x})^2.$$

In calculating $\bar{x}$, S^2 and D for grouped data one uses the formulas

$$\bar{x} = \frac{1}{n}\sum_{k=1}^{r} n_k a_k = \sum_{k=1}^{r} h_k a_k,$$

$$S^2 = \sum_{k=1}^{r} n_k(a_k - \bar{x})^2 = \sum_{k=1}^{r} n_k a_k^2 - n\bar{x}^2$$

or

$$D^2 = \sum_{k=1}^{r} h_k a_k^2 - \bar{x}^2,$$

where r is the number of grouped intervals and the a_k are their means. If the material is grouped into intervals which are too large, then these calculations are too rough. Sometimes, in such cases it is useful to resort to special refinements of the classification. However, it only makes sense to introduce these refinements when definite probabilistic assumptions are satisfied.

Regarding the joint distribution of two or more attributes see **Correlation (in statistics); Regression.**

The connection between statistical and probabilistic distributions. Parameter estimators. Testing probabilistic hypotheses. Above, only certain selected simple modes of statistical description, which form a fairly extensive discipline with a well-developed system of ideas and techniques of calculation, were presented. Modes of statistical description, however, are of interest not just by themselves, but as a means of obtaining, from statistical material, inferences on the laws to which the phenomena studied are subject, and for obtaining inferences on the grounds leading in each individual case to various observed statistical distributions.

For example, the data drawn in Figs. 1, 2 and 3 was collected with the aim of establishing the precision in the manufacturing of parts with design diameter equal to 13.40 mm under normal variations in manufacture. The simplest assumption, which may in this case be based upon some theoretical consideration, is that the diameters of the individual parts can be considered as a random variable X subject to the normal probability distribution

$$P\{X < x\} = \frac{1}{(2\pi)^{1/2}\sigma} \int_{-\infty}^{x} e^{-(t-a)^2/2\sigma^2} \, dt. \qquad (1)$$

If this assumption is true, then the parameters a and σ^2

— the mean and the variance of the probability distribution — can be fairly precisely estimated by the corresponding characteristics of the statistical distribution (since the number of observations $n = 200$ is sufficiently large). As an estimator of the theoretical variance it is preferred not to use the statistical variance

$$D^2 = \frac{S^2}{n},$$

but the **unbiased estimator**

$$s^2 = \frac{S^2}{n-1}.$$

For the theoretical mean-square deviation σ there does not exist a single (suitable for any probability distribution) expression of an unbiased estimator. As an estimator (in general, biased) for σ it is most common to use s. The accuracy of the estimators $\bar{x}$ and s for a and σ is clarified by the corresponding variances, which, in the case of a normal distribution (1), have the form

$$\sigma_{\bar{x}}^2 = \frac{\sigma^2}{n} \sim \frac{s^2}{n},$$

$$\sigma_{s^2}^2 = \frac{2\sigma^4}{n-1} \sim \frac{2s^4}{n},$$

$$\sigma_s^2 \sim \frac{\sigma^2}{2n} \sim \frac{s^2}{2n},$$

where the sign $\sim$ denotes 'approximate equality for large n'. Thus, if one agrees to add to the estimators $\pm$ their mean-square deviation, one has for large n, under the assumption of a normal distribution (1),

$$a = \bar{x} \pm \frac{s}{\sqrt{n}}, \quad \sigma = s \pm \frac{s}{\sqrt{2n}}. \tag{2}$$

The sample size $n = 200$ is sufficient for the use in these formulas of laws from the theory of large samples.

For more information on the estimation of the parameters of theoretical probability distributions see **Statistical estimation; Confidence estimation.**

All rules based on probability theory for the statistical estimation of parameters and hypotheses testing operate only at a definite **significance level** $\omega < 1$, that is, they may lead to false results with probability $\alpha = 1 - \omega$. For example, if, under the assumption of a normal distribution and known theoretical variance σ^2, an interval estimator of a based on $\bar{x}$ is produced by the rule

$$\bar{x} - \frac{k\sigma}{\sqrt{n}} < a < \bar{x} + \frac{k\sigma}{\sqrt{n}},$$

then the probability of an error will be equal to α, which is related to k through

$$\alpha = \frac{2}{\sqrt{2\pi}} \int_k^\infty e^{-x^2/2} \, dx.$$

The question of a rational choice of the significance level under given concrete conditions (for example, in the development of rules for statistical quality control

in mass production) is very essential. In this connection the desire to apply only rules with a very high (close to 1) significance level faces the situation that for a restricted number of observations such rules only allow inferences with poor precision (it may not be possible to establish the inequality of probabilities even given a noticeable inequality of the frequencies, etc.).

Further problems in mathematical statistics. The above-mentioned methods of parameter estimation and hypotheses testing are based on the assumption that the number of observations required to attain a given precision in the conclusions is determined in advance (before carrying out the sampling). However, frequently an a priori determination of the number of observations is inconvenient, since by not fixing the number of trials in advance, but by determining it during the experiment, it is possible to decrease the expected number of trials. This situation was first observed in the example of choosing between one of two hypotheses in a sequence of independent trials. The corresponding procedure (first proposed in connection with problems of statistical sampling) is as follows: at each step decide, by the results of the observations already carried out, whether to a) conduct the next trial, or b) stop the trials and accept the first hypothesis, or c) stop the trials and accept the second hypothesis. With an appropriate choice of the quantitative characteristics such a procedure can secure (with the same precision in the calculations) a reduction in the average number of observations to almost half that of the fixed size sampling procedure (see **Sequential analysis**). The development of the methods of sequential analysis led, on the one hand, to the study of controlled stochastic processes (cf. **Controlled stochastic process**) and, on the other, to the appearance of **statistical decision theory**. This theory arises because the results of sequentially carrying out observations serve as a basis for the adoption of certain decisions (intermediate — to continue the trial, and final — when the trials are stopped). In problems on parameter estimation the final decisions are numbers (the values of the estimators), in problems on hypotheses testing they are the accepted hypothesis. The aim of the theory is to give rules for the acceptance of decisions which minimise the mean loss or risk (the risk depends on the probability distributions of the results of the observations, on the final decision, on the expense of conducting the trials, etc.).

Questions on the expedient distribution of effort in carrying out a statistical analysis of phenomena are considered in the theory of design of experiments, which plays a major part in modern mathematical statistics.

Side by side with the development and elaboration of the general ideas of mathematical statistics there have evolved various specialized branches such as **dispersion analysis; covariance analysis; multi-dimensional statistical analysis; the statistical analysis of stochastic processes; and factor analysis**. New considerations in regression analysis have appeared (see also **Stochastic approximation**). A major part in problems of mathematical statistics is played by the **Bayesian approach** to statistical problems.

Historical information. The first elements of mathematical statistics can already be found in the writings of the originators of probability theory — J. Bernoulli, P. Laplace and S. Poisson. In Russia the methods of mathematical statistics in the application to demography and actuarial work were developed by V.Ya. Bunyakovskiĭ (1846). Of key importance for all subsequent development of mathematical statistics was the work of the classical Russian school of probability theory in the second half of the 19-th and beginning of the 20-th centuries (P.L. Chebyshev, A.A. Markov, A.M. Lyapunov, and S.N. Bernshteĭn). Many questions of statistical estimation theory were essentially devised on the basis of the theory of errors and the method of least squares (C.F. Gauss and Markov). The work of A. Quételet, F. Galton and K. Pearson has great significance, but in terms of utilizing the achievements of probability theory they lagged behind that of the Russian school. Pearson widely expanded the work on the formation of tables of functions necessary for applying the methods of mathematical statistics. This important work was continued in many scientific centres (in the USSR it was carried out by E.E. Slutskiĭ, N.V. Smirnov and L.N. Bol'shev). In the creation of small sample theory, the general theory of statistical estimation and hypotheses testing (free of assumptions on the presence of a priori distributions), and sequential analysis, the role of the Anglo-American school (Student, the pseudonym of W.S. Gosset, R.A. Fisher, Pearson, and J. Neyman), whose activity began in the 1920's, was very significant. In the USSR noteworthy results in the field of mathematical statistics were obtained by V.I. Romanovskiĭ, A.N. Kolmogorov and Slutskiĭ, to whom belongs important work on the statistics of dependent stationary series, Smirnov, who laid the foundations of the theory of **non-parametric methods in statistics**, and Yu.V. Linnik, who enriched the analytical apparatus of mathematical statistics with new methods. On the basis of mathematical statistics, statistical methods of research and investigation in **queueing theory**, physics, hydrology, climatology, stellar astronomy, biology, medicine, etc., were particularly intensively developed.

See also the references to the articles on branches of mathematical statistics.

References

[1] SMIRNOV, N.V. and DUNIN-BARKOVSKIĬ, I.V.: *Mathematische Statistik in der Technik*, Deutsch. Verlag Wissenschaft., 1969 (translated from the Russian).

[2] BOL'SHEV, L.N. and SMIRNOV, N.V.: *Tables of mathematical statistics*, Libr. of mathematical tables, 46, Nauka, Moscow, 1983 (in Russian). Processed by L.S. Bark and E.S. Kedova.

[3] WAERDEN, B.L. VAN DER: *Mathematische Statistik*, Springer, 1957.

[4] CRAMÉR, H.: *Mathematical methods of statistics*, Princeton Univ. Press, 1946.

[5] WALD, A.: *Statistical decision functions*, Wiley, 1950.

[6] KENDALL, M.G. and STUART, A.: *The advanced theory of statistics*, 1. Distribution theory, Griffin, 1977.

[7] KENDALL, M.G. and STUART, A.: *The advanced theory of statistics*, 2. Inference and relationship, Griffin, 1979.

[8] KENDALL, M.G. and STUART, A.: *The advanced theory of statistics*, 3. Design and analysis and time series, Griffin, 1983.

A.N. Kolmogorov
Yu.V. Prokhorov

AMS 1980 Subject Classification: 62-XX

MATHEMATICAL SYMBOLS - Conventional signs used for the written notation of mathematical notions and reasoning. For example, the notion 'the square root of the number equal to the ratio of the length of the circumference of a circle to its diameter' is denoted briefly by $\sqrt{\pi}$, while the statement 'the ratio of the length of the circumference of the circle to its diameter is greater than three and ten seventy-firsts and less than three and one seventh' is written as

$$3\frac{10}{71} < \pi < 3\frac{1}{7}.$$

The development of mathematical notation was intimately bound up with the general evolution of mathematical concepts and methods.

The first mathematical symbols were signs for the depiction of numbers — **ciphers**, the appearance of which apparently preceded the introduction of written language. The most ancient systems of numbering (see **Numbers, representations of**) — the Babylonian and the Egyptian — date back to around 3500 B.C.

The first mathematical symbols for arbitrary quantities appeared much later (from the 5-4 centuries B.C.) in Greece. Arbitrary quantities (areas, volumes, angles) were represented by the lengths of lines and the product of two such quantities was represented by a rectangle with sides representing the respective factors. In Euclid's *Elements* (3-th century B.C), quantities are denoted by two letters, the initial and final letters of the corresponding segment, and sometimes by one letter. Dating from Archimedes (287 - 213 B.C.), the latter device became standard. This mode of notation could potentially have developed into a calculus of letters. In the mathematics of classical Antiquity, however, no operations were carried out on letters and such a letter calculus did not materialize.

The rudiments of letter notation and calculus appeared in the post-Hellenistic era, thanks to the liberation of algebra from its geometric setting. Diophantus (probably 3-th century A.D.) denoted the unknown (x) and its powers by the following symbols:

x	x^2	x^3	x^4	x^5	x^6
ς'	$\delta^{\tilde{\nu}}$	$\kappa^{\tilde{\nu}}$	$\delta\delta^{\tilde{\nu}}$	$\delta\kappa^{\tilde{\nu}}$	$\kappa\kappa^{\tilde{\nu}}$

($\delta^{\tilde{\nu}}$ — from the Greek term $\delta\acute{\nu}\nu\alpha\mu\iota\varsigma$, denoting the square of the unknown; $\kappa^{\tilde{\nu}}$ — from the Greek $\kappa\acute{\nu}\beta os$, cube). Diophantus wrote coefficients to the right of the unknown or its powers, e.g. $3x^5$ was denoted by $\delta\kappa^{\tilde{\nu}}\bar{\gamma}$ (where $\bar{\gamma}=3$). Terms to be added together were simply juxtaposed, while subtraction required the special symbol Λ; equality was denoted by the letter ι (from the Greek $'\iota\sigma os$, equal). For example, Diophantus would have written the equation

$$(x^3+8x)-(5x^2+1) = x$$

as follows:

$$x^{\tilde{\nu}}\,\bar{\alpha}\,\varsigma'\,\bar{\eta}\,\Lambda\,\delta^{\tilde{\upsilon}}\,\bar{\varepsilon}\,\mu^o\,\bar{\alpha}\,\iota\,\varsigma'\,\bar{\alpha}$$

(here $\bar{\alpha}=1$, $\bar{\eta}=8$, $\bar{\epsilon}=5$, and $\mu^o\bar{\alpha}$ means that the unit $\bar{\alpha}$ is not to be multiplied by a power of the unknown).

Several centuries later, the Indians, who had developed a numerical algebra, introduced various mathematical symbols for several unknowns (abbreviations for the names of colours, which denoted the unknowns), the square, the square root, and the subtrahend. Thus, the equation

$$3x^2+10x-8 = x^2+1$$

was written in Brahmaputra's notation (7-th century) as follows:

$$\underline{\text{ya va}}\ 3\ \underline{\text{ya}}\ 10\ \underline{\text{ru}}\ \overset{\cdot}{8}$$
$$\underline{\text{ya va}}\ 1\ \underline{\text{ya}}\ 0\ \underline{\text{ru}}\ 1$$

(ya — from $\underline{\text{yavat}}$ — $\underline{\text{tavat}}$, unknown; va — from $\underline{\text{varga}}$, squared number; $\underline{\text{ru}}$ — from $\underline{\text{rupa}}$, a rupee coin — free term; a dot above a number denotes subtraction).

The creation of modern algebraic symbols dates to the 14 - 15-th centuries; it was conditioned by achievements in practical arithmetic and the study of equations. Symbols for various operations and for powers of an unknown quantity appeared spontaneously in different countries. Many decades — sometimes centuries — elapsed until a specific symbol became accepted as convenient for calculations. Thus, at the end of the 15-th century N. Chuquet and L. Pacioli were using the symbols $\tilde{p}$ and $\tilde{m}$ (from the Latin plus and minus) for addition and subtraction, respectively, while German mathematicians introduced the modern $+$ (probably an abbreviation for the Latin et) and $-$. As late as the 17-th century, one could count about ten different symbols for multiplication:

$\square$, $*$; „ , , , ; $\int$; $\cdot$; $\times$.

The history of the radical sign is instructive. Following Leonardo Pisano (1220), and up to the 17-th century, the symbol R_x (from the Latin radix = root) was widely employed for 'square root'. Chuquet denoted square, cube, etc., roots by $R_x 2$, $R_x 3$, etc. In a German manuscript of ca. 1480 the square root is denoted by a dot before the number, the cube root by three dots, and the fourth root by two dots. By 1525 one can already find the symbol $\sqrt{}$ (Ch. Rudolff). For higher-order roots, some scholars simply repeated this symbol; others wrote a suitable letter after the symbol (an abbreviation of the name of the exponent), and still others inscribed a suitable figure in a circle or between parentheses or square brackets in order to distinguish it from the number under the radical sign (the horizontal line over the radicand was introduced by R. Descartes, 1637). Only at the beginning of the 18-th century did it become customary to write the exponent above the opening of the radical sign; the first appearance of this convention, though, was much earlier (A. Girard, 1629). Thus, the evolution of the radical sign extended over almost 500 years.

Mathematical symbols for an unknown quantity and its powers were highly diverse. During the 16-th and early 17-th centuries, more than ten rival notations were current for just one square of an unknown; among these were ce (from census — the Latin term serving as translation for the Greek term $\delta\acute{\upsilon}\nu\alpha\mu\iota\varsigma$), Q (for quadratum), $\mathcal{Z}$, $\dfrac{\overset{ii}{}}{1}$, A (2), $1^{\underline{2}}$, A^{ii}, aa, a^2, etc. G. Cardano (1545) would have written the equation

$$x^3+5x = 12$$

as follows:

1. cubus $\dot{p}$. ς. positionibus æquantur 1 2

(cubus = cube, positio = unknown, æquantur = equals).

The same equation, written by M. Stifel (1544), would have been:

$$1c\!\ae + \varsigma\,\mathfrak{X}.\,\text{æqu.}\,1\,2$$

by R. Bombelli (1572):

$$1\overset{\mathit{3}}{\smile}\text{p.}\ 5\overset{\mathit{1}}{\smile}\text{eguale à}\,1\,2$$

($\overset{3}{\smile}$ = cube of the unknown, $\overset{1}{\smile}$ = unknown, eguale à = equals);

by F. Viète (1591):

$$1\,C + \varsigma\,N,\ \text{æquatur}\,1\,2$$

(C = cubus = cube, N = numerus = number);

and by T. Harriot (1631):

$$aaa + 5.a = 12$$

The 16-th and early 17-th centuries saw the first appearance and use of the equality sign and brackets;

square brackets (Bombelli, 1550), parentheses (N. Tartaglia, 1556), and curly brackets (Viète, 1593).

A significant step forward in the development of mathematical notation was Viète's introduction (1591) of capital letters of the Latin alphabet to denote both arbitrary constant quantities and unknowns; consonants, such as B, D, . . . , were reserved for constants, and vowels A, E, . . . , for unknowns. This made it possible for the first time to write down algebraic equations with arbitrary coefficients and to operate with them. For example, Viète's equation

$$\text{A cubus} + \text{B plano in A } 3, \text{ æquatur D solido}$$

(cubus = cube, planus = plane, i.e. B is a two-dimensional constant; solidus = solid (three-dimensional); the dimensionality was indicated to ensure homogeneity of the different terms) stands for the following equation in our notation:

$$x^3 + 3Bx = D.$$

Viète, then, was the creator of algebraic formulas. Descartes (1637) gave algebraic notation its modern appearance, denoting unknowns by the last letters of the alphabet x, y, z, and arbitrary given quantities by the first letters a, b, c. Descartes is also to be credited with the modern notation for powers. As his notation offered considerable advantages over its predecessors, it rapidly gained universal recognition.

The further development of mathematical symbols was intimately connected with the invention of **infinitesimal calculus**, though the basis had already been prepared to a considerable extent in algebra. I. Newton, in his method of fluxions and fluents (1666 and later), introduced symbols for successive fluxions (derivatives) of a quantity x: $\dot{x}, \ddot{x}, \dddot{x}$, and the symbol o for an infinitesimal increment. Somewhat earlier J. Wallis (1655) had proposed the symbol ∞ for infinity. The creator of the modern notation for the differential and integral calculus was G. Leibniz. In particular, it was he who invented the modern differentials dx, d^2x, d^3x and the integral

$$\int y \, dx.$$

It is worth emphasizing the essential advantage of Leibniz' integral symbol over Newton's proposal, namely the incorporation of the x. Leibniz's notation $\int y \, dx$, while hinting at the actual process of constructing an integral sum, also includes explicit indication of the integrand and the variable of integration. As a result, the notation $\int y \, dx$ is also suited for writing formulas for transformation of variables and is readily used for multiple and line integrals. Newton's notation $'x$ does not directly offer such possibilities. Similar remarks hold concerning Leibniz's differential signs as against Newton's signs for fluxions and infinitesimal increments.

L. Euler deserves the credit for a considerable proportion of modern mathematical notation. He introduced the first generally accepted symbol for a variable operation, the function symbol fx (from the Latin functio = function; 1734). Somewhat earlier, the symbol ϕx had been used by J. Bernoulli (1718). After Euler, the symbols for many individual functions (including the trigonometric functions) became standard. Euler was also the first to use the notations e (the base of the natural logarithms, 1736), π (probably from the Greek $\pi\epsilon\rho\iota\phi\acute{\epsilon}\rho\epsilon\iota\alpha$ = circumference, 1736), and the imaginary unit $i = \sqrt{-1}$ (from the French 'imaginaire', 1777, published in 1794), which soon gained universal acceptance.

During the 19-th century, the role of notation became even more important; as new fields of mathematics were opened up, scholars endeavored to standardize the basic symbols. Some widely employed modern symbols appeared only at that time: the absolute value $|x|$ (K. Weierstrass, 1841), the vector $\bar{r}$ (A. Cauchy, 1853), the determinant

$$\begin{vmatrix} a_1 & a_2 \\ b_1 & b_2 \end{vmatrix}$$

(A. Cayley, 1841), and others. Many of the new theories of the 19-th century, such as the tensor calculus, could not have been developed without suitable notation. A characteristic phenomenon in this respect was the increase in the relative proportion of symbols denoting relations, such as the congruence $\equiv$ (C.F. Gauss, 1801), membership $\in$, isomorphism $\cong$, equivalence $\sim$, etc. Symbols for variable relations appeared with the advent of mathematical logic, which makes particularly extensive use of mathematical symbols.

From the point of view of mathematical logic, mathematical symbols can be classified under the following main headings: A) symbols for objects, B) symbols for operations, C) symbols for relations. For example, the symbols 1, 2, 3, 4 denote numbers, i.e., the objects studied in arithmetic. The symbol for the addition operation, $+$, standing on its own, does not denote any object; it takes an objective content only when the numbers to be added are specified: $1+3$ denotes the number 4. The symbol $>$ (greater) denotes a relation between numbers. A relation symbol assumes a definite content only when the objects that can stand in that specific relation are specified. One further, fourth, group of symbols may be added: D) auxiliary symbols, which determine the order in which the basic symbols are to be combined. A good example of this type of symbol is provided by parentheses, which indi-

TABLE 1 Symbols for individual objects

Symbol	Meaning	Introduced by	Year
∞	infinity	J. Wallis	1655
e	base of the natural logarithms	L. Euler	1736
π	ratio of the length of a	H. Jones	1706
	circumference to the diameter	L. Euler	1736
i	square root of -1	L. Euler	1777
		(publ. 1794)	
i, j, k	unit vectors	W. Hamilton	1853
$\Pi(\alpha)$	angle of parallelism	N.I. Lobachevskii	1835

TABLE 2 Symbols for variable objects

Symbol	Meaning	Introduced by	Year
x, y, z	unknown or variable quantities	R. Descartes	1637
$\bar{r}$	vector	A.L. Cauchy	1853

TABLE 4 Symbols for variable operations

Symbol	Meaning	Introduced by	Year
ϕx	function	J. Bernoulli	1718
fx	function	L. Euler	1734

TABLE 5 Symbols for individual relations

Symbol	Meaning	Introduced by	Year
$=$	equality	R. Recorde	1557
$>$	greater than	T. Harriot	1631
$<$	smaller than	T. Harriot	1631
$\equiv$	congruence	C.F. Gauss	1801
$\parallel$	parallel	W. Oughtred	1677
		(posthumously publ.)	
$\perp$	perpendicular	P. Hérigone	1634

TABLE 3 Symbols for individual operations

Symbol	Meaning	Introduced by	Year
$+$ $-$	addition	German mathematicians	end of the 15-th century
$\times$	multiplication	W. Oughtred	1631
$\cdot$	multiplication	G. Leibniz	1698
$:$	division	G. Leibniz	1684
$a^2, \ldots, a^n$	powers	R. Descartes	1637
$\sqrt{\ }, \sqrt[3]{\ }, \ldots$	roots	I. Newton / K. Rudolff / A. Girard	1676 / 1525 / 1629
Log	logarithm	J. Kepler	1624
log	logarithm	B. Cavalieri	1632
sin	sine	L. Euler	1748
cos	cosine	L. Euler	1748
tg	tangent	L. Euler	1753
tan	tangent	L. Euler	1753
arc sin	arcsine	J. Lagrange	1772
Sh	hyperbolic sine	V. Ricatti	1757
Ch	hyperbolic cosine	V. Ricatti	1757
$dx, ddx, \ldots, d^2x, d^3x, \ldots$	differentials	G. Leibniz (publ. 1684)	1675
$\int y\, dx$	integral	G. Leibniz (publ. 1684)	1675
d/dx	derivative	G. Leibniz	1675
$f', y', f'x$	derivative	J. Lagrange	1770-1179
Δx	difference, increment	L. Euler	1755
$\partial/\partial x$	partial derivative	A. Legendre	1786
$\int_a^b f(x)\, dx$	definite integral	J. Fourier	1819-1820
Σ	sum	L. Euler	1755
Π	product	C.F. Gauss	1812
$!$	factorial	Ch. Kramp	1808
$\lvert x \rvert$	absolute value	K. Weierstrass	1841
lim	limit	S. l'Huillier	1786
$\lim_{n=\infty}$	limit	W. Hamilton	1853
$\lim_{n\to\infty}$	limit	various mathematicians	beginning of the 20-th century
ζ	zeta-function	B. Riemann	1857
Γ	gamma-function	A. Legendre	1808
B	beta-function	J. Binet	1839
Δ	delta, Laplace operator	-R. Murphy	1833
∇	nabla, Hamilton operator	W. Hamilton	1853

cate the order in which arithmetical operations are to be carried out.

The symbols of each of the three main groups A), B), C) are of two kinds: 1) individual symbols for definite objects, operations and relations; and 2) general symbols for 'variable' or 'unknown' objects, operations and relations. Examples of symbols of the first kind are the following (see also the table in this article):

A_1) The notation for the natural numbers 1, 2, 3, 4, 5, 6, 7, 8, 9; the transcendental numbers e and π; the imaginary unit $i = \sqrt{-1}$; etc.

B_1) The signs for the arithmetical operations, $+, -, \cdot, \times, :$; root extraction $\sqrt{}$, $(\cdot)^{1/n}$, differentiation d/dx, the Laplace operator

$$\Delta = \frac{\partial^2}{\partial x^2} + \frac{\partial^2}{\partial y^2} + \frac{\partial^2}{\partial z^2}.$$

This subgroup also contains the individual symbols sin, tan, log, etc.

C_1) Equality and inequality signs, $=, >, <, \neq$, the symbols denoting parallel ($\parallel$) and perpendicular ($\perp$), etc.

Symbols of the second kind denote arbitrary objects, operations and relations of a certain class, or objects, operations and relations resulting from some previously mentioned conditions. For example, in the written identity

$$(a+b)(a-b) = a^2 - b^2$$

the letters a and b denote arbitrary numbers; when one is studying the functional dependence

$$y = x^2$$

the letters x and y denote arbitrary numbers standing in the given relation; in the solution of the equation

$$x^2 - 1 = 0$$

x denotes any number satisfying the equation (by solving the equation, one knows that there are only two numbers satisfying the condition: $+1$ and -1).

From a logical point of view it is quite legitimate to call all symbols of this kind *variable symbols*, as is customary in mathematical logic (the 'domain of variation' of the variable may prove to consist of a single object; it may even be 'empty' — e.g. in the case of equations with no solutions). Further examples of this kind of signs are:

A_2) Symbols for points, straight lines, planes, and more complex geometrical figures, denoted in geometry by letters.

B_2) Notations such as f, F, ϕ for functions and notations in operator calculus, when one letter L may be used to denote, say, an arbitrary operator of the form

$$L[y] = a_0 + a_1 \frac{dy}{dx} + \cdots + a_n \frac{d^n y}{dx^n}.$$

Symbols for 'variable relations' are less common; they find application only in mathematical logic and in comparatively abstract, primarily axiomatic, branches of mathematics.

References

[1] CAJORI, F.: *A history of mathematical notations*, 1-2, Open Court, La Salle, 1952-1974.

Editorial comments. In fact, the notation π was borrowed by Euler from H. Jones.

References

[A1] KLINE, M.: *Mathematical thought from ancient to modern times*, Oxford Univ. Press, 1972.
[A2] BOYER, C.: *A history of mathematics*, Wiley, 1968.

AMS 1980 Subject Classification: 00A25

MATHEMATICAL THEORY OF COMPUTATION

Editorial comments. The last twenty years have witnessed most vigorous growth in areas of mathematical study connected with computers and computer science. The enormous development of computers and the resulting profound changes in scientific methodology have opened new horizons for the science of mathematics at a speed without parallel during the long history of mathematics.

The following two observations should be kept in mind. First, various developments in mathematics have directly initiated the 'beginning' of computers and computer science. Secondly, advances in computer science have induced very vigorous developments in certain branches of mathematics. More specifically, the second of these observations refers to the growing importance of discrete mathematics — and only the very beginning of the influence of discrete mathematics is witnessed now (1990).

Because of reasons outlined above, mathematics plays a central role in the foundations of computer science. A number of significant research areas can be listed in this connection. It is interesting to note that these areas also reflect the historical development of computer science.

1) The classical computability theory initiated by the work of K. Gödel, A. Tarski, A. Church, E. Post, A. Turing, and S.C. Kleene occupies a central position. This area is rooted in mathematical logic (cf. **Computable function**).

2) In classical formal language and automata theory, the central notions are those of an **automaton**, a grammar (cf., e.g., **Grammar, generative**), and a language (cf., e.g., **Alphabet**). Apart from developments in area 1), the work of N. Chomsky on the foundations of natural languages, as well as the work of Post concerning rewriting systems, should be mentioned here. It is, however, fascinating to observe that the modern theory of formal languages and rewriting systems was initiated by the work of the Norwegian mathematician A. Thue at the beginning of this century! (Cf. **Formal languages and automata; L-systems**.)

3) An area initiated in the nineteen-sixties is complexity theory. In it the performance of an algorithm is investigated. The central notions are those of a tractable and an intractable problem. This area is gaining in importance because of

several reasons, one of them being the advances in area 4). (Cf. **Complexity theory; Algorithm, computational complexity of an.**)

4) Quite recent developments concerning the security of computer systems have increased the importance of cryptography to a great extent. Moreover, the idea of public-key cryptography is of specific theoretical interest and has drastically changed the ideas concerning what is doable in communication systems. (Cf. **Cryptography.**)

Areas 1) through 4) constitute the core of the mathematical theory of computation. Many other important areas dealing with the mathematical foundations of computer science (e.g., semantics and the theory of correctness of programming languages, the theory of data structures, and the theory of data bases) have been developed.

All the areas listed above comprise a fascinating part of contemporary mathematics that is very dynamic in character, full of challenging problems requiring most interesting and ingenious mathematical techniques.

The basic question in the theory of computing can be formulated in any of the following ways: What is computable? For which problem can one construct effective mechanical procedures that solve every instance of the problem? Which problems possess algorithms for their solution?

Fundamental developments in mathematical logic during the 1930-s showed the existence of unsolvable problems: No **algorithm** can possibly exist for the solution of the problem. Thus, the existence of such an algorithm is a logical impossibility — its non-existence has nothing to do with ignorance. This state of affairs led to the present formulation of the basic question in the theory of computing. Previously, people always tried to construct an algorithm for every precisely formulated problem until (if ever) the correct algorithm was found. The basic question is of definite practical significance: One should not try to construct algorithms for an unsolvable problem. (There are some notorious examples of such attempts in the past.)

A model of computation is necessary for establishing unsolvability. If one wants to show that no algorithm for a specific problem exists, one must have a precise definition of an algorithm. The situation is different in establishing solvability: it suffices to exhibit some particular procedure that is effective in the intuitive sense. (The terms 'algorithm' and 'effective procedure' are used synonymously.)

One is now confronted with the necessity of formalizing a notion of a model of computation that is general enough to cover all conceivable computers, as well as the intuitive notion of an algorithm. Some initial observations are in order.

Assume that the algorithms that need formalization compute functions mapping the set of non-negative integers into the same set. Although this is not important at this point, one could observe that this assumption is no essential restriction of generality. This is due to the fact that other input and output formats can be encoded into non-negative integers.

After having somehow defined a general model of computation, denoted by MC, one observes that each specific instance of the model possesses a *finitary description*; that

is, it can be described in terms of a formula or finitely many words. By enumerating these descriptions, one obtains an enumeration $MC_1, MC_2, \ldots$, of all specific instances of the general model of computation. In this enumeration, each MC_i represents some particular algorithm for computing a function from non-negative integers into non-negative integers. Denote by $MC_i(j)$ the value of the function computed by MC_i for the argument value j.

Define a function $f(x)$ by

$$f(x) = MC_x(x) + 1. \tag{A1}$$

Clearly, the following is an algorithm (in the intuitive sense) to compute the function $f(x)$. Given an input x, start the algorithm MC_x with the input x and add one to the output.

However, is there any specific algorithm among the formalized MC-models that would compute the function $f(x)$? The answer is no, and the argument is an indirect one. Assume that MC_t would give rise to such an algorithm, where t is some natural number. Hence, for all x,

$$f(x) = MC_t(x). \tag{A2}$$

A contradiction now arises by substituting the value t for the variable x in both (A1) and (A2).

This contradiction, referred to as the *dilemma of diagonalization*, shows that independently of the model of computation — indeed, no MC-model was specified — there will be algorithms not formalized by the model.

There is a simple and natural way to avoid the dilemma of diagonalization. So far the MC_i-algorithms are assumed to be defined everywhere: For all inputs j, the algorithm MC_i produces an output. This assumption is unreasonable from many points of view, one of which is computer programming; one cannot be sure that every program produces an output for every input. Therefore, one should allow also the possibility that some of the MC_i-algorithms enter an infinite loop for some inputs j and, consequently, do not produce any output for such j. Moreover, the set of such values j is not known a priori.

Thus, some algorithms in the list

$$MC_1, MC_2, \ldots, \tag{A3}$$

produce an output only for some of the possible inputs; that is, the corresponding functions are not defined for all non-negative integers. The dilemma of diagonalization does not arise after the inclusion of such partial functions among the functions computed by the algorithms of (A3). Indeed, the argument presented above does not lead to a contradiction because MC_t is not necessarily defined.

The general model of computation, now referred to as a **Turing machine**, was introduced quite a long time before the advent of electronic computers. Turing machines constitute by far the most widely-used general model of computation. Other general models are Markov algorithms (cf. **Normal algorithm**), Post systems (cf. **Post canonical system**), grammars, and L-systems. Each of these models leads to a list such as (A3), where partial functions are also included. All models are also equivalent in the sense that they define the same set of solvable problems or computable functions.

Only the general question of characterizing the class of solvable problems has been considered here. This question was referred to as basic in the theory of computing. It led to a discussion of general models of computation.

More specific questions in the theory of computing deal with the complexity of solvable problems. Is a problem P_1 more difficult than P_2 in the sense that every algorithm for P_1 is more complex (for instance, in terms of time or memory space needed) than a reasonable algorithm for P_2? What is a reasonable classification of problems in terms of complexity? Which problems are so complex that they can be classified as intractable in the sense that all conceivable computers require an unmanageable amount of time for solving the problem?

Undoubtedly, such questions are of crucial importance from the point of view of practical computing. A problem is not yet settled if it is known to be solvable or computable and remains intractable at the same time. As a typical example, many recent results in cryptography are based on the assumption that the factorization of the product of two large primes is impossible in practice. More specifically, if one knows a large number n consisting of, for example, 200 digits and if one also knows that n is the product of two large primes, it is still practically impossible, in general, to find the two primes. This assumption is reasonable because the problem described is intractable, at least in view of the factoring algorithms know at present. Of course, from a merely theoretical point of view where complexity is not considered, such a factoring algorithm can be trivially constructed.

Such specific questions lead to more specific models of computing. The latter are obtained either by imposing certain restrictions on Turing machines or else by some direct construction. Of particular importance is the finite automaton (cf. also **Automaton, finite**). It is a model of a strictly finitary computing device: The automaton is not capable of increasing any of its resources during the computation.

It is clear that no model of computation is suitable for all situations; modifications and even entirely new models are needed to match new developments. Theoretical computer science by now has a history long enough to justify a discussion about good and bad models. The theory is mature enough to produce a great variety of different models of computation and prove some interesting properties concerning them. Good models should be general enough so that they are not too closely linked with any particular situation or problem in computing — they should be able to lead the way. On the other hand, they should not be too abstract. Restrictions on a good model should converge, step by step, to some area of real practical significance. Typical examples are certain restrictions of abstract grammars especially suitable for considerations concerning parsing. The resulting aspects of parsing are essential in compiler construction.

To summarize: A good model represents a well-balanced abstraction of a real practical situation — not too far from and not too close to the real thing.

Formal languages constitute a descriptive tool for models of computation, both in regard to input-output format and the mode of operation. Formal language theory is by its very essence an interdisciplinary area of science; the need for a formal grammatical description arises in various scientific disciplines, ranging from linguistics to biology.

References

[A1] SALOMAA, A.: *Formal languages*, Acad. Press, 1973.
[A2] ROZENBERG, G. and SALOMAA, A.: *The mathematical theory of L systems*, Acad. Press, 1980.
[A3] KUICH, W. and SALOMAA, A.: *Semirings, automata, languages*, Springer, 1986.
[A4] SALOMAA, A.: *Computation and automata*, Cambridge Univ. Press, 1985.

G. Rozenberg

A. Salomaa

AMS 1980 Subject Classification: 68CXX, 03DXX, 68DXX, 68FXX

MATHEMATICS - The science of quantitative relations and spatial forms in the real world. Being inseparably connected with the needs of technology and natural science, the accumulation of quantitative relations and spatial forms studied in mathematics is continuously expanding; so this general definition of mathematics becomes ever richer in content.

The clear recognition of the independent position of mathematics as a separate science became possible only after the collection of a fairly large store of factual material, and arose first in Ancient Greece in the 6-th - 5-th centuries B.C.. The development of mathematics up to that time is naturally referred to as the *period of the origin of mathematics*, and the 6-th - 5-th centuries B.C. as the beginning of the *period of elementary mathematics*. During these first two periods mathematical investigation dealt almost exclusively with very restricted questions concerning fundamental ideas, already there in the very early stages of historical development in relation with very simple aspects of economic life. The first problems of mechanics and physics were already in this collection of fundamental mathematical ideas.

In the 17-th century new questions in natural science and technology compelled mathematicians to concentrate their attention on the creation of methods to allow the mathematical study of motion, the processes of variation of quantities and the transformation of geometrical figures. With the use of variable quantities in analytic geometry and the creation of differential and integral calculus, the *period of the mathematics of variable quantities* began.

Further expansion of the circle of quantitative relations and spatial forms studied in mathematics led, at the beginning of the 19-th century, to the need to consciously regard the very process of expansion as a topic of mathematical research by setting the problem of systematically studying, from a fairly general point of view, the possible types of quantitative relations and spatial forms. The creation of the 'imaginary' geometry of Lobachevskiĭ was the first significant step in this direction. The development of this type of research introduced such an important new feature into mathematics that mathematics in the 19-th and 20-th centuries is naturally referred to as the special *period of modern mathematics*.

1. The origin of mathematics. Consideration of the objects in the very early stages in the development of cultures leads to the creation of the simplest ideas of the arithmetic of natural numbers. Based on an elaborate system of verbal calculation, written systems of calculation arose and slowly the methods of doing the four arithmetic operations over the natural numbers were perfected. The demands of measurement (quantity of grain, length of a road, etc.) led to the emergence of names and notation for the simplest fractions and to the elaboration of methods for performing arithmetic operations on fractions. In this way material was accumulated which gradually added up to that most ancient mathematical science: **arithmetic**. The measurement of area and volume, the needs of building technology and, somewhat later, astronomy motivated the development of the rudiments of **geometry**. These processes occurred in many nations, largely independently and in parallel. Of special significance for the later development of science was the accumulation of arithmetic and geometric knowledge in Egypt and Babylon. In Babylon, on the basis of the techniques which were developed for arithmetic calculation, **algebra** appeared also in connection with the needs of astronomy, and the rudiments of **trigonometry** appeared.

2. The period of elementary mathematics. Only after the accumulation of a larger amount of concrete material in the form of un-coordinated methods of arithmetic calculation and methods for determining area and volume, did mathematics arise as an independent science with a clear understanding of the originality of its method and the necessity for a systematic development of its basic concepts and assumptions in a fairly general form. In the application to arithmetic and algebra this process had already begun in Babylon. However, this new trend, comprising the systematic and logical succession of the construction of the foundations of mathematics, was fully defined in Ancient Greece. The system of exposition of elementary geometry created by the Ancient Greeks remained for more than two thousand years the standard for the deductive construction of a mathematical theory. From arithmetic gradually grew **number theory**. The systematic study of magnitudes and measurements was created. The process of forming (in connection with the problems of measurement of magnitude) the notion of a real number (see **Number**) turned out to be very protracted. The problem was that the idea of an irrational or a negative number was related to more complicated

mathematical abstractions which, in contrast to the concepts of a natural number, a fraction or a geometric figure, have no fairly sound support in prescientific ordinary human experience. The creation of algebra as a literal calculus was completed only at the end of this period. The period of elementary mathematics ended (in Western Europe at the beginning of the 17-th century), when the emphasis of mathematical interests shifted to the domain of mathematics of variable quantities.

3. The period of creation of the mathematics of variable quantities. With the 17-th century an essentially new period in the development of mathematics began. The circle of quantitative relations and spatial forms of mathematics studied now was no longer exhausted by numbers, quantities and geometric figures. On this basis there resulted the explicit introduction into mathematics of the ideas of motions and change. Already algebra contained the idea of dependence between variables in a latent form (the value of a sum depends on the values of the terms, etc.). However, in order to include quantitative relations in the process of variation it was necessary that the very dependence between the variables be made an independent object of study. Therefore, in the first scheme the notion of a **function** was put forward, which later played the same role of fundamental and independent object of study as the notion of quantity or number had played earlier. The study of variable quantities and functional dependence leads to the fundamental ideas of mathematical analysis, introducing explicitly into mathematics the idea of the infinite, the notions of a limit, a derivative, a differential, and an integral. Infinitesimal analysis was born; in the first place in the form of the **differential calculus** and **integral calculus**, allowing one to relate finite variations of variable quantities to their behaviour in an immediate vicinity of their individual values. The basic laws of mechanics and physics were described by differential equations, and the problem of investigating these equations comes to the foreground as one of the major problems of mathematics. The search for unknown functions defined by conditions of another kind (conditions of maxima or minima of certain related quantities) forms the topic of the calculus of variations (cf. **Variational calculus**). In this way, side-by-side with equations in which the unknowns are numbers, equations emerge in which functions are the unknowns and have to be determined.

The subject of geometry is also significantly expanded with the penetration into geometry of the ideas of motions and transformations of figures. Geometry begins to study motions and transformations for their own sake. For example, in **projective geometry** one of the basic objects of study is the set of projective transformations of the plane or space. However, the conscious development of these ideas dates only from the end of the 18-th and the beginning of the 19-th centuries. Much earlier, with the advent of **analytic geometry** in the 17-th century, the relation of geometry to the remainder of mathematics was essentially changed; a universal method was found for transferring questions of geometry into the language of algebra and analysis and for solving them neatly by algebraic and analytic methods. On the other hand, the broad possibility of sketching (illustrating) algebraic and analytic facts by geometric means was discovered, for example, in the graphical illustration of functional dependence.

4. Modern mathematics. All the divisions of mathematical analysis created in the 17-th and 18-th centuries continued to develop with great intensity in the 19-th and 20-th centuries. The circle of applications to problems of science and technology was greatly expanded at this time. However, in addition to this quantitative growth, at the end of the 18-th and the beginning of the 19-th centuries a number of essentially new features were observed in the development of mathematics.

The enormous amount of factual material which had been accumulated in the 17-th and 18-th centuries led to the demand for a deep logical analysis and unification of it from a new point of view. In essence, the relationship between mathematics and natural science was no less close but was now increased in complexity. The majority of new theories arose not just as a result of the immediate needs of natural science and technology, but also from internal requirements of mathematics itself. Such, in essence, was the development of the theory of functions of a complex variable (cf. **Functions of a complex variable, theory of**), which occupied a central position in mathematical analysis at the beginning and middle of the 19-th century. Another remarkable example of a theory arising as a result of the internal development of mathematics was **Lobachevskiĭ geometry**.

In more immediate and continuous relation to the needs of mechanics and physics the formation of vector and tensor calculus arose. The translation of vector and tensor notions into infinite-dimensional quantities resulted in the framework of **functional analysis** and is closely connected with the requirements of modern physics.

In this way, as a result of both the internal requirements of mathematics and the new needs of natural science, the circle of quantitative relations and spatial forms studied in mathematics was greatly expanded: relations between elements of arbitrary groups, operations in function spaces, the whole diversity of forms of spaces of any number of dimensions, etc. are now parts of mathematics.

The essential novelty of this stage in the development of mathematics, beginning in the 19-th century, is that questions concerning the necessary expansion of the circle of ideas in the study of quantitative relations and spatial forms themselves became the subject of conscious and active interest for mathematicians. If before, for example, the introduction of negative and complex numbers and the exact formulation of their rules of operation required protracted effort, now the development of mathematics required elaboration of methods for the deliberate and planned creation of new geometric and algebraic systems.

The extraordinary expansion of the subject of mathematics in the 19-th century attracted vigorous attention to the question of its 'foundations', that is, a critical review of its initial conditions (axioms), the construction of rigorous systems of definitions and proofs and also a critical consideration of the logical methods used in these proofs. The standard requirement of logical rigour imposed on the practical work of mathematicians in the development of individual mathematical theories was added only at the end of the 19-th century. A deep and careful analysis of the requirement of logical rigour in proofs, the construction of mathematical theories, questions of algorithmic solvability and unsolvability of mathematical problems, comprise the subject of **mathematical logic**.

At the beginning of the 19-th century a new significant expansion of the domain of application of mathematical analysis was initiated. Up to this time the fundamental divisions of physics demanding the greatest mathematical apparatus were mechanics and optics. Now were added electrodynamics and the theories of magnetism and thermodynamics. The important division of the mechanics of continuous media underwent broad development. The mathematical needs of technology also grew rapidly. As the fundamental apparatus of the new areas of mechanics and mathematical physics, the theories of ordinary and partial differential equations and of the equations of mathematical physics were strongly developed (cf. **Differential equation, ordinary; Differential equation, partial; Mathematical physics, equations of**).

The theory of differential equations served as the starting point for the investigation of the topology of manifolds. Here were obtained the first 'combinatorial', 'homological' and 'homotopical' methods of **algebraic topology**. Another direction in topology arose, based on **set theory** and **functional analysis**, and has led to the systematic construction of the theory of general topological spaces (cf. **Topological space**).

An essential completion to the methods of differential equations in the study of the character and solution of technical problems are the methods of **probability theory**. If at the beginning of the 19-th century the principal use of probabilistic methods was in ballistics and the theory of errors, then at the end of the 19-th and the beginning of the 20-th centuries probability theory had many applications, due to the creation of the theory of random processes and the development of the apparatus of **mathematical statistics.**

The theory of numbers, a collection of diverse results and ideas, was developed in various directions in the 19-th century as a structured theory (see **Algebraic number theory; Analytic number theory; Diophantine approximations**).

The emphasis of algebraic research shifted to new areas of algebra: the theory of groups, rings, fields, and general algebraic structures. On the interface between algebra and geometry grew the theory of continuous groups, the methods of which later permeated into all new areas of mathematics and natural science.

Elementary and projective geometry attracted the attention of mathematicians mainly from the point of the study of logical and axiomatic foundations. But the fundamental areas of geometry on which most of the significant scientific force was concentrated became **differential geometry; algebraic geometry** and **Riemannian geometry.**

As a result of the systematic construction of mathematical analysis based on the strictly arithmetic theory of irrational numbers and on set theory, the theory of functions of a real variable arose (cf. **Functions of a real variable, theory of**).

The practical application of the results of pure mathematical research required obtaining an answer to a given problem in numerical form. However, even after exhaustive theoretical analysis of a problem this often turns out to be a very difficult matter. The numerical methods of analysis and algebra that arose at the end of the 19-th and the beginning of the 20-th centuries have grown with the manufacture and use of computers into an independent branch of mathematics: **computational mathematics.**

These fundamental distinctive features of modern mathematics and the basic directions of research within the divisions of mathematics are listed as at the beginning of the 20-th century. To a large extent this division into branches has been preserved in spite of the tremendous growth of mathematics in the 20-th century. However, the demand for the development of mathematics itself, the 'mathematization' of various domains of science, the penetration of mathematical methods into many spheres of practical activity, and the rapid progress in computational techniques has led to a mixing of the basic efforts of mathematicians over the branches of mathematics and to the appearance of a whole series of new mathematical disciplines (see, for example, **Automata, theory of; Information theory; Games, theory of; Operations research**, and also **Cyber-** **netics; Mathematical economics**). On the basis of problems in the theory of control systems (cf. **Control system**), combinatorial analysis, graph theory, coding theory, and **discrete analysis** arose. Questions on optimal (in some sense) control of physical or mechanical systems described by differential equations has led to the creation of the mathematical theory of optimal control (cf. **Optimal control, mathematical theory of**).

Studies in the domain of general control problems and related areas of mathematics, in conjunction with the progress in computational techniques, provided a basis for the automatization of new spheres of human activity.

References

[1] KOLMOGOROV, A.N.: 'Mathematics', in *Great Soviet Encyclopedia*, Vol. 26, Moscow, 1954 (in Russian).
[2] VINOGRADOV, I.M.: 'Mathematics and scientific progress', in *Lenin and Modern Science*, Vol. 2, Moscow, 1970 (in Russian).
[3] HILBERT, D. and BERNAYS, P.: *Grundlagen der Mathematik*, 1-2, Springer, 1968-1970.
[4] ALEKSANDROV, A.D., KOLMOGOROV, A.N. and LAVRENT'EV, M.A. (EDS.): *Mathematics, its content, methods and meaning*, 1-6, Amer. Math. Soc., 1962 (translated from the Russian).
[5] *The history of mathematics from ancient times up to the beginning of the 19-th century*, 1-3, Moscow, 1970-1972 (in Russian).
[6] *Mathematics of the 19-th century. Mathematical logic. Algebra. Number theory. Probability theory*, Moscow, 1978 (in Russian).
[7] *Mathematics of the 19-th century. Geometry. The theory of analytic functions*, Moscow, 1981 (in Russian).
[8] STRUIK, D.J.: *A concise history of mathematics*, Dover, reprint, 1967.
[9] MARDZHANISHVILI, K.K.: 'Mathematics in the Academy of Science of the USSR', *Vestnik Akad. Nauk SSSR* **6** (1974).
[10] WEYL, H.: 'A half-century of mathematics', *Amer. Math. Monthly* **58**, no. 8 (1951), 523-553.

From the article by A.N. Kolmogorov [1].

Editorial comments.

References

[A1] KLINE, M.: *Mathematical throught from ancient to modern times*, Oxford Univ. Press, 1972.
[A2] BOURBAKI, N.: *Eléments d'histoire des mathématiques*, Hermann, 1960.
[A3] MACLANE, S.: *Mathematics, form and function*, Springer, 1986.
[A4] LAKATOS, I.: *Proofs and refutations*, Cambridge Univ. Press, 1979.
[A5] HARDY, G.: *A mathematician's apology*, Cambridge Univ. Press, 1977.
[A6] YOUNG, L.C.: *Mathematicians and their times*, North-Holland, 1981.
[A7] NEWMAN, J.R.: *The world of mathematics*, 1-3, Simon & Schuster, 1956.
[A8] BROWDER, F.E. (ED.): *Mathematical developments arising from Hilbert problems*, Proc. Symp. Pure Math., 28, Amer. Math. Soc., 1976.
[A9] DIEUDONNÉ, J.: *Abrégé d'histoire des mathématiques: 1700-1900*, 1-2, Hermann, 1978.
[A10] DIEUDONNÉ, J.: *Panorama des mathématiques pures: le choix bourbachique*, Gauthier-Villars, 1977.
[A11] BOCHNER, S.: *The role of mathematics in the rise of science*, Princeton Univ. Press, 1981.
[A12] COURANT, R. and ROBBINS, H.: *What is mathematics?*, Oxford Univ. Press, 1980.

AMS 1980 Subject Classification: 00A25, 01-XX

MATHIEU EQUATION - The following ordinary differential equation with real coefficients:

$$\frac{d^2u}{dz^2} + (a + b\cos 2z)u = 0, \quad z \in \mathbf{R}.$$

It was introduced by E. Mathieu [1] in the investigation of the oscillations of an elliptic membrane; it is a particular case of a **Hill equation**.

A **fundamental system of solutions** of the Mathieu equation has the form

$$u_1(z) = e^{\alpha z}\phi(z), \quad u_2(z) = u_1(-z), \tag{*}$$

for $\alpha \neq ni$, n an integer, where $\phi(z)$ is a π-periodic function and the **Lyapunov characteristic exponent** α is either real or purely imaginary. For $\operatorname{Im}\alpha = 0$ one of the solutions grows unboundedly, whereas the other tends to zero as $z \to +\infty$ (instability zones in the plane of the parameters a, b); for $\operatorname{Re}\alpha = 0$ these solutions are both bounded (stability zones). On the boundary of these zones (the case excluded in (*)) one of the functions of the fundamental system of solutions is either π-periodic or 2π-periodic (the latter is called a Mathieu function, cf. **Mathieu functions**), while the second is obtained from the first through multiplication by z. The instability zones have the form of curvilinear triangles with vertices at the points $a = n^2$, $b = 0$, $n = 0, 1, \ldots$ (see [2], [4]).

The Mathieu equation is known also in a different form (see [3]).

References

[1] MATHIEU, E.: *Course de physique mathématique*, Paris, 1873.
[2] STRETT, M.J.O.: *Lamésche- Mathieusche- und verwandter Funktionen in Physik und Technik*, Springer, 1932.
[3] KAMKE, E.: *Differentialgleichungen: Lösungsmethoden und Lösungen*, 1. Gewöhnliche Differentialgleichungen, Chelsea, reprint, 1971.
[4] YAKUBOVICH, V.A. and STARZHINSKIĬ, V.M.: *Linear differential equations with periodic coefficients and their applications*, 1-2, Wiley, 1975.

V.M. Starzhinskiĭ

Editorial comments. The operator involved in the Mathieu equation is called a *Mathieu operator*. In various applications, especially in solid state theory, a discrete analogue, the *discrete Mathieu operator*, defined by

$$(M_{A,\alpha,\nu}g)(n) = g(n+1) + 2A\cos(2\pi n\alpha - \nu)g(n) + g(n-1),$$

$$A, \alpha, \nu \in \mathbf{R},$$

is important. If α is rational this is a periodic operator, otherwise it is almost periodic. Let $\operatorname{Spec}(A, \alpha, \nu)$ be the spectrum of $M_{A,\alpha,\nu}$ on $l_2(\mathbf{Z})$ and let

$$\operatorname{Spec}(A, \alpha) = \bigcup_\nu \operatorname{Spec}(A, \alpha, \nu).$$

The spectrum $\operatorname{Spec}(1, \alpha)$ as a function of α gives a figure in the plane with remarkable combinatorial regularity and Cantor set like properties. It is known as *Hofstadter's butterfly* [A1]. M. Kac conjectured (the *Martini problem*) that $\operatorname{Spec}(A, \alpha, \nu)$ is a Cantor set for all irrational α, $A \neq 0$, $\nu \in \mathbf{R}$; another conjecture states that the Lebesgue measure of $\operatorname{Spec}(1, \alpha)$ is zero for all irrational α. For some detailed results on these spectra for rational α and a survey of this problem area cf. [A2]. A selection of noteworthy papers on these matters as well as results for the continuous analogues is [A3] - [A5].

References

[A1] HOFSTADTER, D.: 'The energy levels of Bloch electrons in rational and irrational magnetic fields', *Phys. Rev.* **B14** (1976), 2239-2249.
[A2] MOUCHÉ, P.M.M. VAN: *Sur les régions interdites du spectre de l'opérateur périodique et discret de Mathieu*, Math. Inst. Univ. Utrecht, 1988. Thesis.
[A3] BÉLISSARD, J. and SIMON, B.: 'Cantor spectrum for the almost Mathieu potential', *J. Funct. Anal.* **48** (1982), 408-419.
[A4] BÉLISSARD, J., LIMA, R. and TESTAREL, D.: 'Almost periodic Schrödinger operators', in L. Streit (ed.): *Mathematics and Physics, lectures on recent results*, Vol. 1, World Scientific, 1985, pp. 1-64.
[A5] SIMON, B.: 'Almost periodic Schrödinger operators, a review', *Adv. Appl. Math.* **3** (1982), 463-490.
[A6] MEIXNER, J. and SCHÄFKE, F.W.: *Mathieu functions and spheroidal functions and their mathematical foundations: further studies*, Springer, 1980.

AMS 1980 Subject Classification: 34B30

MATHIEU FUNCTIONS - The 2π-periodic solutions of the **Mathieu equation**

$$\frac{d^2u}{dz^2} + (a + 16q\cos 2z)u = 0, \quad z \in \mathbf{R},$$

which exist only when the point (a, q) in the parameter plane lies on the boundary of the stability zones. A Mathieu function is even or odd, and is unique up to a factor; the second linearly-independent solution grows linearly in z for $|z| \to \infty$, provided $q \neq 0$. The even Mathieu functions are the eigen functions of the integral equation

$$G(z) = \lambda \int_{-\pi}^{\pi} e^{k\cos z\cos t}G(t)\,dt, \quad k = \sqrt{32q}.$$

An analogous equation is satisfied by the odd Mathieu functions. The notation for Mathieu functions is:

$$ce_0(z, q), ce_1(z, q), \ldots; \quad se_1(z, q), se_2(z, q), \ldots.$$

For $q \to 0$ these functions reduce to the **trigonometric system**

$$1, \cos z, \ldots; \quad \sin z, \sin 2z, \ldots,$$

and they possess the same orthogonality properties on the interval $(-\pi, \pi)$. The Mathieu functions admit Fourier-series expansions which converge for small $|q| \leq r_n$; the coefficients of these series are convergent power series in q, for example,

$$ce_0(z, q) = 1 +$$

$$+ \sum_{n=1}^{\infty} \left[2^{n+1}\frac{q^n}{(n!)^2} - \frac{n(3n+4)2^{n+3}q^{n+2}}{((n+1)!)^2} + O(q^{n+4}) \right] \cos 2nz.$$

References

[1] WHITTAKER, E.T. and WATSON, G.N.: *A course of modern analysis*, Cambridge Univ. Press, 1952.
[2] BATEMAN, H. and ERDÉLYI, A.: *Higher transcendental functions*.

Automorphic functions, 3, McGraw-Hill, 1955.

[3] SANSONE, G.: *Equazioni differenziali nel campo reale*, 1, Zanichelli, 1948.

[4] STRETT, M.J.O.: *Lamésche-, Mathieusche- und verwandte Funktionen in Physik und Technik* , Springer, 1932.

[5] MAC-LACHLAN, N.W.: *Theory and application of Mathieu functions*, Clarendon Press, 1947.

M.V. Fedoryuk

AMS 1980 Subject Classification: 34B30, 33A55

MATHIEU GROUP - A **finite group** isomorphic to one of the five groups discovered by E. Mathieu [1]. The series of Mathieu groups consists of the groups denoted by

$$M_{11}, M_{12}, M_{22}, M_{23}, M_{24}.$$

They are representable as permutation groups (cf. **Permutation group**) on sets with 11, 12, 22, 23, and 24 elements, respectively. The groups M_{12} and M_{24} are five-fold transitive. M_{11} is realized naturally as the **stabilizer** in M_{12} of an element of the set on which M_{12} acts; similarly, M_{23} and M_{22} are stabilizers of elements of M_{24} and M_{23}, respectively. The Mathieu groups have the respective orders

$$7\,920,\ 95\,040,\ 443\,520,\ 10\,200\,960,\ 244\,823\,040.$$

When considering a Mathieu group, one often uses (see [2]) its representation as the group of automorphisms of the corresponding **Steiner system** $S(l, m, n)$, i.e. of the set of n elements in which there is distinguished a system of

$$\binom{m}{l}^{-1}\binom{n}{l}$$

subsets, called blocks, consisting of m elements of the set, and such that every set of l elements is contained in one and only one block. An *automorphism of a Steiner system* is defined as a permutation of the set of its elements which takes blocks into blocks. The list of Mathieu groups and corresponding Steiner systems for which they are automorphism groups is as follows: M_{11} — $S(4, 5, 11)$; M_{12} — $S(5, 6, 12)$; M_{22} — $S(3, 6, 22)$; M_{23} — $S(4, 7, 23)$; M_{24} — $S(5, 8, 24)$.

The Mathieu groups were the first (and for over 80 years the only) known sporadic finite simple groups (cf. also **Sporadic simple group**).

References

[1A] MATHIEU, E.: 'Mémoire sur l'étude des fonctions de plusieures quantités, sur la manière de les formes et sur les substitutions qui les laissant invariables', *J. Math. Pures Appl.* **6** (1861), 241-323.

[1B] MATHIEU, E.: 'Sur la fonction cinq fois transitive des 24 quantités', *J. Math. Pures Appl.* **18** (1873), 25-46.

[2A] WITT, E.: 'Die 5-fach transitiven Gruppen von Matthieu', *Abh. Math. Sem. Univ. Hamburg* **12** (1938), 256-264.

[2B] WITT, E.: 'Ueber Steinersche Systeme', *Abh. Math. Sem. Univ. Hamburg* **12** (1938), 265-275.

[3] MAZUROV, V.D.: 'Finite groups', *Itogi Nauk. i Tekhn. Algebra Topol. Geom.* **14** (1976), 5-56 (in Russian).

S.P. Strunkov

Editorial comments. For more information (e.g. charac-

ter tables and maximal subgroups) see [A1].

References

[A1] CONWAY, J.H., ET AL.: *Atlas of finite groups*, Clarendon Press, 1985.

AMS 1980 Subject Classification: 20D08, 20B25

MATRIX - A rectangular array

$$\left\| \begin{array}{ccc} a_{11} & \cdots & a_{1n} \\ \cdots & \cdots & \cdots \\ a_{m1} & \cdots & a_{mn} \end{array} \right\|, \tag{1}$$

consisting of m rows and n columns, the entries a_{ij} of which belong to some set K. (1) is called also an $(m \times n)$-*dimensional matrix over* K, or a *matrix of dimensions* $m \times n$ *over* K. Let $M_{m,n}(K)$ denote the set of all $(m \times n)$-dimensional matrices over K. If $m = n$, then (1) is called a *square matrix of order* n. The set of all square matrices of order n over K is denoted by $M_n(K)$.

Alternative notations for matrices are:

$$\left[\begin{array}{ccc} a_{11} & \cdots & a_{1n} \\ \cdots & \cdots & \cdots \\ a_{m1} & \cdots & a_{mn} \end{array} \right], \left[\begin{array}{ccc} a_{11} & \cdots & a_{1n} \\ \cdots & \cdots & \cdots \\ a_{m1} & \cdots & a_{mn} \end{array} \right] \text{ and } \| a_{ij} \|.$$

In the most important cases the role of K is played by the field of real numbers, the field of complex numbers, an arbitrary field, a ring of polynomials, the ring of integers, a ring of functions, or an arbitrary associative ring. The operations of addition and multiplication defined on K are carried over naturally to matrices over K, and in this way one is led to the *matrix calculus* — the subject matter of the theory of matrices.

The notion of a matrix arose first in the middle of the nineteenth century in the investigations of W. Hamilton, and A. Cayley. Fundamental results in the theory of matrices are due to K. Weierstrass, C. Jordan and G. Frobenius. I.A. Lappo-Danilevskiĭ has developed the theory of analytic functions of several matrix variables and has applied it to the study of systems of linear differential equations.

Operations with matrices. Let K be an associative ring and let $A = \| a_{ij} \|, B = \| b_{ij} \| \in M_{m,n}(K)$. Then the *sum of the matrices* A and B is, by definition,

$$A + B = \| a_{ij} + b_{ij} \|.$$

Clearly, $A + B \in M_{n,m}(K)$ and addition of matrices is associative and commutative. The *null matrix* in $M_{m,n}(K)$ is the matrix 0, all entries of which are zero. For every $A \in M_{m,n}(K)$,

$$A + 0 = 0 + A = A.$$

Let $A = \| a_{ij} \| \in M_{m,k}(K)$ and $B = \| b_{ij} \| \in M_{k,n}(K)$. The *product of the two matrices* A and B is defined by the rule

$$AB = \| c_{\mu\nu} \| \in M_{m,n}(K),$$

where

$$c_{\mu\nu} = \sum_{j=1}^{k} a_{\mu j} b_{j\nu}.$$

The product of two elements of $M_n(K)$ is always defined and belongs to $M_n(K)$. Multiplication of matrices is associative: If $A \in M_{m,k}(K)$, $B \in M_{k,n}(K)$ and $C \in M_{n,p}(K)$, then

$$(AB)C = A(BC)$$

and $ABC \in M_{m,p}(K)$. The distributivity rule also holds: For $A \in M_{m,n}(K)$ and $B, C \in M_{n,m}(K)$,

$$A(B+C) = AB+AC, \quad (B+C)A = BA+CA. \tag{2}$$

In particular, (2) holds also for $A, B, C \in M_n(K)$. Consequently, $M_n(K)$ is an associative ring. If K is a ring with an identity, then the matrix

$$E_n = \left\| \begin{array}{ccc} 1 & \cdots & 0 \\ \cdots & \cdots & \cdots \\ 0 & \cdots & 1 \end{array} \right\|$$

is the identity of the ring $M_n(K)$:

$$E_n A = A E_n = A$$

for all $A \in M_n(K)$. Multiplication of matrices is not commutative: If $n \geqslant 2$, for every associative ring K with an identity there are matrices $A, B \in M_n(K)$ such that $AB \neq BA$.

Let $\alpha \in K$, $A = \| a_{ij} \| \in M_{m,n}(K)$; the *product of the matrix A by the element* (*number, scalar*) α is, by definition, the matrix $\alpha A = \| \alpha a_{ij} \|$. Then

$$(\alpha+\beta)A = \alpha A + \beta A, \quad \alpha(\beta A) = (\alpha\beta)A, \quad \alpha(A+B) = \alpha A + \alpha B.$$

Let K be a ring with an identity. The matrix e_{ij} is defined as the element of $M_{m,n}(A)$ the only non-zero entry of which is the entry (i, j), which equals 1, $1 \leqslant i \leqslant m$, $1 \leqslant j \leqslant n$. For every $A = \| a_{ij} \| \in M_{m,n}(K)$,

$$A = \sum_{i=1}^{m} \sum_{j=1}^{n} a_{ij} e_{ij}.$$

If K is a field, then $M_{m,n}(K)$ is an mn-dimensional **vector space** over K, and the matrices e_{ij} form a basis in this space.

Block matrices. Let $m = m_1 + \cdots + m_k$, $n = n_1 + \cdots + n_l$, where m_μ and n_ν are positive integers. Then a matrix $A \in M_{m,n}(K)$ can be written in the form

$$A = \left\| \begin{array}{ccc} A_{11} & \cdots & A_{1l} \\ \cdots & \cdots & \cdots \\ A_{k1} & \cdots & A_{kl} \end{array} \right\|, \tag{3}$$

where $A_{\mu\nu} \in M_{m_\mu, n_\nu}(K)$, $\mu = 1, \ldots, k$, $\nu = 1, \ldots, l$. The matrix (3) is called a *block matrix*. If $B \in M_{n,p}(K)$, $p = p_1 + \cdots + p_t$, $p_i > 0$, and B is written in the form

$$B = \left\| \begin{array}{ccc} B_{11} & \cdots & B_{1t} \\ \cdots & \cdots & \cdots \\ B_{l1} & \cdots & B_{lt} \end{array} \right\|, \quad B_{ij} \in M_{n_i, p_j}(K),$$

then

$$AB = C = \left\| \begin{array}{ccc} C_{11} & \cdots & C_{1t} \\ \cdots & \cdots & \cdots \\ C_{k1} & \cdots & C_{kt} \end{array} \right\|, \quad C_{\mu j} = \sum_{i=1}^{l} A_{\mu i} B_{ij}.$$

For example, if $n = kl$, then $M_n(K)$ may be regarded as $M_k(\Sigma)$, where $\Sigma = M_l(K)$.

The matrix $A \in M_n(K)$ of the form

$$\left\| \begin{array}{cccc} A_1 & 0_{12} & \cdots & 0_{1k} \\ 0_{21} & A_2 & \cdots & 0_{2k} \\ \cdots & \cdots & \cdots & \cdots \\ 0_{k1} & 0_{k2} & \cdots & A_k \end{array} \right\|,$$

where $A_i \in M_{n_i}(K)$ and $0_{ij} \in M_{n_i, n_j}(K)$ is the null matrix, is denoted by $\mathrm{diag}[A_1, \ldots, A_k]$ and is called *block diagonal*. The following holds:

$$\mathrm{diag}[A_1, \ldots, A_k] + \mathrm{diag}[B_1, \ldots, B_k] =$$
$$= \mathrm{diag}[A_1+B_1, \ldots, A_k+B_k],$$

$$\mathrm{diag}[A_1, \ldots, A_k]\mathrm{diag}[B_1, \ldots, B_k] =$$
$$= \mathrm{diag}[A_1 B_1, \ldots, A_k B_k],$$

provided that the orders of A_i and B_i coincide for $i = 1, \ldots, k$.

Square matrices over a field. Let K be a field, let $A \in M_n(K)$ and let $\det A$ be the **determinant** of the matrix A. A is said to be *non-degenerate* (or *non-singular*) if $\det A \neq 0$. A matrix $A^{-1} \in M_n(K)$ is called the *inverse* of A if $AA^{-1} = A^{-1}A = E_n$. The invertibility of A in $M_n(K)$ is equivalent to its non-degeneracy, and

$$A^{-1} = \| c_{ij} \|, \quad c_{ij} = \frac{A_{ji}}{\det A},$$

where A_{ij} is the **cofactor** of the entry a_{ji}, $\det(A^{-1}) = (\det A)^{-1}$. For $A, B \in M_n(K)$,

$$AB = E_n \Leftrightarrow BA = E_n.$$

The set of all invertible elements of $M_n(K)$ is a group under multiplication, called the **general linear group** and denoted by $\mathrm{GL}(n, K)$. The powers of a matrix A are defined as follows

$$A^0 = E_n;$$
$$A^k = A^{k-1}A \quad \text{for } k > 0,$$

and if A is invertible, then $A^{-k} = (A^{-1})^k$. For the polynomial

$$f(x) = \alpha_0 + \alpha_1 x + \cdots + \alpha_t x^t, \quad f(x) \in K[x],$$

the matrix polynomial

$$f(A) = \alpha_0 E_n + \alpha_1 A + \cdots + \alpha_t A^t$$

is defined.

Every matrix from $M_n(K)$ gives rise to a linear transformation of the n-dimensional vector space V over K. Let $v_1, \ldots, v_n$ be a basis in V and let $\sigma: V \to V$ be a linear transformation of V. Then σ is uniquely determined by the set of vectors

$$u_1 = \sigma(v_1), \ldots, u_n = \sigma(v_n).$$

Moreover,

$$\left.\begin{array}{c} \sigma(v_1) = v_1 a_{11} + \cdots + v_n a_{n1}, \\ \cdots\cdots\cdots\cdots \\ \sigma(v_n) = v_1 a_{1n} + \cdots + v_n a_{nn}, \end{array}\right\} \quad (4)$$

where $a_{ij} \in K$. The matrix $A = \| a_{ij} \|$ is called the *matrix of the transformation* σ in the basis $v_1, \ldots, v_n$. For a fixed basis, the matrix $A + B$ is the matrix of the linear transformation $\sigma + \tau$, while AB is the matrix of $\sigma\tau$ if B is the matrix of the linear transformation τ. Equality (4) may be written in the form

$$[\sigma(v_1), \ldots, \sigma(v_n)] = [v_1, \ldots, v_n]A.$$

Suppose that $w_1, \ldots, w_n$ is a second basis in V. Then $[w_1, \ldots, w_n] = [v_1, \ldots, v_n]T$, $T \in \mathrm{GL}(n, K)$, and $T^{-1}AT$ is the matrix of the transformation σ in the basis $w_1, \ldots, w_n$. Two matrices $A, B \in M_n(K)$ are *similar* if there is a matrix $T \in \mathrm{GL}(n, K)$ such that $B = T^{-1}AT$. Here, also, $\det A = \det T^{-1}AT$ and the ranks of the matrices A and B coincide. The linear transformation σ is called *non-degenerate*, or *non-singular*, if $\sigma(V) = V$; σ is non-degenerate if and only if its matrix is non-degenerate. If V is regarded as the space of columns $M_{n,1}(K)$, then every linear transformation in V is given by left multiplication of the columns $v \in V$ by some $A \in M_n(K)$: $\sigma(v) = Av$, and the matrix of σ in the basis

$$v_1 = \left\|\begin{array}{c} 1 \\ 0 \\ . \\ . \\ 0 \end{array}\right\|, \ldots, v_n = \left\|\begin{array}{c} 0 \\ . \\ . \\ 0 \\ 1 \end{array}\right\|$$

coincides with A. A matrix $A \in M_n(K)$ is *singular* (or *degenerate*) if and only if there is a column $v \in M_{n,1}(K)$, $v \neq 0$, such that $Av = 0$.

Transposition and matrices of special form. Let $A = \| a_{ij} \| \in M_{m,n}(K)$. Then the matrix $A^T = \| a_{ij}^T \| \in M_{n,m}(K)$, where $a_{ij}^T = a_{ji}$, is called the *transpose* of A. Alternative notations are ${}^t A$ and A'. Let $A = \| a_{ij} \| \in M_{m,n}(\mathbf{C})$. Then $\overline{A} = \| \overline{a_{ij}} \|$, where $\overline{a_{ij}}$ is the complex conjugate of the number a_{ij}, is called the *complex conjugate* of A. The matrix $A^* = \overline{A}^T$, where $A \in M_n(\mathbf{C})$, is called the *Hermitian conjugate* of A. Many matrices used in applications are given special names:

name of the matrix	defining condition
symmetric	$A^T = A$
skew-symmetric	$A^T = -A$
orthogonal	$A^T = A^{-1}$
Hermitian	$A^* = A$
unitary	$A^* = A^{-1}$
normal	$A^* A = AA^*$
unipotent	$(A - E_n)^n = 0$
stochastic	$A = \| a_{ij} \| \in M_n(\mathbf{C}^n)$, $a_{ij} \geqslant 0$, $\sum_{j=1}^{n} a_{ij} = 1$, $i = 1, \ldots, n$
doubly-stochastic	A and A^T are stochastic
(0, 1)-matrix	every entry of A is either 0 or 1

Polynomial matrices. Let K be a field and let $K[x]$ be the ring of all polynomials in the variable x with coefficients from K. A matrix over $K[x]$ is called a *polynomial matrix*. For the elements of the ring $M_n(K[x])$ one introduces the following *elementary operations*: 1) multiplication of a row or column of a matrix by a non-zero element of the field K; and 2) addition to a row (column) of another row (respectively, column) of the given matrix, multiplied by a polynomial from $K[x]$. Two matrices $A, B \in M_n(K[x])$ are called *equivalent* ($A \sim B$) if B can be obtained from A through a finite number of elementary operations.

Let

$$N = \mathrm{diag}[f_1(x), \ldots, f_r(x), 0, \ldots, 0] \in M_n(K[x]),$$

where a) $f_i(x) \neq 0$; b) $f_j(x)$ is divisible by $f_i(x)$ for $j > i$; and c) the coefficient of the leading term in $f_i(x)$ is equal to 1. Then N is called a *canonical polynomial matrix*. Every equivalence class of elements of the ring $M_n(K[x])$ contains a unique canonical matrix. If $A \sim N$, where

$$N = \mathrm{diag}[f_1(x), \ldots, f_r(x), 0, \ldots, 0]$$

is a canonical matrix, then the polynomials

$$f_1(x), \ldots, f_r(x)$$

are called the *invariant factors* of A; the number r is identical with the **rank** of A. A matrix $A \in M_n(K[x])$ has an inverse in $M_n(K[x])$ if and only if $A \sim E_n$. The last condition is in turn equivalent to $\det A \in K \setminus 0$. Two matrices $A, B \in M_n(K[x])$ are equivalent if and only if

$$B = PAQ,$$

where $P, Q \in M_n(K[x])$, $P \sim Q \sim E_n$.

Let $A \in M_n(K)$. The matrix

$$xE_n - A \in M_n(K[x])$$

is called the *characteristic matrix* of A and $\det(xE_n - A)$ is called the *characteristic polynomial* of A. For every polynomial of the form

$$f(x) = \alpha_0 + \alpha_1 x + \cdots + \alpha_{n-1} x^{n-1} + x^n \in K[x]$$

there is an $F \in M_n(K)$ such that

$$\det(xE_n - F) = f(x).$$

Such is, for example, the matrix

$$F = \begin{Vmatrix} 0 & 0 & \cdots & 0 & -\alpha_0 \\ 1 & 0 & \cdots & 0 & -\alpha_1 \\ 0 & 1 & \cdots & 0 & -\alpha_2 \\ \cdots & \cdots & \cdots & \cdots & \cdots \\ 0 & 0 & \cdots & 1 & -\alpha_{n-1} \end{Vmatrix} \qquad (*)$$

The characteristic polynomials of two similar matrices coincide. However, the fact that two matrices have identical characteristic polynomials does not necessarily entail the fact that the matrices are similar. A *similarity criterion* is: Two matrices $A, B \in M_n(K)$ are similar if and only if the polynomial matrices $xE_n - A$ and $xE_n - B$ are equivalent. The set of all matrices from $M_n(K)$ having a given characteristic polynomial $f(x)$ is partitioned into a finite number of classes of similar matrices; this set reduces to a single class if and only if $f(x)$ does not have multiple factors in $K[x]$.

Let $A \in M_n(K)$, $v \in M_{n,1}(K)$, $v \neq 0$, and suppose that $Av = \lambda v$, where $\lambda \in K$. Then v is called an *eigen vector* of A and λ is called an *eigen value* of A. An element $\lambda \in K$ is an eigen value of a matrix A if and only if it is a root of the characteristic polynomial of A. The set of all columns $u \in M_{n,1}(K)$ such that $Au = \lambda u$ for a fixed eigen value λ of A is a subspace of $M_{n,1}(K)$. The dimension of this subspace equals the *defect* (or *deficiency*) d of the matrix $\lambda E_n - A$ ($d = n - r$, where r is the rank of $\lambda E_n - A$). The number d does not exceed the multiplicity of the root λ, but need not coincide with it. A matrix $A \in M_n(K)$ is similar to a diagonal matrix if and only if it has n linearly independent eigen vectors. If for an $A \in M_n(K)$,

$$\det(xE_n - A) = (x - \lambda_1)^{n_1} \cdots (x - \lambda_t)^{n_t}, \quad \lambda_j \in K,$$

and the roots λ_j are distinct, then the following holds: A is similar to a diagonal matrix if and only if for each λ_j, $j = 1, \dots, t$, the defect of $\lambda_j E_n - A$ coincides with n_j. In particular, every matrix with n distinct eigen values is similar to a diagonal matrix. Over an algebraically closed field every matrix from $M_n(K)$ is similar to some triangular matrix from $M_n(K)$. The *Hamilton–Cayley theorem*: If $f(x)$ is the characteristic polynomial of a matrix A, then $f(A)$ is the null matrix.

By definition, the *minimum polynomial* of a matrix $A \in M_n(K)$ is the polynomial $m(x) \in K[x]$ with the properties: α) $m(A) = 0$; β) the coefficient of the leading term equals 1; and γ) if $0 \neq \psi(x) \in K[x]$ and the degree of $\psi(x)$ is smaller than the degree of $m(x)$, then $\psi(A) \neq 0$. Every matrix has a unique minimum polynomial. If $g(x) \in K[x]$ and $g(A) = 0$, then the minimum polynomial $m(x)$ of A divides $g(x)$. The minimum polynomial and the characteristic polynomial of A coincide with the last invariant factor, and, respectively, the product of all invariant factors, of the matrix $xE_n - A$.

The minimum polynomial of A equals

$$\frac{\det(xE_n - A)}{d_{n-1}(x)},$$

where $d_{n-1}(x)$ is the **greatest common divisor** of the minors (cf. **Minor**) of order $n - 1$ of the matrix $xE_n - A$. A matrix $A \in M_n(K)$ is similar to a diagonal matrix over the field K if and only if its minimum polynomial is a product of distinct linear factors in the ring $K[x]$.

A matrix $A \in M_n(K)$ is called *nilpotent* if $A^k = 0$ for some integer k. A matrix A is nilpotent if and only if $\det(xE_n - A) = x^n$. Every nilpotent matrix from $M_n(K)$ is similar to some triangular matrix with zeros on the diagonal.

References

[1] VOEVODIN, V.V.: *Linear algebra*, Moscow, 1974 (in Russian).
[2] GANTMAKHER, F.R.: *The theory of matrices*, Chelsea, reprint, 1977 (translated from the Russian).
[3] KOSTRIKIN, A.I.: *Introduction to algebra*, Springer, 1982 (translated from the Russian).
[4] KUROSH, A.G.: *Higher algebra*, Mir, 1972 (translated from the Russian).
[5] MAL'TSEV, A.I.: *Foundations of linear algebra*, Freeman, 1963 (translated from the Russian).
[6] MISHINA, A.P. and PROSKURYAKOV, I.B.: *Higher algebra*, Moscow, 1965 (in Russian).
[7] TYSHKEVICH, A.R.I. and FEDENKO, A.S.: *Linear algebra and analytic geometry*, Minsk, 1976 (in Russian).
[8] BELLMAN, R.: *Introduction to matrix analysis*, McGraw-Hill, 1970.
[9] BOURBAKI, N.: *Elements of mathematics. Algebra: Algebraic structures. Linear algebra*, 1, Addison-Wesley, 1974, Chapt. 1;2 (translated from the French).
[10] LANCASTER, P.: *Theory of matrices*, Acad. Press, 1969.
[11] MARCUS, M. and MINC, H.: *A survey of matrix theory and matrix inequalities*, Allyn & Bacon, 1964.

D.A. Suprunenko

Editorial comments. The result on canonical polynomial matrices quoted above has a natural generalization to matrices over principal ideal domains. An $(m \times n)$-matrix A over a principal ideal domain R of the form

$$\begin{Vmatrix} d_1 & 0 & \cdot & \cdot & \cdot & 0 \\ 0 & \cdot & & & & \cdot \\ \cdot & & d_r & & & \cdot \\ \cdot & & & 0 & & \cdot \\ \cdot & & & & \cdot & \cdot \\ 0 & \cdot & \cdot & \cdot & & 0 \end{Vmatrix} \qquad (A1)$$

with d_i divisible by d_{i+1}, $i = 1, \dots, r-1$, is said to be in *Smith canonical form*. Every matrix A over a principal ideal domain R is equivalent to one in Smith canonical form in the sense that there are an $(m \times m)$-matrix P and an $(n \times n)$-matrix Q such that P and Q are invertible in $M_m(R)$ and $M_n(R)$, respectively, and such that PAQ is in Smith canonical form.

A matrix of the form (A1) is said to be in *companion form*, especially in linear systems and control theory where the theory of (polynomial) matrices finds many applications.

References

[A1] COHN, P.M.: *Algebra*, 1, Wiley, 1974, Sect. 10.6.

[A2] WOLOVICH, W.A.: *Linear multivariable systems*, Springer, 1974.
[A3] KALMAN, R.E., FALB, P.L. and ARBIB, M.A.: *Topics in mathematical system theory*, Prentice-Hall, 1969.

AMS 1980 Subject Classification: 15-XX

MATRIX ALGEBRA, *algebra of matrices* - A subalgebra of the full matrix algebra F_n of all $(n \times n)$-dimensional matrices over a field F. The operations in F_n are defined as follows:

$$\lambda a = \| \lambda a_{ij} \|, \quad a+b = \| a_{ij}+b_{ij} \|,$$

$$ab = c = \| c_{ij} \|, \quad c_{ij} = \sum_{\nu=1}^{n} a_{i\nu}b_{\nu j}$$

where $\lambda \in F$, and $a = \| a_{ij} \|$, $b = \| b_{ij} \| \in F_n$. The algebra F_n is isomorphic to the algebra of all endomorphisms of an n-dimensional vector space over F. The dimension of F_n over F equals n^2. Every associative algebra with an identity (cf. **Associative rings and algebras**) and of dimension over F at most n is isomorphic to some subalgebra of F_n. An associative algebra without an identity and with dimension over F less than n can also be isomorphically imbedded in F_n. By *Wedderburn's theorem*, the algebra F_n is simple, i.e. it has only trivial two-sided ideals. The centre of the algebra F_n consists of all scalar $(n \times n)$-dimensional matrices over F. The group of invertible elements of F_n is the **general linear group** $\mathrm{GL}(n, F)$. Every **automorphism** h of F_n is inner:

$$h(x) = txt^{-1}, \quad x \in F_n, \quad t \in \mathrm{GL}(n, F).$$

Every irreducible matrix algebra (cf. also **Irreducible matrix group**) is simple. If a matrix algebra A is absolutely reducible (for example, if the field F is algebraically closed), then $A = F_n$ for $n > 1$ (*Burnside's theorem*). A matrix algebra is semi-simple if and only if it is completely reducible (cf. also **Completely-reducible matrix group**).

Up to conjugation, F_n contains a unique maximal nilpotent subalgebra — the algebra of all upper-triangular matrices with zero diagonal entries. In F_n there is an r-dimensional commutative subalgebra if and only if

$$r \leqslant \left[\frac{n^2}{4} \right] + 1$$

(*Schur's theorem*). Over the complex field $\mathbf{C}$ the set of conjugacy classes of maximal commutative subalgebras of $\mathbf{C}_n$ is finite for $n < 6$ and infinite for $n > 6$.

In F_n one has the standard identity of degree $2n$:

$$\sum_{\sigma \in S_{2n}} (\mathrm{sgn}\, \sigma) x_{\sigma(1)} \cdots x_{\sigma(2n)} = 0,$$

where S_{2n} denotes the **symmetric group** and $\mathrm{sgn}\, \sigma$ the sign of the permutation σ, but no identity of lower degree.

References
[1] WEYL, H.: *The classical groups, their invariants and representations*, Princeton Univ. Press, 1946.
[2] JACOBSON, N.: *Structure of rings*, Amer. Math. Soc., 1956.
[3] HERSTEIN, I.: *Noncommutative rings*, Math. Assoc. Amer., 1968.
[4] WAERDEN, B.L. VAN DER: *Algebra*, 1-2, Springer, 1967-1971 (translated from the German).
[5] SUPRUNENKO, D.A. and TYSHKEVICH, R.I.: *Commutable matrices*, Minsk, 1966 (in Russian).

D.A. Suprunenko

Editorial comments. A frequently used notation for F_n is $M_n(F)$.

Wedderburn's theorem on the structure of semi-simple rings says that any semi-simple ring R is a finite direct product of full matrix rings $M_{n_i}(F_i)$ over skew-fields F_i, and conversely every ring of this form is semi-simple. Further, the F_i and n_i are uniquely determined by R.

The *Wedderburn–Artin theorem* says that a right Artinian simple ring is a total matrix ring (E. Artin, 1928; proved for finite-dimensional algebras by J.H.M. Wedderburn in 1907). A far-reaching generalization of this is the Jacobson density theorem, cf. **Associative rings and algebras** and [A1].

References
[A1] COHN, P.M.: *Algebra*, 2, Wiley, 1977, Sect. 10.2.

AMS 1980 Subject Classification: 16A40, 16A42, 16A46

MATRIX DIFFERENTIAL EQUATION - An equation in which the unknown is a matrix of functions appearing in the equation together with its derivative.

Consider a linear matrix differential equation of the form

$$X' = A(t)X, \quad t \in \mathbf{R}, \tag{1}$$

where $A(t)$ is an $(n \times n)$-dimensional matrix function with locally Lebesgue-integrable entries, and let $X(t)$ be an absolutely-continuous solution of equation (1) satisfying the condition $X(t_0) = I$, where I is the identity matrix. Then the vector function $x(t) = X(t)h$, $h \in \mathbf{R}^n$, is a solution of the linear system

$$x' = A(t)x \tag{2}$$

satisfying the condition $x(t_0) = h$. Conversely, if $h_1, \ldots, h_n \in \mathbf{R}^n$ and $x_i(t)$ is a solution of the system (2) satisfying the condition $x_i(t_0) = h_i$, $i = 1, \ldots, n$, then the matrix $X(t)$ with as columns the solutions $x_i(t)$ is a solution of the matrix differential equation (1). If, in addition, the vectors $h_1, \ldots, h_n$ are linearly independent, then $\det X(t) \neq 0$ for all $t \in \mathbf{R}$.

Equation (1) is a particular case of the following matrix differential equation (arising in the theory of stability)

$$X' = A(t)X - XB(t) + C(t). \tag{3}$$

The solution of (3) with initial condition $X(t_0) = X_0$ is given by the formula

$$X(t) = U(t, t_0)X_0 V(t, t_0) + \int_{t_0}^{t} U(t, s)C(s)V(s, t)\, ds,$$

where $U(t, s)$ is the solution of (1) with the condition $X(s, s) = I$, and $V(t, s)$ is the solution of the matrix differential equation $X' = B(t)X$ with the condition $X(s, s) = I$.

In various applied problems (theories of stabilization, optimal control, filtration of control system, and others) an important role is played by the so-called *matrix Riccati differential equation*

$$X' = A(t)X - XB(t) + C(t) + XD(t)X.$$

Thus, if the matrix Riccati equation

$$X' = -(F(t) + \lambda I)^T X - X(F(t) + \lambda I) - I + XG(t)G^T(t)X,$$

where ^{T} stands for transposition, has for $\lambda \geqslant 0$ a bounded solution $X(t)$ on the line $\mathbf{R}$, and if for all $h \in \mathbf{R}^n$, all $t \in \mathbf{R}$, and some $\epsilon > 0$, the inequality $h^T X(t)h \geqslant \epsilon h^T h$ holds, then every solution of the controllable system

$$x' = F(t)x + G(t)u, \quad x \in \mathbf{R}^n, \quad u \in \mathbf{R}^m,$$

closed by the feedback law $u = -G^T(t)X(t)x/2$, satisfies the inequality

$$|x(t)| \leqslant M |x(s)| e^{-\lambda(t-s)}, \quad s \leqslant t,$$

where $|\cdot|$ is the Euclidean norm and M does not depend on s.

References

[1] LAPPO-DANILEVSKY, I.A.: *Théorie des systèmes des équations différentielles linéaires*, Chelsea, reprint, 1953 (translated from the Russian).
[2] DALETSKIĬ, YU.L. and KREĬN, M.G.: *Stability of solutions of differential equations in Banach space*, Amer. Math. Soc., 1974 (translated from the Russian).
[3] ATKINSON, F.V.: *Discrete and continuous boundary problems*, Acad. Press, 1964.
[4] REID, W.T.: *Riccati differential equations*, Acad. Press, 1972.
[5] ZAKHAR-ITKIN, M.KH.: 'The matrix Riccati differential equation and the semi-group of linear fractional transformations', *Russian Math. Surveys* **28**, no. 3 (1973), 89-131. (*Uspekhi Mat. Nauk* **28**, no. 3 (1973), 83-120)

E.L. Tonkov

Editorial comments.

References

[A1] HALE, J.K.: *Ordinary differential equations*, Wiley, 1969.

AMS 1980 Subject Classification: 34-XX

MATRIX FACTORIZATION METHOD, *matrix forward-backward substitution method* - A method for solving finite-difference systems that approximate boundary value problems for systems of ordinary differential equations in one-dimensional problems, and for elliptic equations in two-dimensional problems.

The solution of the three-point difference scheme

$$A_i Y_{i-1} - C_i Y_i + B_i Y_{i+1} = -F_i, \quad i = 1, \dots, N-1,$$

where $Y_i = \{y_{1,i}, \dots, y_{n,i}\}$ is an unknown grid vector, F_i is the right-hand side vector and A_i, B_i, C_i are given square matrices, under the boundary conditions

$$-C_0 Y_0 + B_0 Y_1 = -F_0, \quad A_N Y_{N-1} - C_N Y_N = -F_N,$$

is sought for, as in the scalar case, in the form

$$Y_i = R_{i+1} Y_{i+1} + Q_{i+1}, \quad i = 0, \dots, N-1. \qquad (*)$$

The coefficients (the matrix R_{i+1} and the vector Q_{i+1}) are determined from the recurrence relations ('forward substitution')

$$R_{i+1} = (C_i - A_i R_i)^{-1} B_i,$$

$$Q_{i+1} = (C_i - A_i R_i)^{-1}(A_i Q_i + F_i), \quad i = 1, \dots, N-1,$$

while R_1 and Q_1 are given by the left boundary condition:

$$R_1 = C_0^{-1} B_0, \quad Q_1 = C_0^{-1} F_0.$$

The Y_i are calculated by formula (*) ('backward substitution'), and

$$Y_N = (C_N - A_N R_N)^{-1}(A_N Q_N + F_N).$$

There is stability of this method to rounding errors under the conditions

$$\| C_0^{-1} B_0 \| < 1, \quad \| C_N^{-1} A_N \| < 1,$$

$$\| C_i^{-1} B_i \| + \| C_i^{-1} A_i \| < 1, \quad i = 1, \dots, N-1,$$

which implies that $\| R_i \| < 1, i = 1, \dots, N$ (see [1]). A different form of the stability conditions is also available (see [2], [3]). The matrix factorization method is applied also to two-point difference schemes (see [3]). A variant in which inversion of matrices is replaced by orthogonalization is also used (see [4]).

References

[1] SAMARSKIĬ, A.A.: *Theorie der Differenzverfahren*, Akad. Verlagsgesell. Geest u. Portig K.-D., 1984 (translated from the Russian).
[2] OGNEVA, V.V.: 'The "sweep" method for the solution of difference equations', *USSR Comp. Math. Math. Phys.* **7**, no. 4 (1967), 113-126. (*Zh. Vychisl. Mat. i Mat. Fiz.* **7**, no. 4 (1967), 803-812)
[3] SAMARSKIĬ, A.A. and NIKOLAEV, E.S.: *Numerical methods for grid equations*, 1-2, Birkhäuser, 1989 (translated from the Russian).
[4] GODUNOV, S.K.: 'A method of orthogonal successive substitution for the solution of systems of difference equations', *USSR Comp. Math. Math. Phys.* **2**, no. 6 (1962), 1151-1165. (*Zh. Vychisl. Mat. i Mat. Fiz.* **2**, no. 6 (1962), 972-982)
[5] WACHSPRESS, E.L.: *Iterative solution of elliptic systems and applications to the neutron diffusion equations of reactor physics*, Prentice-Hall, 1966.

T.A. Germogenova

Editorial comments.

References

[A1] BABUSKA, I., PRAGER, M. and VITASEK, E.: *Numerical processes in differential equations*, Interscience, 1966.

AMS 1980 Subject Classification: 39-XX, 65-XX

MATRIX GAME - A **two-person zero-sum game** in which each player possesses only a finite number of pure strategies. If player I possesses m strategies and player II possesses n strategies, then the matrix game can be given by an $(m \times n)$-matrix $A = \| a_{ij} \|$, where $a_{ij}, i = 1, \dots, m, j = 1, \dots, n$, is the payoff of player I if (s)he chooses strategy i while player II chooses strategy j. According to the general optimality principle in two-person zero-sum games (see also **Minimax principle**), player I tends to choose a strategy i_0 on which

$$\max_i \min_j a_{ij} = \underline{v}$$

is attained, whereas player II tends to choose a strategy j_0 on which

$$\min_j \max_i a_{ij} = \bar{v}$$

is attained. If $\underline{v} = \bar{v}$, then the pair (i_0, j_0) is called a saddle point (cf. $\overline{\textbf{Saddle point in game theory}}$) of the game; the number $a_{i_0 j_0}$ is called the value of the game, and the strategies i_0, j_0 are optimal pure strategies. If $\underline{v} \neq \bar{v}$ (i.e. a pure strategy solution does not exist), then always $\underline{v} < \bar{v}$. In this case one must seek the optimal strategies of the players among their mixed strategies (cf. **Strategy (in game theory)**). Let $X \subset \mathbf{R}^m$ (respectively, $Y \subset \mathbf{R}^n$) be the set of mixed strategies of player I (respectively, player II). Then players I and II will tend to choose the strategies x^* and y^* on which

$$\underline{v}^* = \max_{x \in X} \min_{y \in Y} xAy^T$$

and

$$\bar{v}^* = \min_{y \in Y} \max_{x \in X} xAy^T$$

are attained, respectively (the superscript T denotes transposition). The main theorem in the theory of matrix games (*von Neumann's minimax theorem*) asserts that $\underline{v}^* = \bar{v}^* = v$, i.e. for every matrix game there are optimal mixed strategies x^*, y^* and a value v of the game.

For the *numerical solution of matrix games* (i.e. for finding optimal strategies and the value of the game) one most frequently uses the fact that the problem of solving a matrix game can be reduced to a **linear programming** problem. A less efficient approach is the *Brown — Robinson iterative method*, which amounts to 'playing' fictitiously the matrix game so that at each step the players choose the best pure strategies under an 'accumulated' mixed strategy of the opponent. Matrix games in which one of the players has only two strategies are readily solved using graphical methods.

Matrix games serve as mathematical models of many of the simplest conflict situations in the areas of economics, mathematical statistics, war science, and biology. In applications, the role of one of the players is sometimes assigned to 'nature', by which one understands the totality of external circumstances which are unknown to the decision maker (the other player).

References

[1] DRESHER, J.M.: *Games of strategy: theory and practice*, Prentice-Hall, 1961.
[2] NEUMANN, J. VON and MORGENSTERN, O.: *Theory of games and economic behavior*, Princeton Univ. Press, 1947.
[3] OWEN, H.G.: *Game theory*, Saunders, 1968.
[4] VOROB'EV, N.N.: *Game theory. Lectures for economists and system scientists*, Springer, 1977 (translated from the Russian).

A.A. Korbut

Editorial comments. The theory of matrix games is divided into zero-sum and non-zero-sum games. *Two-person non-zero-sum matrix games* are usually referred to as bimatrix games, cf. **Bimatrix game**.

References

[A1] KARLIN, S.: *Matrix games, programming and mathematical economics*, 1-2, Addison-Wesley, 1959.

AMS 1980 Subject Classification: 90D05

MATRIX GROUP - A **group** of square $(n \times n)$-dimensional matrices (cf. **Matrix**) with entries from an associative ring with identity (cf. **Associative rings and algebras**), with the usual multiplication of matrices. See **Linear group**.

AMS 1980 Subject Classification: 20GXX, 20HXX

MATRIX OF TRANSITION PROBABILITIES - The matrix $P_t = \| p_{ij}(t) \|$ of **transition probabilities** in time t for a homogeneous **Markov chain** $\xi(t)$ with at most a countable set of states S:

$$p_{ij}(t) = P\{\xi(t) = j \mid \xi(0) = i\}, \quad i, j \in S.$$

The matrices $\| p_{ij}(t) \|$ of a Markov chain with discrete time or a regular Markov chain with continuous time satisfy the following conditions for any $t > 0$ and $i, j \in S$:

$$p_{ij}(t) \geq 0, \quad \sum_{j \in S} p_{ij}(t) = 1,$$

i.e. they are stochastic matrices (cf. **Stochastic matrix**), while for irregular chains

$$p_{ij}(t) \geq 0, \quad \sum_{j \in S} p_{ij}(t) \leq 1,$$

such matrices are called *sub-stochastic*.

By virtue of the basic (Chapman — Kolmogorov) property of a homogeneous Markov chain,

$$p_{ij}(s + t) = \sum_{k \in S} p_{ik}(s) p_{kj}(t),$$

the family of matrices $\{P_t : t > 0\}$ forms a **multiplicative semi-group**; if the time is discrete, this semi-group is uniquely determined by P_1.

A.M. Zubkov

Editorial comments.

References

[A1] CHUNG, K.L.: *Elementary probability theory with stochastic processes*, Springer, 1974.

AMS 1980 Subject Classification: 60J35

MATRIX RING, *full matrix ring* - The ring of all square matrices of a fixed order over a ring R. The ring of $(n \times n)$-dimensional matrices over R is denoted by R_n or $M_n(R)$. Throughout this article R is an associative ring with identity (cf. **Associative rings and algebras**).

The ring R_n is isomorphic to the ring $\operatorname{End} M$ of all endomorphisms of the free right R-module M, possessing a basis with n elements. The matrix $E_n = \mathrm{diag}[1, \ldots, 1]$ is the identity in R_n. An associative ring A with identity 1 is isomorphic to R_n if and only if there is in A a set of n^2 elements e_{ij}, $i, j = 1, \ldots, n$, subject to the following conditions:

1) $e_{ij}e_{kl} = \delta_{jk}e_{il}$, $\sum_{i=1}^{n} e_{ii}e_{ii} = 1$;

2) the centralizer of the set of elements e_{ij} in A is isomorphic to R.

The centre of R_n coincides with $Z(R)E_n$, where $Z(R)$ is the centre of R; for $n > 1$ the ring R_n is non-commutative.

The multiplicative group of the ring R_n (the group of all invertible elements), called the **general linear group**, is denoted by $\mathrm{GL}(n, R)$. A matrix from R_n is invertible in R_n if and only if its columns form a basis of the free right module of all $(n \times 1)$-dimensional matrices over R. If R is commutative, then the invertibility of a matrix a in R_n is equivalent to the invertibility of its determinant, $\det a$, in R. The equality $(R_m)_n = R_{mn}$ holds.

The ring R_n is simple if and only if R is simple, for the two-sided ideals in R_n are of the form k_n, where k is a two-sided ideal in R. An **Artinian ring** is simple if and only if it is isomorphic to a matrix ring over a skew-field (the *Wedderburn−Artin theorem*). If $J(R)$ denotes the **Jacobson radical** of the ring R, then $J(M_n(R)) = M_n(J(R))$. Consequently, every matrix ring over a semi-simple ring R is semi-simple. If R is regular (i.e. if for every $a \in R$ there is a $b \in R$ such that $aba = a$), then so is R_n. If R is a ring with an invariant basis number, i.e. the number of elements in a basis of each free R-module does not depend of the choice of the basis, then R_n also has this property. The rings R and R_n are equivalent in the sense of Morita (see **Morita equivalence**): The category of R-modules is equivalent to the category of R_n-modules. However, the fact that projective R-modules are free does not necessarily entail that projective R_n-modules are free too. For instance, if R is a field and $n > 1$, then there exist finitely-generated projective R_n-modules which are not free.

References

[1] FAITH, C.: *Algebra: rings, modules, and categories*, 1, Springer, 1973.
[2] LAMBEK, J.: *Lectures on rings and modules*, Blaisdell, 1966.
[3] BOKUT', L.A.: *Associative rings*, 1, Novosibirsk, 1977 (in Russian).

D.A. Suprunenko

Editorial comments.

References

[A1] COHN, P.M.: *Algebra*, 1-2, Wiley, 1974-1977.

AMS 1980 Subject Classification: 15-XX, 20GXX, 20HXX

MATRIX SUMMATION METHOD - One of the methods for summing series and sequences using an infinite matrix. Employing an infinite matrix $\| a_{nk} \|$, $n, k = 1, 2, \ldots$, a given sequence $\{s_n\}$ is transformed into the sequence $\{\sigma_n\}$:

$$\sigma_n = \sum_{k=1}^{\infty} a_{nk}s_k.$$

If the series on the right-hand side converges for all $n = 1, 2, \ldots$, and if the sequence σ_n has a limit s for $n \to \infty$:

$$\lim_{n \to \infty} \sigma_n = s,$$

then the sequence $\{s_n\}$ is said to be *summable by the method determined by the matrix* $\| a_{nk} \|$, or simply *summable by the matrix* $\| a_{nk} \|$, and the number s is referred to as its limit in the sense of this summation method. If $\{s_n\}$ is regarded as the sequence of partial sums of a series

$$\sum_{k=1}^{\infty} u_k, \tag{1}$$

then this series is said to be *summable to the sum s by the matrix* $\| a_{nk} \|$.

A matrix summation method for series can be also defined directly by transforming the series (1) into a sequence $\{\gamma_n\}$:

$$\gamma_n = \sum_{k=1}^{\infty} g_{nk}u_k, \tag{2}$$

where $\| g_{nk} \|$ is a given matrix. In this case the series (1) is said to be *summable to the sum s* if, for all $n = 1, 2, \ldots$, the series on the right-hand side in (2) converges and

$$\lim_{n \to \infty} \gamma_n = s.$$

Less often used are matrix summation methods defined by a transformation of a series (1) into a series

$$\sum_{n=1}^{\infty} \alpha_n, \tag{3}$$

where

$$\alpha_n = \sum_{k=1}^{\infty} \alpha_{nk}u_k,$$

or by a transformation of a sequence $\{s_n\}$ into a series

$$\sum_{n=1}^{\infty} \beta_n, \tag{4}$$

where

$$\beta_n = \sum_{k=1}^{\infty} \beta_{nk}s_k, \quad n = 1, 2, \ldots,$$

which use matrices $\| \alpha_{nk} \|$ and $\| \beta_{nk} \|$, respectively. In these cases the series (1) with the partial sums s_n is summable to the sum s if the series (3) converges to s or, respectively, if the series (4) converges to s.

The matrix of a summation method all entries of which are non-negative is called a *positive matrix*. Among the matrix summation methods one finds, for example, the **Voronoï summation method**, the **Cesàro summation methods**, the **Euler summation method**, the **Riesz summation method** (R, p_n), the **Hausdorff summation method**, and others (see also **Summation methods**).

References

[1] HARDY, G.H.: *Divergent series*, Clarendon, 1949.
[2] COOKE, R.G.: *Infinite matrices and sequence spaces*, MacMillan, 1950.
[3] KANGRO, G.P.: 'Theory of summability of sequences and series', *J. Soviet Math.* **5**, no. 1 (1976), 1-45. (*Itogi Nauk. i Tekhn. Mat. Anal.* **12** (1974), 5-70)
[4] BARON, S.A.: *Introduction to the theory of summability of series*, Tartu, 1966 (in Russian).

I.I. Volkov

AMS 1980 Subject Classification: 40C05

MATROID

MATROID - A combinatorial abstraction of a linear algebra. A matroid is specified by a set V of elements and a family $\mathscr{E} = \{E_1, E_2, \ldots\}$ of subsets of V, called *independent subsets*, which satisfy the following axioms: 1) the empty set is independent; 2) any subset of an independent set is independent; 3) for every subset $A \subseteq V$, all the independent subsets of the matroid which are contained in A and are maximal with respect to inclusion relative to A have the same number of elements.

Examples. 1) The set V of rows of an arbitrary rectangular matrix and the family $\mathscr{E}$ of all subsets of V consisting of linearly independent rows form a matroid. 2) Let $\mathscr{L} = \{L_1, L_2, \ldots\}$ be the set of all skeleton forests (see **Tree**) of a graph G, and let $R(L_i)$ be the set of edges of the forest L_i, $i = 1, 2, \ldots$. Then the set V of edges of the graph G and the family $\mathscr{E} = \{R(L_i): i = 1, 2, \ldots\}$ form a matroid. 3) Let G be a bipartite graph (cf. **Graph, bipartite**) with parts W', W''. A subset $V \subseteq W'$ of vertices for which there is a matching P of the graph G such that every vertex $v \in V$ is incident to some edge of P is called a *partial transversal*. The set W' and the set of all transversals of the graph G form a so-called *transversal matroid*.

A matroid can be also specified by a set V of elements and a family $C = \{C_1, C_2, \ldots\}$ of non-empty subsets $C_i \subseteq V$, called *circuits*, which satisfy the following axioms: no proper subset of a circuit is a circuit; and if $v \in C_i \cap C_j$, then $(C_i \cup C_j) \setminus \{v\}$ contains a circuit. The independent subsets of this matroid are the subsets $E \subseteq V$ that do not contain circuits.

If G is a graph, then the set of its edges and the family of its simple circuits form what is called a *graphic matroid*. If for the circuits of a matroid one takes the cocycles (cuts, cf. **Graph, connectivity of a**) of the graph G, the resulting matroid is called a *cographic matroid*. The matroids of the last two types are also termed *cyclic* and *cocyclic*. The notion of a 'matroid' is used in graph theory and combinatorics in the proof of some assertions on **covering and packing** of matchings.

References

[1] WHITNEY, H.: 'On the abstract properties of linear dependence', *Amer. J. Math.* **57** (1935), 509-533.
[2] TUTTE, W.T.: 'Lectures on matroids', *J. Res. Nat. Bur. Standards Sec. B* **69**, no. 1-2 (1965), 1-47.

A.A. Sapozhenko

Editorial comments. The example of the matroid consisting of the rows of a (rectangular) matrix gave rise to the name.

In matroid theory, most often the underlying set V is supposed to be finite. And, in fact, it is not particularly clear what the right definitions are in the infinite case (cf. [A1], Chapt. 20). For a finite V the third independence axiom (given the other 2) is equivalent to 3') If $E_1, E_2 \in \mathscr{E}$ and $|E_1| = |E_2| + 1$, then there is a $v \in E_1 \setminus E_2$ such that $E_2 \cup \{v\} \in \mathscr{E}$.

There are many axiom systems for matroids. In addition to those based on the ideas of independent subsets and circuits there are axiom systems based on the idea of a rank function, the idea of a basis, the idea of a hyperplane, or the idea of a closure operation.

A maximal independent set is called a basis (and a minimal dependent set is called a *circuit* or *cycle* of the matroid). The maximal cardinality of an independent set contained in a subset A of V is called the *rank* $\rho(A)$. Given A, the set $\sigma(A) = \{x \in V: \rho(A \cup \{x\}) = \rho(A)\}$ is called the *closure* of A; one calls A *closed* if and only if $A = \sigma(A)$. A maximal (proper) closed subset H of V is called a *hyperplane*.

For a finite V the 'basis axiomatization' is as follows. A non-empty collection $\mathscr{B}$ of subsets of V is the set of bases of a matroid if and only if for all $B_1, B_2 \in \mathscr{B}$ and $x \in B_1 \setminus B_2$ there is an $y \in B_2 \setminus B_1$ such that $((B_1 \cup \{y\}) \setminus \{x\}) \in \mathscr{B}$.

A *closure operation* on a set V is a mapping $A \mapsto \sigma(A)$ of subsets of V to subsets of V such that $\sigma(\sigma(A)) = \sigma(A)$, $A \subset \sigma(A)$, $A \subset B \Rightarrow \sigma(A) \subset \sigma(B)$. Such a closure operation defines a matroid if for all $A \subset V$, $p, q \in V$, $p \notin \sigma(A)$ one has $p \in \sigma(A \cup \{q\}) \Rightarrow q \in \sigma(A \cup \{p\})$ (the *exchange axiom*). The corresponding independent subsets are defined by: $A \in \mathscr{E}$ if and only if $x \in A \Rightarrow x \notin \sigma(A \setminus \{x\})$.

Given any matroid $M = (V, \mathscr{E})$, there is a *dual matroid* M^*, which is most simply defined in terms of bases as follows: If $\mathscr{B}$ is the set of all bases of M, then the set $\{V \setminus B: B \in \mathscr{B}\}$ is the set of bases of M^*. The duality theory has important applications; as a striking example, one could mention the following result of H. Whitney: A graph G with associated graphic matroid M (determined by the circuits, respectively forests, of G as above) is planar if and only if the dual matroid M^* is also graphic.

Matroids are also studied from a more geometric point of view, under the name 'combinatorial geometries' (cf. also **Combinatorial geometry**).

Finally, it is also possible to define matroids in an algorithmic way. Let $M = (V, \mathscr{E})$ be a set system satisfying conditions 1) and 2) above and consider a *weighting* on M, i.e. a mapping $w: V \to \mathbf{R}$ into the real numbers satisfying $w(v) \geqslant 0$ for all $v \in V$. One extends w to the power set of V by putting $w(A) = \sum_{v \in A} w(v)$ for each subset A of V. It is required to find a subset $E \in \mathscr{E}$ of maximal weight (among all subsets in $\mathscr{E}$). Then M is a matroid if and only if the *greedy algorithm* solves this problem for every weighting w. One orders the elements of V according to weight, say $w(v_1) \geqslant \cdots \geqslant w(v_n)$ and determines E (starting from the empty set) recursively as follows: In step *i*, the element v_i is added to E unless this results in a set no longer contained in $\mathscr{E}$. Due to this algorithmic property, matroids are an extremely important tool in combinatorial optimization.

Matroids are also used in the stratification of Grassmann manifolds, in the analysis of higher-dimensional splines and *p*-adic curves, and in many other areas.

An important related concept is that of an *oriented matroid*, which is an abstraction of a linear algebra over an ordered field, and is used in the study of convex polytopes.

The seminal paper that started the whole theory of matroids is [1] (also included in [A8]).

References

[A1] WELSH, D.J.A.: *Matroid theory*, Acad. Press, 1976.
[A2] LAWLER, E.L.: *Combinatorial optimization: Networks and matroids*, Holt, Rinehart & Winston, 1976.
[A3] PAPADIMITRIOU, C.H. and STEIGLITZ, K.: *Combinatorial optimization. Algorithms and complexity*, Prentice-Hall, 1982.
[A4] WHITE, N. (ED.): *Theory of matroids*, Cambridge Univ. Press, 1986.
[A5] WHITE, N. (ED.): *Combinatorial geometries*, Cambridge Univ. Press, 1987.
[A6] WHITE, N. (ED.): *Combinatorial geometries: Advanced theory*, Cambridge Univ. Press, 1986.
[A7] AIGNER, M.: *Kombinatorik II. Matroide und Transversaltheorie*, Springer, 1976.
[A8] KUNG, J.P.S.: *A source book in matroid theory*, Birkhäuser, 1986.

AMS 1980 Subject Classification: 05B35

MATSUSHIMA CRITERION - The homogeneous space G/H, where G is an affine reductive **algebraic group** (cf. also **Affine group; Reductive group**) defined over an algebraically closed field k and H is a closed subgroup of G, is an affine algebraic variety if and only if H is a reductive group. This criterion was first found by Y. Matsushima [1] in the case where k is the complex field. Later, proofs were given that are valid for every algebraically closed field of characteristic zero (see [2], [3], [4]). In the case where the characteristic of k is positive, the proof of the criterion was obtained only after the proof of the **Mumford hypothesis** (see [5], [6]).

References

[1] MATSUSHIMA, Y.: 'Espaces homogènes de Stein des groupes de Lie complexes', *Nagoya Math. J.* **16** (1960), 205-218.
[2] BIAŁYNICKI-BIRULA, A.: 'On homogeneous affine spaces of linear algebraic groups', *Amer. J. Math.* **85** (1963), 577-582.
[3] LUNA, D.: 'Slices étales', *Bull. Soc. Math. France* **33** (1973), 81-105.
[4] BOREL, A. and HARISH-CHANDRA: 'Arithmetic subgroups of algebraic groups', *Ann. of Math.* **75** (1962), 485-535.
[5] NISNEVICH, E.A.: 'Affine homogeneous spaces and finite subgroups of arithmetic groups over function fields', *Funct. Anal. Appl.* **11**, no. 1 (1977), 64-65. (*Funktsional. Anal. i Prilozhen.* **11**, no. 1 (1977), 73-74)
[6] RICHARDSON, R.W.: 'Affine coset spaces of reductive algebraic groups', *Bull. London Math. Soc.* **9** (1977), 38-41.

V.L. Popov

AMS 1980 Subject Classification: 14LXX

MAUPERTUIS PRINCIPLE - A principle of least action, the first verbal formulation of which was given by P. Maupertuis. Originally (1744) Maupertuis deduced, from this principle, the laws of reflection and refraction of light, being compatible, in his words, 'with the important principle by which nature, when realizing its actions, always goes along the simplest path' (see [1]), and then (1746) he published his universal law of motion and equilibrium: 'A general principle: When a change occurs in nature, the quantity of action necessary for this change is least possible. The quantity of action is the product of the masses of the bodies by their speeds and by the distance over which they move' (see [2]). The universality of this principle was justified by Maupertuis with an obscure reasoning of metaphysical nature using teleological arguments, which in the subsequent discussion of his principle gave rise to strong objections by a number of his contemporaries. In addition to the laws of propagation of light, Maupertuis deduced from the principle only the known laws of collisions of bodies and the equilibrium of a lever. In the opinion of J.L. Lagrange 'the applications mentioned were of too special a nature for it to be possible to construct from them a proof of a general principle; in addition, they were somewhat indeterminate and arbitrary, which imparts some unreliability to the conclusions which could be made on their basis as well as to the principle itself' (see [3]).

The first mathematical idea of a principle of least action, for the special case of an isolated body, is due to L. Euler. He showed (1744) that under the actions of central forces bodies describe trajectories for which the integral $\int v \, ds$ takes a minimum or maximum (see [4]); here v is the velocity and ds is the arc element.

For the general case of motion of any system of bodies, interacting in any way so that the total mechanical energy of the system is conserved, the principle of least action was established by Lagrange (1760). Starting from the laws of mechanics he proved that the sum of the products of the masses by the integrals of the velocities multiplied by the elements of the traversed paths, is always a maximum or a minimum (see [5]), that is,

$$\delta \sum_i m_i \int v_i \, ds_i = 0.$$

'This principle, together with the principle of vis viva and developed by the rules of variational calculus, gives immediately all the equations necessary for the solution of each problem; hence there arises a method, as simple as it is general, for solving problems concerning the motions of bodies' (see [3]).

It is accepted, mathematically, to write the principle of least action in Lagrange's form (**Lagrange principle**), in the form of equality (14) (see **Variational principles**

of classical mechanics for the equations related to the equation numbers given here). By eliminating the time from (14) using the energy integral (13) C.G.J. Jacobi (1837) presented the principle of least action in the form (16) (see also **Jacobi principle**).

References

[1] MAUPERTUIS, P.: *Accord de différentes lois de la Nature qui avaient jusqu'ici paru incompatibles*, 1744.

[2] MAUPERTUIS, P.: 'Les lois de mouvement et du répos déduites d'un principe métaphysique', *Mem. Acad. R. Sci. et Belles Lettres Berlin* (1746), 267-294.

[3] LAGRANGE, J.L.: *Mécanique analytique*, Paris, 1788. (Also: Oeuvres, Vol. 11.)

[4] EULER, L.: *Methodus inveniendi lineas curvas maximi minimive proprietate gandentes sive solutio problematis isoperimetrici latissimo sensu accepti*, Lausannae et Genevae, 1744.

[5] LAGRANGE, J.L.: 'Essai d'une nouvelle méthode pour déterminer les maxima et les minima des formules intégrales indéfinies', in J.-A. Serret (ed.): *Oeuvres*, Vol. 1, G. Olms, reprint, 1973, pp. 335-362.

[6] JACOBI, C.: 'Note sur l'intégration des équations différentielles de la dynamique', *C.R. Acad. Sci. Paris* **5** (1837), 61-67.

V.V. Rumyantsev

Editorial comments.

References

[A1] ARNOL'D, V.I.: *Mathematical methods of classical mechanics*, Springer, 1978 (translated from the Russian).

[A2] GANTMACHER, F. [F. GANTMAKHER]: *Lectures in analytical mechanics*, Mir, 1975 (translated from the Russian).

[A3] WHITTAKER, E.T.: *Analytical dynamics of particles and rigid bodies*, Dover, reprint, 1944.

AMS 1980 Subject Classification: 49H05, 70H30, 70GXX, 70FXX

MAURER — CARTAN FORM - A left-invariant 1-form on a Lie group G, i.e. a differential form ω of degree 1 on G satisfying the condition $l_g^* \omega = \omega$ for any left translation $l_g: x \to gx$, g, $x \in G$. The Maurer — Cartan forms on G are in one-to-one correspondence with the linear forms on the tangent space $T_e(G)$ at the point e; specifically, the mapping which sends each Maurer — Cartan form ω to its value $\omega_e \in T_e(G)^*$ is an isomorphism of the space of Maurer — Cartan forms onto $T_e(G)^*$. The differential of a Maurer — Cartan form ω is a left-invariant 2-form on G, defined by the formula

$$d\omega(X, Y) = -\omega([X, Y]), \qquad (1)$$

where X, Y are arbitrary left-invariant vector fields on G. Suppose that $X_1, \ldots, X_n$ is a basis in $T_e(G)$ and let ω_i, $i = 1, \ldots, n$, be Maurer — Cartan forms such that

$$(\omega_i)_e(X_j) = \delta_{ij}, \quad j = 1, \ldots, n.$$

Then

$$d\omega_i = -\sum_{j,k=1}^{n} c_{jk}^i \omega_j \wedge \omega_k, \qquad (2)$$

where c_{jk}^i are the structure constants of the Lie algebra $\mathfrak{g}$ of G consisting of the left-invariant vector fields on G, with respect to the basis $\tilde{X}_1, \ldots, \tilde{X}_n$ determined by

$$(\tilde{X}_i)_e = X_i, \quad i = 1, \ldots, n.$$

The equalities (2) (or (1)) are called the *Maurer — Cartan equations*. They were first obtained (in a different, yet equivalent form) by L. Maurer [1]. The forms ω_i were introduced by E. Cartan in 1904 (see [2]).

Let $x_1, \ldots, x_n$ be the canonical coordinates in a neighbourhood of the point $e \in G$ determined by the basis $X_1, \ldots, X_n$. Then the forms ω_i are written in the form

$$\omega_i = \sum_{j=1}^{n} A_{ij}(x_1, \ldots, x_n) \, dx_j,$$

in which the matrix

$$A(x_1, \ldots, x_n) = (A_{ij}(x_1, \ldots, x_n))$$

is calculated by the formula

$$A(x_1, \ldots, x_n) = \frac{1 - e^{-\operatorname{ad} X}}{\operatorname{ad} X},$$

where $X = \sum_{i=1}^{n} x_i \tilde{X}_i$ and ad is the adjoint representation of the Lie algebra $\mathfrak{g}$.

Furthermore, let θ be the $\mathfrak{g}$-valued 1-form on G which assigns to each tangent vector to G the unique left-invariant vector field containing this vector (θ is called the *canonical left differential form*). Then

$$\theta = \sum_{i=1}^{n} \tilde{X}_i \omega_i$$

and

$$d\theta + \frac{1}{2}[\theta, \theta] = 0,$$

which is yet another way of writing the Maurer — Cartan equations.

References

[1] MAURER, L.: *Sitzungsber. Bayer. Akad. Wiss. Math. Phys. Kl. (München)* **18** (1879), 103-150.

[2] CARTAN, E.: 'Sur la structure des groupes infinis de transformations', *Ann. Ecole Norm.* **21** (1904), 153-206.

[3] CHEVALLEY, C.: *Theory of Lie groups*, 1, Princeton Univ. Press, 1946.

[4] BOURBAKI, N.: *Elements of mathematics. Lie groups and Lie algebras*, Addison-Wesley, 1975 (translated from the French).

[5] HELGASON, S.: *Differential geometry, Lie groups, and symmetric spaces*, Acad. Press, 1978.

A.L. Onishchik

AMS 1980 Subject Classification: 22E15, 22E60, 53C30

MAXIMAL AND MINIMAL EXTENSIONS of a symmetric operator A - The operators $\overline{A}$ (the closure of A, cf. **Closed operator**) and A^* (the adjoint of A, cf. **Adjoint operator**), respectively. All closed symmetric extensions of A occur between these. Equality of the maximal and minimal extensions is equivalent to the self-adjointness of A (cf. **Self-adjoint operator**) and is a necessary and sufficient condition for the uniqueness of a self-adjoint extension.

A.I. Loginov
V.S. Shul'man

Editorial comments.

References

[A1] REED, M. and SIMON, B.: *Methods of modern mathematical physics*, 1. Functional analysis, Acad. Press, 1972, Chapt. 8.

AMS 1980 Subject Classification: 47B25

MAXIMAL AND MINIMAL OPERATORS - The **maximal and minimal extensions** of an operator defined by a given differential expression on the subspace of functions of compact support. The domains of definition of the maximal and minimal operators can be concretely described in a number of cases, for example, for an ordinary differential operator, for elliptic operators and for differential operators with constant coefficients.

References

[1] BEREZANSKIY, YU.M. [YU.M. BEREZANSKIĬ]: *Expansion in eigenfunctions of selfadjoint operators*, Amer. Math. Soc., 1968 (translated from the Russian).

A.I. Loginov
V.S. Shul'man

AMS 1980 Subject Classification: 34B25, 47E05, 47F05

MAXIMAL COMPACT SUBGROUP *of a* **topological group** G - A compact subgroup (cf. **Compact group**) $K \subset G$ which is not contained as a proper subgroup in any compact subgroup of G. For example, $K = SO(n)$ for $G = SL(n, \mathbf{R})$, $K = \{e\}$ for a solvable simply-connected Lie group G.

In an arbitrary group G maximal compact subgroups need not exist (for example, if $G = GL(V)$, where V is an infinite-dimensional Hilbert space), and if they do exist there may be non-isomorphic ones among them.

Maximal compact subgroups of Lie groups have been studied most. If G is a connected Lie group, then any compact subgroup of G is contained in some maximal compact subgroup (in particular, maximal compact subgroups must exist) and all maximal compact subgroups of G are connected and conjugate to each other. The space of the group G is diffeomorphic to $K \times \mathbf{R}^n$, therefore most of the topological questions about Lie groups reduce to the corresponding questions for compact Lie groups (cf. **Lie group, compact**).

References

[1] CARTAN, E.: 'La géometrie des groupes de transformations', *J. Math. Pures Appl.* **6** (1927), 1-119.
[2] HELGASON, S.: *Differential geometry, Lie groups, and symmetric spaces*, Acad. Press, 1978.

V.V. Gorbatsevich

AMS 1980 Subject Classification: 22C05, 22-01

MAXIMAL CORRELATION COEFFICIENT - A measure of dependence of two random variables X and Y, defined as the least upper bound of the values of the correlation coefficients between the real random variables $\phi_1(X)$ and $\phi_2(Y)$, which are functions of X and Y such that $E\phi_1(X) = E\phi_2(Y) = 0$ and $D\phi_1(X) = D\phi_2(Y) = 1$:

$$\rho^*(X, Y) = \sup E[\phi_1(X)\phi_2(Y)].$$

If this least upper bound is attained at $\phi_1 = \phi_1^*(X)$ and $\phi_2 = \phi_2^*(Y)$, then the maximal correlation coefficient between X and Y is equal to the **correlation coefficient** of $\phi_1^*(X)$ and $\phi_2^*(Y)$. The maximal correlation coefficient has the property: $\rho^*(X, Y) = 0$ is necessary and sufficient for the independence of X and Y. If there is a linear correlation between the variables, then the maximal correlation coefficient coincides with the usual correlation coefficient.

References

[1] SARMANOV, O.V.: 'The maximum correlation coefficient (symmetric case)', *Dokl. Akad. Nauk SSSR* **120**, no. 4 (1958), 715-718 (in Russian).
[2] SARMANOV, O.V.: *Dokl. Akad. Nauk SSSR* **53**, no. 9 (1946), 781-784.
[3] PROHOROV, YU.V. [YU.V. PROKHOROV] and ROZANOV, YU.A.: *Probability theory, basic concepts. Limit theorems, random processes*, Springer, 1969 (translated from the Russian).

I.O. Sarmanov

Editorial comments. See also Canonical correlation.

References

[A1] GEBELEIN, H.: 'Das statistische Problem der Korrelation als Variations- und Eigenwertproblem und sein Zusammenhang mit der Ausgleichungrechnung', *Z. Angew. Math. Mech.* **21** (1941), 364-379.
[A2] KOYAK, R.: 'On measuring internal dependence in a set of random variables', *Ann. Statist.* **15** (1987), 1215-1229.

AMS 1980 Subject Classification: 62H20

MAXIMAL ERGODIC THEOREM - If T is an **endomorphism** of a measure space (X, μ), if $f \in L_1(X, \mu)$ and if E is the set of $x \in X$ for which

$$\sup_{n > 0} \sum_{i=0}^{n} f(T^i x) \geq 0,$$

then

$$\int_E f \, d\mu \geq 0.$$

The maximal ergodic theorem is due to K. Yosida and S. Kakutani [1], who showed that it can play a central role in the proof of the **Birkhoff ergodic theorem** (G.D. Birkhoff himself, instead of the maximal ergodic theorem, used somewhat different arguments). In later proofs of generalizations of Birkhoff's theorem (and also in related questions on the decomposition of the phase space into conservative and dissipative parts under conditions such that these generalizations make sense) a generalized maximal ergodic theorem is used in a similar way. There is a generalization of the maximal ergodic theorem due to E. Hopf and a simple proof of this generalization was given by A. García (see [2]). See also [3] and the references in **Birkhoff ergodic theorem**.

References

[1] YOSIDA, K. and KAKUTANI, S.: 'Birkhoff's ergodic theorem and the maximal ergodic theorem', *Proc. Imp. Acad. Tokyo* **15** (1939), 165-168.

[2] NEVEU, J.: *Mathematical foundations of the calculus of probability*, Holden-Day, 1965 (translated from the French).

[3] VERSHIK, A.M. and YUZVINSKIĬ, S.A.: 'Dynamical systems with invariant measure', *Progress in Math.* **8** (1970), 151-215. *(Itogi Nauk. Anal.* (1967), 133-187)

D.V. Anosov

Editorial comments. A variety of ergodic theorems (including historical remarks) can be found in [A1].

References

[A1] KRENGEL, U.: *Ergodic theorems*, De Gruyter, 1985.

AMS 1980 Subject Classification: 28DXX, 47A35

MAXIMAL IDEAL - A maximal element in the **partially ordered set** of proper ideals of a corresponding algebraic structure. Maximal ideals play an essential role in ring theory. Every ring with identity has maximal left (also right and two-sided) ideals. The quotient module $M = R/I$ of R regarded as a left (respectively, right) R-module relative to a left (respectively, right) maximal ideal I is irreducible (cf. **Irreducible module**); a homomorphism ϕ of R into the field of endomorphisms of M is a representation of R. The kernel of all such representations, that is, the set of elements of the ring which are mapped to zero by all representations, is called the *Jacobson radical* of R; it coincides with the intersection of all maximal left (also, all right) ideals.

In the ring $R = C[a, b]$ of continuous real-valued functions on a closed interval $[a, b]$, the set of functions vanishing at some fixed point x_0 is a maximal ideal. Such ideals exhaust all maximal ideals of R. This relation between the points of the interval and the maximal ideals has resulted in the construction of various theories for representing rings as rings of functions on a topological space.

The **Zariski topology** on the set of prime ideals (cf. **Prime ideal**) $\operatorname{Spec} R$ of a ring R has weak separation properties (that is, there are non-closed points). A similar topology in the non-commutative case can be introduced on the set $\operatorname{Spec} R$ of primitive ideals (cf. **Primitive ideal**), which are the annihilators of irreducible R-modules. The set of maximal ideals, and in the non-commutative case, of maximal primitive ideals, forms a subspace $\operatorname{Specm} R \subset \operatorname{Spec} R$ which satisfies the T_1-separation axiom.

References

[1] JACOBSON, N.: *Structure of rings*, Amer. Math. Soc., 1956.

V.E. Govorov

Editorial comments. Maximal ideals also play an important role in the structure and representation theory of lattices (particularly distributive lattices). In a **distributive lattice**, as in a commutative ring, all maximal ideals are prime; the converse implication holds in a **Boolean algebra**, and indeed a distributive lattice in which all prime ideals are maximal is necessarily Boolean. As with rings, the set $\operatorname{Specm} L$ of maximal ideals of a distributive lattice L can be topologized as a subspace of the space $\operatorname{Spec} L$ of all prime ideals, and it is a compact T_1-space; moreover, every compact T_1-space arises in this way. A distributive lattice L is said to be *normal* if, given elements $a, b \in L$ with $a \vee b = 1$, there exist $c, d \in L$ with $a \vee d = c \vee b = 1$ and $c \wedge d = 0$. Normal distributive lattices can be characterized as those for which every prime ideal is contained in a unique maximal ideal, or equivalently as those for which there is a continuous **retraction** of $\operatorname{Spec} L$ onto $\operatorname{Specm} L$; they have the property that $\operatorname{Specm} L$ is a Hausdorff space. For a topological space X, the lattice $\mathcal{O}(X)$ of open subsets of X is normal if and only if X is a **normal space**; if X is a T_1-space, then $\operatorname{Specm} \mathcal{O}(X)$ yields a T_1 compactification of X, which coincides with the Stone–Čech compactification if X is normal (see **Wallman compactification**).

The construction of maximal ideals in arbitrary rings or lattices generally requires an appeal to Zorn's lemma (see **Axiom of choice** or **Zorn lemma**), and indeed the *maximal ideal theorem* for many classes of rings or lattices (i.e. the assertion that every non-trivial ring or lattice in the class has a maximal ideal) has been shown to be equivalent in ZF set theory to the axiom of choice. This applies to the class of all (commutative) unique factorization domains, and of all Heyting algebras (see **Brouwer lattice**); however, for the classes of principal ideal domains, of Brouwer lattices, and of normal distributive lattices, the maximal ideal theorem is equivalent to the 'prime ideal theorem' for the corresponding class, and is strictly weaker than the axiom of choice.

References

[A1] JOHNSTONE, P.T.: *Stone spaces*, Cambridge Univ. Press, 1983.

In the theory of semi-groups (cf. **Semi-group**) maximal ideals play a lesser role than minimal ideals (cf. **Minimal ideal**). If M is a maximal two-sided ideal of a semi-group S, then either $M = S \setminus \{a\}$, where a is some indecomposable element of S (that is, $a \in S^2 \setminus S$), or M is a *prime ideal* (that is, for any two ideals A and B, $AB \subseteq M$ implies $A \subseteq M$ or $B \subseteq M$). This implies that every maximal two-sided ideal in S is prime if and only if $S^2 = S$. In a semi-group S with a maximal two-sided ideal a prime ideal $P \neq S$ is maximal if (and, obviously, only if) P contains the intersection I of all maximal two-sided ideals of S. The Rees quotient semi-group S/I is an 0-**direct union** of semi-groups each of which is either 0-simple or two-element nilpotent.

Sometimes a semi-group S with proper left ideals may have a largest such ideal L^* (that is, containing all other proper left ideals). This, for example, is the case when S has a right identity. In that case, if $S \setminus L^*$ is not a singleton, then it is a sub-semi-group. In a periodic semi-group S the existence of L^* implies that L^* is a (largest proper) two-sided ideal. Another example is given by subgroups with separating group part (see **Invertible element**) which is not a group.

References

[1A] SCHWARZ, S.: 'On maximal ideals in the theory of semigroups I', *Czechoslovak. Math. J.* **3** (1953), 139-153 (in Russian). English abstract.

[1B] SCHWARZ, S.: 'On maximal ideals in the theory of semigroups II', *Czechoslovak. Math. J.* **4** (1953), 365-383 (in Russian). English abstract.

[2] SCHWARZ, S.: 'Prime ideals and maximal ideals in semigroups', *Czechoslovak. Math. J.* **19** (1969), 72-79.

[3] GRILLET, P.A.: 'Intersections of maximal ideals in semigroups', *Amer. Math. Monthly* **76** (1969), 503-509.

L.N. Shevrin

AMS 1980 Subject Classification: 13CXX, 13A15, 16A66, 20M12, 06D05, 04A25

MAXIMAL INVARIANT - An **invariant statistic** that takes different values on the different orbits generated by a group of one-to-one measurable transformations of the **sampling space**. Thus, if $(\mathfrak{X}, \mathfrak{B})$ is a sampling space and $G=\{g\}$ is a group of one-to-one $\mathfrak{B}$-measurable transformations of $\mathfrak{X}$ onto itself, then an invariant statistic $T(x)$ is a maximal invariant if $T(x_2)=T(x_1)$ implies that $x_2=gx_1$ for some $g \in G$. For example, if $\mathfrak{X}=\mathbf{R}^n$, $x=(x_1,\ldots,x_2)^T$, $G=\{\Gamma\}$ is the group of orthogonal transformations $\mathbf{R}^n \to \mathbf{R}^n$, and $y=\Gamma x$, then the statistic $T(x)=\sum x_i^2$ is a maximal invariant. Any invariant statistic is a function of the maximal invariant.

Maximal invariants are used for the construction of invariant tests (cf. **Invariant test**).

References

[1] LEHMANN, E.L.: *Testing statistical hypotheses*, Wiley, 1959.

[2] ZACKS, S.: *The theory of statistical inference*, Wiley, 1975.

[3] KLIMOV, G.P.: *Invariant inferences in statistics*, Moscow, 1973 (in Russian).

M.S. Nikulin

AMS 1980 Subject Classification: 62-01

MAXIMAL SPECTRAL TYPE - The type of the *maximal spectral measure* μ (i.e. its equivalence class) of a **normal operator** A acting on a Hilbert space H. This measure is defined (up to equivalence) by the following condition. Let $E(\lambda)$ be the **resolution of the identity** in the spectral representation of the normal operator $A=\int\lambda\,dE(\lambda)$, and let $E(\Lambda)=\int_\Lambda dE(\lambda)$ (where Λ denotes a Borel set) be the associated 'operator-valued' measure. Then $E(\Lambda)=0$ precisely for those Λ for which $\mu(\Lambda)=0$. Any $x \in H$ has an associated *spectral measure* $\mu_x(\Lambda)=(x, E(\Lambda)x)$; in these terms the definition of μ implies that for any x the measure μ_x is absolutely continuous with respect to μ and there is an x_0 for which μ_{x_0} is equivalent to μ (that is, x_0 has maximal spectral type). If H is separable, then a measure μ with these properties always exists, but if H is not separable, then there is no such measure and A does not have maximal

spectral type. This complicates the theory of unitary invariants of normal operators in the non-separable case.

References

[1] PLESNER, A.I.: *Spectral theory of linear operators*, F. Ungar, 1969 (translated from the Russian).

D.V. Anosov

AMS 1980 Subject Classification: 47B15

MAXIMAL SUBGROUP - A proper **subgroup** of a group G which is not contained in any other proper subgroup of G, that is, a maximal element in the set of proper subgroups of G ordered by inclusion. There exist groups without maximal subgroups, for example, a **group of type** p^∞.

A generalization of the concept of a maximal subgroup is that of a *subgroup maximal with respect to some property* σ, i.e. a subgroup H_0 of G with the property σ and such that no other proper subgroup H of G has σ and contains H_0.

References

[1] KARGAPOLOV, M.I. and MERZLYAKOV, YU.I.: *Fundamentals of the theory of groups*, Springer, 1979 (translated from the Russian).

N.N. Vil'yams

Editorial comments. A *proper subgroup* of a group G is a subgroup H of G satisfying $H \neq G$.

References

[A1] HALL, M., JR.: *The theory of groups*, Macmillan, 1959.

AMS 1980 Subject Classification: 20E28

MAXIMAL TERM OF A SERIES - The term of a convergent **series** of numbers or functions with positive terms the value of which is not less than the values of all other terms of this series.

Applying this idea to the study of **power series**

$$\sum_{k=0}^{\infty} c_k(z-a)^k$$

in one complex variable z with positive radius of convergence R, $0<R\leqslant\infty$, one has in mind the maximal term $\mu(r)$ of the series

$$\sum_{k=0}^{\infty} |c_k|r^k, \quad 0<r=|z-a|<R.$$

Thus,

$$|c_k|r^k \leqslant \mu(r), \quad k=0, 1, \ldots.$$

The index $\nu(r)$ of the maximal term $\mu(r)$ is called the *central index*:

$$\mu(r) = |c_{\nu(r)}|r^{\nu(r)}.$$

If there are several terms in modulus equal to $\mu(r)$, then the central index is taken to be the largest of the indices of these terms. The function

$$y = \ln\mu(e^x), \quad -\infty<x<\infty,$$

is non-decreasing and convex; the function $\nu(r)$ is a

step-function, increases at discontinuity points in natural numbers and is everywhere continuous from the right.

References

[1] VALIRON, G.: *Les fonctions analytiques*, Paris, 1954.
[2] WITTICH, H.: *Neuere Untersuchungen über eindeutige analytische Funktionen*, Springer, 1955.

E.D. Solomentsev

Editorial comments.

References

[A1] PÓLYA, G. and SZEGÖ, G.: *Problems and theorems in analysis*, 2, Springer, 1976, Part IV, Chapt. 1 (translated from the German).

AMS 1980 Subject Classification: 30B10, 40-01

MAXIMAL TORUS - 1) A *maximal torus of a linear algebraic group* G is an algebraic subgroup of G which is an **algebraic torus** and which is not contained in any larger subgroup of that type. Now let G be connected. The union of all maximal tori of G coincides with the set of all semi-simple elements of G (see **Jordan decomposition**) and their intersection coincides with the set of all semi-simple elements of the centre of G. Every maximal torus is contained in some **Borel subgroup** of G. The centralizer of a maximal torus is a **Cartan subgroup** of G; it is always connected. Any two maximal tori of G are conjugate in G. If G is defined over a field k, then there is a maximal torus in G also defined over k; its centralizer is also defined over k.

Let G be a **reductive group** defined over a field k. Consider the maximal subgroups among all algebraic subgroups of G which are k-split algebraic tori. The maximal k-split tori thus obtained are conjugate over k. The common dimension of these tori is called the k-*rank* of G and is denoted by $\mathrm{rk}_k G$. A maximal k-split torus need not, in general, be a maximal torus, that is, $\mathrm{rk}_k G$ is in general less than the *rank* of G (which is equal to the dimension of a maximal torus in G). If $\mathrm{rk}_k G = 0$, then G is called an *anisotropic group over* k, and if $\mathrm{rk}_k G$ coincides with the rank of G, then G is called a *split group over* k. If k is algebraically closed, then G is always split over k. In general, G is split over the separable closure of k.

Examples. Let k be a field and let $\bar{k}$ be an algebraic closure. The group $G = \mathrm{GL}_n(\bar{k})$ of non-singular matrices of order n with coefficients in $\bar{k}$ (see **Classical group**; **General linear group**) is defined and split over the prime subfield of k. The subgroup of all diagonal matrices is a maximal torus in G.

Let the characteristic of k be different from 2. Let V be an n-dimensional vector space over $\bar{k}$ and F a non-degenerate quadratic form on V defined over k (the latter means that in some basis $e_1, \ldots, e_n$ of V, the form $F(x_1 e_1 + \cdots + x_n e_n)$ is a polynomial in $x_1, \ldots, x_n$ with coefficients in k). Let G be the group of all non-singular linear transformations of V with determinant 1 and preserving F. It is defined over k. Let V_k be the linear hull over k of $e_1, \ldots, e_n$; it is a k-form of V. In V there always exists a basis $f_1, \ldots, f_n$ such that

$$F(x_1 f_1 + \cdots + x_n f_n) = x_1 x_n + x_2 x_{n-1} + \cdots + x_p x_{n-p+1},$$

where $p = n/2$ if n is even and $p = (n+1)/2$ if n is odd. The subgroup of G consisting of the elements whose matrix in this basis takes the form $\| a_{ij} \|$, where $a_{ij} = 0$ for $i \neq j$ and $a_{ii} a_{n-i+1,n-i+1} = 1$ for $i = 1, \ldots, p$, is a maximal torus in G (thus the rank of G is equal to the integer part of $n/2$). In general, this basis does not belong to V_k. However, there always is a basis $h_1, \ldots, h_n$ in V_k in which the quadratic form can be written as

$$F(x_1 h_1 + \cdots + x_n h_n) =$$
$$= x_1 x_n + \cdots + x_q x_{n-q+1} + F_0(x_{q+1}, \ldots, x_{n-q}), \quad q > p,$$

where F_0 is a quadratic form which is anisotropic over k (that is, the equation $F_0 = 0$ only has the zero solution in k, see **Witt decomposition**). The subgroup of G consisting of the elements whose matrix in the basis $h_1, \ldots, h_n$ takes the form $\| a_{ij} \|$, where $a_{ij} = 0$ for $i \neq j$, $a_{ii} a_{n-i+1,n-i+1} = 1$ for $i = 1, \ldots, q$ and $a_{ii} = 1$ for $i = q+1, \ldots, n-q$, is a maximal k-split torus in G (so $\mathrm{rk}_k G = q$ and G is split if and only if q is the integer part of $n/2$).

Using maximal tori one associates to a reductive group G a **root system**, which is a basic ingredient for the classification of reductive groups. Namely, let $\mathfrak{g}$ be the Lie algebra of G and let T be a fixed maximal torus in G. The adjoint representation of T in $\mathfrak{g}$ is rational and diagonalizable, so $\mathfrak{g}$ decomposes into a direct sum of weight spaces for this representation. The set of non-zero weights of this representation (considered as a subset of its linear hull in the vector space $X(T) \otimes_{\mathbf{Z}} \mathbf{R}$, where $X(T)$ is the group of rational characters of T) turns out to be a (reduced) root system. The **relative root system** is defined in a similar way: If G is defined over k and S is a maximal k-split torus in G, then the set of non-zero weights of the adjoint representation of S in $\mathfrak{g}$ forms a root system (which need not be reduced) in some subspace of $X(S) \otimes_{\mathbf{Z}} \mathbf{R}$. See also **Weyl group**; **Semi-simple group**.

References

[1] BOREL, A.: *Linear algebraic groups*, Benjamin, 1969.
[2] BOREL, A. and MOSTOW, G.D. (EDS.): *Algebraic groups and discontinuous subgroups*, Proc. Symp. Pure Math., 9, Amer. Math. Soc., 1966.

Editorial comments. For k-forms see **Form of an (algebraic) structure**.

See especially the article by A. Borel in [2].

2) A *maximal torus of a connected real Lie group G* is a connected compact commutative Lie subgroup T of G

not contained in any larger subgroup of the same type. As a Lie group T is isomorphic to a direct product of copies of the multiplicative group of complex numbers of absolute value 1. Every maximal torus of G is contained in a maximal compact subgroup of G; any two maximal tori of G (as any two maximal compact subgroups) are conjugate in G. This, in a well-known sense, reduces the study of maximal tori to the case when G is compact.

Now let G be a compact group. The union of all maximal tori of G is G and their intersection is the centre of G. The Lie algebra of a maximal torus T is a maximal commutative subalgebra in the Lie algebra $\mathfrak{g}$ of G, and each maximal commutative subalgebra in $\mathfrak{g}$ can be obtained in this way. The centralizer of a maximal torus T in G coincides with T. The adjoint representation of T in $\mathfrak{g}$ is diagonalizable and all non-zero weights of this representation form a **root system** in $X(T) \otimes_{\mathbf{Z}} \mathbf{R}$, where $X(T)$ is the group of characters of T. This is a basic ingredient for the classification of compact Lie groups.

References
[1] PONTRYAGIN, L.S.: *Topological groups*, Princeton Univ. Press, 1958 (translated from the Russian).
[2] ZHELOBENKO, D.P.: *Compact Lie groups and their representations*, Amer. Math. Soc., 1973 (translated from the Russian).
[3] HELGASON, S.: *Differential geometry, Lie groups, and symmetric spaces*, Acad. Press, 1978.

V.L. Popov

Editorial comments.

References
[A1] BOURBAKI, N.: *Groupes et algèbres de Lie*, Masson, 1982, Chapt. IX.
[A2] BRÖCKER, TH. and TOM DIECK, T.: *Representations of compact Lie groups*, Springer, 1985.

AMS 1980 Subject Classification: 20-01, 20GXX, 22-01

MAXIMIN - The mixed extrema

$$\sup_{x \in X} \inf_{y \in Y} F(x, y), \quad \max_{x \in X} \min_{y \in Y} F(x, y), \text{ etc.} \qquad (*)$$

A maximin can be interpreted (for example, in decision theory, **operations research** or game theory, cf. **Games, theory of**) as the greatest gain among those that can be attained by decision making under the worst conditions; it is, thereby, a guaranteed gain. Therefore, decision making oriented on a maximin may reasonably be regarded as optimal.

The value of a maximin does not exceed the value of the corresponding **minimax**. Conditions for their equality are very important in game theory (see **Minimax principle**). Such conditions are, for example, the presence of a linear structure in X, the convexity of X and the concavity of the function F relative to $x \in X$ for each $y \in Y$ (or, the linearity of X and convexity of Y, and convexity of F relative to $y \in Y$ for each $x \in X$).

Finding a maximin as a mathematical operation formally consists in the successive calculation of extrema, that is, in the solution of standard ('single-criterion') optimal programming problems, and, therefore, involves no conceptual complications. However, even when Y is 'well arranged' and the function F is uniformly continuous on X, the function which associates to x the value of y at which the extremum $\inf_{y \in Y} F(x, y)$ is attained (or 'almost attained') can turn out to be 'badly arranged' and, in particular, can be a discontinuous function of x. In these cases the calculation of a maximum (*) analytically is difficult and it must be found by numerical methods (see **Maximin, numerical methods**). The above also applies to finding the minimax.

N.N. Vorob'ev

Editorial comments. See also (references [1] and [2] to) Maximin, numerical methods.

AMS 1980 Subject Classification: 49B40, 90DXX

MAXIMIN CRITERION, *maximin test* - A statistical test for testing a compound hypothesis H_0: $\theta \in \Theta_0 \subset \Theta$ against the compound alternative H_1: $\theta \in \Theta_1 = \Theta \setminus \Theta_0$, whose minimal power (cf. **Power of a statistical test**) is maximal in the class of all statistical tests intended for testing H_0 against H_1 and having the same size α, $0 < \alpha < 1$. In statistically testing H_0 against H_1 a maximin **invariant test** exists if the problem itself is invariant relative to some group of transformations G, and there is a **uniformly most-powerful test** in the class of corresponding invariant tests (cf. **Hunt−Stein theorem**). A maximin test exists, in general, if the family of probability distributions determined by the null and competing hypotheses H_0 and H_1 is absolutely continuous relative to some σ-finite measure.

References
[1] LEHMANN, E.L.: *Testing statistical hypotheses*, Wiley, 1959.
[2] HAJEK, J. and SIDAK, Z.: *Theory of rank tests*, Acad. Press, 1967.

M.S. Nikulin

AMS 1980 Subject Classification: 62F03, 62G10, 62C20

MAXIMIN, NUMERICAL METHODS - The branch of computational mathematics devoted to the solution of maximin (minimax) problems. Problems of calculating maximins and minimaxes often arise in **operations research** and in game theory (cf. **Games, theory of**), for example, from the **minimax principle** or from the **principle of the largest sure result**.

First there is the problem of calculating the maximin

$$\sup_{x \in X} \inf_{y \in Y} F(x, y), \qquad (1)$$

which arises, for example, in the solution of two-person

zero-sum games with complete information. A natural generalization of it is the problem of finding a maximin with 'constrained' variables:

$$\sup_{x \in X} \inf_{y \in B(x)} F(x, y), \qquad (2)$$

where the set $B(x)$ is often given in the form

$$B(x) = \{y \in Y : \phi(x, y) \geqslant 0\} \neq \varnothing$$

for each $x \in X$. This problem is fundamental in the theory of two-person games with exchange of information (see, for example, **Game with a hierarchy structure**). Iteration of the first problem leads to a multiple or sequential maximin problem:

$$\sup_{x_1 \in X_1} \inf_{y_1 \in Y_1} \cdots \sup_{x_n \in X_n} \inf_{y_n \in Y_n} F(x_1, y_1, \ldots, x_n, y_n), \qquad (3)$$

which is related to the solution of certain dynamical games. Stochastic problems of calculating a maximin and also minimax problems of optimal control are of interest.

The basis of most methods for solving minimax problems is the **gradient method** or the method of penalty functions (cf. **Penalty functions, method of**). In the first of these, (1) is regarded as a problem in optimal programming:

$$\sup_{x \in X} f(x), \qquad (4)$$

where

$$f(x) = \inf_{y \in Y} F(x, y). \qquad (5)$$

The construction of numerical methods for solving (4) - (5) is connected with directional differentiability of the minimum function (5).

If $X \subset \mathbf{R}^n$, if Y is a compact set in $\mathbf{R}^n$ and if $\partial f(x) / \partial g$ is the derivative in the direction $g \in \mathbf{R}^n$ (see [1] - [3]),

$$\left. \begin{aligned} \frac{\partial f(x)}{\partial g} &= \min_{y \in Y(x)} \left[\frac{\partial}{\partial x} F(x, y), g \right], \\[2mm] Y(x) &= \{y \in Y : F(x, y) = f(x)\}, \end{aligned} \right\} \qquad (6)$$

then for a finite set Y formula (6) allows one to construct an iterative sequence $x_1, x_2, \ldots$, in which $x_{k+1} = x_k + \alpha_k g_k$ for $k = 0, 1, \ldots$, and which, under certain additional restrictions, converges to a point satisfying the necessary condition for a maximin.

In the method of penalty functions the problem (1), for a function $F(x, y)$ that is continuous on the product of a compact set $X \subset \mathbf{R}^n$ and the unit cube $Y \subset \mathbf{R}^m$, reduces to finding

$$\lim_{c \to \infty} \max_{x, u \in X} \mathscr{L}(x, u, c),$$

where

$$\mathscr{L}(x, u, c) = u - c \int_Y (\min(0, F(x, y) - u))^2 \, dy \qquad (7)$$

(u is an auxiliary variable). Here, if a pair $(u(c), c(c))$ realizes the maximum

$$\max_{u, x \in X} \mathscr{L}(x, u, c),$$

then any limit point (u^*, x^*) of the sequence $\{(u(c_k), x(c_k)) : c_k \to \infty\}$ gives

$$u^* = \max_{x \in X} \min_{y \in Y} F(x, y)$$

as the maximin of (1) and one of the optimal strategies x^* for which

$$u^* = \min_{y \in Y} F(x^*, y)$$

(see [4]). Thereby the solution of (1), up to arbitrary accuracy, reduces to the search for the maximum of the function (7) for sufficiently large values of the penalty c. To avoid difficulties connected with the calculation of the integrals in (7), the stochastic gradient method is used (see [5], [8]):

$$\left. \begin{aligned} x_{k+1} &= x_k + \alpha_k \xi_k(c_k), \\[2mm] u_{k+1} &= u_k + \alpha_k \eta_k(c_k), \end{aligned} \right\} \quad k = 1, 2, \ldots, \qquad (8)$$

where $(\xi_k(c_k), \eta_k(c_k))$ is the stochastic gradient of $\mathscr{L}(x, u, c)$, that is, a random variable whose mathematical expectation is

$$\left[\frac{\partial}{\partial x} \mathscr{L}(x_k, u_k, c_k), \frac{\partial}{\partial u} \mathscr{L}(x_k, u_k, c_k) \right].$$

Under certain conditions on the sequences $\{\alpha_k\}$ and $\{c_k\}$ and the function $F(x, y)$, for any first approximation (x_1, y_1) the sequence defined by (8) converges with probability 1 to the set of solutions of the maximin problem (1).

To avoid large values of the penalty c in (7), the so-called 'method of discrepancies' is used (see [6]), which is another method of transforming the problem (1). Here u^*, the maximin of (1), is defined as the maximum value of u for which

$$\min_{x \in X} \Phi(x, u) = 0, \qquad (9)$$

where

$$\Phi(x, y) = \int_Y (\min(0, F(x, y) - u))^2 \, dy.$$

Here the value x^* that realizes

$$\min_{x \in X} \Phi(x, u^*)$$

is an optimal strategy for (1). In order to find the largest root u^* of (9) it is possible to use gradient methods for minimizing Φ relative to x.

The transformation methods based on the method of penalty functions permit the approximate reduction of the minimax problem with constrained variables (2) to a maximum problem (see [7], [8]).

The extremal problems obtained by reducing minimax problems to maximum problems are very complicated and their solution by known methods entails great, sometimes even insurmountable, difficulties for modern electronic computers. In particular, this makes the problem of finding a sequential maximin (3) very difficult. Together with these general approaches to the solution of minimax problems there are a number of methods orientated towards particular classes of problems; for example, to problems of the theory of two-person games and transmission of information (see [6]).

References

[1] DEM'YANOV, V.F. and MALOZEMOV, V.N.: *Introduction to minimax*, Wiley, 1974 (translated from the Russian).

[2] DANSKIN, J.M.: *The theory of max-min and its application to weapons allocation problems*, Springer, 1967.

[3] DEM'YANOV, V.F. and VASIL'EV, L.V.: *Non-differentiable optimization*, Optim. Software, 1985 (translated from the Russian).

[4] GERMEÏER, YU.B.: *Introduction to the theory of Operations Research*, Moscow, 1971 (in Russian).

[5] ERMOL'EV, YU.M.: *Methods of stochastic programming*, Moscow, 1976 (in Russian).

[6] GERMEÏER, YU.B.: *Games with non-conflicting interests*, Moscow, 1976 (in Russian).

[7] ERESHKO, F.I. and ZLOBIN, A.S.: 'An algorithm for centralized resource distribution among active subsystems', *Ekonomika i Mat. Metody* 13, no. 4 (1977), 703-713 (in Russian).

[8] FEDOROV, V.V.: *Numerical methods of maximin*, Moscow, 1979 (in Russian).

V.V. Fedorov

AMS 1980 Subject Classification: 90C99

MAXIMIN PRINCIPLE - See **Minimax principle**.

AMS 1980 Subject Classification: 90C99, 49-XX

MAXIMIZATION AND MINIMIZATION OF FUNCTIONS *of a finite number of variables* - The problem of finding the extrema of a function $f(x)$, $x = (x^1, \ldots, x^n) \in X \subseteq \mathbf{R}^n$. This means

1) finding $\bar{f} = \sup_{x \in X} f(x)$ or $\underline{f} = \inf_{x \in X} f(x)$;

2) searching maximum or minimum points if $\bar{f}$ or $\underline{f}$ are attained on the admissible set (see **Maximum and minimum of a function**);

3) the construction of *maximizing* or *minimizing* *sequences* $\{x_i\}$ such that:

$$\lim_{i \to \infty} f(x_i) = \bar{f}, \quad \lim_{j \to \infty} f(x_j) = \underline{f},$$

if $\bar{f}$ or $\underline{f}$ is not attained on X.

The investigation of the extrema of functions of discrete arguments is referred to as **discrete programming** or **integer programming**. Below only maximization and minimization for functions of continuous arguments is explained.

The classical (indirect) methods of maximization and minimization apply only to smooth functions. They use necessary conditions for an extremum in order to locate stationary points. Zeros of the derivatives $\partial^\alpha f / \partial x^\alpha$, $\alpha = 1, \ldots, n$, are usually computed in practice by some successive approximation method (see [3]). On the other hand, each problem of solving a finite set of functional equations of the form

$$\phi_m(x^1, \ldots, x^n) = 0, \quad m \leq n,$$

can be interpreted as a maximization or minimization problem of some function, for example, of the function

$$f(x) = \phi_1^2(x) + \cdots + \phi_m^2(x) \to \min,$$

and one of the specific methods for maximizing and minimizing can be applied.

Direct methods for maximizing and minimizing functions are based on direct comparison of the values of f at two or more points.

The practical search of extrema uses iterative algorithms of the form:

$$x_{i+1} = \hat{X}(i, x_i, x_{i-1}, \ldots, x_{i-j}),$$

where i is the index of the iteration and $\hat{X}(\cdot)$ is some operator. Here it is usually assumed that

a) the algorithm converges in some sense, most often in the sense

$$x_\infty = \bar{x} \ (x_\infty = \underline{x}) \text{ or } f(x_\infty) = \bar{f} \ (f(x_\infty) = \underline{f});$$

b) the iteration procedure is local, that is, $j \ll i$ ($j = o(i)$ as $i \to \infty$); the algorithm 'remembers' the values of x only for iterations in some neighbourhood of the current position x_i. For $j = 0$ a memoryless simple Markov computational process is obtained.

The operator $X(\cdot)$ may be deterministic, in deterministic methods, or may contain stochastic parameters. In computational practice stochastic methods are often combined with deterministic methods, for example, in the **coordinate-wise descent method** the direction of descent may be determined randomly. The probabilistic characteristics of the stochastic parameters, in turn, may be changed from iteration to iteration (search with adaptation and 'self-learning' random search).

Various deterministic methods are widely used in combination, including sequential and parallel computation of an extremum by several methods, composition of algorithms of the form $\hat{X} = \hat{X}_2(\hat{X}_1(\cdot))$, etc. For example, the *Levenberg−Marquardt method*

$$x_{i+1} = x_i - (\alpha_i \nabla \nabla f(x_j) + \beta_i I)^{-1} \nabla f(x_i),$$

which coincides with the **gradient method** for $\alpha_i = 0$ and with the **Newton method** for $\beta_i = 0$.

One-dimensional optimization, that is, maximizing and minimizing a function $f(x)$, $x \in \mathbf{R}^1$, in addition to its interest by itself, is a necessary stage in most applicable methods. Specific one-dimensional methods include, for example, the **Fibonacci method**; the **dichotomy method** (division by halves) and the **parabola method**. Methods for maximizing and minimizing functions in several variables are the **gradient method**, the method of steepest descent (cf. **Steepest descent, method of**), the **coordinate-wise descent method**, the **simplex method**, the **scanning method**, the method of conjugate gradients, the method of the heavy sphere (cf. **Conjugate**

gradients, method of; Heavy sphere, method of the), the **adjustment method**, and others.

The algorithms of most of the methods listed fall into the scheme of the *descent (ascent) method*:

$$x_{i+1} = x_i \mp \kappa_i y_i,$$

where $f(x_{i+1}) \leqslant f(x_i)$ or $f(x_{i+1}) \geqslant f(x_i)$ for all i (the relaxation condition). These are distinguished either by the choice of a descent direction y_i or by the method of moving along the descent vector, defined by the *step factor* κ.

Cut methods have been developed for functions whose contours are 'valleys with steep slopes' (see **Minimization methods for functions depending strongly on a few variables**). Ordinary (non-cut) methods when applied here give a zig-zag relaxational path, requiring excessive amounts of machine time to calculate the extremum.

The comparative effectiveness of the methods is estimated by many, even contradictory, criteria. E.g., accuracy of the solution, speed of approach to the solution, reliability of the method, preparation time from the problem to the calculation, convergence of the algorithm, etc. The domain of applicability of each of the approved methods is very limited.

For testing the methods a set of standard *test functions*, characteristic of different function classes, has been collected (see [1]). Convergence of methods for maximizing and minimizing functions has been extensively studied (see [6], [8]). However, convergence is a property which is neither necessary nor sufficient for the effective termination of a calculation.

All methods above lead to a local extremum if the first approximation belongs to the domain of attraction of this extremum. The detection of a global extremum is guaranteed only for convex and related unimodal functions. The theory of finding a global extremum is still (1989) in the initial phase of development (see **Multi-extremum problem**). Another area of maximizing and minimizing of functions which is developing is optimization of non-smooth functions (see [4], [13], [16]). In particular, the problem of minimizing the maxima of a family of functions generally leads to non-smooth functions (see **Maximin, numerical methods**). Apparently, all of the commonly-used methods for optimization have an interesting physical, economic or biological meaning. The corresponding research is only just developing (see [9]) and leads to the creation of new methods (see also **Continuous analogues of iteration methods**). If the values of the functions being considered are defined statistically with random noise, then one of the methods of **stochastic approximation** is applied to find extrema. Here one is bordering on the **design of experiments**.

Experimental methods of maximizing and minimizing functions use the reproduction of various physical processes in the search for extrema. A related area is simulation on analogue computers (see [17]). In spite of the convenience and cheapness of using the simplest automatic optimizers, they do not guarantee high accuracy of calculation.

Graphical methods are suitable only for obtaining rough estimates and for the construction of a first approximation for iterative methods.

If the admissible set is given by functional conditions

$$\phi_m(x) \leqslant 0$$

(constraints and restrictions, conditional extremum), then methods of **mathematical programming** are applied in the search of extrema. This problem can be reduced to a sequence of problems on unconstrained extrema by the use of penalty and barrier functions (see **Penalty functions, method of**).

References

[1] AOKI, M.: *Introduction to optimization techniques*, Macmillan, 1971.

[2] BAKHVALOV, N.S.: *Numerical methods: analysis, algebra, ordinary differential equations*, Mir, 1977 (translated from the Russian).

[3] BEREZIN, I.S. and ZHIDKOV, N.P.: *Computing methods*, Pergamon, 1973 (translated from the Russian).

[4] VASIL'EV, F.P.: *Lectures on methods of solving extremal problems*, Moscow, 1974 (in Russian).

[5] GUPAL, A.M.: *Stochastic methods of solving non-smooth extremal problems*, Kiev, 1979 (in Russian).

[6] KARMANOV, V.G.: *Mathematical programming*, Moscow, 1975 (in Russian).

[7] MOISEEV, N.N., IVANILOV, YU.P. and STOLYAROVA, E.M.: *Methods of optimization*, Moscow, 1978 (in Russian).

[8] PSHENICHNYĬ, B.N. and DANILIN, YU.M.: *Numerical methods in extremum problems*, Moscow, 1975 (in Russian).

[9] RAZYMIKHIN, B.S.: *Physical models and methods of the theory of equilibrium in programming and economics*, Reidel, 1984 (translated from the Russian).

[10] PASTRIGIN, L.A.: *Extremal control systems*, Moscow, 1974 (in Russian).

[11] SAUL'EV, V.K. and SAMOĬLOVA, I.I.: 'Approximation methods for the unconstrained optimization of functions of several variables', *J. Soviet Math.* **4**, no. 6 (1975), 681-705. (*Itogi Nauk. i Tekhn. Mat. Anal.* **11** (1973), 91-128)

[12] WILDE, D.J.: *Extremum searching methods*, Prentice-Hall, 1965.

[13] FEDOROV, V.V.: *Numerical maximin methods*, Moscow, 1979 (in Russian).

[14] ORTEGA, J.M. and RHEINBOLDT, C.: *Iterative solution of non-linear equations in several variables*, Acad. Press, 1970.

[15] *The current state of the theory of Operations Research*, Moscow, 1979 (in Russian).

[16] LEMARECHAL, C. and MIFFLIN, R. (EDS.): *Nonsmooth optimization*, Pergamon, 1978.

[17] DIXON, L.C.W. and SZEGÖ, G.P. (EDS.): *Towards global optimisation*, 1-2, North-Holland, 1975-1978.

[18] EVTUSHENKO, YU.G.: 'Numerical methods for finding global extrema (case of a non-uniform mesh)', *USSR Comp. Math. Math. Phys.* **11**, no. 6 (1971), 38-54. (*Zh. Vychisl. Mat. i Mat. Fiz.* **11**, no. 6 (1971), 1390-1403)

Yu.P. Ivanilov

V.V. Okhrimenko

Editorial comments.

References

[A1] ROCKAFELLAR, R.T.: *The theory of subgradients and its applications to problems of optimization. Convex and nonconvex functions*, Heldermann, 1981.

[A2] LAARHOVEN, P.J.M. VAN: *Theoretical and computational aspects of simulated annealing*, CWI Tract, 51, Centre Math. Computer Sci., Amsterdam, 1988.

[A3] LUENBERGER, D.G.: *Linear and nonlinear programming*, Addison-Wesley, 1984.

[A4] (CONDOR), COMMITTEE ON THE NEXT DECADE IN OPERATIONS RESEARCH: 'Operations Research: the next decade', *Oper. Research* 36, no. 4 (1988), 619-637.

[A5] DANIEL, J.W.: *The approximate minimization of functionals*, Prentice-Hall, 1971.

[A6] BUCKLEY, A.G. and GOFFIN, J.L.: *Algorithms for constrained minimization of smooth nonlinear functions*, North-Holland, 1982.

AMS 1980 Subject Classification: 49-01, 65-01, 26-01, 90CXX

MAXIMUM AND MINIMUM OF A FUNCTION - A largest, respectively smallest, value of a real-valued function. A point of the domain of definition of a real-valued function at which a maximum or minimum is attained is called a maximum or minimum point, respectively (see **Maximum and minimum points**). If some point is an absolute (local) maximum or minimum point, strict or non-strict, then the value of the function at that point is correspondingly called an absolute (local), strict or non-strict, maximum or minimum. A continuous function on a compact set always takes maximum and minimum values on that set.

All together, the maxima and minima of a function are called its extrema or extremal values.

L.D. Kudryavtsev

AMS 1980 Subject Classification: 26-XX

MAXIMUM AND MINIMUM POINTS - Points in the domain of definition of a real-valued function at which it takes its greatest and smallest values; such points are also called *absolute maximum* and *absolute minimum points*. If f is defined on a topological space X, then a point x_0 is called a *local maximum (local minimum) point* if there is a neighbourhood $U \subseteq X$ of x_0 such that x_0 is an absolute maximum (minimum) point for the restriction of f to this neighbourhood. One distinguishes between *strict* and *non-strict maximum (minimum) points* (both absolute and local). For example, a point $x_0 \in \mathbf{R}$ is called a non-strict (strict) local maximum point of f if there is a neighbourhood U of x_0 such that for all $x \in U$, $f(x) \leqslant f(x_0)$ $(f(x) < f(x_0)$, $x \neq x_0)$.

For functions defined on finite-dimensional domains there are conditions and tests, in terms of differential calculus, for a given point to be a local maximum (minimum) point. Let f be defined in a neighbourhood of a point x_0 of the real line. If x_0 is a non-strict local maximum (minimum) point and if the derivative $f'(x_0)$ exists, then the latter is equal to zero.

If a function f is differentiable in a neighbourhood of x_0 except, possibly, at x_0 itself where it is continuous, and if the derivative f' is of constant sign on each side of x_0 in this neighbourhood, then for x_0 to be a strict local maximum (local minimum) point it is necessary and sufficient that the derivative changes sign from plus to minus, that is, $f'(x) > 0$ for $x < x_0$ and $f'(x) < 0$ for $x > x_0$ (respectively, from minus to plus; $f'(x) < 0$ for $x < x_0$ and $f'(x) > 0$ for $x > x_0$). However, it is not possible to speak of the change of sign of the derivative at x_0 for every function f that is differentiable in a neighbourhood of x_0.

If f has m derivatives at x_0 and if $f^{(k)}(x_0) = 0$, $k = 1, \ldots, m - 1$, $f^{(m)}(x_0) \neq 0$, then for x_0 to be a strict local maximum point it is necessary and sufficient that m be even and that $f^{(m)}(x_0) < 0$, and for a local minimum that m be even and $f^{(m)}(x_0) > 0$.

Let $f(x_1, \ldots, x_n)$ be defined in an n-dimensional neighbourhood of a point $x^{(0)} = (x_1^{(0)}, \ldots, x_n^{(0)})$ and let it be differentiable at this point. If $x^{(0)}$ is a non-strict local maximum (minimum) point, then the differential of f at this point is equal to zero. This condition is equivalent to all first-order partial derivatives of f being zero at this point. If the function has continuous second-order partial derivatives at $x^{(0)}$, if all its first-order derivatives are equal to zero at $x^{(0)}$, and if the second-order differential at $x^{(0)}$ is a negative-definite (positive-definite) quadratic form, then $x^{(0)}$ is a strict local maximum (minimum) point. Conditions for maximum and minimum points of differentiable functions are known when restrictions are imposed on the variation of the arguments in the domain: coupling equations must be satisfied. Necessary and sufficient conditions for a maximum (minimum) of a real-valued function with a more complicated structure of its domain of definition have been obtained in special areas of mathematics; for example, in **convex analysis** and **mathematical programming** (see also **Maximization and minimization of functions**). Maximum and minimum points of functions on manifolds are studied in **variational calculus in the large**, and maximum and minimum points for functions on function spaces, that is, for functionals, are studied in **variational calculus**. There are also various numerical approximation methods for finding maximum and minimum points.

References

[1] IL'IN, V.A. and POZNYAK, E.G.: *Fundamentals of mathematical analysis*, 1, Mir, 1982 (translated from the Russian).

[2] KUDRYAVTSEV, L.D.: *A course of mathematical analysis*, 1-2, Moscow, 1981 (in Russian).

[3] NIKOL'SKII, S.M.: *A course of mathematical analysis*, 1, Mir,

1977 (translated from the Russian).

[4] BAKHVALOV, N.S.: *Numerical methods: analysis, algebra, ordinary differential equations*, Mir, 1977 (translated from the Russian).

L.D. Kudryavtsev

Editorial comments.

References

[A1] STROMBERG, K.R.: *An introduction to classical real analysis*, Wadsworth, 1981.

AMS 1980 Subject Classification: 26-XX

MAXIMUM-ENTROPY SPECTRAL ESTIMATOR,

auto-regressive spectral estimator - An estimator $f_q^*(\lambda)$ for the **spectral density** $f(\lambda)$ of a discrete-time stationary stochastic process such that 1) the first q values of the auto-correlations are equal to the sample auto-correlations calculated from the observational data, and 2) the **entropy** of the Gaussian stochastic process with spectral density $f_q^*(\lambda)$ is maximized subject to condition 1). If N sample values x_t, $t = 1, \ldots, N$, are known from observing a realization of a real stationary process X_t having spectral density $f(\lambda)$, then the maximum-entropy spectral estimator $f_q^*(\lambda)$ is defined by the relations

$$\int_{-\pi}^{\pi} \cos k\lambda f_q^*(\lambda) \, d\lambda = r_k^* \equiv \tag{1}$$

$$\equiv N^{-1} \sum_{j=1}^{N-k} x_j x_{j+k}, \quad k = 0, \ldots, q,$$

$$\int_{-\pi}^{\pi} \log f_q^*(\lambda) \, d\lambda = \max, \tag{2}$$

where the sign $\equiv$ denotes 'equal by definition'. The maximum-entropy spectral estimator has the form

$$f_q^*(\lambda) = \frac{\sigma^2}{2\pi |\, 1 + \beta_1 \exp(i\lambda) + \cdots + \beta_q \exp(iq\lambda) \,|^2}, \tag{3}$$

where the coefficients $\beta_1, \ldots, \beta_q$ and σ^2 are given by the $q + 1$ equations (1) (see, e.g., [1], [9], [10]). Formula (3) shows that the maximum-entropy spectral estimator coincides with the so-called auto-regressive spectral estimator (introduced in [2], [3]). The positive integer q here plays a role related to that played by the reciprocal width of a spectral window in the case of non-parametric estimation of the spectral density by periodogram smoothing (see **Spectral window**; **Statistical problems in the theory of stochastic processes**). There are several methods for estimating the optimal value of q from given observations (see, for example, [1], [4], [5], [8]). The values of the coefficients $\beta_1, \ldots, \beta_q, \sigma^2$ can be found using a solution of the Yule−Walker equations

$$r_k^* + \sum_{j=1}^{q} \beta_j r_{|k-j|}^* = 0, \quad k = 1, \ldots, q, \tag{4}$$

$$r_0^* + \sum_{j=1}^{q} \beta_j r_j^* = \sigma^2; \tag{5}$$

there are also other, numerically more convenient, methods for calculating these coefficients (see, e.g., [1], [4] - [6], [10]).

In the case of small sample size or spectral densities of complex form, maximum-entropy spectral estimators and parametric spectral estimators (cf. **Spectral estimator, parametric**), which generalize them, possess definite advantages over non-parametric estimators of $f(\lambda)$: they usually have a more regular form and possess better resolving power, that is, they permit one to better distinguish close peaks of the graph of the spectral density (see [1], [4] - [7]). Therefore maximum-entropy spectral estimators are widely used in the applied **spectral analysis of a stationary stochastic process**.

References

[1] CHILDERS, D.G. (ED.): *Modern spectrum analysis*, IEEE Press, 1978.
[2] PARZEN, E.: 'An approach to empirical time series analysis', *Radio Sci.* **68** (1964), 937-951.
[3] AKAIKE, H.: 'Power spectrum estimation through autoregressive model fitting', *Ann. Inst. Stat. Math.* **21**, no. 3 (1969), 407-419.
[4] HAYKIN, S. (ED.): *Nonlinear methods of spectral analysis*, Springer, 1979.
[5] KAY, S.M. and MARPL, S.L.: 'Spectrum analysis — a modern perspective', *Proc. IEEE* **69**, no. 11 (1981), 1380-1419.
[6] 'Spectral estimation', *Proc. IEEE* **70**, no. 9 (1982).
[7] PISARENKO, V.F.: 'Sampling properties of maximum entropy spectral estimation', in *Numerical Seismology*, Moscow, 1977, pp. 118-149 (in Russian).
[8] GOOYER, J.G. DE, ABRAHAM, B., GOULD, A. and ROBINSON, L.: 'Methods for determining the order of an autoregressive-moving average process: A survey', *Internat. Stat. Rev.* **55** (1985), 301-329.
[9] PRIESTLEY, M.B.: *Spectral analysis and time series*, 1-2, Acad. Press, 1981.
[10] PAPOULIS, A.: *Probability, random variables and stochastic processes*, McGraw-Hill, 1984.

A.M. Yaglom

AMS 1980 Subject Classification: 62M15, 62G05, 62G10

MAXIMUM-LIKELIHOOD METHOD - One of the

fundamental general methods for constructing estimators of unknown parameters in statistical estimation theory.

Suppose one has, for an observation X with distribution P_θ depending on an unknown parameter $\theta \in \Theta \subseteq \mathbf{R}^k$, the task to estimate θ. Assuming that all measures P_θ are absolutely continuous relative to a common measure ν, the *likelihood function* is defined by

$$L(\theta) = \frac{dP_\theta}{d\nu}(X).$$

The maximum-likelihood method recommends taking as an estimator for θ the statistic $\hat{\theta}$ defined by

$$L(\hat{\theta}) = \max_{\theta \in \Theta} L(\theta).$$

$\hat{\theta}$ is called the *maximum-likelihood estimator*. In a broad class of cases the maximum-likelihood estimator is the solution of a *likelihood equation*

$$\frac{\partial}{\partial\theta_i}\log L(\theta) = 0, \quad i=1,\ldots,k, \quad \theta=(\theta_1,\ldots,\theta_k). \quad (1)$$

Example 1. Let $X=(X_1,\ldots,X_n)$ be a sequence of independent random variables (observations) with common distribution P_θ, $\theta\in\Theta$. If there is a density

$$f(x,\theta) = \frac{dP_\theta}{dm}(x)$$

relative to some measure m, then

$$L(\theta) = \prod_{j=1}^{n} f(X_j,\theta)$$

and the equations (1) take the form

$$\sum_{j=1}^{n}\frac{\partial}{\partial\theta_i}\log f(X_j,\theta) = 0, \quad i=1,\ldots,k. \quad (2)$$

Example 2. In Example 1, let P_θ be the **normal distribution** with density

$$\frac{1}{\sigma\sqrt{2\pi}}\exp\left\{-\frac{(x-a)^2}{2\sigma^2}\right\},$$

where $x\in\mathbf{R}^1$, $\theta=(a,\sigma^2)$, $-\infty<a<\infty$, $\sigma^2>0$. Equations (2) become

$$\frac{1}{\sigma^2}\sum_{j=1}^{n}(X_j-a) = 0,$$

$$\frac{1}{2\sigma^4}\sum_{j=1}^{n}(X_j-a)^2 - \frac{n}{2\sigma^2} = 0;$$

and the maximum-likelihood estimator is given by

$$\hat{a} = X = \frac{1}{n}\sum_{j=1}^{n}X_j, \quad \hat{\sigma}^2 = \frac{1}{n}\sum_{j=1}^{n}(X_j-\overline{X})^2.$$

Example 3. In Example 1, let X_j take the values 0 and 1 with probabilities $1-\theta$, θ, respectively. Then

$$L(\theta) = \prod_{j=1}^{n}\theta^{X_j}(1-\theta)^{1-X_j},$$

and the maximum-likelihood estimator is $\hat{\theta}=\overline{X}$.

Example 4. Let the observation $X=X_t$ be a **diffusion process** with **stochastic differential**

$$dX_t = \theta a_t(X_t)+dW_t, \quad X_0 = 0, \quad 0\leqslant t\leqslant T,$$

where W_t is a **Wiener process** and θ is an unknown one-dimensional parameter. Here (see [3]),

$$\log L(\theta) = \theta\int_0^T a_t(X_t)\,dX_t - \frac{\theta^2}{2}\int_0^T a_t^2(X_T)\,dt,$$

$$\hat{\theta} = \frac{\int_0^T a_t(X_t)\,dX_t}{\int_0^T a_t^2(X_t)\,dt}.$$

There are no definitive reasons for optimality of the maximum-likelihood method and the widespread belief in its efficiency is partially based on the great success with which it has been applied to numerous concrete problems, and partially on rigorously established asymptotic optimality properties. For example, in Example 1, under broad assumptions, $\hat{\theta}_n\to\theta$ with P_θ-probability 1. If the Fisher information

$$I(\theta) = \int\frac{|f_\theta'(x,\theta)|^2}{f(x,\theta)}m(dx)$$

exists, then the difference $\sqrt{n}(\hat{\theta}_n-\theta)$ is asymptotically normal with parameters $(0, I^{-1}(\theta))$, and $\hat{\theta}_n$, in a well-defined sense, has an asymptotically-minimal mean-square deviation from θ (see [4], [5]).

References

[1] CRAMÉR, H.: *Mathematical methods of statistics*, Princeton Univ. Press, 1946.
[2] ZACKS, S.: *The theory of statistical inference*, Wiley, 1975.
[3] LIPSTER, R.SH. and SHIRYAEV, A.N.: *Statistics of random processes*, Springer, 1977 (translated from the Russian).
[4] IBRAGIMOV, A.I. and HAS'MINSKII, R.Z. [R.Z. KHAS'MINSKIĬ]: *Statistical estimation: asymptotic theory*, Springer, 1981 (translated from the Russian).
[5] LEHMANN, E.L.: *Theory of point estimation*, Wiley, 1983.

I.A. Ibragimov

AMS 1980 Subject Classification: 62FXX

MAXIMUM-MODULUS PRINCIPLE - A theorem expressing one of the basic properties of the modulus of an **analytic function**. Let $f(z)$ be a regular analytic, or holomorphic, function of n complex variables $z=(z_1,\ldots,z_n)$, $n\geqslant 1$, defined on an (open) domain D of the complex space $\mathbf{C}^n$, which is not a constant, $f(z)\neq\text{const}$. The *local formulation of the maximum-modulus principle* asserts that the modulus of $f(z)$ does not have a local maximum at a point $z^0\in D$, that is, there is no neighbourhood $V(z^0)$ of z^0 for which $|f(z)|\leqslant|f(z^0)|$, $z\in V(z^0)$. If in addition $f(z^0)\neq 0$, then z^0 also cannot be a local minimum point of the modulus of $f(z)$. An equivalent *global formulation of the maximum-modulus principle* is that, under the same conditions as above, the modulus of $f(z)$ does not attain its least upper bound

$$M = \sup\{|f(z)|: z\in D\}$$

at any $z^0\in D$. Consequently, if $f(z)$ is continuous in a finite closed domain D, then M can only be attained on the boundary of D. These formulations of the maximum-modulus principle still hold when $f(z)$ is a holomorphic function on a connected complex (analytic) manifold, in particular, on a **Riemann surface** or a Riemann domain (cf. **Riemannian domain**) D.

The maximum-modulus principle has generalizations in several directions. First, instead of $f(z)$ being holomorphic, it is sufficient to assume that $f(z)=u(z)+iv(z)$ is a (complex) **harmonic function**. Another generalization is connected with the fact that for a holomorphic function $f(z)$ the modulus $|f(z)|$ is a **logarithmically-subharmonic function**. If $f(z)$ is a bounded holomorphic function in a finite domain $D\subset\mathbf{C}^n$ and if

$$\limsup \{ | f(z) | : z \to \zeta, \ z \in D \} \leqslant M$$

holds for all $\zeta \in \partial D$, except at some set $E \subset \partial D$ of outer **capacity** zero (in $\mathbf{R}^{2n} = \mathbf{C}^n$), then $| f(z) | \leqslant M$ everywhere in D. See also **Two-constants theorem; Phragmén − Lindelöf theorem**.

The maximum-modulus principle can also be generalized to holomorphic mappings. Let $f : D \to \mathbf{C}^m$ be a **holomorphic mapping** of an (open) domain $D \subset \mathbf{C}^n$, $n \geqslant 1$, into $\mathbf{C}^m$, that is, $f = (f_1, \dots, f_m)$, $m \geqslant 1$, where f_j, $j = 1, \dots, m$, are holomorphic functions on D, $f(z) \not\equiv$ const and $\| f \| = \sqrt{| f_1 |^2 + \cdots + | f_m |^2}$ is the Euclidean norm. Then $\| f(z) \|$ does not attain a local maximum at any $z^0 \in D$. The maximum-modulus principle is valid whenever the principle of preservation of domain is satisfied (cf. **Preservation of domain, principle of**).

References

[1] Stoĭlov, S.: *The theory of functions of a complex variable*, 1-2, Moscow, 1962 (in Russian; translated from the Rumannian).
[2] Vladimirov, V.S.: *Methods of the theory of functions of many complex variables*, M.I.T., 1966 (translated from the Russian).
[3] Shabat, B.V.: *Introduction to complex analysis*, 1-2, Moscow, 1976 (in Russian).

E.D. Solomentsev

Editorial comments. This principle is also called the *maximum principle*, cf. [A2].

References

[A1] Burckel, R.B.: *An introduction to classical complex analysis*, 1, Acad. Press, 1979.
[A2] Ahlfors, L.V.: *Complex analysis*, McGraw-Hill, 1979.

AMS 1980 Subject Classification: 30C80, 32A10

MAXIMUM PRINCIPLE, *discrete* - The Pontryagin maximum principle for discrete-time control processes. For such a process the maximum principle need not be satisfied, even if the **Pontryagin maximum principle** is valid for its continuous analogue, obtained by replacing the finite difference operator $x_{t+1} - x_t$ by the differential dx / dt. For example, consider the optimal control problem

$$\max J(x_{T+1}), \tag{1}$$

$$x_{t+1} = f_t(x_t, u_t), \tag{2}$$

$$u_t \in U, \ t = 0, \dots, T, \tag{3}$$

$$x_0 = a. \tag{4}$$

This can be treated as a standard problem on an extremum in the presence of constraints. Then the condition for optimality of a trajectory $\{x_t^*, u_t^*\}$ can be obtained with the help of the **Lagrange function**

$$L = J(x_{T+1}) + \sum_{t=0}^{T} (p_{t+1} f_t(x_t, u_t) - p_{t+1} x_{t+1}).$$

Here

$$p_{t+1} f_t(x_t, u_t) = H_t,$$

by analogy with the continuous case, is called the **Hamilton function**. Let $J, f_t, t = 0, \dots, T$, be differentiable

functions with respect to all variables and let U be a bounded closed set. Then for a solution $\{x_t^*, u_t^*\}$ of (1) - (4) to be optimal it is necessary that there exist **Lagrange multipliers** $\{p_{t+1}^*\}$ such that $(x_t^*, u_t^*, p_{t+1}^*)$ is a stationary point of the Lagrange function, that is, at this point

$$\frac{\partial L}{\partial x} = 0; \quad \frac{\partial L}{\partial p} = 0; \quad \delta_n L = \frac{\partial L}{\partial u} \delta u \leqslant 0$$

for all admissible variations of the control δu. The first condition leads to the dynamical equations of the discrete process (2) and the initial condition (4). The second leads to the boundary condition and the conjugate system for the impulses $\{p_{t+1}\}$:

$$p_{T+1} = \frac{\partial J}{\partial x_{T+1}},$$

$$p_t = p_{t+1} \frac{\partial f}{\partial x_t}, \quad t = T, \dots, 0.$$

The third condition leads to a condition on the first variation of the Hamilton function:

$$\delta_n H_t = \frac{\partial H_t}{\partial u_t} \delta u_t \leqslant 0. \tag{5}$$

However, (5) does not imply that the Hamilton function attains its maximum at the optimal control,

$$H_t(x_t^*, u_t^*, p_{t+1}^*) = \max_{u \in U} H_t(x_t^*, u_t^*, p_{t+1}^*),$$

over all controls satisfying the constraint (3); it shows that u_t^* is a stationary point of the Hamilton function. If the first variation of the Hamilton function, $(\partial H_t / \partial u_t) \delta u_t$, is zero (this holds, in particular, when u_t^* is an interior point or when there are admissible variations δu_t of the control at u_t^* orthogonal to $\partial H_t / \partial u_t$), then the nature of the stationary point is determined by the successive terms in the expansion:

$$H_t(u_t^* + \epsilon \delta u_t) - H_t(u_t^*) = \epsilon \frac{\partial H_t}{\partial u_t} \delta u_t < 0.$$

Examples have been constructed in which u_t^* is a local maximum, a local minimum and even a saddle point of the Hamilton function. Therefore, in general, the maximum principle does not hold for discrete systems.

For systems which are linear in the phase variables,

$$x_{t+1} = A(u_t) x_t + \phi(u_t),$$

or in the controls,

$$x_{t+1} = g(x_t) + B(x_t) u_t,$$

under the additional condition of linearity of the objective $J = c_{T+1} x_{T+1}$ in the first case, or convexity of U in the second case, the maximum principle is satisfied (see [1] - [5]).

If the optimal control problem for a linear discrete system is treated as a **linear programming** problem (see [6], [7]), then its dual discrete-time dynamical problem may be obtained. The conjugate system for the impulses gives the dynamical equation for the dual problem. On an optimal trajectory not only objective but also the Hamilton functions of dual dynamical systems coincide.

References

[1] FAN, L.-Ts. and WANG, Ch.-S.: *The discrete maximum principle*, Wiley, 1964.
[2] PROPOÏ, A.I.: *Elements of the theory of optimal processes*, Moscow, 1973 (in Russian).
[3] PSHENICHNYĬ, B.H.: *Necessary conditions for an extremum*, M. Dekker, 1971 (translated from the Russian).
[4] BOLTYANSKIĬ, V.G.: *Optimal control of discrete systems*, Wiley, 1978 (translated from the Russian).
[5] GABASOV, R. and KIRILLOVA, F.M.: 'Extending L.S. Pontryagin's maximum principle to discrete systems', *Automation and Remote Control* **27** (1966), 1878-1882. (*Avtomatika i Telemekhanika* **11** (1966), 46-51)
[6] IVANILOV, YU.P.: 'Conjugate (dual) linear dynamic optimization problems and procedures for their computation', *Prikl. Mat. i Programmirov.* **4** (1971), 31-40 (in Russian).
[7] IVANILOV, YU.P. and PROPOÏ, A.I.: 'On linear dynamic programming problems', *Soviet Math. Dokl.* **12**, no. 3 (1971), 926-930. (*Dokl. Akad. Nauk SSSR* **198**, no. 5 (1971), 1011-1014)

Yu.P. Ivanilov

Editorial comments. For the maximum principle for analytic functions see **Maximum-modulus principle**.

References

[A1] BRYSON, A.E. and HO, Y.C.: *Applied optimal control*, Ginn, 1969.

AMS 1980 Subject Classification: 49D05, 90C99

MAXWELL DISTRIBUTION - The **probability distribution** with probability density

$$
p(x) = \begin{cases} \sqrt{\dfrac{2}{\pi}}\,\dfrac{x^2}{\sigma^3}e^{-x^2/2\sigma^2}, & x \geqslant 0, \\[2mm] 0, & x < 0, \end{cases} \qquad (*)
$$

depending on a parameter $\sigma > 0$. The **distribution function** of the Maxwell distribution has the form

$$
F(x) = \begin{cases} 2\Phi\!\left(\dfrac{x}{\sigma}\right) - \sqrt{\dfrac{2}{\pi}}\,\dfrac{x}{\sigma}e^{-x^2/2\sigma^2} - 1, & x \geqslant 0, \\[2mm] 0, & x < 0, \end{cases}
$$

where $\Phi(x)$ is the standard **normal distribution** function. The Maxwell distribution has positive coefficient of skewness; it is unimodal, the unique mode occurring at $x = \sqrt{2}\,\sigma$. The Maxwell distribution has finite moments of all orders; the mathematical expectation and variance are equal to $2\sigma\sqrt{2/\pi}$ and $(3\pi - 8)\sigma^2/\pi$, respectively.

If X_1, X_2 and X_3 are independent random variables having the normal distribution with parameters 0 and σ^2, then the random variable $\sqrt{X_1^2 + X_2^2 + X_3^2}$ has a Maxwell distribution with density (*). In other words, a Maxwell distribution can be obtained as the distribution of the length of a random vector whose Cartesian coordinates in three-dimensional space are independent and normally distributed with parameters 0 and σ^2. The Maxwell distribution with $\sigma = 1$ coincides with the distribution of the square root of a variable having the χ^2-distribution with three degrees of freedom (see also **Rayleigh distribution**). The Maxwell distribution is widely known as the velocity distribution of particles in statistical mechanics and physics. The distribution was first defined by J.C. Maxwell (1859) as the solution of the problem on the distribution of velocities of molecules in an ideal gas.

References

[1] FELLER, W.: *An introduction to probability theory and its applications*, 2, Wiley, 1971.

A.V. Prokhorov

AMS 1980 Subject Classification: 82A40, 82A05, 60EXX, 62EXX

MAXWELL EQUATIONS - The equations of an electromagnetic field in material surroundings; established in the 1860's by J.C. Maxwell on the basis of the experimental evidence at that time of the laws of electric and magnetic phenomena.

In classical electrodynamics the electromagnetic field in a medium is described by four vector fields; the electric field strength **E**, the electric displacement **D**, the magnetic field strength **H**, and the magnetic flux density **B**. These are continuously-differentiable functions of the radius vector **r** of a point of the 3-dimensional space and of the time t.

The Maxwell equations are a system of inhomogeneous partial differential equations of the first order for the fields **E**, **D**, **H**, and **B**, which, in SI-units, takes the form

$$
-\frac{\partial \mathbf{D}}{\partial t} + \operatorname{rot}\mathbf{H} = \mathbf{J}, \qquad (1a)
$$

$$
\frac{\partial \mathbf{B}}{\partial t} + \operatorname{rot}\mathbf{E} = 0, \qquad (1b)
$$

$$
\operatorname{div}\mathbf{B} = 0, \qquad (1c)
$$

$$
\operatorname{div}\mathbf{D} = \rho, \qquad (1d)
$$

where the terms $\rho(t, \mathbf{r})$ — a given volume density of electric charge in the medium — and $\mathbf{J}(t, \mathbf{r})$ — the volume density of electric current (the charge passing in unit time through a unit area perpendicular to the direction of motion of the charge) — are the sources of the flow. The Maxwell equations may also be described in integral form:

$$
\begin{aligned}
\oint_C \mathbf{H}\,dl &= \int_S \left[\mathbf{J} + \frac{\partial \mathbf{D}}{\partial t}\right]d\mathbf{s}, \\
\oint_C \mathbf{E}\,dl &= -\int_S \frac{\partial \mathbf{B}}{\partial t}\,d\mathbf{s}, \\
\oint_{\partial V} \mathbf{B}\,d\mathbf{s} &= 0, \\
\oint_{\partial V} \mathbf{D}\,d\mathbf{s} &= 4\pi\int_V \rho\,dV,
\end{aligned} \qquad (2)
$$

where S is any two-sided surface with closed boundary curve C and ∂V is the closed boundary surface of any bounded domain C in $\mathbf{R}^3$; further, $d\mathbf{l}$ is the elementary arc length along C, $d\mathbf{s}$ is the elementary oriented surface area, and ∂V is the elementary volume on S; C, $d\mathbf{s}$ and $d\mathbf{l}$ form a right-hand system, while on ∂V the element $d\mathbf{s}$ points away from V.

The fields $\mathbf{E}$, $\mathbf{D}$, $\mathbf{H}$, $\mathbf{B}$, and $\mathbf{J}$ are not independent. In a large number of material media, $\mathbf{D}$ and $\mathbf{J}$ depend only on $\mathbf{E}$, and $\mathbf{B}$ depends only on $\mathbf{H}$, that is, the following functional dependencies hold:

$$\mathbf{D} = \mathbf{D(E)}, \quad \mathbf{J} = \mathbf{J(E)}, \quad \mathbf{B} = \mathbf{B(H)}, \tag{3}$$

called the *equations of state* or the *constitutive equations of the medium*. Within the limits of classical macroscopic electrodynamics the equations of state (3) must be given in addition (postulated or determined from experimental data) and then the equations for the two remaining independent vector fields $\mathbf{E}$ and $\mathbf{H}$ become closed. The concrete form of the equations of state (3) is determined by the electric and magnetic properties of the given medium and its states. In general, in (3), the vector fields $\mathbf{D}$, $\mathbf{J}$ and $\mathbf{B}$, at a point $\mathbf{r}$ at time t, may depend non-linearly on the values of $\mathbf{E}$ and $\mathbf{H}$, respectively, at all points of the medium (non-local case) and at all times prior to, by the physical principle of causality, the given time t (a medium with after-effects or memory). Most media of practical interest are characterized by local linear dependency of $\mathbf{D}$ and $\mathbf{J}$ on $\mathbf{E}$, and $\mathbf{B}$ on $\mathbf{H}$, and, in this case, the Maxwell equations turn out to be linear; however, in applications, more complicated cases are met (for example, in non-linear optics).

The equations of states (3) may be deduced, in principle, from microscopic electrodynamics and taking account of the motions of different parts of the medium, their individual microscopic characteristics (the values of electrical charge, mass) and their interactions. The values of the macroscopic fields $\mathbf{E}$, $\mathbf{H}$, $\mathbf{D}$, and $\mathbf{B}$ are then defined as the volume-average values of the microscopic fields created by the individual motions of the charged particles in the medium and for them the Maxwell equations hold.

On the boundary surface between two different media the following boundary conditions must be satisfied:

$$[\mathbf{n}\times\mathbf{H}_2]-[\mathbf{n}\times\mathbf{H}_1] = \mathbf{J}_S,$$

$$[\mathbf{n}\times\mathbf{E}_2]-[\mathbf{n}\times\mathbf{E}_1] = 0,$$

$$(\mathbf{n}\mathbf{D}_2)-(\mathbf{n}\mathbf{D}_1) = \sigma,$$

$$(\mathbf{n}\mathbf{B}_2)-(\mathbf{n}\mathbf{B}_1) = 0,$$

where $\mathbf{J}_S$ is the surface current density, σ is the surface charge density, $\mathbf{n}$ is a unit vector normal to the boundary surface, and the subscripts 1 and 2 refer to the values of the fields on either side of the boundary.

A consequence of the Maxwell equations is the *continuity equation*

$$\frac{\partial \rho}{\partial t}+\operatorname{div}\mathbf{J} = 0,$$

expressing the law of conservation of electrical charge.

The Maxwell equations are invariant under **Lorentz transformation**. In the pseudo-Euclidean 4-dimensional space-time with coordinates $x_1=x$, $x_2=y$, $x_3=z$, and $x_4=ict$ one introduces two anti-symmetric 4-dimensional tensors F_{kl} and G_{kl} ($k,l=1,2,3,4$) with components

$$\left.\begin{aligned}F_{12} = B_z, \quad F_{23} = B_x, \quad F_{31} = B_y, \quad F_{4k} = \frac{i}{c}E_k,\\ G_{12} = H_z, \quad G_{23} = H_x, \quad G_{31} = H_y, \quad G_{4k} = icD_k,\\ k = 1,2,3,4,\end{aligned}\right\} \tag{4}$$

and a 4-dimensional current vector j_k ($k=1,2,3,4$), whose spatial components $j_1=j_x, j_2=j_y, j_3=j_z$ coincide with the components of the current $\mathbf{j}$ and whose fourth component $j_4=ic\rho$ is proportional to the charge density, then the Maxwell equations (1) may be written in relativistic covariant form:

$$\frac{\partial F_{kl}}{\partial x_m}+\frac{\partial F_{lm}}{\partial x_k}+\frac{\partial F_{mk}}{\partial x_l} = 0, \quad k,l,m=1,2,3,4, \tag{5}$$

and

$$\sum_{l=1}^{4}\frac{\partial G_{kl}}{\partial x_l} = j_k, \quad k=1,2,3,4. \tag{6}$$

In these equations, $c=299\,792\,458\,m/s$ is the speed of light in vacuum. The equations (5) are the 4-dimensional form of (1b) and (1c), and the equations (6) are the 4-dimensional form of (1a) and (1d).

For an electromagnetic field in vacuum, $\mathbf{D}\equiv\epsilon_0\mathbf{E}$ and $\mathbf{B}\equiv\mu_0\mathbf{H}$, and consequently $G_{kl}=F_{kl}/\mu_0$, where $\mu_0=4\pi\cdot10^{-7}\,H/m$ is the permeability of vacuum and $\epsilon_0=1/\mu_0c^2$ is the permitivity of vacuum, and the electromagnetic field is described by just one tensor F_{kl}. If one introduces a 4-dimensional electromagnetic vector potential A_k, $k=1,2,3,4$, whose spatial components $A_1=A_x, A_2=A_y, A_3=A_z$ form the so-called 3-dimensional vector potential $\mathbf{A}(t,\mathbf{r})$ and whose fourth time component $A_4=(i/c)\phi$ is proportional to the scalar potential $\phi(t,\mathbf{r})$, then the components of the anti-symmetric electromagnetic field tensor F_{kl} can be expressed in terms of the components A_k by

$$F_{kl} = \frac{\partial A_k}{\partial x_l}-\frac{\partial A_l}{\partial x_k}, \quad k,l=1,2,3,4. \tag{7}$$

Because of (7), the equations (5) are satisfied identically and the equations (6) take the form

$$\sum_{l=1}^{4}\left[\frac{\partial^2 A_k}{\partial x_l^2}-\frac{\partial}{\partial x_k}\frac{\partial A_l}{\partial x_l}\right] = -\mu_0 j_k, \quad k=1,2,3,4, \tag{8}$$

that is, are inhomogeneous wave equations for the A_k.

The introduction of A_k allows the Maxwell equations to be written in the simple form (8). However, the potential A_k is not uniquely defined, which reflects the invariance of the Maxwell equations in the form (8) relative to gauge transformations. This non-uniqueness in the definition of A_k can be removed (see **Gauge transformation**).

According to (4) and (7) the physically observable fields **E** and **H** can be expressed in terms of the vector potential **A** and the scalar potential ϕ:

$$\mathbf{H} = \frac{1}{\mu_0}\operatorname{rot}\mathbf{A}, \quad \mathbf{E} = -\operatorname{grad}\phi - \frac{\partial\mathbf{A}}{\partial t}.$$

When the electromagnetic field in vacuum is free from sources, the Maxwell equations (1) and (8) become homogeneous and it is possible to obtain from them homogeneous wave equations for the electric and magnetic field strength:

$$\frac{1}{c^2}\frac{\partial^2}{\partial t^2}\mathbf{E} - \Delta\mathbf{E} = 0, \quad \frac{1}{c^2}\frac{\partial^2}{\partial t^2}\mathbf{H} - \Delta\mathbf{H} = 0,$$

where Δ is the Laplacian (cf. **Laplace operator**) and c is the speed of propagation of electromagnetic waves in vacuum.

The Maxwell equations for an electromagnetic field are used only in the classical theory. Thus, when the variable electric and magnetic field have very high frequencies and very small wavelengths (comparable with the dimensions of the atoms), significant quantum effects arise and the theory of an electromagnetic field and its sources must be built on the basis of quantum electrodynamics.

References

[1] MAXWELL, J.C.: *A treatise on electricity and magnetism*, Clarendon Press, 1873.
[2] TAMM, I.E.: *Fundamentals of the theory of electricity*, Mir, 1979 (translated from the Russian).
[3] LANDAU, L.D. and LIFSHITZ, E.M.: *The classical theory of fields*, Pergamon, 1951 (translated from the Russian).
[4] LANDAU, L.D. and LIFSHITZ, E.M.: *Electrodynamics of continuous media*, Pergamon, 1960 (translated from the Russian).

V.D. Kukin

Editorial comments. For a historical survey see [A3].

References

[A1] JACKSON, J.D.: *Classical electrodynamics*, Wiley, 1962.
[A2] STATTON, J.A.: *Electromagnetic theory*, McGraw-Hill, 1941.
[A3] WHITTAKER, E.T.: *A history of the theories of aether and electricity*, 1-2, Nelson, 1951-1953.

AMS 1980 Subject Classification: 78A25

MAYER PROBLEM - One of the fundamental problems in the calculus of variations (cf. **Variational calculus**) on a conditional extremum. The Mayer problem is the following: Find a minimum of the functional

$$J(y) = g(x_1, y(x_1), x_2, y(x_2)), \quad g\colon \mathbf{R}\times\mathbf{R}^n\times\mathbf{R}\times\mathbf{R}^n \to \mathbf{R},$$

in the presence of differential constraints of the type

$$\phi(x, y, y') = 0, \quad \phi\colon \mathbf{R}\times\mathbf{R}^n\times\mathbf{R}^n \to \mathbf{R}^m, \quad m < n,$$

and boundary conditions

$$\psi(x_1, y(x_1), x_2, y(x_2)) = 0, \quad \psi\colon \mathbf{R}\times\mathbf{R}^n\times\mathbf{R}\times\mathbf{R}^n \to \mathbf{R}^p,$$

$$p < 2n + 2.$$

For details see **Bolza problem**.

The Mayer problem is named after A. Mayer, who studied necessary conditions for its solution (at the end of the 19-th century).

I.B. Vapnyarskiĭ

AMS 1980 Subject Classification: 49B10

MEAN CURVATURE *of a surface* Φ^2 *in 3-dimensional Euclidean space* $\mathbf{R}^3$ - Half of the sum of the principal curvatures (cf. **Principal curvature**) k_1 and k_2, calculated at a point A of this surface:

$$H(A) = \frac{k_1 + k_2}{2}.$$

For a hypersurface Φ^n in the Euclidean space $\mathbf{R}^{n+1}$, this formula is generalized in the following way:

$$H(A) = \frac{k_1 + \cdots + k_n}{n},$$

where k_i, $i = 1, \ldots, n$, are the principal curvatures of the hypersurface, calculated at a point $A \in \Phi^n$.

The mean curvature of a surface in $\mathbf{R}^3$ can be expressed by means of the coefficients of the first and second fundamental forms of this surface:

$$H(A) = \frac{1}{2}\frac{LG - 2MF + NE}{EG - F^2},$$

where E, F, G are the coefficients of the **first fundamental form**, and L, M, N are the coefficients of the **second fundamental form**, calculated at a point $A \in \Phi^2$. In the particular case where the surface is defined by an equation $z = f(x, y)$, the mean curvature is calculated using the formula:

$$H(A) =$$
$$= \frac{\left[1 + \left(\frac{\partial f}{\partial y}\right)^2\right]\frac{\partial^2 f}{\partial x^2} - 2\frac{\partial f}{\partial x}\frac{\partial f}{\partial y}\frac{\partial^2 f}{\partial x\partial y} + \left[1 + \left(\frac{\partial f}{\partial x}\right)^2\right]\frac{\partial^2 f}{\partial y^2}}{\left[1 + \left(\frac{\partial f}{\partial x}\right)^2 + \left(\frac{\partial f}{\partial y}\right)^2\right]^{3/2}},$$

which is generalized for a hypersurface Φ^n in $\mathbf{R}^{n+1}$, defined by the equation $x_{n+1} = f(x_1, \ldots, x_n)$, as follows:

$$H(A) = \frac{\sum_{i=1}^{n}\left[1 + p^2 - \left(\frac{\partial f}{\partial x_i}\right)^2\right]\frac{\partial^2 f}{\partial x_i^2} - \sum_{i,j=1}^{n}\frac{\partial f}{\partial x_i}\frac{\partial f}{\partial x_j}\frac{\partial^2 f}{\partial x_i\partial x_j}}{(1 + p^2)^{3/2}},$$

where

$$p^2 = |\operatorname{grad}f|^2 = \left(\frac{\partial f}{\partial x_1}\right)^2 + \cdots + \left(\frac{\partial f}{\partial x_n}\right)^2.$$

L.A. Sidorov

Editorial comments. For an *m*-dimensional submanifold *M* of an *n*-dimensional Euclidean space of codimension $n-m>1$, the mean curvature generalizes to the notion of the *mean curvature normal*

$$\nu_p = \frac{1}{m} \sum_{j=1}^{n-m} [\mathrm{Tr}\, A(e_j)] e_j,$$

where $e_1, \ldots, e_{n-m}$ is an orthonormal frame of the normal space (cf. **Normal space (to a surface)**) of *M* at *p* and $A(e_j): T_pM \to T_pM$ (T_pM denotes the tangent space to *M* at *p*) is the *shape operator* of *M* at *p* in the direction e_j, which is related to the second fundamental tensor *V* of *M* at *p* by $<A(e_j)(X), Y> = <V(X, Y), e_j>$.

References

[A1] BERGER, M. and GOSTIAUX, B.: *Differential geometry*, Springer, 1988 (translated from the French).
[A2] DO CARMO, M.: *Differential geometry of curves and surfaces*, Prentice-Hall, 1976.
[A3] BLASCHKE, W. and LEICHTWEISS, K.: *Elementare Differentialgeometrie*, Springer, 1973.
[A4] CHEN, B.-Y.: *Geometry of submanifolds*, M. Dekker, 1973.

AMS 1980 Subject Classification: 53A05, 53A07

MEAN-SQUARE APPROXIMATION OF A FUNCTION - An approximation of a function $f(t)$ by a function $\phi(t)$, where the error measure $\mu(f; \phi)$ is defined by the formula

$$\mu_\sigma(f; \phi) = \int_a^b [f(t) - \phi(t)]^2 \, d\sigma(t),$$

where $\sigma(t)$ is a non-decreasing function on $[a, b]$ different from a constant.

Let

$$u_1(t), u_2(t), \ldots, \qquad (*)$$

be an **orthonormal system** of functions on $[a, b]$ relative to the distribution $d\sigma(t)$. In the case of a mean-square approximation of the function $f(t)$ by linear combinations $\sum_{k=1}^n \lambda_k u_k(t)$, the minimal error for every $n = 1, 2, \ldots,$ is given by the sums

$$\sum_{k=1}^n c_k(f) u_k(t),$$

where $c_k(f)$ are the **Fourier coefficients** of the function $f(t)$ with respect to the system (*); hence, the best method of approximation is linear.

References

[1] GONCHAROV, V.L.: *The theory of interpolation and approximation of functions*, Moscow, 1954 (in Russian).
[2] SZEGÖ, G.: *Orthogonal polynomials*, Amer. Math. Soc., 1975.

N.P. Korneĭchuk
V.P. Motornyĭ

Editorial comments. Cf. also Approximation in the mean; Approximation of functions; Approximation of functions, linear methods; Best approximation; Best approximation in the mean; Best linear method.

References

[A1] CHENEY, E.W.: *Introduction to approximation theory*, McGraw-Hill, 1966.

[A2] NATANSON, I.P.: *Constructive theory of functions*, 1-2, F. Ungar, 1964-1965 (translated from the Russian).

AMS 1980 Subject Classification: 41-01, 41A45, 42A16

MEASURABLE DECOMPOSITION, *measurable partition*, of a **measure space** (M, μ) - A partition (cf. **Decomposition of a space**) ξ of the space into disjoint subsets (called the elements of the partition) that can be obtained as the partition into level sets of some **measurable function** (with numerical values) on *M*. This definition can be restated in terms of 'intrinsic' properties of the partition (see [1]). In accordance with the general tendency in questions of measure theory to ignore sets of measure zero, a measurable partition is often understood to be a partition measurable modulo 0, that is, equivalent modulo 0 to some measurable partition (two partitions ξ and η of a measure space *M* are equivalent modulo 0 if there exists a set *N* of measure 0 such that the partitions of $M \setminus N$ consisting of the intersections with $M \setminus N$ of the elements of ξ and η coincide).

Although the definition given makes sense for any measure space, in fact measurable partitions are almost always considered for Lebesgue spaces (cf. **Lebesgue space**) (and sometimes for spaces having the properties of the latter to some extent, for example, for spaces with perfect measures; see [2] and [3] and **Perfect measure**), since in these spaces measurable partitions have a number of 'good' properties. Thus, in this case there exists a system of conditional measures (or, as used to be said in earlier times [1], a canonical system of measures) belonging to measurable partitions. This is a system of measures $\mu(\cdot \mid C)$ on the elements *C* of the partition ξ which enables one to consider integration with respect to μ as repeated integration: first, integration is performed over the *C* and with respect to appropriate $\mu(\cdot \mid C)$, and then it is necessary to integrate the result, which can be looked upon as a function on the quotient space M / ξ, with respect to the natural measure μ_ξ on the latter space (by definition, M / ξ has the elements of ξ as points, while its measurable subsets are those with measurable pre-images under the natural projection $\pi: M \to M / \xi$; the measure is defined to be $\mu_\xi(A) = \mu(\pi^{-1}(A))$). Interpreting (M, μ) as a space of elementary events in probability theory, one can say that the system of conditional measures is an 'improvement' of the conditional probability, closely connected with the specific Lebesgue space; for arbitrary spaces of elementary events, the conditional probability cannot, in general, be interpreted using some set of measures on the elements of some partition.

Non-measurable (and non-measurable modulo 0) partitions are by no means always 'pathological' objects, like non-measurable sets or functions. For example, the partition of the phase space of an ergodic dynamical system into its trajectories can have a completely 'classical' origin; it is simply that its properties are different from those of a measurable partition.

References

[1] ROKHLIN, V.A.: 'On mean notions of measure theory', *Mat. Sb.* **25**, no. 1 (1949), 107-150 (in Russian).
[2] GNEDENKO, B.V. and KOLMOGOROV, A.N.: *Limit distributions for sums of independent random variables*, Addison-Wesley, 1954 (translated from the Russian).
[3] SAZONOV, V.V.: 'On perfect measures', *Izv. Akad. Nauk. SSSR Ser. Mat.* **26**, no. 3 (1962), 391-414 (in Russian).

D.V. Anosov

Editorial comments. A useful result in ergodic theory concerning Lebesgue spaces and measurable partitions is the *Rokhlin—Halmos theorem*. In its strong form it states: Let $\pi=(A_1,\ldots,A_r)$ be a finite measurable partition of a Lebesgue space $(M,\mathscr{A},\mu)$ and let T be an **automorphism** of this space. Then, for any $\epsilon>0$ and any $n\in\mathbf{N}$ there exists an $E\in\mathscr{A}$ such that $E, TE,\ldots,T^{n-1}E$ are pairwise disjoint, $\mu(\bigcup_{i=0}^{n-1}T^iE)>1-\epsilon$ and $\mu(E\cap A_j)=\mu(E)\mu(A_j)$, $j=1,\ldots,r$. (The weak form results by taking the trivial partition (M).)

References

[A1] CORNFELD, I.P. [I.P. KORNFEL'D], FOMIN, S.V. and SINAĬ, YA.G.: *Ergodic theory*, Springer, 1982, Chapt. 10, § 5.

AMS 1980 Subject Classification: 28A99

MEASURABLE FLOW *in a measure space* (M,μ) - A family $\{T^t\}$ (t runs over the set of real numbers $\mathbf{R}$) of automorphisms of the space such that: 1) $T^t(T^s(x))=T^{t+s}(x)$ for all $t, s\in\mathbf{R}$, $x\in M$; and 2) the mapping $M\times\mathbf{R}\to M$ taking (x, t) to $T^t x$ is measurable (a measure is introduced on $M\times\mathbf{R}$ as the direct product of the measure μ in M and the Lebesgue measure in $\mathbf{R}$). 'Automorphisms' here are to be understood in the strict sense of the word (and not modulo 0), that is, the T^t must be bijections $M\to M$ carrying measurable sets to measurable sets of the same measure. In using automorphisms modulo 0, it turns out to be expedient to replace condition 2) by a condition of a different character, which leads to the concept of a **continuous flow**. Measurable flows are used in **ergodic theory**.

D.V. Anosov

AMS 1980 Subject Classification: 28D05, 28D10

MEASURABLE FUNCTION - 1) Originally, a measurable function was understood to be a function $f(x)$ of a real variable x with the property that for every a the set E_a of points x at which $f(x)<a$ is a (Lebesgue-) **measurable set**. A measurable function on an interval $[x_1, x_2]$ can be made continuous on $[x_1, x_2]$ by changing its values on a set of arbitrarily small measure; this is the so-called C-property of measurable functions (N.N. Luzin, 1913, cf. also **Luzin C-property**).

2) A measurable function on a space X is defined relative to a chosen system A of measurable sets in X. If A is a σ-ring, then a real-valued function f on X is said to be a *measurable function* if

$$R_f\cap E_a \in A$$

for every real number a, where

$$E_a = \{x\in X: f(x)<a\},$$
$$R_f = \{x\in X: f(x)\neq 0\}.$$

This definition is equivalent to the following: A real-valued function f is measurable if

$$R_f\cap\{x\in X: f(x)\in B\} \in A$$

for every **Borel set** B. When A is a σ-algebra, a function f is measurable if E_a (or $\{x\in X: f(x)\in B\}$) is measurable. The class of measurable functions is closed under the arithmetical and lattice operations; that is, if f_n, $n=1, 2,\ldots,$ are measurable, then f_1+f_2, f_1f_2, $\max(f_1,f_2)$, $\min(f_1,f_2)$ and af (a real) are measurable; $\varlimsup f_n$ and $\varliminf f_n$ are also measurable. A complex-valued function is measurable if its real and imaginary parts are measurable. A generalization of the concept of a measurable function is that of a **measurable mapping** from one **measurable space** to another.

References

[1] HALMOS, P.: *Measure theory*, v. Nostrand, 1950.
[2] DUNFORD, N. and SCHWARTZ, J.T.: *Linear operators*, 1, Interscience, 1958.
[3] KOLMOGOROV, A.N. and FOMIN, S.V.: *Elements of the theory of functions and functional analysis*, 1-2, Graylock, 1957-1961 (translated from the Russian).

V.V. Sazonov

AMS 1980 Subject Classification: 28A20

MEASURABLE MAPPING - A mapping f of a **measurable space** $(X_1,\mathscr{A}_1)$ to a measurable space $(X_2,\mathscr{A}_2)$ such that

$$f^{-1}(A) = \{x: f(x)\in A\} \in \mathscr{A}_1 \quad\text{for each } A\in\mathscr{A}_2.$$

In the case where $\mathscr{A}_1$ is a σ-algebra and $(X_2,\mathscr{A}_2)$ is the real line with the σ-algebra $\mathscr{A}_2$ of Borel sets (cf. **Borel set**), the concept of a measurable mapping reduces to that of a **measurable function** (however, when $\mathscr{A}_1$ is only a σ-ring, the definition of a measurable function is usually modified in accordance with the requirements of integration theory). The superposition of measurable mappings is measurable. If $\mathscr{A}_1$ and $\mathscr{A}_2$ are rings and $f^{-1}(B)\in\mathscr{A}_1$ for each B in some class of sets $B\in\mathscr{A}_2$ such that the ring generated by it is the whole of $\mathscr{A}_2$, then f is measurable. The analogous assertions hold in the case of σ-rings, algebras and σ-algebras. If $(X_1,\mathscr{A}_1)$ and $(X_2,\mathscr{A}_2)$ are topological spaces with the σ-algebras of Borel sets, then every continuous mapping from X_1 to X_2 is measurable. Let X be a topological space, let $\mathscr{A}$ be the σ-algebra of Borel sets and let μ be a finite non-negative *regular measure* on $\mathscr{A}$ (regularity means that $\mu(A)=\sup\{\mu(F): F\subset A, F$ closed$\}$). Suppose further that Y is a separable metric space, $\mathscr{B}$ is the σ-algebra of Borel sets, and let f be a measurable mapping from $(X, \mathscr{A})$ to $(Y, \mathscr{B})$. Then for any $\epsilon>0$ there is a closed subset $F\subset X$ such that $\mu(X\setminus F)<\epsilon$ and f is continuous on F (*Luzin's theorem*).

References

[1] HALMOS, P.: *Measure theory*, v. Nostrand, 1950.
[2] NEVEU, J.: *Mathematical foundations of the calculus of probabilities*, Holden-Day, 1965 (translated from the French).
[3] BOURBAKI, N.: *Elements of mathematics. Integration*, Addison-Wesley, 1975, Chapt. 6; 7; 8 (translated from the French).
[4] DUNFORD, N. and SCHWARTZ, J.T.: *Linear operators. General theory*, 1, Interscience, 1958.

V.V. Sazonov

AMS 1980 Subject Classification: 28A20

MEASURABLE SET - A subset of a **measurable space** $(X, \mathscr{A})$ belonging to $\mathscr{A}$, where $\mathscr{A}$ is a ring or σ-ring of subsets of X. The concept arose and was developed in the process of the solution and generalization of the measurement of areas (lengths, volumes) of various sets; that is, the problem of the extension of area (length, volume) as an additive function of polygons (segments, polyhedra) to a wider system of sets. A measurable set was defined to be a set in the system to which the extension can be realized; this extension is said to be the measure. Thus were defined the **Jordan measure**, the **Borel measure** and the **Lebesgue measure**, with sets measurable according to Jordan, Borel and Lebesgue, respectively. The solution of the problem of extending any fixed measure in $\mathbf{R}^n$ led to the **Radon measure** (Lebesgue−Stieltjes measure) and sets measurable with respect to the Radon (Lebesgue−Stieltjes) measure. The measurable sets connected with a measure defined on an abstract set are the sets on which the measure under discussion is defined.

References

[1] KOLMOGOROV, A.N. and FOMIN, S.V.: *Elements of the theory of functions and functional analysis*, 1-2, Graylock, 1957-1961 (translated from the Russian).
[2] HALMOS, P.: *Measure theory*, v. Nostrand, 1950.

A.P. Terekhin

Editorial comments.

References

[A1] HEWITT, E. and STROMBERG, K.: *Real and abstract analysis*, Springer, 1965.

AMS 1980 Subject Classification: 28A05

MEASURABLE SPACE - A set X with a distinguished ring or σ-ring $\mathscr{A}$ (in particular, an algebra or a σ-algebra) of subsets of X.

Examples: $\mathbf{R}^n$ with the ring of Jordan-measurable sets (see **Jordan measure**); $\mathbf{R}^n$ with the σ-ring of sets of

finite **Lebesgue measure**; a topological space E with the σ-algebra of Borel sets (cf. **Borel set**).

References

[1] HALMOS, P.: *Measure theory*, v. Nostrand, 1950.

V.V. Sazonov

AMS 1980 Subject Classification: 28A05

MEASURE, *measure of a set* - A notion that generalizes those of the length of segments, the area of figures and the volume of bodies, and that corresponds intuitively to the mass of a set for some mass distribution throughout the space. The notion of the measure of a set arose in the theory of functions of a real variable in connection with the study and improvement of the notion of an **integral**.

Definition and general properties. Let X be a set and let $\mathscr{E}$ be a class of subsets of X. A non-negative (not necessarily finite) set function λ defined on $\mathscr{E}$ is called *additive, finitely additive* or *countably additive* if

$$\lambda\left[\bigcup_{i=1}^{n} E_i\right] = \sum_{i=1}^{n}\lambda(E_i)$$

whenever

$$E_i \in \mathscr{E}, \quad \bigcup_{i=1}^{n} E_i \in \mathscr{E}, \quad E_i \cap E_j = \varnothing, \, i\neq j,$$

for, respectively, $n=2$, n arbitrary finite, and $n\leqslant\infty$.

A collection $\mathscr{P}$ of subsets of X is called a *semi-ring of sets* if

1) $\varnothing\in\mathscr{P}$;
2) $E_1, E_2\in\mathscr{P}$ imply $E_1\cap E_2\in\mathscr{P}$;
3) $E, E_1\in\mathscr{P}$, $E_1\subset E$ imply that E is representable as $E=\bigcup_{i=1}^{n}E_i$, $E_i\cap E_j=\varnothing$ for $i\neq j$, $E_i\in\mathscr{P}$, $i=1,\ldots,n$, $n<\infty$.

A collection $\mathscr{R}$ of subsets of X is called a *ring of sets* if

1) $\varnothing\in\mathscr{R}$;
2) $E_1, E_2\in\mathscr{R}$ imply $E_1\cup E_2\in\mathscr{R}$, $E_1\setminus E_2\in\mathscr{R}$.

An example of a semi-ring is: $X=\mathbf{R}^k$, $\mathscr{P}$ is the collection of all intervals of the form

$$\{x=(x_1,\ldots,x_k)\in\mathbf{R}^k: a_i\leqslant x_i<b_i, \, i=1,\ldots,k\},$$

where $a_i, b_i\in\mathbf{R}$ for $i=1,\ldots,k$. The collection of all possible finite unions of such intervals is a ring.

A collection $\mathscr{S}$ of subsets of X is called a σ-*ring* if

1) $\varnothing\in\mathscr{S}$;
2) $E_1, E_2\in\mathscr{S}$ imply $E_1\setminus E_2\in\mathscr{S}$;
3) $E_i\in\mathscr{S}$, $i=1,2,\ldots$, implies $\bigcup_{i=1}^{\infty}E_i\in\mathscr{S}$.

Every σ-ring is a ring; every ring is a semi-ring.

A *finitely-additive measure* is a non-negative finitely-additive set function m such that $m(\varnothing)=0$. The domain of definition $\mathscr{E}_m$ of a finitely-additive measure may be a semi-ring, a ring or a σ-ring. In the definition of a finitely-additive measure on a ring or on a σ-ring the condition of finite additivity can be weakened to additivity, which leads to the same notion.

If m is a finitely-additive measure, if the sets $E, E_1, \ldots, E_n$ belong to its domain of definition, and if $E \subset \bigcup_{i=1}^{n} E_i$, then

$$m(E) \leqslant \sum_{i=1}^{n} m(E_i).$$

Let m_1 be a finitely-additive measure with domain $\mathscr{E}_{m_1}$. A finitely-additive measure m_2 with domain $\mathscr{E}_{m_2}$ is called an *extension* of m_1 if $\mathscr{E}_{m_1} \subset \mathscr{E}_{m_2}$ and $m_2(E) = m_1(E)$ for all $E \in \mathscr{E}_{m_1}$.

Every finitely-additive measure m defined on a semiring $\mathscr{P}$ admits a unique extension to a finitely-additive measure m' on the smallest ring $\mathscr{R}(\mathscr{P})$ containing $\mathscr{P}$. This extension is defined as follows: Every $E \in \mathscr{R}(\mathscr{P})$ is representable as $E = \bigcup_{i=1}^{n} E_i$, $E_i \in \mathscr{P}$, $E_i \cap E_j = \varnothing$, $i \neq j$, and one sets

$$m'(E) = \sum_{i=1}^{n} m(E_i).$$

A finitely-additive measure that has the property of countable additivity is called a *measure*. Examples of measures: Let X be an arbitrary non-empty set, let $\mathscr{E}_\mu$ be a σ-ring, a ring or a semi-ring of subsets of X, let $\{x_1, x_2, \ldots\}$ be a countable subset of X, and let $p_1, p_2, \ldots$, be non-negative numbers. Then the function

$$\mu(E) = \sum_n p_n \delta_{x_n}(E),$$

where $\delta_x(E) = 1$ if $x \in E$ and $\delta_x(E) = 0$ if $x \notin E$, is a measure defined on $\mathscr{E}_\mu$. The measures δ_x are called *elementary*, *degenerate* or *Dirac measures* (sometimes, *Dirac masses*). Not every finitely-additive measure is a measure. For example, if X is the set of rational points of the segment $[0, 1]$, $\mathscr{P}$ is the semi-ring of all possible intersections of subintervals of $[0, 1]$ with X, and for every $a, b, 0 \leqslant a \leqslant b \leqslant 1$,

$$m((a, b) \cap X) = m([a, b) \cap X) = m((a, b] \cap X) =$$
$$= m([a, b] \cap X) = b - a,$$

then m is finitely additive, but not countably additive on $\mathscr{P}$.

A (finitely-additive) measure m with domain $\mathscr{E}_m$ is said to be *finite* (respectively, *σ-finite*) if $m(E) < \infty$ for all $E \in \mathscr{E}_m$ (respectively, if for every $E \in \mathscr{E}_m$ there is a sequence of sets $\{E_i\}$ in $\mathscr{E}_m$ such that $E \subset \bigcup_{i=1}^{\infty} E_i$ and $m(E_i) < \infty$, $i = 1, 2, \ldots$). A (finitely-additive) measure m is said to be *totally finite* (*totally σ-finite*) if it is finite (respectively, σ-finite) and $X \in \mathscr{E}_m$.

A pair $(X, \mathscr{S})$, where X is a set and $\mathscr{S}$ is a σ-ring of subsets of X such that $\bigcup_{E \in \mathscr{S}} E = X$, is called a **measurable space**. A triple $(X, \mathscr{S}, \mu)$, where $(X, \mathscr{S})$ is a measurable space and μ is a measure on $\mathscr{S}$, is called a **measure space**. A space with a totally-finite measure μ normalized by the condition $\mu(X) = 1$ is called a **probability space**. In abstract measure theory, where the basic notions are a measurable space $(X, \mathscr{S})$ or a measure space $(X, \mathscr{S}, \mu)$, the elements of $\mathscr{S}$ are also referred to as *measurable sets* (cf. also **Measurable set**).

Properties of measure spaces. Let $\{E_i\}$ be an arbitrary sequence of measurable sets. Then

1) $\mu(\liminf_{i \to \infty} E_i) \leqslant \liminf_{i \to \infty} \mu(E_i)$;

2) if $\mu(\bigcup_{i=i_0}^{\infty} E_i) < \infty$ for some i_0, then

$$\mu\left[\limsup_{i \to \infty} E_i\right] \geqslant \limsup_{i \to \infty} \mu(E_i);$$

3) if $\lim_{i \to \infty} E_i$ exists and the condition in 2) is satisfied, then

$$\mu\left[\lim_{i \to \infty} E_i\right] = \lim_{i \to \infty} \mu(E_i).$$

A finitely-additive measure m defined on a ring $\mathscr{R}$ is a measure if and only if

$$m\left[\lim_{i \to \infty} E_i\right] = \lim_{i \to \infty} m(E_i)$$

for every monotone increasing sequence $\{E_i\}$ of elements of $\mathscr{R}$ such that $\bigcup_{i=1}^{\infty} E_i \in \mathscr{R}$.

Let $(X_1, \mathscr{S}_1, \mu_1)$ be a measure space, let $(X_2, \mathscr{S}_2)$ be a measurable space and let T be a *measurable mapping* from X_1 into X_2, i.e.

$$T^{-1}(E) = \{x \in X_1 : Tx \in E\} \in \mathscr{S}_1$$

for all $E \in \mathscr{S}_2$. The *measure generated by the mapping T* (denoted here by μT^{-1}) is the measure on $\mathscr{S}_2$ defined by

$$\mu T^{-1}(E) = \mu(T^{-1}E).$$

Let $(X, \mathscr{S}, \mu)$ be a measure space and let $X_1 \subset X$. Define μ_{X_1} on the sets E from the σ-ring $\mathscr{S} \cap X_1 = \{E \cap X_1 : E \in \mathscr{S}_1\}$ by

$$\mu_{X_1}(E) = \inf_{E \subset F \in \mathscr{S}} \mu(F).$$

Then $(X_1, \mathscr{S} \cap X_1, \mu_{X_1})$ is a measure space; μ_{X_1} is called the *restriction of the measure μ to X_1*.

An *atom* of the space $(X, \mathscr{S}, \mu)$ (or of the measure μ) is any set $E \in \mathscr{S}$ of positive measure such that if $F \subset E$, $F \in \mathscr{S}$, then either $\mu(F) = 0$ or $\mu(F) = \mu(E)$. A measure space without atoms is called *non-atomic* or *continuous* (in this case μ is also called *non-atomic* or *continuous*). If $(X, \mathscr{S}, \mu)$ is a space with a non-atomic σ-finite measure and $E_1 \in \mathscr{S}$, then for every α with $0 \leqslant \alpha \leqslant \mu(E_1)$ (possibly $\alpha = \infty$) there is an element $E_2 \in \mathscr{S}$ such that $E_2 \subset E_1$ and $\mu(E_2) = \alpha$.

A measure space $(X, \mathscr{S}, \mu_1)$ (or the measure μ) is said to be *complete* if $E \in \mathscr{S}$, $F \subset E$, $\mu(E) = 0$ imply $F \in \mathscr{S}$. Every measure space $(X, \mathscr{S}, \mu)$ can be completed by adjoining to $\mathscr{S}$ all the sets of the form $E \cup N$ with $E \in \mathscr{S}$, $N \subset N'$, $N' \in \mathscr{S}$, $\mu(N') = 0$, and putting for such sets $\bar{\mu}(E \cup N) = \mu(E)$. The class of sets of the indicated form is a σ-ring, and $\bar{\mu}$ is a complete measure on it. The sets of null measure are called *null sets*. If the set of points of X at which a property Q is not satisfied is a null set, then property Q is said to hold **almost-everywhere**.

Extension of measures. A measure μ_2 is an extension of a measure μ_1 if μ_2 is an extension of μ_1 in the class of finitely-additive measures (see above). Every measure defined on a semi-ring $\mathscr{P}$ admits a unique extension to a measure on the ring $\mathscr{R}(\mathscr{P})$ generated by $\mathscr{P}$ (the extension is realized in the same way as in the case of finitely-additive measures). Further, every measure μ defined on a ring $\mathscr{R}$ can be extended to a measure μ' on the σ-ring $\mathscr{S}(\mathscr{R})$ generated by $\mathscr{R}$; if μ is σ-finite, then μ' is unique and σ-finite. The value of μ' on any set $E \in \mathscr{S}(\mathscr{R})$ can be given by the formula

$$\mu'(E) = \inf\left\{\sum_{i=1}^{\infty}\mu(E_i): E_i \in \mathscr{R},\ i=1,2,\ldots,\ E \subset \bigcup_{i=1}^{\infty}E_i\right\}. \quad (*)$$

A class of subsets of X is called *hereditary* if it contains, together with any set in the class, all its subsets. An **outer measure** is a set function m^*, defined on a hereditary σ-ring $\mathscr{H}$ (i.e. a class of sets which is simultaneously hereditary and a σ-ring), which has the following properties:

1) $0 \leqslant m^*(E) \leqslant \infty$, $m^*(\varnothing)=0$;
2) $E \subset F$ implies $m^*(E) \leqslant m^*(F)$;
3) $m^*(\bigcup_{i=1}^{\infty}E_i) \leqslant \sum_{i=1}^{\infty}m^*(E_i)$.

Given a measure μ on the ring $\mathscr{R}$ one can construct an outer measure μ^* on the hereditary σ-ring $\mathscr{H}(\mathscr{R})$ generated by $\mathscr{R}$ ($\mathscr{H}(\mathscr{R})$ consists of all sets that can be covered by a countable union of elements of $\mathscr{R}$) by means of the formula

$$\mu^*(E) = \inf\left\{\sum_{i=1}^{\infty}\mu(E_i): E_i \in \mathscr{R},\ i=1,2,\ldots,\ E \subset \bigcup_{i=1}^{\infty}E_i\right\}.$$

The outer measure μ^* is called the *outer measure induced by the measure* μ.

Let m^* be an outer measure on a hereditary σ-ring $\mathscr{H}$ of subsets of X. A set $E \in \mathscr{H}$ is called m^*-*measurable* if

$$m^*(A) = m^*(A \cap E) + m^*(A \cap (X \setminus E))$$

for every $A \in \mathscr{H}$. The collection $\overline{\mathscr{S}}$ of m^*-measurable sets is a σ-ring which contains all sets of null outer measure. The set function $\overline{m}$ on $\overline{\mathscr{S}}$ defined by the equality $\overline{m}(E)=m^*(E)$ is a complete measure and is called the *measure induced by the outer measure* m^*.

Suppose that μ is a measure on a ring $\mathscr{R}$ and that μ^* is the outer measure on $\mathscr{H}(\mathscr{R})$ induced by μ. Let $\overline{\mathscr{S}}$ and $\overline{\mu}$ denote the collection of μ^*-measurable sets and the measure on $\overline{\mathscr{S}}$ induced by μ^*, respectively. Then $\overline{\mu}$ is an extension of μ, and since $\mathscr{S}(\mathscr{R}) \subset \overline{\mathscr{S}}$ it follows that the function μ' on $\mathscr{S}(\mathscr{R})$ given by formula $(*)$ is also a measure extending μ. If the original measure μ

on $\mathscr{R}$ is σ-finite, then the space $(X, \overline{\mathscr{S}}, \overline{\mu})$ is the completion of the space $(X, \mathscr{S}(\mathscr{R}), \mu')$ (see $(*)$). If μ is given on the σ-ring $\mathscr{S}$, then the induced outer measure μ^* on the hereditary σ-ring $\mathscr{H}(\mathscr{S})$ generated by $\mathscr{S}$ is given by the formula

$$\mu^*(E) = \inf\{\mu(F): E \subset F,\ F \in \mathscr{S}\}.$$

Alongside with the outer measure μ^*, one defines the *inner measure induced by the measure* μ on $\mathscr{S}$. It is defined as

$$\mu_*(E) = \sup\{\mu(F): E \supset F,\ F \in \mathscr{S}\},\ E \in \mathscr{H}(\mathscr{S}).$$

For every set $E \in \mathscr{H}(\mathscr{S})$ a *measurable kernel* E' and a *measurable envelope* E'' are defined as elements of $\mathscr{S}$ such that $E' \subset E \subset E''$ and $\mu(F')=\mu(F'')=0$ for all $F', F'' \in \mathscr{S}$ such that $F' \subset E \setminus E'$, $F'' \subset E'' \setminus E$. A measurable kernel exists always, while a measurable envelope exists whenever E has σ-finite outer measure; moreover, $\mu_*(E)=\mu(E')$ and $\mu^*(E)=\mu(E'')$. Let μ be a measure on a ring $\mathscr{R}$ and let μ' be its extension to the σ-ring $\mathscr{S}(\mathscr{R})$ generated by $\mathscr{R}$. The inner measure μ'_* on the subsets E of finite μ-measure can be expressed in terms of the outer measure μ^* (and hence μ):

$$\mu'_*(A) = \mu(E)-\mu^*(E \setminus A),\ A \subset E.$$

Furthermore, a set F belonging to the hereditary σ-ring $\mathscr{H}(\mathscr{R})$ with finite outer μ^*-measure is μ^*-measurable if and only if $\mu^*(F)=\mu'_*(F)$. In case the original measure μ on $\mathscr{R}$ is totally finite, one has the following necessary and sufficient condition for the μ^*-measurability of a set $E \subset X$:

$$\mu(X) = \mu^*(E)+\mu^*(X \setminus E).$$

For totally-finite measures on $\mathscr{R}$ this condition is frequently taken as the definition of μ^*-measurability of the set E.

If $(X, \mathscr{S}, \mu)$ is a space with a σ-finite measure and $X_1, \ldots, X_n$ is a finite collection of elements of the hereditary σ-ring $\mathscr{H}(\mathscr{S})$ generated by $\mathscr{S}$, then on the σ-ring $\widetilde{\mathscr{S}}$ generated by $\mathscr{S}$ and the sets $X_1, \ldots, X_n$ one can define a measure $\widetilde{\mu}$ which agrees with μ on $\mathscr{S}$.

Jordan, Lebesgue and Lebesgue$-$Stieltjes measures. An example of an extension of a measure is provided by the Lebesgue measure in $\mathbf{R}^k$. The intervals of the form

$$I = \{(x_1, \ldots, x_k): a_i \leqslant x_i < b_i,\ i=1, \ldots, k\}$$

form a semi-ring $\mathscr{P}$ in $\mathbf{R}^k$. For each such interval, let

$$\lambda(I) = \prod_{i=1}^{k}(b_i-a_i)$$

($\lambda(I)$ coincides with the volume of I). The function λ is σ-finite and countably additive on $\mathscr{P}$ and admits a unique extension to a measure λ' on the σ-ring $\mathscr{S}$ generated by $\mathscr{P}$; $\mathscr{S}$ is identical with the σ-ring of Borel sets (cf. **Borel set**) (or Borel-measurable sets) in $\mathbf{R}^k$. The measure λ' was first defined by E. Borel in 1898 (see

Borel measure). The completion $\bar{\lambda}$ of λ' (defined on $\overline{\mathscr{S}}$) is called the *Lebesgue measure*, and was introduced by H. Lebesgue in 1902 (see **Lebesgue measure**). A set belonging to the domain $\overline{\mathscr{S}}$ of $\bar{\lambda}$ is called *Lebesgue measurable*. A bounded set $E \subset \mathbf{R}^k$ belongs to $\overline{\mathscr{S}}$ if and only if $\lambda(I) = \lambda^*(E) + \lambda^*(I \setminus E)$, where $I \in \mathscr{P}$ is some interval containing E; in this case $\bar{\lambda}(E) = \lambda^*(E)$. A set $E \subset \mathbf{R}^k$ belongs to $\overline{\mathscr{S}}$ if and only if for some sequence $\{r_n\}$, $r_n > 0$, $n = 1, 2, \ldots$, such that $r_n \to \infty$, one has $E \cap B_{r_n} \in \overline{\mathscr{S}}$ for all n, where $B_r = \{x \in \mathbf{R}^k : \|x\| \leqslant r\}$. The cardinality of the family of all Borel sets in $\mathbf{R}^k$ is c (the cardinality of the continuum), whereas the cardinality of the family of all Lebesgue-measurable sets is 2^c, so that the inclusion $\mathscr{S} \subset \overline{\mathscr{S}}$ is strict, i.e. there exist Lebesgue-measurable sets that are not Borel measurable.

The Lebesgue measure $\bar{\lambda}$ is invariant under linear orthogonal transformations A of $\mathbf{R}^k$ as well as under translations by elements $x \in \mathbf{R}^k$, i.e. $\bar{\lambda}(AE + x) = \bar{\lambda}(E)$ for all $E \in \mathscr{S}$.

Using the **axiom of choice** one can show that there exist sets which are not Lebesgue measurable. On the straight line, for example, such a set can be obtained by picking one point in each coset in $\mathbf{R}$ of the additive subgroup of rational numbers (*Vitali's example*).

Historically the Borel and Lebesgue measures in $\mathbf{R}^k$ were preceded by the measure defined by C. Jordan in 1892 (see **Jordan measure**). The idea of the definition of the Jordan measure is very close to that of the classic definition of area and volume, which goes back to ancient Greece. Thus, a set $E \subset \mathbf{R}^k$ is called *Jordan measurable* if there exist two sets, representable as finite unions of disjoint rectangles, one contained in E and the other containing E, such that the difference of their volumes (defined in an obvious manner) is arbitrarily small. The *Jordan measure* of such a set is the infimum of the volumes of finite unions of rectangles covering E. A Jordan-measurable set is also Lebesgue measurable, and its Jordan and Lebesgue measures are equal. The domain of the Jordan measure is merely a ring, and not a σ-ring, which restricts considerably its domain of applicability.

The Lebesgue measure is a particular case of the more general Lebesgue−Stieltjes measure. The latter is defined by means of a real-valued function F on $\mathbf{R}^k$ with the properties:

1) $-\infty < F < \infty$;

2) $\Delta_{b_1 - a_1} \cdots \Delta_{b_k - a_k} F(a_1, \ldots, a_k) \geqslant 0$ for $a_i < b_i$, $i = 1, \ldots, k$, where $\Delta_{b_i - a_i}$ is the difference operator with step $b_i - a_i$ taken at the point a_i with respect to the i-th coordinate;

3) $F(a_1, \ldots, a_k) \uparrow F(b_1, \ldots, b_k)$ as $a_i \uparrow b_i$, $i = 1, \ldots, k$.

Given such a function F, the measure μ_F of the interval

$$I = \{(x_1, \ldots, x_k): a_i \leqslant x_i < b_i, \ i = 1, \ldots, k\}$$

is defined by the formula

$$\mu_F(I) = \Delta_{b_1 - a_1} \cdots \Delta_{b_k - a_k} F(a_1, \ldots, a_k).$$

It turns out that μ_F is countably additive on the semi-ring of all such intervals and that it admits an extension to the σ-algebra of Borel sets; the completion of this extension yields what is called the *Lebesgue−Stieltjes measure* corresponding to F. For the particular choice

$$F(x_1, \ldots, x_k) = x_1 \cdots x_k$$

one obtains the Lebesgue measure.

Measures in product spaces. By definition, the *product of two measurable spaces* $(X_1, \mathscr{S}_1)$, $(X_2, \mathscr{S}_2)$ is the measurable space consisting of the set $X_1 \times X_2 = \{(x_1, x_2): x_1 \in X_1, x_2 \in X_2\}$ (the product of X_1 and X_2) and the σ-ring $\mathscr{S}_1 \times \mathscr{S}_2$ of subsets of X (the product of the σ-rings $\mathscr{S}_1$ and $\mathscr{S}_2$) generated by the semi-ring $\mathscr{P}$ of sets of the form

$$E_1 \times E_2 = \{(x_1, x_2): x_1 \in E_1, x_2 \in E_2\},$$

where $E_1, E_2 \in \mathscr{S}$. If $(X_1, \mathscr{S}_1, \mu_1)$ and $(X_2, \mathscr{S}_2, \mu_s)$ are measure spaces, the formula

$$\mu(E_1 \times E_2) = \mu_1(E_1)\mu_2(E_2), \quad E_1 \in \mathscr{S}_1, \ E_2 \in \mathscr{S}_2,$$

defines a measure on $\mathscr{P}$; if μ_1 and μ_2 are σ-finite, μ extends uniquely to a measure on $\mathscr{S}_1 \times \mathscr{S}_2$, denoted by $\mu_1 \times \mu_2$. The measure $\mu_1 \times \mu_2$ and the space $(X_1 \times X_2, \mathscr{S}_1 \times \mathscr{S}_2, \mu_1 \times \mu_2)$ are called, respectively, the *product of the measures* μ_1 and μ_2, and the *product of the measure spaces* $(X_1, \mathscr{S}_1, \mu_1)$ and $(X_2, \mathscr{S}_2, \mu_2)$. The completion of the product of the Lebesgue measure in $\mathbf{R}^k$ and the Lebesgue measure in $\mathbf{R}^l$ is the Lebesgue measure in $\mathbf{R}^{k+l}$. Analogously one defines the product of an arbitrary finite number of measure spaces.

Let $(X_i, \mathscr{S}_i, \mu_i)$, $i \in I$, be an arbitrary family of measure spaces such that $\mu_i(X_i) = 1$, $i \in I$. The *product space* $X = \prod_{i \in I} X_i$ is, by definition, the set of all functions on I such that the value at each $i \in I$ is an element $x_i \in X_i$. A *measurable rectangle* in X is any set of the form $\prod_{i \in I} E_i$, where $E_i \in \mathscr{S}_i$ and only finitely many sets E_i are different from X_i. The family of measurable rectangles forms a semi-ring $\mathscr{P}$. The σ-ring generated by $\mathscr{P}$ is denoted by $\prod_{i \in I} \mathscr{S}_i$ and is called the *product of the σ-rings* $\mathscr{S}_i$. Now, let μ be the function on $\mathscr{P}$ defined by $\mu(E) = \prod_{i \in I} \mu_i(E_i)$ for $E = \prod_{i \in I} E_i$. The function μ thus defined is a measure which admits a unique extension to a measure on $\prod_{i \in I} \mathscr{S}_i$, denoted by $\prod_{i \in I} \mu_i$. The measure space $(\prod_{i \in I} X_i, \prod_{i \in I} \mathscr{S}_i, \prod_{i \in I} \mu_i)$ is called the *product of the spaces* $(X_i, \mathscr{S}_i, \mu_i)$, $i \in I$.

The product of an arbitrary number of measure

spaces is a particular case of the following, more general, scheme, which plays an important role in probability theory. Let $(X_i, \mathscr{S}_i)$, $i \in I$, be a family of measurable spaces (each $\mathscr{S}_i$ is a σ-algebra), and suppose that for each finite subset $I_1 \subset I$ there is given a probability measure μ_{I_1} on the measurable spaces $(\prod_{i \in I_1} X_i, \prod_{i \in I_1} \in \mathscr{S}_i)$ (the product of measures corresponds to the case that $\mu_{I_1} = \prod_{i \in I_1} \mu_i$ for all finite $I_1 \subset I$). Suppose further that each two measures μ_{I_1}, μ_{I_2} are *compatible* in the sense that if $I_1 \subset I_2$ and p_{21} is the projection of $\prod_{i \in I_2} X_i$ onto $\prod_{i \in I_1} X_i$, then $\mu_{I_1}(E) = \mu_{I_2} p_{21}^{-1}(E)$ for all $E \in \prod_{i \in I_1} \mathscr{S}_i$ (by definition, p_{21} is the mapping of $\prod_{i \in I_2} X_i$ onto $\prod_{i \in I_1} X_i$ such that $(p_{21}(x))_i = x_i$ for all $i \in I_1$). The following question arises: Is there a probability measure on $\prod_{i \in I} \mathscr{S}_i$ such that $\mu_{I_1}(E) = \mu p^{-1}(E)$ for every finite $I_1 \subset I$ and every $E \in \prod_{i \in I_1} \mathscr{S}_i$, where p denotes the projection of $\prod_{i \in I} X_i$ onto $\prod_{i \in I_1} X_i$? It turns out that such a measure does not always exist, and additional conditions must be imposed to guarantee its existence. One such condition is perfectness of the measures μ_i (corresponding to the one-point sets $i \in I$). The notion of a **perfect measure** was first introduced by B.V. Gnedenko and A.N. Kolmogorov [6]. A space $(X, \mathscr{S}, \mu)$ with a totally-finite measure, as well as the measure μ itself, is called *perfect* if for every $\mathscr{S}$-measurable real-valued function f on X there is a Borel set $B \subset f(X)$ such that $\mu(f^{-1}(B)) = \mu(X)$. The perfectness assumption eliminates a series of 'pathological' phenomena that arise in general measure theory.

Measures in topological spaces. The study of measures in topological spaces is usually concerned with measures defined on sets connected in some way or another with the topology of the underlying space. One of the typical approaches is the following. Let X be an arbitrary topological space and let $\mathscr{Z}$ be the class of subsets of the form $f^{-1}(F)$, where f is a continuous real-valued function on X and $F \subset \mathbf{R}^1$ is a closed set. Let $\mathfrak{A}$ be the algebra generated by the class $\mathscr{Z}$ and let $\mathscr{B}$ be the σ-algebra generated by $\mathscr{Z}$ ($\mathscr{B}$ is called the σ-*algebra of Baire sets*, cf. also **Algebra of sets**). Now let $\mathscr{M}$ be the class of totally-finite finitely-additive measures m on $\mathfrak{A}$ that are regular in the sense that

$$m(E) = \sup\{m(Z): Z \subset E, \ Z \in \mathscr{Z}\}$$

for all $E \in \mathfrak{A}$. In $\mathscr{M}$ one distinguishes the subclasses $\mathscr{M}_\sigma$, $\mathscr{M}_\tau$ and $\mathscr{M}_t$ formed by the (finitely-additive) measures possessing additional smoothness properties. By definition, $\mu \in \mathscr{M}_\sigma$ if $\mu(Z_n) \downarrow 0$ for every sequence $Z_n \downarrow \varnothing$, $Z_n \in \mathscr{Z}$ (this property is equivalent to the countable additivity of μ; the measures from $\mathscr{M}_\sigma$ admit

unique extensions to $\mathscr{B}$ and hereafter it is assumed that they are given on $\mathscr{B}$); $\mu \in \mathscr{M}_\tau$ if $\mu(\mathscr{Z}_\alpha) \downarrow 0$ for every net $Z_\alpha \downarrow \varnothing$, $Z_\alpha \in \mathscr{Z}$; and $\mu \in \mathscr{M}_t$ if for every $\epsilon > 0$ there is a compact set K such that $\mu(E) < \epsilon$ whenever $E \subset X \setminus K$, $E \in \mathfrak{A}$.

The inclusions $\mathscr{M} \supset \mathscr{M}_\sigma \supset \mathscr{M}_\tau \supset \mathscr{M}_t$ hold. The elements of $\mathscr{M}_\sigma$ are called *Baire measures*.

There is an intimate connection between the measures belonging to $\mathscr{M}$ and the linear functionals on the space $C(X)$ of bounded continuous functions on X. Namely, the formula

$$\Lambda(f) = \int_X f \, dm$$

establishes a one-to-one correspondence between the finitely-additive measures $m \in \mathscr{M}$ and the non-negative linear functionals Λ on $C(X)$ (non-negative means that $\Lambda(f) \geq 0$ whenever $f(x) \geq 0$, $x \in X$). Moreover, for every set $Z \in \mathscr{Z}$,

$$m(Z) = \inf\{\Lambda(f): \chi_Z \leq f \leq 1\},$$

where χ_Z is the indicator function of Z. This correspondence takes the measures from $\mathscr{M}_\sigma$ into σ-*smooth functionals* Λ (i.e. functionals Λ with the property that $\Lambda(f_n) \to 0$ if $f_n \downarrow 0$ in $C(X)$), the measures from $\mathscr{M}_\tau$ into τ-*smooth functionals* Λ (i.e. functionals such that $\Lambda(f_\alpha) \to 0$ for every net $f_\alpha \downarrow 0$ in $C(X)$), and the measures from $\mathscr{M}_t$ into *dense functionals* Λ (i.e. with the property that $\Lambda(f_\alpha) \to 0$ for every net f_α in $C(X)$ such that $\| f_\alpha \| \leq 1$ for all α and $f_\alpha \to 0$ uniformly on compact subsets; here $\| \cdot \|$ is the uniform norm).

The space $\mathscr{M}$ is usually endowed with the weak topology w, in which a basis of neighbourhoods consists of the sets of the form

$$U(m_0; f_1, \ldots, f_n, \epsilon) =$$

$$= \left\{ m: \left| \int_X f_k \, (dm - dm_0) \right| < \epsilon, \ k = 1, \ldots, n, f_1, \ldots, f_n \in C(X) \right\}.$$

With the topology w, $\mathscr{M}$ is a completely-regular Hausdorff space. Convergence in the topology w is usually denoted by the symbol $\Rightarrow$. For the convergence of a net m_α to m: $m_\alpha \Rightarrow m$, it is necessary and sufficient that $m_\alpha(X) \to m(X)$ and $\limsup m_\alpha(Z) \leq m(Z)$ for all $Z \in \mathscr{Z}$. Another necessary and sufficient condition for the convergence $m_\alpha \Rightarrow m$ is that $m_\alpha(E) \to m(E)$ for all $E \in \mathfrak{A}$ such that there are $Z_1, Z_2 \in \mathscr{Z}$ with $X \setminus E \subset Z_1$, $E \subset Z_2$, and $m(Z_1 \cap Z_2) = 0$. If the space X is completely regular and Hausdorff, then $\mathscr{M}_\tau$ is metrizable if and only if X is metrizable. If X is metrizable, then $\mathscr{M}_\tau$ admits a metric in which it is separable if and only if X is separable, and it admits a metric in which it is complete if and only if X has a complete metric. If X is metrizable, then $\mathscr{M}_\sigma$ is metrizable if and only if it is metrizable by the **Lévy – Prokhorov metric**.

The space $\mathscr{M}_\sigma$ is sequentially closed in $\mathscr{M}$

(*Aleksandrov's theorem*). A set $A \subset \mathcal{M}$ is called *tight* if $\sup\{m(X): m \in A\} < \infty$ and if for every $\epsilon > 0$ there is a compact set K such that $m(E) < \epsilon$ for all $E \subset X \setminus K$, $m \in A$ and $E \in \mathfrak{A}$. If $A \subset \mathcal{M}_\sigma$ is tight, then A is relatively compact in $\mathcal{M}_\sigma$; conversely, if X is metrizable and topologically complete, then $A \subset \mathcal{M}_\sigma$ is relatively compact, and if every measure in A is concentrated on some separable subset of X, then A is tight (*Prokhorov's theorem*).

Under certain conditions the elements of $\mathcal{M}_\sigma$ can be extended to Borel measures, i.e. measures defined on the σ-algebra of Borel sets (see **Borel set**; **Borel measure**). For example, if X is a normal countably-paracompact Hausdorff space, then every measure $\mu \in \mathcal{M}_\sigma$ admits a unique extension to a regular Borel measure. If X is completely regular and Hausdorff, then every τ-smooth (tight) Baire measure admits a unique extension to a τ-smooth (tight) Borel measure.

The *support* of a Baire (Borel) measure is the smallest set $Z \in \mathcal{Z}$ (respectively, the smallest closed set) the measure of which is equal to the measure of the whole space. Every τ-smooth measure has a support.

Often, when measures in topological spaces (especially in locally compact Hausdorff spaces) are considered, it is assumed that the Borel and Baire measures are given on less-wide classes of sets, more precisely — on σ-rings generated by compact sets and, respectively, compact G_δ-sets.

Let G be a locally compact Hausdorff topological group. A *left Haar measure* on G is a measure defined on the σ-ring generated by all compact subsets that does not vanish identically and is such that $\mu(xE) = \mu(E)$ for all $x \in G$ and E in the domain of μ. A *right Haar measure* is defined in the same manner but with the condition $\mu(xE) = \mu(E)$ replaced by $\mu(Ex) = \mu(E)$. On any group of the type considered a left Haar measure exists and is unique (up to a multiplicative positive constant). Every left Haar measure is regular in the sense that $\mu(E) = \sup\{\mu(K): K \subset E\}$, where K are compact sets. The right Haar measure has analogous properties. The Lebesgue measure on $\mathbf{R}^k$ is a particular case of the **Haar measure**. See also **Measure in a topological vector space**.

Isomorphism of measure spaces. Let $(X, \mathcal{S}, \mu)$ be a measure space. Call two sets $E, E' \in \mathcal{S}$ μ-*equal* (written $E = E'$ $[\mu]$) if $\mu(E \Delta E') = 0$ (where $E \Delta E'$ denotes the symmetric difference of E and E', cf. **Symmetric difference of sets**). Denote by $\mathcal{S}_\mu$ the class of sets $\mathcal{S}$ with this equality relation. In $\mathcal{S}_\mu$ the set-theoretic operations, performed a finite (or countable) number of times are correctly defined: for example, if $E_1 = E_1'$ $[\mu]$ and $E_2 = E_2'$ $[\mu]$, then $E_1 \cup E_2 = E_1' \cup E_2'$ $[\mu]$. The measure μ is carried over, in an obvious manner, to $\mathcal{S}_\mu$.

Let $\tilde{\mathcal{S}}_\mu$ be the subset of $\mathcal{S}_\mu$ consisting of the sets of finite measure. The function $\rho(E, E') = \mu(E \Delta E')$ on $\tilde{\mathcal{S}}_\mu \times \tilde{\mathcal{S}}_\mu$ is a metric. The measure space $(X, \mathcal{S}, \mu)$ is said to be *separable* if the space $\tilde{\mathcal{S}}_\mu$ with metric ρ is separable. If $(X, \mathcal{S}, \mu)$ is a space with a σ-finite measure and the σ-ring $\mathcal{S}$ is countably generated (i.e. there is a countable family $\{E_n\} \subset \mathcal{S}$ such that $\mathcal{S}$ is the smallest σ-ring that contains this family), then the metric space $\mathcal{S}_\mu$ is separable.

Two measure spaces, $(X_1, \mathcal{S}_1, \mu_1)$ and $(X_2, \mathcal{S}_2, \mu_2)$ are said to be *isomorphic* if there is a one-to-one mapping ϕ of $(\mathcal{S}_1)_{\mu_1}$ onto $(\mathcal{S}_2)_{\mu_2}$ such that

$$\phi(E \setminus F) = \phi(E) \setminus \phi(F), \quad \phi(E \cup F) = \phi(E) \cup \phi(F)$$

and

$$\mu_1(E) = \mu_2(\phi(E)) \quad \text{for all } E, F \in (\mathcal{S}_1)_{\mu_1}.$$

Now, let $(X, \mathcal{S}, \mu)$ be an arbitrary space with a totally-finite measure. There is a partition of X into disjoint sets $X_n \in \mathcal{S}$, $n = 1, 2, \ldots$, such that the restriction of μ to X_n is isomorphic either to a measure concentrated at one point or to a measure which is equal, up to a positive factor, to the direct product $\prod_{i \in I}(U_i, \mathcal{U}_i, u_i)$, where $U_i = \{0, 1\}$, $u_i(\{0\}) = u_i(\{1\}) = 1/2$, and the set I may have arbitrary cardinality (the *Maharan–Kolmogorov theorem*). If $(X, \mathcal{S}, \mu)$ is separable, non-atomic and $\mu(X) = 1$, then it is isomorphic to the space $\prod_{i \in I}(U_i, \mathcal{U}_i, u_i)$ with I countable, which in turn is isomorphic to the unit interval with the Lebesgue measure.

Side by side with the theory of measures regarded as functions on subsets of some set, the theory of measures as functions on the elements of a Boolean ring (or on a **Boolean algebra**) has been developed; these theories are in many respects parallel. Another widespread construction of measures goes back to W. Young and P. Daniell (see [12]). Theories dealing with measures with real or complex values, or with values belonging to some algebraic structure, were developed in addition to the theory of positive measures.

References

[1] SAKS, S.: *Theory of the integral*, Hafner, 1952 (translated from the Polish).

[2] HALMOS, P.: *Measure theory*, v. Nostrand, 1950.

[3] DUNFORD, N. and SCHWARTZ, J.T.: *Linear operators. General theory*, 1, Interscience, 1958.

[4] KOLMOGOROV, A.N. and FOMIN, S.V.: *Elements of the theory of functions and functional analysis*, 1-2, Graylock, 1957-1961 (translated from the Russian).

[5] NEVEU, J.: *Mathematical foundations of the calculus of probabilities*, Holden-Day, 1965 (translated from the French).

[6] GNEDENKO, B.V. and KOLMOGOROV, A.N.: *Limit distributions for sums of independent random variables*, Addison-Wesley, 1954 (translated from the Russian).

[7] VARADARAJAN, V.S.: 'Measures on topological spaces', *Mat. Sb.* **55**, no. 1 (1961), 35-100 (in Russian).

[8] PARTHASARATHY, K.R.: *Probability measures on metric spaces*, Acad. Press, 1967.

[9] BILLINGSLEY, P.: *Convergence of probability measures*, Wiley, 1968.

[10] SIKORSKI, R.: *Boolean algebras*, Springer, 1969.
[11] VLADIMIROV, D.A.: *Boolesche Algebren*, Akad. Verlag, 1978 (translated from the Russian).
[12] BOURBAKI, N.: *Elements of mathematics. Integration*, Addison-Wesley, 1975 , p. Chapt.6;7;8 (translated from the French).
[13] DIESTEL, J. and UHL, J.: *Vector measures*, Amer. Math. Soc., 1977.

V.V. Sazonov

Editorial comments. Properties 1) and 2) listed under the heading 'Properties of measure spaces' are usually called *Fatou's lemma*, cf. **Fatou theorem**.

The procedure for extending a measure, as described under the heading 'Extension of measures', is due to C. Carathéodory, and one often speaks of *Carathéodory extension*, with the accompanying phrases *Carathéodory extension theorem* and *Carathéodory outer (inner) measure* (cf. **Carathéodory measure**).

Recall that a ring (respectively, a σ-ring) $\mathscr{A}$ of subsets of a set X such that $A \in \mathscr{A}$ implies $X \setminus A \in \mathscr{A}$, is called a *Boolean algebra* or an *algebra* (respectively, a *σ-algebra* or a *σ-field*, cf. also **Algebra of sets**). Usually, in a measure space $(X, \mathscr{S}, \mu)$ the σ-ring $\mathscr{S}$ can be proved to be a σ-field (this holds, in particular, if $\mu(X) < \infty$).

The phrase 'totally (σ-) finite' is seldom used.

Borel has given very nice ideas in order to construct the measure λ', but Lebesgue was the first to give a satisfactory construction of it, as a byproduct of the construction of $\overline{\lambda}$.

A product space is also often written as a (kind of) tensor product: $(X_1 \times X_2, \mathscr{S}_1 \otimes \mathscr{S}_2, \mu_1 \otimes \mu_2)$.

A family of measurable spaces $(X_i, \mathscr{S}_i)_i$ with compatible probability measures on each finite product is called a *projective system of measure spaces*, and the corresponding probability measure on $\prod X_i$, if it exists, is called the *projective limit*; it exists if I is countable (the *Ionescu–Tulcea theorem*, cf. [5]).

Suppose that X is a topological space and $\mathscr{S}$ is its Borel σ-field; then $(X, \mathscr{S}, \mu)$ is perfect for every finite measure μ if X is a *Polish space* or, more generally, a *Luzin space* (in which case $(X, \mathscr{S})$ is often called a *standard measurable space*) or, still more generally, a *Suslin space* (in which case $(X, \mathscr{S})$ is sometimes called a *Blackwell measurable space*) (cf. (the editorial comments to) **Descriptive set theory**).

The converse part of *Prokhorov's theorem* is not true when X is the space of rational numbers, or, more generally, when X is a **Luzin space** which is not Polish. See [A1].

In the abstract setting, whenever (μ_n) is a sequence of finite measures on $(X, \mathscr{S})$, where $\mathscr{S}$ is a σ-field, such that

$$m(A) = \lim_n \mu_n(A)$$

exists for any $A \in \mathscr{S}$, then m is also a measure (the *Vitali–Hahn–Saks theorem*, cf. [3] or [5]).

References

[A1] PREISS, D.: 'Metric spaces in which Prokhorov's theorem is not valid', *Z. Wahrscheinlichkeitstheor. Verw. Gebiete* **27** (1973), 109-116.
[A2] COHN, D.: *Measure theory*, Birkhäuser, 1980.
[A3] HEWITT, E. and ROSS, K.A.: *Abstract harmonic analysis*, I, Springer, 1979.
[A4] HEWITT, E. and STROMBERG, K.R.: *Real and abstract analysis*, Springer, 1965.

AMS 1980 Subject Classification: 28CXX, 28AXX

MEASURE IN A TOPOLOGICAL VECTOR SPACE - A term used to designate a measure given in a **topological vector space** when one wishes to stress those properties of the measure that are connected with the linear and topological structure of this space. A general problem encountered in the construction of a measure in a topological vector space is that of extending a **pre-measure** to a measure. Let E be a (real or complex) locally convex space and let $\mathfrak{A}(E)$ be the algebra of its cylindrical sets (cf. **Cylinder set**). Suppose that a pre-measure is defined on $\mathfrak{A}(E)$. It is required to extend this pre-measure to a countably-additive measure defined on the σ-algebra $\mathfrak{B}_0(E)$ — the smallest σ-algebra containing $\mathfrak{A}(E)$. $\mathfrak{B}_0(E)$ is the smallest of all σ-algebras (weakly Borel, Borel, etc.) that are naturally connected with the topology of E; for a large class of spaces E these σ-algebras coincide. In the particular, yet most important, case when the space $E = V'$, i.e. it is the dual of some locally convex space V, endowed with the weak-* topology (so that $E' = V$), in order that a pre-measure μ in V' admits an extension to a measure, it suffices that its characteristic functional (*Fourier transform*)

$$\chi(\phi) = \int_V \exp\{ix(\phi)\} \, d\mu(x), \quad x \in V', \quad \phi \in V,$$

be continuous in the so-called *Sazonov topology* on the space V (i.e. in the topology generated by all continuous Hilbert semi-norms in V) and in a number of cases — for example, if V is a **Fréchet space** — it is necessary that the characteristic functional be continuous in the original topology of V. For instance, if V is a **nuclear space**, the Sazonov topology is identical with the original topology, and every pre-measure in V' with a continuous characteristic functional extends to a measure. In the case of a pre-measure defined on a Hilbert space H the sufficient condition for its extendability to a measure formulated above is also necessary. In addition to this general criterion for the extendability of pre-measures to measures, partial results of this type, applicable to specific classes of measures (or classes of spaces), are available. For example, a *Gaussian pre-measure* on V', where V' is a locally convex space (i.e. a pre-measure the restriction of which to any σ-algebra $\mathfrak{B}(L) \subset \mathfrak{A}(V'), L \subset V$, $\dim L < \infty$, is a Gaussian distribution with correlation functional $B(\phi_1, \phi_2)$, $\phi_1, \phi_2 \in V$) extends to a measure if there exists a convex neighbourhood of zero in V with **ϵ-entropy**, in the metric defined by the inner product $<\phi_1, \phi_2> = B(\phi_1, \phi_2)$, smaller than two.

For a sequence of (probability) measures in a dual

space V' to converge weakly it suffices that the characteristic functionals of these measures converge pointwise (this condition is also necessary) and that they be equicontinuous at zero in the Sazonov topology in V, and it is necessary that these functionals be equicontinuous in the original topology of V. In the case of a Hilbert space V, conditions necessary and sufficient for the weak compactness of a family of measures in V are known. These are also expressible in terms of their characteristic functionals. The following aspects have been investigated (1982) only for Gaussian measures: the problem of the quasi-invariance of a measure in a topological vector space (see **Quasi-invariant measure**) with respect to some set of translations (the set of quasi-invariance) of this space (it is known that for a number of infinite-dimensional vector spaces the set of quasi-invariance of a non-zero measure does not necessarily coincide with the whole space); and criteria for the absolute continuity of one measure with respect to another. The study of measures in topological vector spaces is mainly connected with integrals over trajectories (cf. **Integral over trajectories**), and also with the theory of generalized random fields, and is to a high degree stimulated by the applications of these theories in physics and mechanics.

References

[1] DUNFORD, N. and SCHWARTZ, J.T.: *Linear operators. General theory*, 1, Interscience, 1958.
[2] BOURBAKI, N.: *Elements of mathematics. Integration*, Addison-Wesley, 1975, Chapt. 6; 7; 8 (translated from the French).
[3] GIKHMAN, I.I. and SKOROKHOD, A.V.: *The theory of stochastic processes*, 1, Springer, 1974 (translated from the Russian).
[4] GEL'FAND, I.M. and VILENKIN, N.YA.: *Generalized functions*, 4. Applications of harmonic analysis, Acad. Press, 1964 (translated from the Russian).
[5] SUDAKOV, V.N.: 'Geometric problems in the theory of infinite-dimensional probability distributions', *Proc. Steklov Inst. Math.* **141** (1976). (*Trudy Mat. Inst. Steklov.* **141** (1976))
[6] SMOLYANOV, O.G. and FOMIN, S.V.: 'Measures on linear topological spaces', *Russian Math. Surveys* **31**, no. 4 (1976), 1-53. (*Uspekhi Mat. Nauk* **31**, no. 4 (1976), 3-56)

R.A. Minlos

Editorial comments.

References

[A1] SCHWARTZ, L.: *Radon measures on arbitrary topological spaces and cylindrical measures*, Oxford Univ. Press, 1973.
[A2] VAKHANIA, N.N.: *Probability distributions on linear spaces*, North-Holland, 1981.
[A3] VAKHANIA, N.N., TARIELADZE, V.I. and CHOBANYAN, S.A.: *Probability distributions on Banach spaces*, Reidel, 1987 (translated from the Russian).

AMS 1980 Subject Classification: 28C15

MEASURE OF IRRATIONALITY *of a real number* ξ - The function

$$L(\xi, H) = \min \, | h_1 \xi + h_0 |,$$

where the minimum is over all pairs h_0, h_1 of integral rational numbers such that

$$| h_0 |, | h_1 | \leqslant H, \quad | h_0 | + | h_1 | \neq 0.$$

The concept of the measure of irrationality is a particular case of those of the measure of linear independence and the measure of transcendency (cf. **Linear independence, measure of; Transcendency, measure of**). The measure of irrationality indicates how 'well' the number ξ can be approximated by rational numbers. For all real irrational numbers one has

$$L(\xi, H) < \frac{1}{\sqrt{5}} \frac{1}{H},$$

but for any $\epsilon > 0$ and almost-all (in the sense of the Lebesgue measure) real numbers ξ,

$$L(\xi, H) > \frac{C}{H^{1+\epsilon}},$$

where $C = C(\epsilon, \xi) > 0$. However, for any function ϕ with $\phi(H) \to 0$ as $H \to \infty$ and $\phi(H) > 0$, there exists a number ξ_ϕ such that for all $H \geqslant 1$,

$$0 < L(\xi_\phi, H) < \phi(H).$$

References

[1] KHINCHIN, A.YA.: *Continued fractions*, Univ. Chicago Press, 1964 (translated from the Russian).

A.I. Galochkin

AMS 1980 Subject Classification: 10-XX

MEASURE-PRESERVING TRANSFORMATION of a measure space $(X, \mathfrak{A}, \mu)$.

Editorial comments. A measurable mapping $T: X \to X$ such that $\mu(T^{-1}(A)) = \nu(A)$ for every $A \in \mathfrak{A}$; μ is called an *invariant measure* for T. A measurable mapping $T: X \to Y$ between measure spaces $(X, \mathfrak{A}, \mu)$ and $(Y, \mathfrak{B}, \nu)$ such that $\mu(T^{-1}(B)) = \nu(B)$ for every $B \in \mathfrak{B}$ is usually called a *measure-preserving mapping*. A surjective measure-preserving transformation T of a measure space $(X, \mathfrak{A}, \mu)$, i.e., T maps X onto itself, is often called an *endomorphism* of $(X, \mathfrak{A}, \mu)$; an endomorphism which is bijective and whose inverse is also measure preserving is called an **automorphism** of $(X, \mathfrak{A}, \mu)$.

Measure-preserving transformations arise, for example, in the study of classical dynamical systems (cf. (measurable) **Cascade; Measurable flow**). In that case the transformation is first obtained as a continuous (or smooth) transformation of some, often compact, topological space (or manifold), and the existence of an invariant measure is proved. An example is Liouville's theorem for a **Hamiltonian system** (cf. also **Liouville theorem**).

For further information and references see **Ergodic theory**.

AMS 1980 Subject Classification: 28D05

MEASURE SPACE (X, A, μ) - A **measurable space** (X, A) with a **measure** μ given on A (i.e. a countably-additive function μ with values in $[0, \infty]$ for which $\mu(\varnothing) = 0$; the latter property follows from additivity if the measure is finite, i.e. does not take the value ∞, or even if there is some $Y \in A$ with $\mu(Y) < \infty$). The notation (X, A, μ) is often shortened to (X, μ) and one says that μ is a measure on X; sometimes the notation is shortened to X. The basic case is when A is a σ-algebra (cf. **Algebra of sets**) and X can be represented as $\bigcup_{n=1}^{\infty} X_i$ with $X_n \in A$ and $\mu(X_n) < \infty$. In this case the measure is called *(totally) σ-finite* (while if $\mu(X) < \infty$, then it is called *(totally) finite*). Such is, e.g., the Lebesgue measure on $\mathbf{R}$ (cf. **Lebesgue space**). However, sometimes non-σ-finite measures are encountered, such as, e.g., the k-dimensional **Hausdorff measure** on $\mathbf{R}^n$ for $k < n$. One may also encounter modifications in which μ takes values in $(-\infty, \infty)$, or complex or vector values, as well as cases when μ is only finitely additive.

References

[1] HALMOS, P.: *Measure theory*, v. Nostrand, 1950.
[2] DUNFORD, N. and SCHWARTZ, J.T.: *Linear operators. General theory*, 1, Interscience, 1958.

$D.V.\ Anosov$

AMS 1980 Subject Classification: 28AXX

MECHANICAL QUADRATURE, METHOD OF, *method of mechanical cubature* - A method for solving integral equations, based on replacing an integral by a sum using quadrature (cubature) formulas. Consider the equation

$$x(t) = \int_{\Omega} K(t, s)x(s)\,ds + y(t), \qquad (1)$$

where $\Omega \subset \mathbf{R}^n$ is a bounded open domain. Using a quadrature (cubature) process

$$\int_{\Omega} z(s)\,ds = \sum_{j=1}^{n} \alpha_{jn} z(s_{jn}) + \phi_n(z)$$

one forms the system of linear equations

$$x_{in} = \sum_{j=1}^{n} \alpha_{jn} K(s_{in}, s_{jn})x_{jn} + y(s_{in}), \quad i = 1, \ldots, n, \qquad (2)$$

where $x_{in} \approx x(x_{in})$, $i = 1, \ldots, n$.

Let the absolute term y and the kernel K be continuous on $\bar{\Omega}$ and $\bar{\Omega} \times \bar{\Omega}$, respectively ($\bar{\Omega}$ is the closure of Ω), and let (1) have a unique solution $x(t)$. Let $\phi_n(z) \to 0$ as $n \to \infty$ for any continuous function $z(t)$ on $\bar{\Omega}$. Then for sufficiently large n the system (2) is uniquely solvable and

$$c_1 \epsilon_n \leqslant \max_{1 \leqslant i \leqslant n} |x_{in} - x(s_{in})| \leqslant c_2 \epsilon_n, \quad n \geqslant n_0,$$

where c_1 and c_2 are positive constants and

$$\epsilon_n = \max_{1 \leqslant i \leqslant n} |\phi_n(K(s_{in}, s)x(s))| \to 0$$

as $n \to \infty$.

A mechanical quadrature method can be applied for the solution of non-linear integral equations [3] and eigen value problems for linear operators. The method converges even for a certain class of equations with discontinuous kernels [4].

References

[1] KRYLOV, V.I., BOBKOV, V.V. and MONASTYRNYĬ, P.I.: *Compu-

tational methods*, 2, Moscow, 1977 (in Russian).
[2] BEREZIN, I.S. and ZHIDKOV, N.P.: *Computing methods*, Pergamon, 1973 (translated from the Russian).
[3] KRASNOSEL'SKIĬ, M.A., ET AL.: *Approximate solution of operator equations*, Wolters-Noordhoff, 1972 (translated from the Russian).
[4] VAĬNIKKO, G.M.: 'On the convergence of the method of mechanical quadratures for integral equations with discontinuous kernels', *Sib. Math. J.* **12**, no. 1 (1971), 29-38. (*Sibirsk. Mat. Zh.* **12**, no. 1 (1971), 40-53)
[5] MICHLIN, S.G. [S.G. MIKHLIN] and PRÖSSDORF, S.: *Singular integral operators*, Springer, 1986 (translated from the German).

$G.M.\ Va\breve{\imath}nikko$

Editorial comments.

References

[A1] BRUNNER, H. and HOUWEN, P.J. VAN DER: *The numerical solution of Volterra equations*, North-Holland, 1986.
[A2] BAKER, C.T.H.: *The numerical treatment of integral equations*, Clarendon Press, 1977.
[A3] ENGELS, H.: *Numerical quadrature and cubature*, Acad. Press, 1980.
[A4] ATKINSON, K.E.: *A survey of numerical methods for the solution of Fredholm integral equations of the second kind*, SIAM, 1976.

AMS 1980 Subject Classification: 65D32

MEDIAN *of a trapezium* - A segment joining the mid-points of the two sides of the trapezium. The median is parallel to the bases of the trapezium and is equal to half the sum of their lengths.

$BSE-3$

Editorial comments. See also **Median (of a triangle)**.

AMS 1980 Subject Classification: 51N10

MEDIAN (IN STATISTICS) - One of the numerical characteristics of probability distributions, a particular case of a **quantile**. For a real-valued random variable X with distribution function F, a median is defined as a number m such that $F(m) \leqslant 1/2$ and $F(m+0) \geqslant 1/2$. Every random variable has at least one median. If $F(x) = 1/2$ for all x in a closed interval, then every point of this interval is a median. If F is a strictly-monotone function, then the median is unique. In the symmetric case, if the median is unique, it is identical with the **mathematical expectation**, provided that the latter exists. The fact that a median always exists is used for centering random variables (see, for instance, **Lévy inequality**). In mathematical statistics, to estimate the median of a distribution in terms of independent results of observations $X_1, \ldots, X_n$ one uses a so-called sample median — a median of the corresponding order statistics (cf. **Order statistic**) $X_{(1)}, \ldots, X_{(n)}$, that is, of the quantity $X_{(k+1)}$ if $n = 2k+1$ is odd, and of $[X_{(k)} + X_{(k+1)}]/2$ if $n = 2k$ is even.

References

[1] LOÈVE, M.: *Probability theory*, Springer, 1977.
[2] CRAMÉR, H.: *Mathematical methods of statistics*, Princeton Univ. Press, 1946.

$A.V.\ Prokhorov$

AMS 1980 Subject Classification: 62-XX

MEDIAN (OF A TRIANGLE) - A straight line (or its segment contained in the triangle) which joins a vertex of the triangle with the midpoint of the opposite side. The three medians of a triangle intersect at one point, called the *centre of gravity*, the *centroid* or the *barycentre of the triangle*. This point divides each median into two parts with ratio 2:1 if the first segment is the one that starts at the vertex.

P.S. Modenov

Editorial comments. J. Hjelmslev has shown that also in hyperbolic geometry (cf. **Lobachevskiĭ geometry**) the meridians of a triangle intersect at a point.

References

[A1] COXETER, H.S.M.: *Introduction to geometry*, Wiley, 1989.

AMS 1980 Subject Classification: 51M05, 51M15

MEDIANT of two fractions a/b and c/d with positive denominators - The fraction $(a+c)/(b+d)$. The mediant of two fractions is positioned between them, i.e. if $(a/b) \leqslant (c/d)$, $b, d > 0$, then

$$\frac{a}{b} \leqslant \frac{a+c}{b+d} \leqslant \frac{c}{d}.$$

A finite sequence of fractions in which each intermediary term is the mediant of its two adjacent fractions is called a **Farey series**. The mediant of two adjacent convergent fractions of the continued-fraction expansion of a real number α is positioned between α and the convergent fraction of lower order (cf. also **Continued fraction**). Thus, if P_n/Q_n and P_{n+1}/Q_{n+1} are convergent fractions of orders n and $n+1$ in the continued-fraction expansion of α, then

$$\left| \alpha - \frac{P_n}{Q_n} \right| > \left| \frac{P_n + P_{n+1}}{Q_n + Q_{n+1}} - \frac{P_n}{Q_n} \right| = \frac{1}{Q_n(Q_n + Q_{n+1})}.$$

References

[1] KHINCHIN, A.YA.: *Continued fractions*, Univ. Chicago Press, 1964 (translated from the Russian).

V.I. Nechaev

Editorial comments.

References

[A1] HARDY, G.H. and WRIGHT, E.M.: *An introduction to the theory of numbers*, Clarendon Press, 1979.

AMS 1980 Subject Classification: 10A32, 10F20

MEHLER − FOCK TRANSFORM, *Mehler − Fok transform* - The **integral transform**

$$F(x) = \int_0^\infty P_{i\tau - 1/2}(x) f(\tau) \, d\tau, \quad 1 \leqslant x < \infty, \tag{1}$$

where $P_\nu(x)$ is the Legendre function of the first kind (cf. **Legendre functions**). If $f \in L[0, \infty)$, the function $|f'(\tau)|$ is locally integrable on $[0, \infty)$ and $f(0) = 0$, then the following inversion formula is valid:

$$f(\tau) = \tau \tanh \pi\tau \int_1^\infty P_{i\tau - 1/2}(x) F(x) \, dx. \tag{2}$$

The *Parseval identity*. Consider the Mehler − Fock transform and its inverse defined by the equalities

$$G(\tau) = \int_1^\infty \sqrt{\tau \tanh \pi\tau} \, P_{i\tau - 1/2}(x) g(x) \, dx,$$

$$g(x) = \int_0^\infty \sqrt{\tau \tanh \pi\tau} \, P_{i\tau - 1/2}(x) G(\tau) \, d\tau.$$

If $g_i(x)$, $i = 1, 2$, are arbitrary real-valued functions satisfying the conditions

$$g_i(x) x^{-1/2} \ln(1+x) \in L(1, \infty), \quad g_i(x) \in L_2(1, \infty),$$

then

$$\int_0^\infty G_1(\tau) G_2(\tau) \, d\tau = \int_1^\infty g_1(x) g_2(x) \, dx.$$

The generalized Mehler − Fock transform and the corresponding inversion formula are:

$$F(x) = \int_0^\infty P_{i\tau - 1/2}^{(k)}(x) f(\tau) \, d\tau, \tag{3}$$

and

$$f(\tau) = \frac{1}{\pi} \tau \sinh \pi\tau \, \Gamma\left[\frac{1}{2} - k + i\tau\right] \Gamma\left[\frac{1}{2} - k - i\tau\right] \times \tag{4}$$

$$\times \int_1^\infty P_{i\tau - 1/2}^{(k)}(x) F(x) \, dx,$$

where $P_\nu^{(k)}(x)$ are the associated Legendre functions of the first kind. For $k = 0$ formulas (3) and (4) reduce to (1) and (2); for $k = 1/2$, $y = \cosh \alpha$, formulas (3) and (4) lead to the **Fourier cosine transform**, and for $k = -1/2$, $y = \cosh \alpha$ to the **Fourier sine transform**. The transforms (1) and (2) were introduced by F.G. Mehler [1]. The basic theorems were proved by V.A. Fock [V.A. Fok].

References

[1] MEHLER, F.G.: 'Ueber eine mit den Kugel- und Cylinderfunctionen verwandte Function und ihre Anwendung in der Theorie der Electricitätsvertheilung', *Math. Ann.* **18** (1881), 161-194.

[2] FOK, V.A.: 'On the representation of an arbitrary function by an integral involving Legendre functions with complex index', *Dokl. Akad. Nauk SSSR* **39** (1943), 253-256 (in Russian).

[3] DITKIN, V.A. and PRUDNIKOV, A.P.: 'Operational calculus', *Progress in Math.* **1** (1968), 1-75. (*Itogi Nauk. Mat. Anal. 1966* (1967), 7-82)

Yu.A. Brychkov
A.P. Prudnikov

Editorial comments.

References

[A1] SNEDDON, I.N.: *The use of integral transforms*, McGraw-Hill, 1972.

AMS 1980 Subject Classification: 44A15

MEHLER QUADRATURE FORMULA - The quadrature formula for the segment $[-1, 1]$ and the weight $1/\sqrt{1-x^2}$ which gives the highest algebraic degree of accuracy. It has the form

$$\int_{-1}^{1} \frac{1}{\sqrt{1-x^2}} f(x) \, dx \approx \frac{\pi}{N} \sum_{k=1}^{N} f\left[\cos \frac{2k-1}{2N} \pi\right]. \tag{1}$$

The nodes are the roots of the Chebyshev polynomial

$$T_N(x) = \cos N \arccos x;$$

the coefficients are identical and equal to π/N. The algebraic degree of accuracy equals $2N-1$. Formula (1) was established by F.G. Mehler [1].

The quadrature formula of highest algebraic degree of accuracy for the weight $1/\sqrt{1-x^2}$ and with $2N+1$ nodes for which N fixed nodes coincide with the nodes of the quadrature formula (1), has the form

$$\int_{-1}^{1} \frac{1}{\sqrt{1-x^2}} f(x)\,dx \approx \qquad (2)$$

$$\approx \frac{\pi}{2N}\left[\frac{f(-1)+f(1)}{2} + \sum_{k=1}^{2N-1} f\left(\cos\frac{k\pi}{2N}\right)\right].$$

Formula (2) is used to improve the accuracy of the approximate value of the integral obtained by means of formula (1); since the values of the integrand at the nodes of formula (1) have already been computed, it is necessary to compute its values at $N+1$ supplementary nodes only. Formula (2) represents also the quadrature formula of highest algebraic degree of accuracy with the weight $1/\sqrt{1-x^2}$ for which the fixed nodes are the end points of $[-1, 1]$, and hence the other nodes of which are the roots of the orthogonal polynomial of degree $2N-1$ on $[-1, 1]$ with weight $\sqrt{1-x^2}$, i.e. of the Chebyshev polynomial $U_{2N-1}(x)$ of the second kind. The algebraic degree of accuracy of the quadrature formula (2) is $4N-1$.

Formula (1) is sometimes referred to as *Hermite's quadrature formula*.

References

[1] MEHLER, F.G.: 'Bemerkungen zur Theorie der mechanischen Quadraturen', *J. Reine Angew. Math.* **63** (1864), 152-157.
[2] KRYLOV, N.M.: *Approximate calculation of integrals*, Macmillan, 1962 (translated from the Russian).

I.P. Mysovskikh

Editorial comments. The quadrature formula above is more commonly referred to as *Gauss–Chebyshev quadrature* (see **Gauss quadrature formula**). It may be viewed as being based on Hermite (oscillatory) interpolation given the weight function $1/\sqrt{1-x^2}$. Hermite's quadrature formula is a Gauss-type quadrature formula with weight e^{-x^2} and integration interval $(-\infty, \infty)$.

References

[A1] DAVIS, P.J. and RABINOWITZ, P.: *Methods of numerical integration*, Acad. Press, 1975.
[A2] HILDERBRAND, F.B.: *Introduction to numerical analysis*, MacGraw-Hill, 1974.

AMS 1980 Subject Classification: 65D32

MEIER THEOREM - Let $f(z)$ be a meromorphic function in the unit disc $D = \{z \in \mathbf{C}: |z| < 1\}$; then all points of the circle $\Gamma = \{z \in \mathbf{C}: |z| = 1\}$ except, possibly, for a set of the first category on Γ, are either Plessner points or Meier points. By definition, a point $e^{i\theta}$ on Γ is a *Plessner point* for $f(z)$ if the angular cluster set $C_A(e^{i\theta}, f)$ is total (i.e., coincides with the whole extended complex plane $\overline{\mathbf{C}}$) for every angle Δ between pairs of chords through $e^{i\theta}$. The point $e^{i\theta}$ is said to be a *Meier point* (or to have the *Meier property*) if: 1) the complete **cluster set** $C(e^{i\theta}; f)$ of $f(z)$ at $e^{i\theta}$ is subtotal, i.e. does not coincide with the whole extended complex plane $\overline{\mathbf{C}}$; and 2) the set of all limit values along arbitrary chords of the disc D drawn at the point $e^{i\theta}$ is identical to $C(e^{i\theta}; f)$. The theorem was proved by K. Meier [1].

Meier's theorem is the analogue, in terms of the category of a set, of the **Plessner theorem**, which is formulated in terms of measure theory. A sharpening of Meier's theorem is given in [3].

References

[1] MEIER, K.: 'Ueber die Randwerte der meromorphen Funktionen', *Math. Ann* **142** (1961), 328-344.
[2] COLLINGWOOD, E.F. and LOHWATER, A.J.: *The theory of cluster sets*, Cambridge Univ. Press, 1966.
[3] GAVRILOV, V.I. and KANATNIKOV, A.N.: 'Characterization of the set $M(f)$ for meromorphic functions', *Soviet Math. Dokl.* **18**, no. 2 (1977), 15-17. (*Dokl. Akad. Nauk SSSR* **233**, no. 1 (1977))

E.D. Solomentsev

AMS 1980 Subject Classification: 30D40

MEIJER TRANSFORM - The **integral transform**

$$F(x) = \int_{0}^{\infty} e^{-xt/2}(xt)^{-\mu-1/2} W_{\mu+1/2,\nu}(xt) f(t)\,dt,$$

where $W_{\mu,\nu}(x)$ is the Whittaker function (cf. **Whittaker functions**). The corresponding inversion formula is

$$f(t) = \lim_{\lambda \to +\infty} \frac{1}{2\pi i} \frac{\Gamma(1-\mu+\nu)}{\Gamma(1+2\nu)} \times$$

$$\times \int_{\beta-i\lambda}^{\beta+i\lambda} e^{xt/2}(xt)^{\mu-1/2} W_{\mu-1/2,\nu}(xt) F(x)\,dx.$$

For $\mu = \pm\nu$ the Meijer transform becomes the **Laplace transform**; for $\mu = -1/2$ it becomes the K_ν-transform

$$F(x) = \frac{1}{\sqrt{\pi}} \int_{0}^{\infty} e^{-xt/2}(xt)^{1/2} K_\nu\left(\frac{xt}{2}\right) f(t)\,dt,$$

where $K_\nu(x)$ is the **Macdonald function**.

The *Varma transform*

$$F(x) = \int_{0}^{\infty} (xt)^{\nu-1/2} e^{-xt/2} W_{\mu,\nu}(xt) f(t)\,dt$$

reduces to a Meijer transform.

The *Meijer K-transform* (or the *Meijer–Bessel transform*) is the integral transform

$$F(x) = \sqrt{\frac{2}{\pi}} \int_{0}^{\infty} K_\nu(xt) \sqrt{xt}\, f(t)\,dt.$$

If the function f is locally integrable on $(0, \infty)$, has bounded variation in a neighbourhood of the point

$t = t_0 > 0$, and if the integral

$$\int_0^\infty e^{-\beta t} | f(t) | \, dt, \quad \beta > \alpha \geqslant 0,$$

converges, then the following inversion formula is valid:

$$\frac{f(t_0+0)+f(t_0-0)}{2} =$$

$$= \lim_{\lambda \to \infty} \frac{1}{i\sqrt{2\pi}} \int_{\beta-i\lambda}^{\beta+i\lambda} I_\nu(t_0 x)(t_0 x)^{1/2} F(x) \, dx.$$

For $\nu = \pm 1/2$ the Meijer K-transform turns into the Laplace transform.

The Meijer transform and Meijer K-transform were introduced by C.S. Meijer in [1] and, respectively, [2].

References

[1] MEIJER, C.S.: 'Eine neue Erweiterung der Laplace Transformation I', *Proc. Koninkl. Ned. Akad. Wet.* **44** (1941), 727-737.

[2A] MEIJER, C.S.: 'Ueber eine neue Erweiterung der Laplace Transformation I', *Proc. Koninkl. Ned. Akad. Wet.* **43** (1940), 599-608.

[2B] MEIJER, C.S.: 'Ueber eine neue Erweiterung der Laplace Transformation II', *Proc. Koninkl. Ned. Akad. Wet.* **43** (1940), 702-711.

[3] BRYCHKOV, YU.A. and PRUDNIKOV, A.P.: *Integral transforms of generalized functions*, Moscow, 1977 (in Russian).

[4] DITKIN, V.A. and PRUDNIKOV, A.P.: 'Operational calculus', *Progress in Math.* **1** (1968), 1-75. (*Itogi Nauk. Mat. Anal. 1966* (1967), 7-82)

Yu.A. Brychkov
A.P. Prudnikov

AMS 1980 Subject Classification: 44A15

MELLIN TRANSFORM - The integral transform

$$M(p) = \int_0^\infty f(t) t^{p-1} \, dt, \quad p = \sigma + i\tau.$$

The substitution $t = e^{-z}$ reduces it to the **Laplace transform**. The Mellin transform is used for solving a specific class of planar problems for harmonic functions in a sectorial domain, of problems in elasticity theory, etc.

The *inversion theorem*. Suppose that $\tau^{\sigma-1} f(\tau) \in L(0, \infty)$ and that the function $f(\tau)$ has bounded variation in a neighbourhood of the point $\tau = t$. Then

$$\frac{f(t+0)-f(t-0)}{2} = \frac{1}{2\pi i} \lim_{\lambda \to \infty} \int_{\sigma-i\lambda}^{\sigma+i\lambda} M(s) t^{-s} \, ds.$$

The *representation theorem*. Suppose that the function $M(\tau + iu)$ is summable with respect to u on $(-\infty, +\infty)$ and has bounded variation in a neighbourhood of the point $u = t$. Then

$$\frac{M(\sigma+i(t+0))+M(\sigma+i(t-0))}{2} = \lim_{\lambda \to \infty} \int_{1/\lambda}^{\lambda} f(x) x^{\sigma+it-1} \, dx,$$

where

$$f(x) = \frac{1}{2\pi i} \int_{\sigma-i\infty}^{\sigma+i\infty} M(s) x^{-s} \, ds.$$

References

[1] MELLIN, H.: 'Ueber die fundamentelle Wichtigkeit des Satzes von Cauchy für die Theorie der Gamma- und hypergeometrischen Funktionen', *Acta Soc. Sci. Fennica* **21**, no. 1 (1896), 1-115.

[2] MELLIN, H.: 'Ueber den Zusammenhang zwischen linearen Differential- und Differenzengleichungen', *Acta Math.* **25** (1902), 139-164.

[3] TITCHMARSH, E.C.: *Introduction to the theory of Fourier integrals*, Oxford Univ. Press, 1948.

[4] DITKIN, V.A. and PRUDNIKOV, A.P.: *Transformations intégrales et calcul opérationnel*, Mir, 1978 (translated from the Russian).

P.I. Lizorkin

Editorial comments. If $M(p)$ denotes the Mellin transform of $f(t)$, then the *Parseval equality* takes the form:

$$\int_0^\infty | f(t) |^2 x^{2k-1} \, dx = \frac{1}{2\pi} \int_{-\infty}^{+\infty} | M(k+iy) |^2 \, dy$$

if $f(t) t^{k-1/2} \in L_2(0, \infty)$.

The Mellin transform also serves to link **Dirichlet series** with automorphic functions (cf. **Automorphic function**); in particular, the inversion formula plays a role in the proof of a functional equation for Dirichlet series similar to that for the Riemann zeta-function. Cf. [A1] - [A5].

References

[A1] HECKE, E.: 'Ueber die Bestimmung Dirichletscher Reihen durch ihre Funktionalgleichung', *Math. Ann.* **112** (1936), 664-699.

[A2] WEIL, A.: 'Ueber die Bestimmung Dirichletscher Reihen durch ihre Funktionalgleichung', *Math. Ann.* **168** (1967), 149-156.

[A3] WEIL, A.: 'Zeta functions and Mellin transforms', in *Algebraic Geometry (Bombay Coll., 1968)*, Oxford Univ. Press & Tata Inst. Fundam. Res., 1968, pp. 409-426.

[A4] OGG, A.: *Modular forms and Dirichlet series*, Benjamin, 1969.

[A5] SHIMURA, G.: *Introduction to the arithmetic theory of automorphic functions*, Princeton Univ. Press & Iwanami-Shoten, 1971, § 3.6, pp 89-94.

AMS 1980 Subject Classification: 44A15

MEMORYLESS CHANNEL - A **communication channel** for which the statistical properties of the output signal at a time t are determined only by the input signal transmitted at this moment t of time (and consequently do not depend on the signal transmitted prior to or after the moment t). More precisely, a discrete-time communication channel whose input and output signals are given by random sequences $\eta = (\eta_1, \eta_2, \ldots)$ and $\tilde\eta = (\tilde\eta_1, \tilde\eta_2, \ldots)$ with values in spaces (Y, S_Y) and $(\tilde Y, S_{\tilde Y})$, respectively, is called a memoryless channel if for any natural number n and any sets $\tilde A_1, \ldots, \tilde A_n$, $\tilde A_k \in S_{\tilde Y}$, $k = 1, \ldots, n$, the equality

$$P\{\tilde\eta_1 \in \tilde A_1, \ldots, \tilde\eta_n \in \tilde A_n \,|\, \eta^n\} =$$

$$= P\{\tilde\eta_1 \in \tilde A_1 \,|\, \eta_1\} \cdots P\{\tilde\eta_n \in \tilde A_n \,|\, \eta_n\}$$

holds, where $\eta^n = (\eta_1, \ldots, \eta_n)$. If furthermore the conditional probabilities $P\{\tilde\eta_k \in \tilde A_k \,|\, \eta_k\}$ do not depend on k, then the channel is called a *homogeneous memoryless channel*.

If one denotes by C_n the transmission rate of the channel (cf. **Transmission rate of a channel**) for a segment of length n of a homogeneous memoryless channel, then $C_n = nC_1$. If Y and $\tilde{Y}$ are finite (or countable) sets, a homogeneous memoryless channel is completely determined by the matrix of transition probabilities $\{q(y, \tilde{y}): y \in Y, \tilde{y} \in \tilde{Y}\}$, where

$$q(y, \tilde{y}) = P\{\tilde{\eta}_k = \tilde{y} \mid \eta_k = y\}, \quad k = 1, 2, \ldots.$$

For references see [1], [3] - [5] cited in **Communication channel**.

R.L. Dobrushin
V.V. Prelov

AMS 1980 Subject Classification: 62B10, 94A40

MENELAUS THEOREM - A theorem on the relations between the lengths of the segments on the sides of a triangle determined by an intersecting straight line. It asserts that if the given line intersects the sides of a triangle ABC (or their extensions) at the points C', A' and B', then

$$\frac{AC'}{BC'} \cdot \frac{BA'}{CA'} \cdot \frac{CB'}{AB'} = 1.$$

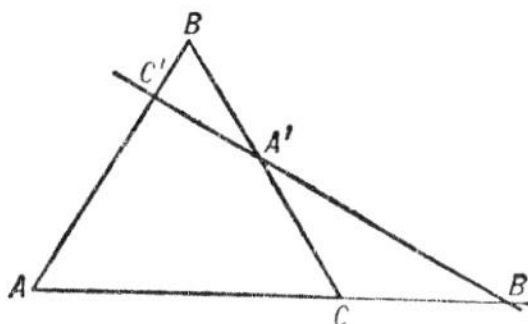

Menelaus' theorem is a particular case of the **Carnot theorem**; it can be generalized to the case of a **polygon**. Thus, suppose that a straight line l intersects the edges $A_1 A_2, \ldots, A_{n-1} A_n, A_n A_1$ of a polygon $A_1 \cdots A_n$ at the respective points $a_1, \ldots, a_n$. Then the following relation is valid:

$$\frac{A_1 a_1}{A_2 a_1} \cdots \frac{A_{n-1} a_{n-1}}{A_n a_{n-1}} \cdot \frac{A_n a_n}{A_n a_n} = 1.$$

The theorem was proved by Menelaus (first century) and apparently it was known to Euclid (third century B.C.).

P.S. Modenov

Editorial comments.

References

[A1] WAERDEN, B.L. VAN DER: *Science awakening*, 1, Noordhoff & Oxford Univ. Press, 1961, p. 275.

AMS 1980 Subject Classification: 51M15

MENGER CURVE - An example of a curve containing the topological image of any curve (and, in addition, of every one-dimensional separable metrizable space). For this reason it is referred to as a *universal curve*. It was constructed by K. Menger [1] (for Menger's construction see **Line (curve)**). The Menger curve is topologically characterized [3] as a one-

dimensional locally connected metrizable continuum K without locally separating points (i.e. for every connected neighbourhood O of any point $x \in K$ the set $O \setminus \{x\}$ is connected) and also without non-empty open subsets imbeddable in the plane.

References

[1] MENGER, K.: *Kurventheorie*, Teubner, 1932.
[2] PARKHOMENKO, A.S.: *What kind of curve is that?*, Moscow, 1954 (in Russian).
[3] ANDERSON, R.: 'One-dimensional continuous curves and a homogeneity theorem', *Ann. of Math.* **68** (1958), 1-16.

B.A. Pasynkov

AMS 1980 Subject Classification: 54B99, 54F45, 54F50

MEN'SHOV EXAMPLE OF A ZERO-SERIES - The first non-trivial example of a **trigonometric series** that converges to zero in the complement of a **perfect set** of measure zero; constructed by D.E. Men'shov [1]. A series with this property is called a *zero-series*. A problem naturally connected with this notion is that of the uniqueness of a trigonometric series of a function (see **Uniqueness set**).

References

[1] MEN'SHOV, D.E.: 'Sur l'unicité du développement trigonométrique', *C.R. Acad. Sci. Paris* **163** (1916), 433-436.
[2] BARY, N.K. [N.K. BARI]: *A treatise on trigonometric series*, Pergamon, 1964 (translated from the Russian).

M.I. Voĭtsekhovskiĭ

Editorial comments.

References

[A1] ZYGMUND, A.: *Trigonometric series*, 1-2, Cambridge Univ. Press, 1988.

AMS 1980 Subject Classification: 42A63

MEN'SHOV — RADEMACHER THEOREM - A theorem on the almost-everywhere convergence of orthogonal series: If a system of functions $\{\phi_n(t)\}_{n=1}^{\infty}$ is orthonormal on a segment $[a, b]$ and if

$$\sum_{n=1}^{\infty} a_n^2 \log^2 n < \infty,$$

then the series

$$\sum_{n=1}^{\infty} a_n \phi_n(t) \qquad (*)$$

converges almost-everywhere on $[a, b]$. This result has been proved independently by D.E. Men'shov [1] and H. Rademacher [2]. Men'shov showed also that this assertion is sharp in the following sense. If a monotone increasing sequence $\omega(n)$ satisfies the condition $\omega(n) = o(\log^2 n)$, then one can find an orthogonal series $(*)$, diverging everywhere, the coefficients of which satisfy the condition

$$\sum_{n=1}^{\infty} a_n^2 \omega(n) < \infty.$$

References

[1] MEN'SHOV, D.E.: 'Sur la séries de fonctions orthogonales (I)', *Fund. Math.* **4** (1923), 82-105.
[2] RADEMACHER, H.: 'Einige Sätze über Reihen von allgemeinen Orthogonalfunktionen', *Math. Ann.* **87** (1922), 112-138.
[3] ALEXITS, G.: *Konvergenzprobleme der Orthogonalreihen*, Deutsch. Verlag Wissenschaft., 1960.

B.I. Golubov

Editorial comments.

References

[A1] ZYGMUND, A.: *Trigonometric series*, 1-2, Cambridge Univ. Press, 1988.

AMS 1980 Subject Classification: 42C15

MERCER THEOREM - The bilinear series

$$\sum_m \frac{\phi_m(s)\overline{\phi_m(t)}}{\lambda_m}$$

of a Hermitian positive-definite continuous kernel $K(s, t)$ on $D \times D$ (cf. **Integral equation with symmetric kernel**; **Kernel of an integral operator**), where D is the closure of a bounded domain in $\mathbf{R}^n$, converges absolutely and uniformly in $D \times D$ to $K(s, t)$. Here the λ_m are the characteristic numbers of the kernel $K(s, t)$ and the $\phi_m(s)$ are the corresponding orthonormalized eigen functions. If a kernel K satisfies the conditions of Mercer's theorem, then the integral operator $T: L_2(D) \to L_2(D)$,

$$Tf(s) = \int_D K(s, t)f(t)\,dt = \sum_m \frac{1}{\lambda_m}(f, \phi_m)\phi_m$$

is nuclear (cf. **Nuclear operator**) and its **trace** $\sum_m 1/\lambda_m$ can be calculated by the formula

$$\sum_m \frac{1}{\lambda_m} = \int_D K(s, s)\,ds.$$

Mercer's theorem can be generalized to the case of a bounded discontinuous kernel.

The theorem was proved by J. Mercer [1].

References

[1] MERCER, J.: *Philos. Trans. Roy. Soc. London Ser. A* **209** (1909), 415-446.
[2] MERCER, J.: 'Functions of positive and negative type, and their connection with the theory of integral equations', *Proc. Roy. Soc. London Ser. A* **83** (1908), 69-70.
[3] PETROVSKIĬ, I.G.: *Lectures on the theory of integral equations*, Graylock, 1957 (translated from the Russian).
[4] TRICOMI, F.G.: *Integral equations*, Interscience, 1957.
[5] KRASNOSEL'SKIĬ, M.A., ET AL.: *Integral operators in spaces of summable functions*, Noordhoff, 1976 (translated from the Russian).

V.B. Korotkov

Editorial comments.

References

[A1] GOHBERG, I. and GOLDBERG, S.: *Basic operator theory*, Birkhäuser, 1977.
[A2] ZAANEN, A.C.: *Linear analysis*, North-Holland, 1956.

AMS 1980 Subject Classification: 47A70, 45B05

MERGELYAN THEOREM - A theorem on the possibility of uniform approximation of functions of one complex variable by polynomials. Let K be a compact subset of the complex z-plane $\mathbf{C}$ with a connected complement. Then every function f continuous on K and holomorphic at its interior points can be approximated uniformly on K by polynomials in z.

This theorem was proved by S.N. Mergelyan (see [1], [2]); it is the culmination of a large number of studies on approximation theory in the complex plane and has many applications in various branches of complex analysis.

In the case where K has no interior points this result was proved by M.A. Lavrent'ev [3]; the corresponding theorem in the case where K is a compact domain with a connected complement is due to M.V. Keldysh [4] (cf. also **Keldysh—Lavrent'ev theorem**).

Mergelyan's theorem has the following consequence. Let K be an arbitrary compact subset of $\mathbf{C}$. Let a function f be continuous on K and holomorphic in its interior. Then in order that f be uniformly approximable by polynomials in z it is necessary and sufficient that f admits a holomorphic extension to all bounded connected components of the set $\mathbf{C} \setminus K$.

The problem of polynomial approximation is a particular case of the problem of approximation by rational functions with poles in the complement of K. Mergelyan found also several sufficient conditions for rational approximation (see [2]). A complete solution of this problem (for compacta $K \subset \mathbf{C}$) was obtained in terms of analytic capacities (cf. **Analytic capacity**), [5].

Mergelyan's theorem touches upon a large number of papers concerning polynomial, rational and holomorphic approximation in the space $\mathbf{C}^n$ of several complex variables. Here only partial results for special types of compact subsets have been obtained up till now.

References

[1] MERGELYAN, S.N.: 'On the representation of functions by series of polynomials on closed sets', *Transl. Amer. Math. Soc.* **3** (1962), 287-293. (*Dokl. Akad. Nauk SSSR* **78**, no. 3 (1951), 405-408)
[2] MERGELYAN, S.N.: 'Uniform approximation to functions of a complex variable', *Transl. Amer. Math. Soc.* **3** (1962), 294-391. (*Uspekhi Mat. Nauk* **7**, no. 2 (1952), 31-122)
[3] LAVRENTIEFF, M.A. [M.A. LAVRENT'EV]: *Sur les fonctions d'une variable complexe représentables par des series de polynômes*, Hermann, 1936.
[4] KELDYSH, M.V.: 'Sur la réprésentation par des séries de polynômes des fonctions d'une variable complexe dans des domaines fermés', *Mat. Sb.* **16**, no. 3 (1945), 249-258.
[5] VITUSHKIN, A.G.: 'The analytic capacity of sets in problems of approximation theory', *Russian Math. Surveys* **22**, no. 6 (1967), 139-200. (*Uspekhi Mat. Nauk* **22**, no. 6 (1967), 141-199)
[6] *Some questions in approximation theory*, Moscow, 1963 (in Russian; translated from the English).
[7] GAMELIN, T.W.: *Uniform algebras*, Prentice-Hall, 1969.

E.M. Chirka

Editorial comments. Another important forerunner of Mergelyan's theorem was the *Walsh theorem*: the case where K is the closure of a *Jordan domain* (a set with boundary consisting of Jordan curves, cf. **Jordan curve**).

An interesting proof of Mergelyan's theorem, based on functional analysis, is due to L. Carlesson, see [A1].

For analogues of Mergelyan's theorem in $\mathbf{C}^n$, see [A2]. See also **Approximation of functions of a complex variable**.

References

[A1] CARLESSON, L.: 'Mergelyan's theorem on uniform polynomial approximation', *Math. Scand.* 15 (1964), 167-175.
[A2] CHIRKA, E.M. and KHENKIN, G.M.: 'Boundary properties of holomorphic functions of several complex variables', *J. Soviet Math.* 5, no. 5 (1976), 612-687. (*Itogi Nauk. i Tekhn. Sovrem. Probl. Mat.* 4 (1975), 13-142)
[A3] GAIER, D.: *Lectures on complex approximation*, Birkhäuser, 1987 (translated from the German).

AMS 1980 Subject Classification: 30E10, 32A20

MEROMORPHIC FUNCTION *of one complex variable* in a domain $\Omega \subset \mathbf{C}$ (or on a Riemann surface Ω) - A holomorphic function in a domain $\Omega \setminus \{a_1, a_2, \ldots\}$ which has at every singular point a_ν a pole (cf. **Pole (of a function)**, i.e. a_ν is an isolated point of the set $\{a_1, a_2, \ldots\}$, which has no limit points in Ω, and $\lim_{z \to a_\nu} |f(z)| = \infty$). The collection $M(\Omega)$ of all meromorphic functions in Ω is a field with respect to the usual pointwise operations followed by redefinition at the removable singularities.

The quotient ϕ/ψ of two arbitrary holomorphic functions in Ω, $\psi \not\equiv 0$, is a meromorphic function in Ω. Conversely, every meromorphic function in a domain $\Omega \subset \mathbf{C}$ (or on a non-compact Riemann surface Ω) can be expressed as $f = \phi/\psi$, $\psi \not\equiv 0$, where ϕ, ψ are holomorphic and have no common zeros in Ω. It follows that on a non-compact Riemann surface Ω the field $M(\Omega)$ coincides with the field of fractions of the ring $O(\Omega)$ of holomorphic functions in Ω.

Every meromorphic function $f \in M(\Omega)$ defines a continuous mapping $\tilde{f}$ of the domain Ω into the Riemann sphere $\mathbf{C} \cup \{\infty\}$, which is a **holomorphic mapping** relative to the standard complex structure on $\mathbf{C} \cup \{\infty\} \approx \mathbf{C}P^1$. Conversely, every holomorphic mapping $\tilde{f} : \Omega \to \mathbf{C} \cup \{\infty\}$, $\tilde{f} \not\equiv \infty$, defines a meromorphic function f in Ω: The set of poles of f coincides with the discrete set $\tilde{f}^{-1}(\infty)$ and $f(z) = \tilde{f}(z) \in \mathbf{C}$ if $z \in \Omega \setminus \tilde{f}^{-1}(\infty)$. Thus, the meromorphic functions of one variable may be identified with the holomorphic mappings ($\not\equiv \infty$) into the Riemann sphere.

The basic problems in the theory of meromorphic functions are those concerning the existence (and construction) of meromorphic functions with prescribed singularities.

I. One is given a (closed) discrete subset $\{a_1, a_2, \ldots\} \subset \Omega$ and, at each point a_ν, the principal part of a Laurent expansion (cf. **Laurent series**)

$$f_\nu(z) = \sum_{j=1}^{n_\nu} c_{\nu j}(z - a_\nu)^{-j};$$

it is required to find a meromorphic function $f \in M(\Omega)$ with these principal parts, i.e. a holomorphic function f in $\Omega \setminus \{a_1, a_2, \ldots\}$ such that $f - f_\nu$ is holomorphic in a neighbourhood of a_ν for each ν. If the number of points a_ν is finite, then (in a domain $\Omega \subset \mathbf{C}$) the problem is trivially solved by the function $\sum f_\nu$. In the general case this problem is solved by the *Mittag-Leffler theorem*: On every non-compact Riemann surface there exists a meromorphic function with given principal parts f_ν, $\nu = 1, 2, \ldots$. On a compact Riemann surface (for instance, a torus) this problem has in general no solution — supplementary conditions concerning the compatibility of the principal parts must be imposed.

The second basic problem is conveniently formulated in the language of divisors (cf. **Divisor**), i.e. of mappings $D : \Omega \to \mathbf{Z}$ such that for every compactum $K \subset \Omega$ the number of points $z \in K$ at which $D(z) \neq 0$ is finite (the number $D(z)$ is called the multiplicity of D at z). Divisors can explicitly be written as formal sums $\sum k_\nu a_\nu$, where $a_\nu \in \Omega$ are the points at which $D(a_\nu)(= k_\nu) \neq 0$; in the case of finitely many terms the number $\sum k_\nu$ ($= \deg D$) is called the *degree of the divisor D*. For a meromorphic function f its divisor (f) is equal to zero everywhere apart from the zeros and poles of f, at which the multiplicity is set equal to the order of the zero or of the pole (poles have negative orders).

II. At the points of a (closed) discrete subset $\{a_1, a_2, \ldots\} \subset \Omega$ one is given 'multiplicities' — integers $k_\nu \neq 0$. It is required to find a meromorphic function with zeros and poles of the respective multiplicities, i.e. a holomorphic function f in $\Omega \setminus \{a_1, a_2, \ldots\}$ such that $f(z)(z - a_\nu)^{-k_\nu}$ is holomorphic and does not vanish in a neighbourhood of the point a_ν, $\nu = 1, 2, \ldots$. In the case of finitely many points a_ν (and $\Omega \subset \mathbf{C}$) such a function is, for example, $f(z) = \prod(z - a_\nu)^{k_\nu}$. In the general case the problem is solved by *Weierstrass' theorem*: On a non-compact Riemann surface Ω, for every given divisor D there is a meromorphic function f with divisor (f) equal to D. For a compact Riemann surface Ω the holomorphic mapping into the Riemann sphere defined by a non-constant meromorphic function f is a branched covering, and hence the function f takes every value the same number of times; in particular, the number of zeros of f equals the number of its poles (multiplicities taken into account). Therefore, the condition $\deg(f) = 0$ is necessary in order that problem II admits a solution on a compact Riemann surface. In general, it is not sufficient; a necessary and sufficient condition for the existence of a meromorphic function with a given divisor is given by *Abel's theorem* (see [2]).

Let D be a divisor on a compact Riemann surface Ω. The functions $f \in M(\Omega)$ satisfying the condition $(f) + D \geq 0$ form a finite-dimensional linear space O_D (over $\mathbb{C}$); if $\deg D < 0$, then $O_D = \{0\}$.

The **Riemann—Roch theorem** asserts that

$$\dim O_D - \dim O_{K-D} = \deg D - g + 1,$$

where K and g are the so-called canonical divisor and, respectively, the genus of the Riemann surface Ω. From this relation one can obtain many existence theorems (if $\dim O_{K-D} + \deg D > g$, then $\dim O_D \geq 2$, and hence O_D contains non-constant meromorphic functions). For example, on every compact Riemann surface Ω of genus g there is a meromorphic function which realizes a branched covering $f : \Omega \to \mathbb{C} \cup \{\infty\}$ with at most $g + 1$ sheets.

An important place in the theory of meromorphic functions of one complex variable is occupied by **value-distribution theory** (Nevanlinna theory), which studies the distribution of the roots of the equations $f(z) = a$, $a \in \mathbb{C} \cup \{\infty\}$, when approaching the boundary of the domain.

Meromorphic functions of several complex variables. Let Ω be a domain in $\mathbb{C}^n$ (or an n-dimensional complex manifold) and let $P \subset \Omega$ be a (complex-) analytic subset of codimension one (or empty). A holomorphic function f defined on $\Omega \setminus P$ is called a meromorphic function in Ω if for every point $p \in P$ one can find an arbitrarily small neighbourhood U of p in Ω and functions ϕ, ψ holomorphic in U without common non-invertible factors in $O(U)$, such that $f = \phi / \psi$ in $U \setminus P$. The set $P = P_f$ is called the **polar set** of the meromorphic function f. Its subset $N = N_f$, defined locally by the condition $\phi = \psi = 0$, is called the *set of (points of) indeterminacy* of f; N is an analytic subset of Ω of (complex) codimension ≥ 2. At each point $p \in N$ the function f is essentially undefined: The limiting values of $f(z)$ for $z \to p$, $z \in \Omega \setminus P$, fill up the Riemann sphere $\mathbb{C} \cup \{\infty\}$. On the other hand, at the points of $P \setminus N$ the limit $\lim_{z \to p} f(z) = \infty$ exists, and upon redefining $f(p) = \infty$ if $p \in P \setminus N$, one obtains a holomorphic mapping of $\Omega \setminus N$ into the Riemann sphere. Conversely, if N is an arbitrary (possibly empty) complex-analytic subset of Ω of codimension ≥ 2, then every holomorphic mapping $f : \Omega \setminus N \to \mathbb{C} \cup \{\infty\}$ defines a meromorphic function on Ω that is equal to f on $\Omega \setminus P$, where $P = N \cup f^{-1}(\infty)$ is either an analytic subset of Ω of codimension 1 or is empty. Thus, a meromorphic function f in Ω can be defined as a holomorphic mapping into the Riemann sphere defined in the complement of an analytic subset $N_f \subset \Omega$ of codimension ≥ 2.

A third, completely localized, definition of meromorphic functions (equivalent to the one given above) is stated in the language of sheaves. Let O be the **sheaf** of germs of holomorphic functions on Ω, and for each point $z \in \Omega$ let M_z denote the field of fractions of the ring O_z (the stalk of the sheaf O over z). Then $M = \bigcup M_z$ is naturally endowed with the structure of a sheaf of fields, called the *sheaf of germs of meromorphic functions* in Ω. A meromorphic function in Ω is defined as a global section of M, i.e. a continuous mapping $f : z \to f_z$ such that $f_z \in M_z$ for all $z \in \Omega$. The sets P_f and N_f are defined as follows: If $f_z = \phi_z / \psi_z$, $\phi_z, \psi_z \in O_z$, $\psi_z \neq 0$, then one may assume that ϕ_z and ψ_z are mutually prime, i.e. they have no common non-invertible factors in O_z; then $z \in P_f$ if $\psi_z(z) = 0$, while $z \in N$ if $\psi_z(z) = \phi_z(z) = 0$. The value at a point $z \notin P$ of the meromorphic function f thus defined is $\phi_z(z) / \psi_z(z)$.

As in the one-dimensional case, the collection of all meromorphic functions in Ω forms a field $M(\Omega)$ with respect to the pointwise algebraic operations with a subsequent redefinition at the removable singularities.

The closure $Z = Z_f$ of the zero set of a meromorphic function f, i.e. of the set $\{z \in \Omega \setminus P_f : f(z) = 0\}$, is an analytic subset of Ω of codimension one (or empty); the set of indeterminacy is $N = Z \cap P$. On Z_f and P_f one can define the order (multiplicity) of the zeros (or poles) of the meromorphic function f. If p is a regular point of the analytic set $Z \cup P$, then in some neighbourhood U of p the set $(Z \cup P) \cap U$ is connected and is given by an equation $g = 0$, $g \in O(U)$, where $dg \neq 0$ throughout U. Hence there is a maximal integer $k(p)$ such that the function $fg^{-k(p)}$ admits a holomorphic extension to U; this number is called the *order (of the zero p if $p \in Z$, and of the pole p if $p \in N$)* of the meromorphic function f at the point p. The function $k(p)$ is locally constant on the set of regular points of $Z \cup P$. Therefore one can attach to each meromorphic function in Ω its divisor $(f) = \sum k_\nu A_\nu$, where A_ν are the irreducible components of $Z_f \cup P_f$ and k_ν is the multiplicity (order) of f at the regular points of $Z \cup P$ that belong to A_ν (alternative notations: $(f) = D_f = \Delta_f$, etc.). On a compact complex manifold a meromorphic function is uniquely defined by its divisor, up to a multiplicative constant.

The problems solved in the one-dimensional case by the Mittag-Leffler and Weierstrass theorems are known in the higher-dimensional case as the first (additive) and the second (multiplicative) **Cousin problems**. Due to the complicated structure of the polar set P_f, the notion of a principal part of a meromorphic function is not defined in general, and accordingly the Cousin problems are formulated as follows.

I. Suppose that an open covering $\{U_\nu\}$ of the manifold Ω and in each U_ν a meromorphic function f_ν are given; it is required to find a meromorphic function $f \in M(\Omega)$ such that $f - f_\nu \in O(U_\nu)$ for all ν.

II. For a given divisor $D = \sum k_\nu A_\nu$ on Ω, find a meromorphic function $f \in M(\Omega)$ such that $(f) = D$.

The conditions of solvability of these problems in the higher-dimensional case are considerably more stringent than in the one-dimensional case.

The problem of representing a meromorphic function as a quotient of two holomorphic functions is called the *Poincaré problem*. The *strong Poincaré problem* is to represent a meromorphic function as a quotient of holomorphic functions the germs of which at each point $z \in \Omega$ are mutually prime in O_z. The Poincaré problem is unsolvable on a compact connected complex manifold if there are non-constant meromorphic functions on it. However, this problem is solvable in every domain $\Omega \subset \mathbf{C}^n$ and, in fact, in an arbitrary domain on a **Stein manifold** (see [7]). The solvability of the strong Poincaré problem follows from that of the Cousin II problem (the converse is not true).

Functions $f_1, \ldots, f_k \in M(\Omega)$ are said to be *algebraically dependent* if there is a polynomial $F \not\equiv 0$ in k variables with complex coefficients such that $F(f_1(z), \ldots, f_k(z)) \equiv 0$ in the common domain of definition of the functions f_ν. The maximal number of algebraically-independent meromorphic functions on Ω is called the *transcendence degree* of the field $M(\Omega)$. On a compact complex manifold this number does not exceed the (complex) dimension of the manifold (*Siegel's theorem*); furthermore, the field $M(\Omega)$ has a finite number of generators (see [6]).

On concrete complex manifolds, meromorphic functions may have supplementary properties. For instance, in the complex projective space $\mathbf{C}P^n$ the set of indeterminacy of any non-constant meromorphic function is not empty. Every meromorphic function on a projective algebraic variety is rational, i.e. is expressible as a quotient p/q of homogeneous polynomials in homogeneous coordinates. On algebraic varieties the field $M(\Omega)$ is quite rich. On the other hand, there exist complex manifolds (for example, some non-algebraic tori) on which every meromorphic function is constant. Higher-dimensional generalizations of the Riemann$-$Roch theorem are less effective, and existence theorems for various classes of meromorphic functions can only be obtained for some classes of complex manifolds.

See also **Weierstrass theorem**; **Mittag-Leffler theorem**; **Riemann$-$Roch theorem**.

References

[1] NEVANLINNA, R.: *Analytic functions*, Springer, 1970 (translated from the German).
[2] FORSTER, O.: *Lectures on Riemann surfaces*, Springer, 1981 (translated from the German).
[3] HAYMAN, W.: *Meromorphic functions*, Clarendon Press, 1964.
[4] SHABAT, B.V.: *Introduction to complex analysis*, 1-2, Moscow, 1976 (in Russian).
[5] HÖRMANDER, L.: *An introduction to complex analysis in several variables*, North-Holland, 1973.
[6] SHAFAREVICH, I.R.: *Basic algebraic geometry*, Springer, 1977 (translated from the Russian).
[7] KAJIWARA, J. and SAKAI, E.: 'Generalization of Levi$-$Oka's theorem concerning meromorphic functions', *Nagoya Math. J.* **29** (1967), 75-84.

E.M. Chirka

Editorial comments. *Abel's theorem on meromorphic functions* states the following. Let S be a compact Riemann surface. A necessary and sufficient condition for a differential D to be the divisor of a meromorphic function is that there is a singular 1-chain γ (cf. **Integration on manifolds**) of which D is the boundary, $\partial \gamma = D$, such that $\int_\gamma \phi = 0$ for all differentials of the first kind on S (cf. **Abelian differential** and **Integration on manifolds**). Cf. also **Jacobi variety** for another formulation of Abel's theorem.

References

[A1] SPRINGER, G.: *Introduction to Riemann surfaces*, Addison-Wesley, 1957, Sect. 10-7.
[A2] FARKAS, H.M. and KRA, I.: *Riemann surfaces*, Springer, 1980, Sect. III.6.

AMS 1980 Subject Classification: 30D30, 32A20

MEROMORPHIC MAPPING *of complex spaces* - A generalization of the notion of a **meromorphic function**. Let X and Y be complex spaces (cf. **Complex space**), let A be an open subset of X such that $X \setminus A$ is a nowhere-dense analytic subset (cf. **Analytic set**) and suppose that an **analytic mapping** $f: A \to Y$ has been given. Then f is called a *meromorphic mapping* of X into Y if the closure Γ_f of the graph A^* of f in $X \times Y$ is an analytic subset of $X \times Y$ and if the projection $\pi: \Gamma_f \to X$ is a proper mapping (cf. also **Proper morphism**). The set Γ_f is called the *graph of the meromorphic mapping f*. The mapping $\pi: \Gamma_f \to X$ is surjective and defines a bijective mapping of the set of irreducible components. If $A_0^f \subset X$ denotes the largest open subset to which f can be extended as an analytic mapping, then $I_f = X \setminus A_0^f$ is a nowhere-dense analytic subset of X, called the *set of indeterminacy* of f. The set $\pi^{-1}(A_0^f) = A_0^{f*}$ is open and dense in Γ_f; also, $A^* \subseteq A_0^{f*}$ and $\Gamma_f \setminus A_0^{f*}$ is analytic and nowhere dense in Γ_f. The restriction $\pi: A_0^{f*} \to A_0^f$ is an isomorphism of analytic spaces. If X is a normal complex space (cf. **Normal analytic space**), then $\mathrm{codim}\, I_f \geq 2$ and $\dim_z \pi^{-1}(x) > 0$ if and only if $z \in \pi^{-1}(x)$ and $x \in I_f$. If X is not normal, $\pi^{-1}(x)$ may consist of a finite number of points, even if $x \in I_f$. In the case $Y = \mathbf{C}P^1$ the notion of a meromorphic mapping reduces to that of a meromorphic function.

Let $f: X \to Y$, $g: Y \to Z$, $k: X \to Z$ be meromorphic mappings of complex spaces. One says that the composite $g \circ f$ of the mappings f and g is defined and equals k if there is an open dense subset U of X such that $U \subseteq A_0^f$, $f(U) \subset A_0^g$, $U \subset A_0^k$, and $k|_U = g \circ f|_U$. A meromorphic mapping $f: X \to Y$ is called *bimeromorphic* if there is a meromorphic mapping $g: Y \to X$ such that $f \circ g = 1|_Y$ and $g \circ f = 1|_X$. Composition of two bimeromorphic mappings $X \to Y$ and $Y \to Z$ is always defined.

References

[1] ANDREOTTI, A. and STOLL, W.: *Analytic and algebraic dependence of meromorphic functions*, Springer, 1971.
[2] REMMERT, R.: 'Holomorphe und meromorphe Abbildungen komplexer Räume', *Math. Ann.* **133**, no. 3 (1957), 328-370.

D.A. Ponomarev

Editorial comments.

References

[A1] WHITNEY, H.: *Complex analytic varieties*, Addison-Wesley, 1972, Sect. 6.3.

AMS 1980 Subject Classification: 32HXX, 32A20

MERSENNE NUMBER - A prime number of the form $M_n = 2^n - 1$, where $n = 1, 2, \ldots$. Mersenne numbers were considered in the seventeenth century by M. Mersenne. The numbers M_n can be prime only for prime values of n. For $n = 2, 3, 5, 7$ one obtains the prime numbers $M_n = 3, 7, 31, 127$. However, for $n = 11$ the number M_n is composite. For prime values of n larger than 11, among the M_n one encounters both prime and composite numbers. The fast growth of the numbers M_n makes their study difficult. By considering concrete numbers M_n it has been shown, for example, that M_{31} (L. Euler, 1750) and M_{61} (I.M. Pervushin, 1883) are Mersenne numbers. Computers were used to find other very large Mersenne numbers, among them M_{11213}. The existence of an infinite set of Mersenne numbers is still an open problem (1989). This problem is closely related with the problem on the existence of perfect numbers.

References

[1] HASSE, H.: *Vorlesungen über Zahlentheorie*, Springer, 1950.
[2] BUKHSHTAB, A.A.: *Number theory,* Moscow, 1966 (in Russian).

B.M. Bredikhin

Editorial comments. Presently (1989) it is known that for the following n the Mersenne number M_n is prime: 2, 3, 5, 7, 13, 17, 19, 31, 61, 89, 107, 127, 521, 607, 1279, 2203, 2281, 3217, 4253, 4423, 9689, 9941, 11213, 19937, 21701, 23209, 44497, 86243, 132049, 216091. See [A1].

The *Lucas test* provides a very simple method to establish primality of these numbers. This test consists of the following (cf. [A2]). Define $S_1 = 4$ and $S_{k+1} = S_k^2 - 2$ for $k \geq 1$. Then M_n is prime if and only if M_n divides S_{n-1} (and n is a prime number).

References

[A1] RIESEL, H.: *Prime numbers and computer methods for factorisation*, Birkhäser, 1986.
[A2] SHANKS, D.: *Solved and unsolved problems in number theory*, Chelsea, reprint, 1978.

AMS 1980 Subject Classification: 10A40

META-ABELIAN GROUP, *metabelian group* - A **solvable group** of derived length two, i.e. a group whose **commutator subgroup** is Abelian. The family of all metabelian groups is a variety (see **Variety of groups**) defined by the identity

$$[[x, y], [z, t]] = 1.$$

Special interest is attached to finitely-generated metabelian groups. These are all residually finite (see **Residually-finite group**) and satisfy the maximum condition (see **Chain condition**) for normal subgroups. An analogous property is shared by a generalization of these groups — the finitely-generated groups for which the quotient by an Abelian normal subgroup is polycyclic (see **Polycyclic group**).

In the Russian mathematical literature, by a metabelian group one sometimes means a **nilpotent group** of nilpotency class 2.

References

[1] KUROSH, A.G.: *The theory of groups*, 1-2, Chelsea, 1955-1956 (translated from the Russian).
[2] KARGAPOLOV, M.I. and MERZLYAKOV, YU.I.: *Fundamentals of the theory of groups*, Springer, 1979 (translated from the Russian).

A.L. Shmel'kin

Editorial comments.

References

[A1] ROBINSON, D.J.S.: *A course in the theory of groups*, Springer, 1980.

AMS 1980 Subject Classification: 20D10, 20F16

META-LANGUAGE - A logico-mathematical language used to create a **meta-theory**. In a wider sense, a meta-language is a non-formalized language that is used to formulate statements of **meta-mathematics**.

A.G. Dragalin

Editorial comments.

References

[A1] GRZEGORCZYK, A.: *An outline of mathematical logic*, Reidel, 1974.

AMS 1980 Subject Classification: 03-XX, 03BXX

META-LOGIC - A logic used in the discussion of a formal axiomatic theory within the frame of some **meta-theory**. In the foundations of mathematics one often imposes specific demands on the meta-theory, related to the rejection of some common mathematical abstractions, with the aim of improving the philosophical acceptability of the meta-theory. Examples of such abstractions subject to criticism are the **abstraction of actual infinity**, the abstraction of reasoning corresponding to the appearance of antinomies (cf. **Antinomy**), etc. This, in general, leads to the use of a logic different from the classical one, e.g. **modal logic**, or intuitionistic logic if the meta-theory is built within the frame of **intuitionism**.

On the other hand, in **proof theory** intuitionistic and other non-classical logical theories are studied by traditional mathematical means without any restrictions, e.g. by means of set theory. In this case, classical logic acts as the meta-logic.

A.G. Dragalin

AMS 1980 Subject Classification: 03-XX, 03BXX

META-MATHEMATICS - The totality of mathematical theories used in the study of formal theories (calculi). The meta-mathematics related to the study of a given formal theory forms the so-called **meta-theory** of this formal theory.

In a narrow sense, the term 'meta-mathematics' is also used as a synonym for **proof theory**.

A.G. Dragalin

Editorial comments.

References

[A1] RASIOWA, H. and SIKORSKY, R.: *The mathematics of metamathematics*, PWN, 1963.

AMS 1980 Subject Classification: 03-XX, 03BXX

META-THEOREM - A statement about a formal axiomatic theory, obtained within a definite **meta-theory**.

A.G. Dragalin

Editorial comments. In category theory (cf. **Category**), the term 'meta-theorem' has acquired a more specific meaning: it refers to an assertion of the form 'if P is any statement (expressed in an appropriate formal language) about categories of a given type (e.g. Abelian categories or regular categories), then the validity of P in some particular category (e.g. the category of Abelian groups or the category of sets) implies its validity in all categories of the given type'. Results of this kind are generally deduced from imbedding theorems which assert that any category of the given type can be imbedded (in a structure-preserving way) into (a power of) the particular category under consideration.

References

[A1] FREYD, P.: *Abelian categories*, Harper & Row, 1964.
[A2] FREYD, P. and SCEDROV, A.: *Geometric logic*, North-Holland, 1989.

AMS 1980 Subject Classification: 03-XX, 03BXX, 18B15

META-THEORY - The totality of mathematical methods and means intended for the description and definition of some formal axiomatic theory, as well as for the investigation of its means. A meta-theory is an important component of the formalization method, which is one of the most central methods in mathematical logic.

The essence of this method may briefly be described as follows. Suppose that one is interested in some meaningful mathematical theory T_1. This may be a complicated theory, the **semantics** of which is intuitively not sufficiently clear (e.g. it may be set theory, mathematical analysis, second-order arithmetic, etc.). One is interested in whether T_1 is non-contradictory or whether some mathematical principle (e.g. the **axiom of choice**) is compatible with T_1. In order to clarify this question one constructs at first a precise logico-mathematical language Ω such that all statements in T_1 of interest can be described by formulas in Ω. Then the logical principles that are used in the theory to derive new facts are formalized as axioms and derivation rules; these are used to derive new formulas in the language Ω from the axioms and formulas already derived. Thus, a *formal system* (or, in other words, a *formal axiomatic theory*, a *calculus*) $\tilde{T}_1$ arises that exactly describes some fragment of the meaningful theory T_1 of interest. It is essential here that in stating $\tilde{T}_1$ one does not use an exhaustive penetration into the, possibly very complicated, semantics of $\tilde{T}_1$. The calculus $\tilde{T}_1$ is constructed by simple laws as a purely symbolic system and for understanding its operation one need not penetrate into the sense of the formulas that are derived in it.

Such an approach opens, first, the possibility of mathematically formulating the problems of interest, related to the derivability of certain formulas in $\tilde{T}_1$, and, secondly, of investigating $\tilde{T}_1$ by means of some meaningful theory T_2. In this case, $\tilde{T}_1$ is called the *object theory*, and T_2 its *meta-theory*.

From the point of view of the foundations of mathematics it is important that T_2 is related to more reliable theories than T_1, so that investigating $\tilde{T}_1$ by means of T_2 can be regarded as a real clarification and justification of the unclear part of the semantics of T_1 using the more convincing theory T_2. In this connection special preference has been given to sufficiently reliable meta-theories reflecting finite parts of mathematics, as well as to theorems constructed within the frame of intuitionism or constructive mathematics. Moreover, outside the foundations of mathematics this restriction is superfluous. If one is interested not in the question of intuitive clarity of T_1 but only in the simple fact of derivability or non-derivability of certain formulas in $\tilde{T}_1$, it is natural to investigate $\tilde{T}_1$ by means of any previously established mathematical theory T_2 that is convincing for the investigator, without imposing a priori restrictions.

One may further analogously investigate the meta-theory T_2, by constructing a formal system $\tilde{T}_2$ and by studying $\tilde{T}_2$ by means of some *meta-meta-theory* T_3. The investigations in **proof theory** are of this nature.

References

[1] KLEENE, S.C.: *Introduction to metamathematics*, North-Holland, 1951.

A.G. Dragalin

Editorial comments.

References

[A1] RASIOWA, H. and SIKORSKY, R.: *The mathematics of metamathematics*, PWN, 1963.

[A2] SUPPES, P.: *Introduction to logic*, v. Nostrand, 1957.

AMS 1980 Subject Classification: 03-XX, 03BXX

METACYCLIC GROUP - A group having a cyclic **normal subgroup** such that the quotient group by this normal subgroup is also cyclic (cf. **Cyclic group**). Every finite group of square-free order (i.e. the order is not divisible by the square of a natural number) is metacyclic. Polycyclic groups (cf. **Polycyclic group**) are a generalization of metacyclic groups.

A.L. Shmel'kin

Editorial comments. Sometimes, the term metacyclic is reserved for the more special class of groups whose derived group and derived quotient group are both cyclic.

References

[A1] HALL, M., JR.: *The theory of groups*, Macmillan, 1959.

AMS 1980 Subject Classification: 20E99

METEOROLOGY, MATHEMATICAL PROBLEMS IN - Problems in the area of the physics, the chemistry and the biology of the atmosphere, solved with the aid of mathematical methods. The majority of mathematical problems in meteorology are characterized by their complexity and the huge volume of information processing that is involved. Therefore for the solution of these problems, alongside analytic methods, numerical simulation on computers is widely used.

Mathematical problems of the physics of the atmosphere are primarily problems in the hydro-thermodynamics of a stratified fluid with distinctive features, created by the rotation, inhomogeneities of the relief and the surface temperature of the Earth. The theoretical basis of the mathematical models are the laws of conservation of mass, momentum and energy, which, together with the laws of thermodynamics and chemistry, describe the processes occurring in the atmosphere and the interactions of the atmosphere with the oceans and the Earth's surface. In mathematical terms this is a system of non-linear partial differential equations, which must be solved under the assumption that the external source of energy is the Sun. Besides the functions describing the state and behaviour of the atmosphere (temperature, pressure, density, wind velo-

cities, etc.), these equations contain a number of parameters. By parameters one usually means the coefficients of the equations, and in non-stationary problems — the initial values of the functions, the state of the atmosphere, the characteristics of the Earth's surface, the external sources, etc. The initial conditions are determined as a result of measurements in the real physical system 'atmosphere−Earth'. The process of mathematical modelling consists of several stages: a qualitative analysis of the mathematical model (well-posedness of the problem, its solvability in physically reasonable functions, etc.), the construction of its discrete analogue, the development of computational algorithms and programs for the computer realization of the discrete models, an analysis of the sensitivity of the model to variations of the parameters, an estimate of the parameters based on a priori information and measurements, etc. The structure of the mathematical model depends on the space-time scales of the processes being studied in the atmosphere and the method of describing them.

Numerical simulation is one of the fundamental means for the solution of problems in weather forecasting and the theory of the climate. Of particularly great significance is the problem of numerical simulation in the study of climatic variations under the influence of natural and antropogenic factors and under an estimate of the influence of the activity of man on the environment. The selection and justification of the physical models for a given class of problems is closely connected with research into the fundamental questions of stability, predictability and sensitivity of the physical system consisting of the atmosphere, the oceans, the snow cover, the continents, and the biosphere, which is usually called the climatic system. Predictability determines the possibility of a sufficiently deterministic approach to the prediction of physical processes and, at the same time, determines the possibility of constructing mathematical models for the description and prediction of the behaviour of the climatic system, or part of it. Sensitivity characterizes the degree of sensitivity of the system with respect to variations of the external influences and the internal parameters. If the influence of anthropogenic factors is interpreted as perturbations to the system, then an estimate of their influence can be considered as one of the applied aspects of sensitivity theory.

The problem of short-term weather forecasting (from several hours up to several days) is to find a non-stationary solution of the system of non-linear differential equations of the hydro-thermodynamics of the atmosphere with given boundary values and initial conditions. In problems of long-term weather forecasting certain generalized, or averaged, characteristics of the behaviour of the atmosphere are determined. Numerical experiments of the general circulation of the atmosphere consist of integrating the corresponding equations over a long period of time under idealized initial conditions. Finding a stationary solution, or a solution with an annual period, is a numerical experiment with a climatic background.

Atmospheric processes have a wave structure. The construction of different types of atmospheric waves is accomplished by methods of non-linear mechanics using asymptotic series expansions in powers of a small parameter. The mathematical theory of atmospheric waves for linearized models is well developed. Among the solutions of the hydrodynamic equations several classes of waves are isolated (acoustic, gravitational, Rossby waves). Among the non-linear waves only individual examples have been studied (Gerstner gravitational waves, Rossby non-linear waves, and others). To clarify the structure of atmospheric motions, numerical solution methods are widely used to solve eigen value and eigen function problems for the hydrothermodynamic operators and their discrete analogues. The mathematical problem of atmospheric acoustics is the study of propagation of waves in stratified media. Here analytic (the WKB-method, etc.) and numerical methods are used. A major role, particularly in connection with the study and simulation of the dynamics of cyclones and fronts in the atmosphere, is taken by the research into hydrodynamic instability of atmospheric waves and the interaction of waves of various scales.

In atmospheric optics specific inverse problems in the class of conditionally well-posed problem arise. A typical example is the problem of recovering the parameters of the atmosphere relative to data from remote sounding by satellites. The process of multiple scattering of light in the atmosphere has been investigated by various asymptotic and numerical methods. For the solution of the radiative-transfer equation in the atmosphere, numerical methods are used. Specially effective is the **Monte-Carlo method**.

Many theoretical and applied problems of meteorology are connected with the problem of atmospheric turbulence. The spectrum of scales of turbulence in the atmosphere is extremely broad. Turbulence plays a determining role in the interaction of the atmosphere with the ocean and with the Earth's surface, in the diffusion of atmospheric impurities, in bumps of airplanes and other aircrafts, in the vibration of buildings under the pressure of the wind, in fluctuations of light and radio signals from terrestrial and cosmic sources, etc. There are several approaches to the creation of mathematical models of turbulence and to methods for parametrizing it.

In solving meteorological problems, a number of problems arise which are typical of the complicated problems of mathematical physics. First of all, in the preparation of the input data (objective analysis, statistical processing of **time series** on a network of measurements, space-time analysis and the compatibility of meteorological fields), and also the use of methods of sensitivity theory and optimization to identify the parameters of the model with respect to the measurement data. Mathematical models in meteorology have a large number of degrees of freedom, and therefore the problem of reduction arises (for example, by parametrization or by taking account of the use of informative generalized variables).

Related to the mathematical problems in meteorology are problems connected with the study and estimation of the influence of human activity on the atmosphere. These are problems of simulating the microclimate of towns and industrial areas with due regard for anthropogenic factors, of estimating the pollution of the atmosphere by industrial waste, of estimating the influence of variations in the Earth's surface on the dynamics and thermal behaviour of the atmosphere, etc. The problem of choosing an effective economic policy is formalized by methods of optimization theory, applied to meteorological problems. In particular, the question of optimal siting of economic complexes with due regard for sanitarily permissible norms of environmental pollution can be formalized mathematically as a constrained variational problem.

A complex of mathematical problems, including practically all those listed above, arises in connection with the problem of monitoring or realizing a procedure for tracking atmospheric processes on local and global scales.

References

[1] BELOV, P.N.: *Practical methods of numerical weather forecasting*, Leningrad, 1967 (in Russian).
[2] BLINOVA, E.N.: 'A hydrodynamic theory of pressure and temperature waves and centers of atmospheric action', *Dokl. Akad. Nauk SSSR* **39**, no. 7 (1943), 284-287 (in Russian).
[3] GANDIN, L.S.: *Objective analysis of meteorological fields*, Leningrad, 1963 (in Russian).
[4] GOLITSYN, G.S.: *Introduction to the dynamics of planetary atmospheres*, Leningrad, 1973 (in Russian).
[5] KIBEL', I.A.: *Introduction to the hydrodynamical methods of short-term weather forecasting*, Moscow, 1957 (in Russian).
[6] KONDRAT'EV, K.YA.: *Actinometry*, Leningrad, 1965 (in Russian).
[7] LAĬKHTMAN, D.L.: *The physics of a boundary layer in the atmosphere*, Leningrad, 1970 (in Russian).
[8] LORENZ, E.N.: *The nature and theory of the general circulation of the atmosphere*, Techn. Papers 115, WMO, 1967.
[9] MALKEVICH, M.S.: *Optimal research of the atmosphere by satellites*, Moscow, 1973 (in Russian).
[10] MARCHUK, G.I.: *Numerical methods in weather forecasting*, Leningrad, 1967 (in Russian).
[11] MATVEEV, L.T.: *The foundations of general meteorology. The physics of the atmosphere*, Leningrad, 1965 (in Russian).
[12] MONIN, A.S. and YAGLOM, A.M.: *Statistical hydromechanics*,

1-2, Moscow, 1965-1967 (in Russian).

[13] *Non-linear systems of hydrodynamic type*, Moscow, 1974 (in Russian).
[14] THOMPSON, P.D.: *Numerical weather analysis and prediction*, Macmillan, 1961.
[15] ECKART, C.: *Hydrodynamics of the ocean and atmosphere*, Moscow, 1962 (in Russian; translated from the English).
[16] YUDIN, M.I.: *New methods and problems in short-term weather forecasting*, Leningrad, 1963 (in Russian).

G.I. Marchuk

Editorial comments. The development of numerical schemes for handling a large amount of information and for solving large systems of coupled differential equations forms the main contribution of mathematics to meteorology. It started with a numerical integration procedure by L.F. Richardson [A9] and has come to its present development at national and international institutes like the National Centre for Atmospheric Research at Boulder (Colorado, USA) and the European Centre for Medium-Range Weather Forecasts at Reading (UK), see [A3]. Recently, methods have been developed to make the data of observations consistent with the model for which the data is meant [A1]. From meteorology mathematicians have learned about an essentially non-linear phenomenon that may occur in systems of three of more coupled ordinary differential equations. That is the possible presence of a **strange attractor**, which was first observed by E.N. Lorenz [A7] in a heavily truncated spectral model of the local vertical circulation of the atmosphere. Spectral models of the global circulation have since then been studied intensively, see [A4] and [A2] for a survey. The general theory of the physics of the atmosphere is found in [A5] and [A6], while the introduction of J. Pedlosky [A8] is more mathematically oriented. The references [A10] and [A11] deal with boundary-layer phenomena and turbulence in relation with meteorological problems.

References

[A1] BENGTSSON, L., GHIL, M. and KÄLLÉN, E. (EDS.): *Dynamic meteorology: data assimilation methods*, Springer, 1981.
[A2] DE SWART, H.E.: 'Low-order spectral models of the atmospheric circulation. A survey', *Acta Applic. Math.* 11 (1988), 49-96.
[A3] *Seminar: Numerical methods for weather prediction*, European Centre for Medium-Range Weather forecasts, Berkshire, Reading, 1984.
[A4] GHIL, M. and CHILDRESS, S.: *Topics in geophysical fluid dynamics: Atmospheric dynamics, dynamo theory and climate dynamics*, Springer, 1987.
[A5] GILL, A.: *Atmosphere-ocean dynamics*, Acad. Press, 1982.
[A6] HOLTON, J.R.: *An introduction to dynamic meteorology*, Acad. Press, 1979.
[A7] LORENZ, E.N.: 'Deterministic nonperiodic flow', *J. Atmos. Sci.* 20 (1963), 130-141.
[A8] PEDLOSKY, J.: *Geographical fluid dynamics*, Springer, 1979.
[A9] RICHARDSON, L.F.: *Weather prediction by numerical process*, Cambridge Univ. Press, 1922.
[A10] STULL, R.B.: *Introduction to boundary layer meteorology*, Kluwer, 1988.
[A11] TENNEKES, H. and LUMLEY, J.L.: *A first course in turbulence*, M.I.T., 1972.
[A12] HALTINER, G.J. and WILLIAMS, R.J.: *Numerical weather prediction*, Wiley, 1980.

AMS 1980 Subject Classification: 86A10

METHOD OF BOUNDARY INTEGRATION, *method of contour integration* - An important method in the geometric theory of functions of a complex variable, enabling one to obtain various inequalities which express extremal properties of univalent and multivalent functions, as well as identities between the mapping functions of a domain (fundamental domain functions) in the theory of conformal mapping. The method makes essential use of properties of functions conformally mapping a given domain onto various canonical domains. By means of such mappings one can construct domain functions possessing the following *boundary property*: On each boundary component of the domain, the values of the function differ by an additive constant from the complex conjugate values corresponding to another such function. The method of boundary integration consists basically of the following.

Some integral is considered, taken over the entire boundary of a given domain (the boundary can generally be taken to consist of a finite number of simple closed analytic curves). This integral is chosen so that the integrand contains factors with the boundary property indicated above and such that, after using this property, an integral is obtained that can be calculated by means of the residue theorem (cf. **Contour integration, method of; Cauchy integral theorem**). If from other considerations either the value of the original integral is known or its sign, then one obtains as a result some relation between the functions used or some inequality connecting them. Often the boundary integral to which the above method can be applied appears as the result of a transformation according to Green's formula of a non-negative double integral, namely, the integral of the square of the modulus of the derivative of some function that is regular in the given domain. Hence the connection between the method of boundary integration and the **area method**. By means of the method of boundary integration, results have been obtained concerning: **distortion theorems** for univalent conformal mappings between multiply-connected domains (see [1], [2]); necessary and sufficient conditions on the coefficients of univalent functions (see [3]); and identities relating the fundamental domain functions in the theory of conformal mapping (see [4]).

The method of boundary integration is also employed in the study of univalent functions in the following form. Suppose, for example, that B is a domain in the w-plane with boundary C consisting of a finite number of simple closed analytic curves; suppose that $S(w)$ is a function that is harmonic in the entire w-plane except for a finite number of points of B; and suppose that $p(w)$ is a function with the following property: The difference $S(w)-p(w)$ is harmonic in the domain B, continuous in the closed domain, and $p(w)|_C = 0$. Then

$$\int_C S \frac{\partial p}{\partial n} \, ds \leq 0,$$

where $\partial / \partial n$ denotes differentiation along the outward normal to B. If $\sigma(w)$ and $q(w)$ are analytic functions for which $S = \operatorname{Re} \sigma$, $p = \operatorname{Re} q$, then the above inequality can be written in the form

$$\operatorname{Re}\left\{ \frac{1}{i} \int_C (\sigma - q) \sigma' \, dw \right\} \leq 0.$$

The integral in this inequality can be calculated by the residue theorem. By choosing various functions $S(w)$ and $p(w)$ that are suitably related to the functions under investigation, one can thus obtain various new inequalities for univalent functions (see [5] - [7]).

The method of boundary integration is also successfully employed in the study of non-univalent conformal mappings. Thus, there have been established by this method a number of new extremal properties of functions that are meromorphic in a multiply-connected domain and satisfy certain additional properties (see [8]); and generalizations have been obtained for multiply-connected domains, for the case of several poles and for functions that are p-valent in an appropriate generalized sense, of Goluzin's area theorem for functions that are p-valent in the disc (see [9]). The domain functions referred to above are closely related to Bergman kernel functions (cf. **Bergman kernel function**), and results obtained by the method of boundary integration are often expressed in terms of them. Hence there is also a connection between the method of boundary integration and the theory of orthonormal systems of analytic functions.

References

[1] GRUNSKY, H.: 'Neue Abschätzungen zur konformen Abbildung ein- und mehrfachzusammenhängender Bereiche', *Schriften Math. Sem. Inst. Angew. Math. Univ. Berlin* (1932), 95-140.

[2] GOLUZIN, G.M.: 'On distortion theorems for conformal mappings of multiply-connected domains', *Mat. Sb.* 2, no. 1 (1937), 37-63 (in Russian).

[3] GRUNSKY, H.: 'Koeffizientenbedingungen für schlicht abbildende meromorphe Funktionen', *Math. Z.* 45 (1939), 29-61.

[4] GARABEDIAN, P.R. and SCHIFFER, M.: 'Identities in the theory of conformal mapping', *Trans. Amer. Math. Soc.* 65, no. 2 (1949), 187-238.

[5] NEHARI, Z.: 'Some inequalities in the theory of functions', *Trans. Amer. Math. Soc.* 75, no. 2 (1953), 256-286.

[6] ALENITSYN, YU.E.: 'On univalent functions in multiply-connected domains', *Mat. Sb.* 39, no. 3 (1956) (in Russian).

[7] ALENICYN, YU.E. [YU.E. ALENITSYN]: 'Univalent functions without common values in a multiply connected domain', *Proc. Steklov Inst. Math.* 94 (1968), 1-18. (*Trudy Mat. Inst. Steklov.* 94 (1968), 4-18)

[8] MESCHKOWSKI, H.: 'Beiträge zur Theorie der Orthonomalsysteme', *Mat. Ann.* 127 (1954), 107-129.

[9] ALENICYN, YU.E. [YU.E. ALENITSYN]: 'Area theorems for functions analytic in a finitely connected domain', *Math. USSR Izv.* 7, no. 5 (1973), 1130-1151. (*Izv. Akad. Nauk SSSR Ser. Mat.* 37, no. 5 (1973), 1132-1154)

[10] GOLUZIN, G.M.: *Geometric theory of functions of a complex variable*, Amer. Math. Soc., 1969 (translated from the Russian).

Yu.E. Alenitsyn

Editorial comments. The *mapping functions* of a domain, or *fundamental domain functions*, are univalent conformal mappings of the given domain onto certain standard domains, e.g. from a simply-connected domain onto the unit disc (cf. **Riemann theorem**) or from a doubly-connected domain onto a standard annulus, [10].

AMS 1980 Subject Classification: 30CXX

METHOD OF CHARACTERISTICS - A method for the numerical integration of equations of hyperbolic type. In the domain of hyperbolicity there is a linear combination of initial equations in which there occur only interior derivatives along characteristic surfaces. Then the equations to be solved simplify substantially. In the method of characteristics the solution is computed on a characteristic grid, which is constructed in the process of calculation, and so the domain of dependency of the solution can be determined exactly. For the method of characteristics the existence of a solution and its convergence have been proved. The widest application of the method of characteristics is in the solution of problems of the mechanics of continuous media (see [1]). For example, the equations in characteristic form

$$\frac{\partial J}{\partial t} - \frac{p}{\rho^2} \frac{\partial \rho}{\partial t} + u \frac{\partial J}{\partial t} - \frac{p}{\rho^2} \frac{\partial \rho}{\partial x} = 0, \tag{1}$$

$$\frac{\partial p}{\partial t} + (u \pm c) \frac{\partial p}{\partial x} \pm \rho c \left[\frac{\partial u}{\partial t} + (u \pm c) \frac{\partial u}{\partial x} \right] = 0 \tag{2}$$

are a linear combination of the traditional equations of gas dynamics: continuity, momentum and energy. Here and below ρ is the density, u is the velocity, J is the intrinsic energy of a unit mass, $p = p(\rho, J)$ is the pressure, T is the temperature, x is the spatial coordinate, and t is the time. The Cauchy problem asks for a solution in the domain $t > 0$ for given data on the line $t = 0$. Entropy is the name for an integral $S(\rho, J) = \mathrm{const}$ of the equation

$$dJ - \frac{p(\rho, J)}{\rho^2} \, d\rho = 0.$$

Then (1) has the form

$$\frac{\partial S}{\partial t} + u \frac{\partial S}{\partial x} = 0. \tag{1'}$$

On the left-hand sides of (1') and (2) there stand the derivatives dS/dt, dp/dt, du/dt, taken in the directions

$$\frac{dx}{dt} = u \tag{3}$$

and

$$\frac{dx}{dt} = u \pm c, \tag{4}$$

the so-called *characteristics*. The system (1), (2) has three families of real characteristics. Along the characteristics (3) the relation

$$dS = 0$$

holds, and along the characteristic (4) the relations

$$dp \pm \rho c \, du = 0 \tag{5}$$

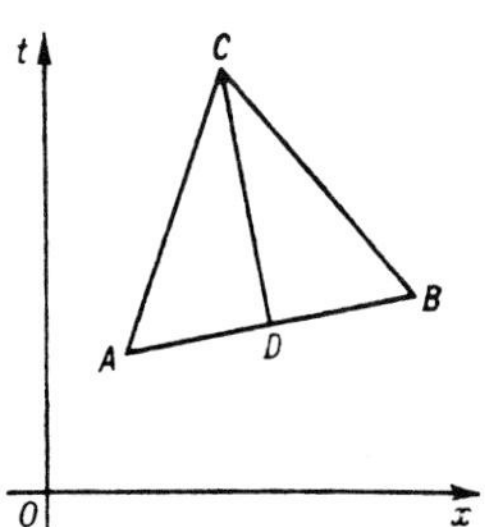

hold. Through the point A (see Fig.) there passes a characteristic (4) on the side of increasing t,

$$\frac{x - x_A}{t - t_A} = u_A + c_A.$$

Through a point close to A on the right, say B, there passes a characteristic (4) of another family

$$\frac{x - x_B}{t - t_B} = u_B - c_B,$$

where C is the point of intersection of the characteristics. When one replaces the differential relations (5), which hold along characteristics, by differences, one obtains the algebraic system

$$(p_C - p_A) + \rho_A c_A (u_C - u_A) = 0,$$

$$(p_C - p_B) - \rho_B c_B (u_C - u_B) = 0,$$

from which p_C and u_C can be determined. From C one draws the characteristic (3),

$$\frac{x - x_C}{t - t_C} = u_C,$$

up to the intersection with AB at D. The value of the entropy S at D is determined by means of interpolation between the points A and B (here $S_C = S_D$). From the equations

$$p(\rho_C, J_C) = p_C,$$

$$S(\rho_C, J_C) = S_C$$

one can find the values of the intrinsic energy J_C and the density ρ_C at C. When the data at the two points A and D are known, one can find the solution at C for large values of t. This procedure of computations is repeated for every pair of points. Then, by using new points C instead of the original A and B, one takes the next step in t. The calculation is made for the required values of t. However, since the equations of gas dynam-

ics are non-linear, the calculation may come to a stop at a certain moment of time if the characteristics of one family touch each other or intersect.

The difference scheme thus described is of the first order of accuracy (an analogue of Euler's method of polygonal lines for solving ordinary differential equations). An increase in accuracy can be achieved by repetitions of the calculations, etc.

Using the method of characteristics one can solve stationary multi-dimensional problems in a domain of hyperbolicity (for gas dynamics — of supersonic flow). One can also determine the position of secondary shock waves at places where the characteristics of a single family intersect or touch. By the method of characteristics one can tackle only problems with a small number of discontinuities, since under an accumulation of singularities the computations become tedious. A computation by the method of characteristics consists of a number of elementary problems: the calculation of an interior point, of a point on the shock wave, or on a body around which the flow is taking place, etc.

One can construct numerical schemes for the method of characteristics, which make it possible to carry out the calculations in 'layers', the *grid-characteristic method* (see [2]).

References

[1] An attempt at calculating plane and axially-symmetric supersonic flows of a gas by the method of characteristics, Moscow, 1961 (in Russian).

[2] MAGOMEDOV, K.M. and KHOLODOV, A.S.: 'The constructions of difference schemes for hyperbolic equations based on characteristic relations', *USSR Math. Math. Comp.* **9**, no. 2 (1969), 158-176. (*Zh. Vyshch. Mat. i Mat. Fiz.* **9**, no. 2 (1969), 373-386)

Yu.M. Davydov

Editorial comments. The method of characteristics goes back to J. Massau (see [A2]).

References

[A1] AMES, W.F.: *Numerical methods for partial differential equations*, Acad. Press, 1977.

[A2] MASSAU, J.: *Mémoire sur l'intégration graphique des équations aux derivées partielles*, F. Mayer-van Loo, Ghent, 1899.

AMS 1980 Subject Classification: 65M25, 35A25

METHOD OF EXTENSIONS AND RESTRICTIONS, *method of prolongations and restrictions* - A method for studying various differential-geometric structures (cf. **Differential-geometric structure**) on smooth manifolds and their submanifolds. At the basis of this method there lies a differential-algebraic criterion for an operation that allows one to associate in an invariant (coordinate-free) way to a given structure structures intrinsically related to it, among them their differential invariants (cf. **Differential invariant**). Historically this method arose as a consequence of the **moving-frame**

method as an invariant method for studying submanifolds of homogeneous spaces or of spaces with a connection. Subsequently the method of prolongations and restrictions was extended to the geometry of arbitrary fibre spaces (cf. **Fibre space**). In distinction from the aim of the moving-frame method — to construct a canonical field of frames and differential invariants of the unknown structure by means of subsequent restriction of corresponding principal fibre spaces — the method of prolongations and restrictions has as its aim the construction of invariants and invariantly associated structures without restricting the principal fibres of frames. The process of canonization of a frame is included in the method of prolongations and restrictions.

Let G be a **Lie group** and let $K(G)$ be the class of G-spaces with a left action of G as transformation group on them. A G-restriction is a smooth surjective mapping

$$f : X \to Y, \quad X, Y \in K(G),$$

such that for any $g \in G$ the following diagram is commutative:

$$\begin{array}{ccc} & f & \\ X & \to & Y \\ l_g \downarrow & & \downarrow l_g^1 \\ X & \to & Y \\ & f & \end{array}$$

Here l_g and l_g^1 are the transformations of the G-spaces X and Y, respectively, determined by g. In this case one says that Y is a *restriction* of X by means of f, or that X is a *prolongation* of Y. The class $K(G)$ becomes a category with the G-restrictions as morphisms.

Examples of G-restrictions.

1) Let $T(p, q) \in K(GL(n, \mathbf{R}))$ be the space of tensors of type $(p, q), p, q \geqslant 1$. The contraction mapping

$$T(p, q) \to T(p-1, q-1)$$

is a restriction. The complete contraction of tensors of $T(p, p)$,

$$T(p, p) : T(p, p) \to \mathbf{R},$$

is an example of a restriction invariant.

2) If $X, Y \in K(G)$, then $X \times Y$ restricts by means of pr_X and pr_Y, respectively, to X and Y. In other words, $X \times Y$ is a prolongation of both X and Y.

The concept of a restriction can be naturally generalized to classes of fibre spaces associated with principal fibre bundles. Let $\pi : P(M, H) \to M$ be a principal fibre bundle with structure group H, acting on P from the right, and let $F \in K(H)$ be a left H-space. Fibre spaces associated with P by objects from $K(P)$ are spaces of the type

$$F(P) = (P \times F)/H,$$

where factorization is by the following right action of H on $P \times F$:

$$(\xi, Y)h = (\xi h, h^{-1}Y), \quad (\xi, Y) \in P \times F, \quad h \in H.$$

The space $F(P) \in K(P)$ is a fibre bundle over the base

M with typical fibre F. The element $y \in F(P)$ determined by a pair $(\xi, Y) \in P \times F$ is written as $y = \xi Y$. If $F, \Phi \in K(H)$ and $f : F \to \Phi$ is an H-restriction mapping, then, by construction, $F(P)$ and $\Phi(P)f$ induce a fibrewise surjective mapping $\tilde{f} : F(P) \to \Phi(P)$, called a P-restriction. The P-restriction $\tilde{f}$ is defined by

$$\tilde{f}(\xi Y) = \xi f(Y), \quad \xi \in P, \quad Y \in F.$$

Thus, the class $K(P)$ of fibre bundles associated with P is a category with P-restrictions $\tilde{f}$ as morphisms. The correspondence $F \mapsto F(P)$, $f \mapsto \tilde{f}$ is a bijective functor from the category $K(H)$ to the category $K(P)$. Hence it is sufficient to study the restriction operation in the category of H-spaces.

If $s : M \to F(P)$ is a section of a fibre bundle $F(P)$ (a *field of geometric objects* of type F), then the P-restriction $\tilde{f} : F(P) \to \Phi(P)$ associates the section $\tilde{s} = \tilde{f} \circ s$ of the restricted fibre bundle $\Phi(P)$ to s. In other words, the field of geometric objects $s(x)$, $x \in M$, restricts the field of geometric objects $\tilde{f} \circ s(x)$. If $s(x)$ is the structure object of a G-structure, then the study of the G-structures and its invariants reduces largely to the search for restricting geometric objects. In the latter process, an important role is played by differential criteria for restrictions, formulated in terms of structure differential forms of fibre spaces forming the base of the method of restrictions and prolongations.

References

[1] LAPTEV, G.F.: 'Differential geometry of imbedded manifolds. Group-theoretical method of differential-geometric investigation', *Trudy Moskov. Mat. Obshch.* **2** (1953), 275-382 (in Russian).
[2] LAPTEV, G.F.: *Proc. 3-th All-Union Mat. Congress Moscow, 1956*, Vol. 3, Moscow, 1958, pp. 409-418.

L.E. Evtushik

Editorial comments.

References

[A1] JENSEN, G.: *Higher order contact of submanifolds of homogeneous spaces*, Lecture note in math., 610, Springer, 1977.
[A2] KOBAYASHI, S. and NOMIZU, K.: *Foundations of differential geometry*, 1-2, Interscience, 1979.

AMS 1980 Subject Classification: 55R05, 58A10, 55R10, 53C10

METHOD OF STEEPEST DESCENT (FOR INTEGRALS) - See **Saddle point method.**

AMS 1980 Subject Classification: 41A60, 45M05

METHOD OF UNDETERMINED COEFFICIENTS IN DEVELOPING NUMERICAL ALGORITHMS - A special method for constructing algorithms based on the requirement that the algorithms should be exact or have an error of prescribed order of accuracy on some set of problems.

A typical example of problems that can be solved by the method of undetermined coefficients is the following (see [1], [2]). (It can also be solved by other methods.) Let the values $f(P_1), \ldots, f(P_N)$ of a function be known. It is required to construct a formula for the approximation of the function

$$f(P) \approx g(f(P_1), \ldots, f(P_N), P);$$

to construct a formula for computing the derivative:

$$Df(P) \approx g_D(f(P_1), \ldots, f(P_N), P);$$

and to construct a formula for computing the integral:

$$I(f) = \int_\Omega f(P)p(P)\,dP \approx g_1(f(P_1), \ldots, f(P_n)).$$

To solve the last of these problems one is given some form of an approximate solution, for example, linear:

$$g_1 = \sum_{n=1}^{N} C_n f(P_n),$$

and one determines the coefficients C_n from the requirement that the approximate formula $I(f) \approx g_1$ should be exact for functions of a certain collection; for example, for those of the form

$$f(P) = \sum_{m=1}^{M} a_m \omega_m(P),$$

where the $\omega_m(P)$ are fixed and the a_m are arbitrary. As a rule one takes $M = N$. For the equality

$$I\left[\sum_{m=1}^{M} a_m \omega_m \right] = \sum_{n=1}^{N} C_n \left[\sum_{m=1}^{M} a_m \omega_m(P_n) \right]$$

to hold for all a_m it suffices that

$$I(\omega_m) = \sum_{n=1}^{N} C_n \omega_m(P_n), \quad m = 1, \ldots, M.$$

From this one determines the C_n (if this is possible).

Sometimes one is given a more complicated form of dependence. For example, frequently it is known that the function in question has a good approximation by functions of the form

$$g(a_1, \ldots, a_M, P),$$

where the a_m are known. The parameters a_m are chosen from the system of equations

$$g(a_1, \ldots, a_M, P_n) = f(P_n), \quad n = 1, \ldots, N.$$

In the case of formulas for numerical integration, often the coordinates of the nodes of integration stand out as unknown parameters. For example, in the **Gauss quadrature formula**

$$I(f) = \int_a^b f(x)p(x)\,dx \approx \sum_{n=1}^{N} C_n f(P_n)$$

one considers as free parameters the coordinates of the nodes P_n; owing to this one succeeds in constructing quadratures that are exact for polynomials of degree $2N-1$. In the construction of approximations of differential equations by the method of undetermined coefficients it is required that under substitution in a finite-difference scheme the solutions of the problem give compatible (**discrepancy**) values of prescribed order of smallness in relation to the step of the grid. Such a device lies at the basis of ways for constructing Runge−Kutta methods and finite-difference methods (see [1], [2]).

The method of undetermined coefficients is also used in constructing approximations of partial differential equations (see [3]).

References
[1] BEREZIN, I.S. and ZHIDKOV, N.P.: *Computing methods*, Pergamon, 1973 (translated from the Russian).
[2] BAKHVALOV, N.S.: *Numerical methods: analysis, algebra, ordinary differential equations*, Mir, 1977 (translated from the Russian).
[3] GODUNOV, S.K. and RYABEN'KII, V.S.: *The theory of difference schemes*, North-Holland, 1964 (translated from the Russian).

N.S. Bakhvalov

Editorial comments. The techniques of curve and surface fitting may be considered as a form of, what is here called, the method of undetermined coefficients.

AMS 1980 Subject Classification: 00A25, 65-XX

METRIC, *distance on a set* X - A function ρ with non-negative real values, defined on the Cartesian product $X \times X$ and satisfying for any $x, y \in X$ the conditions:

1) $\rho(x, y) = 0$ if and only if $x = y$ (the *identity axiom*);

2) $\rho(x, y) + \rho(y, z) \geqslant \rho(x, z)$ (the *triangle axiom*);

3) $\rho(x, y) = \rho(y, x)$ (the *symmetry axiom*).

A set X on which it is possible to introduce a metric is called *metrizable* (cf. **Metrizable space**). A set X provided with a metric is called a **metric space**.

Examples. 1) On any set there is the *discrete metric*

$$\rho = 0 \; (x = y) \quad \text{and} \quad \rho = 1 \; (x \neq y).$$

2) In the space $\mathbf{R}^n$ various metrics are possible, among them are:

$$\rho(x, y) = \sqrt{\sum (x_i - y_i)^2};$$

$$\rho(x, y) = \sup_i |x_i - y_i|;$$

$$\rho(x, y) = \sum |x_i - y_i|;$$

here $\{x_i\}, \{y_i\} \in \mathbf{R}^n$.

3) In a Riemannian space a metric is defined by a **metric tensor**, or a quadratic differential form (in some sense, this is an analogue of the first metric of example 2)). For a generalization of metrics of this type see **Finsler space**.

4) In function spaces on a (countably) compact space X there are also various metrics; for example, the *uniform metric*

$$\rho(f, g) = \sup_{x \in X} |f(x) - g(x)|$$

(an analogue of the second metric of example 2)), and the *integral metric*

$$\rho(f, g) = \int_X |f - g|\,dx.$$

5) In normed spaces over $\mathbf{R}$ a metric is defined by the norm $|\cdot|$:
$$\rho(x,y) = |x-y|.$$

6) In the space of closed subsets of a metric space there is the **Hausdorff metric**.

If, instead of 1), one requires only:

1') $\rho(x,y)=0$ if $x=y$ (so that from $\rho=0$ it does not always follows that $x=y$), the function ρ is called a *pseudo-metric* [2], [3], or finite écart [4].

A metric (and even a pseudo-metric) makes the definition of a number of additional structures on the set X possible. First of all a topology (see **Topological space**), and in addition a uniformity (see **Uniform space**) or a proximity (see **Proximity space**) structure. The term metric is also used to denote more general notions which do not have all the properties 1) - 3); such are, for example, an **indefinite metric**, a **symmetry on a set**, etc.

References
[1] ALEKSANDROV, P.S.: *Introduction to set theory and general topology*, Moscow, 1977 (in Russian).
[2] KELLEY, J.L.: *General topology*, Springer, 1975.
[3] KURATOWSKI, K.: *Topology*, 1, Acad. Press, 1966 (translated from the French).

M.I. Voĭtsekhovskiĭ

Editorial comments. Potentially, any metric space (X, ρ) has a second metric $\sigma \geqslant \rho$ naturally associated: the *intrinsic* or *internal metric*. Potentially, because the definition may give $\sigma(x, y) = \infty$ for some pairs of points x, y. One defines the *length* (which may be ∞) of a continuous path $f : [0, 1] \to X$ by $L(f) = \lim_{\epsilon \to 0} \sup L_\epsilon(f)$, where $L_\epsilon(f)$ is the infimum of all finite sums $\sum \rho(x_i, x_{i+1})$ with $\{x_i\}$ a finite subset of $[0, 1]$ which is an ϵ-net (cf. **Metric space**) and is listed in the natural order. Then $\sigma(x, y)$ is the infimum of the lengths of paths f with $f(0)=x$, $f(1)=y$, but $\sigma(x, y) = \infty$ if there is no such path of finite length.

No reasonable topological restriction on (X, ρ) suffices to guarantee that the intrinsic 'metric' (or *écart*) σ will be finite-valued. If σ is finite-valued, suitable compactness conditions will assure that minimum-length paths, i.e. paths from x to y of length $\sigma(x, y)$, exist. When every pair of points x, y is joined by a path (non-unique, in general) of length $\sigma(x, y)$, the metric is often called *convex*. (This is much weaker than the surface theorists' **convex metric**.) The main theorem in this area is that every locally connected metric **continuum** admits a convex metric [A1], [A2].

References
[A1] BING, R.H.: 'Partitioning a set', *Bull. Amer. Math. Soc.* **55** (1949), 1101-1110.
[A2] MOÏSE, E.E.: 'Grille decomposition and convexification', *Bull. Amer. Math. Soc.* **55** (1949), 1111-1121.

AMS 1980 Subject Classification: 54E35, 46A06

METRIC CONNECTION - A **linear connection** in a vector bundle $\pi : X \to B$, equipped with a bilinear form in the fibres, for which parallel displacement along an arbitrary piecewise-smooth curve in B preserves the form, that is, the scalar product of two vectors remains constant under parallel displacement. If the bilinear form is given by its components $g_{\alpha\beta}$ and the linear connection by a matrix 1-form ω_α^β, then this connection is metric if
$$dg_{\alpha\beta} = g_{\gamma\beta}\omega_\alpha^\gamma + g_{\alpha\gamma}\omega_\beta^\gamma.$$

In the case of a non-degenerate symmetric bilinear form, i.e. $g_{\alpha\beta} = g_{\beta\alpha}$ and $\det | g_{\alpha\beta} | \neq 0$, the metric connection is called a **Euclidean connection**. In the case of a non-degenerate skew-symmetric bilinear form, the metric connection is called a *symplectic connection* in the vector bundle.

Under projectivization of a vector bundle, when the symmetric bilinear form generates some projective metric in each fibre (as in a projective space), the role of the metric connection is played by the projective-metric connection.

Ü. Lumiste

Editorial comments. In the case of a positive-definite bilinear form, the metric connection is also called a **Riemannian connection**.

References
[A1] KLINGENBERG, W.: *Riemannian geometry*, de Gruyter, 1982.
[A2] KOBAYASHI, S. and NOMIZU, K.: *Foundations of differential geometry*, Wiley (Interscience), 1969.

AMS 1980 Subject Classification: 53B05, 53B10

METRIC DIMENSION - A numerical characteristic of a compact set, defined in terms of coverings of 'standard measure', the number of which defines the metric dimension. Let F be a compact set, and let $N_F(\epsilon)$ be the minimal number of sets with diameter not exceeding ϵ that are needed in order to cover F. This function, depending on the metric in F, takes integer values for all $\epsilon > 0$, and increases without bound as $\epsilon \to 0$; it is called the *volume function* of F. The *metric order* of the compact set F is the number
$$k = \varlimsup \left[-\frac{\ln N_F(\epsilon)}{\ln \epsilon} \right].$$

This quantity is not yet a topological invariant. Thus, the metric order of a curve in the sense of Jordan (cf. **Line (curve)**) with the Euclidean metric is equal to 1, but for a curve in the sense of Jordan passing through a perfect totally-disconnected set in $\mathbf{R}^n$ of positive measure, this value is equal to n. However, the greatest lower bound of the metric orders for all metrics on F (called the *metric dimension*) is equal to the **Lebesgue dimension** (the *Pontryagin — Shnirel'man theorem*, 1931, see [1]).

References
[1] HUREVICZ, W. and WALLMAN, G.: *Dimension theory*, Princeton Univ. Press, 1948.

M.I. Voĭtsekhovskiĭ

Editorial comments. Metric dimension makes sense for

non-compact separable metrizable spaces (using totally bounded metrics), and the Pontryagin–Shnirel'man theorem extends to them. This was shown by E. Szpilrajn-Marczewski. See [A2].

There are also other types of metric-dependent dimension functions.

One example is the **Hausdorff dimension**.

Another example is obtained by modifying the definition of the covering dimension dim (see **Dimension**): If (X, d) is a metric space, one defines $\mu\dim(X, d)$ by $\mu\dim(X, d) \leq n$ if and only if for every $\epsilon > 0$ there is an open covering $\mathfrak{U}$ of X with $\operatorname{mesh}\mathfrak{U} \leq n+1$ and $\operatorname{ord}\mathfrak{U} < \epsilon$. Here $\operatorname{mesh}\mathfrak{U} = \sup\{\operatorname{diam}(U): U \in \mathfrak{U}\}$ and $\operatorname{ord}\mathfrak{U} \leq n+1$ means that no point of X is an element of more than $n+1$ elements of $\mathfrak{U}$. One can show that $\mu\dim(X, d) \leq \dim X \leq 2\mu\dim(X, d)$ and that these inequalities are best possible, see [A1].

References

[A1] ENGELKING, R.: *Dimension theory*, PWN & North-Holland, 1978.

[A2] NAGATA, J.-I.: *Modern dimension theory*, Interscience, 1965.

AMS 1980 Subject Classification: 54F45

METRIC ENTROPY of a dynamical system - One of the most important invariants in **ergodic theory**. Basic is the concept of the entropy $h(S)$ of an endomorphism S (see **Metric isomorphism**) of a Lebesgue space (X, μ). For any finite **measurable decomposition** (*measurable partition*) ξ the limit

$$h(S, \xi) = \lim_{n \to \infty} \frac{1}{n} H(\xi_S^n),$$

$$\xi_S^n = \xi \vee S^{-1}\xi \vee \cdots \vee S^{-n+1}\xi$$

(the entropy of ξ in unit time relative to S) exists, where $H(\xi)$ is the entropy (cf. **Entropy of a measurable decomposition**) of ξ, and $\xi \vee \eta$ is the partition whose elements are the intersections of the elements of ξ and η. (This definition carries over verbatim to ξ with $H(\xi) < \infty$; by another method $h(S, \xi)$ can be defined for any measurable ξ.) The entropy $h(S)$ is defined as the least upper bound of the $h(S, \xi)$ over all possible finite measurable ξ. (It may be ∞; the use of all ξ with $H(\xi) < \infty$ or of all measurable ξ yields the same entropy.)

Originally the entropy was defined by A.N. Kolmogorov somewhat differently (see [1]); the version given above came later (see [2]). In the basic case of an **aperiodic automorphism** of a Lebesgue space the definitions are ultimately equivalent [3].

It turns out that $h(S^n) = nh(S)$, and if S is an automorphism, then $h(S^{-1}) = h(S)$. Therefore, the entropy of a **cascade** $\{S^n\}$ is naturally taken to be $h(S)$. For a **measurable flow** $\{S_t\}$ it turns out that $h(S_t) = |t| h(S_1)$. Therefore the entropy of a flow is naturally taken to be $h(S_1)$. The definition of the entropy for other transformation groups with an invariant measure is somewhat different. (It does not reduce

to the entropy of a single transformation in the group; see [5], [6].) There are modifications of the entropy for the case of an infinite invariant measure [7]; another modification is the *A-entropy* (where $A = \{k_n\}$ is an ascending sequence of natural numbers), which is obtained when ξ_S^n is replaced by

$$S^{-k_1}\xi \vee \cdots \vee S^{-k_n}\xi$$

and lim by $\overline{\lim}$ (see [8]).

The entropy is a metric isomorphism invariant of dynamical systems and is fundamentally different from the earlier-known invariants, which are basically connected with the **spectrum of a dynamical system**. In particular, by means of the entropy of Bernoulli automorphisms (cf. **Bernoulli automorphism**; see [1]) it was first established that there exist non-isomorphic ergodic systems with the same continuous spectrum (which contrasts with the situation for a discrete spectrum). In a wider setting the role of the entropy is related to the fact that a new trend arose in ergodic theory: the *entropy theory* of dynamical systems (see [3], [4], and **Ergodic theory**).

The entropy provides a tool for characterizing the rate of **mixing** of sets of small measure (more accurately, the collection of those that form the partition). Side-by-side with this 'global' role, the entropy also plays a 'local' role, which is established by *Breiman's ergodic theorem* (an *individual ergodic theorem of information theory*): For ergodic S and almost-all x,

$$\frac{1}{n} |\log \mu(C_{\xi_x}(x))| \to h(S, \xi) \quad \text{for } n \to \infty,$$

where $C_\eta(x)$ is the element of the partition η containing x and the logarithm is taken to the same base as in the definition of H (see [9], [4]). (Breiman's theorem is true for ξ with $H(\xi) < \infty$ [10], but, generally speaking, not for countable ξ with $H(\xi) = \infty$ [11]; there are variants for non-ergodic S (see [4], [12]) and an infinite μ [13]. A weaker assertion on the convergence in the sense of l_1 has been proved for a certain general class of transformation groups [6].)

For smooth dynamical systems with a smooth invariant measure a connection has been established between the entropy and the **Lyapunov characteristic exponent** of the equations in variations (see [14] - [16]).

The name 'entropy' is explained by the analogy between the entropy of dynamical systems and that in **information theory** and statistical physics, right up to the fact that in certain examples these entropies are the same (see, for example, [4], [17]). The analogy with statistical physics was one of the stimuli for introducing in ergodic theory (even in a not-purely metric context and for topological dynamical systems, cf. **Topological dynamical system**) new concepts such as 'Gibbsian measures', the 'topological pressure' (an analogue to the free energy) and the 'variational principle' for the latter (see the references to *Y*-system; **Topological entropy**).

References

[1A] KOLMOGOROV, A.N.: 'A new metric invariant of transitive dynamical systems, and Lebesgue space automorphisms', *Dokl. Akad. Nauk SSSR* **119**, no. 5 (1958), 861-864 (in Russian).

[1B] KOLMOGOROV, A.N.: 'On entropy per unit time as a metric invariant of automorphisms', *Dokl. Akad. Nauk SSSR* **124**, no. 4 (1959), 754-755 (in Russian).

[2] SINAĬ, YA.G.: 'On the notion of entropy of dynamical systems', *Dokl. Akad. Nauk SSSR* **124**, no. 4 (1959), 768-771 (in Russian).

[3] ROKHLIN, V.A.: 'Lectures on the entropy theory of transformations with invariant measure', *Russian Math. Surveys* **22**, no. 5 (1967), 1-52. (*Uspekhi Mat. Nauk* **22**, no. 5 (1967), 3-56)

[4] BILLINGSLEY, R.: *Ergodic theory and information*, Wiley, 1965.

[5] SAFONOV, A.V.: 'Information parts in groups', *Math. USSR. Izv.* **22** (1984), 393-398. (*Izv. Akad. Nauk SSSR Ser. Mat.* **47**, no. 2 (1983), 421-426)

[6] KIEFFER, J.C.: 'A generalized Shannon–McMillan theorem for the action of an amenable group on a probability space', *Ann. of Probab.* **3**, no. 6 (1975), 1031-1037.

[7] KRENGEL, V.: 'Entropy of conservative transformations', *Z. Wahrscheinlichkeitstheor. Verw. Geb.* **7**, no. 3 (1967), 161-181.

[8] KUSHNIRENKO, A.G.: 'Metric invariants of entropy type', *Russian Math. Surveys* **22**, no. 5 (1967), 53-61. (*Uspekhi Mat. Nauk* **22**, no. 5 (1967), 37-65)

[9A] BREIMAN, L.: 'The individual ergodic theorem of information theory', *Ann. Math. Stat.* **28**, no. 3 (1957), 809-811.

[9B] BREIMAN, L.: 'Correction to ' The individual ergodic theorem of information theory", *Ann. Math. Stat.* **31**, no. 3 (1960), 809-810.

[10] CHUNG, K.L.: 'A note on the ergodic theorem of information theory', *Ann. Math. Stat.* **32**, no. 3 (1961), 612-614.

[11] PITSKEL, B.S.: 'Nonuniform distribution of entropy for processes with a countable set of states', *Probl. Peredatsi Inform.* **12**, no. 2 (1976), 98-103 (in Russian).

[12] IONESCO-TULCEA, A.: 'Contributions to information theory for abstract alphabets', *Arkiv for Mat.* **4**, no. 2-3 (1961), 235-247.

[13] KLIMKO, E.M. and SUCHESTON, L.: 'On convergence of information in spaces with infinite invariant measure', *Z. Wahrscheinlichkeitstheor. Verw. Geb.* **10**, no. 3 (1968), 226-235.

[14] MILLIONSHCHIKOV, V.M.: 'A formula for the entropy of smooth dynamical systems', *Differential Eq.* **12** (1976), 1527-1530. (*Differents. Uravnen.* **12**, no. 12 (1976), 2188-2192)

[15] PESIN, YA.B.: 'Characteristic Lyapunov exponents, and smooth ergodic theory', *Russian Math. Surveys* **32**, no. 4 (1977), 55-114. (*Uspekhi Mat. Nauk* **32**, no. 4 (1977), 55-112)

[16] MAÑÉ, R.: 'A proof of Pesin's formula', *Ergod. Th. and Dynam. Syst.* **1**, no. 1 (1981), 95-102.

[17] ROBINSON, D.W. and RUELLE, D.: 'Mean entropy of states in classical statistical mechanics', *Comm. Math. Phys.* **5**, no. 4 (1967), 288-300.

D.V. Anosov

Editorial comments. Instead of *A*-entropy the term *sequence entropy* is used in the English literature. See e.g. [A1], § 4.11. For several useful recent references concerning the computation of entropy, see [A2].

References

[A1] WALTERS, P.: *An introduction to ergodic theory*, Springer, 1982.

[A2] WOJTKOWSKI, M.P.: 'Measure theoretic entropy of the system of hard spheres', *Ergod. Th. and Dynam. Syst.* **8** (1988), 133-153.

AMS 1980 Subject Classification: 28D20, 58F11, 94A17

METRIC ISOMORPHISM *of two measure spaces* $(X_1, \mathfrak{B}_1, \mu_1)$ and $(X_2, \mathfrak{B}_2, \mu_2)$ - A bijective mapping $f: X_1 \to X_2$ for which images and inverse images of measurable sets are measurable and have the same measure (here $\mathfrak{B}_i$ is some Boolean σ-algebra or σ-ring of subsets of the set X_i, called measurable, and μ_i is a given measure on $\mathfrak{B}_i$). There is the more general notion of a (*metric*) *homomorphism* of these spaces, that is, a mapping $f: X_1 \to X_2$ such that inverse images of measurable sets are measurable and have the same measure. For $(X_1, \mathfrak{B}_1, \mu_1) = (X_2, \mathfrak{B}_2, \mu_2)$, instead of an isomorphism or a homomorphism one speaks of a (*metric*) *automorphism* or an *endomorphism*.

In correspondence with the usual tendency in measure theory to ignore sets of measure zero, there is (and is primarily used) a 'modulo 0' version of all these ideas. For example, let $N_i \subset X_i$, $\mu_i(N_i) = 0$, $i = 1, 2$, and let $f: X_1 \setminus N_1 \to X_2 \setminus N_2$ be a metric isomorphism; then it is said that f is an isomorphism modulo 0 of the original measure spaces (the stipulation 'modulo 0' is often omitted).

For a number of objects given in X_i (subsets, functions, transformations, and systems of these) one can give a meaning to the assertion that under a metric isomorphism f these objects transform into each other. It is then said that f is a metric isomorphism of the corresponding objects. It is also possible to speak of their being metrically isomorphic modulo 0. This means that for certain N_i of measure zero the corresponding objects O_i may be considered as objects O_i' in $X_i \setminus N_i$ (for transformations this means that the $X_i \setminus N_i$ are invariant relative to these transformations, whereas for subsets and functions this makes sense for any N_i: take the intersection of the considered subsets with $X_i \setminus N_i$ or the restrictions of the functions to $X_i \setminus N_i$) and that f is a metric isomorphism of the objects O_i'. A class of all objects metrically isomorphic modulo 0 to each other is called a (*metric*) *type*; two objects of this class are said to have the same type.

Associated with $(X_i, \mathfrak{B}_i, \mu_i)$ are the Hilbert spaces $L_2(X_i, \mathfrak{B}_i, \mu_i)$ in which, in addition to the usual Hilbert space structure, there is also the operation of ordinary multiplication of functions (defined, it is true, not everywhere, since the product of L_2 functions is not always in L_2), and the Boolean measure σ-algebras $\mathfrak{M}_i$, obtained from $\mathfrak{B}_i$ by identifying sets with symmetric difference of measure zero (that is, factorizing with respect to the ideal of sets of measure zero). A metric isomorphism f modulo 0 induces an isomorphism of the Boolean measure algebras $\mathfrak{M}_i$ and a unitary isomorphism of the Hilbert spaces $L_2(X_i, \mu_i)$ which is also *multiplicative*, that is, takes a product (whenever defined) to the product of the images of the multiplicands. If $(X_i, \mathfrak{B}_i, \mu_i)$ is a **Lebesgue space**, then the converse is true: Every isomorphism of the Boolean measure σ-algebras $\mathfrak{M}_i$, or every multiplicative unitary isomorphism of the spaces $L_2(X_i, \mu_i)$, is induced by some metric isomorphism modulo 0.

References

[1] ROKHLIN, V.A.: 'On mean notions in measure theory', *Mat. Sb.* **25**, no. 1 (1949), 107-150 (in Russian).

D.V. Anosov

Editorial comments. See also **Ergodic theory** for additional references. As a rule, the adjective 'metric' is not anymore used and one simply speaks of an *isomorphism of measure spaces*, a *homomorphism of measure spaces*, etc.

References

[A1] HALMOS, P.: *Measure theory*, v. Nostrand, 1950.

AMS 1980 Subject Classification: 28D05

METRIC PROJECTION, *operator of best approximation* - A many-valued mapping $P_M: x \to P_M x$, associating to each element x of a metric space $X = (X, \rho)$ the set

$$P_M x = \{m \in M: \rho(x, m) = \rho(x, M)\}$$

of elements of best approximation (cf. **Element of best approximation**) from the set $M \subset X$. If M is a **Chebyshev set**, then the metric projection is a single-valued mapping. The problem of constructing an element of best approximation is often solved approximately, that is, an element is determined in the set

$$P_M^t x = \{m \in M: \rho(x, m) \leqslant t + \rho(x, M)\},$$

where $t > 0$ is sufficiently small. From the properties of the mapping $P_M^t: x \to P_M^t x$ it is sometimes possible to obtain properties of the set M. E.g., if for any element x of a normed space X a number $t = t(x) > 0$ exists such that $P_M^t x$ is convex (connected), then M is convex (respectively, connected).

From the point of view of applications it is useful to know whether the metric projection has such properties as linearity, continuity, uniform continuity, etc. A metric projection on a Chebyshev subspace of a normed space is, in general, not linear. If the metric projection on each subspace of fixed dimension is single-valued and linear, then X is linearly isometric to an inner-product space. The metric projection on a non-empty **approximately-compact set** in a metric space is upper semi-continuous; in particular, in a normed space the metric projection onto a finite-dimensional Chebyshev subspace is continuous; the metric projection may be not lower semi-continuous if the subspace is not Chebyshev. There exists a reflexive strictly-convex space and an infinite-dimensional subspace on which the metric projection is discontinuous. The metric projection on any closed convex set M in a Hilbert space satisfies a Lipschitz condition:

$$\| P_M x - P_M y \| \leqslant K \| x - y \|,$$

with constant $K = 1$.

The continuity property of a metric projection and its generalizations have found applications in ill-posed problems, in the convexity problem for Chebyshev sets, in the construction of elements of best approximation, etc.

References

[1] SINGER, I.M.: *The theory of best approximation and functional analysis*, SIAM, 1974.
[2] VLASOV, L.P.: 'Approximative properties of sets in normed linear spaces', *Russian Math. Surveys* **28**, no. 6 (1973), 3-66. (*Uspekhi Mat. Nauk* **28**, no. 6 (1973), 3-66)
[3] BERDYSHEV, V.I.: 'Uniform continuity of the metric projection and of the S-projection', in *Theory of Approximation of Functions. Internat. Conf., Kaluga 1975*, Moscow, 1977, pp. 37-41 (in Russian).

V.I. Berdyshev

AMS 1980 Subject Classification: 41A65, 41A50

METRIC SPACE - A set X together with a **metric** ρ. The set-theoretic approach to the study of figures (spaces) is based on the study of the relative position of their elementary constituents. A fundamental characteristic of the relative position of points of a space is the distance between them. This approach leads to the idea of a metric space, first suggested by M. Fréchet [2] in connection with the discussion of function spaces. It turns out that sets of objects of very different types carry natural metrics. As metric spaces one may consider sets of states, functions and mappings, subsets of Euclidean spaces, and Hilbert spaces. Metrics are important in the study of convergence (of series, functions) and for the solution of questions concerning approximation.

The development of the theory of metric spaces has proceeded in the following main directions.

General theory of metric spaces. Here one studies properties of metric spaces which are invariant relative to isometries: one-to-one and onto mappings which preserve distance (cf. **Isometric mapping**). Such properties include completeness, boundedness, total boundedness, and widths. Properties of this type are called *metric*.

Topological theory of metric spaces. Its subject is the properties of metric spaces which are preserved under homeomorphisms (cf. **Homeomorphism**). Among these are compactness, separability, connectedness, the Baire property, and zero dimensionality. Properties of this type are called *topological*.

Theory of spaces on which a metric has been given that is compatible with some additional algebraic structure (for example, a vector space or a group). Here one is concerned with Euclidean spaces, pre-Hilbert and Hilbert spaces (cf. **Hilbert space**) (of any weight), Banach spaces, Banach algebras, Banach lattices (cf. **Banach space; Banach algebra; Banach lattice**), and countably-normed spaces (cf. **Countably-normed space**). The facts available here are essentially connected with the discussion of important properties of metrics or norms, but the content, on the whole, belongs to the corresponding domains of algebra and functional analysis.

The discussion of particular metrics plays an important role in investigations in non-Euclidean geometries, differential geometry, mechanics, and physics. Here a central place is occupied by the notion of a Riemannian metric in a Riemannian space (see **Riemannian geometry**). A broader approach to the study of the surfaces and figures that arise in differential geometry is related to the concept of a G-space, resulting from the addition of certain conditions to the metric axioms (see **Geodesic geometry**), which creates a basis for the discussion of geodesics in G-spaces by ensuring their existence and nice properties. Typical here is the abandoning of the methods of differential calculus. In this connection it turns out that much of differential geometry is not connected with differentiability conditions, but is determined only by geometric axioms. Geodesic geometry is of interest not only as a generalization of Riemannian geometry, but also as an attempt to investigate geometric objects more geometrically, without using complicated computations.

In each set X a metric ρ_T can be defined by the following rule: $\rho_T(x, y) = 0$ if $x = y$, and $\rho_T(x, y) = 1$ if $x \neq y$. This is called the *trivial metric*. Each metric ρ on a set X gives rise, in a natural way, to a topology $\mathcal{T}_\rho$ on X. The concept of a **topological space** axiomatizes the relation of absolute nearness of a point to a set, whereas the concept of a metric space formalizes the notion of relative nearness of points. The *distance* $\rho(x, A)$ of a point x to a set A in a metric space (X, ρ) is defined as $\inf\{\rho(x, y): y \in A\}$. A point x is said to be absolutely near to a set A if $\rho(x, A) = 0$. The *closure* $[A]$ of A in (X, ρ) is the set of all points of X absolutely near to A. The topology in X uniquely associated with this operation is called the *topology generated in X by the metric ρ*. The trivial metric leads to the trivial topology, in which all sets are closed.

In research on metric spaces (particularly on their topological properties) the idea of a convergent sequence plays an important role. This is explained by the fact that the topology of a metric space can be completely described in the language of sequences.

Let $\xi = \{x_n: n = 1, 2, \ldots\}$ be a sequence of points in a metric space (X, ρ). It is said to *converge* to a point $x \in X$ if for each $\epsilon > 0$ there is an integer N such that $\rho(x, x_n) < \epsilon$ for all $n > N$. The sequence ξ is called *fundamental* if for each $\epsilon > 0$ there is an integer N such that $\rho(x_n, x_m) < \epsilon$ for all $m, n > N$.

An important metric property is completeness. A metric space (X, ρ) is called *complete* if each fundamental sequence in it converges to a point of it. The space (X, ρ_T) is always complete. Completeness of a metric space is not a topological property: A metric space homeomorphic to a complete metric space may be non-complete, for example, the real line **R** with the usual metric $\rho(x, y) = |x - y|$ is homeomorphic to the interval $(0, 1) = \{x \in \mathbf{R}: 0 < x < 1\}$ with the same metric; however, the first metric space is complete and the second is not.

Examples of complete metric spaces are Euclidean and Banach spaces. An important property of complete metric spaces, preserved under homeomorphisms, is the **Baire property**, on the strength of which each complete metric space without isolated points is uncountable. Therefore the usual topological space of the rational numbers is not generated by any complete metric. However, each metric space may be represented as a subset of some complete metric space by the standard construction of completion. Two fundamental sequences $\xi = \{x_n\}$ and $\eta = \{y_n\}$ in a metric space (X, ρ) are called *equivalent* if

$$\lim_{n \to \infty} \rho(x_n, y_n) = 0.$$

Let $\tilde{X}$ be the resulting collection of equivalence classes. A metric $\tilde{\rho}$ is introduced on $\tilde{X}$ by the rule: If $a', a'' \in \tilde{X}$ and $a' \ni \xi = \{x_n'\}$, $a'' \ni \xi'' = \{x_n''\}$, then

$$\tilde{\rho}(a', a'') = \lim_{n \to \infty} \rho(x_n', x_n'').$$

For $x \in X$, let $i(x) = \{x_n\}$, where $x_n = x$ for all n. Then $(\tilde{X}, \tilde{\rho})$ is a complete metric space and $i: X \to \tilde{X}$ is an isometry of (X, ρ) onto an everywhere-dense subset in $(\tilde{X}, \tilde{\rho})$ ($(\tilde{X}, \tilde{\rho})$ is called the *completion* of (X, ρ)).

Related to the discussion of completion is the **Lav'rent'ev theorem** on the extension of homeomorphisms. This theorem implies that the property of a metric space being a G_δ-set in its completion is a topological invariant (in contrast with the non-invariance of metric completeness itself relative to homeomorphisms).

Two metrics ρ_1 and ρ_2 on a set X are called *topologically equivalent* if the topologies $\mathcal{T}_{\rho_1}$ and $\mathcal{T}_{\rho_2}$ generated by them coincide. On a finite set all metrics are equivalent; they generate the discrete topology. The *Aleksandrov $-$ Hausdorff theorem*: A metric ρ on a set X is topologically equivalent to some complete metric if and only if X is a G_δ-set in the completion of (X, ρ). In particular, the space of irrational numbers with the usual metric, relative to which it is not complete, is homeomorphic to the complete metric Baire space whose points are the infinite sequences $\{n_i\}$ of natural numbers, with distance given by $\rho(\{n_i\}, \{m_i\}) = 1/k$, where k is such that $n_k \neq m_k$ and $n_i = m_i$ for all $i < k$.

The following example of a complete metric space is important: The space $C[0, 1]$ of all continuous functions on $[0, 1]$, with metric defined by the rule

$$\rho(f, g) = \max\{\,|\,f(x) - g(x)\,| : x \in [0, 1]\}$$

for all $f, g \in C[0, 1]$. The space $C[0, 1]$ is separable — there is a countable everywhere-dense set in it (cf. **Separable space**). It turns out that each separable metric space is isometric to some subset of $C[0, 1]$ (the *Banach–Mazur theorem*). This result means that all metrics which generate separable topologies are obtained by restricting the natural metric on the set of continuous functions.

A subset Y of a complete metric space (X, ρ), equipped with the same metric ρ (more precisely, its restriction to $Y \times Y$), is a complete metric space if and only if Y is closed in (X, ρ).

There is a fundamental connection between the ideas of completeness and compactness for a metric space. *Compactness* of a metric space (X, ρ) is equivalent to any of the following conditions: 1) any sequence in (X, ρ) contains a convergent subsequence; 2) each countable open covering (cf. **Covering (of a set)**) of (X, ρ) contains a finite subcovering; 3) in any open covering of (X, ρ) there is a finite subcovering; 4) each decreasing sequence of non-empty closed sets in (X, ρ) has a non-empty intersection; and 5) every closed discrete subset of (X, ρ) is finite.

The simplest examples of compact metric spaces are: finite discrete spaces, any interval (together with its end points), a square, a circle, and a sphere. In general, a subset of the Euclidean space E^n, with the usual metric, is compact if and only if it is closed and bounded.

The conditions listed are not all equivalent outside the class of metric spaces (see **Compact space**). H. Lebesgue (1911) established that for each open covering γ of a compact metric space (X, ρ) there is a number $\delta > 0$ such that each set $A \subset X$ of diameter $\leqslant \delta$ is contained in some element of γ. This implies the fundamental property of compact metric spaces: Every continuous mapping of such a space into an arbitrary metric space is uniformly continuous (cf. **Uniform continuity**). Further, a metric space is compact if and only if each real-valued continuous function on it is bounded (and attains its least and greatest values).

Each compact metric space is complete, but the converse is false; the simplest example is an infinite discrete space with the trivial metric. However, the following characteristic property holds: A metric space is compact if and only if every metric space homeomorphic to it is complete.

It is intuitively clear that compactness implies, besides completeness, some sort of boundedness; this is confirmed by the consideration of compact subsets of E^n. In general, a metric space (X, ρ) is called bounded if there is a real number a such that $\rho(x, y) < a$ for all $x, y \in X$. Every compact metric space is bounded. The space $(X, \rho_{\mathcal{T}})$ is complete and bounded, but not compact if X is infinite; thus, completeness and boundedness together are not sufficient for compactness in the class of metric spaces. In general, each metric on a set is topologically equivalent to some bounded metric, which is complete if the given metric is complete. In this connection, there is the important notion of total boundedness. A metric space (X, ρ) is called *totally bounded* if for each $\epsilon > 0$ there is a finite set $A_\epsilon \subset X$ such that $\rho(x, A_\epsilon) < \epsilon$ for all $x \in X$. The set A_ϵ here is called an *ϵ-net* in (X, ρ). A metric space (X, ρ) is compact if and only if it is complete and totally bounded, and (X, ρ) is totally bounded if and only if it is isometric to a subset of some compact metric space. More precisely, total boundedness of a metric space is equivalent to compactness of its completion $(\tilde{X}, \tilde{\rho})$. Each subspace of a totally-bounded metric space is totally bounded. All totally-bounded metric spaces (in particular, all compact metric spaces) are separable and have a countable base. Compactness, in general, is not inherited by subsets; a set $A \subset X$ is relatively compact in a metric space (X, ρ) if the closure of A in (X, ρ) is a compact metric space. If (X, ρ) is complete, then relative compactness of a set $A \subset X$ in (X, ρ) is equivalent to total boundedness of A equipped with the metric ρ.

An important role in functional analysis is played by a criterion for compactness of a set A of continuous functions on $[0, 1]$ in the metric space $C[0, 1]$. This criterion is the following (the *Arzelà–Ascoli theorem*): A set A is relatively compact in $C[0, 1]$ if and only if: 1) there is a number M such that $|\,f(x)\,| < M$ for all $x \in [0, 1]$ and all $f \in A$; and 2) for each $\epsilon > 0$ there is a $\delta > 0$ such that $|\,f(x') - f(x'')\,| < \epsilon$ for all $f \in A$ and all $x', x'' \in [0, 1]$ for which $|\,x' - x''\,| < \delta$.

A mapping f of a metric space (X, ρ) into itself is called a *contraction* if there is a real number $\lambda < 1$ such that

$$\rho(f(x), f(y)) \leqslant \lambda \rho(x, y)$$

for all $x, y \in X$. An important theorem for complete metric spaces is the *contracting- (contraction-) mapping principle* (cf. also **Contraction-mapping principle**): For each such mapping of a (non-empty) complete metric space (X, ρ) into itself there is precisely one **fixed point**.

The topological theory of metric spaces is significantly simpler than the general theory of topological spaces. Below the most important topological properties of metric spaces (X, ρ) are given. Here one has in mind properties of the topology $\mathcal{T}_\rho$ that is generated by the metric ρ.

Each metric space is normal and even collectionwise normal (cf. **Normal space**). This permits the extension of continuous real-valued functions from closed subsets of a metric space to the whole space. A stronger result is: For each closed subset Y of a metric space (X, ρ) there is a linear mapping ϕ of the space of all continuous real-valued functions on (Y, ρ) to the space of all continuous real-valued functions on (X, ρ) such that (for any f) $\phi(f)$ is an extension of f and

$$\sup\{\,|\,f(x)\,|: x \in Y\} = \sup\{\,|\,\phi(f)(x)\,|: x \in X\}$$

(*Dugundji's theorem*). This theorem is related to *Hausdorff's theorem on the extension of metrics*: If a closed subspace Y of a metrizable space X is already metrizable with a metric ρ_1 (generating the topology on Y as a subspace of X), then it is possible to extend ρ_1 to a metric ρ on the whole of X, generating the original topology on X. Similar results are valid for totally-bounded metrics and complete metrics.

Research into the topological properties of metric spaces is, to a large extent, based on the following *theorem of A.H. Stone*: A metric space is *paracompact*, that is, any open covering γ has an open locally finite refinement λ (locally finite means that each point has a neighbourhood intersecting only a finite number of elements of λ, cf. also **Paracompact space**). The *Nagata–Smirnov metrization criterion* (see **Metrizable space**) is based on the paracompactness of metric spaces.

For a metric space there are important theorems on the equivalence of topological properties which are distinct in general topological spaces. Thus, the following cardinal-valued invariants coincide: the density, the character, the weight, the Suslin number, and the Lindelöf number (cf. also **Cardinal characteristic**). For metric spaces countable compactness, pseudo-compactness and compactness are equivalent. For metric spaces the dimensions dim (the covering dimension) and Ind (the large inductive dimension) coincide, and for separable metric spaces the small inductive dimension ind coincides with dim and Ind (see **Dimension theory**).

Each metric space (X, ρ) is star normal; any open covering γ of (X, ρ) has an open star refinement λ, that is, for each point $x \in X$ there is a $U \in \gamma$ containing every $V \in \lambda$ for which $x \in V$. Related to this theorem is the following *metrizability criterion* (*Stone–Arkhangel'skiĭ*): A **regular space** is metrizable by a totally-bounded metric if the space has a countable base. But even a countable regular space need not be metrizable. The simplest example is obtained by adjoining to the discrete natural numbers one exterior point from its **Stone–Čech compactification**. The criterion for metrizability of a metric space X by a complete metric is unexpected: It is necessary and sufficient that X be a G_δ-set in some (and then in any) Hausdorff compactification of X. However, Hausdorff compactifications of metric spaces carry complete information on the topology of the latter; this is clear from *Čech's theorem*: Metric spaces are homeomorphic if and only if their Stone–Čech compactifications are homeomorphic.

A metric space need not have a countable base, but it always satisfies the **first axiom of countability**: it has a countable base at each point. In addition, each compact set in a metric space has a countable base. Moreover, in each metric space there is a base such that each point of the space belongs to only countably many of its elements — a point-countable base, but this property is weaker than metrizability, even for paracompact Hausdorff spaces. Regular separable spaces satisfying the first axiom of countability need not be metrizable.

The condition for metrizability of a separated topological group is easily found: It is necessary and sufficient that the space of the group satisfies the first axiom of countability; there are then both left-invariant and right-invariant metrics generating the topology.

Connected with each metric space (X, ρ), in a standard way, there is another metric space, namely the space $F(X)$ of its non-empty bounded closed subsets with the **Hausdorff metric**, which is defined as follows:

$$\rho_H(A, B) = \max(\sup\{\rho(a, B): a \in A\}, \sup\{\rho(b, A): b \in B\}).$$

The space (X, ρ) is isometric to a closed subspace of $(F(X), \rho_H)$. If ρ is complete, then ρ_H is complete. But topological equivalence of two metrics ρ' and ρ'' given on X does not imply, in general, that the corresponding Hausdorff metrics ρ'_H and ρ''_H are topologically equivalent.

A continuous image of a metric space need not be homeomorphic to any metric space, even when the Hausdorff **separation axiom** is satisfied. This also applies to quotient spaces of metric spaces. For example, if in the plane one shrinks a fixed line to a point, taking as individual elements of the decomposition all the points of the plane not on the line, then one obtains a non-metrizable separable space, at whose special point the first axiom of countability is not satisfied. There is a general criterion for the metrizability of a quotient space of a metric space (see [6]). In particular, the quotient space associated with a **continuous decomposition** of a metric space into compact sets is always metrizable. Every Hausdorff space which is a continuous image of a compact metric space is metrizable and compact. This is a particular case of a general proposition on the non-increase of the weight of a topological space under a continuous mapping into a compact set. But even when the image Y of a metric space (X, ρ) is metrizable, the metric realizing the metrization of Y need not be obtained from ρ by means of any formula. Instead of a metric on $Y \times Y$, with respect to ρ it is natural to define a function d by means of the rule: For $y', y'' \in Y$, $d(y', y'')$ is equal to the distance in the sense of ρ between the inverse images of the points y', y'' under the mapping under discussion. Often (for example, if Y is the decomposition space of a metric space into compact sets) d is closely related to the topology of Y and is *symmetric*. This means that $d(y',y'')=d(y'',y')$ for all $y',y''\in Y$, $d(y',y'')=0$ if and only if $y'=y''$. The symmetric relation d thus defined almost never satisfies the triangle axiom (cf. **Metric**), but if the decomposition into compact sets is continuous, then d has topological properties which successfully replace the triangle axiom and guarantee the metrizability of the image by a 'true' metric. A topological space which is the image of a metric space under a continuous open and closed mapping is itself homeomorphic to a metric space. However, under continuous open mappings, metrizability is not always preserved: All spaces satisfying the first axiom of countability, and only they, are the images of metric spaces under continuous open mappings.

Among the generalizations of metric spaces the most important are pseudo-metric spaces, spaces with a symmetric relation and spaces with 0-metrics [7]. These are defined axiomatically by a natural weakening of the axioms of a metric space. However, the distance here, as usual, is expressed by a non-negative real number. One may consider generalized metrics with values in ordered semi-groups, semi-fields, etc. (see [8]). In this way a generalized metrization of an arbitrary completely-regular space can be obtained.

A fundamental generalization of the concept of a metric space is the notion of a **uniform space**. Further, there are purely topological extensions of the class of metric spaces, among which are the important classes of spaces with a uniform base, Moore spaces, feathered and paracompact feathered spaces, and lattice spaces (cf. **Moore space**; **Feathered space**). The class of paracompact spaces is too broad a generalization of the class of metric spaces, because paracompactness is not even preserved under finite products. On the contrary, the class of paracompact feathered spaces is a successful simultaneous generalization of the class of spaces homeomorphic to metric spaces and the class of compact spaces. In another direction the idea of a metric generalizes to a κ-metric and a δ-metric [4]. The concept of a statistical metric space, introduced by K. Menger, turns out to be topologically equivalent to the idea of a space with a symmetric relation.

References

[1] ALEKSANDROV, P.S.: *Introduction to set theory and general topology*, Moscow, 1977 (in Russian).
[2] FRÉCHET, M.: 'Sur quelques points du calcul fonctionnel', *Rend. Circ. Mat. Palermo* **22** (1906), 1-74.
[3] ARKHANGEL'SKIĬ, A.V. and PONOMAREV, V.I.: *Fundamentals of general topology: problems and exercises*, Reidel, 1984 (translated from the Russian).
[4] SHCHEPIN, E.V.: 'Topology of limit spaces of uncountable inverse spectra', *Russian Math. Surveys* **31**, no. 5 (1976), 155-191. (*Uspekhi Mat. Nauk* **31**, no. 5 (1976), 191-226))
[5] ENGELKING, R.: *General topology*, Heldermann, 1989.
[6] ARKHANGEL'SKIĬ, A.V.: 'Factor-mappings of metric spaces', *Soviet Math. Dokl.* **5**, no. 2 (1964), 368-371. (*Dokl. Akad. Nauk SSSR* **155**, no. 2 (1964), 247-250))
[7] NEDEV, S.I.: 'o-metrizable spaces', *Trans. Moscow Math. Soc.* **24** (1971), 213-247. (*Trudy Moskov. Mat. Obshch.* **24** (1971), 201-236))
[8] ANTONOVSKIĬ, M.YA., BOLTYANSKIĬ, V.G. and SARYMSAKOV, T.A.: 'An outline of the theory of topological semi-fields', *Russian Math. Surveys* **21**, no. 4 (1966), 163-192. (*Uspekhi Mat. Nauk* **21**, no. 4 (1966), 185-218))
[9] NEDEV, S.I. and CHOBAN, M.M.: 'o-metrics and proximity spaces. Metrization of proximity spaces', *Serdica* **1** (1975), 12-28 (in Russian).

A.V. Arkhangel'skiĭ

Editorial comments. The trivial metric is also called the *discrete metric*. A **fundamental sequence** is also called a *Cauchy sequence*. Star-normal spaces are also called *fully normal*.

There are fairly obvious numerical invariants of metric spaces such as width (diameter) and (various kinds of) dimension. A rather more hidden numerical invariant is the *Gross dispersion number* or *rendezvous number*, whose existence and uniqueness is guaranteed by the following theorem, [A12]. Let (X, d) be a compact connected metric space; then there is a unique number $a(X, d)\in\mathbf{R}$ such that for all $n\in\mathbf{N}$ and all sets of n points $x_1,\dots,x_n\in X$ there is a point $y\in X$ such that

$$\frac{1}{n}\sum_{i=1}^{n} d(x_i, y) = a(X, d). \tag{A1}$$

Some examples are as follows. If (X, d) is a ball of radius $1/2$ in Euclidean n-space, then $a(X, d)=1/2$; if S^n is the n-dimensional sphere of unit diameter in Euclidean $(n+1)$-space, then, [A15],

$$a(S^n, d) = \frac{1}{\sqrt{\pi}}\Gamma\left[\frac{2n+1}{2}\right]^{-1} 2^{n-1}\Gamma\left[\frac{n+1}{2}\right]^2,$$

where $\Gamma(x)$ is the **gamma-function**; if X is an equilateral triangle in $\mathbf{R}^2$, then $a(X, d)=1/3+\sqrt{2}/6$. The theorem guaranteeing the existence of $a(X, d)$ generalizes in two directions. First, $d: X\times X\to\mathbf{R}$ can be replaced by any symmetric function $f: X\times X\to\mathbf{R}$ (where symmetric means $f(x, y)=f(y, x)$), [A13], and further the average on the left of (A1) can be replaced by an integral, [A14]. Thus, for a compact connected Hausdorff space X and a symmetric function $f: X\times X\to\mathbf{R}$ there exists a unique real number $a(X, f)$ such that for any regular Borel probability measure μ on X there is a point $y\in X$ such that

$$a(X, f) = \int_X f(x, y)\,d\mu(x).$$

The metric spaces on which every continuous function (to any metric space, or just to the real line) is uniformly continuous are studied in [A2]. The simplest description is this: There is a compact subset C such that the complement of any neighbourhood of C is discrete.

Beyond the completion of a metric space (X, ρ) is the *injective envelope* (I, d). In general, X is not dense in I (e.g., if X is a circle, I is infinite dimensional), but I is an *essential extension* of X; this means that a non-expansive mapping f from I to any metric space M whose restriction to X preserves all distances must preserve all distances in I

(non-expansive means that $d(f(x), f(y)) \leq d(x, y)$ for all $x, y \in I$). I is characterized, up to a unique isometry, as an essential extension of X which has no further essential extension. This is equivalent to injectivity in the sense of extendability of mappings, and also to the following Helly-type property: Any family of spherical neighbourhoods $O(x_\alpha, \epsilon_\alpha)$ satisfying the consistency conditions $d(x_\alpha, x_\beta) \leq \epsilon_\alpha + \epsilon_\beta$, for all α, β, has a common point. Because of this equivalence, the injective metric spaces are also called *hyperconvex*. See [A1], [A5].

The injective envelope of a real Banach space is itself a real Banach space in a unique compatible way. This result is known only from a highly non-constructive proof combining H. Cohen's construction of relative injective envelopes in the category of real Banach spaces [A3], the *Aronszajn−Panitchpakdi theorem* that an injective real Banach algebra is an injective metric space [A1], and the *Mazur−Ulam theorem* that every isometry of real Banach spaces is affine [A9]. Compare [A6].

Injective spaces support a much stronger fixed-point theorem than the contraction-mapping theorem. Every non-expansive mapping of a bounded injective metric space into itself has a fixed point (the *Sine−Soardi theorem*). However, the extensive development of fixed-point theory of non-expansive mappings has been done mostly in the important special case of convex subsets of Banach spaces. A survey of it, up to 1980, is in [A8]. Compare [A7].

It should be noted that the *Stone−Arkhangel'skiĭ metrization criterion* involves A.H. Stone, who also proved the paracompactness of metric spaces. In the **Stone−Čech compactification**, it is M.H. Stone.

It was noted above that two topologically-equivalent metrics on a space X do not, in general, give topologically-equivalent Hausdorff metrics on the hyperspace $F(X)$. In fact, they do so if and only if they are uniformly equivalent [A4].

References

[A1] ARONSZAJN, N. [N. ARONSZAĬN] and PANITCHPAKDI, P.: 'Extensions of uniformly continuous transformations and hyperconvex metric spaces', *Pacific J. Math.* 6 (1956), 405-439.

[A2] ATSUJI, M.: 'Uniform continuity of continuous functions of metric spaces', *Pacific J. Math.* 8 (1958), 11-16.

[A3] COHEN, H.: 'Injective envelopes of Banach spaces', *Bull. Amer. Math. Soc.* 70 (1964), 723-726.

[A4] HAMMOND SMITH, D.: 'Hyperspaces of a uniformizable space', *Proc. Cambridge Philos. Soc.* 62 (1966), 25-28.

[A5] ISBELL, J.: 'Six theorems about injective metric spaces', *Comment. Math. Helv.* 39 (1964), 65-76.

[A6] ISBELL, J.: 'Three remarks on injective envelopes of Banach spaces', *J. Math. Anal. Appl.* 66 (1969), 301-306.

[A7] ISTRAŢESCU, V.I.: *Fixed point theory*, Reidel, 1981.

[A8] KIRK, W.A.: 'Fixed point theory for nonexpansive mappings', in *Fixed Point Theory*, Lecture notes in math., Vol. 886, Springer, 1981, pp. 484-505.

[A9] MAZUR, S. and ULAM, S.: 'Sur les transformations isométriques d'espaces vectoriels'', *C.R. Acad. Sci. Paris* 194 (1932), 946-948.

[A10] DUGUNDJI, J.: *Topology*, Allyn & Bacon, 1966.

[A11] KELLEY, J.L.: *General topology*, Springer, 1975.

[A12] GROSS, O.: 'The rendezvous value of a metric space', in M. Drester, L.S. Shapley and A.W. Tucker (eds.): *Advances in Game Theory*, Princeton Univ. Press, 1964, pp. 49-53.

[A13] STADJE, W.: 'A property of compact connected spaces', *Arch. Math.* 36 (1981), 275-280.

[A14] CLEARY, J., MORRIS, S.A. and YOST, D.: 'Numerical geometry — numbers for shapes', *Amer. Math. Monthly* 93 (1986), 260-275.

[A15] MORRIS, S.A. and NICKOLAS, P.: 'On the average distance property of compact spaces', *Arch. Math.* 40 (1983), 459-463.

AMS 1980 Subject Classification: 54E35, 46A06, 53B20, 53C70

METRIC TENSOR, *basic tensor, fundamental tensor* - A twice covariant symmetric tensor field $g = g(X, Y)$ on an n-dimensional differentiable manifold M^n, $n \geq 2$. The assignment of a metric tensor on M^n introduces a scalar product $<X, Y>$ of contravariant vectors $X, Y \in M_p^n$ on the tangent space M_p^n of M^n at $p \in M^n$, defined as the bilinear function $g_p(X, Y)$, where g_p is the value of the field g at the point p. In coordinate notation:

$$<X, Y> = g_{ij}(p) X^i Y^j, \quad X = \{X^i\}, \quad Y = \{Y^j\}, \quad 0 \leq i, j \leq n.$$

The metric in M_p^n with this scalar product is regarded as infinitesimal for the metric of the manifold M^n, which is expressed by the choice of the quadratic differential form

$$ds^2 = g_{ij}(p) \, dx^i \, dx^j \qquad (*)$$

as the square of the differential of the arc length of curves in M^n, going from p in the direction $dx^1, \ldots, dx^n$. With respect to its geometric meaning the form (*) is called the *metric form* or *first fundamental form* on M^n, corresponding to the metric tensor g. Conversely, if a symmetric quadratic form (*) on M^n is given, then there is a twice covariant tensor field $g(X, Y) = g_{ij} X^i Y^j$ associated with it and whose corresponding metric form is g. Thus, the specification of a metric tensor g on M^n is equivalent to the specification of a metric form on M^n with a quadratic line element of the form (*). The metric tensor completely determines the intrinsic geometry of M^n.

The collection of metric tensors g, and the metric forms defined by them, is divided into two classes, the degenerate metrics, when $\det(g_{ij}) = 0$, and the non-degenerate metrics, when $\det(g_{ij}) \neq 0$. A manifold M^n with a degenerate metric form (*) is called isotropic. Among the non-degenerate metric tensors, in their turn, are distinguished the *Riemannian metric tensors*, for which the quadratic form (*) is positive definite, and the *pseudo-Riemannian metric tensors*, when (*) has variable sign. A Riemannian (pseudo-Riemannian) metric introduced on M^n via a Riemannian (pseudo-Riemannian) metric tensor defines on M^n a Riemannian (respectively, pseudo-Riemannian) geometry.

Usually a metric tensor, without special indication, means a Riemannian metric tensor; but if one wishes to stress that the discussion is about Riemannian and not about pseudo-Riemannian metric tensors, then one speaks of a proper Riemannian metric tensor. A proper Riemannian metric tensor can be introduced on any paracompact differentiable manifold.

References

[1] EISENHART, L.P.: *Riemannian geometry*, Princeton Univ. Press, 1949.
[2] RASHEWSKI, P.K. [P.K. RASHEVSKIĬ]: *Riemannsche Geometrie und Tensoranalyse*, Deutsch. Verlag Wissenschaft., 1959 (translated from the Russian).
[3] GROMOLL, D., KLINGENBERG, W. and MEYER, W.: *Riemannsche Geometrie im Grossen*, Springer, 1968.

I.Kh. Sabitov

Editorial comments.

References

[A1] KLINGENBERG, W.: *Riemannian geometry*, de Gruyter, 1982.

AMS 1980 Subject Classification: 53B20

METRIC THEORY OF FUNCTIONS - The branch of the theory of functions of a real variable (cf. **Functions of a real variable, theory of**) in which properties of functions are studied on the basis of the idea of the **measure** of a set.

The research of many 19-th century mathematicians created a new mathematical discipline — the theory of functions of a real variable. At the end of the 19-th century there was a clear crystallization of certain problems which required solution: the problem of the measure of a set, the length of curves and the area of surfaces, the representation of functions by series (in particular, by trigonometric series), the primitive and the integral, relations between integration and differentiation, term-by-term integration of series, etc. The solution of these problems had a general mathematical significance; the most prominent mathematicians worked in this direction, which, in particular, explained the vigorous development of the metric theory of functions in the first third of the 20-th century. The foundations of this theory of functions were laid by E. Borel, R. Baire, H. Lebesgue, and others.

In 1902 Lebesgue introduced the extraordinarily important idea of the measure of a set (**Lebesgue measure**). On the basis of this idea he created a theory of the integral (**Lebesgue integral**). These two fundamental ideas — measure and integral — formed the foundation of the metric theory of functions, which is concerned with the study of properties of functions, derivatives, integrals, series of functions (in particular, trigonometric and general orthogonal series), areas of surfaces, etc.

Many basic properties of the Lebesgue measure and integral were established at the beginning of the 20-th century by Lebesgue himself (countable additivity of the measure and the integral, passage to the limit under the integral sign, differentiation of an indefinite integral, etc.). In addition, Lebesgue gave numerous applications of his research to various questions in mathematical analysis (area of surfaces, various properties of trigonometric series, singular integrals, etc.). G. Vitali (1904) independently discovered a measure identical to the Lebesgue measure, and, somewhat later, W.H. Young (1905) also constructed an integral and a measure equivalent to the Lebesgue integral and measure. However, they did not develop their theories and, in this period, made no significant applications of them. The origin of the development of the metric theory of functions in Russia must be associated with the beginning of the 20-th century. However, the first important results in this area were obtained by Russian mathematicians in the second decade of the 20-th century (D.F. Egorov, N.N. Luzin), when a new important centre of research into the metric theory of functions was formed. The founder and leader of the school of the metric theory of functions in the USSR was Luzin.

The development of the metric theory of functions can be characterized by two major directions. The first direction was to study on the basis of the measure of a set and the Lebesgue integral, and also their generalizations, both general properties of functions, integrals, trigonometric series, orthogonal series, etc., and more concrete subtle properties of them, the development and study of which by classical analytic methods was almost unattainable. This direction is, precisely, the metric theory of functions. The second, no less important direction, was the penetration of methods of the metric theory of functions into other divisions of mathematics, and also the creation, based on its ideas, of other new areas of mathematics which, in their turn, had a stimulating effect on the theory of functions itself.

On the basis of the metric theory of functions a detailed study of **boundary properties of analytic functions** was initiated and the **metric theory of numbers**, whose methods are inseparably connected with the metric theory of functions, was created. The influence of the theory of functions on the creation of functional analysis was great.

Below, certain typical results from the metric theory of functions are given, each of which signified the solution of some central question and entailed numerous subsequent investigations. Thus, in 1911, Egorov proved that every convergent sequence of measurable functions is uniformly convergent if some set of arbitrarily small measure is neglected (see **Egorov theorem**). Luzin (1912) established that every measurable function becomes continuous if some set of arbitrarily small measure is neglected (see **Luzin C-property**). These two results are difficult to overestimate since, on the one hand, they have shown a definite interrelation between basic ideas of the metric theory of functions and classical analysis, and, on the other hand, they are the foundation on which much research in the theory of functions has often been conducted.

Luzin proved (1915) the existence of primitives in the class of measurable functions. Namely, he proved that for every finite measurable function f there is a continuous function F such that $F'=f$ almost-everywhere. On the basis of this result he solved the Dirichlet problem in the class of measurable functions. This was the beginning of an active development of the Moscow school concerning the theory of functions of a complex variable and, in particular, the beginning of the study of boundary properties of analytic functions.

The Lebesgue integral does not solve the problem of finding a primitive F from the finite exact derivative $F'=f$. This problem was solved by A. Denjoy (1912) on the basis of a special integral (the narrow Denjoy integral) which naturally generalizes the Lebesgue integral and does not contradict it (see **Denjoy integral**). In 1916 Denjoy and A.Ya. Khinchin constructed an even more general integral, named the Denjoy integral in the wide sense; it is related to **approximate differentiability**.

At the beginning of the 3-rd decade of the 20-th century D.E. Men'shov and H. Rademacher established that the sequence $\{\log^2 n\}$ is a Weyl multiplier for the almost-everywhere convergence of series in any orthogonal system. In addition (and this is particularly important and fundamental), Men'shov proved (1923) that the above result is not true if the sequence $\{\log^2 n\}$ is replaced by any sequence $\{\omega(n)\}$ with $\omega(n)=o(\log^2 n)$ as $n\to\infty$. These results became the foundation on which various investigations into the theory of convergence and summability of orthogonal series have been and still are being built.

In 1926 Kolmogorov constructed an example of an everywhere-divergent trigonometric Fourier series of a summable function. However, the problem of almost-everywhere convergence of the trigonometric Fourier series of a function in L_2, proposed by Luzin in 1914, remained unsolved for a long time. A positive solution to this problem was given by L. Carleson only in 1966 (see **Carleson theorem**). P. du Bois-Reymond, Lebesgue and Ch. de la Vallée-Poussin solved the problem of the recovery of the coefficients of a trigonometric series which converges to a summable function. In the 1940's Denjoy constructed an integral by which the problem of the recovery of coefficients in a sum of an arbitrary everywhere-convergent trigonometric series could be solved. At the same time Men'shov proved that every finite measurable function is representable by some almost-everywhere convergent trigonometric series.

G. Cantor, Young, N.K. Bari, etc., have laid the foundations of the theory of uniqueness of trigonometric series.

References

[1] LEBESGUE, H.: *Leçons sur l'intégration et la récherche des fonctions primitives*, Gauthier-Villars, 1928.
[2] LUZIN, N.N.: *The integral and trigonometric series*, Moscow, 1951 (in Russian).
[3] BARY, N.K. [N.K. BARI]: *A treatise on trigonometric series*, Pergamon, 1964 (translated from the Russian).
[4] SAKS, S.: *Theory of the integral*, Hafner, 1952 (translated from the Polish).
[5] HALMOS, P.: *Measure theory*, v. Nostrand, 1950.
[6] PAPLAUSKAS, A.B.: *Trigonometric series from Euler to Lebesgue*, Moscow, 1966 (in Russian).
[7] PESIN, I.N.: *Classical and modern integration theories*, Acad. Press, 1970 (translated from the Russian).
[8] MEN'SHOV, D.E. and UL'YANOV, P.L.: 'On the representation of functions by series', *Vestnik Moskov. Univ. Ser. Mat. Mekh.* **5** (1967), 24-36 (in Russian).
[9] KOLMOGOROV, A.N.: 'The theory of functions of a real variable', in *Science in the USSR during fifteen years: Mathematics*, Moscow-Leningrad, 1932 (in Russian).
[10] 'The metric theory of functions of a real variable', in *Mathematics in the USSR during thirty years*, Moscow-Leningrad, 1948 (in Russian).
[11] LOZINSKIĬ, S.M. and NATANSON, I.P.: 'Metric and constructive theory of functions of a real variable', in *Mathematics in the USSR during forty years*, Vol. 1, Moscow, 1959 (in Russian).
[12] UL'YANOV, P.L.: 'The metric theory of functions', in *The history of Soviet Mathematics*, Vol. 3, Kiev, 1968, pp. 530-568 (in Russian).
[13] TATALYAN, A.A.: 'Representation and uniqueness problems in the theory of orthogonal series', *J. Soviet Math.* **1**, no. 6 (1973), 635-700. (*Itogi Nauk. Mat. Anal.* (1970), 5-64)

P.L. Ul'yanov

Editorial comments. For the notion of Weyl multiplier see **Multiplier theory**.

AMS 1980 Subject Classification: 28A20

METRIC THEORY OF NUMBERS - The branch of **number theory** which studies and metrically (that is, based on **measure** theory) characterizes sets of numbers with fixed arithmetic properties. Metric number theory is closely connected with probability theory, which sometimes proves an opportunity to use its methods and results in the analysis of number-theoretic models.

Many problems touching on arithmetic properties of individual numbers also have a metric formulation; for example, alongside questions on the uniform distribution of the sequence of fractional parts $\{\alpha n\}$, $n=1,2,\ldots$, for $\alpha=\sqrt{2}$ or $\log 3$, it is possible to pose the question: What is the Lebesgue measure of those α from $(0, 1)$ for which this sequence is uniformly distributed? Such a metric generalization of a problem often turns out to be very useful and provides an opportunity to represent a phenomenon in the large. Sometimes one can fairly easily prove, based on metric arguments, the existence of numbers with definite arithmetic properties, whereas a direct construction of these numbers is complicated (Borel normal numbers, cf. **Normal number**, numbers with certain approximation properties, etc.).

The most significant achievements in metric number theory are related to the metric theory of Diophantine approximation (cf. **Diophantine approximation, metric theory of**), to the theory of **uniform distribution** of sequences of numbers, to the theory of continued fractions (cf. **Continued fraction**), and to other branches of number theory.

One of the first theorems in metric number theory is *Borel's theorem* (E. Borel, 1909): When written in an arbitrary fixed integer base g, almost-all (in the sense of Lebesgue measure) real numbers α in the interval $(0, 1)$ are normal (cf. **Normal number**). In an equivalent formulation, this theorem asserts that the fractional parts $\{\alpha g^n\}$, $n = 1, 2, \ldots$, are uniformly distributed in the interval $(0, 1)$. Borel's theorem has been generalized and extended by many mathematicians. It turned out to be fruitful to take the view that the 'digits' $0, \ldots, g-1$ in the base g expansion of α (in the g-ary system) are independent random variables. Founded explicitly or implicitly on this situation, and by applying methods developed in probability theory for finding the asymptotic distributions of sums of independent or weakly-dependent random variables, the basic questions regarding the distribution of the 'digits' $0, \ldots, g-1$, and arbitrary groups of 'digits', in the g-ary expansion of numbers 'randomly' chosen from $(0, 1)$, were solved. For example, putting

$$\alpha = \frac{a_1}{g} + \frac{a_2}{g^2} + \cdots,$$

where the a_i are numbers from $0, \ldots, g-1$, it turns out that $a_i = a_i(\alpha)$ can be considered as independent random variables defined on $(0, 1)$ with the Lebesgue measure as probability measure. If a is any number from $0, \ldots, g-1$, and $k_n(\alpha)$ denotes the number of $i \leq n$ for which $a_i(\alpha) = a$ for a given α, then

$$\frac{(k_n(\alpha) - n/g)g}{\sqrt{n(g-1)}}$$

is asymptotically normally distributed, that is, for any real x the measure of the set of α's for which

$$k_n(\alpha) - \frac{n}{g} < \frac{x}{g}\sqrt{n(g-1)}$$

tends, as $n \to \infty$, to the limit

$$\frac{1}{\sqrt{2\pi}} \int_{-\infty}^{x} e^{-t^2/2}\, dt.$$

H. Weyl (1916) proved that if a_n, $n = 1, 2, \ldots$, is an arbitrary increasing sequence of natural numbers, then for almost-all α the fractional parts $\{\alpha a_n\}$ are uniformly distributed in the interval $(0, 1)$. On the assumption that the a_n are the values of a function defined on the infinite interval $(1, \infty)$ and satisfying special analytic conditions, this theorem has a more precise version concerning the 'quality' of the uniform distribution. J. Koksma [8] proved a general theorem on the distribution of the fractional parts (cf. **Distribution modulo one**) of a function in two variables $f(\alpha, n)$, where α is a real number taking almost-all values from the interval $(1, \infty)$, and $n = 1, 2, \ldots$. For example, the fractional parts $\{\alpha^n\}$, for almost-all $\alpha > 1$, are uniformly distributed in the interval $(0, 1)$.

Apart from questions connected with Borel normal numbers, one of the basic objects in metric number theory, from the start of its development, was the metric theory of continued fractions. Let α be a real number in the interval $(0, 1)$, let $\alpha = [0, a_1, \ldots]$ be its **continued fraction** expansion, let $r_n(\alpha) = [a_n, a_{n+1}, \ldots]$, and let $q_n(\alpha)$ be the denominator of the n-th convergent $[0, a_1, \ldots, a_n]$. A.Ya. Khinchin established (1935) that for almost-all α, as $n \to \infty$,

$$(a_1 \cdots a_n)^{1/n} \to K = \prod_{k=1}^{\infty} \left[1 + \frac{1}{k(k+1)}\right]^{\ln k / \ln 2} =$$

$$= 2.68545 \cdots,$$

and there is an absolute constant γ such that for almost-all α, $(q_n(\alpha))^{1/n} \to \gamma$ as $n \to \infty$. P. Lévy found $\gamma = \exp(\pi^2 / 12 \ln 2)$. In addition, Khinchin [5] used his results on the metric properties of continued fractions to prove a theorem on the approximation of numbers by rationals. Let f be a positive continuous function of a positive argument x such that $xf(x)$ is a non-increasing function. Then the inequality

$$\left| \alpha - \frac{p}{q} \right| < \frac{f(q)}{q} \qquad (*)$$

has, for almost-all α, an infinite set of solutions in integers p and q ($q > 0$) if for some $c > 0$ the integral

$$\int_{c}^{\infty} f(x)\, dx$$

diverges; conversely, the inequality (*) has, for almost-all α, at most a finite number of solutions in integers p and q ($q > 0$) if the integral converges for all $c > 0$.

This theorem has been transcended and generalized from various points of view. It became the starting point of an intensive development of the metric theory of Diophantine approximation. P. Erdös [7], completing a series of papers of his predecessors, obtained the following result. A necessary and sufficient condition under which for almost-all α an infinite number of $q_i(\alpha)$ are contained in an arbitrary sequence $n_1 < n_2 < \cdots$ is

$$\sum_{i=1}^{\infty} \frac{\phi(n_i)}{n_i^2} = \infty,$$

where $\phi(n)$ is the **Euler function**. Under the same condition, for almost-all α the inequality

$$\left|\alpha - \frac{m}{n_i}\right| < \frac{\epsilon}{n_i^2}, \quad (m, n_i) = 1,$$

where m is an integer and $\epsilon > 0$ is arbitrary, has an infinite number of solutions. These results are close to the conjecture (1982): If $n_1 < n_2 \cdots$ is an arbitrary sequence of integers and $\delta_i > 0$ are arbitrary, then the inequality

$$\left|\alpha - \frac{m}{n_i}\right| < \frac{\delta_i}{n_i}, \quad (m, n_i) = 1,$$

has an infinite number of solutions for almost-all α if and only if

$$\sum_{i=1}^{\infty} \frac{\delta_i \phi(n_i)}{n_i} = \infty.$$

R.O. Kuz'min proved (1928) that for any $x \in (0, 1)$ the measure $m_n(x)$ of the set of α's for which $r_n(\alpha) - a_n < x$, is equal to

$$\frac{\ln(1+x)}{\ln 2} + O\left(e^{-\lambda \sqrt{n}}\right),$$

where $\lambda > 0$ is an absolute constant. The asymptotic relation

$$m_n(x) \sim \frac{\ln(1+x)}{\ln 2} \quad (n \to \infty)$$

was known to C.F. Gauss, but he did not publish it, and he mentions in one of his letters to P. Laplace that it would be very desirable to estimate the difference $m_n(x) - \ln(1+x)/\ln 2$. Kuz'min's estimate $O(e^{-\lambda \sqrt{n}})$ was improved by Lévy (1929) to $O(e^{-\lambda n})$. Kuz'min's method was the source of many other metric theorems on continued fractions.

The modern treatment of metric questions connected with Borel normal numbers and the theory of continued fractions uses ideas of **ergodic theory**. This is based on the fact that the mappings $\alpha \to \{\alpha g\}$ and $\alpha \to \{1/\alpha\}$ of the interval $(0, 1)$ into itself, which are closely connected with the g-ary fraction and the continued fraction respectively, preserve a measure and are ergodic: the first preserves the Lebesgue measure, the second the measure $\mu(A)$ defined for every measurable set A in $(0, 1)$ by

$$\mu(A) = \frac{1}{\ln 2} \int_A \frac{dx}{1+x}.$$

From this point of view the Gauss–Kuz'min theorem, without estimates on the remainder, follows immediately from Birkhoff's **individual ergodic theorem**. Arguments from ergodic theory turn out to be useful even in order to obtain estimates of the remainder in certain limit theorems. For example, the results of Khinchin admit the improvement (see [10])

$$(a_1 \cdots a_n)^{1/n} = K + O(n^{-1/2}(\ln n)^{3/2+\epsilon})$$

$$(q_n(\alpha))^{1/n} = \exp\left[\frac{\pi^2}{12 \ln 2}\right] + O(n^{1/2}(\ln n)^{2+\epsilon}),$$

where $\epsilon > 0$ is arbitrary. The ideas of ergodic theory are useful in many other problems of the metric theory of numbers (linear Diophantine equations, the distribution

of values of matrix exponential functions, the Jacobi–Perron algorithm, and others).

In certain cases the metric characteristics of sets of numbers, based on the Lebesgue measure, turn out to be too coarse, and then more precise characteristics are used, for example, the **Hausdorff dimension**. Such an approach is particularly useful in the theory of Diophantine approximations and in the theory of transcendental numbers. For example, it has been shown [6] that for any fixed n and $w \geqslant n+1$ the set of real numbers x for which the inequality

$$|x - a| < h_\alpha^{-w}$$

has an infinite number of solutions in algebraic numbers α of degree n and height h_α, has Hausdorff dimension $(n+1)/w$. If $w < (n+1)$, then the corresponding inequality has an infinite number of solutions for almost-all x, at the same time it has been suggested that this is true for all x (*Wirsing's conjecture*). Similar results are known for complex numbers. They are directly related to the fundamental classification questions for transcendental numbers [3].

From the metric point of view, not only problems concerning real and complex numbers have been analyzed, but also problems on p-adic numbers, adèles, formal power series, etc., and, in general, elements of arbitrary spaces in which a measure can be introduced and in which some 'arithmetic' problem can be posed. In particular, for p-adic numbers there are analogues of many metric theorems from the theory of uniform distribution and the theory of Diophantine approximations of real numbers, despite the fact that the domain of p-adic numbers differs in its metric and topology (see [3], [9]).

The metric approach is effective for the solution of 'ill-posed' problems, when shortage of information on the object of research is compensated by the assumption of a 'random' choice of this object from some suitable set of objects. Here, of course, one will not succeed in investigating the primary object, which is sometimes essentially impossible because of lack of information concerning it, but one may conclude that 'almost-all' objects of the set considered have a certain property. For example, let $A^* = \{1 < a_1^* < a_2^* < \cdots\}$ be a sequence of natural numbers increasing less rapidly than some power, that is, $a_i^* < i^\mu$, μ a constant. One poses the question: Does there exist a number r such that any natural number can be represented as a sum of at most r terms of A^*? Clearly, the information given on A^* is not sufficient to solve this problem.

Let $\mathfrak{A}$ be the set of sequences $A = \{1 < a_1 < a_2 < \cdots\}$ of integers, where each a_i is 'randomly' chosen from the interval $[a_i^*, a_{i+1}^*)$. Lebesgue measure can be defined on $\mathfrak{A}$ and it can be proved that for almost-all A the required number exists and that $r < \infty$ (see [4]).

The connection between 'global' results in the metric theory of numbers and their 'individual' realizations is very interesting and profound. Despite the fact that almost-all elements of some set have a given property, it can be very difficult to establish that a given concrete element of the set has the property. For example, Borel conjectured that numbers such as $\sqrt{2}$, e, π, $\log 3$, etc., are normal; and, although almost-all numbers are normal, up till now (1989) it is not known whether any of these numbers is normal. In many cases it is easier to prove a metric theorem than to prove a similar 'individual' theorem. However, this does not mean that there are no deep problems in the metric theory of numbers, since many problems in the metric theory of numbers are closely connected with certain 'individual' problems which are sometimes discovered fairly quickly. On the other hand, the solution of 'individual' problems reveals their relationship to the metric problems.

The ideas of the metric theory of numbers play a fundamental role in many areas of **analytic number theory**, in particular when integration with respect to some measure is involved. In these cases the conclusions of certain metric theorems may not be the aim of research, but the metric understanding is used as an intermediate stage in the arguments. These theorems may be a main principle underlying the arguments; however, the final formulation of the result will not contain any metric notions. An example of the systematic use of this type of argument is the Hardy−Littlewood−Vinogradov method, in which an essential role is played by metric properties of approximations of numbers by rational fractions (the 'major' and 'minor' arcs of Hardy−Littlewood). This situation allowed I.M. Vinogradov to formulate his own theorems on estimates of Weyl sums as certain metric theorems [1]. In addition, Vinogradov's method of estimating Weyl sums has a clearly expressed metric character, establishing a connection between the 'global' estimate of the integral and the 'individual' estimate of a concrete sum. Examples of this type are rare in number theory.

References

[1] VINOGRADOV, I.M.: *The method of trigonometric sums in the theory of numbers*, Interscience, 1954 (translated from the Russian).

[2] POSTNIKOV, A.G.: *Ergodic problems in the theory of congruences and of Diophantine approximations*, Amer. Math. Soc., 1967 (translated from the Russian).

[3] SPRINDZHUK, V.G.: *Mahler's problem in metric number theory*, Amer. Math. Soc., 1969 (translated from the Russian).

[4] SPRINDZHUK, V.G.: 'Ueber die Darstellung ganzer Zahlen durch eine beschränkte Zahl von Zufälligen Summanden', *Izv. Akad. Nauk BSSR Ser. Fiz. Mat. Nauk* 1 (1970), 5-14.

[5] KHINCHIN, A.YA.: *Continued fractions*, Univ. Chicago Press, 1964 (translated from the Russian).

[6] BAKER, A. and SCHMIDT, W.: 'Diophantine approximation and Hausdorff dimension', *Proc. London Math. Soc. (3)* 21 (1970), 1-11.

[7] ERDÖS, P.: 'On the distribution of the convergents of almost all real numbers', *J. Number Theory* 2 (1970), 425-441.

[8] KOKSMA, J.F.: *Diophantische Approximationen*, Springer, 1936.

[9] LUTZ, E.: *Sur les approximations diophantiennes linéaires p-adiques*, Hermann, 1955.

[10] PHILIPP, W.: 'Some metrical theorems in number theory', *Pacific J. Math.* 20, no. 1 (1967), 109-127.

[11] CIGLER, J. and HELMBERG, G.: 'Neuere Entwicklungen der Theorie der Gleichverteilung', *Jhber. Deutsch. Math. Verein.* 64, no. 1 (1961), 1-50.

V.G. Sprindzhuk

Editorial comments. For a number of striking applications of measure-theoretic and ergodic-theoretic ideas in number theory see also [A1].

References

[A1] FURSTENBERG, H.: *Recurrence in ergodic theory and combinatorial number theory*, Princeton Univ. Press, 1981.

AMS 1980 Subject Classification: 10KXX, 10FXX, 10LXX

METRIC TRANSITIVITY *of a dynamical system* $\{T_t\}$ *with a (quasi-) invariant measure* μ - The property of $\{T_t\}$ that any measurable subset A of the phase space W which is invariant relative to T_t (in the sense that it coincides with its complete inverse images $(T_t)^{-1}A$) either has measure zero or coincides with W up to a set of measure zero. A formally stronger version of this property is obtained if in the definition instead of invariant sets one considers sets which are invariant modulo 0 (a set A differing for each t from $(T_t)^{-1}A$ by a set of measure zero). These two versions are equivalent if μ is a σ-finite measure and if the 'time' t runs through a locally compact group with a countable base (see the proof for flows in [1]), but for arbitrary transformation groups this is not the case (see the example, given for another reason, in [2]).

Instead of metric transitivity one may also speak of *metric indecomposability*, or **ergodicity**. If several (quasi-) invariant measures are considered in a given system and μ is one of them, then instead of speaking of metric transitivity (ergodicity) relative to μ one speaks of metric transitivity (ergodicity) of μ. Of course, *decomposability of a normalized invariant measure* μ means something different, i.e. the possibility to represent μ as $a_1\mu_1 + a_2\mu_2$, where μ_i are normalized invariant measures different from μ and $a_i > 0$. Indecomposability of μ is equivalent to the second (stronger) version of metric transitivity (see [2], [3]).

If there is a topology on W, then under natural assumptions (see the discussion for flows in [1]) metric transitivity implies that for almost-all $w \in W$ the trajectory of w is everywhere dense (in this sense, metric transitivity implies **topological transitivity**). The converse is false.

Starting with J. von Neumann [4], a number of results were obtained concerning the decomposition of non-ergodic systems into ergodic components. For the traditional, purely metric version (W a **Lebesgue space** and only one (quasi-) invariant measure involved) see [5], [6], [11]. N.M. Krylov and N.N. Bogolybov [7] obtained stronger results for topological flows and cascades on a metric compactum W, of a twofold kind (see also [8], [9]): 1) a decomposition of normalized invariant measures, compatible with the topology, into ergodic measures; and 2) a 'geometric' realization of all these decompositions at once by means of ergodic sets (cf. **Ergodic set**). Regarding the generalization of these results to other transformation groups, it is known that 1) is valid under broader assumptions than 2) (see [2], [3]). There are also similar results for dynamical systems (at present only cascades and flows) in appropriate measurable spaces (cf. **Measurable space**) and (or) for quasi-invariant measures (see [9] - [10]).

Metric transitivity of a partition ξ *of a measure space* (W, μ) is the property that any measurable subset $A \subset W$ consisting entirely of elements of ξ either has measure zero or coincides with W up to a set of measure zero. Instead of metric transitivity one also speaks of *absolute non-measurability* of ξ. Metric transitivity of a dynamical system generated by a group of transformations, whose trajectories form a partition of the phase space, coincides with metric transitivity of this partition. If the transformations forming the dynamical system are not invertible, then their metric transitivity also reduces to that of some partition; however, its description is more complicated: the element containing some point consists of all images of the point and all inverse images of these images.

References

[1]   HOPF, E.: *Ergodentheorie*, Springer, 1937.
[2]   FOMIN, S.V.: 'On measures invariant under certain groups of transformations', *Izv. Akad. Nauk SSSR Ser. Mat.* **14**, no. 3 (1950), 261-274 (in Russian).
[3]   BOGOLYUBOV, N.N.: 'On some ergodic properties of continuous transformation groups', in *Selected Works*, Vol. 1, Kiev, 1969, pp. 561-569 (in Russian).
[4A] NEUMANN, J. VON: 'Zur Operatorenmethode in der klassischen Mechanik', *Ann. of Math.* **33**, no. 3 (1932), 587-642.
[4B] NEUMANN, J. VON: 'Zusätze zur Arbeit 'Zur Operatorenmethode in der klassischen Mechanik'', *Ann. of Math.* **33**, no. 4 (1933), 789-792.
[5]   ROKHLIN, V.A.: 'Selected topics from the metric theory of dynamical systems', *Transl. Amer. Math. Soc. Ser. 2* **49** (1966), 171-240. (*Uspekhi Mat. Nauk* **4**, no. 2 (1949), 57-128)
[6]   ROKHLIN, V.A.: 'On the decomposition of a dynamical system into transitive components', *Mat. Sb.* **25**, no. 2 (1949), 235-249 (in Russian).
[7]   BOGOLYUBOV, N.N. and KRYLOV, N.M.: 'La theorie générale de la mesure dans son application à l'étude des systèmes dynamiques de la mécanique non-linéaire', *Ann. of Math. (2)* **38** (1937), 65-113.
[8]   NEMYTSKIĬ, V.V. and STEPANOV, V.V.: *Qualitative theory of differential equations*, Princeton Univ. Press, 1960 (translated from the Russian).
[9]   OXTOBY, J.: 'Ergodic sets', *Bull. Amer. Math. Soc.* **58** (1952), 116-136.
[10] KIFER, YU.I. and PIROGOV, S.A.: 'The decomposition of quasi-ergodic measures into ergodic components', *Uspekhi Mat. Nauk* **27**, no. 5 (1972), 239-240 (in Russian).

D.V. Anosov

Editorial comments. The term 'ergodic' is mostly used for the first version of metric transitivity; for the second (stranger) version sometimes the term 'irreducible' is used; see e.g. [A4]. The two notions coincide if for every measurable subset A of W which is invariant mod 0 there exists an invariant measurable subset A_1 of W such that $\mu(A \Delta A_1) = 0$. For conditions such that this holds, see also (the proof of) Lemma 1 in Chapt. 1, § 2 of [A8], and e.g. Lemma 3.3 in [A6]. A simple example where the two notions do not coincide is in [A5], p. 84.

What is called above 'indecomposability' is usually expressed by saying that μ is an *extreme point* of the (convex) set $J(W)$ of all invariant probability measures on W (a **probability measure** is a normalized measure). Proofs of the fact that an invariant probability measure is indecomposable (i.e., is an extreme point of $J(W)$) if and only if it is irreducible (according to the above definition), under various conditions for W or $\{T_t\}$, can also be found in [A1] and in [A6]; also the proof of theorem 6.10(iii) in [A7] can easily be adapted to the case of a general semi-group $\{T_t\}$ and arbitrary W.

As to the decomposition of the phase space into ergodic components, see also [A6] and [A2]. Such decompositions lead to representations of invariant measures as integrals of ergodic measures; if the phase space W is compact, this is an easy consequence of Choquet theory: see [A5], p. 82. For such a representation in the context of random transformations, see Appendix A.1 in [A3].

References

[A1] BLUM, J.R. and HANSON, D.L.: 'On invariant probability measures I', *Pacific J. Math.* 10 (1960), 1125-1129.
[A2] FARRELL, R.H.: 'Representation of invariant measures', *Illinois J. Math.* 6 (1962), 447-467.
[A3] KIFER, YU.: *Ergodic theory of random transformations*, Birkhäuser, 1986 (translated from the Russian).
[A4] MACKEY, G.W.: 'Point realizations of transformation groups', *Illinois J. Math.* 6 (1962), 327-335.
[A5] PHELPS, R.R.: *Lectures on Choquet's theorem*, v. Nostrand, 1966.
[A6] VARADARAJAN, V.S.: 'Groups of automorphisms of Borel spaces', *Trans. Amer. Math. Soc.* 109 (1963), 191-220.
[A7] WALTERS, P.: *An introduction to ergodic theory*, Springer, 1982.
[A8] CORNFELD, I.P. [I.P. KORNFEL'D], FOMIN, S.V. and SINAĬ, YA.G.: *Ergodic theory*, Springer, 1982 (translated from the Russian).

AMS 1980 Subject Classification: 28D15, 58F11

METRIZABLE SPACE - A space whose topology is generated by some **metric** via the rule: a point belongs to the closure of a set if and only if it lies at zero distance from the set. If such a metric exists, then it is not unique, unless the space is empty or consists of one point only. In particular, the topology of each metrizable space is generated by a bounded metric. In a metrizable space strong separation axioms (cf. **Separation axiom**) are satisfied: it is normal and even collectionwise normal. Every metrizable space is paracompact. All metrizable spaces satisfy the **first axiom of countability**. But none of the named conditions, nor any collection of them, is sufficient for a space to be metrizable. A sufficient condition for metrizability was found by P.S. Urysohn (1923): Every **normal space** (and even every **regular space**, A.N. Tikhonov, 1925) with a countable **base** is metrizable. The first general criterion for metrizability of a space was proposed in 1923 by P.S. Aleksandrov and Urysohn (see [1]). On its basis two subsequent, more precise, criteria for metrizability were developed: 1) a space is metrizable if and only if it is collectionwise normal and has a countable refining set of open coverings; 2) a space is metrizable if and only if has a countable fundamental set of open coverings and satisfies the T_1 separation axiom (the *Stone—Arkhangel'skiĭ criterion*). Here a set ξ of open coverings of a space X is called *fundamental* if for each point $x \in X$ and each of its neighbourhoods O_x there is a covering $\gamma \in \xi$ and a neighbourhood O_{1x} of x such that every element of γ intersecting O_{1x} is contained in O_x. These criteria are connected with the property of *unrestricted divisibility* and with the following fundamental property of *full normality of metrizable spaces*. Every open covering γ of a metrizable space X can be refined to an open covering γ' such that for any $x \in X$ there is an $U \in \gamma$ for which $\bigcup \{W \in \gamma' : x \in W\} \subset U$.

Based on another important idea — local finiteness — there is an important general criterion for metrizability. The *Nagata—Smirnov criterion*: A space X is metrizable if and only if it is regular and has a base decomposing into a countable set of locally finite families (cf. **Locally finite family**) of sets. *Bing's criterion* is similar, but instead of locally finite uses discrete families of sets (cf. **Discrete family of sets**). Convenient versions of the above metrizability criteria are related to the notions of a uniform base and a regular base. A base $\mathscr{B}$ of a space X is called *regular* (*uniform*) if for every point $x \in X$ and any of its neighbourhoods O_x there is a neighbourhood O_{1x} of this point such that the number of elements of the base $\mathscr{B}$ simultaneously intersecting O_{1x} and the complement of O_x is finite (respectively, if the set $\{U \in \mathscr{B} : U \ni x, U \subset O_x\}$ is finite). A space X is metrizable if and only if it is collectionwise normal and has a uniform base. Finally, for a T_1-space to be metrizable it is necessary and sufficient that it has a regular base. Regular bases are convenient in that they reveal the mechanism of paracompactness of arbitrary metrizable spaces: In order to inscribe a locally finite open covering inside any open covering γ of a space X with a regular base $\mathscr{B}$, it is sufficient to

take the collection of all maximal elements of the family

$$\mathscr{B}_\gamma = \{U \in \mathscr{B} : \text{there is a } W \in \gamma \text{ such that } U \subset W\}.$$

Metrizability criteria become very simple in a number of special classes of spaces. Thus, for a compactum X to be metrizable, any of the following four conditions is necessary and sufficient: a) X has a countable base; b) X has a point-countable base; c) there is a countable network (cf. **Net (of sets in a topological space)**; **Network**) in X; or d) the diagonal in $X \times X$ is a G_δ-set. For the metrizability of the space of a topological group it is necessary and sufficient that the group satisfies the first axiom of countability; moreover, the space is then metrizable by an invariant metric (for example, with respect to left multiplication).

A characteristic property of a metrizable space is the coincidence of a number of cardinality properties. In particular, in a metrizable space the Suslin number, the Lindelöf number, the density, the character, the spread, and the weight all coincide. The non-coincidence of these numbers is an indication of the non-metrizability of the corresponding space.

Not every metrizable space is metrizable by a complete metric: an example is the space of rational numbers. A space is metrizable by a complete metric if and only if it is metrizable and is a set of type G_δ in some compact space containing it. An important topological property of a space metrizable by a complete metric is the *Baire property*: The intersection of any countable family of everywhere-dense open sets is everywhere dense.

Very close to metrizable spaces in their properties are the so-called *Moore spaces*, i.e. completely-regular spaces having a countable refining family of open coverings, and lattice spaces.

A broad range of generalizations of the idea of a metrizable space is obtained if the metric axioms are varied, weakening them in some way or other, and by considering the topologies generated by such v-metrics. In this way *symmetrizable spaces* are obtained by abandoning the triangle axiom. Moore spaces fit into this scheme. Another important generalization of the idea of metrizability is related to the discussion of 'metrics' with values in semi-fields and other algebraic structures of a general nature.

References

[1] ARKHANGEL'SKIĬ, A.V. and PONOMAREV, V.I.: *Fundamentals of general topology: problems and exercises*, Reidel, 1984 (translated from the Russian).
[2] ENGELKING, R.: *General topology*, Heldermann, 1989.
[3] ANTONOVSKIĬ, M.YA., BOLTYANSKIĬ, V.G. and SARYMSAKOV, T.A.: *Metric spaces over semi-fields*, Tashkent, 1961 (in Russian).

A.V. Arkhangel'skiĭ

Editorial comments. The topology of a metrizable space is described here in terms of the closure operation. It is somewhat more common to use the open balls: If for $x \in X$ and $\epsilon > 0$, $B(x, \epsilon)$ denotes the set of points at distance less than ϵ from x (the ϵ-ball around x), then one calls a set U open if and only if for every $x \in U$ there is an $\epsilon > 0$ such that $B(x, \epsilon) \subseteq U$.

Metrizability criterion 1) is due to R.H. Bing [A1]. A countable refining family of open coverings is also called a *development*, see (the comments to) **Moore space** for the definition and more information.

A fundamental collection is sometimes called *locally starring*. A fully-normal space is also called a *star-normal space* (every open covering admits a *star refinement*), especially in the (translated) Russian literature. A regular topological space is fully normal if and only if it is paracompact.

Unrestricted divisibility is a property defined in [A4]; it is analogous to full normality, which is defined above, but not apparently equivalent.

For a survey (and, in part, a survey of surveys) of generalized metrizable spaces, see [A3]. For metrizable spaces, see [A5], especially pages 244-360.

References

[A1] BING, R.H.: 'Metrization of topological spaces', *Canad. J. Math.* 3 (1951), 175-186.

[A2] BURKE, D.K.: 'Covering properties', in J. Barwise (ed.): *Handbook of Set-Theoretic Topology*, North-Holland, 1984, pp. 347-422.

[A3] GRUENHAGE, G.: 'Generalized metric spaces', in J. Barwise (ed.): *Handbook of Set-Theoretic Topology*, North-Holland, 1984, pp. 423-501.

[A4] ALEKSANDROV, P.S. and PONOMAREV, V.I.: 'Some classes of n-dimensional spaces', *Sib. Mat. Zh.* 1 (1960), 3-13 (in Russian).

[A5] NAGATA, J.-I.: *Modern general topology*, North-Holland, 1985.

AMS 1980 Subject Classification: 54E35

MEUSNIER THEOREM - If γ is a curve lying on a surface and P is a point on γ, then the curvature k of γ at P, the curvature k_N of the normal section of the surface by the plane N passing through the tangent to γ at P, and the angle α between the osculating plane of γ at P and the normal plane N, satisfy the relation

$$k_N = k \cos \alpha.$$

In particular, the curvature of every inclined section of the surface can be expressed in terms of the curvature of the normal section with the same tangent.

This theorem was proved by J. Meusnier in 1779 (and was published in [1]).

References

[1] MEUSNIER, J.: *Mém. prés. par div. Etrangers. Acad. Sci. Paris* **10** (1785), 477-510.

D.D. Sokolov

Editorial comments.

References

[A1] DO CARMO, M.: *Differential geometry of curves and surfaces*,

Prentice-Hall, 1976, p. 142.

[A2] BLASCHKE, W. and LEICHTWEISS, K.: *Elementare Differentialgeometrie*, Springer, 1973.

AMS 1980 Subject Classification: 53A05

MICRO-BUNDLE - A mapping $p: E \to X$ which is a **retraction** (that is, there is a $g: X \to E$ for which $pg = 1_X$) and which is locally trivial in the sense that for each $x \in X$ there is a neighbourhood U of $g(x)$ in E which can be represented as a direct product $U = V \times \mathbf{R}^n$, with $p \,|\, _U$ the projection onto V. If for each such neighbourhood U there is fixed a piecewise-linear structure in each fibre $(p \,|\, _U)^{-1}(x)$, if, moreover, the projection of U on $\mathbf{R}^n$ is piecewise linear and for two neighbourhoods U_1 and U_2 and any $x \in p(U_1) \cap p(U_2)$ the structures on $(p \,|\, _{U_1})^{-1}(x)$ and $(p \,|\, _{U_2})^{-1}(x)$ coincide in a neighbourhood of $g(x)$, then the micro-bundle is called *piecewise linear*. Other structures may be introduced similarly.

The notion of a micro-bundle was introduced in order to define an analogue of the **tangent bundle** for a topological or piecewise-linear manifold N. Namely, here $E = N \times N$, $p(x, y) = y$ and $g(x) = (x, x)$. Each topological micro-bundle is equivalent to a unique locally trivial bundle with fibres $\mathbf{R}^n$ of corresponding dimension, that is, there is a homeomorphism h of some neighbourhood W of $g(X)$ in E into a neighbourhood $\overline{W}$ of the zero section of some bundle $\overline{p}: E \to X$ with fibre $\mathbf{R}^n$. This fact is also true for piecewise-linear micro-bundles. Despite the fact that, because of this theorem, the notion of a micro-bundle has lost its theoretical interest, it is still used in concrete problems.

A.V. Chernavskiĭ

Editorial comments.

References

[A1] MILNOR, J.: 'Microbundles, Part I', *Topology* 3, Suppl. 1 (1964), 53-80.

[A2] KIRBY, R.C. and SIEBENMANN, L.C.: *Foundational essays on topological manifolds, smoothings, and triangulations*, Princeton Univ. Press, 1977.

AMS 1980 Subject Classification: 57N55

MICROLOCAL ANALYSIS

Editorial comments. Microlocal analysis considers (generalized, hyper-) functions, operators, etc. in the 'microlocal' range. Here, 'microlocal' means seeing the matter more locally than usual by introducing the (cotangential) direction at every point. In Fourier analysis it corresponds to viewing things locally in both x and ξ. In view of the uncertainty principle, this is possible only by considering the objects modulo regular parts. This idea was first used in the study of pseudo-differential operators by P.D. Lax, S. Mizohata, L. Hörmander, etc. V.P. Maslov has enriched the theory by the introduction of a canonical structure. M. Sato has constructed the sheaf of micro-functions on the cotangent sphere bundle S^*M of the base space M as the basic object of microlocal analysis.

Micro-analyticity of hyperfunctions. A hyperfunction $f \in \mathscr{B}(M)$ is said to be *micro-analytic* at $(x_0, \xi_0) \in S^*M$ if on a neighbourhood of x_0 it admits **analytic continuation** into the half-space $<\mathrm{Im}\, z, \xi_0> < 0$, in the sense that it admits a boundary-value representation $\sum_{j=1}^{N} F_j(x + i\Gamma_j 0)$ such that $\Gamma_j \cap \{<\mathrm{Im}\, z, \xi_0> < 0\} \neq \varnothing$ for every j. This is equivalent to saying that near x_0, $f = \mathscr{F}^{-1}(g) + h$, where h is the germ of a real-analytic function and g is a Fourier hyperfunction (cf. **Hyperfunction**) exponentially decreasing in a conic neighbourhood of ξ_0. The set of points $(x_0, \xi_0) \in \Omega \times S^{n-1}$ at which f is not micro-analytic is called the *singular spectrum* of f, and is denoted by S.S. f. By definition,

$$\text{S.S. } F(x + i\Gamma 0) \subset \Omega \times (\Gamma^\circ \cap S^{n-1}),$$

where $\Gamma^\circ = \{\xi \in \mathbf{R}^n : <y, \xi> \geqslant 0 \text{ for all } y \in \Gamma\}$ is the dual cone of Γ. Conversely, a hyperfunction satisfying this estimate can be written in the form $F(x + i\Gamma 0)$.

Operations and the singular spectrum. The following inclusions hold:

$$\text{S.S.}(fg) \subset$$
$$\subset \{(x, \xi + \eta) : (x, \xi) \in \text{S.S. } f, (x, \eta) \in \text{S.S. } g\} \bigcup$$
$$\bigcup \{(x, \xi) : x \in \text{supp } f, (x, \xi) \in \text{S.S. } g,$$
$$\text{or } x \in \text{supp } g, (x, \xi) \in \text{S.S. } f\};$$
$$\text{S.S. } f(\Phi(\tilde{x})) \subset (\Phi^{-1} \times {}^t d\Phi)(\text{S.S. } f).$$

Here, the operations are legitimate if the vector 0 does not appear in the ξ-component of the right-hand side. In particular, under coordinate transformation the singular spectrum behaves like a subset of $S^*\mathbf{R}^n$. Restriction $f(x, 0) = f(x, t)|_{t=0}$ is possible if S.S. $f \cap S^*_{\{t=0\}}M = \varnothing$, in which case f is said to contain t as *real-analytic parameter at $t = 0$*, and then

$$\text{S.S. } f(x, 0) \subset \rho_*(\text{S.S. } f)|_{t=0},$$

where ρ denotes projection on the dx components. The dual assertion is:

$$\text{S.S. } \int f(x, t)\, dx \subset$$
$$\subset \{(t, \theta) : (x, t, 0, \theta) \in \text{S.S. } f \text{ for some } x\}.$$

The combination of these assertions gives a convolution, dual to the product:

$$\text{S.S.}(f * g) \subset$$
$$\subset \{(x + y, \xi) : (x, \xi) \in \text{S.S. } f, (y, \xi) \in \text{S.S. } g\}.$$

Let $P(x, \partial)$ be a **linear differential operator** with real-analytic coefficients and let $\text{Char } P = \{(x, \xi) : P_m(x, \xi) = 0\}$ be its **characteristic manifold**. Then

$$\text{S.S. } P(x, \partial)u(x) \subset \text{S.S. } u(x) \subset$$
$$\subset \text{S.S. } P(x, \partial)u(x) \bigcup \text{Char } P$$

(*Sato's fundamental theorem*). Hence Cauchy data for solutions can be specified on a non-characteristic manifold. *Holmgren's uniqueness theorem* holds with these data. More generally, for any hyperfunction u, $0 \in \text{supp } u \subset \{x_1 \geqslant 0\}$ implies $(0, \pm dx_1) \in \text{S.S. } u$ (the *Kashiwara–Kawai Holmgren-type theorem*); further, the

fibre E of S.S. u at 0 has the form $\rho^{-1}\rho(E \setminus \{\pm dx_1\}) \bigcup \{\pm dx_1\}$, where $\rho : S^{n-1} \setminus \{\pm dx_1\} \to S^{n-2}$ denotes projection to the equator $\xi_1 = 0$ (the *watermelon theorem*).

Decomposition of singular spectra. One has

$$\delta(x) = \int_{S^{n-1}} W(x, \omega)\, d\omega,$$

$$W(x, \omega) = \frac{(n-1)!}{(-2\pi i)^n} \frac{\det(\text{grad}_\omega \psi(x, \omega))}{(\phi(x, \omega) + i0)^n},$$

where the *twisted phase* $\phi(x, \omega)$ is a real-analytic function of (x, ω) which is of positive type in x (that is, $\mathrm{Re}\, \phi(x, \omega) = 0$ implies $\mathrm{Im}\, \phi(x, \omega) \geqslant 0$), is homogeneous of degree 1 in ω, and $\phi(0, \omega) = 0$, $\text{grad}_x \phi(0, \omega) = \omega$; and the vector $\psi(x, \omega)$ is such that $<\psi(x, \omega), x> = \phi(x, \omega)$. This is a generalization of the classical *Radon decomposition*, in which

$$W(x, \omega) = \frac{(n-1)!}{(-2\pi i)^n} \frac{1}{(x\omega + i0)^n}.$$

The component, regarded as a hyperfunction of x, has a singular spectrum with only one direction ω. Via convolution it gives a similar decomposition of general hyperfunctions. If $\phi(x, \omega)$ also satisfies $\phi(x, \omega) \neq 0$ for $x \neq 0$, then the singular spectrum of the component as a hyperfunction of x is precisely one point $(0, \omega)$; this is useful in applications. Typical examples are:

$$W(x, \omega) = \frac{(n-1)!}{(-2\pi i)^n} \frac{(1 - ix\omega)^{n-1} - (1 - ix\omega)^{n-2}(x^2 - (x\omega)^2)}{(x\omega + i(x^2 - (x\omega)^2) + i0)^n}$$

(*Kashiwara example*);

$$W(x, \omega) = \frac{(n-1)!}{(-2\pi i)^n} \frac{1 + i\alpha x\omega}{(x\omega + i\alpha x^2 + i0)^n}, \quad \alpha > 0,$$

(*Bony example*). For such a decomposition f is micro-analytic at (x_0, ξ_0) if and only if $f *_x W(x, \omega)$ is real analytic in (x, ω) at (x_0, ω_0).

The *Fourier–Bros–Iagolnitzer transform* of a hyperfunction f (the *FBI-transform* of f) is

$$\int_{\mathbf{R}^n} e^{i\lambda\phi(x, y, \xi, \alpha)} f(y)\, dy,$$

where $\phi(x, y, \xi, \alpha)$ is a real-analytic function satisfying: 1) for $x = y = \alpha$ one has $\phi = 0$ and $\phi_x = -\phi_y = \xi$; and 2) $\mathrm{Im}\, \phi \geqslant C(|x - \alpha|^2 + |y - \alpha|^2)$ for some $C > 0$. A typical example of such a ϕ is

$$\phi(x, y, \xi, \alpha) = (x - y)\xi + i((x - \alpha)^2 + (y - \alpha)^2).$$

A hyperfunction f is micro-analytic at (x_0, ξ_0) if and only if for some (equivalently, any) modification of f with compact support its FBI-transform is exponentially decreasing with respect to λ uniformly in (x, ξ, α) in a neighbourhood of (x_0, ξ_0, x_0). Integration of the inversion formula over the radial variable gives the formula

$$= \frac{2^n \Gamma(3n/2)}{(-2\pi i)^{3n/2}} \times$$
$$\times \int_{\mathbf{R}^n \times S^{n-1}} \frac{1 + (i/2)(x - y)\omega}{((x - y)\omega + i((x - \alpha)^2 + (y - \alpha)^2) + i0)^{3n/2}}\, d\alpha\, d\omega.$$

This supplies a partition of unity of the sheaf $\mathscr{B}/\mathscr{A}$. All these arguments are compatible with the corresponding theory of (analytic) **wave front sets** for distributions.

The *sheaf* $\mathscr{C}$ of *micro-functions* on S^*M is the sheaf associated with the pre-sheaf

$$S^*M \supset \Omega \times \Delta \mapsto \mathscr{C}(\Omega \times \Delta) =$$
$$= \mathscr{B}(\Omega) / \{ f \in \mathscr{B}(\Omega): \text{S.S. } f \cap \Omega \times \Delta = \varnothing \}.$$

The sequence of sheaves on M:

$$0 \to \mathscr{A} \to \mathscr{B} \xrightarrow{\text{sp}} \pi_* \mathscr{C} \to 0$$

is exact. Here $\pi: S^*M \to M$ denotes projection. By definition, S.S. $f = \text{supp sp}[f]$ for a hyperfunction f. The sheaf $\mathscr{C}$ is flabby, which implies the possibility of arbitrary modification of hyperfunctions preserving the singular spectrum. Analytic pseudo-differential operators (cf. **Pseudo-differential operator**) and *micro-differential operators* naturally act on $\mathscr{C}$ as sheaf homomorphisms. They act isomorphically at a non-characteristic point. Canonical transforms induce ring isomorphisms of the sheaf of pseudo-differential operators. Using these, a simple characteristic system of pseudo-differential equations can be locally reduced to the direct sum of copies of de Rham, Cauchy–Riemann and Lewy–Mizohata equations (the *fundamental structure theorem*).

The sheaf $\mathscr{C}$ is constructed from the sheaf $\mathscr{O}$ by *Sato micro-localization*: Let M be a real-analytic manifold and X a complex neighbourhood of it. Let ${}^M\tilde{X} = (X \setminus M) \amalg S_M X$ be the real blowing-up, $\iota: X \setminus M \to X$, $j: X \setminus M \to {}^M\tilde{X}$ the canonical inclusions, and let DM be the subset of the fibre product of $S_M X$ and $S^*_M X$ over M defined by $<\xi, \eta> \leq 0$; let ϕ, τ be the canonical projections to the factors. Let ω be the *orientation* sheaf of M. Then

$$\mathscr{C} = \mathrm{R}\tau_* \pi^{-1} \mathrm{R}\Gamma_{S_M X}(j_* \iota^{-1} \mathscr{O}) \otimes \omega[n] =$$
$$= \mathrm{R}^{n-1}\tau_* \pi^{-1} \mathscr{H}^1_{S_M X}(j_* \iota^{-1} \mathscr{O}) \otimes \omega$$

(the *fundamental vanishing theorem*). This argument can be generalized to the *second micro-localization* of micro-functions with respect to a holomorphic parameter or to micro-localization of any sheaf.

For additional references see also **Hyperfunction**.

References

[A1] SATO, M., KAWAI, T. and KASHIWARA, M.: 'Microfunctions and pseudo-differential equations', in H. Komatsu (ed.): *Hyperfunctions and Pseudo-differential Equations, Part II*, Lecture notes in math., Vol. 287, Springer, 1973, pp. 263-529.
[A2] KASHIWARA, M.: *Systems of micro-differential equations*, Birkhäuser, 1983 (translated from the French).
[A3] SJÖSTRAND, J.: 'Singularités analytiques microlocales', *Astérisque* 95 (1982).
[A4] LAURENT, Y.: *Théorie de la deuxième microlocalisation dans le domaine complexe*, Birkhäuser, 1985.
[A5] KASHIWARA, M. and SCHAPIRA, P.: 'Microlocal study of sheaves', *Astérisque* 128 (1986).

A. Kaneko

AMS 1980 Subject Classification: 32A35, 46F15, 58G07

MIKHAĬLOV CRITERION - All roots of a polynomial

$$P(z) = z^n + a_{n-1}z^{n-1} + \cdots + a_0$$

with real coefficients have strictly negative real part if and only if the complex-valued function $z = P(i\omega)$ of a real variable $\omega \in [0, \infty)$ describes a curve (the *Mikhaĭlov hodograph*) in the complex z-plane which starts on the positive real semi-axis, does not hit the origin and successively generates an anti-clockwise motion through n quadrants. (An equivalent condition is: The radius vector $P(i\omega)$, as ω increases from 0 to $+\infty$, never vanishes and monotonically rotates in a positive direction through an angle $n\pi/2$.)

This criterion was first suggested by A.V. Mikhaĭlov [1]. It is equivalent to the **Routh–Hurwitz criterion**; however, it is geometric in character and does not require the verification of determinant inequalities (see [2], [3]). The Mikhaĭlov criterion gives a necessary and sufficient condition for the asymptotic stability of a linear differential equation of order n,

$$x^{(n)} + a_{n-1}z^{(n-1)} + \cdots + a_0 x = 0,$$

with constant coefficients, or of a linear system

$$\dot{x} = Ax, \quad x \in \mathbf{R}^n,$$

with a constant matrix A, the characteristic polynomial of which is $P(z)$ (see [4]).

Mikhaĭlov's criterion is one of the frequency criteria for the stability of linear systems of automatic control (closely related to, for example, the **Nyquist criterion**). A generalization of Mikhaĭlov's criterion is known for systems of automatic control with delay, for impulse systems (see [5]), and there is also an analogue of Mikhaĭlov's criterion for non-linear control systems (see [6]).

References

[1] MIKHAĬLOV, A.V.: *Avtomat. i Telemekh.* **3** (1938), 27-81.
[2] CHEBOTAREV, N.G. and MEĬMAN, N.N.: 'The Routh–Hurwitz problem for polynomials and entire functions', *Trudy Mat. Inst. Steklov.* **76** (1949) (in Russian).
[3] LAVRENT'EV, M.A. and SHABAT, B.V.: *Methoden der komplexen Funktionentheorie*, Deutsch. Verlag Wissenschaft., 1967 (translated from the Russian).
[4] DEMIDOVICH, B.P.: *Lectures on the mathematical theory of stability*, Moscow, 1967 (in Russian).
[5] GNOENSKIĬ, L.S., KAMENSKIĬ, G.A. and EL'SGOL'TS, L.E.: *Mathematical foundations of the theory of control systems*, Moscow, 1969 (in Russian).
[6] BLAQUIÈRE, A.: *Mécanique non-lineaire*, Gauthier-Villars, 1960.

N.Kh. Rozov

Editorial comments. Recently a generalization of stability criteria with respect to roots of polynomials has been found. It is named after V.L. Kharitonov [A1]. The generalization is that the coefficients a_i in the polynomial $p(z) = z^n + a_{n-1}z^{n-1} + \cdots + a_0$ take values in given intervals $[a_i^-, a_i^+]$, $i = 0, \ldots, n-1$. The problem addressed by Kharitonov is whether all polynomials $p(z)$ with $a_i \in [a_i^-, a_i^+]$ are strictly stable. It turns out that the stability of only four specific polynomials has to be investigated in order to answer this question.

References

[A1] KHARITONOV, V.L.: 'Asymptotic stability of an equilibrium position of a family of systems of linear differential equations', *Differential Eq.* **14**, no. 11 (1978), 1483-1485. (*Differentsial'nye Uravnen.* **14**, no. 11 (1978), 2086-2088)

[A2] BARMISH, B.R.: 'New tools for robustness analysis', in *IEEE Proc. 27th Conf. Decision and Control, Austin, Texas, December 1988*, IEEE, 1988, pp. 1-6.

[A3] LASALLE, S. and LEFSCHETZ, S.: *Stability by Liapunov's direct method*, Acad. Press, 1961.

AMS 1980 Subject Classification: 34D99, 65H05, 12D10, 93DXX

MILNE METHOD - A finite-difference method for the solution of the **Cauchy problem** for systems of first-order ordinary differential equations:

$$y' = f(x,y), \ y(a) = b.$$

The method uses the finite-difference formula

$$y_i - y_{i-2} = 2hf(x_{i-1}, y_{i-1}),$$

where

$$x_i = a + ih, \ i = 0, 1, \ldots.$$

In computations using this formula it is necessary, by some other means, to find an additional initial value $y_1 \approx y(x_1)$. The Milne method has second-order accuracy and is Dahlquist stable, that is, all solutions of the homogeneous difference equation $y_i - y_{i-2} = 0$ are bounded uniformly with respect to h for $i = 0, \ldots, [(A-a)/h]$, for any fixed interval $[a, A]$. For Dahlquist stability it is sufficient that the simple roots of the characteristic polynomial of the left-hand side of the difference equation do not exceed one in modulus, and that the multiple roots are strictly less than one in modulus. In the case given, the characteristic polynomial $\rho(\lambda) = \lambda^2 - 1$ has roots $\lambda = \pm 1$ and, consequently, the stability condition holds. However, in solving a system of equations $y' = Ay$ with a matrix A having negative eigen values, a rapid growth in the calculation error occurs.

The *predictor-corrector Milne method* uses a pair of finite-difference formulas:

a predictor

$$\bar{y}_i = y_{i-4} + \frac{4h}{3}(2f_{i-3} + f_{i-2} + 2f_{i-1}), \ i = 4, 5, \ldots,$$

and a corrector

$$\bar{\bar{y}}_i = y_{i-2} + \frac{h}{3}(f_i + 4f_{i-1} + f_{i-2}), \ i = 4, 5, \ldots,$$

where

$$f_i = f(x_i, y_i), \ \bar{f}_i = f(x_i, \bar{y}_i).$$

An approximate expression for the error is given by the quantity

$$\epsilon_i = \frac{1}{29}(\bar{y}_i - \bar{\bar{y}}_i).$$

Additional initial values $y_j \approx y(x_j)$, $j = 1, 2, 3$, are calculated by some other means, for example, by the **Runge−Kutta method**, which has fourth-order accuracy. The method was proposed in [3].

References

[1] BAKHVALOV, N.S.: *Numerical methods: analysis, algebra, ordinary differential equations*, Mir, 1977 (translated from the Russian).

[2] DEMIDOVICH, B.P., MARON, I.A. and SHUVALOVA, E.Z.: *Numerische Methoden der Analysis*, Deutsch. Verlag Wissenschaft., 1968 (translated from the Russian).

[3] MILNE, W.E.: *Numerical solution of differential equations*, Dover, reprint, 1970.

V.V. Pospelov

Editorial comments. In the Western literature, the method here called 'Milne method' is called the (explicit) *midpoint rule*. Instead, the corrector appearing in the 'predictor-corrector Milne method' is called the *Milne method* or a *Milne device*. This method is direct generalization of the Simpson quadrature rule to differential equations. The terminology 'Dahlquist stability' is nowadays seldom used in the English literature. More frequently, one uses 'zero-stability', 'root-stability', or just 'stability'.

References

[A1] HENRICI, P.: *Discrete variable methods in ordinary differential equations*, Wiley, 1962.

AMS 1980 Subject Classification: 65L99, 39A99

MILNE PROBLEM - A problem in **radiative transfer theory** concerning the single-velocity kinetic transport equation of quanta or particles in a half-space. The integral equation of the Milne problem with a source at infinity under zero incident flux of the radiation was first introduced by E. Milne [1] for the case of isotropic scattering of quanta, diffusing without absorption in a stellar atmosphere.

The *Milne equation* takes the form

$$B(x) = \frac{1}{2}\int_0^\infty B(t)E_1(|x - t|)\,dt. \tag{1}$$

Here $B(x)$ is the radiation (or particle) density and

$$E_1(x) = \int_0^1 \frac{e^{-x/\mu}}{\mu}\,d\mu$$

is the exponential integral function $(E_1(x) = -\mathrm{Ei}(-x))$.

In neutron physics the Milne problem is used for the formulation of approximate boundary conditions for solutions of the equations of a diffusion approximation in a bounded domain; in this connection one takes into account neutron capture by the medium, anisotropic scattering and the curvature of the boundary.

Here the Milne problem is to solve the integro-differential equation

$$\mu\frac{\partial\psi}{\partial x} + \psi(x, \mu) = \frac{c}{2}\int_{-1}^{+1} p(\mu, \mu')\psi(x, \mu')\,d\mu'$$

with boundary conditions on the boundary of the half-space occupied by the matter, with a vacuum

$$\psi(0, \mu) = 0 \ \text{for} \ 0 < \mu \leqslant 1, \tag{2}$$

where c is the mean number of secondary neutrons, colliding once with a nucleus ($c < 1$ for scattering and absorbing neutrons by the medium), and $p(\mu, \mu')$ is the indicatrix of the scattering ($p(\mu, \mu')=1$ for isotropic scattering). The spherical or cylindrical Milne problem on the distribution of neutrons in the space outside absorbing spheres or cylinders is stated analogously.

The solution of the Milne problem is conveniently given by applying the **Laplace transform** to the integro-differential transfer equation (see [3]) and using the **Wiener–Hopf method** to solve the functional equations thus obtained.

In order to solve the Milne problem it was suggested that expansion relative to generalized eigen functions and methods for solving singular integral equations be used (see [4]). The solution of the Milne problem for $p(\mu, \mu')=1$, $c<1$ is sought in the form

$$\psi(x, \mu) = \psi_{0-}(x, \mu) + a_{0+}\psi_{0+}(x, \mu) + \int_0^1 A(\nu)\psi_\nu(x, \mu)\, d\nu,$$

where

$$\psi_\nu(x, \mu) = \phi_\nu(\mu)e^{x/\nu},$$

$$\phi_\nu(\mu) = \frac{c\nu}{2}P\frac{1}{\nu-\mu} + \lambda(\nu)\delta(\nu-\mu)$$

are the eigen functions of the continuous spectrum, P denotes the Cauchy principal value, $\delta(\nu-\mu)$ is the Dirac δ-function, and

$$\lambda(\nu) = 1 - \nu c\,\text{Artanh}\,\nu, \quad \psi_{0\pm} = \phi_{0\pm}(\mu)e^{\mp x/\nu_0}.$$

The discrete eigen values $\pm\nu_0$ are the roots of the characteristic equation

$$\nu c\,\text{Artanh}\frac{1}{\nu} = 1.$$

The eigen functions of the discrete spectrum take the form

$$\phi_{0\pm}(\mu) = \pm\frac{c\nu_0}{2}\frac{1}{\pm\nu_0-\mu}.$$

The system of eigen functions $\phi_{0+}(\mu)$ and $\phi_\nu(\mu)$, $0\leqslant\nu\leqslant 1$, turns out to be complete in the space of generalized functions on the interval $0\leqslant\mu\leqslant 1$ and they are orthogonal with respect to a weight $W(\mu)$ which is the solution of a singular integral equation (see [4]).

The boundary condition (2) of the Milne problem gives ($\mu\geqslant 0$):

$$-\phi_{0-}(\mu) = a_{0+}\phi_{0+}(\mu) + \int_0^1 A(\nu)\phi_\nu(\mu)\, d\nu,$$

that is, a_{0+} and $A(\nu)$ are defined as the coefficients in the expansion of the function

$$\phi_{0-}(\mu) = \frac{c\nu_0}{2}\frac{1}{\nu_0+\mu}.$$

The asymptotic density of the neutrons,

$$B_{ac}(x) = 2\pi\int_{-1}^{+1}[\psi_{0-}(x, \mu) + a_{0+}\psi_{0+}(x, \mu)]\, d\mu =$$

$$= 4\pi e^{-x_0/\nu_0}\sinh\frac{x+x_0}{\nu_0},$$

vanishes for

$$x = -x_0 = \frac{\nu_0}{2}(\ln a_{0+} - i\pi).$$

For $c=0$, $p(\mu, \mu')=1$, the *Hopf constant* $x_0 = 0.710\,446$.

References
[1] MILNE, E.A.: *Mon. Notices Roy. Astron. Soc.* **81** (1921), 361-375.
[2] HOPF, E.: *Mathematical problems of radiative equilibrium*, Cambridge Univ. Press, 1934.
[3] SNEDDON, I.: *Fourier transforms*, McGraw-Hill, 1951.
[4] CASE, K.M. and ZWEIFEL, P.F.: *Linear transport theory*, Addison-Wesley, 1967.

V.A. Chuyanov

Editorial comments.

References
[A1] GREENBERG, W., MEE, C. VAN DER and PROTOPOPESCU, V.: *Boundary value problems in abstract kinetic theory*, Birkhäuser, 1987.
[A2] CERCIGNANI, C.: *The Boltzmann equation and its applications*, Springer, 1988.
[A3] DAVISON, B. and SYKES, J.B.: *Neutron transport theory*, Clarendon Press, 1957.
[A4] WILLIAMS, M.M.R.: *The slowing down and thermalization of neutrons*, North-Holland, 1966.

AMS 1980 Subject Classification: 82A70, 82A75, 85A25

MILNOR SPHERE - A smooth manifold homeomorphic (and piecewise-linearly isomorphic), but not diffeomorphic, to the sphere S^n. The first example of such a manifold was constructed by J. Milnor in 1956 (see [1]); the same example was the first example of homeomorphic but not diffeomorphic manifolds.

Construction of a Milnor sphere. Any compact smooth oriented closed manifold, homotopically equivalent to S^n, $n\geqslant 5$, is homeomorphic (and even piecewise-linearly isomorphic) to S^n (see **Poincaré conjecture**, generalized; *h*-**cobordism**). The index of a closed smooth almost **parallelizable manifold** of dimension $4k$ is divisible by a number σ_k which exponentially increases with k. For any k there is a parallelizable manifold P^{4k} of index 8 (namely, the *plumbing construction* of Milnor) whose boundary $M=\partial P$ is, for $k>1$, a homotopy sphere (see [2], [6]). If M were diffeomorphic to the sphere S^{4k-1}, then the manifold W^{4k} obtained from P^{4k} by the addition of a cone over the boundary would be a smooth almost parallelizable closed manifold of index 8. Thus M is a Milnor sphere.

There are other examples of Milnor spheres (see [5]).

Classification of Milnor spheres. In the sequel the term 'Milnor sphere' will be used also for the standard sphere S^n. There are 28 distinct (non-diffeomorphic) 7-dimensional Milnor spheres.

The set of all smooth structures on the piecewise-linear sphere is equivalent to the set of elements of the group $\pi_i(\text{PL}/O)$. The latter group is trivial for $i<7$, so in the PL-case any Milnor sphere of dimension less than 7 is diffeomorphic to the standard sphere.

Let θ_n be the set of classes of h-cobordant n-dimensional smooth manifolds which are homotopically equivalent to S^n. The operation of **connected sum** transforms this set into a group, where the zero is the h-cobordism class of S^n. For $n>5$ the elements of θ_n are in one-to-one correspondence with the diffeomorphism classes of n-dimensional Milnor spheres. To calculate the groups θ_n, $n>5$, one specifies (see [3]) a trivialization of the stable normal bundle (a framing) of the Milnor sphere M^n. This is possible since M^n is stably parallelizable. The framed manifold obtained defines an element of the stable homotopy group $\Pi_n = \lim_i \pi_{i+n}(S^i)$. This element depends, in general, on the choice of the framing ($\theta_n \to \Pi_n$ is a 'multi-valued mapping'). Let $\theta_n(\partial \pi)$ be the subgroup in θ_n consisting of Milnor spheres that bound parallelizable manifolds. This multi-valued mapping induces a homomorphism $\alpha: \theta_n / \theta_n(\partial \pi) \to \operatorname{Coker} J_n$, where $J_n: \pi_n(SO) \to \Pi_n$ is the stationary **Whitehead homomorphism** and α is an isomorphism. The calculation of the group $\theta_n / (\theta_n(\partial \pi))$ reduces to the problem of calculating Π_n and $\theta_n(\partial \pi)$ (unsolved, 1989), which is done by means of surgery (cf. **Morse surgery**) of the manifold (preserving the boundary). Let $[M^n] \in \theta_n(\partial \pi)$, that is, $M^n = \partial W^{n+1}$ and W^{n+1} is parallelizable. If W is a contractible manifold, then after cutting out from W a small disc, the manifold M is h-cobordant to S^n, that is, $[M^n] = 0 \in \theta_n$. If n is even, then it is possible to modify W by means of surgery so that the new manifold W_1 with $\partial W_1 = M$ is contractible (here one requires parallelizability of W and $n \geqslant 4$). Thus $\theta_{2n}(\partial \pi) = 0$.

The case $n+1 = 4k$. If the index $\sigma(W)$ of W is 0, then W can be transformed by surgery into a contractible manifold, so that in this case M is a standard sphere. If $M = \partial W$ and $M_1 = \partial W_1$, then $M \# (-M_1) = \partial(W \# (-W))$ and $\sigma(W \# (-W_1)) = \sigma(W) - \sigma(W_1)$ (here $A \# B$ is the connected sum or the boundary connected sum of two manifolds A and B). If $\sigma(W) = \sigma(W_1)$, then $[M] = [M_1]$, so that the invariant $\sigma(W)$ defines an element $[M] \in \theta_n$. If $[M] = 0 \in \theta_{4k-1}(\partial \pi)$ and $M = \partial W$, then $\sigma(W)$ is divisible by σ_k. Conversely, for any $k>1$ there is a smooth closed manifold B^{4k} with $\sigma(B^{4k}) = \sigma_k$; therefore, if $M = \partial W$ and $\sigma(W) = n\sigma_k$, then $M = \partial(W \# (-nB^{4k}))$, where $W \# (-nB^{4k})$ is parallelizable and $\sigma(W \# (-nB^{4k})) = 0$. The element $[M] \in \theta_{4k-1}(\partial \pi)$ is completely determined by the residue of $\sigma(W)$ modulo σ_k, and different residues determine different manifolds. Since $\sigma(W)$ takes any value divisible by 8, $\operatorname{ord} \theta_{4k-1}(\partial \pi) = \sigma_k / 8$. E.g., $\theta_7(\partial \pi) = \mathbf{Z}_{28}$, and $\operatorname{Coker} J_7 = 0$, so $\theta_7 = \mathbf{Z}_{28}$.

The case $n = 4k+1$. Let $M = \partial W^{4k+2}$. If the **Kervaire**

invariant of W is zero, that is, $\psi(W) = 0$, then W can be converted by surgery into a contractible manifold, that is, $[M] = 0$. Now let $\psi(W) \neq 0$. Since for $4k + 2 \neq 2^i - 2$ there is no smooth closed almost-parallelizable (which in dimension $4k + 2$ is equivalent to stably-parallelizable) manifold with Kervaire invariant not equal to zero, M is not diffeomorphic to S^{4k+1}. In this case $\theta_{4k+1}(\partial \pi) \neq 0$, that is, $\theta_{4k+1}(\partial \pi) = \mathbf{Z}_2$. For $4k + 2 = 2^i - 2$ and those i for which there is a manifold with non-zero Kervaire invariant, $M \approx S^{4k+1}$, that is, $\theta_{4k+1}(\partial \pi) = 0$, but the question of describing all such i has not been solved (1989). However, for $i \leqslant 6$ the answer is positive. Thus $\theta_{4k+1}(\partial \pi)$ is $\mathbf{Z}_2$ or 0.

There is another representation of a Milnor sphere. Let W be an algebraic variety in $\mathbf{C}^{n+1}$ with equation

$$z_1^{a_1} + \cdots + z_{n+1}^{a_{n+1}} = 0$$

and let S_ϵ be the $(2n+1)$-dimensional sphere of (small) radius ϵ with centre at the origin. For suitable values of a_k, $M = W \cap S_\epsilon$ is a Milnor sphere (see [4]). For example, for $n = 4$ and $a_1 = 6k - 1$, $a_2 = 3$, $a_3 = a_4 = a_5 = 2$ and $k = 1, \ldots, 28$, all 28 7-dimensional Milnor spheres are obtained.

References

[1] MILNOR, J.W.: 'On manifolds homeomorphic to the 7-sphere', *Ann. of Math.* **64** (1956), 399–405.
[2] KERVAIRE, M.A. and MILNOR, J.W.: 'Bernoulli numbers, homotopy groups, and a theorem of Rohlin', in *Proc. Internat. Congress Mathematicians, Cambridge 1958*, Cambridg. Univ. Press, 1960, pp. 454–458.
[3] KERVAIRE, M.A. and MILNOR, J.W.: 'Groups of homotopy spheres', *Ann. of Math.* **77** (1963), 504–537.
[4] MILNOR, J.: *Singular points of complex hypersurfaces*, Princeton Univ. Press, 1968.
[5] MILNOR, J. and STASHEFF, J.: *Characteristic classes*, Princeton Univ. Press, 1974.
[6] BROWDER, W.: *Surgery on simply-connected manifolds*, Springer, 1972.

Yu.B. Rudyak

Editorial comments. The general problem of constructing different smooth structures on a topological manifold has received much attention since the above article was written (around 1982). In particular, it has been proven that $\mathbf{R}^4$ has different smooth structures (but not $\mathbf{R}^n$ for $n \neq 4$). general reference is [A1].

References

[A1] FREED, D.S. and UHLENBECK, K.K.: *Instantons and four-manifolds*, Springer, 1984.

AMS 1980 Subject Classification: 57R20, 57N15, 58A99

MINIMAL DISCREPANCY METHOD, *minimal residual method* - An iteration method for the solution of an operator equation

$$Au = f \tag{1}$$

with a self-adjoint positive-definite bounded operator A acting in a Hilbert space H, and with a given element

$f \in H$. The formula for the minimal discrepancy method takes the form

$$u^{k+1} = u^k - \alpha_k(Au^k - f), \quad k = 0, 1, \dots, \tag{2}$$

where the parameter

$$\alpha_k = \frac{(A\xi^k, \xi^k)}{(A\xi^k, A\xi^k)} \tag{3}$$

is chosen in each step $k \geq 0$ from the condition for maximal minimization of the norm of the discrepancy (or residual vector) $\xi^{k+1} = Au^{k+1} - f$; that is, it is required that

$$\| \xi^k - \alpha_k A\xi^k \| = \inf_\alpha \| \xi^k - \alpha A\xi^k \|. \tag{4}$$

If the spectrum of A belongs to an interval $[m, M]$ on the real line, where $m \leq M$ are positive numbers, then the sequence of approximations $\{u^k\}$ in the method (2) - (3) converges to a solution u^* of (1) with the speed of a geometric progression with multiplier $q = (M - m)/(M + m) < 1$.

Different ways of defining the scalar product in H lead to different iteration methods. In particular, for special scalar products the formulas for the minimal discrepancy method coincide with the formulas for the method of steepest descent (cf. **Steepest descent, method of**) and the method of minimal errors (see [2]).

The condition for the convergence of the minimal discrepancy method may be weakened in comparison with that given above if it is considered on certain subsets of H. For example, if the minimal discrepancy method is considered only in real spaces, then it is possible to drop the requirement that A be self-adjoint (see [3] - [5]).

References

[1] KRASNOSEL'SKIĬ, M.A. and KREĬN, S.G.: 'An iteration process with minimal residuals', *Mat. Sb.* **31** (1952), 315-334 (in Russian).
[2] KRASNOSEL'SKIĬ, M.A.: *Approximate solution of operator equations*, Wolters-Noordhoff, 1972 (translated from the Russian).
[3] SAMARSKIĬ, A.A.: *Theorie der Differenzverfahren*, Akad. Verlagsgesell. Geest u. Portig K.-D., 1984 (translated from the Russian).
[4] MARCHUK, G.I. and KUZNETSOV, YU.A.: *Iterative methods and quadratic functionals*, Novosibirsk, 1972 (in Russian).
[5] MARCHUK, G.I. and KUZNETSOV, YU.A.: *Sur les méthodes numériques en sciences physique et economique*, Dunod, 1974 (translated from the Russian).

Yu.A. Kuznetsov

AMS 1980 Subject Classification: 47A50, 65J10

MINIMAL FUNCTIONAL CALCULUS, *minimal predicate calculus* - The calculus of predicates given by all axiom schemes of the **minimal propositional calculus** and by the usual quantifier axiom schemes and deduction rules, that is,

$$\forall x\, A(x) \supset A(t), \quad A(t) \supset \exists x\, A(x)$$

(t an arbitrary term), **modus ponens** and

$$\frac{C \supset A(a)}{C \supset \forall x\, A(x)}, \quad \frac{A(a) \supset C}{\exists x\, A(x) \supset C}$$

(provided the variable a does not occur in $A(x)$ and C).

References

[1] CHURCH, A.: *Introduction to mathematical logic*, 1, Princeton Univ. Press, 1956.

S.K. Sobolev

AMS 1980 Subject Classification: 03B05

MINIMAL IDEAL - A minimal element of the **partially ordered set** of ideals of a given type of some algebraic system. Since the ordering on the set of ideals is defined by the inclusion relation, a minimal ideal is an ideal not containing ideals of the same type different from itself. For multi-operator groups (in particular, rings) and for lattices, in contrast to semi-groups, it is always assumed that the partially ordered set of ideals does not contain the zero ideal. If no class of ideals is specifically mentioned, a minimal ideal is taken to be minimal in the set of all (non-zero) two-sided ideals.

A minimal two-sided ideal, if one exists, in a **semi-group** S is unique and is the smallest two-sided ideal: it is called the *kernel of the semi-group S*. Not every semi-group has a kernel (for example, an infinite **monogenic semi-group**) but, for example, the kernel exists in any finite semi-group. The kernel is an ideally-simple semi-group (see **Simple semi-group**). If the kernel of a semi-group S is a **group**, then S is called a *homogroup*. A semi-group S is a homogroup if and only if there is an element z in S which is divisible on the left and right by any element of S (that is, $z \in xS \cap Sx$ for any $x \in S$); in this case the kernel consists of all such elements. For example, every finite commutative semi-group is a homogroup.

If a semi-group S has a minimal left ideal L, then for any $x \in S$ the product Lx is also a minimal left ideal, moreover, every minimal left ideal can be obtained in this way. Every minimal left ideal is a left simple semi-group. In a semi-group with minimal left ideals every left ideal contains a minimal left ideal, and the union of all minimal left ideals (which are pairwise disjoint) is the kernel of the semi-group. If a semi-group S has a minimal left ideal L and a minimal right ideal R, then $R \cap L = RL$ is a subgroup in S and $L = Se$, $R = eS$, where e is the identity of this subgroup; the product LR coincides with the kernel of S and is, in this case, a **completely-simple semi-group**.

For semi-groups with a zero, the interest is in the consideration of non-zero ideals, and a minimal element in the corresponding partially ordered set of ideals is called a *0-minimal (left, right, two-sided) ideal*. The properties of 0-minimal ideals are in many ways similar to the properties of minimal ideals, with some natural restrictions. For example, a 0-minimal two-sided ideal is not necessarily unique and need not be a 0-simple semi-group; it may be a semi-group with zero multiplication (see **Nilpotent semi-group**). The union of all 0-minimal left ideals (respectively, 0-minimal right ideals) is a semi-group with zero, called its *left* (respectively, *right*) *socle* (by definition, the **socle** is equal to zero if there are no corresponding 0-minimal ideals in the semi-group). A semi-group coincides with its left and right socles if and only if it is an *O*-direct union of completely-simple semi-groups and semi-groups with zero multiplication.

The consideration of minimal ideals and 0-minimal ideals plays an essential role in the structure theory of a number of important classes of semi-groups (see, for example, **Completely-simple semi-group**; **Regular semi-group**, and also [1], §§ 2.5, 2.7, Chapt. 6, §§ 7.7, 8.2, 8.3; [2], Chapt. V).

L.N. Shevrin

Rings (like semi-groups) need not have minimal ideals (the simplest example is the ring of integers) and a minimal ideal in a ring, if it exists, need not be unique. The sum of all (left, right, two-sided) minimal ideals in a ring is called the (*left, right, two-sided*) *socle of the ring*. An **Artinian ring**, obviously, has a non-zero socle. The presence of minimal ideals in a **primitive ring** makes it close to a matrix ring in the following sense: A primitive ring with a non-zero socle is isomorphic to a dense subring of the ring of all linear transformations of some vector space over a skew-field, containing all transformations of finite rank [3].

V.E. Govorov

References

[1] CLIFFORD, A.H. and PRESTON, G.B.: *The algebraic theory of semigroups*, 1-2, Amer. Math. Soc., 1961-1967.
[2] LYAPIN, E.S.: *Semigroups*, Amer. Math. Soc., 1974 (translated from the Russian).
[3] JACOBSON, N.: *Structure of rings*, Amer. Math. Soc., 1956.

Editorial comments. In commutative ring theory, and also in the theory of distributive lattices, the study of minimal prime ideals (i.e. minimal elements in the ordered set of prime ideals) plays an important part. To a large extent this is simply the order-theoretic dual of the study of maximal ideals of these structures (see **Maximal ideal**), but the parallel is not exact — for instance, the space of minimal prime ideals of a distributive lattice is always Hausdorff but need not be compact, whereas the space of maximal ideals is always compact but need not be Hausdorff.

References

[A1] HENRIKSEN, M. and JERISON, M.: 'The space of minimal prime ideals of a commutative ring', *Trans. Amer. Math. Soc.* 115 (1965), 110-130.
[A2] SIMMONS, H.: 'Reticulated rings', *J. Algebra* 66 (1980), 169-192.
[A3] SUN, S.-H.: 'A localic approach to minimal prime spectra', *Math. Proc. Cambridge Philos. Soc.* 103 (1988), 47-53.

AMS 1980 Subject Classification: 06B10, 13A15, 20M12, 16A66

MINIMAL ITERATION METHOD - A method for solving linear algebraic equations $Ax=b$, in which the solution x is represented as a linear combination of basis vectors which are orthogonal in some metric connected with the matrix of the system.

In the case of a symmetric matrix A, the orthogonal system of vectors $p_0, \dots, p_{n-1}$ is constructed using the three-term recurrence formula

$$p_{k+1} = Ap_k - \alpha_k p_k - \beta_k p_{k-1}, \quad k=1, \dots, n-2, \qquad (1)$$

$p_1 = Ap_0 - \alpha_0 p_0$, p_0 an arbitrary vector, where

$$\alpha_k = \frac{(Ap_k, p_k)}{(p_k, p_k)}, \quad k=0, \dots, n-2,$$

$$\beta_k = \frac{(p_k, p_k)}{(p_{k-1}, p_{k-1})}, \quad k=1, \dots, n-2.$$

The solution of the system $Ax=b$ is found by the formula $x = \sum_{k=0}^{n-1} c_k p_k$, and the coefficients c_k are given as the solutions of the system

$$\left. \begin{array}{l} c_{i-1} + \alpha_i c_i + \beta_{i+1} c_{i+1} = \dfrac{(b, p_i)}{(p_i, p_i)}, \quad i=1, \dots, n-2, \\[2mm] \alpha_0 c_0 + \beta_1 c_1 = \dfrac{(b, p_0)}{(p_0, p_0)}, \\[2mm] c_{n-2} + \alpha_{n-1} c_{n-1} = \dfrac{(b, p_{n-1})}{(p_{n-1}, p_{n-1})}. \end{array} \right\} \qquad (2)$$

If the orthogonalization algorithm is degenerate, that is, if $p_r = 0$ for $r < n$, one has to choose a new initial vector $p_0^{(1)}$, orthogonal to $p_0, \dots, p_{r-1}$ and one has to complete the system of basis vectors to a complete system.

In the case of a non-symmetric matrix a bi-orthogonal algorithm is used.

If A is symmetric and positive definite, then constructing an A-orthogonal system $p_0, \dots, p_{n-1}$ by formula (1) with

$$\alpha_k = \frac{(Ap_k, Ap_k)}{(Ap_k, p_k)}, \quad \beta_k = \frac{(Ap_k, p_k)}{(Ap_{k-1}, p_{k-1})}$$

enables one to avoid solving the auxiliary system (2) and gives an explicit expression for the coefficients c_k: $c_k = (b, p_k)/(Ap_k, p_k)$. Here, to the method of A-minimal iteration one can add the iteration

$$x_{k+1} = x_k + c_{k+1} p_{k+1}, \quad k=0, \dots, n-2, \quad x_0 = c_0 p_0,$$

where $x = x_{n-1}$. This modification of the method does not require a repeated use of all the vectors $p_0, \dots, p_{k-1}$. A minimal iteration method is used also for the solution of the complete eigen value problem and for finding the inverse matrix.

References

[1] LANCZOS, C.: 'An iteration method for the solution of the eigenvalue problem of linear differential and integral operators', *Res. Nat. Bur. Stand.* 45, no. 4 (1950), 255-288.
[2] FADDEEV, D.K. and FADDEEVA, V.N.: *Computational methods of linear algebra*, Freeman, 1963 (translated from the Russian).

E.S. Nikolaev

AMS 1980 Subject Classification: 65F10

MINIMAL MODEL - An **algebraic variety** which is minimal relative to the existence of birational morphisms into non-singular varieties. More precisely, let B be the class of all birationally-equivalent non-singular varieties over an algebraically closed field k, the fields of functions of which are isomorphic to a given finitely-generated extension K over k. The varieties in the class B are called *projective models* of this class, or projective models of the field K/k. A variety $X \in B$ is called a *relatively minimal model* if every **birational morphism** $f: X \to X_1$, where $X_1 \in B$, is an isomorphism. In other words, a relatively minimal model is a minimal element in B with respect to the partial order defined by the following domination relation: X_1 dominates X_2 if there exists a birational morphism $h: X_1 \to X_2$. If a relatively minimal model is unique in B, then it is called the *minimal model.*

In each class of birationally-equivalent curves there is a unique (up to an isomorphism) non-singular projective curve. So each non-singular projective curve is a minimal model. In the general case, if B is not empty, then it contains at least one relatively minimal model. The non-emptiness of B is known (thanks to theorems about **resolution of singularities**) for varieties of arbitrary dimension in characteristic 0 for and for varieties of dimension $n \leqslant 3$ in characteristic $p > 5$.

The basic results on minimal models of algebraic surfaces are included in the following.

1) A non-singular projective surface X is a relatively minimal model if and only if it does not contain exceptional curves of the first kind (see **Exceptional subvariety**).

2) Every non-singular complete surface has a birational morphism onto a relatively minimal model.

3) In each non-empty class B of birationally-equivalent surfaces, except for the classes of rational and ruled surfaces, there is a (moreover, unique) minimal model.

4) If B is the class of ruled surfaces (cf. **Ruled surface**) with a curve C of genus $g > 0$ as base, then all relatively minimal models in B are exhausted by the geometric ruled surfaces $\pi: X \to C$.

5) If B is the class of rational surfaces, then all relatively minimal models in B are exhausted by the projective plane P^2 and the series of minimal rational ruled surfaces $F_n = P(\mathcal{O}_{P^1} + \mathcal{O}_{P^1}(n))$ for all integers $n \geqslant 2$ and $n = 0$.

There is (see [6], [7]) a generalization of the theory of minimal models of surfaces to regular two-dimensional schemes. Minimal models of rational surfaces over an arbitrary field have been described (see [2]).

References

[1] SHAFAREVICH, I.R.: 'Algebraic surfaces', *Proc. Steklov Inst. Math.* **75** (1975). (*Trudy Mat. Inst. Steklov.* **75** (1975))
[2] ISKOVSKIKH, V.A.: 'Minimal models of rational surfaces over arbitrary fields', *Math. USSR Izv.* **14**, no. 1 (1980), 17-39. (*Izv. Akad. Nauk SSSR Ser. Mat.* **43**, no. 1 (1979), 19-43)
[3] SHAFAREVICH, I.R.: *Basic algebraic geometry*, Springer, 1977 (translated from the Russian).
[4] BOMBIERI, E. and HUSEMOLLER, D.: 'Classification and embeddings of surfaces', in R. Hartshorne (ed.): *Algebraic Geometry, Arcata 1974*, Proc. Symp. Pure Math., Vol. 29, Amer. Math. Soc., 1975, pp. 329-420.
[5] HARTSHORNE, R.: *Algebraic geometry*, Springer, 1977.
[6] LICHTENBAUM, S.: 'Curves over discrete valuation rings', *Amer. J. Math.* **90**, no. 2 (1968), 380-405.
[7] SHAFAREVICH, I.R.: *Lectures on minimal models and birational transformations of two-dimensional schemes*, Tata Inst. Fundam. Res., 1966.

V.A. Iskovskikh

Editorial comments. Since 1982 important progress has been made (over the field of complex numbers) in the theory of minimal models for higher-dimensional varieties, and especially for varieties of dimension 3. It has turned out to be necessary to allow a mild type of singularities, namely so-called terminal and canonical singularities. For the precise (very technical) definitions see the references below. (Terminal singularities are special canonical singularities, and for surfaces a point with a terminal (respectively, canonical) singularity is in fact smooth (respectively, a rational double point).) Allowing terminal singularities, the *'minimal model problem'* (i.e. the existence of a minimal model in a class of birational equivalence) has been solved by S. Mori for varieties of dimension three; in particular, for non-uniruled 3-dimensional algebraic varieties [A2]. A new phenomenon in the higher-dimensional case is also the non-uniqueness of minimal models. References [A1], [A2] and [A4] are good surveys of this new theory.

References

[A1] KOLLÁR, J.: 'The structure of algebraic threefolds: an introduction to Mori's program', *Bull. Amer. Math. Soc.* **17** (1987), 211-273.
[A2] MORI, S.: 'Flip theorem and the existence of minimal models for 3-folds', *J. Amer. Math. Soc.* **1** (1988), 117-253.
[A3] MORI, S.: 'Classification of higher-dimensional varieties', in *Algebraic Geometry*, Proc. Symp. Pure Math., Vol. 46, Part 1, Amer. Math. Soc., 1987, pp. 165-171.
[A4] WILSON, P.M.H.: 'Toward a birational classification of algebraic varieties', *Bull. London Math. Soc.* **19** (1987), 1-48.
[A5] KOLLÁR, J.: 'Minimal models of algebraic threefolds: Mori's program', *Sém. Bourbaki* **712** (1989).
[A6] KAWAMATA, Y., MATSUDA, K. and MATSUKI, K.: 'Introduction to the minimal model problem', in T. Oda (ed.): *Algebraic Geometry, Sendai 1985*, North-Holland & Kinokuniya, 1987, pp. 283-360.

AMS 1980 Subject Classification: 14E30

MINIMAL NORMAL SUBGROUP - A non-trivial **normal subgroup** H such that between it and the identity subgroup there are no other normal subgroups of the group. Not all groups have a minimal normal subgroup. If the group is finite, then any minimal normal subgroup of it is a direct product of isomorphic simple groups. If a minimal normal subgroup exists and is unique, then it is called a *monolith* (sometimes, a *socle*), and the group itself is called a *monolithic group.*

References

[1] KUROSH, A.G.: *The theory of groups*, 1-2, Chelsea, 1960 (translated from the Russian).

A.L. Shmel'kin

Editorial comments. I.e. a minimal normal subgroup is a non-trivial normal subgroup that is minimal in the set of all such subgroups, ordered by inclusion.

References

[A1] ROBINSON, D.J.S.: *A course in the theory of groups*, Springer, 1980.

AMS 1980 Subject Classification: 20E28

MINIMAL PROPERTY *of the partial sums of an orthogonal expansion* - For any function $f \in L_2[a, b]$, any orthonormal system $\{\phi_k\}_{k=1}^{\infty}$ on $[a, b]$ and for any n, the equality

$$\inf_{\{a_k\}_{k=1}^{n}} \int_a^b \left| f(x) - \sum_{k=1}^{n} a_k \phi_k(x) \right|^2 dx =$$

$$= \int_a^b | f(x) - S_n(f, x) |^2 dx$$

holds, where

$$S_n(f, x) = \sum_{k=1}^{n} c_k(f)\phi_k(x)$$

is the n-th partial sum of the expansion of f with respect to the system $\{\phi_k\}$, that is,

$$c_k(f) = \int_a^b f(x)\phi_k(x)\,dx.$$

The minimum is attained precisely at the sum $S_n(f, x)$ and

$$\int_a^b | f(x) - S_n(f, x) |^2 dx =$$

$$= \int_a^b f^2(x)\,dx - \sum_{k=1}^{n} | c_k(f) |^2, \quad n = 1, 2, \dots.$$

Bessel's inequality, Parseval's equality for complete systems and also certain other basic properties of orthogonal expansions essentially are corollaries of this equality (cf. **Bessel inequality**; **Parseval equality**; **Complete system of functions**; **Orthogonal series**; **Orthogonal system**).

References

[1] KOLMOGOROV, A.N. and FOMIN, S.V.: *Elements of the theory of functions and functional analysis*, 1-2, Graylock, 1957-1961 (translated from the Russian).
[2] KACZMARZ, S. and STEINHAUS, H.: *Theorie der Orthogonalreihen*, Chelsea, reprint, 1951.

A.A. Talalyan

Editorial comments.

References

[A1] YOSIDA, K.: *Functional analysis*, Springer, 1978, Sect. III.4.

AMS 1980 Subject Classification: 42C15, 46CXX

MINIMAL PROPOSITIONAL CALCULUS, *minimal calculus of expressions* - The logical calculus obtained from the **positive propositional calculus** Π by the addition of a new connective $\neg$ (negation) and the axiom scheme

$$(A \supset B) \supset ((A \supset \neg B) \supset \neg A),$$

which is called the *law of reductio ad absurdum*.

The minimal propositional calculus is distinguished by the fact that in it not every formula is deducible from 'false', that is, from a formula of the form $\neg A \& A$. The minimal propositional calculus can be obtained from the calculus Π in another way by adding to the language instead of the connective $\neg$ a new propositional constant $\perp$ (falsehood) without the addition of new axiom schemes. Here the formula $A \supset \perp$ serves as the negation $\neg A$ of a formula A.

References

[1] CHURCH, A.: *Introduction to mathematical logic*, 1, Princeton Univ. Press, 1956.

S.K. Sobolev

AMS 1980 Subject Classification: 03B05

MINIMAL SET - 1) A *minimal set in a Riemannian space* is a generalization of a **minimal surface**. A minimal set is a k-dimensional closed subset X_0 in a Riemannian space M^n, $n > k$, such that for some subset Z of k-dimensional **Hausdorff measure** zero the set $X_0 \setminus Z$ is a differentiable k-dimensional minimal surface (that is, is an extremum of the k-dimensional volume functional Λ^k, defined on k-dimensional surfaces imbedded in M^n). The notion of a 'minimal set' amalgamates several mathematical ideas called upon to serve in the so-called multi-dimensional **Plateau problem** (cf. also **Plateau problem, multi-dimensional**).

A.T. Fomenko

2) A *minimal set in a topological dynamical system* $\{S_t\}$ is a non-empty closed invariant (that is, consisting wholly of trajectories) subset F of the phase space W of the system which does not have proper closed invariant subsets. The latter is equivalent to saying that each trajectory in F is everywhere dense in F. The notion of a minimal set was introduced by G.D. Birkhoff (see [1]) for the case of a flow (the 'time' t running through the real numbers). He proved (see [1], [2]) that if F is a compact minimal set and $w \in F$, then for any neighbourhood U of w the set of those t for which $S_t w \in U$ is relatively dense in $\mathbf{R}$ (that is, there is an l such that in each 'time interval' $[s, s+l]$ of length l there is at least one t with $S_t w \in U$); conversely, if W is a complete metric space and a point w has the above property, then the closure of its trajectory $\{S_t w\}$ is a compact minimal set (the same is true for a **cascade**; regarding more general groups of transformations see, for example, [3] and [4]). Birkhoff called this property of w (and its trajectory) *recurrence*; another terminology, suggested by W.H. Gottschalk and G.A. Hedlund [3], is

also used, in which this property is called *almost-periodicity of the point w*. If $F = W$, then the dynamical system itself is called *minimal*.

If a trajectory has compact closure, then it contains a minimal set F (for semi-groups of continuous transformations $\{S_t\}$, with non-negative real or integer t, an analogue of this result holds, where in F the transformations S_t are even invertible [5]). However, research into the limit behaviour of the trajectories of a dynamical system does not reduce to the study of only the minimal sets of the latter. A minimal set of a smooth flow of class C^2 on a two-dimensional closed surface S has a very simple structure: it is either a point, a closed trajectory or the whole surface, which is then a torus (*Schwarz's theorem*, [6]). In the general case the structure of a minimal set can be very complicated (in this connection, in addition to what is said in [2] - [4] it must be said that minimality of a dynamical system places no restrictions on its ergodic properties with respect to any of its invariant measures, [7]). Minimal sets are the fundamental objects of study in **topological dynamics**.

References

[1] BIRKHOFF, G.D.: *Dynamical systems*, Amer. Mat. Soc, 1927.
[2] NEMYTSKIĬ, V.V. and STEPANOV, V.V.: *Qualitative theory of differential equations*, Princeton Univ. Press, 1960 (translated from the Russian).
[3] GOTTSCHALK, W.H. and HEDLUND, G.A.: *Topological dynamics*, Amer. Math. Soc., 1955.
[4] BRONSTEĬN, I.U.: *Extensions of minimal transformations groups*, Noordhoff & Sythoff, 1979 (translated from the Russian).
[5] LEVITAN, B.M. and ZHIKOV, V.V.: *Almost-periodic functions and differential equations*, Cambridge Univ. Press, 1982 (translated from the Russian).
[6] HARTMAN, P.: *Ordinary differential equations*, Wiley, 1964.
[7] KATOK, A.B., SINAĬ, YA.G. and STEPIN, A.M.: 'Theory of dynamical systems and general transformation groups with invariant measure', *J. Soviet Math.* 7 (1977), 974-1065. (*Itogi. Nauk. i Tekhn. Mat. Anal.* 13 (1975), 129-262)

D.V. Anosov

Editorial comments. The terminology around the notions of recurrence and almost periodicity of points in a topological dynamical system is confusing. These are two mainstreams of nomenclature, represented by [1], [2], [A8] on the one hand, and by [3], [A1], [A2] on the other. The type of point mentioned above, namely, a point w such that for every neighbourhood U of w the set $\{t: S_t w \in U\}$ is relatively dense in R, is called *almost periodic* in [3], [4], [A1], [A2], and *almost recurrent* in [2] and [A8]. (In [2], [A8], almost periodicity has another meaning.) Formally, the notion of a recurrent point as defined in [1], [2], [A8] is different: see **Recurrent point**; a recurrent point is always [3]-almost periodic (i.e., almost-recurrent), but not conversely. In a dynamical system on a complete metric space the two notions coincide. (In [3] the notion of a recurrent point is used in the meaning of 'positively and negatively Poisson stable'.) What Birkhoff proved was the equivalence of *recurrence* of a point w (according to the terminology of [1], [2]) and the property that w has a compact minimal orbit closure, provided the phase space is a complete metric space. Using the terminology of [3] one can show: If w has a compact minimal orbit closure, the w is an almost-periodic point (no conditions on the phase space); conversely, an almost-periodic point has a minimal orbit closure, which is compact if the phase space is locally compact and Hausdorff (no metrizability assumed).

The classification of compact minimal sets in topological dynamics is a largely unsolved problem. Only for special classes something can be said (cf. **Distal dynamical system**); see [4], [A2] and [A1]. Unsolved is also the problem as to which (compact) Hausdorff spaces can be the phase space of a minimal flow or a minimal cascade. In this respect, Schwarz's theorem, mentioned above, gives a partial solution for compact surfaces; for a generalization, see [A4]. A Klein bottle cannot be minimal under a continuous flow (the **Kneser theorem**, see also [A6]), neither can the real projective plane (see [A5]). Still open is *Gottschalk's conjecture* (a particular case of *Seifert's conjecture*): S^3 cannot be the phase space of a minimal flow; see Appendix II of [A7] for references (the Seifert conjecture states that any smooth flow on S^3 has a periodic orbit; there is a C^1-counterexample, [A9]). For results about cascades, see [A3], [A10].

References

[A1] AUSLANDER, J.: *Minimal flows and their extensions*, North-Holland, 1988.
[A2] ELLIS, R.: *Lectures on topological dynamics*, Benjamin, 1969.
[A3] GLASNER, G. and WEISS, B.: 'On the construction of minimal skew products', *Israel J. Math.* 34 (1979), 321-336.
[A4] GUTIERREZ, C.: 'Smoothing continuous flows on two-manifolds and recurrences', *Ergod. Th. Dynam. Sys.* 6 (1986), 17-44.
[A5] LAM, P.-F.: 'Inverses of recurrent and periodic points under homomorphisms of dynamical systems', *Math. Systems Theory* 6 (1972), 26-36.
[A6] MARKLEY, N.G.: 'The Poincaré − Bendixson theorem for the Klein bottle', *Trans. Amer. Math. Soc.* 135 (1969), 159-165.
[A7] MARKUS, L.: *Lectures in differentiable dynamics*, Amer. Math. Soc., 1980.
[A8] SIBIRSKY, K.S. [K.S. SIBIRSKIĬ]: *Introduction to topological dynamics*, Noordhoff, 1975 (translated from the Russian).
[A9] SCHWEITZER, P.A.: 'Counterexamples to the Seifert conjecture and opening closed leaves of foliations', *Amer. of Math.* (2) 100 (1974), 386-400.
[A10] FAHTI, A. and HERMAN, M.: 'Existence de diffeomorphismes minimaux', *Astérisque* 49 (1977), 37-59.

AMS 1980 Subject Classification: 54H20, 53A10, 53B20, 58F25

MINIMAL SIMPLE GROUP - A non-Abelian **simple group** all proper subgroups of which are solvable (cf. **Solvable group**). A complete description of the finite minimal simple groups has been obtained (see [1], [2]), together with the classification of all finite groups whose local subgroups (that is, normalizers of p-subgroups) are solvable. Namely, a finite minimal simple group is isomorphic to one of the following projective special linear groups:

PSL$(2, 2^p)$, p any prime;
PSL$(2, 3^p)$, p any odd number;
PSL$(2, p)$, $p \neq 3$ a prime satisfying $p^2 + 1 \equiv 0 \pmod 5$;
PSL$(3, 3)$; or
the **Suzuki group** Sz(2^p), p any odd prime. In particular, every finite minimal simple group is generated by two elements.

References

[1] THOMPSON, J.G.: 'Nonsolvable finite groups all of whose local subgroups are solvable', *Bull. Amer. Math. Soc.* **74** (1968), 383-437.

[2A] THOMPSON, J.G.: 'Nonsolvable finite groups all of whose local subgroups are solvable II', *Pacific J. Math.* **33** (1970), 451-536.

[2B] THOMPSON, J.G.: 'Nonsolvable finite groups all of whose local subgroups are solvable III', *Pacific J. Math.* **39** (1971), 483-534.

[2C] THOMPSON, J.G.: 'Nonsolvable finite groups all of whose local subgroups are solvable IV', *Pacific J. Math.* **48** (1973), 511-592.

[2D] THOMPSON, J.G.: 'Nonsolvable finite groups all of whose local subgroups are solvable V', *Pacific J. Math.* **50** (1974), 215-297.

[2E] THOMPSON, J.G.: 'Nonsolvable finite groups all of whose local subgroups are solvable VI', *Pacific J. Math.* **51** (1974), 573-630.

S.P. Strunkov

AMS 1980 Subject Classification: 20D08

MINIMAL SUFFICIENT STATISTIC - A statistic X which is a **sufficient statistic** for a family of distributions $\mathscr{P} = \{P_\theta: \theta \in \Theta\}$ and is such that for any other sufficient statistic Y, $X = g(Y)$, where g is some measurable function. A sufficient statistic is minimal if and only if the sufficient σ-algebra it generates is minimal, that is, is contained in any other sufficient σ-algebra.

The notion of a $\mathscr{P}$-*minimal sufficient statistic* (or σ-*algebra*) is also used. A sufficient σ-algebra $\mathscr{B}_0$ (and the corresponding statistic) is called $\mathscr{P}$-minimal if $\mathscr{B}_0$ is contained in the completion $\overline{\mathscr{B}}$, relative to the family of distributions $\mathscr{P}$, of any sufficient σ-algebra $\mathscr{B}$. If the family $\mathscr{P}$ is dominated by a σ-finite measure μ, then the σ-algebra $\mathscr{B}_0$ generated by the family of densities

$$\left\{ p_\theta(\omega) = \frac{dp}{d\mu}(\omega): \theta \in \Theta \right\}$$

is sufficient and $\mathscr{P}$-minimal.

A general example of a minimal sufficient statistic is given by the canonical statistic $T = (T_1, \ldots, T_n)$ of an exponential family

$$p_\theta(\omega) = C(\theta) \exp \sum_j Q_j(\theta) T_j(\omega).$$

References

[1] BARRA, J.-R.: *Mathematical bases of statistics*, Acad. Press, 1981 (translated from the French).

[2] SCHMETTERRER, L.: *Introduction to mathematical statistics*, Springer, 1974 (translated from the German).

A.S. Kholevo

Editorial comments.

References

[A1] LEHMANN, E.: *Theory of point estimation*, Wiley, 1983.

[A2] LEHMANN, E.: *Testing statistical hypotheses*, Wiley, 1986.

AMS 1980 Subject Classification: 62B05

MINIMAL SURFACE - A surface for which the **mean curvature** H is zero at all points.

The first research on minimal surfaces goes back to J.L. Lagrange (1768), who considered the following variational problem: Find a surface of least area stretched across a given closed contour. Assuming that the required surface is given in the form $z = z(x, y)$, Lagrange showed that $z(x, y)$ must satisfy the so-called *Euler − Lagrange equation*

$$(1 + q^2)\frac{\partial^2 z}{\partial x^2} - 2pq\frac{\partial^2 z}{\partial x \partial y} + (1 + p^2)\frac{\partial^2 z}{\partial y^2} = 0, \qquad (1)$$

$$p = \frac{\partial z}{\partial x}, \quad q = \frac{\partial z}{\partial y}.$$

Later G. Monge (1776) discovered that the condition for minimality of a surface leads to the condition $H = 0$, and therefore surfaces with $H = 0$ are called 'minimal'. In reality, however, it is necessary to distinguish the notions of a minimal surface and a surface of least area, since the condition $H = 0$ is only a necessary condition for minimality of area, which follows from the vanishing of the first variation of the surface area among all surfaces of class C^2 with the given boundary. To verify that in this class even a relative (local) minimum is attained, it is necessary to investigate the second variation of the surface area.

The theory of minimal surfaces has a rich history; it was taken up by almost-all prominent mathematicians of the 19-th and 20-th centuries. Its problems stimulated the development of many neighbouring domains of mathematics. The first general method of integration of the Euler − Lagrange equations was suggested by Monge (1784) and A. Legendre (1787) in the form of the so-called *Monge formula*, obtained from the complex characteristic of (1):

$$x = A(t) + A_1(\tau), \quad y = B(t) + B_1(\tau), \quad z = C(t) + C_1(\tau),$$

where t and τ are complex variables and $A(t), \ldots, C_1(\tau)$ are holomorphic functions satisfying the conditions

$$A'^2(t) + B'^2(t) + C'^2(t) = 0, \quad A_1'^2(\tau) + B_1'^2(\tau) + C_1'^2(\tau) = 0.$$

However, these methods, by virtue of the insufficient development of complex function theory at that time, remained without application for a long time. (In 1832 S. Poisson wrote that it is difficult to derive any benefit from Monge's formula since it had been complicated by the introduction of complex variables.)

New results on minimal surfaces began to appear only from the beginning of the 1830's. Poisson announced (1832) his solution of the Lagrange variational problem when the boundary of the surface is near to a plane curve. Shortly afterwards a third minimal surface, the **Scherk surface** (H. Scherk, 1834), was added to the already known minimal surfaces: the **catenoid** (L. Euler, 1774, J. Meusnier, 1776) and the

234

helicoid (Meusnier, 1776). In 1842 E. Catalan proved that the helicoid is the unique ruled minimal surface; in 1844 the **Björling problem** was raised and solved; in the 1850's, in a series of papers, O. Bonnet gave new proofs of the facts known at that time on the theory of minimal surfaces and found other properties of minimal surfaces (the uniqueness of the catenoid as a minimal surface of revolution, the conformality of spherical Gauss mappings of minimal surfaces (cf. also **Spherical map**), etc.).

In 1866 the *Weierstrass formula* was discovered:

$$x = x_0 + \mathrm{Re} \int_0^w (f^2 - g^2)\, dw,$$

$$y = y_0 + \mathrm{Re} \int_0^w i(f^2 + g^2)\, dw,$$

$$z = z_0 + 2\mathrm{Re} \int_0^w fg\, dw,$$

representing a simply-connected minimal surface $S(x, y, z)$ by holomorphic functions $f(w)$ and $g(w)$, defined in the disc or in the whole plane of variation of the intrinsic **isothermal coordinates** (u, v), $w = u + iv$. These formulas are equivalent to, or contain as particular cases, all the other then known parametric representations of minimal surfaces (B. Riemann (1860), A. Enneper (1864), K.M. Peterson (1866), and others) and give a regular minimal surface if and only if f and g do not have common zeros. In the case of common zeros of f and g, the surface is a so-called *generalized minimal surface* with a degenerate metric at the common zeros of f and g:

$$ds^2 = (|f|^2 + |g|^2)(du^2 + dv^2).$$

At such a point, a branching point (cf. **Branching point (of a minimal surface)**) of the minimal surface appears. Using the Weierstrass formula it became possible to explicitly construct and study many concrete minimal surfaces; in particular, algebraic minimal surfaces are obtained for algebraic functions f and g.

In 1874 H.A. Schwarz obtained a representation of a minimal surface in isothermal coordinates (u, v) in the form

$$r(w) = r(u + iv) = \mathrm{Re}\left[F(w) - i \int_0^w n(w)\, dF(w) \right],$$

where $F(w)$ and $n(w)$ are three-dimensional vectors with holomorphic coordinates, coinciding respectively with $r(u, 0)$ and the unit normal $n(u, 0)$ to the minimal surface. For $\mathrm{Im}\, w = 0$ this formula gives an explicit solution to the Björling problem and allows the extension to minimal surfaces of the Schwarz symmetry principle.

In the same decade S. Lie (1878) developed his interpretation of the Monge formula, assigning to each minimal surface with a harmonic radius vector $r(w)$ a complex-analytic curve:

$$z = R(w) \in C^3, \quad \mathrm{Re}\, R(w) = \frac{1}{2} r(w),$$

and representing the minimal surface as a translation surface of the curve $R(w)$ and its complex conjugate curve $\overline{R(w)}$; this served as a starting point for the establishment of a fruitful connection between minimal surfaces and analytic curves. Thus, the work of K. Weierstrass, Lie, Riemann, Schwarz, and others resulted, at the end of the 19-th century, in the wide use of methods and results of complex function theory in the theory of minimal surfaces.

In his experiments J. Plateau (1849) arrived at the physical realization of minimal surfaces in the form of soap films, stretched on wire frames of various shapes. His experiments revived the interest in the old problem of finding minimal surfaces with a given boundary contour. This problem came to be called the **Plateau problem**. The first solution of it was obtained for various polygonal contours; in this connection, in particular, the **Riemann−Schwarz surface** was discovered.

In 1816, in the theory of minimal surfaces there appeared the novel *Gergonne problem*: To find a minimal surface if part of its boundary is given and the remainder is situated in some preassigned surface; this problem came to be called the problem on a *minimal surface with free boundary*. The first results towards its solution also related to cases when the given part of the boundary consists of line segments and the remainder lies in given planes (for more details see [1], [2]).

At the beginning of the 20-th century there was an intensified interest in the study of minimal surfaces 'in the large'. The Dirichlet problem for the Euler−Lagrange equation was studied (A. Korn (1909), S.N. Bernshteĭn (1910)) and Bernshteĭn proved his theorem on minimal surfaces (1916, cf. also **Bernstein theorem**); in general, starting with Bernshteĭn's work on the theory of minimal surfaces, methods of the theory of partial differential equations came to be used. H. Liebmann (1919) established a connection between minimal surfaces and infinitesimal deformations of spheres. The crowning achievement of the first half of the 20-th century in the theory of minimal surfaces was the complete solution of Plateau's problem, initially for simply-connected surfaces, and then for two-dimensional surfaces of arbitrary topological type in a Euclidean or a Riemannian space (see [1] - [5]). The results on minimal surfaces were extended to the case of more general variational problems (generalizing the variational problems for a minimal surface), more general differential equations (generalizing the Euler−Lagrange equations for a minimal surface) and to a more general class of surfaces (for example, with constant mean curvature) (see [3], [6], [7]).

From the available results and the leading research on minimal surfaces several directions can be selected.

1. The *intrinsic geometry of minimal surfaces*. Not every Riemannian manifold of non-positive curvature can be isometrically immersed in a Euclidean space E^n, $n \geqslant 3$, as a minimal surface. The criterion for the existence of a minimal surface in E^3 with a given metric is given in the following *theorem of Ricci*: For a given metric ds^2 to be isometric to the metric of some minimal surface in E^3 it is necessary and sufficient that its curvature K be non-positive and that at the points where $K < 0$ the metric $d\sigma^2 = \sqrt{-K} \, ds^2$ be Euclidean.

There is also a necessary and sufficient condition that a given two-dimensional metric be the metric of a minimal surface in a Euclidean space E^n, $n > 3$. In particular, a metric with the Ricci condition (i.e. satisfying the conditions of Ricci's theorem) can be the metric of a minimal surface only in E^3 or E^6, and there is also a complete description of all minimal surfaces with the same metric: they make up a so-called *associated family of isometric minimal surfaces*; for instance, in E^3 all minimal surfaces isometric to a given minimal surface F with $X_k = \operatorname{Re} \Phi_k(w)$, $w = u + iv$, where (u, v) are isothermal coordinates, $\Phi_k(w)$ is a holomorphic function and $1 \leqslant k \leqslant 3$, form the family of associated minimal surfaces F_α with $x_k = \operatorname{Re}(\Phi_k(w)e^{-i\alpha})$, $F_0 = F$. So the catenoid and helicoid are two associated (hence isometric) minimal surfaces, with parameters $\alpha = 0$ and $\alpha = \pi/2$. In the problem of immersion as a minimal surface of a metric ds^2 in a Euclidean space E^n, $n > 3$, one can indicate lower and upper bounds on n using a generalized Ricci condition. Moreover, in terms of certain classes of holomorphic mappings this problem has a complete solution. There is also a number of results on the intrinsic nature of the metrics of two-dimensional minimal surfaces in a sphere S^n, $n \geqslant 3$ (see [8]). For the metrics of multi-dimensional minimal surfaces a necessary and sufficient condition is known only in the case of hypersurfaces [16]. Isometric deformations of minimal surfaces into a surface of another extrinsic structure have also not been very much studied, not even for E^3.

Another circle of problems in the intrinsic geometry of minimal surfaces is formed by isoperimetric inequalities (cf. **Isoperimetric inequality; Isoperimetric inequality, classical**), which, however, in their most exact and interesting forms depend on the exterior structure of the minimal surface (see [9]).

2. *Local properties of minimal surfaces*. Here one may mention results on the analyticity of minimal surfaces (the definition of a minimal surface requires only that it be of class C^2) and the solution of a number of variational problems, the result that a sufficiently small domain of a minimal surface realizes absolutely minimal area among all surfaces with the same boundary as the domain in question, theorems on the removability of isolated or low-order singularities of the solutions of the equations of a minimal surface and their generalizations, the study of the structure of generalized minimal surfaces in a neighbourhood of branching points, research on singularities of minimal surfaces, and the solution of so-called elliptic variational problems in multi-dimensional Euclidean and Riemannian spaces (see [2], [3], [6], [10], [12]).

3. The *study of concrete minimal surfaces* or of minimal surfaces with preassigned boundary properties, plane sections, spherical Gauss mappings, etc. For example, the theorem on the convexity of the horizontal sections of doubly-connected minimal surfaces with convex horizontal boundaries; the theorem on the coincidence with the catenoid of a complete minimal surface situated between two planes $z = C_1$ and $z = C_2$, $-\infty \leqslant C_1 < C_2 \leqslant \infty$, and having star-shaped sections $z = \mathrm{const}$, the detailed study of the classical minimal surfaces of Enneper, Scherk, Riemann $-$ Schwarz and others. Special mention must be given to the research on minimal cones in $\mathbf{R}^n$, $n \geqslant 4$, leading to the construction of a counter-example to Bernstein's theorem in E^n, $n \geqslant 9$, and to the finding of an example of an irregular solution of the Plateau problem for hyperplanes in E^8. Concrete examples of various minimal surfaces have also begun to be studied in multi-dimensional Riemannian spaces (see [1], [2], [6], [12], [13]).

4. The *development of the theory of minimal surfaces* in E^n, $n \geqslant 3$, by analogy with functions of a complex variable in $\mathbf{C}^1$. Here one may mention results on boundary properties of minimal surfaces, the theorem on the analytic continuation of minimal surfaces across regular arcs of the boundary, theorems on the smoothness of a minimal surface in dependence on the smoothness of its boundary (for example, if the boundary is of class $C^{n,\alpha}$, $n \geqslant 1$, $0 < \alpha < 1$, then the minimal surface is of the same class), and work on the construction for minimal surfaces of an analogue of Nevanlinna value-distribution theory for meromorphic functions (see [2], [6], [13], [15]).

5. *Work on the Euler $-$ Lagrange theorem*. Side-by-side with the study 'in the large' of solutions of this equation and its generalization for n-dimensional minimal surfaces $(x_1, \ldots, x_n, f_1(x), \ldots, f_k(x))$ in E^{n+k}, $k \geqslant 1$,

$$\sum_{i,j=1}^{n} g^{ij} \frac{\partial^2 f}{\partial x^i \partial x^j} = 0 \qquad (2)$$

(where $f = (f_1, \ldots, f_k)$ and g^{ij} are the contravariant components of the metric tensor g_{ij}), a large amount of work has been devoted to the local behaviour of the solutions of the Euler $-$ Lagrange equation and its gen-

236

eralization (2). There are the questions on the removability of singularities for codimension $k=1$, the existence of irregular solutions for large codimension, and, particularly, the investigation of the Dirichlet problem. This problem can be treated in another way as the Plateau problem of determining a minimal surface with given contour and with the additional condition of single-valued projection of the minimal surface into the plane containing the boundary contour. Such a treatment, in many cases, allows one to obtain a conclusion of the Dirichlet problem on the basis of known results on the Plateau problem (for example, the existence and uniqueness of the solution of the Dirichlet problem in E^3 for a convex domain D is a corollary of Radó's theorem on the Plateau problem).

In the general case it has been established that for the solvability of the Dirichlet problem in codimension 1 for any continuous boundary function, it is necessary and sufficient that the mean curvature vector of the boundary ∂D is directed into the domain D (for $n=2$ this means that D is convex); here the solution of the problem is also unique. Regarding non-convex domains it has been proved, e.g., that for any non-convex domain D with Jordan boundary ∂D it is possible to find a continuous function $\phi(p)$, $p \in \partial D$, such that the corresponding Dirichlet problem is unsolvable. In large codimensions the Dirichlet problem is unsolvable even when the domain is a ball in E^n.

The Dirichlet problem has also been discussed in the exterior of D. Here, new facts have been obtained (for example, uniqueness of the solution is lost, solvability is not obtained everywhere, etc.). There are also results on the Dirichlet problem with incomplete or partially unbounded boundary values (see [2], [6] - [8], [14]).

6. *Complete minimal surfaces.* By complete minimal surfaces are meant minimal surfaces which are complete as metric spaces relative to their intrinsic metric (cf. **Internal metric**). Complete minimal surfaces may be compact (without boundary) and non-compact, or open. In research on complete minimal surfaces interest is basically directed to the study of connections between the global metric, geometric and topological properties of the surfaces.

Most progress has been made in the study of two-dimensional complete minimal surfaces in E^n, $n \geqslant 3$, where most of the results have been obtained by applying methods of complex function theory. In particular, it has been shown that there is a complete minimal surface in E^3 of any preassigned genus g and connectivity k; theorems have been obtained on the relations between the integral of the curvature, $\int\int K\,dS$, and topological and conformal types of a complete minimal surface (for example, $\int\int K\,dS \leqslant 2\pi(\chi-k)$, where χ is the Euler characteristic of the surface and k is the

number of components of the boundary; if $|\int\int K\,dS| < \infty$, then the surface S is conformally equivalent to a Riemann sphere with a finite number of punctured points; if a complete minimal surface S is infinitely connected or of conformally hyperbolic type, then its total curvature is infinite and the normals to S attain all directions an infinite number of times, with the possible exception of a set of directions of capacity zero, etc.). The structure of spherical Gauss mappings has been studied (for example, a complete minimal surface in E^3 is either a plane or its spherical image does not contain a set of capacity zero, a complete minimal surface in E^n, $n \geqslant 3$, is either a plane or its image under a generalized spherical Gauss mapping intersects an everywhere-dense set of hyperplanes, for any $n \leqslant 4$ there exist complete minimal surfaces in E^3 whose spherical images do not contain precisely n preassigned points, etc.); surfaces have been found which are completely determined by their total curvature and topological type (these are the catenoid and the **Enneper surface**); the exterior unboundedness in E^n, $n \geqslant 3$, of a complete minimal surface with bounded total curvature has been shown (see the references in [2], [8], [18], [21]).

Results on compact minimal surfaces are, on the whole, concerned with complete minimal surfaces situated in spheres $S^n \subset E^{n+1}$. The interest in these minimal surfaces is explained, apart from the difficulties in the case of a general Riemannian space, by the presence of important relationships between minimal cones in E^{n+1} and minimal surfaces in S^n (each hypercone in E^{n+1}, definable by its vertex 0 and its intersection with a sphere S^n with centre at 0, is minimal if and only if its intersection with S^n is a minimal surface in S^n). The questions analyzed here are, on the whole, the same as for complete minimal surfaces in E^n. For example, it has been shown that any compact two-dimensional manifold, with the exception of the projective plane, can be realized as a complete minimal surface in S^3; in S^3 an analogue of Bernstein's theorem has been obtained: If the normals to a complete minimal surface lie in an open hemisphere, then the minimal surface is the equatorial hypersphere; other criteria have been found for a complete minimal surface in S^n to be a hypersphere; the possible forms and the questions of uniqueness of complete minimal surfaces in S^n, in dependence on the value of their scalar curvature, have been investigated, etc. (see [2], [6], [8], [13]).

7. The construction of a different kind of *generalized minimal surface*, and also of equations and variational problems the solutions of which preserve the properties of a minimal surface. Here the classical theory of minimal surfaces is directly bordering on work about surfaces with given mean curvature, about quasi-

conformal spherical Gauss mappings, and on work about quasi-linear elliptic equations with many of the properties of the Euler–Lagrange equation in two or more variables. More remote generalizations have been made in theories connecting minimal surfaces with sets which minimize integral flows or various Hausdorff measures, etc. (see [2], [6], [7], [10], [11]).

8. *Work on the Plateau problem* for two-dimensional and multi-dimensional minimal surfaces (see **Plateau problem; Plateau problem, multi-dimensional**).

From this list of problems and results it is clear that the range of questions in the theory of minimal surfaces is very large, and correspondingly diverse are the methods applied in them. In classical investigations, on the whole, the methods of differential geometry, complex function theory and differential equations were applied, whereas now there is a growing use of the methods of topology, measure theory and functional analysis, particularly in research on minimal surfaces in multi-dimensional spaces.

Since the time of publication and, especially, preparation of this article, important progress has been made in all questions mentioned above, see [18] - [22]; in addition one could mention the development of computer-aided investigations on minimal surfaces; see, e.g., [17].

References

[1] COURANT, R.: *Dirichlet's principle, conformal mapping, and minimal surfaces*, Interscience, 1950.

[2] NITSCHE, J.C.C.: *Vorlesungen über Minimalflächen*, Springer, 1975.

[3] RADÓ, T.: *On the problem of Plateau*, Chelsea, reprint, 1951.

[4] DOUGLAS, J.: 'Solution of the problem of Plateau', *Trans. Amer. Math. Soc.* **33** (1931), 263-321.

[5] MORREY, C.: 'The problem of Plateau on a Riemannian manifold', *Ann. of Math.* **49** (1948), 807-851.

[6] NITSCHE, J.C.C.: 'On new results in the theory of minimal surfaces', *Bull. Amer. Math. Soc.* **71** (1965), 195-270.

[7] OSSERMAN, R.: *A survey of minimal surfaces*, v. Nostrand Reinhold, 1969.

[8A] OSSERMAN, R.: 'Minimal varieties', *Bull. Amer. Math. Soc.* **75** (1969), 1092-1120.

[8B] OSSERMAN, R.: 'Global properties of minimal surfaces in E^3 and E^n', *Ann. of Math.* (2) **80** (1964), 340-364.

[9] OSSERMAN, R.: 'The isoperimetric inequality', *Bull. Amer. Math. Soc.* **84** (1978), 1182-1238.

[10] FEDERER, H.: *Geometric measure theory*, Springer, 1969.

[11] MORREY, C.: *Multiple integrals in the calculus of variations*, Springer, 1966.

[12] FOMENKO, A.T.: 'Minimal compacta in Riemannian manifolds and Reïfenberg's conjecture', *Math. USSR Izv.* **6**, no. 5 (1972), 1037-1066. (*Izv. Akad. Nauk SSSR Ser. Mat.* **36**, no. 5 (1972), 1049-1079)

[13] LAWSON, H.: 'Complete minimal surfaces in S^3', *Ann. of Math.* **92** (1970), 335-374.

[14] LAWSON, H. and OSSERMAN, R.: 'Non-existence, non-uniqueness and irregularity of solutions to the minimal surface system', *Acta. Math.* **139** (1977), 1-17.

[15] VIKARUK, A.YA.: 'Analogs of Nevanlinna's theorems for minimal surfaces', *Math. USSR Sb.* **29**, no. 4 (1976), 497-520. (*Mat. Sb.* **100**, no. 4 (1976), 555-579)

[16] CHERN, S.-S. and OSSERMAN, R.: 'Remarks on the Riemannian metric of a minimal submanifold', in E. Looijenga, D. Siersma and F. Takens (eds.): *Geometry Symp. Utrecht 1980*, Lecture notes in math., Vol. 894, Springer, 1981, pp. 49-90.

[17] HOFFMAN, D.: 'The computer-aided discovery of new embedded minimal surfaces', *Math. Intell.* **9**, no. 3 (1987), 8-21.

[18] MEEUS, W.H., III: 'A survey of the geometric results in the classical theory of minimal surfaces', *Bol. Soc. Brasil. Mat.* **12**, no. 1 (1981), 29-86.

[19] BOMBIERI, E. (ED.): *Seminar of minimal submanifolds*, Anal. Math. Stud., 103, Princeton Univ. Press, 1983.

[20] YAU, S.-T.: 'Minimal surfaces and their role in differential geometry', in T.J. Willmore and N.J. Hitchin (eds.): *Global Riemannian geometry*, Horwood, 1984, pp. 99-103.

[21] DAO CHONG TKHI and FOMENKO, A.T.: *Minimal surfaces and Plateau's problem*, Amer. Math. Soc., Forthcoming (translated from the Russian).

[22] FOMENKO, A.T.: *Variational problems in topology*, Kluwer, 1990.

I.Kh. Sabitov

AMS 1980 Subject Classification: 53A10, 53C42, 49F10, 49F20, 58E12

MINIMAX - A mixed extremum

$$\inf_{y\in Y}\sup_{x\in X} F(x,y), \quad \min_{y\in Y}\max_{x\in X} F(x,y),$$

etc. (see also **Maximin**); it can be interpreted (for example, in decision theory, **operations research** or statistics) as the least of the losses which cannot be prevented by decision making under the given circumstances.

N.N. Vorob'ev

Editorial comments. Cf. Minimax statistical procedure for an interpretation in statistics, [A1] for minimaxima in game theory, and [A2], [A3] for a discussion of minimaxima and maximinima in decision theory. Minimax (and maximin) considerations also occur in other parts of mathematics, for instance in approximation theory, [A4].

References

[A1] SZÉP, J. and FORGO, F.: *Introduction to the theory of games*, Reidel, 1985, Sect. 9.1.

[A2] THIERAUF, R.J. and GROSSE, R.A.: *Decision making through operations research*, Wiley, 1970, Chapt. 3.

[A3] SENGUPTA, J.K.: *Stochastic optimization and economic models*, Reidel, 1986, Chapt. IV.

[A4] RIVLIN, TH.J.: *An introduction to the approximation of functions*, Dover, reprint, 1981.

AMS 1980 Subject Classification: 49B40, 90BXX, 90DXX

MINIMAX ESTIMATOR - A statistical estimator obtained as a result of the application of the notion of a **minimax statistical procedure** in the problem of statistical estimation.

Example 1. Let a random variable X be subject to the binomial law with parameters n and θ, where θ, $0<\theta<1$, is unknown. The statistic

$$t = \frac{X}{n}\frac{\sqrt{n}}{1+\sqrt{n}} + \frac{1}{2(1+\sqrt{n})}$$

is a minimax estimator for the parameter θ with respect to the loss function

$$L(\theta, t) = (\theta - t)^2.$$

Example 2. Let $X_1, \ldots, X_n$ be independent random variables subject to the same probability law, with a continuous probability density $f(x - \theta)$, $|x| < \infty$, $|\theta| < \infty$. The *Pitman estimator*

$$t = t(X_1, \ldots, X_n) = X_{(1)} - \frac{\int_{-\infty}^{\infty} x f(x) \prod_{i=2}^{n} f(x + Y_{(i)}) \, dx}{\int_{-\infty}^{\infty} f(x) \prod_{i=2}^{n} f(x + Y_{(i)}) \, dx}$$

is a minimax estimator for the unknown shift parameter θ relative to the loss function $L(\theta, t) = (\theta - t)^2$, where $X_{(1)} \leqslant \cdots \leqslant X_{(n)}$ are the order statistics (cf. **Order statistic**) obtained from the sample $X_1, \ldots, X_n$ and $Y_{(i)} = X_{(i)} - X_{(1)}$. In particular, if $X_1, \ldots, X_n \sim N(\theta, 1)$, then $t = (X_1 + \cdots + X_n)/n$.

References

[1] ZACKS, S.: *The theory of statistical inference*, Wiley, 1971.
[2] COX, D.R. and HINKLEY, D.: *Theoretical statistics*, Chapman & Hall, 1974.

M.S. Nikulin

AMS 1980 Subject Classification: 62F11, 62C20

MINIMAX PRINCIPLE - An optimality principle for a **two-person zero-sum game**, expressing the tendency of each player to obtain the largest sure pay-off. The minimax principle holds in such a game $\Gamma = \langle A, B, H \rangle$ if the equality

$$v = \max_{a \in A} \min_{b \in B} H(a, b) = \min_{b \in B} \max_{a \in A} H(a, b) \qquad (*)$$

holds, that is, if there are a value of the game, equal to v, and optimal strategies for both players.

For a **matrix game** and for certain classes of infinite two-person zero-sum games (see **Infinite game**) the minimax principle holds if mixed strategies are used. It is known that (*) is equivalent to the inequalities (see **Saddle point in game theory**):

$$H(a, b^*) \leqslant H(a^*, b^*) \leqslant H(a^*, b)$$

for all $a \in A$, $b \in B$, where a^* and b^* are the strategies on which the external extrema in (*) are attained. Thus, the minimax principle expresses mathematically the intuitive conception of stability, since it is not profitable for either player to deviate from his optimal strategy a^* (respectively, b^*). At the same time the minimax principle guarantees to player I (II) a gain (loss) of not less (not more) than the value of the game. An axiomatic characterization of the minimax principle for matrix games has been given (see [1]).

References

[1] VILKAS, E.: 'Axiomatic definition of the value of a matrix game', *Theory Probabl. Appl.* **8** (1963), 304-307. (*Teor. Veroyatnost. i Primenen.* **8**, no. 3 (1963), 324-327)

E.B. Yanovskaya

Editorial comments. The fact that the minimax principle holds if mixed strategies are allowed is called the *minimax theorem*; it is due to J. von Neumann.

References

[A1] NEUMANN, J. VON and MORGENSTERN, O.: *Theory of games and economic behavior*, Princeton Univ. Press, 1947.

AMS 1980 Subject Classification: 90D05

MINIMAX PROPERTY *of eigen values* - A special type of relationship connecting the eigen values of a completely-continuous **self-adjoint operator** A (cf. also **Completely-continuous operator**) with the maximum and minimum values of the associated quadratic form (Ax, x). Let A be a completely-continuous self-adjoint operator on a Hilbert space H. The spectrum of A consists of a finite or countable set of real eigen values λ_n having unique limit point zero. The root subspaces corresponding to the non-zero eigen values consist of eigen vectors and are finite dimensional; the eigen subspaces associated with distinct eigen values are mutually orthogonal; A has a complete system of eigen vectors. The spectral decomposition of A (cf. **Spectral decomposition of a linear operator**) has the form: $A = \sum \lambda_i P_i$, where λ_i are the distinct eigen values, P_i are the projection operators onto the corresponding eigen spaces, and the series converges in the operator norm. The norm of A coincides with the maximum modulus of the eigen values and with $\max \{ |(Ax, x)| : x \in H, |x| = 1 \}$; the maximum is attained at the corresponding eigen vector.

Let $\lambda_1^+ \geqslant \lambda_2^+ \geqslant \cdots$ be the positive eigen values of A, where each eigen value is repeated as often as its multiplicity. Then

$$\left. \begin{aligned} \lambda_1^+ &= \max_{x \in H} \frac{(Ax, x)}{|x|^2}, \\ \lambda_{n+1}^+ &= \min_{y_1, \ldots, y_n} \max_{\substack{(x, y_i) = 0 \\ i = 1, \ldots, n}} \frac{(Ax, x)}{|x|^2}, \quad n > 1, \end{aligned} \right\} \qquad (1)$$

where $x, y_1, \ldots, y_n$ are arbitrary non-zero vectors in H. Similar relations hold for the negative eigen values $\lambda_1^- \geqslant \lambda_2^- \geqslant \cdots$:

$$\left. \begin{aligned} \lambda_1^- &= \min_{x \in H} \frac{(Ax, x)}{|x|^2}, \\ \lambda_{n+1}^- &= \max_{y_1, \ldots, y_n} \min_{\substack{(x, y_i) = 0 \\ i = 1, \ldots, n}} \frac{(Ax, x)}{|x|^2}, \quad n > 1. \end{aligned} \right\} \qquad (2)$$

Relations (1) and (2) are applied for finding the eigen values of integral operators with a symmetric kernel. If A and B are completely-continuous self-adjoint operators, $A \leqslant B$ (that is, $(Ax, x) \leqslant (Bx, x)$), λ_n and μ_n the sequences of their positive eigen values, listed in decreasing order, where each value is repeated as often as its multiplicity, then $\lambda_n \leqslant \mu_n$.

References
[1] DUNFORD, N. and SCHWARTZ, J.T.: *Linear operators*, 2. Spectral theory, Wiley, 1988.

A.I. Loginov

AMS 1980 Subject Classification: 47A10, 47A70, 47B15

MINIMAX STATISTICAL PROCEDURE - One of the versions of optimality in mathematical statistics, according to which a statistical procedure is pronounced optimal in the minimax sense if it minimizes the maximal risk. In terms of decision functions (cf. **Decision function**) the notion of a minimax statistical procedure is defined as follows. Let a random variable X take values in a sampling space $(\mathfrak{X}, \mathfrak{B}, P_\theta)$, $\theta \in \Theta$, and let $\Delta = \{\delta\}$ be the class of decision functions which are used to make a decision d from the decision space D on the basis of a realization of X, that is, $\delta(\cdot) \colon \mathfrak{X} \to D$. In this connection, the loss function $L(\theta, d)$, defined on $\Theta \times D$, is assumed given. In such a case a statistical procedure $\delta^* \in \Delta$ is called a *minimax procedure* in the problem of making a statistical decision relative to the loss function $L(\theta, d)$ if for all $\delta \in \Delta$,

$$\sup_{\theta \in \Theta} \mathsf{E}_\theta L(\theta, \delta^*(X)) \leqslant \sup_{\theta \in \Theta} \mathsf{E}_\theta L(\theta, \delta(X)),$$

where

$$\mathsf{E}_\theta L(\theta, \delta(X)) = R(\theta, \delta) = \int_{\mathfrak{X}} L(\theta, \delta(X)) \, dP_\theta(x)$$

is the risk function associated to the statistical procedure (decision rule) δ; the decision $d^* = \delta^*(x)$ corresponding to an observation x and the minimax procedure δ^* is called the *minimax decision*. Since the quantity

$$\sup_{\theta \in \Theta} \mathsf{E}_\theta L(\theta, \delta(X))$$

shows the expected loss under the procedure $\delta \in \Delta$, δ^* being maximal means that if δ^* is used to choose a decision d from D, then the largest expected risk,

$$\sup_{\theta \in \Theta} R(\theta, \delta^*),$$

will be as small as possible.

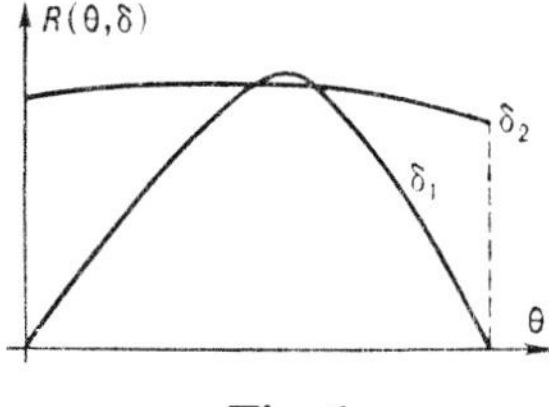

Fig. 1.

The minimax principle for a statistical procedure does not always lead to a reasonable conclusion (see Fig. 1); in this case one must be guided by δ_1 and not by δ_2, although

$$\sup_{\theta \in \Theta} R(\theta, \delta_1) > \sup_{\theta \in \Theta} R(\theta, \delta_2).$$

The notion of a minimax statistical procedure is useful in problems of statistical decision making in the absence of a priori information regarding θ.

References
[1] LEHMANN, E.: *Testing statistical hypotheses*, Wiley, 1986.
[2] ZACKS, S.: *The theory of statistical inference*, Wiley, 1971.

M.S. Nikulin

AMS 1980 Subject Classification: 62C20

MINIMIZATION METHODS FOR FUNCTIONS DEPENDING STRONGLY ON A FEW VARIABLES - Numerical methods for finding minima of functions of several variables. Suppose one is given a function that is bounded from below and twice continuously differentiable,

$$J(x) = J(x_1, \ldots, x_m),$$

and that is known to take its least value at a vector $x^* = (x_1^*, \ldots, x_m^*)^T$ (where T denotes transposition). One wishes to construct a sequence of vectors $\{x_n\}$, $x_n = (x_{1n}, \ldots, x_{mn})^T$, such that

$$\lim_{n \to \infty} J(x_n) = J(x^*).$$

There are many methods for obtaining such a sequence. However, the majority of algorithms suffer from a sharp deterioration when the level surfaces $J(x) = \text{const}$ of the function to be minimized have a structure which is far from spherical. In this case a domain Q in which the norm of the gradient vector

$$J'(x) = \left[\frac{\partial J}{\partial x_1}, \ldots, \frac{\partial J}{\partial x_m} \right]^T$$

is essentially smaller than in the rest of the space is called a *bottom of a valley*, and the function itself is called a *valley function*. If the dimension of the space of arguments of the function to be minimized exceeds two, the structure of the level surfaces of the valley function may be very complicated. $(m - k)$-dimensional valleys occur, where k varies from 1 to $m - 1$. In three-dimensional space, for example, one and two-dimensional valleys are possible.

Functions of valley type are locally characterized by an ill-conditioned matrix of second derivatives (*Hesse matrix* or *Hessian*)

$$J''(x) = \left\| \frac{\partial^2 J(x)}{\partial x_i \partial x_j} \right\|, \quad i, j = 1, \ldots, m,$$

which leads to strong variations of $J(x)$ along directions coinciding with eigen vectors of the Hesse matrix corresponding to large eigen values, and to small variations along directions corresponding to small eigen values.

Most known methods of optimization allow a fairly rapid descent to the bottom Q of the valley, leading sometimes to a considerable reduction in the values of

$J(x)$ as compared with its value at the starting point (descent into the bottom of the valley). However, the process then slows down sharply, and practically stops at some point of Q which may be far from the required minimum point.

A twice continuously-differentiable function $J(x)$ is called a *valley function* (see [1]) if there exists a domain $G \subset \mathbf{R}^m$ in which the eigen values of the Hesse matrix $J''(x)$, in order of decreasing modulus, satisfy, at any point $x \in G$, the inequality

$$0 < \left| \min_i \lambda_i(x) \right| \ll \lambda_1(x). \tag{1}$$

The degree of steepness of the valley is characterized by the number

$$S = \frac{\lambda_1}{\left| \min_{\lambda_i \neq 0} \lambda_i \right|}. \tag{2}$$

If the eigen values of $J''(x)$ in G satisfy the inequalities

$$| \lambda_m(x) | \leqslant \cdots \leqslant | \lambda_{m-r+1}(x) | \ll \lambda_{m-r}(x) \leqslant \cdots \leqslant \lambda_1(x),$$

then the number r is called the *dimension of the valley* of the function $J(x)$ at $x \in G$ (see [1]).

The system of differential equations describing the trajectory of descent of $J(x)$,

$$\frac{dx}{dt} = -J'(x), \quad x(0) = x_0, \tag{3}$$

is a **stiff differential system**.

In particular, when $J(x)$ is strictly convex and the Hesse matrix is positive definite (all its eigen values are strictly positive), the inequalities (1) coincide with the well-known requirement for ill-conditioning of the Hesse matrix:

$$k(J''(x)) = \frac{\max_i \lambda_i(x)}{\min_i \lambda_i(x)} \gg 1.$$

In this case the spectral condition number is the same as the degree of steepness of the valley.

The **coordinate-wise descent method** (see [2])

$$J(x_{1,k+1}, \ldots, x_{i-1,k+1}, x_{i,k+1}, x_{i+1,k}, \ldots, x_{m,k}) = \tag{4}$$

$$= \min_y J(x_{1,k+1}, \ldots, x_{i-1,k}, y, x_{i+1,k}, \ldots, x_{m,k}),$$

$$k = 0, 1, \ldots,$$

in spite of its simplicity and universality, is effective in the valley situation only in the rare case when the orientation of the valleys is along the coordinate axes.

A modernization of the method (4) was proposed in [2], consisting of a rotation of the coordinate axes so that one of the axes lies along $x_k - x_{k-1}$, after which the search begins at the $(k+1)$-st step. Such an approach leads to one of the axes having a tendency to align along a generator of the bottom of the valley, allowing one to perform successfully the minimization of functions with one-dimensional valleys in several cases. The method is unsuitable for multi-dimensional valleys.

The scheme of the method of steepest descent (cf. **Steepest descent, method of**) is given by the difference equation

$$x_{k+1} = x_k - h_k J'_k, \quad J'_k = J'(x_k), \tag{5}$$

where h_k is chosen by the condition

$$J(x_{k+1}) = \min_{h>0} J(x_k - h J'_k).$$

For strictly-convex valley functions, in particular for quadratic ones

$$J(x) = \frac{1}{2} x^T D x - b^T x, \tag{6}$$

the sequence $\{x_k\}$ constructed by algorithm (5) converges geometrically to the minimum point x^* of the function (see [3]):

$$\| x_k - x^* \| \leqslant C q^k,$$

where $C = \text{const}$ and

$$q = \frac{k(J''(x^*)) - 1}{k(J''(x^*)) + 1}.$$

Since $k(J''(x)) \gg 1$ for a valley function, $q \simeq 1$ and convergence is effectively absent.

A similar picture can be seen for the simple gradient scheme (see [4]; **Gradient method**)

$$x_{k+1} = x_k - h J'_k, \quad J_{k+1} = J(x_{k+1}), \quad h = \text{const.} \tag{7}$$

Acceleration of its convergence is based on using the results of the previous iterations to make the bottom of the valley more precise. The gradient method (7) could be used (see [4], [5]) with computation of the ratio $q = \| J'_k \| / \| J'_{k-1} \|$ at each iteration. When it becomes steady near the constant value $q = 1$, a large accelerating step is performed in accordance with the expression

$$x_{k+1} = x_k - \frac{h}{1-q} J'_k.$$

Then descent by the gradient method is continued until the next accelerating step.

Various versions of the method of parallel tangents (see [4] - [6]) are based on carrying out the accelerating step along the direction $x_{k+2} - x_k$ given by the points x_k, x_{k+2} in the gradient method. In the 'heavy-sphere' method (see [4]; **Heavy sphere, method of the**) the next approximation has the form

$$x_{k+1} = x_k - \alpha J'_k + \beta(x_k - x_{k-1}).$$

In the *valley method* (see [7]) local descents are carried out by the gradient method (7) from two arbitrarily chosen starting points, and then an accelerating step is made in a direction given by two of the points at the bottom of the valley.

These methods are not much more complicated than the gradient method (7), and they are based on it. Accelerated convergence is obtained for a one-dimensional valley. In more general cases of multi-dimensional valleys, where convergence of these methods slows down sharply, one has to turn to the more powerful methods of quadratic approximation, based on Newton's method (cf. **Newton method**)

$$x_{k+1} = x_k - (J_k'')^{-1} J_k', \quad J_k'' = J''(x_k). \tag{8}$$

The minimum of the function (6) satisfies the system of linear equations

$$D\mathbf{x} = \mathbf{b}, \tag{9}$$

and under the condition of absolute accuracy of all computations, for a quadratic function Newton's method is independent of the degree of steepness of the valley (2) and the dimension of the valley, and leads to the minimum in one step. In reality, with a large condition number $k(D)$ and a restricted number of digits in computation, the problem of solving (9) may be ill-posed, and small changes in the elements of the matrix D and the vector $\mathbf{b}$ can lead to large variations in $\mathbf{x}^*$.

With moderate degrees of steepness in a convex situation, Newton's method often turns out to be preferable, with respect to speed of convergence, to other methods, such as the gradient method.

A large class of quadratic (quasi-Newtonian) methods are based on the use of **conjugate directions** (see [2], [3], [8]). For the case of minimization of a convex function these algorithms are very effective, since having a quadratic termination they do not require a computation of the matrix of second derivatives.

Sometimes (see [8]) the iteration is performed according to the scheme

$$x_{k+1} = x_k - (\beta_k E + J_k'')^{-1} J_k', \tag{10}$$

where E is the identity matrix. The scalars β_k are chosen so that the matrix $J_k'' + \beta_k E$ is positive definite and

$$\| x_{k+1} - x_k \| \leqslant \epsilon_k.$$

There are several similar approaches (see [8]) based on obtaining strictly positive definite approximations to the Hesse matrix. For the minimization of valley functions such algorithms turn out to be inefficient because of the difficulty in choosing the parameters β_k, ϵ_k, etc. The choice of these parameters is based on information on the size of the eigen values of smallest modulus of the Hesse matrix, and for real calculations and a high degree of steepness this information is strongly distorted.

A more appropriate generalization of Newton's method for the case of minimization of valley functions is based on the continuous principle of optimization. The function $J(x)$ is juxtaposed to a differential system (3), which is integrable by the system method (see **Stiff differential system**). The minimization algorithm takes the form

$$\left.\begin{aligned} x_{k+1} &= x_k - \Phi(2^N h_k^0) J_k', \\[4pt] \Phi(2^N h_k^0) &= \int_0^{2Nh_k^0} \exp(-J_k'' \tau)\, d\tau, \\[4pt] J_{k+1} &= \min_N J(x_k - \Phi(2^N h_k^0) J_k'), \\[4pt] h_k^0 &\leqslant \frac{1}{\| J_k' \|}, \\[4pt] \Phi(h_k^0) &= h_k^0 \left[E - \frac{h_k^0}{2} J_k'' + \frac{(h_k^0)^2}{6} (J_k'')^2 - \cdots \right], \\[4pt] \Phi(2^{s+1} h_k^0) &= \Phi(2^s h_k^0)[2E - J_k'' \Phi(2^s h_k^0)], \\[4pt] s &= 0, \ldots, N-1. \end{aligned}\right\} \tag{11}$$

In [1], an algorithm for the minimization of valley functions has been proposed based on the use of properties of stiff systems. Let the function $J(x)$ be approximated in a neighbourhood of x_0 by the quadratic function (6). The matrix D and the vector $\mathbf{b}$ are computed by, for example, finite-difference approximation. From the representation of the elements of the matrix

$$d_{ij} = \sum_{s=1}^{m} u_{is} u_{js} \lambda_s,$$

where $u_s = (u_{1s}, \ldots, u_{ms})^T$, $s = 1, \ldots, m$, is an orthonormal basis of eigen vectors of D, it follows that imprecise measurement of these elements distorts the information about the smallest eigen values of an ill-conditioned matrix, and hence leads to ill-posedness of the problem of minimization of (6).

Moreover, the system of differential equations of descent for the valley function (6),

$$\frac{dx}{dt} = -Dx + b, \quad x(0) = x_0,$$

has a solution in which, by virtue of condition (1), the terms with factors $\exp(-\lambda_1 t)$ have an influence only on a small initial interval of length $\tau_{ps} = O(\lambda_1^{-1})$.

In other words, the components of the vector $x(t)$ satisfy the equation

$$x^T(t) u_1 - \lambda_1^{-1} b^T u_1 = (x_0^T u_1 - \lambda_1^{-1} b^T u_1) \exp(-\lambda_1 t),$$

which quickly goes over to the stationary relation

$$\sum_{i=1}^{m} u_{i1} \bar{x}_i + \lambda_1^{-1} \sum_{i=1}^{m} b_i u_{i1} = 0, \tag{12}$$

where $\bar{x}_i$ are the components of a vector satisfying (12). This property is used in the algorithm. Expressing the j-th component of the vector $\bar{\mathbf{x}}$ to which the maximal component of the vector u_1 corresponds in terms of the remaining components, one replaces $J(x)$ by a new function with argument of dimension $(m-1)$:

$$J\left[\bar{x}_1, \ldots, \bar{x}_{i-1}, \right. \tag{13}$$

$$\left. -u_{j1}^{-1} \sum_{\substack{i=1 \\ i \neq j}}^{m} (u_{i1} \bar{x}_i - \lambda_1^{-1} b_i u_{i1}), x_{j+1}, \ldots, x_m \right] =$$

$$= \bar{J}(\bar{x}_1, \ldots, \bar{x}_{j-1}, \bar{x}_{j+1}, \ldots, \bar{x}_m).$$

One then finds a new matrix $\bar{D}$ of order $(m-1)$ and a new vector $\bar{\mathbf{b}}$ from the function (13) by means of finite-difference approximation. Here it is not so much the reduction of the dimension of the search space that is important, as the reduction of the degree of steepness, so that in the minimization of the new function in the subspace orthogonal to the vector u_1, the largest eigen value no longer influences the computation. The requirement that $\bar{D}$ and $\bar{\mathbf{b}}$ are obtained in terms of the function (13), and not in terms of D and $\mathbf{b}$, is a very important point here. The coefficients of relation (12) are found by the power method as coefficients of an arbitrary equation of the system

$$D^k \mathbf{x} = D^{k-1} \mathbf{b}.$$

If the degree of steepness is not reduced or is reduced insignificantly, then the process of elimination of coordinates of the vector $\mathbf{x}$ is continued recursively until the necessary reduction is obtained.

References

[1] RAKITSKIĬ, YU.V., USTINOV, S.M. and CHERNORUSTSKIĬ, I.G.: *Computational methods for the solution of stiff systems*, Moscow, 1979 (in Russian).
[2] CÉA, J.: *Optimisation: théorie et algorithmes*, Dunod, 1971.
[3] PSHENICHNYĬ, B.N. and DANILIN, YU.M.: *Numerical methods in extremal problems*, Mir, 1978 (translated from the Russian).
[4] POLYAK, B.T.: 'Methods for minimizing functions of several variables', *Ekonom. i Mat. Metody* 3, no. 6 (1967), 881-902 (in Russian).
[5] FADDEEV, D.K. and FADDEEVA, V.N.: *Computational methods of linear algebra*, Freeman, 1963 (translated from the Russian).
[6] WILDE, D.J.: *Optimum seeking methods*, Prentice Hall, 1964.
[7] GEL'FAND, I.M. and TSETLIN, M.L.: 'The principle of nonlocal search in automatic optimization systems', *Soviet Phys. Dokl.* 6 (1961), 192-194. (*Dokl. Akad. Nauk SSSR* 137, no. 2 (1961), 295-298)
[8] GILL, P.E. and MURRAY, W. (EDS.): *Numerical methods for constrained optimization*, Acad. Press, 1974.

Yu.V. Rakitskiĭ

Editorial comments.

References

[A1] LUENBERGER, D.G.: *Linear and nonlinear programming*, Addison-Wesley, 1984.

AMS 1980 Subject Classification: 49DXX, 65L99, 90C30

MINIMIZATION OF AN AREA - The problem of finding the minimum of the area $A(F)$ of a **Riemann surface** to which a given domain B of the z-plane is mapped by a one-to-one regular function F of a given class R, that is, the problem of finding

$$\min_{F \in R} A(F) = \min_{F \in R} \iint_B |F'(z)|^2 \, d\sigma \qquad (*)$$

($d\sigma$ is the surface element). The integral in (*), taken over B, is understood as the limit of integrals over domains B_n, $n = 1, 2, \ldots$, which *exhaust the domain B*,

that is, are such that $\bar{B}_n \subset B$, $B_n \subset B_{n+1}$ and such that any closed set $E \subset B$ lies in B_n from some n onwards.

When R is the class of functions $F(z)$, $F(0) = 0$, $F'(0) = 1$, regular in a given simply-connected domain B containing $z = 0$ and having more than one boundary point, the minimum A of the areas $A(F)$ of the images of B in the class R is given by the unique function univalently mapping B onto the full disc $|z| < r$, where r is the conformal radius of B at $z = 0$ (cf. **Conformal radius of a domain**); moreover, $A = \pi r^2$.

The problem of finding the minimal area of the image of a multiply-connected domain has also been considered (see [1]).

References

[1] GOLUZIN, G.M.: *Geometric theory of functions of a complex variable*, Amer. Math. Soc., 1969 (translated from the Russian).

E.G. Goluzina

AMS 1980 Subject Classification: 30C75

MINIMIZATION OF THE LABOUR OF CALCULATION, *minimization of the computational work* - A branch of modern computational mathematics, devoted to the design and investigation of methods which allow one to find an approximate solution, with preassigned accuracy $\epsilon > 0$, of a problem P from a class $\{P\}$ with the least computational cost (least volume of calculation). This branch of computational mathematics can be related to the more general branch concerned with the problem of optimizing methods (see, for example, [1], [2]) in which alongside the problem of minimizing the computational work one considers the dual problem of finding, among the approximation methods requiring roughly the same amount of calculation, a method with maximal accuracy (least error). The latter problem is characteristic, for example, of problems in numerical integration in which the number of points of integration, which measures the computational work, is fixed.

Let $\epsilon \geq 0$ (usually $\epsilon > 0$, for $\epsilon = 0$ a precise solution of P is sought) be the required accuracy for the solution of a problem P from a given class $\{P\}$ of related problems, and let m be an admissible method for finding the solution of any problem from $\{P\}$; the set of such methods will be denoted by $\{m\}$. Let a number $W_m(P, \epsilon) > 0$ characterize the computational cost when by using the method m a solution of P is to be found with accuracy ϵ, and put

$$W_m(\{P\}, \epsilon) = \sup_{P \in \{P\}} W_m(P, \epsilon).$$

Then the minimization problem is to find a method m_0 such that

$$W_{m_0}(\{P\}, \epsilon) = \inf_{m \in \{m\}} W_m(\{P\}, \epsilon),$$

that is, in essence one seeks an optimal method for

solving not just a fixed problem P, but a whole class of problems (optimization for a class). Mostly the minimization of the computational work is done in the asymptotic sense as $\epsilon \to 0$, $N \to \infty$, where N is a parameter defining the 'dimension' of the problem to be solved.

A method m_0 is called *order optimal* if the computational cost in it is at most finitely many times a quantity $\underline{W}(\{P\}, \epsilon)$ which is a lower estimate of the computational cost in any possible method; a method m_0 is called *logarithmically optimal* if

$$\ln W_{m_0}(\{P\}, \epsilon) = (1+o(1))\ln \underline{W}(\{P\}, \epsilon).$$

The computational work $W_m(P, \epsilon)$ itself is usually characterized by the number of arithmetic operations generated by the method m in attaining the accuracy ϵ in its realization on a conventional computer. To some extent this simplification is justified, since in many methods logical operations on real computers are performed in at most finitely many arithmetic operations. For admissible methods, as a rule, their stability with respect to rounding errors is required. It is important that the asymptotic growth of the amount of machine memory used in these methods is not too large.

To make this problem of minimizing the computational work concrete, the description of the class $\{P\}$ must be made precise and the metric space H which is used to define the accuracy of a solution of the resulting problem must be stated.

The solution of many such concrete minimization problems is not only of theoretical interest, but has also a great applied significance, allowing one to solve problems on computers with comparatively small expense of machine time. This is particularly important both for problems which require a large amount of calculations, characteristic for example of multi-dimensional problems in mathematical physics (see [2] - [8]), and for problems similar to the problems of calculating elementary functions and finding discrete Fourier transforms (see [1], [9]), which are standard and repeatedly used for the solution of other, more complex, problems. The problem of minimization of the computational work is fairly complicated and in many cases only a partial solution has been obtained.

In practice, for the solution of a problem of relatively small dimension and with no great accuracy, sometimes methods with poor asymptotic characteristics may require less machine time. Often, if the computational cost is acceptable, then the choice of the method is guided first of all by considerations of simplicity and reliability.

Suppose that the original problem is finite dimensional and that there are methods which give a precise solution in a finite number of arithmetic operations, if these operations are performed without rounding

errors. As examples one can take the problem of calculating the value of a polynomial $p_N(x)$ of degree N for given values of x with $|x| \leqslant 1$, the multiplication of two square matrices of order N, the solution of a given system of algebraic equations $Ax = b$ with a square matrix of order N, and the problem of finding the discrete Fourier transform (see [1], [9]):

$$v_n = \sum_{k=0}^{N-1} u_k e^{-2\pi n i / N}, \quad n = 0, \ldots, N-1, \tag{1}$$

where i is the imaginary unit, the vector $u = (u_0, \ldots, u_{n-1})$ is given and a vector $v \equiv (v_0, \ldots, v_{n-1})$, $N = 2^r$, $r = 0, 1, \ldots$, is sought. No concrete restrictions are imposed on the form of $p_N(x)$, $x \in [-1, 1]$, A, b, the number r, and the vector u, and, therefore, in each of these problems the admissible class $\{P\}$ consists of all problems of such form. In similar problems N takes the role of a parameter and special consideration is given to the behaviour of the computational cost $W_m(P, 0) \equiv W_m(P)$ as $N \to \infty$. For the first of these problems Horner's method, writing $p_N(x)$ in the form

$$a_0 + x(\cdots + x(a_{N-1} + x a_N) \cdots),$$

allows one to calculate $p_N(X)$ using N multiplications and N additions. It has been proved (see [9]) that this method is optimal: There is no method which would require fewer additions and subtractions or fewer multiplications and divisions; the stability is acceptable if $\sum_i |a_i|$ is not large (see [1]).

For the second and third problems there are a number of methods giving solutions as $N \to \infty$ in $W(N) \sim N^3$ arithmetic operations (see [1]) and they are actually applied in practice. The least computational cost among all methods known at present is attained in a method with estimate $W(N) \sim N^\alpha$, $\alpha \approx 2.5$ (see [9], [10]). This method is rather complicated and for a number of reasons is at present interesting only from the theoretical point of view. It is not known (1989), whether there is even a logarithmically-optimal method. There is a conjecture that there exists a logarithmically-optimal method with $W(N)$ arithmetic operations, where

$$\ln W(N) = 2 \ln N + o(\ln N).$$

For problem (1), the subject of harmonic analysis, the simplest methods require $\sim N^2$ arithmetic operations with complex numbers. In 1965 a method was suggested which allows one to find the vector in $W(N) \sim N \ln N$ arithmetic operations (see [1], [9]); this method is called the *method of the fast discrete Fourier transform*. This method is logarithmically optimal; it is widely used in practice. There are a large number of similar minimization problems, solved and unsolved (see [9], [10]); order-optimal or logarithmically-optimal estimates of the computational cost for finding the solution of similar problems may be considered as an indicator of their complexity.

Minimization of the computational work for the solution of grid systems of equations, which arise either in difference methods or in projection-grid methods (finite-element methods) for the approximate solution of boundary value problems for equations and systems of equations of strongly-elliptic type, has a special theoretical and applied significance, and, usually, is accomplished asymptotically as $N \to \infty$, where N is the number of unknowns in the system, and as $\epsilon \to 0$. For the solution of the simplest difference analogues of certain boundary value problems for the Poisson equation in a rectangle or a parallelopiped, certain direct methods have been applied successfully which are logarithmically optimal and require $O(N \ln N)$ arithmetic operations (see [3], [5], [8], [11], [12]). In the case of a rectangle (see [13]) an order-optimal method is known which requires $O(N)$ operations. Using the same iteration method one can obtain a logarithmically-optimal estimate of the type

$$W = O(N \ln N |\ln \epsilon|),$$

where ϵ is the accuracy of the solution of the system in some metric, for a fairly broad class of discrete boundary value problems in a parallelopipedal grid for linear and non-linear strongly-elliptic systems in certain ideal domains (see [2] - [8], [14]). (For example, in a plane Q they may be generated from a finite number of rectangles with sides parallel to the coordinate axes; and in a three-dimensional space Q they may be decomposed by a finite number of plane sections parallel to the given coordinate planes into parallelopipeds with boundaries parallel to the coordinate planes.) For more complicated domains Q the use of grids, topologically equivalent to the above grid domains Q, and certain types of projection-grid methods often allows one to obtain systems of equations which are as effectively solvable as in the case of ideal domains (see [3], [8], [15], [16]). Here the right-hand sides of these systems can be arbitrary vectors; if it is taken into account that they arise in special form as functionals on sufficiently smooth functions, then methods have been constructed with computational cost $O(N \log N)$ and even $O(N)$, under the condition that $\epsilon \sim N^{-\alpha}$, $\alpha > 0$. Methods of the latter type, for the solution of a problem on a given grid, use a sequence of similar problems on sparser grids (see [8], [16]).

There are methods which allow one to find with accuracy ϵ the minor eigen values and corresponding eigen functions for certain grid analogues of elliptic eigen value problems, at an expense of $O(N \ln N |\ln \epsilon|)$, or even $O(N \ln N)$ (for $\epsilon \sim N^{-\alpha}$,

$\alpha > 0$), operations (see [8], [15], [16]).

Let the problem P be to solve a well-posed operator equation

$$L(u) = f, \qquad (2)$$

where the operator L acts from H to F, and H and F are infinite-dimensional Banach spaces. Let $\{P\}$ be the class of such problems with different f for which the solution to (2) belongs to some compact set U. Usually U is given by one condition:

$$\| u \|_{H'} \leq R, \qquad (3)$$

where H' is a Banach space imbedded in H. If $u \in U$ is sought with accuracy $\epsilon > 0$ in the norm of H, then often information estimates, uniform with respect to all $u \in U$, are known for the least dimension $N \equiv N(\epsilon)$ of a vector $v_N \equiv (v_1, \ldots, v_N)$, the assignment of which allows one to obtain an element

$$\hat{v}_N \equiv \sum_{i=1}^{N} v_i \psi_i(z) \in H \quad \text{with} \quad \| u - \hat{v}_N \|_H \leq \epsilon \qquad (4)$$

(see [1], [2], [8], [17], [18]). These lower estimates for $N(\epsilon)$ give obvious lower estimates for the required computational cost in any admissible method. In a number of boundary value problems for strongly-elliptic systems variants of the projection-grid method have been constructed which lead to algebraic systems of equations

$$L_N(\bar{u}_N) = \bar{f}_N \qquad (5)$$

in N unknowns, for which, first, order-optimal iteration methods of approximate solution are known and, secondly, the corresponding functions $\hat{u}_N$ of the vectors $\bar{u}_N$ (see (4)) are ϵ-approximations in H of the solution of (2), where $N \leq kN(\epsilon)$ and k is a constant not depending on ϵ (see [7], [8], [16]). If the effort in forming (5) is not taken into account, these methods lead to order-minimal estimates for the computational cost. For example, for a second-order elliptic equation, where $H = W_2^1(\Omega)$, $H' = W_2^2(\Omega)$ ($W_2^k(\Omega)$ are the Sobolev spaces [19]), if Ω is a domain in the plane, the estimate $\sim \epsilon^{-2}$ for the expense of the number of arithmetic operations is obtained. The computational work in forming (5) can often also be estimated as $O(N \ln N)$ if information on the given functions in the corresponding spaces is used. In particular, in the example f must be considered as an element of $F = W_2^{-1}(\Omega)$, but not of $L_2(\Omega)$ (see [16]). A similar kind of estimate has been obtained for certain differential eigen value problems (see [8], [15]).

Questions on the minimization of the computational work for the solution of integral equations, ordinary differential equations and non-stationary partial differential equations have also been discussed (see, for example, [1] - [4], [20] - [22]).

References

[1] BAKHVALOV, N.S.: *Numerical methods: analysis, algebra, ordinary differential equations*, Mir, 1977 (translated from the Russian).

[2] TRAUB, J.F. and WOZNIAKOWSKI, H.: *A general theory of optimal algorithms*, Acad. Press, 1980.

[3] MARCHUK, G.I.: *Methods of numerical mathematics*, Springer, 1975 (translated from the Russian).

[4] SAMARSKII, A.A.: *Theorie der Differenzverfahren*, Akad. Verlagsgesell. Geest u. Portig K.-D., 1984 (translated from the Russian).

[5] SAMARSKII, A.A. and NIKOLAEV, E.S.: *Numerical methods for grid equations*, 1-2, Birkhäuser, 1989 (translated from the Russian).

[6] GLOWINSKI, R.: *Numerical methods for nonlinear variational problems*, Springer, 1984.

[7] KORNEEV, V.G.: *Schemes of the method of finite elements for high orders of accuracy*, Leningrad, 1977 (in Russian).

[8] D'YAKONOV, A.G.: *Minimization of computational work. Asymptotically optimal algorithms for elliptic problems*, Moscow, 1989 (in Russian).

[9] BORODIN, A. and MUNRO, I.: *The computational complexity of algebraic and numeric problems*, Amer. Elsevier, 1975.

[10] TRAUB, J.F. (ED.): *Analytic computational complexity*, Acad. Press, 1976.

[11] CONCUS, P. and GOLUB, G.: 'Use of fast direct methods for the efficient numerical solution of nonseparable elliptic equations', *SIAM J. Numer. Anal.* **10** (1973), 1103-1120.

[12] BARKER, R.J.: 'Advanced techniques for the direct numerical solution of Poisson's equation', in *Mathematical Models and Numerical Methods*, Banach Center Publ., Vol. 3, PWN, 1978, pp. 255-268.

[13] BANK, R.E. and ROSE, D.J.: 'Extrapolated fast direct algorithms for elliptic boundary value problems', in *Algorithms and Complexity. Proc. Symp. Carnegie-Mellon Univ.*, Acad. Press, 1976, pp. 201-249.

[14] MARCHUK, G.I., KUZNETSOV, YU.A. and MATSOKIN, A.M.: 'Fictitious domain and domain decomposition methods', *Soviet J. Numer. Anal. Math. Modelling* **1**, no. 1 (1986), 3-35.

[15] D'YAKONOV, E.G.: 'Effective methods for solving eigenvalue problems with fourth-order elliptic operators', *Soviet J. Anal. Math. Modelling* **1**, no. 1 (1986), 59-82.

[16] HACKBUSK, W. and TROTTENBERG, U. (EDS.): *Multigrid methods*, Lecture notes in math., 960, Springer, 1982.

[17] *Theoret. Foundations and Construction of Numerical Algorithms for Problems in Math. Physics*, Moscow, 1979, pp. 45-118 (in Russian).

[18] TIKHOMIROV, V.M.: *Some questions in approximation theory*, Moscow, 1976 (in Russian).

[19] SOBOLEV, S.L.: *Applications of functional analysis in mathematical physics*, Amer. Math. Soc., 1963 (translated from the Russian).

[20] EMEL'YANOV, K.V. and IL'IN, A.M.: 'Number of arithmetical operations necessary for the approximate solution of Fredholm integral equations of the second kind', *USSR Comp. Math. Math. Phys.* **7**, no. 4 (1967), 259-266. (*Zh. Vychisl. Mat. i Mat. Fiz.* **7**, no. 4 (1967), 905-910)

[21] IVANOV, V.V.: 'Optimal algorithms for the numerical solution of singular integral equations', in *Continuum Mechanics and Related Problems in Anal.*, Moscow, 1972, pp. 209-219 (in Russian).

[22] AKOPYAN, YU.R. and OGANESYAN, L.A.: 'A variational-difference method for solving two-dimensional linear parabolic equations', *USSR Comp. Math. Math. Phys.* **17**, no. 1 (1977), 101-110. (*Zh. Vychisl. Mat. Mat. Fiz.* **17**, no. 1 (1977), 109-118)

E.G. D'yakonov

Editorial comments.

References

[A1] TRAUB, J.F. and WOZNIAKOWSKI, H.: *Information, uncertainty, complexity*, Addison-Wesley, 1983.

AMS 1980 Subject Classification: 90C99, 68E99, 65K99

MINIMIZING SEQUENCE - A sequence of elements y_n from a set M for which the corresponding sequence of function values $\phi(y_n)$ tends to the greatest lower bound of ϕ on M, that is,

$$\lim_{n \to \infty} \phi(y_n) = \inf_{y \in M} \phi(y).$$

The compactness of a minimizing sequence, that is, the existence of a subsequence converging to an element of M, in combination with the lower semi-continuity of ϕ, guarantees the existence of an optimal element

$$y^n \in M, \quad \phi(y^n) = \min_{y \in M} \phi(y).$$

In approximation theory, a minimizing sequence $\{y_n\} \in M$ for a given element x of a metric space $X = (X, \rho)$ is a sequence for which

$$\rho(x, y_n) \to \rho(x, M) = \inf\{\rho(x, y) : y \in M\}.$$

See **Approximately-compact set**.

Yu.N. Subbotin

AMS 1980 Subject Classification: 41A65

MINIMIZING SEQUENCE FOR AN OPERATOR - A sequence of elements $\{z_n\}$, $z_n \in Z$, minimizing a continuous functional $I[z]$, $z \in Z$:

$$I[z_n] \to \inf_{z \in Z} I[z], \quad n \to \infty.$$

Minimization problems for functionals are usually divided into two groups. To the first group belong problems of finding the minimal value of a functional, for which it is irrelevant at which element z the minimum is attained. In this case the values of the functional associated with any minimizing sequence may be used as an approximate solution. The other group of problems concerns the search for an element z^* at which the functional $I[z]$ attains its least value:

$$\inf_{z \in Z} I[z] = I[z^*] = I^*. \tag{1}$$

In this connection, there are minimizing sequences which do not converge to an element z^*.

Let the minimization problem (1) have a unique solution z^* and let $\{z_n\}$ be a minimizing sequence, that is, a sequence such that

$$\lim_{n \to \infty} I[z_n] = I^*. \tag{2}$$

The minimization problem (1) is called *stable* if every minimizing sequence (2) converges to the element $z^* \in Z$.

In the solution of stable problems a minimizing sequence is obtained by constructing a sequence of iterates so that relative to z_n (the n-th iterate) a 'direction' y_n is found, and an element

$$z_{n+1} = z_n - f(\rho_n)y_n$$

is chosen from the set of elements $z_n - f(\rho)y_n$ minimizing the function $I[z_n - f(\rho)y_n]$ of the variable ρ.

Methods for constructing minimizing sequences for stable problems (1) are divided into three families. In the first, derivatives are not used; these are the direct methods. The second family uses the first derivatives of the functional; these methods are usually called descent methods. The third group of methods consists of algorithms which use second derivatives of the functional.

In non-stable minimization problems for functionals, regularization methods are applied to construct a sequence $\{z_n\}$ converging to the element z^*.

References
[1] TIKHONOV, A.N. and ARSENIN, V.YA.: *Solution of ill-posed problems*, Winston, 1977 (translated from the Russian).
[2] CÉA, J.: *Optimisation. Théorie et algorithmes*, Dunod, 1971.

Yu.V. Rakitskiĭ

AMS 1980 Subject Classification: 40A05, 41A65, 65J99

MINIMUM - See **Maximum and minimum of a function; Maximum and minimum points.**

AMS 1980 Subject Classification: 26A24

MINKOWSKI GEOMETRY - The geometry of a finite-dimensional **normed space**, that is, an affine space with a *Minkowski metric* — a metric invariant under parallel translation — in which the role of the unit sphere is played by a given centrally-symmetric **convex body**.

V.A. Zalgaller

Editorial comments. Note that the geometry of special relativity theory (cf. **Space-time**), which is not of this type, is also called Minkowski geometry.

References
[A1] BENSON, R.V.: *Euclidean geometry and convexity*, McGraw-Hill, 1966,Chapt. 6.

AMS 1980 Subject Classification: 51B20, 52A99

MINKOWSKI HYPOTHESIS *on the product of inhomogeneous linear forms* - A statement according to which for real linear forms

$$L_j(\bar{x}) = a_{j1}x_1 + \cdots + a_{jn}x_n, \quad 1 \leqslant j \leqslant n,$$

in n variables $x_1, \ldots, x_n$, with a non-zero determinant Δ, and any real $\alpha_1, \ldots, \alpha_n$, there are integers $x_1, \ldots, x_n$ such that the inequality

$$\prod_{j=1}^{n} |L_j(\bar{x}) - \alpha_j| \leqslant 2^{-n} |\Delta| \tag{*}$$

holds. This hypothesis was proved by H. Minkowski (1918) in case $n=2$. A proof of the hypothesis is known (1982) for $n \leqslant 5$, and (*) has been proved for $n > 5$ under certain additional restrictions (see [2]).

References
[1] CASSELS, J.W.S.: *An introduction to the geometry of numbers*, Springer, 1972.

[2] SKUBENKO, B.F.: 'A proof of Minkowski's conjecture on the product of n linear inhomogeneous forms in n variables for $n \leqslant 5$', in *Investigations in number theory*, Zap. Nauchn. Sem. Leningrad. Otdel. Mat. Inst. Steklov, Vol. 33, 1973, pp. 6-36 (in Russian).

E.I. Kovalevskaya

Editorial comments. See also **Geometry of numbers.**

References
[A1] GRUBER, P.M. and LEKKERKERKER, C.G.: *Geometry of numbers*, North-Holland, 1987.
[A2] ERDÖS, P., GRUBER, P.M. and HAMMER, J.: *Lattice points*, Longman, 1989.

AMS 1980 Subject Classification: 10EXX, 52A43, 52A45

MINKOWSKI INEQUALITY - 1) The *proper Minkowski inequality*: For real numbers $x_i, y_i \geqslant 0$, $i=1, \ldots, n$, and for $p > 1$,

$$\left[\sum_{i=1}^{n}(x_i + y_i)\right]^{1/p} \leqslant \left[\sum_{i=1}^{n} x_i^p\right]^{1/p} + \left[\sum_{i=1}^{n} y_i^p\right]^{1/p}. \tag{1}$$

This was derived by H. Minkowski [1]. For $p < 1$, $p \neq 0$, the inequality is reversed (for $p < 0$ one must have $x_i, y_i > 0$). In each case equality holds if and only if the rows $\{x_i\}$ and $\{y_i\}$ are proportional. For $p=2$ Minkowski's inequality is called the *triangle inequality*. Minkowski's inequality can be generalized in various ways (also called *Minkowski inequalities*). Below some of them are listed.

2) *Minkowski's inequality for sums.* Let $x_{ij} \geqslant 0$ for $i=1, \ldots, n$ and $j=1, \ldots, m$ and let $p > 1$. Then

$$\left[\sum_{i=1}^{n}\left[\sum_{j=1}^{m} x_{ij}\right]^p\right]^{1/p} \leqslant \sum_{j=1}^{m}\left[\sum_{i=1}^{n} x_{ij}^p\right]^{1/p}. \tag{2}$$

The inequality is reversed for $p < 1$, $p \neq 0$, and for $p < 0$ it is assumed that $x_{ij} > 0$. In each case equality holds if and only if the rows $\{x_{i1}\}, \ldots, \{x_{im}\}$ are proportional. There are also generalizations of (1) for multiple and infinite sums. However, when limit processes are used special attention is required in formulating the case of possible equality (see [2]).

Inequalities (1) and (2) are homogeneous with respect to $\sum$ and therefore have analogues for various means, for example, if $M_\phi(x_i) = \phi^{-1}\{\sum\phi(x_i)\}$, where $\phi(t) = \log t$, then

$$M_\phi\left[\frac{x_i + y_i}{2}\right] \leqslant \frac{1}{2}M_\phi(x_i) + \frac{1}{2}M_\phi(y_i);$$

for more details see [2].

3) *Minkowski's inequality for integrals* is similar to (2) and also holds because of the homogeneity with respect to $\int$. Let f, g be integrable functions in a domain $X \subset \mathbf{R}^n$ with respect to the volume element dV. Then for $p > 1$,

$$\left[\int_X |f+g|^p \, dV\right]^{1/p} \leqslant \tag{3}$$

$$\leqslant \left[\int_X |f|^p \, dV \right]^{1/p} + \left[\int_X |g|^p \, dV \right]^{1/p}.$$

A generalization of (3) to more general functions can be obtained naturally. A further generalization is: If $k > 1$, then

$$\left[\int \left[\int f(x,y) \, dy \right]^k dx \right]^{1/k} \leqslant \int \left[\int f^k(x,y) \, dx \right]^{1/k} dy,$$

where equality holds only if $f(x,y) = \phi(x)\psi(y)$.

4) *Other inequalities of Minkowski type*:

a) *for products*: If $x_i, y_i \geqslant 0$, then

$$\prod_{i=1}^n (x_i + y_i)^{1/n} \geqslant \left[\prod_{i=1}^n x_i \right]^{1/n} + \left[\prod_{i=1}^n y_i \right]^{1/n};$$

b) *Mahler's inequality*: Let $F(x)$ be a generalized norm on E^n and $G(y)$ its polar function; then

$$(x,y) \leqslant F(x)G(y),$$

where $(\cdot, \cdot)$ is the **inner product**;

c) *for determinants*: If A, B are non-negative Hermitian matrices over **C**, then

$$(\det(A+B))^{1/n} \geqslant (\det A)^{1/n} + (\det B)^{1/n}.$$

5) Finally, connected with the name of Minkowski are other inequalities; in particular, in convex analysis and number theory. For example, the **Brunn−Minkowski theorem**.

References

[1] MINKOWSKI, H.: *Geometrie der Zahlen*, Chelsea, reprint, 1953.
[2] HARDY, G.H., LITTLEWOOD, J.E. and PÓLYA, G.: *Inequalities*, Cambridge Univ. Press, 1934.
[3] BECKENBACH, E.F. and BELLMAN, R.: *Inequalities*, Springer, 1961.
[4] MARCUS, M. and MINC, H.: *Survey of matrix theory and matrix inequalities*, Allyn & Bacon, 1964.

M.I. Voĭtsekhovskiĭ

Editorial comments. A *generalized norm* on E^n is a function F for which: 1) $F(x) > 0$ for $x \neq 0$; 2) $F(tx) = tF(x)$ for $t \geqslant 0$; and 3) $F(x) + F(y) \geqslant F(x+y)$. The *polar form* (or *polar function*) G of the generalized norm F is defined by:

$$G(y) = \max_x \frac{(x,y)}{F(x)},$$

where $(\cdot, \cdot)$ is the inner product.

References

[A1] BULLEN, P.S., MITRINOVIĆ, D.S. and VASIĆ, P.M.: *Means and their inequalities*, Reidel, 1988.

AMS 1980 Subject Classification: 26D15

MINKOWSKI PROBLEM - Does there exist a closed convex hyperplane F for which the **Gaussian curvature** $K(\xi)$ is a given function of the unit outward normal ξ? This problem was posed by H. Minkowski [1], to whom is due a generalized solution of the problem in the sense that it contains no information on the nature of regularity of F, even if $K(\xi)$ is an analytic function. He proved that if a continuous positive function $K(\xi)$, given on the hypersphere S, satisfies the condition

$$\int_S \xi \frac{ds}{K(\xi)} = 0, \tag{1}$$

then there exists a closed convex surface F, which is moreover unique (up to a parallel translation), for which $K(\xi)$ is the Gaussian curvature at a point with outward normal ξ.

A regular solution of Minkowski's problem has been given by A.V. Pogorelov in 1971 (see [2]); he also considered certain questions in geometry and in the theory of differential equations bordering on this problem. Namely, he proved that if $K(\xi)$ is of class C^m, $m \geqslant 3$, then the surface F is of class $C^{m+1,\alpha}$, $\alpha > 0$, and if $K(\xi)$ is analytic, then F also turns out to be analytic.

A natural generalization of Minkowski's problem is the solution of the question of the existence of convex hypersurfaces with given elementary symmetric principal curvature functions $\phi_\nu(\xi)$ of any given order ν, $\nu \leqslant n = \dim F$. In particular, for $\nu = 1$ this is Christoffel's problem on the recovery of a surface from its mean curvature. A necessary condition for the solvability of this generalized Minkowski problem, analogous to (1), has the form

$$\int_S \xi \phi_\nu(\xi) \, dS = 0.$$

However, this condition is not sufficient (A.D. Aleksandrov, 1938, see [3]). There are examples of sufficient conditions:

$$\int_S \xi \Phi_\nu(\xi) \, dS = 0,$$

$$\left[1 - \frac{1}{n} \right]^{1/2(\nu-1)} \max(\Phi_{\nu,t} - \Phi_{\nu,t}'') < \Phi_{\nu,t}(\xi),$$

where $\Phi_{\nu,t} = (\phi_{\nu,t}/C_n^\nu)^{1/n}$, $\phi_{\nu,t} = t\phi_\nu + 1 - t$, $0 \leqslant t \leqslant 1$. Here the regularity of F is as in the Minkowski problem. Using approximations these results turn out to be valid even for functions $\phi(\xi)$ which are non-negative, symmetric and concave.

References

[1] MINKOWSKI, H.: 'Volumen und Oberfläche', *Math. Ann.* **57** (1903), 447-495.
[2] POGORELOV, A.V.: *The Minkowski multidimensional problem*, Winston, 1978 (translated from the Russian).
[3] BUSEMANN, H.: *Convex surfaces*, Interscience, 1958.

M.I. Voĭtsekhovskiĭ

Editorial comments.

References

[A1] POGORELOV, A.V.: *Extrinsic geometry of convex surfaces*, Amer. Math. Soc., 1973 (translated from the Russian).
[A2] BERGER, M. and GOSTIAUX, B.: *Differential geometry*, Springer, 1988 (translated from the French).
[A3] SCHNEIDER, R.: 'Boundary structure and curvature of convex bodies', in J. Tölke and J.M. Wills (eds.): *Contributions to Geometry*, Birkhäuser, 1979, pp. 13-59.

AMS 1980 Subject Classification: 53A05, 52A20

MINKOWSKI SPACE - A four-dimensional **pseudo-Euclidean space** of signature $(1, 3)$, suggested by H.

Minkowski (1908) as a geometric model of space-time in the special theory of relativity (see [1]). Corresponding to each event there is a point of Minkowski space, three coordinates of which represent its coordinates in the three-dimensional space; the fourth coordinate is ct, where c is the velocity of light and t is the time of the event. The space-time relationship between two events is characterized by the so-called square interval:

$$s^2 = c^2(\Delta t)^2 - (\Delta x)^2 - (\Delta y)^2 - (\Delta z)^2.$$

The interval in Minkowski space plays a role similar to that of distance in Euclidean geometry. A vector with positive square interval is called a *time-like vector*, one with negative square interval, a *space-like vector*, one with square interval zero, a *null* or *isotropic vector*. A curve with a time-like tangent vector at each point is called a *time-like curve*. *Space-like* and *isotropic curves* are similarly defined. An event at a given moment of time is called a *world point*; a set of world points describing the development of some process or phenomenon through time is called a *world line*. If a vector joining neighbouring world points is time-like, then there is a frame of reference in which the events project to one and the same point of three-dimensional space. The time separating the events in this frame of reference is equal to $\Delta t = \tau = s/c$, where τ is the so-called *proper time*. There is no frame of reference in which these events can be simultaneous (that is, have the same time coordinate t). If the vector joining the world points of two events is space-like, then there is a frame of reference in which these two events occur simultaneously; they are not connected by a causal relation; the modulus of the square interval defines the spatial distance between the points (events) in this frame of reference. A tangent vector to a world line is a time-like vector. The tangent vector to a light ray is an isotropic vector.

The motions of Minkowski space, that is, the interval-preserving transformations, are the Lorentz transformations (cf. **Lorentz transformation**).

A generalization of Minkowski space is the **pseudo-Riemannian space** used in the construction of the theory of gravitation.

References

[1A] MINKOWSKI, H.: 'Raum und Zeit', *Phys. Z. Sowjetunion* **10** (1909), 104- (Reprint in: Lorentz − Einstein − Minkowski, Teubner, 1922).
[1B] MINKOWSKI, H.: 'Das Relativitätsprinzip', *Jahresber. Deutsch. Math. Verein* **24** (1915), 372-.
[2] LANDAU, L.D. and LIFSHITZ, E.M.: *The classical theory of fields*, Addison-Wesley, 1962 (translated from the Russian).
[3] FOCK, V.A. [V.A. FOK]: *The theory of space, time and gravitation*, Macmillan, 1954 (translated from the Russian).
[4] RASHEWSKI, P.K. [P.K. RASHEVSKIĬ]: *Riemannsche Geometrie und Tensoranalyse*, Deutsch. Verlag Wissenschaft., 1959 (translated from the Russian).

[5] SYNGE, J.L.: *Relativity: the general theory*, North-Holland, 1960.

D.D. Sokolov

Editorial comments. See in particular [A3], pp. 10-11 and Chapt. 3, and [A4] - [A6] for material on the pseudo-Riemannian spaces used in gravitational theories.

References

[A1] CHOQUET-BRUHAT, Y., DEWITT-MORETTE, C. and DILLARD-BLEIK, M.: *Analysis, manifolds and physics*, North-Holland, 1977 (translated from the French).
[A2] WEINBERG, S.: *Gravitation and cosmology*, Wiley, 1972.
[A3] BESSE, A.: *Einstein manifolds*, Springer, 1987.
[A4] HAWKING, S. and ELLIS, G.: *The large scale structure of space-time*, Cambridge Univ. Press, 1973.
[A5] MISNER, C., THORNE, K. and WHEELER, J.: *Gravitation*, Freeman, 1973.
[A6] O'NEILL, B.: *Semi-riemannian geometry*, Acad. Press, 1983.

AMS 1980 Subject Classification: 51N25, 53A07, 53B30, 83A05

MINKOWSKI THEOREM - 1) *Minkowski's theorem on convex bodies* is the most important theorem in the geometry of numbers, and is the basis for the existence of the **geometry of numbers** as a separate division of number theory. It was established by H. Minkowski in 1896 (see [1]). Let $\mathfrak{R}$ be a closed convex body, symmetric with respect to the origin 0 and having volume $V(\mathfrak{R})$. Then every point lattice Λ of determinant $d(\Lambda)$ for which

$$V(\mathfrak{R}) \geqslant 2^n d(\Lambda)$$

has a point in $\mathfrak{R}$ distinct from 0.

An equivalent formulation of Minkowski's theorem is:

$$\Delta(\mathfrak{R}) \geqslant 2^{-n} V(\mathfrak{R}),$$

where $\Delta(\mathfrak{R})$ is the critical determinant of the body $\mathfrak{R}$ (see **Geometry of numbers**). A generalization of Minkowski's theorem to non-convex bodies is Blichfeldt's theorem (see **Geometry of numbers**). The theorems of Minkowski and Blichfeldt enable one to estimate from above the arithmetic minima of distance functions.

References

[1] MINKOWSKI, H.: *Geometrie der Zahlen*, Chelsea, reprint, 1953.

A.V. Malyshev

Editorial comments. A refinement of Minkowski's theorem employing Fourier series was given by C.L. Siegel. A different refinement is Minkowski's theorem on successive minima (see **Geometry of numbers**). These refinements have applications in algebraic number theory and in Diophantine approximation. For a collection of other conditions which guarantee the existence of lattice points in a convex body see [A2].

2) *Minkowski's theorem on linear forms*: The system of inequalities

$$\left| \sum_{j=1}^{n} a_{1j} x_j \right| \leqslant c_1, \quad \left| \sum_{j=1}^{n} a_{ij} x_j \right| < c_i, \quad 2 \leqslant i \leqslant n,$$

where a_{ij}, c_i are real numbers, has an integer solution $(x_1, \ldots, x_n) \neq 0$ if $c_1 \cdots c_n \geqslant |\det a_{ij}|$. This was established by H. Minkowski in 1896 (see [1]). Minkowski's theorem on linear forms is a corollary of the more general theorem of Minkowski on a convex body (see part 1).

References

[1] MINKOWSKI, H.: *Geometrie der Zahlen*, Chelsea, reprint, 1953.
[2] MINKOWSKI, H.: *Diophantische Approximationen*, Chelsea, reprint, 1957.
[3] CASSELS, J.W.S.: *An introduction to the geometry of numbers*, Springer, 1972.

E.I. Kovalevskaya

Editorial comments. The problem when the first inequality in Minkowski's theorem on linear forms can be replaced by strict inequality was solved by G. Hajós.

References

[A1] GRUBER, P.M. and LEKKERKERKER, C.G.: *Geometry of numbers*, North-Holland, 1987.
[A2] ERDÖS, P., GRUBER, P.M. and HAMMER, J.: *Lattice points*, Longman, 1989.

AMS 1980 Subject Classification: 10E05, 52A20, 52A43, 52A45

MINOR *of order* k - A **determinant** of a **matrix** whose entries are located in a given matrix at the intersection of k distinct columns and k distinct rows. If the row indices and column indices are the same, then the minor is called *principal*, and if they are the first k rows and columns, then it is called a *corner*. A *basic minor* of a matrix is any non-zero minor of maximal order. In order that a non-zero minor be basic it is necessary and sufficient that all minors *bordering* it (that is, minors of an order higher by one and containing it) are equal to zero. The system of rows (columns) of a matrix related to a basic minor form a maximal linearly independent subsystem of the system of all rows (columns) of the matrix.

V.N. Remeslennikov

Editorial comments. Instead of 'minor of order k' one also uses 'minor of degree k'. Sometimes a minor is not understood to be a determinant (as defined above), but the corresponding submatrix (the notion of 'bordering' uses this interpretation).

AMS 1980 Subject Classification: 15A15

MINOR OF A GRAPH

Editorial comments. Let G be a **graph** (possibly with loops and multiple edges). A *minor* of G is any graph obtained by a succession of the following operations:

i) deletion of a single edge;
ii) contraction of a single edge;
iii) removal of an isolated vertex.

The *graph minor theorem* of N. Robertson and P.D. Seymour says the following. Given an infinite sequence $G_1, G_2, \ldots$, of finite graphs, there are indices $i < j$ such that G_i is (isomorphic to) a minor of G_j. This was conjectured by K. Wagner (*Wagner's well-quasi-ordering conjecture*).

Let P be any property of a graph which is closed under taking minors. For instance, the property of being imbeddable in a compact Riemann surface. Let L be the set of all minor-minimal isomorphism classes of graphs not possessing property P. Then the *graph minor theorem* says that L is finite. This implies that property P is characterized by saying that a graph has property P if and only if it does not contain a minor from the finite set L. In the case of the property 'planar' the set L is given by the two graphs $K_{3,3}$ and K_5, cf. **Kuratowski graph**.

The finite forms of the graph minor theorem are as follows:

a) For all k there is an n such that for n-tuples of graphs $G_1, \ldots, G_n$ with $|G_i| \leqslant k + i$ there is a pair $i < j$ with G_i isomorphic to a minor of G_j. Here $|G_i|$ denotes the sum of the number of vertices and the number of edges of G_i.

b) For all k there is an n such that for all n-tuples of graphs $G_1, \ldots, G_n$ with $|G_i| \leqslant i$ there are $i_1 < \cdots < i_k$ such that G_{i_j} is isomorphic to a minor of $G_{i_{j+1}}$ for all $j = 1, \ldots, k-1$.

Wagner's conjecture does not hold, in general, for infinite graphs, in the sense that there is a sequence of infinite graphs $G_1, G_2, \ldots$, such that for all $i \neq j$, G_i is not a minor of G_j. The conjecture is open (1988) for countable graphs.

A consequence (application) is that each graph property that is closed under taking minors can be tested in polynomial time. So, for instance, planarity can be tested in polynomial time. Further, there exists a *polynomial-time algorithm* for the following problem: Given a graph G and vertices $r_1, s_1, \ldots, r_k, s_k$ of G, find paths $P_1, \ldots, P_k$ so that P_i connects r_i and s_i for all $i = 1, \ldots, k$ and such that the $P_1, \ldots, P_k$ are vertex disjoint. Such problems arise in VLSI design.

References

[A1] FRIEDMAN, H., ROBERTSON, N. and SEYMOUR, P.D.: 'The mathematics of the graph minor theorem', *Contempory Math.* **65** (1987), 229-261.
[A2] ROBERTSON, N. and SEYMOUR, P.D.: 'Graph minors — a survey', in I. Anderson (ed.): *Surveys in Combinatorics 1985*, Cambridge Univ. Press, 1985, pp. 153-171.
[A3] THOMAS, R.: 'A counter-example to 'Wagner's conjecture' for infinite graphs', *Math. Proc. Cambridge Philos. Soc.* **103** (1988), 55-57.

AMS 1980 Subject Classification: 05CXX, 05C10

MINORANT - See **Majorant and minorant**.

AMS 1980 Subject Classification: 26A99, 41-XX

MINUTE - The unit of measurement of plane angles equal to $1/60$ part of a **degree**, denoted by the sign $'$. A minute is divided into 60 seconds ($60''$). A *metric minute* is $1/10000$ part of a right angle, denoted by the symbol ^{c}. A metric minute is divided into 100 metric seconds (100^{cc}).

AMS 1980 Subject Classification: 51-XX, 00A22

MITTAG-LEFFLER FUNCTION - An entire function $E_\rho(z)$ of a complex variable z, introduced by G. Mittag-Leffler [1] as a generalization of the exponential function:

$$E_\rho(z) = \sum_{k=0}^{\infty} \frac{z^k}{\Gamma(1+k/\rho)}, \quad 1 \le \rho < \infty.$$

Since the Mittag-Leffler function and the more general *functions of Mittag-Leffler type*

$$E_\rho(z;\mu) = \sum_{k=0}^{\infty} \frac{z^k}{\Gamma(\mu+k/\rho)}, \quad \mu, \rho \in \mathbf{C},$$

are widely used in integral representations and transforms of analytic functions, their properties, in particular asymptotic properties, have been studied in great detail (see [2], [3]).

References

[1] MITTAG-LEFFLER, G.: 'Sur la représentation analytique d'une branche uniforme d'une fonction monogène', *Acta Math.* **29** (1905), 101-181.
[2] DZHRBASHYAN, M.M.: *Integral transforms and representation of functions in a complex domain*, Moscow, 1966 (in Russian).
[3] GOL'DBERG, A.A. and OSTROVSKIĬ, I.V.: *Distribution of the values of meromorphic functions*, Moscow, 1970 (in Russian).

E.D. Solomentsev

Editorial comments.

References

[A1] CARTWRIGHT, M.L.: *Integral functions*, Cambridge Univ. Press, 1962.

AMS 1980 Subject Classification: 30D15

MITTAG-LEFFLER STAR - The same as the **star of a function element**.

AMS 1980 Subject Classification: 30BXX

MITTAG-LEFFLER SUMMATION METHOD - A **semi-continuous summation method** for summing series of numbers and functions, defined by a sequence of functions

$$g_k(\delta) = \frac{1}{\Gamma(1+\delta k)}, \quad \delta > 0, \quad k = 0, 1, \ldots,$$

where $\Gamma(x)$ is the gamma-function. A series

$$\sum_{k=0}^{\infty} u_k$$

is *summable by the Mittag-Leffler method* to a sum s if

$$\lim_{\delta \to 0} \sum_{k=0}^{\infty} \frac{u_k}{\Gamma(1+\delta k)} = s$$

and if the series under the limit sign converges. The method was introduced by G. Mittag-Leffler [1] primarily for the series

$$\sum_{k=0}^{\infty} z^k.$$

A Mittag-Leffler summation method is regular (see **Regular summation methods**) and is used as a tool for the **analytic continuation** of functions. If $f(z)$ is the principal branch of an analytic function, regular at zero and represented by a series

$$\sum_{k=0}^{\infty} a_k z^k$$

for small z, then this series is summable by the Mittag-Leffler method to $f(z)$ in the whole star of the function $f(z)$ (cf. **Star of a function element**) and, moreover, uniformly in any closed bounded domain contained in the interior of the star.

For summation methods defined by transformations of sequences by semi-continuous matrices $a_k(\omega)$ of the type

$$a_k(\omega) = \frac{c_{k+1}\omega^{k+1}}{E(\omega)},$$

where

$$E(\omega) = \sum_{k=0}^{\infty} c_k \omega^k$$

is an entire function, Mittag-Leffler considered the case when

$$E(\omega) = \sum_{k=0}^{\infty} \frac{\omega^k}{\Gamma(1+ak)}.$$

A matrix $a_k(\omega)$ with such an entire function is called a *Mittag-Leffler matrix*.

References

[1] MITTAG-LEFFLER, G.: *Atti del IV Congresso Internationale*, Vol. 1, Rome, 1908, pp. 67-85.
[2] MITTAG-LEFFLER, G.: 'Sur la représentation analytique d'une branche uniforme d'une fonction monogène', *Acta Math.* **29** (1905), 101-181.
[3] HARDY, G.H.: *Divergent series*, Clarendon Press, 1949.
[4] COOKE, R.G.: *Infinite matrices and sequence spaces*, MacMillan, 1950.

I.I. Volkov

Editorial comments. The function $E(\omega)$ considered by Mittag-Leffler is called a **Mittag-Leffler function**.

AMS 1980 Subject Classification: 40G99

MITTAG-LEFFLER THEOREM - 1) The *Mittag-Leffler theorem on expansion of a meromorphic function* (see [1], [2]) is one of the basic theorems in analytic function theory, giving for meromorphic functions an analogue of the expansion of a rational function into the simplest partial fractions. Let $\{a_n\}_{n=1}^{\infty}$ be a sequence of distinct complex numbers,

$$|a_1| \le |a_2| \le \cdots, \quad \lim_{n \to \infty} a_n = \infty,$$

and let $\{g_n(z)\}$ be a sequence of rational functions of the form

$$g_n(z) = \sum_{k=1}^{l_n} \frac{c_{nk}}{(z-a_n)^k}, \tag{1}$$

so that a_n is the unique pole of the corresponding function $g_n(z)$. Then there are meromorphic functions $f(z)$ in the complex z-plane $\mathbf{C}$ having poles at a_n, and only there, with given principal parts (1) of the **Laurent series** corresponding to the points a_n. All these functions $f(z)$ are representable in the form of a *Mittag-*

Leffler expansion

$$f(z) = h(z) + \sum_{n=1}^{\infty} [g_n(z) + p_n(z)], \qquad (2)$$

where $p_n(z)$ is a polynomial chosen in dependence of a_n and $g_n(z)$ so that the series (2) is uniformly convergent (after the removal of a finite number of terms) on any compact set $K \subset \mathbf{C}$ and $h(z)$ is an arbitrary entire function.

The Mittag-Leffler theorem implies that any given meromorphic function $f(z)$ in $\mathbf{C}$ with poles a_n and corresponding principal parts $g_n(z)$ of the Laurent expansion of $f(z)$ in a neighbourhood of a_n can be expanded in a series (2) where the entire function $h(z)$ is determined by $f(z)$. G. Mittag-Leffler gave a general construction of the polynomials $p_n(z)$; finding the entire function $h(z)$ relative to a given $f(z)$ is sometimes a more difficult problem. To obtain (2) it is possible to apply methods of the theory of residues (cf. **Residue of an analytic function**, see also [3] - [5]).

A generalization of the quoted theorem, also due to Mittag-Leffler, states that for any domain D of the extended complex plane $\overline{\mathbf{C}}$, any sequence $\{a_n\}$ of points $a_n \in D$ all limit points of which are in the boundary ∂D, and corresponding principal parts (1), there is a function $f(z)$, meromorphic in D, having poles at $\{a_n\}$, and only there, with the given principal parts (1). In this form the Mittag-Leffler theorem generalizes to open Riemann surfaces D (see [7]); for the existence of meromorphic functions on compact Riemann surfaces with given singularities see **Abelian differential; Differential on a Riemann surface; Riemann—Roch theorem**. The Mittag-Leffler theorem is also true for abstract meromorphic functions $g_n, f: D \to F$, $D \subset \overline{\mathbf{C}}$, with values in a Banach space F (see [8]).

Another generalization of the Mittag-Leffler theorem states that for any sequence $\{a_n\} \subset \mathbf{C}$, $|a_1| \leqslant |a_2| \leqslant \cdots$, $\lim a_n = \infty$, and corresponding functions

$$g_n(z) = \sum_{k=1}^{\infty} \frac{c_{nk}}{(z - a_n)^k}$$

that are entire functions of the variable $w_n = 1/(z - a_n)$, there is a single-valued analytic function $f(z)$ having singular points at a_n, and only there, and with principal parts $g_n(z)$ (see [3]).

For analytic functions of several complex variables a generalization of the Mittag-Leffler problem on the construction of a function with given singularities is the first (additive) Cousin problem (cf. **Cousin problems**). In this connection the following equivalent statement of the Mittag-Leffler theorem is often useful. Let $\Omega = \bigcup_j \Omega_j$, where the Ω_j are open sets in $\mathbf{C}$, and let there be given meromorphic functions g_j, respectively, on the sets Ω_j, where the differences $g_j - g_k$ are regular functions on the intersections $\Omega_j \cap \Omega_k$ for all j and k. Then there is on Ω a meromorphic function f such that the differences $f - g_j$ are regular on Ω_j for all j (see [5], [6]).

2) For the *Mittag-Leffler theorem on the expansion of single-valued branches of an analytic function in a star* see **Star of a function element**.

References

[1] MITTAG-LEFFLER, G.: 'En metod att analytisk framställa en funktion at rationel karacte...', *Öfversigt Kongl. Vetenskap-Akad. Förhandlinger* **33** (1876), 3-16.
[2] MITTAG-LEFFLER, G.: 'Sur la répresentation analytique des fonctions monogènes uniformes d'une variable indépendante', *Acta Math.* **4** (1884), 1-79.
[3] GOURSAT, E.: *Cours d'analyse mathématique*, Gauthier-Villars, 1927.
[4] MARKUSHEVICH, A.I.: *The theory of functions of a complex variable*, 2, Chelsea, 1977 (translated from the Russian).
[5] SHABAT, B.V.: *Introduction to complex analysis*, 1-2, Moscow, 1976 (in Russian).
[6] HÖRMANDER, L.: *An introduction to complex analysis in several variables*, North-Holland, 1973.
[7] BEHNKE, H. and SOMMER, F.: *Theorie der analytischen Funktionen einer komplexen Veränderlichen*, Springer, 1972.
[8] SCHWARTZ, L.: *Analyse mathématique*, 2, Hermann, 1967.

E.D. Solomentsev

Editorial comments.

References

[A1] CONWAY, J.B.: *Functions of one complex variable*, Springer, 1978.
[A2] HEINS, M.: *Complex function theory*, Acad. Press, 1968.

AMS 1980 Subject Classification: 30D30, 32C35

MIXED AND BOUNDARY VALUE PROBLEMS FOR HYPERBOLIC EQUATIONS AND SYSTEMS - Problems of finding solutions of (systems of) partial differential equations of hyperbolic type that satisfy specific conditions on the boundaries of their domains (or on parts of it) (see **Boundary conditions; Initial conditions**).

A boundary value problem for (a system of) hyperbolic equations in some domain D of a Euclidean space $\mathbf{R}^{n+1}$ is called a *mixed* or *initial boundary value problem* if the desired solution, as well as the boundary conditions, must also satisfy initial conditions, or if the support ∂D of the boundary data consists of both characteristic and non-characteristic oriented manifolds.

For second-order hyperbolic equations, the support of the initial data in a mixed problem is the spatially-oriented part of ∂D. On the time-oriented part of ∂D, the boundary conditions are usually of the same type as for parabolic equations (see **Mixed and boundary value problems for parabolic equations and systems**).

Let Ω be a domain in $\mathbf{R}^n$ of points $x = (x_1, \ldots, x_n)$ with a sufficiently smooth boundary $\partial \Omega$, and let

$$D = \{(x, x_0): x \in \Omega, \ 0 < x_0 < T\},$$
$$S = \{(x, x_0): x \in \partial \Omega, \ 0 < x_0 < T\},$$

$\Omega_0 = \{(x, x_0): x \in \Omega,\ x_0 = 0\},\ \Omega_T = \{(x, x_0): x \in \Omega,\ x_0 = T\}.$

Consider in D the second-order linear hyperbolic equation

$$u_{x_0 x_0} - a^{ij}(x, x_0)u_{x_i x_j} + a^i(x, x_0)u_{x_i} + \qquad (1)$$

$$+ a^0(x, x_0)u_{x_0} + a(x, x_0)u = f(x, x_0),$$

where summation is from 1 to n over repeated subscripts i, j and the form $a^{ij}(x, x_0)\xi_i\xi_j$ must be positive definite.

The basic mixed problems for equation (1) can be combined into the following: One has to find a solution $u = u(x, x_0)$ of equation (1) in D that satisfies the initial conditions

$$u\big|_{x_0 = 0} = \tau(x),\ \ u_{x_0}\big|_{x_0 = 0} = \nu(x) \qquad (2)$$

on Ω_0 and one of the boundary conditions

$$u\big|_S = \phi_1(x, x_0), \qquad (3)$$

$$\frac{\partial u}{\partial N}\bigg|_S = \phi_2(x, x_0), \qquad (4)$$

$$\left[\frac{\partial u}{\partial N} + \sigma(x, x_0)u\right]\bigg|_S = \phi_s(x, x_0) \qquad (5)$$

on S, where N is a co-normal with respect to the operator $a^{ij}(\partial/\partial x_i)(\partial/\partial x_j)$. Problems (2), (3); (2), (4); and (2), (5) are usually called the *first, second* and *third mixed problem* for equation (1), respectively.

Under fairly general assumptions on the coefficients of (1) and on the boundary ∂D, and also on the given functions, one can prove the existence and uniqueness of both regular and generalized solutions of all three mixed problems. The structural and differentiability properties of these solutions in the closed domain $\overline{D}$ in dependence on the smoothness of the boundary have also been investigated [8]. When $n = 1$, the solutions of the mixed problems can be obtained in explicit form.

Mixed problems have been investigated for a wide class of linear and non-linear hyperbolic equations and systems (see **Quasi-linear hyperbolic equations and systems**). There is a satisfactory theory of mixed problems for strictly-hyperbolic equations and systems of the form

$$u_{x_0 x_0} = \sum_{|\alpha| \leqslant m} a_\alpha(x, x_0) \frac{\partial^{|\alpha|} u}{\partial x_1^{\alpha_1} \cdots \partial x_n^{\alpha_n}} + f(x, x_0)$$

with initial data on the spatially-oriented part of the boundary of D lying in the plane $x_0 = 0$. There has also been a successful study of mixed problems for hyperbolic equations and systems in case the supports of the initial or the boundary conditions are surfaces of degenerate type or of the order of these equations (see **Degenerate partial differential equation**). Most substantial results have been obtained for second-order linear equations of the form

$$u_{x_0 x_0} - \frac{\partial}{\partial x_j}[a^{ij}(x, x_0)u_{x_i}] + b^i(x, x_0)u_{x_i} +$$

$$+ b^0(x, x_0)u_{x_0} + c(x, x_0)u = f(x, x_0)$$

with coefficients satisfying the condition

$$a^{ij}\xi_i\xi_j \geqslant \lambda_0 \mu_0 x_0 (b^i \xi_i)^2 - \mu_0 a^{ij}_{x_0} \xi_i \xi_j \quad \text{(for all } \xi \in \mathbf{R}^n),$$

where λ_0 and μ_0 are positive constants, and especially for equations of the form

$$y^m u_{xx} - u_{yy} + a(x, y)u_x + b(x, y)u_y + \qquad (6)$$

$$+ c(x, y)u = f(x, y),$$

$$x_1 = x,\ \ x_0 \equiv y \geqslant 0,\ \ m = \text{const.}$$

Mixed problems with interior or exterior boundary conditions (see **Exterior and interior boundary value problems**) result from mathematical models of many processes in the theory of scattering of waves on obstructions. For example, the Sommerfeld radiation condition (see **Radiation conditions**) results from the problem of finding a solution u of the wave equation

$$\Box u \equiv u_{x_0 x_0} - \cdots - u_{x_n x_n} = 0$$

for all points $x \in \mathbf{R}^n$ lying outside a bounded domain $\Omega \subset \mathbf{R}^n$, given that the derivative of u in the direction of the exterior normal to $\partial \Omega$ vanishes for any moment of time $x_0 \geqslant 0$, where the initial conditions correspond to a planar wave running from infinity in the direction of the x_1-axis.

The basic boundary value problems for hyperbolic equations and systems are the Goursat and the Darboux – Picard problems and their multi-dimensional analogues (see **Goursat problem; Cauchy characteristic problem**, and also [1]).

The Goursat and Darboux – Picard problems, and various generalizations of them, have been thoroughly investigated for second-order hyperbolic equations and systems with split principal parts of the form

$$u_{xx} - u_{yy} + a(x, y)u_x + b(x, y)u_y + c(x, y)u = f(x, y),$$

where a, b and c are given real $(m \times m)$-matrices, and f is a given and u the desired m-dimensional vector. Substantial results have also been obtained for a fairly wide class of systems of second-order hyperbolic equations with non-split principal parts in the absence of parabolic degeneracy. It has been shown that the characteristic Goursat problem

$$u_i(x, x) = \phi_i(x),\ \ u_i(x, -x) = \psi_i(x),\ \ x \geqslant 0,\ \ i = 1, 2,$$

for the hyperbolic system

$$\frac{\partial^2 u_i}{\partial x^2} - \frac{\partial^2 x_i}{\partial y^2} + 2\frac{\partial^2 u_j}{\partial x \partial y} = 0,\ \ i = 1, 2,\ \ i + j = 3,$$

with two independent variables x, y does not have a unique solution, and the effect of the lowest terms on the well-posedness of this problem has been found [3]. The influence of the nature of parabolic degeneracy on the well-posedness of both the local and the non-local boundary value problem for degenerate hyperbolic equations and systems has been thoroughly investigated [3]. This applies in particular to basic (local) boundary value problems for linear degenerate hyperbolic equations of the form

$$u_{yy} - k(x,y)u_{xx} + a(x,y)u_x + b(x,y)u_y +$$
$$+ c(x,y)u = f(x,y)$$

in bounded domains with an arbitrary piecewise-smooth boundary, and the influence of the order of non-characteristic degeneracy on the well-posedness of the Darboux problem and the non-equivalence of characteristics as supports of boundary conditions has been described (see [10]).

As to finding multi-dimensional analogues of the Darboux and Tricomi problems (see **Mixed-type differential equation**), much work has been done on non-local boundary value problems, and especially on problems with a shift (see [9]), for hyperbolic equations with a given condition on the characteristic parts of the boundary that connect pointwise the values of the desired solution and its (fractional) derivatives or integrals of some specific order.

Many boundary value problems with a shift that have been studied with great completeness and generality in the case of equation (1) are encompassed by the following formulation. In the domain bounded by the characteristics

$$\Gamma_0: x - \frac{2}{m+2}y^{(m+2)/2} = 0, \quad \Gamma_1: x + \frac{2}{m+2}y^{(m+2)/2} = 1$$

and the interval I: $0<x<1$ on the line $y=0$, one has to find a (sufficiently smooth) solution $u(x,y)$ of equation (1) that satisfies on I the local condition

$$\lim_{y\downarrow 0}\left[A_1(x)\frac{\partial u}{\partial x} + A_2\frac{\partial u}{\partial y} + A_3 u\right] = A_4(x), \quad x\in I, \qquad (7)$$

and on $\Gamma_0 \bigcup \Gamma_1$ the non-local condition

$$B_1(x)D_{0x}^\alpha u[\theta_0(x)] + B_2(x)D_{1x}^\alpha u[\theta_1(x)] + \qquad (8)$$

$$+ \lim_{y\downarrow 0}\left[a_1(x)\frac{\partial u}{\partial x} + a_2(x)\frac{\partial u}{\partial y} + a_3(x)u\right] = B_3(x), \quad x\in I.$$

Here A_i, B_i, a_i are given functions, D_{jx}^ϵ is the fractional integro-differential operator of order $|\epsilon|$ defined by

$$D_{jx}^\epsilon \phi(x) = \frac{1}{\Gamma(-\epsilon)}\int_j^x |x-t|^{-\epsilon-1}\phi(t)\,dt$$

if $\epsilon<0$, and by

$$D_{jx}^\epsilon \phi(x) = \frac{d^{[\epsilon]+1}}{dx^{[\epsilon]+1}}D^{\epsilon-[\epsilon]-1}\phi(x), \quad j=0,1,$$

if $\epsilon>0$, where $\Gamma(z)$ is the gamma-function, $[\epsilon]$ is the integer part of ϵ and $\theta_j(x)$ is the point of intersection of the characteristic emanating from $x\in I$ with the characteristic Γ_j of equation (1):

$$u[\theta_j(x)] = u(\operatorname{Re}\theta_j, \operatorname{Im}\theta_j).$$

A detailed study has been made of boundary value problems with a shift for an equation of the form (1) which, in characteristic coordinates ξ and η, reduces to an Euler−Darboux−Poisson equation

$$(\xi-\eta)u_{\xi\eta} - \beta u_\xi + \beta' u_\eta = 0.$$

A special case of the problems (7), (8) is the *Darboux problem*, which asks for a (sufficiently smooth) solution $u(x,y)$ of equation (6) satisfying the (local) boundary conditions

$$u[\theta_0(x)] = \psi(x), \quad u(x,0) = \tau(x), \quad x\in I,$$

or

$$u[\theta_1(x)] = \phi(x), \quad u_y(x,0) = \nu(x), \quad x\in I.$$

The condition $m<2$, or the *Gellerstedt condition* $a(x,y)=O(1)y^p$, $p>(m/2)-1$, $m\geqslant 2$, is essential in order that the Darboux problem be well-posed (see [3], [10]).

A qualitatively new multi-dimensional analogue of the Darboux problem is the *Bitsadze problem*, which in the case of the wave equation $\square u=f(x,x_0)$ can be posed in the following way (see [1]). In the domain $D\subset\mathbf{R}^{n+1}$ bounded by a part S_0 of the plane $x_n=0$ and the two characteristic surfaces

$$S_1: x_0 = |x| = \sqrt{x_1^2 + \cdots + x_n^2}, \quad x_n \leqslant 0,$$

and

$$S_2: 1-x_0 = |x|, \quad x_n \leqslant 0,$$

one has to find a solution of the equation satisfying $u|_{S_0}=0$, $u|_{S_2}=0$ or $u_{x_n}|_{S_0}=0$, $u|_{S_2}=0$.

Other multi-dimensional analogues have been studied, both of the Darboux problem and of non-local boundary value problems (of the type (2), (3)) for hyperbolic equations in special domains, for which the non-characteristic part of the boundary is usually a spatially-oriented surface. Most complete results have been obtained in the case of the Euler−Darboux−Poisson equation $x_0\square u=ku_{x_0}$.

For an equation of the form

$$u_{xy} + A(x,y)u_x + B(x,y)u_y + c(x,y)u = f(x,y)$$

much attention has been devoted to the following non-local problem. In the domain $\{(x,y): 0<x<h, 0<y<T\}$ one has to find a (sufficiently smooth) solution $u(x,y)$ of equation (9) given that, for all $y\in[0,T]$,

$$u(x,0) = \phi(x),\ 0\leqslant x\leqslant h, \quad \frac{\partial}{\partial y}\int_0^{x_0} u(x,y)\,dx = \tau(y),$$

or

$$u(x,0) = \phi(x),\ 0\leqslant x\leqslant h, \quad \frac{\partial}{\partial y}\sum_{i=1}^q \alpha_j(y)u(x^j,y) = \tau(y),$$

where $x_0,\ldots,x_q$ are given points in the interval $[0,h]$.

A new aspect in the theory of boundary value problems is the study of third-order hyperbolic equations of the form

$$u_{xxy} + A(x,y)u_{xx} + a(x,y)u_x + b(x,y)u_y + c(x,y)u = \quad (10)$$
$$= f(x,y),$$

which lie at the basis of mathematical models for many processes and phenomena in the theory of mass or heat transfer in porous media. An extensive theory has been developed of both local and non-local linear boundary value problems for hyperbolic equations of the form (10), and in particular, an analogue of the **Riemann method** has been devised.

For linear symmetric hyperbolic systems of the first order (see **Linear hyperbolic partial differential equation and system**), a study has been made, within the limits of the theory of systems of first-order equations, of boundary value problems with admissible boundary conditions on $\partial\Omega$ (see [6]).

The **Dirichlet problem** is not, in general, well-posed for hyperbolic equations and systems in arbitrary domains. By methods of energy estimates and integral equations the well-posedness of this problem has been established for a wide class of second-order hyperbolic equations in special cylindrical domains.

References

[1] BITSADZE, A.V.: *Certain classes of partial differential equations*, Moscow, 1981 (in Russian).

[2] VLADIMIROV, V.S.: *Equations of mathematical physics*, Mir, 1984 (translated from the Russian).

[3] GELLERSTEDT, S.: 'Sur une équation linéaire aux derivées partielles du type mixte', *Ark. Mat. Astr. Fys.* **25A**, no. 29 (1937), 1-23.

[4] GODUNOV, S.K.: *The equations of mathematical physics*, Moscow, 1979 (in Russian).

[5] DZHURAEV, T.D.: *Boundary value problems for equations of mixed and mixed-composite type*, Tashkent, 1979 (in Russian).

[6] DEZIN, A.A.: *General questions in the theory of boundary value problems*, Moscow, 1980 (in Russian).

[7] LAVRENT'EV, M.M., ROMANOV, V.G. and VASIL'EV, V.G.: *Multidimensional inverse problems for differential equations*, Springer, 1970 (translated from the Russian).

[8] LADYZHENSKAYA, O.A.: *A mixed problem for a hyperbolic equation*, Moscow, 1953 (in Russian).

[9] NAKHUSHEV, A.M.: 'Certain boundary value problems for hyperbolic equations and for equations of mixed type', *Differential Eq.* **5**, no. 1 (1969), 37-57. (*Differentsial'nye Uravn.* **5**, no. 1 (1969), 44-59)

[10] NAKHUSHEV, A.M.: 'Darboux's problem for degenerate hyperbolic equations', *Differential Eq.* **7**, no. 1 (1971), 41-48. (*Differentsial'nye Uravn.* **7**, no. 1 (1971), 49-56)

[11] SALAKHITDINOV, M.S.: *Equations of mixed-composite type*, Tashkent, 1974 (in Russian).

[12] TIKHONOV, A.N. and SAMARKSIĬ, A.A.: *Differentialgleichungen der mathematischen Physik*, Deutsch. Verlag Wissenschaft., 1959 (translated from the Russian).

[13] SHKHANUKOV, M.KH.: 'Boundary-value problems for a third-order equation occurring in the modelling of water filtration in porous media', *Differential Eq.* **18**, no. 4 (1982), 509-517. (*Differentsial'nye Uravn.* **18**, no. 4 (1982), 689-698)

A.M. Nakhushev

Editorial comments. Concerning the terminology 'mixed' see (the editorial comment to) **Mixed and boundary value problems for parabolic equations and systems.**

AMS 1980 Subject Classification: 35LXX

MIXED AND BOUNDARY VALUE PROBLEMS FOR PARABOLIC EQUATIONS AND SYSTEMS -
Problems of finding solutions

$$u(x,t) = (u_1(x,t), \ldots, u_m(x,t))$$

in a domain D of a Euclidean space $\mathbf{R}^{n+1}$ (with points $(x,t) = (x_1, \ldots, x_n, t)$) of a parabolic system of equations or, when $m=1$, of a parabolic equation satisfying additional conditions on some part of the boundary ∂D of the domain D.

Let Ω be a domain in $\mathbf{R}^n$ with sufficiently smooth boundary $\partial\Omega$ and let D be a cylinder $\{x \in \Omega: 0 < t < T\}$ with lateral surface $\Gamma = \{x \in \partial\Omega: 0 < t < T\}$, lower base $\Omega_0 = \{x \in \Omega: t = 0\}$ and upper base $\Omega_T = \{x \in \Omega: t = T\}$. The mixed Petrovskiĭ problem for a linear parabolic system

$$u_t + \sum_{|\alpha| < 2p} A_\alpha(x,t)D_x^\alpha u = f(x,t), \quad (x,t) \in D, \qquad (1)$$
$$f(x,t) = (f_1(x,t), \ldots, f_m(x,t)),$$

in the cylinder D consists of finding solutions of this system satisfying the initial conditions

$$u\big|_{\Omega_0} = \phi(x), \qquad (2)$$

where $\phi(x) = (\phi_1(x), \ldots, \phi_m(x))$, and the boundary condition

$$B\left[x, t, \frac{\partial}{\partial x}\right]u\bigg|_\Gamma = \psi(x,t), \qquad (3)$$

where $\psi(x,t) = (\psi_1(x,t), \ldots, \psi_p(x,t))$ and $B(x,t,\partial/\partial x)$ is a rectangular matrix with

$$B_{ij}\left[x, t, \frac{\partial}{\partial x}\right] = \sum_{|\alpha| < q_{i,j}} b_\alpha^{i,j}(x,t)D_x^\alpha,$$
$$i = 1, \ldots, m; \quad j = 1, \ldots, p.$$

Suppose that the system is uniformly parabolic.

A classical solution of the mixed problem (1) - (3) is a vector function $U(x,t)$ belonging to

$$C_{x,t}^{2p;1}(D) \cap C_{x,1}^{q;0}(D \cup \Gamma) \cap C(D \cup \Gamma \cup \bar\Omega_0),$$

where $q = \max q_{i,j}$ for $1 \leq i \leq m$, $1 \leq j \leq p$, and satisfying (1) in D and conditions (2) and (3) on Ω_0 and Γ, respectively. Sometimes one considers more general solutions than this. In particular, one may drop the requirement that the solution be continuous at the points of $\bar\Gamma \cap \bar\Omega_0$ and replace it by the condition that it is bounded in D.

If the complementarity (or Lopatinskiĭ) condition holds (and if, for the sake of simplicity, Ω is assumed to be bounded), then for sufficiently smooth data (the coefficients in (1) and (3) and the vector functions f, ϕ and ψ) and under certain compatibility conditions, a classical solutions exists and is unique.

The basic mixed problems for a general linear uniformly-parabolic second-order equation

$$u_t - Lu \equiv u_t - \sum_{i,j=1}^n (a_{ij}(x,t)u_{x_i})_{x_j} + \qquad (1')$$

$$-\sum_{i=1}^{n} b_i(x,t)u_{x_i} - c(x,t)u = f(x,t), \quad (x,t)\in D,$$

$$a_{ij}(x,t) = a_{ji}(x,t), \quad i,j=1,\ldots,n,$$

are those of finding solutions of (1') that satisfy the initial condition

$$u\,|_{\Omega_0} = \phi(x) \tag{2'}$$

and one of the boundary conditions

$$u\,|_\Gamma = \psi(x,t), \tag{4}$$

the *first mixed problem*,

$$\left.\frac{\partial u}{\partial \nu}\right|_\Gamma = \psi(x,t), \tag{5}$$

the *second mixed problem*, or

$$\left.\left[\frac{\partial u}{\partial N} + \sigma(x,t)u\right]\right|_\Gamma = \psi(x,t), \tag{6}$$

the *third mixed problem*, where N is a co-normal of the elliptic operator L.

Each of these problems satisfies the complementarity condition and, consequently, when the data are sufficiently smooth and the compatibility conditions hold, each has a classical solution. This solution can be obtained by the method of potentials, the method of finite differences, the **Galerkin method**, or, in the case when the functions $a_{i,j}$ $(i,j=1,\ldots,n)$, c and σ do not depend on t and $b_i\equiv0$, $i=1,\ldots,n$, by the **Fourier method**. For example, in order to solve the first mixed problem for equation (1) it is sufficient to require that the coefficients of the equation belong to the Hölder space $C^\alpha(\overline{D})$ for some $\alpha>0$, that the coefficients $a_{i,j}(x,t)$ have derivatives $\partial a_{i,j}/\partial x_i$ in $C^\alpha(\overline{D})$, $i,j=1,\ldots,n$, that $f(x,t)$ belongs to $C^\alpha(\overline{D})$, that ϕ and ψ are continuous on $\overline{\Omega}_0$ and $\overline{\Gamma}$, respectively, and that $\phi\,|_{\partial\Omega}=\psi(x,0)$. For this it is sufficient that the boundary $\partial\Omega$ of Ω satisfies the following condition: For any point $x^0\in\partial\Omega$ there is a closed sphere S having a unique point in common with Ω, namely, the point x^0: $S\cap\Omega=x^0$. Under certain conditions on the lateral surface (that it contains no characteristic points, i.e. points of contact with the planes $t=\text{const}$), an analogous statement also holds in the case of a non-cylindrical domain D.

Existence theorems for the basic mixed problems for equation (1') also hold under other conditions on the given functions and the domain Ω. For example, in the case of the first mixed problem in a cylindrical domain D, for the homogeneous **heat equation** with continuous functions ϕ and ψ satisfying the compatibility condition $\phi\,|_{\partial\Omega}=\psi(x,0)$, a solution exists provided that Ω is such that the **Dirichlet problem** for the **Laplace equation** is solvable in Ω (there is a classical solution) for an arbitrary continuous boundary function.

Let the coefficients a_{ij}, b_i and c be measurable and bounded in D, and let σ be measurable and bounded

on Γ. Further, let $f\in L_2(D)$, $\phi\in L_0(\Omega)$ and, in the case of the first mixed problem, let ψ be the trace on Γ of some function from the **Sobolev space** $W_2^{1,0}(D)$, while in the case of the third (or second) mixed problem, let ψ belong to $L_2(\Gamma)$.

A function $u(x,t)$ belonging to $W_2^{1,0}(D)$ and with trace on Γ equal to ψ: $u\,|_\Gamma=\psi$, is called a *generalized solution of the first mixed problem* (1'), (2'), (4) if it satisfies the integral identity

$$\int_D \left[-uv_t + \sum_{i,j=1}^{n} a_{i,j}u_{x_i}v_{x_j} - \left(\sum_{i=1}^{n} b_i u_{x_i} + cu\right)v\right] dx\,dt =$$
$$= \int_D fv\,dx\,dt + \int_{\Omega_0}\phi v\,dx$$

for all v in the Sobolev space $W_2^1(D)$ for which $v\,|_\Gamma=0$, $v\,|_{\Omega_T}=0$.

A function $u(x,t)$ belonging to $W_2^{1,0}(D)$ is called a *generalized solution of the third (second, if $\sigma\equiv0$) mixed problem* (1), (2), (6) if it satisfies the integral identity

$$\int_D \left[-uv_t + \sum_{i,j=1}^{n} a_{i,j}u_{x_i}v_{x_j} - \left(\sum_{i=1}^{n} b_i u_{x_i} + cu\right)v\right] dx\,dt +$$
$$+ \int_\Gamma \sigma uv\,dS =$$
$$= \int_D fv\,dx\,dt + \int_{\Omega_0}\phi v\,dx + \int_\Gamma \psi v\,dS$$

for all v in W_2^1 such that $v\,|_{\Omega_T}=0$.

A generalized solution of each of these problems exists and is unique; moreover, if $f\in L_p(D)$ is continuous in D, for sufficiently large p, then it even satisfies a Hölder condition for some exponent $\alpha>0$. By increasing the smoothness of the given functions and the boundary of the domain subject to the compatibility conditions, the smoothness of the generalized solution increases. For example, consider the heat equation with $\phi\equiv0$ and $\psi\equiv0$, and let $\partial\Omega$ be a sufficiently smooth surface. Then the generalized solution of the first mixed problem belongs to $W_2^{2(s+1),s+1}(D)$, provided that $f\in W_2^{2s,s}(D)$ and the compatibility conditions

$$f\,|_{\partial\Omega_0} = (\Delta f + f_t)\,|_{\partial\Omega_0} = \cdots = \left.\sum_{i=0}^{s-1}\Delta^i \frac{\partial^{s-1-i}f}{\partial t^{s-1-i}}\right|_{\partial\Omega_0} \tag{7}$$

hold.

In particular, if $f\in L_2(D)$, then the solution belongs to $W_2^{2,1}(D)$ when $(x,t)\in D$, it satisfies the heat equation and its trace on Ω_0 is equal to zero. If $f\in W_2^{2s,s}$ for sufficiently large s and the compatibility conditions (7) hold, then by virtue of the **imbedding theorems**, the generalized solution is classical. An analogous statement holds for generalized solutions of the basic mixed problems for equation (1') when the coefficients are sufficiently smooth.

Let $\Omega=\mathbf{R}^n$. The problem of finding a solution in the strip $D=\mathbf{R}^n\times(0,T)$ for the parabolic system (1) satisfying the initial condition (2) on the characteristic $\Omega_0 = \{x \in \mathbf{R}^n, \ t = 0\}$ is called the **Cauchy problem** for (1). A classical solution of the Cauchy problem (1), (2) is a vector function $u(x, t)$ belonging to $C^{2p, 1}(D) \cap C(D \cup \Omega_0)$ and satisfying (1) in D and (2) on Ω. If the right-hand side $f(x, t)$ belongs to the Hölder space $C^{\alpha}(\overline{D})$ for some $\alpha > 0$, and the coefficients are sufficiently smooth in $\overline{D}$ (they and their derivatives are bounded), then for any bounded continuous initial vector function $\phi(x)$ on $\mathbf{R}^n$ there is a bounded solution of the Cauchy problem on D, and this bounded solution is unique.

The condition of boundedness can be replaced by the condition of 'not too rapid growth'. For example, the following is true for second-order equations. Let the coefficients of equation (1'),

$$a_{i,j}(x, t), \ b_i(x, t), \ c(x, t) \text{ and } \frac{\partial a_{i,j}(x, t)}{\partial x_i}$$

belong to the Hölder space $C^{\alpha}(\overline{D})$ for some $\alpha > 0$. Further, let $\phi(x)$ be continuous in $\mathbf{R}^n$ and let $f(x, t)$ be continuous in $\overline{D}$, locally Hölder continuous in x uniformly for $t \in [0, T]$ (for some exponent $\alpha > 0$), and such that

$$|\phi(x)| \leqslant Ce^{h|x|^2}, \ x \in \mathbf{R}^n,$$

$$|f(x, t)| \leqslant Ce^{h|x|^2}, \ (x, t) \in D.$$

Then for sufficiently small T (depending on h), there is a solution of the Cauchy problem (1'), (2') in the strip $D = \mathbf{R}^n \times (0, T)$. It can be written in the form

$$u(x, t) = \int_{\mathbf{R}^n} \Gamma(x, t; \xi, 0)\phi(\xi) \, d\xi +$$

$$+ \int_0^t \int_{\mathbf{R}^n} \Gamma(x, t; \xi, \tau) f(\xi, \tau) \, d\xi \, d\tau,$$

where $\Gamma(x, t; \xi, \tau)$ is a fundamental solution of (1'), and satisfies the estimate

$$|u(x, t)| \leqslant C_1 e^{h|x|^2} \tag{8}$$

for some positive constants C_1 and k. Condition (8) guarantees the uniqueness of the solution of the Cauchy problem.

In the case of an equation with constant coefficients it is possible to find a condition of type (8) on the growth of the solution that is necessary and sufficient for its uniqueness. For example, for a solution of the Cauchy problem for the heat equation to be unique in the class of functions satisfying the inequality

$$|u(x, t)| \leqslant Ce^{|x|h(|x|)},$$

where $h(|x|)$ is a positive continuous function on $[0, \infty)$, it is necessary and sufficient that the integral $\int_0^{\infty} dr / h(r)$ diverges.

For parabolic equations one can also consider problems without initial conditions (the Fourier problem). For example, one can ask for the solution of the homo-geneous heat equation in the cylinder $D = \{x \in \Omega, \ -\infty < t < +\infty\}$, where Ω is a bounded domain with sufficiently smooth boundary $\partial\Omega$, satisfying the boundary condition

$$u(x, t)|_{x \in \partial\Omega} = \psi(x, t).$$

If ψ is continuous and bounded, then there is a bounded solution of the Fourier problem, and this is the unique bounded solution.

For parabolic equations and systems it is also possible to consider the first mixed problem in a non-cylindrical domain D in the case when the lateral surface contains characteristic points (points of contact with the planes $t = $const). In particular, it is possible to consider the **Dirichlet problem**, where boundary conditions are given on the entire boundary $\partial\Omega$. Under specific conditions on the set of characteristic points and on the order of contact of the characteristic points of $\partial\Omega$ with a characteristic plane, the Dirichlet problem has a unique solution (in the space $W_2^{B^0}$). For example, suppose (for the sake of simplicity) that $D \subset \mathbf{R}^2$ is a strictly-convex domain and that the equation of the boundary in a neighbourhood of the upper characteristic point (x^0, t^0) has the form $x - x^0 = \phi_1(t)$ when $x \leqslant x^0$, and $x - x^0 = \phi_2(t)$ when $x \geqslant x^0$ $(t^0 - \delta < t \leqslant t^0)$. Then the divergence of both integrals

$$\int_{t^0 - \delta}^{t^0} |\phi_i(t)|^{-2p} \, dt, \ i = 1, 2,$$

guarantees the existence and uniqueness of a solution of the Dirichlet problem for a second-order parabolic equation. This condition is also necessary in this class of equations.

References

[1] VLADIMIROV, V.S.: *Equations of mathematical physics*, Mir, 1984 (translated from the Russian).

[2] IL'IN, V.A.: 'The solvability of mixed problems for hyperbolic and parabolic equations', *Russian Math. Surveys* **15**, no. 2 (1960), 85-142. (*Uspekhi Mat. Nauk* **15**, no. 2 (1960), 97-154)

[3] IL'IN, A.M., KALASHNIKOV, A.S. and OLEĬNIK, O.A.: 'Linear equations of the second order of parabolic type', *Russian Math. Surveys* **17**, no. 3 (1962), 1-143. (*Uspekhi Mat. Nauk* **17**, no. 3 (1962), 3-146)

[4] KRUZHKOV, S.N.: 'A priori bounds and some properties of solutions of elliptic and parabolic equations', *Mat. Sb.* **65**, no. 4 (1964), 522-570 (in Russian).

[5] LADYZHENSKAYA, O.A., SOLONNIKOV, V.A. and URAL'TSEVA, N.N.: *Linear and quasi-linear equations of parabolic type*, Amer. Math. Soc., 1968 (translated from the Russian).

[6] LADYZHENSKAYA, O.A.: *The boundary value problems of mathematical physics*, Springer, 1985 (translated from the Russian).

[7] LADYZHENSKAYA, O.A.: 'On uniqueness of a solution of the Cauchy problem for a linear parabolic equation', *Mat. Sb.* **27**, no. 2 (1950), 175-184 (in Russian).

[8] MIKHAĬLOV, V.P.: 'The Dirichlet problem for a parabolic equation I', *Mat. Sb.* **61**, no. 1 (1963), 40-64 (in Russian).

[9] MIKHAĬLOV, V.P.: 'The Dirichlet problem for a parabolic equation II', *Mat. Sb.* **62**, no. 2 (1963), 140-159 (in Russian).

[10] NASH, J.: 'Continuity of solutions of parabolic and elliptic equations', *Amer. J. Math.* **80** (1958), 931-954.

[11] PETROVSKIĬ, I.G.: *Partial differential equations*, Saunders, 1967 (translated from the Russian).

[12] PETROVSKIĬ, I.G.: 'On Cauchy's problem for systems of linear partial differential equations in a nonanalytic function domain', *Bull. Moskov. Gos. Univ. Ser. A.* **1**, no. 7 (1938), 1-72 (in Russian).

[13] PETROVSKIĬ, I.G.: 'Zur ersten Randwertaufgabe der Wärmeleitungsgleichung', *Comp. Math.* **1** (1934), 383-419.

[14] SOBOLEV, S.L.: *Partial differential equations of mathematical physics*, Pergamon, 1964 (translated from the Russian).

[15] SOLONNIKOV, V.A.: 'Boundary value problems of mathematical physics III', *Proc. Steklov Inst. Math.* **83** (1965). (*Trudy Mat. Inst. Steklov* **83** (1965))

[16] TIKHONOV, A.N. and SAMARSKIĬ, A.A.: *Differentialgleichungen der mathematischen Physik*, Deutsch. Verlag Wissenschaft., 1959 (translated from the Russian).

[17] TIKHONOV, A.N.: *Byull. Moskov. Univ. (A)* **1**, no. 9 (1938), 1-43.

[18] TIKHONOV, A.N.: *Mat. Sb.* **42**, no. 2 (1935), 199-216.

[19] FRIEDMAN, A.: *Partial differential equations of parabolic type*, Prentice-Hall, 1964.

[20] EĬDEL'MAN, S.D.: *Parabolic systems*, North-Holland, 1969 (translated from the Russian).

V.P. Mikhaĭlov

Editorial comments. In the current literature the initial boundary value problems (1'), (2'), (4) - (6) are not referred to as 'mixed'. Sometimes expressions like *Cauchy−Dirichlet* or *Cauchy−Neumann* are used. Quite often, by a problem with Dirichlet data for a parabolic equation is meant a problem in which such data are prescribed on the parabolic boundary. Besides the first, second and third kind of boundary data, higher-order problems are also considered. For further comments and more references see **Linear parabolic partial differential equation and system**.

AMS 1980 Subject Classification: 35KXX

MIXED AUTOREGRESSIVE MOVING-AVERAGE PROCESS, *autoregressive moving-average process, ARMA process* - A wide-sense stationary **stochastic process** $X(t)$ with discrete time $t = 0, \pm 1, \ldots$, the values of which satisfy a difference equation

$$X(t) + a_1 X(t-1) + \cdots + a_p X(t-p) = \quad (1)$$
$$= Y(t) + b_1 Y(t-1) + \cdots + b_q Y(t-q),$$

where $EY(t) = 0$, $EY(t)Y(s) = \sigma^2 \delta_{ts}$, δ_{ts} being the Kronecker delta (i.e. $Y(t)$ is a white noise process with spectral density $\sigma^2/2\pi$), p and q are non-negative integers, and $a_1, \ldots, a_p, b_1, \ldots, b_q$ are constant coefficients. If all roots of the equation

$$\phi(z) = 1 + a_1 z + \cdots + a_p z^p = 0$$

are of modulus distinct from 1, then the stationary autoregressive moving-average process $X(t)$ exists and has **spectral density**

$$f(\lambda) = \frac{\sigma^2}{2\pi} \frac{|\psi(e^{i\lambda})|^2}{|\phi(e^{i\lambda})|^2},$$

where $\psi(z) = 1 + b_1 z + \cdots + b_q z^q$. However, for the solution of equation (1) with given initial values $X(t_0 - 1), \ldots, X(t_0 - p)$ to tend to the stationary process $X(t)$ as $t - t_0 \to \infty$, it is necessary that all roots of

the equation $\phi(z) = 0$ be situated outside the unit disc $|z| \leq 1$ (see [1] and [2], for example).

The class of Gaussian autoregressive moving-average processes coincides with the class of stationary processes that have a spectral density and are one-dimensional components of multi-dimensional Markov processes (see [3]). Special cases of autoregressive moving-average processes are auto-regressive processes (when $q = 0$, cf. **Auto-regressive process**) and moving-average processes (when $p = 0$, cf. **Moving-average process**).

Generalizations of autoregressive moving-average processes are the autoregressive integrated moving-average processes introduced by G.E.P. Box and G.M. Jenkins (see [1]) and often used in applied problems. These are non-stationary processes with stationary increments such that the increments of some fixed order form an autoregressive moving-average process.

References

[1] BOX, G.E.P. and JENKINS, G.M.: *Time series analysis. Forecasting and control*, 1-2, Holden-Day, 1976.

[2] ANDERSON, T.W.: *The statistical analysis of time series*, Wiley, 1971.

[3] DOOB, J.L.: 'The elementary Gaussian processes', *Ann. Math. Stat.* **15** (1944), 229-282.

A.M. Yaglom

Editorial comments. The class of autoregressive moving-average processes is of interest because they represent stationary processes with a rational spectral density.

The problem of representing a stationary process as an autoregressive moving-average process is known in the Western literature as the *stochastic realization problem*; see [A2], [A4] for references on this problem.

Autoregressive moving-average processes are used by statisticians [A3], econometricians [A1] and engineers [A5].

References

[A1] AOKI, M.: *Notes on economic time series analysis: system theory perspectives*, Lecture notes in economics and math. systems, 220, Springer, 1983.

[A2] FAURRE, P., CLERGET, M. and GERMAIN, F.: *Opérateurs rationnels positifs*, Dunod, 1979.

[A3] HANNAN, E.J.: *Multiple time series*, Wiley, 1970.

[A4] LINDQUIST, A. and PICCI, G.: 'Realization theory for multivariate stationary Gaussian processes', *SIAM J. Control Optim.* **23** (1985), 809-857.

[A5] LJUNG, L. and SÖDERSTRÖM, T.: *Theory and practice of recursive identification*, M.I.T., 1983.

AMS 1980 Subject Classification: 60G10

MIXED GROUP - A **group** that contains both elements of infinite order and elements of finite order other than the unit element (see **Order** of an element of a group).

O.A. Ivanova

Editorial comments. This terminology is not very common.

References

[A1] KUROSH, A.G.: *The theory of groups*, 1-2, Chelsea, 1955-

1956 (translated from the Russian).

AMS 1980 Subject Classification: 20FXX

MIXED INTEGRAL EQUATION - An **integral equation** that, in the one-dimensional case, has the form

$$\phi(x) - \lambda \int_a^b K(x, s)\phi(s)\, ds - \lambda \sum_{j=1}^m K_1(x, s_j)\phi(s_j) = \tag{1}$$
$$= f(x),$$

where ϕ is the unknown and f is a given continuous function on $[a, b]$, $s_j \in [a, b]$, $j = 1, \ldots, m$, are given points, and K, K_1 are given continuous functions on the rectangle $[a, b] \times [a, b]$. If

$$K_1(x, s_j) = a_j K(x, s_j),$$

where the a_j are positive constants, then (1) can be written as

$$\phi(x) - \lambda^{\bullet} \int_a^b K(x, s)\phi(s)\, ds = f(x), \quad x \in [a, b], \tag{2}$$

where the new integration symbol, with ψ an arbitrary finite integrable function, is defined by (see [1]):

$$^{\bullet}\int_a^b \psi(s)\, ds = \int_a^b \psi(s)\, ds + \sum_{j=1}^m a_j \psi(s_j).$$

The theory of Fredholm equations (cf. **Fredholm equation**) and, in the case of a symmetric kernel, the theory of integral equations with symmetric kernel (cf. **Integral equation with symmetric kernel**), is valid for equation (2).

In the case of multi-dimensional mixed integral equations, the unknown function can be part of the integrands of integrals over manifolds of different dimensions. For example, in the two-dimensional case the integral equation may have the form

$$\phi(x) - \lambda \iint_D K_1(x, y)\phi(y)\, d\sigma_y + \lambda \int_\Gamma K_2(x, y)\phi(y)\, ds_y +$$
$$+ \lambda \sum_{j=1}^m K_3(x, y_j)\phi(y_j) = f(x), \quad x \in D,$$

where D is some domain in the plane, Γ is its boundary, and y_j are fixed points in $D \cup \Gamma$. This equation may also be written as

$$\phi(x) - \lambda \iint_{D \cup \Gamma} K(x, y)\phi(y)\, d\omega_y = f(x),$$

if the function K and the volume element $d\omega_y$ are correspondingly defined. In this case, moreover, the theory of Fredholm integral equations remains valid.

References

[1] KNESER, A.: 'Belastete Integralgleichungen', *Rend. Circolo Mat. Palermo* **37** (1914), 169-197.
[2] LICHTENSTEIN, L.: 'Bemerkungen über belastete Integralgleichungen', *Studia Math.* **3** (1931), 212-225.
[3] GUNTER, N.M.: 'Sur le problème des "Belastete Integralgleichungen"', *Studia Math.* **4** (1933), 8-14.
[4] SMIRNOV, V.I.: *A course of higher mathematics*, 4, Addison-

Wesley, 1964 (translated from the Russian).

B.V. Khvedelidze

AMS 1980 Subject Classification: 45H05

MIXED PROBLEM - A problem involving (systems of) partial differential equations, with **initial conditions** and **boundary conditions**, and also problems with data on both characteristic and non-characteristic oriented manifolds (see **Mixed and boundary value problems for hyperbolic equations and systems; Mixed and boundary value problems for parabolic equations and systems; Mixed-type differential equation**). The term mixed problem is also applied to boundary value problems for elliptic equations when on various parts of the boundary conditions of different kinds are given (see **Boundary value problem, elliptic equations**).

A.M. Nakhushev

AMS 1980 Subject Classification: 35-XX, 35M05

MIXED PRODUCT $(\mathbf{a}, \mathbf{b}, \mathbf{c})$ *of vectors* $\mathbf{a}$, $\mathbf{b}$, $\mathbf{c}$ - The **inner product** of the vector $\mathbf{a}$ by the **vector product** of $\mathbf{b}$ and $\mathbf{c}$:

$$(\mathbf{a}, \mathbf{b}, \mathbf{c}) = (\mathbf{a}, [\mathbf{b}, \mathbf{c}]).$$

See **Vector algebra**.

Editorial comments. The phrase *scalar triple product* also occurs instead of mixed product.

AMS 1980 Subject Classification: 15-XX, 15A71, 15A63

MIXED-TYPE DIFFERENTIAL EQUATION - A partial differential equation which is of varying type (elliptic, hyperbolic or parabolic) in its domain of definition. A linear (or quasi-linear) differential equation of the second order with two unknown variables,

$$Au_{xx} + 2Bu_{xy} + Cu_{yy} = f(x, y, u, u_x, u_y) \tag{1}$$

and with coefficients defined in the domain Ω is an equation of mixed type if the discriminant $\Delta = AC - B^2$ of the characteristic form

$$A\, dy^2 + 2B\, dx\, dy + C\, dx^2 = Q$$

takes the value zero in Ω but is not identically zero there.

The curve δ defined by the equation $\Delta = 0$ is called the *parabolic line* of equation (1) or the *line of degeneracy (change) of type*.

If Δ does not change sign when the point (x, y) crosses the parabolic line δ in Ω, then equation (1) is a degenerate equation of elliptic-parabolic ($\Delta \geqslant 0$) or hyperbolic-parabolic ($\Delta \leqslant 0$) type (see **Degenerate partial differential equation**).

Under certain smoothness conditions on A, B, C, and δ, there is a non-singular real transformation of the independent variables sending equation (1) (in the case where the sign of Δ alternates in the neighbourhood of a chosen point of δ where $A^2 + B^2 \neq 0$) to one of the following canonical forms (the notation for the independent variables is preserved):

$$y^{2m+1} u_{xx} + u_{yy} = F(x, y, u, u_x, u_y), \qquad (2)$$

$$u_{xx} + y^{2m+1} u_{yy} = F(x, y, u, u_x, u_y). \qquad (3)$$

The equations (2) and (3) are equations of mixed (elliptic-hyperbolic) type in any domain containing a segment of the line of degeneracy $y = 0$.

The domain Ω of definition of an equation of mixed type is sometimes called a *mixed domain*, and boundary value problems in mixed domains are called *mixed boundary value problems*. The part Ω^+ (Ω^-) of a mixed domain Ω where the equation is of elliptic (hyperbolic) type is called the *domain of ellipticity (hyperbolicity)*.

Many problems of an applied nature reduce to finding specific solutions of equations of mixed type; in particular, problems of plane transonic flow of a compressible medium, and problems in the theory of envelopes.

An equation (1) of mixed type is called an *equation of the first kind* (*second kind*) if the characteristic form $Q \neq 0$ ($Q = 0$) everywhere on the parabolic line. The *Chaplygin equation*

$$k(y) u_{xx} + u_{yy} = 0, \qquad (4)$$

where $k(y)$ is a continuously-differentiable monotone function such that $y k(y) > 0$ when $y \neq 0$, is a typical example of an equation of mixed type of the first kind. When $k(y) = y$, equation (4) is usually called the **Tricomi equation**.

An important model for an equation of mixed type (with a discontinuous coefficient in front of one of the higher derivatives) is the *equation of Lavrent'ev — Bitsadze*

$$(\operatorname{sign} y) \cdot u_{xx} + u_{yy} = 0. \qquad (5)$$

One of the basic boundary value problems for equations of mixed type (of the first kind) is the *Tricomi problem*, which is as follows for an equation of the form (2). Let Ω be a finite simply-connected domain in the Euclidean plane with independent variables x and y, bounded by a simple *Jordan curve* σ with end points $A(0, 0)$, $B(1, 0)$ lying in the half-plane $y > 0$, and by the parts AC and BC of the characteristics of equation (2) going through the point $C(1/2, y_C)$, $y_C < 0$. The Tricomi problem is to find a solution $u(x, y)$ of equation (2) that is continuous in the closure $\overline{\Omega}$ of Ω and takes given values on the curve $\sigma \cup AC$.

In the theory of the Tricomi problem, an essential role is played by the *Bitsadze extremum principle*, which, in the case of equation (5), states that a solution $u(x, y)$ of equation (5) in the class $C(\overline{\Omega}) \cap C^1(\Omega)$ which vanishes on the characteristic AC: $x + y = 0$, $0 \leqslant x \leqslant 1/2$, in the closure $\overline{\Omega}^+$ of the domain of ellipticity $\Omega^+ = \Omega \cap \{y > 0\}$, attains its extremum on the curve σ.

This principle, which guarantees the uniqueness and stability of a solution of the Tricomi problem (and also provides a basic estimate to prove existence by the alternating method), can be generalized to a very wide class of linear and quasi-linear equations of mixed type. In particular, it applies to Chaplygin equations (and Tricomi equations) whenever $k(y)$ is twice continuously differentiable and $5 k'^2 \geqslant 4 k k'$ for $y < 0$. The Bitsadze extremum principle also holds for the equation

$$\operatorname{sign} y \cdot |y|^{\alpha} u_{xx} + u_{yy} = 0, \quad \alpha = \text{const} > 0. \qquad (6)$$

The solution of the Tricomi problem for equation (6) in the corresponding mixed domain Ω can be written in explicit form when the elliptic part σ of the boundary of this domain coincides with the so-called *normal contour σ_0*:

$$x^2 + \left[\frac{2}{\alpha + 2} \right]^2 y^{\alpha + 2} = \frac{1}{4}.$$

In the general case, under specific conditions on the curve σ and on the class in which the solutions are sought, the Tricomi problem for equation (6) reduces to an equivalent **singular integral equation** which is (by virtue of the uniqueness condition) unconditionally solvable. The method of integral equations can also be applied to prove existence of a solution of the Tricomi problem, and of other mixed problems, for more general equations of the form

$$\operatorname{sign} y \cdot |y|^{\alpha} u_{xx} + u_{yy} = F(x, y, u, u_x, u_y)$$

with power degeneracy of order α.

Function-theoretical methods and functional analysis, in particular the use of a priori estimates, have made it possible to extend significantly the class of equations of mixed type and mixed domains for which existence and uniqueness of a (generalized) solution can be proved, both for the Tricomi problems and for various other mixed problems.

An important generalization of the Tricomi problem is the *general mixed Bitsadze problem*, which, in the case of equation (5), can be posed as follows. Let Ω be a simply-connected mixed domain bounded by a simple Jordan curve σ lying in the half-plane $y > 0$ with end points $A(0, 0)$, $B(1, 0)$ and (smooth) monotone curves Γ_0 and Γ_1 through these points, meeting at a point $C(x_1, y_1)$, $y_1 < 0$. It is assumed that Γ_0 and Γ_1 lie in the domain bounded by the characteristics $x + y = 0$, $x - y = 1$ and the interval AB: $0 \leqslant x \leqslant 1$ on the x-axis. Let B_0 and B_1 denote the points of intersection of the characteristics $x - y = x_0$ and $x + y = x_0$ with the curves Γ_0 and Γ_1, where x_0 is any fixed point of the semi-interval $x_1 + y_1 < x_1 \leqslant x_1 - y_1$, and let γ_0 and γ_1 denote the parts of Γ_0 and Γ_1 lying between A, B_0, and B, B_1, respectively. The general mixed Bitsadze problem consists of finding a regular solution (when $y \neq 0$, $x \pm y \neq x_0$) of equation (5) in Ω which is continuous in $\overline{\Omega}$, has continuous first derivatives in Ω when $x = -y \neq x_0$, and satisfies given boundary conditions on σ, γ_0 and γ_1. The uniqueness and existence of a solution of this problem, both for equation (5) and for more general equations, can be proved under certain geometric conditions on the boundary of Ω, especially on the curve σ. The general mixed Bitsadze problem can be regarded as completely solved in the special case when Γ_1 coincides ($x_0 = 1$) with the characteristic BC through the point B. An important consequence of the fact that the general mixed Bitsadze problem is correctly posed, in the case of equation (5) for example, is that the **Dirichlet problem** for mixed domains of the form Ω is incorrectly posed, whatever the size and form of the domain Ω^- of hyperbolicity.

For a fairly large class of linear equations

$$k(y)u_{xx} + u_{yy} + au_x + bu_y + cu = f$$

it is known that the coefficient $a(x, y)$ has a substantial influence on the correctness of posing the Dirichlet problem in corresponding mixed domains of the form Ω.

Another type of mixed problem is the *Frankl problem*. Let Ω be a simply-connected domain with the following boundary: the interval $A'A$: $-1 \leqslant y \leqslant 1$ of the line $x = 0$, a smooth curve σ with end points at $A(0, 1)$ and $B(a, 0)$ and lying in the quadrant $x > 0, y > 0$, the interval CB: $a_1 \leqslant x \leqslant a$ of the line $y = 0$, and the characteristic through $A'(0, -1)$ and $C(a_1, 0)$ of the equation of mixed type under consideration (e.g. equation (4)). The Frankl problem consists of finding a solution $u(x, y)$ of the equation of mixed type in Ω, given the value of $u(x, y)$ on $\sigma \bigcup CB$ and the conditions

$$\frac{\partial u}{\partial x} = 0, \quad u(0, y) - u(0, -y) = f(y),$$

$$-1 \leqslant y \leqslant 1, \quad x = 0,$$

on $A'A$. This problem has been investigated chiefly for model equations of mixed type and has been completely solved for equation (5) in the case where the curve σ: $x = x(s), y = y(s)$ is such that $dy/ds \geqslant 0$, where s is the arc length of σ measured from the point $B(a, 0)$.

Basic boundary value problems have been formulated for equations of mixed type of the first kind and, adapted with appropriate modifications, for equations of mixed type of the second kind. These modifications are necessary because the Dirichlet problem for elliptic equations with characteristic degeneracy is not always correctly posed.

In the formulation of the boundary value problems for equation (1) in mixed domains, a new aspect is introduced if the line δ of change of type is also a line

of degeneracy of the order of equation, which occurs, for example, in the case of the equation

$$y^{2p}u_{xx} + yu_{yy} + \beta u_y = 0, \tag{7}$$

where p is a natural number and β is a constant such that $1 - 2p \leqslant 2\beta < 1$.

For equations (5), (6), (7), there are, in addition to the above, a number of essentially new boundary value problems. These are chiefly characterized by the fact that the entire boundary $\sigma \bigcup AC \bigcup BC$ of Ω (where the Tricomi problem is posed) carries the following boundary conditions: the Dirichlet conditions, for example, on Σ and on $AC \bigcup BC$, with some non-local condition pointwise connecting the values of the desired solution or a (fractional) derivative of it of a certain order. In particular, these problems include a simple example of a correctly-posed self-adjoint mixed boundary value problem.

Boundary value problems have also been studied for equations (and systems) of mixed type in domains containing in their interiors several lines of degeneracy of type, or one single closed parabolic line.

Analogues of the Tricomi problem have been studied for certain classes of equations and systems of mixed type in two independent variables, and for equations of higher order.

Significant difficulties arise in the search for well-posed problems for equations of mixed type with many variables. Nevertheless, several important results have been obtained also in this direction. For the equation

$$(\text{sign } z) \cdot u_{xx} + u_{yy} + u_{zz} = f(x, y, z), \tag{8}$$

which is a simple model of an equation of mixed type having $z = 0$ as a time-like plane of degeneracy of type, the following problem is known to be correctly posed. Let Ω be a finite simply-connected three-dimensional domain, bounded by a piecewise smooth surface $z = f(x, y) \geqslant 0$ and by the characteristic surfaces

$$S_1 : x + x_0 = \sqrt{y^2 + z^2},$$

$$S_2 : x - x_0 = \sqrt{y^2 + z^2}$$

of equation (8). One has to find a continuously differentiable function in Ω, satisfying equation (8) in Ω for $z \neq 0$, that vanishes on σ and on one of the characteristic surfaces S_1, S_2. Existence of a weak solution and uniqueness of a strong solution for this problem have been proved for the more general equation

$$(\text{sign } x_n) \cdot u_{x_n x_n} + \Delta_x u = f(x_0, x), \quad x = (x_1, \ldots, x_n),$$

where Δ_x is the **Laplace operator** in the variables $x_1, \ldots, x_n$.

For the equation

$$x_0^{2m} \Delta_x u - x_0 u_{x_0 x_0} + \left[m - \frac{1}{2} \right] u_{x_0} = 0 \tag{9}$$

with part of the space-like hyperplane of degeneracy

both of type and of order $x=0$ contained in the mixed domain Ω, boundary value problems of a special form have been studied. Here the part of $\partial\Omega$ lying in the half-space $x_0<0$ carries data $u(x_0,x)$, and the part lying in the half-space $x_0>0$ (the *characteristic conoid* of equation (9)) carries certain integral averages of $u(x_0,x)$.

Other model equations of mixed type in bounded and unbounded three-dimensional domains have been studied, including the equations

$$z^{2m+1}u_{xx}+u_{yy}+u_{zz} = 0,$$

$$z^{2m+1}(u_{xx}+u_{yy})+u_{zz} = 0.$$

There is also a uniqueness criterion of the solution of the Dirichlet problem for a large class of self-adjoint equations of mixed type in cylindrical domains.

References

[1] BERS, L.: *Mathematical aspects of subsonic and transonic gas dynamics*, Wiley, 1958.

[2] BITSADZE, A.V.: *Equations of the mixed type*, Pergamon, 1964 (translated from the Russian).

[3] BITSADZE, A.V.: 'On the theory of equations of mixed type whose order is degenerate on the line of change of type', in *Continuum mechanics and related problems of analysis*, Moscow, 1972, pp. 47-52 (in Russian).

[4] BITSADZE, A.V. and NAKHUSEV, A.M.: 'Correct formulation of problems for equations of mixed type in multidimensional domains', *Soviet Math. Dokl.* **13**, no. 4 (1972), 857-860. (*Dokl. Akad. Nauk. SSSR* **205**, no. 1 (1972), 9-12)

[5] VEKUA, I.N.: *Generalized analytic functions*, Pergamon, 1962 (translated from the Russian).

[6] KARATOPAKLIEV, G.D.: 'A class of equations of mixed type', *Diff. Equations* **5**, no. 1 (1969), 171-176. (*Differentsial'nye Uravn.* **5**, no. 1 (1969), 199-205)

[7] KELDYSH, M.V.: 'On certain cases of degeneracy on the boundary of a domain for equations of elliptic type', *Dokl. Akad. Nauk SSSR* **77** (1951), 181-183 (in Russian).

[8] SALAKHITDINOV, M.S.: 'Certain boundary value problems for equations of mixed type', *Izv. Akad. Nauk UzbSSR, Ser. Fiz.-Mat. Nauk* **1** (1969), 27-33 (in Russian).

[9] SMIRNOV, M.M.: *Equations of mixed type*, Amer. Math. Soc., 1978 (translated from the Russian).

[10] SOLDATOV, A.P.: 'A problem in function theory', *Diff. Equations* **9**, no. 2 (1973), 248-253. (*Differentsial'nye Uravn.* **9**, no. 2 (1973), 325-332)

[11] TRICOMI, F.: *Atti Accad. Naz. Lincei, Ser. 5* **14** (1932), 134-247.

[12] FRANKL, F.I.: *Selected work on gas dynamics*, Moscow, 1973 (in Russian).

[13] FRIEDRICHS, K.O.: 'Symmetric positive linear differential equations', *Comm. Pure Appl. Math.* **11** (1958), 333-418.

[14] GELLERSTEDT, S.: 'Quelques problèmes mixtes pour l'équation $y^m z_{xx}+z_{yy}=0$', *Ark. Mat. Astr. Fysik* **26A**, no. 3 (1937), 1-32.

[15] GERMAIN, P. and BADER, R.: 'Sur le problème de Tricomi', *C.R. Acad. Sci. Paris* **232** (1951), 463-465.

A.M. Nakushev

Editorial comments. More recent workers tend to favour functional-analytic methods, see [A1]. For a constructive approach, using Fourier integral operator methods, see [A2].

References

[A1] SCHNEIDER, M.: 'Ueber Differentialgleichungen zweiter Ordnung vom gemischten Typ im R^3', *Math. Nachr.* **66** (1975), 57-66.

[A2] GROOTHUIZEN, R.J.P.: *Mixed elliptic-hyperbolic partial differential operators: a case-study in Fourier integral operators*, CWI Tract, 16, CWI, Amsterdam, 1985.

AMS 1980 Subject Classification: 35M05

MIXED-VOLUME THEORY - A branch of the theory of convex bodies concerned with the functionals that arise in the study of linear combinations of bodies (see **Addition of sets**).

The volume V of a linear combination $\sum_{i=1}^{r}\lambda_i K_i$ of convex bodies K_i in a Euclidean space $\mathbf{R}^n$ with coefficients $\lambda_i\geq0$ is a homogeneous polynomial of degree n in $\lambda_1,\ldots,\lambda_r$:

$$V\left[\sum_{i=1}^{r}\lambda_i K_i\right] = \sum_{i_1=1}^{r}\cdots\sum_{i_n=1}^{r}V_{i_1\cdots i_n}\lambda_{i_1}\cdots\lambda_{i_n}. \qquad (*)$$

The coefficients $V_{i_1\cdots i_n}$ are assumed to be symmetric with respect to permutations of the subscripts and are denoted by $V(K_{i_1},\ldots,K_{i_n})$, since they depend only on the bodies $K_{i_1},\ldots,K_{i_n}$. These coefficients are called the *mixed volumes* of the bodies $K_{i_1},\ldots,K_{i_n}$.

The significance of this theory lies in the universality of the concept of mixed volumes: by substituting concrete bodies $K_1,\ldots,K_{n-1}$ into $V(K,K_1,\ldots,K_{n-1})$, one can obtain various quantities related to a body K. These include: its volume, its surface area, the surface integral of the elementary symmetric function of its principal curvatures (in the case of a C^2-smooth body), and also the corresponding characteristics of its projections to i-dimensional planes, $0<i<n$. A special case of the expression $(*)$ is the *Steiner formula* for volumes of parallel bodies in $\mathbf{R}^3$:

$$V_\epsilon = V+S\epsilon+\pi B\epsilon^2+\frac{4}{3}\pi\epsilon^3,$$

where V is the volume, S the surface area, B the total mean curvature of the original body, and V_ϵ the volume of an ϵ-neighbourhood of it. The mixed volume $V(K_1,\ldots,K_n)$ is invariant under parallel translation of any K_i, is monotonic (with respect to inclusion of bodies), continuous and non-negative; $V(K_1,\ldots,K_n)>0$ if and only if it is possible to choose a segment in every K_i such that these segments are linearly independent (see [1]).

If K' is the projection of K to a hypersurface orthogonal to a segment e of length one, then

$$V(K_1,\ldots,K_{n-1},e) = nV(K'_1,\ldots,K'_{n-1}).$$

The volume of the projection of K to a p-dimensional subspace is called its *p-th cross-sectional measure* (or *p-quermass*). Establishing relations between the mean values $W_p(K)$ of these measures is one of the concerns of **integral geometry**. Up to a multiplicative constant, the functionals $W_p(K)$ coincide with the p-th integral curvatures:

$$V_p(K) = V(K, \ldots, K, U, \ldots, U)$$

(p occurrences of K, $n-p$ occurrences of U), where U is the unit sphere. For a C^2-smooth strictly convex body K, the mixed volume $V_p(K)$, $0<p<n$, is equal to the integral of the p-th elementary symmetric function D_p of the principal radii of curvature, regarded as a function of the normal to the sphere S^{n-1}. In the case of a general convex body, $V_p(K)$ is the total value of the measure μ_p on S^{n-1}, defined below and called the *curvature function*. (In the smooth case, D_p is the density of μ_p.) Just as the volume of the body K is $1/n$-th of the integral of its **support function** $K(u)$ with respect to its *surface function*, i.e. the surface area of the image on S^{n-1} under the spherical mapping (cf. **Spherical map**), the mixed volume of n bodies can be written as the integral of the support function $K_1(u)$ of one of them with respect to some measure $\mu(\omega)=\mu(K_2, \ldots, K_n, \omega)$ on S^{n-1} that depends on the other bodies, called the *mixed surface function* of $K_2, \ldots, K_n$:

$$V(K_1, \ldots, K_n) = \frac{1}{n} \int_{S^{-1}} K_1(u)\, d\mu.$$

The curvature function $\mu_p(\omega)$ is defined by the equation

$$\mu_p(\omega) = \mu(K, \ldots, K, U, \ldots, U, \omega)$$

(p occurrences of K, $n-p-2$ occurrences of U).

The main content of the theory of mixed volumes is formed by inequalities between mixed volumes (see [2], [3]). They include the *Minkowski inequality*

$$V^n(K, L, \ldots, L) \geqslant V(K)V^{n-1}(L)$$

and the *quadratic Minkowski inequality*

$$V^2(K, L, \ldots, L) \geqslant V(L)V(K, K, L, \ldots, L).$$

Both of these are closely connected with the **Brunn−Minkowski theorem**, which is true not only for convex bodies. The *Aleksandrov−Fenchel inequality* generalizes these, and has the following modification (see [2]):

$$V^m(K_1, \ldots, K_m, L_1, \ldots, L_{n-m}) \geqslant$$
$$\geqslant \prod_{i=1}^{m} V(K_i, \ldots, K_i, L_1, \ldots, L_{n-m}).$$

In particular,

$$V^n(K_1, \ldots, K_n) \geqslant V(K_1) \cdots V(K_n).$$

A complete system of inequalities characterizing the mixed volume $V(K_1, \ldots, K_n)$ has been obtained in the case of two bodies, and more general inequalities have been established (see [4]).

Many geometric inequalities, such as the classical isoperimetric inequality (cf. **Isoperimetric inequality, classical**) and several of its refinements, are special cases of inequalities for mixed volumes of convex bodies. The extremum of one of the functionals $V_p(K)$ when some other such functional is fixed is attained for a sphere. Inequalities in the theory of mixed volumes are used in proving the uniqueness of the solution of the generalized **Minkowski problem** (see [2]), in establishing stability in Minkowski problems (see [5]) and Weil problems (see [6]), and in the solution of the van der Waerden problem on permanents (see [7]). Infinite-dimensional analogues of the concepts of the theory of mixed volumes have found application in the theory of Gaussian stochastic processes (see [7]).

The theory of mixed volumes has deep connections with algebraic geometry. Given a polynomial $f(z_1, \ldots, z_n)$ in n complex variables, one can define its Newton polyhedron $\mathrm{Nw}(f)$ in the following way. To each monomial $z_1^{a_1} \cdots z_n^{a_n}$ occurring in f with a non-zero coefficient, one assigns the point $(a_1, \ldots, a_n)\in\mathbf{R}^n$, and one defines $\mathrm{Nw}(f)$ as the convex hull of these points. The typical number of solutions of the system of polynomial equations $f_1 = \cdots = f_n = 0$ is equal to the mixed volume of the polyhedron $\mathrm{Nw}(f_1 \cdots f_n)$ divided by $n!$. Among other things, this allows one to give an algebraic proof of the Aleksandrov−Fenchel inequality (see [10]).

In the theory of mixed volumes, a convex body is identified with its support function. This is extended to differences of these functions, and then to arbitrary continuous functions on a sphere (see [2], [9]). Using the analogous decomposition of the vector of the centre of gravity of a body $\sum_{i=1}^{r}\lambda_i K_i$ multiplied by its volume, one can define the so-called *mixed direction vectors*, which are vector analogues of mixed volumes. The centres of gravity of a body K coincide up to a multiplicative constant with the mixed direction vectors of K and the sphere U (see [11]).

References

[1] MINKOWSKI, H.: 'Theorie der konvexen Körpern, insbesonder der Begründung ihres Oberflächenbegriffs', in *Gesammelte Abh.*, Vol. 2, Teubner, 1911, pp. 131-229.
[2A] ALEKSANDROV, A.D.: 'Zur Theorie gemischter Volumina von konvexen Körpern I. Verallgemeinerungen einiger Begriffe der Theorie von konvexen Körpern', *Mat. Sb.* **2**, no. 5 (1937), 947-972.
[2B] ALEKSANDROV, A.D.: 'Zur Theorie gemischter Volumina von konvexen Körpern II. Neue Ungleichungen zwischen den gemischten Volumina und ihre Anwendungen', *Mat. Sb.* **2**, no. 6, 1205-1238.
[2C] ALEKSANDROV, A.D.: 'Zur Theorie gemischter Volumina von konvexen Körpern III. Die Erweiterung zweier Lehrsätze Minkowski's über die konvexen Polyedern auf die beliebigen konvexen Körper', *Mat. Sb.* **3**, no. 1 (1938), 27-46.
[2D] ALEKSANDROV, A.D.: 'Zur Theorie gemischter Volumina von konvexen Körpern IV. Die gemischten Diskriminanten und die gemischten Volumina', *Mat. Sb.* **3**, no. 2, 227-251.
[3] BUSEMANN, H.: *Convex surfaces*, Interscience, 1958.
[4] SHEPHARD, G.: 'Inequalities between mixed volumes of convex sets', *Mathematika* **7**, no. 14 (1960), 125-138.
[5] DISKANT, V.I.: 'Stability of solutions of Minkowski's equation', *Sib. Math. J.* **14**, no. 3 (1973), 466-469. (*Sibirsk. Mat. Zh.* **14**, no. 3 (1973), 669-673)

[6] VOLKOV, YU.A.: 'Estimate of deformation of a convex surface as a function of the change in internal metric', *Ukrain. Geom. Sb.* **5-6** (1968), 44-69 (in Russian).

[7] KNUTH, D.E.: 'A permanent inequality', *Amer. Math. Monthly* **88** (1981), 731-740.

[8] SUDAKOV, V.N.: 'Gaussian random processes and measures of solid angles in Hilbert space', *Soviet Math. Dokl.* **12**, no. 2 (1971), 412-415. (*Dokl. Akad. Nauk SSSR* **197**, no. 1 (1971), 43-45)

[9] BUSEMANN, H., EWALD, G. and SHEPARD, G.: 'Convex bodies and convexity on Grassmann cones I-IV', *Math. Ann.* **151**, no. 1 (1963), 1-41.

[10] KHOVANSKIĬ, A.G.: 'Algebra and mixed volumes', in Yu.D. Burago and V.A. Zalgaller (eds.): *Geometric Inequalities*, Springer, 1988, pp. 182-207 (translated from the Russian).

[11] SCHNEIDER, R.: 'Krümmungsschwerpunkte konvexer Körper I', *Abh. Math. Sem. Univ. Hamburg* **37** (1972), 112-132.

Yu.D. Burago

Editorial comments.

References

[A1] SCHNEIDER, R.: 'On the Aleksandrov—Fenchel inequality', *Ann. New York Acad. Sci.* **440** (1985), 132-141.

[A2] MCMULLEN, P. and SCHNEIDER, R.: 'Valuations on convex bodies', in P.M. Gruber and J.M. Wills (eds.): *Convexity and Its Applications*, Birkhäuser, 1983, pp. 170-247.

AMS 1980 Subject Classification: 52A20, 52A40

MIXING - A property of a dynamical system (a **cascade** $\{S^n\}$ or a **flow (continuous-time dynamical system)** $\{S_t\}$) having a finite **invariant measure** μ, in which for any two measurable subsets A and B of the phase space W, the measure

$$\mu((S^n)^{-1}A \cap B),$$

or, respectively,

$$\mu((S_t)^{-1}A \cap B),$$

tends to

$$\frac{\mu(A)\mu(B)}{\mu(W)}$$

as $n \to \infty$, or, respectively, as $t \to \infty$. If the transformations S and S_t are invertible, then in the definition of mixing one may replace the pre-images of the original set A with respect to these transformations by the direct images $S^n A$ and $S_t A$, which are easier to visualize. If a system has the mixing property, one also says that the system is mixing, while in the case of a mixing cascade $\{S^n\}$, one says that the endomorphism S generating it in the measure space (W, μ) also is mixing (has the property of mixing).

In ergodic theory, properties related to mixing are considered: *multiple mixing* and *weak mixing* (see [1]; in the old literature the latter is often called mixing in the wide sense or simply mixing, while mixing was called mixing in the strong sense). A property intermediate between mixing and weak mixing has also been discussed [2]. All these properties are stronger than **ergodicity**.

There is an analogue of mixing for systems having an infinite invariant measure [3].

References

[1] HALMOS, P.R.: *Lectures on ergodic theory*, Math. Soc. of Japan, 1956.

[2] FURSTENBERG, H. and WEISS, B.: 'The finite multipliers of infinite ergodic transformations', in N.G. Markley, J.C. Martin and W. Perrizo (eds.): *The Structure of Attractors in Dynamical Systems*, Lecture notes in math., Vol. 668, Springer, 1978, pp. 127-132.

[3] KRENGEL, U. and SUCHESTON, L.: 'On mixing in infinite measure spaces', *Z. Wahrscheinlichkeitstheor. Verw. Geb.* **13**, no. 2 (1969), 150-164.

D.V. Anosov

Editorial comments. For a cascade $\{S^n\}$ on a finite measure space (W, μ) the notion of weak mixing as defined above is equivalent to the property that the cascade generated by $S \times S$ on the measure space $(W \times W, \mu \otimes \mu)$, where $\mu \otimes \mu$ denotes the product measure, is ergodic (cf. **Ergodicity**; **Metric transitivity**). See [1].

For topological dynamical systems the notions of strong and weak mixing have been defined as well [A2]. A flow on a topological space W is said to be *topologically weakly mixing* whenever the flow $\{S_t \times S_t\}$ on $W \times W$ (with the usual product topology) is topologically ergodic; equivalently: whenever for every choice of four non-empty open subsets U_i, V_i ($i = 1, 2$) of W where exists a t such that $S_t U_i \cap V_i \neq \varnothing$ for $i = 1, 2$. On compact spaces the weakly mixing minimal flows are the minimal flows that have no non-trivial equicontinuous factors; see [A1], p. 133. A flow $\{S_t\}$ on a space W is said to be *topologically strongly mixing* whenever for every two non-empty open subsets U and V of W there exists a value t_0 such that $S_t U \cap V \neq \varnothing$ for all $|t| \geq t_0$. For example, the geodesic flow on a complete two-dimensional Riemannian manifold of constant negative curvature is topologically strongly mixing; see [A3], 13.49. For cascades, the definitions are analogous.

References

[A1] AUSLANDER, J.: *Minimal flows and their extensions*, North-Holland, 1988.

[A2] FURSTENBERG, H.: 'Disjointness in ergodic theory, minimal sets and a problem in diophantine approximations', *Math. Systems Th.* **1** (1967), 1-49.

[A3] GOTTSCHALK, W.H. and HEDLUND, G.A.: *Topological dynamics*, Amer. Math. Soc., 1955.

AMS 1980 Subject Classification: 28DXX, 58F99, 54H20

MÖBIUS FUNCTION - An **arithmetic function** of natural argument: $\mu(1) = 1$, $\mu(n) = 0$ if n is divisible by the square of a prime number, otherwise $\mu(n) = (-1)^k$, where k is the number of prime factors of the number n. This function was introduced by A. Möbius in 1832.

The Möbius function is a **multiplicative arithmetic function**; $\sum_{d \mid n} \mu(d) = 0$ if $n > 1$. It is used in the study of other arithmetic functions; it appears in inversion formulas (see, e.g. **Möbius series**). The following estimate is known for the mean value of the Möbius function [2]:

$$\frac{1}{x}\left|\sum_{n \leqslant x}\mu(n)\right| \leqslant \exp\{-c\ln^{3/5}x(\ln\ln x)^{-1/5}\},$$

where c is a constant. The fact that the mean value tends to zero as $x\to\infty$ implies an asymptotic law for the **distribution of prime numbers** in the natural series.

References

[1] VINOGRADOV, I.M.: *Elements of number theory*, Dover, 1954 (translated from the Russian).
[2] WALFISZ, A.: *Weylsche Exponentialsummen in der neueren Zahlentheorie*, Deutsch. Verlag Wissenschaft., 1963.

N.I. Klimov

Editorial comments. The multiplicative arithmetic functions form a **group** under the *convolution product* $(f\star g)(n)=\sum_{d\mid n}f(d)g(n/d)$. The Möbius function is in fact the inverse of the constant multiplicative function E (defined by $E(n)=1$ for all $n\in\mathbb{N}$) under this convolution product. From this there follows many 'inversion formulas', cf. e.g. Möbius series.

References

[A1] HARDY, G.H. and WRIGHT, E.M.: *An introduction to the theory of numbers*, Clarendon Press, 1979.

AMS 1980 Subject Classification: 10A20

MÖBIUS PLANE, *circular plane, inversive plane* - A plane whose elements form two disjoint sets: the set of points and the set of circles, and which is endowed with a symmetric incidence relation (relating points and circles). The incidence relation satisfies the following axioms:

1) any three distinct points are incident to one and only one circle;

2) given a point A on a circle γ and a point B not on γ, there is a unique circle through B whose common point with γ is A;

3) there exist at least four distinct points not incident to a circle. Every circle is incident to at least three distinct points.

From a Möbius plane one can obtain an affine plane if one calls one of its points the ideal point, and calls the circles incident to this point straight lines.

In three-dimensional projective space $\mathbf{P}_3$ the points of an **ovoid** o and the planes that intersect the ovoid at more than one point form, with the incidence relation inherited from $\mathbf{P}_3$, a Möbius plane $M(o)$ (see [1]). A Möbius plane is called egg-like if it is isomorphic to $M(o)$ for some ovoid o. Among the egg-like Möbius planes the best known is the model $M(S)$, where S is the sphere in three-dimensional Euclidean space, i.e. the plane isomorphic to $M(c)$, where c is a non-ruled quadric in three-dimensional projective space over the field of real numbers.

A Möbius plane is said to be *finite* if it has a finite number of points and circles. Each circle in a Möbius plane has the same number of points, and through each point of the plane pass the same number of circles. By definition, the *order of the plane* is the number of points on a circle minus one. A Möbius plane of order n contains n^2+1 points and $n(n^2+1)$ circles; through each point of the plane $n(n+1)$ circles pass. The following model of the Möbius plane of order $n=p^h$ is the best known. The points of the plane are the elements of the **Galois field** $\mathrm{GF}(p^{2h})$ and the ideal point $\{\infty\}$; the circles of the plane are the images of the set $K=\mathrm{GF}(p^h)\bigcup\{\infty\}$ under the group of permutations of the form

$$x \to \frac{x^\alpha a+c}{x^\alpha b+d},\quad a,b,c,d\in\mathrm{GF}(p^{2h}),\quad ad\neq bc,$$

$$\alpha\in\mathrm{Aut}\,\mathrm{GF}(p^{2h}).$$

A necessary condition for the existence of a Möbius plane of order n is the existence of a finite affine plane of the same order. The uniqueness of the Möbius plane of order $n=2,3,4,5,7,11$ has been proven [5]. If a Möbius plane of order n contains a proper subplane of order m, then $m\equiv n\ (\mathrm{mod}\,2)$ and $m^2+m\leqslant n$ (see [2]).

The classification of Möbius planes has been carried out (see [3], [4]). The planes are named after A. Möbius (1855), who laid the foundations of the theory of circles.

References

[1] DEMBOWSKI, P.: *Finite geometries*, Springer, 1968, p. 254.
[2] DEMBOWSKI, P. and HUGHES, D.R.: 'On finite inversive planes', *J. London Math. Soc.* **40** (1965), 171-182.
[3] HERING, C.H.: 'Eine Klassifikation der Möbius-Ebenen', *Math. Z.* **87** (1965), 252-262.
[4] KRIER, N.: 'The Hering classification of Möbius planes', in *Proc. Internat. Conf. Projective Planes*, Washington State Univ. Press, 1973, pp. 157-163.
[5] ISTOMINA, L.I.: 'Uniqueness of the circle plane of order 11', *Perm Univ. Uchen. Zap.* **156** (1976), 81-83 (in Russian).

V.V. Afanas'ev

Editorial comments. For more recent developments, see the chapters by I.M. Yaglom, J.F. Rigby, J.B. Wilker, and N.W. Johnson in [A1].

References

[A1] DAVIS, C., GRÜNBAUM, B. and SHERK, F.A. (EDS.): *The geometric vein*, Springer, 1980.

AMS 1980 Subject Classification: 51E99, 51M99

MÖBIUS SERIES - A series of functions of the form

$$F(x) = \sum_{s=1}^{\infty}\frac{f(x^s)}{s^n}.\qquad (*)$$

These series were investigated by A. Möbius [1], who found for a series (*) the inversion formula

$$f(x) = \sum_{s=1}^{\infty}\mu(s)\frac{F(x^s)}{s^n},$$

where $\mu(s)$ is the **Möbius function**. Möbius considered also *inversion formulas* for finite sums running over the divisors of a natural number n:

$$F(n) = \sum_{d \mid n} f(d), \quad f(n) = \sum_{d \mid n} \mu(d) F\left[\frac{n}{d}\right].$$

Another inversion formula: If $P(n)$ is a totally-multiplicative function (cf. **Multiplicative arithmetic function**) for which $P(1)=1$, and $f(x)$ is a function defined for all real $x>0$, then

$$g(x) = \sum_{n \leqslant x} P(n) f\left[\frac{x}{n}\right]$$

implies

$$f(x) = \sum_{n \leqslant x} \mu(n) P(n) g\left[\frac{x}{n}\right].$$

References

[1] MÖBIUS, A.: 'Ueber eine besondere Art der Umkehrung der Reihen', *J. Reine Angew. Math.* **9** (1832), 105-123.
[2] VINOGRADOV, I.M.: *Elements of number theory*, Dover, 1954 (translated from the Russian).
[3] PRACHAR, K.: *Primzahlverteilung*, Springer, 1957.

B.M. Bredikhin

Editorial comments. All these (and many other) inversion formulas follow from the basic property of the Möbius function that it is the inverse of the unit arithmetic function $E(n) \equiv 1$ under the convolution product, cf. (the editorial comments to) **Möbius function** and **Multiplicative arithmetic function**.

AMS 1980 Subject Classification: 10A20, 10A22

MÖBIUS STRIP - A non-orientable surface with **Euler characteristic** zero whose boundary is a closed curve. The Möbius strip can be obtained by identifying two opposite sides AB and CD of a rectangle $ABCD$ so that the points A and B are matched with the points C and D, respectively (see Fig.).

In the Euclidean space E^3 the Möbius strip is a one-sided surface (see **One-sided and two-sided surfaces**).

The Möbius strip was considered (in 1858 - 1865) independently by A. Möbius and I. Listing.

A.B. Ivanov

AMS 1980 Subject Classification: 57N12

MODAL LOGIC - The domain of logic in which along with the usual statements *modal statements* are considered, that is, statements of the type 'it is necessary that $\cdots$', 'it is possible that $\cdots$', etc. In mathematical logic various formal systems of modal logic have been considered, interrelations between these systems have been revealed, and their interpretations have been studied.

Elements of modal logic were in essence already known to Aristotle (4-th century B.C.) and became part of classical philosophy. Modal logic was formalized for the first time by C.I. Lewis [1], who constructed five propositional systems of modal logic, given in the literature the notations S1 - S5 (their formulations are given below). Other systems of modal logic were then constructed and investigated. The great variety of systems of modal logic is explained by the fact that the ideas of 'possible' and 'necessary' can be made precise in various ways; in addition, there are various ways to treat complex modalities (cf. **Modality**) of the type 'necessarily possible', and 'interrelations' of modality with the logical connectives. The majority of systems of modal logic which have been studied are based on classical logic; however, systems based on intuitionistic logic have also been discussed (see, for example, [6]).

Below, several of the most widely-studied propositional systems of modal logic are described. The language of each of these systems is obtained from the language of classical propositional calculus P by the addition of the new one-place connectives (*modal operators*) $\Box$ (necessary) and $\Diamond$ (possible). Since in almost all these systems the relation

$$\Diamond A \equiv \neg \Box \neg A \qquad (*)$$

holds, initially one modal operator is chosen, for example $\Box$, and the other is defined by (*).

The *system* S1. Axiom schemes:

1) all formulas of the form $\Box A$, where A is a formula derivable in P;

2) $\Box(\Box A \supset A)$;

3) $\Box(\Box(A \supset B) \& \Box(B \supset C) \supset \Box(A \supset C))$.

Derivation rules:

I) $\dfrac{A \quad A \supset B}{B}$ (**modus ponens**);

II) $\dfrac{\Box A}{A}$;

III) $\dfrac{\Box(A \supset B) \quad \Box(B \supset A)}{\Box(\Box A \supset \Box B)}$.

The *system* S2: S1 + $\{\Box(\Box A \supset \Box(A \vee B))\}$.

The *system* S3: S2 + $\{\Box(\Box(A \supset B) \supset \Box(\Box A \supset \Box B))\}$.

The *system* K. Axiom schemes:

1) the axiom schemes of the propositional calculus P;

2) $\Box(A \supset B) \supset (\Box A \supset \Box B)$.

Derivation rules: modus ponens and

$$\frac{A}{\Box A} \quad (\Box-\text{prefix}).$$

The *system* T: K + $\{\Box A \supset A\}$ = S2 + $\{$rule of $\Box$-prefix$\}$.

The *system* B: T + $\{A \supset \Box \Diamond A\}$.

The *system* S4: S3 + $\{\Box(\Box A \supset \Box \Box A)\}$ = T +

$\{\Box A \supset \Box \Box A\}$.

The *system* S5: S4 + $\{\Box (A \supset \Box \Diamond A)\}$.

Among the above systems, S4 has important significance since the intuitionistic propositional calculus I can be interpreted in it, that is, with respect to every propositional (non-modal) formula A it is possible to construct a formula A^* of modal logic such that

$$I \vdash A \Leftrightarrow S4 \vdash A^*.$$

In this connection the *system of Grzegorczyk* is of particular interest (see [5]):

$$G = S4 + \{\Box (\Box (\Box (A \supset \Box A) \supset A) \supset A)\};$$

for it the *transference theorem* is true: For any set of axiom schemes Γ and any formula A,

$$I + \Gamma \vdash A \Leftrightarrow S4 + \Gamma^* \vdash A^* \Leftrightarrow G + \Gamma^* \vdash A^*,$$

where $\Gamma^* = \{B^* : B \in \Gamma\}$; moreover, G is the strongest system with this property. This theorem makes it possible to transfer a property (for example, completeness or decidability) from an extension of the system S4 (or G) to an **intermediate logic**.

For each propositional system of modal logic S it is possible to consider the corresponding predicate system, which is obtained by the addition of object variables, predicate symbols and the quantifiers $\forall$, $\exists$ (or one of these) to the language of S. The usual axiom schemes and derivation rules for quantifiers are added. In addition one sometimes also adds axioms which describe the actions of modal operators on quantifiers such as, for example, the *Barcan formula*:

$$\forall x \, \Box A(x) \supset \Box \forall x \, A(x).$$

The algebraic interpretation of a system of modal logic is given by some algebra (also called a *matrix*)

$$\mathfrak{M} = <M, D; \&^*, \vee^*, \supset^*, \neg^*, \Box^*>,$$

where M is the set of truth values (cf. **Truth value**), D is a set of *distinguished truth values*, $D \subset M$, and $\&^*$, $\vee^*$, $\supset^*$, $\neg^*$, $\Box^*$ are the operations in M corresponding to the connectives $\&$, $\vee$, $\supset$, $\neg$, $\Box$. A formula is called *generally valid* in M if for every valuation of its propositional variables by elements of M it takes a distinguished value. A system of modal logic S is called *complete* relative to a class of algebras $\mathcal{K}$ if a formula is derivable in S if and only if it is generally valid in every algebra in the class $\mathcal{K}$. For example, the system S4 is complete relative to the class of so-called finite topological Boolean algebras (see [3]). In general, a system S is called *finitely approximable* if it is complete relative to finite algebras. If a system is finitely axiomatizable and finitely approximable, then it is *decidable*, that is, the problem of recognition of derivability is algorithmically decidable. A matrix $\mathfrak{M}$ is called *characteristic* or *adequate* for a system S if S is complete relative to $\{\mathfrak{M}\}$. None of the above-mentioned proposi-

tional systems of modal logic has a finite adequate matrix, but each of them is finitely approximable and therefore decidable. On the other hand, every extension of S5 has a finite adequate matrix with one distinguished value. The property of finite approximability also holds for all extensions of the system

$$S4.3 = S4 + \{\Box (\Box A \supset \Box B) \vee \Box (\Box B \supset \Box A)\}.$$

An important tool in the study of modal logics are **Kripke models**, having the form (W, R, θ), where W is a set (called the set of 'worlds', 'situations'), R is a relation on W and θ is a valuation of the propositional variables by subsets of W. For $s, t \in W$ the relation sRt can be treated as 'the world t is possible in the world s'. The pair (W, R) is called a *Kripke structure*, or *frame* (the term *scale* is also used). A formula A is *generally valid in the frame* (W, R) if for each valuation θ formula A is true in the Kripke model (W, R, θ). A system S is called *Kripke complete* relative to a class of Kripke structures if the S-derivable formulas are exactly the formulas which are generally valid in all Kripke structures in the class $\mathcal{K}$. For example, the system T is Kripke complete relative to the class of structures (W, R), where R is a reflexive relation; S4 is Kripke complete relative to a structure with a reflexive and transitive relation. Among the finitely-axiomatizable extensions of S4 there are extensions which are not Kripke complete (see [7]).

For predicate systems of modal logic the Kripke models have the form (W, R, D, ψ, θ), where $D = \{D_s\}_{s \in W}$, D_s is a universe for the world s, ψ is an interpretation of the predicate symbols in D, and θ is a valuation associating to object variables some elements of the set $\bigcup_{s \in W} D_s$. For systems containing the Barcan formula, it is also necessary to require

$$sRt \Leftrightarrow D_t \subseteq D_s.$$

Kripke models, as a rule, have a more easily visualized structure than algebraic models; therefore they are often more convenient for the study of different systems of modal logic.

References
[1] LEWIS, C.I. and LANGFORD, C.H.: *Symbolic logic*, Dover, reprint, 1930.
[2] FEYS, R.: *Modal logic*, Moscow, 1974 (in Russian).
[3] RASIOWA, E. and SIKORSKI, R.: *The mathematics of metamathematics*, Polska Akad. Nauk, 1963.
[4] MINTS, G.A.: 'On some calculi of model logic', *Proc. Steklov Inst. Math.* **98** (1968), 97-124. (*Trudy Mat. Inst. Steklov.* **98** (1968), 88-111)
[5] GRZEGORCZYK, A.: 'Some relational systems and the corresponding topological spaces', *Fund. Math.* **60**, no. 2 (1967), 223-231.
[6] BULL, R.A.: 'A model extension of intuitionist logic', *Notre Dame J. Formal Logic* **6** (1965), 142-146.
[7] FINE, K.: 'An incomplete logic containing S4', *Theoria* **40**, no. 1 (1974), 23-29.

[8] GABBAY, D.M.: 'Deducability results in non-classical logics I', *Ann. Math. Logic* **8**, no. 3 (1975), 237-295.

S.K. Sobolev

Editorial comments. A formula A *holds* at a world $s \in W$ if and only if (inductively) either: 1) more precisely, 'being true' of a formula in a Kripke model is defined as follows: A is a propositional variable and $s \in \theta(A)$; 2) A is $\neg B$ and B does not hold at s; 3) A is $B \vee C$ and at least one of B, C holds at s; or (and this the distinctive clause) 4) A is $\Box B$ and B holds at each t for which sRt.

General validity of a modal formula A on a frame (W, R) expresses a monadic universal second-order condition on (W, R). (For, propositional variables are related to subsets of W, hence *monadic*; and in the condition 'for each θ and s, A holds at s' the part 'A holds at s' is first-order expressible — as can be seen immediately from the above definition.) Therefore, modal logic, through its *Kripke semantics*, can be considered as part of second-order logic.

Questions like: which modal formulas have a first-order equivalent (on a given class of frames)?, and: which (monadic universal) second-order formulas can be modally expressed?, belong to the *correspondence theory of modal logic*. Cf., e.g., [A1], [A2].

In *provability logic* the modal expression $\Box A$ is interpreted as 'A is provable'. A basic result here is *Solovay's completeness theorem*, which states that the theorems of *Löb's modal logic* (the extension of S4 with the scheme $\Box(\Box A \rightarrow A) \rightarrow \Box A$, expressing the generalization of Gödel's second incompleteness theorem known as *Löb's theorem*) are exactly those modal formulas with the following property: Every arithmetical instance of it (where $\Box$ is replaced by the formalized provability predicate of formal (Peano) arithmetic) is a theorem of formal arithmetic; cf. [A4].

References

[A1] BENTHEM, J. VAN: *Modal logic and classical logic*, Bibliopolis, 1983.
[A2] BENTHEM, J. VAN: 'Correspondence theory', in D. Gabbay and F. Guenthner (eds.): *Handbook of Philos. Logic*, Vol. II, Reidel, 1984, pp. 167-242.
[A3] BULL, R.A. and SEGERBERG, K.: 'Basic modal logic', in D. Gabbay and F. Guenthner (eds.): *Handbook of Philos. Logic*, Vol. II, Reidel, 1984, pp. 1-88.
[A4] SMORYNSKI, C.: *Self-reference and modal logic*, Springer, 1985.
[A5] HUGHES, G.E. and CRESSWELL, M.J.: *An introduction to modal logic*, Methuen, 1968.

AMS 1980 Subject Classification: 03B45

MODALITY - A property of a judgement characterizing its degree of certainty. Different modalities and the interelations between them are studied in **modal logic**. The modalities 'necessary' and 'possible' have been studied in logic since Aristotle (4-th century B.C.) who, however, did not impart a precise meaning to them. These modalities are called *fundamental* and are denoted by the symbols $\Box$ and $\Diamond$ (or L and M), respectively. Various combinations of the fundamental modalities and the negation $\neg$ are also modalities. The *dual of a modality Q* is the modality $\bar{Q}$ obtained from Q by replacing each occurrence of $\Box$ by $\Diamond$, and conversely. In most systems of modal logic, for a modality Q and its dual $\bar{Q}$ the equivalence

$$Q \neg A \Leftrightarrow \neg \bar{Q} A \qquad (*)$$

holds.

In principle, it is possible to form an infinite number of combinations of $\Box$, $\Diamond$ and $\neg$; however, in concrete systems the number of pairwise inequivalent modalities often remains bounded (in view of (*) and the presence of axioms simplifying certain modalities, or reducing one modality to another). For example, in the system S3 there are exactly 40 different modalities, and in S4 there are only 12:

$$\Box A, \quad \Box \Diamond A, \quad \Box \Diamond \Box A, \quad \neg \Box A, \quad \neg \Box \Diamond A, \quad \neg \Box \Diamond \Box A,$$

and their duals. In S5 there are only 4 modalities: $\Box A$, $\Diamond A$, $\neg \Box A$, $\neg \Diamond A$. On the other hand, in the system of modal logic T, and also in S1 and S2, the number of modalities is infinite and, even more, there are no reductions of modalities; that is, two positive (not containing $\neg$) modalities, Q_1 and Q_2 are equivalent if and only if $Q_1 = Q_2$.

Sometimes the term modality refers to such notions (formalized in corresponding theories) as 'truth', 'provability', 'disprovability', and is also connected with the temporal 'will be', 'always was', etc.

For references see **Modal logic.**

S.K. Sobolev

AMS 1980 Subject Classification: 03B45

MODE - One of the numerical characteristics of the **probability distribution** of a **random variable.** For a random variable with density $p(x)$ (cf. **Density of a probability distribution**), a mode is any point x_0 where $p(x)$ is maximal. A mode is also defined for discrete distributions: If the values x_k of a random variable X with distribution $p_k = P\{x = x_k\}$ are arranged in increasing order, then a point x_m is called a mode if $p_m \geqslant p_{m-1}$ and $p_m \geqslant p_{m+1}$.

Distributions with one, two or more modes are called, respectively, *unimodal* (*one-peaked* or *single-peaked*), *bimodal* or *multimodal*. The most important in probability theory and mathematical statistics are the unimodal distributions (cf. **Unimodal distribution**). Along with the **mathematical expectation** and the **median (in statistics)** the mode acts as a measure of location of the values of a random variable. For distributions which are unimodal and symmetric with respect to some point a, the mode is equal to a and to the median and to the mathematical expectation, if the latter exists.

A.V. Prokhorov

Editorial comments.

References

[A1] MOOD, A.M. and GRAYBILL, F.A.: *Introduction to the theory of statistics*, McGraw-Hill, 1963.

[A2] BREIMAN, L.: *Statistics with a view towards applications*, Houghton Mifflin, 1973.

AMS 1980 Subject Classification: 62E10, 60EXX

MODEL FOR CALCULATIONS, *computational model* - A typical problem used as a model for investigating and developing numerical methods for some class of problems. For example, in the theory of **quadrature** the problem of calculating integrals of functions satisfying a condition $|f^{(n)}| \leqslant A$ is considered. The processing of methods for the solution of the **Cauchy problem** for systems of ordinary differential equations historically was done by investigating the properties of the methods on models from a sequence of increasing complexity (with integration interval $[0, X]$):

1) the equation $y' = 0$;

2) the equation $y' = my$, $|m|X$ of order 1 (models 1) and 2) correspond to the problem of integration on small time intervals of systems with smooth solutions);

3a) the equation $y' = my$, $m < 0$, $|m|X \gg 1$; this model corresponds to the problem of integration on large time intervals of systems with stable solutions;

3b) the equation $y' = x^\lambda$; a model of an equation with singularities in the derivatives of solutions;

4) the system $y_1' = m_1 y_1$, $y_2' = m_2 y_2$, $0 > m_1 > m_2$, $|m_2|X \gg |m_1|X$, $|m_1|X$ of order 1; a model of so-called stiff differential systems (cf. **Stiff differential system**), in which one component varies relatively slowly and the other rapidly.

References

[1] BAKHVALOV, N.S.: *Numerical methods: analysis, algebra, ordinary differential equations*, Mir, 1977 (translated from the Russian).

N.S. Bakhvalov

AMS 1980 Subject Classification: 65-XX

MODEL (IN LOGIC) - An interpretation of a **formal language** satisfying certain axioms (cf. **Axiom**). The basic formal language is the *first-order language L_Ω* of a given signature Ω including predicate symbols R_i, $i \in I$, function symbols f_j, $j \in J$, and constants c_k, $k \in K$. A *model of the language L_Ω* is an **algebraic system** of signature Ω.

Let Σ be a set of closed formulas in L_Ω. A model for Σ is a model for L_Ω in which all formulas from Σ are true. A set Σ is called *consistent* if it has at least one model. The class of all models of Σ is denoted by $\operatorname{Mod}\Sigma$. Consistency of a set Σ means that $\operatorname{Mod}\Sigma \neq \varnothing$.

A class $\mathscr{K}$ of models of a language L_Ω is called *axiomatizable* if there is a set Σ of closed formulas of L_Ω such that $\mathscr{K} = \operatorname{Mod}\Sigma$. The set $T(\mathscr{K})$ of all closed formulas of L_Ω that are true in each model of a given class $\mathscr{K}$ of models of L_Ω is called the **elementary theory** of $\mathscr{K}$. Thus, a class $\mathscr{K}$ of models of L_Ω is axiomatizable if and only if $\mathscr{K} = \operatorname{Mod} T(\mathscr{K})$. If a class $\mathscr{K}$ consists of models isomorphic to a given model, then its elementary theory is called the *elementary theory of this model*.

Let **A** be a model of L_Ω having universe A. One may associate to each element $a \in A$ a constant c_a and consider the first-order language $L_{\Omega A}$ of signature ΩA which is obtained from Ω by adding the constants c_a, $a \in A$. $L_{\Omega A}$ is called the *diagram language of the model* **A**. The set $O(\mathbf{A})$ of all closed formulas of $L_{\Omega A}$ which are true in **A** on replacing each constant c_a by the corresponding element $a \in A$ is called the *description* (or *elementary diagram*) of **A**. The set $D(\mathbf{A})$ of those formulas from $O(\mathbf{A})$ which are atomic or negations of atomic formulas is called the *diagram of A*.

Along with models of first-order languages, models of other types (infinitary logic, **intuitionistic logic**, many-sorted logic, second-order logic, **many-valued logic**, and **modal logic**) have also been considered.

For references see **Model theory**.

D.M. Smirnov

Editorial comments. English usage prefers the word 'structure' where Russian speaks of a 'model of a language' or an 'algebraic system'; 'model' is reserved for structures satisfying a given theory (set of closed formulas).

AMS 1980 Subject Classification: 03CXX

MODEL, MATHEMATICAL - See **Mathematical model**.

AMS 1980 Subject Classification: 00A25

MODEL, REGULAR - An interpretation of a formal system in which all axioms are true or have the value 'true' for all values of their parameters and in which the deduction rules preserve the property of taking the value 'true'. This definition concerns two-valued logical calculi. If the logic is multi-valued and some of the values are distinguished, then in the definition of a regular model one has to say 'taking distinguished values' instead of 'taking the value 'true''. If the above property of a model is not satisfied, the model is said to be *non-regular*. Models of formal systems obtained by adjoining to restricted predicate calculus some set T of axioms are also called *models of the axiom system T* or models for T. Any interpretation of such a formal system is called an **algebraic system** or simply a **model (in logic)**.

V.N. Grishin

Editorial comments. In English usage, one normally distinguishes between *structures* for a language, which may or may not satisfy the axioms of a theory under consideration, and *models*, which necessarily do so; thus the term 'regular model' is redundant, and is not used in this sense. See also (the editorial comments to) **Algebraic system**; **Model (in logic)**.

For another use of the phrase 'regular model' see [A1].

References

[A1] SHOENFIELD, J.R.: *Mathematical logic*, Addison-Wesley, 1967.

AMS 1980 Subject Classification: 03CXX

MODEL THEORY - The part of mathematical logic studying mathematical models (cf. **Model (in logic)**).

The origins of model theory go back to the 1920's and 1930's, when the following two fundamental theorems were proved.

Theorem 1 (*Gödel compactness theorem*). If each finite subcollection of a collection T of propositions in a first-order language is consistent, then the whole collection T is consistent (see [1]).

Theorem 2 (*Löwenheim—Skolem theorem*). If a collection of propositions in a first-order language of signature Ω has an infinite model, then it has a model of any infinite cardinality not less than the cardinality of Ω.

Theorem 1 has had extensive application in algebra. On the basis of this theorem, A.I. Mal'tsev created a method of proof of local theorems in algebra (see **Mal'tsev local theorems**).

Let A be an **algebraic system** of signature Ω, let $|A|$ be the underlying set of A, let $X \subseteq |A|$, let $<\Omega, X>$ denote the signature obtained from Ω by the addition of symbols for distinguished elements c_a for all $a \in X$, and let (A, X) denote the algebraic system of signature $<\Omega, X>$ which is an enrichment of A in which for each $a \in X$ the symbol c_a is interpreted by the element a. The set $O(A)$ of all closed formulas of the signature $<\Omega, |A|>$ in a first-order language which are true in the system $(A, |A|)$ is called the *elementary diagram of the algebraic system A* (or the *description of the algebraic system A*), and the set $D(A)$ of those formulas from $O(A)$ which are either atomic or the negation of an atomic formula is called the *diagram of A*. An algebraic system B is called an *elementary extension of A* if $|A| \subseteq |B|$ and if $(B, |A|)$ is a model for $O(A)$. In this case A is called an *elementary subsystem of B*. For example, the set of rational numbers with the usual order relation is an elementary subsystem of the system of real numbers with the usual order relation.

A subsystem A of an algebraic system B of signature Ω is an elementary subsystem of B if and only if for each closed formula $(\exists v)\Phi(v)$ in the first-order language of signature $<\Omega, |A|>$ which is true in $(B, |A|)$ there is an $a \in |A|$ such that $\Phi(c_a)$ is true in $(B, |A|)$. It follows at once from this criterion that the union of an increasing chain of elementary subsystems is an elementary extension of each of these subsystems. If a closed $\forall\exists$-formula in a first-order language is true in every system of an increasing chain of systems, then it is true in the union of the chain (see [1], [8]).

Let the signature Ω contain a one-place relation symbol U. One says that a model A of a theory T of signature Ω has *type* (α, β) if the cardinality of $|A|$ is equal to α and if the cardinality of $U(A) = \{a \in |A|: A \vDash U(a)\}$ is equal to β. *Vaught's theorem*: If an elementary theory T of countable signature has a model of type (α, β) where $\alpha > \beta$, then T has a model of type $(\aleph_1, \aleph_0)$ (see [7], [8], [10]). Under the assumption that the generalized **continuum hypothesis** holds, an **elementary theory** of countable signature has models of types $(\aleph_{\alpha+2}, \aleph_{\alpha+1})$ for each α, if it has a model of type $(\aleph_1, \aleph_0)$ (see [10]). Under the same assumption the theory $\mathrm{Th}(A)$, where the signature of A is $< +, \cdot, 0, 1, <, U, >$, $|A|$ is the set of all real numbers, $U(A)$ the set of all integers, and $+, \cdot, 0, 1, <$ are defined in the usual way, does not have a model of type $(\aleph_2, \aleph_0)$.

Let (A, P) denote the enrichment of the algebraic system A by a predicate P, and let $<\Omega, P>$ be the signature obtained from Ω by the addition of the predicate symbol P. In many cases it is important to understand when in each member of a class $\mathscr{K}$ of algebraic systems of signature $<\Omega, P>$ the predicate P is given by a formula in the first-order language of the signature Ω. A partial answer to this question is given by *Beth's definability theorem*: There exists a formula $\Phi(x)$ in the first-order language of signature Ω such that the formula $(\forall x)(\Phi(x) \leftrightarrow P(x))$ is true for all members of an axiomatizable class $\mathscr{K}$ of signature $<\Omega, P>$ if and only if the set $\{P: (A, P) \in \mathscr{K}\}$ contains at most one element for each algebraic system A of signature Ω (see [2], [8]).

Much research in model theory is connected with the study of properties preserved under operations on algebraic systems. The most important operations include homomorphism, direct and filtered products.

A statement Φ is said to be *stable* with respect to homomorphisms if the truth of Φ in an algebraic system A implies the truth of Φ in all epimorphic images of A. A formula Φ in a first-order language is called *positive* if Φ does not contain negation and implication signs. It has been proved (see [1], [8]) that a statement Φ in a first-order language is stable relative to homomorphisms if and only if Φ is equivalent to a positive statement. A similar theorem holds for the language $L_{\omega_1\omega}$.

A formula $\Phi(x_1, \ldots, x_n)$ in a first-order language of signature Ω is called a *Horn formula* it it can be obtained by conjunction and quantification from formulas of the form $(\Phi_1 \& \cdots \& \Phi_s) \to \Phi$, $\neg(\Phi_1 \& \cdots \& \Phi_s)$, where $\Phi_1, \ldots, \Phi_s$ are atomic formulas in the first-order language of Ω. Examples of Horn formulas are identities and quasi-identities. Central in the theory of ultraproducts is the *theorem of J. Łos*: Every formula in a first-order language is stable with respect to any **ultrafilter** (see [1]). A formula in a first-order language is conditionally stable with respect to any filter if and only if it is equivalent to a Horn formula. There is the following theorem (see [9]): Two algebraic systems A and B of signature Ω are elementarily equivalent if and only if there is an ultrafilter D on a set I such that A^I / D and B^I / D are isomorphic. The cardinality of a filtered product is countably infinite if for each natural number n the number of factors of cardinality n is finite. If for each natural number n the set of indices for which the corresponding factors have cardinality n does not belong to D, then the cardinality of the ultraproduct with respect to a non-principal ultrafilter D on a countable set I is equal to that of the continuum. For each infinite set I of cardinality α there is a filter D on I such that for each filter D_1 on I containing D, and each infinite set A, the cardinality of A^I / D_1 is not less than 2^α (see [1]).

Many applications have been found for the *Ehrenfeucht — Mostrowski theorem* on the existence of models with a large number of automorphisms (see [3]): For any totally ordered set X in an axiomatizable class $\mathcal{X}$ of algebraic systems containing an infinite system, there is a system A such that $X \subseteq |A|$ and such that each order-preserving one-to-one mapping of X onto X can be extended to an automorphism of A.

The major notions in model theory are those of universal, homogeneous and saturated systems. Let A and B be algebraic systems of a signature Ω. A mapping f of a set $X \subseteq |A|$ into a set $Y \subseteq |B|$ is called *elementary* if for each formula $\Phi(x_1, \ldots, x_n)$ in the first-order language of the signature Ω and any $a_1, \ldots, a_n \in X$ the equivalence $A \vDash \Phi(a_1, \ldots, a_n) \Leftrightarrow B \vDash \Phi(fa_1, \ldots, fa_n)$ holds. A system A is called α-*universal* if for every system B that is elementarily equivalent to A and of cardinality not exceeding α, there is an elementary mapping from $|B|$ into $|A|$. A system A is called α-*homogeneous* if for every set $X \subseteq |A|$ of cardinality less than α, every elementary mapping from X into $|A|$ can be extended to an elementary mapping of $|A|$ onto $|A|$ (that is, to an **automorphism** of A). A system A of signature Ω is called α-*saturated* if for every set $X \subseteq |A|$ of cardinality less than α and every collection Σ of formulas in the first-order language of the signature $<\Omega, X>$ not containing free variables other than x_0, Σ finitely satisfiable in (A, X) implies that Σ is satisfiable in (A, X). A system is called *universal* (respectively, *homogeneous* or *saturated*) if A is α-universal (respectively, α-homogeneous or α-saturated), where α is the cardinality

of $|A|$. A system is saturated if and only if it is simultaneously universal and homogeneous. Two elementary equivalent saturated systems of the same cardinality are isomorphic (see [3]). All uncountable models of elementary theories which are categorical in uncountable cardinalities (cf. **Categoricity in cardinality**) and of countable signature are saturated (*Morley's theorem*, see [3], [8]). A large number of examples of α-saturated systems is given by ultraproducts. For example, if D is a non-principal ultrafilter on a countable set I, then $\prod_{i \in I} A_i / D$ is an $\aleph_1$-saturated system for any algebraic system A_i $(i \in I)$ of a countable signature Ω.

The basic problems of model theory are the study of the expressive possibilities of a formalized language and the study of classes of structures defined by means of such a language. Some important properties of stable theories have been found, and the classes of categorical and superstable theories have been studied in even more detail.

The basic apparatus for the study of stable theories is the classification of formulas and locally consistent sets of formulas in these theories. Such a classification can be obtained by means of ascribing to formulas their ranks. Such ranks are usually ordinals and the ranking functions are given with the help of special topologies and other means. The study of ranking functions and their improvements is a rich source of information on the theories.

In the study of classes of models one is concerned with the number of distinct models, up to isomorphism, of a theory of a given cardinality, the existence of special models, for example, simple, minimal, saturated, homogeneous, universal, etc., and one creates means for constructing them.

The classical examples of application of methods of model theory are the papers of A. Robinson and his school, which developed an independent science — **non-standard analysis**; from the work of Mal'tsev and his school the applications of model-theoretic methods to topological algebra have been developed; the latest results on the properties of stable theories have been used in the study of concrete algebraic questions.

The above problems arose also in the study of various non-elementary languages, for example, obtained by the addition of new quantifiers, the introduction of infinite expressions, modalities, etc.

References

[1] MAL'TSEV, A.I.: *Algebraic systems*, Springer, 1973 (translated from the Russian).
[2] ROBINSON, A.: *Introduction to model theory and to the metamathematics of algebra*, North-Holland, 1963.
[3] TAĬTSLIN, M.A.: *Model theory*, Novosibirsk, 1970 (in Russian).
[4] ERSHOV, YU.L.: *Decidability problems and constructive models*, Moscow, 1980 (in Russian).

[5] PALYUTIN, YU.A.: *Mathematical logic*, Moscow, 1979 (in Russian).

[6] ERSHOV, YU.L., LAVROV, I.A., TAĬMANOV, A.D. and TAĬTSLIN, M.A.: 'Elementary theories', *Russian Math. Surveys* **20**, no. 4 (1965), 35-105. (*Uspekhi Mat. Nauk* **20**, no. 4 (1965), 37-108)

[7] MAL'TSEV, A.I.: 'Some problems in the theory of classes of models', in *Proc. 4-th All-Union Math. Congress*, Vol. 1, Leningrad, 1963, pp. 169-198 (in Russian).

[8] CHANG, C.C. and KEISLER, H.: *Model theory*, North-Holland, 1973.

[9] SACKS, G.E.: *Saturated model theory*, Benjamin, 1972.

[10] VAUGHT, R.L.: 'Denumerable models of complete theories', in *Infinitistic methods. Proc. Symp. Foundations of Math. Warsaw, 1959*, Pergamon, 1961, pp. 303-321.

[11] MORLEY, M. and VAUGHT, R.: 'Homogeneous universal models', *Math. Scand.* **11**, no. 1 (1962), 37-57.

[12] MORLEY, M.: 'Categoricity in power', *Trans. Amer. Math. Soc.* **114**, no. 2 (1965), 514-538.

[13] SHELAH, S.: *Classification theory and the number of non-isomorphic models*, North-Holland, 1978.

[14] BELL, J.L. and SLOMSON, A.B.: *Models and ultraproducts: an introduction*, North-Holland, 1971.

A.D. Taĭmanov
M.A. Taĭtslin

Editorial comments. Theorem 1 was proved by K. Gödel in [A1].

Let $\mathfrak{A}_i = <A_i, \{R_n(i)\}>$, $i \in I$, be a collection of *relational structures* of the same *type* (*algebraic systems of the same signature*). (I.e. the A_i are sets, the $R_n(i)$ relations.) Let F be an **ultrafilter** on the index set I. Then the *ultraproduct* $(\prod_i A_i)/F$ is a relational structure of the same type. (Cf. (the editorial comments to) **Ultrafilter** for the definition of ultraproduct.)

A precise formulation of Łos' theorem is now as follows (cf. [A8] or [A9]). Let $f = <f_1, \ldots, f_n, \ldots>$ be a sequence of elements from $\prod A_i$, i.e. $f_n = (f_n(i))_{i \in I} \in \prod A_i$ for all n. Let f/F be the sequence $<f_1/F, \ldots, f_n/F, \ldots>$ of elements from $(\prod A_i)/F$ and let $f(i)$ be the sequence $f(i) = <f_1(i), \ldots, f_n(i)>$. Then for any formula ϕ from the language L for which the $\mathfrak{A}_i$ are interpretations,

$$(\prod_i A_i)/F \underset{f/F}{\vDash} \phi \Leftrightarrow \left\{ i \in I: \mathfrak{A}_i \underset{f(i)}{\vDash} \phi \right\} \in F.$$

Łos' theorem is also called the *fundamental theorem on ultraproducts*.

The theorem to the effect that two algebraic systems $\mathfrak{A}$ and $\mathfrak{B}$ are equivalent if and only if they have isomorphic ultrapowers is known as the *Keisler ultrapower theorem*.

One of the basic applications of logic to algebra is the work by J. Ax and S. Kochen [A2].

See [A3] for the model theory of infinitary languages; [A4], [A5] for stability theory; and [A6], [A7] for categorical model theory.

The Gödel compactness theorem and the Löwenheim−Skolem theorem are in the Russian literature sometimes known as the *Gödel−Mal'tsev theorem* and the *Löwenheim−Skolem−Mal'tsev theorem*, respectively.

References

[A1] GÖDEL, K.: 'Die Vollständigkeit der Axiome des logischen Funktionenkalküls', *Monatshefte für Math. und Physik* **37** (1930), 344-360.

[A2] AX, J. and KOCHEN, S.: 'Diophantine problems over local fields III: decidable fields', *Ann. of Math.* **83** (1966), 437-456.

[A3] DICKMANN, M.A.: *Large infinitary languages*, North-Holland, 1975.

[A4] PILLAY, A.: *An introduction to stability theory*, Oxford Univ. Press, 1985.

[A5] BALDWIN, T.: *Fundamentals of stability theory*, Springer, 1988.

[A6] MALEKAI, M. and REGES, G.E.: *First order categorical logic*, Springer, 1977.

[A7] LAMBEK, J. and SCOTT, P.: *Higher order categorical logic*, Cambridge Univ. Press, 1986.

[A8] BELL, J.L. and SLOMSON, A.B.: *Models and ultraproducts*, North-Holland, 1969.

[A9] COMFORT, W.W. and NEGREPONTIS, S.: *The theory of ultrafilters*, Springer, 1974, §11.

AMS 1980 Subject Classification: 03CXX

MODIFICATION *of an analytic space* - An analytic mapping $f: X \to Y$ of analytic spaces such that for certain analytic sets $S \subset X$ and $T \subset Y$ of smaller dimensions, the conditions

$$f: X \setminus S \to Y \setminus T \text{ is an isomorphism}$$

and

$$f(S) = T$$

hold. A modification is also called a *contraction* of S onto T. An example of a modification is a **monoidal transformation**.

See also **Exceptional analytic set; Exceptional subvariety**.

References

[1] BEHNKE, H. and STEIN, K.: 'Modifikation komplexer Mannigfaltigkeiten und Riemannschen Gebiete', *Math. Ann.* **124**, no. 1 (1951), 1-16.

A.L. Onishchik

Editorial comments.

References

[A1] HARTSHORNE, R.: *Algebraic geometry*, Springer, 1977.

AMS 1980 Subject Classification: 14E15, 32C45

MODULAR CURVE - A complete **algebraic curve** $X_{\tilde{\Gamma}}$ uniformized by a subgroup $\tilde{\Gamma}$ of finite index in the **modular group** Γ; more precisely, a modular curve is a complete algebraic curve obtained from a quotient space $H/\tilde{\Gamma}$, where H is the upper half-plane, together with a finite number of parabolic points (the equivalence classes relative to $\tilde{\Gamma}$ of the rational points of the boundary of H). The best known examples of subgroups $\tilde{\Gamma}$ of finite index in Γ are the congruence subgroups containing a principal congruence subgroup $\Gamma(N)$ of level N for some integer $N > 1$, represented by the matrices

$$A \in \mathrm{SL}_2(\mathbf{Z}), \quad A \equiv \begin{bmatrix} 1 & 0 \\ 0 & 1 \end{bmatrix} \bmod N$$

(see **Modular group**). The least such N is called the *level of the subgroup* $\tilde{\Gamma}$. In particular, the subgroup $\Gamma_0(N)$ represented by matrices which are congruent $\bmod N$ to upper-triangular matrices has level N. Corresponding to each subgroup $\tilde{\Gamma}$ of finite index there is a covering of the modular curve $X_{\tilde{\Gamma}} \to X_{\Gamma}$, which ramifies only over the images of the points $z=i$, $z=(1+i\sqrt{3})/2$, $z=\infty$. For a congruence subgroup $\tilde{\Gamma}$ the ramification of this covering allows one to determine the genus of $X_{\tilde{\Gamma}}$ and to prove the existence of subgroups $\tilde{\Gamma}$ of finite index in Γ which are not congruence subgroups (see [4], Vol. 2, [2]). The genus of $X_{\Gamma(N)}$ is 0 for $N \leqslant 2$ and equals

$$1 + \frac{N^2(N-6)}{24} \prod_{p \mid N}(1 - p^{-2}),$$

p a prime number, for $N > 2$. A modular curve is always defined over an algebraic number field (usually over $\mathbf{Q}$ or a cyclic extension of it). The rational functions on a modular curve lift to modular functions (of a higher level) and form a field; the automorphisms of this field have been studied (see [2]). A holomorphic differential form on a modular curve $X_{\tilde{\Gamma}}$ is given on H by a differential $f(z)\,dz$ (where $f(z)$ is a holomorphic function) which is invariant under the transformations $z \to \gamma(z)$ of $\tilde{\Gamma}$; here $f(z)$ is a cusp form of weight 2 relative to $\tilde{\Gamma}$. The **zeta-function** of a modular curve is a product of the Mellin transforms (cf. **Mellin transform**) of modular forms and, consequently, has a meromorphic continuation and satisfies a functional equation. This fact serves as the point of departure for the Langlands–Weil theory on the relationship between modular forms and Dirichlet series (see [7], [8]). In particular, there is a hypothesis that each **elliptic curve** over $\mathbf{Q}$ (with conductor N) can be uniformized by modular functions of level N. The homology of a modular curve is connected with modular symbols, which allows one to investigate the arithmetic of the values of the zeta-function of a modular curve in the centre of the critical strip and to construct the p-adic zeta-function of a modular curve (see [1]).

A modular curve parametrizes a family of elliptic curves, being their moduli variety (see [7], Vol. 2). In particular, for $\tilde{\Gamma} = \Gamma(N)$ a point z of $H/\Gamma(N)$ is in one-to-one correspondence with a pair consisting of an elliptic curve E_z (analytically equivalent to a complex torus $\mathbf{C}/(\mathbf{Z}+\mathbf{Z}z)$) and a point of order N on E_z (the image of z/N).

Over each modular curve $X_{\tilde{\Gamma}}$ there is a natural algebraic fibre bundle $E_{\tilde{\Gamma}} \to X_{\tilde{\Gamma}}$ of elliptic curves if $\tilde{\Gamma}$ does not contain -1, compactified by degenerate curves above the parabolic points of $X_{\tilde{\Gamma}}$. Powers $E_{\tilde{\Gamma}}^{(w)}$, where $w \geqslant 1$ is an integer, are called *Kuga varieties* (see [3], [5]). The zeta-functions of $E_{\tilde{\Gamma}}^{(w)}$ are related to the Mellin transforms of modular forms, and their homology to the periods of modular forms (see [3], [7]).

The rational points on a modular curve correspond to elliptic curves having rational points of finite order

(or rational subgroups of points); their description (see [6]) made it possible to solve the problem of determining the possible torsion subgroups of elliptic curves over $\mathbf{Q}$.

The investigation of the geometry and arithmetic of modular curves is based on the use of groups of automorphisms of the projective limit of the curves $X_{\tilde{\Gamma}}$ with respect to decreasing $\tilde{\Gamma}$, which (in essence) coincides with the group $\mathrm{SL}_2(A)$ over the ring A of rational adèles. On each modular curve $X_{\tilde{\Gamma}}$ this gives a non-trivial ring of correspondences $R_{\tilde{\Gamma}}$ (a Hecke ring), which has applications in the theory of modular forms (cf. **Modular form**, [3]).

References

[1] MANIN, YU.I.: 'Parabolic points and zeta-functions of modular curves', *Math. USSR Izv.* **6**, no. 1 (1972), 19-64. (*Izv. Akad. Nauk SSSR Ser. Mat.* **36**, no. 1 (1972), 19-66)
[2] SHIMURA, G.: *Introduction to the arithmetic theory of automorphic functions*, Math. Soc. Japan, 1971.
[3] ŠOKUROV, V.V. [V.V. SHOKUROV]: 'Holomorphic differential forms of higher degree on Kuga's modular varieties', *Math. USSR Sb.* **30**, no. 1 (1976), 119-142. (*Mat. Sb.* **101**, no. 1 (1976), 131-157)
[4] KLEIN, F. and FRICKE, R.: *Vorlesungen über die Theorie der elliptischen Modulfunktionen*, 1-2, Teubner, 1890-1892.
[5] KUGA, M. and SHIMURA, G.: 'On the zeta function of a fibre variety whose fibres are abelian varieties', *Ann. of Math.* **82** (1965), 478-539.
[6] MAZUR, B. and SERRE, J.-P.: 'Points rationnels des courbes modulaires $X_0(N)$ (d'après A. Ogg)', in *Sem. Bourbaki 1974/1975*, Lecture notes in math., Vol. 514, Springer, 1976, pp. 238-255.
[7] *Modular functions of one variable. 1-6*, Lecture notes in math., 320; 349; 350; 476; 601; 627, Springer, 1973-1977.
[8] WEIL, A.: 'Ueber die Bestimmung Dirichletscher Reihen durch Funktionalgleichungen', *Math. Ann.* **168** (1967), 149-156.

A.A. Panchishkin
A.N. Parshin

Editorial comments.

References

[A1] KATZ, N.M. and MAZUR, B.: *Arithmetic moduli of elliptic curves*, Princeton Univ. Press, 1985.

AMS 1980 Subject Classification: 10D99, 14H45

MODULAR FORM of one complex variable, *elliptic modular form* - A function f on the upper half-plane $H = \{z \in \mathbf{C}: \operatorname{Im} z > 0\}$ satisfying for some fixed k the *automorphicity condition*

$$f\left(\frac{az+b}{cz+d}\right) = (cz+d)^k f(z) \qquad (1)$$

for any element

$$\begin{bmatrix} a & b \\ c & d \end{bmatrix} \in \mathrm{SL}_2(\mathbf{Z})$$

($\mathrm{SL}_2(\mathbf{Z})$ is the group of integer-valued matrices with determinant $ad-bc=1$), and such that

$$f(z) = \sum_{n=0}^{\infty} a_n q^n,$$

where $q = \exp(2\pi i z)$, $z \in H$, $a_n \in \mathbf{C}$. The integer $k \geqslant 0$ is called the *weight of the modular form f*. If $a_0 = 0$, then f is called a *parabolic modular form*. There is also [8] a definition of modular forms for all real values of k.

An example of a modular form of weight $k \geqslant 4$ is given by the Eisenstein series (see [4])

$$G_k(z) = \sum_{m_1, m_2 \in \mathbf{Z}}^{*} (m_1 + m_2 z)^{-k},$$

where the asterisk means that the pair $(m_1, m_2) = (0, 0)$ is excluded from summation. Here $G_k(z) \equiv 0$ for odd k and

$$G_k(z) = \frac{2(2\pi i)^k}{(k-1)!} \left[-\frac{B_k}{2k} + \sum_{n=1}^{\infty} \sigma_{k-1}(n) q^n \right],$$

where $\sigma_{k-1}(n) = \sum_{d \mid n} d^{k-1}$ and B_k is the k-th Bernoulli number (cf. **Bernoulli numbers**).

The set of modular forms of weight k is a complex vector space, denoted by M_k; in this connection, $M_k M_l \subset M_{k+l}$. The direct sum $\oplus_{k=0}^{\infty} M_k$ forms a **graded algebra** isomorphic to the ring of polynomials in the independent variables G_4 and G_6 (see [3]).

For each $z \in H$ the **complex torus** $\mathbf{C}/(\mathbf{Z} + \mathbf{Z}z)$ is analytically isomorphic to the **elliptic curve** given by the equation

$$y^2 = 4x^3 - g_2(z)x - g_3(z), \tag{2}$$

where $g_2(z) = 60G_4(z)$, $g_3(z) = 140G_6(z)$. The **discriminant** of the cubic polynomial on the right-hand side of (2) is a parabolic modular form of weight 12:

$$\frac{1}{2^4}(g_2^3 - 27g_3^2) = \frac{(2\pi)^{12}}{2^4} q \prod_{m=1}^{\infty} (1 - q^m)^{2k} = \frac{(2\pi)^{12}}{2^4} \sum_{n=1}^{\infty} \tau(n) q^n,$$

where $\tau(n)$ is the **Ramanujan function** (see [1]).

For each integer $N \geqslant 1$ *modular forms of higher level* N have been introduced, satisfying (1) only for elements

$$\begin{bmatrix} a & b \\ c & d \end{bmatrix}$$

of a congruence subgroup $\tilde{\Gamma}$ of level N of the modular group. In this case, related to the modular form f is the holomorphic differential $f(z)(dz)^{k/2}$ on the modular curve $X_{\tilde{\Gamma}}$. A well-known example of a modular form of higher level is the **theta-series** $f(z)$ associated to an integer-valued positive-definite quadratic form $F(x_1, \ldots, x_m)$:

$$f(z) = \sum_{x_1, \ldots, x_m \in \mathbf{Z}} \exp(2\pi i F(x_1, \ldots, x_m)),$$

which is a modular form of higher level and of weight $k = m/2$. In this example a_n is the integer equal to the number of solutions of the Diophantine equation $F(x_1, \ldots, x_m) = n$.

The theory of modular forms allows one to obtain an estimate, and sometimes a precise formula, for numbers of the type a_n (and congruences, such as the Ramanujan congruence $\tau(n) \equiv \sum_{d \mid n} d^{11} \pmod{691}$), and also to investigate their divisibility properties (see [7]). Best estimates for numbers of the type a_n have been obtained (see [2]).

Important arithmetic applications of modular forms are related to the Dirichlet series

$$L_f(s) = \sum_{n=1}^{\infty} a_n n^{-s},$$

i.e. the **Mellin transform** of f. Such Dirichlet series have been the subject of detailed study (estimates of coefficients, analyticity properties, the functional equation, Euler product expansion) in view of the presence of a non-trivial ring of correspondences R on a modular curve. For a curve X_{Γ} this ring is generated by the correspondence $T_n(z) = \sum_{\gamma} \gamma(z)$, where γ runs through the set of all representatives of the elements of the quotient set

$$SL(2, \mathbf{Z}) \setminus \{A \in M_2(\mathbf{Z}) : \det A = n\}.$$

The correspondences induce linear operators (*Hecke operators*) acting on the space of modular forms. They are self-adjoint relative to the Peterson scalar product (see [3], [7]). Modular forms which are eigen functions of the Hecke operators are characterized by the fact that their Mellin transforms have Euler product expansions.

Another direction in the theory of modular forms is related to the study of modular curves and the associated fibrations, the Kuga varieties (cf. **Modular curve**), and also to the theory of infinite-dimensional representations of algebraic adèle groups. Here the theory of modular forms of one variable was successfully transferred to the case of several variables (see [6]). A survey of the number-theoretic applications of modular forms is given in [5].

References
[1] HURWITZ, A. and COURANT, R.: *Vorlesungen über allgemeine Funktionentheorie und elliptische Funktionen*, 1, Springer, 1964, Chapt. 8.
[2] DELIGNE, P.: 'La conjecture de Weil I', *Publ. Math. IHES* **43** (1974), 273-307.
[3] LANG, S.: *Introduction to modular forms*, Springer, 1976.
[4] SERRE, J.-P.: *A course in arithmetic*, Springer, 1973 (translated from the French).
[5] FENKO, O.M.: 'Applications of the theory of modular forms to number theory', *J. Soviet Math.* **14**, no. 4 (1977), 1307-1362. (*Itogi Nauk. i Tekhn. Algebra Topol. Geom.* **15** (1977), 5-91)
[6] *Modular functions of one variable. 1-6*, Lecture notes in math., 320; 349; 350; 476; 601; 627, Springer, 1973-1977.
[7] OGG, A.: *Modular forms and Dirichlet series*, Benjamin, 1969.
[8] RANKIN, R.: *Modular forms and functions*, Cambridge Univ. Press, 1977.

A.A. Panchishkin

Editorial comments. A parabolic modular form is also called a *cusp form*.

AMS 1980 Subject Classification: 10D12

MODULAR FUNCTION, *elliptic modular function*, of one complex variable - An **automorphic function** of a complex variable $z = x + iy$, associated with the group Γ of all fractional-linear transformations γ of the form

$$z \to \gamma(z) = \frac{az+b}{cz+d}, \quad ad - bc = 1, \tag{1}$$

where a, b, c, d are real integers (this group is called the **modular group**). The transformations of Γ transform the real axis into itself and the domain of definition of a modular function can be regarded as being the upper half-plane $\{z: \operatorname{Im} z > 0\}$. The group Γ is generated by the two transformations $z \to z+1$, $z \to -1/z$.

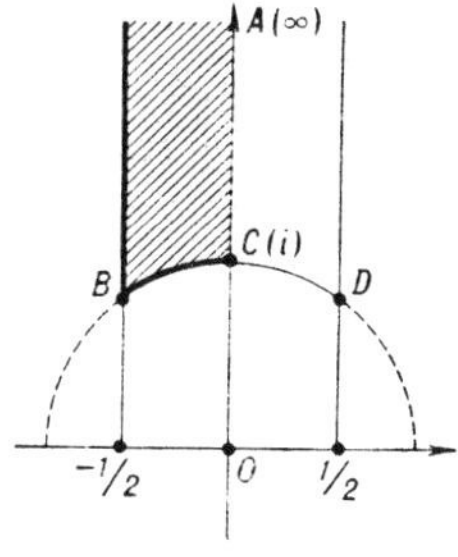

Fig. 1.

A fundamental domain G of the modular group is depicted in Fig. 1; this is the curvilinear quadrangle $ABCDA$ with vertices $A(\infty)$, $B(-1/2+i\sqrt{3}/2)$, $C(i)$, $D(1/2+i\sqrt{3}/2)$ two sides of which, AB and DA, are segments of the lines $x=-1/2$ and $x=1/2$, respectively, and BD is an arc of the circle $|z|=1$. AB and BC are included in G, CD and DA are not. The images of G under all possible mappings of Γ cover the half-plane $\operatorname{Im} z > 0$ without intersections.

The study of modular functions began in the 19-th century in connection with the study of elliptic functions and preceded the appearance of the general theory of automorphic functions. In the theory of modular functions the following **theta-series** are used as basic modular forms:

$$g_2(z) = 60 \sum_{m_1, m_2 \in \mathbb{Z}}^{*} \frac{1}{(m_1+m_2 z)^4} =$$

$$= (2\pi)^4 \left\{ \frac{1}{12} + 20 \sum_{n=1}^{\infty} \frac{n^3 h^{2n}}{1-h^{2n}} \right\},$$

$$g_3(z) = 140 \sum_{m_1, m_2 \in \mathbb{Z}}^{*} \frac{1}{(m_1+m_2 z)^6} =$$

$$= (2\pi)^6 \left\{ \frac{1}{216} - \frac{7}{3} \sum_{n=1}^{\infty} \frac{n^5 h^{2n}}{1-h^{2n}} \right\},$$

$$\Delta(z) = g_2^3(z) - 27 g_3^2(z),$$

where $h = e^{i\pi z}$ and the asterisk means that the null pair $(m_1, m_2) = (0, 0)$ is omitted. According to the terminology of K. Weierstrass these are *relative invariants*, playing a major role in his theory of elliptic functions (see **Weierstrass elliptic functions**), and Δ is also called the *discriminant*. From the point of view of the theory of automorphic functions (cf. **Automorphic function; Automorphic form**) these are automorphic forms of weights 2, 3 and 6, respectively, associated with the modular group. The fundamental modular form has the form

$$J(z) = \frac{g_2^3(z)}{\Delta(z)} = 1 + \frac{27 g_3^2(z)}{\Delta(z)}. \qquad (2)$$

$J(z)$ is also called the *absolute invariant*. It is regular in the upper half-plane and in the interior of the fundamental domain G it takes each finite value, except 0 and 1, precisely once; in addition, $J(-1/2+i\sqrt{3}/2)=0$, $J(i)=1$.

The modular function $J(z)$ plays a major role in the theory of elliptic functions, allowing one to determine the periods $2\omega_1$, $2\omega_3$ with respect to given Weierstrass relative invariants $a=g_2$, $b=g_3$, $a^3-27b^2 \neq 0$, and, consequently, to construct all Weierstrass elliptic functions. If τ is the unique solution in the fundamental domain of the equation

$$J(\tau) = \frac{a^3}{(a^3-27b^2)},$$

then for $a \neq 0$, $b \neq 0$ one has $\omega_1^2 = a/b$, $\omega_3 = \omega_1 \tau$; for $a=0$ one has $\tau = -1/2 + i\sqrt{3}/2$, and ω_1 is determined by the equation

$$\omega_1^6 = \frac{140}{b} \sum_{m_1, m_2 \in \mathbb{Z}}^{*} \frac{1}{(m_1+m_2\tau)^6}, \quad \omega_3 = \omega_1 \tau;$$

for $b=0$ one has $\tau=i$, and ω_1 is determined by the equation

$$\omega_1^4 = \frac{60}{a} \sum_{m_1, m_2 \in \mathbb{Z}}^{*} \frac{1}{(m_1+m_2\tau)^4}, \quad \omega_3 = \omega_1 \tau.$$

For the construction of **Jacobi elliptic functions**, instead of $J(z)$ it is more convenient to use

$$\lambda(z) = k^2(z) = 1 - \prod_{n=1}^{\infty} \left[\frac{1-h^{2n-1}}{1+h^{2n+1}} \right]^8, \quad h = e^{2\pi i z}, \qquad (3)$$

also called a modular function. By the same token, $\lambda(z)$ is an automorphic function only relative to the subgroup Γ_2 of Γ, where Γ_2 consists of all transformations of the form (1) in which (as an extra condition) a and d are odd numbers and b and c are even. The fundamental domain G_2 of Γ_2 is depicted in Fig. 2; this is the curvilinear quadrangle $ABOCA$ with vertices $A(\infty)$, $B(-1)$, 0, $C(1)$, two sides of which, AB and CA, are segments of the lines $x=-1$ and $x=1$, respectively, and BO and OC are arcs of the circles $|z+1/2|=1/2$ and $|z-1/2|=1/2$, respectively. The parts of the boundary to the left of the imaginary axis are included and OC and CA are not included.

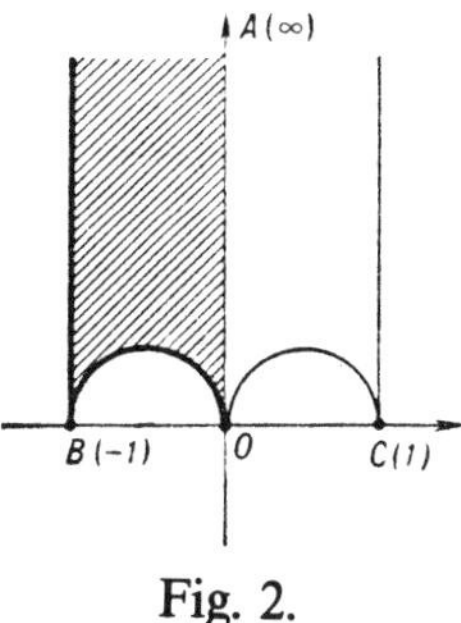

Fig. 2.

The function $\lambda(z)$ is also regular in the upper half-plane

$\operatorname{Im} z > 0$. In the interior of G_2 it takes each finite value, except 0 and 1, precisely once; in addition, $\lambda(\infty) = 0$ and $\lambda(0) = 1$. For the construction of a Jacobi elliptic function of given modulus k the value $\tau = \omega_3 / \omega_1$, or $h = e^{\pi i z}$, uniquely defined by the equation $\lambda(\tau) = k^2$, is required. In practice, in the normal case $0 < k < 1$, one first determines $\epsilon = (1 - \sqrt{k'}) / 2(1 + \sqrt{k'})$, where $k' = \sqrt{(1 - k^2)}$, and then constructs a solution of this equation in the form of a series $h = \epsilon + 2\epsilon^5 + 15\epsilon^9 + 150\epsilon^{13} + O(\epsilon^{17})$. The modular functions $\lambda = \lambda(z)$ and $J(z)$ are related by

$$J(z) = \frac{4}{27} \cdot \frac{\lambda^2 - \lambda + 1}{\lambda^2 (1 - \lambda)^2}.$$

The modular function $w = J(z)$ gives the most convenient representation of the conformal classes of Riemann surfaces of elliptic functions (cf. **Riemann surfaces, conformal classes of**), when the genus $g = 1$ and the Euler characteristic $\chi = 0$. Corresponding to each w there is a solution $\tau = \omega_3 / \omega_1$ of $J(\tau) = w$, which determines a conformal class and the corresponding field of elliptic functions. For example, $w = 0$ corresponds to a period parallelogram in the form of a rhombus with angles $120°$ and $60°$, and $w = 1$ corresponds to a square. Modular functions have also been applied in the study of **conformal mapping; boundary properties of analytic functions** and cluster sets (cf. **Cluster set**). The modular function $w = J(z)$ gives a conformal mapping of the left half of the fundamental domain G (Fig. 1), that is, the curvilinear triangle $ABCA$, onto the upper half-plane $\operatorname{Im} w > 0$, where B, C and A are mapped to 0, 1 and ∞, respectively. The modular function $w = \lambda(z)$ conformally maps the curvilinear triangle $ABOA$ (Fig. 2) onto the upper half-plane, where B, O and A are mapped to 0, 1 and ∞, respectively.

In geometric questions it is often more convenient to take the unit disc as the domain of the modular functions. The modular group (1) is then replaced by the modular group of automorphisms of the unit disc. For example, it is convenient to apply the fractional-linear transformation

$$\zeta = \zeta(z) = \frac{i(z - e^{i\pi/3})}{z - e^{-i\pi/3}},$$

which maps the upper half-plane $\operatorname{Im} z > 0$ onto the unit

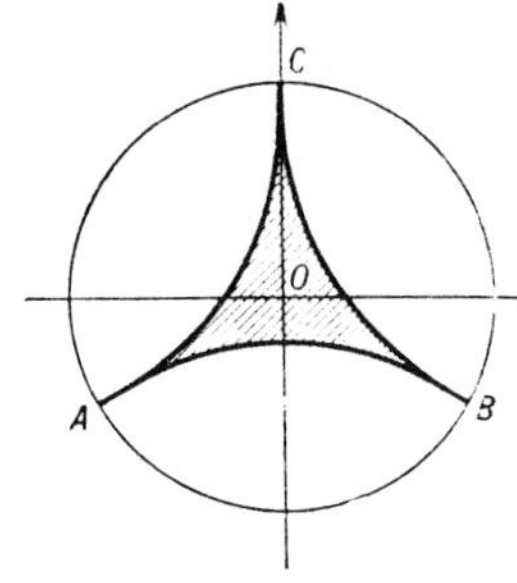

Fig. 3.

disc $|\zeta| < 1$, where 0, 1 and ∞ are mapped to $A(e^{-i5\pi/6})$, $B(e^{-i\pi/6})$ and $C(i)$, respectively, on the unit circle $|\zeta| = 1$ (Fig. 3).

Then the composite function $w = \mu(\zeta) = \lambda(z(\zeta))$ is a modular function that is regular in the unit disc and takes there all values except 0, 1 and ∞. It conformally maps the curvilinear triangle $ABCA$ (Fig. 3) onto the upper half-plane $\operatorname{Im} z > 0$. It is precisely this modular function $\mu(\zeta)$ that is used in the proof of the **Picard theorem** and in a number of geometric questions.

References

[1] HURWITZ, A. and COURANT, R.: *Vorlesungen über allgemeine Funktionentheorie und elliptische Funktionen*, 1, Springer, 1964, Chapt. 8.
[2] AKHIEZER, N.I.: *Elements of the theory of elliptic functions*, Moscow, 1970 (in Russian).
[3] FORD, L.R.: *Automorphic functions*, Chelsea, reprint, 1951.
[4] KLEIN, F. and FRICKE, R.: *Vorlesungen über die Theorie der elliptischen Modulfunktionen*, 1-2, Teubner, 1890-1892.

E.D. Solomentsev

Editorial comments. When considering the upper half-plane and the modular group acting on it, the point $i\infty$ and the rational points on the real axis are often referred to as cusps.

More generally, consider the group $GL_2(\mathbb{C})$ of invertible complex (2×2)-matrices, $g = \left(\begin{smallmatrix} a & b \\ c & d \end{smallmatrix}\right)$, and the corresponding *fractional-linear transformations*

$$z \mapsto \frac{az + b}{cz + d}, \quad z \in \mathbb{C} \bigcup \{\infty\}. \qquad (A1)$$

The fractional-linear transformation (A1) is called *parabolic* if it is $\neq$ id and the associated matrix has two equal eigen values (cf. also **Fractional-linear mapping**). This is equivalent to saying that the Jordan canonical form is of the form $\left(\begin{smallmatrix} \lambda & 1 \\ 0 & \lambda \end{smallmatrix}\right)$, or, if $\det(g) = 1$ is also imposed, that $\operatorname{Tr}(g) = \pm 2$. Now let Γ' be some discrete subgroup of $GL_2(\mathbb{R})$. A point $x \in \mathbb{R} \bigcup \{\infty\}$ is called a *cusp* of Γ' if there is a parabolic element of Γ' which has x as a fixed point.

The cusps of $\Gamma = SL_2(\mathbb{Z}) \subset GL_2(\mathbb{R})$ are precisely the points of $\mathbb{Q} \bigcup \{\infty\}$. To aid visualization, cf. Fig. 1, the point ∞ is written as $i\infty$ (= the point A in Fig. 1).

Let H^* be the *extended upper half-plane* $H \bigcup \mathbb{Q} \bigcup \{i\infty\}$, where $H = \{z \in \mathbb{C} : \operatorname{Im} z > 0\}$. The action of Γ on H naturally extends to H^*, and all the points of $\mathbb{Q} \bigcup \{\infty\}$ form one orbit. The translates of the fundamental region G of Fig. 1 form a tesselation of H (or of H^*), called the *modular tesselation*. Each translate γG, $\gamma \in \Gamma$, is called a *modular triangle*. In the special points γB, $B = -1/2 + i\sqrt{3}/2$, $\gamma \in \Gamma$, six modular triangles meet; in the special points γC, $C = i$, $\gamma \in \Gamma$, two modular triangles meet; and in a cusp (a point of $\mathbb{Q} \bigcup \{i\infty\}$) countably infinite many modular triangles meet (at angle 0; whence the terminology 'cusp').

The modular functions form a field. Indeed, this is the field $\mathbb{C}(J)$, where J is the *fundamental modular function* (2) above.

Let Γ_1 be a subgroup of finite index in Γ. The quotient H^* / Γ_1 can be given a natural complex structure making it a compact Riemann surface, cf. e.g. [A1], Chapt. IV, § 6. This is a natural compactification of H/Γ_1. For $\Gamma_1 = \Gamma$ one finds the Riemann sphere (of genus zero). For the principal congruence subgroups

$$\Gamma(N) = \left\{ \begin{bmatrix} a & b \\ c & d \end{bmatrix} \in \Gamma: a \equiv d \equiv 1 \bmod N,\ b \equiv c \equiv 0 \bmod N \right\},$$

the resulting quotients $H^*/\Gamma(N)$, the *modular curves* $X(N)$, for $N = 2, \ldots, 12$ have genus 0, 0, 0, 0, 1, 3, 5, 10, 13, 26, 25, respectively. For the general formula cf. **Modular curve**.

A *modular function for a subgroup* of finite index Γ_1 of Γ is a complex meromorphic function f on H^* such that $f(S(\tau)) = f(\tau)$ for $\tau \in H^*$, $S \in \Gamma_1$, and such that at a rational cusp $-d/c$, $(c, d) = 1$, f admits an expansion of the form

$$f(\tau) = \sum_{n \geqslant n_0} c_n \exp\left[\frac{2\pi i}{K} A(\tau) n\right],$$

for some $A = \left(\begin{smallmatrix} a & b \\ c & d \end{smallmatrix}\right) \in \Gamma$, natural number K and $n_0 \in \mathbf{Z}$. This is valid for $\tau \in H$ with $\operatorname{Im} A(\tau)$ large enough. This last condition reflects the requirement that f also defines a meromorphic function on the compactification H^*/Γ_1 of H/Γ_1, cf. **Automorphic function**. In case $\Gamma_1 = \Gamma$ this last requirement takes the following form: There is an $a > 0$ such that for $\operatorname{Im} \tau > a$, $\tau \in H$, $f(\tau)$ has an expansion of the form

$$f(\tau) = \sum_{n \geqslant n_0} b_n \exp(2\pi i n \tau).$$

References

[A1] SCHOENEBERG, B.: *Elliptic modular functions*, Springer, 1974.
[A2] SHIMURA, G.: *Introduction to the arithmetic theory of automorphic functions*, Princeton Univ. Press, 1971.
[A3] RANKIN, R.A.: *Modular forms and functions*, Cambridge Univ. Press, 1977.
[A4] LANG, S.: *Elliptic functions*, Addison-Wesley, 1973.

AMS 1980 Subject Classification: 10D12

MODULAR GROUP - The group Γ of all fractional-linear transformations γ of the form

$$z \to \gamma(z) = \frac{az+b}{cz+d}, \quad ad - bc = 1, \tag{1}$$

where a, b, c, d are rational integers. The modular group can be identified with the quotient group $SL_2(\mathbf{Z})/\{\pm E\}$, where

$$E = \begin{bmatrix} 1 & 0 \\ 0 & 1 \end{bmatrix},$$

and is a **discrete subgroup** in the **Lie group** $PSL_2(\mathbf{R}) = SL_2(\mathbf{R})/\{\pm E\}$. Here $SL_2(\mathbf{R})$ (respectively, $SL_2(\mathbf{Z})$) is the group of matrices

$$\begin{bmatrix} a & b \\ c & d \end{bmatrix},$$

with a, b, c, d real numbers (respectively, integers) and $ad - bc = 1$. The modular group is a **discrete group of transformations** of the complex upper half-plane $H = \{z = x + iy: y > 0\}$ (sometimes called the *Lobachevskiĭ plane* or *Poincaré upper half-plane*) and has a presentation with generators $T: z \to z + 1$ and $S: z \to -1/z$, and relations $S^2 = (ST)^3 = 1$, that is, it is

the free product of the cyclic group of order 2 generated by S and the cyclic group of order 3 generated by ST (see [2]).

Interest in the modular group is related to the study of modular functions (cf. **Modular function**) whose Riemann surfaces (cf. **Riemann surface**) are quotient spaces of H/Γ, identified with a fundamental domain G of the modular group. The compactification $X_\Gamma = (H/\Gamma) \bigcup \infty$ is analytically isomorphic to the complex projective line, where the isomorphism is given by the fundamental modular function $J(z)$. The fundamental domain G has finite Lobachevskiĭ area:

$$\int_G y^{-2}\, dx\, dy = \frac{\pi}{3},$$

that is, the modular group is a **Fuchsian group** of the first kind (see [3]). For the lattice $L = \mathbf{Z} + \mathbf{Z}z$, $z \in H$, the lattice $L_1 = \mathbf{Z} + \mathbf{Z}\gamma(z)$,

$$\gamma = \begin{bmatrix} a & b \\ c & d \end{bmatrix} \in \Gamma,$$

is equivalent to L, that is, can be obtained from L by multiplying the elements of the latter by a non-zero complex number λ, $\lambda = (cz + d)^{-1}$.

Corresponding to each lattice there is a complex torus $\mathbf{C}/L$ that is analytically equivalent to a non-singular cubic curve (an **elliptic curve**). This gives a one-to-one correspondence between the points of the quotient space H/Γ, classes of equivalent lattices and the classes of (analytically) equivalent elliptic curves (see [3]).

The investigation of the subgroups of the modular group is of interest in the theory of modular forms and algebraic curves (cf. **Algebraic curve**; **Modular form**). The *principal congruence subgroup $\Gamma(N)$ of level $N \geqslant 1$* (N an integer) is the group of transformations $\gamma(z)$ of the form (1) for which $a \equiv d \equiv 1 \pmod{N}$, $c \equiv b \equiv 0 \pmod{N}$. A subgroup $\tilde{\Gamma} \subset \Gamma$ is called a *congruence subgroup* if $\tilde{\Gamma} \supset \Gamma(N)$ for some N; the least such N is called the *level* of $\tilde{\Gamma}$. Examples of congruence subgroups of level N are: the group $\Gamma_0(N)$ of transformations (1) with c divisible by N, and the group $\Gamma_1(N)$ of transformations (1) with $a \equiv d \equiv 1 \pmod{N}$ and $c \equiv 0 \pmod{N}$. The **index** of $\Gamma(N)$ in the modular group is $(N^3/2)\prod_{p \mid N}(1 - p^{-2})$ if $N > 2$, p is a prime number, and 6 if $N = 2$; thus, each congruence subgroup has finite index in the modular group.

Corresponding to each subgroup $\tilde{\Gamma}$ of finite index in the modular group there is a complete algebraic curve $X_{\tilde{\Gamma}}$ (a **modular curve**), obtained from the quotient space $H/\tilde{\Gamma}$ and the covering $X_{\tilde{\Gamma}} \to X_\Gamma$. The study of the branches of this covering allows one to find generators and relations for the congruence subgroup $\tilde{\Gamma}$, the genus of $X_{\tilde{\Gamma}}$ and to prove that there are subgroups of finite index in the modular group which are not congruence subgroups (see [3], [8], [7], Vol. 2). The study of presentations of the modular group was initiated in work (see [4], [6]) connected with the theory of modular forms. Such presentations were intensively studied within the theory of automorphic forms (see [7] and **Automorphic form**). Many results related to the modular group have been transferred to the case of arithmetic subgroups of algebraic Lie groups (cf. **Arithmetic group; Lie algebra, algebraic**).

References

[1] HURWITZ, A. and COURANT, R.: *Vorlesungen über allgemeine Funktionentheorie und elliptische Funktionen*, 1, Springer, 1964, Chapt. 8.

[2] SERRE, J.-P.: *A course in arithmetic*, Springer, 1973 (translated from the French).

[3] SHIMURA, G.: *Introduction to the arithmetic theory of automorphic functions*, Math. Soc. Japan, 1971.

[4] HECKE, E.: 'Analytische Arithmetik der positiven quadratischen Formen', in *Mathematische Werke*, Vandenhoeck & Ruprecht, 1959, pp. 789-918.

[5] KLEIN, F. and FRICKE, R.: *Vorlesungen über die Theorie der elliptischen Modulfunktionen*, 1-2, Teubner, 1890-1892.

[6] KLOOSTERMAN, H.D.: 'The behaviour of general theta functions under the modular group and the characters of binary modular congruence groups I, II', *Ann. of Math.* **47** (1946), 317-375; 376-417.

[7] *Modular functions of one variable. 1-6*, Lecture notes in math., 320; 349; 476; 601; 627, Springer, 1973-1977.

[8] RANKIN, R.: *Modular forms and functions*, Cambridge Univ. Press, 1977.

A.A. Panchishkin

AMS 1980 Subject Classification: 10D07, 20H05

MODULAR IDEAL - A right (left) **ideal** J of a ring R having the following property: There is at least one element e in R such that for all x from R the difference $x - ex$ belongs to J (respectively, $x - xe \in J$). The element e is called a *left (right) identity modulo the ideal J*. In a ring with identity every ideal is modular. Every proper modular right (left) ideal can be imbedded in a maximal right (left) ideal, which is automatically modular. The intersection of all maximal modular right ideals of an associative ring coincides with the intersection of all maximal left modular ideals and is the **Jacobson radical** of the ring. Modular ideals are also called *regular ideals*.

References

[1] JACOBSON, N.: *Structure of rings*, Amer. Math. Soc., 1956.

K.A. Zhevlakov

AMS 1980 Subject Classification: 16A99

MODULAR LATTICE, *Dedekind lattice* - A **lattice** in which the *modular law* is valid, i.e. if $a \leqslant c$, then $(a+b)c = a + bc$ for any b. This requirement amounts to saying that the identity $(ac+b)c = ac + bc$ is valid. Examples of modular lattices include the lattices of subspaces of a linear space, of normal subgroups (but

not all subgroups) of a group, of ideals in a ring, etc. A lattice with a **composition sequence** is a modular lattice if and only if there exists on it a *dimension function d*, i.e. an integer-valued function such that $d(x+y) + d(xy) = d(x) + d(y)$ and such that if the interval $[a, b]$ is prime, it follows that $d(b) = d(a) + 1$. If $w = a_1^{(1)} \cdots a_{m_1}^{(1)} = a_1^{(2)} \cdots a_{m_2}^{(2)}$, if none of the elements $a_i^{(k)}$ can be represented as a product of elements other than itself and if

$$a_1^{(k)} \cdots a_{i-1}^{(k)} a_{i+1}^{(k)} \cdots a_{mk}^{(k)} \not\leqslant a_i^{(k)},$$

then $m_1 = m_2$ and for any $a_i^{(1)}$ it is possible to find an element $a_j^{(2)}$ such that

$$w = a_1^{(1)} \cdots a_{i-1}^{(1)} a_j^{(2)} a_{i+1}^{(1)} \cdots a_{m_1}^{(1)},$$

[3], [6]. Non-zero elements $a_1, \ldots, a_n$ of a modular lattice with a zero 0 are said to be *independent* if $(a_1 + \cdots + a_{i-1} + a_{j+1} + \cdots + a_n)a_i = 0$ for all i. This definition makes it possible to generalize many properties of systems of linearly independent vectors [3], [5], [6]. If $a_1, \ldots, a_n$ are independent, their sum is denoted by $a_1 \oplus \cdots \oplus a_n$. *Ore's theorem*: If a modular lattice has a composition sequence and if

$$1 = a_1^{(1)} \oplus \cdots \oplus a_{m_1}^{(1)} = a_1^{(2)} \oplus \cdots \oplus a_{m_2}^{(2)},$$

none of the elements $a_i^{(k)}$ being representable as a sum of two independent elements, then $m_1 = m_2$ and for each $a_i^{(1)}$ it is possible to find an element $a_j^{(2)}$ such that

$$1 = a_1^{(1)} \oplus \cdots \oplus a_{i-1}^{(1)} \oplus a_j^{(2)} \oplus a_{i+1}^{(1)} \oplus \cdots \oplus a_{m_1}^{(1)},$$

[3], [6]. In the case of completely modular lattices (cf. also **Complete Dedekind lattice**), which must satisfy certain additional requirements, the theorems on independent elements and direct decompositions may be applied to infinite sets as well [4], [5]. *Complemented modular lattices* have been studied; these are modular lattices with a 0 and a 1 in which for each element x there exists at least one element y (said to be a *complement of the element x*) such that $x + y = 1$, $xy = 0$. A complemented modular lattice which has a composition sequence, is isomorphic to the modular lattice of all subspaces of a finite-dimensional linear space over some skew-field. A complemented completely modular lattice L is isomorphic to the modular lattice of all subspaces of a linear (not necessarily finite-dimensional) space over some skew-field if and only if the following conditions are met: a) if $0 \neq a \in L$, it is possible to find an **atom** $p \leqslant a$; b) if p is an atom and $p \leqslant \sup A$, where $A \subseteq L$, then $p \leqslant \sup F$ for some finite set $F \subseteq A$; c) if p, q are distinct atoms, it is possible to find a third atom $r \leqslant p + q$; and d) there exist at least three independent atoms. The last condition d) may be replaced by the requirement that the **Desargues assumption** be valid [2]. A further generalization of this result, which leads to regular rings [7], [5], is connected

with the theory of von Neumann algebras. For a modular lattice with a composition sequence the presence of complements is equivalent to the representability of the unit as a sum of atoms.

Modular lattices are (in the Soviet Union) also called Dedekind lattices, in honour of R. Dedekind, who was the first to formulate the modular law and established a number of its consequences [1].

References

[1] DEDEKIND, R.: 'Ueber die von drei Moduln erzeugte Dualgruppe', *Math. Ann.* **53** (1900), 371-403.
[2] BAER, R.: *Linear algebra and projective geometry*, Acad. Press, 1952.
[3] BIRKHOFF, G.: *Lattice theory*, Colloq. Publ., 25, Amer. Math. Soc., 1973.
[4] KUROSH, A.G.: *The theory of groups*, 1-2, Chelsea, 1955-1956 (translated from the Russian).
[5] SKORNYAKOV, L.A.: *Complemented modular lattices and regular rings*, Oliver & Boyd, 1964 (translated from the Russian).
[6] SKORNYAKOV, L.A.: *Elements of lattice theory*, Hindushtan Publ. Comp., 1977 (translated from the Russian).
[7] NEUMANN, J. VON: *Continuous geometry*, Princeton Univ. Press, 1960.

L.A. Skornyakov

Editorial comments. Modular lattices satisfying the Desargues assumption are called *Desarguesian lattices*. Complemented completely modular lattices satisfying the identity

$$a\sum_{i\in I} b_i = \sum_{i\in I} ab_i$$

whenever the set $\{b_i: i\in I\}$ is (upwards) directed, and the dual of the latter condition, are called *continuous geometries* [5], [7].

AMS 1980 Subject Classification: 06CXX

MODULE - An **Abelian group** with a ring of operators. A module is a generalization of a (linear) **vector space** over a **field** K, when K is replaced by a **ring**.

Let a ring A be given. An additive Abelian group M is called a *left A-module* if there is a mapping $A\times M\to M$ whose value on a pair (a,m), for $a\in A$, $m\in M$, written am, satisfies the axioms:

1) $a(m_1+m_2)=am_1+am_2$;

2) $(a_1+a_2)m=a_1m+a_2m$;

3) $a_1(a_2m)=(a_1a_2)m$. If A has a unit, then it is usual to require in addition that for any $m\in M$, $1m=m$. A module with this property is called *unitary* or *unital* (cf. **Unitary module**).

Right A-modules are defined similarly; axiom 3) is replaced by $(ma_1)a_2=m(a_1a_2)$. Any right A-module can be considered as a left A^{opp}-module over the ring A^{opp} anti-isomorphic to A; hence, corresponding to any result about right A-modules there is a result about left A^{opp}-modules, and conversely. When A is commutative, any left A-module can be considered as a right A-module and the distinction between left and right modules disappears. Below only left A-modules are discussed.

The simplest examples of modules (finite Abelian groups; they are **Z**-modules) were known already to C.F. Gauss as class groups of binary quadratic forms. The general notion of a module was first encountered in the 1860's till 1880's in the work of R. Dedekind and L. Kronecker, devoted to the arithmetic of algebraic number and function fields. At approximately the same time research on finite-dimensional associative algebras, in particular, group algebras of finite groups (B. Pierce, F. Frobenius), led to the study of ideals of certain non-commutative rings. At first the theory of modules was developed primarily as a theory of ideals of a ring. Only later, in the work of E. Noether and W. Krull, it was observed that it was more convenient to formulate and prove many results in terms of arbitrary modules, and not just ideals. Subsequent developments of the theory of modules were connected with the application of methods and ideas of the theory of categories (cf. **Category**), in particular, methods of **homological algebra**.

Examples of modules. 1) Any Abelian group M is a module over the ring of integers **Z**. For $a\in\mathbf{Z}$ and $m\in M$ the product am is defined as the result of adding m to itself a times.

2) When A is a field, the notion of a unitary A-module is exactly equivalent to the notion of a linear vector space over A.

3) An n-dimensional vector space V over a field K (provided with coordinates) can be considered as a module over the ring $M_n(K)$ of all $(n\times n)$-matrices with coefficients in K. For $v\in V$ and $X\in M_n(K)$ the product Xv is defined as multiplication of the matrix X by the column of coordinates of the vector v.

4) An associative ring (cf. **Associative rings and algebras**) A is a left A-module. Multiplication of elements of the ring by elements of the module is ordinary multiplication in A.

5) The set of differential forms on a smooth manifold X has the natural structure of a module over the ring of all smooth functions on X.

6) Connected with any Abelian group M is the associative ring with identity, End(M), of all endomorphisms of M. The group M has a natural End(M)-module structure.

If there is an A-module structure on M, for some ring A, then the mapping $m\to am$ is an endomorphism of M for any $a\in A$. Associating with the element $a\in A$ the endomorphism of M that it generates, one obtains a homomorphism ϕ of A into End(M). Conversely, any homomorphism $\phi:A\to\text{End}(M)$ defines the structure of an A-module on M. Thus, the specification of an A-module structure on an Abelian group M is equivalent to the specification of a homomorphism of rings $\phi:A\to\text{End}(M)$. Such a homomorphism is also called a *representation of the ring A*, and M is called a *representation module*. Connected with any representation ϕ is a two-sided ideal $\mathrm{Ann}(M) = \mathrm{Ker}\,\phi$, consisting of the $a \in A$ such that $am = 0$ for all $m \in M$. This ideal is called the *annihilator of the module* M. When $\mathrm{Ann}(M) = 0$, the representation is called *faithful* and M is called a *faithful module* (or *faithful representation*).

It is obvious that a module M can also be considered as a module over the quotient ring $A / \mathrm{Ann}(M)$. In particular, although the definition of a module does not assume the associativity of A, the ring $A / \mathrm{Ann}(M)$ is always associative. Therefore, in the majority of cases the discussion may be restricted to modules over associative rings. Everywhere below, unless stated otherwise, A is assumed to be associative.

G-modules. Let G be a **group**. An additive Abelian group M is called a *left G-module* if there is a mapping $G \times M \to M$ whose value at a pair (g, m), where $g \in G$, $m \in M$, is written as gm, and where for any $g \in G$ the mapping $m \to gm$ is an endomorphism of M; for any $g_1, g_2 \in G$, $m \in M$, $(g_1 g_2)m = g_1(g_2 m)$; and for all $m \in M$, $1m = m$, where 1 is the identity of G. For any $g \in G$ the mapping $m \to gm$ is an automorphism of the group M.

Right G-modules may be defined similarly.

Examples of G-modules. 1) Let K be a **Galois extension** of a field k with **Galois group** G. Then the additive and multiplicative groups of K have the natural structure of G-modules. If k is an algebraic number field, then other G-modules are: the additive group of the ring of integers of K, the group of units of K, the group of divisors and the **divisor class group** of K, etc. A module over a Galois group is called a *Galois module*.

2) Let an extension of an Abelian group M be given, that is, an **exact sequence** of groups

$$1 \to M \to F \to G \to 1,$$

where M is an Abelian **normal subgroup** of F and G is an arbitrary group. Then M can be given the natural structure of a G-module by putting, for $g \in G$, $m \in M$, $gm = \bar{g}m\bar{g}^{-1}$, where $\bar{g}$ is an inverse image of g in F.

When the group operation in the Abelian group M is written multiplicatively (for example, if M is the multiplicative group of a field), the notation m^g is also used instead of gm, that is, the action of G is written exponentially.

Let a G-module M be given. By associating with an element $g \in G$ the automorphism $m \to gm$ of M, a homomorphism of G into the group of invertible elements of the ring $\mathrm{End}(M)$ is obtained. Conversely, any homomorphims of G into the group of invertible elements of $\mathrm{End}(M)$ gives M the structure of a G-module.

The notions of a module over a ring and a G-module are closely connected. Namely, any G-module M can be regarded as a module over the group ring (cf. **Group algebra**) $\mathbf{Z}G$ if the action of G on M is extended linearly, that is, if one puts

$$\left[\sum a_i g_i\right] m = \sum a_i(g_i m),$$

where $a_i \in \mathbf{Z}$, $g_i \in G$, $m \in M$. Conversely, given a unitary $\mathbf{Z}G$-module structure on M, M may be regarded to be a G-module.

When M is simultaneously a K-module over a commutative ring K and a G-module, where the action of the elements of G on M commutes with the action of the elements of K, then M may be given the structure of a KG-module by linearly extending the action from G to KG. For example, if V is a linear vector space over a field K, then the specification of a KG-module structure on V is equivalent to giving a representation of G in V.

Using the standard involution $g \to g^{-1}$ in G, any left G-module M can be made into a right G-module by putting $mg = g^{-1}m$ for $m \in M$, $g \in G$. Similarly, any right G-module can be made into a left G-module.

Modules over a Lie algebra. Let $\mathfrak{g}$ be a **Lie algebra** over a commutative ring K and let M be a K-module. The specification of a $\mathfrak{g}$-module structure on M consists of the specification of a K-endomorphism $m \to gm$ of the group M for each $g \in \mathfrak{g}$, where the axiom

$$[g_1, g_2]m = g_1(g_2 m) - g_2(g_1 m)$$

holds for $g_1, g_2 \in \mathfrak{g}$, $m \in M$. This definition differs from that of an A-module given earlier. Giving a $\mathfrak{g}$-module structure on M is equivalent to giving a Lie algebra homomorphism of $\mathfrak{g}$ into the Lie algebra of the ring $\mathrm{End}(M)$. A module over a Lie algebra $\mathfrak{g}$ may also be regarded as a module in the usual sense over the **universal enveloping algebra** of $\mathfrak{g}$.

Constructions in the theory of modules. Starting from a given A-module it is possible to obtain new A-modules by a number of standard constructions. Thus, with any module M is associated the **lattice** of its submodules. For example, if A is considered as left module over itself, then its left submodules are precisely the left ideals in A. A number of important types of modules are defined in terms of the lattice of submodules. For example, the condition for termination of a descending (ascending) chain of submodules defines Artinian modules (respectively, Noetherian modules, cf. **Artinian module**; **Noetherian module**). The condition for absence of non-trivial submodules, that is, submodules other than 0 or M, defines irreducible or simple modules (cf. **Irreducible module**).

For a module M and any submodule N, the quotient group M / N can be given the structure of an A-module. This module is called the *quotient module of M over N*.

A homomorphism of A-modules is defined as an Abelian group **homomorphism** $f: M \to N$ commuting with multiplication by elements of A, that is, $f(am) = af(m)$ for all $m \in M$, $a \in A$. If two homomorphisms $f_1, f_2: M \to N$ are given, then their sum, defined by $(f_1 + f_2)(m) = f_1(m) + f_2(m)$, is again a homomorphism of A-modules. This addition gives an Abelian group structure to the set $\mathrm{Hom}_A(M, N)$ of all homomorphisms of M into N. For any homomorphism $f: M \to N$ the submodules $\mathrm{Ker}\, f$ (the kernel of f) and $\mathrm{Im}\, f$ (the image of f), and also the quotient modules $\mathrm{Coker}\, f = N / \mathrm{Im}\, f$ (the cokernel of f) and $\mathrm{Coim}\, f = M / \mathrm{Ker}\, f$ (the coimage of f) are defined. The modules $\mathrm{Im}\, f$ and $\mathrm{Coim}\, f$ are canonically isomorphic and therefore usually identified. For example, for any left ideal J of A the quotient module A / J is defined. The module A / J is irreducible if and only if J is a maximal left ideal (cf. **Maximal ideal**). If M is an irreducible A-module not annihilating the ring A, then M is isomorphic to A / J for some maximal left ideal J.

For any family of A-modules $\{M_i\}$, where i runs through some index set J, the direct sum and direct product of $\{M_i\}$ exist in the category of A-modules. Here an element of the direct product $\prod_{i \in J} M_i$ may be interpreted as a vector $(\dots, m_i, \dots)$ the components of which are indexed by J and where for each i, $m_i \in M_i$. The sum of such vectors and their multiplication by elements of the ring are defined componentwise. The direct sum $\sum_{i \in J} M_i$ of the family $\{M_i\}$ can be interpreted as the submodule of the direct product consisting of the vectors all components of which, except for finitely many, are equal to zero.

For a projective (inductive) system of A-modules the projective (inductive) limit of this system can be naturally equipped with the structure of an A-module. The direct product and direct sum may be considered as special cases of the notions of a projective and an inductive limit.

Generators and relations. Let X be a subset of an A-module M. The *submodule* generated by X is the intersection of the submodules of M which contain X. If this submodule coincides with M, then X is called a *family (system) of generators of the module M*. A module admitting a finite family of generators is called a *finitely-generated module*. For example, in a Noetherian ring any ideal is a finitely-generated module. A direct sum of a finite number of finitely-generated modules is again finitely generated. Any quotient module of a finitely-generated module is also finitely generated. For the construction of a system of generators for a module M, *Nakayama's lemma* often turns out to be useful: For any ideal $\mathfrak{A}$ contained in the radical of a ring A the condition $\mathfrak{A}M = M$ implies $M = 0$. In particular, under the conditions of Nakayama's lemma elements $m_1, \dots, m_r$ form a system of generators for M if their images generate the module $M / \mathfrak{A}M$. This is used particularly often when A is a **local ring** and $\mathfrak{A}$ is the maximal ideal in A.

Let M be a module with system of generators $\{x_i\}_{i \in J}$. Then a mapping $\phi: y_i \to x_i$ defines an epimorphism of the free A-module F with generators $\{y_i\}_{i \in I}$ onto M (F can be defined as the set of formal finite sums $\sum a_i y_i$, $a_i \in A$, and ϕ is extended from the generators to F by linearity). The elements of $R = \mathrm{Ker}\, \phi$ are called *relations between the generators* $\{x_i\}$ of M. If M can be represented as a quotient module of a finitely-generated free module F so that the module of relations R is also finitely generated, then M is called a *finitely-presented module*. For example, over a Noetherian ring A any finitely-generated module is finitely presented. In general, being finitely generated does not imply being finitely presented.

Change of rings. There are standard constructions which allow an A-module M to be considered as a module over some other ring. For example, let $\phi: B \to A$ be a homomorphism of rings. Then, putting $bm = \phi(b)m$, M can be considered as a B-module. The resulting B-module is said to be obtained by *base change* or, in particular in the case that B is a subring of A, by *restriction of scalars*. If M is a unitary A-module and ϕ takes the identity to the identity, M becomes a unitary B-module.

Let a ring homomorphism $\phi: A \to B$ and an A-module M be given. Then B may be given the structure of a (B, A)-module (cf. **Bimodule**) by putting $ba = b\phi(a)$ for $b \in B$, $a \in A$, and the left B-module $B \otimes_A M$ can be considered. One says that this module is obtained from M by *extension of scalars*.

The category of modules. The class of all modules over a given ring A with homomorphisms of modules as morphisms forms an **Abelian category**, denoted, for instance, by A-mod or Mod_A. The most important functors defined on this category are Hom (homomorphism) and $\otimes$ (tensor product). The functor Hom takes values in the category of Abelian groups and associates to a pair of A-modules M, N the group $\mathrm{Hom}_A(M, N)$. For $f: M_1 \to M$ and $\phi: N \to N_1$ the mappings

$$f': \mathrm{Hom}_A(M, N) \to \mathrm{Hom}_A(M_1, N)$$

and

$$\phi': \mathrm{Hom}_A(M, N) \to \mathrm{Hom}_A(M, N_1)$$

are defined in the obvious way; that is, the functor Hom is contravariant in its first argument and covariant in the second. When M or N carry a bimodule structure, the group $\mathrm{Hom}_A(M, N)$ has an additional module structure. If N is an (A, B)-module, $\mathrm{Hom}_A(M, N)$ is a right B-module and if M is an

(A, B)-module, then $\text{Hom}_A(M, N)$ is a left B-module.

The functor $\otimes_A$ takes a pair M, N, where M is a right A-module and N is a left A-module, to the tensor product $M \otimes_A N$ of M and N over A. This functor takes values in the category of Abelian groups and is covariant with respect to both M and N. When M or N is a bimodule, the group $M \otimes_A N$ may be equipped with an additional structure. Namely, if M is a (B, A)-module, $M \otimes_A N$ is a B-module, and if N is an (A, B)-module, then $M \otimes_A N$ is a right B-module. The study of the functors Hom and $\otimes$, and also of their derived functors, is one of the fundamental problems of homological algebra.

Many important types of modules can be characterized in terms of Hom and $\otimes$. Thus, a **projective module** M is defined by the requirement that the functor $\text{Hom}_A(M, X)$ (as a functor in X) is exact (cf. **Exact functor**). Similarly, an **injective module** N is defined by the requirement of exactness of $\text{Hom}_A(X, N)$ (in X). A **flat module** M is defined by the requirement of exactness of the functor $M \otimes_A X$.

A module over a given ring A can be considered from two points of view.

A) Modules can be studied from the point of view of their intrinsic structure. The fundamental problem here is the complete classification of modules, that is, the construction for each module of a system of invariants which characterizes the module up to an isomorphism, and, given a set of invariants, the ability to construct a module with those invariants. For certain types of rings such a description is possible. For example, if M is a finitely-generated module over the group ring KG of a finite group G, where K is a field of characteristic coprime with the order of G, then M is representable as a finite direct sum of irreducible submodules (M is completely reducible, cf. **Completely-reducible module**). This representation is unique up to an isomorphism (the choice of the irreducible modules is, in general, not unique). All irreducible submodules also admit a simple description: All of them are contained in the **regular representation** of G and are in one-to-one correspondence with the irreducible characters of the group. Modules over principal ideal rings and over Dedekind rings also have a simple description. Namely, any finitely-generated module M over a **principal ideal ring** A is isomorphic to a finite direct sum of modules of the form $A / \mathfrak{A}_i$, where $\mathfrak{A}_i$ are ideals of A (possibly null), and where $\mathfrak{A}_1 \subseteq \cdots \subseteq \mathfrak{A}_m \neq A$. The ideals $\mathfrak{A}_i$ are uniquely determined by this last condition. Thus, the set of invariants $\{\mathfrak{A}_i\}$ completely determines M. If M is a finitely-generated module over a **Dedekind ring** A, then $M = M_1 \oplus M_2$, where M_2 is a torsion module (periodic module) and M_1 is a **torsion-free module** (the choice of M_1 is not unique). The module M_2 is annihi-

lated by some ideal $\mathfrak{A}$ of A and, consequently, is a module over the principal ideal ring $A / \mathfrak{A}$ and admits the description given above; M_1 is representable in the form $(\oplus^n A) \oplus \mathfrak{B}$, where $\mathfrak{B}$ is an ideal of A and $\oplus^n$ is the n-fold direct sum. The module M_1 is, up to an isomorphism, determined by two invariants: the number n and the class of $\mathfrak{B}$ in the ideal class group.

B) Another approach to the study of modules consists of studying the category A-mod and in considering a given module M as an object of this category. Such a study is the object of **homological algebra** and **algebraic K-theory**. On this route many important and deep results have been found.

Often, modules which carry some extra structure are considered. Thus one considers graded modules, filtered modules, topological modules, modules with a **sesquilinear form**, etc. (cf. **Graded module**; **Topological module**; **Filtered module**).

References

[1] BOURBAKI, N.: *Elements of mathematics. Algebra: Algebraic structures. Linear algebra*, 1, Addison-Wesley, 1974, Chapt. 1; 2 (translated from the French).
[2] BOURBAKI, N.: *Elements of mathematics. Commutative algebra*, Addison-Wesley, 1972 (translated from the French).
[3] BOURBAKI, N.: *Elements of mathematics. Lie groups and Lie algebras*, Addison-Wesley, 1975 (translated from the French).
[4] LANG, S.: *Algebra*, Addison-Wesley, 1984.
[5] WAERDEN, B.L. VAN DER: *Algebra*, 1-2, Springer, 1967-1971 (translated from the German).
[6] KOSTRIKIN, A.I.: *An introduction to algebra*, Moscow, 1977 (in Russian).
[7] JACOBSON, N.: *Structure of rings*, Amer. Math. Soc., 1956.
[8] HERSTEIN, I.: *Noncommutative rings*, Math. Assoc. Amer., 1968.
[9] FAITH, C.: *Algebra: rings, modules and categories*, I-II, Springer, 1981-1976.
[10] CARTAN, H. and EILENBERG, S.: *Homological algebra*, Princeton Univ. Press, 1956.
[11] MACLANE, S.: *Homology*, Springer, 1963.
[12] BASS, H.: *Algebraic K-theory*, Benjamin, 1968.
[13] MILNOR, J.: *Introduction to algebraic K-theory*, Princeton Univ. Press, 1971.

L.V. Kuz'min

AMS 1980 Subject Classification: 13CXX, 16-XX

MODULES, CATEGORY OF - The **category** mod-R whose objects are the right unitary modules over an arbitrary associative ring R with identity, and whose morphisms are the homomorphisms of R-modules. This category is the most important example of an **Abelian category**. Moreover, for every small Abelian category there is a full exact imbedding into some category of modules.

If $R = \mathbf{Z}$, the ring of integers, then mod-R is the category of Abelian groups, and if $R = D$ is a skewfield, then mod-R is the category of vector spaces over D.

The properties of mod-R reflect a number of important properties of the ring R (see **Homological classification of rings**). Connected with this category is a number of important homological invariants of the ring; in particular, its **homological dimension**. The *centre* of mod-R (that is, the set of natural transformations of the identity functor of the category) is isomorphic to the centre of R.

In ring theory, homological algebra and algebraic K-theory, various subcategories of the category of modules are discussed; in particular, the subcategory of finitely-generated projective R-modules and the associated K-functors (see **Algebraic K-theory**). By analogy with **Pontryagin duality**, dualities between full subcategories of the category of modules have been studied; in particular between subcategories of finitely-generated modules. For example, it has been established that if R and S are Noetherian rings and if there is duality between finitely-generated right R-modules and finitely-generated left S-modules, then there is a bimodule $_S U_R$ such that the given duality is equivalent to the duality defined by the functors

$$\mathrm{Hom}_R(-, U) \text{ and } \mathrm{Hom}_S(-, U),$$

the ring of endomorphisms $\mathrm{End}\, U_R$ is isomorphic to S, $\mathrm{End}_S U$ is isomorphic to R, the bimodule U is a finitely-generated injective cogenerator (both as an R-module and an S-module), and the ring R is semiperfect (cf. **Semi-perfect ring**). The most important class of rings, arising in the consideration of duality of modules, is the class of quasi-Frobenius rings (cf. **Quasi-Frobenius ring**). A left **Artinian ring** R is quasi-Frobenius if and only if the mapping

$$M \to \mathrm{Hom}_R(M, R)$$

defines a duality between the categories of finitely-generated left and right R-modules.

References

[1] BASS, H.: *Algebraic K-theory*, Benjamin, 1968.
[2] BUCUR, I. and DELEANU, A.: *Introduction to the theory of categories and functors*, Wiley, 1968.
[3] FAITH, C.: *Algebra: rings, modules and categories*, 1-2, Springer, 1973-1976.

A.V. Mikhalev

Editorial comments. A duality given by a bimodule U as described above is called a U-duality or Morita duality; cf. also (the comments to) Morita equivalence.

AMS 1980 Subject Classification: 18E99

MODULI OF A RIEMANN SURFACE - Numerical characteristics (parameters) which are one and the same for all conformally-equivalent Riemann surfaces, and in their totality characterize the conformal equivalence class of a given Riemann surface. Here two Riemann surfaces R_1 and R_2 are called *conformally equivalent* if there is a conformal mapping from R_1 onto R_2. For example, the conformal classes of compact Riemann surfaces of topological genus $g > 1$ are characterized by $6g - 6$ real moduli; a Riemann surface of torus type ($g = 1$) is characterized by 2 moduli; an n-connected plane domain, considered as a Riemann surface with boundary, is characterized by $3n - 6$ moduli for $n \geq 3$. About the structure of the moduli space of a Riemann surface see **Riemann surfaces, conformal classes of**.

A necessary condition for the conformal equivalence of two plane domains is that they have the same connectivity. According to the **Riemann theorem**, all simply-connected domains with more than one boundary point are conformally equivalent to each other; each such domain can be conformally mapped onto one canonical domain, usually taken to be the unit disc. For n-connected domains, $n \geq 2$, a precise equivalent of this Riemann mapping theorem does not exist: It is impossible to give any fixed domain whatever onto which it is possible to univalently and conformally map all domains of a given order of connectivity. This has led to a more flexible definition of a canonical n-connected domain, which reflects the general geometric structure of this domain, but does not fix its moduli (see **Conformal mapping**).

Each doubly-connected domain D of the z-plane with non-degenerate boundary continua can be conformally mapped onto some circular annulus $r < |w| < R$, $0 < r < R < \infty$. The ratio R/r of the radii of the boundary circles of this annulus is a conformal invariant and is called the *modulus of the doubly-connected domain D*. Let D be an n-connected domain, $n \geq 3$, with a non-degenerate boundary. D can be conformally mapped onto some n-connected circular domain Δ, which is a circular annulus $r < |w| < R$ with $n - 2$ discs with bounding circles $C_k = \{w: \ |w - w_k| = r_k\}$, $k = 1, \ldots, n - 2$, removed; the circles C_k, $k = 1, \ldots, n - 2$, lie in the annulus $r < |w| < R$ and pairwise do not have points in common. Here it can be assumed that $R = 1$ and $w_1 > 0$. Then Δ depends on $3n - 6$ real parameters: the $n - 1$ numbers $r, r_1, \ldots, r_{n-2}$ and the $2n - 5$ real parameters defining the centres w_k of the circles C_k, $k = 1, \ldots, n - 2$. These $3n - 6$ real parameters can be taken as *moduli of the n-connected domain D* in the case $n \geq 3$.

As moduli of n-connected domains D it is also possible to take any other μ real parameters ($\mu = 1$ if $n = 2$, and $\mu = 3n - 6$ if $n \geq 3$) which determine a conformal mapping of D onto some canonical n-connected domain of another shape.

References

[1] SPRINGER, G.: *Introduction to Riemann surfaces*, Addison-Wesley, 1957.
[2] BERS, L.: 'Uniformization, moduli, and Kleinian groups', *Bull. London Math. Soc.* **4** (1972), 257-300.
[3] GOLUZIN, G.M.: *Geometric theory of functions of a complex variable*, Amer. Math. Soc., 1969 (translated from the Russian).
[4] COURANT, R.: *Dirichlet's principle, conformal mapping and minimal surfaces*, Interscience, 1950.

G.V. Kuz'mina

E.D. Solomentsev

AMS 1980 Subject Classification: 32G15

MODULI PROBLEM - The classical problem of the rationality or uni-rationality of the moduli variety of algebraic curves of genus g.

Riemann surfaces of genus g (up to isomorphism) depend on $3g-3$ complex parameters — the moduli (see **Moduli of a Riemann surface**). The set of classes of non-singular projective curves of genus g over an algebraically closed field k has the structure of a quasi-projective algebraic variety M_g (see [3] - [5]).

The manifolds M_g for $g=0$ and 1 have a simple structure: M_0 consists of one point, and M_1 is isomorphic to the affine line A^1. Therefore the moduli problem refers to curves of genus $g \geqslant 2$ and is formulated as follows: Is the moduli variety M_g of curves of genus $g \geqslant 2$ rational, or at least uni-rational? The rationality of M_g has been established only for $g=2$ (see [2], where M_2 is explicitly described).

A general method for proving uni-rationality of M_g has been constructed [6]. By this method, in particular, the uni-rationality of M_g for all $g \leqslant 10$ has been proved. The uni-rationality of M_{11}, M_{12} and M_{13} has also been proved.

The moduli problem frequently receives a broader interpretation (see, for example, [5]): It refers to the whole complex of problems associated with the existence of moduli spaces of certain algebraic objects (varieties, vector bundles, endomorphisms, etc.), with the study of their various algebraic-geometric properties and with compactification techniques for moduli spaces (see **Moduli theory**).

References

[1] HARTSHORNE, R.: *Algebraic geometry*, Springer, 1977.
[2] IGUSA, J.: 'Arithmetic variety of moduli for genus two', *Ann. of Math.* **72**, no. 3 (1960), 612-649.
[3] MUMFORD, D.: *Geometric invariant theory*, Springer, 1965.
[4] MUMFORD, D.: 'Stability of projective varieties', *l'Enseign. Math. (2)* **23**, no. 1-2 (1977), 39-110.
[5] POPP, H.: *Moduli theory and classification theory of algebraic varieties*, Springer, 1977.
[6] SEVERI, F.: 'Sulla classificazione delle curve algebriche e sul teorema d'esistenza di Riemann', *Atti R. Accad. Naz. Lincei Rend.* **24** (1915), 877-888.

V.A. Iskovskikh

Editorial comments. It is now known that M_g is of general type (cf. **General-type algebraic surface**) for $g \geqslant 24$ and has positive **Kodaira dimension** for $g=23$ (cf. [A1], [A2]); thus M_g is not uni-rational for $g \geqslant 23$. For $g=11, 12, 13$, M_g is uni-rational. Also, M_{15} has negative Kodaira dimension. The nature of M_g for $g=16, \ldots, 22$ is still (1989) unknown.

References

[A1] CHANG, M. and RAN, Z.: 'Unirationality of the moduli space of curves of genus 11, 13 (and 12)', *Invent. Math.* **76** (1984), 41-54.
[A2] EISEBUD, D. and HARRIS, J.: 'The Kodaira dimension of the moduli space of curves of genus $g \geqslant 23$', *Invent. Math.* **90** (1987), 359-387.
[A3] HARRIS, J. and MUMFORD, D.: 'On the Kodaira dimension of the moduli space of curves', *Invent. Math.* **67** (1982), 23-86.
[A4] MORI, S. and MUKAI, S.: 'Uniruledness of the moduli space of curves of genus 11', in M. Reynard and T. Shioda (eds.): *Algebraic Geometry*, Lecture notes in math., Vol. 1016, Springer, 1983, pp. 334-353.
[A5] SERNESI, E.: 'L'unirazionalità della varietà dei moduli delle curvi di genere dodici', *Ann. Scuola Norm. Sup. Pisa (IV)* **VIII** (1981), 405-439.
[A6] SHEPHERD-BARRON, N.: 'The rationality of certain spaces associated to trigonal curves', in S.J. Bloch (ed.): *Algebraic Geometry*, Proc. Symp. Pure Math., Vol. 46.1, Amer. Math. Soc., 1987, pp. 165-171.
[A7] SHEPHERD-BARRON, N.: 'Invariant theory for S_5 and the rationality of M_6', *Compos. Math.* **70** (1989), 13-25.
[A8] HARRIS, J.: 'Curves and their moduli', in S.J. Block and S.J. Bloch (eds.): *Algebraic Geometry*, Proc. Symp. Pure Math., Vol. 46.1, Amer. Math. Soc., 1985, pp. 99-143.
[A9] EISENBUD, D. and HARRIS, J.: 'Limit linear series', *Bull. Amer. Math. Soc.* **10** (1984), 277-280.

AMS 1980 Subject Classification: 14D20, 32G13

MODULI THEORY - A theory studying continuous families of objects in algebraic geometry.

Let A be a class of objects in algebraic geometry (varieties, schemes, vector bundles, etc.) on which an equivalence relation R has been given. The fundamental classification problem (the description of the set of classes A/R) has the following two parts: 1) the description of discrete invariants, which usually allow a partition of A/R into a countable number of subsets, the objects of which already continuously depend on parameters; 2) the assignment and study of algebro-geometric structures on the parameter sets. The second part forms the matter of moduli theory.

Moduli theory arose in the study of elliptic functions: There is a continuous family of different fields of elliptic functions (or of their models — isomorphic elliptic curves over $\mathbf{C}$), parametrized by the complex numbers. B. Riemann, who introduced the term 'moduli', showed that an algebraic function field over $\mathbf{C}$ (or their models, compact Riemann surfaces) of genus $g \geqslant 2$ depends on $3g-3$ continuous complex parameters — the *moduli*.

Basic concepts in moduli theory. Let S be a **scheme** (a complex or algebraic space). A *family of objects parametrized by the scheme S* (or, as is often said, 'over S' or 'with basis S') is a set of objects

$$\{X_s : s \in S, \; X_s \in A\},$$

equipped with an additional structure compatible with the structure of the base S. This structure, in each concrete case, is given explicitly. A *functor of families* is a contravariant functor $\mathcal{M}$ from the category of the schemes (or spaces) into the category of sets defined as follows: $\mathcal{M}(S)$ is the set of classes of isomorphic families over S. To every morphism $f: T \to S$ is associated a mapping $f^*: \mathcal{M}(S) \to \mathcal{M}(T)$ which assigns to a family over S the pullback, or induced, family over T.

Let M be an object in the category of schemes (complex or algebraic spaces) and let h_M be a functor of the points in this category, that is, $h_M = \mathrm{Hom}(S, M)$. If the functor of families $\mathcal{M}$ is representable, that is, $\mathcal{M} = h_M$ for some M, then there exists a universal family with base $\mathcal{M}$, and M is called a *fine moduli scheme* (respectively, *fine complex moduli space* or *fine algebraic moduli space*). The functor $\mathcal{M}$ is representable in very few cases. Therefore the notion of a coarse moduli scheme was introduced. M is called a *coarse moduli scheme* if there is a morphism of functors $\phi: \mathcal{M} \to h_M$ with the properties: a) if $S = \mathrm{Spec}\, K = \mathrm{pt}$ is one point (where K is an algebraically closed field), then the mapping $\phi: \mathcal{M}(\mathrm{pt}) \to h_M(\mathrm{pt})$ is bijective; in other words, the set of geometric points of the scheme M is in a natural one-to-one correspondence with the set of equivalence classes of parametrized objects; and b) for each scheme N and morphism of functors $\psi: \mathcal{M} \to h_N$ there is a unique morphism $\chi: h_M \to h_N$ such that $\psi = \chi \circ \phi$. Coarse schemes of complex and algebraic moduli spaces are similarly defined.

Although a coarse moduli scheme uniquely parametrizes the class of objects defined by given discrete invariants, the natural family over it (in contrast to the family over a fine moduli scheme) does not have the strong universality property. A coarse moduli scheme (space) already exists in a fairly large number of cases.

Examples. 1) *Moduli of algebraic curves.* Let $A/R = \overline{\mathfrak{M}}_g$ (respectively $\mathfrak{M}_g$) be the set of classes of isomorphic projective non-singular curves (respectively, stable curves) of genus $g \geqslant 2$ over an algebraically closed field K. A family over S is a smooth (flat) proper morphism of schemes $f: X \to S$ whose fibres are smooth (stable) curves of genus g. Then there is a coarse (but not a fine) moduli scheme M_g (respectively $\overline{M}_g$), which is a quasi-projective (projective), irreducible, normal variety over K (see [3], [5], [6]).

2) *Moduli of algebraic curves with level n structure* (with *Jacobian rigidity*). Let $f: X \to S$ be a smooth family of projective curves (respectively, a flat family of stable curves) of genus $g \geqslant 1$, let n be an integer invertible on S, and let $R^1 f_*(\mathbf{Z}/n\mathbf{Z})$ be the first direct image of the constant sheaf $\mathbf{Z}/n\mathbf{Z}$ in the étale topology. Then $R^1 f_*(\mathbf{Z}/n\mathbf{Z})$ is locally free, has rank $2g$ and is equipped with a locally non-degenerate symplectic form with values in $\mathbf{Z}/n\mathbf{Z}$, up to an invertible element in $\mathbf{Z}/n\mathbf{Z}$. A Jacobian structure of level n on X is an assignment of a symplectic isomorphism

$$R^1 f_*(\mathbf{Z}/n\mathbf{Z}) \xrightarrow{\sim} (\mathbf{Z}/n\mathbf{Z})^{2g}.$$

Let $\mathcal{M}_{g,n}$ (respectively, $\overline{\mathcal{M}}_{g,n}$) be the functor of families of smooth (stable) curves of genus $g \geqslant 2$ with a Jacobian rigidity of level n. Then for $n \geqslant 3$ the functor $\mathcal{M}_{g,n}$ (respectively, $\overline{\mathcal{M}}_{g,n}$) is represented by a quasi-projective (projective) scheme $M_{g,n}$ (respectively, $\overline{M}_{g,n}$) over $\mathrm{Spec}\, \mathbf{Z}[1/n, \xi_n]$, where ξ_n is the inverse image of an n-th root of unity, that is, there is a fine moduli scheme $M_{g,n}$ (respectively, $\overline{M}_{g,n}$) for the smooth (stable) curves of genus $g \geqslant 2$ over a field of characteristic coprime with n, equipped with a Jacobian rigidity of level n. For sufficiently large n the scheme $M_{g,n}$ is smooth [5].

3) *Polarized algebraic varieties.* A polarized family is a pair $(X/S, \mathfrak{P}/S)$, where X/S is a smooth family of varieties, i.e. a smooth proper morphism $f: X \to S$, and $\mathfrak{P}/S$ is the class of the relatively ample invertible sheaf $\mathscr{L}_{X/S}$ in $\mathrm{Hom}(S, \mathrm{Pic}\, X/S)$ modulo $\mathrm{Hom}(S, \mathrm{Pic}^0 X/S)$, where $\mathrm{Pic}\, X/S$ is the relative Picard scheme and $\mathrm{Pic}^0 X/S$ is the connected component of its zero section. In this case a functor of the polarized families $\mathcal{M}_h$, with a given Hilbert polynomial h, is constructed. Without additional restrictions this functor is not representable. The existence of a coarse moduli space is known (1989) only in individual cases.

For polarized algebraic varieties the idea of rigidity of level n also exists.

4) *Vector bundles.* There are also results on moduli spaces for vector bundles of rank n over an algebraic variety X. In this case a family over S is a vector bundle over $X \times S$. Cf. [7], [10] - [14] for a description of results and more detail.

Local and global theory. The local theory arose as the theory of deformations of complex structures (see **Deformation** 1) and 2)). The fundamental methods of the global theory are those of the theory of representable functors and geometric invariant theory, the theory of algebraic stacks, and the algebraization of formal moduli.

The method of construction of a global moduli space goes back to the classical theory of invariants (cf. **Invariants, theory of**). It is as follows. A sufficiently large family $X \to H$ is constructed which contains representatives of all equivalence classes of the objects in questions, and so that the equivalence relation on H reduces to the action of an algebraic group G. Then the theory of actions of algebraic groups on algebraic varieties (schemes, spaces) is exploited with the aim of clarifying conditions for the existence of the quotient H/G in the corresponding category. The basic tool in the construction of the family $X \to H$ is the theory of Hilbert schemes (cf. **Hilbert scheme**). In such an approach the difficulty in constructing the family $X \to H$ is reduced to the problem on a simultaneous immersion of the objects in question into a projective space. An important result on the possibility of such a simultaneous immersion is Matsusaka's theorem. Then the difficult problem remains of the existence of the quotient H/G. Here one has the notions of categorical and geometric quotients. The construction of a coarse moduli space reduces to the problem of the existence of geometric quotients; here the idea of stability of points, corresponding to the idea of orbits in general position, is used. Results concerning actions of reductive groups on algebraic varieties over fields of characteristic 0 have been extended to the case of fields of characteristic $p > 0$.

Another approach to global moduli theory is the method of algebraic stacks, that is, a method of globalization of local deformation theory. The first step in the investigation of the representability of a global functor of families in this approach is the establishment of the algebraizability of a formal versal deformation for each object X_0. The difficulty in the construction of a global moduli space is that not every factorization of the base of the family with respect to an equivalence relation is a separated space. In such cases the object representing the functor $\mathcal{M}$ is replaced by an algebraic stack, the study of the properties of which gives some information on the moduli space.

One of the approaches to global moduli theory over $\mathbf{C}$ is the theory of period mappings (cf. **Period mapping**). The fundamental object here is the classifying space D of polarized Hodge structures (cf. **Hodge structure**) of weight k for given Hodge numbers. For a family $X \to S$ of polarized algebraic varieties over $\mathbf{C}$ the periods define a mapping of S onto the corresponding classifying space D of Hodge structures. The moduli problem reduces to the study of conditions for the period mapping to be bijective. The presence of (global) injectivity for the period mapping is the so-called *local-global Torelli problem*. Along this route the existence of coarse moduli spaces has been proved for curves, Abelian varieties and $K3$-surfaces.

The compactification problem for a moduli variety M is that of finding a natural and complete (projective or compact, in the theory over the field $\mathbf{C}$) variety $\overline{M}$ containing M as a dense open subset, and also the description and geometric interpretation of the boundary $\overline{M} \setminus M$. In example 1) the natural compactification of the coarse moduli variety M_g of curves of genus $g \geq 2$ is the projective moduli variety $\overline{M}_g$ of stable curves. For polarized Abelian varieties over $\mathbf{C}$ several means for compactifying moduli varieties are known.

References

[1] ARTIN, M.: 'Algebraization of formal moduli I', in *Global Analysis*, Univ. Tokyo Press, 1969, pp. 21-71.
[2] ARTIN, M.: 'Versal deformations and algebraic stacks', *Invent. Math.* **27** (1974), 165-189.
[3] DELIGNE, P. and MUMFORD, D.: 'The irreducibility of the space of curves of given genus', *Publ. Math. IHES* **36** (1969), 75-109.
[4] DIEUDONNÉ, J. and CARRELL, J.: *Invariant theory, old and new*, Acad. Press, 1971.
[5] MUMFORD, D.: *Geometric invariant theory*, Springer, 1965.
[6] MUMFORD, D.: 'Stability of projective varieties', *l'Enseign. Math. (2)* **23** (1977), 39-110.
[7] GIESEKER, D.: 'Global moduli for varieties of general type', *Invent. Math.* **43** (1977), 233-282.
[8] GIESEKER, D.: 'On the moduli of vector bundles on an algebraic surface', *Ann. of Math.* **106** (1977), 45-60.
[9] MATSUSAKA, T.: 'Polarized varieties with a given Hilbert polynomial', *Amer. J. Math.* **94**, no. 4 (1972), 1027-1077.
[10] NEWSTEAD, P.E.: *Lectures on introduction to moduli problems and orbit spaces*, Springer, 1978.
[11] OKONEK, C., SCHNEIDER, M. and SPINDLER, H.: *Vector bundles on complex projective spaces*, Birkhäuser, 1980.
[12] POPP, H.: *Moduli theory and classification theory of algebraic varieties*, Springer, 1977.
[13] SESHADRI, C.S.: 'Spaces of unitary vector bundles on a compact Riemann surface', *Ann. of Math.* **85** (1967), 302-336.
[14] TYURIN, A.N.: 'The geometry of moduli of vector bundles', *Russian Math. Surveys* **29**, no. 6 (1974), 57-88. (*Uspekhi Mat. Nauk* **29**, no. 6 (1974), 59-88)

V.A. Iskovskikh

Editorial comments. Much progress has been made recently in the study of the moduli spaces of algebraic curves and Abelian varieties; see the appendices of [A2]. Among the most important ones are the question of the compactification of M_g (see the appendix to Chapt. 5 in [A1], which is an enlarged edition of [5]) and the proof that M_g is of general type for $g \geq 24$ [A2].

G. Faltings has constructed a compactification of the moduli space A_g of principally-polarized Abelian varieties over $\mathbf{Z}$, see [A3].

References

[A1] MUMFORD, D. and FOGARTY, J.: *Geometric invariant theory*, Springer, 1982.
[A2] HARRIS, J.: 'Curves and their moduli', in S.J. Bloch (ed.): *Algebraic Geometry*, Proc. Symp. Pure Math., Vol. 46.1, Amer. Math. Soc., 1985, pp. 99-143.
[A3] FALTINGS, G.: 'Arithmetische Kompaktifizierung des Modulraums der abelschen Varietäten', in F. Hirzebruch, et al. (ed.): *Arbeitstagung Bonn 1984*, Lecture notes in math., Vol. 1111, Springer, 1985, pp. 321-383.

AMS 1980 Subject Classification: 14D20, 32G13

MODULUS - A numerical characteristic of various mathematical objects. Usually the value of a modulus is a non-negative real number, an element of $\mathbf{R}^+$, having certain characteristic properties, conditioned by properties of the set Ω of objects under discussion. The notion of a modulus figures in various branches of mathematics, although sometimes under other names — **absolute value**; **norm**, etc. All of them, in essence, are generalizations of the idea of the **absolute value** of a real or complex number (but the term modulus usually means a generalization of special form). Here the function $\Omega \to \mathbf{R}^+$ turns out to be a morphism of some structure in Ω onto one of the (algebraic) structures in $\mathbf{R}^+$, among which the most important ones are the order, the addition and the multiplication. In this connection the basic properties of the absolute value must be preserved (see below: α) - ϵ)). In more abstract situations it is natural to use an ordered semi-ring instead of $\mathbf{R}^+$ (this conception of a modulus is satisfied by, for example, a **measure**, a **capacity**, a **mass**, etc.). Finally, the term modulus denotes numerical characteristics of other objects, such as, for example, the moduli of a plane domain, the **modulus of an annulus**, the **moduli of a Riemann surface**, and the modulus of continuity or smoothness (cf. **Continuity, modulus of; Smoothness, modulus of**) (and even moduli in the theory of elasticity (compression, shear)). However, in all these cases it is possible to introduce a value functionally depending on the modulus and more adequately reflecting the nature of the objects under discussion (for example, for a family of curves, the **extremal length** instead of the modulus).

Examples. 1) The modulus of an element x of a **semi-ordered space** P is the number

$$|x| = x^+ + x^-,$$

where x^+ (x^-) is the positive (negative) part of x. Here, as for real numbers,

$\alpha)$ $|x| \geqslant x, -x;$ $|x| = |-x|;$
$\beta)$ $|x| = 0 \Leftrightarrow x = 0$ (0 is the zero in P).

2) The modulus of an element x of a separable **pre-Hilbert space** H, in particular, a finite-dimensional vector space, is the number

$$|x| = <x, x>^{1/2},$$

where $<\cdot, \cdot>$ is the **inner product** in H. This is a norm in H and thus

$\gamma)$ $|x+y| \leqslant |x| + |y|;$
$\delta)$ $|\lambda x| = |\lambda| |x|,$ λ a scalar.

3) The modulus of an element x of a **locally compact skew-field** is the number

$$|x| = \frac{\mu(xS)}{\mu(S)} \ (x \neq 0) \ \text{ or } \ 0 \ (x = 0),$$

where μ is a **Haar measure** on the additive group of K and S is a measurable subset. Here, as for numbers from **R, C, H**,

$\epsilon)$ $|xy| = |x| |y|.$

A generalization of this idea is the **modulus of an automorphism**.

4) The modulus of an endomorphism A of a vector space V over a field K (a special case is the modulus of an automorphism) is the number $\text{mod}_V(A)$, which turns out to be simply equal to $\text{mod}_K(\det A) = |\det A|,$ where $|\cdot|$ is the modulus of Example 3).

M.I. Voĭtsekhovskiĭ

MODULUS OF A FAMILY OF CURVES - A

conformally-invariant numerical characteristic of the family of curves. It is the value reciprocal to the **extremal length** of this family.

MODULUS OF AN ANNULUS - The value recipro-

cal to the **extremal length** of the family of closed curves in the annulus $r_1 \leqslant |z| \leqslant r_2$ which separate the boundary circles; the modulus is equal to

$$\frac{1}{2\pi} \ln \frac{r_2}{r_1}.$$

By a **conformal mapping** onto an associated annulus K, the *modulus m_G of an annular domain* G can be obtained. It turns out that $m_G = D(u)/2\pi$, where $D(u)$ is the **Dirichlet integral** of the real part of the function u mapping G onto K: $\{1 < |w| < e^{m_G}\}$. (Thus, a given **annular domain** is mapped onto an annulus with a fixed ratio of the radii of the boundary circles. This fact can be taken as another definition of the modulus of an annulus, its generalization leads to the idea of the modulus of a plane domain.)

A generalization of the modulus of an annular domain is the *modulus m_γ of a prime end* (cf. **Cluster set; Limit elements**) γ of an open **Riemann surface** R relative to a neighbourhood. Depending on whether m_γ is finite or infinite, the prime end has hyperbolic or parabolic type and R either does or does not have a **Green function**.

For a simply-connected domain D of hyperbolic type the so-called *reduced modulus m_{z_0}* relative to $z_0 \in D$ is defined as the limit

$$\lim_{r \to 0} \left[m(r) + \frac{1}{2\pi} \ln r \right],$$

where $m(r)$ is the modulus of the annular domain $D(r) = D \cap \{|z - z_0| > r\}$. It turns out that $m_{z_0} = (\ln R)/2\pi$, where R is the conformal radius (cf. **Conformal radius of a domain**) of D relative to $z_0 \in D$.

M.I. Voĭtsekhovskiĭ

MODULUS OF AN AUTOMORPHISM - A positive

real number associated to an automorphism of a locally compact group. Let G be such a group and let α be an automorphism of G, regarded as a topological group. Then the modulus of α is defined by

$$\text{mod}_G(\alpha) = \frac{\mu(\alpha S)}{\mu(S)},$$

where μ is left-invariant **Haar measure** on G and S is any compact subset of G with positive measure (indeed,

$\mathrm{mod}_G(\alpha)$ does not depend on S). If G is compact or discrete, then $\mathrm{mod}_G(\alpha) \equiv 1$, since for a compact group one can put $S = G$, and for a discrete group one can take $S = \{1\}$, where 1 is the identity element of G.

If α and β are two automorphism of G, then

$$\mathrm{mod}_G(\alpha \cdot \beta) = \mathrm{mod}_G(\alpha) \, \mathrm{mod}_G(\beta).$$

If Γ is a topological group which acts continuously on G by automorphisms, then the associated homomorphism $\pi : \Gamma \to \mathrm{Aut}\, G$ defines a continuous homomorphism $\mathrm{mod}_G \circ \pi : \Gamma \to \mathbf{R}_+^*$, where $\mathbf{R}_+^*$ is the multiplicative group of positive real numbers. In particular, if $\Gamma = G$ and $\pi(g)(x) = gxg^{-1}$, then $\pi \circ \mathrm{mod}_G : G \to \mathbf{R}_+^*$ is a continuous homomorphism. This homomorphism is trivial if and only if the left-invariant Haar measure on G is simultaneously right invariant. Groups satisfying the latter condition are called *unimodular*.

If K is a locally compact skew-field, then each non-zero element $a \in K$ defines an automorphism $\mu(a)$ of the additive group of K via multiplication by a. The function $\mathrm{mod}_K \circ \mu : K \setminus \{0\} \to \mathbf{R}_+^*$ is used in the study of the structure of locally compact skew-fields.

References

[1] BOURBAKI, N.: *Elements of mathematics. Integration*, Addison-Wesley, 1975 , Chapt. 6; 7; 8 (translated from the French).
[2] WEIL, A.: *L'intégration dans les groupes topologiques et ses applications*, Hermann, 1940.
[3] WEIL, A.: *Basic number theory*, Springer, 1974.

L.V. Kuz'min

AMS 1980 Subject Classification: 43-XX, 22A10

MODULUS OF AN ELLIPTIC INTEGRAL - The parameter k which enters into the expression of the **elliptic integral** in Legendre normal form. For example, in the incomplete elliptic integral of the first kind,

$$F(\phi, k) = \int_0^\phi \frac{dt}{\sqrt{1 - k^2 \sin^2 t}}. \qquad (*)$$

The number k^2 is sometimes called the *Legendre modulus*, $k' = \sqrt{(1 - k^2)}$ is called the *complementary modulus*. In applications the *normal case* $0 < k < 1$ usually holds; here the sharp angle θ for which $\sin\theta = k$ is called the *modular angle*. The modulus k also enters into the expression of the **Jacobi elliptic functions**, which arise from the inversion of elliptic integrals of the form (*).

E.D. Solomentsev

Editorial comments.

References

[A1] BOWMAN, F.: *Introduction to elliptic functions*, Dover, reprint, 1961.

AMS 1980 Subject Classification: 33A25

MODUS PONENS, *law of detachment, rule of detachment* - A **derivation rule** in formal logical systems. The rule of modus ponens is written as a scheme

$$\frac{A \quad A \supset B}{B},$$

where A and B denote formulas in a formal logical system, and $\supset$ is the logical connective of **implication**. Modus ponens allows one to deduce B from the premise A (the *minor premise*) and $A \supset B$ (the *major premise*). If A and $A \supset B$ are true in some interpretation of the formal system, then B is true. Modus ponens, together with other derivation rules and axioms of a formal system, determines the class of formulas that are derivable from a set of formulas M as the least class that contains the formulas from M and the axioms, and closed with respect to the derivation rules.

Modus ponens can be considered as an operation on the derivations of a given formal system, allowing one to form the derivation of a given formula B from the derivation α of A and the derivation β of $A \supset B$.

V.N. Grishin

Editorial comments. The more precise Latin name of the law of detachment is *modus ponendo ponens*. In addition there is *modus tollendo ponens*, which is written as the scheme

$$\frac{\neg B \quad A \vee B}{A},$$

where $\neg$ stands for negation and $\vee$ denotes the logical 'or'.

References

[A1] SUPPES, P.: *Introduction to logic*, v. Nostrand, 1957.
[A2] GRZEGORCZYK, A.: *An outline of mathematical logic*, Reidel, 1974.

AMS 1980 Subject Classification: 03BXX

MOMENT - A numerical characteristic of a **probability distribution**. The *moment of order k* ($k > 0$ an integer) of a random variable X is defined as the mathematical expectation $\mathsf{E}X^k$, if it exists. If F is the **distribution function** of the random variable X, then

$$\mathsf{E}X^k = \int_{-\infty}^{\infty} x^k \, dF(x). \qquad (*)$$

For the definition of a moment in probability theory, a direct analogy is used with the corresponding idea which plays a major role in mechanics: Formula (*) is defined as the moment of a mass distribution. The first-order moment (a statistical moment in mechanics) of a random variable X is the mathematical expectation $\mathsf{E}X$. The value $\mathsf{E}(X - a)^k$ is called the *moment of order k relative to a*, $\mathsf{E}(X - \mathsf{E}X)^k$ is the *central moment of order k*. The second-order central moment $\mathsf{E}(X - \mathsf{E}X)^2$ is called the **dispersion** (or *variance*) $\mathsf{D}X$ (the moment of inertia in mechanics). The value $\mathsf{E}|X|^k$ is called the **absolute moment** of order k (absolute moments are also defined for non-integral k). The moments of the **joint distribution** of random variables $X_1, \ldots, X_n$ (see **Multi-dimensional distribution**) are defined similarly:

For any integers $k_i \geq 0$, $k_1 + \cdots + k_n = k$, the mathematical expectation $E(X_1^{k_1} \cdots X_n^{k_n})$ is called a *mixed moment of order k*, and $E(X_1 - EX_1)^{k_1} \cdots (X_n - EX_n)^{k_n}$ is called a *central mixed moment of order k*. The mixed moment $E(X_1 - EX_1)(X_2 - EX_2)$ is called the **covariance** and is one of the basic characteristics of dependency between random variables (see **Correlation (in statistics)**). Many properties of moments (in particular, the inequality for moments) are consequences of the fact that for any random variable X the function $g(k) = \log E|X|^k$ is convex with respect to k in each finite interval on which the function is defined; $(E|X|^k)^{1/k}$ is a non-decreasing function of k. The moments EX^k and $E(X-a)^k$ exist if and only if $E|X|^k < \infty$. The existence of $E|X|^{k_0}$ implies the existence of all moments of orders $k \leq k_0$. If $E|X_i|^k < \infty$ for all $i = 1, \ldots, n$, then the mixed moments $EX_1^{k_1} \cdots X_n^{k_n}$ exist for all $k_i \geq 0$, $k_1 + \cdots + k_n \leq k$. In some cases, for the definition of moments, the so-called *moment generating function* is useful — the function $M(t)$ with the moments of the distribution as coefficients in its power-series expansion; for integer-valued random variables this function is related to the **generating function** $P(s)$ by the relation $M(t) = P(e^t)$. If $E|X|^k < \infty$, then the **characteristic function** $f(t)$ of the random variable X has continuous derivatives up to order k inclusively, and the moment of order k is the coefficient of $(it)^k / k!$ in the expansion of $f(t)$ in powers of t,

$$EX^k = (-i)^k \frac{d^k}{dt^k} f(t)\bigg|_{t=0}.$$

If the characteristic function has a derivative of order $2k$ at zero, then $E|X|^{2k} < \infty$.

For the connection of moments with semi-invariants see **Semi-invariant**. If the moments of a distribution are known, then it is possible to make some assertions about the probabilities of deviation of a random variable from its mathematical expectation in terms of inequalities; the best known are the **Chebyshev inequality in probability theory**

$$P\{|X - EX| \geq \epsilon\} \leq \frac{DX}{\epsilon^2}, \quad \epsilon > 0,$$

and its generalizations.

Problems of determining a probability distribution from its sequence of moments are called moment problems (cf. **Moment problem**). Such problems were first discussed by P.L. Chebyshev (1874) in connection with research on limit theorems. In order that the probability distribution of a random variable X be uniquely defined by its moments $\alpha_k = EX^k$ it is sufficient, for example, that *Carleman's condition* be satisfied:

$$\sum_{k=1}^{\infty} \frac{1}{\alpha_{2k}^{1/2k}} = \infty.$$

A similar result even holds for moments of random vectors.

The use of moments in the proof of limit theorems is based on the following fact. Let F_n, $n = 1, 2, \ldots$, be a sequence of distribution functions, all moments $\alpha_k(n)$ of which are finite, and for each integer $k \geq 1$ let

$$\alpha_k(n) \to \alpha_k \text{ as } n \to \infty,$$

where α_k is finite. Then there is a sequence F_{n_i} that weakly converges to a distribution function F having α_k as its moments. If the moments determine F uniquely, then the sequence F_n weakly converges to F. Based on this is the so-called method of moments (cf. **Moments, method of (in probability theory)**), used, in particular, in mathematical statistics for the study of the deviation of empirical distributions from theoretical ones, and for the statistical estimation of parameters of a distribution (on sampling moments as well as estimators for moments of a certain distribution see **Empirical distribution**).

A.V. Prokhorov

Editorial comments. The moment generating function is defined by $M(t) = Ee^{tX}$.

References

[A1] FELLER, W.: *An introduction to probability theory and its applications*, 2, Wiley, 1971.

AMS 1980 Subject Classification: 60E99

MOMENT PROBLEM - One of the **interpolation** problems in the real or complex domain.

The first precise formulation of the original version of the *moment problem in the real domain* is due to T.J. Stieltjes (1894). He proposed and solved the following problem in connection with the study of continued fractions (cf. **Continued fraction**): Given a sequence of real numbers $\{\mu_n\}$, $n = 0, 1, \ldots$, determine a bounded and non-decreasing function $\psi(x)$ on $[0, +\infty)$ such that

$$\int_0^\infty x^n \, d\psi = \mu_n, \quad n = 0, 1, \ldots. \tag{1}$$

As in every interpolation problem, the solution of (1) consists of two parts.

Problem A. Let $\mathfrak{M}$ be the set of all sequences of real numbers $\{\mu_n\}$ for which the infinite system of equations (1) has at least one solution ψ with the above properties; determine necessary and sufficient (constructive) conditions which must be satisfied by the numbers μ_n, $n = 0, 1, \ldots$, in order that $\{\mu_n\} \in \mathfrak{M}$.

Problem B. Determine the set of all solutions in the class of bounded non-decreasing functions ψ on $[0, +\infty)$ satisfying the infinite system (1) for given μ_n, $n = 0, 1, \ldots$, $\{\mu_n\} \in \mathfrak{M}$.

The left-hand sides of (1) were called 'moments' by Stieltjes. He borrowed the terminology from mechanics. If $d\psi(x)$ is interpreted as the mass on $[x, x + dx]$, then the integral $\int_0^X d\psi(t)$ is the mass on $[0, X]$. The integrals (1) for $n=1$ and $n=2$ are then, respectively, the first (static) and second (inertial) moments with respect to the origin $x=0$ of the total mass $\int_0^\infty d\psi(x)$ (this corresponds to $n=0$ in (1)) on $[0, \infty)$. Generalizing this idea, Stieltjes called the integral

$$\int_0^\infty x^n \, d\psi(x)$$

the *moment of order* n (relative to $x=0$) of the given mass $\int_0^\infty d\psi(x)$ with ψ as distribution on $[0, +\infty)$.

Stieltjes related the solution of the moment problem in the following way to the 'natural' continued fraction associated with the integral

$$I(z, \psi) = \int_0^\infty \frac{d\psi(x)}{z+x} \sim \frac{\mu_0}{z} - \frac{\mu_1}{z^2} + \frac{\mu_2}{z^3} - \frac{\mu_3}{z^4} + \cdots, \quad (2)$$

more precisely, to the formal series

$$\sum_{n=0}^\infty (-1)^n \frac{\mu_n}{z^{n+1}}.$$

Corresponding to the integral $I(z, \psi)$ there is a continued fraction:

$$I(z, \psi) \sim \frac{1 \, |}{| \, a_1 z} + \frac{1 \, |}{| \, a_2} + \frac{1 \, |}{| \, a_3 z} + \frac{1 \, |}{| \, a_4} + \cdots, \quad (3)$$

and also a 'closely related' continued fraction

$$I(z, \psi) \sim \frac{\lambda_1 \, |}{| \, z+c_1} - \frac{\lambda_2 \, |}{| \, z+c_2} - \frac{\lambda_3 \, |}{| \, z+c_3} - \cdots. \quad (4)$$

The continued fraction (4) is obtained from (3) by 'reductions' of the form

$$z - \frac{\alpha \, |}{| \, 1} - \frac{\beta \, |}{| \, z-\gamma} = z - \alpha - \frac{\alpha\beta}{z-(\beta+\gamma)}.$$

Making use of the theory of continued fractions, Stieltjes proved that in a certain sense a necessary and sufficient condition for the solvability of (1) (which is equivalent to $\{\mu_n\} \in \mathfrak{M}$) is the positivity of all a_n in (3), which, in turn, is a consequence of the positivity of λ_n and c_n in (4). In terms of μ_n these conditions are equivalent to the positivity of the determinants

$$\Delta = \det \| \mu_{i+j} \|_{i,j=0}^n$$

and

$$\Delta_n^{(1)} = \det \| \mu_{i+j+1} \|_{i,j=0}^n.$$

The moment problem (1) is called *well-posed* or *determined* for a given sequence $\{\mu_n\}$, $\{\mu_n\} \in \mathfrak{M}$, if the system (1) has a unique solution ψ. On the other hand, it has been shown that if the system (1) has more than one solution for a given μ_n, $\{\mu_n\} \in \mathfrak{M}$, then it has an infinite number of solutions.

Example: the two functions

$$\psi_1(x) = \int_0^x t^{-\ln t} \, dt$$

and

$$\psi_2(x) = \int_0^x t^{-\ln t}[1 - \theta \sin(\pi \ln t)] \, dt, \quad \theta \in [0, 1],$$

have the same moments

$$\int_0^\infty x^n \, d\psi_1(x) = \int_0^\infty x^n \, d\psi_2(x)$$

for all $n = 0, 1, \ldots.$.

Stieltjes effectively constructed certain solutions of (1), which, of course, all coincide in a well-known sense if (1) is well-posed. When the moment problem (1) is *ill-posed* or *undetermined*, the Stieltjes solutions have a number of extremal properties. Stieltjes subsequently showed that (1) is well-posed or ill-posed depending on the convergence or divergence of the continued fraction (3) (which is equivalent to the divergence or convergence of the series $\sum_{n=0}^\infty a_n$). Here the fraction (3) may be convergent to $I(z, \psi)$, whereas the series

$$\sum_{n=0}^\infty \frac{(-1)^n \mu_n}{z^{n+1}}$$

may, at the same time, diverge for all $z \in \mathbf{C}$.

Preceding the work of Stieltjes [1], the moment problem in the real domain was considered in a less general and less precise formulation; such as, for example, in a series of papers by P.L. Chebyshev [2] and A.A. Markov [3]. They mainly investigated the following problem: Give a description of the properties of a class U of functions defined on $(-\infty, +\infty)$ such that the relations

$$p(x) \in U$$

and

$$\int_{-\infty}^{+\infty} x^n p(x) \, dx = \int_{-\infty}^{+\infty} x^n e^{-x^2} \, dx, \quad n = 0, 1, \ldots, \quad (5)$$

lead to the identity

$$p(x) = e^{-x^2}.$$

In other words, the question here concerns a maximally complete and constructive characterization of the uniqueness class U of the interpolation problem (5). The solution of the moment problem (5) plays a major role in probability theory and mathematical statistics. Also of major significance are the polynomials $\omega_n(x)$, the dominators of the successive approximations (that is, the approximants) of the continued fraction (4). The study of the properties of the polynomials $\{\omega_n(x)\}$ later initiated a broad field of research into the theory of **orthogonal polynomials**.

H. Hamburger (1920) generalized the moment problem (1) to the case of the whole real line $\mathbf{R} = (-\infty, +\infty)$. Here the consideration of negative values of x introduced a number of peculiarities and was non-trivial. Hamburger, making essential use of Helly's selection principle (cf. **Helly theorem**), aimed at obtaining necessary and sufficient conditions for the solvability of the system

$$\int_{-\infty}^{+\infty} x^n \, d\psi(x) = \mu_n, \quad n = 0, 1, \ldots, \tag{6}$$

thereby completely solving the problem of convergence of the continued fractions (3) and (4) generated by (6). The union of problems A and B in relation to (6) is called the *moment problem of equation* (6). Hamburger obtained a criterion for the existence of a unique solution of the moment problem for (6). In this connection, the moment problem for (6) may be ill-posed, whereas at the same time the corresponding moment problem (1) (with the same μ_n) may be well-posed (have a unique solution). R. Nevanlinna (1922) gave a solution to the moment problem (6) using the integrals

$$I(z, \psi) = \int_{-\infty}^{+\infty} \frac{d\psi(x)}{z - x}, \quad x \in \mathbf{C} \setminus \mathbf{R},$$

and studied properties of these solutions. He made an important observation about the so-called 'extremal solution' of the moment problem (6).

M. Riesz (1921) obtained solutions of the moment problem (6) based on the theory of quasi-orthogonal polynomials. These consist of linear combinations of the form $A_n \omega_n(x) + A_{n-1} \omega_{n-1}(x)$, where A_k are constants and $\omega_k(x)$ is the dominator of the k-th approximant of the continued fraction (4) associated with (6). He observed a close connection between the solutions of the moment problem (6) and the validity of Parseval's formula for the system of orthogonal polynomials $\{\omega_k(x)\}$. T. Carleman (1923 - 1926) established connections between the moment problem (6), the theory of quasi-analytic functions and the theory of quadratic forms in a countable set of variables. He also obtained the most general criterion for the well-posedness of the moment problem (6). F. Hausdorff (1923) obtained a criterion for the solvability of the moment problem (6) ($\Leftrightarrow \{\mu_n\} \in \mathfrak{M}$) under the condition that the function $\psi(x)$ in (6) is a constant outside a given interval. He effectively constructed the solution $\psi(x)$ of (6) (which, under the assumption given above, is always unique); this provides an opportunity to obtain criteria for additional properties of solutions $\psi(x)$ of (6) (continuity, differentiability, etc.). Carleman and subsequently M.H. Stone (1932) fully investigated (6) based on results in the theory of Jacobi quadratic forms and the theory of singular integral equations. E.K. Haviland (1935) and H. Cramér (1937) extended Riesz's theory of (6) to the multi-dimensional case.

Numerous different generalizations of the moment problem have also been considered. Mainly these are variants (or a combination of variants) of the following two themes.

Replacement of the powers x^n in the integrals (6) by 'moment' sequences of functions $\{\phi_n(x)\}$ of another

form, and replacement of the left-hand sides of (6) by other kinds of integrals (for example, the case when $d\psi(x)$ is replaced by $\phi(x)\,dx$, where $\phi(x) \in L_p$, $p \geq 1$, has been studied) or even by operators acting in abstract spaces.

Thus, with respect to the first theme, one has the so-called *trigonometric moment problem*, which is the following: Given an infinite sequence of numbers $\{c_n\}_{n=-\infty}^{\infty}$, determine a function $\psi(x)$, non-decreasing on $[-\pi, \pi]$, satisfying

$$\frac{1}{2\pi} \int_{-\pi}^{\pi} e^{inx} \, d\psi(x) = c_n, \quad n = 0, \pm 1, \ldots, \tag{7}$$

that is, solve problems A and B for the system (7).

Precise formulations of certain results concerning the theory of moment problems in the real domain are given below. Let $\mathbf{R}^n$ be the n-dimensional Euclidean space. A set function $\Phi(e)$, defined on the family $\mathscr{B}$ of all Borel sets in $\mathbf{R}^n$, is called a distribution if $\Phi(e) \geq 0$ for all $e \in \mathscr{B}$ and if

$$\sum_{i=1}^{\infty} \Phi(e_i) = \Phi\left[\sum_{i=1}^{\infty} e_i\right]$$

whenever $e_i \cap e_j = \varnothing$, $i \neq j$, where $e_i \in \mathscr{B}$ for all $i, j = 1, 2, \ldots$.

The *spectrum* $\sigma(\Phi)$ of a distribution Φ is the set of all points $x = (x_1, \ldots, x_n) \in \mathbf{R}^n$ such that $\Phi(G) > 0$ for an arbitrary open set $G \subset \mathbf{R}^n$ containing x. Let

$$\{\mu_{i_1 \cdots i_n}\}, \quad i_1, \ldots, i_n = 0, 1, \ldots, \tag{8}$$

be an n-fold infinite sequence of real numbers. The question is: What are necessary and sufficient conditions to be satisfied by the numbers (8) in order that there is a distribution Φ, with spectrum $\sigma(\Phi)$ contained in a given closed set F, which is a solution of the system

$$\int_{\mathbf{R}^n} t_1^{i_1} \cdots t_n^{i_n} \, d\Phi = \mu_{i_1 \cdots i_n}, \quad i_1, \ldots, i_n = 0, 1, \ldots \tag{9}$$

(problem A for (9)). Problem B for (9) is formulated similarly. The union of problems A and B for (9) is called the *F-moment problem*. The F-moment problem is well-posed if its solution is in some way unique. Otherwise the F-moment problem (9) is called ill-posed.

Theorem. A necessary and sufficient condition that the F-moment problem (9) has a solution in $\mathbf{R}^2$ is that the condition

$$\sum a_i b_j \mu_{ij} \geq 0$$

holds for any polynomial

$$P(u, v) = \sum a_i b_j u^i v^j$$

taking non-negative values for all $(u, v) \in F$.

This theorem is the basis for obtaining solvability conditions (that is, for the solution of problem A) for different versions of (9). Here are some of them.

Theorem 1. In order that the moment problem (6)

(with $F=\mathbf{R}$) have a solution it is necessary that

$$\Delta_n = \det \| \mu_{i+j} \|_{i,j=0}^{n} \geqslant 0, \quad n=0, 1, \ldots.$$

For the existence of a solution to the moment problem (6) having a spectrum which is not a finite number of points, it is necessary and sufficient that

$$\Delta_n > 0, \quad n=0, 1, \ldots.$$

For the existence of a solution to the moment problem (6) having a spectrum consisting of precisely $k+1$ different points, it is necessary and sufficient that

$$\Delta_0, \ldots, \Delta_k > 0, \quad \Delta_{k+1} = \Delta_{k+2} = \cdots = 0.$$

In the latter case the moment problem (6) is always well-posed.

Theorem 2. In order that the moment problem (1) (with $F=[0, \infty)$) is solvable it is necessary that

$$\Delta_n = \det \| \mu_{i+j} \|_{i,j=0}^{n} \geqslant 0$$

and

$$\Delta_n^{(1)} = \det \| \mu_{i+j+1} \|_{i,j=0}^{n} \geqslant 0, \quad n=0, 1, \ldots.$$

For the existence of a solution to the moment problem (1) having a spectrum which is not a finite number of points, it is necessary and sufficient that

$$\Delta_n > 0 \quad \text{and} \quad \Delta_n^{(1)} > 0, \quad n=0, 1, \ldots.$$

Necessary and sufficient conditions have also been obtained for the existence of a solution to the moment problem (1) having a spectrum $\sigma(\Phi)$ consisting of precisely $k+1$ points different from $x=0$. The conditions are similar to those given in the final part of Theorem 1.

Theorem 3. A necessary and sufficient condition that the Hausdorff moment problem in $\mathbf{R}$,

$$\int_0^1 x^n \, d\Phi = \mu_n, \quad n=0, 1, \ldots, \quad F=[0, 1],$$

has a solution, is that $\Delta^k \mu_v \geqslant 0$ for all $k, v = 0, 1, \ldots$ (here Δ^k denotes the k-th difference operator).

Theorem 4. A necessary and sufficient condition that the Hausdorff moment problem in $\mathbf{R}^2$,

$$\int_0^1 \int_0^1 u^i v^j \, d\Phi = \mu_{ij}, \quad i,j=0, 1, \ldots, \quad F=[0, 1]\times[0, 1],$$

has a solution, is that

$$\Delta_1^n \Delta_2^m \mu_{ij} \geqslant 0, \quad n, m, i, j=0, 1, \ldots.$$

Theorem 5. The moment problem (6) is well-posed if

$$\sum_{n=0}^{\infty} \frac{1}{\mu_{2n}^{1/2n}} = +\infty. \qquad (10)$$

Necessary and sufficient conditions are known (see, for example, [4]) which must be satisfied by μ_n in order that the moment problem (6) (the moment problem (1)) be well-posed; however, these conditions are less simple than the sufficient condition (10) and their formulation is somewhat cumbersome.

The *moment problem in the complex domain* is the name of a wide class of interpolation problems described as follows. Let D be an open simply-connected domain in the complex plane $\mathbf{C}$, $\infty \notin D$; let $A(D)$ be the space of analytic functions in D with topology defined by uniform convergence on arbitrary compact sets $K \subset D$; let $A^*(D)$ be the space of all functions $\gamma(z)$ analytic in a neighbourhood $V^\infty = V^\infty(\gamma)$ of the point at infinity for which $\gamma(\infty)=0$ and $\operatorname{supp}\gamma \subset D$ (the latter is another way of saying that the set of singularities of $\gamma \in A^*(D)$ lies in D). The topology in $A^*(D)$ is defined by uniform convergence on one of the curves of the family of simple closed Jordan curves $\{\Gamma_\alpha\} \subset D$ having the property: For any compact set $K \subset D$ there is a $\Gamma_{\alpha_0} = \Gamma_{\alpha_0}(K) \in \{\Gamma_\alpha\}$ such that $K \subset \operatorname{int}\Gamma_{\alpha_0}(D)$ (here $\operatorname{int}\Gamma_\alpha$ denotes the open simply-connected domain with boundary Γ_{α_0} lying inside Γ_{α_0}). It is well known that the spaces $A(D)$ and $A^*(D)$ are dual.

The moment problem in a complex domain is as follows. Given an integer $p>1$, functions $0 \neq A_s(z) \in A(D)$, $s=0, \ldots, p-1$, a univalent function $W(z) \in A(D)$, and a set of p sequences of complex numbers

$$\alpha_p = \{\{a_{ns}\}: n=0, 1, \ldots; s=0, \ldots, p-1\},$$

can one find a function $\gamma(z) \in A^*(D)$ for which

$$\frac{1}{2\pi} \int_\Gamma [W(z)]^{np} A_s(z) \gamma(z) \, dz = a_{ns}, \qquad (11)$$

$$n = 0, 1, \ldots; \quad s = 0, \ldots, p-1,$$

where

$$\operatorname{supp}\gamma \subset \operatorname{int}\Gamma \subset \Gamma \subset D?$$

In general, it is not true for every given collection α_p that the infinite system (11) has at least one solution $\gamma(z) \in A^*(D)$. Therefore a collection α_p is called D-admissible if there is (at least one) $\gamma(z) \in A^*(D)$ satisfying (11).

Problem A. Determine necessary and sufficient conditions (of a constructive nature) for the D-admissibility of a collection α_p.

Problem B. Let α_p be D-admissible. The question is: How can one determine the complete set of functions $\gamma(z) \in A^*(D)$ satisfying (11) with respect to given numbers a_{ns} in the right-hand side of (11)?

The union of problems A and B is called a moment problem in the complex domain. Problem B, for the case $p=1$ and $A_0(z)=1$, was first treated in 1937 by A.O. Gel'fond [6]; he discussed whether, in principle, problem B can be solved (for $p=1$ and $A_0(z)=1$ the system (11) always has a unique solution for D-admissible right-hand sides a_{ns}). Numerous special cases of problems A and B have been investigated (see [7] - [10]). Using tools from the theory of boundary value problems allows one to attain (see [11] - [14]) a fairly complete investigation of the moment problem in the complex domain.

A domain $G \subset \mathbf{C}$ is called $2\pi/p$-invariant, $G \in \mathrm{Inv}(2\pi/p)$, if $\exp(2\pi i/p)G \equiv G$.

An exhaustive solution to the moment problem in a complex domain D under natural assumptions concerning the functions $A_s(z)$, $s = 0, \ldots, p-1$, has been given in [10], when $W(D) = G \in \mathrm{Inv}(2\pi/p)$, as well as for a domain whose image $W(D)$ can be imbedded in some domain $G \in \mathrm{Inv}(2\pi/p)$. The theory of boundary value problems can be fruitfully used to obtain a complete solution by quadrature of problem B for domains of the types indicated. In particular, for $p = 1$ every domain G belongs to the class $\mathrm{Inv}(2\pi/p)$. Thus necessary and sufficient conditions for the uniqueness of the solution of the system (11) have been found for domains D whose W-images cannot be imbedded. These domains are important in applications. Here there are two essentially different cases: $0 \in W(D)$ and $0 \notin W(D)$ (in the latter, the question of the uniqueness of the solution to (11) has been exhaustively studied on the assumption that $n = 0, \pm 1, \ldots$). Several versions of the moment problem (11) are possible with regard to the behaviour of the corresponding functions on Γ.

A number of well known interpolation problems reduce to the moment problem in the complex domain by means of the Borel transformation and its generalizations (see **Comparison function** and **Borel transform**), for example:

$$F^{(n)}(hn) = a_n; \quad F^{(n)}(\omega^n) = a_n;$$

$$F(\omega^n) = a_n; \quad \Delta^n F(hn) = a_n;$$

$$F^{(np+l_s)}(\alpha_s) = a_{ns};$$

$$n = 0, 1, \ldots, \quad s = 0, \ldots, p-1, \quad l_s = 0, 1;$$

$$\Delta^{2n+l_s} F(\alpha_s + 2hn) = a_{ns}, \quad s = 0, 1; \quad l_s = 0, 1; \quad n = 0, 1, \ldots .$$

In addition, many theorems on integer-valued functions reduce to very specific cases of problem A.

References

[1A] STIELTJES, T.J.: 'Recherches sur les fractions continues', *Ann. Fac. Sci. Univ. Toulouse* **8** (1894), 1-122.
[1B] STIELTJES, T.J.: 'Recherches sur les fractions continues', *Ann. Fac. Sci. Univ. Toulouse* **9** (1895), 1-47.
[2] CHEBYSHEV, P.L.: *Oevres*, 1-2, Chelsea, No date.
[3] MARKOV, A.A.: *Selected works on the theory of continued fractions*, Moscow-Leningrad, 1948 (in Russian).
[4] SHOHAT, J.A. and TAMARKIN, J.D.: *The problem of moments*, Amer. Math. Soc., 1950.
[5] AKHIEZER, N.I.: *The classical moment problem and related questions in analysis*, Hafner, 1965 (translated from the Russian).
[6] GELFOND, A.O. [A.O. GEL'FOND]: *Differenzenrechnung*, Deutsch. Verlag Wissenschaft., 1958 (translated from the Russian).
[7] BUCK, R.C.: 'Interpolation series', *Trans. Amer. Math. Soc.* **64** (1948), 283-298.
[8] BUCK, R.C.: 'Integral valued entire functions', *Duke Math. J.* **15** (1948), 879-891.
[9] BUCK, R.C.: 'On admissibility of sequences and a theorem of Pólya', *Comment. Mat. Helv.* **27** (1953), 75-80.
[10] LOKHIN, I.F.: 'An interpolation problem for entire functions', *Mat. Sb.* **35**, no. 2 (1954), 223-230 (in Russian).
[11] KAZ'MIN, YU.A.: 'On a general problem in the theory of interpolation', *Soviet Math. Dokl.* **11** (1970), 1357-1361. (*Dokl. Akad. Nauk SSSR* **194**, no. 6 (1970), 1251-1254)
[12] KAZ'MIN, YU.A.: 'On the moment problem in the complex domain', *Soviet Math. Dokl.* **13** (1972), 833-837. (*Dokl. Akad. Nauk SSSR* **204**, no. 6 (1972), 1309-1312)
[13] KAZ'MIN, YU.A.: 'The general moment problem in the complex domain. Uniqueness theorems', *Soviet Math. Dokl.* **13** (1972), 868-872. (*Dokl. Akad. Nauk SSSR* **205**, no. 1 (1972), 19-22)
[14] BIEBERBACH, L.: *Analytische Fortsetzung*, Springer, 1955.
[15] GAKHOV, F.D.: *Boundary value problems*, Pergamon, 1966 (translated from the Russian).
[16] KREĬN, M.G. and NUDEL'MAN, A.A.: *The Markov moment problem and extremal problems*, Amer. Math. Soc., 1977 (translated from the Russian).

Yu.A. Kaz'min

Editorial comments. The classical moment problem is connected with a large number of fundamental theoretical and applied topics, including function theory, spectral decomposition of operators, positive definiteness, probability, approximation theory, electrical and mechanical inverse problems, prediction of stochastic processes, and the design of algorithms for signal-processing VLSI chips. A survey of some of these ramifications is given in [A23].

As is obvious from the main article above, there is an intimate connection between moment problems and continued fractions; from there it is but a very small step to the field of rational **approximation** and **interpolation**, viz. Padé and Hermite–Padé approximation (cf., e.g., **Padé approximation**).

The vast literature on the last mentioned subject contains many papers connected with and having effect on several types of moment problems, cf. contributions in [A3] - [A12]; also, applications in physics should be mentioned, see [A13] - [A15].

Moreover, in the last 15 years the study of different types of extensions of the moment problem in the setting of the theory of continued fractions and linear analysis has intensified; cf. [A16] - [A20].

Finally, [A1] - [A2] are important from a historical point of view.

References

[A1] GROMMER, J.: 'Ganze transzendente Funktionen mit lauter reellen Nullstellen', *J. Reine Angew. Math.* **144** (1914), 212-238.
[A2A] HAMBURGER, H.: 'Ueber eine Erweiterung des Stieltjesschen Momentenproblems I', *Math. Ann.* **81** (1920), 235-319 .
[A2B] HAMBURGER, H.: 'Ueber eine Erweiterung des Stieltjesschen Momentenproblems II', *Math. Ann.* **82** (1921), 120-164.
[A2C] HAMBURGER, H.: 'Ueber eine Erweiterung des Stieltjesschen Momentenproblems III', *Math. Ann.* **82** (1921), 168-187.
[A3] GRAVES, P.R. (ED.): *Padé Approximants and their Applications (Canterbury, 1972)*, Acad. Press, 1973.
[A4] SAFF, E.B. and VARGA, R.S. (EDS.): *Padé and Rational Approximation (Tampa, 1976)*, Acad. Press, 1977.
[A5] WUYTACK, L. (ED.): *Padé Approximation and its Applications (Antwerp, 1979)*, Lecture notes in math., 765, Springer, 1979.
[A6] BRUIN, M.G. DE and ROSSUM, H. VAN (EDS.): *Padé Approxi-*

mation and its Applications (Amsterdam, 1980), Lecture notes in math., 888, Springer, 1981.

[A7] GILEWICZ, J. (ED.): *Proc. first French-Polish Meeting on Padé Approximation and Convergence Acceleration Techniques (Warszaw, 1981)*, CPT-81/PE 1354, CNRS, 1982.

[A8] WERNER, H. and BÜNGER, H.J. (EDS.): *Padé Approximation and its Applications (Bad Honnef, 1983)*, Lecture notes in math., 1071, Springer, 1984.

[A9] GRAVES-MORRIS, P.R., SAFF, E.B. and VARGA, R.S. (EDS.): *Rational Approximation and Interpolation (Tampa, 1983)*, Lecture notes in math., 1105, Springer, 1984.

[A10] BREZINSKI, C., DRAUX, A. and MAGNUS, A.P., ET AL. (EDS.): *Polynomes Orthogonaux et Applications (Bar-le-Duc, 1984)*, Lecture notes in math., 1171, Springer, 1985.

[A11] GILEWICZ, J., PINDOR, M. and SIEMASKO, W. (EDS.): *Rational Approximation and its Application in Mathematics and Physics (Lańcut, 1985)*, Lecture notes in math., 1237, Springer, 1987.

[A12] CUYT, A. (EDS.): *Nonlinear Numerical Methods and Rational Approximation (Antwerp, 1987)*, Reidel, 1988.

[A13] ANTOLIN, J. and CRUZ, A.: *J. Phys.* G12 (1986), 297.

[A14] CORCORAN, C.T. and LANGHOFF, P.W.: 'Moment-theory approximations for nonnegative spectral densities', *J. Math. Phys.* 18 (1977), 651-657.

[A15] LANGHOFF, P.W.: B.J. Dalton et al. (ed.): *Moment methods in many Fermion systems*, Plenum, Forthcoming.

[A16] JONES, W.B., THRON, W.J. and WAADELAND, H.: 'A strong Stieltjes moment problem', *Trans. Amer. Math. Soc.* 261 (1980), 503-528.

[A17] JONES, W.B. and THRON, W.J.: 'Survey of continued fraction methods in solving moment problems and related topics', in W.B. Jones, W.J. Thron and H. Waaderland (eds.): *Analytic Theory of Continued Fractions*, Lecture notes in math., Vol. 932, Springer, 1982, pp. 4-36.

[A18] JONES, W.B., THRON, W.J. and NJASTAD, O.: 'Orthogonal Laurent polynomials and the strong Hamburger moment problem', *J. Math. Anal. Applic.* 98 (1984), 528-554.

[A19] JONES, W.B., NJASTAD, O. and THRON, W.J.: 'Continued fractions associated with the trigonometric and other strong moment problems', (To appear).

[A20] JONES, W.B., NJASTAD, O. and THRON, W.J.: 'Perron–Carathéodory continued fractions', (To appear).

[A21] AKHIEZER, N.I. and KRIEN, M.: *Some questions in the theory of moments*, Amer. Math. Soc., 1962 (translated from the Russian).

[A22] LANDAU, H.J.: 'The classical moment problem: Hilbertian proofs', *J. Funct. Anal.* 38 (1980), 255-272.

[A23] LANDAU, H.J. (ED.): *Moments in mathematics*, Amer. Math. Soc., 1987.

[A24] NATANSON, I.P.: *Constructive theory of functions*, 1-2, F. Ungar, 1964-1965 (translated from the Russian).

AMS 1980 Subject Classification: 41A05, 42A70, 44A60

MOMENTS, METHOD OF - The same as the **Galerkin method.**

AMS 1980 Subject Classification: 65N30, 49D15

MOMENTS, METHOD OF (IN PROBABILITY THEORY) - A method for determining a **probability distribution** by its moments (cf. **Moment**). Theoretically the method of moments is based on the uniqueness of the solution of the **moment problem**: If $\alpha_0 = 1, \alpha_1, \alpha_2, \ldots$, are constants, then under what conditions does there exist a unique distribution P such that

$$\alpha_n = \int x^n P(dx)$$

are the moments of P for all n? There are various types of sufficient conditions for a distribution to be uniquely determined by its moments, for example, the *Carleman condition*

$$\sum_{n=1}^{\infty} \frac{1}{\alpha_{2n}^{1/2n}} = \infty.$$

The use of the method of moments in the proof of limit theorems in probability theory and mathematical statistics is based on the correspondence between moments and the convergence of distributions: If F_n is a sequence of distribution functions with finite moments $\alpha_k(n)$ of any order $k \geqslant 1$, and if $\alpha_k(n) \to \beta_k$, as $n \to \infty$, for each k, then the β_k are the moments of a distribution function F; if F is uniquely determined by its moments, then as $n \to \infty$, the F_n converge weakly to F. The method of moments in the case of convergence to a **normal distribution** was first treated by P.L. Chebyshev (1887), and a proof of the **central limit theorem** by the method of moments was accomplished by A.A. Markov (1898).

The method of moments in mathematical statistics is one of the general methods for finding statistical estimators of unknown parameters of a probability distribution from results of observations. The method of moments was first used to this end by K. Pearson (1894) to solve the problem of the approximation of an empirical distribution by a system of Pearson distributions (cf. **Pearson distribution**). The procedure in the method of moments is this: The moments of the empirical distribution are determined (the sample moments), equal in number to the number of parameters to be estimated; they are then equated to the corresponding moments of the probability distribution, which are functions of the unknown parameters; the system of equations thus obtained is solved for the parameters and the solutions are the required estimates. In practice the method of moments often leads to very simple calculations. Under fairly general conditions the method of moments allows one to find estimators that are asymptotically normal, have mathematical expectation that differs from the true value of the parameter only by a quantity of order $1/n$ and standard deviation that deviates by a quantity of order $1/\sqrt{n}$. However, the estimators found by the method of moments need not be best possible from the point of view of efficiency: their variance need not be minimal. For a normal distribution the method of moments leads to estimators that coincide with the estimators of the **maximum-likelihood method**, that is, with asymptotically-unbiased asymptotically-efficient estimators.

References
[1] PROKHOROV, YU.V. and ROZANOV, YU.A.: *Probability theory,*

Springer, 1969 (translated from the Russian).

[2] CRAMÉR, H.: *Mathematical methods of statistics*, Princeton Univ. Press, 1946.

[3] KENDALL, M.G. and STUART, A.: *The advanced theory of statistics. Distribution theory*, 1, Griffin, 1987.

A.V. Prokhorov

AMS 1980 Subject Classification: **60E99**

MONGE — AMPÈRE EQUATION - A second-order partial differential equation of the form

$$rt - s^2 = ar + 2bs + ct + \phi,$$

$$r = \frac{\partial^2 z}{\partial x^2}, \quad s = \frac{\partial^2 z}{\partial x \partial y}, \quad t = \frac{\partial^2 z}{\partial y^2},$$

with coefficients depending on variables x, y, the unknown function $z(x, y)$, and its first derivatives

$$p = \frac{\partial z}{\partial x}, \quad q = \frac{\partial z}{\partial y}.$$

The type of a Monge — Ampère equation depends on the sign of the expression

$$\Delta = \phi + ac + b^2.$$

If $\Delta > 0$, then the Monge — Ampère equation is of elliptic type, if $\Delta < 0$ it is of hyperbolic type and if $\Delta = 0$ it is of parabolic type. A Monge — Ampère equation is invariant under contact transformations (cf. **Contact transformation**). In particular, the transformation

$$\xi = p, \quad \eta = q, \quad \zeta = z - px - qy$$

transforms the equation

$$rt - s^2 = \phi(p, q)$$

into the equation

$$\frac{\partial^2 \zeta}{\partial x^2} \frac{\partial^2 \zeta}{\partial y^2} - \left[\frac{\partial^2 z}{\partial x \partial y} \right]^2 = \frac{1}{\phi(\xi, \eta)}.$$

The development of the theory of Monge — Ampère equations is mainly connected with the solution of various geometric problems which, when formulated analytically, reduce to the discussion of such equations. For example, the construction of a surface with a given line element reduces to the solution of the **Darboux equation**, which is a Monge — Ampère equation. In the case of a semi-geodesic parametrization (with line element $ds^2 = du^2 + c^2 \, dv^2$) this equation takes the form

$$r \left[t + cc_u p - \frac{c_v}{c} q \right] - \left[s - \frac{c_u}{c} q \right]^2 +$$

$$+ \frac{c_{uu}}{c}(c^2 - c^2 p^2 - q^2) = 0.$$

The type of the Darboux equation depends on the sign of the **Gaussian curvature** $k = -c_{uu}/c$. For a hyperbolic Darboux equation (negative Gaussian curvature) the characteristics are asymptotic lines. An application of the **Cauchy — Kovalevskaya theorem** to Darboux's equation gives a theorem about the existence of a surface with a given line element whose coefficients are analytic functions.

The particularly high level of development of the theory of elliptic Monge — Ampère equations is due to the introduction of the idea of a **generalized solution** and the application of geometric methods for their solution. In the simplest case of the equation $rt - s^2 = \phi$, a generalized solution is defined as a convex function $z(x, y)$ satisfying the equality

$$\iint\limits_{M^*} dp \, dq = \iint\limits_{M} \phi(x, y, z, p, q) \, dx \, dy,$$

where M is an arbitrary Borel set in the xy-plane in the domain in which the solution is to be discussed, and M^* is the so-called normal image of M, which consists of those points of the pq-plane for which p and q are the angular coefficients of the support planes to the surface $z = (x, y)$ at points (x, y, z) projecting into M. A regular (twice-differentiable) solution is also a generalized solution. A generalized solution in the case that ϕ is a continuous positive function is smooth; however, the second derivatives need not exist. The construction of a generalized solution essentially reduces to the purely geometric problem of constructing an infinite convex polyhedron with given directions for the finite faces and given functions on these faces. In particular, if the right-hand side ϕ of the equation depends only on x and y, then this function is the area of the face. Passage to the limit of these polyhedra gives the graph of a generalized solution of the equation. For generalized solutions fairly general existence and uniqueness theorems have been obtained for the solution of the Dirichlet problem. In particular, there is the following theorem. The **Dirichlet problem** for the equation

$$rt - s^2 = \phi(x, y, z, p, q)$$

in a convex domain G is solvable for any continuous boundary values if the curvature of the curve bounding G is positive, if ϕ is continuous, positive, non-decreasing in z, and, as $p^2 + q^2 \to \infty$, has order of growth not exceeding $p^2 + q^2$. For a given direction of convexity this solution is unique.

An essential result in the theory of elliptic Monge — Ampère equations is the *theorem on regularity of generalized solutions*. For the simplest equation

$$rt - s^2 = \phi(x, y, z, p, q)$$

this theorem reads: For $\phi, \phi_z > 0$ every generalized solution with a regular right-hand side (ϕ) is regular. Namely, if ϕ is k times differentiable ($k \geqslant 3$), then the generalized solution is $k + 1$ times differentiable. If ϕ is analytic, then the generalized solution is analytic.

Among the general elliptic Monge — Ampère equations the *strongly-elliptic Monge — Ampère equations* are the ones most studied. These are the equations for which $\phi > 0$ and the quadratic form

$$a\xi^2 + 2b\xi\eta + c\eta^2$$

is non-negative. The basic results, quoted above for the simplest Monge—Ampère equations, have been extended to the strongly-elliptic case. The idea of a generalized solution has been introduced, and under very general conditions the existence and uniqueness of the solution to the Dirichlet problem as well as regularity of a generalized solution, depending on the regularity of the coefficients of the equation, has been proved. Darboux's equation for a line element with positive curvature is, in general, not strongly elliptic. A generalized solution for it is defined as the z-coordinate of a surface realizing the given line element. Darboux's equation has a generalized solution in any domain that is convex in the sense of the given metric (the geodesic curvature of the boundary should be positive). In any such domain, under sufficiently general conditions, the solvability of the Dirichlet problem has been proved. A generalized solution is regular if the coefficients of the line element are regular. The generalized solution is analytic if the coefficients are analytic.

Important results have been obtained for elliptic Monge—Ampère equations on manifolds homeomorphic to the sphere. In particular, two classical problems reduce to such equations: the **Weyl problem** and the **Minkowski problem**. The solutions of these problems, obtained by passage to the limit over polyhedra, are generalized. Regularity of these solutions is obtained from theorems on the regularity of generalized solutions.

Monge—Ampère equations were considered by G. Monge (1784) and A. Ampère (1820).

References

[1] GOURSAT, E.: *Cours d'analyse mathématique*, 3, Gauthier-Villars, 1923, Part 1.
[2] POGORELOV, A.V.: *On Monge—Ampère equations of elliptic type*, Kharkhov, 1960 (in Russian).
[3] POGORELOV, A.V.: *Extrinsic geometry of convex surfaces*, Amer. Math. Soc., 1972 (translated from the Russian).

A.V. Pogorelov

Editorial comments. In 1950 - 1970 the study of *boundary value problems for general two-dimensional Monge—Ampère equations* was a most active research area.

Necessary and sufficient conditions for solvability of the Dirichlet problem and other boundary value problems were obtained for generalized and regular solutions of general two-dimensional elliptic Monge—Ampère equations. Also completely investigated was the regularity of generalized solutions for the most important classes of two-dimensional elliptic Monge—Ampère equations (the Darboux equation, equations $rt - s^2 = f(x, y, z, p, q)$ for which $f, f_z > 0$, and strongly-elliptic equations), under the condition that the prescribed data are sufficiently regular. The *sharp uniqueness* and *non-uniqueness theorems* were obtained for generalized and regular solutions of these boundary value problems. Profound applications to global problems of classical differential geometry were done on the basis of these results.

The global theorems of non-existence of regular solutions (mainly for the Darboux equation) were established for two-dimensional hyperbolic Monge—Ampère equations. A few existence theorems for regular solutions of the Cauchy problem were proved for two-dimensional hyperbolic Monge—Ampère equations. The solution of a few important global problems for saddle surfaces was obtained as an application of these results.

From 1970 up to the present time (1989) the most active research area shifts to the theory of elliptic solutions for real and complex n-dimensional Monge—Ampère equations, $n \geq 2$.

Real multi-dimensional Monge—Ampère equations. New fundamental difficulties arise at the transition from two-dimensional Monge—Ampère equations to the multi-dimensional case. Necessary and sufficient conditions were obtained for generalized and regular elliptic solutions of n-dimensional Monge—Ampère equations, $n \geq 2$. Completely investigated is the regularity of certain generalized elliptic equations under the condition that the prescribed data are sufficiently regular. The sharp uniqueness and non-uniqueness theorems were obtained for generalized and regular solutions of the corresponding boundary value problems. The connections between variational problems and the Dirichlet problem for elliptic solutions of Monge—Ampère equations were also established.

Profound applications to global problems of differential geometry, non-linear elliptic partial differential equations and the calculus of variations were obtained on the basis of these results.

Complex multi-dimensional Monge—Ampère equations. Various important investigations were made for elliptic solutions of n-dimensional complex Monge—Ampère equations. The solvability of the Dirichlet problem for regular elliptic solutions of such equations was obtained. Deep connections with Kähler and other classical complex manifolds were established, and many interesting applications to differential geometry and topology were obtained.

References

[A1] NIRENBERG, L.: 'The Weyl and Minkowski problems in differential geometry in the large', *Comm. Pure Appl. Math.* 6 (1953), 103-156.
[A2] BAKELMAN, I.J.: 'Generalized solutions of Monge—Ampère equations', *Dokl. Akad. Nauk SSSR* 114 (1957), 1143-1145 (in Russian).
[A3] ALEKSANDROV, A.D.: 'The Dirichlet problem for the equation $\det(z_{ij}) = \phi(z_1, \ldots, z_n, z, x_1, \ldots, x_n)$', *Vestnik Leningradsk. Gosudarstv. Univ.*, no. 1 (1958) (in Russian).
[A4] BAKELMAN, I.J.: *Geometric methods for solving elliptic equations*, Moscow, 1965 (in Russian).
[A5] POGORELOV, A.V.: *Monge—Ampère equations of elliptic type*, Noordhoff, 1964 (translated from the Russian).
[A6] EFIMOV, N.V.: 'The appearance of singularities on surfaces of negative curvature', *Mat. Sb.* 64, no. 2 (1964), 286-320 (in Russian).
[A7] EFIMOV, N.V.: 'Surfaces with a slowly changing negative curvature', *Russian Math. Surveys* 21, no. 5 (1966), 1-55. (*Uspekhi Mat. Nauk.* 21, no. 5 (1966), 3-58)
[A8] POZNYAK, E.G.: 'A regular realization of two-dimensional

metrics with negative curvature', *Ukr. Geom. Sb.* 3 (1966), 78-92 (in Russian).

[A9] Pogorelov, A.V.: *The Minkowski multidimensional problem*, Wiley, 1978 (translated from the Russian).

[A10] Cheng, S. and Yau, S.T.: 'On the regularity of the solutions of the Monge−Ampère equations $\det(u_{ij}) = F(x, u)$', *Comm. Pure Appl. Math.* 30 (1977), 41-68.

[A11] Yau, S.T.: 'Survey on partial differential equations in differential geometry', in S.T. Yau (ed.): *Sem. Differential Geom.*, Princeton Univ. Press, 1982, pp. 669-706.

[A12] Yau, S.T.: 'On the Ricci curvature of a compact Kähler manifold and the complex Monge−Ampère equation I', *Comm. Pure Appl. Math.* 31 (1978), 339-411.

[A13] Bakelman, I.J.: 'Applications of the Monge−Ampère operators to the Dirichlet problem for quasilinear elliptic equations', in S.T. Yau (ed.): *Sem. Differential Geom.*, Princeton Univ. Press, 1982, pp. 239-258.

[A14] Aubin, T.: *Nonlinear analysis on manifolds. Monge−Ampère equations*, Springer, 1982.

[A15] Bakelman, I.J.: 'Variational problems and elliptic Monge−Ampère equations', *J. Differential Geom.* 18 (1983), 669-699.

[A16] Caffarelli, L., Nirenberg, L. and Spruck, J.: 'The Dirichlet problem for nonlinear second order elliptic equations, I: Monge−Ampère equations', *Comm. Pure Appl. Math.* 37 (1984), 369-402.

[A17] Caffarelli, L., Kohn, J.J., Nirenberg, L. and Sprick, J.: 'The Dirichlet problem for nonlinear second order elliptic equations, II: Complex Monge−Ampère equations and uniformly elliptic equations', *Comm. Pure Appl. Math.* 38 (1985), 261-301.

[A18] Bakelman, I.J.: 'Generalized elliptic solutions of the Dirichlet problem for n-dimensional Monge−Ampère equations', in F.E. Browder (ed.): *Nonlinear Functional Anal. Applic.*, Proc. Symp. Pure Math., Vol. 45.1, Amer. Math. Soc., 1966, pp. 73-102.

[A19] Gilbarg, D. and Trudinger, N.: *Elliptic partial differential equations of the second order*, Springer, 1984.

[A20] Krylov, N.V.: *Nonlinear elliptic and parabolic equations of the second order*, Kluwer, 1987 (translated from the Russian).

[A21] Minkowski, H.: 'Volumen und Oberflächen', *Math. Ann.* 57 (1903), 447-495.

[A22] Weyl, H.: 'Ueber die Starrheit der Eiflächen und konvexen Flächen durch ihr Linienelement', *Vierteljahrschrift naturforsch. Gesellschaft Zürich* 61 (1916), 40-72.

[A23] Lewy, H.: 'On differential geometry in the large 1', *Trans. Amer. Math. Soc.* 43 (1938), 258-270.

[A24] Calabi, E.: 'Improper affine hyperspheres of convex type and a generalization of a theorem by K. Jörgens', *Michigan Math. J.* 5 (1958), 105-126.

[A25] Serrin, J.: 'The Dirichlet problem for the minimal surface equation in higher dimension', *J. Reine Angew. Math.* 223 (1968), 170-187.

[A26] Bakelman, I.J.: 'Geometric problems in quasilinear elliptic equations', *Russian Math. Surveys* 25, no. 3 (1970), 45-109. (*Uspekhi Mat. Nauk* 25, no. 3 (1970), 49-112)

[A27] Cheng, S.Y. and Yau, S.T.: 'On the regularity of the solution of the n-dimensional Minkowski problem', *Comm. Pure Appl. Math.* 29 (1976), 495-516.

[A28] Lions, P.L.: 'Sur les équations de Monge−Ampère', *Manuscripta Math.* 41 (1983), 1-43.

AMS 1980 Subject Classification: 53A10, 35M05, 35Q20

Monge cone, *directing cone* - The **envelope** of the tangent planes to the integral surface at a point (x_0, y_0, z_0) of a partial differential equation

$$F(x, y, z, p, q) = 0, \qquad (*)$$

where $p = \partial z / \partial x$, $q = \partial z / \partial y$. If F is a non-linear function in p and q, then the general case holds: The tangent planes form a one-parameter family of planes passing through a fixed point; their envelope is a cone. If F is a linear function in p and q, then a bundle of planes passing through a line is obtained, that is, the Monge cone degenerates to the so-called *Monge axis*. The directions of the generators of the Monge cone corresponding to some point (x_0, y_0, z_0) are called *characteristic directions*. A line on the integral surface which is tangent at each point to a corresponding generator of the Monge cone is called a *characteristic line*, a *characteristic*, a *focal curve*, or a *Monge curve*.

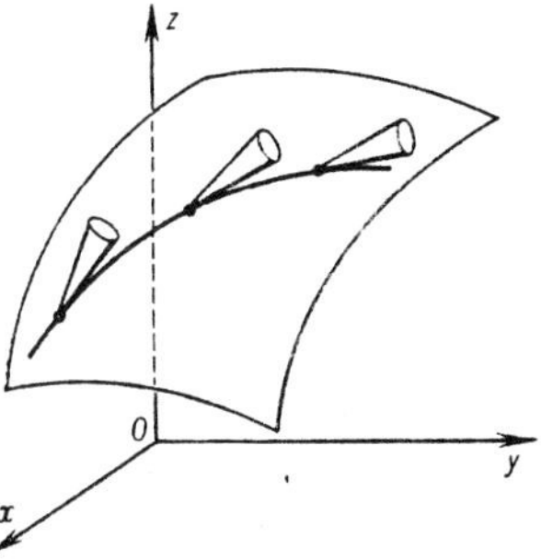

The geometric interpretation (see Fig.) of equation (*), as a field of directing cones, was given by G. Monge (1807).

A.B. Ivanov

Editorial comments. See also **Monge equation**; **Characteristic**; **Characteristic strip** and the references therein.

AMS 1980 Subject Classification: 35A30, 35F20

Monge equation - A differential equation of the form

$$F(x, y, z, dx, dy, dz) = 0.$$

G. Monge (see [1]) studied these equations in connection with the construction of a geometric theory of first-order partial differential equations. A particular case of a Monge equation is a **Pfaffian equation**.

For example, consider a first-order partial differential equation for an unknown function z of two independent variables x, y:

$$\Phi\left(x, y, z, \frac{\partial z}{\partial x}, \frac{\partial z}{\partial y}\right) = 0,$$

then the directions of the generators of the **Monge cone** (characteristic directions) at each point have to satisfy the Monge equation, which can be written in the form

$$M\left(x, y, z, \frac{dy}{dx}, \frac{dz}{dx}\right) = 0.$$

This is an ordinary differential equation in two unknown functions, that is, it is the simplest case of an **underdetermined system.** Often an arbitrary underdetermined system of ordinary differential equations, in which the number of equations is less than the number of unknown functions, is called a Monge equation.

References

[1] MONGE, G.: *Application de l'analyse à la géometrie*, Paris, 1850.
[2] COURANT, R. and HILBERT, D.: *Methods of mathematical physics. Partial differential equations*, 2, Interscience, 1965 (translated from the German).
[3] RASHEVSKIĬ, P.K.: *Geometric theory of partial differential equations*, Moscow-Leningrad, 1947 (in Russian).

N.Kh. Rozov

AMS 1980 Subject Classification: 35A30, 34A34

MONODROMIC FUNCTION *in a domain D of the complex plane* - A single-valued continuous function in D (except, perhaps, for poles). The term 'monodromic function' was applied by A.L. Cauchy in connection with the necessity of subdividing the class of analytic functions (cf. **Analytic function**) into monodromic functions and many-valued analytic functions; at present (1989) the term has gone out of use.

See also **Monodromy theorem**.

E.D. Solomentsev

AMS 1980 Subject Classification: 30A05

MONODROMY GROUP of an ordinary linear differential equation or system of equations - The group of $(n \times n)$-matrices associated with the n-th order system

$$\dot{x} = A(t)x, \qquad (*)$$

defined as follows. Let the matrix $A(t)$ be holomorphic in a domain $G \subset \mathbf{C}$, let $t_0 \in G$ and let $X(t)$ be the **fundamental matrix** of the system (*) given in a small neighbourhood of t_0. If $\gamma \subset G$ is a closed curve with initial point t_0, then by analytic continuation along γ, $X(t) \to X(t)C_\gamma$, where C_γ is a constant $(n \times n)$-matrix. If two curves γ_1, γ_2 are homotopic in G, then $C_{\gamma_1} = C_{\gamma_2}$; if $\gamma = \gamma_1 \gamma_2$, then $C_\gamma = C_{\gamma_1} C_{\gamma_2}$. The mapping $\gamma \to C_\gamma$ is a homomorphism of the **fundamental group** of G:

$$\pi_1(G, t_0) \to \mathrm{GL}(n, \mathbf{C}),$$

where $\mathrm{GL}(n, \mathbf{C})$ is the group of $(n \times n)$-matrices with complex entries; the image of this homomorphism is called the *monodromy group* $M(t_0, G)$ of (*). In this connection,

$$M(t_1, G) = T^{-1} M(t_0, G)T,$$

where T is a constant matrix. The monodromy group has been computed for the equations of Euler and Papperitz (see [1], [2]).

References

[1] GOLUBEV, V.V.: *Vorlesungen über Differentialgleichungen im Komplexen*, Deutsch. Verlag Wissenschaft., 1958 (translated from the Russian).
[2] INCE, E.L.: *Ordinary differential equations*, Dover, reprint, 1956.

M.V. Fedoryuk

Editorial comments. Cf. also **Monodromy matrix** and **Monodromy operator**. If $\gamma(s)$ is a closed differentiable curve in G with initial point t_0, then $Y(s) = X(\gamma(s))$ satisfies a matrix equation $\dot{Y}(s) = \dot{\gamma}(s)A(\gamma(s))Y(s)$ and C_γ is the monodromy matrix of this system of linear differential equations with periodic coefficients.

AMS 1980 Subject Classification: 34A15

MONODROMY MATRIX - A constant $(n \times n)$-matrix $X(\omega)$ which is the value at $t = \omega$ of the **fundamental matrix** $X(t)$, normalized at zero, of a linear system of differential equations

$$\dot{x} = A(t)x, \quad t \in \mathbf{R}, \quad x \in \mathbf{R}^n,$$

with an ω-periodic matrix $A(t)$ that is summable on each compact interval in $\mathbf{R}$.

Yu.V. Komlenko

Editorial comments.

References

[A1] HALE, J.K.: *Ordinary differential equations*, Wiley, 1969.

AMS 1980 Subject Classification: 34A15

MONODROMY OPERATOR - A bounded linear operator $U(T)$ which associates to the initial value $x(0) = x_0$ the value at time T of the solution of a differential equation in a Banach space, $\dot{x} = A(t)x$, where $A(t)$ is a bounded operator, depending on t, that is continuous and periodic with period T: $x(T) = U(T)x_0$. For every solution, $x(t + T) = U(T)x(t)$. In the finite-dimensional case $U(T)$ corresponds to the **monodromy matrix**. The position of the spectrum of the monodromy operator influences the existence of periodic solutions of the equation, the behaviour of the solution at infinity, the reducibility of the equation to an equation with constant coefficients, and the presence of exponential dichotomy. The problems of existence and uniqueness of periodic solutions of an inhomogeneous equation $\dot{x} = A(t)x + f(t)$ with periodic $A(t)$ and $f(t)$ have also been solved in terms of the spectrum of the monodromy operator.

See also **Qualitative theory of differential equations in Banach spaces**.

S.G. Kreĭn

Editorial comments.

References

[A1] DALETSKIĬ, YU.L. and KREIN, M.G.: *Stability of solutions of differential equations in Banach space*, Amer. Math. Soc., 1974 (translated from the Russian).

AMS 1980 Subject Classification: 34GXX, 47E05

MONODROMY THEOREM - A sufficient criterion for the single-valuedness of a **branch of an analytic function**. Let D be a simply-connected domain in the complex space $\mathbf{C}^n$, $n \geqslant 1$. Now, if an analytic function element $\Sigma(z^0;r)$, with centre $z^0 \in D$, can be analytically continued along any path in D, then the branch of an analytic function $f(z)$, $z=(z_1,\ldots,z_n)$, arising by this **analytic continuation** is single-valued in D. In other words, the branch of the analytic function $f(z)$ defined by the simply-connected domain D and the element $\Sigma(z^0;r)$ with centre $z^0 \in D$ must be single-valued. Another equivalent formulation is: If an element $\Sigma(z^0;r)$ can be analytically continued along all paths in an arbitrary domain $D \subset \mathbf{C}^n$, then the result of this continuation at any point $z^* \in D$ (that is, the element $\Sigma(z^*;r^*)$ with centre z^*) is the same for all homotopic paths in D joining z^0 to z^*.

The monodromy theorem is valid also for analytic functions $f(z)$ defined in domains D on Riemann surfaces or on Riemann domains. See also **Complete analytic function**.

References

[1] MARKUSHEVICH, A.I.: *Theory of functions of a complex variable*, 2, Chelsea, 1977 (translated from the Russian).
[2] STOÏLOV, S.: *The theory of functions of a complex variable*, 1-2, Moscow, 1962 (in Russian; translated from the Rumannian).
[3] VLADIMIROV, V.S.: *Methods of the theory of functions of several complex variables*, M.I.T, 1966 (translated from the Russian).

E.D. Solomentsev

Editorial comments.

References

[A1] CONWAY, J.B.: *Functions of one complex variable*, Springer, 1978.

AMS 1980 Subject Classification: 30B40, 32D15

MONODROMY TRANSFORMATION - A transformation of the fibres (or of their homotopy invariants) of a fibre space corresponding to a path in the base. More precisely, let $p:E \to B$ be a locally trivial **fibre space** and let $\gamma:[0,1] \to B$ be a path in B with initial point $a=\gamma(0)$ and end-point $b=\gamma(1)$. A trivialization of the fibration $\gamma^*(E)$ defines a homeomorphism T_γ of the fibre $p^{-1}(a)$ onto the fibre $p^{-1}(b)$, $T_\gamma: p^{-1}(a) \to p^{-1}(b)$. If the trivialization of $\gamma^*(E)$ is modified, then T_γ changes into a homotopically-equivalent homeomorphism; this also happens if γ is changed to a homotopic path. The homotopy type of T_γ is called the *monodromy transformation* corresponding to a path γ. When $a=b$, that is, when γ is a loop, the monodromy transformation T_γ is a homeomorphism of $F=p^{-1}(a)$ into itself (defined, yet again, up to a homotopy). This mapping, and also the homomorphisms induced by it on the homology and cohomology spaces of F, is also called a monodromy transformation. The correspondence of γ with T_γ gives a representation of the **fundamental group** $\pi_1(B,a)$ on $H^*(F)$.

The idea of a monodromy transformation arose in the study of multi-valued functions (see **Monodromy theorem**). If $S \to P^1(\mathbf{C})$ is the **Riemann surface** of such a function, then by eliminating the singular points of the function from the Riemann sphere $P^1(\mathbf{C})$, an unbranched covering is obtained. The monodromy transformation in this case is also called a *covering* or *deck transformation*.

The monodromy transformation arises most frequently in the following situation. Let $D=\{z \in \mathbf{C}: |z|<1\}$ be the unit disc in the complex plane, let X be an **analytic space**, let $f:X \to D$ be a proper holomorphic mapping (cf. **Proper morphism**), let X_t be the fibre $f^{-1}(t)$, $t \in D$, $D^* = D \setminus \{0\}$, and let $X^* = f^{-1}(D^*)$. Diminishing, if necessary, the radius of D, the fibre space $f:X^* \to D^*$ can be made locally trivial. The monodromy transformation T associated with a circuit around 0 in D is called the *monodromy of the family* $f:X \to D$ at $0 \in D$, it acts on the (co)homology spaces of the fibre X_t, where $t \in D^*$. The most studied case is when X and the fibres X_t, $t \neq 0$, are smooth. The action of T on $H^*(X_t, \mathbf{Q})$, in this case, is *quasi-unipotent* [4], that is, there are positive integers k and N such that $(T^k - 1)^N = 0$. The properties of the monodromy display many characteristic features of the degeneracy of the family $f:X \to D$. The monodromy of the family $f:X \to D$ is closely related to mixed Hodge structures (cf. **Hodge structure**) on the cohomology spaces $H^*(X_0)$ and $H^*(X_t)$ (see [5] - [7]).

When the singularities of $f:X \to D$ are isolated, the monodromy transformation can be localized. Let x be a singular point of f (or, equivalently, of X_0) and let B be a sphere of sufficiently small radius in X with centre at x. Diminishing, if necessary, the radius of D, a local trivialization of the fibre space $B \cap f^{-1}(D^*)$ can be defined. It is compatible with the trivialization of the fibre space $\partial B \cap f^{-1}(D) \to D$ on the boundary. This gives a diffeomorphism T of the manifold of 'vanishing cycles' $V_t = B \cap X_t$ into itself which is the identity on ∂V_t, and which is called the *local monodromy* of f at x. The action of the monodromy transformation on the cohomology spaces $H^*(V_t)$ reflects the singularity of f at x (see [1], [2], [7]). It is known that the manifold V_t is homotopically equivalent to a bouquet of μ n-dimensional spheres, where $n+1 = \dim X$ and μ is the Milnor number of the germ of f at x.

The simplest case is that of a Morse singularity when, in a neighbourhood of x, f reduces to the form $f = z_0^2 + \cdots + z_n^2$ (cf. **Morse lemma**). In this case $\mu = 1$, and the interior V_t^0 of V_t is diffeomorphic to the tangent bundle of the n-dimensional sphere S^n. A *vanishing cycle* δ is a generator of the cohomology group with compact support $H_c^n(V_t^0, \mathbf{Z}) \cong \mathbf{Z}$, defined up to sign. In general, if $f:X \to D$ is a proper holomorphic mapping (as above, having a unique Morse singularity at x), then a cycle δ_x *vanishing at x* is the image of a cycle $\delta \in H_c^n(V_t^0)$ under the natural mapping

$H^n_c(V^0_t) \to H^n(X_t)$. In this case the specialization homomorphism $r^*_t: H^i(X_0) \to H^i(X_t)$ is an isomorphism for $i \neq n, n+1$, and the sequence

$$0 \to H^n(X_0) \to H^n(X_t) \xrightarrow{(\,,\delta_x)} \mathbf{Z} \to$$
$$\to H^{n+1}(X_0) \to H^{n+1}(X_t) \to 0$$

is exact. The monodromy transformation T acts trivially on $H^i(X_t)$ for $i \neq n$ and its action on $H^n(X_t)$ is given by the *Picard–Lefschetz formula*: For $z \in H^n(X_t)$,

$$T_z = z \pm (z, \delta_x)\delta_x.$$

The sign in this formula and the values of (δ_x, δ_x) are collected in the table.

$n \bmod 4$	0	1	2	3
$\pm$	$-$	$-$	$+$	$+$
(δ_x, δ_x)	2	0	-2	0

A monodromy transformation preserves the intersection form on $H^n(X_t)$.

Vanishing cycles and monodromy transformations are used in the *Picard–Lefschetz theory*, associating the cohomology space of a projective complex manifold and its hyperplane sections. Let $X \subset P^N$ be a smooth manifold of dimension $n+1$, and let $\{X_t\}$, $t \in P^1$, be a pencil of hyperplane sections of X with basic set (axis of the pencil) $Y \subset X$; let the following conditions be satisfied: a) Y is a smooth submanifold in X; b) there is a finite set $S \subset P^1$ such that X_t is smooth for $t \in P^1 \setminus S$; and c) for $s \in S$ the manifold X_s has a unique non-degenerate quadratic singular point x_s, where $x_s \in Y$. Pencils with these properties (*Lefschetz pencils*) always exist. Let $\sigma: \overline{X} \to X$ be a monoidal transformation with centre on the axis Y of the pencil, and let $f: \overline{X} \to P^1$ be the morphism defined by the pencil $\{X_t\}$; here $f^{-1}(t) \cong X_t$ for all $t \in P^1$. Let a point $0 \in P^1 \setminus S$ be fixed; then the monodromy transformation gives an action of $\pi_1(P^1 \setminus S, 0)$ on $H^i(X_0)$ (non-trivial only for $i = n$). To describe the action of the monodromy on $H^n(X_0)$ one chooses points s', situated near $s \in S$, and paths γ'_s leading from 0 to s'. Let $\gamma_s \in \pi_1(P^1 \setminus S, 0)$ be the loop constructed as follows: first go along γ'_s, then once round s and, finally, return along γ'_s to 0. In addition, let δ_s be a cycle vanishing at x_s (more precisely, take a vanishing cycle in $H^n(X_{s'})$ and transfer it to $H^n(X_0)$ by means of the monodromy transformation corresponding to the path γ'_s). Finally, let $E \subset H^n(X_0, \mathbf{Q})$ be the subspace generated by the vanishing cycles δ_s, $s \in S$ (the *vanishing cohomology space*). Then the following hold.

1) $\pi_1(P^1 \setminus S, 0)$ is generated by the elements γ_s, $s \in S$;

2) the action of γ_s is given by the formula

$$\gamma_s(z) = z \pm (z, \delta_s)\delta_s;$$

3) the space $E \subset H^n(X_0)$ is invariant under the action of the monodromy group $\pi_1(P^1 \setminus S, 0)$;

4) the space of elements in $H^n(X_0)$ that are invariant relative to monodromy coincides with the orthogonal complement of E relative to the intersection form on $H^n(X_0)$, and also with the images of the natural homomorphisms $H_n(\overline{X}) \to H_n(X_0)$ and $H^n(X) \to H^n(X_0)$;

5) the vanishing cycles $\pm \delta_s$ are conjugate (up to sign) under the action of $\pi_1(P^1 \setminus S, 0)$;

6) the action of $\pi_1(P^1 \setminus S, 0)$ on E is absolutely irreducible.

The formalism of vanishing cycles, monodromy transformations and the Picard–Lefschetz theory has also been constructed for l-adic cohomology spaces of algebraic varieties over any field (see [3]).

References

[1] ARNOL'D, V.I.: 'Normal forms of functions in neighbourhoods of degenerate critical points', *Russian Math. Surveys* **29**, no. 2 (1974), 10-50. (*Uspekhi Mat. Nauk* **29**, no. 2 (1974), 11-49)

[2] MILNOR, J.: *Singular points of complex hypersurfaces*, Princeton Univ. Press, 1968.

[3] DELIGNE, P. and KATZ, N. (EDS.): *Groupes de monodromie en géométrie algébrique. SGA 7^{II}*, Lecture notes in math., 340, Springer, 1973.

[4] CLEMENS, C.H.: 'Picard–Lefschetz theorem for families of nonsingular algebraic varieties acquiring ordinary singularities', *Trans. Amer. Math. Soc.* **136** (1969), 93-108.

[5] SCHMID, W.: 'Variation of Hodge structure: the singularities of the period mapping', *Invent. Math.* **22** (1973), 211-319.

[6] STEENBRINK, J.: 'Limits of Hodge structures', *Invent. Math.* **31** (1976), 229-257.

[7] STEENBRINK, J.H.M.: 'Mixed Hodge structure on the vanishing cohomology', in P. Holm (ed.): *Real and Complex Singularities, Oslo, 1976. Proc. Nordic Summer School*, Sythoff & Noordhoff, 1977, pp. 524-563.

[8] LEFSCHETZ, S.: *L'analysis situs et la géométrie algébrique*, Gauthier-Villars, 1924.

[9] LEFSCHETZ, S.: 'A page of mathematical autobiography', *Bull. Amer. Math. Soc.* **74**, no. 5 (1968), 854-879.

V.I. Danilov

AMS 1980 Subject Classification: 55R05, 55R55, 14D05, 14E20

MONOGENEITY SET - The set of all derived numbers (Dini derivatives, cf. **Dini derivative**) of a given function of a complex variable at a given point. More precisely, let E be a set in the complex plane $\mathbf{C}$, let ζ be a non-isolated point of E and let $f(z)$ be a complex function of $z \in E$. A complex number a (proper or equal to ∞) is called a *derived number* (or *Dini derivative*) of f at ζ relative to E if there is a sequence $z_n \in E$ with the properties: $z_n \neq \zeta$, $z_n \to \zeta$,

$$\frac{f(z_n) - f(\zeta)}{z_n - \zeta} \to a \quad \text{as } n \to \infty.$$

The set $\mathfrak{M}(\zeta, f, E)$ of all derived numbers of f at ζ relative to E is called the *monogeneity set* of f at ζ relative to E (see [1]). The set $\mathfrak{M}(\zeta, f, E)$ consists of a unique finite point a if and only if $f(z)$ is a **monogenic function** at ζ relative to E and $f'_E(\zeta) = a$. The set $\mathfrak{M}(\zeta, f, E)$ is always closed, and for each closed set A in the extended complex plane $\overline{\mathbf{C}}$, each set $E \subset \mathbf{C}$ and each finite non-isolated point ζ of E, there is a function $f(z)$, $z \in E$, such that $\mathfrak{M}(\zeta, f, E) = A$. If ζ is an interior point of E, then for any function $f(z)$ that is continuous in a neighbourhood of ζ, the set $\mathfrak{M}(\zeta, f, E)$ is closed and connected (a **continuum**) in $\overline{\mathbf{C}}$ and, conversely, for any continuum $K \subset \overline{\mathbf{C}}$ there is a function $f(z)$, continuous in a neighbourhood of ζ, for which $\mathfrak{M}(\zeta, f, E) = K$. If $f(z) = f(x + iy) = u(x, y) + iv(x, y)$ is differentiable with respect to the set of real variables (x, y) at an interior point $\zeta = \xi + i\eta$ of E, then $\mathfrak{M}(\zeta, f, E)$ is the circle $\gamma(r, c) = \{w : |w - c| = r\}$ (possibly degenerating into a point, $r = 0$) with centre $c = \partial f(\zeta)/\partial z$ and radius $r = |\partial f(\zeta)/\partial \overline{z}|$, where

$$\frac{\partial f}{\partial z} = \frac{1}{2}\left[\frac{\partial f}{\partial x} - i\frac{\partial f}{\partial y}\right] = \frac{1}{2}\left[\frac{\partial u}{\partial x} + \frac{\partial v}{\partial y}\right] + \frac{i}{2}\left[\frac{\partial v}{\partial x} - \frac{\partial u}{\partial y}\right],$$

$$\frac{\partial f}{\partial \overline{z}} = \frac{1}{2}\left[\frac{\partial f}{\partial x} + i\frac{\partial f}{\partial y}\right] = \frac{1}{2}\left[\frac{\partial u}{\partial x} - \frac{\partial v}{\partial y}\right] + \frac{i}{2}\left[\frac{\partial v}{\partial x} + \frac{\partial u}{\partial y}\right]$$

are the so-called *formal derivatives*. The converse is also true: Each circle is the monogeneity set for some function f, differentiable with respect to (x, y), at a given interior point ζ of E.

If $f(z)$ is continuous in a domain G, then at almost every $\zeta \in G$ the set $\mathfrak{M}(\zeta, f, G)$ is either a circle $\gamma(r, c)$, $0 \leqslant r < \infty$, or is $\overline{\mathbf{C}}$ (see [2]). In the general case of an arbitrary (not necessarily measurable) set E and an arbitrary (not necessarily measurable) finite function $f(z)$, $z \in E$, at almost every point $\zeta \in E$ one of the following three cases holds: a) $\mathfrak{M}(\zeta, f, E) = \gamma(r, c)$, $c \in \mathbf{C}$, $0 \leqslant r < \infty$; b) $\mathfrak{M}(\zeta, f, E) = \overline{\mathbf{C}}$; or c) $\mathfrak{M}(\zeta, f, E) = \gamma(r, c) \bigcup \infty$, $c \in \mathbf{C}$, $0 \leqslant r < \infty$. Here, a) holds at almost every differentiability point of $f(z) = f(x + iy)$ with respect to $(x, y) \in E$ and one of the first two cases holds at almost every continuity point of $f(z)$. Each of the cases a) - c) may be realized individually at almost every point $\zeta \in E$.

For some natural generalizations to the multi-dimensional case see [4].

References

[1] FEDOROV, V.S.: 'The work of N.N. Luzin on the theory of functions of a complex variable', *Uspekhi Mat. Nauk* **7**, no. 2 (1952), 7-16 (in Russian).
[2] TROKHIMCHUK, YU.YU.: *Continuous mappings and monogeneity conditions*, Moscow, 1963 (in Russian).
[3] DOLZHENKO, E.P.: 'On the derived numbers of complex functions', *Izv. Akad. Nauk SSSR Ser. Mat.* **26** (1962), 347-360 (in Russian).
[4] BONDAR, A.V.: 'Continuous operator conformal mappings', *Ukr. Math. J.* **32**, no. 3 (1980), 207-212. (*Ukrain. Mat. Zh.* **32**, no. 3 (1980), 314-322)

$E.P.\ Dolzhenko$

AMS 1980 Subject Classification: 30A05

MONOGENIC FUNCTION - A function of a complex variable having a finite derivative. More precisely, a function $f(z)$ defined on a set E of the complex plane $\mathbf{C}$ is called *monogenic* (with respect to E) at a finite non-isolated point $\zeta \in E$ if it has a finite derivative $f'_E(\zeta)$ with respect to $z \in E$ at this point:

$$f'_E(\zeta) = \lim_{\substack{z \to \zeta \\ z \in E}} \frac{f(z) - f(\zeta)}{z - \zeta}.$$

A function which is monogenic at every non-isolated point of E is called *monogenic on E*.

If $E = G$ is a domain in $\mathbf{C}$, then a monogenic function on G is called an **analytic function** on G.

If E is not a domain, then a monogenic function on E, in general, will no longer have the typical properties of an analytic function. However, if a set E, which itself is not a domain, is sufficiently 'massive' near a majority of its points (more precisely, if the complement of E in $\mathbf{C}$ is sufficiently thin near a majority of $\zeta \in E$), then a monogenic function on E has, in a weakened form, many of the properties of an analytic function. In attempts to gain a deep understanding of connections between properties of analytic functions, the notion of an analytic function has been generalized in various directions: by generalizing the domain of definition, by generalizing the very notion of the derivative itself, by weakening the Cauchy−Riemann conditions, by weakening the condition of Morera's theorem, etc. (see [13], Chapt. 6). To this end, for 'thin' sets E, for example, for a segment $E = [a, b] \subset (-\infty, \infty)$, the so-called quasi-analytic classes (cf. **Quasi-analytic class**) of functions have been picked out. Functions defined on nowhere-dense sufficiently massive compact sets E, which are close to analytic functions in the sense that they can be uniformly approximated to any accuracy by analytic functions, or, what is the same thing, by rational functions of z, have also been studied.

Below some results in the areas listed are given.

1. Functions monogenic in a domain. If $f(z) = f(x + iy) = u(x, y) + iv(x, y)$ (here $u(x, y)$ and $v(x, y)$ are real-valued functions), then for $f(z)$ to be analytic in a domain G it is sufficient (and necessary) that at every point $x + iy \in G$ the following two conditions are simultaneously satisfied: 1) $u(x, y)$ and $v(x, y)$ have total differentials $du(x, y)$ and $dv(x, y)$ with respect to the set of real variables (x, y); and 2) the Cauchy−Riemann conditions are satisfied:

$$\frac{\partial u(x, y)}{\partial x} = \frac{\partial v(x, y)}{\partial y}, \quad \frac{\partial u(x, y)}{\partial y} = -\frac{\partial v(x, y)}{\partial x}.$$

The conditions of this theorem can be weakened and generalized. For example, it has been shown that the requirement of existence of the total differentials for u and v can be replaced by the essentially weaker condition of boundedness of u and v in G and, while preserving the requirement of the existence of the first partial derivatives of u and v everywhere in G, requiring only that the Cauchy–Riemann equations be satisfied almost-everywhere in G (in the sense of the Lebesgue measure in the plane) (see [6]).

Let $L(z)$ be any pair of distinct lines intersecting at z, let $R(z)$ be any triple of pairwise non-collinear rays emanating from z and let $E(z)$ be any measurable set having z as a Lebesgue **density point**. Analyticity of a continuous function $f(z)$ in G is guaranteed by each of the following conditions individually (see [3] - [5]): a) $f'_{L(z)} \neq \infty$ exists everywhere in G; b) $f'_{R(z)} \neq \infty$ exists everywhere in G; or c) $f'_{E(z)} \neq \infty$ exists everywhere in G. Here each of the sets $L(z)$, $R(z)$ and $E(z)$ has its own definition at each point. It must be noted that $f'_{E(z)}(z)$ depends on f and z but not on $E(z)$; the derivative

$$f'_{as}(z) = f'_{E(z)}(z)$$

is called an *asymptotic (approximate)* derivative, and a function having a asymptotic derivative at a point z (respectively, in a domain G) is called *asymptotically monogenic* at z (in G). Instead of condition b) it suffices to require the existence of a limit for

$$\arg \frac{f(z')-f(z)}{z'-z} \quad \text{as } z' \to z, \ z' \in R(z).$$

2. Functions monogenic on nowhere-dense sets. E. Borel [8] constructed a connected **perfect set** (a **continuum**) E without interior points and an increasing sequence $E_1 \subset E_2 \subset \cdots \subset E$ of perfect sets such that the area $\text{mes}_2 E$ of E is positive,

$$\text{mes}_2 \left[E \setminus \bigcup_{n=1}^{\infty} E_n \right] = 0,$$

and such that monogeneity of $f(z)$ on E (or even on each set E_n, $n = 1, 2, \ldots$) implies that $f(z)$ is infinitely differentiable with respect to $z \in E_{n_0}$ for each fixed n_0. One could give comparatively general sufficient conditions on E so that $f(z)$ has this property, or the weaker property of k times differentiability with respect to $z \in E_n$, where k is a natural number given in advance. Sufficient conditions on the nowhere-dense continuum E have also been found so that for monogenic functions on it, analogues of Cauchy's integral theorem, Taylor series expansions and various forms of uniqueness theorems hold. One example of the latter: If two monogenic functions $f_1(z)$ and $f_2(z)$ on E coincide on some **portion** of E (that is, on the non-empty intersection of E with an open disc), then $f_1(z) \equiv f_2(z)$ on E (see [9]).

3. Functions close to analytic functions. As is well known, any function $f(z)$ that is analytic in a domain G is the limit of a sequence of rational functions $r_n(z)$ which converge to $f(z)$ uniformly on each compact set $K \subset G$. If one considers a sufficiently massive nowhere-dense continuum E and the class $R(E)$ of functions

$f(z)$, $z \in E$, for which there is a sequence of rational functions $r_n(z, f)$ uniformly converging to $f(z)$, then one obtains yet another generalization of the notions of a domain and a function $f(z)$ analytic on E. If E is a nowhere-dense set in $\mathbf{C}$, then one can always find a function $f \in R(E)$ not having a derivative $f'_E(z)$ at any $z \in E$ (see [10]). However, for every m, $2 \leq m \leq \infty$, it is possible to give conditions on E under which every $f \in R(E)$ has derivatives $f^{(k)}_{E_n}(z)$, where $1 \leq k < m$, where E_n is closed, $E_n \subset E$ and $\text{mes}_2(E \setminus E_n) < 1/n$ (see [11]). M.V. Keldysh has constructed an example of a nowhere-dense continuum E on which the uniqueness theorem in the form quoted at the end of Subsection 2 holds for functions $f_1, f_2 \in R(E)$ (see also [12]). (Concerning the monogenic properties of functions from $R(E)$ in terms of functional analysis see [12], Chapt. I, Sect. 17.) The dependence of the monogenic properties of a function $f \in R(E)$ on the speed of its approximation by rational functions has also been investigated.

References
[1] FEDOROV, V.S.: 'The work of N.N. Luzin on the theory of functions of a complex variable', *Uspekhi Mat. Nauk* **7**, no. 2 (1952), 7-16 (in Russian).
[2] BOHR, H.: 'Ueber streckentreue und konforme Abbildung', *Math. Z.* **1** (1918), 403-420.
[3] MEN'SHOV, D.E.: 'Sur la généralisation des conditions de Cauchy–Riemann', *Fund. Math.* **25** (1935), 59-97.
[4] MEN'SHOV, D.E.: *Les conditions de monogénéité*, Hermann, 1936.
[5] MEN'SHOV, D.E.: 'On asymptotic monogeneity', *Mat. Sb.* **1** (1936), 189-210 (in Russian).
[6] TOLSTOV, G.P.: 'On curvilinear and repeated integrals', *Trudy Mat. Inst. Steklov.* **35** (1950) (in Russian).
[7] TROKHIMCHUK, YU.YU.: *Continuous mappings and monogeneity conditions*, Moscow, 1963 (in Russian).
[8] BOREL, E.: *Leçons sur les fonctions monogènes uniformes d'une variable complexe*, Gauthier-Villars, 1917.
[9] SELEZNEV, A.I.: 'On functions monogenic on nowhere closed sets and sets of type F_σ', *Dokl. Akad. Nauk SSSR* **108**, no. 4 (1956), 591-594 (in Russian).
[10] DOLZHENKO, E.P.: 'Construction on a nowhere dense continuum of a nowhere differentiable function which can be expanded in a series of rational functions', *Dokl. Akad. Nauk SSSR* **125**, no. 5 (1959), 970-973 (in Russian).
[11] DOLZHENKO, E.P.: 'Approximation on closed regions and zero-sets', *Soviet Math. Dokl.* **3**, no. 2 (1962), 472-475. (*Dokl. Akad. Nauk SSSR* **143**, no. 4 (1962), 771-774)
[12] MEL'NIKOV, M.S. and SINANYAN, S.O.: 'Aspects of approximation theory for functions of one complex variable', *J. Soviet Math.* **5**, no. 5 (1976), 688-752. (*Itogi Nauk. i Tekhn. Sovrem. Probl. Mat.* **4** (1975), 143-250)
[13] BERMANT, A.F. and MARKUSHEVICH, A.I.: 'Theory of functions of a complex variable', in *Mathematics in the USSR during thirty years: 1917-1947*, Moscow-Leningrad, 1948, pp. 319-481 (in Russian).
[14] TELYAKOVSKIĬ, D.S.: 'Generalization of the Looman–Men'shov theorem', *Math. Notes* **19**, no. 4 (1986), 296-301. (*Mat. Zametki* **39**, no. 4 (1986), 539-549)

E.P. Dolzhenko

AMS 1980 Subject Classification: 30A05

MONOGENIC SEMI-GROUP, *cyclic semi-group* - A

semi-group generated by one element. The monogenic semi-group generated by an element a is usually denoted by $<a>$ (sometimes by $[a]$) and consists of all powers a^k with natural exponents. If all these powers are distinct, then $<a>$ is isomorphic to the additive semi-group of natural numbers. Otherwise $<a>$ is finite, and then the number of elements in it is called the *order of the semi-group* $<a>$, and also the *order of the element a*. If $<a>$ is infinite, then a is said to have *infinite order*. For a finite monogenic semi-group $A=<a>$ there is a smallest number h with the property $a^h=a^k$, for some $k>h$; h is called the *index of the element a* (and also *the index of the semi-group A*). In this connection, if d is the smallest number with the property $a^h=a^{h+d}$, then d is called the *period* of a (of A). The pair (h, d) is called the *type* of a (of A). For any natural numbers h and d there is a monogenic semi-group of type (h, d); two finite monogenic semi-groups are isomorphic if and only if their types coincide. If (h, d) is the type of a monogenic semi-group $A=<a>$, then $a, \ldots, a^{h+d-1}$ are distinct elements and, consequently, the order of A is $h+d-1$; the set

$$G = \{a^h, \ldots, a^{h+d-1}\}$$

is the largest subgroup and smallest ideal in A; the identity e of the group G is the unique idempotent in A, where $e=a^{ld}$ for any l such that $ld \geqslant h$; G is a **cyclic group**, a generator being, for example, ae. An idempotent of a monogenic semi-group is a unit (zero) in it if and only if its index (respectively, period) is equal to 1; this is equivalent to the given monogenic semi-group being a **group** (respectively, a **nilpotent semi-group**). Every sub-semi-group of the infinite monogenic semi-group is finitely generated.

References

[1] CLIFFORD, A.H. and PRESTON, G.B.: *The algebraic theory of semigroups*, 1, Amer. Math. Soc., 1961.
[2] LYAPIN, E.S.: *Semigroups*, Amer. Math. Soc., 1974 (translated from the Russian).

L.N. Shevrin

AMS 1980 Subject Classification: 20M14

MONOID - A term used as an abbreviation for the phrase '**semi-group** with identity'. Thus, a *monoid* is a set M with an associative binary operation, usually called *multiplication*, in which there is an element e such that $ex=x=xe$ for any $x \in M$. The element e is called the *identity* (or *unit*) and is usually denoted by 1. In any monoid there is exactly one identity. If the operation given on the monoid is commutative, it is often called *addition* and the identity is called the *zero* and is denoted by 0.

Examples of monoids. 1) The set of all mappings of an arbitrary set S into itself is a monoid relative to the operation of successive application (composition) of mappings. The identity mapping is the identity. 2) The set of endomorphisms of a **universal algebra** A is a monoid relative to composition; the identity is the identity endomorphism. 3) Every **group** is a monoid.

Every semi-group P without an identity can be imbedded in a monoid. For this it suffices to take a symbol 1 not in P and give a multiplication on the set $P \cup \{1\}$ as follows: $1 \cdot 1 = 1$, $1 \cdot x = x = x \cdot 1$ for any $x \in P$, and on elements from P the operation is as before. Every monoid can be represented as the monoid of all endomorphisms of some universal algebra.

An arbitrary monoid can also be considered as a **category** with one object. This allows one to associate with a monoid M its dual (opposite, adjoint) monoid M^{op}. The elements of both monoids coincide, but the product of x and y in M^{op} is put equal to the product yx in M.

The development of the theory of monoids and adjoint functors has shown the utility of the definition of a monoid in so-called *monoidal categories*. Suppose given a category $\mathfrak{M}$ equipped with a bifunctor $\otimes : \mathfrak{M} \times \mathfrak{M} \to \mathfrak{M}$, an object Z and natural isomorphisms

$$\alpha_{ABC} : (A \otimes B) \otimes C \to A \otimes (B \otimes C),$$

$$\lambda_A : Z \otimes A \to A, \quad \rho_A : A \otimes Z \to A,$$

satisfying coherence conditions. An object M is called a monoid in the category $\mathfrak{M}$ if there are morphisms $\mu : M \otimes M \to M$ and $\epsilon : Z \to M$ such that the following diagrams are commutative:

$$
\begin{array}{ccccc}
(M \otimes M) \otimes M & \xrightarrow{\mu \otimes 1_M} & M \otimes M & \xrightarrow{\mu} & M \\
{\scriptstyle \alpha_{MMM}} \downarrow & & & & \| \\
M \otimes (M \otimes M) & \xrightarrow[1_M \otimes \mu]{} & M \otimes M & \xrightarrow[\mu]{} & M
\end{array}
$$

$$
\begin{array}{ccccc}
Z \otimes M & \xrightarrow{\epsilon \otimes 1_M} & M \otimes M & \xleftarrow{1_M \otimes \epsilon} & M \otimes Z \\
& {\scriptstyle \lambda_M} \searrow & \downarrow{\scriptstyle \mu} & \swarrow {\scriptstyle \rho_M} & \\
& & M & &
\end{array}
$$

If $\mathfrak{M}$ is taken to be the category of sets (cf. **Sets, category of**), $\otimes$ the **Cartesian product**, Z a one-point set, and the isomorphisms α, λ and ρ are chosen in the natural way $(\alpha((a, b), c)=(a, (b, c))$, $\lambda(z, a)=a=\rho(a, z))$, then the second definition of a monoid turns out to be equivalent to the original definition.

References

[1] CLIFFORD, A.H. and PRESTON, G.B.: *The algebraic theory of semigroups*, 1-2, Amer. Math. Soc., 1961-1967.
[2] MACLANE, S.: *Categories for the working mathematician*, Springer, 1971.

M. Sh. Tsalenko

Editorial comments. For monoidal categories, and particularly the *coherence conditions* that the isomorphisms α_{ABC}, λ_A must satisfy cf. [1], Chapt. 7, Sects. 1-2.

AMS 1980 Subject Classification: 20M99, 18D10

MONOIDAL TRANSFORMATION, *blowing up, σ-process* - A special kind of **birational morphism** of an algebraic variety or bimeromorphic morphism of an analytic space. For example, let X be an algebraic variety (or an arbitrary scheme), and let $D \subset X$ be a closed subvariety given by a sheaf of ideals J. The *monoidal transformation of X with centre D* is the X-scheme $X^1 = \text{Proj}(\oplus_{n \geqslant 0} J^n)$ — the projective spectrum of the graded sheaf of $\mathcal{O}_X$-algebras $\oplus_{n \geqslant 0} J^n$. If $f: X^1 \to X$ is the structure morphism of the X-scheme X^1, then the sheaf of ideals $f^*(J) = J \cdot \mathcal{O}_{X^1}$ on X^1 (defining the exceptional subscheme $f^{-1}(D)$ on X^1) is invertible. This means that $f^{-1}(D)$ is a **divisor** on X^1; in addition, f induces an isomorphism between $X^1 \setminus f^{-1}(D)$ and $X \setminus D$. A monoidal transformation $f: X^1 \to X$ of a scheme X with centre D is characterized by the following universal property [1]: The sheaf of ideals $f^*(J)$ is invertible and for any morphism $g: X_1 \to X$ for which $g^*(J)$ is invertible there is a unique morphism $h: X_1 \to X^1$ such that $g = f \circ h$.

A monoidal transformation of an algebraic or analytic space X with as centre a closed subspace $D \subset X$ can be defined and characterized in the same way.

An important class of monoidal transformations are the *admissible monoidal transformations*, which are distinguished by the condition that D is non-singular and X is a *normally flat scheme along D*. The latter means that all sheaves J^n / J^{n+1} are flat $(\mathcal{O}_X / J)$-modules. The importance of admissible monoidal transformations is explained by the fact that they do not worsen the singularities of the variety. In addition, it has been proved (see [1]) that a suitable sequence of admissible monoidal transformations improves singularities, which allows one to prove the theorem on the resolution of singularities for an algebraic variety over a field of characteristic zero.

Admissible monoidal transformations of non-singular varieties are particularly simple to construct. If $f: X_1 \to X$ is a monoidal transformation with a non-singular centre $D \subset X$, then X_1 is again non-singular and the exceptional subspace $f^{-1}(D)$ is canonically isomorphic to the projectivization of the conormal sheaf to D in X. In the special case when D consists of one point, the monoidal transformation consists of blowing up this point into the whole projective space of tangent directions. For the behaviour of various invariants of non-singular varieties (such as Chow rings, cohomology spaces, the K-functor, and Chern classes) under admissible monoidal transformations see [2] - [5].

References

[1] HIRONAKA, H.: 'Resolution of an algebraic variety over a field of characteristic zero I, II', *Ann. of Math.* (2) **79** (1964), 109-326.

[2] BERTHELOT, P., GROTHENDIECK, A. and ILLUSIE, L., ET AL. (EDS.): *Théorie des intersections et théorème de Riemann − Roch,*

SGA 6, Lecture notes in math., 225, Springer, 1971.

[3] GROTHENDIECK, A., ET AL. (EDS.): *Cohomologie l-adique et fonctions L. SGA 5,* Springer, 1977.

[4] PORTEOUS, I.: 'Blowing up Chern classes', *Proc. Cambridge Philos. Soc.* **56**, no. 2 (1960), 118-124.

[5] MANIN, YU.I.: 'Lectures on the K-functor in algebraic geometry', *Russian Math. Surveys* **24** (1969), 1-89. (*Uspekhi Mat. Nauk* **24**, no. 5 (1969), 3-86)

V.I. Danilov

Editorial comments. The word 'σ-process' appeared for the first time in [A1].

References

[A1] HOPF, H.: 'Schlichte Abbildungen und lokale Modifikationen 4-dimensionaler komplexer Mannigfaltikeiten', *Comm. Math. Helv.* **29** (1954), 132-156.

AMS 1980 Subject Classification: 14B25

MONOMIAL - The simplest form of an algebraic expression, a **polynomial** containing only one term.

Like polynomials (see **Ring of polynomials**), monomials can be considered not only over a field but also over a ring. A monomial over a commutative ring A in a set of variables $\{x_i\}$, where i runs through some index set I, is a pair (a, ν), where $a \in A$ and ν is a mapping of the set I into the set of non-negative integers, where $\nu(i) = 0$ for all but a finite number of i. A monomial is usually written in the form

$$a x_{i_1}^{\nu(i_1)} \cdots x_{i_n}^{\nu(i_n)},$$

where $i_1, \ldots, i_n$ are all the indices for which $\nu(i) > 0$. The number $\nu(i)$ is called the *degree of the monomial in the variable x_i*, and the sum $\sum_{i \in I} \nu(i)$ is called the *total degree of the monomial*. The elements of the ring can be regarded as monomials of degree 0. A monomial with $a = 1$ is called *primitive*. Any monomial with $a = 0$ is identified with the element $0 \in A$.

The set of monomials over A in the variables $\{x_i\}$, $i \in I$, forms a commutative semi-group with identity. Here the product of two monomials (a, ν) and (b, κ) is defined as $(ab, \nu + \kappa)$.

Let B be a commutative A-algebra. Then the monomial $a x_{i_1}^{\nu(i_1)} \cdots x_{i_n}^{\nu(i_n)}$ defines a mapping of B^n into B by the formula $(b_1, \ldots, b_n) \to ab_1^{\nu(i_1)} \cdots b_n^{\nu(i_n)}$.

Monomials in non-commuting variables are sometimes considered. Such monomials are defined as expressions of the form

$$a x_{i_1}^{\nu(i_1)} \cdots x_{i_n}^{\nu(i_n)},$$

where the sequence of (not necessarily distinct) indices $i_1, \ldots, i_n$ is fixed.

References

[1] LANG, S.: *Algebra,* Addison-Wesley, 1974.

L.V. Kuz'min

AMS 1980 Subject Classification: 12E05

MONOMIAL MATRIX - A square **matrix** over an

associative ring with identity, in each row and column of which there is exactly one non-zero element. If the non-zero entries of a monomial matrix are equal to 1, then the matrix is called a *permutation matrix*. Any monomial matrix is the product of a permutation matrix and a diagonal matrix.

D.A. Suprunenko

AMS 1980 Subject Classification: 15A33

MONOMIAL REPRESENTATION *of a finite group G* - A representation ρ of G in a finite-dimensional vector space V (cf. **Representation of a group**) such that in some basis $(e)=\{e_1,\ldots,e_n\}$ of V the matrix $M^{(e)}_{\rho(g)}$ of $\rho(g)$, for any $g\in G$, has only one non-zero element in each row and each column. Sometimes, by a monomial representation is meant the regular matrix representation $g\to M^{(e)}_{\rho(g)}$ (cf. also **Regular representation**). A monomial representation is a special case of an imprimitive representation (see **Imprimitive group**). Namely, the set of one-dimensional subspaces in V generated by the vectors $e_1,\ldots,e_n$ is a system of imprimitivity for ρ. Conversely, if for some representation ϕ of G in a vector space W there is a system of imprimitivity consisting of one-dimensional subspaces, then ϕ is a monomial representation. Let H be any subgroup of G; an example of a monomial representation is the representation of G induced from a one-dimensional representation of H (see **Induced representation**). Such a representation is also called an *induced monomial representation* (see [1]). Not every monomial representation ρ is induced (however, if ρ is irreducible, then it will be induced). The definition of a monomial representation given above arose in the classical theory of representations of finite groups. Often, however, this definition can be modified — any representation of a group G induced from a one-dimensional representation of some subgroup H of it being called a monomial representation. In this form the definition of a monomial representation makes sense not just for finite groups and their finite-dimensional representations, but also, for example, for Lie groups and their representations in Hilbert spaces (the subgroup H is then assumed to be closed). For a fairly large class of groups the construction of the monomial representations turns out to be sufficient for the description of all irreducible unitary representations. Namely, groups all irreducible unitary representations (cf. **Irreducible representation**; **Unitary representation**) of which are monomial are called *monomial groups*, or *M-groups*. They include all finite nilpotent groups and all connected nilpotent Lie groups. All finite *M*-groups, and also all monomial Lie groups, are solvable (see [2]).

References

[1] CURTIS, C.W. and REINER, I.: *Representation theory of finite groups and associative algebras*, Interscience, 1962.
[2] KIRILLOV, A.A.: *Elements of the theory of representations*, Springer, 1976 (translated from the Russian).

V.I. Popov

Editorial comments. For a fuller account on the state of *M*-groups as known 20 years ago see [A1].

References

[A1] HUPPERT, B.: *Endliche Gruppen*, I, Springer, 1967, Chapt. V, § 18.
[A2] FEIT, W.: *Characters of finite groups*, Benjamin, 1967.

AMS 1980 Subject Classification: 20C99

MONOMIAL SUBSTITUTIONS, GROUP OF - The subgroup of the group $GL(m, \mathbf{Z}[H])$ of all invertible matrices of order m over the integral group ring $\mathbf{Z}[H]$ (see **Group algebra**) of a group H, consisting of all matrices which precisely contain one non-zero element of H in each row and column. Each such matrix, having a non-zero element $h_{ij}\in H$ in place (i,j), corresponds to a monomial substitution, that is, a mapping $\psi: u_i\to h_{ij}u_j$, where $j=j(i)$, $i=1,\ldots,m$, and $u_i\to u_j$ is a permutation of the finite set $S=\{u_1,\ldots,u_m\}$. The product of such mappings is given by the formula

$$\psi_1\psi_2: u_i \to (h_{ij}h_{ik})u_k$$

($\psi_1: u_i\to h_{ij}u_j$, $\psi_2: u_j\to h_{ik}u_k$), and corresponds to the product of the matrices associated with ψ_1 and ψ_2. Any group G containing H as a subgroup of index m can be isomorphically imbedded in a group of monomial substitutions. The group of monomial substitutions is isomorphic to the (unrestricted) **wreath product** of H with the **symmetric group** $S(m)$ of degree m.

References

[1] KARGAPOLOV, M.I. and MERZLYAKOV, YU.I.: *Fundamentals of the theory of groups*, Springer, 1979 (translated from the Russian).
[3] HALL, M., JR.: *The theory of groups*, Macmillan, 1959.

N.N. Vil'yams

AMS 1980 Subject Classification: 20H99

MONOMORPHISM *in a category* - A morphism $\mu: A\to B$ of a **category** $\mathfrak{K}$ for which $\alpha\mu=\beta\mu$ (α,β from $\mathfrak{K}$) implies that $\alpha=\beta$ (in other words, μ can be cancelled on the right). An equivalent definition of a monomorphism is: For any object X of a category $\mathfrak{K}$ the mapping of sets induced by μ,

$$\text{Hom}(X,A) \to \text{Hom}(X,B),$$

must be injective. The product of two monomorphisms is a monomorphism. Each left divisor of a monomorphism is a monomorphism. The class of all objects and all monomorphisms of an arbitrary category $\mathfrak{K}$ forms a subcategory of $\mathfrak{K}$ (usually denoted by Mon $\mathfrak{K}$).

In the category of sets (cf. **Sets, category of**) monomorphisms are the injections (cf. **Injection**). Dual to the notion of a monomorphism is that of an **epimorphism**.

References

[1] TSALENKO, M.SH. and SHUL'GEĬFER, E.G.: *Fundamentals of category theory*, Moscow, 1974 (in Russian).
[2] BUCUR, I. and DELEANU, A.: *Introduction to the theory of categories and functors*, Wiley, 1968.

O.A. Ivanova

Editorial comments. In the first definition above, composition of morphisms is written in 'diagram order' (that is, $\alpha\mu$ means 'α followed by μ'). If, as is frequently done, the opposite convention is employed, then monomorphisms are morphisms which can be cancelled on the left.

References

[A1] MACLANE, S.: *Categories for the working mathematician*, Springer, 1971.

AMS 1980 Subject Classification: 18A20

MONOSPLINE - The difference between the function x^n and a polynomial **spline** $S_{n-1}(x)$ of degree $n-1$. Monosplines arise in the study of quadrature formulas for differentiable functions.

Yu.N. Subbotin

Editorial comments.

References

[A1] JOHNSON, R.S.: 'On monosplines of least deviation', *Trans. Amer. Math. Soc.* 96 (1960), 458-477.
[A2] SCHOENBERG, I.J.: 'Monosplines and quadrature formulae', in T.N.E. Greville (ed.): *Theory and Applications of Spline Functions*, Acad. Press, 1969, pp. 157-207.
[A3] SCHUMAKER, L.L.: *Spline functions: basic theory*, Wiley, 1981.
[A4] ZHENSYKBAEV, A.A.: 'Monosplines of minimal norm and the best quadrature formulae', *Russ. Math. Surveys* 36, no. 4 (1981), 121-180. (*Uspekhi Mat. Nauk* 36, no. 4 (1981), 107-159)
[A5] ZHENSYKBAEV, A.A.: 'On monosplines with nonnegative coefficients', *J. Approximation Theory* 55 (1988), 172-182.

AMS 1980 Subject Classification: 41A15, 65D07

MONOTONE BOOLEAN FUNCTION - A **Boolean function** $f(x_1, \ldots, x_n)$, $n = 0, 1, \ldots$, having the following property: If for some sets $\tilde{\alpha} = (\alpha_1, \ldots, \alpha_n)$ and $\tilde{\beta} = (\beta_1, \ldots, \beta_n)$, $\alpha_i, \beta_j \in \{0, 1\}$, the condition $\alpha_i \leq \beta_i$ holds for all i (one then writes $\tilde{\alpha} \prec \tilde{\beta}$), then $f(\tilde{\alpha}) \leq f(\tilde{\beta})$. For example, the function $f(x_1, x_2) = x_1 \oplus x_2$ (addition modulo 2) is not monotone since $(0, 1) \prec (1, 1)$ but $1 = f(0, 1) > f(1, 1) = 0$.

Examples of monotone Boolean functions are: The constants 0 and 1, the identity function $f(x) = x$, the disjunction $x_1 \vee x_2$, the conjunction $x_1 \& x_2$, etc. Examples of non-monotone Boolean functions are: the negation $\bar{x}$, the implication $x_1 \rightarrow x_2$, etc. Any function obtained by composition of monotone Boolean functions is itself monotone. In other words, the class of all monotone Boolean functions is closed. Moreover, the class of all monotone Boolean functions is one of the five maximal (pre-complete) classes in the set of all Boolean functions, that is, there is no closed class of

Boolean functions containing all monotone Boolean functions and distinct from the class of monotone functions and the class of all Boolean functions. The reduced **disjunctive normal form** of any monotone Boolean function distinct from 0 and 1 does not contain negations of variables. The set of functions $\{0, 1, x_1 \vee x_2, x_1 \& x_2\}$ is a complete system (and, moreover, a basis) in the class of all monotone Boolean functions.

For the number $\psi(n)$ of monotone Boolean functions depending on n variables, it is known that

$$\psi(n) = 2^{\left[\binom{n}{[n/2]}\right](1+\epsilon(n))},$$

where $0 < \epsilon(n) < c(\log n)/n$ and c is a constant (see [2]).

The complexity of realization of the class of monotone Boolean functions by diagrams of functional elements and by switching circuits (cf. **Diagram of functional elements; Contact scheme**) has a lower value than the complexity of realization of arbitrary Boolean functions (see **Synthesis problems**). Certain discrete extremal problems reduce to the *problem of evaluating monotone Boolean functions*. In this problem it is required, knowing that a function $f(x_1, \ldots, x_n)$ is a monotone Boolean function, to clarify its value on all sets using fewest possible questions of the form: 'What is the value of $f(\alpha_1, \ldots, \alpha_n)$ on a certain set $\tilde{\alpha} = (\alpha_1, \ldots, \alpha_n)$?' An algorithm has been suggested, [3], which for the evaluation of an arbitrary monotone Boolean function requires at most

$$\binom{n}{[n/2]} + \binom{n}{[n/2]+1}$$

questions. On the other hand, there is no evaluation algorithm that would distinguish the function

$$f(x_1, \ldots, x_n) = \begin{cases} 0 & \text{if } \sum_{n=1}^{n} x_i \leq \left[\dfrac{n}{2}\right], \\ 1 & \text{if } \sum_{i=1}^{n} x_i \geq \left[\dfrac{n}{2}\right]+1 \end{cases}$$

among all other monotone Boolean functions in fewer than

$$\binom{n}{[n/2]} + \binom{n}{[n/2]+1}$$

questions.

A generalization of the idea of a monotone Boolean function is that of *monotone function of k-valued logic*. If an arbitrary partial order S is given on the set $E_k = \{0, \ldots, k-1\}$ (written as $\leq_S$), then, by definition, for any two sets $\tilde{\alpha} = (\alpha_1, \ldots, \alpha_n)$ and $\tilde{\beta} = (\beta_1, \ldots, \beta_n)$, $\alpha_i, \beta_j \in E_k$, $\tilde{\alpha} \prec_S \tilde{\beta}$ means that $\alpha_i \leq_S \beta_i$ for all i. A function of k-valued logic $f(x_1, \ldots, x_n)$ (that is, defined on and taking values in E_k) is called *monotone relative* to S if for any sets $\tilde{\alpha} = (\alpha_1, \ldots, \alpha_n)$ and $\tilde{\beta} = (\beta_1, \ldots, \beta_n)$, the condition $\tilde{\alpha} \prec_S \tilde{\beta}$ implies $f(\tilde{\alpha}) \leq_S f(\tilde{\beta})$. The class of all functions that are monotone relative to some partial order S on E_k is always a closed class; it is a pre-complete class in k-valued logic if and only if there is precisely one maximal element and precisely one minimal element in S [4]. The number $\psi_S(n)$ of functions of k-valued logic that depend on n variables and that are monotone relative to S satisfies, as $n \to \infty$,

$$\log_2 \psi_S(n) \sim C(S)\frac{k^n}{\sqrt{n}},$$

where $C(S)$ is a constant that is effectively computable relative to a given partial order S (see [5]).

References

[1] YABLONSKIĬ, S.V.: *Introduction to discrete mathematics*, Moscow, 1986 (in Russian).
[2] KLEITMAN, D.: 'On Dedekind's problem: the number of monotone Boolean functions', *Proc. Amer. Math. Soc.* **21** (1969), 677-682.
[3] HANSEL, G.: 'Sur le nombre des fonctions booleennes monotones de n variables', *C.R. Acad. Paris* **262** (1966), 1080-1090.
[4] MARTYNYUK, V.V.: 'Investigation of certain classes of functions in many-valued logics', *Probl. Kibernet.* **3** (1960), 49-60 (in Russian).
[5] ALEKSEEV, V.B.: 'On the number of monotone k-valued functions', *Probl. Kibernet.* **28** (1974), 5-24 (in Russian).
[6] ALEKSEEV, V.B.: 'On the number of k-valued monotonic functions', *Soviet Math. Dokl.* **14**, no. 1 (1973), 87-91. (*Dokl. Akad. Nauk SSSR* **208**, no. 3 (1973), 505-508)

V.B. Alekseev

Editorial comments. In 1985 superpolynomial lower bounds have been proved for the size of monotone circuit realizations of explicit monotone Boolean functions. This solves a problem which had been open since the early days of circuit theory, and which still is open for general Boolean circuits. The first result in this direction was obtained by A.A. Razborov, [A1].

References

[A1] RAZBOROV, A.A.: 'Lower bounds for the monotone complexity of some Boolean functions', *Soviet Math. Dokl.* **31** (1985), 354-357. (*Dokl. Akad. Nauk SSSR* **281**, no. 4 (1985), 798-801)
[A2] SAVAGE, J.E.: *The complexity of computing*, Wiley, 1976.

AMS 1980 Subject Classification: 06E99

MONOTONE FUNCTION - A function of one variable, defined on a subset of the real numbers, whose increment $\Delta f(x) = f(x') - f(x)$, for $\Delta x = x' - x > 0$, does not change sign, that is, is either always negative or always positive. If $\Delta f(x)$ is strictly greater (less) than zero when $\Delta x > 0$, then the function is called *strictly monotone* (see **Increasing function; Decreasing function**). The various types of monotone functions are represented in the Table.

If at each point of an interval f has a derivative that does not change sign (respectively, is of constant sign),

$\Delta f(x) \geqslant 0$	Increasing (non-decreasing)	
$\Delta f(x) \leqslant 0$	Decreasing (non-increasing)	
$\Delta f(x) > 0$	Strictly increasing	
$\Delta f(x) < 0$	Strictly decreasing	

then f is *monotone* (*strictly monotone*) *on this interval*.

The idea of a monotone function can be generalized to functions of various classes. For example, a function $f(x_1, \ldots, x_n)$ defined on $\mathbf{R}^n$ is called monotone if the condition $x_1 \leqslant x_1', \ldots, x_n \leqslant x_n'$ implies that everywhere either $f(x_1, \ldots, x_n) \leqslant f(x_1', \ldots, x_n')$ or $f(x_1, \ldots, x_n) \geqslant f(x_1', \ldots, x_n')$ everywhere. A monotone function in the **algebra of logic** is defined similarly.

A monotone function of many variables, increasing or decreasing at some point, is defined as follows. Let f be defined on the n-dimensional closed cube Q^n, let $x_0 \in Q^n$ and let $E_t = \{x: f(x) = t, x \in Q^n\}$ be a **level set** of f. The function f is called *increasing* (respectively, *decreasing*) at x_0 if for any t and any $x' \in Q^n \setminus E_t$ not separated in Q^n by E_t from x_0, the relation $f(x') < t$ (respectively, $f(x') > t$) holds, and for any $x'' \in Q^n \setminus E_t$ that is separated in Q^n by E_t from x_0, the relation $f(x'') > t$ (respectively, $f(x'') < t$) holds. A function that is increasing or decreasing at some point is called *monotone at that point.*

L.D. Kudryavtsev

AMS 1980 Subject Classification: 26A48

MONOTONE MAPPING - An isotone mapping or an antitone mapping.

AMS 1980 Subject Classification: 06A10

MONOTONE OPERATOR - One of the notions in **non-linear functional analysis**.

Let E be a **Banach space**, E^* its dual, and let (y, x) be the value of a linear functional $y \in E^*$ at an element $x \in E$. An operator A, in general non-linear and acting from E into E^*, is called *monotone* if

$$\mathrm{Re}(Ax_1 - Ax_2, x_1 - x_2) \geqslant 0 \qquad (1)$$

for any $x_1, x_2 \in E$. An operator A is called *semi-continuous* if for any $u, v, w \in E$ the numerical function $(A(u + tv), w)$ is continuous in t. An example of a semi-continuous monotone operator is the gradient of a convex Gâteaux-differentiable functional. Many functionals in variational calculus are convex and hence generate monotone operators; they are useful in the solution of non-linear integral equations and were in fact first applied there.

Various applications of monotone operators in questions regarding the solvability of non-linear equations are based on the following theorem (see [1], [2]). Let E be a reflexive Banach space (cf. **Reflexive space**) and let A be a semi-continuous monotone operator with the property of coerciveness:

$$\lim_{\|u\| \to \infty} \frac{\mathrm{Re}(Au, u)}{\|u\|} = \infty.$$

Then for any $f \in E$ the equation $Au = f$ has at least one solution.

An operator A defined on a set $D \subset E$ with values in E^* is called *monotone* on D if (1) holds for any

$x_1, x_2 \in D$, and it is called *maximal monotone* if it is monotone on D and has no monotone proper (strict) extension.

Research into equations with monotone operators has been stimulated to a large extent by problems in the theory of quasi-linear elliptic and parabolic equations. For example, boundary value problems for quasi-linear parabolic equations lead to equations of the form

$$\Lambda x + Ax = f \tag{2}$$

in a suitable Banach space E. The same equation also arises naturally in the investigation of the **Cauchy problem** for an abstract **evolution equation** with a non-linear operator in Banach spaces. If E is reflexive and A is a bounded, semi-continuous and coercive operator with dense domain of definition in E, then (2) is solvable for any $f \in E^*$. The idea of monotonicity has also been applied in the problem of almost-periodic solutions of non-linear parabolic equations.

References

[1] BROWDER, F.: 'Non-linear parabolic boundary value problems of arbitrary order', *Bull. Amer. Math. Soc.* **69** (1963), 858-861.

[2] MINTY, G.J.: 'On a "monotonicity" method for the solution of non-linear problems in Banach spaces', *Proc. Nat. Acad. Sci. USA* **50** (1963), 1038-1041.

[3] VAĬNBERG, M.M. and KACHUROVSKIĬ, R.I.: 'On the variational theory of non-linear operators and equations', *Dokl. Akad. Nauk SSSR* **129**, no. 6 (1959), 1199-1202 (in Russian).

[4] VAĬNBERG, M.M.: *Variational method and method of monotone operators in the theory of nonlinear equations*, Wiley, 1973 (translated from the Russian).

[5] LIONS, J.L.: *Quelques méthodes de résolution des problèmes aux limites non-linéaires*, Dunod, 1969.

[6] LEVITAN, B.M. and ZHIKOV, V.V.: *Almost-periodic functions and differential equations*, Cambridge Univ. Press, 1982 (translated from the Russian).

[7] KACHUROVSKIĬ, R.I.: 'Nonlinear monotone operators in Banach spaces', *Russian Math. Surveys* **23**, no. 2 (1968), 117-165. (*Uspekhi Mat. Nauk* **23**, no. 2 (1968), 121-168)

V.V. Zhikov

AMS 1980 Subject Classification: 47H05

MONOTONE SEQUENCE - A sequence $\{x_n\}$ such that for all $n = 1, 2, \ldots$, either

$x_n < x_{n+1}$ (a strictly-increasing sequence);

$x_n \leqslant x_{n+1}$ (a non-decreasing sequence);

$x_n > x_{n+1}$ (a strictly-decreasing sequence); or

$x_n \geqslant x_{n+1}$ (a non-increasing sequence).

Editorial comments. See also Increasing sequence; Decreasing sequence.

AMS 1980 Subject Classification: 40-XX

MONTE-CARLO METHOD, *method of statistical trials* - A numerical method based on simulation by random variables and the construction of statistical estimators for the unknown quantities. It is usually supposed that the Monte-Carlo method originated in 1949 (see [1]) when, in connection with work on the construction of atomic reactors, J. von Neumann and S. Ulam suggested using the apparatus of probability theory in the computer solution of applied problems. The Monte-Carlo method is named after the town of Monte-Carlo, famous for its casinos.

Simulation by random variables with given distributions. As a rule such a simulation is achieved by a transformation of one or more independent values of a random number α, uniformly distributed in the interval $(0, 1)$. The sequence of 'sample' values of α is usually obtained on a computer by number-theoretic algorithms, of which the most widely used is the so-called *method of residues*, for example, in the form

$$u_0 = 1, \quad u_n \equiv u_{n-1} 5^{2p+1} \pmod{2^m}, \quad \alpha_n = u_n \cdot 2^{-m}.$$

Here m is the order of the mantissa of the computer and

$$p = \max\{q: 5^{2q+1} < 2^m\}.$$

Numbers of this type are called *pseudo-random numbers*; they are used in statistical testing and in solving typical problems (see [2] - [6]). The length of the period of the above version of the method of residues is 2^{m-2}. Physical generators, tables of random numbers and quasi-random numbers are also used in the Monte-Carlo method. There are Monte-Carlo methods with a small number of playing parameters (see [7]).

The standard method for simulating a discrete random variable with distribution $P\{\xi = x_k\} = p_k$, $k = 0, 1, \ldots$, is as follows: Put $\xi = x_m$ if the chosen value of α satisfies

$$\sum_{k=0}^{m-1} p_k \leqslant \alpha < \sum_{k=0}^{m} p_k.$$

The standard method for simulating a continuous random variable (sometimes called the *method of inverse functions*) is to use the, easily verified, representation $\xi = F^{-1}(\alpha)$, where F is the distribution function with given density f. Sometimes **randomization** of the simulation is useful (in other words, the *method of superposition*), based on the expression

$$f(x) = \sum_k p_k f_k(x);$$

here one first chooses a number m with distribution $P\{m = k\} = p_k$, and then obtains a sample value ξ from the distribution with density f_m. In other methods of randomization certain parameters of a deterministic method for solving the problem are considered as random variables (see [7] - [9]).

Another, more general, method for simulating a continuous random variable is the *method of exclusion* (*method of selection*), at the basis of which lies the following result: If (ξ, η) is uniformly distributed in a domain $G = \{(x, y): 0 \leqslant y \leqslant g(x)\}$, then $f_\xi(x) = g(x)/\overline{G}$.

In the method of exclusion, choose a point (ξ_0, η) uniformly in a domain $G_1 \supset G$ and put $\xi = \xi_0$ if $(\xi_0, \eta) \in G$; otherwise repeat the selection of (ξ_0, η), etc. For example, if $a \leqslant \xi \leqslant b$ and $g(x) = cf(x) \leqslant R$, one can take $\xi_0 = a + (b-a)\alpha_1$, $\eta = R\alpha_2$. The average number of operations in the method of exclusion is proportional to $\overline{G_1}/\overline{G}$.

For many random variables, special representations of the form $\xi = \phi(\alpha_1, \ldots, \alpha_n)$ have been obtained. For example, the random variables

$$\sqrt{-2 \ln \alpha_1} \cdot \cos 2\pi\alpha_2 \quad \text{and} \quad \sqrt{-2 \ln \alpha_2} \cdot \sin 2\pi\alpha_2$$

have standard normal distributions and are independent; the random variable $\ln(\alpha_1 \cdots \alpha_n)$ has the gamma-distribution with parameter n; the random variable $\max(\alpha_1, \ldots, \alpha_n)$ is distributed with density nx^{n-1}, $0 \leqslant x \leqslant 1$; the random variable $\exp\{\sum_{k=1}^{n} (\ln \alpha)/(p+k+1)\}$ has the beta-distribution with parameters p, n (see [3] - [6]).

The standard algorithm for simulating a continuous random vector $\vec{\xi} = (\xi_1, \ldots, \xi_n)$ is to successively choose the values of its components from conditional distributions corresponding to the representation

$$f_{\vec{\xi}}(x_1, \ldots, x_n) = f_1(x_1)f_2(x_2 \mid x_1) \cdots f_n(x_n \mid x_1, \ldots, x_{n-1}).$$

The method of exclusion extends to the multi-dimensional case without change; it is only necessary, in its formulation, to regard ξ, ξ_0 and x as vectors. A multi-dimensional normal vector can be simulated by using a special linear transformation of a vector of independent standard normal random variables. Special methods have also been developed for the approximate simulation of stationary Gaussian processes (see, for example, [3], [6]).

If, in a Monte-Carlo method calculation, random variables defined by a real phenomenon are simulated, then the calculation is a direct simulation (imitation) of this phenomenon. Computer simulations have been worked out for: the processes of transport, scattering and reproduction of particles: neutrons, gamma-quanta, photons, electrons, and others (see, for example, [11] - [18]); simulations of the evolution of ensembles of molecules for the solution of various problems in classical and quantum statistical physics (see, for example, [10] - [18]); simulation of queueing and industrial processes (see, for example [2], [6], [18]); simulation of various random processes in technology, biology; etc. (see [18]). Simulation algorithms are usually carefully developed; for example, they tabulate complicated functions, modify standard procedures, etc. None-the-less, a direct simulation often cannot provide the necessary accuracy in the estimates of the required variables. Many methods have been developed for increasing the effectiveness of a simulation.

Monte-Carlo algorithms for estimating multiple integrals. Suppose it is required to estimate an integral $J = \int h(x) \, dx$ with respect to the Lebesgue measure in an s-dimensional Euclidean space X and let $f_\xi(x)$ be a probability density such that J can be written as:

$$J = \int_X f_\xi(x) \frac{h(x)}{f_\xi(x)} \, dx = \mathsf{E}\zeta,$$

where $\zeta = h(\xi)/f_\xi(\xi)$. By computer simulation of ξ it is possible to obtain N sample values $x_1, \ldots, x_N$. By the law of large numbers,

$$J \approx J_N = \frac{1}{N} \sum_{k=1}^{N} \frac{h(x_k)}{f(x_k)}.$$

Simultaneously it is possible to estimate the mean-square error in J_N, that is, the quantity $\sigma_N = (\mathsf{D}\zeta/N)^{1/2}$, and to approximately construct a suitable confidence interval for J. By the choice of the density f it is possible to arrange for estimates with possibly smaller variance. For example, if $0 \leqslant m_1 \leqslant h/f \leqslant m_2 < +\infty$, then $\mathsf{D}\zeta \leqslant (m_2 - m_1)^2/4$ and if $f = h/J$, then $\mathsf{D}\zeta = 0$. The corresponding algorithm is called *essential sampling* (*choice by importance*). Another common modification — the *method of selection of principal part* — occurs when a function $h_0 \approx h$ is determined whose integral is known. It is sometimes useful to combine Monte-Carlo methods and classical quadratures (cf. **Quadrature**) in so-called *random quadrature formulas*, the basic idea of which is that the nodes and coefficients in any quadrature sum (for example, in interpolation) are chosen randomly from a distribution, providing an **unbiased estimator** of the integral [3]. Particular cases of these formulas are: the so-called *method of stratified sampling*, in which the nodes are chosen one in each part of a fixed partition of the domain of integration and the coefficients are proportional to the corresponding volumes; and the so-called *method of symmetric sampling*, which, in the case of integration over $(0, 1)$, is defined by the expression (see [10])

$$2J = \mathsf{E}\left[\frac{h(\xi) + h(1-\xi)}{f_\xi(\xi)}\right].$$

Here the order of the speed of convergence of the Monte-Carlo method is increased and, in certain cases, becomes best possible in the class of problems being considered.

In general, the domain of integration is partitioned into parallelopipeds. In each parallelopiped the value of the integral is calculated as the average of the value at a random point and the point symmetric to it relative to the centre of the parallelopiped.

A number of modifications of Monte-Carlo methods are based on the (perhaps formal) representation of the required value as a double integral

$$J = \int_X \int_Y f(x, y)h(x, y) \, dx \, dy = \mathsf{E}\zeta,$$

where $\zeta = h(\xi, \eta)$ and the vector (ξ, η) is distributed with density $f(x, y)$. It is known that $E(\zeta) = EE(\zeta | \xi)$ and that

$$D\zeta = DE(\zeta | \xi) + ED(\zeta | \xi) = A_1 + A_2, \qquad (1)$$

where $E(\zeta | \xi)$ is the conditional mathematical expectation and $D(\zeta | \xi)$ is the conditional variance of ζ given a fixed value of ξ. Formula (1) is widely used in Monte-Carlo methods. In particular, it shows that $DE(\zeta | \xi) < D\zeta$, that is, analytic averaging over any variable increases the accuracy of the Monte-Carlo method. However, in this connection, the amount of computation may be significantly increased. The computer time necessary to achieve a given accuracy is proportional to $tD\zeta$, where t is the average time it takes to obtain one value of ζ. The *method of splitting* is optimal with respect to this criterion. Its simplest version uses the 'unbiased' estimator

$$\zeta_n = \frac{1}{n} \sum_{k=1}^{n} h(\xi, \eta_k),$$

where $\eta_1, \ldots, \eta_n$ are conditionally-independent and distributed as η for a fixed value of ξ. Using (1) it is possible to obtain an optimal value

$$n = \left[\frac{A_2}{A_1} \frac{t_1}{t_2} \right]^{1/2},$$

where t_1, t_2 are the average computing times corresponding to the samples ξ, η (see, for example, [4]).

If the integrand depends on a parameter, it is expedient to use the *method of dependent trials*, that is, to estimate the integral for various values of the parameter at one and the same random points [20]. An important property of the Monte-Carlo method is the comparatively relatively weak dependence of the mean-square error σ_N on the number of measurements; moreover, the order of convergence relative to the number of points N is always one and the same: $N^{-1/2}$. This allows one to estimate (after a preliminary transformation of the problem) integrals of very high, and even infinite, multiplicity. For example, methods have been worked out for the estimation of Wiener integrals [19].

Monte-Carlo algorithms for solving integral equations of the second kind. Suppose it is required to estimate a linear functional $J_h = (\phi, h)$, where $\phi = K\phi + f$, where the integral operator K with kernel $k(x', x)$ satisfies a condition providing convergence of the **Neumann series**: $\| K^{n_0} \| < 1$. A **Markov chain** $\{x_n\}$ is defined by an initial density $\pi(x)$ and a transition density $p(x', x) = p(x' \rightarrow x)$; the probability of termination of the chain at x' is equal to

$$g(x') = 1 - \int p(x', x) \, dx.$$

Let N be the random index of the last state. Further, let a functional of the trajectories of the chain with

expectation J_h be defined. Most often the so-called *collision estimator*

$$\xi = \sum_{n=0}^{N} Q_n h(x_n)$$

is used, where

$$Q_0 = \frac{f(x_0)}{\pi(x_0)}, \quad Q_n = Q_{n-1} \frac{k(x_{n-1}, x_n)}{p(x_{n-1}, x_n)}.$$

If $p(x', x) \neq 0$ for $k(x', x) \neq 0$ and $\pi(x) \neq 0$ for $f(x) \neq 0$, then under certain additional conditions

$$E\xi = \sum_{n=0}^{\infty} (K^n f, h) = (\phi, h) = \int_X \phi(x) h(x) \, dx$$

(see [3] - [5]). The possibility of attaining a small variance in the case of constant sign is ensured by the following result: If

$$\pi(x) = \frac{f(x) \phi^*(x)}{(f, \phi^*)} \quad \text{and} \quad p(x', x) = \frac{k(x', x) \phi^*(x)}{[K^* \phi^*](x')},$$

where $\phi^* = K^* \phi^* + h$, then $D\xi = 0$ and $E\xi = J_h$ (see [4]). If a suitable Markov chain is simulated on a computer, statistical estimates of linear functionals in the solution of an integral equation of the second kind can be obtained. This gives the possibility of locally estimating the solution on the basis of the representation $\phi(x) = (\phi, h_x) + f(x)$, where $h_x(x') = k(x', x)$. In a number of cases, alongside Monte-Carlo methods, number-theoretic methods are applied in order to solve these problems (see [21]). A Monte-Carlo method for estimating the first eigen value of an integral operator has been realized by an iteration method on the basis of the relation [22]

$$E[Q_n h(x_n)] = (K^n f, h).$$

All these results can be almost automatically extended to systems of linear algebraic equations of the form $x + Hx = h$ (see [23]).

Modifications of Monte-Carlo methods in radiative transport theory. (See [11] - [17].) The density of the average number of particle collisions in a phase space with coordinates $\mathbf{r}$ and velocities ω satisfies an integral equation of the second kind; its kernel in the single-velocity case has the form

$$\frac{\sigma_s(\mathbf{r}') g(\mu) \exp(-\tau(\mathbf{r}', \mathbf{r})) \sigma(\mathbf{r})}{\sigma(\mathbf{r}') 2\pi |\mathbf{r}' - \mathbf{r}|^2} \delta \left[\omega - \frac{\mathbf{r} - \mathbf{r}'}{|\mathbf{r} - \mathbf{r}'|} \right].$$

Here $\sigma_s(\mathbf{r})$ is the scattering coefficient (section), $\sigma(\mathbf{r})$ is the relaxation coefficient, $g(\mu)$ is the scattering indicatrix, and $\tau(\mathbf{r}', \mathbf{r})$ is the optical length of a path from $\mathbf{r}'$ to $\mathbf{r}$ (see [3], [4]). For the construction of estimates with small variances one uses, for example, asymptotic solutions of the adjoint transport equation [4]; the simplest algorithm of this type is the so-called exponential transformation (see [4], [11]). A modification of a local estimate of the flow of particles has been developed (see [3], [4], [11] - [13], [17], [18]). Using a simulation of a Markov chain (for example, the physical transport process in a certain medium) it is possible to simultaneously obtain dependent estimators of functionals for different values of the parameters; by differentiating the 'weight' Q_n it is sometimes possible to obtain unbiased estimators of the corresponding derivatives (see [4], [12]). This provides an opportunity to use Monte-Carlo methods in solving certain inverse problems [12]. For the solution of problems in transport theory 'bifurcation' of the trajectory and analytic averaging is effective [11]. The simulation of trajectories of particles in compound media can sometimes be essentially simplified by the method of a maximal section (see [3] - [5]).

Monte-Carlo algorithms for solving elliptic equations. These are constructed on the basis of the corresponding integral relations. For example, the standard five-point difference approximation for the Laplace equation has the form of the formula for the complete mathematical expectation of a symmetric random walk over a grid with absorption at the boundary (see, for example, [2], [3]). A continuous analogue of this formula is the relation

$$u(P) = \frac{\int\limits_{S(P)} u(\mathbf{r}(s))\,ds}{\int\limits_{S(P)} ds}, \qquad (2)$$

where the integral is taken over the surface of a sphere lying entirely within the given domain and with centre at P. Formula (2) and other similar relations provide an opportunity to use isotropic 'random walk on spheres' when solving elliptic and parabolic equations (see [24], [4]). Monte-Carlo methods are effective, for example, for estimating the solution of multidimensional boundary value problems at a point.

A simulation of Markov branching processes allows one to construct estimates of the solution of certain non-linear equations, for example, the Boltzmann equation in the theory of rarefied gases [3].

References

[1] NEUMANN, J. VON: *Nat. Bureau Standard Appl. Math. Series* **12** (1951), 36-38.

[2] BUSLENKO, N.P., ET AL.: *The methods of statistical trials (Monte-Carlo method)*, Moscow, 1962 (in Russian).

[3] ERMAKOV, S.M.: *Die Monte-Carlo Methode und verwandte Fragen*, Deutsch. Verlag Wissenschaft., 1975 (translated from the Russian).

[4] MIKHAĬLOV, G.A.: *Some questions in the theory of Monte-Carlo methods*, Novosibirsk, 1971 (in Russian).

[5] SOBOL', I.M.: *Numerical Monte-Carlo methods*, Moscow, 1973 (in Russian).

[6] POLLYAK, YU.G.: *Probabilistic simulation on computers*, Moscow, 1971 (in Russian).

[7] BAKHVALOV, N.S.: 'On optimal convergence estimates for quadrature processes and integration methods of Monte-Carlo type on function classes', in *Numerical Methods for Solving Differential and Integral Equations and Quadrature Formulas*, Moscow, 1964, pp. 5-63 (in Russian).

[8] BAKHVALOV, N.S.: 'An estimate of the remainder term in quadrature formula', *Comp. Math. Math. Phys.* **1**, no. 1 (1961), 68-82. (*Zh. Vychisl. Mat. i Mat. Fiz.* **1**, no. 1 (1961), 64-77)

[9] BAKHVALOV, N.S.: 'Approximate computation of multiple integrals', *Vestnik Moskov. Gos. Univ. Ser. Mat. Mekh. Astronom. Fiz. Khim.* **4** (1959), 3-18 (in Russian).

[10] HAMMERSLEY, J.M. and HANDSCOMB, D.C.: *Monte-Carlo methods*, Methuen, 1964.

[11] *Monte-Carlo methods and problems of radiative transport*, Moscow, 1967.

[12] MARCHUK, G.I., ET AL.: *The Monte-Carlo method in atmospheric optics*, Novosibirsk, 1976 (in Russian).

[13] SPANIER, J. and GELBARD, E.: *Monte-Carlo principles and neutron transport problems*, Addison-Wesley, 1969.

[14] CHAVCHANIDZE, V.V.: *Izv. Akad. Nauk SSSR Ser. Fiz.* **19**, no. 6 (1955), 629-638.

[15] *The penetration of radiation through non-uniformity in protection*, Moscow, 1968 (in Russian).

[16] FRANK-KAMENETSKIĬ, A.D.: *Atomnaya Energiya* **16**, no. 2 (1964), 119-122.

[17] KALOS, M.H.: *Nuclear Sci. and Eng.* **33** (1968), 284-290.

[18] *Monte-Carlo methods and their application. Reports III All-Union Conf. Monte-Carlo Methods*, Novosibirsk, 1971.

[19] GELFAND, I.M., FROLOV, A.S. and CHENTSOV, N.N.: 'The computation of continuous integrals by the Monte-Carlo method', *Izv. Vuz. Ser. Mat.* **5** (1958), 32-45 (in Russian).

[20] FROLOV, A.S. and CHENTSOV, N.N.: 'On the calculation of definite integrals dependent on a parameter by the Monte-Carlo method', *USSR Comp. Math. Math. Phys.* **2**, no. 4 (1962), 802-807. (*Zh. Vychisl. Mat. i. Mat. Fiz.* **2**, no. 4 (1962), 714-717)

[21] KOROBOV, N.M.: *Number-theoretic methods in applied analysis*, Moscow, 1963 (in Russian).

[22] VLADIMIROV, V.S.: 'Monte-Carlo methods as applied to the calculation of the lowest eigenvalue and the associated eigenfunction of a linear integral equation', *Theor. Probab. Appl.* **1**, no. 1 (1956), 101-116. (*Teor. Veroyatnost. i. Primenen.* **1**, no. 1 (1956), 113-130)

[23] CURTISS, J.H.: 'Monte-Carlo methods for the iteration of linear operators', *J. Math. Phys.* **32**, no. 4 (1954), 209-232.

[24] MULLER, M.E.: 'Some continuous Monte-Carlo methods for the Dirichlet problem', *Ann. Math. Stat.* **27**, no. 3 (1956), 569-589.

G.A. Mikhaĭlov

Editorial comments. For an up-to-date account of the use of random processes in solving (numerically) the classical equations of mathematical physics cf. [A2].

References

[A1] BRATLEY, P., FOX, B.L. and SCHRAGE, L.E.: *A guide to simulation*, Springer, 1987.

[A2] ERMAKOV, S.M., NEKRUTKIN, V.V. and SIPIN, A.S.: *Random processes for the classical equations of mathematical physics*, Kluwer, 1989 (translated from the Russian).

AMS 1980 Subject Classification: 65C05

MONTEL SPACE - A **barrelled space** (in particular, a **Fréchet space**) in which each closed bounded set is compact. The space $H(G)$ of all holomorphic functions in a domain G, with the topology of uniform convergence on compact sets, is a Fréchet space and, in view of a theorem of Montel (cf. **Montel theorem**, 2), every bounded sequence of holomorphic functions is relatively compact in $H(G)$, so $H(G)$ is a Montel space. The space $C^\infty(\Omega) = \mathcal{E}(\Omega)$ of all infinitely-differentiable functions in a domain $\Omega \subset \mathbf{R}^n$, the space $D(\Omega)$ of all functions of compact support and the space $S(\mathbf{R}^n)$ of differentiable functions that are rapidly decreasing at infinity, are also Montel spaces in their natural topologies.

A Montel space is reflexive (cf. **Reflexive space**). The strong dual of a Montel space is a Montel space; in particular, the spaces of generalized functions $\mathscr{E}'(\Omega)$, $D'(\Omega)$ and $S'(\Omega)$ are Montel spaces. A **normed space** is a Montel space if and only if it is finite-dimensional.

References

[1] BOURBAKI, N.: *Elements of mathematics. Topological vector spaces*, Addison-Wesley, 1977 (translated from the French).
[2] ROBERTSON, A.P. and ROBERTSON, W.: *Topological vector spaces*, Cambridge Univ. Press, 1964.
[3] EDWARDS, R.E.: *Functional analysis: theory and applications*, Holt, Rinehardt, Winston, 1965.

S.G. Kreĭn

Editorial comments.

References

[A1] JARCHOW, H.: *Locally convex spaces*, Teubner, 1981.
[A2] KÖTHE, G.: *Topological vector spaces*, 1, Springer, 1969.
[A3] SCHAEFER, H.H.: *Topological vector spaces*, Springer, 1971.

AMS 1980 Subject Classification: 46A14

MONTEL THEOREM - 1) *Montel's theorem on the approximation of analytic functions by polynomials*: If D is an open set in the complex z-plane not containing $z=\infty$ and $f(z)$ is a single-valued function, analytic at each point $z\in D$, then there is a sequence of polynomials $\{P_n(z)\}$ converging to $f(z)$ at each $z\in D$. This theorem is one of the basic results in the theory of **approximation of functions of a complex variable**; it was obtained by P. Montel [1].

2) *Montel's theorem on compactness conditions for a family of holomorphic functions* (*principle of compactness*, see [2]): Let $\Phi=\{f(z)\}$ be an infinite family of holomorphic functions in a domain D of the complex z-plane, then Φ is pre-compact, that is, any subsequence $\{f_k(z)\}\subset\Phi$ has a subsequence converging uniformly on compact subsets of D, if Φ is uniformly bounded in D. This theorem can be generalized to a domain D in $\mathbf{C}^n$, $n\geqslant 1$ (see **Compactness principle**).

3) *Montel's theorem on conditions for normality of a family of holomorphic functions* (*principle of normality*, see [2]): Let $\Phi=\{f(z)\}$ be an infinite family of holomorphic functions in a domain D of the complex z-plane. If there are two distinct values a and b that are not taken by any of the functions $f(z)\in\Phi$, then Φ is a **normal family**, that is, any sequence $\{f_k(z)\}\subset\Phi$ has a sequence uniformly converging on compact subsets of D to a holomorphic function or to ∞. The conditions of this theorem can be somewhat weakened: It suffices that all $f(z)\in\Phi$ do not take one of the values, say a, and that the other value b is taken at most m times, $1\leqslant m<\infty$. This theorem can be generalized to a domain D in $\mathbf{C}^n$, $n\geqslant 1$.

References

[1] MONTEL, P.: *Leçons sur les séries de polynomes à une variable complexe*, Gauthier-Villars, 1910.
[2] MONTEL, P.: *Leçons sur les familles normales de fonctions analytiques et leurs applications*, Gauthier-Villars, 1927.

E.D. Solomentsev

Editorial comments.

References

[A1] MARKUSHEVICH, A.I.: *Theory of functions of a complex variable*, 3, Sect. 11; 1, Sect. 86; 3, Sect. 50, Chelsea, 1977 (translated from the Russian).

AMS 1980 Subject Classification: 30E10, 30D45

MOORE SPACE - A topological space M with a unique non-trivial reduced homology group:

$$\tilde{H}_k(M) = G; \ \tilde{H}_i(M) = 0, \ i\neq k.$$

If $K(\mathbf{Z}, n)$ is the **Eilenberg−MacLane space** of the group of integers $\mathbf{Z}$ and $M_k(G)$ is the Moore space with $H_k(M_k(G))=G$, then

$$\lim_{N\to\infty} \left[\Sigma^{N+k}X, K(\mathbf{Z}, N+n)\wedge M_k(G) \right] \cong H^n(X, G),$$

that is, $\{K(\mathbf{Z}, n)\wedge M_k(G)\}$ is the spectrum of the cohomology theory $H^*(\ , G)$. This allows one to extend the idea of cohomology with arbitrary coefficients to a generalized cohomology theory. For any spectrum E, the spectrum $E\wedge M_k(G)$ defines a cohomology theory $(E\wedge M_k(G))^*$, called the E^*-*cohomology theory with coefficient group* G. For the definition of generalized homology theories with coefficients in a group G, the so-called *co-Moore space* $M^k(G)$ is used, which is characterized by

$$\tilde{H}^k(M^k(G)) = G, \ \tilde{H}^i(M^k(G)) = 0, \ i\neq k.$$

For example, the group $\pi_i(X, G)=[M^k(G), X]$ is called the *homotopy group of the space X with coefficients in G*. However, the space $M^k(G)$ does not exist for all pairs (G, k). If G is a finitely-generated group, then $M^k(G)$ does exist.

References

[1] MOORE, J.C.: 'On homotopy groups of spaces with a single non-vanishing homotopy group', *Ann. of Math.* **59**, no. 3 (1954), 549-557.

A.F. Kharshiladze

Editorial comments. For a construction of a Moore space as a CW-complex with one zero cell and further only cells in dimensions n and $n+1$, cf. [A1]. The Eilenberg−MacLane space $K(G, n)$ can be obtained from the Moore space $M(G, n)$ by killing the higher homotopy groups.

In general topology, a *Moore space* is a **regular space** with a development. (A *development* is a sequence $\{\mathscr{U}_n\}$ of open coverings such that for every x and every open set O containing x there is an n such that

$$\text{St}(x, \mathscr{U}_n) = \bigcup\{U\in\mathscr{U}_n: x\in U\} \subseteq O;$$

in other words, $\{\text{St}(x, \mathscr{U}_x)\}_n$ is a neighbourhood base at x.)

The idea of a development can be found in [A4] (Axiom 1). Moore spaces are generalizations of metric spaces and one can show that collectionwise normal Moore spaces are metrizable [A2]. The question whether every normal Moore space is metrizable generated lots of research; its solution is described in [A3].

References

[A1] GRAY, B.: *Homotopy theory*, Acad. Press, 1975, §17.
[A2] BING, R.H.: 'Metrization of topological spaces', *Canad. J. Math.* **3** (1951), 175-186.
[A3] FLEISSNER, W.G.: 'The normal Moore space conjecture and large cardinals', in K. Kunen and J. Vaughan (eds.): *Handbook of Set-Theoretic Topology*, North-Holland, 1984, pp. 733-760.
[A4] MOORE, R.L.: *Foundations of point set theory*, Amer. Math. Soc., 1962.

AMS 1980 Subject Classification: 54E30, 55P20

MORDELL CONJECTURE - A conjecture on the finiteness of the set of rational points on an **algebraic curve** of genus $g>1$. Advanced by L.J. Mordell [1] for the case when the ground field K is the field of rational numbers. At present (1982) Mordell's conjecture is taken to be the assertion of the finiteness of the set of rational points $X(L)$ of an irreducible algebraic curve X of genus $g>1$ defined over a field K of finite type over the field of rational numbers $\mathbf{Q}$ in any finite extension L/K. A reduction of Mordell's conjecture to the most difficult case when K is an algebraic number field has been obtained (see [3]). A number of special results related to Mordell's conjecture are known. Thus, it has been proved [2] that $X(K)$ is finite if the rank of the group of K-automorphisms from X into an elliptic curve Y is larger than the rank of the group $Y(K)$. The finiteness of $X(K)$ has been established, [7], for the broad class of modular curves (cf. **Modular curve**) and their fields of definition K. An estimate for the growth of the height,

$$h(x_n) \geqslant \exp(an+b),$$

of the rational points $x_n \in X(K)$ has been found, [8], showing that they are situated more 'sparsely' than on the curves of genus $g \leqslant 1$. It has also been proved that Mordell's conjecture is a consequence of Shafarevich's conjecture on the finiteness of the number of algebraic curves having a given genus $g>1$, a given field of definition (a finite extension of $\mathbf{Q}$) and a given set of points of bad reduction (see [4], and also **Siegel theorem** on integer points).

The geometric analogue of Mordell's conjecture is the assertion of the finiteness of the number of sections of a bundle

$$f: V \to B,$$

where V is a non-singular projective surface, B is a curve and the general fibre of f is an irreducible curve of genus $g>1$. This assertion is true if the bundle is non-constant, that is, if it is not a direct product after a certain covering of the base B, and if the characteristic of the ground field k is equal to zero (see [3], [6]). For constant bundles it is possible only to assert the finiteness of the number of classes consisting of sections that are algebraically equivalent as curves on V. If the characteristic of k is positive, the geometric analogue of Mordell's conjecture is false [4].

References

[1] MORDELL, L.J.: 'On the rational solutions of the indeterminate equation of the third and fourth degrees', *Proc. Cambridge Philos. Soc.* **21** (1922), 179-192.
[2] DEM'YANENKO, V.A.: 'Rational points of a class of algebraic curves', *Izv. Akad. Nauk SSSR Ser. Mat.* **30**, no. 6 (1966), 1373-1396 (in Russian).
[3] MANIN, YU.I.: 'Rational points of algebraic curves over function fields', *Transl. Amer. Math. Soc. (2)* **50** (1966), 189-234. (*Izv. Akad. Nauk SSSR Ser. Mat.* **27**, no. 6 (1963), 1395-1440)
[4] PARSHIN, A.N.: 'Quelques conjectures de finitude en géométrie diophantienne', in *Actès Congres International des Mathematiciens, Nice*, Vol. 1, Gauthier-Villars, 1971, pp. 467-471.
[5] GRAUERT, H.: 'Modell's Vermutung über rationale Punkte auf algebraischen Kurven und Funktionenkörpern', *Publ. Math. IHES*, no. 25 (1965), 131-149.
[6] LANG, S.: *Diophantine geometry*, Interscience, 1962.
[7] MAZUR, B.: 'Rational points on modular curves', in *Modular functions of one variables V*, Lecture notes in math., Vol. 601, Springer, 1977, pp. 107-148.
[8] MUMFORD, D.: 'A remark on Mordell's conjecture', *Amer. J. Math.* **87**, no. 4 (1965), 1007-1016.

A.N. Parshin

Editorial comments. The article above was originally written in 1982. Since then much has happened, and the field of arithmetic algebraic geometry has taken on additional prominence.

The Mordell conjecture over a number field has been proved in 1983 by G. Faltings. In his fundamental paper [A1], opening a new chapter in arithmetic algebraic geometry, Faltings proves at the same time two other famous conjectures, namely the *Tate conjecture* concerning endomorphisms of Abelian varieties over number fields (cf. **Tate conjectures**) and the *Shafarevich conjecture* concerning the finiteness of isomorphism classes of curves of genus g (respectively, Abelian varieties) defined over a number field and with good reduction outside a finite number of given places. The relation of the Mordell conjecture with the Shafarevich conjecture was pointed out by A.N. Parshin in 1970 [4]. Important for the work of Faltings were ideas of S.Yu. Arakelov, Yu.G. Zarkhin and L. Szpiro; essential in Faltings' proof is the theory of heights on the moduli space of Abelian varieties. For a good introduction to the proof of the Mordell conjecture see [A5].

References

[A1] FALTINGS, G.: 'Endlichkeitssätze für abelsche Varietäten über Zahlkörper', *Invent. Math.* **73** (1983), 349-366. Erratum: Ibid **75** (1984), 381.
[A2] FALTINGS, G.: 'Die Vermutungen von Tate und Mordell', *Jhber. Deutsch. Math. Verein.* **86** (1984), 1-13.
[A3] DELIGNE, P.: 'Preuve et conjectures de Tate et de Shafarevich', *Sém. Bourbaki* **616** (1983).
[A4] SZPIRO, L.: 'Le conjecture de Mordell', *Sém. Bourbaki* **619** (1983).
[A5] FALTINGS, G. and WÜSTHOLZ, G.: *Rational points*, Vieweg, 1984.

AMS 1980 Subject Classification: 14H99

MORERA THEOREM - If a (single-valued) function $f(z)$ of a complex variable z in a domain D is continuous and if its integral over any closed rectifiable contour $\Gamma \subset D$ is equal to zero, that is, if

$$\int_\Gamma f(z)\,dz = 0, \quad \Gamma \subset D, \qquad (*)$$

then $f(z)$ is an **analytic function** in D. This theorem was obtained by G. Morera [1].

The conditions of Morera's theorem can be weakened by restricting the requirement on vanishing integrals (*) to those taken over the boundary $\Gamma = \partial\Delta$ of any triangle Δ that is compactly contained in D, i.e. such that $\Delta \subset \bar{\Delta} \subset D$. Morera's theorem is an (incomplete) converse of the **Cauchy integral theorem** and is one of the basic theorems in the theory of analytic functions.

Morera's theorem can be generalized to functions of several complex variables. Let $f(z) = f(z_1, \ldots, z_n)$ be a function of the complex variables $z_1, \ldots, z_n$, $n \geqslant 1$, continuous in a domain D of $\mathbf{C}^n$ and such that its integral vanishes when taken over the boundary ∂T_ν of any *prismatic domain* compactly contained in D of the form

$$T_\nu = [a_1, z_1] \times \cdots \times [a_{\nu-1}, z_{\nu-1}] \times \Delta_\nu \times$$
$$\times [a_{\nu+1}, z_{\nu+1}] \times \cdots \times [a_n, z_n], \quad \nu = 1, \ldots, n,$$

where $[a_\mu, z_\mu]$, $\mu \neq \nu$, $\mu = 1, \ldots, n$, are line segments in the planes $\mathbf{C}(z_\mu)$ with end points a_μ and z_μ, and Δ_ν is a triangle in the plane $\mathbf{C}(z_\nu)$. Then $f(z)$ is a holomorphic function in D.

References

[1] MORERA, G.: 'Un teorema fondamentale nella teorica delle funzioni di una variabili complessa', *Rend. R. Ist. Lomb. Sci. Lettere* **19** (1886), 304-308.
[2] MARKUSHEVICH, A.I.: *Theory of functions of a complex variable*, 1, Chelsea, 1977 (translated from the Russian).
[3] SHABAT, B.V.: *Introduction to complex analysis*, 1-2, Moscow, 1976 (in Russian).
[4] VLADIMIROV, V.S.: *Methods of the theory of functions of several complex variables*, M.I.T, 1966 (translated from the Russian).

E.D. Solomentsev

Editorial comments.

References

[A1] REMMERT, R.: *Funktionentheorie*, 1, Springer, 1984.
[A2] CONWAY, J.B.: *Functions of one complex variable*, Springer, 1978.

AMS 1980 Subject Classification: 30A05, 32A30

MORITA EQUIVALENCE - An equivalence relation on the class of all rings (cf. **Ring**) defined as follows: Two rings R and S are called *Morita equivalent* if the categories of left (right) R- and S-modules are equivalent. The most important examples of Morita-equivalent rings are: a ring R and the ring of all $(n \times n)$-matrices over it. In order that there is Morita equivalence between two rings R and S it is necessary

and sufficient that in the **category** of left R-modules there is a finitely-generated projective generator U such that its ring of endomorphisms is isomorphic to S. The left R-module A is put in correspondence with the left S-module $\mathrm{Hom}_R(U, A)$. Among the properties preserved by transition to a Morita-equivalent ring are the properties of being: Artinian, Noetherian, primary, simple, classically semi-simple, regular, self-injective, hereditary, and primitive.

Alongside with Morita equivalence one considers *Morita duality*, relating some subcategories of the categories of left R-modules and right S-modules (mostly the subcategories of finitely-generated modules). However, the very existence of such a duality places definite restrictions on the rings R and S. In particular, for $R = S$ this implies that R is a **quasi-Frobenius ring**.

The general concept of Morita equivalence was developed by K. Morita [1].

References

[1] MORITA, K.: *Sci. Reports Tokyo Kyoiku Dajkagu A* **6** (1958), 83-142.
[2] BASS, H.: *Algebraic K-theory*, Benjamin, 1968.
[3] FAITH, C.: *Algebra: rings, modules and categories*, I-II, Springer, 1981-1976.
[4] COHN, P.: *Morita equivalence and duality*, London, 1976.

L.A. Skornyakov

Editorial comments. For generating objects of categories see also **Generator of a category**.

Let $\mathscr{C}$ and $\mathscr{D}$ be categories. A *duality* is a pair of contravariant functors $T: \mathscr{C} \to \mathscr{D}$ and $S: \mathscr{D} \to \mathscr{C}$ such that $ST \simeq \mathrm{id}_\mathscr{C}$, $TS \simeq \mathrm{id}_\mathscr{D}$, where $\simeq$ denotes natural equivalence (functorial isomorphism) and $\mathrm{id}_\mathscr{C}$ is the identity functor on $\mathscr{C}$.

Let A and B be rings and let $\mathscr{C}$ and $\mathscr{D}$ be full subcategories of the categories of right A-modules Mod_A and left B-modules $_B\mathrm{Mod}$, respectively (cf. **Module**). Let U be a (B, A) bimodule. A duality (T, S) between $\mathscr{C}$ and $\mathscr{D}$ is called a *U-duality* or *Morita duality* if T and S are, respectively, naturally equivalent to $\mathrm{Hom}_A(\ , U)$ and $\mathrm{Hom}_B(\ , U)$. A theorem of Morita says that if $\mathscr{C}$ and $\mathscr{D}$ are Abelian full subcategories with $A \in \mathscr{C}$ and $B \in \mathscr{D}$, then any duality (T, S) between $\mathscr{C}$ and $\mathscr{D}$ is a U-duality with $U = TA$.

AMS 1980 Subject Classification: 16-XX, 13-XX

MORPHISM *of a category* - A term used to denote the elements of an arbitrary **category** which play the role of mappings of one set into another, homomorphisms of groups, rings, algebras, continuous mappings of topological spaces, etc. A morphism of a category is an undefined concept. Each category consists of elements of two classes, called the *class of objects* and the *class of morphisms*, respectively. The class of morphisms of a category $\mathfrak{R}$ is usually denoted by $\mathrm{Mor}\,\mathfrak{R}$.

Any morphism α of a category $\mathfrak{R}$ has a uniquely defined domain (source) A and a uniquely defined codomain (target) B. All morphisms with common domain A and codomain B form a subset $H_{\Re}(A, B)$ of Mor $\Re$. The fact that α has domain A and codomain B can be written in the usual way: $\alpha \in H_{\Re}(A, B)$ or, using arrows, $\alpha: A \to B$, $A \to^{\alpha} B$, etc.

The division of the elements of a category into morphisms and objects is meaningful only within the context of a fixed category, since the morphisms of one category may be the objects of another and conversely. The morphisms of any category form a system that is closed under a partial binary operation — multiplication. Depending on the properties of morphisms relative to this operation, special classes of morphisms can be distinguished, for example, the classes of monomorphisms, epimorphisms, bimorphisms, isomorphisms, null (zero) morphisms, normal monomorphisms, normal epimorphisms, etc. (cf. **Monomorphism; Epimorphism; Bimorphism; Isomorphism; Normal monomorphism; Normal epimorphism**).

$M.Sh.\ Tsalenko$

Editorial comments.

References
[A1] MITCHELL, B.: *Theory of categories*, Acad. Press, 1965.

AMS 1980 Subject Classification: 18A05

MORSE FUNCTION - A smooth function with certain special properties. Morse functions arise and are used in **Morse theory**.

Let W be a smooth complete (in some Riemannian metric) Hilbert manifold (for example, finite dimensional) whose boundary ∂W is a disconnected union (possibly empty) of manifolds V_0 and V_1. A Morse function for the triple $(W; V_0, V_1)$ is a smooth (of Fréchet class C^2) function $f: W \to [a, b]$, $-\infty < a, b < +\infty$ (or $f: W \to [a, \infty]$ for $V_1 = \varnothing$), such that:

1) $f^{-1}(a) = V_0, f^{-1}(b) = V_1$;

2) all critical points (cf. **Critical point**) of f lie in $W \setminus \partial W = f^{-1}(a, b)$ and are non-degenerate;

3) *condition C of Palais$-$Smale* is fulfilled (see [2], [3]). I.e. on any closed set $S \subset W$ where f is bounded and the greatest lower bound of $x \to \| df(x) \|$ is zero, there is a critical point of f.

For example, if f is a proper function, that is, all sets $f^{-1}[c, d]$, $-\infty < c, d \leqslant \infty$, are compact (possible only for dim $W < \infty$), then F satisfies condition C. A Morse function attains a (global) minimum on each connected component of W. If V is a finite-dimensional manifold, then for $k \geqslant 2$ the set of Morse functions of class C^k is a set of the second category (and, if W is compact, even a dense open set) in the space of all functions

$$f: (W; V_0, V_1) \to ([a, b], a, b)$$

in the C^k-topology.

References
[1] MORSE, M.: *The calculus of variations in the large*, Amer. Math. Soc., 1934.
[2] PALAIS, R.S.: 'Morse theory on Hilbert manifolds', *Topology* **2** (1963), 299-340.
[3] SMALE, S.: 'Morse theory and a nonlinear generalization of the Dirichlet problem', *Ann. of Math.* **80** (1964), 382-396.

$M.M.\ Postnikov$
$Yu.B.\ Rudyak$

Editorial comments. There exist generalizations to Morse functions on stratified spaces (cf. (the editorial comments to) **Morse theory** and [A1]) and to equivariant Morse functions (cf. [A2] and [A3]).

References
[A1] GORESKI, M. and MACPHERSON, R.: *Stratified Morse theory*, Springer, 1988.
[A2] WASSERMAN, A.: 'Morse theory for *G*-manifolds', *Bull. Amer. Math. Soc.* **71** (1965), 384-388.
[A3] WASSERMAN, A.: 'Equivariant differential topology', *Topology* **8** (1969), 127-150.

AMS 1980 Subject Classification: 58C25, 57R45

MORSE INDEX - A number associated with a **critical point** of a smooth function on a manifold or of a geodesic on a Riemannian (or Finsler) manifold.

1) The Morse index of a critical point p of a smooth function f on a manifold M is equal, by definition, to the negative index of inertia of the Hessian of f at p (cf. **Hessian of a function**), that is, the dimension of the maximal subspace of the tangent space TM_p of M at p on which the Hessian is negative definite. This definition makes sense also for twice (Fréchet) differentiable functions on infinite-dimensional Banach spaces. The only difference is that the value $+\infty$ is admissible for the index. In this case it is expedient to introduce the idea of the *co-index* of a critical point p of f as the positive index of inertia of the Hessian (the second Fréchet differential) of f at p.

2) Let V_0 and V_1 be smooth submanifolds of a complete Riemannian space M. For a piecewise-smooth path $\omega: [0, 1] \to M$ with $\omega(i) \in V_i$, $i = 0, 1$, transversal to V_0 and V_1 at its end-points $\omega(0)$ and $\omega(1)$, the analogue of a tangent space is the vector space $T_\omega = T_{\omega, V_0, V_1}$ of all piecewise-smooth vector fields W along ω for which $W(\omega(i)) \in (TV_i)_{\omega(i)}$, $i = 0, 1$. For any geodesic $\gamma: [0, 1] \to M$ with $\gamma(i) \in V_i$, orthogonal at its end-points $\gamma(0)$ and $\gamma(1)$ to V_0 and V_1, respectively, the second variation $\delta^2 E$ of the action functional (see **Morse theory**) defines a symmetric bilinear functional $E_{\bullet\bullet}$ on T_γ (the analogue of the Hessian). The Morse index of the geodesic is equal, by definition, to the negative index of inertia of this functional. The null space N_γ of $E_{\bullet\bullet}$ on T_γ (the set of $X \in T_\gamma$ at which $E_{\bullet\bullet}(X, Y) = 0$ for all $Y \in T_\gamma$) consists exactly of the Jacobi fields (cf. **Jacobi vector field**) $J \in T_\gamma$. If $N_\gamma \neq 0$, the geodesic is called (V_0, V_1)-*degenerate*, and dim N_γ is called the *order of degeneracy* of the geodesic.

The case when V_1 is a point $q \in M$ is considered below. Let ν be the **normal bundle** to $V = V_0$ in M and let $\nu(p)$ be its fibre over $p \in V$. The restriction of the **exponential mapping** $TM \to M$ defines a mapping $\exp: \nu \to M$. A geodesic $\gamma(t) = \exp(t\xi)$, $\xi \in \nu(p)$, $0 \leqslant t \leqslant 1$, is $(V, \exp\xi)$-degenerate if and only if the kernel of the differential $d_\xi \exp: T\nu_\xi \to TM_{\exp\xi}$ of $\exp$ at ξ is not null; in this connection, the dimension of the kernel is equal to the order of degeneracy of the geodesic γ. A point $s = \gamma(t_0)$, $0 < t_0 \leqslant 1$, is called a *focal point* of V along γ if the geodesic $\gamma': t \to \gamma(t/t_0)$ is (V, s)-degenerate; the order of degeneracy of γ is called the *multiplicity of the focal point s*. By the **Sard theorem**, the set of focal points has measure zero, so a typical geodesic is non-degenerate. If V also consists of one point $p \in M$ ($p = q$ is not excluded), then a focal point is called *adjoint to p* along γ. The *Morse index theorem* [1] asserts that the Morse index of a geodesic is finite and equal to the number of focal points $\gamma(t)$ of V, $0 < t < 1$, taking account of multiplicity.

References

[1] MORSE, M.: *The calculus of variations in the large*, Amer. Math. Soc., 1934.
[2] AMBROSE, W.: 'The index theorem in Riemannian geometry', *Ann. of Math.* 73 (1961), 49-86.

M.M. Postnikov
Yu.B. Rudyak

Editorial comments. There is a natural generalization of the Morse index of geodesics to **variational calculus**, which runs as follows. Let f be a real-valued smooth function on an open subset Z of $[0, 1] \times TM$ and let R be a smooth sub-manifold of $M \times M$. Let C be the space of smooth curves $\omega: [0, 1] \to M$ for which the 1-jet lies in Z and $(\omega(0), \omega(1)) \in R$. Then C is a Banach manifold, on which one has the smooth functional

$$F: \omega \mapsto \int_0^1 f\left[t, \omega(t), \frac{d\omega}{dt}(t)\right] dt.$$

One then considers the Morse index of F at critical curves ω; it is finite if the Hessian of $v \mapsto f(t, x, v)$ is positive definite at $x = \omega(t)$, $v = (d\omega/dt)(t)$, $t \in [0, 1]$ (*Legendre's condition*, cf. **Legendre condition**).

References

[A1] MILNOR, J.: *Morse theory*, Princeton Univ. Press, 1963.
[A2] KLINGENBERG, W.: *Riemannian geometry*, de Gruyter, 1982.
[A3] KLINGENBERG, W.: *Lectures on closed geodesics*, Springer, 1978.

AMS 1980 Subject Classification: 58C27, 58E05

MORSE INEQUALITIES - Inequalities following from Morse theory and relating the number of critical points (cf. **Critical point**) of a Morse function on a manifold to its homology invariants.

Let f be a **Morse function** on a smooth n-dimensional manifold M (without boundary) having a finite number of critical points. Then the **homology group** $H_\lambda(M)$ is finitely generated and is therefore determined by its rank, $r_\lambda = \mathrm{rk}(H_\lambda(M))$, and its torsion rank, $t_\lambda = t(H_\lambda(M))$ (the *torsion rank of an Abelian group A* with a finite number of generators is the minimal number of cyclic groups in a direct-sum decomposition of which a maximal torsion subgroup of A can be imbedded). The Morse inequalities relate the number m_λ of critical points of f with **Morse index** λ to these ranks, and have the form:

$$r_\lambda + t_\lambda + t_{\lambda-1} \leqslant m_\lambda, \quad \lambda = 0, \ldots, n;$$

$$\sum_{i=0}^{\lambda} (-1)^{\lambda-i} r_i \leqslant \sum_{i=0}^{\lambda} (-1)^{\lambda-i} m_i, \quad \lambda = 0, \ldots, n.$$

For $\lambda = n$ the last Morse inequality is always an equality, so that

$$\sum_{i=0}^{n} (-1)^i m_i = \chi(M),$$

where $\chi(M)$ is the **Euler characteristic** of M.

The Morse inequalities also hold for Morse functions of a triple (W, V_0, V_1), it suffices to replace the groups $H_\lambda(M)$ by the relative homology groups $H_\lambda(W, V_0)$.

According to the Morse inequalities, a manifold having 'large' homology groups does not admit a Morse function with a small number of critical points. It is remarkable that the estimates in the Morse inequalities are sharp: On a closed simply-connected manifold of dimension $n \geqslant 6$ there is a Morse function for which the Morse inequalities are equalities (*Smale's theorem*, see [2]). In particular, on any closed manifold that is homotopically equivalent to the sphere S^n, with $n \geqslant 6$, there is a Morse function with two critical points; hence it follows immediately (see **Morse theory**) that M is homeomorphic to S^n (see **Poincaré conjecture**). A similar application of Smale's theorem allows one to prove theorems on h- and s-cobordism.

An analogue of the Morse inequalities holds for a Morse function $f: X \to \mathbf{R}$ on an infinite-dimensional Hilbert manifold, and they relate (for any regular values $a, b \in \mathbf{R}$, $a < b$, of f) the numbers $m_\lambda(a, b)$ of critical points of finite index λ lying in $f^{-1}[a, b]$, with the rank $r_\lambda(a, b)$ and torsion rank $t_\lambda(a, b)$ of the group $H_\lambda(M^b, M^a)$, where $M^c = f^{-1}(-\infty, c]$. Namely,

$$r_\lambda(a, b) + t_\lambda(a, b) + t_{\lambda-1}(a, b) \leqslant m_\lambda,$$

$$\sum_{i=0}^{\lambda} (-1)^{\lambda-i} r_i(a, b) \leqslant \sum_{i=0}^{\lambda} (-1)^{\lambda-i} m_i;$$

$$\lambda = 0, 1, \ldots.$$

For λ large enough the latter inequality becomes an equality.

References

[1] MORSE, M.: *The calculus of variations in the large*, Amer. Math. Soc., 1934.
[2] SMALE, S.: 'Generalized Poincaré's conjecture in dimensions greater than four', *Ann. of Math.* 74 (1961), 391-466.

M.M. Postnikov
Yu.B. Rudyak

Editorial comments. Another version of the Morse inequalities can be stated as follows, cf. [A1].

For a Morse function f one introduces the quantity

$$M_t(f) = \sum_p t^{\lambda(p)}$$

where the sum is taken over the critical points p of f and $\lambda(p)$ is the index of p relative to f. In the compact case this sum is finite, since the critical points are discrete. The polynomial $M_t(f)$, which is also called the *Morse polynomial* of f, has the Poincaré polynomial of the manifold W as a lower bound in the following sense. Let

$$P_t(W) = \sum t^k \dim H_k(W; K),$$

where the homology is taken relative to some fixed coefficient field K. Then the following Morse inequality holds: For every non-degenerate f there exists a polynomial $Q_t(f) = q_0 + q_1 t + \cdots$ with non-negative coefficients such that

$$M_t(f) - P_t(f) = (1 + t) \cdot Q_t(f).$$

References

[A1] BOTT, R.: 'Lectures on Morse theory, old and new', *Bull. Amer. Math. Soc.* 7, no. 2 (1982), 331-358.
[A2] MILNOR, J.: *Morse theory*, Princeton Univ. Press, 1963.
[A3] PALAIS, R.S.: 'Morse theory on Hilbert manifolds', *Topology* 2 (1963), 299-340.

AMS 1980 Subject Classification: 58C27, 58E05

MORSE LEMMA - The statement describing the structure of the **germ** of a twice continuously-differentiable function. Let $f: \mathbf{R}^n \to \mathbf{R}^1$ be a function of class C^2, having the point $0 \in \mathbf{R}^n$ as a non-degenerate **critical point**. Then in some neighbourhood U of 0 there is a local coordinate system (a chart) $x_1, \ldots, x_n$, with centre at 0, so that for all $x \in U$,

$$f(x) = f(0) - x_1^2 - \cdots - x_\lambda^2 + x_{\lambda+1}^2 + \cdots + x_n^2.$$

Here the number λ, $0 \leqslant \lambda \leqslant n$, is the **Morse index** of the critical point 0 of f. An analogue of the Morse lemma for functions $f: \mathbf{C}^n \to \mathbf{C}$ is also true, namely: If f is holomorphic in a neighbourhood of a non-degenerate critical point (in other terminology, a saddle point, see **Saddle point method**) $0 \in \mathbf{C}^n$, then in some neighbourhood U of 0 there is a local coordinate system $z_1, \ldots, z_n$ such that

$$f(x) = f(0) + z_1^2 + \cdots + z_n^2.$$

The Morse lemma also holds for functions $f: E \to \mathbf{R}$ on a separable (infinite-dimensional) Hilbert space E. Let f be twice (Fréchet) differentiable in some neighbourhood of a non-degenerate critical point $0 \in E$. Then there are a convex neighbourhood of zero $V \subset E$, a convex neighbourhood of zero $U \subset E$ and a diffeomorphism (a chart) $\theta: U \to V$ with $\theta(0) = 0$, such that for all $x \in U$,

$$f(x) = \| P(\theta(x)) \|^2 - \| (I - P)(\theta(x)) \|^2,$$

where $P: E \to E$ is a continuous orthogonal projection and I is the identity operator. Here the dimension $\dim \mathrm{Im}(I - P)$ coincides with Morse index of the critical point $0 \in E$ of f and the dimension $\dim \mathrm{Im} P$ coincides with its co-index.

References

[1] MORSE, M.: *The calculus of variations in the large*, Amer. Math. Soc., 1934.

M.M. Postnikov
Yu.B. Rudyak

Editorial comments. There exist generalizations of the Morse lemma to the following cases:

Equivariant Morse lemma. Consider a holomorphic function $f: \mathbf{C}^k \to \mathbf{C}$ that is invariant with respect to the linear action of a compact subgroup G on $\mathbf{C}^k$. If f has at 0 a critical point with critical value 0, then it can be reduced to its quadratic part by a G-invariant change of independent variables, biholomorphic at the point 0.

An analogous 'equivariant Morse lemma' is true in the real-analytic and the differentiable context. Cf. [A1] and [A2].

Morse lemma depending on parameters. Let $f: \mathbf{R}^n \times \mathbf{R}^s \to \mathbf{R}$ be a real-valued differentiable function defined in a neighbourhood of $(0, 0)$. Let $(x, \lambda) \in \mathbf{R}^n \times \mathbf{R}^s$. Assume that $Df_x(0, 0) = 0$ and that $A = D^2 f_{xx}(0, 0)$ is non-singular. Then there exist coordinates (z, λ) in a neighbourhood of $(0, 0)$ such that

$$f(x, \lambda) = f(x(\lambda), \lambda) + \frac{1}{2} <Az, z>.$$

In this formula $x(\lambda)$ is the local solution of the equations $Df_x(x, \lambda) = 0$ and $x(0) = 0$. The proof is a modification of that in the case without parameters. A good reference is [A3], p. 502.

References

[A1] ARNOL'D, V.I.: 'Wave front evolution and the equivariant Morse lemma', *Comm. Pure Appl. Math.* 29 (1976), 557-582.
[A2] ARNOL'D, V.I., GUSEIN-ZADE, S.M. and VARCHENKO, A.N.: *Singularities of differentiable maps*, 1, Birkhäuser, 1985 (translated from the Russian).
[A3] HÖRMANDER, L.: *The analysis of linear partial operators*, 3. Pseudo-differential operators, Springer, 1985.
[A4] PALAIS, R.S.: 'Morse theory on Hilbert manifolds', *Topology* 2 (1963), 299-340.

AMS 1980 Subject Classification: 58C25, 58E05

MORSE — SMALE SYSTEM, *Morse — Smale dynamical system* - A smooth **flow (continuous-time dynamical system)** $\{S_t\}$ or **cascade** (discrete-time dynamical system) $\{S^n\}$ (generated by a diffeomorphism S, which is called a *Morse — Smale diffeomorphism* in this case) on a compact (usually closed) differentiable m-dimensional manifold M^m, having the following properties:

1) The system has a finite number of periodic trajectories (including fixed points in the case of a cascade) and (in the case of a flow) equilibrium states.

2) Each periodic trajectory listed in 1) has **local structural stability** (usually the definition requires an equivalent property of the corresponding linearized system). This guarantees the existence of stable and unstable invariant manifolds W^s and W^u for each such

trajectory (if the trajectory is stable, or totally unstable, W^u, respectively W^s, reduces to the trajectory itself); the dimension of W^u is called its *index*. The index generalizes the notion of the **Morse index** of a non-degenerate critical (stationary) point w_0 of a smooth function $f: M \to \mathbf{R}$, since the latter coincides with the index of w as an equilibrium point of the **gradient dynamical system**

$$\dot{w} = -\nabla f(w), \tag{1}$$

where the gradient is taken with respect to any Riemannian metric on M.

3) The invariant manifolds of the trajectories listed in 1) intersect transversely (that is, if $w \in W_1^s \cap W_2^u$, then for the tangent spaces $T_w W_1^s + T_w W_2^u = T_w M$).

4) All remaining trajectories tend, as $t \to \pm\infty$ or as $n \to \pm\infty$, to one of the trajectories listed in 1).

5) If M has a boundary, then some condition must be placed on the behaviour of the system near the boundary. For flows (up to now, only this case has been considered) it is usual to require that the phase velocity vector always be transversal to the boundary.

Morse—Smale systems are structurally stable systems (cf. **Rough system**, [1]). It was precisely in connection with the study of the latter that special cases of Morse—Smale systems were first discussed — flows in plane domains (see the more detailed account in [2]) and cascades on the circle (see [4] - [6]). In the general case Morse—Smale systems were introduced by S. Smale, who considered Morse—Smale systems on a closed M, for which the following *Morse—Smale inequalities* were proved. For a cascade, let m_i be the number of periodic points of index i, and for a flow it denotes the sum of the number of equilibrium positions of index i and the number of periodic trajectories of indices i and $i+1$. Then for $i = 0, \dots, m$,

$$\sum_{j=0}^{i} (-1)^j m_{i-j} \geq \sum_{j=0}^{i} (-1)^j b_{i-j}, \tag{2}$$

where b_i is the i-th **Betti number** of M (if some of the W^u, W^s introduced in 2) are non-orientable, then the Betti number is taken over a field of characteristic two). For $i = m$, (2) becomes an equality.

The inequalities (2) generalize the usual **Morse inequalities** for a smooth function $f: M \to \mathbf{R}$ with non-degenerate critical points. Namely, the Morse inequalities can be obtained by an application of (2) to the system (1) (which, in reality, need not be a Morse—Smale system, so minor additional arguments are required, see [7], [8]).

The question of when there is a Morse—Smale diffeomorphism in a given isotopy class (see [9], [10]) and, for an M^m with Euler characteristic zero, the analogous question with respect to a homotopy class of non-singular vector fields (here for $m \geq 4$ the answer is

always positive, see [11]) has been investigated. For flows with $m = 2$ (see [12], [13]) and certain special types of flows with $m \geq 3$ (see [14], [15]) it has been clarified which topological invariants determine the topological equivalence of two Morse—Smale systems. (In the two-dimensional case this question has been solved for a broader class of flows (see [3], [16]), and the case $m = 1$ is trivial.)

References

[1] PALIS, J. and SMALE, S.: 'Structural stability theorems', in S.S. Chern and S. Smale (eds.): *Global analysis*, Proc. Symp. Pure Math., Vol. 14, Amer. Math. Soc., 1970, pp. 223-232.

[2] ANDRONOV, A.A., LEONTOVICH, E.A., GORDON, I.I. and MAIER, A.G.: *Theory of bifurcations of dynamic systems on a plane*, Israel Progr. Sci. Transl., 1971 (translated from the Russian).

[3] ANDRONOV, A.A., LEONTOVICH, E.A., GORDON, I.I. and MAIER, A.G.: *Qualitative theory of second-order dynamic systems*, Wiley, 1973 (translated from the Russian).

[4] MAIER, A.G.: 'A structurally stable map of the circle onto itself', *Uchen. Zap. Gor'k. Gos. Inst.* **12** (1939), 215-229 (in Russian).

[5] PLISS, V.A.: 'On the structural stability of differential equations on the torus', *Vestnik Leningrad. Univ. Ser. Mat.* **15**, no. 13 (1960), 15-23 (in Russian).

[6A] ARNOL'D, V.I.: 'Small denominators I. Mapping the circle onto itself', *Transl. Amer. Math. Soc. (2)* **46** (1965), 213-284. (*Izv. Akad. Nauk SSSR Ser. Mat.* **25**, no. 1 (1961), 21-86)

[6B] ARNOL'D, V.I.: 'Correction to 'Small dominators I'', *Izv. Akad. Nauk SSSR Ser. Mat.* **28**, no. 2 (1964), 479-480 (in Russian).

[7] SMALE, S.: 'Morse inequalities for dynamical systems', *Bull. Amer. Math. Soc.* **66** (1960), 43-49.

[8] SMALE, S.: 'On gradient dynamical systems', *Ann. of Math. (2)* **74** (1961), 199-206.

[9] SHUB, M.: 'Morse—Smale diffeomorphisms are unipotent in homology', in M.M. Peixoto (ed.): *Dynamical Systems (Proc. Conf. Salvador, 1971)*, Acad. Press, 1973, pp. 489-491.

[10] SHUB, M. and SULLIVAN, D.: 'Homology theory and dynamical systems', *Topology* **14** (1975), 109-132.

[11] ASIMOV, D.: 'Homotopy of non-singular vector fields to structurally stable ones', *Ann. of Math.* **102**, no. 1 (1975), 55-65.

[12] PEIXOTO, M.: 'Sur la classification des équations différentielles', *C.R. Acad. Sci. Paris* **272** (1971), A262-A265.

[13] PEIXOTO, M.: 'Dynamical systems', in M. Peixoto (ed.): *Dynamical Systems (Proc. Conf. Salvador, 1971)*, Acad. Press, 1973, pp. 389-419.

[14] UMANSKIĬ, YA.L.: 'The scheme of a 3-dimensional Morse—Smale dynamical system without closed trajectories', *Soviet Math. Dokl.* **17** (1976), 1479-1482. (*Dokl. Akad. Nauk SSSR* **230**, no. 6 (1976), 1286-1289)

[15] PILYUGIN, S.YU.: 'Phase diagrams that determine Morse—Smale systems without periodic trajectories on spheres', *Diff. Eq.* **14**, no. 2 (1978), 170-177. (*Diff. Uravnen.* **14**, no. 2 (1978), 245-254)

[16] NEUMANN, D. and O'BRIEN, T.: 'Global structure of continuous flows on 2-manifolds', *J. Diff. Eq.* **22**, no. 1 (1976), 89-110.

D.V. Anosov

Editorial comments. Condition (2) above is often called *hyperbolicity* of the periodic trajectories and invariant points (cf. **Hyperbolic set**). Condition (4), together with condition (1), is often formulated as: The non-wandering set (cf. **Non-wandering point**) consists only of a finite number of periodic trajectories and invariant points (each of which is, by (2), hyperbolic).

References

[A1] SMALE, S.: 'Diffeomorphisms with many periodic points', in *Differential and Combinatorial Topology (A Symp. in honour of M. Morse)*, Princeton Univ. Press, 1965, pp. 63-80.
[A2] SHUB, M.: 'Dynamical systems, filtrations and entropy', *Bull. Amer. Math. Soc.* **80** (1974), 27-41.
[A3] FRANTS, J. and SHUB, M.: 'The existence of Morse–Smale diffeomorphisms', *Topology* **20** (1981), 273-290.
[A4] MALLER, M.: 'Fitted diffeomorphisms of non-simply connected manifolds', *Topology* **19** (1980), 395-410.
[A5] PALIS, J.: 'On Morse–Smale dynamical systems', *Topology* **8** (1969), 385-405.

AMS 1980 Subject Classification: 58F09

MORSE SURGERY, *surgery* - A transformation of smooth manifolds to which the level manifold of a smooth function is subjected on passage through a non-degenerate **critical point**; it is the most important construction in the topology of manifolds.

Let V be a smooth n-dimensional manifold (without boundary) in which a $(\lambda-1)$-dimensional sphere $S^{\lambda-1}$ is (smoothly) imbedded. Suppose that the **normal bundle** of $S^{\lambda-1}$ in V is trivial, that is, a closed **tubular neighbourhood** T of $S^{\lambda-1}$ in V decomposes into a direct product $T=S^{\lambda-1}\times D^{n-\lambda+1}$, where $D^{n-\lambda+1}$ is a disc of dimension $n-\lambda+1$. After having chosen such a decomposition, remove the interior of T from V. A manifold is obtained whose boundary decomposes into a product $S^{\lambda-1}\times S^{n-\lambda}$ of spheres. But the manifold $D^{\lambda}\times S^{n-\lambda}$ has precisely the same boundary. Identifying the boundaries of $V\setminus\operatorname{Int} T$ and $D^{\lambda}\times S^{n-\lambda}$ by a diffeomorphism preserving the product structure of $S^{\lambda-1}\times S^{n-\lambda}$, a manifold V' without boundary is again obtained. It is called the *result of surgery* of V along $S^{\lambda-1}$.

In the realization of a surgery it is necessary to give a decomposition of the neighbourhood T of $S^{\lambda-1}$ into a direct product, that is, a *trivialization* of the normal bundle of $S^{\lambda-1}$ in V; in this connection different trivializations (riggings) may give essentially distinct (even homotopically) manifolds V'.

The number λ is called the *index of the surgery*, and the pair $(\lambda, n-\lambda+1)$ its *type*. If V' is obtained from V by a surgery of type $(\lambda, n-\lambda+1)$, then V is obtained from V' by a surgery of type $(n-\lambda+1, \lambda)$. For $\lambda=0$, V' is a disjoint union of V (which may be empty) and S^n. The construction of a surgery may also be carried out for piecewise-linear and topological manifolds.

Example. For $V=S^2$ and $\lambda=2$ the result of surgery is a disjoint union of spheres, and for $\lambda=1$ a torus. For $V=S^3$ and $\lambda=2$ the product $S^1\times S^2$ is obtained. The case $V=S^3$ and $\lambda=1$ is more complicated: if $S^{\lambda-1}=S^1$ is imbedded in S^3 in the standard way (as a great circle), then, depending on the choice of the trivialization of the normal bundle, all lens spaces (cf. **Lens space**) are obtained; if, however, a knotting of S^1 is allowed, then a still larger set of three-dimensional manifolds is obtained.

If V is the boundary of an $(n+1)$-dimensional manifold M, then V' will be the boundary of the manifold M' obtained from M by glueing a handle of the index of M'. In particular, if f is a smooth function on M and if $a<b$ are numbers such that $f^{-1}[a, b]$ is compact and contains a unique non-degenerate critical point p, then $V^b=f^{-1}(b)$ is obtained from $V^a=f^{-1}(a)$ by a surgery of index λ, where λ is the **Morse index** of p. In a more general form, any surgery V' of a manifold V of index λ defines a **bordism** (*cobordism*) $(W; V, V')$ (obtained from the product $V\times[0, 1]$ by glueing a handle of index λ to its 'right-hand boundary' $V\times\{1\}$), and on the triple $(W; V, V')$ there is a **Morse function** with a unique critical point of index λ; moreover, any bordism $(W; V, V')$ on which there is such a Morse function is obtained in this way. Hence (and from the existence of Morse functions on triples) it follows that two manifolds are bordant if one can be obtained from the other by a sequence of surgeries.

With the known precautions on the treatment of orientations, the result of a surgery on an oriented manifold is again an oriented manifold. In general, for any structure series (B, ϕ) (see (B, ϕ)-**structure**) it is possible to define the idea of (B, ϕ)-*surgery*; in this connection, two manifolds are (B, ϕ)-bordant if they are connected by a finite sequence of (B, ϕ)-surgeries.

The important role of surgery in the topology of manifolds is explained by the fact that it allows one to 'delicately' (without infringing on the various properties of manifolds) kill 'superfluous' homotopy groups (the operation usually used to this end in homotopy theory, i.e. the operation of 'glueing' cells, instantaneously leads out of the class of manifolds). In practice, all theorems on the classification of structures on manifolds are based on the question: Given a mapping $f: M\to X$ of a closed manifold M into a CW-complex X, when does there exist a bordism $(W; M, N)$ and a mapping $F: W\to X$ such that $F|_M=f$ and $F|_N: N\to X$ is a homotopy equivalence. The natural route to the solution of this problem is to annihilate the kernels of the homomorphisms $f_*: \pi_i(M)\to\pi_i(X)$ by a sequence of surgeries (where π_i are the homotopy groups). If this succeeds, the resulting mapping will be a homotopy equivalence. The study of the corresponding obstructions (lying in the so-called Wall groups, see [4] and **Wall group**) is one of the most important stimuli in the development of **algebraic K-theory**.

References

[1] MORSE, M.: *The calculus of variations in the large*, Amer. Math. Soc., 1934.
[2] NOVIKOV, S.P.: 'Homotopically equivalent smooth manifolds I', *Transl. Amer. Math. Soc.* **48** (1965), 271-396. (*Izv. Akad. Nauk SSSR Ser. Mat.* **28**, no. 2 (1964), 365-474)

[3] KERVAIRE, M.A. and MILNOR, J.W.: 'Groups of homotopy spheres: I', *Ann. of Math.* **77** (1963), 504-537.

[4] MISHCHENKO, A.S.: 'Hermitian *K*-theory. The theory of characteristic classes and methods of functional analysis', *Russian Math. Surveys* **31**, no. 2 (1976), 71-138. (*Uspekhi Mat. Nauk* **31**, no. 2 (1976), 69-134)

M.M. Postnikov
Yu.B. Rudyak

Editorial comments. There are clear relations between surgery and Morse theory, as indicated above, which is why in the Russian literature the term Morse surgery is frequently used. In the Western literature one simply speaks of surgery. The construction was invented by J.W. Milnor ([A4]).

References

[A1] BROWDER, W.B.: *Surgery on simply-connected manifolds*, Springer, 1972.

[A2] WALL, C.T.-C.: *Surgery on compact manifolds*, Acad. Press, 1970.

[A3] WALL, C.T.-C.: 'Surgery on non-simply connected manifolds', *Ann. of Math.* **84** (1966), 217-276.

[A4] MILNOR, J.: 'A procedure for killing the homotopy groups of differentiable manifolds', in *Differential Geometry*, Proc. Symp. Pure Math., Vol. 3, Amer. Math. Soc., 1961, pp. 39-55.

[A5] MILNOR, J.: *Morse theory*, Princeton Univ. Press, 1963.

[A6] HIRSCH, M.: *Differential topology*, Springer, 1976.

AMS 1980 Subject Classification: 58E05, 57R65

MORSE THEORY - The common name for three different theories based on ideas of M. Morse [1] and describing the relation between algebraic-topological properties of topological spaces and extremal properties of functions (functionals) on them. Morse theory is a branch of **variational calculus in the large** (*calculus of variations in the large*); however, the latter is broader: for example, it includes the theory of categories (cf. **Category (in the sense of Lyusternik—Shnirel'man))**.

1) Morse theory of critical points (cf. **Critical point**) of smooth functions f on a smooth manifold M (briefly, Morse theory 1) is divided into two parts: local and global. The local part is related to the idea of a critical point of a smooth function, the **Hessian of a function** at its critical point, the **Morse index** of a critical point, etc. The basic result is the **Morse lemma**, which describes the structure of a smooth function in a neighbourhood of a non-degenerate critical point.

The study of smooth functions in neighbourhoods of degenerate points does not properly belong to Morse theory, it does rather belong to the separate theory of **singularities of differentiable mappings**.

The basic results in global Morse theory are as follows. Let f be a function on a smooth manifold M. If the set $f^{-1}(a)$ does not contain a critical point of f and does not intersect the boundary of M, then $M^a = f^{-1}(-\infty, a)$ is a smooth manifold with boundary $f^{-1}(a)$. If the set $f^{-1}[a, b]$ is compact, does not intersect the boundary of M and does not contain a critical point of f, then there is a smooth isotopy $h_t: M \to M$,

$0 \leqslant t \leqslant 1$ (realized by shifting along the trajectories of the gradient of f), such that $h_0 = \mathrm{id}_M$ and h_1 diffeomorphically maps M^b onto M^a. In particular, M^b is diffeomorphic to M^a and the inclusion $M^a \subset M^b$ is a homotopy equivalence.

If $f^{-1}[a, b]$ is compact, does not intersect the boundary of M and contains precisely one critical point $p \in f^{-1}(a, b)$ with Morse index λ, then M^b is diffeomorphic to a manifold obtained from M^a by glueing a handle of index λ (see **Morse surgery**). In particular, if p is the unique global minimum point of f, then for small $\epsilon > 0$ the set $M^{f(p)+\epsilon}$ is diffeomorphic to the disc D^n, where $n = \dim M$. Hence it follows that if M is a closed smooth manifold having a function with precisely two critical points (both non-degenerate), then M is obtained by glueing two smooth discs along their common boundary and, therefore, it is homeomorphic (but in general not diffeomorphic) to the sphere S^n.

Since glueing a handle of index λ is homotopically equivalent to glueing a cell of dimension λ, the following *fundamental theorem of Morse theory* 1 follows immediately: Corresponding to each **Morse function** f on a smooth manifold M (without boundary) is a **CW-complex** homotopically equivalent to M; its cells are in bijective correspondence with the critical points of f and the dimension of a cell is equal to the index of the corresponding critical point. The **Morse inequalities** are an immediate consequence of this theorem. An analogous theorem is valid for a Morse function on a triple $(W; V_0, V_1)$.

2) Morse theory of geodesics on a Riemannian manifold (briefly, Morse theory 2) describes the homotopy type of the **loop space** ΩM of a smooth manifold M with a Riemannian metric g_{ij}. Its aim is to transfer the results of Morse theory 1 to this space (more correctly, to a suitable model of it). The role of f is played here by an action functional E (sometimes called energy functional, [5]), defined on the space $\mathrm{PS}(M)$ of piecewise-smooth paths $\omega: t \to \omega(t)$, $0 \leqslant t \leqslant 1$, whose value on a path $\omega \in \mathrm{PS}(M)$ is defined, in local coordinates $x^1, \ldots, x^n$, by the formula

$$E(\omega) = \int_0^1 g_{ij} \, dx^i \, dx^j.$$

In the initial construction of Morse theory the length functional

$$L(\omega) = \int_0^1 \sqrt{g_{ij} \, dx^i \, dx^j}$$

was considered, but for many technical reasons E turned out to be preferable. At the same time the extremals of E (that is, paths $\omega \in \mathrm{PS}(M)$ for which the linear functional E_* defined by the variation δE of E is zero on T_ω) coincide with the geodesics of the metric g_{ij} (the extremals of the functional L) in their natural parametrization.

Let p and q be two (not necessarily distinct) points of M, and let $\Omega^{PS}(M; p, q) \subset PS(M)$ be the space of piecewise-smooth paths joining p to q. For each $l \in \mathbf{R}$, put

$$\Omega_l = \Omega_l^{PS}(M; p, q) = E^{-1}[0, l] \cap \Omega^{PS}(M; p, q).$$

If M is complete, then $\Omega_l^0 = E^{-1}[0, l) \cap \Omega_l$ (the interior of Ω_l) is a **deformation retract** of a smooth manifold B whose points are 'polygonal geodesics' with a fixed number of links, joining p to q (so that, in particular, B contains all geodesics from Ω_l^0). Here $E' = E|_B : B \to \mathbf{R}$ is a smooth function; for any $a < l$ the set $B^a = (E')^{-1}[0, a]$ is compact and is a deformation retract of Ω_a; the critical points of E' coincide with the extremals of the functional $E : \Omega_l^0 \to \mathbf{R}$ and are geodesics of length $< \sqrt{l}$ joining p and q; the Morse indices of the critical points of E' are equal to the Morse indices of the corresponding geodesics; the null space N_γ of E_{**} on a geodesic $\gamma \in \Omega_l^0$ is finite dimensional and isomorphic to the null space of the Hessian of E' at the corresponding critical point; in particular, if p and q are not conjugate on any geodesic γ joining them, then E' is a Morse function. Applying Morse theory 1, passing to the limit as $l \to \infty$ and noting that $\Omega^{PS}(M; p, q)$ is homotopically equivalent to the space $\Omega(M; p, q)$ of all continuous paths joining p to q, one obtains the following *fundamental theorem of Morse theory* 2: Let M be a complete Riemannian manifold and let p and q be two points not conjugate on any geodesic joining them. The space $\Omega(M; p, q)$ of all paths joining p and q is homotopically equivalent to a CW-complex all cells of dimension λ of which are in bijective correspondence with the geodesics of index λ joining p to q. Since the homotopy type of $\Omega(M; p, q)$ does not depend on the choice of p and q, this theorem gives, in particular, a description of the homotopy type of the loop space ΩM.

It is known [10] that for a non-contractible manifold M the space ΩM has non-trivial homology groups in arbitrarily high dimension. By the fundamental theorem of Morse theory 2 it follows that non-conjugate points in a complete Riemannian non-contractible manifold are joined by infinitely many geodesics (by the example of the sphere it is clear, in general, that these geodesics may be segments of one periodic geodesic).

In the description of the homotopy type given by the fundamental theorem, Jacobi fields (cf. **Jacobi equation** and **Jacobi vector field**) (implicitly) appear, therefore Morse theory establishes a connection between the curvature of a manifold and its topology. For example, if M is a complete simply-connected Riemannian manifold of non-positive curvature in all two-dimensional directions, then any Jacobi field vanishing at two points

of a geodesic is identically zero. Therefore the loop space ΩM of such a manifold has the type of a zero-dimensional CW-complex, and consequently (in view of the simple connectedness of M) is contractible. Therefore M is contractible, that is, is homotopically equivalent to $\mathbf{R}^n$. A more precise use of Morse theory shows that M is even diffeomorphic to $\mathbf{R}^n$ (see [3], [5]).

The application of Morse theory to the topology of Lie groups has turned out to be very effective [2]. For example, for any simply-connected Lie group G the space ΩG has the homotopy type of a CW-complex with only odd-dimensional cells. The apotheosis here is the **Bott periodicity theorem**, which plays a fundamental role in K-theory and, consequently, in the whole of differential topology. Let U be the limit of the sequence of nested unitary groups $\cdots \subset U_n \subset U_{n+1} \subset \cdots$ and let O be the limit of the sequence of nested orthogonal groups $\cdots \subset O_n \subset O_{n+1} \subset \cdots$. Bott's periodicity theorem asserts that there are homotopy equivalences $\Omega^2 U \sim U$, $\Omega^8 O \sim O$, where Ω^n is the n-th iterate of the functor of passing to the loop space. This theorem allows one to calculate the homotopy groups $\pi_i U$ and $\pi_i O$ and, consequently, the homotopy groups $\pi_i U_n$ and $\pi_j O_n$ for $i < 2n$, $j < n$.

More theory 2 generalizes also to the case when instead of points p, q smooth submanifolds V_0, V_1 of M are considered. The action functional is studied on the space $\Omega^{PS}(M; V_0, V_1)$ of all piecewise-smooth paths $\omega : t \to \omega(t)$, $0 \leqslant t \leqslant 1$, $\omega(i) \in V_i$, $i = 0, 1$, that are transversal at the end-points to V_0 and V_1, and a relation between the extremals of this functional and the homotopy type of $\Omega(M; V_0, V_1)$ has been established. The corresponding fundamental theorem is analogous to the above-mentioned fundamental theorem of Morse theory 2; the difficulty is in the geometric interpretation of the Morse index of a geodesic.

3) The natural development of Morse theory 2 is Morse theory for critical points of smooth functions on Banach (infinite-dimensional) manifolds — Morse theory 3, which is no longer an analogue, but a direct generalization of Morse theory 1. At present (1989) Morse theory 3 is at an initial stage and has been constructed only in a very preliminary context under very strong (and clearly not necessary) conditions on the model Banach space (on separable- and Hilbert-type spaces), when no specifically functional-analytic difficulties arise [9], although there have been attempts at a construction of Morse theory 3 in fairly general situations. Therefore, in its modern form, Morse theory 3 is an almost verbatim re-iteration of Morse theory 1. The only difference worth mentioning is that in Morse theory 3 the compactness of $f^{-1}[a, b]$ is replaced by condition C of Palais$-$Smale (see **Morse function**), which, besides, is not satisfied in all situations of

interest. In addition, although it is possible to glue to a Banach manifold a handle of infinite index, in view of the homotopic triviality of infinite-dimensional spheres this handle has no effect on the homotopy type. Therefore only critical points of finite index occur in the fundamental theorem of Morse theory 3.

References

[1] MORSE, M.: *The calculus of variations in the large*, Amer. Math. Soc., 1934.

[2] MILNOR, J.: *Morse theory*, Princeton Univ. Press, 1963.

[3] MILNOR, J.: *Lectures on the h-cobordism theorem*, Princeton Univ. Press, 1965.

[4] SEIFERT, H. and THRELFALL, W.: *Variationsrechnung im Groszen (Morsesche Theorie)*, Teubner, 1938.

[5] GROMOLL, D., KLINGENBERG, W. and MEYER, W.: *Riemannsche Geometrie im Grossen*, Springer, 1968.

[6] BISHOP, R.L. and CRITTENDEN, R.J.: *Geometry of manifolds*, Acad. Press, 1964.

[7] POSTNIKOV, M.M.: *Introduction to Morse theory*, Moscow, 1971 (in Russian).

[8] POSTNIKOV, M.M.: *The variational theory of geodesics*, Saunders, 1967 (translated from the Russian).

[9] EELLS, J.: 'A setting for global analysis', *Bull. Amer. Math. Soc.* **72** (1966), 751-807.

[10A] SERRE, J.-P.: 'Homologie singulière des espaces fibrés I', *C.R. Acad. Sci. Paris* **231** (1950), 1408-1410.

[10B] SERRE, J.-P.: 'Homologie singulière des espaces fibrés. Applications', *Ann. of Math. (2)* **54** (1951), 425-505.

[10C] SERRE, J.-P.: 'Homologie singulière des espaces fibrés II', *C.R. Acad. Sci. Paris* **232** (1951), 31-33.

[10D] SERRE, J.-P.: 'Homologie singulière des espaces fibrés III', *C.R. Acad. Sci. Paris* **232** (1951), 142-144.

M.M. Postnikov
Yu.B. Rudyak

Editorial comments. A useful survey of Morse theory is [A1]; historical remarks can be found in [A2] and [A3], Sect. 1.7.

There is an analogue (generalization) of (finite-dimensional) smooth Morse theory for suitable spaces with singularities, called *stratified Morse theory*. Let X be a compact Whitney-stratified space contained in a smooth manifold M (cf. (the editorial comments to) **Stratification**). Let f be the restriction to X of a smooth real-valued function on M. A critical point of f is any critical point of f restricted to a stratum of X. In particular, all the zero-dimensional strata are critical points. The proper smooth function f is called a *Morse function on the stratified space* X if:

a) all critical values of f are distinct;

b) at each critical point p of f, the restriction of f to the stratum containing p has a non-degenerate critical point at p;

c) the differential of f at a critical point p does not annihilate any limit of tangent spaces to any stratum S other than the stratum containing p.

It follows that the set of critical points is discrete in X and that the critical values are discrete in R. If $M = \mathbf{R}^m$, the distance function on X from a point $q \in \mathbf{R}^m$ is a Morse function for almost-all q. The Morse functions also form an open dense set in the space of all proper smooth functions with the appropriate topology.

For each $c \in \mathbf{R}$, let $X_c = \{x \in X : f(x) \leqslant c\}$. Then (for a Morse function f on X) one has the following analogue of smooth finite-dimensional Morse theory. As c varies in the open interval between two adjacent critical values, the topological type of X_c does not vary, and as c crosses a critical value c_0 (from below), the topological type of $X_{c+\epsilon}$, with ϵ sufficiently small, is obtained from that of $X_{c-\epsilon}$ by glueing in a suitable (stratified) space A along a subspace B. The major difference is that the pair (A, B) can be far more complicated than the pair $(D^\lambda \times D^{n-\lambda}, \partial D^\lambda \times D^{n-\lambda})$, where D^i is the i-dimensional solid ball, of the smooth theory. Also, the pair (A, B) is not determined by a single integer. **Intersection homology** plays an analogous role vis à vis stratified Morse theory as ordinary homology does with respect to the smooth theory, in that if (A, B) is the pair belonging to the critical point $p \in S$, then the intersection homology group $IH_i(A, B)$ vanishes for all i except $i = n - s + \lambda_p$, where $s = \dim S$, $n = \dim X$ and λ_p is the **Morse index** of f restricted to S at p.

There are two other important generalizations of ordinary finite-dimensional Morse theory:

The non-isolated case. This applies to functions with non-degenerate critical manifolds. One assumes that f restricted to the normal direction is non-degenerate. Cf. [A5].

The equivariant case. This applies to functions which are equivariant under the action of a Lie group. Cf. [A1]. There are applications e.g. to Yang−Mills theory (cf. **Yang−Mills field**) in 2 dimensions. Cf. [A4].

References

[A1] BOTT, R.: 'Lectures on Morse theory, old and new', *Bull. Amer. Math. Soc.* **7**, no. 2 (1982), 331-358.

[A2] BOTT, R.: 'Marston Morse and his mathematical works', *Bull. Amer. Math. Soc.* **3**, no. 3 (1980), 907-950.

[A3] GORESKI, M. and MACPHERSON, R.: *Stratified Morse theory*, Springer, 1988.

[A4] ATIYAH, M. and BOTT, R.: 'The Yang−Mills equations over Riemann surfaces', *Phil. Trans. R. Soc. London A* **308** (1982), 523-615.

[A5] BOTT, R.: 'Non-degenerate critical manifolds', *Ann. of Math. (2)* **60** (1954), 248-261.

[A6] KLINGENBERG, W.: *Lectures on closed geodesics*, Springer, 1978.

[A7] SMALE, S.: 'Morse theory and a non-linear generalization of the Dirichlet problem', *Ann. of Math.* **80** (1964), 382-346.

[A8] KLINGENBERG, W.: *Riemannian geometry*, de Gruyter, 1982.

AMS 1980 Subject Classification: 57R45, 58E05, 49F15, 58C27

MOST-POWERFUL TEST - A statistical test that has the greatest power of all tests with the same **significance level**. Suppose that the results of observations are to be used to test a hypothesis H_0 against a simple alternative H_1; let α be the (given) admissible probability of an error of the first kind — the error arising from the rejection of the hypothesis H_0 being tested, according to a statistical test devised to test H_0 against H_1, when H_0 is actually true. In the theory of statistical hypotheses testing, the best test of all those intended for testing H_0 against H_1 and offering the same probability of an error of the first kind, or, equivalently, having the same significance level α, is the test that has the highest power. In other words, the best test is the one that offers the highest probability of rejecting the hypothesis H_0 being tested when the alternative H_1 is true. This best test is called the most-powerful test of level α among all tests of level α intended for testing a simple hypothesis H_0 against a simple alternative H_1. Since the power of a statistical test is one minus the probability of an error of the second kind, the error arising from accepting H_0 when it is in fact false, the concept of a most-powerful test is frequently formulated in terms of probabilities of errors of the first and second kinds: A most-powerful test is a test intended for testing a simple hypothesis against a simple alternative and offering the least probability of an error of the second kind among all tests with given probability of the error of the first kind. The problem of constructing most-powerful tests for simple hypotheses is solved by the **Neyman−Pearson lemma**, according to which the **likelihood-ratio test** is a most-powerful test.

If the competing hypotheses H_0 and H_1 are compound, the problem of devising most-powerful tests is formulated in terms of a **uniformly most-powerful test**, if such a test exists.

References
[1] LEHMANN, E.L.: *Testing statistical hypotheses*, Wiley, 1986.
[2] NEYMAN, J. and PEARSON, E.: 'On the problem of the most efficient tests of statistical hypotheses', *Philos. Trans. Royal Soc. London A* **231** (1933), 289-337.

M.S. Nikulin

AMS 1980 Subject Classification: 62F03

MOTION - A transformation of a space which preserves the geometrical properties of figures (dimension, shape, etc.). The concept of a motion has been formulated as an abstraction of real displacements of solid bodies in Euclidean space. Motion is sometimes accepted as the basic concept in the axiomatic construction of geometry.

A *motion of Euclidean space* is a transformation of the space which preserves the distances between points. It is called *proper* (a *motion of the first kind*) or *improper* (a *motion of the second kind*), depending on whether or not the **orientation** of the space is preserved.

Proper motions of a plane are analytically expressed in an orthogonal coordinate system (x, y) by the formulas

$$\tilde{x} = x \cos\phi - y \sin\phi + a,$$
$$\tilde{y} = x \sin\phi + y \cos\phi + b,$$

which show that the totality of all proper motions on a plane depends on three parameters a, b, ϕ. The first two parameters characterize the **parallel displacement** of the plane by the vector (a, b), while the parameter ϕ describes the **rotation** of the plane around the coordinate origin. A proper motion is the product (composite)

of rotation around the coordinate origin through an angle ϕ and parallel displacement by a vector (a, b). Any proper motion can be represented as a parallel displacement or as a rotation around some point.

Improper motions are expressed by the formulas

$$\tilde{x} = x \cos\phi + y \sin\phi + a,$$
$$\tilde{y} = x \sin\phi - y \cos\phi + b,$$

which show that an improper motion is the product of a proper motion by a **symmetry** with respect to some straight line. Any improper motion is the product of a parallel displacement in some direction and a symmetry with respect to a straight line having the same direction.

A proper motion in three-dimensional space is either a rotation around an axis or a parallel displacement, or else can be represented as the product of a rotation around an axis and a parallel displacement in the direction of the axis (helical motion). An improper motion is either a symmetry with respect to a plane or can be represented as the product of a symmetry with respect to a plane and a rotation around an axis normal to that plane, or as the product of a symmetry with respect to a plane by a displacement in the direction of a vector parallel to this plane.

In a rectangular coordinate system (x, y, z) in space, the analytic expression of a motion is:

$$\tilde{x} = a_{11}x + a_{12}y + a_{13}z + a,$$
$$\tilde{y} = a_{21}x + a_{22}y + a_{23}z + b,$$
$$\tilde{z} = a_{31}x + a_{32}y + a_{33}z + c,$$

where the elements of the matrix $\| a_{ij} \|$ satisfy the following orthogonality conditions:

$$a_{1r}a_{1s} + a_{2r}a_{2s} + a_{3r}a_{3s} = \delta_{rs}$$

($\delta_{rs} = 1$ if $r = s$, $\delta_{rs} = 0$ if $r \neq s$). A motion is proper or improper depending on whether the determinant of this matrix is 1 or -1. Cf. **Orthogonal transformation**.

In n-dimensional Euclidean spaces a motion is analytically expressed, in a similar manner, in rectangular coordinates, using an orthogonal matrix $A = \| a_{ij} \|$, as:

$$a_{1r}a_{1s} + \cdots + a_{nr}a_{ns} = \delta_{rs},$$
$$s = 1, \ldots, n.$$

A *motion of a Riemannian space* is a one-to-one continuously-differentiable mapping of class C^s, $s \geq 2$, under which the lengths of the respective lines are preserved. The line element of the space,

$$ds^2 = g_{\alpha\beta}(x^1, \ldots, x^n) \, dx^\alpha \, dx^\beta,$$

where the summation is over the indices α, β, is invariant under a motion. Analytically, a motion is defined by formulas which express the coordinates of a point $\tilde{M}(\tilde{x}^i)$ after transformation in terms of the coordinates of $M(x^i)$, the initial point, using functions $f^i(x)$ of class C^s:

$$\tilde{x}^\alpha = f^\alpha(x^1, \ldots, x^n), \quad \alpha = 1, \ldots, n. \qquad (*)$$

The condition of invariance of the line element means that

$$g_{\alpha\beta}(\tilde{x}^1, \ldots, \tilde{x}^n)\frac{\partial \tilde{x}^\alpha}{\partial x^i}\frac{\partial \tilde{x}^\beta}{\partial x^j} = g_{ij}(x^1, \ldots, x^n).$$

The concept of a motion in the Riemannian spaces of the general theory of relativity is of importance: In strong asymmetric gravitational fields solid bodies may become displaced to a very limited extent only; cases in which all motion is impossible may also occur, in which case the metric does not remain invariant under any transformation of space or, in other words, any displacement is accompanied by a deformation of the body.

A *motion of a space with an affine connection* is a one-to-one continuously-differentiable mapping of class C^s, $s \geq 2$, under which any field of parallel vectors along any smooth curve becomes a field of parallel vectors of the transformed curve. Under a motion the object of affine connection $\Gamma^\alpha_{\beta\gamma}(x)$ (cf. **Geometric objects, theory of**) is mapped to itself. Conversely, any mapping (*) which maps the affine connection to itself is a motion.

The motions constitute a group of transformations (cf. **Group of motions**). They are the simplest transformations of a space.

References

[1] EGOROV, I.P.: *Lectures on the axiomatics of Weyl in non-Euclidean geometry*, Ryazan', 1973 (in Russian).
[2] EGOROV, I.P.: *Motions in a space with an affine connection*, Kazan', 1965 (in Russian).
[3] BAZYLEV, V.T., DUNICHEV, K.I. and IVANITSKAYA, V.P.: *Geometry*, 1, Moscow, 1974 (in Russian).

I.P. Egorov

Editorial comments. A motion of a Euclidean space is thus an **isometric mapping** of the space onto itself. For detailed information on groups of motions admitted by various kinds of Riemannian manifolds see [A3].

References

[A1] EVES, H.: *A survey of geometry*, Allyn & Bacon, 1972.
[A2] KOBAYASHI, S. and NOMIZU, K.: *Foundations of differential geometry*, 1-2, Interscience, 1963-1969.
[A3] KOBAYASHI, S.: *Transformation groups in differential geometry*, Springer, 1972.

AMS 1980 Subject Classification: 51N25, 51-XX, 53B05, 53B20

MOTIVES, THEORY OF - A generalization of the various **cohomology** theories of algebraic varieties. The theory of motives systematically generalizes the idea of using the **Jacobian** of an algebraic curve X as a replacement for the cohomology group $H^1(X, \mathbf{Q})$ in the classical theory of correspondences, and the use of this theory in the study of the **zeta-function** of a curve X over a finite field. The theory of motives is universal in the sense that every geometric cohomology theory, of the type of the classical singular cohomology for algebraic varieties over $\mathbf{C}$ with constant coefficients, every l-adic cohomology theory for various prime numbers l different from the characteristic of the ground field, every crystalline cohomology theory, etc. (see **Weil cohomology**) are functors on the category of motives.

Let $V(k)$ be the category of smooth projective varieties over a field k and let $X \to C(X)$ be a contravariant functor of global **intersection theory** from $V(k)$ into the category of commutative Λ-algebras, where Λ is a fixed ring. For example, $C(X)$ is the **Chow ring** of classes of algebraic cycles (cf. **Algebraic cycle**) on X modulo a suitable (rational, algebraic, numerical, etc.) equivalence relation, or $C(X) = K(X)$ is the Grothendieck ring, or $C(X) = H^{ev}(X)$ is the ring of cohomology classes of even dimension, etc. The category $V(k)$ and the functor $X \to C(X)$ enable one to define a new category, the category of correspondences $CV(k)$, whose objects are varieties $X \in V(k)$, denoted by $\overline{X}$, and whose morphisms are defined by the formula

$$\mathrm{Hom}(\overline{X}, \overline{Y}) = C(X \times Y)$$

with the usual composition law for correspondences (see [1]). Let the functor C take values in the category of commutative graded Λ-algebras $A(\Lambda)$. Then $CV(k)$ will be the Λ-additive category of graded correspondences. Moreover, $CV(k)$ will have direct sums and tensor products.

The category whose objects are the varieties from $V(k)$ and whose morphisms are correspondences of degree 0 is denoted by $CV^0(k)$. A natural functor from $V(k)$ into $CV^0(k)$ has been defined, and the functor C extends to a functor T from $CV^0(k)$ to $A(\Lambda)$. The category $CV^0(k)$, like $CV(k)$, is not Abelian. Its pseudo-Abelian completion, the category $M_C^+(k)$, has been considered. It is obtained from $CV^0(k)$ by the formal addition of the images of all projections p. More precisely, the objects of $M_C^+(k)$ are pairs $(\overline{X}, p)$, where $\overline{X} \in CV^0(k)$ and $p \in \mathrm{Hom}(\overline{X}, \overline{X})$, $p^2 = p$, and $H((\overline{X}, p), (\overline{Y}, q))$ is the set of correspondences $f: \overline{X} \to \overline{Y}$ such that $f \circ p = = q \circ f$ modulo a correspondence g with $g \circ p = p \circ g = 0$. The category $CV^0(k)$ is imbedded in $M_C^+(k)$ by means of the functor $\overline{X} \to (\overline{X}, \mathrm{id})$. The natural functor $h: V(k) \to M_C^+(k)$ is called the *functor of motive cohomology spaces* and $M_C^+(k)$ is called the *category of effective motives.*

Let $p = (1 \times e)$, where e is the class of any rational point on the projective line P^1, and let $L = (P^1, p)$. Then

$$h(P^n) = 1 \oplus L \oplus L^{\otimes_2} \oplus \cdots \oplus L^{\otimes_n}.$$

If $X = P(E)$ is the projectivization of a locally free sheaf

E of rank r on Y, then

$$h(X) = \bigoplus_{i=0}^{r-1}(h(Y)\otimes L^{\otimes i}).$$

Motives of a monodial transformation with a non-singular centre, motives of curves (see [1]), motives of Abelian manifolds (see [2]), and motives of Weil hyper-surfaces have also been calculated.

The *category of motives* $M_C(k)$ is obtained from $M_C^+(k)$ by the formal addition of negative powers of the motives L. By analogy with *l-adic cohomology*, $T = L^{\otimes -1}$ is called the *Tate motive*. Tensor multiplication with T is called *twisting by the Tate motive*. Twisting enables one to define the level of a motive as in an l-adic cohomology theory. Any functor of the Weil cohomology factors through the functor $h\colon V(k)\to M_C(k)$. There is the conjecture that $M_C(k)$ does not, in some sense, depend on the intersection theory of C, and that the functor $X\to h(X)$ is itself a (universal) theory for the Weil cohomology. This conjecture is closely related to the standard Grothendieck conjectures (see [5]) on algebraic cycles (at present, 1982, not proved).

References

[1] MANIN, YU.I.: 'Correspondences, motives and monoidal transformations', *Math. USSR Sb.* **6**, no. 4 (1968), 439-470. (*Mat. Sb.* **77**, no. 4 (1968), 475-507)
[2] SHERMENEV, A.M.: 'The motif of an abelian variety', *Uspekhi Mat. Nauk* **26**, no. 2 (1971), 215-216 (in Russian).
[3] DEMAZURE, M.: 'Motives des variétés algébrique', in *Sem. Bourbaki, Exp. 365*, Lecture notes in math., Vol. 180, Springer, 1971, pp. 19-38.
[4] KLEIMAN, S.L.: 'Motives', in P. Holms (ed.): *Algebraic Geom. Proc. 5-th Nordic Summer School Math. Oslo, 1970*, Wolters-Noordhoff, 1972, pp. 53-96.
[5] KLEIMAN, S.L.: 'Algebraic cycles and the Weil conjectures', in *Dix exposés sur la cohomologie des schémas*, North-Holland, 1968, pp. 359-386.

V.A. Iskovskikh

Editorial comments. The theory of motives has been created by A. Grothendieck in the 1960-s. Although the above-mentioned *standard conjectures on algebraic cycles* have not yet (1989) been proved, the theory of motives has played an important role in various recent developments, for instance: i) as a guide for the *Deligne—Hodge theory* ([A1]); ii) in the study of absolute Hodge cycles on Abelian varieties ([A2]), where a variant of the notion of a motive has been used; iii) in the study of Chow groups on certain varieties over a finite field ([A3]); and iv) in work on the *Beilinson's conjectures* on special values of L-functions (see [A4]).

References

[A1] DELIGNE, P.: 'Theory de Hodge I ', in *Actès de Congrès Internat. des Mathématiciens, Nice 1970*, Vol. 1, Gauthier-Villars, 1971, pp. 425-430.
[A2] DELIGNE, P., MILNE, J.S., OGUS, A. and SHIH, K. (EDS.): *Hodge cycles, motives and Shimura varieties*, Lecture notes in math., 900, Springer, 1980.
[A3] SOULÉ, C.: 'Groupes de Chow et K-theory des variétés sur un corps fini', *Math. Ann.* **268** (1984), 317-345.
[A4] RAPOPORT, M., SCHAPPACHER, N. and SCHNEIDER, P. (EDS.):

Beilinson's conjectures on special values of L-functions, Acad. Press, 1988.

AMS 1980 Subject Classification: 55N20, 14F99, 14C17

MOUFANG LOOP - A **loop** in which the following (equivalent) identities hold:

$$x(y\cdot xz) = (xy\cdot x)z,$$
$$(zx\cdot y)x = z(x\cdot yx),$$
$$xy\cdot zx = x(yz\cdot x).$$

These loops were introduced and studied by R. Moufang [1]. In particular, she proved the following theorem, showing that the loops of this class are close to groups: If the elements a, b and c of a Moufang loop satisfy the associativity relation $ab\cdot c = a\cdot bc$, then they generate an associative subloop, that is, a **group** (*Moufang's theorem*). A corollary of this theorem is the *di-associativity of a Moufang loop*: Any two elements of the loop generate an associative subloop.

For commutative Moufang loops, which are defined by the single identity

$$x^2\cdot yz = xy\cdot xz,$$

the following theorem holds: Every commutative Moufang loop with n generators is centrally nilpotent with nilpotency class not exceeding $n-1$ (see [2]). Central nilpotency is defined analogously to nilpotency in groups (cf. **Nilpotent group**).

If a loop is isotopic (cf. **Isogeny**) to a Moufang loop, then it is itself a Moufang loop, that is, the property of being a Moufang loop is universal. Moreover, isotopic commutative Moufang loops are isomorphic.

References

[1] MOUFANG, R.: 'Zur Struktur von Alternativkörpern', *Math. Ann.* **110** (1935), 416-430.
[2] BRUCK, R.H.: *A survey of binary systems*, Springer, 1958.

V.D. Belousov

AMS 1980 Subject Classification: 20N05

MOULDING SURFACE - A surface generated by the orthogonal trajectories of a one-parameter family of planes. Moulding surfaces have one family of planar lines of curvature that are simultaneously geodesics for the moulding surface. If the family of planes is degenerated into a bundle, then the moulding surface will be a **surface of revolution**. The sections of a moulding surface by planes of the family are called *meridians*, and the orthogonal trajectories are called *parallels* of the moulding surface. All meridians are congruent, so that a moulding surface can be formed by the motion of a planar line L (the meridian), the plane of which moves without sliding along a certain developable surface. This surface is called the *directing surface* of the moulding surface and is one of the sheets of its evolute. If

$\rho(u)$ is the position vector of one parallel position, then the position vector of the moulding surface will be

$$r = \rho(u)+\eta(v)p(u)+\zeta(v)q(u),$$

where $p = v\cos\theta+\beta\sin\theta$, $q = -v\sin\theta+\beta\cos\theta$, v is the principal normal, β is the binormal, x is the torsion of the curve Γ, and $\theta = -\int x\,du$. Its line element is given by:

$$ds^2 = [1+k(\zeta\sin\theta-\eta\cos\theta)]^2\,du^2+(\eta'^2+\zeta'^2)\,dv^2,$$

where $\eta(v)$, $\zeta(v)$ are the equations of L and k is the curvature of Γ.

$$I.Kh.\ Sabitov$$

Editorial comments.

References

[A1] DARBOUX, G.: *Théorie générale des surfaces*, 1, Chelsea, reprint, 1972, Sects. 85-87.

AMS 1980 Subject Classification: 53A05

MOVABLE SINGULAR POINT - A singular point z_0 of the solution $w(z)$ of a differential equation $F(z, w, w')=0$ (F is an analytic function), where $w(z)$ is considered as a function of the complex variable z, which is such that solutions to the same equation with initial data close to the original data have singular points close to z_0 but not coincident with it. The classical example of a movable singular point arises when considering the equation

$$\frac{dw}{dz} = \frac{P(z, w)}{Q(z, w)},$$

where P and Q are holomorphic functions in a certain region of the space $\mathbf{C}^2$. If the surface $\{Q=0\}$ is irreducible and is projected along the Ow-axis on a region $\Omega\subset Oz$, then all points in the region Ω are movable singular points; for the solution with initial condition (z_0, w_0), where

$$Q(z_0, w_0) = 0 \neq P(z_0, w_0),$$

the point z_0 is an **algebraic branch point**.

References

[1] GOLUBEV, V.V.: *Vorlesungen über Differentialgleichungen im Komplexen*, Deutsch. Verlag Wissenschaft., 1958 (translated from the Russian).

$$Yu.S.\ Il'yashenko$$

Editorial comments.

For equations of the form

$$\frac{d^2w}{dz^2} = R\left[\frac{dw}{dz}, w, z\right],$$

where R is rational in dw/dz and w and analytic in z, it is known which equations have only non-movable singularities, cf. **Painlevé equation** and [A1].

References

[A1] INCE, E.L.: *Ordinary differential equations*, Dover, reprint, 1956.

AMS 1980 Subject Classification: 34A20

MOVING-AVERAGE PROCESS - A stochastic process which is stationary in the wide sense and which can be obtained by applying some linear transformation to a process with non-correlated values (that is, to a **white noise** process). The term is often applied to the more special case of a process $X(t)$ in discrete time $t=0, \pm 1, \ldots$, that is representable in the form

$$X(t) = Y(t)+b_1 Y(t-1)+\cdots+b_q Y(t-q), \tag{1}$$

where $EY(t)=0$, $EY(t)Y(s)=\sigma^2\delta_{ts}$, with δ_{ts} the Kronecker delta (so that $Y(t)$ is a white noise process with spectral density $\sigma^2/2\pi$), q is a positive integer, and $b_1, \ldots, b_q$ are constant coefficients. The **spectral density** $f(\lambda)$ of such a process is given by

$$f(\lambda) = \frac{\sigma^2}{2\pi}\,|\,\psi(e^{i\lambda})\,|^2,$$

$$\psi(z) = b_0+b_1 z+\cdots+b_q z^q, \quad b_0=1,$$

and its correlation function $r(k)=EX(t)X(t-k)$ has the form

$$r(k) = \sigma^2\sum_{j=0}^{q-|k|} b_j b_{j+|k|} \quad \text{if } |\,k\,|\leqslant q,$$

$$r(k) = 0 \quad \text{if } |\,k\,|>q.$$

Conversely, if the correlation function $r(k)$ of a stationary process $X(t)$ in discrete time t has the property that $r(k)=0$ when $|\,k\,|>q$ for some positive integer q, then $X(t)$ is a moving-average process of order q, that is, it has a representation of the form (1) where $Y(t)$ is a white noise (see, for example, [1]).

Along with the moving-average process of finite order q, which is representable in the form (1), there are two types of moving-average processes in discrete time of infinite order, namely: *one-sided moving-average processes*, having a representation of the form

$$X(t) = \sum_{j=0}^{\infty} b_j Y(t-j), \tag{2}$$

where $Y(t)$ denotes white noise and the series on the right-hand side converges in mean-square (so that $\sum_{j=0}^{\infty}|\,b_j\,|^2<\infty$), and also more general *two-sided moving-average processes*, of the form

$$X(t) = \sum_{j=-\infty}^{\infty} b_j Y(t-j), \tag{3}$$

where $Y(t)$ denotes white noise and $\sum_{j=-\infty}^{\infty}|\,b_j\,|^2<\infty$. The class of two-sided moving-average processes coincides with that of stationary processes $X(t)$ having spectral density $f(\lambda)$, while the class of one-sided moving-average processes coincides with that of processes having spectral density $f(\lambda)$ such that

$$\int_{-\pi}^{\pi} \log f(\lambda)\,d\lambda > -\infty$$

(see [2], [1], [3]).

A continuous-time stationary process $X(t)$, $-\infty < t < \infty$, is called a one-sided or two-sided moving-average process if it has the form

$$X(t) = \int_0^\infty b(s)\,dY(t-s), \quad \int_0^\infty |\,b(s)\,|^2\,ds < \infty,$$

or

$$X(t) = \int_{-\infty}^\infty b(s)\,dY(t-s), \quad \int_{-\infty}^\infty |\,b(s)\,|^2\,ds < \infty,$$

respectively, where $\mathsf{E}[dY(t)]^2 = \sigma^2\,dt$, that is, $Y'(t)$ is a generalized white noise process. The class of two-sided moving-average processes in continuous time coincides with that of stationary processes $X(t)$ having spectral density $f(\lambda)$, while the class of one-sided moving-average processes in continuous time coincides with that of processes having spectral density $f(\lambda)$ such that

$$\int_{-\infty}^\infty \log f(\lambda)(1+\lambda^2)^{-1}\,d\lambda > -\infty$$

(see [4], [3], [5]).

References

[1] ANDERSON, T.: *The statistical analysis of time series*, Wiley, 1971.
[2] KOLMOGOROV, A.N.: 'Stationary sequences in Hilbert space', in T. Kailath (ed.): *Linear Least-Squares Estimation*, Benchmark Papers in Electric Engin. Computer Sci., Vol. 17, Dowden, Hutchington & Ross, 1977, pp. 66-89 (translated from the Russian).
[3] DOOB, J.L.: *Stochastic processes*, Wiley, 1953.
[4] KARHUNUN, K.: 'Ueber lineare Methoden in der Wahrschein-lichkeitsrechnung', *Ann. Acad. Sci. Fennicae Ser. A. Math. Phys.* 37 (1947).
[5] ROZANOV, YU.A.: *Stationary random processes*, Holden-Day, 1967 (translated from the Russian).

A.M. Yaglom

Editorial comments. Both auto-regressive processes (cf. **Auto-regressive process**) and moving-average processes are special cases of so-called ARMA processes, i.e. auto-regressive moving-average processes (cf. **Mixed autoregressive moving-average process**), which are of great importance in the study of **time series**.

AMS 1980 Subject Classification: 60G07, 60G10, 60G12, 62M10

MOVING-FRAME METHOD - A method in **differential geometry** for the local examination of submanifolds of various homogeneous spaces, in which the starting point is to provide the submanifold itself and all its geometrical objects with the most general possible (mobile) frame (of reference). This method includes making the frame of reference canonical, namely assigning to each point in the submanifold a unique frame of reference in an invariant manner, in order to obtain differential invariants characterizing the submanifold apart from transformations imbedding it in the surrounding **homogeneous space**. This method was proposed in its most general form by E. Cartan [1],

who gave various examples of its application. Subsequently, the method was widely used and developed (see **Method of extensions and restrictions**). The analytic basis of the method is constituted by the invariant linear differential forms of Lie groups and their structure equations, as well as by the theory of representations of Lie groups as transformation groups. In modern geometry, the basic concepts of the method have required refinement, and they have been formulated in terms of the theory of bundles.

Let X_n be an n-dimensional homogeneous space and let G be the r-dimensional **Lie group** of its transformations (G acts from the left). Let $X_n = G/H$ be a representation, where $H \subset G$ is the **isotropy group** (stationary group) of a certain point $x_0 \in X_n$; let (e_k, e_α), $k = 1, \ldots, n$, $\alpha = n+1, \ldots, r$, be a basis of left-invariant vector fields on G such that the e_α restricted to H also constitute a basis of left-invariant vector fields for the Lie subgroup H. The basis (e_k, e_α) corresponds to a dual basis of left-invariant linear differential forms $(\theta^k, \theta^\alpha)$ on the Lie group G. The canonical projection $\pi: G \to X_n$ puts the points $x \in X_n$ into correspondence with the left cosets $\pi(x) = H_x \subset G$ of G with respect to $H = H_{x_0}$, and it introduces the structure of a principal H-bundle with base X_n and structure group H of dimension $r - n$ on the Lie group G. With this representation of G, the vector fields e_α constitute a basis of fundamental vector fields for the bundle $\pi: G \to X_n$, while the vector fields e_k span a certain n-distribution transverse to the fibres of $\pi: G \to X_n$. Correspondingly, the linear differential forms θ^k are a semi-basis of forms of the bundle $\pi: G \to X_n$ and form a completely-integrable subsystem of forms in the system $(\theta^k, \theta^\alpha)$. The fibres $H_x \subset G$ are integral manifolds of maximal dimension for the system of Pfaffian equations $\theta^k = 0$ (cf. **Pfaffian equation**; **Completely-integrable differential equation**).

A *system of frames of reference* in classical differential geometry (Euclidean, affine, projective, etc.) is a set of figures in X_n that is in bijective correspondence with the set of transformations of X_n (or, which is the same, with the set of elements of the **fundamental group** G of that space). Moreover, any frame of reference R from the given system can be obtained from some initial one R_0 by means of only one transformation:

$$L_g: X_n \to X_n, \quad R = L_g(R_0), \quad g \in G.$$

As the main role of the moving frame of reference $L_g(R_0) = R_g$ in relation to the fixed one R_0 is that it enables one to determine any transformation L_g of the homogeneous space X_n, one can identify the set of frames of reference $\{R_g\}$ with the set of elements of the fundamental group G of the space, thus obtaining a notion of abstract frames of reference in any homogeneous space with given fundamental group G.

Let some smooth submanifold $M \subset X_n$ of dimension m be given. *Frames of order zero* for M are elements of the restriction $G(\pi, M) = G|_M \subset G$ of the bundle $\pi: G \to X_n$ to M, as a new basis. This means that the principal bundle $G(\pi, M) \to M$ is imbedded in G and is defined in it as the complete pre-image $\pi^{-1}(M) \subset G$. As the left-invariant forms θ^k and θ^α in the Lie group G satisfy the Maurer–Cartan equations

$$
\left.
\begin{aligned}
d\theta^k &= \tfrac{1}{2} C^k_{lm} \theta^l \wedge \theta^m + C^k_{l\alpha} \theta^l \wedge \theta^\alpha, \\[4pt]
d\theta^\alpha &= \tfrac{1}{2} C^\alpha_{\beta\gamma} \theta^\beta \wedge \theta^\gamma + C^\alpha_{\beta k} \theta^\beta \wedge \theta^k + \tfrac{1}{2} C^\alpha_{lm} \theta^l \wedge \theta^m, \\[4pt]
k, l, m &= 1, \ldots, n; \quad \alpha, \beta, \gamma = n+1, \ldots, r,
\end{aligned}
\right\}
\tag{1}
$$

where C^k_{lm}, $C^k_{l\alpha}$, $C^\alpha_{\beta\gamma}$, $C^\alpha_{\beta k}$, C^α_{lm} are the structure constants of the Lie group, the restrictions of the forms θ^k and θ^α to the subbundle $G(\pi, M)$, i.e. the forms ω^k and ω^α, will be subject to the same equations, but in addition between the forms ω^k one has the linear relations

$$
\omega^p = \Lambda^p_a \omega^a, \quad a = 1, \ldots, m; \; p = m+1, \ldots, n.
\tag{2}
$$

Here the ω^a are forms that remain, along with ω^α, linearly independent in the principal bundle $G(\pi, M) \to M$, while the Λ^p_a are functions also defined on the bundle of frames of order zero of $G(\pi, M) \to M$. The functions Λ^p_a are coordinates in the tangent plane $T_x(M) \subset T_x(X_n)$ of the submanifold $M \subset X_n$, which depend on the point $x \in M$ and the frame

$$
R \in \pi^{-1}(x) = H_x \subset G(\pi, M).
$$

The tangent planes $x \to T_x(M)$ form a section $f: M \to \mathscr{G}_m(M)$ of the Grassmann bundle $\mathscr{G}_m(M) \to M$ of m-planes passing through the points of M. The bundle $\mathscr{G}_m(M) \to M$ is associated to the principal bundle $G(\pi, M) \to M$. The structure of the functions Λ^p_a is characterized by the equations

$$
d\Lambda^p_a + F^p_{a\alpha}(\Lambda) \omega^\alpha = \Lambda^p_{ab} \omega^b,
\tag{3}
$$

the explicit form of which can be obtained by exterior differentiation (cf. **Exterior form**) of (2) by means of (1) and subsequent application of Cartan's lemma (cf. **Cartan lemma**). The functions Λ^p_a and Λ^p_{ab} are the relative coordinates of the one-jet $j^1_x f$ of the section f in relation to the moving frame $R \in \pi^{-1}(x)$ for a point $x \in M$. The geometric object $j^1_x f$ forms also a section $j^1 f: M \to \mathscr{G}^1_m(M)$ of the corresponding bundle $\mathscr{G}^1_m(M) \to (M)$ associated to the principal bundle $G(\pi, M) \to M$. Similarly one obtains the section $j^2 f: M \to \mathscr{G}^2_m(M)$ with coordinates Λ^p_a, Λ^p_{ab}, Λ^p_{abc} of the generating geometric object, and also the subsequent extensions $j^3 f, \ldots, j^q f$, which correspond to differential extensions of (3).

As long as the bundle $\mathscr{G}^a_m(M) \to M$ to which the section $j^q f(M)$ belongs is homogeneous, it is possible to perform a reduction $G^q(\pi, M)$ of the principal bundle $G(\pi, M) \to M$ of frames to a certain subgroup $\tilde{H} \subset H$ defined by Cartan by fixing the relative coordinates Λ^p_a, $\Lambda^p_{ab}, \ldots, \Lambda^p_{a_1 \cdots a_{q+1}}$ for the geometric object $j^q_x f$, which is independent of the point $x \in M$. In this way one defines a partial canonization of the frame of reference. The frames $R \in G^q(\pi, M)$ are called *semi-canonical frames* of order $q+1$ for the given submanifold $M \subset X_n$. If the subsequent continuation gives geometric objects whose isotropy group contains only the identity transformation, it is possible to fix only some of the coordinates of the geometric objects of the section $j^{q+1} f$ which do not depend on the point x, after which the other coordinates of Λ of the geometric object $j^{q+1} f$ depend only on $x \in M$. One thus gets a section $s: M \to G(\pi, M)$ of the bundle of frames of order zero of M. The frame $R = s(x)$ of this section is called the *canonical frame* of the submanifold $M \subset X_n$ or the *accompanying frame* of this submanifold. The above process for the continuation of the equations (3) and the method selected for fixing the function Λ leads to the equations

$$
\omega^p = \Lambda^p_a \omega^a, \quad \omega^\alpha = \Lambda^\alpha_a \omega^a,
\tag{4}
$$

which connect the linear forms ω^k and ω^α in the section $s(M)$. The field of the canonical frame is not constructed unambiguously, being dependent on arbitrarily fixing the relative coordinates of the geometric object $j^{q+1} f$. The only important point is that some of the coefficients in (4) have constant numerical values (preferably the simplest ones), whereas the others form differential invariants for the submanifold $M \subset X_n$ that define it up to transformations in X_n. The canonical frames for the section $s(M)$ are analogues of a classical example, the accompanying Frénet frame (cf. **Frénet trihedron**) for a curve in the Euclidean space, while equations (4) correspond to the Frénet equations (cf. **Frénet formulas**) for the curve. During the canonization of the frame of reference, complications may arise connected with the inhomogeneity of the bundles $\mathscr{G}^q_m(M)$ and the differences in type (in this sense) between the different submanifolds M in X_n, and even in individual parts of them. This is the basis for classifying the various types of points and various classes of submanifolds in X_n. On account of these features, the moving-frame method has played a fruitful part in research on submanifolds in various homogeneous spaces, and it has also indicated a way of developing modern methods of investigating very general differential-geometric structures on smooth manifolds.

References

[1] CARTAN, E.: *La théorie des groupes finis et continus et la géométrie différentielle traitées par la méthode du repère mobile*, Gauthier-Villars, 1937.

[2] FAVARD, J.: *Cours de géométrie différentielle locale*, Gauthier-Villars, 1957.

[3A] CARTAN, H.: *Differential forms*, Karshaw, 1983 (translated from the French).
[3B] CARTAN, H.: *Calcul différentielle*, Herman, 1967.
[4] FINIKOV, S.P.: *Cartan's method of exterior forms in differential geometry*, 1-3, Moscow-Leningrad, 1948 (in Russian).

E.L. Evtushik

Editorial comments. In [A3], Chapt. II, Sect. IV: *La méthode du repère mobile*, Cartan writes as follows. 'We shall resume the projective differentiable study of plane curves by attaching in an intrinsic way, . . . , a moving frame to the current point of the curve and study the properties of the curve by those of the displacements of the frame'.

Let M be n-dimensional differentiable manifold and p a point of M. A frame at p is a basis of the tangent space T_pM at $p \in M$. Given n vector fields $X_1, \ldots, X_n$ on $U \subset M$ such that $X_1(q), \ldots, X_n(q)$ are linearly independent for each $q \in U$, the $X_1(q), \ldots, X_n(q)$ define a *moving frame (repère mobile)* on U. Conversely, every moving frame $p \mapsto F_p \in (T_pM)^n$, i.e. a section of the frame bundle (cf. **Frame**), determines such an n-tuple of vector fields. In Cartan's theory the basic idea is to express everything in terms of an arbitrary moving frame $X_1, \ldots, X_n$ and not just in terms of the 'natural' frames $(\partial/\partial x_1, \ldots, \partial/\partial x_n)$ defined by a (local) system of coordinates. This has been a very fruitful idea, not at least because a moving frame may well exist on a region which cannot even be included in a coordinate system. For example, there is a rather obvious moving frame X_1, X_2 on the whole torus. Similarly, there are the very useful moving frames defined in the main article above on the whole of the homogeneous spaces G/H which are given by suitable left-invariant vector fields.

On a **Riemannian manifold** an *orthonormal moving frame* $X_1, \ldots, X_n$ is one for which the $X_1(p), \ldots, X_n(p)$ form an orthonormal basis of T_pM for all p. An orthonormal moving frame can be obtained from an arbitrary one by Gram−Schmidt orthonormalization.

References

[A1] JENSEN, G.: *Higher order contact of submanifolds of homogeneous spaces*, Lecture notes in math., 610, Springer, 1977.
[A2] KOBAYASHI, S. and NOMIZU, K.: *Foundations of differential geometry*, Wiley, 1969.
[A3] CARTAN, E.: *Théorie des espaces à connexion projective*, Gauthier-Villars, 1937.
[A4] SPIVAK, M.: *Differential geometry*, II, Publish or Perish, 1970, Chapt. VII.

AMS 1980 Subject Classification: 53C30, 53C40, 53A55, 53B25

MULTI-ALGEBRA - A set in which a system of (in general, partial) multi-operations is given. A *partial multi-operation* on a set A is a partial mapping $f: A^n \to A^m$ between Cartesian powers of A, where $n, m \geqslant 0$. Here A^0 means a one-element set. A homomorphism $g: A \to B$ of multi-algebras with the same system of multi-operations is a mapping g such that if f is a multi-operation mapping the n-th power into the m-th, then

$$g^m(f(x_1, \ldots, x_n)) = f(g(x_1), \ldots, g(x_n))$$

for all $x_i \in A$. The concept of a multi-algebra is a generalization of that of a **universal algebra**. At the same time a multi-algebra is a particular case of an **algebraic system**, since a mapping $f: A^n \to A^m$ can be identified with the $(m+n)$-ary relation $(x, f(x))$ on A, $x \in A^n$. Multi-algebras arise most naturally in connection with the functorial approach to universal algebra (see [1]). Namely, let C be a **category** whose objects are the natural numbers including zero, where the object $m+n$ is the direct product of the objects m and n. Then a functor F from C into the category of sets that commutes with direct products is a multi-algebra on the set $F(1) = A$ with system of multi-operations $F(f): A^n \to A^m$, where $f: n \to m$ in C. The homomorphisms in this case are precisely the natural transformations of functors.

References

[1] LAWVERE, F.W.: 'Functorial semantics of algebraic theories', *Proc. Nat. Acad. Sci. USA* **50**, no. 5 (1963), 869-872.
[2] BELOUSOV, V.D.: *Algebraic nets and quasi-groups*, Kishinev, 1971 (in Russian).

V.A. Artamonov

AMS 1980 Subject Classification: 08A55, 18C10

MULTI-CRITERION PROBLEM - A mathematical model for making an optimal decision with respect to several criteria simultaneously. These criteria may reflect evaluations of various qualities of the objects (or processes) about which a decision is to be made, or may be evaluations of some single characteristic from various points of view. The theory of multi-criterion problems belongs to **operations research**.

Formally a multi-criterion problem is given by a set X of 'feasible decisions' and a set of objective functions $f_1, \ldots, f_n$ on X, taking real values. The essence of a multi-criterion problem is finding an optimal decision, that is, an $x \in X$ which, in some sense, maximizes the values of all the functions f_i, $i = 1, \ldots, n$. The existence of a decision maximizing all objective functions is a rare exception. Therefore in the theory of multi-criterion problems the notion of optimality has various and, moreover, non-trivial interpretations. The content of the theory of multi-criterion problems is the development of such concepts of optimality, the proof of their realizability (that is, the existence of decisions which are optimal in the corresponding sense), and the search for these realizations (that is, the actual solution of the problem).

The most direct approach to solving a multi-criterion problem consists in reducing it to an ordinary ('single-criterion') problem of **mathematical programming** by replacing the system of objective functions $f_1, \ldots, f_n$ by an aggregate function $F(f_1, \ldots, f_n)$. In this role there may appear weighted sums $\sum_{i=1}^n \lambda_i f_i$, $\lambda_i \geqslant 0$, weighted maxima $\max_i \lambda_i f_i$, and other combinations of the initial objective functions. This approach is conceptually and technically convenient. Its fundamental shortcoming is the difficulty in satisfying the need for a meaningful comparability of the values of the various objective functions and also the uncertainty (and not infrequent arbitrariness) in the choice of F and, in particular, in the weights λ_i. To establish these one is often advised to resort to expert evaluations.

A particular case of this approach is the selection of a 'main criterion', that is, all weights except a single λ_{i_0} are put equal to zero. Then the multi-criterion problem becomes an ordinary problem of mathematical programming, and the set of optimal solutions of it can be considered as the set of feasible solutions for a new multi-criterion problem with objective functions f_i, $i \neq i_0$.

As the solutions of a multi-criterion problem one can consider solutions which are *Pareto optimal*, that is, solutions which do not allow an improvement with respect to any one criterion, except at the expense of detoriation in another criterion (in other words, an $x \in X$ such that for any $y \in X$, if $f_i(x) < f_i(y)$ then $f_j(y) < f_j(x)$ for some j). The shortcoming of this approach is the plurality of Pareto-optimal solutions. This defect was overcome, following a suggestion of J. Nash, by the method of 'bargaining schemes', which essentially limits the number of selected solutions among the Pareto-optimal solutions. It consists in establishing, on the basis of informal considerations, certain minimal feasible values f_i^0 and subsequently finding a feasible x maximizing $\prod_{i=1}^n (f_i(x) - f_i^0)$ (see also **Arbitration scheme**).

A multi-criterion problem can also be considered as a game (see **Games, theory of**) and its solution can be treated on the basis of various game-theoretic methods. For example, if a feasible solution x is chosen with the aim of maximizing one of the objective functions $f_1, \dots, f_n$, and if it is unknown which one has to be maximized, then it is possible to use a weighted sum of these functions, taking as weights the components of a mixed strategy of 'nature'. A multi-criterion problem can be treated as a **non-cooperative game**, or can include the point of view of cooperative game theory (see **Cooperative game**) as well as the application of related optimality principles.

References

[1] LUCE, R.D. and RAIFFA, H.: *Games and decisions: introduction and critical survey*, Wiley, 1957.

N.N. Vorob'ev

Editorial comments. The method described for selecting a Pareto-optimal solution is also known as *Nash' solution of the bargaining problem*.

Sometimes the terms *vector-valued optimization* or *multiple objective decision problem* are used for multi-criterion problem.

References

[A1] GRAVEN, B.D.: 'Vector valued optimization', in S. Schaible and W.T. Ziemba (eds.): *Generalized Concavity in Optimization and Economics*, Acad. Press, 1981, pp. 661-688.
[A2] CHANKONG, V. and HAIMES, Y.Y.: *Multiobjective decision making: theory and methodology*, North-Holland, 1983.
[A3] LEITMANN, G.: *Cooperative and non-cooperative many players differential games*, Internat. Center Mech. Sci., 190, Springer, 1974.

AMS 1980 Subject Classification: 90A05, 90B50, 90C31, 90D35, 90CXX

MULTI-DIMENSIONAL DISTRIBUTION, *multivariate distribution* - A probability distribution on the σ-algebra of Borel sets of an s-dimensional Euclidean space $\mathbf{R}^s$. One usually speaks of a multivariate distribution as the distribution of a multi-dimensional random variable, or random vector, $X = (X_1, \dots, X_s)$, meaning by this the **joint distribution** of the real random variables $X_1(\omega), \dots, X_s(\omega)$ given on the same space of elementary events Ω ($X_1, \dots, X_s$ may be regarded as coordinate variables in the space $\Omega = \mathbf{R}^s$). A multivariate distribution is uniquely determined by its **distribution function** — the function

$$F(x_1, \dots, x_s) = \mathsf{P}\{X_1 < x_1, \dots, X_s < x_s\}$$

of the real variables $x_1, \dots, x_s$.

As in the one-dimensional case, the most widespread multivariate distributions are the discrete and the absolutely-continuous distributions. In the discrete case a multivariate distribution is concentrated on a finite or countable set of points $(x_{i_1}, \dots, x_{i_s})$ of $\mathbf{R}^s$ such that

$$\mathsf{P}\{X_1 = x_{i_1}, \dots, X_s = x_{i_s}\} = p_{i_1 \cdots i_s} \geq 0,$$

$$\sum_{i_1 \cdots i_s} p_{i_1 \cdots i_s} = 1$$

(see, for example, **Multinomial distribution**). In the absolutely-continuous case almost-everywhere (with respect to Lebesgue measure) on $\mathbf{R}^s$,

$$\frac{\partial^s F(x_1, \dots, x_s)}{\partial x_1 \cdots \partial x_s} = p(x_1, \dots, x_s),$$

where $p(x_1, \dots, x_s) \geq 0$ is the *density of the multivariate distribution*:

$$\mathsf{P}\{X \in A\} = \int_A p(x_1, \dots, x_s) \, dx_1 \cdots dx_s,$$

for any A from the σ-algebra of Borel subsets of $\mathbf{R}^s$, and

$$\int_{\mathbf{R}^s} p(x_1, \dots, x_s) \, dx_1 \cdots dx_s = 1.$$

The distribution of any random variable X_i (and also, for any $m < s$, the distribution of the variables $X_{i_1}, \dots, X_{i_m}$) relative to a multivariate distribution is called a **marginal distribution**. The marginal distributions are completely determined by the given multivariate distribution. When $X_1, \dots, X_s$ are independent, then

$$F(x_1, \ldots, x_s) = F_1(x_1) \cdots F_s(x_s)$$

and

$$p(x_1, \ldots, x_s) = p_1(x_1) \cdots p_s(x_s),$$

where $F_i(x)$ and $p_i(x)$ are, respectively, the marginal distribution functions and densities of the X_i.

The mathematical expectation of any function $f(X_1, \ldots, X_s)$ of $X_1, \ldots, X_s$ is defined by the integral of this function with respect to the multivariate distribution; in particular, in the absolutely-continuous case it is defined by the integral

$$\mathsf{E}f(X_1, \ldots, X_s) =$$
$$= \int_{\mathbf{R}^s} f(x_1, \ldots, x_s) p(x_1, \ldots, x_s) \, dx_1 \cdots dx_s.$$

The characteristic function of a multivariate distribution is the function of $t = (t_1, \ldots, t_s)$ given by

$$\phi(t) = \mathsf{E} e^{itx'},$$

where $tx' = t_1 x_1 + \cdots + t_s x_s$. The fundamental characteristics of a multivariate distribution are the moments (cf. **Moment**): the mixed moments $\mathsf{E}X_1^{k_1} \cdots X_s^{k_s}$ and the central mixed moments $\mathsf{E}(X_1 - \mathsf{E}X_1)^{k_1} \cdots (X_s - \mathsf{E}X_s)^{k_s}$, where $k_1 + \cdots + k_s$ is the order of the corresponding moment. The roles of the expectation and the variance for a multivariate distribution are played by $\mathsf{E}X = (\mathsf{E}X_1, \ldots, \mathsf{E}X_s)$ and the set of second-order central mixed moments, which form the **covariance matrix**. If $\mathsf{E}(X_i - \mathsf{E}X_i)(X_j - \mathsf{E}X_j) = 0$ for all i, j, $i \neq j$, then $X_1, \ldots, X_s$ are called pairwise uncorrelated or orthogonal (the covariance matrix is diagonal). If the rank r of the covariance matrix is less than s, then the multivariate distribution is called a **degenerate distribution**; in this case the distribution is concentrated on some linear manifold in $\mathbf{R}^s$ of dimension $r < n$.

For methods of investigating dependencies between $X_1, \ldots, X_s$ see **Correlation; Regression**.

A.V. Prokhorov

Editorial comments.

References

[A1] JOHNSON, N.L. and KOTZ, S.: *Discrete distributions*, Houghton-Miflin, 1969.
[A2] JOHNSON, N.L. and KOTZ, S.: *Continuous multivariate distributions*, Wiley, 1942.

AMS 1980 Subject Classification: 60E05

MULTI-DIMENSIONAL KNOT - An isotopy class of imbeddings of a sphere into a sphere. More precisely, an n-dimensional knot of codimension q is a pair $K = (S^{n+q}, k^n)$ consisting of an oriented sphere S^{n+q} and an oriented, locally flat, submanifold of it, k^n, homeomorphic to the sphere S^n. Two knots $K_1(S^{n+q}, k_1^n)$ and $K_2 = (S^{n+q}, k_2^n)$ are called equivalent if there is an **isotopy (in topology)** of S^{n+q} which takes k_1^n to k_2^n while preserving the orientation. Depending on

the category (Diff, PL or Top) from which the terms 'submanifold' and 'isotopy' in these definitions are taken, one speaks of smooth, piecewise-linear or topological multi-dimensional knots, respectively. In the smooth case k^n may have a non-standard differentiable structure. An n-dimensional knot of codimension q which is isotopic to the standard imbedding is called a *trivial*, or *unknotted*, *knot*.

The study of multi-dimensional knots of codimension 1 is related to the **Schoenflies conjecture**. Every topological knot of codimension 1 is trivial. This is true for piecewise-linear and smooth knots if $n \neq 3, 4$.

Piecewise-linear and topological multi-dimensional knots of codimension $q \geqslant 3$ are trivial. In the smooth case this is not so. The set of isotopy classes of smooth n-dimensional knots of codimension $q \geqslant 3$ coincides, for $n \geqslant 5$, with the set $\theta^{n+q,n}$ of *cobordism classes of knots*. (Two multi-dimensional knots $K_1 = (S^{n+q}, k_1^n)$ and $K_2 = (S^{n+q}, k_2^n)$ are called *cobordant* if there is a smooth $(n+1)$-dimensional submanifold $W \subset S^{n+q} \times I$ transversal to $\partial(S^{n+q} \times I)$, where $\partial W = (k_1^n \times 0) \bigcup (-k_2^n \times 1)$ and W is an *h-cobordism* between $k_1^n \times 0$ and $k_2^n \times 1$.) The set $\theta^{n+q,n}$ is an Abelian group with respect to the operation of connected sum. In this group the negative of the class of (S^{n+q}, k^n) is the cobordism class of $(-S^{n+q}, -k^n)$, where the minus denotes reversal of orientation. There is a natural homomorphism $\theta^{n+q,n} \to \theta^n$, where θ^n is the group of n-dimensional homotopy spheres; this homomorphism associates the differentiable structure of k^n to the knot (S^{n+q}, k^n). The kernel of this homomorphism, denoted by $\Sigma^{n+q,n}$, is the set of isotopy classes of the standard sphere S^n in S^{n+q}. If $2q > n+3$, then $\Sigma^{n+q,n}$ is trivial. If $2q \geqslant n+3$ and $(n+1) \not\equiv 0 \pmod 4$, then $\theta^{n+q,n}$ and $\Sigma^{n+q,n}$ are finite. When $2q \leqslant n+3$ and $(n+1) \not\equiv 0 \pmod 4$, then $\theta^{n+q,n}$ and $\Sigma^{n+q,n}$ are finitely-generated Abelian groups of rank 1 (see [1], [2]). The set of concordance classes of smooth imbeddings of S^n into S^{n+q} for $q > 2$ has also been calculated (see [3]).

The study of multi-dimensional knots of codimension 2, which will subsequently simply be called knots, proceeds quite similarly in all three categories (Diff, PL, Top). For $n \geqslant 5$ every topological knot may be transformed by an isotopy to a smooth knot. However, there are topological three-dimensional knots in S^5 which are not equivalent, or even cobordant, to smooth knots (see [4]).

The set of isotopy classes of n-dimensional knots (in each category) is an Abelian semi-group with respect to the operation of connected sum. It is known that for $n = 1$ every element in this semi-group is a finite sum of primes, and such a decomposition is unique.

An n-dimensional knot $K = (S^{n+2}, k^n)$ is trivial if and only if $\pi_i(S^{n+2} \setminus k^n) = \pi_i(S^1)$ for all $i \leqslant [(n+1)/2]$.

An algebraic classification has been given (see [6]) of the knots K for which $\pi_i(S^{n+2}\setminus k^n)=\pi_i(S^1)$, for all $i\leqslant[(n+1)/2]-1$ and n odd (knots of type L): For $n\geqslant5$ the set of isotopy classes of such knots turns out to be in one-to-one correspondence with the set of S-equivalence classes of the **Seifert matrix**. Knots of type L are important from the point of view of applications to algebraic geometry, since they contain all knots obtained by the following construction (see [15]). Let $f(z_1,\ldots,z_{q+1})$ be a complex polynomial of non-zero degree having zero as an isolated singularity and let $f(0)=0$. The intersection k of the hyperplane $V=f^{-1}(0)$ with a small sphere S^{q+1} with centre at zero is a $(q-2)$-connected $(2q-1)$-dimensional manifold. The manifold k is homeomorphic to S^{2q-1} if and only if $|\Delta(1)|=1$, where $\Delta(t)$ is the Alexander polynomial. In this case there thus arises a knot (S^{2q+1},k). Such knots are called *algebraic*; they are all of type L.

The *exterior* of a smooth knot $K=(S^{n+2},k^n)$ is the complement X (of an open tubular neighbourhood) of k^n in S^{n+2}. For $n\geqslant2$, for each n-dimensional knot K there is a knot $\tau(K)$ such that each knot with exterior diffeomorphic to the exterior of K is equivalent to either K or $\tau(K)$. If X_1, X_2 are the exteriors of two smooth n-dimensional knots, $n\geqslant3$, and $\pi_1(X_1)=\pi_1(X_2)=\mathbf{Z}$, then the following statements are equivalent (see [7]): 1) X_1 and X_2 are diffeomorphic; and 2) the pairs $(X_1,\partial X_1)$ and $(X_2,\partial X_2)$ are homotopically equivalent. These results reduce the classification problem for knots to the homotopy classification of pairs $(X,\partial X)$ and the solution of the question: Does the exterior determine the type of a knot, that is, does $K=\tau(K)$ hold? It is known that this equality holds for knots of type L (see [6]) and for knots obtained by the Artin construction and the supertwisting construction (see [8]). However, two-dimensional knots have been found in S^4 for which $K\neq\tau(K)$ (see [9]).

The study of the homotopy type of the exterior of X is complicated because this exterior is not simply connected. If G is the group of the knot (that is, $G=\pi_1(X)$), then $G/[G,G]=\mathbf{Z}$, $H_2(G)=0$, and the weight of G (that is, the minimal number of elements not contained in a proper normal divisor) is equal to 1. For $n\geqslant3$ these properties completely describe the class of groups of n-dimensional knots (see [10]). The groups of one-dimensional and two-dimensional knots have a number of additional properties (see **Knot theory; Two-dimensional knot**).

Since $H^1(X;\mathbf{Z})=\mathbf{Z}$, the exterior X has a unique infinite cyclic covering $p:\tilde{X}\to X$. The homology spaces $H_*(\tilde{X};\mathbf{Z})$ are $\mathbf{Z}[\mathbf{Z}]$-modules. Their **Alexander invariants** are invariants of the knot. For algebraic properties of the modules $H_*(\tilde{X};\mathbf{Z})$ see [10] - [13].

Due to the fact that the group $\mathbf{Z}$ acts without fixed points on an infinite cyclic covering, the $(n+2)$-dimensional non-compact manifold $\tilde{X}$ has a number of the homological properties of compact $(n+1)$-dimensional manifolds. In particular, for the homology of the manifold $\tilde{X}$ with coefficients from a field F there is a non-degenerate pairing

$$H_n(\tilde{X};F)\otimes H_{n+1-k}(\tilde{X};F)\to F,\quad k=1,\ldots,n,$$

with properties resembling the pairing determined by the **intersection index (in homology)** in $(n+1)$-dimensional compact manifolds. There is also a pairing

$$T_k\tilde{X}\otimes T_{n-k}\tilde{X}\to\mathbf{Q}/\mathbf{Z},\quad k=1,\ldots,n-1,$$

similar to the linking coefficients (cf. **Linking coefficient**) in $(n+1)$-dimensional manifolds (see [13]), where $T_j\tilde{X}=\mathrm{Tors}\,H_j(\tilde{X};\mathbf{Z})$. These homology pairings generate invariants of the homotopy type of the pair $(X,\partial X)$. To obtain algebraic invariants, finite-sheeted cyclic branched coverings are also used (see [14]).

The problem of classifying knots of codimension 2 up to cobordism, a coarser equivalence relation than isotopy type, has been completely solved for $n>1$ (see **Cobordism of knots**).

References

[1] HAEFLIGER, A.: 'Knotted $(4k-1)$-spheres in $6k$-space', *Ann. of Math.* **75** (1962), 452-466.
[2] HAEFLIGER, A.: 'Differentiable embeddings of S^n in S^{n+q} for $q>2$', *Ann. of Math.* **83** (1966), 402-436.
[3] LEVINE, J.: 'A classification of differentiable knots', *Ann. of Math.* **82** (1965), 15-50.
[4] CAPPELL, S. and SHANESON, J.: 'Topological knots and knot cobordism', *Topology* **12** (1973), 33-40.
[5] SOSSINSKIĬ, A.B.: 'Decomposition of knots', *Math. USSR Sb.* **10** (1970), 139-150. (*Mat. Sb.* **81**, no. 1 (1970), 145-158)
[6] LEVINE, J.: 'An algebraic classification of some knots of codimension two', *Comment. Math. Helv.* **45** (1970), 185-198.
[7] LASHOF, R. and SHANESON, J.: 'Classification of knots in codimension two', *Bull. Amer. Math. Soc.* **75** (1969), 171-175.
[8] CAPPELL, S.: 'Superspinning and knot complements', in *Topology of Manifolds*, Markham, 1971, pp. 358-383.
[9] CAPPELL, S. and SHANESON, J.: 'There exist inequivalent knots with the same complements', *Ann. of Math.* **103** (1976), 349-353.
[10] KERVAIRE, M.: 'Les noeuds de dimensions supérieures', *Bull. Soc. Math. France* **93** (1965), 225-271.
[11] LEVINE, J.: 'Polynomial invariants of knots of codimension two', *Ann. of Math.* **84** (1966), 537-554.
[12] LEVINE, J.: 'Knot modules', in *Knots, Groups and 3-Manifolds*, Princeton Univ. Press., 1975, pp. 25-34.
[13] FARBER, M.SH.: 'Duality in an infinite cyclic covering and even-dimensional knots', *Math. USSR Izv.* **11** (1974), 749-781. (*Izv. Akad. Nauk SSSR Ser. Mat.* **41** (1977), 794-828)
[14] VIRO, O.YA.: 'Branched coverings of manifolds with boundary and link invariants I', *Math. USSR Izv.* **7** (1973), 1239-1256. (*Izv. Akad. Nauk SSSR Ser. Mat.* **37** (1973), 1242-1258)
[15] MILNOR, J.: *Singular points of complex hypersurfaces*, Princeton Univ. Press, 1968.

M.Sh. Farber

Editorial comments.

References

[A1] MILNOR, J.: 'Infinite cyclic coverings', in J. Hocking (ed.): *Conf. Topology of Manifolds*, Prindle, Weber & Schmidt,

1968, pp. 115-133.

AMS 1980 Subject Classification: 57Q45

MULTI-DIMENSIONAL STATISTICAL ANALYSIS,
multivariate statistical analysis - The branch of
mathematical statistics devoted to mathematical
methods for constructing optimal designs for the collection, systematization and processing of multivariate statistical data, directed towards clarifying the nature and the structure of the correlations between the components of the multivariate attribute in question, and intended for obtaining scientific and practical inferences. By a *multivariate attribute* is meant a p-dimensional vector $\mathbf{x}=(x_1, \ldots, x_p)'$ of components (laws, variables) $x_1, \ldots, x_p$ which may be *quantitative*, that is, measuring in some fixed scale the degree of manifestation of the studied property of an object, it may be *ordering* (or *ordinal*), that is, allowing the objects being analyzed to be ordered relative to the degree of manifestation in them of the studied property, and it may be *classifying* (or *nominal*), that is, allow the collection of objects being investigated, which does not lend itself to ordering, to be separated into homogeneous (relative to the analyzed property) classes. The results of measuring these components,

$$\{\mathbf{x}_i\}_1^n = \{(x_{1i}, \ldots, x_{pi})'\}_1^n \tag{1}$$

for each of n objects of a collection, forms a sequence of multivariate observations, or an initial ensemble of multivariate data, for conducting a multivariate statistical analysis. A significant part of multivariate statistical analysis involves the situation in which $\mathbf{x}$ is interpreted as a multivariate random variable, and the corresponding sequence of observations (1) is a population sample. In this case the choice of a method for processing the initial statistical data and the analysis of their properties is carried out on the basis of assumptions regarding the nature of the multivariate (joint) law of the probability distribution $P(\mathbf{x})$.

The content of multivariate statistical analysis can be conventionally divided into three basic subdivisions: the multivariate statistical analysis of multivariate distributions and their basic characteristics; the multivariate statistical analysis of the nature and structure of the correlations between the components of the multivariate attribute being investigated; and the multivariate statistical analysis of the geometric structure of the set of multi-dimensional observations being investigated.

Multivariate statistical analysis of multivariate distributions and their fundamental characteristics. This branch covers only situations in which the observations (1) being processed have a probabilistic nature, that is, can be interpreted as a sample from a corresponding population. The basic problems of this branch are: the statistical estimation, for the multivariate distributions in question, of their fundamental numerical characteristics and parameters; the investigation of the properties of the statistical estimators used; and the investigation of the probability distributions of a number of statistics that are used to construct statistical tests for the verification of various hypotheses on the nature of the multi-dimensional data being analyzed The fundamental results are related to the particular case when the attribute in question, $\mathbf{x}$, is subject to a multivariate normal law $N_p(\mu, \mathbf{V})$, with density function $f(\mathbf{x}\,|\,\mu, \mathbf{V})$ given by

$$f(\mathbf{x}\,|\,\mu, \mathbf{V}) = \frac{1}{(2\pi)^{p/2}\,|\,\mathbf{V}\,|^{1/2}} \times \tag{2}$$

$$\times \exp\left\{-\frac{1}{2}(\mathbf{x}-\mu)'\mathbf{V}^{-1}(\mathbf{x}-\mu)\right\},$$

where $\mu=(\mu_1, \ldots, \mu_p)'$ is the vector of mathematical expectations (cf. **Mathematical expectation**) of the components of $\mathbf{x}$, that is, $\mu_i=Ex_i$, $i=1, \ldots, p$, and $V=\|\,v_{ij}\,\|_{i,j=1}^p$ is the **covariance matrix** of $\mathbf{x}$, that is, $v_{ij}=E(x_i-\mu_i)(x_j-\mu_j)$ is the covariance of these components of $\mathbf{x}$ (the non-degenerate case rank $\mathbf{V}=p$ is considered; in case rank $\mathbf{V}=p'<p$, all the results remain true, but in a subspace of a smaller dimension p' on which the probability distribution of $\mathbf{x}$ is concentrated).

Thus, if (1) is a sequence of independent observations, forming a random sample from $N_p(\mu, \mathbf{V})$, then the maximum-likelihood estimators for the parameters μ and $\mathbf{V}$ in (2) are, respectively, the statistics (see [1], [2])

$$\hat{\mu} = \frac{1}{n}\sum_{i=1}^n \mathbf{x}_i \tag{3}$$

and

$$\hat{\mathbf{V}} = \frac{1}{n}\sum_{i=1}^n (\mathbf{x}_i-\hat{\mu})(\mathbf{x}_i-\hat{\mu})', \tag{4}$$

where the random vector $\hat{\mu}$ is subject to the p-dimensional normal law $N_p(\mu, \mathbf{V}/n)$ and is statistically independent of $\hat{\mathbf{V}}$, and the joint distribution of the elements of the matrix $\hat{\mathbf{Q}}=n\hat{\mathbf{V}}$ is described by the so-called **Wishart distribution** (see [4]) with density

$$w(\hat{\mathbf{Q}}\,|\,\mathbf{V}; n) =$$

$$= \frac{\hat{\mathbf{Q}}^{(n-p-2)/2}\exp\{-\operatorname{tr}(\mathbf{V}^{-1}\hat{\mathbf{Q}})/2\}}{2^{(n-2)p/2}\pi^{p(p-1)/4}\,|\,\mathbf{V}\,|^{(n-1)/2}\prod_{j=1}^p\Gamma((n-j)/2)}$$

if $\hat{\mathbf{Q}}$ is positive definite, and 0 otherwise.

Within this scheme, the distribution and moments of sampling characteristics of multivariate random variables such as the coefficients of paired, partial and multiple correlations, the generalized variance (i.e., the statistic $|\,\hat{\mathbf{V}}\,|$) and the generalized Hotelling T^2-statistic (cf. **Hotelling T^2-distribution** and [5]) have been investigated. In particular (see [1]), if the sample covariance matrix $\mathbf{S}_n$ is defined as the estimator $\hat{\mathbf{V}}$ made 'unbiased', namely:

$$S_n = \frac{n}{n-1}\hat{\mathbf{V}}, \tag{5}$$

then the distribution of $\sqrt{n}(|\,\mathbf{S}_n\,|\,/\,|\,\mathbf{V}\,|^{-1})$ tends to $N_1(0, 2p)$ as $n \to \infty$, and the random variables

$$\frac{n-p}{p(n-1)}T^2 = \frac{n-p}{p(n-1)}n(\hat{\mu}-\mu)'\mathbf{S}_n^{-1}(\hat{\mu}-\mu) \tag{6}$$

and

$$\frac{n_1+n_2-p-1}{(n_1+n_2-2)p}\tilde{T}^2 = \tag{7}$$

$$= \frac{n_1+n_2-p-1}{(n_1+n_2-2)p}\frac{n_1 n_2}{n_1+n_2}(\hat{\mu}_{n_1}-\hat{\mu}_{n_2})'\mathbf{S}_{n_1+n_2}^{-1}(\hat{\mu}_{n_1}-\hat{\mu}_{n_2})$$

have the **Fisher F-distribution** with degrees of freedom $(p, n-p)$ and (p, n_1+n_2-p-1), respectively. In (7), n_1 and n_2 are the sizes of two independent samples of the form (1) taken from the same population $N_p(\mu, \mathbf{V})$, $\hat{\mu}_{n_i}$ and $\mathbf{S}_{n_i}$ being estimators of the form (3) and (4) - (5), constructed with respect to the i-th sample, and

$$\mathbf{S}_{n_1+n_2} = \frac{1}{n_1+n_2-2}[(n_1-1)\mathbf{S}_{n_1}+(n_2-1)\mathbf{S}_{n_2}]$$

is the common sample covariance matrix constructed with respect to the estimators $\mathbf{S}_{n_1}$ and $\mathbf{S}_{n_2}$.

Multivariate statistical analysis of the nature and structure of correlations between the components of the multivariate attribute in question. This branch unifies the ideas and results used in such methods and models of multivariate statistical analysis as multiple **regression**; multivariate **dispersion analysis** and **covariance analysis**; **factor analysis**; the method of principal components; and the analysis of canonical correlations. The results of this branch may be conventionally divided into two basic types.

1) The construction of best (in a specified sense) statistical estimators for the parameters of these models and the analysis of their properties (more precisely, and in a probabilistic formulation, of their distribution laws, confidence regions, etc.). Thus, let the multivariate attribute x be interpreted as a vector-valued random variable subject to the p-dimensional normal distribution $N_p(\mu, \mathbf{V})$, and let it be partitioned into two subvectors $\mathbf{x}^{(1)}$ and $\mathbf{x}^{(2)}$ of dimensions q and $p-q$, respectively. This defines a corresponding partition of the expectation vector μ and of the theoretical and sample covariance matrices $\mathbf{V}$ and $\hat{\mathbf{V}}$, namely:

$$\mu = \begin{bmatrix}\mu^{(1)}\\\mu^{(2)}\end{bmatrix}, \quad \mathbf{V} = \begin{bmatrix}\mathbf{V}_{11} & \mathbf{V}_{12}\\\mathbf{V}_{21} & \mathbf{V}_{22}\end{bmatrix} \text{ and } \hat{\mathbf{V}} = \begin{bmatrix}\hat{\mathbf{V}}_{11} & \hat{\mathbf{V}}_{12}\\\hat{\mathbf{V}}_{21} & \hat{\mathbf{V}}_{22}\end{bmatrix}.$$

Then (see [1], [2]) the conditional distribution of the subvector $\mathbf{x}^{(1)}$ (under the condition that the second subvector $\mathbf{x}^{(2)}$ takes a fixed value) will also be normal $N_q(\mu^{(1)}+\mathbf{B}(\mathbf{x}^{(2)}-\mu^{(2)}), \Sigma)$. Here the maximum-likelihood estimators $\hat{\mathbf{B}}$ and $\hat{\Sigma}$ of the matrices of regression coefficients $\mathbf{B}$ and covariances Σ, in this classical multivariate model of multiple regression

$$\mathsf{E}(\mathbf{x}^{(1)}\,|\,\mathbf{x}^{(2)}) = \mu^{(1)}+\mathbf{B}(\mathbf{x}^{(2)}-\mu^{(2)}), \tag{8}$$

will be the mutually independent statistics

$$\hat{\mathbf{B}} = \hat{\mathbf{V}}_{12}\hat{\mathbf{V}}_{22}^{-1} \text{ and } \hat{\Sigma} = \hat{\mathbf{V}}_{11}-\hat{\mathbf{V}}_{12}\hat{\mathbf{V}}_{22}^{-1}\hat{\mathbf{V}}_{21},$$

respectively. Here the distribution of $\hat{\mathbf{B}}$ is the normal law $N_{q(p-q)}(\mathbf{B}, \mathbf{V_B})$, and $n\hat{\Sigma}$ has the Wishart distribution with parameters Σ and $n-(p-q)$ (the elements of $\mathbf{V_B}$ are given in terms of the elements of $\mathbf{V}$).

The basic results on the construction of estimators of parameters and in the investigation of their properties in models of factor analysis, of principal components and of canonical correlations are related to the analysis of the probabilistic-statistical properties of the eigen values (characteristic values) and eigen vectors of the various covariance matrices.

In schemes not falling within the limits of the classical normal model or even within the limits of any probabilistic model, the basic results are concerned with the construction of algorithms (and the investigation of their properties) for calculating estimators of parameters which are best from the point of view of some exogeneously given functional of the quality (or adequacy) of the model.

2) The construction of statistical tests for the verification of various hypotheses on the structure of the correlations being investigated. Within the limits of a multivariate normal model (sequences of observations of the form (1) are interpreted as random samples from the corresponding multivariate normal population) statistical tests have been constructed for testing, for example, the following hypotheses.

I. The hypothesis $\mu = \mu^*$, i.e., that the expectation of the variables studied be equal to a specific vector μ^*; this is tested via the Hotelling T^2-statistic by substituting $\mu = \mu^*$ in (6).

II. The hypothesis $\mu^{(1)} = \mu^{(2)}$ of equality of the expectation vectors in two populations (with identical but unknown covariance matrices), based on two samples; this is tested via the statistic $\tilde{T}^2$ (see [7]).

III. The hypothesis $\mu^{(1)} = \cdots = \mu^{(k)} = \mu$ of equality of the expectation vectors in several populations (with identical but unknown covariance matrices), based on samples from them; this is tested via the statistic

$$U_{p,k-1,n-k} = \frac{\left|\sum_{j=1}^{k}\sum_{i=1}^{n_j}(\mathbf{x}_i^{(j)}-\hat{\mu}^{(j)})(\mathbf{x}_i^{(j)}-\hat{\mu}^{(j)})'\right|}{\left|\sum_{j=1}^{k}\sum_{i=1}^{n_j}(\mathbf{x}_i^{(j)}-\hat{\mu})(\mathbf{x}_i^{(j)}-\hat{\mu})'\right|},$$

in which $\mathbf{x}_i^{(j)}$ is the i-th p-dimensional observation in a sample of size n_j, representing the j-th population, and $\hat{\mu}^{(j)}$ and $\hat{\mu}$ are estimators of the form (3), constructed separately with respect to each of the samples and with respect to the joint sample of size $n = n_1 + \cdots + n_k$, respectively.

IV. The hypotheses $\mu^{(1)} = \cdots = \mu^{(k)} = \mu$ and $V_1 = \cdots = V_k = V$ of equivalence of several normal populations, based on samples from them $\{x_{\cdot i}^{(j)}\}_{i=1}^{n_j}$, $j = 1, \ldots, k$; this is tested via the statistic

$$\lambda = \frac{\prod_{j=1}^{k} |\, n_j \hat{V}_j \,|^{(n_j - 1)/2}}{\left| \sum_{j=1}^{k} \sum_{i=1}^{n_j} (x_{\cdot i}^{(j)} - \hat{\mu})(x_{\cdot i}^{(j)} - \hat{\mu})' \right|^{(n-k)/2}},$$

in which the $\hat{V}_j$ are estimators of the form (4) constructed separately with respect to the observations from the j-th sample, $j = 1, \ldots, k$.

V. The hypothesis of mutual independence of the subvectors $x^{(1)}, \ldots, x^{(m)}$ of dimensions $p_1, \ldots, p_m$, respectively, into which the initial p-dimensional vector x has been partitioned, $p_1 + \cdots + p_m = p$; this is tested via the statistic

$$\psi = \frac{|\, n\hat{V} \,|}{\prod_{i=1}^{m} |\, n_i \hat{V}_i \,|},$$

in which $\hat{V}$ and $\hat{V}_i$ are sample covariance matrices of the form (4) for the vector x and its subvectors $x^{(i)}$, respectively.

Multivariate statistical analysis of the geometric structure of the set of multi-dimensional observations being investigated. This branch unifies notions and results of models and schemes such as **discriminant analysis**, mixtures of probability distributions, cluster analysis, taxonomy, and multi-dimensional scaling. The key in all of these schemes is a notion of distance (measure of proximity, measure of similarity) between the elements being analyzed. Here the objects being analyzed may both be real objects, in each of which the values of the components x are fixed — then in the geometrical representation the i-th object will be a point $x_{\cdot i} = (x_{1i}, \ldots, x_{pi})'$ in the corresponding p-dimensional space, as well as the variables x_l, $l = 1, \ldots, p$, themselves — in the geometrical representation the l-th index will be a point $x_l = (x_{l1}, \ldots, x_{ln})$ in the corresponding n-dimensional space.

The methods and results of discriminant analysis (see [1], [2], [7]) are directed to the solution of the following problem. Suppose that the existence of a specific number $k \geq 2$ of populations is known and that there is a sample from each (a 'training sample') known. It is required to construct, on the basis of training samples, the best, in a specified sense, classifying rule which allows one to attribute some new element (an observation x) to its population, when the investigator does not know in advance to which population the element belongs. Usually, a classification rule means a sequence of actions; the calculation of a scalar function of the variables in question, based on which a decision is taken on assigning the element to one of the classes

(the construction of a discriminant function); an ordering of the variables themselves according to their degree of informativeness from the point of view of a proper assignment of elements to classes; and a calculation of the corresponding probabilities of the errors in the classification.

The problem of analysis of a mixture of probability distributions (see [7]) most often (but not always) also arises in connection with the investigation of the 'geometric structure' of some population. Here the idea of the r-th homogeneous class is formalized with the help of a population described by some (as a rule, unimodal) distribution law $P(x \,|\, \theta_r)$, so that the distribution of the general population from which the sample (1) is extracted is described by a mixture of distributions of the form

$$P(x) = \sum_{r=1}^{k} \pi_r P(x \,|\, \theta_r),$$

where π_r is the a priori probability (the specific weight of the elements) of the r-th class in the general population. The problem is to give a 'good' statistical estimation (with respect to a sample $\{x_{\cdot i}\}_1^n$) of the unknown parameters θ_r, π_r, and sometimes even k. This, in particular, allows one to reduce the problem of the classification of the elements to a scheme of discriminant analysis, although in this case training samples are absent.

The methods and results of cluster analysis (classification, taxonomy, pattern recognition 'without a teacher', see [2], [6], [7]) are directed to the solution of the following problem. The geometric structure of the set of elements to be analyzed is given either by the coordinates of the corresponding points (that is, by the matrix $\|\, x_{ij} \,\|$, $i = 1, \ldots, p$, $j = 1, \ldots, n$), or by geometric characteristics of their mutual disposition, for example, by the matrix of pairwise distances $\|\, \rho_{ij} \,\|_{i,j=1}^{n}$. It is required to partition the set of elements being investigated into a comparatively small (known in advance or not) number of classes, so that the elements of a class are at a small distance from each other, and at the same time different classes should, as far as possible, be sufficiently far from each other and could not be partitioned into other subsets equally far from each other.

The problem of multi-dimensional scaling (see [6]) is related to the situation when the set of elements being investigated is given via a matrix of mutual distances $\|\, \rho_{ij} \,\|_{i,j=1}^{n}$ and consists of attributing to each of the elements a given number (p) of coordinates so that the structure of the mutual distances between the elements, measured using these auxiliary coordinates, would on the average differ least from that given. It should be noted that the basic results and methods of cluster analysis and multi-dimensional scaling have usually been developed without any assumptions regarding the probabilistic nature of the initial data.

The merits of multivariate statistical analysis in practice. These consist mainly in processing the following three problems.

The problem of statistical investigation of dependence between the variables being analyzed. Suppose that the set of recorded statistical variables x partitions, according to the meaning of these variables and the final aim of investigation, into a q-dimensional subvector $\mathbf{x}^{(1)}$ of (dependent) variables to be predicted and a $(p-q)$-dimensional subvector $\mathbf{x}^{(2)}$ of predicting (independent) variables. Then it can be said that the problem is to determine, on the basis of a sample (1), a q-dimensional vector-valued function $f(\mathbf{x}^{(2)})$ from the class of acceptable decisions F, which would give the best, in a specific sense, approximation of the behaviour of the subvector $\mathbf{x}^{(1)}$. Depending on the concrete form of the functional of the quality of the approximation and the nature of the variables being analyzed, one arrives at some scheme of multiple regression, variance, covariance, or confluence analysis [8].

The problem of classifying elements. This problem in a *general* (non-rigorous) *formulation* is that the whole set of elements (objects or variables) being analyzed, represented statistically as a matrix $\| x_{ij} \|$, $i=1,\ldots,p$, $j=1,\ldots,n$, or a matrix $\| \rho_{ij} \|$, $i,j=1,\ldots,n$, partitions into a comparatively small number of homogeneous (in a specified sense) groups [7]. Depending on the behaviour of the a priori information and the concrete form of the functional giving the criteria for the quality of the classification, one arrives at some scheme of discriminant analysis, cluster analysis (taxonomy, pattern recognition 'without a teacher'), or splitting mixtures of distributions.

The problem of lowering the dimension of the factor space being investigated and the selection of the most informative variables. This consists of defining a set of a comparatively small number $m \ll p$ of variables $\mathbf{z}=(z_1,\ldots,z_m)'$ in the class of admissible transformations $Z(\mathbf{x})$ of the initial variables $\mathbf{x}=(x_1,\ldots,x_p)$ for which some exogeneously given measure of informativity for an m-dimensional system of tests attains its least upper bound (see [7]). Concretization of the functional giving the *measure of self-informativity* (that is, aimed at a maximal preservation of the information contained in the statistical ensemble (1) relative to the initial attributes themselves) results, in particular, in various schemes of factor analysis and principal components, and in the method of extremal grouping of tests. Functionals giving a *measure of external informativity*, that is, aimed at extracting from (1) maximum information relative to certain other variables or phenomena not directly contained in x, lead to different methods of selecting the most informative variables in schemes of statistical research into dependences and discriminant analysis.

Fundamental mathematical tools in multivariate statistical analysis. These consist of special methods of the theory of systems of linear equations and matrices (the method of solution of simple and generalized problems on eigen values and vectors; simple inversion and pseudo-inversion of matrices; a procedure for the diagonalization of matrices; etc.) and certain optimization algorithms (methods of coordinate-wise descent, conjugate gradients, branch-and-bound, various versions of random scanning and stochastic approximation, etc.).

References

[1] ANDERSON, T.W.: *An introduction to multivariate statistical analysis*, Wiley, 1958.
[2] KENDALL, M.G. and STUART, A.: *The advanced theory of statistics. Design and analysis, and time series*, 3, Griffin, 1983.
[3] BOL'SHEV, L.N.: *Bull. Int. Stat. Inst.* **43** (1969), 425-441.
[4] WISHART, J.: *Biometrika* **20A** (1928), 32-52.
[5] HOTELLING, H.: 'The generalization of student's ratio', *Ann. Math. Statist.* **2** (1931), 360-378.
[6] KRUSKAL, J.B.: 'Multidimensional scaling by optimizing goodness of fit to a nonmetric hypothesis', *Psychometrika* **29** (1964), 1-27.
[7] AÏVAZYAN, S.A., BUKHSHTABER, V.M., YENYUKOV, I.S. and MESHALKIN, L.D.: *Applied statistics: classification and reduction of dimensionality*, Moscow, 1989 (in Russian).
[8] AÏVAZYAN, S.A., YENYUKOV, I.S. and MESHALKIN, L.D.: *Applied statistics: study of relationships*, Moscow, 1985 (in Russian).

S.A. Aïvazyan

Editorial comments.

References

[A1] GNANADESIKAN, R.: *Methods for statistical data analysis of multivariate observations*, Wiley, 1977.
[A2] SCHERVISH, M.J.: *Stat. Science* **2** (1987), 396-433.
[A3] FARRELL, R.: *Techniques of multivariate calculation*, Springer, 1976.
[A4] EATON, M.L.: *Multivariate statistics: A vector space approach*, Wiley, 1983.
[A5] MUIRHEAD, R.J.: *Aspects of multivariate statistical theory*, Wiley, 1982.

AMS 1980 Subject Classification: 62HXX

MULTI-DIMENSIONAL VARIATIONAL PROBLEM, *variational problem involving partial derivatives* - A problem in the calculus of variations (cf. **Variational calculus**) in which it is required to determine an extremum of a functional depending on a function of several independent variables. Ordinary variational problems, in which functionals of functions of one independent variable are considered, may be called one-dimensional variational problems, in this sense.

An example of a *two-dimensional variational problem* is the problem of determining a function of two independent variables, $u(x,y)$, which, together with its first-order partial derivatives, is continuous and yields an extremum of the functional

$$I(u) = \int\int_D F(x, y, u, u_x, u_y)\, dx\, dy \qquad (1)$$

under the boundary condition

$$u(x, y)|_l = u_0(x, y), \qquad (2)$$

where l is a closed contour bounding a domain D, $u_0(x, y)$ is a given function and $F(x, y, u, u_x, u_y)$ is a twice continuously-differentiable function jointly in its arguments. Let $u(x, y)$ be a solution of the problem (1), (2). Substitution of a comparison function $u(x, y) + \alpha\eta(x, y)$, where $\eta(x, y)|_l = 0$ and α is a numerical parameter, into (1), differentiation with respect to α and equating $\alpha = 0$, gives the following expression for the first variation of the functional:

$$\delta I = \int\int_D (F_u\eta + F_{u_x}\eta_x + F_{u_y}\eta_y)\, dx\, dy. \qquad (3)$$

If $u(x, y)$ has continuous second-order derivatives, then it is easy to show that a necessary condition for δI to vanish is:

$$F_u - \frac{\partial}{\partial x}F_{u_x} - \frac{\partial}{\partial y}F_{u_y} = 0. \qquad (4)$$

Equation (4) is called the *Euler−Ostrogradski equation* (sometimes the *Ostrogradski equation*). This equation must be satisfied by a function $u(x, y)$ which gives an extremum of (1) under the boundary conditions (2). The Euler−Ostrogradski equation is analogous to the **Euler equation** for one-dimensional problems. In expanded form, (4) is a second-order partial differential equation.

In the case of a triple integral and a function $u(x, y, z)$ depending on three independent variables, the Euler−Ostrogradski equation takes the form:

$$F_u - \frac{\partial}{\partial x}F_{u_x} - \frac{\partial}{\partial y}F_{u_y} - \frac{\partial}{\partial z}F_{u_z} = 0.$$

The following condition is an analogue of the **Legendre condition**. In order that $u(x, y)$ gives at least a weak extremum of (1) it is necessary that at each interior point of D,

$$F_{u_xu_x}F_{u_yu_y} - F_{u_xu_y}^2 \geq 0.$$

For a minimum necessarily $F_{u_xu_x} \geq 0$, and for a maximum $F_{u_xu_x} \leq 0$.

Discontinuous multi-dimensional variational problems have also been considered (see [4]).

References

[1] COURANT, R. and HILBERT, D.: *Methods of mathematical physics. Partial differential equations*, 2, Interscience, 1965 (translated from the German).

[2] SMIRNOV, V.I.: *A course of higher mathematics*, 4, Addison-Wesley, 1964 (translated from the Russian).

[3] AKHIEZER, N.I.: *The calculus of variations*, Blaisdell, 1962 (translated from the Russian).

[4] KERIMOV, M.K.: 'On two-dimensional continuous problems of variational calculus', *Trudy Tbilis. Mat. Inst. Akad. Nauk GruzSSR* **18** (1951), 209-219 (in Russian).

I.B. Vapnyarskiĭ

AMS 1980 Subject Classification: 49B21

MULTI-EXTREMUM PROBLEM - An extremum problem having several, or an unknown number of, local extrema.

The problem of finding a global extremum of a function $f(x)$, $x = (x^1, \ldots, x^n) \in X \subset \mathbf{R}^n$, $\overline{X}$ compact, has been solved for the basic classes of unimodal functions (first of all for convex and related functions, see **Convex programming**). For multimodal functions, even for smooth and slowly varying ones, at present (1989) there are no methods which are guaranteed to calculate a global extremum, with the exception of scanning along trajectories which form an everywhere-dense set in the admissible set X. In practice, time-consuming scanning is combined with algorithms for finding a local extremum: by scanning and a priori reduction of $f(x)$ (estimates of derivatives, functional equations and inequalities, etc.) one delineates a *domain of attraction* for each local extremum and *dead zones*, where a concrete local algorithm loses effectiveness (for example, in the **gradient method** the dead zones are neighbourhoods of saddle points and 'bottoms of valleys'). The extrema are then estimated or looked for by local methods and compared with each other.

For functions satisfying a Lipschitz condition

$$(\forall x, x' \in X)(\exists l < \infty)\ |f(x) - f(x')| \leq l\,\|x - x'\|,$$

a more universal method is that of *covering* (over a non-uniform grid), which is, moreover, easily realized in the multi-dimensional case.

For finding local extrema close to global ones in an approximation, empirical 'heuristic' methods are used, which may be called *quasi-global*:

1. Algorithms of the heavy-sphere type (cf. **Heavy sphere, method of the**), **minimization methods for functions depending strongly on a few variables**, or methods of 'galloping' through a local extremum of 'fitting' domains of attraction.

2. Random running through the values of $f(x)$ by a **Monte-Carlo method**.

3. Modification of a local method by the introduction of stochastic parameters. For example, the trajectory of a randomized gradient method

$$\dot{x} = \pm\nabla f(x) + \xi(t),$$

where $\xi(t)$ is a white noise process, can, under certain conditions, move from a point of local to a point of global extremum.

4. 'Hardening' a function by approximating it by a sequence of unimodal functions. The approximation can be carried out with respect to the values of $f(x)$ at certain points, by expansion with respect to the parameters in an analytic expression of the function, by majorants or minorants, etc.

5. The search for a local extremum of an averaged function $\overline{f}(x)$ (a *sliding average*),

$$\overline{f(x)} = \int_X p(x, \xi) f(\xi)\, d\xi, \qquad (1)$$

$$p(x, \xi) \geqslant 0; \quad \int_X p(x, \xi)\, d\xi = 1.$$

Examples are:

$$\overline{f(x)} = \frac{1}{2T} \int_{x-T}^{x+T} f(x)\, d\xi, \quad x \in \mathbf{R}^1, \quad x \pm T \in X;$$

$$\overline{f(x)} = \frac{\int_X e^{-(x-\xi)^2} f(\xi)\, d\xi}{\int_X e^{-(x-\xi)^2}\, d\xi}, \quad x \pm \xi \in X.$$

The physical meaning of this averaging is a 'smoothing' of the initial function and a 'filtering off' of its oscillating terms.

If $\overline{f(x)} \cong \mathrm{const}(x)$ (a rapidly-oscillating function $f(x)$), then before averaging it can be 'detected' with the help of a non-linear transformation, for example,

$$\overline{\overline{f(x)}} = \int_X p(x, \xi)\, |\, f(\xi)\,|\, d\xi. \qquad (2)$$

As a weight function $p(x, \xi)$ for averaging it is possible to take

$$p_n(x, \xi) = \frac{|\, f(x-\xi)\,|^n}{\int_X |\, f(x)\,|^n\, dx}, \quad n \geqslant 0,$$

$$\operatorname*{ess\,sup}_{M\ \ X} |\, f(x)\,| = \lim_{n \to \infty} \left[\frac{\int_X |\, f(x)\,|^n\, dM_x}{\int_X dM_x} \right].$$

The latter can also be used for obtaining estimates for essential maxima and minima.

Formulas (1) and (2) can be interpreted as mathematical expectations for functions of a random variable ξ, distributed with probability density $p(x, \xi)$. Therefore the given multi-extremum problem can be treated as a problem in **stochastic programming** and methods of the latter can be applied to it.

After finding the extremum of an approximating or averaged function (*outline search*), a *detailed search* for an extremum of the initial function $f(x)$ can be carried out in a neighbourhood of the point found.

Global optimization of individual functions can be carried out by particular specific procedures. For example, by constructing a dynamical system for which the point of global extremum is an asymptotically-stable point of rest.

One of the sources of ideas for new methods of (quasi-) global optimization is the modelling of processes in physical and biological systems.

The cumbersome calculations of the non-local search process can be optimized with respect to a number of indices taking account of limitations on the calculating means, a priori and sequentially accumulated information on the function $f(x)$, the probabilistic characteristics of random factors, etc. One of the approaches which has been tried is based on the theory of statistical decisions.

Apart from the search of global extrema, other multi-extremum problems arise, e.g. the determination of the oscillation of a function, or the listing and finding of all local extrema in a given domain.

For polynomials very effective rules for calculating and separating roots of derivatives have been worked out.

For extremum points of transcendental functions, considerations of the type of the **Rolle theorem** or a **comparison theorem** for the solutions of differential equations may be useful.

The number of stationary points of an analytic function can be estimated using the principle of the argument (cf. **Argument, principle of the**).

If a function has an infinite sequence of extrema, then in practice several of them are calculated directly and the others are obtained by using an asymptotic expansion (example: the Γ-function).

The quasi-classical approximation for differential equations can be regarded as the asymptotic expansion of the solutions of an equation with respect to the number of extrema.

For multi-extremum problems in the infinite-dimensional cases see **Variational calculus in the large**. Discrete analogues are given in **Integer programming** and **Discrete programming**.

For references see **Maximization and minimization of functions**.

Yu.P. Ivanilov
V.V. Okhrimenko

Editorial comments. For global optimization methods based on the Bayes principle of statistical inference cf. [A1]. Other recent texts specifically aimed at global optimization are [A2] - [A5].

The method of *simulated annealing* has attracted attention more recently. Within the classification of this article, it belongs to item 3 (modification of a local method by the introduction of stochastic parameters). See [A6].

References

[A1] MOCKUS, J.: *Bayesian approach to global optimization*, Kluwer, 1989.

[A2] PARDATOS, P.M. and ROSEN, J.B.: *Constrained global optimization: algorithms and applications*, Springer, 1987.

[A3] RINNOOY KAN, A.H.G. and TIMMER, G.T.: 'Stochastic methods for global optimization', *Amer. J. Manag. Sci.* **4** (1984), 7-40.

[A4] PINTER, J. and SZABO, J.: *Global optimization algorithms: theory and some applications*, Springer, 1986.

[A5] ZHIGLYAVSKY, A.A. [A.A. ZHIGLYAVSKIĬ]: *Theory of global random search*, Kluwer, 1990 (translated from the Russian).

[A6] LAARHOVEN, P.J.M. VAN: *Theoretical and computational aspect of simulated annealing*, CWI Tracts, 51, Centre for Math. Comp. Sc., Amsterdam, 1988.

AMS 1980 Subject Classification: 49DXX, 65KXX, 90C30

MULTI-FIELD POTENTIAL - The same as a **multipole potential**.

AMS 1980 Subject Classification: 31B15

MULTI-FUNCTOR, *multi-place functor* - A function of several arguments, defined on categories, taking values in a **category** and giving a one-place **functor** in each argument. More precisely, let n categories $\Re_1, \ldots, \Re_n$ be given. Construct the Cartesian product category $\Re = \bar{\Re}_1 \times \cdots \times \bar{\Re}_n$, where each category $\bar{\Re}_i$ is either $\Re_i$ or the opposite category $\Re_i^*$. A one-place covariant functor F from $\Re$ with values in a category $\mathfrak{C}$ is called an *n-place functor* on $\Re_1, \ldots, \Re_n$ with values in $\mathfrak{C}$. The functor F is covariant in those arguments which correspond to the factors $\Re_i$ in $\Re$, and contravariant in the remaining arguments.

The conditions which must be satisfied by a mapping $F: \Re \to \mathfrak{C}$ are given below (in the case $n = 2$, with the first argument contravariant and the second covariant). The functor $F: \Re_1^* \times \Re_2 \to \mathfrak{C}$ associates to each pair of objects (A, B), $A \in \mathrm{Ob}\,\Re_1$, $B \in \mathrm{Ob}\,\Re_2$, an object $F(A, B) \in \mathrm{Ob}\,\mathfrak{C}$ and to each pair of morphisms (α, β), where

$$\alpha: A \to A_1 \in \mathrm{Mor}\,\Re_1, \quad \beta: B \to B_1 \in \mathrm{Mor}\,\Re_2,$$

a morphism

$$F(\alpha, \beta): F(A_1, B) \to F(A, B_1) \in \mathrm{Mor}\,\mathfrak{C},$$

in such a way that the following conditions are satisfied:

1) $F(1_A, 1_B) = 1_{F(A, B)}$ for any pair of objects A, B;
2) if $\alpha: A \to A_1$, $\alpha_1: A_1 \to A_2$, $\alpha, \alpha_1 \in \mathrm{Mor}\,\Re_1$, $\beta: B \to B_1$, $\beta_1: B_1 \to B_2$, $\beta, \beta_1 \in \mathrm{Mor}\,\Re_2$, then

$$F(\alpha_1 \alpha, \beta_1 \beta) = F(\alpha, \beta_1) F(\alpha_1, \beta).$$

Examples of multi-functors.

A) Let $\Re$ be a category with finite products. Then the product of n objects can be considered as an n-place functor that is covariant in all its arguments, defined on $\Re^n = \Re \times \cdots \times \Re$ (n times) and taking values in $\Re$. Similar functors can be constructed for coproducts, etc.

B) Let $\Re$ be an arbitrary category. Associate with each pair of objects A, B from $\Re$ the set of morphisms $H_\Re(A, B)$ and with each pair of morphism $\alpha: A \to A_1$, $\beta: B \to B_1$ the mapping $H_\Re(\alpha, \beta): H_\Re(A_1, B) \to H_\Re(A, B_1)$ given as follows: if $\phi: A_1 \to B$ then $H_\Re(\alpha, \beta)(\phi) = \beta \phi \alpha$. This construction gives a two-place functor from $\Re^* \times \Re$ into the category of sets that is contravariant in its first argument and covariant in its second.

If $\Re$ is an **additive category**, then this functor can be regarded as taking values in the category of Abelian groups.

C) Let $\Re$ be a category with finite products. Consider the product as a two-place functor $\times: \Re \times \Re \to \Re$. Then

by combining Examples A) and B) it is possible to construct three-place functors $H_\Re(A, B \times C)$ and $H_\Re(A \times B, C)$. The first functor is naturally equivalent to the functor $H_\Re(A, B) \times H_\Re(A, C)$. If $\mathfrak{C}$ is the category of sets (cf. **Sets, category of**), the second functor is naturally equivalent to the functor $H_\mathfrak{C}(A, H_\mathfrak{C}(B, C))$.

D) Let θ be a **small category** and let $F(\theta, \mathfrak{C})$ be the category of diagrams over the category of sets $\mathfrak{C}$ with scheme θ, that is, the category of one-place covariant functors and their natural transformations. A two-place functor $E: \theta \times F(\theta, \mathfrak{C}) \to \mathfrak{C}$ which is covariant in both arguments is constructed as follows: If $A \in \mathrm{Ob}\,\theta$ and $F \in \mathrm{Ob}\,F(\theta, \mathfrak{C})$, then $E(A, F) = F(A)$; if $\alpha: A \to B \in \mathrm{Mor}\,\theta$ and $\sigma: F \to G$ is a natural transformation, then $E(\alpha, \sigma) = \sigma_B F(\alpha) = G(\alpha) \sigma_A$. The functor E is called the 'evaluation functor'. This functor is naturally equivalent to the functor $\mathrm{Nat}(H_A, F): \theta \times F(\theta, \mathfrak{C}) \to \mathfrak{C}$, which associates with an object $A \in \theta$ and a functor $F: \theta \to \mathfrak{C}$ the set of natural transformations of the **representable functor** H_A into F (*Yoneda's lemma*).

M.Sh. Tsalenko

Editorial comments. A two-place functor is often called a *bifunctor*.

References
[A1] MITCHELL, B.: *Theory of categories*, Acad. Press, 1965.
[A2] MACLANE, S.: *Categories for the working mathematician*, Springer, 1971.

AMS 1980 Subject Classification: 18A99

MULTI-GROUP APPROXIMATION - One of the methods for an approximate description and study of a system of a large number of interacting particles (gas, fluid), used in statistical physics.

The method is based on the assumption that every (or almost-every) configuration of particles can be divided into finite groups of particles, *clusters*, situated sufficiently far apart and almost not interacting with each other. In other words, the system of interacting particles is assumed to be in the form of gas clusters which are not allowed to come too close together. The next step lies in neglecting these restrictions, and also in excluding from the discussion clusters of very large size or of very complicated form (this properly constitutes the multi-group approximation).

The basic assumption that a configuration decomposes into clusters, which underlies this method, is true for low particle density and for short-range forces between them. Whether this assumption holds in other cases is unknown (1989). The method of multi-group approximation, apparently, gives a good approximation only for low densities and becomes too coarse in the case of a phase transition, when clusters of all sizes begin to play a comparably important role.

References
[1] BAND, W.: *J. Chem. Phys.* **7** (1939), 324-326.
[2] FRENKEL, J.: *J. Chem. Phys.* **7** (1939), 200.
[3] MINLOS, R.A. and SINAĬ, YA.G.: 'The phenomenon of 'phase separation' at low temperatures in some lattice models of a gas I', *Math. USSR Sb.* **2**, no. 3 (1967), 335-395. (*Mat. Sb.* **73**, no. 3 (1967), 375-448)
[4] HILL, T.L.: *Statistical mechanics*, McGraw-Hill, 1956.

R.A. Minlos

AMS 1980 Subject Classification: 70-XX, 73-XX, 81-XX, 82-XX, 83-XX, 35Q20

MULTI-OPERATOR GROUP, *group with multiple operators*, Ω-*group* -

A **universal algebra** which is a **group** relative to the addition operation + (which need not be commutative) and in which there is given a system of operations Ω of arity $\geqslant 1$. It is assumed that the zero element 0 of the additive group A is a subalgebra, that is, $0 \cdots 0\omega = 0$ for all $\omega \in \Omega$. Thus, a multi-operator group combines the concepts of a group, a **linear algebra** and a **ring**. An *ideal of an* Ω-*group* is a **normal subgroup** N of A such that

$$-(x_1 \cdots x_n\omega)+(x_1 \cdots x_{i-1}(a+x_i)x_{i+1} \cdots x_n\omega) \in N$$

for all $a \in N$, $x_i \in A$, $\omega \in \Omega$, $1 \leqslant i \leqslant n$. Congruences on a multi-operator group are described by coset classes relative to ideals.

Let A, B and C be Ω-subgroups in an Ω-group G (that is, subalgebras of the universal algebra G), where C is generated by A and B. The *mutual commutator* $[A, B]$ *of the subgroups* A and B is the ideal in C generated by all elements of the form

$$-a-b+a+b,$$

$$-(a_1 \cdots a_n\omega)-(b_1 \cdots b_n\omega)+(a_1+b_1) \cdots (a_n+b_n)\omega,$$

where $a, a_i \in A$, $b, b_i \in B$, $\omega \in \Omega$. Let $G' = [G, G]$. A multi-operator group G is called *Abelian* if $G' = 0$. Inductively one defines ideals $G_{i+1} = [G_i, G]$, where $G_1 = G'$, and $G^{(i+1)} = [G^{(i)}, G^{(i)}]$, where $G^{(1)} = G'$. A multi-operator group G is called *nilpotent* if $G_i = 0$, and *solvable* if $G^{(i)} = 0$ for some i. Many of the properties of the corresponding classes of groups and rings can be transferred to these classes of multi-operator groups. A multi-operator group A is called a *multi-operator (linear)* Ω-*algebra* over a commutative associative ring k with an identity if the addition in A is commutative, if $\Omega_1 = k$, where Ω_1 is the set of unary operations from Ω, and if all operations from Ω are semi-linear over k (see [2] - [6], and **Semi-linear mapping**).

References
[1] HIGGINS, P.J.: 'Groups with multiple operators', *Proc. London Math. Soc.* **6** (1956), 366-416.
[2] KUROSH, A.G.: 'Free sums of multi-operator algebras', *Sibirsk. Mat. Zh.* **1**, no. 1 (1960), 62-70 (in Russian).
[3] KUROSH, A.G.: *Lectures on general algebra*, Chelsea, 1963 (translated from the Russian).
[4] KUROSH, A.G.: *General algebra. Lectures in the academic year 1969-1970*, Mosocw, 1974 (in Russian).
[5] KUROSH, A.G.: 'Multioperator rings and algebras', *Russian Math. Surveys* **24**, no. 1 (1969), 1-13. (*Uspekhi Mat. Nauk.* **24**, no. 1 (1969), 3-15)
[6] BARANOVICH, T.M. and BURGIN, M.S.: 'Linear Ω-algebras', *Russian Math. Surveys* **30**, no. 4 (1975), 65-113. (*Uspekhi Mat. Nauk.* **30**, no. 4 (1975), 61-106)
[7] ARTAMONOV, V.A.: 'Universal algebras', *Itogi Nauk. i Tekhn. Algebra. Topol. Geom.* **14** (1976), 191-248 (in Russian).
[8] *Rings*, Vol. 1, Novosibirsk, 1973, pp. 41-45.

V.A. Artamonov

AMS 1980 Subject Classification: 08A40

MULTI-POLE POTENTIAL, *potential of a multi-pole* -

A **harmonic function** on the domain $\mathbf{R}^n \setminus \{0\}$ in $\mathbf{R}^n$, $n \geqslant 2$, which is a partial derivative of some order $|m| \geqslant 1$ of the main fundamental solution of the **Laplace equation**; that is, a function of the form

$$\frac{\partial^{|m|}}{\partial x_1^{m_1} \cdots \partial x_n^{m_n}} \frac{1}{r^{n-2}} \text{ for } n \geqslant 3,$$

$$\frac{\partial^{|m|}}{\partial x_1^{m_1} \partial x_2^{m_2}} \ln \frac{1}{r} \text{ for } n = 2,$$

$$r = \sqrt{x_1^2 + \cdots + x_n^2}, \quad |m| = m_1 + \cdots + m_n.$$

For brevity, let $n = 3$. For $|m| = 1$ *dipole potentials* have the form

$$-\frac{x_1}{r^3} = -\frac{\cos\alpha}{r^2}, \quad -\frac{x_2}{r^3} = -\frac{\cos\beta}{r^2}, \quad -\frac{x_3}{r^3} = -\frac{\cos\gamma}{r^2},$$

where $\cos\alpha$, $\cos\beta$ and $\cos\gamma$ are the direction cosines of the radius vector of the point of observation (x_1, x_2, x_3). The function $-x_1/r^3$, for example, is interpreted as a dipole potential with moment 1 and axis Ox_1, that is, the limit as $a \to 0+$ of the sum of the Newton potentials of a mass $-1/2a$ placed at $(a, 0, 0)$, and a mass $1/2a$ placed at $(-a, 0, 0)$; otherwise this function can be represented as the magnetic potential of a small magnet placed at the origin along the axis Ox_1. Similarly, the functions $-x_2/r^3$ and $-x_3/r^3$ are dipole potentials with axes Ox_2 and Ox_3, respectively. By taking linear combinations of these functions it is possible to obtain the potential of an arbitrarily oriented dipole with any moment μ. For $|m| = 2$ a *quadrupole potential* arises, obtained by a limit transition from a fixed system of four point masses whose total mass is always equal to zero, etc.

The **Newton potential** $U(x_1, x_2, x_3)$ of a bounded body G of density $\rho = \rho(\xi, \eta, \zeta)$, situated so that $0 \in G$, can be expanded in a series of multi-pole potentials:

$$U(x_1, x_2, x_3) = \tag{1}$$

$$= \iiint_G \frac{\rho(\xi, \eta, \zeta)\, d\xi\, d\eta\, d\zeta}{\sqrt{(x_1 - \xi)^2 + (x_2 - \eta)^2 + (x_3 - \zeta)^2}} =$$

$$= \frac{M}{r} + \sum_{|m|=1}^{\infty} \frac{(-1)^{|m|}}{m_1! m_2! m_3!} a_{m_1 m_2 m_3} \frac{\partial^{|m|}}{\partial x_1^{m_1} \partial x_2^{m_2} \partial x_3^{m_3}} \frac{1}{r},$$

where

$$M = \iiint_G \rho(\xi, \eta, \zeta)\, d\xi\, d\eta\, d\zeta$$

is the total mass of the body G, and the coefficients

$$a_{m_1 m_2 m_3} = \iiint_G \xi^{m_1} \eta^{m_2} \zeta^{m_3} \rho(\xi, \eta, \zeta) \, d\xi \, d\eta \, d\zeta$$

are called *dipole moments* for $|m|=1$, *quadrupole moments* for $|m|=2$, and, in general, *multi-pole moments* for $|m| \geqslant 1$. The series (1) differs from the expansion of $U(x_1, x_2, x_3)$ in spherical functions,

$$U(r, \theta, \lambda) = \frac{M}{r} + \sum_{m=1}^{\infty} \sum_{l=-m}^{m} q_{ml} \frac{Y_{ml}(\theta, \lambda)}{r^{m+1}}, \qquad (2)$$

by rearrangement of the terms; the terms of (2) can also be interpreted as potentials of multi-poles that are oriented in a special way (see [1]). Therefore the coefficients q_{ml} are also often called, respectively, *dipole*, *quadrupole* and, more generally, *multi-pole moments*.

Expansions of the type (1) and (2) are used in the description and approximate representation of scalar and vector potentials, not only in connection with the fundamental solution of the Laplace equation, but also of the **Helmholtz equation** (see [2]).

In the hydrodynamics of planar flows of an ideal incompressible fluid there are also applications of complex multi-pole potentials of the form

$$\frac{\mu e^{i\alpha}}{z^m}, \quad m \geqslant 1,$$

where z is a complex variable and μ and α are, respectively, the moment and angle of orientation of the multi-pole. The dipole potential obtained for $m=1$, $\mu=1$ and $\alpha=0$ can be interpreted as the limit as $a \to 0+$ of the sum of the complex potentials of a source of capacity 1 at $z=a>0$ and a sink of capacity 1 at $z=-a$. The expansion (1) here corresponds to the expansion of the complex potential of the velocity of the flow of a streamlined planar body G, in a neighbourhood of the point at infinity:

$$w = f(z) = \sum_{m=1}^{\infty} \frac{\mu_m e^{i\alpha_m}}{z^m}.$$

Here the action of the streamlined body G is replaced by the resultant action of multi-pole potentials placed at the origin (see [3]).

References
[1] MORSE, F.M. and FESHBACH, H.: *Methods of theoretical physics*, 2, McGraw-Hill, 1953.
[2] JACKSON, J.D.: *Classical electrodynamics*, Wiley, 1962.
[3] MILNE-THOMSON, L.M.: *Theoretical hydrodynamics*, MacMillan, 1950.

E.D. Solomentsev

AMS 1980 Subject Classification: 31B15

MULTI-SHEETED REGION - A region S of a **Riemann surface** R, considered as a covering surface over the complex plane $\mathbf{C}$, such that above each point of its projection $D \subset \mathbf{C}$ there are at least two points of S; a **branch point** of R of order $k-1$ is regarded here as k distinct points. For example, the analytic function

$w = z^2$ is a one-to-one mapping of the disc $D = \{z \in \mathbf{C}: |z| < 1\}$ onto the two-sheeted region (two-sheeted disc) $S = \{w \in R: |w| < 1\}$ of the Riemann surface R of this function; this mapping is conformal everywhere except at the origin.

For analytic functions of several complex variables there arise multi-sheeted Riemann domains (cf. **Riemannian domain**) over the complex space $\mathbf{C}^n$.

E.D. Solomentsev

Editorial comments.

References
[A1] SIEGEL, C.L.: *Topics in complex functions*, 1, Wiley (Interscience), 1988, Chapt. 1, Sect. 4.
[A2] SPRINGER, G.: *Introduction to Riemann surfaces*, Addison-Wesley, 1957.

AMS 1980 Subject Classification: 30B40, 30FXX

MULTI-VALUED FUNCTION - A function which assigns to each element $x \in X$ a subset $f(x)$ of a set Y: $f(x) \subset Y$, where there is at least one such subset of Y consisting of at least two elements. See also **Multi-valued mapping**.

L.D. Kudryavtsev

AMS 1980 Subject Classification: 26CXX, 54C60

MULTI-VALUED MAPPING, *point-to-set mapping* - A mapping $\Gamma: X \to Y$ associating with each element x of a set X a subset $\Gamma(x)$ of a set Y. If for each $x \in X$ the set $\Gamma(x)$ consists of one element, then the mapping Γ is called *single-valued*. A multi-valued mapping Γ can be treated as a single-valued mapping of X into 2^Y, that is, into the set of all subsets of Y.

For two multi-valued mappings $\Gamma_i: X \to Y$, $i=1,2$, their *inclusion* is naturally defined: $\Gamma_1 \subset \Gamma_2$ if $\Gamma_1(x) \subset \Gamma_2(x)$ for all $x \in X$. For any family of multi-valued mappings $\Gamma_\alpha: X \to Y$, $\alpha \in A$, the *union* and *intersection* are defined: $\Gamma = \bigcup_{\alpha \in A} \Gamma_\alpha$ if $\Gamma(x) = \bigcup_{\alpha \in A} \Gamma_\alpha(x)$ for all $x \in X$ and $\Gamma = \bigcap_{\alpha \in A} \Gamma_\alpha$ if $\Gamma(x) = \bigcap_{\alpha \in A} \Gamma_\alpha(x)$ for all $x \in X$. For any family of multi-valued mappings $\Gamma_\alpha: X \to Y_\alpha$, $\alpha \in A$, the multi-valued mapping $\Gamma = \prod_{\alpha \in A} \Gamma_\alpha: X \to \prod_{\alpha \in A} Y_\alpha$ is called the *Cartesian product* of the multi-valued mappings Γ_α if $\Gamma(x) = \prod_{\alpha \in A} \Gamma_\alpha(x)$. A *section of a multi-valued mapping* Γ is a single-valued mapping $f: X \to Y$ such that $f(x) \in \Gamma(x)$ for all $x \in X$. The *graph of a multi-valued mapping* Γ is the set $G(\Gamma) = \{(x,y) \in X \times Y: y \in \Gamma(x)\}$.

A multi-valued mapping Γ of a topological space X into a topological space Y is called *upper semicontinuous* if for every open set $U \subset Y$ the set $\Gamma^+(u) = \{x \in X: \Gamma(x) \subset U, \Gamma(x) \neq \varnothing\}$ is open in X, or equivalently: For any $x \in X$ and any neighbourhood U of $\Gamma(x)$ there is a neighbourhood Ox of x such that $\Gamma(Ox) \subset U$, where $\Gamma(Ox) = \bigcup \{\Gamma(y): y \in Ox\}$. A multi-valued mapping from a topological space X to a topological space Y is called *lower semi-continuous* if for any open set $U \subset Y$ the set $\Gamma^-(U) = \{x \in X: \Gamma(x) \cap U \neq \varnothing\}$ is open in X. If a multi-valued mapping satisfies both properties simultaneously, then it is called a *continuous multi-valued mapping*.

Let Y be a topological vector space. A multi-valued mapping $\Gamma: X \to Y$ is called *convex-compact valued* if $\Gamma(x)$ is a convex compact set for all $x \in X$. For a finite set of multi-valued mappings $\Gamma_i: X \to Y$, $i \in I$, the algebraic sum $\Gamma = \sum_{i \in I} \Gamma_i$ is defined by $\Gamma(x) = \sum_{i \in I} \Gamma_i(x)$. The intersection of any (finite) family of upper semi-continuous (respectively, continuous) multi-valued mappings is upper semi-continuous (respectively, continuous). The Cartesian product of a finite family of upper semi-continuous multi-valued mappings is upper semi-continuous. The algebraic sum of a finite family of upper semi-continuous (convex-compact valued) mappings is upper semi-continuous (convex-compact valued). The intersection and Cartesian product of any family of convex-compact valued mappings is convex-compact valued.

Let X be a **paracompact space** and Y a locally convex metric linear space (cf. **Locally convex space; Linear space; Metric space**). Let $\Gamma: X \to Y$ be a multi-valued mapping which is upper semi-continuous and is such that $\Gamma(x)$ is closed in Y for every $x \in X$. Then the multi-valued mapping Γ admits continuous sections. Let $(X, \mathfrak{A})$ and $(Y, \mathfrak{B})$ be spaces with given σ-algebras $\mathfrak{A}$ and $\mathfrak{B}$; a multi-valued mapping $\Gamma: (X, \mathfrak{A}) \to (Y, \mathfrak{B})$ is called *measurable* if the graph $G(\Gamma)$ belongs to the smallest σ-algebra $\mathfrak{A} \times \mathfrak{B}$ of $X \times Y$ containing all sets of the form $A \times B$, where $A \in \mathfrak{A}$ and $B \in \mathfrak{B}$. If Γ is a measurable multi-valued mapping from $(X, \mathfrak{A})$ to a complete separable metric space $(Y, \mathfrak{B})$, where $\mathfrak{B}$ is the Borel σ-algebra of Y, then Γ has measurable sections f.

References

[1] KURATOWSKI, K.: *Topology*, 1-2, Acad. Press, 1966-1968 (translated from the French).

V.A. Efimov

Editorial comments. A multi-valued mapping is also called a *set-valued* or *many-valued mapping*. Sections are also called *selections*.

Theorems which prove that certain kinds of multi-valued mappings admit selections are called *selection theorems*. The measurable selection theorem stated in the last sentence of the main article above is known as *von Neumann's measurable choice theorem*. A number of selection theorems and some applications are discussed in [A4].

References

[A1] MICHAEL, E.: 'Continuous selections', *Ann. of Math.* **63** (1956), 361-382.
[A2] MICHAEL, E.A.: 'A survey of continuous selections', in W.M. Fleischmann (ed.): *Set-valued mappings. Proc. Conf. SUNY Buffalo, 1969*, Lecture notes in math., Vol. 171, Springer, 1970, pp. 54-58.
[A3] PRZESLAWSKI, K. and YOST, D.: 'Continuity properties of selectors and Michael's Theorem', *Mich. Math. J.* **36** (1989), 113-134.
[A4] PARTHASARATHY, T.: *Selection theorems and their applications*, Springer, 1972.

AMS 1980 Subject Classification: 54C60, 04A05

MULTI-VALUED REPRESENTATION *of a connected topological group G* - An ordinary representation π of a connected topological group G_1 (cf. **Representation of a topological group**) such that G is isomorphic (as a topological group) to a quotient group of G_1 relative to a discrete normal subgroup N which is not contained in the kernel of π. A multi-valued representation is called *n-valued* if $\pi(N)$ contains exactly n elements. By identifying the elements of G with the elements of G_1/N one obtains for the sets $\pi(g)$, $g \in G = G_1/N$, the relations $\pi(e) \ni 1$, $\pi(g_1 g_2) = \pi(g_1)\pi(g_2)$, $g_1, g_2 \in G$. Multi-valued representations of connected, locally path-connected topological groups G exist only for non-simply-connected groups. The most important example of a multi-valued representation is the **spinor representation** of the complex orthogonal group $SO(n, \mathbf{C})$, $n \geqslant 2$; this representation is a two-valued representation of $SO(n, \mathbf{C})$ and is determined by a faithful representation of the universal covering group of $SO(n, \mathbf{C})$.

References

[1] KIRILLOV, A.A.: *Elements of the theory of representations*, Springer, 1976 (translated from the Russian).
[2] NAĬMARK, M.A.: *Theory of group representations*, Springer, 1982 (translated from the Russian).

A.I. Shtern

AMS 1980 Subject Classification: 22A25

MULTIGRAPH - A **graph** in which multiple edges are allowed.

Editorial comments. Multigraphs are very useful for modelling the molecular structure of chemical compounds, see [A1]. Multigraphs in which loops are also allowed (called *pseudo-graphs* in [A2]) are called graphs by the Canadian School of graph-theorists (see [A3]). See also **Graph**.

References

[A1] BALABAN, A.T. (ED.): *Chemical application of graph theory*, Acad. Press, 1976.
[A2] HARARY, F.: *Graph theory*, Addison-Wesley, 1969.
[A3] BONDY, J.A. and MURTY, U.S.R.: *Graph theory with applications*, Macmillan, 1976.

AMS 1980 Subject Classification: 05C99

MULTIHARMONIC FUNCTION - A **harmonic function** such that the **Laplace operator** acting on separate groups of independent variables vanishes. More precisely: A function $u = u(x_1, \ldots, x_n)$, $n \geqslant 2$, of class C^2 in a domain D of the Euclidean space $\mathbf{R}^n$ is called a multiharmonic function in D if there exist natural numbers $n_1, \ldots, n_k$, $n_1 + \cdots + n_k = n$, $n \geq k \geq 2$, such that the following identities hold throughout D:

$$\sum_{\nu=1}^{n_1} \frac{\partial^2 u}{\partial x_\nu^2} = 0,$$

$$\sum_{\nu=n_1+1}^{n_1+n_2} \frac{\partial^2 u}{\partial x_\nu^2} = 0, \ldots, \sum_{\nu=n_1+\cdots+n_{k-1}+1}^{n} \frac{\partial^2 u}{\partial x_\nu^2} = 0.$$

An important proper subclass of the class of multiharmonic functions consists of the pluriharmonic functions (cf. **Pluriharmonic function**), for which $n = 2m$, $n_j = 2$, $j = 1, \ldots, m$, i.e. $k = m$, and which also satisfy certain additional conditions.

References

[1] STEIN, E.M. and WEISS, G.: *Introduction to harmonic analysis on Euclidean spaces*, Princeton Univ. Press, 1971.

E.D. Solomentsev

Editorial comments. Multiharmonic functions are also called *multiply harmonic* functions. It was shown by P. Lelong that if u is *separately harmonic*, that is, u is harmonic as a function of $x_{n_j+1}, \ldots, x_{n_{j+1}}$ $(j = 0, \ldots, k;$ $n_0 = 0)$ while the other variables remain fixed, then u is multiharmonic. A different proof is due to J. Siciak. See [A1].

References

[A1] HERVÉ, M.: *Analytic and plurisubharmonic functions*, Lecture notes in math., 198, Springer, 1971.

AMS 1980 Subject Classification: 31C10, 31C15

MULTILINEAR ALGEBRA - The branch of algebra dealing with multilinear mappings (cf. **Multilinear mapping**) between modules (in particular, vector spaces). The first sections of multilinear algebra were the theory of bilinear and quadratic forms, the theory of determinants, and the Grassmann calculus that extends this (see **Exterior algebra**; **Bilinear form**; **Quadratic form**; **Determinant**). A basic role in multilinear algebra is played by the concepts of a **tensor product**, a **tensor on a vector space** and a **multilinear form**. The applications of multilinear algebra to geometry and analysis are related mainly to **tensor calculus** and differential forms (cf. **Differential form**).

A.L. Onishchik

Editorial comments.

References

[A1] MARCUS, M.: *Finite dimensional multilinear algebra*, I-II, M. Dekker, 1973.

[A2] BOURBAKI, N.: *Elements of mathematics. Algebra: Algebraic structures. Linear algebra*, 1, Addison-Wesley, 1974, Chapt. 1; 2 (translated from the French).

AMS 1980 Subject Classification: 15A69

MULTILINEAR FORM, *n-linear form*, *on a unitary A-module E* - A **multilinear mapping** $E^n \to A$ (here A is a commutative associative ring with a unit, cf. **Associative rings and algebras**). A multilinear form is also called a *multilinear function* (*n-linear function*). Since a multi-

linear form is a particular case of a multilinear mapping, one can speak of symmetric, skew-symmetric, alternating, symmetrized, and skew-symmetrized multilinear forms. For example, the **determinant** of a square **matrix** of order n over A is a skew-symmetrized (and therefore alternating) n-linear form on A^n. The n-linear forms on E form an A module $L_n(E, A)$, which is naturally isomorphic to the module $(\otimes^n E)^*$ of all linear forms on $\otimes^n E$. In the case $n = 2$ ($n = 3$), one speaks of bilinear forms (cf. **Bilinear form**) (respectively, trilinear forms).

The n-linear forms on E are closely related to n-times covariant tensors, i.e. elements of the module $T^n(E^*) = \otimes^n E^*$. More precisely, there is a linear mapping

$$\gamma_n : T^n(E^*) \to L_n(E, A),$$

such that

$$\gamma_n(u_1 \otimes \cdots \otimes u_n)(x_1, \ldots, x_n) = u_1(x_1) \cdots u_n(x_n)$$

for any $u_i \in E^*$, $x_i \in E$. If the module E is free (cf. **Free module**), γ is injective, while if E is also finitely generated, γ is bijective. In particular, the n-linear forms on a finite-dimensional vector space over a field are identified with n-times covariant tensors.

For any forms $u \in L_n(E, A)$, $v \in L_m(E, A)$ one can define the tensor product $u \otimes v \in L_{n+m}(E, A)$ via the formula

$$u \otimes v(x_1, \ldots, x_{n+m}) = u(x_1, \ldots, x_n)v(x_{n+1}, \ldots, x_{n+m}).$$

For symmetrized multilinear forms (cf. **Multilinear mapping**), a symmetrical product is also defined:

$$(\sigma_n u) \vee (\sigma_m v) = \sigma_{n+m}(u \otimes v),$$

while for skew-symmetrized multilinear forms there is an exterior product

$$(\alpha_n u) \wedge (\alpha_m v) = \alpha_{n+m}(u \otimes v).$$

These operations are extended to the module $L_*(E, A) = \oplus_{n=0}^{\infty} L(E, A)$, where $L_0(E, A) = A$, $L_1(E, A) = E^*$, to the module of symmetrized forms $L_\sigma(E, A) = \oplus_{n=0}^{\infty} \sigma_n L_n(E, A)$ and to the module of skew-symmetrized forms $L_\alpha(E, A) = \oplus_{n=0}^{\infty} \alpha_n L_n(E, A)$ respectively, which transforms them into associative algebras with a unit. If E is a finitely-generated free module, then the mappings γ_n define an isomorphism of the **tensor algebra** $T(E^*)$ on $L_*(E, A)$ and the **exterior algebra** $\Lambda(E^*)$ on the algebra $L_\alpha(E, A)$, which in that case coincides with the algebra of alternating forms. If A is a field of characteristic 0, then there is also an isomorphism of the symmetric algebra $S(E^*)$ on the algebra $L_\sigma(E, A)$ of symmetric forms.

Any multilinear form $u \in L_n(E, A)$ corresponds to a function $\omega_n(u) : E \to A$, given by the formula

$$\omega_n(u)(x) = u(x, \ldots, x), \quad x \in E.$$

Functions of the form $\omega_n(u)$ are called *forms of degree n* on E; if E is a free module, then in coordinates relative to an arbitrary basis they are given by homogeneous polynomials of degree n. In the case $n=2$ ($n=3$) one obtains quadratic (cubic) forms on E (cf. **Quadratic form; Cubic form**). The form $F=\omega(u)$ completely determines the symmetrization $\sigma_n u$ of a form $u \in L_n(E, A)$:

$$\sigma_n u(x_1, \ldots, x_n) =$$

$$= \sum_{r=1}^{n} (-1)^{n-r} \sum_{i_1 < \cdots < i_r} F(x_{i_1} + \cdots + x_{i_r}).$$

In particular, for $n=2$,

$$(\sigma_2 u)(x, y) = F(x+y) - F(x) - F(y).$$

The mappings γ_n and ω_n define a homomorphism of the algebra $S(E^*)$ on the algebra of all polynomial functions (cf. **Polynomial function**) $P(E)$, which is an isomorphism if E is a finitely-generated free module over an infinite integral domain A.

References

[1] BOURBAKI, N.: *Elements of mathematics. Algebra: Algebraic structures. Linear algebra*, 1, Addison-Wesley, 1974, Chapt. 1; 2 (translated from the French).

[2] BOURBAKI, N.: *Elements of mathematics. Algebra: Modules. Rings. Forms*, 2, Addison-Wesley, 1975, Chapt. 4; 5; 6 (translated from the French).

[3] LANG, S.: *Algebra*, Addison-Wesley, 1984.

A.L. Onishchik

AMS 1980 Subject Classification: 15A69, 15A63, 10C10

MULTILINEAR MAPPING, *n-linear mapping, multilinear operator* - A mapping f of the direct product $\prod_{i=1}^{n} E_i$ of unitary modules E_i (cf. **Unitary module**) over a commutative associative ring A with a unit into a certain A-module F which is linear in each argument, i.e. which satisfies the condition

$$f(x_1, \ldots, x_{i-1}, ay + bz, x_{i+1}, \ldots, x_n) =$$

$$= af(x_1, \ldots, x_{i-1}, y, x_{i+1}, \ldots, x_n) +$$

$$+ bf(x_i, \ldots, x_{i-1}, z, x_{i+1}, \ldots, x_n)$$

$$(a, b \in A; \ y, z \in E_i, \ i = 1, \ldots, n).$$

In the case $n=2$ ($n=3$) one speaks of a **bilinear mapping** (respectively, a **trilinear mapping**). Each multilinear mapping

$$f : \prod_{i=1}^{n} E_i \to F$$

defines a unique linear mapping $\bar{f}$ of the tensor product $\otimes_{i=1}^{n} E_i$ into F such that

$$\bar{f}(x_1 \otimes \cdots \otimes x_n) = f(x_1, \ldots, x_n), \ x_i \in E_i,$$

where the correspondence $f \mapsto \bar{f}$ is a bijection of the set of multilinear mappings $\prod_{i=1}^{n} E_i \to F$ into the set of all linear mappings $\otimes_{i=1}^{n} E_i \to F$. The multilinear mappings $\prod_{i=1}^{n} E_i \to F$ naturally form an A-module.

On the A-module $L_n(E, F)$ of all n-linear mappings $E^n \to F$ there acts the **symmetric group** S_n:

$$(sf)(x_1, \ldots, x_n) = f(x_{s(1)}, \ldots, x_{s(n)}),$$

where $s \in S_n$, $f \in L_n(E, F)$, $x_i \in E$. A multilinear mapping f is called *symmetric* if $sf=f$ for all $s \in S_n$, and *skew-symmetric* if $sf = \epsilon(s)f$, where $\epsilon(s) = \pm 1$ in accordance with the sign of the permutation s. A multilinear mapping is called *sign-varying* (or *alternating*) if $f(x_1, \ldots, x_n)=0$ when $x_i = x_j$ for some $i \neq j$. Any alternating multilinear mapping is skew-symmetric, while if in F the equation $2y=0$ has the unique solution $y=0$ the converse also holds. The symmetric multilinear mappings form a submodule in $L_n(E, F)$ that is naturally isomorphic to the module of linear mappings $L(S^n E, F)$, where $S^n E$ is the n-th symmetric power of E (see **Symmetric algebra**). The alternating multilinear mappings form a submodule that is naturally isomorphic to $L(\Lambda^n E, F)$, where $\Lambda^n E$ is the n-th exterior power of the module E (see **Exterior algebra**). The multilinear mapping $\alpha_n f = \sum_{s \in S_n} sf$ is called the *symmetrized multilinear mapping defined by f*, while the multilinear mapping $\sigma_n f = \sum_{s \in S_n} \epsilon(s) sf$ is called the *skew-symmetrized mapping defined by f*. Symmetrized (skew-symmetrized) multilinear mappings are symmetric (respectively, alternating), and if in F the equation $n!y=c$ has a unique solution for each $c \in F$, then the converse is true. A sufficient condition for any alternating multilinear mapping to be a skew-symmetrization is that the module E is free (cf. **Free module**). For references see **Multilinear form**.

A.L. Onishchik

AMS 1980 Subject Classification: 15A63, 15A69

MULTIMODAL DISTRIBUTION, *multipeaked distribution* - A probability distribution with several modes (cf. **Mode**) or, what is the same, with a density (cf. **Density of a probability distribution**) having several relative maxima corresponding to these modes. In contrast to unimodal distributions (cf. **Unimodal distribution**), multimodal distributions are comparatively rare (in practice) and, as a rule, arise as mixtures of probability distributions.

A.V. Prokhorov

AMS 1980 Subject Classification: 62EXX, 60EXX

MULTINOMIAL COEFFICIENT - The coefficient

$$\frac{n!}{n_1! \cdots n_m!}, \ n_1 + \cdots + n_m = n,$$

of $x_1^{n_1} \cdots x_m^{n_m}$ in the expansion of the polynomial $(x_1 + \cdots + x_m)^n$. In combinatorics, the multinomial coefficient expresses the following: a) the number of possible permutations of n elements of which n_1 are of one form, n_2 of another form, $\ldots, n_m$ of the m-th form; b) the number of ways of locating n different elements in m different cells in which cell i contains n_i elements, $i=1,\ldots,m$, without taking the order of the elements in any cell into account.

Particular cases of multinomial coefficients are the **binomial coefficients**.

References

[1] HALL, M.: *Combinatorial theory*, Wiley, 1986.
[2] RIORDAN, J.: *An introduction to combinatorial analysis*, Wiley, 1967.

S.A. Rukova

AMS 1980 Subject Classification: 05A10

MULTINOMIAL DISTRIBUTION, *polynomial distribution* - The joint distribution of random variables $X_1,\ldots,X_k$ that is defined for any set of non-negative integers $n_1,\ldots,n_k$ satisfying the condition $n_1+\cdots+n_k=n$, $n_j=0,\ldots,n$, $j=1,\ldots,k$, by the formula

$$P\{X_1=n_1,\ldots,X_k=n_k\} = \frac{n!}{n_1!\cdots n_k!}p_1^{n_1}\cdots p_k^{n_k}, \quad (*)$$

where $n,p_1,\ldots,p_k$ $(p_j\geqslant 0, \sum p_j=1)$ are the parameters of the distribution. A multinomial distribution is a multivariate discrete distribution, namely the distribution for the random vector $(X_1,\ldots,X_k)$ with $X_1+\cdots+X_k=n$ (this distribution is in essence $(k-1)$-dimensional, since it is degenerate in the Euclidean space of k dimensions). A multinomial distribution is a natural generalization of a **binomial distribution** and coincides with the latter for $k=2$. The name of the distribution is given because the probability (*) is the general term in the expansion of the multinomial $(p_1+\cdots+p_k)^n$. The multinomial distribution appears in the following probability scheme. Each of the random variables X_i is the number of occurrences of one of the mutually exclusive events A_j, $j=1,\ldots,k$, in repeated independent trials. If in each trial the probability of event A_j is p_j, $j=1,\ldots,k$, then the probability (*) is equal to the probability that in n trials the events $A_1,\ldots,A_k$ will appear $n_1,\ldots,n_k$ times, respectively. Each of the random variables X_j has a binomial distribution with mathematical expectation np_j and variance $np_j(1-p_j)$.

The random vector $(X_1,\ldots,X_k)$ has mathematical expectation $(np_1,\ldots,np_k)$ and covariance matrix $B=\|b_{ij}\|$, where

$$b_{ij} = \begin{cases} np_i(1-p_i), & i=j, \\ -np_ip_j, & i\neq j, \end{cases} \quad i,j=1,\ldots,k$$

(the rank of the matrix B is $k-1$ because $\sum_{i=1}^k n_i=n$). The characteristic function of a multinomial distribution is

$$f(t_1,\ldots,t_k) = \left[p_1e^{it_1}+\cdots+p_ke^{it_k}\right]^n.$$

For $n\to\infty$, the distribution of the vector $(Y_1,\ldots,Y_k)$ with normalized components

$$Y_i = \frac{X_i-np_i}{\sqrt{np_i(1-p_i)}}$$

tends to a certain multivariate **normal distribution**, while the distribution of the sum

$$\sum_{i=1}^k (1-p_i)Y_i^2$$

(which is used in mathematical statistics to construct the 'chi-squared' test) tends to the 'chi-squared' distribution with $k-1$ degrees of freedom.

References

[1] CRAMÉR, H.: *Mathematical methods of statistics*, Princeton Univ. Press, 1946.

A.V. Prokhorov

Editorial comments.

References

[A1] JOHNSON, N.L. and KOTZ, S.: *Discrete distributions*, Wiley, 1969.

AMS 1980 Subject Classification: 60E99, 62E99

MULTIPLE *of a natural number a* - A **natural number** that is divisible by a without remainder (cf. **Division**). A number n divisible by each of the numbers $a, b, \ldots, m$ is called a *common multiple* of these numbers. Among all common multiples of two or more numbers, one (distinct from zero) is the smallest (the *lowest common multiple*) and the others are multiples of the lowest common multiple. If the **greatest common divisor** d of two numbers a and b is known, the lowest common multiple m is found from the formula $m=ab/d$.

Editorial comments.

References

[A1] VINOGRADOV, I.M.: *Elements of number theory*, Dover, reprint, 1954 (translated from the Russian).

AMS 1980 Subject Classification: 10A05

MULTIPLE COMPARISON - The problem of testing hypotheses with respect to the values of scalar products $\theta^T\cdot c$ of a vector $\theta=(\theta_1,\ldots,\theta_k)^T$, the coordinates of which are unknown parameters, with a number of given vectors $c=(c_1,\ldots,c_k)^T$. In statistical research the multiple comparison problem often arises in **dispersion analysis** where, as a rule, the vectors c are chosen so that $c_1+\cdots+c_k=0$, and the scalar product $\theta^T\cdot c$ itself, in this case, is called a **contrast**. On the assumption that $\theta_1,\ldots,\theta_k$ are unknown mathematical expectations of one-dimensional normal laws, J.W. Tukey and H. Scheffé proposed the *T-method* and the *S-method*, respectively, for the simultaneous estimation of contrasts, which are the fundamental methods in the problem of constructing confidence intervals for contrasts.

References

[1] SCHEFFÉ, H.: *The analysis of variance*, Wiley, 1959.
[2] KENDALL, M. and STUART, A.: *The advanced theory of statistics. Design and analysis, and time series*, 3, Griffin, 1983.

M.S. Nikulin

Editorial comments.

References

[A1] MILLER, R.: *Simultaneous statistical inference*, McGraw-Hill, 1966.

AMS 1980 Subject Classification: 62J10

MULTIPLE-CORRELATION COEFFICIENT - A measure of the linear dependence between one random variable and a certain collection of random variables. More precisely, if $(X_1, \ldots, X_k)$ is a random vector with values in $\mathbf{R}^k$, then the multiple-correlation coefficient between X_1 and $X_2, \ldots, X_k$ is defined as the usual **correlation coefficient** between X_1 and its best linear approximation $\mathsf{E}(X_1 \mid X_2, \ldots, X_k)$ relative to $X_2, \ldots, X_k$, i.e. as its **regression** relative to $X_2, \ldots, X_k$. The multiple-correlation coefficient has the property that if $\mathsf{E}X_1 = \cdots = \mathsf{E}X_k = 0$ and if

$$X_1^* = \beta_2 X_2 + \cdots + \beta_k X_k$$

is the regression of X_1 relative to $X_2, \ldots, X_k$, then among all linear combinations of $X_2, \ldots, X_k$ the variable X_1^* has largest correlation with X_1. In this sense the multiple-correlation coefficient is a special case of the canonical correlation coefficient (cf. **Canonical correlation coefficients**). For $k=2$ the multiple-correlation coefficient is the absolute value of the usual correlation coefficient ρ_{12} between X_1 and X_2. The multiple-correlation coefficient between X_1 and $X_2, \ldots, X_k$ is denoted by $\rho_{1 \cdot (2 \cdots k)}$ and is expressed in terms of the entries of the correlation matrix $R = \| \rho_{ij} \|$, $i, j = 1, \ldots, k$, by

$$\rho_{1 \cdot (2 \cdots k)}^2 = 1 - \frac{|R|}{R_{11}},$$

where $|R|$ is the determinant of R and R_{11} is the **cofactor** of $\rho_{11} = 1$; here $0 \leqslant \rho_{1 \cdot (2 \cdots k)} \leqslant 1$. If $\rho_{1 \cdot (2 \cdots k)} = 1$, then, with probability 1, X_1 is equal to a linear combination of $X_2, \ldots, X_k$, that is, the **joint distribution** of $X_1, \ldots, X_k$ is concentrated on a hyperplane in $\mathbf{R}^k$. On the other hand, $\rho_{1 \cdot (2 \cdots k)} = 0$ if and only if $\rho_{12} = \cdots = \rho_{1k} = 0$, that is, if X_1 is not correlated with any of $X_2, \ldots, X_k$. To calculate the multiple-correlation coefficient one can use the formula

$$\rho_{1 \cdot (2 \cdots k)}^2 = 1 - \frac{\sigma_{1 \cdot (2 \cdots k)}^2}{\sigma_1^2},$$

where σ_1^2 is the variance of X_1 and

$$\sigma_{1 \cdot (2 \cdots k)}^2 = \mathsf{E}[X_1 - (\beta_2 X_2 + \cdots + \beta_k X_k)]^2$$

is the variance of X_1 with respect to the regression.

The sample analogue of the multiple-correlation coefficient $\rho_{1 \cdot (2 \cdots k)}$ is

$$r_{1 \cdot (2 \cdots k)} = \sqrt{1 - \frac{s_{1 \cdot (2 \cdots k)}^2}{s_1^2}},$$

where $s_{1 \cdot (2 \cdots k)}^2$ and s_1^2 are estimators of $\sigma_{1 \cdot (2 \cdots k)}^2$ and σ_1^2 based on a sample of size n. To test the hypothesis of no relationship, the sampling distribution of $r_{1 \cdot (2 \cdots k)}$ is used. Given that the sample is taken from a multivariate normal distribution, the variable $r_{1 \cdot (2 \cdots k)}^2$ has the beta-distribution with parameters $((k-1)/2, (n-k)/2)$ if $\rho_{1 \cdot (2 \cdots k)} = 0$; if $\rho_{1 \cdot (2 \cdots k)} \neq 0$, then the distribution of $r_{1 \cdot (2 \cdots k)}^2$ is known, but is somewhat complicated.

References

[1] CRAMÉR, H.: *Mathematical methods of statistics*, Princeton Univ. Press, 1946.
[2] KENDALL, M. and STUART, A.: *The advanced theory of statistics. Inference and relationships*, 2, Griffin, 1979.

A.V. Prokhorov

Editorial comments. For the distribution of $r_{1 \cdot (2 \cdots k)}^2$ if $\rho_{1 \cdot (2 \cdots k)} \neq 0$ see [A2], Chapt. 10.

References

[A1] ANDERSON, T.W.: *An introduction to multivariate statistical analysis*, Wiley, 1958.
[A2] EATON, M.L.: *Multivariate statistics: A vector space approach*, Wiley, 1983.
[A3] MUIRHEAD, R.J.: *Aspects of multivariate statistical theory*, Wiley, 1982.

AMS 1980 Subject Classification: 62H20

MULTIPLE INTEGRAL - A definite integral of a function of several variables. There are several different concepts of a multiple integral (Riemann integral, Lebesgue integral, Lebesgue—Stieltjes integral, etc.).

The multiple Riemann integral is based on the concept of a **Jordan measure** μ. Let E be a Jordan-measurable set in the n-dimensional Euclidean space $\mathbf{R}^n$, let μ_n be the n-dimensional Jordan measure and let $\tau = \{E_i\}_{i=1}^k$ be a partition of E, i.e. a system of Jordan-measurable sets E_i such that $\bigcup_{i=1}^k E_i = E$ and $\mu_n(E_i \cap E_j) = 0$, $i \neq j$, $i, j = 1, \ldots, n$. The quantity

$$\delta_\tau = \max_{i=1, \ldots, k} d(E_i),$$

where $d(E_i)$ is the diameter of E_i, is called the *mesh of the partition* τ. If $f(x)$, $x = (x_1, \ldots, x_n)$, is a function defined on E, then any sum of the type

$$\sigma_\tau = \sigma_\tau(f; \xi^{(1)}, \ldots, \xi^{(k)}) = \sum_{i=1}^k f(\xi^{(i)}) \mu_n(E_i),$$

$$\xi^{(i)} \in E_i \in \tau,$$

is called a *Riemann integral sum of the function f*. If f has the property that $\lim_{\delta_\tau \to 0} \sigma_\tau$ exists, independently of the specific sequence of partitions, then this limit is called the *n-tuple Riemann integral of f* over E, and is denoted by

$$\int_E f(x)\,dx$$

or

$$\int\limits_E \cdots \int f(x_1,\ldots,x_n)\,dx_1 \cdots dx_n.$$

The function f itself is then said to be *Riemann integrable* or, more briefly, *R-integrable*.

When $n=1$, the set E over which the integration takes place is usually an interval and τ is a partition consisting exclusively of intervals (see **Riemann integral**). Hence both the set over which the integration is performed and the elements of the partition are Jordan-measurable sets of a very special form — intervals. That is why not all the properties of functions which are R-integrable on an interval are valid for functions which are R-integrable on arbitrary Jordan-measurable sets. For example, since any function defined on a set of Jordan measure zero is R-integrable on that set, it follows that R-integrable functions need not be bounded. This is impossible for R-integrable functions on intervals. If one wishes R-integrability of a function on some set to imply that the function is bounded, certain additional conditions must be imposed on the set; for example, one might require that the set have arbitrarily fine partitions all elements of which have positive Jordan measure. The class defined by this condition includes all Jordan-measurable open sets and their closures, in particular all Jordan-measurable open domains and their closures. These are precisely the sets for which multiple Riemann integrals are most often used. When $n=2$ $(n=3)$, a multiple integral is called a *double (triple) integral* (cf. also **Double integral**).

Since a multiple Riemann integral can be evaluated only over Jordan-measurable sets (if $n=2$ such a set is also called squarable; if $n=3$ it is also called cubable), double (triple) Riemann integrals are considered only on sets (usually domains or closures of domains) with boundaries of Jordan area (volume) zero.

The Riemann integral of a bounded function of n variables $(n \geq 1)$ possesses the usual properties of an **integral** (linearity, additivity with respect to the set of integration, preservation of non-strict inequalities under integration, integrability of the product of integrable functions, etc.).

A multiple Riemann integral can be reduced to a **repeated integral**. Let $x=(x',x'')\in\mathbf{R}^n$,

$$x' = (x_1,\ldots,x_m) \in \mathbf{R}^m,$$

$$x'' = (x_{m+1},\ldots,x_n) \in \mathbf{R}^{n-m}, \quad E \subset \mathbf{R}^n,$$

where E is a Jordan-measurable set in $\mathbf{R}^n$, $E(x_0')=E\cap\{x'=x_0'\}$ is the intersection of E with the $(n-m)$-dimensional hyperplane $x'=x_0'$, $E_{x''}$ is the projection of E on the hyperplane $\mathbf{R}^m=\{x:\,x''=0\}$, with $E(x')$ and $E_{x''}$ measurable in the sense of the $(n-m)$-dimensional and m-dimensional Jordan measure, respectively. If f is an R-integrable function on E and if for all $x'\in E_{x''}$ the $(n-m)$-multiple integrals of the restrictions of f to the set $E(x')$ exist, then the repeated integral

$$\int\limits_{E_{x''}} dx' \int\limits_{E(x')} f(x',x'')\,dx'',$$

where the outer integral is an m-tuple Riemann integral, exists, and

$$\int\limits_E f(x)\,dx = \int\limits_E \int f(x',x'')\,dx'\,dx'' =$$

$$= \int\limits_{E_{x''}} dx' \int\limits_{E(x')} f(x',x'')\,dx''.$$

For $n=3$ this implies the following formulas: 1) If $E\subset\mathbf{R}^3_{xyz}$, if E_{xy} is the projection of E on the xy-plane, and if $\phi(x,y)$ and $\psi(x,y)$, $x,y\in E_{xy}$, are functions with graphs bounded by the set E in the z-direction, i.e.

$$E = \{(x,y,z):\,(x,y)\in E_{xy},\,\phi(x,y)\leq z \leq \psi(x,y)\},$$

then

$$\int\int\limits_E\int f(x,y,z)\,dx\,dy\,dz =$$

$$= \int\limits_{E_{xy}} dx\,dy \int\limits_{\phi(x,y)}^{\psi(x,y)} f(x,y,z)\,dz.$$

2) Let the projection of E on the x-axis be an interval $[a,b]$, and let $E(x)$ be the intersection of E with the plane through the point x parallel to the yz-plane; then

$$\int\int\limits_E\int f(x,y,z)\,dx\,dy\,dz = \int\limits_a^b dx \int\int\limits_{E(x)} f(x,y,z)\,dy\,dz.$$

In case G is a Jordan-measurable domain in the space $\mathbf{R}^n_x$ and ϕ is also continuously differentiable on the closure $\overline{G}$ of G into $\mathbf{R}^n$, one has the following formula for substitution of variables in the integral of a function f which is integrable on $\Gamma=\phi(G)$:

$$\int\limits_{\phi(G)} f(x)\,dx = \int\limits_G f(\phi(t))\,|J(t)|\,dt, \tag{1}$$

where $J(t)$ is the **Jacobian** of the mapping ϕ.

The geometrical meaning of the multiple Riemann integral of a function of n variables is connected with the concept of the $(n+1)$-dimensional Jordan measure μ_{n+1}: If f is integrable on a set $E\subset\mathbf{R}^n_x$, $f(x)\geq 0$ on E and if

$$A = \{(x,y):\,x\in E,\,0\leq y \leq f(x)\} \subset \mathbf{R}^{n+1}_{xy},$$

then

$$\int\limits_E f(x)\,dx = \mu_{n+1}(A). \tag{2}$$

A *multiple Lebesgue integral* is the **Lebesgue integral** of a function of several variables; the definition is based on the concept of the **Lebesgue measure** in the n-dimensional Euclidean space. A multiple Lebesgue integral can be reduced to a repeated integral (see **Fubini theorem**). For continuously-differentiable one-to-one mappings of domains, formula (1) for substitution of variables holds, as well as formula (2), which conveys the geometrical meaning of the multiple Lebesgue integral, with μ_{n+1} now being interpreted as the $(n+1)$-dimensional Lebesgue measure.

The concept of a multiple integral carries over to functions integrable on a subset A of the product $X \times Y$ of two sets X and Y, on each of which a σ-finite complete non-negative measure, μ_x and μ_y, respectively, has been given; in this situation integration over A involves the measure μ which is the product of μ_x and μ_y.

For functions of several variables one also has a concept of an improper multiple integral (see **Improper integral**). The concept of a multiple integral is also applied to indefinite integrals of functions of several variables: An indefinite multiple integral is a set function

$$F(E) = \int\int_E f(x)\,dx,$$

where E is a measurable set. For example, if f is Lebesgue integrable on some set, then it is the **symmetric derivative** of its indefinite integral $F(E)$ almost-everywhere on that set. In this sense (in analogy to the case of functions of one variable), the evaluation of an indefinite integral is the operation inverse to differentiation of set functions.

References

[1] IL'IN, V.A. and POZNYAK, E.G.: *Fundamentals of mathematical analysis*, 2, Mir, 1982 (translated from the Russian).
[2] KOLMOGOROV, A.N. and FOMIN, S.V.: *Elements of the theory of functions and functional analysis*, 1-2, Graylock, 1957-1961 (translated from the Russian).
[3] NIKOL'SKIĬ, S.M.: *A course of mathematical analysis*, 2, Mir, 1977 (translated from the Russian).

L.D. Kudryavtsev

Editorial comments.

References

[A1] APOSTOL, T.M.: *Mathematical analysis*, Addison-Wesley, 1957.
[A2] BARTLE, R.G.: *The elements of real analysis*, Wiley, 1976.
[A3] SMITH, K.T.: *Primer of modern analysis*, Bogden & Quiley, 1971.
[A4] STROMBERG, K.R.: *Introduction to classical real analysis*, Wadsworth, 1981.

AMS 1980 Subject Classification: 26B15

MULTIPLE POINT of a planar curve $F(x, y)=0$ - A singular point at which the partial derivatives of order up to and including n vanish, but where at least one partial derivative of order $n+1$ does not vanish. For example, if $F(x_0, y_0)=0$, $F'_x(x_0, y_0)=0$, $F''_{yy}(x_0, y_0)$ does not vanish, the multiple point $M(x_0, y_0)$ is called a **double point**; if the first and second partial derivatives vanish at $M(x_0, y_0)$, but at least one third derivative does not, the multiple point is called a *triple point*; etc.

A.B. Ivanov

Editorial comments.

References

[A1] COOLIDGE, J.L.: *Algebraic plane curves*, Dover, reprint, 1959.
[A2] HILBERT, D. and COHN-VOSSEN, S.: *Geometry and the imagination*, Chelsea, reprint, 1952, p. 173 (translated from the German).
[A3] GRIFFITHS, P. and HARRIS, S.: *Principles of algebraic geometry*, Wiley, 1978.
[A4] FULTON, W.: *Algebraic curves*, Benjamin, 1969.

AMS 1980 Subject Classification: 14H20, 53A04

MULTIPLE RECURSION - A form of **recursion** using several variables simultaneously. The set of values of these variables is ordered lexicographically. This definition subsumes numerous concrete recursive descriptions. If in such a description the unknown function is not substituted into itself, then it results in a **primitive recursion**. In general, multiple recursion leads outside the limits of the set of primitive recursive functions, since by a double recursion (performed with respect to two variables) it is possible to construct a function which is universal for the primitive recursive functions (similarly, for k-recursive functions there is a $(k+1)$-multiple universal function); cf. **Universal function**. All possible forms of multiple recursion can be reduced to the following *normal form*:

$$\phi(n_1, \ldots, n_k) = 0 \quad \text{if } n_1, \ldots, n_k = 0,$$

$$\phi(n_1+1, \ldots, n_k+1) = \beta(n_1, \ldots, n_k, \phi_1, \ldots, \phi_k),$$

where

$$\phi_i = \phi(n_1+1, \ldots, n_{i-1}+1, n_i,$$
$$\gamma^{(i)}(n_1, \ldots, n_k, \phi(n_1+1, \ldots, n_{k-1}+1, n_k)), \ldots,$$
$$\ldots, \gamma^{(i)}_{k-1}(n_1, \ldots, n_k, \phi(n_1+1, \ldots, n_{k-1}+1, n_k))).$$

References

[1] PETER, R.: *Recursive functions*, Acad. Press, 1967 (translated from the German).

N.V. Belyakin

Editorial comments. It is more common to speak of *simultaneous recursion*, rather than multiple recursion; cf., e.g., [A1].

References

[A1] KLEENE, S.C.: *Introduction to metamathematics*, North-Holland, 1951.
[A2] ROGERS, JR., H.: *Theory of recursive functions and effective computability*, McGraw-Hill, 1967.

AMS 1980 Subject Classification: 03D20

MULTIPLE SEQUENCE, *k-tuple sequence*, of elements of a given set X - A mapping of the k-th power $\mathbf{N}^k$ of the set of natural numbers $\mathbf{N}$ into the set X. An *element* (or *term*) of a multiple sequence $f: \mathbf{N}^k \to X$ is an ordered set of $k+1$ elements $(n_1, \ldots, n_k, x)$, where $x = f(n_1, \ldots, n_k)$, $x \in X$, $(n_1, \ldots, n_k) \in \mathbf{N}^k$, i.e. $n_j \in \mathbf{N}$ $(j=1, \ldots, k)$. The element is also denoted by $x_{n_1 \cdots n_k}$.

L.D. Kudryavtsev

AMS 1980 Subject Classification: 04A99, 10A99, 40-01

MULTIPLE SERIES, *s-tuple series* - An expression of the form

$$\sum_{m,n,\ldots,p=1}^{\infty} u_{m,n,\ldots,p},$$

consisting of the terms from a table $\| u_{m,n,\ldots,p} \|$. Each term of the table is indexed by $s \geqslant 2$ indices $m, n, \ldots, p$, which run independently of each other over the set of natural numbers. The theory of multiple series is analogous to the theory of **double series**. See also **Absolutely convergent series**.

References

[1] FICHTENHOLZ, G.M.: *Differential- und Integralrechnung, 2,* Deutsch. Verlag Wissenschaft., 1964.

E.G. Sobolevskaya

AMS 1980 Subject Classification: 40B05

MULTIPLICATION *of numbers* - One of the basic arithmetic operations. Multiplication consists in assigning to two numbers a, b (called the *factors*) a third number c (called the *product*). Multiplication is denoted by the sign $\times$ or $\cdot$; in notations using letters the sign is, as a rule, omitted.

Multiplication of positive integers is defined in the following way by means of addition: the product of a number a by a number b is the number c equal to the sum of b summands each of which is equal to a; thus,

$$ab = a + \cdots + a \quad (b \text{ times}).$$

The number a is called the *multiplicand* and b the *multiplier*. Multiplication of two positive rational numbers m/n and p/q is defined by the equation

$$\frac{m}{n} \cdot \frac{p}{q} = \frac{mp}{nq}$$

(cf. **Fraction**). The product of two negative fractions is positive, and that of a positive and a negative fraction is negative, where the modulus of the product in both cases is equal to the product of the moduli of the factors. The product of irrational numbers is defined as the limit of the products of rational approximations of them. Multiplication of two complex numbers $\alpha = a + bi$ and $\beta = c + di$ is given by the formula

$$\alpha\beta = (a + bi)(c + di) = ac - bd + (ad + bc)i,$$

or, in trigonometric form $(\alpha = r_1(\cos\phi_1 + i\sin\phi_1), \beta = r_2(\cos\phi_2 + i\sin\phi_2))$, by

$$\alpha\beta = r_1 r_2(\cos(\phi_1 + \phi_2) + i\sin(\phi_1 + \phi_2)).$$

Multiplication of numbers is commutative, associative and distributive on the left and right relative to addition (cf. **Commutativity; Associativity; Distributivity**). Further, $a \cdot 0 = 0$, $a \cdot 1 = a$.

In general algebra, multiplication may be any **algebraic operation** (n-ary, $n > 2$); most frequently a binary operation (cf. **Groupoid**). In certain cases this operation is a generalization of the usual multiplication of numbers. E.g. multiplication of quaternions, multiplication of matrices and multiplication of transformations. However, properties of the multiplication of numbers (e.g. commutativity) may be lost in these cases.

O.A. Ivanova

AMS 1980 Subject Classification: 10-XX, 00A25

MULTIPLICATIVE ARITHMETIC FUNCTION - An **arithmetic function** of one argument, $f(m)$, satisfying the condition

$$f(mn) = f(m) \cdot f(n) \qquad (*)$$

for any pair of coprime integers m, n. It is usually assumed that $f(m)$ is not identically zero (which is equivalent to the condition $f(1) = 1$). A multiplicative arithmetic function is called *strongly multiplicative* if $f(p^\alpha) = f(p)$ for all prime numbers p and all natural numbers α. If (*) holds for any two numbers m, n, and not just for coprime numbers, then f is called *totally multiplicative*; in this case $f(p^\alpha) = [f(p)]^\alpha$.

Examples of multiplicative arithmetic functions. The function $\tau(m)$, the number of natural divisors of a natural number m; the function $\sigma(m)$, the sum of the natural divisors of the natural number m; the **Euler function** $\phi(m)$; and the **Mobius function** $\mu(m)$. The function $\phi(m)/m$ is a strongly-multiplicative arithmetic function, a power function $f(m) = m^s$ is a totally-multiplicative arithmetic function.

I.P. Kubilyus

Editorial comments. The convolution product

$$(f \star g)(n) = \sum_{d \mid n} f(d) g\left[\frac{n}{d}\right]$$

yields a **group** structure on the multiplicative functions. The unit element is given by the function e, where $e(1) = 1$ and $e(n) = 0$ for all $n > 1$. Another standard multiplicative function is the constant function E ($E(n) = 1$ for all $n \in \mathbf{N}$) and its inverse μ, the **Möbius function**. Note that $\phi = \mu \star N_1$, where $N_1(n) = n$ for all n, and that $\tau = E \star E$, $\sigma = E \star N_1$.

Formally, the **Dirichlet series** of a multiplicative function f has an **Euler product**:

$$\sum_{n=1}^{\infty} \frac{f(n)}{n^s} = \prod_{p} \left[1 + \frac{f(p)}{p^s} + \cdots + \frac{f(p^k)}{p^{ks}} + \cdots\right],$$

whose form simplifies considerably if f is strongly or totally multiplicative.

References

[A1] HARDY, G.H. and WRIGHT, E.M.: *An introduction to the theory of numbers*, Clarendon Press, 1960, Chapts. XVI-XVII.

AMS 1980 Subject Classification: 10A20

MULTIPLICATIVE ERGODIC THEOREM, *Oseledets's multiplicative ergodic theorem, Oseledec's multiplicative ergodic theorem*

Editorial comments. Consider a linear homogeneous

system of differential equations

$$\dot{x} = A(t)x, \quad x(0; x_0) = x_0 \in \mathbf{R}^n, \quad t \geq 0. \qquad (A1)$$

The *Lyapunov exponent* of a solution $x(t; x_0)$ of (A1) is defined as

$$\lambda(x_0) = \limsup_{t \to \infty} t^{-1} \log \| x(t; x_0) \|.$$

A more general setting (*Lyapunov exponents for families of system of differential equations*) for discussing Lyapunov exponents and related matters is as follows. Let $\Phi = (\Phi_t)_{t \in \mathbf{R}}$ be a **measurable flow** on a measure space (E, μ). For all $e \in E$, let V_e be an n-dimensional vector space. (Think, for example, of a vector bundle $T \to E$.) A *cocycle* $C(t, e)$ associated with the flow Φ is a measurable function on $\mathbf{R} \times E$ that assigns to (t, e) an invertible linear mapping $V_e \to V_{\Phi_t(e)}$ such that

$$C(t+s, e) = C(t, \Phi_s(e))C(s, e). \qquad (A2)$$

I.e. if the collection of vector spaces V_e is viewed as an n-dimensional vector bundle over E, then $C(t, \cdot)$ defines an isomorphism of vector bundles $\tilde{\Phi}_t$ over Φ_t,

$$
\begin{array}{ccc}
& \tilde{\Phi}_t & \\
V & \to & V \\
\downarrow & & \downarrow \\
E & \to & E \\
& \Phi_t &
\end{array}
$$

and condition (A2) simply says that $\tilde{\Phi}_{t+s} = \tilde{\Phi}_t \circ \tilde{\Phi}_s$. So $\tilde{\Phi}$ is a flow on V that lifts Φ. $\tilde{\Phi}$ is sometimes called the *skew product flow* defined by Φ and C. This set-up is sufficiently general to discuss Lyapunov exponents for non-linear flows, and even stochastic non-linear flows and such things as products of random matrices. If $E = \{e\}$, $\Phi_t = \mathrm{id}$, the classical situation (A1) reappears. Let $\dot{x} = f(x)$ be a differential equation on a manifold M. Take $V = TM$, the tangent bundle over M. Let Φ_t be the flow on M defined by $\dot{x} = f(x)$. The associated cocycle is defined by the differential $d\Phi_t$ of Φ_t,

$$C(t, m) = d\Phi_t(m): T_m M \to T_{\Phi_t(m)} M.$$

For a skew product flow $\tilde{\Phi}$ on V the Lyapunov exponent at $e \in E$ in the direction $v \in V_e$ is defined by

$$\lambda(e, v) = \limsup_{t \to \infty} t^{-1} \log \| C(t, e)v \|.$$

The multiplicative ergodic theorem of V.I. Oseledets [A1] now is as follows. Let $\tilde{\Phi}$ be a skew product flow and assume that there is an invariant probability measure ρ on (E, μ) for Φ, i.e. $\Phi_t \rho = \rho$ for all $t \in \mathbf{R}$. Suppose, moreover, that

$$\int_E \sup_{-1 < t < 1} \log^+ \| C^{\pm 1}(t, e) \| \, d\rho < \infty.$$

Then there exists a measurable Φ-invariant set $E_0 \subseteq E$ of ρ-measure 1 such that for all $x \in E_0$ there are $l(e)$ numbers $\lambda_e^l < \cdots < \lambda_1^1$, $l(e) \leq d$, and corresponding subspaces $0 \subset W_e^l \subset \cdots \subset W_e^1 = V_e$ of dimensions $d_e^l < \cdots < d_e^1 = d$ such that for all $i = 1, \ldots, l(e)$,

$$\lim_{t \to \infty} t^{-1} \log \| C(t, e)v \| = \lambda_e^i \Leftrightarrow v \in W_e^i \setminus W_e^{i+1}.$$

If moreover ρ is ergodic for Φ_t, i.e. all Φ_t-invariant subsets have ρ-measure 0 or 1, then the $l(e)$, λ_e^i, d_e^i are constants independent of e (or E_0). However, the spaces W_e^i may still depend on $e \in E_0$ (if the bundle V is a trivial bundle so that all the V_e can be identified). The set $\{\lambda_1, \ldots, \lambda_l\}$ is called

the *Lyapunov spectrum* of the flow. For more details and applications cf. [A2], [A3].

References
[A1] OSELEDEC, V.I. [V.I. OSELEDETS]: 'A multiplicative ergodic theorem. Lyapunov characteristic numbers for dynamical systems', *Trans. Moscow Math. Soc.* **19** (1968), 197-231. (*Trudy Moskov. Mat. Obshch.* **19** (1968), 179-210)
[A2] KLIEMANN, W.: 'Analysis of nonlinear stochastic systems', in W. Schuhlen and W. Wedig (eds.): *Analysis and estimation of stochastic mechanical systems*, Springer, 1988, pp. 43-102.
[A3] ARNOLD, L. and WIHSTUTZ, V. (EDS.): *Lyapunov exponents*, Springer, 1986.

AMS 1980 Subject Classification: 28DXX, 58F11

MULTIPLICATIVE GROUP *of a skew-field* - The group of all elements of the given **skew-field** except the zero element and with the operation of multiplication in the skew-field. The multiplicative group of a field is Abelian.

O.A. Ivanova

Editorial comments. The finite multiplicative subgroups of skew-fields of finite non-zero characteristic are cyclic, and this is not the case in characteristic zero. There are only a finite number of even groups and an infinite number of odd groups, and the minimal order is 63. The classification is given in [A1]. There exists a similar problem for proving a kind of *Tits alternative*: Any finite normal subgroup of the multiplicative group of a skew-field contains a free non-cyclic group or is a finitely-solvable group and has an extension to a linear group over a skew-field. Some cases are known, e.g., [A2].

References
[A1] AMITSUR, S.A.: 'Finite subgroups of division rings', *Trans. Amer. Math. Soc.* **80** (1955), 361-396.
[A2] LICHTMAN, A.I.: 'Free subgroups in linear groups over some skew fields', *J. of Algebra* **105** (1987), 1-28.
[A3] SCOTT, W.R.: *Group theory*, Prentice Hall, 1964, Chapt. 14, p. 426.

AMS 1980 Subject Classification: 16A39

MULTIPLICATIVE LATTICE - A **complete lattice** $L = \langle L, \vee, \wedge \rangle$ with an additional commutative and associative binary operation, called multiplication (and denoted by $\cdot$) such that the largest element of the lattice acts as the multiplicative identity and such that

$$a \cdot \left[\bigvee_{\alpha \in J} b_\alpha \right] = \bigvee_{\alpha \in J} a \cdot b_\alpha$$

for any $a, b_\alpha \in L$ and an arbitrary index set J. The theory of multiplicative lattices arose as a result of the application of lattice-theoretic methods in the study of lattices of ideals of commutative rings (see [2]) and therefore the majority of concepts and results have an analogue (or an application) in commutative rings (see [1]).

Let L be a multiplicative lattice and let $a, b \in L$; then one defines $a:b = \vee \{x: x \in L, x \cdot b \leq a\}$. An element $e \in L$ is called $\vee$-principal (respectively, $\wedge$-principal) if

$(a \vee (b \cdot e)){:}e = (a{:}e) \vee b$ (respectively, $(a \wedge (b{:}e)) \cdot e = (a \cdot e) \wedge b$) for any $a, b \in L$; an element which is simultaneously $\vee$-principal and $\wedge$-principal is called *principal*. A *Noether lattice* is a modular multiplicative lattice satisfying the ascending chain condition and in which each element is a union of principal elements (cf. also **Modular lattice**). A complete lattice is called a *module over a multiplicative lattice L* if for any $\lambda \in L$, $a \in M$ a product $\lambda a \in M$ is defined, where

$$(\lambda_1 \lambda_2)a = \lambda_1(\lambda_2 a), \quad 1a = a, \quad 0_L a = 0_M,$$

$$\left[\bigvee_{\alpha \in I} \lambda_\alpha \right] \left[\bigvee_{\beta \in J} a_\beta \right] = \bigvee_{\alpha, \beta} \lambda_\alpha a_\beta$$

(here $\lambda_1, \lambda_2, \lambda_\alpha \in L$, $a, a_\beta \in M$, 1 is the largest element in L and $0_L, 0_M$ are the zeros in L and M, respectively).

The best studied class of multiplicative lattices are the Noether lattices. Here it is possible to distinguish the following directions. 1) Questions of the representation of a Noether lattice as the lattice of ideals of a suitable **Noetherian ring** (it is known that the lattice of ideals of a Noetherian ring is a Noether lattice; however, there are Noether lattices that cannot even be imbedded in the lattice of ideals of a Noetherian ring [3]). 2) The study of Noetherian modules over a multiplicative lattice. 3) The study of concepts and properties from the theory of ideals of Noetherian rings that translate to Noether lattices (the notions of prime and primary elements, of dimension, of a proper maximal element, and of semi-local and local lattices). The distributive regular Noether multiplicative lattices have been described [4] (cf. also **Distributive lattice**; **Regular lattice**). A theory of localization and associated prime elements has been constructed for a class of multiplicative lattices considerably broader than Noether lattices, including lattices of ideals of arbitrary commutative rings.

References

[1] BOURBAKI, N.: *Elements of mathematics. Commutative algebra*, Addison-Wesley, 1972 (translated from the French).
[2] DILWORTH, R.P.: 'Abstract commutative ideal theory', *Pacific J. Math.* **12** (1962), 481-498.
[3] BOGART, K.: 'Nonimbeddable Noether lattices', *Proc. Amer. Math. Soc.* **22**, no. 1 (1969), 129-133.
[4A] BOGART, K.: 'Structure theorems for regular local Noether lattices', *Michigan J. Math.* **15**, no. 2 (1968), 167-176.
[4B] BOGART, K.: 'Distributive Noether lattices', *Michigan J. Math.* **16**, no. 3 (1969), 215-223.
[5] FOFANOVA, T.S.: 'General theory of lattices', in *Ordered Sets and Lattices*, Vol. 3, Saratov, 1975, pp. 22-40 (in Russian).

T.S. Fofanova

Editorial comments. For a fixed $a \in L$, the two functions $x \mapsto a{:}x$, $a \cdot x{:} L \to F$ form a covariant Galois connection. If $a \cdot b = a \wedge b$, then L is a complete *Heyting algebra* with $a{:}b = a \Rightarrow b$.

The study of multiplicative lattices began with a celebrated paper of M. Ward and R.P. Dilworth [A1], who called them *residuated lattices*; this name focuses attention on the *residuation* operation $a{:}b$ characterized by

$$c \leqslant a{:}b \quad \text{if and only if} \quad c \cdot b \leqslant a.$$

The existence of this operation, given the multiplication, is equivalent to the distributive law given in the main article above. Many authors (e.g. [A2]) have sought to dispense with the requirement that the multiplication be commutative (in this case, both left and right distributive laws must be assumed, and there are two different residuation operations) and/or that the top element of the lattice be a unit for multiplication; some have been dropped the associativity of the multiplication (cf. [A3]), but relatively little work has been done on the non-associative case. Recently much interest has been shown in the *quantales* introduced by C.J. Mulvey [A4], where the multiplication is non-commutative but is generally assumed to be idempotent (i.e. to satisfy the identity $a \cdot a = a$) and to have the top element as a one-sided unit. In the commutative case, the assumption of idempotency forces the multiplication to coincide with the lattice-theoretic meet (and forces the underlying lattice to be a frame, cf. **Locale**); thus idempotent quantales may be viewed as a 'non-commutative' generalization of topological spaces, cf. [A5], [A6]. These concepts find application in the representation theory of non-commutative C^*-algebras (see C^*-**algebra**; the lattice of closed right ideals of a C^*-algebra is an example of a quantale).

References

[A1] WARD, M. and DILWORTH, R.P.: 'Residuated lattices', *Trans. Amer. Math. Soc.* **45** (1939), 335-354.
[A2] DILWORTH, R.P.: 'Non-commutative residuated lattices', *Trans. Amer. Math. Soc.* **46** (1939), 426-444.
[A3] BIRKHOFF, G.: *Lattice theory*, Amer. Math. Soc., 1967.
[A4] MULVEY, C.J.: '&', *Suppl. ai Rend. Circ. Mat. Palermo (2)* **12** (1986), 99-104.
[A5] NIEFIELD, S.B. and ROSENTHAL, K.I.: 'Constructing locales from quantales', *Math. Proc. Cambridge Philos. Soc.* **104** (1988), 215-234.
[A6] BORCEUX, F. and VAN DEN BOSSCHE, G.: 'An essay in non-commutative topology', *Topology Appl.* **31** (1989), 203-223.
[A7] BLYTH, T.S. and JANOWITZ, M.F.: *Residuation theory*, Pergamon, 1972.
[A8] KEIMEL, K.: 'A unified theory of minimal prime ideals', *Acta. Math. Acad. Scient. Hung.* **23** (1972), 51-69.

AMS 1980 Subject Classification: 06B99, 06F05

MULTIPLICATIVE SEMI-GROUP *of an associative ring* - The **semi-group** formed by the elements of the given associative ring relative to multiplication. A non-associative ring is, relative to multiplication, only a **groupoid**; it is called the *multiplicative groupoid* of the ring.

O.A. Ivanova

AMS 1980 Subject Classification: 16-XX

MULTIPLICATIVE SYSTEM - An **orthonormal system** of functions $\{\phi_n\}$ on $[a, b]$ satisfying the conditions:

1) for any two functions ϕ_k and ϕ_l the system $\{\phi_n\}$ contains their product $\phi_m(x) = \phi_k(x)\phi_l(x)$;

2) for each function ϕ_k the system $\{\phi_n\}$ contains the function $\phi_m(x)=1/\phi_k(x)$.

Examples of multiplicative systems are the exponential system $\{e^{i2\pi nx}\}_{n=-\infty}^{\infty}$, which is orthogonal on $[0,1]$, and the **Walsh system** of functions.

References

[1] KACZMARZ, S. and STEINHAUS, H.: *Theorie der Orthogonalreihen*, Chelsea, reprint, 1951.
[2] HARMUTH, H.F.: *Transmission of information by orthogonal functions*, Springer, 1972.
[3] ZEEK, R.W. and SHOWALTER, A.E. (EDS.): *Applications of Walsh functions. Proc. Symp. Washington, April 1971*, Univ. Maryland, 1971.

A.V. Efimov

AMS 1980 Subject Classification: 42C05

MULTIPLICITY OF A MODULE *with respect to an ideal*

Editorial comments. Let A be a commutative ring with unit. A module M over A is said to be of *finite length n* if there is a sequence of submodules (a *Jordan–Hölder sequence*) $M_0 \subset \cdots \subset M_n$ such that each of the quotients M_i/M_{i+1}, $i=0,\ldots,n-1$, is a simple A-module. (The number n does not depend on the sequence chosen, by the **Jordan–Hölder** theorem.) Now let M be an A-module of finite type and $\mathfrak{a}$ an ideal contained in the radical of A and such that $M/\mathfrak{a}M$ is of finite length, and let $M\neq 0$ be of Krull dimension d. (The *Krull dimension of a module M* is equal to the dimension of the ring $A/\mathfrak{q}(M)$ where $\mathfrak{q}(M)$ is the *annihilator* of M, i.e. $\mathfrak{q}(M)=\{a\in A:\ aM=0\}$.) Then there exists a unique integer $e_A(\mathfrak{a};M)$ such that

$$\text{length}_A(M/\mathfrak{a}^{n+1}M) = e_A(\mathfrak{a};M)\frac{n^d}{d!}+(\text{lower degree terms})$$

for n large enough. The number $e_A(\mathfrak{a};M)$ is called the multiplicity of M with respect to $\mathfrak{a}$. The multiplicity of an ideal $\mathfrak{a}$ is $e(\mathfrak{a})=e_A(\mathfrak{a};A)$. Thus, the multiplicity of the maximal ideal $\mathfrak{m}$ of a local ring A of dimension d is equal to $(d-1)!$ times the leading coefficient of the Hilbert–Samuel polynomial of A, cf. **Local ring**.

There are some mild terminological discrepancies in the literature with respect to the Hilbert–Samuel polynomial. Let $\psi(n)=\text{length}_A(M/\mathfrak{a}^{n+1}M)$ and $\chi(n)=\text{length}_A(\mathfrak{a}^n M/\mathfrak{a}^{n+1}M)$. Then both $\psi(n)$ and $\chi(n)$ are sometimes called *Hilbert–Samuel functions*. For both $\psi(n)$ and $\chi(n)$ there are polynomials in n (of degree d and $d-1$, respectively) such that $\psi(n)$ and $\chi(n)$ coincide with these polynomials for large n. Both these polynomials occur in the literature under the name *Hilbert–Samuel polynomial*.

For a more general set-up cf. [A1].

The *multiplicity of a local ring A* is the multiplicity of its maximal ideal $\mathfrak{m}$, $e_A(\mathfrak{m};A)$.

References

[A1] BOURBAKI, N.: *Algèbre commutative*, Masson, 1983, Chapt. 8, § 4: Dimension.
[A2] NAGATA, M.: *Local rings*, Interscience, 1962, Chapt. III, § 23.
[A3] MUMFORD, D.: *Algebraic geometry* , I. Complex projective varieties, Springer, 1976, Appendix to Chapt. 6.
[A4] ZARISKI, O. and SAMUEL, P.: *Commutative algebra*, 2, v. Nostrand, 1960, Chapt. VIII, § 10.

AMS 1980 Subject Classification: 13H15

MULTIPLICITY OF A SINGULAR POINT *of an algebraic variety* -

An integer which is a measure of the singularity of the **algebraic variety** at that point. The multiplicity $\mu(X,x)$ of a variety X at a point x is defined to be the multiplicity of the maximal ideal $\mathfrak{m}$ in the **local ring** $\mathcal{O}_{X,x}$. The multiplicity of X at x coincides with the multiplicity of the **tangent cone** $C(X,x)$ at the vertex, and also with the degree of the special fibre $\sigma^{-1}(x)$ of a blow-up $\sigma: X'\to X$ of X at x, where $\sigma^{-1}(X)$ is considered to be immersed in the projective space $P(\mathfrak{m}/\mathfrak{m}^2)$ (see [3]). One has $\mu(X,x)=1$ if and only if x is a non-singular (regular) point of X. If X is a hypersurface in a neighbourhood of x (i.e. X is given by a single equation $f=0$ in an affine space Z), then $\mu(X,x)$ is identical with the number n such that $f\in\mathfrak{n}^n\setminus\mathfrak{n}^{n+1}$, where $\mathfrak{n}$ is the maximal ideal in the local ring $\mathcal{O}_{Z,x}$. The multiplicity does not change when X is cut by a generic hypersurface through x. If X_d denotes the set of points $x\in X$ such that $\mu(X,x)\geqslant d$, then X_d is a closed subset (a subvariety).

References

[1] MUMFORD, D.: *Algebraic geometry*, Springer, 1976.
[2] SERRE, J.-P.: *Algèbre locale. Multiplicités*, Springer, 1965.
[3] RAMANUJAM, C.P.: 'On a geometric interpretation of multiplicity', *Invent. Math.* **22**, no. 1 (1973), 63-67.

V.I. Danilov

Editorial comments. For the multiplicity of the maximal ideal of a local ring, cf. **Multiplicity of a module**.

AMS 1980 Subject Classification: 14H20, 14J17

MULTIPLICITY OF A WEIGHT M *of a representation ρ of a Lie algebra $\mathfrak{t}$ in a finite-dimensional vector space V* -

The dimension n_M of the weight subspace $V_M\subset V$ corresponding to the weight M (see **Weight of a representation of a Lie algebra**).

Let $\mathfrak{t}$ be a **Cartan subalgebra** of a semi-simple Lie algebra $\mathfrak{g}$ over an algebraically closed field of characteristic zero, and let ρ be the restriction to $\mathfrak{t}$ of a finite-dimensional representation σ of the algebra $\mathfrak{g}$. In this case the space V is the direct sum of the weight subspaces of $\mathfrak{t}$ corresponding to the different weights. These weights and their multiplicities are often called the weights and the multiplicities of the representation σ of the algebra $\mathfrak{g}$.

Suppose that σ is an irreducible representation and let Λ be its highest weight (see **Cartan theorem** on the highest weight vector). Then $n_\Lambda=1$. Various devices are available for computing the multiplicities of weights other than the highest weight. Two of these are classical results in representation theory: Freudenthal's formula and Kostant's formula.

1. *Freudenthal's formula* (see [4], [1]). Let $(,)$ be the natural scalar product on the space $\mathfrak{t}^*$ adjoint to $\mathfrak{t}$, induced by the **Killing form** on $\mathfrak{t}$; let R be the **root system** of the algebra $\mathfrak{g}$ relative to $\mathfrak{t}$ and let $>$ be a partial order relation on $\mathfrak{t}^*$ determined by some fixed system of simple roots $\alpha_1, \ldots, \alpha_r$ in R. Then

$$((\Lambda+\delta, \Lambda+\delta)-(M+\delta, M+\delta))n_M =$$

$$= 2\sum_{\substack{\alpha\in R \\ \alpha>0}}\sum_{k=1}^{\infty} n_{M+k\alpha}(M+k\alpha, \alpha),$$

where $\delta=\sum_{\alpha\in R, \alpha>0}\alpha/2$ and by definition $n_N=0$ if N is not a weight of the representation σ. For any weight $M\neq\Lambda$, the coefficient of n_M on the left of the formula is different from zero. This formula is essentially a recurrence formula: it enables one to express n_M in terms of n_N for $N>M$. Since it is known that $n_\Lambda=1$, Freudenthal's formula yields an effective method for the computation of the multiplicities n_M.

2. *Kostant's formula* (see [5], [1]). Let $\Gamma=\{M\in\mathfrak{t}^*: 2(M, \alpha_i)/(\alpha_i, \alpha_i)\in\mathbf{Z}$ for all $i=1, \ldots, r\}$. This set Γ is a multiplicative subgroup in $\mathfrak{t}^*$ which is invariant under the **Weyl group** W, which acts on $\mathfrak{t}^*$ in a natural way. The element δ — and indeed all weights of the representation σ — are members of Γ. Suppose that for each $M\in\Gamma$ the number $P(M)$ is the number of solutions $\{k_\alpha: \alpha\in R, \alpha>0\}$ of the equation

$$M = \sum_{\substack{\alpha\in R \\ \alpha>0}}k_\alpha\alpha,$$

where $k_\alpha\in\mathbf{Z}$, $k_\alpha>0$ for all α. The function $P(M)$ on Γ is known as the *partition function*. Then

$$n_M = \sum_{S\in W}(\det S)P(S(\Lambda+\delta)-(M+\delta)).$$

Practical application of the above formulas involves cumbersome computations. For semi-simple algebras of rank 2, there are more convenient geometrical rules for evaluating the multiplicity of a weight (see [2]).

References

[1] JACOBSON, N.: *Lie algebras*, Interscience, 1962.
[2] ZHELOBENKO, D.P.: *Lectures on the theory of Lie groups*, Dubna, 1965 (in Russian).
[3] ZHELOBENKO, D.P.: *Compact Lie groups and their representations*, Amer. Math. Soc., 1973 (translated from the Russian).
[4A] FREUDENTHAL, H.: 'Zur Berechnung der Charaktere der halbeinfacher Liescher Gruppen I', *Indag. Math.* **16** (1954), 369-376.
[4B] FREUDENTHAL, H.: 'Zur Berechnung der Charaktere der halbeinfacher Liescher Gruppen II', *Indag. Math.* **16** (1954), 487-491.
[4C] FREUDENTHAL, H.: 'Zur Berechnung der Charaktere der halbeinfacher Liescher Gruppen III', *Indag. Math.* **18** (1956), 511-514.
[5] KOSTANT, B.: 'A formula for the multiplicity of a weight', *Trans. Amer. Math. Soc.* **93** (1959), 53-73.

V.L. Popov

Editorial comments. There is a faster algorithm for computing the full set of weights and multiplicities, due to M. Demazure [A3].

References

[A1] FREUDENTHAL, H. and VRIES, H. DE: *Linear Lie groups*, Acad. Press, 1969.
[A2] HUMPHREYS, J.E.: *Introduction to Lie algebras and representation theory*, Springer, 1972.
[A3] DEMAZURE, M.: 'Une nouvelle formule des charactères', *Bull. Sci. Math. (2)* **98** (1974), 163-172.

AMS 1980 Subject Classification: 17B10

MULTIPLIER GROUP, *multiplicator*, of a group G represented as a quotient group F/R of a free group F - The quotient group

$$R\cap F'/[R, F],$$

where F' is the **commutator subgroup** of F and $[R, F]$ is the mutual commutator subgroup of R and F. The multiplicator of G does not depend on the way in which G is presented as a quotient group of a free group. It is isomorphic to the second homology group of G with integer coefficients. In certain branches of group theory the question of non-triviality of the multiplicator of a group is important.

A.L. Shmel'kin

Editorial comments. The usual name in the Western literature is *Schur multiplier* (or *multiplicator*). It specifically enters in the study of central extensions of G and in the study of *perfect groups* G (i.e. groups G for which $G=[G, G]$, where $[G, G]$ is the **commutator subgroup** of G).

References

[A1] ROBINSON, D.J.S.: *A course in the theory of groups*, Springer, 1980.

AMS 1980 Subject Classification: 20J10, 20E32

MULTIPLIER THEORY

Editorial comments. Given a Fourier series on $[-\pi, \pi)$, $\sum_{n=-\infty}^{\infty}c_n e^{inx}$ say, and a (doubly infinite) sequence $\{\lambda_n\}$, one may form a new Fourier series, $\sum_{n=-\infty}^{\infty}\lambda_n c_n e^{inx}$. The sequence $\{\lambda_n\}$ is called a *Fourier multiplier*. The principal problem about Fourier multipliers is to determine conditions on $\{\lambda_n\}$ which guarantee that, when the old Fourier series corresponds to an element of some space $\mathscr{E}$ of functions or generalized functions (cf. **Generalized function**) on $[-\pi, \pi)$, then the new series corresponds to an element of some other given space $\mathscr{F}$ of functions or generalized functions on $[\pi, \pi)$. Typically, $\mathscr{E}$ and $\mathscr{F}$ are Lebesgue spaces, Sobolev spaces or similar function spaces (cf. **Lebesgue space; Sobolev space**). Particular cases of the problem were first solved by W.H. Young (1913), H. Steinhaus (1915) and S. Sidon (1921), the most significant of these solutions being that $\{\lambda_n\}$ is a multiplier from the space of integrable functions to itself or from the space of continuous functions to itself if and only if $\sum_{n=-\infty}^{\infty}\lambda_n e^{inx}$ is a **Fourier–Stieltjes series**. Equivalently, one can seek to characterize generalized functions ϕ on $[-\pi, \pi)$ with the property that, if $f\in\mathscr{E}$, then the convolution product $\phi\ast f\in\mathscr{F}$; the corresponding Fourier multiplier is the sequence of Fourier coefficients of ϕ.

The analogous problem, of characterizing operators which map one space to another and which correspond to a pointwise multiplication of the Fourier transform by a fixed object, can be posed in the context of Fourier integrals rather than series, and in one or several variables. (Indeed, the theory can even be developed in the general context of locally compact groups.)

The most important results on Fourier multipliers are connected with the theories of singular integral operators and pseudo-differential operators (cf. **Pseudo-differential operator**). The general style of these is exemplified by the theorem that a bounded function m on **R** whose total variation (cf. **Variation of a function**) on each dyadic interval $\pm[2^k, 2^{k+1}]$ is bounded is a Fourier L_p-multiplier (i.e. the associated operator maps $L_p(\mathbf{R})$ into itself) if $1<p<\infty$. Perhaps the other most important result in the field is *C. Fefferman's theorem* that in $\mathbf{R}^n$, where $n\geqslant2$, the characteristic function of the unit ball is a Fourier L_p-multiplier only if $p=2$.

References

[A1] HÖRMANDER, L.: 'Estimates for translation-invariant operators in L^p spaces', *Acta Math.* **104** (1960), 93-139.
[A2] FEFFERMAN, C.: 'The multiplier problem for the ball', *Ann. of Math.* **94** (1971), 330-336.
[A3] ZYGMUND, A.: *Trigonometric series*, 1-2, Cambridge Univ. Press, 1988.

M.G. Cowling
J.F. Price

AMS 1980 Subject Classification: 42A15, 42B15

MULTIPLIERS *of the first and second kinds* - The eigen values of the **monodromy operator** of a canonical equation.

In a complex Hilbert space, equations of the form $\dot{x}=iJH(t)x$, where J and $H(t)$ are self-adjoint operators, $J^2=I$ and $H(t)$ is periodic, are called *canonical*. In the finite-dimensional case the eigen values of the monodromy operator $U(t)$ of this equation are called *multipliers*. If all solutions of a canonical equation are bounded on the entire real axis (the equation is stable), then the multipliers lie on the unit circle. Consider a canonical equation $\dot{x}=i\lambda JH(t)x$ with a real parameter λ; then all multipliers can be divided into two groups: multipliers of the first (second) kind, which move counter-clockwise (clockwise) as λ increases.

A canonical equation is called *strongly stable* if it is stable and remains stable under small variations of $H(t)$. For strong stability it is necessary and sufficient that all multipliers be on the unit circle and that there be no coincident multipliers of different kinds.

The theory of multipliers of the first and second kinds allows one to obtain a number of delicate tests for stability and estimates of the zone of stability for canonical equations. The homotopy classification of stable and unstable canonical equations has been given in terms of multipliers.

References

[1] DALETSKIĬ, YU.L. and KREĬN, M.G.: *Stability of solutions of differential equations in Banach space*, Amer. Math. Soc., 1974 (translated from the Russian).
[2] YAKUBOVICH, V.A. and STARZHINSKIĬ, V.M.: *Linear differential equations with periodic coefficients*, Wiley, 1975 (translated from the Russian).

S.G. Kreĭn

Editorial comments.
References

[A1] KREĬN, M.G.: *Topics in differential and integral equations and operator theory*, Birkhäuser, 1983 (translated from the Russian).
[A2] GOHBERG, I. [I. GOKHBERG], LANCASTER, P. and RODMAN, L.: *Matrices and indefinite scalar products*, Birkhäuser, 1983.

AMS 1980 Subject Classification: 34G99, 47E05, 34D99

MULTIPLY-CONNECTED DOMAIN *in a path-connected space* - A domain D in which there are closed paths not homotopic to zero, or, in other words, whose **fundamental group** is not trivial. This means that there are closed paths in D which cannot be continuously deformed to a point while remaining throughout within D, or, otherwise, a multiply-connected domain D is a domain which is not a **simply-connected domain**.

The *order of connectivity* of a plane domain D in $\mathbf{R}^2$ or $\mathbf{C}=\mathbf{C}^1$ (or in the compactification of these spaces, $\overline{\mathbf{R}}^2$ or $\overline{\mathbf{C}}$) is the number of (homologically) independent one-dimensional cycles, that is, the one-dimensional **Betti number** p^1 of D. If the number k of connected components of the boundary of a plane domain D, considered as a domain in the compactified space $\overline{\mathbf{R}}^2$ or $\overline{\mathbf{C}}$, is finite, then $p^1=k$; otherwise one sets $p^1=\infty$. When $p^1=1$, D is a simply-connected domain, when $p^1<\infty$ it is a *finitely-connected domain* (one also uses such terms as *doubly-connected domain*, *triply-connected domain*, ..., *k-connected domain*), when $p^1=\infty$, D is an *infinitely-connected domain*. All plane finitely-connected domains with equal order of connectivity, k, are homeomorphic to each other. By removing from such a domain D all the points of $k-1$ cuts, that is, Jordan arcs joining pairs of connected components of the boundary, it is always possible to obtain a simply-connected domain $D^*\subset D$. About the conformal types of plane multiply-connected domains see **Riemann surfaces, conformal classes of**.

The topological types of domains in $\mathbf{R}^n$, $n\geqslant3$, or $\mathbf{C}^m$, $m\geqslant2$, are far more diverse and cannot be characterized by a single number. Here, sometimes, the term 'multiply-connected domain' (with various provisos) is used when the fundamental group is trivial but some higher-dimensional **homology group** is not trivial.

E.D. Solomentsev

Editorial comments. For a discussion of non-planar multiply-connected domains see [A1].

There are two rather different concepts which go by the phrase '*multi-connected*' or '*multiply-connected*'. The concept and terminology as described above come from the theory of functions of a complex variable.

On the other hand, in (algebraic) topology one defines an *n-connected space* as a space X such that any mapping from a sphere S^m, $m \leq n$, into X is homotopic to zero. Thus, 0-connectedness is the same as path connectedness.

References

[A1] FRANCIS, G.K.: *A topological picturebook*, Springer, 1987.
[A2] MASSEY, W.: *Algebraic topology, an introduction*, Springer, 1967.
[A3] WHITEHEAD, G.W.: *Elements of homotopy theory*, Springer, 1978.

AMS 1980 Subject Classification: 54D05

MULTIVALENT FUNCTION - A notion which is a natural generalization of that of a **univalent function**. A regular or meromorphic function $f(z)$ in a domain D of the complex z-plane is called *p-valent* in D ($p = 1, 2, \ldots$) if in this domain it takes each of its values at most p times, that is, if the number of roots of the equation $f(z) = w$ in D, for any w, does not exceed p. Geometrically this means that above each point of the w-plane lie at most p points of the **Riemann surface** into which $w = f(z)$ maps D. For $p = 1$, $f(z)$ is univalent in D.

Alongside this most simple class of p-valent functions, a major role in the theory of multivalent functions is played by functions which are p-valent in a certain generalized sense: '*p-valent in the mean*'. Let $f(z)$ be a regular or meromorphic function in a domain D of the z-plane, let $n(w)$ be the number of roots in D of the equation $f(z) = w$ and let p be a positive number. The function $f(z)$ is called *p-valent in the mean over circles* in D if for all $R > 0$:

$$\frac{1}{2\pi} \int_0^{2\pi} n(Re^{i\phi}) \, d\phi \leq p.$$

Geometrically this means that the linear measure of the arc on the Riemann surface to which $w = f(z)$ maps D and projecting to the circle $|w| = R$ does not exceed p times the length of this circle. A function $f(z)$ is called *p-valent in the mean over areas* in D if

$$\int_0^R \left[\int_0^{2\pi} n(\rho e^{i\phi}) \, d\phi \right] \rho \, d\rho \leq p\pi R^2$$

for all $R > 0$. Geometrically this means that the area of a part of the Riemann surface to which $w = f(z)$ maps D and projecting to a disc $|w| < R$ does not exceed p times the area of this disc. From these definitions it follows that a function which is p-valent in some domain is also p-valent in the mean over circles in it, and a function which is p-valent in the mean over circles is p-valent in the mean over areas. A function which is p-valent in the mean may turn out to be infinitely-valent.

Multivalent functions, like univalent ones, have been studied in various ways: from the point of view of distortion characteristics of the domain under the mappings by these functions, from estimates on the coefficients of series of representing these functions, etc. They have many extremal properties, similar to the extremal properties of univalent functions. For example, there are the following generalizations to the case of p-valent functions of two classical results in the theory of univalent functions: the **area principle** and estimates for the second coefficient (see **Bieberbach conjecture**).

If a function

$$F(\zeta) = \sum_{m=1}^{p} \alpha_m \zeta^m + \sum_{n=0}^{\infty} \frac{a_n}{\zeta^n}, \quad \alpha_p \neq 0, \quad |\zeta| > 1, \tag{1}$$

is p-valent and regular in the domain $|\zeta| > 1$, except for a pole at $\zeta = \infty$, then

$$\sum_{n=1}^{\infty} n \, |a_n|^2 \leq \sum_{m=1}^{p} m \, |\alpha_m|^2. \tag{2}$$

If a function

$$f(z) = z^p + a_{p+1} z^{p+1} + \cdots \tag{3}$$

is regular and p-valent in the disc $|z| < 1$, then

$$|a_{p+1}| \leq 2p. \tag{4}$$

Inequalities (2) and (4) are best possible. These two results are related to the earliest fundamental results in the theory of p-valent functions. Inequality (2) has also been proved for functions of the form (1) that are p-valent in the mean over areas in $|\zeta| > 1$, and (4) has been proved for functions of the form (3) that are p-valent in the mean over areas in $|z| < 1$.

Research into the class of p-valent functions was significantly advanced by the possibility of considering it as a subclass of the functions that are p-valent in the mean. Precise analogues of the basic distortion and covering theorems for univalent functions have also been obtained for p-valent functions (see **Distortion theorems**; **Covering theorems**). Namely: For a function $f(z)$ of the form (3) that is p-valent in the mean over circles in the disc $|z| < 1$, one has the sharp estimates:

$$\frac{|z|^p}{(1 + |z|)^{2p}} \leq |f(z)| \leq \frac{|z|^p}{(1 - |z|)^{2p}},$$

$$|f'(z)| < p |z|^{p-1} \cdot \frac{1 + |z|}{(1 - |z|)^{2p+1}}, \quad |z| < 1;$$

the function $f(z)$ takes, in $|z| < 1$, each value w with $|w| < 1/4^p$ exactly p times (a direct analogue of the Koebe covering theorem, cf. **Koebe theorem**). This latter property also holds for functions $f(z)$ of the form (3) that are p-valent in the mean over areas in $|z| < 1$.

For functions that are p-valent in the mean over circles there are a number of best possible results characterizing the growth of their coefficients. Thus, for functions of the form

$$f(z) = \sum_{n=0}^{\infty} a_n z^n \tag{5}$$

that are p-valent in the mean over circles in $|z|<1$, $p>0$, the limit

$$\alpha = \lim_{r \to 1-0} (1-r)^{2p} \max_{|z|=r} |f(z)|$$

exists and is, moreover, finite, and

$$\lim_{n \to \infty} \frac{|a_n|}{n^{2p-1}} = \frac{\alpha}{\Gamma(2p)}$$

for $p>1/4$. Whenever an estimate for α is obtained, a corresponding sharp estimate for the asymptotic growth of the coefficients follows. In particular, if $f(z)$ has the form (3), then the latter equality takes the form

$$\lim_{n \to \infty} \frac{|a_n|}{n^{2p-1}} = \frac{\alpha}{(2p-1)!},$$

where $\alpha<1$, except for the case when $f(z)=z^p(1-ze^{i\theta})^{-2p}$ (θ real). Furthermore, for functions of the form (3) that are p-valent in the mean over circles in $|z|<1$, one has the sharp estimate

$$|a_{p+2}| \leqslant p(2p+1),$$

while for the subclass of p-valent functions of this form there is a sharp estimate for the following coefficient:

$$|a_{p+3}| \leqslant \frac{2}{3}p(p+1)(2p+1).$$

The two latter inequalities are, for multivalent functions, the analogues of the estimates $|a_3| \leqslant 3$ and $|a_4| \leqslant 4$, known for univalent functions (see **Bieberbach conjecture**). Since the extremal functions in the above turn out to be p-valent functions, all of these results are best possible even in the class of p-valent functions.

For functions of the form (5) that are p-valent in the mean over areas in the disc $|z|<1$, the following estimates for their coefficients are known to hold for all $n \geqslant 1$:

$$|a_n| < A(p)\mu_p n^{2p-1} \quad (p>1/4), \tag{6}$$

$$|a_n| < A|a_0|n^{-1/2}\log(n+1) \quad (p=1/4), \tag{7}$$

$$|a_n| < A(p)|a_0|\left[\frac{\log(n+1)}{n}\right]^{1/2} \quad (0<p<1/4), \tag{8}$$

$$||a_{n+1}|-|a_n|| < A(p)\mu_p n^{2p-2} \quad (p \geqslant 1), \tag{9}$$

as well as the estimate

$$\max_{|z|=r} |f(z)| < A(p)\mu_p (1-r)^{-2p} \quad (0<r<1); \tag{10}$$

here $A(p)$ depends only on p and $\mu_p = \max_{0 \leqslant \nu \leqslant p} |a_\nu|$. The order of the quantities in (6), (9) and (10) is best possible.

There is also the following analogue for multivalent functions of a theorem known for univalent mero-

morphic functions: In the class of all functions

$$F(\zeta) = \zeta^p \left[1 + \frac{a_1}{\zeta} + \cdots\right]$$

that are p-valent and regular in $|\zeta|>1$, except for a pole at $\zeta=\infty$, and have at a fixed point $\zeta_0 \neq \infty$ of this domain the expansion

$$F(\zeta) = F(\zeta_0) + \frac{F^{(p)}(\zeta_0)}{p!}(\zeta-\zeta_0)^p + \cdots,$$

the range of values of the functional

$$w = \log \frac{F^{(p)}(\zeta_0)}{p!}$$

is the disc

$$|w| \leqslant -p \log \left[1 - \frac{1}{|\zeta_0|^2}\right].$$

Apart from the above-mentioned fundamental classes of multivalent functions, a significant place in the investigations is held by special classes of multivalent functions, for example, functions which are typically real of order p, p-valent star-like, p-valent convex, p-valent close-to-convex, p-valent bounded, and others, which are generalizations of, respectively, typically-real, star-like, convex, close-to-convex, bounded univalent, and other functions (cf. **Typically-real function; Star-like function; Convex function (of a complex variable)**). A function

$$f(z) = \sum_{n=1}^{\infty} a_n z^n \tag{11}$$

is called *typically real of order p* in $|z|<1$ if it is regular, has real coefficients a_n and if there is a number $\delta=\delta(f)$, $0<\delta<1$, such that for each r in the interval $1-\delta<r<1$ the imaginary part $\mathrm{Im}\{f(z)\}$ changes its sign on the circle $|z|=r$ exactly $2p$ times. Here $f(z)$ may be more than p-valent in $|z|<1$. For its coefficients one has the sharp estimates:

$$|a_n| \leqslant \sum_{k=1}^{p} \frac{2k(n+p)!}{(n^2-k^2)(p-k)!(p+k)!(n-p-1)!} |a_k|, \tag{12}$$

$$n > p.$$

One of the analogues of the Bieberbach conjecture for functions of the form (11) that are regular and p-valent in $|z|<1$ is *Goodman's conjecture* on the validity of (12) for the coefficients a_n. In particular, Goodman's conjecture holds for p-valent typically-real functions of order p in $|z|<1$. It has also been proved to hold for a class of p-valent functions which is a generalization of the class of univalent functions that are convex in the direction of the imaginary axis. Another analogue of the Bieberbach conjecture for p-valent functions is the following *conjecture of Goodman*. Let a function

$$f(z) = z^q + \sum_{n=q+1}^{\infty} a_n z^n, \quad q \geqslant 1,$$

be regular and p-valent, $p=1,2,\ldots,$ in $|z|<1$, and

let it have t zeros $\alpha_1, \ldots, \alpha_t$ in $0 < |z| < 1$, $t \leqslant p - q$. The conjecture is that $|a_n| \leqslant B_n$, $n > q$, where B_n is the n-th coefficient in the expansion

$$\frac{z^q}{(1-z)^{2q}} \left[\frac{1+z}{1-z} \right]^{2(p-q-t)} \frac{1}{(1-z)^{2t}} \times$$

$$\times \prod_{j=1}^{t} \left[1 + \frac{z}{|\alpha_j|} \right] (1 + |\alpha_j|z) \equiv$$

$$\equiv z^q + \sum_{n=q+1}^{\infty} B_n z^n, \quad |z| < 1.$$

For typically-real functions of order p in $|z| < 1$ this inequality has been proved.

The classes of *p-valent star-like* and *p-valent convex functions*, $S(p)$ and $C(p)$ respectively, are defined as follows. A function $f(z)$ belongs to $S(p)$, $p = 1, 2, \ldots$, if it is regular in $|z| < 1$, if $f(0) = 0$ and if there is a number ρ, $0 < \rho < 1$, such that

$$\mathrm{Re}\left\{ \frac{zf'(z)}{f(z)} \right\} > 0,$$

$$\int_0^{2\pi} \mathrm{Re}\left\{ \frac{zf'(z)}{f(z)} \right\} d\theta = 2p\pi, \quad \theta = \arg z,$$

for $\rho < |z| < 1$. A function $f(z)$ belongs to $C(p)$, $p = 1, 2, \ldots$, if it is regular in $|z| < 1$, if $f(0) = 0$ and if there is a ρ such that

$$1 + \mathrm{Re}\left\{ \frac{zf''(z)}{f'(z)} \right\} > 0,$$

$$\int_0^{2\pi} \left[1 + \mathrm{Re}\left\{ \frac{zf''(z)}{f'(z)} \right\} \right] d\theta = 2p\pi, \quad \theta = \arg z,$$

for $\rho < |z| < 1$. A number of sharp estimates has been obtained for functions of these two classes.

The classes $S(p)$ and $C(p)$ turn out to be subclasses of a wider class of *p-valent* functions: the class of *p-valent close-to-convex* functions. A function

$$F(z) = \sum_{n=1}^{\infty} a_n z^n$$

that is regular in $|z| < 1$ is called *p-valent close-to-convex* if it satisfies one of the following conditions:

A) there is a function $f(z) \in S(p)$ and a number ρ, $0 < \rho < 1$, such that

$$\mathrm{Re}\left\{ \frac{zF'(z)}{f(z)} \right\} > 0 \quad (\rho < |z| < 1); \tag{13}$$

B) $F(z)$ is regular on $|z| = 1$ and there is a function $f(z) \in S(p)$, also regular on $|z| = 1$, such that (13) is satisfied on $|z| = 1$.

For functions $F(z)$ of this class sharp upper and lower bounds for $|F'(z)|$ have been found and (12) has been proved: for $n = p + 1$ for all functions of this class, and for $n \geqslant p + 1$ for the functions in this class with real coefficients. Sharp estimates generalizing certain results for bounded univalent functions have been

obtained for bounded functions that are *p*-valent in the corresponding generalized sense. Thus, the *radius of p-valency* has been found in the class of functions regular and bounded in a disc: If $f(z)$ is regular and bounded in modulus by one in the disc $|z| < 1$ and normalized by the conditions $f(0) = 0$, $f'(0) = a_1$, $0 < |a_1| < 1$, then the radius ρ of the largest disc $|z| < \rho$ in which it is *p*-valent is given by the equation

$$\frac{(p+1)\rho^p(1-\rho^2)}{1-\rho^{2p+2}} = |a_1|, \quad 0 < \rho < 1.$$

This theorem generalizes, to the case $p > 1$, *Landau's theorem on the radius of univalency* of functions that are regular and bounded in the disc $|z| < 1$.

Various sufficient conditions are known for a function that is regular in a domain to be *p*-valent in it. For example, if $f(z)$ is regular in a convex domain D and if there are a real number θ and an integer k, $0 \leqslant k \leqslant p - 1$, such that

$$\mathrm{Re}\left\{ e^{i\theta} \frac{d^p}{dz^p} [z^k f(z)] \right\} > 0, \quad z \in D,$$

then $f(z)$ is *p*-valent in D.

Multivalent functions have also been studied in multiply-connected domains. In this case many estimates can be expressed in terms of functions mapping the given multiply-connected domain into a canonical Riemann surface, and in terms of the **Bergman kernel function**. The first basic result related to the question of the existence of conformal mappings of a multiply-connected domain onto a multi-sheeted canonical surface is the following *theorem of Grunsky*: Let D be a finitely-connected domain in the z-plane with $z = \infty$ as an interior point and with boundary components which are not points, and let $Q_p(z)$ be a given polynomial of degree $p \geqslant 1$; then for any given θ, $0 \leqslant \theta < \pi$, there is a unique function $\Phi_\theta(z)$, regular in D except for a pole at $z = \infty$, whose principal part at $z = \infty$ (including the free term) coincides with $Q_p(z)$ and which associates with each boundary component of D a rectilinear segment of slope θ with the real axis. In other words, the function $w = \Phi_\theta(z)$ maps D into the whole p-sheeted w-plane with parallel slits of slope θ. The existence of conformal mappings of a given finitely-connected domain onto other canonical multi-sheeted surfaces has also been proved; extremal properties similar to certain extremal properties of univalent functions have been established for multivalent functions. It has been shown that the most general class of multivalent functions, meromorphic in a finitely-connected domain, for which the area theorem holds has a simple geometric characterization.

The basic methods of research of multivalent functions are the **method of boundary integration**, the **symmetrization method** and the method of quadratic differentials (cf. **Quadratic differential**).

Variational methods in the theory of multivalent functions are less effective than in the theory of univalent functions.

References

[1] GOLUZIN, G.M.: 'On p-valent functions', *Mat. Sb.* **8**, no. 2 (1940), 277-283 (in Russian). German abstract.

[2] HAYMAN, W.K.: *Multivalent functions*, Cambridge Univ. Press., 1958.

[3] JENKINS, J.A.: *Univalent functions and conformal mapping*, Springer, 1958.

[4] PETHE, K.: 'Estimation du coefficient a_{p+3} de la fonction p-valente dans le cercle unité', *Bull. Acad. Polon. Sci.* **20**, no. 3 (1972), 219-220. Russian and English abstracts.

[5] LIVINGSTON, A.E.: 'p-valent close-to-convex functions', *Trans. Amer. Math. Soc.* **115**, no. 3 (1965), 161-179.

[6] LEACH, R.J.: 'Coefficient estimates for certain multivalent functions', *Pacific J. Math.* **74**, no. 1 (1978), 133-142.

[7A] KRZYZ, J.: 'On the derivative of bounded p-valent functions', *Ann. Univ. Mariae Curie-Sklodowska Sect. A* **12**, no. 2 (1958), 23-28. Russian and Polish abstracts.

[7B] KRZYZ, J.: 'Distortion theorems for bounded p-valent functions', *Ann. Univ. Mariae Curie-Sklodowska Sect. A* **12**, no. 3 (1958), 29-38. English and Polish abstracts.

[8] OZAKI, S.: *Sci. Rep. Tokyo Bunrika Daigaku A* **2**, no. 40 (1935), 167-188.

[9] ALENITSYN, YU.E.: 'Area theorems for functions analytic in a finitely connected domain', *Math. USSR Izv.* **7**, no. 5 (1973), 1129-1151. (*Izv. Akad. Nauk SSSR Ser. Mat.* **37**, no. 5 (1973), 1132-1154)

[10] SINGH, S.K.: *Math. Student* **30**, no. 1-2 (1973), 79-90.

[11] GOODMAN, A.W.: 'Open problems on univalent and multivalent functions', *Bull. Amer. Math. Soc.* **74**, no. 6 (1968), 1035-1050.

Yu.E. Alenitsyn

AMS 1980 Subject Classification: 30C55

MUMFORD HYPOTHESIS - The hypothesis that each **semi-simple algebraic group** G is *geometrically reductive*, i.e., has the following property: For any **rational representation** of G in a finite-dimensional vector space V and any non-zero vector $v \in V$ fixed by G, there is a G-invariant homogeneous polynomial f of positive degree on V for which $f(v) \neq 0$.

This hypothesis was stated by D. Mumford [1] (in a different, but equivalent form) with the aim of finding a property of semi-simple groups, defined over an algebraically closed field of arbitrary characteristic, which, from the point of view of the geometric theory of invariants (cf. **Invariants, theory of**), would serve as a substitute for the classical property of complete reducibility of rational linear representations of semi-simple groups defined over fields of characteristic zero (this latter property not holding for ground fields of positive characteristic). It would allow the removal of restrictions on the characteristic of the ground field in a number of central results in the geometric theory of invariants, such as the theorem on finite generation of the algebra of invariants of reductive groups of automorphisms of an algebra of finite type over a field (see **Hilbert theorem** on invariants).

If the characteristic of the ground field k is zero, then a proof of Mumford's hypothesis is given by Weyl's classical theorem on complete reducibility of rational representations of semi-simple groups (see [2]): in this case the invariant line $L = kv$ in V has an invariant complement Γ (an invariant hyperplane such that $L \cap \Gamma = 0$), and f can be taken to be the linear form giving the equation of Γ. When k has positive characteristic p, Mumford's hypothesis generalizes the fact that there is an invariant homogeneous hypersurface Γ in V for which $L \cap \Gamma = 0$ (with the degree of Γ equal to p^n for some integer n).

Mumford's hypothesis is also equivalent to the assertion that for any regular action of a semi-simple group G on an affine algebraic variety X and for any two closed non-intersecting invariant subsets X_1 and X_2 in X there is an invariant regular function h on X for which $h(X_1) = 0$ and $h(X_2) = 1$ (i.e., X_1 and X_2 can be separated by regular invariants, see [3]).

Mumford's hypothesis was first proved in [4]; the proof was extended in [5] to the general case of reductive group schemes over a field.

The proof of Mumford's hypothesis, together with the results of [6] and [10], allow one, first, to give a final form to the generalization of Hilbert's theorem on invariants: If R is an algebra of finite type over an algebraically closed field k, G is a reductive group, acting as an automorphism group on $\mathbf{R}^n$, and R^G is the subalgebra of all G-invariant elements in R, then R^G is also an algebra of finite type over k; and, secondly, to establish that a **linear algebraic group** over a field of arbitrary characteristic is geometrically reductive if and only if it is reductive (cf. **Reductive group**). Mumford's hypothesis has applications in the geometric theory of invariants and in moduli theory (see [7] - [9]).

References

[1] MUMFORD, D.: *Geometric invariant theory*, Springer, 1965.

[2] FOGARTY, J.: *Invariant theory*, Benjamin, 1969.

[3] DIEUDONNÉ, J. and CARRELL, J.B.: *Invariant theory: old and new*, Acad. Press, 1971.

[4] HABOUSH, W.J.: 'Reductive groups are geometrically reductive', *Ann. of Math.* **102** (1975), 67-83.

[5] SESHADRI, C.S.: 'Geometric reductivity over arbitrary base', *Adv. Math.* **26** (1977), 225-274.

[6] NAGATA, M.: 'Invariants of a group in an affine ring', *J. Math. Kyoto Univ.* **3** (1964), 369-377. With appendix by M. Miyanishi.

[7] SESHADRI, C.S.: 'Mumford's conjecture for GL(2) and applications', in *Algebraic Geometry. Papers presented at the Bombay Colloq. 1968*, Oxford Univ. Press, 1969, pp. 347-371.

[8] POPP, H.: *Moduli theory and classification theory of algebraic varieties*, Springer, 1977.

[9] SESHADRI, C.S.: 'Theory of moduli', in R. Hartshorne (ed.): *Algebraic geometry. Arcata 1974*, Proc. Symp. Pure Math., Vol. 29, Amer. Math. Soc., 1975, pp. 263-304.

[10] NAGATA, M. and MIYATA, T.: 'Note on semi-reductive groups', *J. Math. Kyoto Univ.* **3** (1964), 379-382.

V.L. Popov

AMS 1980 Subject Classification: 14AXX, 20F32, 20F38, 22-XX

MÜNTZ THEOREM, *theorem on the completeness of a system of powers* $\{x^{\lambda_k}\}$ *on an interval* $[a, b]$, $0 < a < b < \infty$ - Let $0 < \lambda_1 < \lambda_2 < \cdots$. In order that for any continuous function f on $[a, b]$ and for any $\epsilon > 0$ there is a linear combination

$$P(x) = \sum_{k=1}^{n} a_k x^{\lambda_k}$$

such that

$$\| f - P \|_C = \max_{a \leqslant x \leqslant b} | f(x) - P(x) | < \epsilon,$$

it is necessary and sufficient that

$$\sum_{k=1}^{\infty} \frac{1}{\lambda_k} = \infty. \qquad (*)$$

In the case of an interval $[0, b]$ one adds the function which is identically equal to 1 to the system $\{x^{\lambda_k}\}$ and condition (*) is, as before, necessary and sufficient for the completeness of the enlarged system. The condition $a \geqslant 0$ is essential: the system $\{x^{2k}\}_{k=0}^{\infty}$ (which satisfies (*)) is not complete on $[-1, 1]$ (an odd function cannot be arbitrarily closely approximated by combinations of even powers).

Condition (*) is necessary and sufficient for the completeness of $\{x^{\lambda_k}\}$, $-1/p < \lambda_1 < \lambda_2 < \cdots$, on $[a, b]$, $a \geqslant 0$, in the metric of L_p, $p > 1$; that is, for each $f \in L_p(a, b)$ and any $\epsilon > 0$ there is a linear combination P such that

$$\| f - P \|_{L_p} = \left| \int_a^b | f(x) - P(x) |^p \, dx \right|^{1/p} < \epsilon.$$

The theorem was proved by H. Müntz [1].

References

[1] MÜNTZ, H.: *Ueber den Approximationssatz von Weierstrass*, Schwarz—Festschrift, Berlin, 1914.

[2] ACHIEZER, N.I. [N.I. AKHIEZER]: *Theory of approximation*, F. Ungar, 1956 (translated from the Russian).

A.F. Leont'ev

Editorial comments. There exists several extensions of the Müntz theorem. First, O. Szász showed that with exponents $\lambda_k \in \mathbf{C}$, $\operatorname{Re} \lambda_k > 0$,

$$\sum \operatorname{Re} \frac{1}{\lambda_k} = \infty \qquad (A1)$$

is necessary and sufficient for completeness of the system $\{x^{\lambda_k}\}$ in $C[a, b]$ or $L_p[a, b]$, $p > 1$, or, equivalently, completeness of $\{e^{\lambda_k z}\}$ in, say, $C_0(-\infty, 0]$. Later, J. Korevaar, A.F. Leont'ev, P. Malliavin, J.A. Siddiqi, and others studied analogous completeness problems on curves $\gamma(x) = x + i\eta(x)$, $-\infty < x \leqslant 0$. Very recently it was shown that if η is piecewise C^1, with $\exp | \eta' | < \infty$, and $\{\lambda_k\}$ satisfies (A1) and is contained in a sufficiently small sector around the positive axis, then $\{e^{\lambda_k z}\}$ spans $C_0[\gamma]$. See [A1], also for further references. Finally, attempts have been made to generalize the Müntz theorem to functions of several variables, see [A2].

References

[A1] KOREVAAR, J. and ZEINSTRA, R.: 'Transformées de Laplace pour les courbes à pente bornée et un résultat correspondant du type Müntz—Szász', *C.R. Acad. Sci. Paris* **301** (1985), 695-698.

[A2] RONKIN, L.I.: 'Some questions of completeness and uniqueness for functions of several variables', *Funct. Anal. Appl.* **7** (1973), 37-45. (*Funkts. Anal. Prilozhen.* **7** (1973), 45-55)

AMS 1980 Subject Classification: 30E10, 42C30, 41A99

MUTUAL KERNELS, *reciprocal kernels* - Two functions $K(x, s)$ and $K_1(x, s)$ of real variables x, s (or, in general, of points P and Q of a Euclidean space), defined on the square $[a, b] \times [a, b]$ and satisfying the condition

$$K_1(x, s) - K(x, s) =$$
$$= \int_a^b K(x, t) K_1(t, s) \, dt = \int_a^b K_1(x, t) K(t, s) \, dt.$$

If a kernel $K_1(x, s)$ reciprocal with $K(x, s)$ exists, then $K_1(x, s)$ is the *resolvent kernel* of the integral **Fredholm equation**

$$\phi(x) - \int_a^b K(x, s) \phi(s) \, ds = f(x). \qquad (*)$$

A.B. Bakushinskiĭ

Editorial comments. Indeed, when $K(x, s)$ and $K_1(x, s)$ are reciprocal kernels, the solution of equation (*) above is given by

$$\phi(x) = f(x) + \int_a^b K_1(x, t) f(t) \, dt.$$

Consider the Fredholm equation

$$\phi(x) = f(x) + \lambda \int_a^b K(x, t) \phi(t) \, dt \qquad (A1)$$

and the iterated kernels $K^{(1)}(x, t) = K(x, t)$,

$$K^{(n)}(x, t) = \int_a^b K^{(n-1)}(x, s) K(s, t) \, ds, \quad n = 2, 3, \ldots.$$

Form the **Neumann series**

$$R(x, t; \lambda) = K^{(1)}(x, t) + \lambda K^{(2)}(x, t) + \cdots =$$
$$= \sum_{n=1}^{\infty} \lambda^{n-1} K^{(n)}(x, t).$$

If $K(x, t)$ is continuous on $[a, b] \times [a, b]$, this series is uniformly convergent for λ small. Then $R(x, t; \lambda)$ satisfies

$$\lambda \int_a^b R(x, t; \lambda) K(t, s) \, dt = R(x, s; \lambda) - K(x, s),$$

and

$$\phi(x) = f(x) + \lambda \int_a^b R(x, t; \lambda) f(t) \, dt$$

solves (A1).

The terminology 'mutual kernels' and 'reciprocal kernels' is rarely used.

References

[A1] SMIRNOV, V.I.: *A course of higher mathematics*, 4, Addison-Wesley, 1964 (translated from the Russian).

[A2] ZABREĬKO, P.P., ET AL.: *Integral equations - a reference text*, Noordhoff, 1975 (translated from the Russian).

[A3] MOISEIWITSCH, B.L.: *Integral equations*, Longman, 1977.

AMS 1980 Subject Classification: 45H05, 45B05

MUTUALLY-PRIME NUMBERS, *coprimes, relatively-prime numbers* - Integers without common (prime) divisors. The **greatest common divisor** of two coprimes a and b is 1, which is usually written as $(a, b) = 1$. If a and b are coprime, there exist numbers u and v, $|u| < |b|$, $|v| < |a|$, such that $au + bv = 1$.

The concept of being coprime may also be applied to polynomials and, more generally, to elements of a **Euclidean ring**.

Editorial comments.

References

[A1] VINOGRADOV, I.M.: *Elements of number theory*, Dover, reprint, 1954 (translated from the Russian).

AMS 1980 Subject Classification: 10A25

MUTUALLY-SINGULAR MEASURES - Two (positive) measures μ and ν, defined on a locally compact space T, such that $\inf\{\mu, \nu\} = 0$.

Two measures μ and ν are mutually singular if and only if there exist in T two disjoint sets M and N such that μ is concentrated on M and ν on N.

References

[1] BOURBAKI, N.: *Elements of mathematics. Integration*, Addison-Wesley, 1975, Chapt.6; 7; 8 (translated from the French).

M.I. Voĭtsekhovskiĭ

Editorial comments.

The second characterization in the main article above holds if μ and ν are σ-additive σ-finite measures on an abstract **measurable space**, and M and N belong to the σ-field.

Mutually-singular measures are also called *singular measures* or *orthogonal measures*.

Instead of 'concentrated on' one also uses 'supported in' (cf. also **Support of a measure**).

References

[A1] HALMOS, P.: *Measure theory*, v. Nostrand, 1950.
[A2] HEWITT, E. and STROMBERG, K.: *Real and abstract analysis*, Springer, 1965.

AMS 1980 Subject Classification: 28A10

N

n-**GROUP** - A generalization of the concept of a **group** to the case of an *n*-ary operation. An *n-group* is a **universal algebra** with one *n*-ary associative operation that is uniquely invertible at each place (cf. **Algebraic operation**). The theory of *n*-groups for $n \geq 3$ substantially differs from the theory of groups (i.e. 2-groups). Thus, if $n \geq 3$, an *n*-group has no analogue of the unit element.

Let $\Gamma(\circ)$ be a group with multiplication operation $\circ$; let $n \geq 3$ be an arbitrary integer. Then an *n*-ary operation ω on the set Γ can be defined as follows:

$$a_1 \cdots a_n \omega = a_1 \circ \cdots \circ a_n.$$

The resulting *n*-group is called the *n-group determined by the group* $\Gamma(\circ)$. Necessary and sufficient conditions for an *n*-group to be of this form are known [1]. Any *n*-group is imbeddable in such an *n*-group (*Post's theorem*).

References

[1] KUROSH, A.G.: *Lectures on general algebra*, Chelsea, 1963 (translated from the Russian).

V.D. Belousov

Editorial comments. The usual notion of a *p-group* (i.e., a group of order a power of *p*) is not to be mixed up with that of an *n*-group in the above sense.

References

[A1] BALCI, D.: *Zur Theorie der topologischen n-Gruppen*, Minerva, Munich, 1981.
[A2] RUSAKOV, S.A.: 'The subgroup structure of Dedekind *n*-ary groups', in *Finite groups (Proc. Gomel. Sem.)*, Minsk, 1978, pp. 81-104 (in Russian).
[A3] RUSAKOV, S.A.: 'On the theory of nilpotent *n*-ary groups', in *Finite groups (Proc. Gomel. Sem.)*, Minsk, 1978, pp. 104-130 (in Russian).

AMS 1980 Subject Classification: 20N15

NAGEL POINT - The point of intersection of the straight lines joining the vertices of a triangle to the points at which the opposite sides are tangent to the escribed circles (see Fig.). Named after Ch. Nagel (1836).

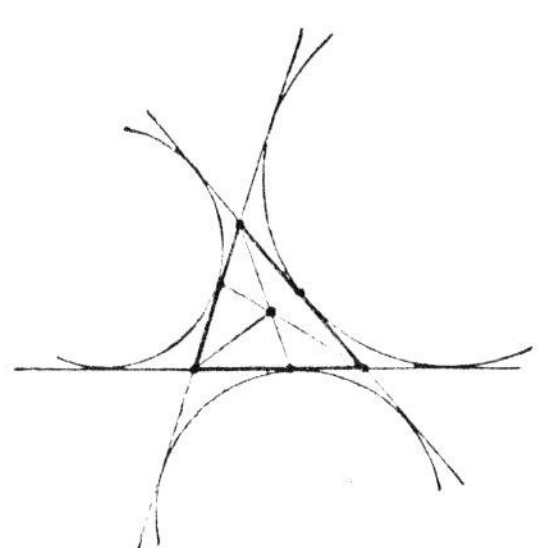

A.B. Ivanov

Editorial comments.

References

[A1] COURT, N.A.: *College geometry*, Barnes & Noble, 1952, p. 160.
[A2] BERGER, M.: *Geometry*, 1, Springer, 1987, Sect. 10.13.30 (translated from the French).

AMS 1980 Subject Classification: 51M05

NAME - A linguistic expression to denote a specified object. The object denoted by a given name is called the *denotation*. In mathematics, names are widely used for specific mathematical objects, for example, e, π for the well-known transcendental numbers, sin for the sine function, $\varnothing$ for the empty set. From such simple names one can form *compound names*, which name an object using the names of two objects. For example, $\sin \pi$ is another name for the number 0. A name not only names the denotation, but it also expresses a definite meaning. Thus, the expressions

$$\lim_{n \to \infty} n^{1/n} \quad \text{and} \quad \sin \frac{\pi}{2}$$

are names for the number 1; however, their meanings are different. If in a compound name, some name occurring in it is replaced by a name having the same denotation, then the denotation of the compound name is unchanged. If in a compound name some name occurring in it is replaced by a *synonym* of it (that is, a name having the same meaning), then the meaning of the compound name is unchanged.

Along with names, one uses in mathematics expressions containing variables. The expressions become names upon replacing the variables by names of objects in the range of values of the variables. Such expressions are called *name forms*. The expressions e^x, $\int_0^x (\sin t)\, dt / t$, where x is a variable for the real numbers, are examples of name forms.

References

[1] CHURCH, A.: *Introduction to mathematical logic*, 1, Princeton Univ. Press, 1956.

V.E. Plisko

Editorial comments.

References

[A1] CARNAP, R.: *Meaning and necessity*, Univ. Chicago Press, 1947.

AMS 1980 Subject Classification: 03A05

NAPIER NUMBER - See e **(number)**.

AMS 1980 Subject Classification: 10A40

NASH THEOREM (IN GAME THEORY) - A theorem on the existence of equilibrium points in a mixed extension of a finite **non-cooperative game**

$$\Gamma = <J, \{S_i\}_{i \in J}, \{H_i\}_{i \in J}>,$$

where J and $\{S_i\}_{i \in J}$ are the finite sets of players and their strategies, respectively, and $H_i: S = \prod_{i \in J} S_i \to \mathbf{R}^1$ is the pay-off function of player $i \in J$ (see also **Games, theory of**). It was established by J. Nash in [1]. Let M_i, $i \in J$, be the set of all probability measures on S_i. Nash' theorem asserts that there is a measure $\mu^* \in M = \prod_{i \in J} M_i$ for which

$$H_i(\mu^*) \geqslant H_i(\mu^* \| \mu_i)$$

for all $\mu_i \in M_i$, $i \in J$, where $\mu^* \| \mu_i$ denotes the measure from M that results from replacing the i-th component of the vector μ^* by μ_i, and $H_i(\mu) = E(H_i, \mu)$. The known proofs of Nash' theorem rely on a fixed-point theorem.

References

[1] NASH, J.: 'Non-cooperative games', *Ann. of Math.* **54** (1951), 286-295.
[2] VOROB'EV, N.N.: *Foundations of game theory. Non-cooperative games*, Moscow, 1984 (in Russian).
[3] VOROB'EV, N.N.: *Game theory. Lectures for economists and system scientists*, Springer, 1977 (translated from the Russian).

E.B. Yanovskaya

AMS 1980 Subject Classification: 90D10

NASH THEOREMS (IN DIFFERENTIAL GEOMETRY) - Two groups of theorems on isometrically imbedded and immersed Riemannian manifolds in a Euclidean space (see also **Immersion of a manifold; Isometric immersion**). The original versions are due to J. Nash [1].

1) *Nash' theorem on C^1-imbeddings and C^1-immersions.* An immersion (imbedding) $f: V^n \to E^m$ of class C^1 of an n-dimensional **Riemannian space** V^n with metric g of class C^0 into an m-dimensional Euclidean space E^m is called *short* if the metric g_f induced by it on V^n is such that the quadratic form $g - g_f$ is positive definite. If V^n has a short immersion (imbedding) in E^m, $m \geqslant n + 1$, then V^n also has an isometric immersion (imbedding) of class C^1 in E^m. Under the restriction $m \geqslant n + 2$ this theorem was proved in [1], and in the form stated above in [2]. This theorem implies, in particular, that if a compact Riemannian manifold V^n has a C^1-imbedding (immersion) in E^m, $m \geqslant n + 1$, then V^n also has an isometric C^1-imbedding (immersion) in E^m. Another consequence of Nash' theorem is that every point of V^n has a sufficiently small neighbourhood that admits an isometric imbedding of class C^1 in E^{n+1}.

2) *Nash' theorem on regular imbeddings.* Every compact Riemannian manifold V^n of class C^r, $3 \leqslant r \leqslant \infty$, has an isometric C^r-imbedding in E^m, where $m = (3n^2 + 11n)/2$. If V^n is not compact, then it has an isometric C^r-imbedding in E^{m_1}, where $m_1 = (3n^2 + 11)(n + 1)/2$.

Nash' theorem on regular imbeddings results from an application of *Nash' implicit-function theorem* on the inversion of a broad class of differential operators. The meaning of this theorem is that when a certain linear algebraic system of equations connected naturally with a differential operator L is solvable and when a reasonable topology is introduced in the image and inverse image, then the operator in question is an open mapping, that is, L is locally invertible near any point of its range. For the equations of an imbedding of a Riemannian manifold in a Euclidean space this reduces to the fact that the first and second derivatives of the mapping $f: V^n \to E^m$ with respect to the intrinsic coordinates of V^n must be linearly independent. Such imbeddings were first considered in [4]; they are called *free*. Nash' implicit-function theorem implies that a compact Riemannian manifold V^n sufficiently close to another one $\overline{V}^n$ having a free imbedding in E^m also has a free imbedding in E^m. This fact and the original method of extension with respect to a parameter lead to Nash' theorem on regular imbeddings (see [3]). By extending Nash' method to non-compact manifolds and analytic imbeddings, and also by a principal refinement of the process of extension with respect to a parameter, it has been proved that every infinitely-differentiable (analytic) Riemannian manifold V^n has an isometric differentiable (analytic) imbedding in E^m, where $m = n(n + 1)/2 + 3n + 5$ (see [5] - [7]).

References

[1] NASH, J.: 'C^1-isometric imbeddings', *Ann. of Math.* **60** (1954), 383-396.

[2] KUIPER, N.: 'On C^1-isometric imbeddings', *Proc. K. Ned. Akad. Wetensch.* **A58**, no. 4 (1955), 545-556.

[3] NASH, J.: 'The imbedding problem for Riemannian manifolds', *Ann. of Math.* **63** (1956), 20-63.

[4] BURSTIN, C.: 'Ein Beitrag zum Problem der Einbettung der Riemannschen Räume in euklidischen Räumen', *Mat. Sb.* **38**, no. 3-4 (1931), 74-85.

[5] NASH, J.: 'Analyticity of the solutions of implicit function problems with analytic data', *Ann. of Math.* **84** (1966), 345-355.

[6] GROMOV, M.L. and ROKHLIN, V.A.: 'Embeddings and immersions in Riemannian geometry', *Russian Math. Surveys* **25**, no. 5 (1970), 1-57. (*Uspekhi Mat. Nauk* **25** (1970), 53-62)

[7] GROMOV, M.L.: 'Isometric imbeddings and immersions', *Soviet Math. Dokl.* **11**, no. 3 (1970), 1206-1209. (*Dokl. Akad. Nauk SSSR* **192** (1970), 794-797)

D.D. Sokolov

Editorial comments. The *Nash theorem in differential topology* says that a compact connected C^∞-manifold without boundary is diffeomorphic to a component of a real algebraic variety.

Let $\pi: X \to V$ be a smooth (i.e. C^∞-) fibration. Denote by $J^r(V, X)$ the space of r-jets (of germs) of smooth sections $f: V \to X$ (cf. **Germ**; **Jet**). The r-th order jet of a section $f: V \to X$ is denoted by $J^r f: V \to J^r(V, X)$. A section $\phi: V \to J^r(V, X)$ is called *holonomic* if there is a C^r-section $f: V \to X$ such that $\phi = J^r f$; ϕ determines f uniquely (if it exists). The *fine topology* on the space $C^0(V, X)$ of C^0-sections $V \to X$ is obtained by taking as a basis the subsets $C^0(V, U)$ where U runs over the open subsets of X. The fine C^r-topology on $C^r(V, X)$ is induced by the imbedding $C^r(V, X) \to C^0(V, J^r(V, X))$, $f \mapsto J^r f$, from the fine C^0-topology to $C^0(V, J^r(V, X))$.

The *Nash approximation theorem* says that an arbitrary Riemannian C^r-metric on V has a fine C^r-approximation by some C^r-metric g' on V that admits C^r-immersions $f': (V, g') \to \mathbf{R}^{2l}$ for some $l = l(n) < \infty$, where $n = \dim V$.

The *Nash−Kuiper theorem* [1], [2] says that an arbitrary differentiable immersion $f_0: V \to \mathbf{R}^q$ for $q > \dim V$ admits a C^1-continuous homotopy of immersions $f_t: V \to \mathbf{R}^q$, $t \in [0, 1]$, to an isometric immersion $f_1: V \to \mathbf{R}^q$.

References

[A1] HIRSCH, M.W.: *Differential topology*, Springer, 1976.

[A2] GROMOV, M.: *Partial differential relations*, Springer, 1986.

AMS 1980 Subject Classification: 53C40, 53B25, 53A07

NATURAL COORDINATE FRAME, *Frénet trihedron, Frénet frame, natural trihedron* - A figure consisting of the tangent, the **principal normal** and the **binormal** of a space curve, and the three planes defined by the pairs of these straight lines. If the edges of the natural frame at a given point of a curve are taken as the axes of a Cartesian coordinate system, the equation of the curve in the natural parametrization (see **Natural parameter**) is, in a neighbourhood of that point,

$$x = s + \cdots, \quad y = \frac{k_1}{2}s^2 + \cdots, \quad z = \frac{k_1 k_2}{6}s^3 + \cdots,$$

where k_1 and k_2 are the **curvature** and **torsion** of the curve at the point.

A.B. Ivanov

Editorial comments. Cf. also **Frénet trihedron**.

References

[A1] KLINGENBERG, W.: *A course in differential geometry*, Springer, 1978 (translated from the German).

[A2] BLASCHKE, W. and LEICHTWEISS, K.: *Elementare Differential-geometrie*, Springer, 1977.

AMS 1980 Subject Classification: 53A04

NATURAL EQUATION *of a curve* - A system of equations

$$k_1 = \phi(s), \quad k_2 = \psi(s),$$

defining the curvature k_1 and torsion k_2 of the curve as functions of the arc length parameter s on the curve. For any regular functions $\phi(s) > 0$ and $\psi(s)$ there exists a curve, unique up to translation in space, with **curvature** $\phi(s)$ and **torsion** $\psi(s)$. A necessary and sufficient condition for a curve to be in a plane is that its torsion vanishes identically. A necessary and sufficient condition for a curve to be a straight line (or a segment of a straight line) is that its curvature vanishes identically.

D.D. Sokolov

Editorial comments. In the article above, ϕ must be positive in order to generate uniqueness of the curve; for existence $\phi(s) \geq 0$ suffices (cf. [A1], Sects. 8.5.8 and 8.6.15).

Instead of 'natural equation' one also finds the phrase '*intrinsic equation of a curve*'. The representation of (certain special) plane curves by means of a relation $k_1 = \phi(s)$ goes back to L. Euler.

References

[A1] BERGER, M. and GOSTIAUX, B.: *Differential geometry*, Springer, 1988 (translated from the French).

[A2] DO CARMO, M.: *Differential geometry of curves and surfaces*, Prentice Hall, 1976.

[A3] O'NEILL, B.: *Elementary differential geometry*, Acad. Press, 1966.

[A4] SPIVAK, M.: *Differential geometry*, 2, Publish or Perish, 1979.

[A5] BLASCHKE, W. and LEICHTWEISS, K.: *Elementare Differential-geometrie*, Springer, 1973.

[A6] STRUIK, D.J.: *Differential geometry*, Addison-Wesley, 1950, Sect. 1-8.

AMS 1980 Subject Classification: 53A04

NATURAL LOGICAL DEDUCTION - A formal deduction approximating as closely as possible the essence of the reasoning usual in mathematics and logic. Criteria for the naturalness and quality of a deduction cannot be specified with complete precision, but they usually concern deductions that can be carried out by the generally accepted rules of logical transformations, that are compact (in particular, do not contain superfluous applications of deduction rules), that are 'glued together' (corresponding part of deductions must be eliminated, for example by extracting auxiliary lemmas), etc.

Originally, formalizations of mathematical and logical theories did not aim at naturalness (see **Logical cal-**

culus); a decisive advance in this direction was made by the calculus of natural deduction (see **Gentzen formal system**), which imitates the form of conventional mathematical argument and allows one to introduce and use assumptions in the usual way. Other quite natural methods are those for handling assumptions in sequent calculi, which have additional advantages — the **subformulation property** — and so form a basis for further advances in the problem of constructing natural logical deductions.

For automating the search for natural logical deductions auxiliary sequent calculi have been proposed [2] that possess the subformulation property but do not allow the transfer of assumptions to the succedent (see **Sequent (in logic)**). It is easier to construct natural logical deductions in terms of deductions in such a calculus. On this basis a method has been developed for the search for natural deductions, taking account of 'likenesses' (that is, of equal subformulas in the structure of the formulas being tested) for shortening the deductions and 'gluing' them together, 'weeding out' superfluous formulas and applications of rules, the possibility of varying the tactics in establishing deducibility, etc. Within the bounds of the logical means of classical **propositional calculus** these methods have led to a computer algorithm; the program finds a natural logical deduction for a given assertion from a given list of hypotheses, and records this deduction as a logico-mathematical text in Russian. Problems related to the search for natural logical deductions are those of correction of hypotheses and strengthening of theorems; these concern methods enabling small corrections to be made in a given formula so that it becomes a theorem or is changed into a stronger theorem, and investigations into criteria for the quality of such corrections.

Developments in the field of natural logical deductions have been mainly concerned with classical logics, but the resulting methods have a more general character.

References

[1] *Mathematical theory of logical deduction*, Moscow, 1967 (in Russian; translated from the English). Collection of translations.
[2] SHANIN, N.A., ET AL.: *An algorithm for computer search for a natural logical deduction in propositional calculus*, Moscow-Leningrad, 1965 (in Russian).
[3] PRAWITZ, D.: *Natural deduction*, Almqvist & Wiksell, 1965.

S. Yu. Maslov

Editorial comments. Cf. also **Derivation, logical**.

References

[A1] SCHÜTTE, K.: *Proof theory*, Springer, 1977.
[A2] GENTZEN, G.: M. Szabo (ed.): *Collected papers*, North-Holland, 1969.

AMS 1980 Subject Classification: 03FXX, 03B10

NATURAL NUMBER - One of the fundamental concepts in mathematics. Natural numbers may be interpreted as the cardinal numbers (cf. **Cardinal number**) of non-empty finite sets. The set $N = \{1, 2, \ldots\}$ of all natural numbers, together with the operations of addition $(+)$ and multiplication $(\cdot)$, forms the natural number system $<N, +, \cdot, 1>$. In this system, both binary operations are associative and commutative and satisfy the distributivity law; 1 is the neutral element for multiplication, i.e. $a \cdot 1 = a$ for any natural number a; there is no neutral element for addition, and, moreover, $a + b \neq a$ for any natural numbers a, b. Finally, the following condition, known as the *axiom of induction*, is satisfied. Any subset of N that contains 1 and, together with any element a also contains the sum $a + 1$, is necessarily the whole of N. See **Natural sequence**; **Arithmetic, formal**.

References

[1] *The history of mathematics from Antiquity to the beginning of the XIX-th century*, 1, Moscow, 1970 (in Russian).
[2] NECHAEV, V.I.: *Number systems*, Moscow, 1975 (in Russian).
[3] DAVENPORT, H.: *The higher arithmetic*, Hutchinson, 1952.

A.A. Bukhshtab
V.N. Nechaev

Editorial comments. A definition more elegant than the definition given above (the one of Frege — Russell) as cardinal numbers is von Neumann's definition, identifying a number with the set of its predecessors: $0 = \emptyset$, $'n + 1' = Sn = \{0, \ldots, n\}$. Here S denotes 'successor'. In this definition 0 is taken to belong to N (this is often done). In this case, 0 is the neutral element for addition.

Cf. also **Natural sequence**.

References

[A1] SCRIBA, C.J.: *The concept of number, a chapter in the history of mathematics, with applications of interest to teachers*, B.I. Mannheim, 1968.

AMS 1980 Subject Classification: 10A99, 04-XX

NATURAL PARAMETER *on a rectifiable curve* - A parameter s for a curve γ with parametric representation $\mathbf{r} = \mathbf{r}(s)$ such that the arc length on the curve between two points $\mathbf{r}(s_1)$ and $\mathbf{r}(s_2)$ is equal to $|s_1 - s_2|$. The parametrization of a curve by the natural parameter is known as its *natural parametrization*. The natural parametrization of a k-times differentiable (analytic) curve with no singular points is also k times differentiable (analytic).

D.D. Sokolov

Editorial comments. See also (the references to) **Natural equation**.

AMS 1980 Subject Classification: 53A04

NATURAL SEQUENCE, *natural number sequence* - The non-empty set $N = \{1, 2, \ldots\}$ in which a *unary operation* S is defined (i.e. S is a single-valued mapping of **N** into itself) satisfying the following conditions (the **Peano axioms**):

1) for any $a \in \mathbf{N}$,
$$1 \neq Sa;$$

2) for any $a, b \in \mathbf{N}$: If
$$Sa = Sb,$$
then
$$a = b;$$

3) any subset of **N** that contains 1 and that together with any element a also contains Sa, is necessarily the whole of **N** (*axiom of induction*).

The element Sa is usually called the *immediate successor* of a. The natural sequence is a **totally ordered set**. It can be proved that the conditions
$$a+1 = Sa, \quad a+Sb = S(a+b),$$
$$a\cdot 1 = a, \quad a\cdot Sb = ab+a,$$
where a and b are arbitrary elements of **N**, define binary operations $(+)$ and $(\cdot)$ on **N**. The system $<\mathbf{N}, +, \cdot, 1>$ is the system of natural numbers (cf. **Natural number**).

References

[1] WAERDEN, B.L. VAN DER: *Algebra*, 1, Springer, 1967 (translated from the German).

A.A. Bukhshtab
V.I. Nechaev

Editorial comments. Often, the natural number sequence is started at 0, cf. also **Natural number**.

The system $(\mathbf{N}, S)$ is the only (up to an isomorphism) system satisfying the Peano axioms.

When saying that $(\mathbf{N}, S)$ is a totally ordered set, one refers to the total order relation $<$ defined by:
$$\neg(a < 1),$$
$$a<Sb \Leftrightarrow a<b \text{ or } a=b.$$

References

[A1] KENNEDY, H.C.: *Selected works of Guiseppe Peano*, Allen & Unwin, 1973.
[A2] LANDAU, E.: *Grundlagen der Analysis*, Akad. Verlagsgesellschaft, 1930.
[A3] MACLANE, S. and BIRKHOFF, G.: *Algebra*, Macmillan, 1967.

AMS 1980 Subject Classification: 10A99

NATURALLY ORDERED GROUPOID - A partially ordered groupoid (cf. **Partially ordered set; Groupoid**) H in which all elements are positive (that is, $a \leqslant ab$ and $b \leqslant ab$ for any $a, b \in H$) and the larger of two elements is always divisible (on both the left and the right) by the smaller, that is, $a<b$ implies that $ax=ya=b$ for some $x, y \in H$. The positive cone of any partially ordered group (cf. **Ordered group**) is a naturally ordered semi-group.

O.A. Ivanova

Editorial comments.

References

[A1] FUCHS, L.: *Partially ordered algebraic systems*, Pergamon, 1963.

AMS 1980 Subject Classification: 06F05

NAVIER — STOKES EQUATIONS - The fundamental equations of motion of a viscous liquid; they are mathematical expressions of the conservation laws of momentum and mass. For a non-stationary flow of a compressible liquid, the Navier — Stokes equations in a Cartesian coordinate system may be written as

$$
\begin{aligned}
\rho\frac{du}{dt} &= X-\frac{\partial p}{\partial x}+\frac{\partial}{\partial x}\left[\mu\left(2\frac{\partial u}{\partial x}-\frac{2}{3}\operatorname{div}\mathbf{w}\right)\right]+ \\
&\quad +\frac{\partial}{\partial y}\left[\mu\left(\frac{\partial u}{\partial y}+\frac{\partial v}{\partial x}\right)\right]+\frac{\partial}{\partial z}\left[\mu\left(\frac{\partial w}{\partial x}+\frac{\partial u}{\partial z}\right)\right], \\
\rho\frac{dv}{dt} &= Y-\frac{\partial p}{\partial y}+\frac{\partial}{\partial y}\left[\mu\left(2\frac{\partial v}{\partial y}-\frac{2}{3}\operatorname{div}\mathbf{w}\right)\right]+ \\
&\quad +\frac{\partial}{\partial z}\left[\mu\left(\frac{\partial v}{\partial z}+\frac{\partial w}{\partial y}\right)\right]+\frac{\partial}{\partial x}\left[\mu\left(\frac{\partial u}{\partial y}+\frac{\partial v}{\partial x}\right)\right], \\
\rho\frac{dw}{dt} &= Z-\frac{\partial p}{\partial z}+\frac{\partial}{\partial z}\left[\mu\left(2\frac{\partial w}{\partial z}-\frac{2}{3}\operatorname{div}\mathbf{w}\right)\right]+ \\
&\quad +\frac{\partial}{\partial x}\left[\mu\left(\frac{\partial w}{\partial x}+\frac{\partial u}{\partial z}\right)\right]+\frac{\partial}{\partial y}\left[\mu\left(\frac{\partial v}{\partial z}+\frac{\partial w}{\partial y}\right)\right], \\
\frac{\partial \rho}{\partial t}&+\frac{\partial(\rho u)}{\partial x}+\frac{\partial(\rho v)}{\partial y}+\frac{\partial(\rho w)}{\partial z} = 0,
\end{aligned}
\tag{1}
$$

where $\mathbf{w}$ is the velocity vector, with projections u, v, w on the coordinate axes x, y, z, respectively; p is the pressure, ρ the density, μ the viscosity coefficient; X, Y, Z are the projections of the mass force vector K on the coordinate axes; and $d\mathbf{w}/dt = \partial\mathbf{w}/\partial t + \mathbf{w}\operatorname{grad}\mathbf{w}$ is the substantive derivative. The derivation of equations (1) is based on Newton's generalized law of friction, according to which the stress in a moving fluid or gas is proportional to the strain rates. For the study of compressible flows one must add to (1) the equation of state, which relates the pressure, the density and the temperature, and the energy equation.

Equations (1), which are the basis of hydrodynamics, were first derived by L. Navier [1] and S.D. Poisson [2], on the basis of considerations involving the action of intermolecular forces. B. Saint-Venant [3] and G.G. Stokes [4] derived the equations under the sole assumption that the normal and tangential stresses are linear functions of the strain rates.

For a flow of an incompressible isothermic liquid ($\rho = \text{const}$), equations (1) can be put in the following vector form:

$$
\left.
\begin{aligned}
\rho\frac{d\mathbf{w}}{dt} &= \mathbf{K}-\operatorname{grad}p+\mu\Delta\mathbf{w}, \\[2ex]
\operatorname{div}\mathbf{w} &= 0.
\end{aligned}
\right\}
\tag{2}
$$

As a rule, when analyzing the Navier—Stokes equations one transforms them to dimension-less form, scaling all quantities figuring therein by dividing by appropriate characteristic units. Thus, for stationary flows of an incompressible liquid, with no mass forces, the Navier—Stokes equations contain one decisive dimension-less parameter, known as the **Reynolds number**:

$$\text{Re} = \frac{\rho V l}{\mu} = \frac{V l}{\nu},$$

where V, l are the characteristic velocity and linear dimension, and ν is the kinematic viscosity.

To study two-dimensional incompressible flows one frequently uses the Helmholtz form of the Navier—Stokes equations:

$$\left. \begin{array}{c} \dfrac{\partial \xi}{\partial t} + \dfrac{\partial \psi}{\partial y} \dfrac{\partial \xi}{\partial x} - \dfrac{\partial \psi}{\partial x} \dfrac{\partial \xi}{\partial y} = \dfrac{1}{\text{Re}} \Delta \xi, \\ \Delta \psi = -\xi, \end{array} \right\} \tag{3}$$

where ψ is the stream function and ξ the rotation, which are related to the velocity projections u and v as follows:

$$u = \frac{\partial \psi}{\partial y}, \quad v = -\frac{\partial \psi}{\partial x}, \quad \xi = \frac{\partial v}{\partial x} - \frac{\partial u}{\partial y}.$$

The fundamental boundary value problems for the stationary Navier—Stokes equations are those associated with the investigation of flows in closed cavities, channels, flows with free surfaces, flows around bodies, jet flows, and wakes behind bodies. In all these problems the equations are integrated over (finite or infinite) domains with boundary conditions determined by physical considerations (adhesion or slipping along the surfaces of bodies, blowing-in or suction through permeable surfaces, conditions governing the external flow far from the body in the flow, conditions on the free boundary, etc.). For non-stationary problems one needs, besides the boundary conditions, suitable initial conditions.

As yet (1989) there exists no rigorous mathematical analysis of the solvability of the boundary value problems of hydro- and aeromechanics for the Navier—Stokes equations. A few results are available in the mathematical theory of the dynamics of viscous incompressible liquids (see **Hydrodynamics, mathematical problems in**).

The efforts of investigators were originally directed towards the determination of exact solutions. For example, in the incompressible case there exist exact solutions for the following stationary flows: in a plane channel with given constant pressure differential (*Poiseuille flow*); between two parallel plane walls, one of which is at rest, while the other is moving in its own plane at constant velocity (*Couette flow*); in a rectilinear tube of circular cross-section with constant pressure drop (*Hagen—Poiseuille flow*). A few self-similar solutions have also been found, including: plane-parallel (and axially-symmetric) flow near a critical point (*Howart flow*); and flows in contracting and expanding channels (*Hamel flow*).

Approximate solutions of the Navier—Stokes equations are based on simplifying assumptions. Worthy of mention here are solutions for very small Reynolds numbers (Re≪1), corresponding to creeping motions, the best-known of which is a *Stokes flow* around a sphere. The limiting case of very large Reynolds numbers leads to the theory of the hydrodynamic boundary layer. The boundary-layer equations have made it possible to solve a broad range of problems of practical importance, using well-developed approximate and numerical methods.

Certain classes of problems in the dynamics of viscous liquids and gases can be solved with the aid of fairly efficient algorithms, based on the use of difference schemes. For example, this is the case for the calculation of laminar flows of viscous incompressible liquids in regions of simple shape (or around bodies of simple shape). The most widely used difference methods in this context are those intended for the equations in Helmholtz form (3), though this system involves certain difficulties stemming from the determination of boundary conditions for ξ. The first results in solving the stationary version of system (3) were obtained using simple explicit five-point schemes and iterative methods (see [6]). The solution of stationary problems in the dynamics of viscous incompressible liquids are largely based on the adjustment method and on the application of explicit and implicit schemes for the system (3). Of the explicit schemes, one uses schemes with two levels in time, with symmetric approximation of the first derivatives by central differences and the solution of the second equation in (3) on each time level by Seidel's method; also used is a three-level scheme in which the convective terms are approximated by a 'cross' scheme and the diffusion terms by the Dufort—Frankel scheme. These schemes are also adequate to yield certain results in the solution of stationary problems for two-dimensional laminar flows in contracting and expanding channels, in a rectilinear cut with moving cover, and also in the non-stationary problem of flow in a channel around a flat plate perpendicular to the direction of the flow.

Implicit schemes are based, as a rule, on the method of fractional steps (see [8]). The general structure of these schemes for equations (3) may be presented, for example, as follows:

$$\frac{\xi^{n+1/2} - \xi^n}{0.5\tau} = L_{1\xi}\xi^{n+1/2} + L_{2\xi}\xi^n,$$

$$\frac{\xi^{n+1} - \xi^{n+1/2}}{0.5\tau} = L_{1\xi}\xi^{n+1/2} + L_{2\xi}\xi^{n+1},$$

$$\frac{\psi^{s+1/2,n+1}-\psi^{s,n+1}}{0.5\sigma}+$$
$$+L_{1\psi}\psi^{s+1/2,n+1}+L_{2\psi}\psi^{s,n+1}=-\xi^{n+1},$$
$$\frac{\psi^{s+1,n+1}-\psi^{s+1/2,n+1}}{0.5\sigma}+$$
$$+L_{1\psi}\phi^{s+1/2,n+1}+L_{2\psi}\psi^{s+1,n+1}=-\xi^{n+1},$$

where $L_{1\xi}$, $L_{2\xi}$, $L_{1\psi}$, $L_{2\psi}$ are one-dimensional difference operators:

$$L_{1\xi}=\frac{\delta^2\xi}{\delta x^2}-\frac{\delta\psi}{\delta y}\frac{\delta\xi}{\delta x},\quad L_{2\xi}=\frac{\delta^2\xi}{\delta y^2}+\frac{\delta\psi}{\delta x}\frac{\delta\xi}{\delta y},$$
$$L_{1\psi}=\frac{\delta^2\psi}{\delta x^2},\quad L_{2\psi}=\frac{\delta^2\psi}{\delta y^2}.$$

In these formulas, τ is the time step; σ the iteration parameter; $s, s+1/2, s+1$ are the iteration indices in the iterative solution of the Poisson equation in (3) on the $(n+1)$-st time level; and $\delta^2/\delta x^2$, $\delta^2/\delta y^2$, $\delta/\delta x$, $\delta/\delta y$ are the difference operators approximating the corresponding second- and first-order derivatives. The first equation of (3) is used to find the values of ξ, the second — to find the values of ψ on the next time level. The approximation of the second derivatives is usually symmetric, while that of the first derivatives in the equation for ξ is done either with symmetric differences or with one-sided differences against the flow, with the sign of the velocity taken into consideration. A highly-recommended method is the scheme developed in [9], together with monotone approximation (see [10]). For the first equation of (3), this scheme is

$$\left.\begin{array}{l}\dfrac{\xi^{n+1}-\xi^{n}}{\tau}+\dfrac{u-|u|}{2}\left[\dfrac{\delta\xi}{\delta x}\right]^{+}+\dfrac{u+|u|}{2}\left[\dfrac{\delta\xi}{\delta x}\right]^{-}+\\[2mm]+\dfrac{v-|v|}{2}\left[\dfrac{\delta\xi}{\delta y}\right]^{+}+\dfrac{v+|v|}{2}\left[\dfrac{\delta\xi}{\delta y}\right]^{-}=\\[2mm]=\dfrac{1}{Re}\left[\dfrac{1}{1+|u|h/2}\dfrac{\delta^2\xi}{\delta x^2}+\dfrac{1}{1+|v|l/2}\dfrac{\delta^2\xi}{\delta y^2}\right],\end{array}\right\}\quad(4)$$

where h and l are the grid steps in the x- and y-directions, and

$$\left[\frac{\delta\xi}{\delta x}\right]^{+}=\frac{\xi^{n+1}_{i+1,j}-\xi^{n+1}_{i,j}}{h},\quad\left[\frac{\delta\xi}{\delta x}\right]^{-}=\frac{\xi^{n+1}_{i,j}-\xi^{n+1}_{i-1,j}}{h},$$
$$\left[\frac{\delta\xi}{\delta y}\right]^{+}=\frac{\xi^{n+1}_{i,j+1}-\xi^{n+1}_{i,j}}{l},\quad\left[\frac{\delta\xi}{\delta y}\right]^{-}=\frac{\xi^{n+1}_{i,j}-\xi^{n+1}_{i,j-1}}{l}.$$

The difference equations (4) are usually transformed to three-diagonal form and solved, together with relations approximating the boundary conditions, by the double-sweep method. When solving stationary problems by the adjustment method one can either successively solve equations (4) (without internal iterations for the determination of ψ) or simultaneously solve the equations of (4), determining ξ and ψ jointly, by a vector double-sweep. The difficulty in the formulation of boundary conditions for equations (3) is that, in the case of the Navier — Stokes equations, the ordinary boundary conditions of adhesion at solid walls yield conditions only for ψ. A formal requirement in the numerical solution of the equation for ξ is the formulation of boundary conditions for the rotation. These conditions can be obtained on each time level, either approximately, on the boundary of the region, or by integrating the equation for ξ only in a region within the basic domain of integration [9].

The equations in the form (2) are less commonly used to study two-dimensional flows; usually, the continuity equation is regularized in some way. Variational-grid methods, in particular the method of finite elements, have found application in solving the equations of the dynamics of viscous fluids in the forms (2) and (3).

Difference methods have been applied to the investigation of a variety of flow problems for viscous incompressible liquids. Among these are flow around an elliptic or circular cylinder (including a rotating cylinder), around a plate of finite thickness (including a plate at an angle of attack), a cylindrical butt-end, a drop, a plane step, etc. Flows have also been studied in a cavity, in plane and cylindrical channels, in a channel with obstacles on the walls, as well as flows with a free surface, and flows of natural, induced and mixed convection.

Calculations for viscous compressible gases, based on the application of difference methods to the full Navier — Stokes equations, involve certain additional difficulties in comparison with calculations for viscous incompressible liquids. The reason is that in the flow of a compressible gas one has not only boundary-layer regions, but also other regions characterized by high gradients of the unknown functions corresponding to shock waves and rarefraction waves in the non-viscous gas flows. The complexity of the Navier — Stokes equations themselves for a viscous compressible gas imposes considerable demands on the speed and memory of the computer. In this case the application of explicit schemes yields simpler algorithms. One explicit scheme recommended for the calculation of stationary flows by the adjustment method is the scheme that, applied as an illustration to the simple equation with constant coefficients

$$\frac{\partial u}{\partial t}=a\frac{\partial u}{\partial x}+\nu\frac{\partial^2 u}{\partial x^2},\qquad(5)$$

may be written as

$$\frac{u^{n+1}_{m}-u^{n}_{m}}{\tau}=a\frac{u^{n}_{m+1}-u^{n+1}_{m-1}}{2h}+\nu\frac{u^{n}_{m+1}-2u^{n+1}_{m}+u^{n+1}_{m-1}}{h^2}.$$

This scheme approximates (5) for a smooth steady-state solution, up to the order $O(h^2)$, and is stable for $\tau<h/a$. The stability condition is independent of ν. Applied to the Navier — Stokes equations, this means that the stability condition is independent of the Rey-

nolds number. Nevertheless, the restriction imposed on the time iteration step in explicit schemes is substantial. Implicit schemes are usually free of this restriction and they are absolutely stable for such linear equations with constant coefficients. Implicit schemes for calculating two-dimensional flows of a viscous gas are used in combination with the method of fractional steps. Various difference schemes of the variable-directions method have recently (1982) been constructed: with complete and incomplete approximation on intermediate levels, divergent and non-divergent (see [11]), and schemes with an improved order of accuracy (higher than the second order) relative to the space step of the grid (see [12]). Numerical simulation of flows of a viscous gas based on the Navier — Stokes equations involves the calculation of flows of a complex structure and the use of sufficiently fine grids. This is impossible, because of limitations on the computer memory, without the use of the method of mutually overlapping regions (see [13]). Difference methods for solving the Navier — Stokes equations have been used to investigate a large number of problems in the dynamics of a viscous gas. Among these are: supersonic flow around blunt bodies (spheres, butt-ends of a circular cylinder), around compound bodies (sphere-cylinder, sphere-cone), and around a wedge or a leading edge of a flat plate. Flows have also been considered in wakes behind bodies of finite dimensions, in nozzles and air intakes, and in cavities with external sub- and supersonic current. Other topics that have been studied are the interaction of a boundary layer with a shock wave, the structure of a shock wave in a plasma, etc. The flows mentioned above are usually assumed to be laminar.

References

[1] NAVIER, L.: *Mém. Acad. Sci.* **7** (1827), 375-394.

[2] POISSON, S.D.: *J. Ecole Polytechn.* **13** (1831), 1-174.

[3] SAINT-VENANT, B.: *Acad. Sci. Paris* **17** (1843).

[4] STOKES, G.G.: *Trans. Cambridge Phil. Soc.* **8** (1843), 287-319.

[5] SCHLICHTING, G.: *Grenzschicht Theorie*, Braun, 1958.

[6] THOM, A. and APELT, C.J.: *Field computations in engineering and physics*, v. Nostrand, 1961.

[7] BRAILOVSKAYA, I.YU., KUSKOVA, T.V. and CHUDOV, L.A.: 'Difference methods for solving the Navier — Stokes equation (a survey)', in *Numerical methods and programming*, Vol. 11, Moscow, 1968, pp. 3-18 (in Russian).

[8] YANENKO, N.N.: *The method of fractional steps; the solution of problems of mathematical physics in several variables*, Springer, 1971 (translated from the Russian).

[9] POLEZHAEV, V.I. and GRYAZNOV, V.L.: *Dokl. Akad. Nauk SSSR* **219**, no. 2 (1974), 301-304.

[10] SAMARSKIĬ, A.A.: 'Monotonic difference schemes for elliptic and parabolic equations in the case of a non-selfadjoint elliptic operator', *USSR Comp. Math. Math. Phys.* **5**, no. 3 (1965), 212-215. (*Zh. Vychisl. Mat. i Mat. Fiz.* **5** (1965), 548-551)

[11] KOVENYA, V.M. and YANENKO, N.N.: *The method of splitting in problems of gas dynamics*, Novosibirsk, 1981 (in Russian).

[12] TOLSTYKH, A.I.: 'Condensation of grid points in the process of solving and using high-accuracy schemes for the numerical integration of viscous gas flows', *USSR Comp. Math. Math.*

Phys. **18**, no. 1 (1978), 134-147. (*Zh. Vychisl. Mat. i Mat. Fiz.* **18**, no. 1 (1978), 139-153)

[13] KOKOSHINSKAYA, N.S., PAVLOV, B.M. and PASKONOV, V.M.: *Numerical investigation of supersonic flow of a viscous gas around bodies*, Moscow, 1980 (in Russian).

[14] TEMAM, R.: *Navier — Stokes equations: theory and numerical analysis*, North-Holland, 1977.

[15] *Numerical investigation of contemporary problems of gas dynamics*, Moscow, 1974 (in Russian).

[16] ROACHE, P.J.: *Computational fluid dynamics*, Hermosa, 1972.

[17] PEYRET, R. and VIVIAND, H.: 'Résolution numérique des équations de Navier — Stokes pour les fluides compressible', in R. Glowinski and J.L. Lions (eds.): *Computing Methods in Applied Sciences and Engineering, 2*, Lect. notes in computer sci., Vol. 11, Springer, 1974, pp. 160-184.

[18] BURGGRAF, O.R.: 'Some recent developments in computation of viscous flow', in A.I. van der Vooren and P.J. Zandbergen (eds.): *Proc. 5th Internat. Conf. Numerical Methods in Fluid Dynamics*, Lect. notes in physics, Vol. 59, Springer, 1976, pp. 52-64.

V.M. Paskonov

Editorial comments. In the last ten years or so considerable progress has been made in some of the questions concerning the relation of the Navier — Stokes equations to finite-dimensional phenomena. This refers e.g. to the dimension of the universal attractor for the 2-dimensional Navier — Stokes equations and upper bounds for the fractal dimension of bounded invariant sets for the 3-dimensional Navier — Stokes equation, cf. [A2].

References

[A1] KREISS, H.O. and LORENZ, J.: *Initial boundary value problems and the Navier — Stokes equations*, Acad. Press, 1989.

[A2] CONSTANTIN, P. and FOIAS, C.: *Navier — Stokes equations*, Univ. Chicago Press, 1988.

AMS 1980 Subject Classification: 35Q10, 76D05, 76D10, 65NXX

NEAR-RING - One of the generalizations of the concept of an associative ring (cf. **Associative rings and algebras**). A near-ring is a **ringoid** over a group, i.e. a **universal algebra** in which an associative multiplication and addition exist; a near-ring is a (not necessarily Abelian) group with respect to addition, and the right distributive property

$$x(y+z) = xy+xz$$

must hold too. A near-ring is also an example of a **multi-operator group**.

Examples of near-rings are the set $M_S(\Gamma)$ of all mappings of a group Γ into itself which commute with the action of a given semi-group S of endomorphisms of Γ. The group operations in $M_S(\Gamma)$ are defined pointwise and multiplication in $M_S(\Gamma)$ is composition of mappings. A near-ring $M_S(\Gamma)$ is an analogue of a ring of matrices. The notions of a sub-near-ring, of an ideal and of a right module over a near-ring are introduced in the usual manner.

Let N_0 (N_c) be the variety of near-rings defined by the identity $0x=0$ $(0x=x)$. Every near-ring A can be decomposed into the sum $A=A_0+A_c$ of sub-near-rings, where $A_0 \in N_0$, $A_c \in N_c$ and $A_0 \cap A_c = 0$. A cyclic right A-module M is called *primitive of type 0* if M is simple; *primitive of type 1* if either $xA = 0$ or $xA = M$ for any $x \in M$; and *primitive of type 2* if M is a simple A_0-module. A near-ring A is called primitive of type ν ($\nu = 0, 1, 2$) if there is a faithful simple A-module Γ of type ν. In this case there is a dense imbedding of A into $M_S(\Gamma)$ for some semi-group S of endomorphisms of Γ. For 2-primitive near-rings A with an identity element and with the minimum condition for right ideals in A_0, the equality $A = M_S(\Gamma)$ holds (an analogue of the Wedderburn $-$ Artin theorem). For every $\nu = 0, 1, 2$, the *Jacobson radical $J_\nu(A)$ of type ν* can be introduced as the intersection of the annihilators of ν-primitive A-modules. The radical $J_{1/2}(A)$ is defined as the intersection of the maximal right module ideals. All four radicals are different, and

$$J_0(A) \subseteq J_{1/2}(A) \subseteq J_1(A) \subseteq J_2(A).$$

It turns out that these radicals posses many properties of the **Jacobson radical** of an associative ring (cf. [4]).

For near-rings an analogue of Ore's theorem on near-rings of fractions [4] holds.

A *distributively-generated near-ring* is a near-ring whose additive group is generated by elements x such that

$$(y + z)x = yx + zx$$

for all y and z in the near-ring. All distributively-generated near-rings generate the variety N_0. For finite distributively-generated near-rings the notions of 1- and 2-primitivity coincide; 1-primitive distributively-generated near-rings have the form $M_0(\Gamma)$ for some group Γ. In a distributively-generated near-ring with the identity

$$(xy - yx)^{n(x,y)} = xy - yx, \quad n(x,y) > 1,$$

multiplication is commutative (cf. [3], [4]).

Every near-ring from N_0 without nilpotent elements is a subdirect product of near-rings without divisors of zero [4]. A near-algebra A can be decomposed into a direct sum of simple near-rings if and only if: a) it satisfies the minimum condition for principal ideals; b) A does not contain ideals with zero multiplication; and c) any annihilator of any minimal ideal is maximal [1].

For near-rings one can prove results similar to those on the structure of regular rings [2] and on near-rings of fractions [5]. Near-rings have applications in the study of permutation groups, block-schemes and projective geometry [4].

References

[1] BELL, H.E.: 'A commutativity theorem for near-rings', *Canad. Math. Bull.* **20**, no. 1 (1977), 25-28.
[2] HEATHERLY, H.E.: 'Regular near-rings', *J. Indian Math. Soc.* **38** (1974), 345-354.
[3] LIGH, S.: 'The structure of certain classes of rings and near

rings', *J. London Math. Soc.* **12**, no. 1 (1975), 27-31.
[4] PILZ, G.: *Near-rings*, North-Holland, 1983.
[5] OSWALD, A.: 'On near-rings of quotients', *Proc. Edinburgh Math. Soc.* **22**, no. 2 (1979), 77-86.
[6] POLIN, S.V.: 'Generalizations of rings', in *Rings*, Vol. 1, Novosibirsk, 1973, pp. 41-45 (in Russian).
[7] MELDRUM, J.D.P.: *Near-rings and their links with groups*, Pitman, 1985.

V.A. Artamonov

AMS 1980 Subject Classification: 16A76

NECESSARY AND SUFFICIENT CONDITIONS - Conditions for the validity of a proposition A without which A cannot possibly be true (necessary conditions), while when these conditions are satisfied, then A must be true (sufficient conditions). Frequently the expression 'necessary and sufficient' is replaced by 'if and only if'. Necessary and sufficient conditions have great significance. In complex mathematical problems the search for necessary and sufficient conditions that are convenient to use, sometimes becomes extremely difficult. In such cases one tries to find sufficient conditions that may be wider, that is, comprise possibly more cases in which the fact of interest still holds, and necessary conditions that may be narrower, that is, comprise possibly fewer cases in which the relevant fact does not hold. In this way, sufficient conditions come close step-by-step to necessary ones.

BSE-3

AMS 1980 Subject Classification: 00A25

NECESSARY SUFFICIENT STATISTIC - See **Minimal sufficient statistic**.

AMS 1980 Subject Classification: 62BXX

NEGATION - The logical operation as a result of which, for a given statement A, the statement 'not A' is obtained. In formal languages, the statement obtained as result of the negation of a statement A is denoted by $\neg A$, $\sim A$, $\bar{A}$, $-A$, A' (these are read: 'not A', 'it is not true that A', 'A does not hold', etc.). Semantically, the negation of a statement A signifies that the assumption A leads to a contradiction (cf. **Contradiction (inconsistency)**). In classical two-valued logic the following **truth table** applies for the operation of negation:

A	$\neg A$
T	F
F	T

V.E. Plisko

AMS 1980 Subject Classification: 03A05, 03B05

NEGATIVE BINOMIAL DISTRIBUTION - A **probability distribution** of a random variable X which takes non-negative integer values $k = 0, 1, \ldots$, in accordance with the formula

$$P\{X=k\} = \begin{bmatrix} r+k-1 \\ k \end{bmatrix} p^r(1-p)^k \qquad (*)$$

for any real values of the parameters $0<p<1$ and $r>0$. The **generating function** and the **characteristic function** of a negative binomial distribution are defined by the formulas

$$P(z) = p^r(1-qz)^{-r}$$

and

$$f(t) = p^r(1-qe^{it})^{-r},$$

respectively, where $q=1-p$. The mathematical expectation and variance are equal, respectively, to $rq=p$ and rq/p^2. The distribution function of a negative binomial distribution for the values $k=0,1,\ldots,$ is defined in terms of the values of the **beta-distribution** function at a point p by the following relation:

$$F(k) = P\{X<k\} = \frac{1}{B(r,k+1)}\int_0^p x^{r-1}(1-x)^k\,dx,$$

where $B(r,k+1)$ is the beta-function.

The origin of the term 'negative binomial distribution' is explained by the fact that this distribution is generated by a **binomial** with a negative exponent, i.e. the probabilities (*) are the coefficients of the expansion of $p^r(1-qz)^{-r}$ in powers of z.

Negative binomial distributions are encountered in many applications of probability theory. For an integer $r>0$, the negative binomial distribution is interpreted as the distribution of the number of failures before the r-th 'success' in a scheme of **Bernoulli trials** with probability of 'success' p; in this context it is usually called a **Pascal distribution** and is a discrete analogue of the **gamma-distribution**. When $r=1$, the negative binomial distribution coincides with the **geometric distribution**. The negative binomial distribution often appears in problems related to the randomization of the parameters of a distribution; for example, if Y is a random variable having, conditionally on λ, a **Poisson distribution** with random parameter λ, which in turn has a gamma-distribution with density

$$\frac{1}{\Gamma(\mu)}x^{\mu-1}e^{-\alpha x}, \quad x>0, \ \mu>0,$$

then the marginal distribution of Y will be a negative binomial distribution with parameters $r=\mu$ and $p=\alpha/(1+\alpha)$. The negative binomial distribution serves as a limiting form of a **Pólya distribution**.

The sum of independent random variables $X_1,\ldots,X_n$ which have negative binomial distributions with parameters p and $r_1,\ldots,r_n$, respectively, has a negative binomial distribution with parameters p and $r_1+\cdots+r_n$. For large r and small q, where $rq\sim\lambda$, the negative binomial distribution is approximated by the Poisson distribution with parameter λ. Many properties of a negative binomial distribution are deter-

mined by the fact that it is a generalized Poisson distribution.

References

[1] FELLER, W.: *An introduction to probability theory and its applications*, 1-2, Wiley, 1957-1971.

A.V. Prokhorov

Editorial comments. See also **Binomial distribution**.

References

[A1] JOHNSON, N.L. and KOTZ, S.: *Distributions in statistics, discrete distributions*, Wiley, 1969.

AMS 1980 Subject Classification: 60E99, 62E15

NEGATIVE CORRELATION - A form of correlative dependence between random variables under which the conditional mean value of one of these diminishes as the value of the other increases. One speaks of negative correlation between variables with a **correlation coefficient** ρ when $\rho<0$. See **Correlation (in statistics)**.

A.V. Prokhorov

AMS 1980 Subject Classification: 62H20

NEGATIVE CURVATURE, SURFACE OF (*in the direct sense*) - A two-dimensional surface in three-dimensional Euclidean space that has negative **Gaussian curvature** $K<0$ at every point. The simplest examples of this are: a one-sheet hyperboloid (Fig. 1a), a hyperbolic paraboloid (Fig. 1b) and a **catenoid**.

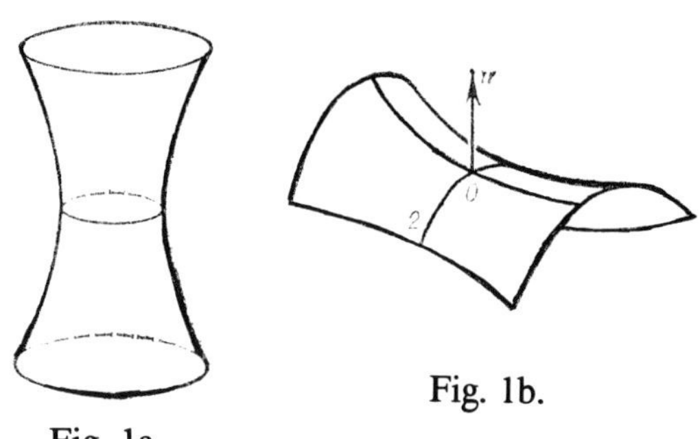

Fig. 1a.

Fig. 1b.

The concept of a surface of negative curvature can be generalized, for example, with respect to the dimension of the surface itself or the dimension and structure of the ambient space.

Surfaces of negative curvature locally have a saddle-like structure. This means that in a sufficiently small neighbourhood of any of its points, a surface of negative curvature resembles a saddle (see Fig. 1b, not considering the behaviour of the surface outside the part of it that has been drawn). The local saddle-like character of the surface is clearly illustrated in this figure, which shows the principal sections of the surface at an arbitrary point O. Let $1/R_1$, $1/R_2$ be their normal curvatures, i.e. the principal curvatures at the point O (cf. **Principal curvature**). According to the classical definition, the Gaussian curvature at O is the number

$K = 1/R_1 R_2$. Since $K < 0$, the principal curvatures have different signs, for which reason the principal sections are convex in opposite directions; in Fig. 1b the section 02 is convex in the direction of the normal n, while the other is convex in the opposite direction, which fits in with the saddle-like character of the surface. The topological structure in the large ('globally') of a surface of negative curvature can be very different. For example, a hyperbolic paraboloid (Fig. 1b) is topologically equivalent to the plane, while a one-sheet hyperboloid (Fig. 1a) is equivalent to a cylindrical tube. Let two one-sheet hyperboloids with parallel axes intersect along a hyperbola (Fig. 2a). Let these hyperboloids not be transparent, and let their invisible parts (i.e. the parts of any of them which lie within another) be disregarded. The surface thus obtained has negative curvature everywhere, apart from the points of the hyperbola mentioned above. At the points of this hyperbola there is, in general, no curvature (in the classical sense), since the hyperbola is an edge of the surface. However, the surface near the edge can, in the given case, be smoothed (see [7]), such that a surface is obtained that has a curvature at all points and that is, moreover, negative (see Fig. 2b). It is topologically equivalent to a torus with two punctures. The topological structure of a surface of negative curvature can be made as complex as desired by the inclusion of a larger number of one-sheet hyperboloids.

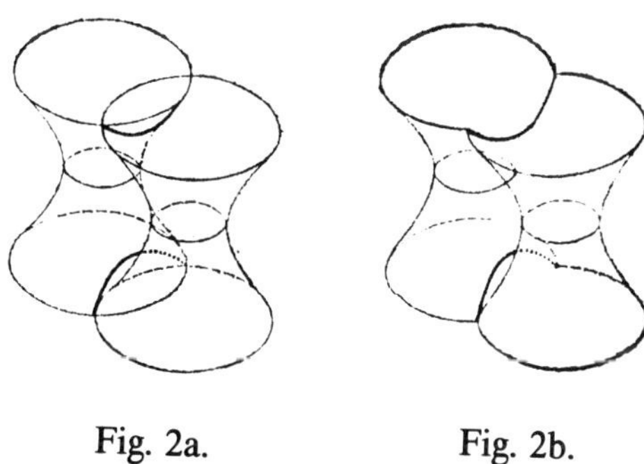

Fig. 2a.   Fig. 2b.

A surface is said to be *complete* if it is a complete metric space with respect to the distance given by its intrinsic geometry (cf. also **Internal metric**). In all examples presented above, complete surfaces were considered. The idea of constructing complete surfaces of negative curvature of various topological types arising from very simple types (as shown above) is attributed to J. Hadamard [1]. A fundamentally different example is proposed in Fig. 3 (see [6], [7]). This surface has negative curvature everywhere, and is completely enclosed in a sphere. It has an infinite number of branches, and a variable point of it moving along the branches can get infinitely close to the boundary of the sphere without ever touching it. The surface is constructed such that a variable point moving along the

branches always follows a path of infinite length. In this way, the completeness of the surface in the sense of its interior metric is guaranteed. This example shows that an intrinsically-complete surface of negative curvature does not necessarily move to infinity in space, like those in Hadamard's examples.

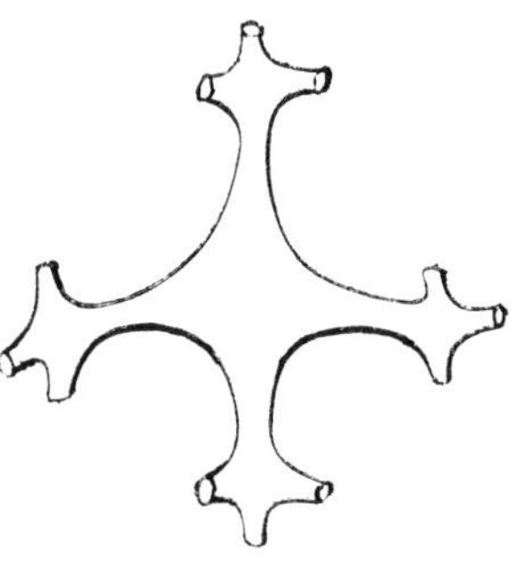

Fig. 3.

In this example, the condition $K < 0$ can generally be observed, allowing for a slight irregularity of the surface. This means that the surface is presumed to be smooth (everywhere of class C^1), but that at certain isolated points it does not belong to C^2; at these points, the curvature must have a limit value. After completing the curvature with those limit values, it is defined everywhere and is continuous. The question of the extension of the surface and its connection with the condition of regularity is complemented by the following theorems [8].

If the Gaussian curvature K of a complete surface of class C^2 satisfies the inequality $-a^2 \leqslant K \leqslant 0$, where $a = \text{const}$, then the surface cannot be contained within a sphere of radius less than $\sqrt{2/3}/a$.

If on a complete surface of class C^2, which is unbounded in the intrinsic sense, the Gaussian curvature $K \to 0$ at infinity, then the surface is not bounded in space (in this theorem, the sign of the curvature may change).

Particular attention has been (and still is) paid to surfaces of constant negative curvature, i.e., they have Gaussian curvature with the same negative value at all points. They are notable for the fact that their intrinsic geometry locally coincides with the geometry on the Lobachevskiĭ plane (see **Lobachevskiĭ geometry**). This means that for figures which lie on a surface of constant negative curvature, precisely the same relations occur as those which hold in the Lobachevskiĭ planimetry (the role of straight lines is played by geodesic lines); for example, the trigonometric formulas for triangles on a surface of constant negative curvature and the corresponding formulas in non-Euclidean Lobachevskiĭ geometry coincide exactly. Thus, on a surface of constant negative curvature, a local model of the Lobachevskiĭ geometry is obtained. A model of this type was constructed by E. Beltrami in 1868 (cf. **Beltrami interpretation**), which made an essential contribution to the recognition of Lobachevskiĭ geometry. Beltrami also proposed an important example of a surface of negative curvature, which he called a *pseudo-sphere* (Fig. 4) and which is formed by the rotation of a tractrix around its asymptote (Fig. 5).

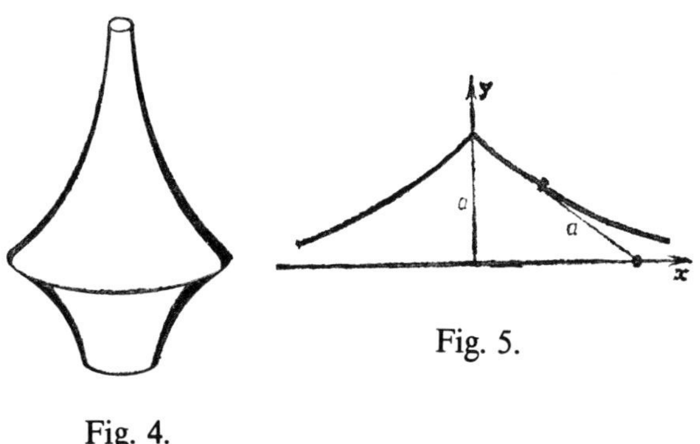

Fig. 5.

Fig. 4.

A characteristic property of a tractrix is that the segment of the tangent from a point of tangency to the asymptote is constant and equal to a number a. The Gaussian curvature of the pseudo-sphere is $K = -1/a^2$. Surfaces of rotation of negative curvature were studied even earlier than Beltrami by F. Minding in 1839. He discovered two forms of periodic surfaces of rotation of constant negative curvature (Fig. 6) which, with the pseudo-sphere, exhaust all possible surfaces of rotation of constant negative curvature (for their equations, see [3], [5]).

Fig. 6.

The construction of a global model of Lobachevskiĭ planimetry on a surface of rotation of constant negative curvature, for example on a pseudo-sphere, is not possible for two reasons. First, there are irregular points on a pseudo-sphere; these form a cuspidal edge (Fig. 4). Secondly, a pseudo-sphere is topologically equivalent to a tube, and therefore, even from the topological point of view, differs from a Lobachevskiĭ plane. However, it is possible to construct a model of the infinite part of the Lobachevskiĭ plane on a regular surface of rotation of constant negative curvature. To this end, an infinite (contractible) domain U is cut out of the pseudo-sphere by a plane perpendicular to the axis. The plane is so formed that it does not contain the cuspidal edge. The

part of the pseudo-sphere in which there is no cuspidal edge is taken as U. Thereafter, only U is considered, and, for the sake of convenience, it is also called a pseudo-sphere; it has no singular points. Let U' be the universal covering surface of U. This concept can be introduced in the following way. A countably infinite number of identical and superposed copies of the pseudo-sphere U is taken, which are distinguished using the integers: $\ldots, U_{-2}, U_{-1}, U_0, U_1, \ldots$. Then they are all cut along a common meridian, and a left and right side of the cut is marked on each of them, such that all left sides are superposed (the same applies also to the right sides). Now, for any number k, $-\infty < k < \infty$, the left side of U_k is joined to the right side of U_{k+1} (similarly, the right side of U_k is joined to the left side of U_{k-1}). A connected surface U' is thus obtained, which is also the universal covering surface of U, and is sometimes called the simply-connected *envelope* of the surface U. A universal covering can be constructed for any surface, but the visual description of its construction will be far less simple than that of a pseudo-sphere. It is essential that the universal covering is always simply connected. It is also important that the local intrinsic geometry of the covered surface can be transferred to the universal covering. For this to happen, it is sufficient to ensure that every point p of the covered surface has a convex neighbourhood $V(p)$ which is covered one-sheetedly by a neighbourhood $V'(p')$ in the covering surface (here p' is a point covering the point p; the correspondence $V \to V'$ is bijective). The geometry of $V(p)$ is transferred to $V'(p')$ identically, i.e. the distance between two points $p'_1, p'_2 \in V'$ is taken to be equal to the distance between their inverse images. Then the global geometry of the universal covering is defined, but, as a rule, it will be completely different from the global geometry of the covered surface.

On the pseudo-sphere U, the meridians, when moving towards infinity, converge; moreover, they are orthogonal to the boundary of U. The boundary of the universal covering is therefore a curve in the Lobachevskiĭ plane, which is the orthogonal trajectory of a pencil of Lobachevskiĭ-parallel (converging) straight lines. This curve is called a *horocycle* in the Lobachevskiĭ geometry. Thus, from the point of view of its intrinsic geometry, the universal covering of the pseudo-sphere U is the interior domain of an horocycle in the Lobachevskiĭ plane. The universal covering U' of the pseudo-sphere U is more convenient for the construction of a model of the Lobachevskiĭ geometry than the pseudo-sphere U itself. E.g., a circle of any radius can be placed on U', but a large circle without overlappings may prove to be too large to be contained in U. However, even U' is not extensive enough to contain some object of the Lobachevskiĭ geometry, such as, for example, a complete straight line. In other words, it is not possible to construct a model of the entire Lobachevskiĭ plane using U'.

In 1900, D. Hilbert posed the question of whether there exists a surface whose intrinsic geometry coincides completely with the geometry of the Lobachevskiĭ plane. Using a simple reasoning, it follows that if such a surface does exist, it must have constant negative curvature and be complete. Moreover, it must be simply connected, since the Lobachevskiĭ plane is simply connected. It would in fact be sufficient, however, to find a complete surface of constant negative curvature (as stated above, the universal covering of such a surface is complete and simply connected). In saying this, one considers only regular surfaces (at least twice continuously differentiable at any point, i.e. of class C^2).

As early as 1901, Hilbert solved this problem (see [2]), and in the negative sense, so that no complete regular surface of constant negative curvature exists in three-dimensional Euclidean space. This theorem has attracted the attention of geometers over a number of decades, and continues to do so today. The reason for this is that a number of interesting questions are related to it and to its proof (see below). The occurrence of singular lines on surfaces of rotation of constant negative curvature (Figs. 4, 6) is not coincidental, but is in line with Hilbert's theorem.

The proof of Hilbert's theorem on surfaces of any topological structure is of special interest, since the concept of a universal covering surface is essential to it. This concept reduces the problem to a surface with the simplest topology, namely, a simply-connected surface. Hilbert's theorem was evidently one of the earliest mathematical statements to explain this concept. Following the publication of Hilbert's first article expounding the proof of his theorem, doubts were expressed on the validity of his proofs. The criticism related to the limitless behaviour of the asymptotic lines, which were studied on the surface in terms of the complexity of the topology of that surface. It was even suggested that Hilbert had refined his proof to make it more accurate. Nevertheless, Hilbert's proof was quite faultless throughout. These criticisms would probably not have arisen if Hilbert had stated clearly that the object of study was a complete simply-connected manifold of constant negative curvature, i.e. the Lobachevskiĭ plane itself, assuming that it can be isometrically and regularly immersed in 3-dimensional Euclidean space.

It is also interesting that *Chebyshev nets* appear unexpectedly in Hilbert's proof. These are nets in which every net quadrangle has opposite sides of equal length (Fig. 7). Moreover, an **asymptotic net** (which is defined and non-degenerate at all points on any surface of

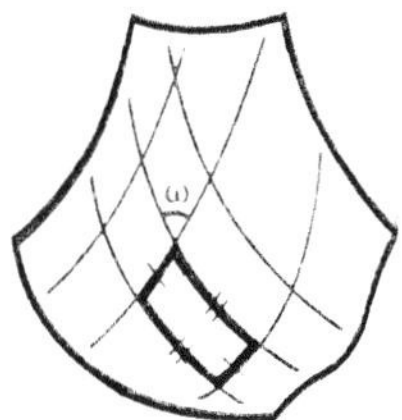

Fig. 7.

negative curvature) is a Chebyshev net in the case of a surface of constant negative curvature. This net can be taken to be a coordinate net: $u = \text{const}$, $v = \text{const}$. The line element then takes the Chebyshev form

$$ds^2 = du^2 + 2\cos\omega \, du \, dv + dv^2, \tag{1}$$

where ω is the angle in the net. Assuming that the surface is regular, complete and simply connected, and taking account of the Chebyshev nature of the net, it can be proved that u, v take all values: $-\infty < u < \infty$, $-\infty < v < \infty$; these exhaust all of the net. In other words, an asymptotic net in the large is homeomorphic to the Cartesian net on the Euclidean coordinate plane (see [9]). On the other hand, assuming that the Gaussian curvature is constant, for example $K = -1$, the equation

$$\frac{\partial^2 \omega}{\partial u \partial v} = \sin\omega \tag{2}$$

is obtained from (1), and the restrictions

$$0 < \omega < \pi \tag{3}$$

follow from geometric considerations. Hilbert demonstrated that equation (2) cannot have a regular solution $\omega = \omega(u, v)$ subject to the condition (3) on the entire coordinate plane, and the complete Lobachevskiĭ plane can therefore not be regularly and isometrically imbedded in 3-dimensional Euclidean space. As explained above, this leads to a more general formulation of this theorem: In 3-dimensional Euclidean space there does not exist a complete regular surface of constant negative curvature (whatever its topological structure).

In various areas of mathematics it happens that one particular statement or one particular question is of great importance in the development of the whole area. In differential geometry, Hilbert's theorem has been, and still remains, one such assertion, as witnessed by the amount of publications devoted to it. Interest in this theorem stems largely from the combination of different ideas and concepts in its proof (for example, the universal covering and the Chebyshev net). A connection with physics was subsequently established.

It turned out that equation (2), if written slightly differently and without restriction (3), proves to express a condition on the potential in the so-called Josephson effect from the theory of superconductivity. Physicists have named equation (2) the 'sine-Gordon' equation.

Using the connection between equation (2) and the theory of surfaces of negative curvature (see [13], [14]), certain solutions of the equation have been found in closed form, which are regular in the entire plane.

S.E. Cohn-Vossen proposed (see [15]) that a far more general theorem than Hilbert's theorem holds; namely, the condition on the Gaussian curvature $K = \text{const} < 0$ can be replaced by the condition $K \leqslant \text{const} < 0$. However, up to the beginning of the 1960's, no theorem had been found which included Hilbert's theorem as a specific case. Not until 1961 (see [16]) was the correctness of Hilbert's theorem proved, i.e. a theorem was proved concerning a surface of variable negative curvature, for which Hilbert's theorem is a limiting case. In other words, the properties of a surface of constant negative curvature, which give rise to the effect of Hilbert's theorem (disruption of regularity) and the effect itself, can be taken in an approximate sense. In the proof, an equation was found in the case of variable curvature, generalizing equation (2):

$$k^2 \frac{\partial}{\partial s_2} \left[\frac{1}{K^2} \frac{\partial \omega}{\partial s_1} \right] = \{ k^2 + A + \alpha \, | \, \text{grad} \, \omega \, | \} \sin \omega, \qquad (4)$$

where ω has its previous sense, $k^2 = -K$, differentiation takes place along the arcs of the asymptotic lines, and A and α are variables whose absolute values allow an intrinsic estimate, i.e. an estimate depending only on the imbedded metric. When $K = \text{const}$, then $A = 0$, $\alpha = 0$ are obtained, and equation (4) becomes equation (2).

In the case of variable curvature the equation

$$L_{s_2} L_{s_1}^{*} (\omega) = M \sin \omega \qquad (5)$$

has been found (see [17]), where L_{s_2} is a linear operator proportional to the differentiation operator along the arc of the asymptotic lines of the second family, and $L_{s_1}^{*}$ is a quasi-linear operator of similar structure. It is essential that the function M, given certain conditions on the metric, has an intrinsic estimate from below by a positive number, thanks to which, and to the structure of the left-hand side, equation (5) is consistent with equation (2). Therefore, given these conditions on the metric, arguments can be put forward which are consistent with those of Hilbert (although far more complex) and which produce a similar result. In this way, a generalization of Hilbert's theorem was first obtained (see [18]). As regards the conditions on the metric mentioned above, they are included in the requirement that the Gaussian curvature K be bounded away from zero and changes slowly. It is proposed that if $K \leqslant -1$, then a slow change of the curvature must be understood in the sense that the first and second derivatives of K along the arc of any geodesic are sufficiently small. Sufficient estimates are made such that in the calculation the above arguments, coming from Hil-

bert, can be used for equation (5). These estimates (see [19]) are expressed using positive constants, denoted by h and Δ. Consequently, metrics with a slowly-changing negative curvature are called (h, Δ)-*metrics* (for their definition see [18] or [19]); moreover, metrics of constant curvature are obtained when $\Delta = +\infty$. Thus, a complete (h, Δ)-metric does not permit a regular isometric immersion in three-dimensional Euclidean space.

The question of generalizing Hilbert's theorem in the form in which it was stated in the 1930's (without restrictions on the character of the behaviour of the curvature, see [12]) along the lines indicated above, i.e. using differential equations of the theory of surfaces, has so far not succeeded at all. However, a solution to this problem has been obtained along entirely different lines, namely, through detailed studies of the boundary properties of the spherical image of a surface of negative curvature. Using this method, it has been proved ([19], [20]) that a complete metric with Gaussian curvature $K \leqslant \text{const} < 0$ does not allow a regular isometric immersion in E^3. This answers the question in the classical formulation. At the same time, the theorem that, in E^3, the equality

$$\sup K = 0 \qquad (6)$$

holds on every complete regular surface of negative Gaussian curvature K, has been proved. Here, C^2 regularity is sufficient. However (see [7]), in the case of $C^{1,1}$ there is a counter-example (in the form of a weakly irregular surface). See [39] for these results.

Apart from (6), other equalities and estimates have been obtained (see [21]), which are universal in the sense that they refer to all surfaces of negative curvature in E^3. Moreover, the method of proving (6) makes it possible to obtain theorems which give sufficient criterions for a mapping from a plane onto a plane with a negative Jacobian to be not only a local, but also a global diffeomorphism.

Despite the results which have now been obtained, studies on (h, Δ)-metrics and other metrics with slowly-changing curvature (for example, q-metrics; see [9]) are still relevant. The situation is that for immersions of metrics with slowly-changing negative curvature, certain effects are created which do not occur in the case of more general metrics of negative curvature (see, for example, the theorem on the simple zone in [9]). As a rough rule, one can say that the slower the change of curvature, the 'harder' it is for the metric of negative curvature to be immersed in E^3. This is particularly noticeable in the case of a metric of constant curvature (see the strengthened Hilbert theorem [9]). As regards differential equations which arise in connection with the question of immersibility of (h, Δ)-metrics, they were essentially used in studies of other questions.

Indeed, a number of strong theorems on the connection between intrinsic and extrinsic properties of surfaces of negative curvature has been proved using these equations; these include theorems on the dependence of the extrinsic regularity of the curvature of a surface of negative curvature on the regularity of its metric (see [7]).

The effect of the regularity of the metric on the extrinsic regularity of a surface with negative curvature is of particular interest, since surfaces of negative curvature form part of a special class of surfaces which can be defined purely geometrically (it is interesting that a purely geometric definition is also possible for the class of convex surfaces). Surfaces of this class are called *saddle-like*. Saddle-like surfaces are, in a certain sense, the antipodes of convex surfaces (see [22] - [25]).

The theorem on the non-existence in Euclidean space of a surface of complete negative curvature $K \leqslant \text{const} < 0$ has a particularly strengthened form in the case of a surface which can be projected injectively onto a plane, i.e. which can be represented as the graph of a function $z = f(x, y)$. In this case, universal estimates for the extension of the surface exist, and its completeness becomes immaterial. The following theorems have been proved (by N.V. Efimov, see [26]):

1) if a surface with Gaussian curvature $K \leqslant -\alpha^2 = \text{const} < 0$ can be regularly projected onto a rectangle with sides a, b, then the inequality $\sqrt{\alpha} \geqslant C_1 + \epsilon$, $\epsilon > 0$, implies $a\sqrt{\alpha} \leqslant C_1 + C_2/\epsilon$, where C_1, C_2 are universal constants (for example, $C_1 = 4\pi$, $C_2 = 12\pi$);

2) if a, b are as above and (for the sake of simplicity) $\alpha = 1$, then $a \leqslant 18.9$.

Let a metric ds^2 with negative Gaussian curvature K be given, and let $k = \sqrt{-K}$. Let the metric ds^2 be immersed in E^3, and let l, m, n be the reduced coefficients of the second fundamental form (reduced means: satisfying the equation $ln - m^2 = -K$). In the immersion problem, l, m and n are unknown functions. The *Riemann invariants* are introduced as new unknown functions:

$$r = \frac{m+k}{n}, \quad s = \frac{m-k}{n}.$$

The system of fundamental equations of the theory of surfaces will then take a symmetric form

$$\left.\begin{array}{l} r_x + sr_y = A_0 + A_1 r + A_2 s + A_3 r^2 + A_4 rs + A_5 r^2 s, \\ s_x + rs_y = A_0 + A_1 s + A_2 r + A_3 s^2 + A_4 rs + A_5 s^2 r, \end{array}\right\} \quad (7)$$

where $A_0, \dots, A_5$ are expressed in terms of only the metric ds^2.

It has already been shown that a complete surface of negative curvature with $K \leqslant \text{const} < 0$ does not permit an isometric regular immersion in E^3. The results obtained on the possibility of an isometric immersion of a two-dimensional manifold of negative curvature in E^3 can be restricted to parts of complete manifolds. The manifold itself is taken to be simply connected. Thus, it has been proved (see [29], [30]) that every geodesic disc on an arbitrary regular manifold of negative curvature $K \leqslant \text{const} < 0$ can be regularly isometrically immersed in E^3. This theorem is obtained as a corollary of a more general theorem: Given certain natural conditions on the regularity of the metric of a complete surface of negative curvature with $K \leqslant \text{const} < 0$, every infinite equidistant strip on this manifold can be isometrically and regularly immersed in E^3. This theorem in turn is obtained as a result of the existence theorem of solutions of hyperbolic systems of quasi-linear equations of the form (7), where $r \neq s$.

An example of a simply-connected manifold of sign-alternating curvature has been constructed on which there is a geodesic disc which cannot be isometrically immersed in E^3 (see [10]). It has been proved [31], given certain restrictions on the complete metric of a simply-connected surface of negative curvature $K \leqslant \text{const} < 0$, that the interior of an arbitrary horocycle of such a manifold can be isometrically immersed in E^3. Thus, for simply-connected manifolds of negative curvature $K \leqslant \text{const} < 0$ it has been proved that isometric immersion is possible for all their compact and non-compact parts which are compatible with parts of the Lobachevskiǐ plane, the isometric immersions of which are known through the works of Minding and Beltrami.

Immersions of a metric with Gaussian curvature $K = \text{const} < 0$ have been studied, i.e. immersions of parts of the Lobachevskiǐ plane. Thus, it has been proved that all polygons, even those of infinite extension, of the Lobachevskiǐ plane can be isometrically immersed in E^3; for new results on this kind of immersions see [40] - [42].

The variety of topological types of complete surfaces of negative curvature in E^3 makes it necessary to consider different subclasses of this too large class, satisfying some supplementary conditions. The subclass of complete surfaces of negative curvature which have a one-to-one spherical (*'single-sheeted'*) mapping is particularly natural (see [33]). If a surface F of this class does not have self-intersections, then it is simply or doubly connected. If such a surface F has a horn, then, using the unboundedness of the saddle horn [34], a number of external geometric properties of this surface can be obtained: F can be defined by an equation $z = f(x, y)$ over a domain obtained from the (x, y)-plane by eliminating a compact convex set; F has a bounding cone $A(F)$ consisting of a ray corresponding to the horn and a convex cone corresponding to the cup of the surface; the closure F^* of the spherical image of F has as its completion to a sphere an open semi-sphere and an open convex set (Figs. 8 - 10).

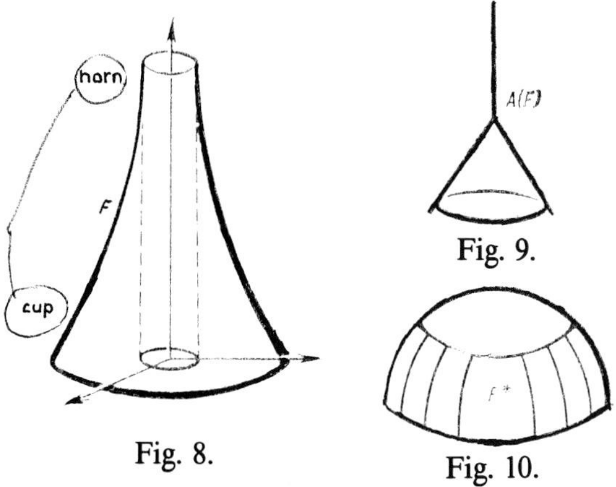

Fig. 8.   Fig. 9.   Fig. 10.

A number of similar questions has been studied for other types of complete surfaces of negative curvature with a one-to-one spherical mapping.

Another class which is of interest because of its properties and the properties of its spherical mapping is the class of *contractible surfaces of negative curvature* (see [36], [37]). An example of this type of surface is the surface defined by an equation $x^2y^2+y^2z^2+z^2x^2=a^2$. In a number of its properties, a contractible surface of negative curvature resembles a closed surface. Its spherical image can be viewed as a Riemann surface with boundary, each component of which is a spherical broken line lying on one of the large circles.

The theory of surfaces of negative curvature in a pseudo-Euclidean space $E^3_{2,1}$ is viewed differently. In this space, surfaces of negative curvature are convex; here the curvature is understood in the usual way, as the curvature of the metric induced by the ambient space. Namely, it is supposed to be negative.

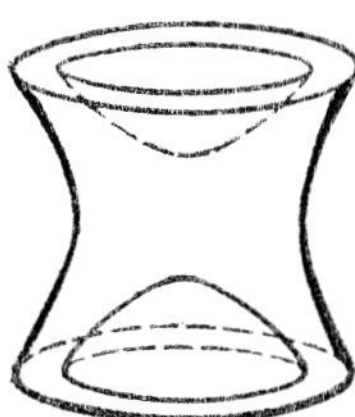

Fig. 11.

The unit sphere of the space $E^3_{2,1}$ consists of three connected components, L_1, L_2, L_3 (see Fig. 11). The components L_1 and L_2 are convex surfaces on which a positive-definite metric of constant negative curvature is induced; these surfaces provide a well-known interpretation of the Lobachevskiĭ geometry. The surface L_3 is saddle-like, and an indefinite metric of constant positive curvature is induced on it. The metric of this surface is known as the two-dimensional de Sitter metric.

Surfaces of negative curvature with a definite metric form a natural broad class of surfaces in $E^3_{2,1}$ generaliz-

ing the properties of the surfaces L_1 and L_2, just as surfaces of positive curvature form in Euclidean space a class generalizing the sphere (see [38]).

The properties of surfaces with a definite metric and negative curvature, the total curvature of which is finite, have been studied exhaustively. The properties of these surfaces are analogous to the properties of convex surfaces with curvature less than 2π in Euclidean space (the class of Olovyanishnikov surfaces). Namely, a complete definite metric of negative curvature and finite total curvature $\tilde{K}$ and a geodesic ray γ on a manifold carrying this metric uniquely determine a convex surface with the given limiting cone of complete curvature $\tilde{K}$, without isotropic generators and with a given generator which is limiting for the ray γ. The regularity of the surfaces obtained is determined by the regularity of the metric.

Complete surfaces of negative curvature with an indefinite metric are also convex. However, their properties are partially compatible with the properties of saddle-like surfaces in Euclidean space. In particular, the estimate $\sup K = 0$ holds for them. These surfaces would have to correspond to the component L_3 of the unit sphere of the space $E^3_{2,1}$, which is not convex, however. Here, both the sign of the Gaussian curvature and the sign of the determinant of the metric form of the surface play a determining role.

References

[1] HADAMARD, J.: 'Les surfaces à courbures opposées et leurs lignes géodesique', *J. Math. Pures Appl.* **4** (1898), 27-73.

[2] HILBERT, D.: 'Ueber Flächen von konstanter Gausscher Krümmung', in *Gesammelte Abhandlungen*, Vol. 2, Springer, 1933, pp. 437-448.

[3] BIANCHI, L.: *Vorlesungen über Differentialgeometrie*, Teubner, 1910 (translated from the Italian).

[4] EFIMOV, N.V.: *Höhere Geometrie*, Deutsch. Verlag Wissenschaft., 1960 (translated from the Russian).

[5] NORDEN, A.P.: *The theory of surfaces*, Moscow, 1956.

[6] ROZENDORN, E.R.: 'The construction of a complete bounded surface with non-positive curvature', *Uspekhi Mat. Nauk* **16**, no. 2 (1961), 149-156 (in Russian).

[7] ROZENDORN, E.R.: 'Weakly irregular surfaces of negative curvature', *Russian Math. Surveys* **21**, no. 5 (1966), 57-112. (*Uspekhi Mat. Nauk* **21**, no. 5 (1966), 59-116)

[8] AMINOV, YU.A.: 'The extrinsic diameter of a surface of negative curvature', *Ukr. Geom. Sb.* **13** (1973), 3-9 (in Russian).

[9] EFIMOV, N.V.: 'Surfaces with slowly changing negative curvature', *Russian Math. Surveys* **21**, no. 5 (1966), 1-56. (*Uspekhi Mat. Nauk* **21**, no. 5 (1966))

[10] POZNYAK, E.G.: 'Isometric immersions of two-dimensional Riemannian metrics in Euclidean space', *Russian Math. Surveys* **28**, no. 4 (1973), 47-77. (*Uspekhi Mat. Nauk* **28**, no. 4 (1973), 47-76)

[11] POZNYAK, E.G. and SHIKIN, E.V.: 'Surfaces of negative curvature', *J. Soviet Math.* **5**, no. 6 (1976), 865-887. (*Itogi Nauk. i Tekhn. Ser. Algebra. Topol. Geom.* **12** (1974), 171-207)

[12] POZNYAK, E.G.: 'Geometric investigations connected with $z_{xy} = \sin z$', *J. Soviet Math.* **13**, no. 5 (1980), 667-686. (*Itogi Nauk. i Tekhn. Probl. Geom.* **8** (1977), 225-241)

[13] GRIBKOV, I.V.: 'Construction of some regular solutions of the "sine-Gordon" equation by means of surfaces of constant

negative curvature', *Vestnik Mosk. Univ. Ser. Mat. Mekh.* **4** (1977), 78-83 (in Russian). English abstract.

[14] GRIBKOV, I.V.: 'Some solutions of the sin-Gordon equation obtained by means of a Bäcklund transformation', *Russian Math. Surveys* **33**, no. 2 (1978), 231-232. (*Uspekhi Mat. Nauk* **33**, no. 2 (1978), 191-192)

[15] COHN-VOSSEN, S.E.: 'Global deformation of surfaces', *Uspekhi Mat. Nauk* **1** (1936), 33-76.

[16] EFIMOV, N.V.: 'The validity of Hilbert's theorem on surfaces of constant negative curvature', *Soviet Math. Dokl.* **2**, no. 1 (1961), 189-191. (*Dokl. Akad. Nauk SSSR* **136**, no. 6 (1961), 1283-1286)

[17] EFIMOV, N.V. and POZNYAK, E.G.: 'Some transformations of the fundamental equations of the theory of surfaces', *Soviet Math. Dokl.* **2**, no. 2 (1961), 225-227. (*Dokl. Akad. Nauk SSSR* **137**, no. 1 (1961), 25-27)

[18] EFIMOV, N.V. and POZNYAK, E.G.: 'Generalization of Hilbert's theorem on surfaces of constant negative curvature', *Soviet Math. Dokl.* **2**, no. 2 (1961), 300-303. (*Dokl. Akad. Nauk SSSR* **137**, no. 3 (1961), 509-512)

[19] EFIMOV, N.V.: 'Imposibility of a complete regular surface in euclidean 3-space whose Gaussian curvature has a negative upper bound', *Soviet Math. Dokl.* **4**, no. 3 (1963), 843-846. (*Dokl. Akad. Nauk SSSR* **150**, no. 6 (1963), 1206-1209)

[20] EFIMOV, N.V.: 'Generation of singularities on surfaces of negative curvature', *Mat. Sb.* **64**, no. 2 (1964), 286-320 (in Russian).

[21] EFIMOV, N.V.: 'Differential criteria for homeomorphism of certain mappings with applications to the theory of surfaces', *Math. USSR Sb.* **5**, no. 4 (1968), 475-488. (*Mat. Sb.* **76**, no. 4 (1968), 499-512)

[22] SHEFEL', S.Z.: *Studies on the geometry of saddle-like surfaces*, Novosibirsk, 1963 (in Russian).

[23] SHEFEL', S.Z.: 'On the intrinsic geometry of saddle surfaces', *Sib. Mat. Zh.* **5**, no. 6 (1964), 1382-1396 (in Russian).

[24] SHEFEL', S.Z.: 'A condition of compactness of a family of saddle surfaces', *Siberian Math. J.* **8**, no. 3 (1967), 528-535. (*Sib. Mat. Zh.* **8**, no. 3 (1967), 705-714)

[25] SHEFEL', S.Z.: 'Saddle surfaces bounded by a rectifiable curve', *Soviet Math. Dokl.* **6**, no. 3 (1965), 684-687. (*Dokl. Akad. Nauk SSSR* **162**, no. 2 (1965), 294-296)

[26] EFIMOV, N.V.: 'Estimates for the dimensions of the domains of regularity of solutions of certain Monge−Ampère equations', *Math. USSR Sb.* **29**, no. 3 (1976), 319-326. (*Mat. Sb.* **100**, no. 3 (1976), 356-363)

[27] EFIMOV, N.V.: 'Study of a single-valued projection of a surface of negative curvature', *Dokl. Akad. Nauk SSSR* **93**, no. 4 (1953), 609-611 (in Russian).

[28] HEINZ, E.: 'Ueber Flächen mit eindeutiger Projektion auf eine Ebene, deren Krümmungen durch Ungleichungen eingeschränkt sind', *Math. Ann.* **129**, no. 5 (1955), 451-454.

[29] POZNYAK, E.G.: 'On the regular realization in the large of two-dimensional metrics of negative curvature', *Soviet Math. Dokl.* **7**, no. 5 (1966), 1288-1291. (*Dokl. Akad. Nauk SSSR* **170**, no. 4 (1966), 786-789)

[30] POZNYAK, E.G.: 'Regular global realization of two-dimensional metrics with negative curvature', *Ukr. Geom. Sb.* **3** (1966), 78-92 (in Russian).

[31] SHIKIN, E.V.: 'On the global isometric imbedding in $\mathbf{R}^3$ of some metrics of nonpositive curvature', *Soviet Math. Dokl.* **15**, no. 2 (1974), 448-451. (*Dokl. Akad. Nauk SSSR* **215**, no. 1 (1974), 61-63)

[32] ROZHDESTVENSKIĬ, B.L.: 'A system of quasi-linear equations in the theory of surfaces', *Dokl. Akad. Nauk SSSR* **143**, no. 1 (1962), 50-52 (in Russian).

[33A] VERNER, A.L.: 'On the extrinsic geometry of elementary complete surfaces with nonpositive curvature I', *Mat. USSR Sb.* **3**, no. 2 (1967), 205-224. (*Mat. Sb.* **74**, no. 2 (1967), 218-240)

[33B] VERNER, A.L.: 'On the extrinsic geometry of elementary complete surfaces with nonpositive curvature II', *Math. USSR Sb.* **4**, no. 1 (1968), 99-123. (*Mat. Sb.* **75**, no. 1 (1967), 112-139)

[33C] VERNER, A.L.: 'Correction to "On the extrinsic geometry of elementary complete surfaces with nonpositive curvature"', *Mat. Sb.* **77**, no. 1 (1968), 136 (in Russian).

[34] VERNER, A.L.: 'Tapering saddle surfaces', *Siberian Math. J.* **11**, no. 4 (1970), 567-581. (*Sib. Mat. Zh.* **11**, no. 4 (1970), 750-769)

[35] VERNER, A.L.: 'The finiteness of the set of branch points of a spherical mapping of a narrowing saddle surface', *Mat. Notes* **12**, no. 3 (1972), 603-605. (*Mat. Zametki* **12**, no. 3 (1972), 281-286)

[36A] SOKOLOV, D.D.: 'The structure of a limiting cone of a convex surface in Euclidean space', *Uspekhi Mat. Nauk* **30**, no. 1 (1975), 261-262 (in Russian).

[36B] SOKOLOV, D.D.: 'On convex surfaces with an indefinite metric', *Russian Math. Surveys* **33**, no. 4 (1978), 265-266. (*Uspekhi Mat. Nauk* **33**, no. 4 (1978), 227-228)

[36C] SOKOLOV, D.D.: 'Convex surfaces having bounded total curvature in pseudo-Euclidean space', *Russian Math. Surveys* **34**, no. 3 (1979), 222-223. (*Uspekhi Mat. Nauk* **34**, no. 3 (1979), 213-214)

[37] ALEKSANDROV, A.D.: *Die innere Geometrie der konvexen Flächen*, Akademie-Verlag, 1955 (translated from the Russian).

[38] POGORELOV, A.V.: *Extrinsic geometry of convex surfaces*, Amer. Math. Soc., 1972 (translated from the Russian).

[39] KLOTZ MILNOR, T.: 'Efimov's theorem about complete immersed surfaces of negative curvature', *Adv. in Math.* **8**, no. 3 (1972), 474-543.

[40] TUNITSKIĬ, D.V.: 'On the regular isometric immersion in E^3 of unbounded domains of negative curvature', *Mat. USSR Sb.* **62**, no. 1 (1989), 121-138. (*Mat. Sb.* **134**, no. 1 (1987), 119-134)

[41] KAĬDASOV, ZH. and SHIKIN, E.V.: 'Isometric immersion in E^3 of a convex domain of the Lobachevskiĭ plane containing two horocycles', *Math. Notes* **39**, no. 4 (1986), 335-340. (*Mat. Zametki* **39**, no. 4 (1986), 612-617)

[42] KAĬDASOV, ZH. and SHIKIN, E.V.: 'Isometric immersion in E^3 of extended strips on manifolds of type L', *Math. Notes* **42**, no. 6 (1987), 962-968. (*Mat. Zametki* **42**, no. 6 (1987), 842-851)

N.V. Efimov

Editorial comments. The Hilbert result that a surface of constant negative curvature cannot be imbedded isometrically in E^3 needs C^2 regularity. The general method of C^1 isometric imbedding, [A1], [A2], applies also here.

References

[A1] KUIPER, N.H.: 'On C^1-isometric imbeddings I', *Indag. Math.* **17**, no. 4 (1955), 545-556.

[A2] KUIPER, N.H.: 'On C^1-isometric imbeddings II', *Indag. Math.* **17**, no. 5 (1955), 683-689.

[A3] BERGER, M. and GOSTIAUX, M.: *Differential geometry: manifolds, curves, and surfaces*, Springer, 1988 (translated from the French).

[A4] KLINGENBERG, W.: *A course in differential geometry*, Springer, 1978 (translated from the German).

[A5] STRUIK, D.J.: *Lectures on classical differential geometry*, Addison-Wesley, 1961.

[A6] POGORELOV, A.V.: *Intrinsic geometry of surfaces*, Amer. Math. Soc., 1973 (translated from the Russian).

[A7] SPIVAK, M.: *A comprehensive introduction to differential geometry*, 3; 5, Publish or Perish, 1975.

[A8] GROMOV, M.: *Partial differential relations*, Springer, 1986.

[A9] GROMOV, M.: 'Embeddings and immersions in Riemannian geometry', *Russian Math. Surveys* **25**, no. 5 (1970), 1-58. (*Uspekhi Mat. Nauk* **25**, no. 5 (1970), 3-62)

AMS 1980 Subject Classification: 53A05, 53A35, 53BXX, 53B25, 53B30, 53C40, 53C45

NEGATIVE EXPONENTIAL DISTRIBUTION - The

same as **exponential distribution**.

AMS 1980 Subject Classification: 60E05, 60J25, 60K05, 62EXX

NEGATIVE HYPERGEOMETRIC DISTRIBUTION -

A **probability distribution** of a random variable X which takes non-negative integer values, defined by the formula

$$P\{X=k\} = \frac{\begin{bmatrix} k+m-n \\ k \end{bmatrix}\begin{bmatrix} N-m-k \\ M-m \end{bmatrix}}{\begin{bmatrix} N \\ M \end{bmatrix}}, \quad 0 \leqslant k \leqslant N-M, \quad (*)$$

where the parameters N, M, m are non-negative integers which satisfy the condition $m \leqslant M \leqslant N$. A negative hypergeometric distribution often arises in a scheme of sampling without replacement. If in the total population of size N, there are M 'marked' and $N-M$ 'unmarked' elements, and if the sampling (without replacement) is performed until the number of 'marked' elements reaches a fixed number m, then the random variable X — the number of 'unmarked' elements in the sample — has a negative hypergeometric distribution (*). The random variable $X+m$ — the size of the sample — also has a negative hypergeometric distribution. The distribution (*) is called a negative hypergeometric distribution by analogy with the **negative binomial distribution**, which arises in the same way for sampling with replacement.

The mathematical expectation and variance of a negative hypergeometric distribution are, respectively, equal to

$$m\frac{N-M}{M+1}$$

and

$$m\frac{(N+1)(N-M)}{(M+1)(M+2)}\left[1-\frac{m}{M+1}\right].$$

When $N, M, N-M \to \infty$ such that $M/N \to p$, $(N-M)/N \to q$, $p+q=1$, the negative hypergeometric distribution tends to the negative binomial distribution with parameters m and p.

The distribution function $F(n)$ of the negative hypergeometric function with parameters N, M, m is related to the **hypergeometric distribution** $G(m)$ with parameters N, M, n by the relation

$$F(n) = 1 - G(m-1).$$

This means that in solving problems in mathematical statistics related to negative hypergeometric distributions, tables of hypergeometric distributions can be used. The negative hypergeometric distribution is used, for example, in **statistical quality control**.

References
[1] BELYAEV, Y.K.: *Probability methods of sampling control*, Moscow, 1975 (in Russian).
[2] BOL'SHEV, L.N. and SMIRNOV, N.V.: *Tables of mathematical statistics*, Libr. of mathematical tables, 46, Nauka, Moscow, 1983 (in Russian). Processed by L.S. Bark and E.S. Kedrova.

A.V. Prokhorov

Editorial comments.

References
[A1] JOHNSON, N.L. and KOTZ, S.: *Distributions in statistics, discrete distributions*, Wiley, 1969.
[A2] PATIL, G.P. and JOSHI, S.W.: *A dictionary and bibliography of discrete distributions*, Hafner, 1968.

AMS 1980 Subject Classification: 60E99, 62E15

NEGATIVE POLYNOMIAL DISTRIBUTION, *negative multinomial distribution* -

The joint **probability distribution** (cf. also **Joint distribution**) of random variables $X_1, \ldots, X_k$ that take non-negative integer values $m = 0, 1, \ldots$, defined by the formula

$$P\{X_1 = m_1, \ldots, X_k = m_k\} =$$
$$= \frac{\Gamma(r+m_1+\cdots+m_k)}{\Gamma(r)m_1!\cdots m_k!}p_0^r p_1^{m_1}\cdots p_k^{m_k}, \quad (*)$$

where $r > 0$ and $p_0, \ldots, p_k$ ($0 < p_i < 1$, $i = 0, \ldots, k$; $p_0 + \cdots + p_k = 1$) are parameters. A negative multinomial distribution is a multi-dimensional **discrete distribution** — a distribution of a random vector $(X_1, \ldots, X_k)$ with non-negative integer components.

The **generating function** of the negative polynomial distribution with parameters $r, p_0, \ldots, p_k$ has the form

$$P(z_1, \ldots, z_k) = p_0^r \left[1 - \sum_{i=1}^{k} z_i p_i\right]^{-r}.$$

A negative multinomial distribution arises in the following multinomial scheme. Successive independent trials are carried out, and in each trial $k+1$ different outcomes with labels $0, \ldots, k$ are possible, having probabilities $p_0, \ldots, p_k$, respectively. The trials continue up to the r-th appearance of the outcome with label 0 (here r is an integer). If X_i is the number of appearances of the outcome with label i, $i = 1, \ldots, k$, during the trials, then formula (*) expresses the probability of the appearance of outcomes with labels $1, \ldots, k$, equal, respectively, $m_1, \ldots, m_k$ times, up to the r-th appearance of the outcome 0. A negative multinomial distribution in this sense is a generalization of a **negative binomial distribution**, coinciding with it when $k = 1$.

If a random vector $(X_0, \ldots, X_k)$ has, conditionally on n, a **multinomial distribution** with parameters $n > 1$, $p_0, \ldots, p_k$ and if the parameter n is itself a random variable having a negative binomial distribution with parameters $r > 0$, $0 < \pi < 1$, then the marginal distribution of the vector $(X_1, \ldots, X_k)$, given the condition $X_0 = r$, is the negative multinomial distribution with parameters $r, p_0(1-\pi), \ldots, p_k(1-\pi)$.

A.V. Prokhorov

Editorial comments.

References

[A1] NEYMAN, J.: *Proceedings of the international symposium on discrete distributions*, Montreal, 1963.

AMS 1980 Subject Classification: 62E15, 60E99

NEGATIVE VARIATION OF A FUNCTION, *negative increment of a function* - One of the two terms whose sum is the complete increment or **variation of a function** on a given interval. Let f be a function of a real variable, defined on an interval $[a, b]$ and taking finite real values.

Let $\Pi = \{a = x_0 < \cdots < x_n = b\}$ be an arbitrary partition of $[a, b]$ and let

$$N_\Pi(f) = -\sum_i{}^{-} [f(x_i) - f(x_{i-1})],$$

where the summation is over those numbers i for which the difference $f(x_i) - f(x_{i-1})$ is non-positive. The quantity

$$N(f) \equiv N(f; [a, b]) = \sup_\Pi N_\Pi(f)$$

is called the *negative variation (negative increment) of the function f* on the interval $[a, b]$. It is always true that $0 \leqslant N(f) \leqslant +\infty$. See also **Positive variation of a function; Variation of a function.**

References

[1] LEBESGUE, H.: *Leçons sur l'intégration et la récherche des fonctions primitives*, Gauthier-Villars, 1928.

B.I. Golubov

AMS 1980 Subject Classification: 26A45

NEGATIVE VECTOR BUNDLE - A holomorphic **vector bundle** (cf. also **Vector bundle, analytic**) E over a **complex space** X that possesses a **Hermitian metric** h such that the function $v \to h(v, v)$ on E is strictly pseudo-convex outside the zero section (this is denoted by $E < 0$). The vector bundle E is negative if and only if the dual vector bundle $E^* > 0$ (see **Positive vector bundle**). If X is a manifold, then the condition of being negative can be expressed in terms of the curvature of the metric h. Any subbundle of a negative vector bundle is negative. A vector bundle E over a complex manifold is said to be *negative in the sense of Nakano* if E^* is positive in the sense of Nakano. A holomorphic vector bundle E over a compact complex space X is said to be *weakly negative* if its zero section possesses a strictly pseudo-convex neighbourhood in E, i.e. if E^* is weakly positive. Every negative vector bundle over X is weakly negative. Negative and weakly negative linear spaces over a space X are also defined in this way.

For references see **Positive vector bundle.**

A.L. Onishchik

AMS 1980 Subject Classification: 32LXX

NEIGHBOURHOOD *of a point x (of a subset A)* of a **topological space** - Any open subset of this space con-

taining the point x (the set A). Sometimes a neighbourhood of the point x (the set A) is defined as any subset of this topological space containing the point x (the set A) in its interior (cf. also **Interior of a set**).

B.A. Pasynkov

AMS 1980 Subject Classification: 54AXX

NEIL PARABOLA - A name for the **semi-cubic parabola**; in honour of W. Neil who found its arc length in 1657.

AMS 1980 Subject Classification: 53A04

NEKRASOV INTEGRAL EQUATION - A non-linear integral equation of the form

$$\phi(x) = \lambda \int_a^b [\phi(y) + R(\lambda, y, \phi(y))] K(x, y)\, dy, \qquad (*)$$

where R and K are known functions, K being symmetric, ϕ is the unknown function, and λ is a numerical parameter. Integral equations of this type were obtained by A.I. Nekrasov (see [1]) in the solution of problems arising in the theory of waves on the surface of a fluid. Under certain conditions Nekrasov has constructed a solution of (*) in the form of a series in powers of a small parameter; its convergence has been proved by the method of majorants.

Sometimes an equation of the type (*) is called a **Hammerstein equation**, although Nekrasov [2] published his investigations before A. Hammerstein [3].

References

[1] NEKRASOV, A.I.: *Collected works*, 1, Moscow, 1961 (in Russian).
[2] NEKRASOV, A.I.: *Izv. Ivanovo-Vozn. Politekhn. Inst.* **6** (1922), 155-171.
[3] HAMMERSTEIN, A.: 'Nichtlineare Integralgleichungen nebst Anwendungen', *Acta Math.* **54** (1930), 117-176.

B.V. Khvedelidze

Editorial comments.

References

[A1] ZABREYKO, P.P. [P.P. ZABREĬKO], ET AL.: *Integral equations — a reference text*, Noordhoff, 1975 (translated from the Russian).

AMS 1980 Subject Classification: 45G10

NÉRON MODEL *of an Abelian variety* - A **group scheme** associated to an **Abelian variety** and having a certain minimality property. If R is a local Henselian discrete valuation ring with residue field k and field of fractions K and if A is an Abelian variety of dimension d over K, then a *Néron model* of A is defined as a smooth commutative group scheme $\mathfrak{A}$ over R whose generic fibre $\mathfrak{A}_K$ is isomorphic to A, while the canonical homomorphism $\mathfrak{A}(R) \to \mathfrak{A}_K(K)$ is an isomorphism. This concept was introduced by A. Néron [1] in the case of a perfect field. In the local case a Néron model exists and is uniquely determined up to an R-isomorphism. A Néron model has the following minimality property: For any smooth R-scheme $\mathfrak{X}$ and any morphism $\phi\colon \mathfrak{X}_K \to \mathfrak{A}_K$ of the generic fibres there exists a unique morphism $\bar{\phi}\colon \mathfrak{X} \to \mathfrak{A}$ of R-schemes induced by ϕ.

If S is a one-dimensional regular Noetherian scheme, η is a generic point of it, $i\colon \eta \to S$ is its canonical imbedding, and A is an Abelian variety over $k(\eta)$, then a Néron model of A is defined as a smooth quasi-projective group scheme $\mathfrak{A}$ over S that represents the sheaf i_*A relative to the flat Grothendieck topology on S (see [4]).

For a generalization of the concept of a Néron model to arbitrary schemes see [3].

References

[1] NÉRON, A.: 'Modèles minimaux des variétés abéliennes sur les corps locaux et globaux', *Publ. Math. IHES* **21** (1964).

[2] MAZUR, B.: 'Rational points of Abelian varieties with values in towers of number fields', *Invent. Math.* **18** (1974), 183-266.

[3] RAYNAUD, M.: 'Modèles de Néron', *C.R. Acad. Sci. Paris Sér. A* **262** (1966), 345-347.

[4] RAYNAUD, M.: 'Caractéristique d'Euler−Poincaré d'un faisceau et cohomologie des variétés abéliennes (d'après Ogg−Shafarévitch et Grothendieck)', in *Dix exposés sur la cohomology des schémas*, North-Holland & Masson, 1968, pp. 12-30.

[5] GROTHENDIECK, A., ET AL. (EDS.): *Groupes de monodromie en géométrie algébrique. SGA 7*, Lecture notes in math., 288, Springer, 1972.

I.V. Dolgachev

Editorial comments.

References

[A1] ARTIN, M.: 'Néron models', in G. Cornell and J. Silverman (eds.): *Arithmetic Geometry*, Springer, 1986, pp. 213-230.

AMS 1980 Subject Classification: 14K99

NÉRON−SEVERI GROUP - The divisor class group under algebraic equivalence on a non-singular projective variety.

Let X be a non-singular projective variety of dimension ≥ 2 defined over an algebraically closed field k, let $D(X)$ be the group of divisors of X and let $D_a(X)$ be the subgroup of divisors that are algebraically equivalent to zero. The quotient group $D(X)/D_a(X)$ is called the *Néron−Severi group* of X and is denoted by $NS(X)$. The *Néron−Severi theorem* asserts that the Abelian group $NS(X)$ is finitely generated.

In the case $k = \mathbf{C}$, F. Severi presented, in a series of papers on the theory of the base (see, for example, [1]), a proof of this theorem using topological and transcendental tools. The first abstract proof (valid for a field of arbitrary characteristic) is due to A. Néron (see [2], [3], and also [4]).

The rank of $NS(X)$ is the algebraic **Betti number** of the group of divisors on X, that is, the algebraic rank of X. This is also called the *Picard number* of the variety X. The elements of the finite torsion subgroup $NS_{\text{tors}}(X)$ are called *Severi divisors*, and the order of this subgroup is called the *Severi number*; the group $NS_{\text{tors}}(X)$ is a birational invariant (see [6]).

There are generalizations of the Néron−Severi theorem to other groups of classes of algebraic cycles (see [1] (classical theory) and [7] (modern theory)).

References

[1] SEVERI, F.: 'La base per le varietà algebriche di dimensione qualunque contenute in una data e la teoria generale delle corrispondénze fra i punti di due superficie algebriche', *Mem. Accad. Ital.* **5** (1934), 239-283.

[2] NÉRON, A.: 'Problèmes arithmétiques et géometriques attachée à la notion de rang d'une courbe algébrique dans un corps', *Bull. Soc. Math. France* **80** (1952), 101-166.

[3] NÉRON, A.: 'La théorie de la base pour les diviseurs sur les variétés algébriques', in 2e *Coll. Géom. Alg. Liège*, G. Thone, 1952, pp. 119-126.

[4] LANG, S. and NÉRON, A.: 'Rational points of abelian varieties over function fields', *Amer. J. Math.* **81** (1959), 95-118.

[5] HARTSHORNE, R.: *Algebraic geometry*, Springer, 1977.

[6] BALDASSARRI, M.: *Algebraic varieties*, Springer, 1956.

[7] DOLGACHEV, I.V. and ISKOVSKIKH, V.A.: 'Geometry of algebraic varieties', *J. Soviet Math.* **5**, no. 6 (1976), 803-864. (*Itogi Nauk. i Tekhn. Algebra. Topol. Geom.* **12** (1974), 77-170)

V.A. Iskovskikh

Editorial comments. A study of the Picard number of a certain number of algebraic varieties has been made by T. Shioda.

The phrase *'theory of the base'* is a somewhat old-fashioned one and refers to the considerations involved in proving that $NS(X)$ is a finitely-generated Abelian group and indicating an explicit minimal set of generators (a *minimal base* in the terminology of Severi), cf. e.g. [A4], Sect. V.7 (for the case of surfaces).

References

[A1] SHIODA, T.: 'On the Picard number of a complex projective variety', *Ann. Sci. Ecole Norm. Sup.* **14** (1981), 303-321.

[A2] SHIODA, T.: 'On the Picard number of a Fermat surface', *J. Fac. Sci. Univ. Tokyo* **28** (1982), 724-734.

[A3] SHIODA, T.: 'An explicit algorithm for computing the Picard number of certain algebraic surfaces', *Amer. J. Math.* **108** (1986), 415-432.

[A4] ZARISKI, O.: *Algebraic surfaces*, Springer, 1935.

AMS 1980 Subject Classification: 14C20

NERVE OF A FAMILY OF SETS α - The **simplicial complex** $K(\alpha)$ with as simplices the finite non-empty subsets of α with non-empty intersection. In particular, the vertices of $K(\alpha)$ are the non-empty elements of α.

M.I. Voĭtsekhovskiĭ

AMS 1980 Subject Classification: 54E60, 55U05

NET - A two-parameter family of curves in a plane, or surfaces in space, that depends linearly on the parameters in the following way. Let F_1, F_2, F_3 be functions of two variables, no one of which is a linear combination of the other two. The family of curves in the plane that are defined by the equation

$$\lambda_1 F_1 + \lambda_2 F_2 + \lambda_3 F_3 = 0$$

for all possible values of the parameters $\lambda_1, \lambda_2, \lambda_3$ (with the exception of $\lambda_1 = \lambda_2 = \lambda_3 = 0$) is a net (in fact, it depends on the ratios $\lambda_1 : \lambda_2 : \lambda_3$). The equation of a net of surfaces in space is written down in a similar way. The three equations $F_1 = 0$, $F_2 = 0$, $F_3 = 0$ give three elements of the net (three curves or surfaces) which determine the whole net.

A net of lines is a two-parameter family of straight lines lying on a surface. In Euclidean geometry a net of lines is the set of all straight lines that pass through a single point. If this point is a finite point, then the net is called *elliptic*, if it is a point at infinity, then the net is called *parabolic*.

A *net of planes* is a set of all planes that pass through a single point (a *non-degenerate net*) or that are parallel to some straight line (a *degenerate net*).

A *net of circles* is a two-parameter family of circles that depends linearly on the parameters. A *non-degenerate net* of circles is the set of all circles relative to which a given point (the *centre* of the net) has a given power. A *degenerate net* of circles is the set of all circles with centres on a certain fixed straight line (the so-called *fundamental straight line*) (see Fig. 1).

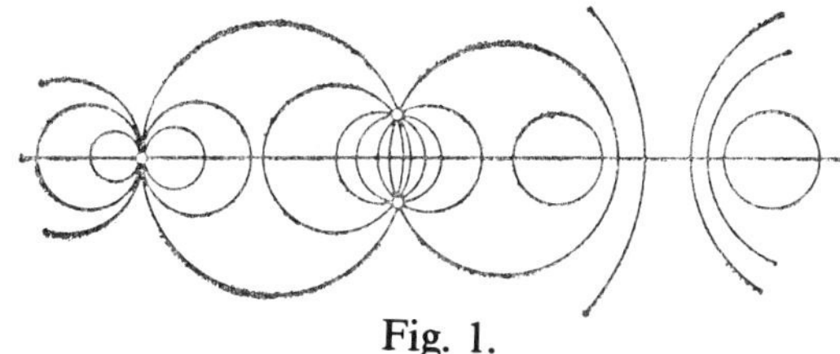

Fig. 1.

If $(0, 0)$ is the centre of a non-degenerate net, then its equation is

$$x^2 + y^2 - 2ax - 2by + p = 0, \quad a^2 + b^2 > p,$$

where a and b are the parameters that define the circle and p is the power of the centre of the net relative to its circles (the power of a degenerate net is assumed to infinite).

There are three types of non-degenerate nets of circles:

1) A *hyperbolic circle net* ($p > 0$), consisting of all circles that are orthogonal to a given circle (the *fundamental circle*) (see Fig. 2).

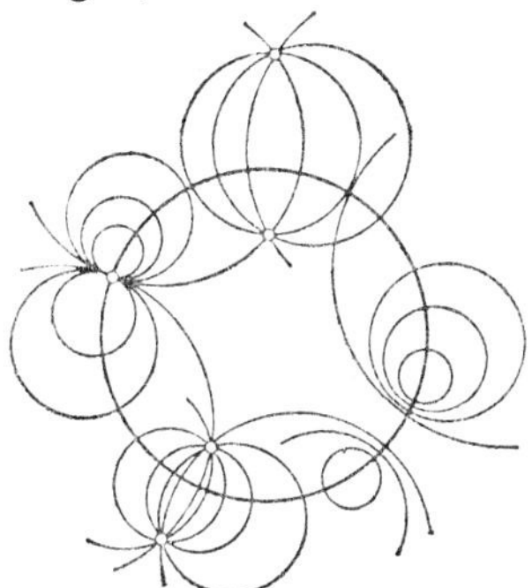

Fig. 2.

2) A *parabolic circle net* ($p = 0$), consisting of all circles passing through a given point (the *centre* of the net) (see Fig. 3).

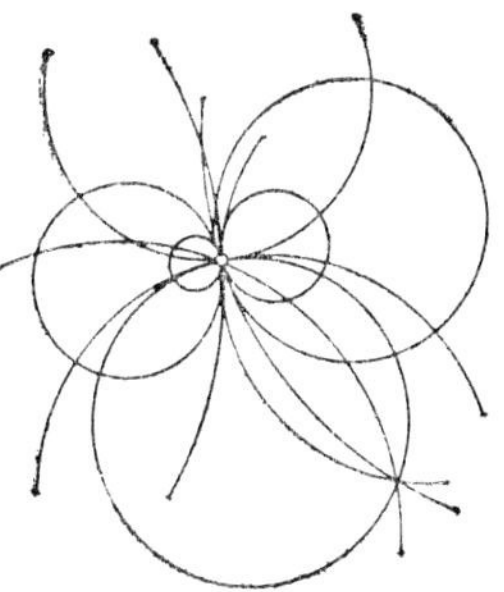

Fig. 3.

3) An *elliptic circle net* ($p < 0$), consisting of circles that intersect a given circle in two diametrically-opposite points of the latter (see Fig. 4).

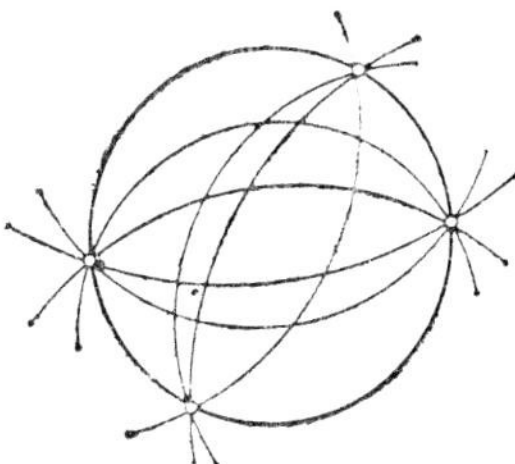

Fig. 4.

The intersection of two nets of circles is a pencil of circles. An elliptic net contains only elliptic pencils, a parabolic net only elliptic and parabolic pencils, and a hyperbolic net contains three types of non-degenerate pencils. A degenerate net contains both degenerate and non-degenerate pencils of all three types.

The intersection of two nets one of which is elliptic can only be an elliptic pencil. The intersection of two nets one of which is parabolic can only be an elliptic or a parabolic pencil. The intersection of two nets one of which is non-degenerate can only be a non-degenerate pencil.

A *net of spheres* is a two-parameter family of spheres that depends linearly on the parameters. A net of spheres consists of the set of spheres relative to which the points of a certain line (the *radical axis*) have the same power (different for different points). The radical axis intersects all the spheres of the net in two points. Depending on whether these points are real (distinct), complex conjugates or coincident, a net of spheres is called *elliptic* (consisting of all the spheres through two common points), *hyperbolic* (consisting of all the spheres orthogonal to two intersecting spheres) or *parabolic* (consisting of all the spheres that touch a given line at a given point). The centres of all the spheres of a net lie on a plane perpendicular to the radical axis.

A net of spheres is the set of all spheres common to

two webs of spheres (cf. **Web of spheres**).

In projective geometry, a net is the set of all lines, respectively, the set of all planes, that pass through a given point.

References

[1] POSTNIKOV, M.M.: *Analytic geometry*, Moscow, 1973 (in Russian).

[2] MODENOV, P.S.: *Analytic geometry*, Moscow, 1969 (in Russian).

A.B. Ivanov

Editorial comments. Instead of the word 'net', for a 2-dimensional linear system of curves or surfaces the word '*bundle*' is also used.

References

[A1] TODD, J.A.: *Projective and analytical geometry*, Pitman, 1947, Chapt. VI.

AMS 1980 Subject Classification: 51N20, 53A04, 53A05

NET (DIRECTED SET) - A mapping of a **directed set** into a (topological) space.

M.I. Voitsekhovskiĭ

Editorial comments. The topology of a space can be described completely in terms of convergence. However, this needs a more general concept of convergence than the concept of convergence of a sequence. What is needed is *convergence of nets*. A net $S: D \to X$ in a topological space X converges to a point $s \in X$ if for each open neighbourhood U of s in X the net S is *eventually in* U. The last phrase means that there is an $m \in D$ such that $S(n) \in U$ for all $n \geq m$ in D.

The theory of convergence of nets is known as *Moore–Smith convergence*, [A1].

References

[A1] KELLEY, J.L.: *General topology*, v. Nostrand, 1955, Chapt. II.

AMS 1980 Subject Classification: 06F99

NET (IN DIFFERENTIAL GEOMETRY) - A system $\Sigma_n = \{\sigma^1, \ldots, \sigma^n\}$ of n families ($n \geq 2$) of sufficiently smooth curves in a domain G of an n-dimensional differentiable manifold M such that: 1) through each point $x \in G$ there passes exactly one curve of each family σ^i; and 2) the tangent vectors to these curves at x form a basis for the tangent space T_x to M at x. The tangent vectors to the curves of one family σ^i belong to a 1-dimensional distribution Δ_1^i defined in G. The conditions for a family of curves to form a net in a certain neighbourhood of a point need not hold when the curves are extended. The curves of σ^i are the integral curves of Δ_1^i. A net $\Sigma_n \subset G$ is defined by specifying n one-dimensional distributions Δ_1^i such that the tangent space T_x at each $x \in G$ is the direct sum of the subspaces Δ_1^i, $i = 1, \ldots, n$. A net $\Sigma_n \subset G$ defines in G $(n-1)$-dimensional distributions Δ_{n-1}^i such that at each point $x \in G$ the subspace $\Delta_{n-1}^i(x) \subset T_x$ is the direct sum of the $n-1$ one-dimensional subspaces $\Delta_1^j(x)$, $j \neq i$.

The following types of nets are distinguished: *holonomic nets*, *partially holonomic nets*, for which certain of the Δ_{n-1}^i are integrable and the remainder are not (such nets are distinguished according to the number of non-integrable distributions), and *non-holonomic nets*, for which all Δ_{n-1}^i are non-integrable.

If a distribution Δ_{n-1}^i, $n > 2$, is integrable and γ^i is an integral curve of Δ_1^i, then through each point $x \in \gamma^i$ there passes an integral manifold of Δ_{n-1}^i that carries a net $\Sigma_{n-1}(x)$ of curves belonging to the families σ^j, $j \neq i$.

A net $\Sigma_n \subset G$ can also be defined by one of the following means: a) by a system of vector fields $X_i \subset \Delta_1^i$; b) by a system of differential 1-forms ω^i such that $\omega^i(X_j) = \delta_j^i$; or c) by the field of an affinor Φ such that $\Phi^n = E$ (E is the identity affinor).

In the study of nets there are three basic problems: the intrinsic properties of nets, the exterior properties and an investigation of diffeomorphisms of nets.

The intrinsic properties of nets are induced by the structure of the manifold that carries the net. For example, a net Σ_n in a space M with an affine connection ∇ is called *geodesic* if all its curves are geodesic. If a Riemannian manifold M with a torsion-free connection in which the metric tensor is covariantly constant carries an orthogonal **Chebyshev net** of the first kind, then M is locally Euclidean. The connection of such nets with parallel transfer of vectors on a surface was established by L. Bianchi (1922). This connection is at the basis of A.P. Norden's definition of a Chebyshev net of the first kind in a space with an affine connection.

The exterior properties of a net are induced by the structure of the ambient space E. For example, suppose that a net Σ_n defined in a domain G on a smooth surface V_n in $(n+k)$-dimensional projective space ($k \geq 1$) is *conjugate*, that is, at each point $x \in G$ the directions $\Delta_1^i(x)$, $\Delta_1^j(x)$ of the tangents to any two curves of Σ_n through x are conjugate (two directions are conjugate if each belongs to the characteristic of the tangent plane T_x when it is displaced in the other direction). If V_n is not contained in a projective space of dimension less than $n+k$, then for $k=1$, V_n carries an infinite set of conjugate nets; for $k=2$ the surface carries, in general, a unique conjugate net, but there are n-dimensional surfaces on which there are no conjugate nets; for $k > 2$ only n-dimensional surfaces of a special structure carry a conjugate net. For $n > 2$ a conjugate net need not be holonomic (see [3]). A special case of a holonomic conjugate net is an *n-conjugate system*: A net Σ_n with the property that the tangents to the curves of each family taken along any curve of any other family form a developable surface. Conjugate systems exist in a projective space of any dimension $n+k$ for $n \geq 2$, $k \geq 0$. Surfaces V_n that carry an n-conjugate system in the

$(n+k)$-dimensional projective space, when $k \geqslant n$, and for which at each point $x \in V_n$ the osculating space (the space of second differentials of the point x) has dimension $2n$, were first considered by E. Cartan [4] under the name 'manifolds of special projective type' (*Cartan surfaces*). The concept of the **Laplace transformation (in geometry)** was extended to such nets (see [5], [6]).

In the study of diffeomorphisms of nets, in terms of known properties of a net $\Sigma_n \subset M$ one describes properties of the net $\phi(\Sigma_n) \subset N$ for a given diffeomorphism $\phi: M \to N$ (for example, under a bending deformation or under a conformal mapping of a surface carrying a net), or one looks for a diffeomorphism ϕ that preserves certain properties of Σ_n. For example, a net Σ_2 on a surface in Euclidean space is called a *rhombic net* (a conformal Chebyshev net) if it admits a conformal mapping onto a Chebyshev net. On every surface of revolution the **asymptotic net** is rhombic.

References

[1] NORDEN, A.P.: *Spaces with an affine connection*, Nauka, Moscow-Leningrad, 1976 (in Russian).

[2] DUBNOV, YA.S. and FUCHS, S.A. [S.A. FUKS]: 'Sur quelques réseaux de l'espace analogues au réseau de Tchebychev', *Dokl. Akad. Nauk SSSR* **28**, no. 2 (1940), 102-105.

[3] BAZYLEV, V.T.: 'Multidimensional nets and their transformations', *Itogi Nauk. Geom. 1963* (1965), 138-164 (in Russian).

[4] CARTAN, E.: 'Sur les variétés de courbure constante d'un espace euclidien ou non euclidien', *Bull. Soc. Math. France* **47** (1919), 125-160.

[5] CHERN, S.S.: 'Laplace transforms of a class of higher dimensional varieties in a projective space of n dimensions', *Proc. Nat. Acad. Sci. USA* **30** (1944), 95-97.

[6] SMIRNOV, R.V.: 'Laplace transforms of p conjugate systems', *Dokl. Akad. Nauk SSSR* **71**, no. 3 (1950), 437-439 (in Russian).

V.T. Bazylev

AMS 1980 Subject Classification: 53BXX, 53CXX, 57R15

NET (IN FINITE GEOMETRY)

Editorial comments. A net is an *incidence structure of points and lines*, i.e. a triple (P, L, I) consisting of a set of points P, a set of lines L and an incidence relation $I \subset P \times L$, such that

1) there exist points, p, and lines, l, and to every point (line) there exist two lines (points) not incident with it;

2) there is at most one line through every two distinct points;

3) if p is not incident with l, then there exist one and only one line m incident with p that does not intersect l.

Thus, a net is a partial plane satisfying the parallelism axiom.

A more elaborate structure in finite geometry also goes by the name net, or *geometrical net*. It consists of a set of n^2 elements and three families of n lines, with each line containing precisely n points. This structure of lines and points is required to satisfy:

a) any two lines of two different families have precisely one point in common;

b) two lines from the same family have no point in common;

c) for each point there is exactly one line from each family passing through that point.

This notion relates to that of a web in differential geometry (cf. **Webs, geometry of**). Pictorially a geometrical net can be seen as an $(n \times n)$-array of points (squares). The first two families of lines form the columns and the rows, and the lines of the third family are subsets with precisely one point in each row and in each column. Giving the squares belonging to the i-th line of the third family the number i, $i = 1, \ldots, n$, results in a **Latin square** (and vice versa).

References

[A1] DEMBOWSKI, P.: *Finite geometries*, Springer, 1968.

[A2] KÁRTESZI, F.: *Introduction to finite geometries*, North-Holland, 1976.

[A3] BATTEN, L.M.: *Combinatorics of finite geometries*, Cambridge Univ. Press, 1986.

AMS 1980 Subject Classification: 51E20, 05B15

NET (OF SETS IN A TOPOLOGICAL SPACE), *network (of sets in a topological space)*

- A family $\mathscr{P}$ of subsets of a topological space X such that for each $x \in X$ and each neighbourhood O_x of x there is an element M of $\mathscr{P}$ such that $x \in M \subset O_x$.

The family of all one-point subsets of a space and every **base** of a space are networks. The difference between a network and a base is that the elements of a network need not be open sets. Networks appear under continuous mappings: If f is a continuous mapping of a topological space X onto a topological space Y and $\mathscr{B}$ is a base of X, then the images of the elements of $\mathscr{B}$ under f form a network $\mathscr{P} = \{fU: U \in \mathscr{B}\}$ in Y. Further, if X is covered by a family $\{X_\alpha: \alpha \in A\}$ of subspaces, then, taking for each $\alpha \in A$ any base $\mathscr{B}_\alpha$ of X_α and amalgamating these bases, a network $\mathscr{P} = \bigcup\{\mathscr{B}_\alpha: \alpha \in A\}$ in X is obtained. Spaces with a countable network are characterized as images of separable metric spaces under continuous mappings.

The minimum cardinality of a network of a space X is called the *network weight*, or *net weight*, of X and is denoted by $\mathrm{nw}(X)$. The net weight of a space never exceeds its weight (cf. **Weight of a topological space**), but, as is shown by the example of a countable space without a countable base, the net weight can differ from the weight. For compact Hausdorff spaces the net weight coincides with the weight. This result extends to locally compact spaces, Čech-complete spaces and feathered spaces (cf. **Feathered space**). Hence, in particular, it follows that weight does not increase under surjective mappings of such spaces. Another corollary: If a feathered space X (in particular, a Hausdorff compactum) is given as the union of a family of cardinality $\leqslant \tau$ of subspaces, the weight of each of which does not exceed τ, supposed infinite, then the weight of X does not exceed τ.

References

[1] ARKHANGEL'SKIĬ, A.V. and PONOMAREV, V.I.: *Fundamentals of general topology: problems and exercises*, Reidel, 1984 (translated from the Russian).
[2] ARKHANGEL'SKIĬ, A.V.: 'An addition theorem for weights of sets lying in bicompacta', *Dokl. Akad. Nauk SSSR* **126**, no. 2 (1959), 239-241 (in Russian).

A.V. Arkhangel'skiĭ

Editorial comments. Most English-language texts (cf. e.g. [A4]) use *network* for the concept called 'net' above. This is because the term 'net' also has a second, totally different, meaning in general topology.

A *net* in a set (topological space) X is an indexed set $\{x_\alpha\}_{\alpha\in\Sigma}$ of points of X, where Σ is a **directed set**. In Russian this is called a **generalized sequence**.

One can build a theory of convergence for nets: *Moore−Smith convergence* (cf. **Moore space**).

References

[A1] ENGELKING, R.: *General topology*, PWN, 1977.
[A2] KELLEY, J.L.: 'Convergence in topology', *Duke Math. J.* 17 (1950), 277-283.
[A3] MOORE, E.H. and SMITH, H.L.: 'A general theory of limits', *Amer. J. Math.* 44 (1922), 102-121.
[A4] NAGATA, J.: *Modern general topology*, North-Holland, 1985.

AMS 1980 Subject Classification: 54A25, 54A99

NETWORK - A generalization of the idea of a **graph**. A network is defined by a pair $(V, \mathscr{E})$, where V is a certain set and $\mathscr{E} = (E_0; E_1, \ldots)$ is a family of collection of elements from V. The elements in the sets $E_i \in \mathscr{E}$ can, in general, be repeated. The elements of V are called the *vertices of the network*, those of E_0 its *poles* and the collections E_i, $i = 1, 2, \ldots$, its *edges*. When the sets of poles is empty and each E_i is a set, a network is a **hypergraph**. If each of the E_i, $i = 1, 2, \ldots$, contains exactly two elements, the network is a *graph with distinguished poles*. Often, a network is regarded as a graph (with or without poles) to the elements of which are ascribed symbols from a certain set. For example, a graph with poles and such that to its edges are ascribed non-negative numbers, called channel capacities, is a *transport network*.

The concept of a network is used in the definition and description of a **control system** and of special classes of control systems (contact schemes, diagrams or functional elements), of transition diagrams of automata, communication networks, etc.

References

[1] YABLONSKIĬ, S.V.: 'Fundamental concepts of cybernetics', *Probl. Kibernet.* **2** (1959), 7-38 (in Russian).
[2] FORD, L.R. and FULKERSON, D.R.: *Flows in networks*, Princeton Univ. Press, 1962.
[3] KUNTZMANN, J.: *Théorie des réseaux (graphes)*, Dunod, 1972.

A.A. Sapozhenko

Editorial comments. For *network in topology* see **Net (of sets in a topological space)**.

Most often, a network is simply defined as a graph $\Gamma = (V, E)$ such that for each $e \in E$ there is specified a (posi-

tive) number c_e, called the *capacity* of the edge e. A *directed network* is a directed graph with for each $e \in E$ a specified capacity c_e. Often c_e is thought of as the maximal amount of something that can be transported over e. But other interpretations are possible and occur. For example, a network of roads with $c_e = $ length of road e for each $e \in E$.

A *transportation network* is a directed network with a specified vertex v_0, the *source*, and a second specified vertex v_1, the *sink*. Usually it is also assumed that there are no incoming arcs at v_0 and no outgoing arcs at v_1. For all $e \in E$, $v \in V$, let

$$n(e, v) = \begin{cases} 1 & \text{if } e \text{ is an incoming edge at } v, \\ -1 & \text{if } e \text{ is an outgoing edge at } v, \\ 0 & \text{otherwise.} \end{cases}$$

A *flow in a transportation network* is a function $\phi: E \to \mathbf{R}$ such that $0 \leqslant \phi(e) \leqslant c_e$ and such that for all $v \in V \setminus \{v_0, v_1\}$ the following *Kirchhoff law* holds

$$\sum_e n(e, v)\phi(e) = 0. \tag{A1}$$

Sometimes an artificial arc, the *return arc*, going from v_1 to v_0, is added with capacity ∞. By defining $\phi(\text{return arc}) = \sum_e n(e, v_1)\phi(e)$, a flow is defined such that the Kirchhoff law holds for all vertices. This is the sole function of the return arc.

The *network flow problem* addresses the problem of calculating those flows for which $\sum_e n(e, v_1)\phi(e)$ is maximal.

In **network planning** one encounters 'transportation networks' in which the edges correspond to tasks which can be performed and where the numbers c_e represent, for example, the time needed to perform task e. Here, finding a 'shortest path' can be important.

An *electrical network* is a directed graph with for each $e \in E$ two real numbers i_e and E_e specified (which are thought of as current and voltage, respectively). More generally, the i_e and E_e may well be functions of time or variables.

The currents i_e are required to satisfy the *Kirchhoff current law*

$$\sum_e n(e, v)i_e = 0,$$

and the voltages E_e are required to satisfy the *Kirchhoff voltage law*, which says that there are numbers E_v, the *potential* at node v, such that

$$\sum_v n(1, v)E_v = E_e.$$

A branch (edge) of a network in which the current is given as an (input) function of time is called a *current source*, and a branch over which there is given an voltage is called a *voltage source*. A branch which is either a voltage or a current source is called a *source branch*. An electrical network with n specified source branches is called an *n-port*.

If the currents in and the voltages across the (non-source) branches are related by

$$E_e = \sum_{e'} z_{ee'} i_{e'}$$

or

$$i_e = \sum_{e'} y_{ee'} E_{e'},$$

where the $z_{ee'}$ or $y_{ee'}$ are linear integro-differential operators, one speaks of a *linear electrical network*. It is called *reciprocal* if $z_{ee'} = z_{e'e}$ or $y_{ee'} = y_{e'e}$. The network is called *passive* if

$$\int_{-\infty}^{t} \sum_{e \in S} i_e(\tau) E_e(\tau)\, d\tau \leqslant 0$$

for all choices of the current sources and voltage sources for which the current/voltage relations are satisfied. Here S denotes the set of source branches.

Under certain non-singularity conditions for the currents in and voltages across the source branches one has the relations

$$i_e = \sum_{e' \in S} Y_{ee'} E_{e'}, \quad e \in S,$$

$$E_e = \sum_{e' \in S} Z_{ee'} i_{e'}, \quad e \in S.$$

The matrices Z and Y are, respectively, called the *port-impedance matrix* and the *port-admittance matrix* of the network.

Synthesis of electrical networks, or *realizability theory*, refers to the problem of finding an electrical network with (partly) given port-impedance and port-admittance matrices. Every passive linear one-port network can be realized using as branches positive *resistors* ($E_e = R_e i_e$), positive *capacitors* ($i_e = c_e(dE_e / dt)$) and positive *inductors* ($E_e = L_e(di_e / dt)$). Every linear passive n-port with $n > 1$ can be synthesized using in addition *ideal transformers* (a pair of branches e_1, e_2 such that $i_2 = -ne_1$, $i_1 = ne_2$) and *ideal gyrators* (a pair of branches e_1, e_2 such that $i_2 = E_1$, $E_2 = -i_1$).

References

[A1] WEINBERG, L.: *Network analysis and synthesis*, McGraw-Hill, 1962.
[A2] BRUNE, O.: 'Synthesis of a finite two-terminal network whose driving-point impedance is a prescribed function of frequency', *J. Math. Phys.* **10** (1931), 191-236.
[A3] PAPADIMITRIOU, CHR. H. and STEIGLITZ, K.: *Combinatorial optimization*, Prentice-Hall, 1982.
[A4] FULKERSON, D.R.: 'Flow networks and combinatorial operations research', in D.R. Fulkerson (ed.): *Studies in graph theory*, Vol. I, Math. Assoc. Amer., 1975, pp. 139-171.
[A5] BERGE, C.: *Graphs*, North-Holland, 1985, Chapt. 5.
[A6] ATEBEKOV, G.I.: *Linear network theory*, Pergamon, 1965.
[A7] SLEPIAN, P.: *Foundations of network analysis*, Springer, 1968.
[A8] BOSE, N.K.: *Applied multidimensional system theory*, v. Nostrand Reinhold, 1982, Chapt. 5.
[A9] BUDAK, A.: *Circuit theory. Fundamentals and applications*, Prentice-Hall, 1978.
[A10] GONDRAN, M. and MINOUX, M.: *Graphs and algorithms*, Wiley, 1986, Chapt. 5 (translated from the French).
[A11] BURR, S.A. (ED.): *The mathematics of networks*, Amer. Math. Soc., 1982.
[A12] COX, C.W.: *Circuits, signals and networks*, MacMillan, 1969.
[A13] CARRÉ, B.: *Graphs and networks*, Clarendon Press, 1979.
[A14] BUSACKER, R.G. and SAATY, T.L.: *Finite graphs and networks*, McGraw-Hill, 1965.

AMS 1980 Subject Classification: 94C15, 90C05, 90C35

NETWORK GRAPH - The graphical representation of a **network model** on the plane.

P.S. Soltan

AMS 1980 Subject Classification: 94C15, 90C05, 90C35

NETWORK MODEL - Interpretations of a program (plan) for realizing a certain complex of interrelated operations in the form of an oriented graph (cf. **Graph, oriented**) without circuits, reflecting the natural order of accomplishing the work in time with certain additional data of the complex of operations (cost, resources, durations, etc.). Usually network models are represented graphically on the plane, and the representation is called a *network graph*. A network model is fundamental to **network planning**, control and calendar planning. Depending on the conditions in the processing of the information, a network model can have other forms of representation: tabular, digital, etc. All forms of representation of a network model are equivalent.

At the basis of a network model is its structure, that is, the graph of the complex of operations. It is defined, as a rule, in the following way. Let $V_1, \ldots, V_n$ be a complex of operations (for example, the erection of a multi-floor building) for which the stages $x_1, \ldots, x_m$ are defined: the first is x_1 and the last is x_m, and depending on the logic of the particular order dictated by the interrelation between the operations, each x_i is an intermediate stage (for example, it is impossible to finish erecting the framework of the tenth floor without completing the appropriate work on the ninth). The intermediate stages, and their number, are to some extent conditional, and are mainly defined by the corresponding stages in the realization of the complex. Let $V = \{v_i\}$ be a complex of operations and let $X = \{x_i\}$ be the set of stages. If one now defines the graph G for which X is the set of vertices, and the operation v_j, $j = 1, \ldots, n$, which begins at stage x_r and ends at stage x_s is an arc, then the resulting oriented graph $G = (X, Y)$ without circuits is the required structure of the network model of the complex of operations in question. In the language of network models, $v_1, \ldots, v_n$ are called *operations*, and the stages $x_1, \ldots, x_m$ are called *events*.

A network model is constructed on the basis of its structure, depending on the aims of the complex of operations. For example, if the aim is to complete the complex of operations in least time with given resources, the network model would include data about the time needed to accomplish each operation v_j; in this case it is said that the network model is constructed according to a *time criterion*. Network models can be constructed in terms of another criterion, or in accordance with several criteria taken simultaneously. Accordingly, a network model is called *one-dimensional* or *multi-dimensional*. One can distinguish *canonical* and *alternative network models* (see [1] and [3]). The first are defined by a fixed structure and the condition that any operation $v_j = (x_r, x_s)$ cannot be started until the operations that end at x_r are finished. The second have a variable structure and allow a certain operation $v_j = (x_r, x_s)$ to begin after the completion of a certain operation with end at x_r. Network models can be *deterministic* or *probabilistic*, depending on the exactness of the criteria or on predictability with a certain probability.

References

[1] *Fundamentals in the development and application of systems of network planning and control*, Moscow, 1974 (in Russian).
[2] *Encyclopaedia of cybernetics*, Kiev, 1974 (in Russian).
[3] LOPATNIKOV, L.I.: *A short mathematical-economic dictionary*, Moscow, 1979 (in Russian).

P.S. Soltan

Editorial comments. For additional references see Network.

AMS 1980 Subject Classification: 94C15, 90C05, 90C35

NETWORK PLANNING, *network method of planning and control* - A method of control in the realization of a certain complex of operations (designs, programs, topics, etc.) on the basis of a **network model** of the complex; it is also known as PERT (see [2]). Network planning allows one to raise significantly the quality of the planning and control in the realization of the complex of operations; in particular, it makes it possible to coordinate efficiently the activity from all sides (organizations) participating in the realization of the complex, to distinguish the most important problems, to assess the most appropriate times to realize designs, to correct plans for realization in good time, etc.

Network planning can be conditionally split into two phases: 1) the construction of a network model of the complex of operations; and 2) the use of this network model for planning and control in the realization of the complex of operations (see [1] - [3]). The construction of a network model of a complex reduces to the representation of the set of stages (events) and the natural order (generally speaking, special) of the operations of the complex as a specially oriented graph, as well as of representing the numerical information necessary (the time required to accomplish each operation, resources, etc.). Depending on the formulated aims, after constructing a network model one proceeds to analyze it to find the best preparations of a plan to achieve these aims. For example, if a network model is constructed in terms of a time orientation, that is, when the whole complex of operations is to be achieved in minimal time for given resources, this analysis reduces to finding the critical paths and determining the least time in which the realization of the complex is possible. This means the following. Let $G = (X, V)$ be the structure of a network model of the complex, where

$$X = \{x_1, \ldots, x_m\} \quad \text{and} \quad V = \{v_1, \ldots, v_m\}$$

are, respectively, the sets of events and operations, and $t(v_j)$ is the time required to accomplish the operation $v_j \in V$. One considers the set P of all paths of the graph G that are maximal with respect to inclusion. In general, there are many such paths in G, and for the simple situation when there is one initial event x_1 and one terminal one x_m, all these paths begin at x_1 and end at x_m. Among the paths of P one can find a path with least length (the *length of a path* $p = (v_{j1}, \ldots, v_{jk})$ is the number $t(p) = t(v_{j1}) + \cdots + t(v_{jk})$). A path $p \in P$ with this property is called a *critical path of the network planning* and its length expresses the fact that the complex of operations cannot be realized in a time less than $t(p)$. Therefore, the method of network planing is also called the *critical path method* (see [1] - [4]).

During the period of the realization of a complex of operations, network planning plays the role of a control mechanism that helps one to process information about the actual state of the operations at a given instant, and to forecast changes and necessary corrections of plans to accomplish the remaining operations.

References

[1] *Fundamentals in the development and application of systems of network planning and control*, Moscow, 1974 (in Russian).
[2] KAUFMANN, A. and DESBAZEILLE, G.: *La méthode du chemin critique*, Dunod, 1964.
[3] *Network planning and control*, Moscow, 1967 (in Russian).
[4] ABRAMOV, S.A., MARINICHEV, M.P. and POLYAKOV, P.D.: *Network methods of planning and control*, Moscow, 1965 (in Russian).
[5] *Encyclopaedia of cybernetics*, Kiev, 1974 (in Russian).
[6] LOPATNIKOV, L.I.: *A short mathematical-economic dictionary*, Moscow, 1979 (in Russian).

P.S. Solatan

Editorial comments. For additional references see also Network.

References

[A1] TAHA, H.A.: *Operations research*, MacMillan, 1982.

AMS 1980 Subject Classification: 94C15, 90C05, 90C35

NEUMANN $\bar{\partial}$-PROBLEM, *Neumann DBAR problem, $\bar{\partial}$-problem, $\bar{\partial}$-Neumann problem, DBAR problem, Neumann problem for the Cauchy$-$Riemann complex*

Editorial comments. A non-coercive boundary problem for the complex Laplacian. Let M be a relatively compact domain of a complex manifold M^1 of dimension $n+1$ with smooth boundary bM. The Cauchy$-$Riemann operator $\bar{\partial}$ (defined on functions on a domain $M \subset \mathbb{C}^{n+1}$ by $\bar{\partial} f = \sum_{i=1}^{n+1} (\partial f / \partial \bar{z}_i) \, d\bar{z}_i$) naturally extends to define the *Dolbeault complex* or *Cauchy$-$Riemann complex*

$$0 \to \Lambda^{p,0}(M) \xrightarrow{\bar{\partial}} \Lambda^{p,1}(M) \xrightarrow{\bar{\partial}} \cdots \xrightarrow{\bar{\partial}} \Lambda^{p,n+1}(M) \to 0,$$

where $\Lambda^{p,q}(M)$ is the space of differential forms of type

(p, q) on M. The holomorphic functions are the solutions of $\bar{\partial} f = 0$ and the inhomogeneous equation $\bar{\partial} f = \phi$ (under the necessary compactibility condition $\bar{\partial}\phi = 0$) is also of interest. For instance, in connection with the *Levi problem*: Given $x \in bM$, is there a holomorphic function on M which blows up at x? Using a general formalism of D.C. Spencer (and general Hilbert space theory), the problem $\bar{\partial} f = \phi$ leads to the $\bar{\partial}$-Neumann problem

$$(\bar{\partial}\bar{\partial}^* + \bar{\partial}^*\bar{\partial})u = \phi. \tag{A1}$$

Here $\bar{\partial}^*$ is the adjoint of $\bar{\partial}$, which is defined by $<\bar{\partial}^* f, g> = <f, \bar{\partial} g>$, where the inner product is given by integration with respect to the volume form determined by a given Hermitian metric on $\overline{M}$. The operator $\Box = \bar{\partial}\bar{\partial}^* + \bar{\partial}^*\bar{\partial}$ is called the *complex Laplacian*. If M is a **Kähler manifold**, then $\Box = \Delta/2$, where Δ is the usual *Laplacian of the de Rham complex*, cf. **de Rham cohomology**.

Strictly speaking, equation (A1) should be written as

$$(\bar{\partial}_q\bar{\partial}_q^* + \bar{\partial}_{q+1}^*\bar{\partial}_{q+1})(u) = \phi, \tag{A2}$$

where $u \in \Lambda^{p,q+1}(M)$, $\bar{\partial}_q: \Lambda^{p,q}(M) \to \Lambda^{p,q+1}(M)$, $\bar{\partial}_q^*: \Lambda^{p,q+1}(M) \to \Lambda^{p,q}(M)$; $q = -1, 0, \ldots, n+1$, $\Lambda^{p,-1}(M) = 0 = \Lambda^{p,n+2}(M)$. Thus equation (A2) comes naturally equipped with the boundary conditions

$$u \in \text{Domain}(\bar{\partial}_q^*), \tag{A3}$$

$$\bar{\partial}_{q+1} u \in \text{Domain}(\bar{\partial}_{q+1}^*). \tag{A4}$$

(The $\bar{\partial}$-*Neumann boundary conditions*.) The operator $\Box$ is elliptic, but the boundary conditions are not. Nevertheless, J.J. Kohn was able to prove existence and to provide a systematic analysis of regularity. A main result is the estimate

$$\| u \|_{s+1} \leqslant A_s \| \Box u \|_s + \| u \|,$$

where $\| \ \|_s$ are Sobolev norms (cf. **Sobolev space**). For more details cf. [A1], [A2]. A great deal of additional and related material can be found in [A1] - [A4].

References

[A1] FOLLAND, G.B. and KOHN, J.J.: *The Neumann problem for the Cauchy–Riemann complex*, Princeton Univ. Press, 1972.
[A2] GREINER, P.C. and SFEIN, E.M.: *Estimates for the $\bar{\partial}$-Neumann problem*, Princeton Univ. Press, 1977.
[A3] TREVES, F.: *Introduction to pseudodifferential and Fourier integral operators*, 1, Plenum, 1980, Sect. III.8.
[A4] KOHN, J.J.: 'Methods of partial differential equations in complex analysis', in R.O. Wells, Jr. (ed.): *Several Complex Variables*, Vol. 1, Amer. Math. Soc., 1977, pp. 215-240.

AMS 1980 Subject Classification: 32F20

NEUMANN FUNCTION - A cylinder function (cf. **Cylinder functions**) of the second kind. The Neumann functions $N_p(x)$ (occasionally the notation $Y_p(x)$ is used) can be defined in terms of the **Bessel functions** as follows:

$$N_p(x) = \lim_{k \to p} \frac{J_k(x)\cos k\pi - J_{-k}(x)}{\sin k\pi}.$$

They are real for positive real x and tend to zero as $x \to \infty$. For large x they have the asymptotic representation

$$N_p(x) \approx \sqrt{\frac{2}{\pi x}} \sin\left[x - \frac{1}{2}p\pi - \frac{\pi}{4} \right].$$

They are connected by the recurrence formulas

$$N_{p-1}(x) + N_{p+1}(x) = \frac{2p}{x} N_p(x),$$

$$N_{p-1}(x) - N_{p+1}(x) = 2N_p'(x).$$

For integers $p = n$:

$$N_{-n}(x) = (-1)^n N_n(x);$$

for small x:

$$N_0(x) \approx -\frac{2}{\pi}\ln\left[\frac{2}{\gamma x} \right], \quad N_n(x) \approx -\frac{(n-1)!}{\pi}\left[\frac{2}{x} \right]^n,$$

where $\ln \gamma = C = 0.5772 \cdots$ is the Euler constant.

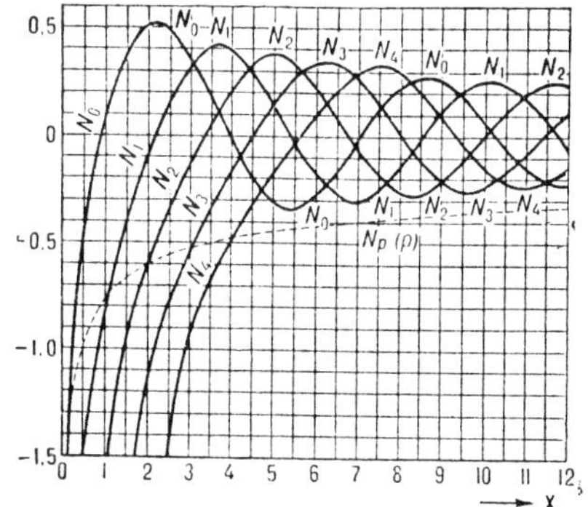

Graphs of Neumann functions.

The Neumann functions of 'half-integral' order $p = (2n+1)/2$ can be expressed in terms of the trigonometric functions; in particular,

$$N_{1/2}(x) = -\sqrt{\frac{2}{\pi x}}\cos x, \quad N_{-1/2}(x) = \sqrt{\frac{2}{\pi x}}\sin x.$$

They were introduced by C.G. Neumann in 1867.

For references see **Cylinder functions**.

V.I. Bityutskov

AMS 1980 Subject Classification: 33A40

NEUMANN PROBLEM - The same as the **second boundary value problem**. It is named after K. Neumann, who was the first to investigate it systematically in 1877.

AMS 1980 Subject Classification: 30-XX, 35-XX

NEUMANN SERIES - 1) A series of the form

$$\sum_{n=0}^{\infty} a_n J_{\nu+n}(z),$$

where $J_{\nu+n}$ is the Bessel function (cylinder function of the first kind, cf. **Bessel functions**) and ν is a (real or complex) number. C.G. Neumann [1] considered the special case when ν is an integer. He showed that if $f(z)$ is an analytic function in a closed disc with centre at the coordinate origin, z is an interior point and C

denotes the boundary of the disc, then

$$f(z) = \sum_{n=0}^{\infty} a_n J_n(z),$$

where

$$a_0 = f(0), \quad a_n = \frac{1}{\pi i} \int_C O_n(t) f(t) \, dt$$

and O_n is a polynomial of degree $n+1$ in $1/t$:

$$O_0(t) = \frac{1}{t},$$

$$O_n(t) = \frac{1}{2t^{n+1}} \int_0^{\infty} e^{-x}[(x+\sqrt{x^2+t^2})^n +$$

$$+ (x - \sqrt{x^2+t^2})^n] \, dx, \quad n \geq 1;$$

it is usually called the *Neumann polynomial* of order n. (Neumann himself called it a *Bessel function of second order*. Nowadays this term is used to denote one of the solutions of the Bessel equation.) Examples of the representation of functions by means of a Neumann series:

$$\cos(z \sin \phi) = J_0(z) + 2 \sum_{n=1}^{\infty} J_{2n}(z) \cos 2n\phi,$$

$$\sin(z \sin \phi) = 2 \sum_{n=1}^{\infty} J_{2n-1}(z) \sin(2n-1)\phi,$$

$$\left(\frac{z}{2}\right)^{\mu} = \sum_{n=0}^{\infty} \frac{(\mu+2n)\Gamma(\mu+n)}{n!} J_{\mu+2n}(z),$$

where μ is an arbitrary number not equal to a non-negative integer and Γ is the **gamma-function**.

2) In the theory of Fredholm integral equations (cf. **Fredholm equation**)

$$\phi(x) - \lambda \int_a^b K(x,s)\phi(s) \, ds = f(x), \quad x \in [a,b], \qquad (1)$$

a *Neumann series* is defined as the expansion of the resolvent $R(x,s;\lambda)$ of the kernel K:

$$R(x,s;\lambda) = \sum_{n=1}^{\infty} \lambda^n K_n(x,s), \qquad (2)$$

where the K_n are the iterated kernels (of K), which are defined by the recurrence formulas

$$K_1(x,s) = K(x,s),$$

$$K_n(x,s) = \int_a^b K_{n-1}(x,t) K(t,s) \, dt, \quad n \geq 2.$$

By means of (2) the solution of (1) for small λ can be represented by

$$\phi(x) = f(x) + \sum_{k=1}^{\infty} \lambda^n \int_a^b K_n(x,s) f(s) \, ds. \qquad (3)$$

The last series is also called a *Neumann series*. In [2] the series (3) is considered in the case of an equation (1) to which the Dirichlet problem in potential theory reduces.

3) Let A be a bounded linear operator mapping a Banach space X into itself, with norm $\|A\| < 1$. Then the operator $I - A$, where I is the identity operator, has a unique bounded inverse $(I-A)^{-1}$, which admits the expansion

$$(I-A)^{-1} = \sum_{n=0}^{\infty} A^n. \qquad (4)$$

In the theory of linear operators this series is called a *Neumann series*. The series (3) can be regarded as a special case of (4).

References

[1] NEUMANN, C.G.: *Theorie der Besselschen Funktionen*, Teubner, 1867.
[2] NEUMANN, C.G.: *Untersuchungen über das logarithmische und Newtonsche Potential*, Teubner, 1877.
[3] WATSON, G.N.: *A treatise on the theory of Bessel functions*, 1, Cambridge Univ. Press, 1952.
[4] KUZ'MIN, R.O.: *Bessel functions*, Moscow-Leningrad, 1035 (in Russian).
[5] YOSIDA, K.: *Functional analysis*, Springer, 1965.
[6] TRICOMI, F.: *Integral equations*, Interscience, 1957.

B.V. Khvedelidze

Editorial comments. The series (4), applied to a specific vector *f*, i.e.

$$\sum_{n=0}^{\infty} A^n f \qquad (\text{A1})$$

may converge also if $\|A\| \geq 1$. For necessary and sufficient conditions for convergence see [A2] (or [A3]).

References

[A1] SMITHIES, F.: *Integral equations*, Cambridge Univ. Press, 1970, Chapt. II.
[A2] SUZUKI, N.: 'On the convergence of Neumann series in Banach space', *Math. Ann.* **220** (1976), 143-146.
[A3] ENGL, H.W.: 'A successive-approximation method for solving equations of the second kind with arbitrary spectral radius', *J. Integral Eq.* **8** (1985), 239-247.
[A4] GOHBERG, I. and GOLDBERG, S.: *Basic operator theory*, Birkhäuser, 1981.
[A5] TAYLOR, A.E. and LAY, D.C.: *Introduction to functional analysis*, Wiley, 1980.
[A6] ZABREYKO, P.P. [P.P. ZABREĬKO], ET.AL.: *Integral equations — a reference text*, Noordhoff, 1975 (translated from the Russian).

AMS 1980 Subject Classification: 33A40, 33A99, 45B05, 47A05, 47A10

NEUTRAL DIFFERENTIAL EQUATION - A differential equation with distributed argument (cf. **Differential equations, ordinary, with distributed arguments**) in which the highest derivative occurs for more than one value of the argument, among them a basic (untransformed) one, and this latter value is the largest of those present in the equation. For example, the equation

$$x'(t) = f(t, x(\alpha(t)), x'(\beta(t))) \qquad (*)$$

is a neutral differential equation when $\alpha(t) \leq t$, $\beta(t) \leq t$.

For a neutral differential equation the initial value problem is solvable; thus, if for (*) with increasing $\beta(t)$ one gives

$$x = \phi(t), \quad t \leq t_0,$$

then for

$$f \in C^{m,n,n}, \quad \alpha, \beta \in C^m, \quad \phi \in C^p, \quad m, n \geqslant 0, \; p \geqslant 1,$$

there exists a (for $n \geqslant 1$ unique) piecewise-smooth solution, which belongs to C^k when $k = 1 + \min\{m, n, p - 1\}$ *compatibility conditions* hold, that is, conditions of the type

$$\phi'(t_0) = f(t_0, \phi(\alpha(t_0)), \phi'(\beta(t_0))).$$

Neutral differential equations are one of the most thoroughly studied classes of equations with distributed arguments. They occur naturally in applied problems that contain in their statement some recurrence property.

A.D. Myshkis

Editorial comments.

References

[A1] HALE, J.K.: *Theory of functional differential equations*, Springer, 1977, Chapt. 12.

AMS 1980 Subject Classification: 34K05

NEUTRON FLOW THEORY, *neutron age theory* - An approximate description of the slowing down of neutrons resulting from their elastic scattering on the nuclei of a medium. It may be used to determine the spatial distribution of neutrons with different energies only in media not containing light nuclei (the permitted mass numbers are $M \gg 2$, i.e. hydrogen and deuterium, for example, are not permitted). The assumption which underlies neutron flow theory is that the slowing down neutrons lose their energy continuously rather than discretely, and that the real behaviour of a large number of individual neutrons is replaced by certain averages; this is legitimate only if, at any given moment after the emergence of the neutrons from their source, their energy scattering is small. On this assumption it is possible to introduce the neutron flow function as an independent variable and employ the neutron flow approximation to obtain the *neutron flow equation* (or *neutron age equation*)

$$\nabla^2 q = \frac{\partial q}{\partial \tau}$$

in the **diffusion approximation**; the form of the equation is the same as that of the *heat equation*. The neutron flow equation is a **parabolic partial differential equation** which describes the slowing down of the neutrons. The unknown function q is the slowing down density, i.e. the number of neutrons in a unit volume which, during their slowing down, cross a given energy value in unit time. The *neutron flow function*, which acts as independent variable τ, is understood to mean the symbolic *age of the neutrons*, which is equal to the time of slowing down of the neutrons (the chronological age) multiplied by the coefficient of diffusion averaged over the slowing down time. The square root of the age of thermal neutrons is said to be the *slowing down length* of a ther-

mal neutron. The physical sense of the concept of 'age' is one-sixth of the mean square of the distance by which the neutron is displaced from the moment of its emergence from a point source (age $\tau = 0$) to the moment under study, corresponding to the age τ. The *age approximation*, which, together with the diffusion approximation, yields the neutron flow equation, is represented by the transition from accurate energy operators which describe the slowing down of the neutron to approximate operators.

The spatial distribution of the neutrons, produced by their diffusion in the course of slowing down, determines the probability of neutron losses in a nuclear reactor, and affects its critical dimensions.

References

[1] GLASSTONE, S. and EDLUND, M.: *The elements of nuclear reactor theory*, New York, 1952.

V.A. Chuyanov

Editorial comments.

References

[A1] WEINBERG, A.M. and WIGNER, E.P.: *The physical theory of neutron chain reactors*, Univ. Chicago Press, 1958.
[A2] GALANIN, A.D.: *Thermal reactor theory*, Pergamon, 1960 (translated from the Russian).
[A3] WILLIAMS, M.M.R.: *The slowing down and thermalization of neutrons*, North-Holland, 1966.

AMS 1980 Subject Classification: 82A75

NEUWIRTH KNOT - A polynomial knot (S^3, k^1) (cf. **Knot theory**) whose group has a finitely-generated commutator subgroup. The complement $S^3 \setminus k^1$ of a Neuwirth knot is a **fibre space** over a circle and the fibre F is a connected surface whose genus is that of the knot. The **commutator subgroup** G' of the group $G = \pi_1(S^3 \setminus k^1)$ of a Neuwirth knot is a free group of rank $2g$, where g is the genus of the knot. The coefficient of the leading term of the Alexander polynomial of a Neuwirth knot (cf. **Alexander invariants**) is 1 and the degree of this polynomial is $2g$. All torus knots (cf. **Torus knot**) are Neuwirth knots. So is every alternating knot whose Alexander polynomial has leading coefficient ± 1.

These knots were introduced by L. Neuwirth (see [1]).

References

[1] NEUWIRTH, L.: *Knot groups*, Princeton Univ. Press, 1965.

M.Sh. Farber

AMS 1980 Subject Classification: 57M25

NEVANLINNA − PICK PROBLEM - Given a class $\mathfrak{H}$ of analytic functions in a domain G of the complex plane (or, in a more general context, of a Riemann surface), to find necessary and sufficient conditions for the solvability in $\mathfrak{H}$ of the interpolation problem

$$f(z_\alpha) = w_\alpha, \tag{1}$$

where $\{z_\alpha\}$ is a subset of G, $\{w_\alpha\}$ is some set of complex numbers, and α usually runs through a countable (sometimes finite, sometimes even uncountable) index set. The classical result of G. Pick [1] and R. Nevanlinna [2] (for finite and countable subsets $\{z_\alpha\} \subset G$, respectively) yields the solution of this problem, for example, in the class B_1 of analytic functions in the unit disc that are bounded by 1 in absolute value. The desired condition here is the non-negativity of the quadratic forms

$$\sum_{j,k=1}^{n} \frac{1-w_j\overline{w_k}}{1-z_j\overline{z_k}}\xi_j\overline{\xi_k}, \quad n\in\mathbf{N}, \quad \xi_j\in\mathbf{C}.$$

The first proof of this proposition (see [3], [4]), as well as quite analogous and similar results for various other function classes $\mathfrak{H}$, relied on algebraic and functional-theoretic methods. Later proofs, based, for example, on reducing the Nevanlinna—Pick problem to a moment problem or obtained from the point of view of the theory of Hilbert spaces, have made it possible to extend the result to uncountable subsets $\{z_\alpha\} \subset G$ and pointed the way to possible generalizations (see [3] - [5]).

A natural development of the Nevanlinna—Pick problem, which necessitated an appeal to functional-analytic methods of investigation, was the question of the solvability of the interpolation problem (1) on a class W of right-hand sides $\{w_\alpha\}$; in this case, as a rule, $\{z_\alpha\}$ is a countable set (a sequence) of points of G, while W may be one of various spaces of sequences of complex numbers. In connection with the class H^∞ of bounded analytic functions in the unit disc and the space l_∞ of bounded sequences, a complete description of the corresponding point sequences $\{z_\alpha\}$ (known as *universal interpolation sequences*) has been obtained (see [6]) in the form of the condition

$$\prod_{\substack{j=1\\j\neq k}}^{\infty}\left|\frac{z_j-z_k}{1-\overline{z_j}z_k}\right| \geqslant \delta > 0, \quad k\in\mathbf{N}. \tag{2}$$

This result played an important role in describing the structure of the maximal ideal space of the algebra H^∞ (see [7]) and was at the same time a starting point for extensive research into the Nevanlinna—Pick problem (in the above generalized formulation) for the **Hardy classes** H^q and the spaces l_p (including weight spaces). It turned out that when $q=\infty$ the solution is independent of p and is given by condition (2), while when $q<\infty$ it necessarily varies when q and p are changed (see [8]). Another generalization of the Nevanlinna—Pick problem is connected with the interpolation problem $\phi_\alpha(f)=w_\alpha$, where $\{\phi_\alpha\}$ is some system of functionals in a class $\mathfrak{H}$. The problem of describing the set $\{\{\phi_\alpha(f)\}: f\in\mathfrak{H}\}$ may be regarded as a generalization of the well-known **coefficient problem** for classes of analytic functions.

References

[1] PICK, G.: 'Ueber die Beschränkungen analytischer Funktionen, welche durch vorgegebene Funktionswerte bewirkt werden', *Math. Ann.* **77** (1916), 7-23.

[2] NEVANLINNA, R.: 'Ueber beschränkte analytische Funktionen', *Ann. Acad. Sci. Fenn. Ser. A* **32**, no. 7 (1929), 1-15.

[3] KREĬN, M.G. and NUDEL'MAN, A.A.: *The Markov moment problem and extremal problems*, Amer. Math. Soc., 1977 (translated from the Russian).

[4] GARNETT, J.B.: *Bounded analytic functions*, Acad. Press, 1981.

[5] SZ.-NAGY, B. and KORÁNYI, A.: 'Rélations d'un problem de Nevanlinna et Pick avec la théorie des opérateurs de l'espace Hilbertien', *Acta Math. Acad. Sci. Hung.* **7** (1956), 295-303.

[6] CARLESON, L.: 'An interpolation problem for bounded analytic functions', *Amer. J. Math.* **80**, no. 4 (1958), 921-930.

[7] SHVEDENKO, S.V.: 'On (H^p, l^r)-interpolation sequences in the unit disk', *Math. USSR Sb.* **46**, no. 4 (1983), 473-492. (*Mat. Sb.* **118**, no. 4 (1982), 470-489)

S.V. Shvedenko

Editorial comments. This classical topic was reactivated and extended to matrix-valued functions in the nineteen-sixties and the beginning of the nineteen-seventies. Connections with operator theory became essential in the new developments (see, e.g., [A1] - [A3]). Applications to problems in control theory, which appeared in the nineteen-eighties, required a revision of the theory and the development of computational methods, in particular, for rational matrix functions (see [A4], [A5]).

See also H^∞ control theory.

References

[A1] ADAMJAN, V.M. [V.M. ADAMYAN], AROV, D.Z. and KREĬN, M.G.: 'Analytic properties of Schmidt pairs for a Hankel operator and the generalized Schur—Tagaki problem', *Math. USSR Sb.* **15** (1971), 31-73. (*Mat. Sb.* **86**, no. 1 (1971), 34-75)

[A2] ROSENBLUM, M. and ROVNYAK, J.: *Hardy classes and operator theory*, Oxford Univ. Press, 1985.

[A3] SARASON, D.: 'Operator-theoretic aspects of the Nevanlinna—Pick interpolation problem', in S.C. Power (ed.): *Operators and Function Theory*, Reidel, 1984, pp. 279-314.

[A4] KIMURA, H.: 'Directional interpolation approach in H^∞-optimization', *IEEE Trans. Autom. Control* **32** (1987), 1085-1093.

[A5] LIMEBEER, D.J.N. and ANDERSON, B.D.O.: 'An interpolation theory approach to H^∞ controller degree bounds', *Linear Algebra Appl.* **98** (1988), 347-386.

[A6] BALL, J.A.: 'Nevanlinna—Pick interpolation: Generalizations and applications', in J.B. Conway (ed.): *Proc. Asymmetric Algebras and Invariant Subspaces. Conf. Indian Univ.*, Pitman, To appear.

[A7] DYM, H.: *J contractive matrix functions, reproducing kernel Hilbert spaces and interpolation*, Amer. Math. Soc., 1989.

[A8] HELTON, J.W.: *Operator theory, analytic functions, matrices, and electrical engineering*, Amer. Math. Soc., 1987.

[A9] DUREN, P.L.: *Theory of H^p spaces*, Acad. Press, 1970.

AMS 1980 Subject Classification: 30E05

NEVANLINNA THEOREMS - Two fundamental theorems, proved by R. Nevanlinna (see [1], [2]), that are basic for the theory of value distribution of meromorphic functions (see **Value-distribution theory**). Let $f(z)$ be a **meromorphic function** on a disc

$$K_R = \{z: |z| < R \leqslant \infty\},$$

where $R=\infty$ means that $f(z)$ is meromorphic in the

entire open complex plane. For every r, $0 \leqslant r < R$, the proximity function of $f(z)$ to a number a is defined by

$$m(r, \infty, f) = \frac{1}{2\pi} \int_0^{2\pi} \ln^+ |f(re^{i\theta})| \, d\theta,$$

$$m(r, a, f) = m\left(r, \infty, \frac{1}{f-a}\right), \quad a \neq \infty,$$

and the counting function of the number of a-points of $f(z)$ by

$$N(r, a, f) = \int_0^r \frac{n(t, a, f) - n(0, a, f)}{t} \, dt + n(0, a, f) \ln r,$$

where $n(t, a, f)$ denotes the number of a-points of $f(z)$, counting multiplicities, in the disc $\{z : |z| \leqslant t\}$, i.e. the number of elements of $f^{-1}(a) \cap \{z : |z| \leqslant t\}$, and $\ln^+ x = \ln x$ for $x \geqslant 1$, $\ln^+ x = 0$ for $0 \leqslant x < 1$.

The function $T(r, f) = m(r, \infty, f) + N(r, \infty, f)$ is called the *Nevanlinna characteristic* of $f(z)$.

Nevanlinna's first theorem. For any function $f(z)$ that is meromorphic on a disc K_R, for any r, $0 \leqslant r < R$, and any complex number a,

$$m(r, a, f) + N(r, a, f) = T(r, f) + \phi(r, a), \tag{1}$$

where

$$|\phi(r, a)| \leqslant \ln^+ |a| + |\ln|c|| + \ln 2.$$

Here c denotes the first non-zero coefficient in the Laurent expansion about zero of the function $f(z) - a$ if $f(0) = a \neq \infty$, and of $f(z)$ itself if $f(0) = \infty$. Thus, for a function whose characteristic $T(r, f)$ increases without limit as $r \to R$, the sum $m(r, a, f) + N(r, a, f)$, considered for different values of a, is equal to the value $T(r, f)$ up to a bounded additive term $\phi(r, a)$. In this sense, all values a are equivalent for any function $f(z)$ that is meromorphic on K_R. For this reason, the theory of value distribution of meromorphic functions concerns itself with questions about the asymptotic behaviour of one term, $m(r, a, f)$ or $N(r, a, f)$, in the invariant sum (1).

Nevanlinna's second theorem shows that, for almost all points a, the principal role in the sum (1) is played by $N(r, a, f)$. The statement of the theorem is as follows.

For any function $f(z)$ that is meromorphic on a disc $K_R = \{z : |z| < R \leqslant \infty\}$, every q, $q \geqslant 3$, and any distinct numbers $\{a_k\}_{k=1}^q$ in the extended complex plane, the relation

$$\sum_{k=1}^q m(r, a_k, f) \leqslant 2T(r, f) - N_1(r, f) + S(r, f) \tag{2}$$

holds, where

$$N_1(r, f) = N(r, \infty, 1/f') + 2N(r, \infty, f) - N(r, \infty, f'),$$

and the term $S(r, f)$ has the following properties:

1) If $R = \infty$, i.e. if $f(z)$ is meromorphic in the entire open complex plane, then

$$S(r, f) = O(\ln(r \cdot T(r, f))),$$

as $r \to \infty$, for all values of r with the possible exception of a set E of finite total measure.

2) If $R < \infty$, then

$$S(r, f) = O\left(\ln\left[\frac{R}{R-r} T(r, f)\right]\right),$$

as $r \to R$, for all values of r with the possible exception of a set E for which

$$\int_E \frac{dr}{R-r} < \infty.$$

The function $N_1(r, f)$ is non-decreasing with increasing r, and therefore the right-hand term in (2) cannot increase as $r \to R$ more rapidly than $2T(r, f)$ outside some exceptional set E.

References
[1] NEVANLINNA, R.: *Analytic functions*, Springer, 1970 (translated from the German).
[2] NEVANLINNA, R.: *Le théorème de Picard—Borel et la théorie des fonctions méromorphes*, Gauthier-Villars, 1929.
[3] WEYL, H. and WEYL, J.: *Meromorphic functions and analytic curves*, Princeton Univ. Press, 1943.
[4] AHLFORS, L.: 'The theory of meromorphic curves', *Acta Soc. Sci. Fennica. Nova Ser. A* **3**, no. 4 (1941), 1-31.
[5] CARTAN, H.: 'Sur les zéros des combinations linéares de p fonctions holomorphes données', *Mathematica (Cluj)* **7** (1933), 5-31.
[6] GRIFFITHS, P. and KING, J.: 'Nevanlinna theory and holomorphic mappings between algebraic varieties', *Acta Math.* **130** (1973), 145-220.
[7] PETRENKO, V.P.: *The growth of meromorphic functions*, Khar'kov, 1978 (in Russian).

V.P. Petrenko

Editorial comments.

References
[A1] HAYMAN, W.K.: *Meromorphic functions*, Oxford Univ. Press, 1964.
[A2] GRIFFITHS, P.A.: *Entire holomorphic mappings in one and several complex variables*, Princeton Univ. Press, 1976.

AMS 1980 Subject Classification: 30D35

NEWTON BINOMIAL, *binomium of Newton* - The formula for the expansion of an arbitrary positive integral power of a **binomial** in a polynomial arranged in powers of one of the terms of the binomial:

$$(z_1 + z_2)^m = \tag{*}$$

$$= z_1^m + \frac{m}{1!} z_1^{m-1} z_2 + \frac{m(m-1)}{2!} z_1^{m-2} z_2^2 + \cdots + z_2^m =$$

$$= \sum_{k=0}^m \binom{m}{k} z_1^{m-k} z_2^k,$$

where

$$\binom{m}{k} = \frac{m!}{k!(m-k)!}$$

are the **binomial coefficients**. For n terms formula (*) takes the form

$$(z_1 + \cdots + z_n)^m =$$

$$= \sum_{k_1 + \cdots + k_n = m} \frac{m!}{k_1! \cdots k_n!} z_1^{k_1} \cdots z_n^{k_n}.$$

For an arbitrary exponent m, real or even complex, the right-hand side of (*) is, generally speaking, a **binomial series**.

The gradual mastering of binomial formulas, beginning with the simplest special cases (formulas for the 'square' and the 'cube of a sum') can be traced back to the eleventh century. I. Newton's contribution, strictly speaking, lies in the discovery of the binomial series.

E.D. Solomentsev

Editorial comments. The coefficients

$$\begin{bmatrix} m \\ k_1 \cdots k_n \end{bmatrix} = \frac{m!}{k_1! \cdots k_n!}, \quad k_1 + \cdots + k_n = m,$$

are called *multinomial coefficients*.

AMS 1980 Subject Classification: 26CXX

Newton–Cotes quadrature formula -

The interpolation **quadrature formula**

$$\int_a^b f(x)\,dx \cong (b-a)\sum_{k=0}^n B_k^{(n)} f(x_k^{(n)})$$

for the computation of an integral over a finite interval $[a, b]$, with nodes $x_k^{(n)} = a + kh$, $k = 0, \ldots, n$, where n is a natural number, $h = (b-a)/n$, and the number of nodes is $N = n + 1$. The coefficients are determined by the fact that the quadrature formula is interpolational, that is,

$$B_k^{(n)} = \frac{(-1)^{n-k}}{k!(n-k)!n} \int_0^n \frac{t(t-1)\cdots(t-n)}{t-k}\,dt.$$

For $n = 1, \ldots, 7, 9$ all coefficients are positive, for $n = 8$ and $n \geqslant 10$ there are both positive and negative ones among them. The *algebraic degree of accuracy* (the number d such that the formula is exact for all polynomials of degree at most d and not exact for x^{d+1}) is n for odd n and $n + 1$ for even n. The simplest special cases of the Newton–Cotes quadrature formula are: $n = 1$, $h = b - a$, $N = 2$,

$$\int_a^b f(x)\,dx \cong \frac{b-a}{2}[f(a) + f(b)],$$

the **trapezium formula**; $n = 2$, $h = (b-a)/2$, $N = 3$,

$$\int_a^b f(x)\,dx \cong \frac{b-a}{6}\left[f(a) + 4f\left(\frac{a+b}{2}\right) + f(b)\right],$$

the **Simpson formula**; $n = 3$, $h = (b-a)/3$, $N = 4$,

$$\int_a^b f(x)\,dx \cong \frac{b-a}{8}\left[f(a) + 3f(a+h) + 3f(a+2h) + f(b)\right],$$

the *'three-eighths'* quadrature formula. For large n the Newton–Cotes formula is seldom used (because of the property of the coefficients for $n \geqslant 10$ mentioned above). One prefers to use for small n the compound Newton–Cotes quadrature formulas, namely, the trapezium formula and Simpson's formula.

The coefficients of the Newton–Cotes quadrature formula for n from 1 to 20 are listed in [3].

The formula first appeared in a letter from I. Newton to G. Leibniz in 1676 (see [1]) and later in the book [2] by R. Cotes, where the coefficients of the formula are given for n from 1 to 10.

References
[1] NEWTON, I.: 'Mathematical principles of natural philosophy', in A.N. Krylov (ed.): *Collected works*, Vol. 7, Moscow-Leningrad, 1936 (in Russian; translated from the Latin).
[2] COTES, R.: *Harmonia Mensurarum*, 1-2, London, 1722.
[3] KRYLOV, V.I. and SHUL'GINA, L.T.: *Handbook on numerical integration*, Moscow, 1966 (in Russian).

I.P. Mysovskikh

Editorial comments. The formulas above are often referred to as *closed Newton–Cotes formulas*, in contrast to *open Newton–Cotes formulas*, which do not include the end points as nodes.

References
[A1] ENGLLS, H.: *Numerical quadrature and cubature*, Acad. Press, 1980.
[A2] BRASS, H.: *Quadraturverfahren*, Vandenhoeck & Ruprecht, 1977.
[A3] DAVIS, P.J. and RABINOWITZ, P.: *Methods of numerical integration*, Acad. Press, 1984.
[A4] STROUD, A.H.: *Numerical quadrature and solution of ordinary differential equations*, Springer, 1974.

AMS 1980 Subject Classification: 65D32

Newton diagram, *Newton polygon* -

A convex polygonal line, introduced by I. Newton in 1669 (see [1]) to determine the exponents of the principal terms of algebraic functions. The process of finding successively the terms of the expansion of an algebraic function with the help of the Newton diagram is called the *method of the Newton diagram*. In more detail it was worked out by V. Puiseux [2] and it is sometimes called the *Puiseux diagram* in mathematical literature. Before Puiseux, an algebraic version of the Newton diagram was studied by J.L. Lagrange [3].

Let $F(x, y)$ be a pseudo-polynomial in y, that is, let

$$F(x, y) = \sum_{s=0}^n F_s(x)y^s,$$

where

$$F_s(x) = x^{\rho_s} \sum_{r=0}^\infty F_{rs}x^{r/p},$$

x and y are complex variables, F_{rs} are complex numbers, p is a natural number, ρ_s are non-negative rational numbers, $F_n(x) \neq 0$, and $F_0(x) \neq 0$. As a rule one assumes that if $F_s(x) \neq 0$, then $F_{0s}(x) \neq 0$, hence, $F_{00} \neq 0$, $F_{0n} \neq 0$. A solution $y = y(\lambda)$ of the equation

$$F(x, y) = 0 \tag{1}$$

is sought for in the form of a series

$$y = y_\epsilon x^\epsilon + y_{\epsilon'} x^{\epsilon'} + \cdots, \tag{2}$$

where $\epsilon < \epsilon' < \cdots$ or, briefly, $y = y_\epsilon x^\epsilon + z$, $z = o(x^\epsilon)$ as

$x \to 0$. To determine the possible values of ϵ and y_ϵ one substitutes (2) in (1), collects terms with equal powers of x, and equates to zero the coefficients of these powers.

The process begins with the term of lowest degree. As long as the exponent ϵ is not yet determined, there is no way of telling which of the resulting terms are lowest in x. However, the terms of lowest order are among the following:

$$F_{00}x^{\rho_0}, \quad F_{0k}y_\epsilon^k x^{\rho_k + k\epsilon}, \quad F_{0n}y_\epsilon^n x^{\rho_n + n\epsilon}, \tag{3}$$

where k ranges over those values $1, 2, \ldots,$ for which $F_k(x) \not\equiv 0$. To annihilate the terms of lowest order one has to choose ϵ so that at least two of the exponents ρ_0, $\rho_k + k\epsilon$, $\rho_n + n\epsilon$ coincide and the remaining ones are to be not smaller. This argument leads to Newton's diagram.

In the plane one takes a rectangular Cartesian coordinate system and plots the points $(0, \rho_0)$, (k, ρ_k) and (n, ρ_n), where k ranges over the same values as in (3). Through the point $(0, \rho_0)$ one draws the line that coincides with the y-axis and then rotates around $(0, \rho_0)$ anti-clockwise until it falls onto one of the plotted points, say (l, ρ_l). The tangent of the angle that the line L passing through $(0, \rho_0)$ and (l, ρ_l) makes with the negative x-axis is one of the values ϵ, since $\rho_0 = \rho_l + l\epsilon$, and $\rho_k + k\epsilon > \rho_l + l\epsilon$ if $(k, \rho_k) \notin L$. Suppose that (s, ρ_s) is the point on L with largest x-coordinate and that L is rotated anti-clockwise around (s, ρ_s) until it falls onto another one of the plotted points, say, (t, ρ_t) with $t > s$. Let L' be the line through (s, ρ_s) and (t, ρ_t). The tangent of the angle between L' and the negative x-axis gives another possible value of ϵ. Continuing these constructions one obtains a convex polygonal line, which is called *Newton's diagram*.

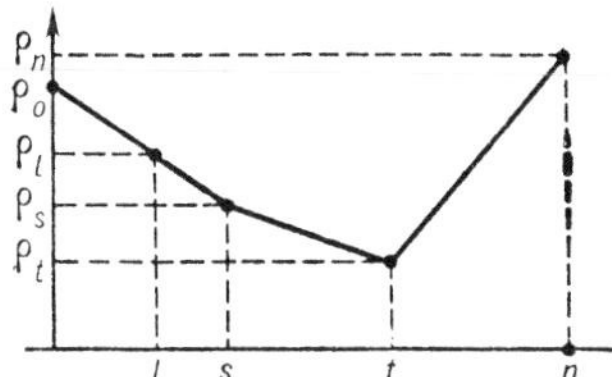

The value of the coefficient of y_s^ϵ is determined as follows. Let (i, ρ_i) and (j, ρ_j) be the extreme points of a segment of the Newton diagram that determines one of the possible values of ϵ. To annihilate the terms of lowest order when (2) is substituted in (1) it is necessary and sufficient that

$$\sideset{}{'}\sum_\epsilon F_{0s}y_\epsilon^s = 0, \tag{4}$$

where the prime in the sum denotes that summation is over those ρ for which $\rho_s + s\epsilon = \rho_i + i\epsilon$. Equation (4) has $j - i$ non-zero roots (including multiplicity), that is, as many as the length of the projection of the relevant

segment of the Newton diagram. Hence it is clear that by the method of the Newton diagram one obtains all n values of the principal term $y_\epsilon x^\epsilon$ in (2). By the same method one determines the next term in the expansion (2), etc. As a result, all n solutions of (1) have the form (2), so-called *Puiseux series* (see **Algebraic function**). The method of the Newton diagram is also applicable to the solution of differential equations.

References

[1] NEWTON, I.: *The mathematical papers of I. Newton*, 1-8, Cambridge Univ. Press, 1967-1981.
[2] PUISEUX, V.: 'Récherches sur les fonctions algébriques', *J. Math. Pure Appl.* **15** (1850), 365-480.
[3] LAGRANGE, J.L.: 'Solution de quelques problèmes d'astronomie sphérique par le moyen des séries', *Nouv. Mém. Acad. Roy. Sci. Belles Lettres Berlin* (1776).
[4] VAĬNBERG, M.M. and TRENOGIN, V.A.: *Theory of branching of solutions of non-linear equations*, Noordhoff, 1974 (translated from the Russian).
[5] *Isaac Newton 1643 - 1727; a collection of articles on the tercentenary of his birth*, Moscow, 1943 (in Russian).
[6] BRUNO, A.D. [A.D. BRYUNO]: *Local methods in nonlinear differential equations*, Springer, 1989 (translated from the Russian).

V.A. Trenogin

AMS 1980 Subject Classification: 30B10, 34-XX

NEWTON INTERPOLATION FORMULA - A form of writing the **Lagrange interpolation formula** by using divided differences:

$$L_n(x) = f(x_0) + (x - x_0)f(x_0; x_1) + \cdots + \tag{1}$$
$$+ (x - x_0) \cdots (x - x_{n-1})f(x_0; \cdots; x_n),$$

where $f(x_0; \cdots; x_k)$ are the divided differences of order k; it was treated by I. Newton in 1687. Formula (1) is called *Newton's interpolation formula for unequal differences*. When the x_i are equidistant, that is, if

$$x_1 - x_0 = \cdots = x_n - x_{n-1} = h,$$

then by introducing the notation $(x - x_0)/h = t$ and expressing the divided differences $f(x_0; \cdots; x_k)$ in terms of the finite differences $f_{k/2}^k$ according to the formula

$$f(x_0; \cdots; x_k) = \frac{f_{k/2}^k}{h^k k!}, \quad k = 0, \ldots, n,$$

one obtains a way of writing the polynomial $L_n(x)$ in the form

$$L_n(x) = L_n(x_0 + th) = \tag{2}$$
$$= f_0 + t f_{1/2}^1 + \frac{t(t-1)}{2!}f_1^2 + \cdots + \frac{t(t-1) \cdots (t-n+1)}{n!}f_{n/2}^n,$$

which is called *Newton's interpolation formula for forward interpolation*. If the same change of variables is made in the interpolation polynomial L_n with nodes $x_0, x_{-1}, \ldots, x_{-n}$, where $x_{-k} = x_0 - kh$,

$$L_n(x) = f(x_0) + (x - x_0)f(x_0; x_{-1}) + \cdots +$$
$$+ (x - x_0) \cdots (x - x_{-n+1})f(x_0; \cdots; x_{-n}),$$

then one obtains *Newton's interpolation formula for backward interpolation*:

$$L_n(x) = L_n(x_0 + th) = \qquad (3)$$

$$= f_0 + t f^1_{-1/2} + \frac{t(t+1)}{2!} f^2_{-1} + \cdots +$$

$$+ \frac{t(t+1) \cdots (t+n-1)}{n!} f^n_{-n/2}.$$

Formulas (2) and (3) are convenient for computing tables of a given function $f(x)$ if the point x is at the beginning or the end of the table, since in this case the addition of one or several nodes caused by the wish to increase the accuracy of the approximation does not lead to a repetition of the whole work done as in computations with Lagrange's formula.

References

[1] BEREZIN, I.S. and ZHIDKOV, N.P.: *Computing methods*, 1, Pergamon, 1973 (translated from the Russian).
[2] BAKHVALOV, N.S.: *Numerical methods: analysis, algebra, ordinary differential equations*, Mir, 1977 (translated from the Russian).

M.K. Samarin

Editorial comments. The *divided differences* $f(x_0; x_1), \ldots, f(x_0; \cdots ; x_n)$ are defined by:

$$f(x_0; x_1) = \frac{f(x_1) - f(x_0)}{x_1 - x_0},$$

$$f(x_0; x_1; x_2) = \frac{1}{x_2 - x_1} \left[\frac{f(x_2) - f(x_0)}{x_2 - x_0} - \frac{f(x_1) - f(x_0)}{x_1 - x_0} \right],$$

$$\cdots \cdots \cdots \cdots \cdots$$

or

$$f(x_0; \cdots ; x_n) = \sum_{i=0}^{n} \prod_{j=0}^{n}{}' \frac{f(x_i)}{x_i - x_j},$$

where the prime in $\prod'$ means that the factor $1/(x_i - x_i)$ is to be omitted. Formula (1) is also known as the *finite Newton series* for a function f.

References

[A1] ATKINSON, K.E.: *An introduction to numerical analysis*, Wiley, 1978.
[A2] DAVIS, P.: *Interpolation and approximation*, Dover, reprint, 1975.
[A3] HILDEBRAND, F.B.: *Introduction to numerical analysis*, McGraw-Hill, 1974.

AMS 1980 Subject Classification: 65D05, 41A05

NEWTON LAWS OF MECHANICS - The three basic laws describing the motion of material bodies under the action of forces applied to them.

First law: If no forces act on a material point (or if the forces applied to it are in equilibrium), then relative to an inertial **reference system** the material point is in a state of rest or uniform rectilinear motion.

Second law: If a force $\mathbf{F}$ acts on a material point, then relative to an inertial reference system the point undergoes an acceleration $\mathbf{a}$ such that its product with the mass m of the point is equal to $\mathbf{F}$:

$$m\mathbf{a} = \mathbf{F}.$$

Third law: Two material points act on each other with forces that are equal in absolute value but opposite in direction along the line joining the two points.

Newton's laws of mechanics cease to be valid for motions of objects of very small dimension (elementary particles) and for motions with velocities close to that of light.

The laws were stated by I. Newton in 1687.

S.M. Targ

Editorial comments.

References

[A1] LINDSAY, R.B. and MARGENAU, H.: *Foundations of physics*, Dover, 1957.
[A2] RUTHERFORD, D.E.: *Classical mechanics*, Oliver & Boyd, 1957.
[A3] TAYLOR, TH.T.: *Mechanics: classical and quantum*, Pergamon, 1976.

AMS 1980 Subject Classification: 70D05, 70A05

NEWTON—LEIBNIZ FORMULA - The formula expressing the value of a definite integral of a given function f over an interval as the difference of the values at the end points of the interval of any primitive (cf. **Integral calculus**) F of the function f:

$$\int_a^b f(x)\,dx = F(b) - F(a). \qquad (*)$$

It is named after I. Newton and G. Leibniz, who both knew the rule expressed by (*), although it was published later.

If f is Lebesgue integrable over $[a, b]$ and F is defined by

$$F(x) = \int_a^x f(t)\,dt + C,$$

where C is a constant, then F is absolutely continuous, $F'(x) = f(x)$ almost-everywhere on $[a, b]$ (everywhere if f is continuous on $[a, b]$) and (*) is valid.

A generalization of the Newton—Leibniz formula is the **Stokes formula** for orientable manifolds with a boundary.

L.D. Kudryavtsev

Editorial comments. The theorem expressed by the Newton—Leibniz formula is called the *fundamental theorem of calculus*, cf. e.g. [A1].

References

[A1] STROMBERG, K.: *An introduction to classical real analysis*, Wadsworth, 1981, p. 318ff.
[A2] RUDIN, W.: *Real and complex analysis*, McGraw-Hill, 1966, p. 165ff.

AMS 1980 Subject Classification: 26A42

NEWTON METHOD, *method of tangents* - A method for the approximation of the location of the roots of a real equation

$$f(x) = 0, \qquad (1)$$

where f is a differentiable function. The successive approximations of Newton's method are computed by the formulas

$$x^{k+1} = x^k - [f'(x^k)]^{-1} f(x^k), \quad k = 0, 1, \ldots. \tag{2}$$

If f is twice continuously differentiable, x^* is a simple root of (1) and the initial approximation x^0 lies sufficiently close to x^*, then Newton's method has quadratic convergence, that is,

$$|x^{k+1} - x^*| \leqslant c |x^k - x^*|^2,$$

where c is a constant depending only on f and the initial approximation x^0.

Frequently, for the solution of (1) one applies instead of (2) the so-called *modified Newton method*:

$$x^{k+1} = x^k - [f'(x^0)]^{-1} f(x^k). \tag{3}$$

Under the same assumptions under which Newton's method has quadratic convergence, the method (3) has linear convergence, that is, it converges with the rate of a **geometric progression** with denominator less than 1.

In connection with solving a non-linear operator equation $A(u) = 0$ with an operator $A : B_1 \to B_2$, where B_1 and B_2 are Banach spaces, a generalization of (2) is the *Newton–Kantorovich method*. Its formulas are of the form

$$u^{k+1} = u^k - [A'(u^k)]^{-1} A(u^k), \quad k = 0, 1, \ldots,$$

where $A'(u^k)$ is the **Fréchet derivative** of A at u^k, which is an invertible operator acting from B_1 to B_2. Under special assumptions the Newton–Kantorovich method has quadratic convergence, and the corresponding modified method has linear convergence (cf. also **Kantorovich process**).

I. Newton worked out his method in 1669.

References

[1] KANTOROVICH, L.V.: 'Functional analysis and applied mathematics', *Nat. Bur. Sci. Rep.* **1509** (1952). (*Uspekhi Mat. Nauk* **3**, no. 6 (1948), 89-185)

[2] KANTOROVICH, L.V. and AKILOV, G.P.: *Funktionalanalysis in normierten Räume*, Akad. Verlag, 1964 (translated from the Russian).

[3] COLLATZ, L.: *Funktionalanalysis und numerische Mathematik*, Springer, 1964.

[4] *Approximate solution of operator equations*, Moscow, 1969 (in Russian).

[5] BAKHVALOV, N.S.: *Numerical methods: analysis, algebra, ordinary differential equations*, Mir, 1977 (translated from the Russian).

Yu.A. Kuznetsov

Editorial comments. The Newton method is also known as the *Newton–Raphson method*, cf., e.g., [A4], Sect. (10.11) for single equations and Sect. (10.13) for systems of n equations.

References

[A1] LAX, P., BURSTEIN, S. and LAX, A.: *Calculus with applications and computing*, New York Univ., 1972.

[A2] DENNIS, J.E., JR. and SCHNABLE, R.: 'Least change secant updates for quasi-Newton methods', *SIAM Review* **21** (1979), 443-459.

[A3] ORTEGA, J.M. and RHEINBOLDT, W.C.: *Iterative solution of nonlinear equations*, Acad. Press, 1970.

[A4] HILDEBRAND, F.B.: *Introduction to numerical analysis*, Dover, reprint, 1987.

AMS 1980 Subject Classification: 26C10, 65D99, 47A62

NEWTON NUMBER - One of the criteria for similarity of mechanical motion, which is obtained from the equation expressing Newton's second law (cf. **Newton laws of mechanics**): $\mathrm{Ne} = 2P/\rho v^2$, where P is the characteristic pressure, ρ is the density and v is the characteristic velocity.

AMS 1980 Subject Classification: 76-XX

NEWTON POTENTIAL *in the broad sense* - A **potential** with Newton kernel $1/|x-y|^{N-2}$, where $|x-y|$ is the distance between two points x and y of the Euclidean space $\mathbf{R}^N$, $N \geqslant 3$, that is, an integral of the form

$$u(x) = \int_S \frac{d\mu(y)}{|x-y|^{N-2}}, \tag{1}$$

where integration is with respect to a certain **Radon measure** μ on $\mathbf{R}^N$ with compact support S. When the measure μ is non-negative, the Newton potential (1) is a superharmonic function in the whole space $\mathbf{R}^n$ (see **Subharmonic function**).

Outside the support S of μ the Newton potential (1) has derivatives of all orders in the coordinates of x and is a regular solution of the **Laplace equation** $\Delta u = 0$, that is, u is a **harmonic function** on the open set CS and is regular at infinity with $u(\infty) = 0$. When μ is absolutely continuous, then u has the form

$$u(x) = \int_D \frac{1}{|x-y|^{N-2}} f(y) \, d\omega(y), \tag{2}$$

where $d\omega$ is the volume element in $\mathbf{R}^N$ and D is a certain bounded domain. If here the *density* f is Hölder continuous in the closed domain D and if the boundary ∂D consists of finitely many closed Lyapunov hypersurfaces (cf. **Lyapunov surfaces and curves**), then u has continuous second-order derivatives inside D and satisfies the **Poisson equation** $\Delta u(x) = -(N-2)2\pi^{N/2} f(x)/\Gamma(N/2)$.

In Newton's work the concept of a 'potential' does not yet occur. The existence of a force function for Newtonian gravitational forces was first proved by J.L. Lagrange in 1773. The terms 'potential function' and 'potential' applied to integrals of the form (2) for $N = 3$ were first used by G. Green in 1828 and C.F. Gauss in 1840. The term 'Newton potential' is sometimes used *in the narrow sense*, applied only to volume potentials of the form (2), and sometimes only to the physically real case of a potential (2) of gravitational forces for $N = 3$, created by masses distributed in D with density $f(y)$.

If an integral of type (2) or (1) is over a hypersurface

$S \subset \mathbf{R}^N$, that is, if

$$u(x) = \int_S \frac{1}{|x-y|^{N-2}} f(y)\, d\sigma(y), \qquad (3)$$

then one speaks of a *simple-layer Newton potential*; it is a regular harmonic function everywhere outside S. If S is a closed Lyapunov hypersurface and the density $f(y)$ is Hölder continuous on S, then the simple-layer Newton potential is continuous everywhere on $\mathbf{R}^N$, and its derivatives are continuous outside S. Moreover, its normal derivative in the direction of the outward normal n_0 to S at $y_0 \in S$ has different limits on approaching S from the inside and the outside. These are expressed by the formulas

$$\lim_{x \to y_0} \frac{du}{dn_0}\bigg|_i = \frac{du(y_0)}{dn_0} + \frac{(N-2)\pi^{N/2}}{\Gamma(N/2)} f(y_0),$$

$$\lim_{x \to y_0} \frac{du}{dn_0}\bigg|_o = \frac{du(y_0)}{dn_0} - \frac{(N-2)\pi^{N/2}}{\Gamma(N/2)} f(y_0),$$

where

$$\frac{du(y_0)}{dn_0} = (N-2)\int_S f(y) \frac{\cos(y-y_0, n_0)}{|y-y_0|^{N-1}}\, d\sigma(y), \quad y_0 \in S,$$

is the so-called direct value of the normal derivative of the simple-layer Newton potential, and $(y-y_0, n_0)$ is the angle between the vector $y-y_0$ and the normal n_0; the normal derivative $du(y_0)/dn_0$ is continuous on S.

A *double-layer Newton potential* has the form

$$v(x) = \int_S f(y) \frac{\cos(y-x, n)}{|y-x|^{N-1}}\, d\sigma(y), \qquad (4)$$

where n is the outward normal to S at $y \in S$. It is also a harmonic function outside S, but upon approaching S it has a discontinuity. Under the same assumptions on S and $f(y)$ it has limits from the inside and the outside of S. These are expressed by the formulas

$$\lim_{x \to y_0} v(x)\bigg|_i = v(y_0) + \frac{\pi^{N/2}}{\Gamma(N/2)} f(y_0),$$

$$\lim_{x \to y_0} v(x)\bigg|_o = v(y_0) - \frac{\pi^{N/2}}{\Gamma(N/2)} f(y_0),$$

where

$$v(y_0) = \int_S f(y) \frac{\cos(y-y_0, n)}{|y-y_0|^{N-1}}\, d\sigma(y)$$

is the so-called direct value of the double-layer Newton potential at $y_0 \in S$. Under somewhat more stringent conditions on S and $f(y)$ the normal derivative of the double-layer Newton potential is, however, continuous on passing through S.

See also **Double-layer potential**; **Potential theory**; **Simple-layer potential**; **Surface potential**.

References

[1] GÜNTER, N.M.: *Potential theory and its applications to basic problems of mathematical physics*, F. Ungar, New-York, 1967 (translated from the French).
[2] SRETENSKIĬ, L.N.: *Theory of the Newton potential*, Moscow-Leningrad, 1946 (in Russian).
[3] LANDKOF, N.S.: *Foundations of modern potential theory*, Springer, 1972 (translated from the Russian).
[4] BRÉLOT, M.: *Eléments de la théorie classique du potentiel*, Sorbonne Univ. Centre Doc. Univ., Paris, 1959.
[5] WERMER, J.: *Potential theory*, Springer, 1974.
[6] KELLOGG, O.D.: *Foundations of potential theory*, F. Ungar, 1929. Re-issue: Springer, 1967.

E.D. Solomentsev

Editorial comments. The 'analogue' in dimension 2 is the logarithmic potential.

References

[A1] GAUSS, C.F.: 'Allgemeine Lehrsätze in Beziehung auf die im verkehrte Verhältnisse des Quadrats der Entfernung wirkenden Anziehungs- und Abstossungskräfte', in *Werke*, Vol. 5, K. Gesellschaft Göttingen, 1876, pp. 195-242.
[A2A] GREEN, G.: 'An essay on the application of mathematical analysis to the theories of electricity and magnetism I', *J. Reine Angew. Math.* 39 (1850), 73-89. Re-issued by Lord Kelvin.
[A2B] GREEN, G.: 'An essay on the application of mathematical analysis to the theories of electricity and magnetism II', *J. Reine Angew. Math.* 44 (1852), 356-374. Re-issued by Lord Kelvin.
[A2C] GREEN, G.: 'An essay on the application of mathematical analysis to the theories of electricity and magnetism III', *J. Reine Angew. Math.* 47 (1854), 161-221. Re-issued by Lord Kelvin.
[A3] LAGRANGE, J.-L.: 'Sur l'équation séculaire de la lune', *Mém. Acad. Roy. Sci. Paris* (1773).

AMS 1980 Subject Classification: 31B10, 31B15, 31B25

NEYMAN METHOD OF CONFIDENCE INTERVALS - One of the methods of **confidence estimation**, which makes it possible to obtain interval estimators (cf. **Interval estimator**) for unknown parameters of probability laws from results of observations. It was proposed and developed by J. Neyman (see [1], [2]). The essence of the method consists in the following. Let $X_1, \ldots, X_n$ be random variables whose joint distribution function $F(x, \theta)$ depends on a parameter $\theta \in \Theta \subset \mathbf{R}^1$, $x = (x_1, \ldots, x_n) \in \mathbf{R}^n$. Suppose, next, that as point estimator of the parameter θ a statistic $T = T(X_1, \ldots, X_n)$ is used with distribution function $G(t, \theta)$, $\theta \in \Theta$. Then for any number P in the interval $0.5 < P < 1$ one can define a system of two equations in θ:

$$G(T, \theta) = \begin{cases} P, \\ 1-P. \end{cases} \qquad (*)$$

Under certain regularity conditions on $F(x, \theta)$, which in almost-all cases of practical interest are satisfied, the system $(*)$ has a unique solution

$$\underline{\theta} = \underline{\theta}(T), \quad \overline{\theta} = \overline{\theta}(T), \quad \overline{\theta}, \underline{\theta} \in \Theta,$$

such that

$$P\{\underline{\theta} < \theta < \overline{\theta} \,|\, \theta\} \geqslant 2P - 1.$$

The set $(\underline{\theta}, \overline{\theta}) \subset \Theta$ is called the *confidence interval* (*confidence estimator*) for the unknown parameter θ with confidence probability $2P - 1$. The statistics $\underline{\theta}$ and $\overline{\theta}$ are called the *lower* and *upper confidence bounds* corresponding to the chosen confidence coefficient P. In turn, the number

$$p = \inf_{\theta \in \Theta} P\{\underline{\theta} < \theta < \overline{\theta} \mid \theta\}$$

is called the *confidence coefficient* of the confidence interval $(\underline{\theta}, \overline{\theta})$. Thus, Neyman's method of confidence intervals leads to interval estimators with confidence coefficient $p \geqslant 2P - 1$.

Example 1. Suppose that independent random variables $X_1, \ldots, X_n$ are subject to one and the same normal law $\Phi(x - \theta)$ whose mathematical expectation θ is not known (cf. **Normal distribution**). Then the best estimator for θ is the **sufficient statistic** $\overline{X} = \sum_{i=1}^{n} X_i / n$, which is distributed according to the normal law $\Phi[\sqrt{n}(x - \theta)]$. Fixing P in $0.5 < P < 1$ and solving the equations

$$\Phi[\sqrt{n}(\overline{X} - \underline{\theta})] = P, \quad \Phi[\sqrt{n}(\overline{X} - \overline{\theta})] = 1 - P,$$

one finds the lower and upper confidence bounds

$$\underline{\theta} = \overline{X} - \frac{1}{\sqrt{n}} \Phi^{-1}(P), \quad \overline{\theta} = \overline{X} - \frac{1}{\sqrt{n}} \Phi^{-1}(1 - P)$$

corresponding to the chosen confidence coefficient P. Since

$$\Phi^{-1}(y) + \Phi^{-1}(1 - y) \equiv 0, \quad y \in [0, 1],$$

the confidence interval for the unknown mathematical expectation θ of the normal law $\Phi(x - \theta)$ has the form

$$\left[\overline{X} - \frac{1}{\sqrt{n}} \Phi^{-1}(P), \overline{X} + \frac{1}{\sqrt{n}} \Phi^{-1}(P) \right],$$

and its confidence coefficient is precisely $2P - 1$.

Example 2. Let μ be a random variable subject to the binomial law with parameters n and θ (cf. **Binomial distribution**), that is, for any integer $m = 0, \ldots, n$,

$$P\{\mu \leqslant m \mid n, \theta\} = \sum_{k=0}^{m} \binom{n}{k} \theta^k (1 - \theta)^{n-k} =$$

$$= I_{1-\theta}(n - m, m + 1), \quad 0 < \theta < 1,$$

where

$$I_x(a, b) = \frac{1}{B(a, b)} \int_0^x t^{a-1}(1 - t)^{b-1} \, dt$$

is the **incomplete beta-function** $(0 \leqslant x \leqslant 1, a > 0, b > 0)$. If the 'success' parameter θ is not known, then to determine the confidence bounds one has to solve, in accordance with Neyman's method of confidence intervals, the equations

$$I_{1-\theta}(n - \mu, \mu + 1) = \begin{cases} P, \\ 1 - P, \end{cases}$$

where $0.5 < P < 1$. From tables of mathematical statistics the roots $\overline{\theta}$ and $\underline{\theta}$ of these equations are determined, which are the upper and lower confidence bounds, respectively, with confidence coefficient P. The coefficient of the resulting confidence interval $(\underline{\theta}, \overline{\theta})$ is precisely $2P - 1$. Obviously, if an experiment gives $\mu = 0$, then $\underline{\theta} = 0$, and if $\mu = n$, then $\overline{\theta} = 1$.

Neyman's method of confidence intervals differs substantially from the Bayesian method (cf. **Bayesian approach**) and the method based on Fisher's fiducial approach (cf. **Fiducial distribution**). In it the unknown parameter θ of the distribution function $F(x, \theta)$ is treated as a constant quantity, and the confidence interval $(\underline{\theta}(T), \overline{\theta}(T))$ is constructed from an experiment in the course of which the value of the statistic T is calculated. Consequently, according to Neyman's method of confidence intervals, the probability for $\underline{\theta} < \theta < \overline{\theta}$ to hold is the **a priori probability** for the fact that the confidence interval $(\underline{\theta}, \overline{\theta})$ 'covers' the unknown true value of the parameter θ. In fact, Neyman's confidence method remains valid if θ is a random variable, because in the method the interval estimator is constructed from carrying out an experiment and consequently does not depend on the **a priori distribution** of the parameter. Neyman's method differs advantageously from the Bayesian and the fiducial approach by being independent of a priori information about the parameter θ and so, in contrast to Fisher's method, is logically sound. In general, Neyman's method leads to a whole system of confidence intervals for the unknown parameter, and in this context arises the problem of constructing an optimal interval estimator having, for example, the properties of being unbiased, accurate or similar, which can be solved within the framework of the theory of statistical hypothesis testing.

References

[1] NEYMAN, J.: 'On the problem of confidence intervals', *Ann. Math. Stat.* **6** (1935), 111-116.
[2] NEYMAN, J.: 'Outline of a theory of statistical estimation based on the classical theory of probability', *Philos. Trans. Roy. Soc. London. Ser. A.* **236** (1937), 333-380.
[3] BOL'SHEV, L.N. and SMIRNOV, N.V.: *Tables of mathematical statistics*, Libr. of mathematical tables, 46, Nauka, Moscow, 1983 (in Russian). Processed by L.S. Bark and E.S. Kedova.
[4] BOLSHEV, L.N.: 'On the construction of confidence limits', *Theor. Probab. Appl.* **10** (1965), 173-177. (*Teor. Veroyatnost. i Primenen.* **10** (1965), 187-192)
[5] LEHMANN, E.L.: *Testing statistical hypotheses*, Wiley, 1986.

M.S. Nikulin

AMS 1980 Subject Classification: 62F25, 62G15

NEYMAN – PEARSON LEMMA - A lemma asserting that in the problem of statistically testing a simple hypothesis H_0 against a simple alternative H_1 the **likelihood-ratio test** is a **most-powerful test** among all statistical tests having one and the same given **significance level**. It was proved by J. Neyman and E.S. Pearson [1]. It is often called the *fundamental lemma of mathematical statistics*. See also **Statistical hypotheses, verification of.**

References

[1] NEYMAN, J. and PEARSON, E.S.: 'On the problem of the most efficient tests of statistical hypotheses', *Philos. Trans. Roy. Soc.*

London Ser. A. **231** (1933), 289-337.
[2] LEHMANN, E.L.: *Statistical hypotheses testing*, Wiley, 1978.

M.S. Nikulin

AMS 1980 Subject Classification: 62F03, 62G10

NEYMAN STRUCTURE - A structure determined by a statistic that is independent of a sufficient statistic. The concept was introduced by J. Neyman (see [1]) in connection with the problem of constructing similar tests (cf. **Similar test**) in the theory of statistical hypothesis testing, and the term 'Neyman structure' is used when referring to the structure of a statistical test if its critical function has Neyman structure. Suppose that in the realization of a random variable X taking values in a sample space $(\mathfrak{X}, \mathfrak{B}, P_\theta)$, $\theta \in \Theta$, it is required to verify a composite hypothesis H_0: $\theta \in \Theta_0 \subset \Theta$ and that for the family $\{P_\theta: \theta \in \Theta_0\}$ there exists a **sufficient statistic** T with distribution in the family $\{P_e^T: e \in \Theta_0\}$. Then any statistical test of level α intended for testing H_0 has Neyman structure if its critical function ϕ satisfies the condition:

$$\mathsf{E}\{\phi(X) \mid T = t\} = \alpha \qquad (1)$$

almost everywhere with respect to the measure P_θ^T, $\theta \in \Theta_0$. Evidently, if a statistical test has Neyman structure, then it is similar (cf. **Similar test**) relative to the family $\{P_\theta: \theta \in \Theta_0\}$, since

$$\mathsf{E}_\theta\{\phi(X)\} = \mathsf{E}_\theta\{\mathsf{E}\{\phi(X) \mid T = t\}\} = \alpha$$

for all $\theta \in \Theta_0$.

The validity of (1) essentially reduces the problem of testing the composite hypothesis H_0 to that of testing H_0 as a simple hypothesis for every fixed value t of the sufficient statistic T.

Example. Suppose that two independent random variables X_1 and X_2 are subject to Poisson laws with unknown parameters λ_1 and λ_2 (cf. **Poisson distribution**) and that the hypothesis H_0: $\lambda_1 = \lambda_2$ is to be tested against the alternative H_1: $\lambda_1 \neq \lambda_2$. Thanks to the independence of X_1 and X_2 the statistic $T = X_1 + X_2$ is subject to the Poisson law with parameter $\lambda_1 + \lambda_2$ and the conditional distributions of X_1 and X_2 under the condition $T = t$ are binomial with parameters t, $\lambda_1/(\lambda_1 + \lambda_2)$ and t, $\lambda_2/(\lambda_1 + \lambda_2)$, respectively, that is,

$$\mathsf{P}\{X_i = k \mid T = t\} = \binom{t}{k}\left[\frac{\lambda_i}{\lambda_1 + \lambda_2}\right]^k \left[1 - \frac{\lambda_i}{\lambda_1 + \lambda_2}\right]^{t-k}, \qquad (2)$$

$$k = 0, \ldots, t.$$

When H_0 is valid, then T is sufficient for the unknown common value $\lambda = \lambda_1 = \lambda_2$, and from (2) it follows that when H_0 holds, then the conditional distribution of X_1 for a fixed value of the sufficient statistic $T = t$ is binomial with parameters t and $1/2$, that is, under H_0,

$$\mathsf{P}\{X_1 = k \mid T = t\} = \binom{t}{k}\left[\frac{1}{2}\right]^t, \quad k = 0, \ldots, t.$$

Thus, in this case the problem of testing the composite hypothesis H_0 reduces to that of testing the simple hypothesis H_0', according to which the conditional distribution of X_1 (for a fixed sum $X_1 + X_2 = t$) is binomial with parameters t and $1/2$. For testing H_0' one can use, for example, the **sign test**.

The concept of a Neyman structure is of great significance in the problem of testing composite statistical hypotheses, since among the tests having Neyman structure there frequently is a **most-powerful test**. E. Lehmann and H. Scheffé have shown that a statistical test for testing a composite hypothesis H_0: $\theta \in \Theta_0$ has Neyman structure relative to a sufficient statistic T if and only if the family $\{P_\theta^T: \theta \in \Theta_0\}$ induced by T is boundedly complete. On the basis of the concept of a Neyman structure general methods have been worked out for the construction of similar tests. See **Distributions, complete family of; Similar test**.

References

[1] NEYMAN, J.: 'Current problems of mathematical statistics', in *Proc. Internat. Congress Mathematicians Amsterdam, 1954*, Vol. 1, Noordhoff & North-Holland, 1957, pp. 349-370.
[2] LEHMANN, E.L.: *Testing statistical hypotheses*, Wiley, 1986.
[3] LINNIK, YU.V.: *Statistical problems with nuisance parameters*, Moscow, 1966 (in Russian).

M.S. Nikulin

AMS 1980 Subject Classification: 62F03, 62G10

NICOMEDES CONCHOID - A plane **algebraic curve** of order 4 whose equation in Cartesian rectangular coordinates has the form

$$(x^2 + y^2)(y - a)^2 - l^2 y^2 = 0;$$

and in polar coordinates

$$\rho = \frac{a}{\sin \psi} \pm l.$$

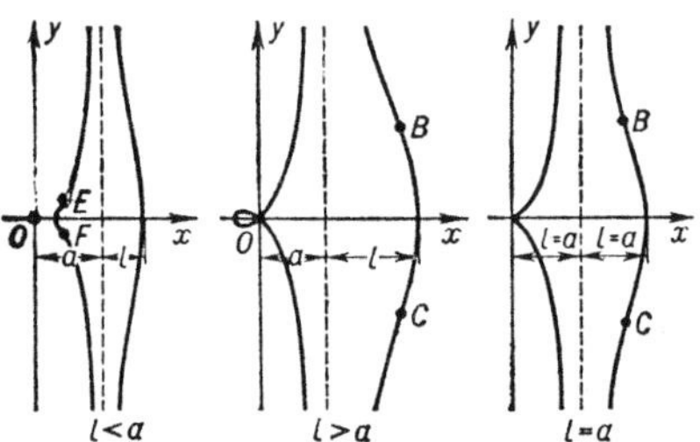

Outer branch (see Fig.). Asymptote $x = a$. Two points of inflection, B and C.

Inner branch. Asymptote $x = a$. The coordinate origin is a double point whose character depends on the values of a and l. For $l < a$ it is an isolated point and, in addition, the curve has two points of inflection, E and F; for $l > a$ it is a **node**; for $l = a$ it is a **cusp**. The curve is a **conchoid** of the straight line $x = a$.

The curve is named after Nicomedes (third century B.C.), who used it to solve the problem of trisecting an angle.

References

[1] SAVELOV, A.A.: *Plane curves*, Moscow, 1960 (in Russian).

D.D. Sokolov

Editorial comments.

References

[A1] LAWRENCE, J.D.: *A catalog of special plane curves*, Dover, reprint, 1972.

AMS 1980 Subject Classification: 14H45, 53A04

NIKOL'SKIĬ SPACE - A **Banach space** $H_p^r(\Omega)$ consisting of functions defined on an open set Ω of an n-dimensional Euclidean space $\mathbf{R}^n$ and having certain difference-differentiability properties characterized by a vector $r = (r_1, \ldots, r_n)$, $r_i > 0$, $i = 1, \ldots, n$, in the L_p-metric, $1 \leq p \leq \infty$. The concept was introduced by S.M. Nikol'skiĭ.

The Nikol'skiĭ space $H_p^r(\Omega)$ can be described in terms of properties of the partial derivatives of order r_i^* in the variable x_i, where $r_i = r_i^* + \alpha_i$, r_i^* is an integer, $0 < \alpha_i \leq 1$, $i = 1, \ldots, n$; if $\Delta_{h_i}^s$ denotes the difference of order $s = 1, 2, \ldots$, and of step h_i with respect to x_i of a function f, then

$$f \in H_p^r(M_1, \ldots, M_n; \Omega), \quad M_i > 0,$$

if and only if f has in Ω generalized partial derivatives

$$f_{x_i}^{(r_i^*)} = \frac{\partial^{r_i^*} f}{\partial x_i^{r_i^*}},$$

$i = 1, \ldots, n$, and if for $0 < \alpha_i < 1$,

$$\| \Delta_{h_i}^1 f_{x_i}^{(r_i^*)} \|_{L_p(\Omega_{|h_i|})} \leq M_i | h_i |^{\alpha_i},$$

while for $\alpha_i = 1$,

$$\| \Delta_{h_i}^2 f_{x_i}^{(r_i^*)} \|_{L_p(\Omega_{2|h_i|})} \leq M_i | h_i |,$$

where Ω_η is the set of points $x \in \Omega$ that are distant by more than $\eta > 0$ from the boundary of Ω and h_i is arbitrary.

The space $H_p^r(\Omega) = H_p^{r_1, \ldots, r_n}(\Omega)$ is defined as the union of all $H_p^r(M_1, \ldots, M_n; \Omega)$ for all $M_i > 0$, $i = 1, \ldots, n$.

If $\Omega \neq \varnothing$, then for any $r_i > 0$, $i = 1, \ldots, n$, $1 \leq p \leq \infty$, the Nikol'skiĭ space $H_p^{r_1, \ldots, r_n}(\Omega)$ is not empty and contains functions that do not belong to $H_p^{r_1, \ldots, r_{i-1}, r_i + \epsilon, r_{i+1}, \ldots, r_n}$ for any $\epsilon > 0$ and any $i = 1, \ldots, n$.

When $p = \infty$, the r_i are not integers and the relevant derivatives are continuous, then a Nikol'skiĭ space is a **Hölder space**. The concept of a Nikol'skiĭ space generalizes to the case of functions that are defined on sufficiently smooth manifolds (see [2]).

There is a description of the Nikol'skiĭ space $H_p^{r_1, \ldots, r_n}(\Omega)$ in terms of properties of the differences of the partial derivatives of orders less than r_i^*; in particular, of those of sufficiently high order of the function itself.

Let $H_p^r(\Omega)$ be an isotropic space, that is,

$r_1 = \cdots r_n = r$. If the domain Ω is such that any function f of class $H_p^r(\Omega)$ can be extended with preservation of the class to the whole space $\mathbf{R}^n$, that is, in such a way that the extended function belongs to $H_p^r(\mathbf{R}^n)$ (this is always the case when the boundary of the domain is sufficiently smooth), then $f \in H_p^r(\Omega)$ if and only if for any non-negative integers k and s such that $0 < r - s < k$ the function f has partial derivatives $f^{(s)}$ of all orders s and there is a constant $M > 0$ such that

$$\| \Delta_h^k f^{(s)} \|_{L_p(\Omega_{k|h|})} \leq M | h |^{r-s}, \tag{1}$$

where $h = (h_1, \ldots, h_n)$ and $\Delta_h^k f^{(s)}$ is the difference of order k of $f^{(s)}$ with vectorial step h. Condition (1) is equivalent to the analogous condition for the modulus of continuity of $f^{(s)}$: There is an $M > 0$ such that

$$\Omega^k(f^{(s)}, \delta) \leq M \delta^{r-s}, \quad \delta > 0,$$

where

$$\Omega^k(f^{(s)}, \delta) = \sup_{|h| = 1} \sup_{0 < t \leq \delta} \| \Delta_{th}^k f^{(s)} \|_{L_p(\Omega_{kt})} \leq M \delta^{r-s}.$$

If M_f, for $f \in H_p^r(\Omega)$, denotes the infimum of all M for which (1) holds for all $h \in \mathbf{R}^n$ and all partial derivatives of an admissible order s, then

$$\| f \| = \| f \|_{L_p(\Omega)} + M_f$$

is a norm in $H_p^r(\Omega)$ and the norms obtained for distinct admissible pairs k, s are equivalent.

A Nikol'skiĭ space consisting of functions defined on the whole space $\mathbf{R}^n$ can be characterized in terms of best approximations of the functions in this space by entire functions of exponential type. Let $E_{v_1, \ldots, v_n}(f)_p$ be the best approximation (error) in the $L_p(\mathbf{R}^n)$-metric of an $f \in L_p(\mathbf{R}^n)$ by entire functions $q_{v_1, \ldots, v_n}(x_1, \ldots, x_n) \in L_p(\mathbf{R}^n)$ of exponential type and of order v_i in x_i, $i = 1, \ldots, n$. The following direct and inverse theorems of Bernshteĭn, Jackson and Zygmund type hold for Nikol'skiĭ functions.

If $f \in H_p^{r_1, \ldots, r_n}(M_1, \ldots, M_n; \mathbf{R}^n)$, then for any $v_i > 0$,

$$E_{v_1, \ldots, v_n}(f) \leq c \sum_{i=1}^n \frac{M_i}{v_i^{r_i}} \tag{2}$$

(the constant $c > 0$ does not depend on f).

Conversely, if (2) holds for a function $f \in L_p(\mathbf{R}^n)$ for $v_i = a_i^k$, $k = 0, 1, \ldots$, $a_i > 1$, $i = 1, \ldots, n$, and if q is an entire function of order 1 in each variable $x_1, \ldots, x_n$ for which

$$\| f - q \|_{L_p(\mathbf{R}^n)} \leq c \sum_{i=1}^n M_i$$

(which exist for $k = 0$, by (2)), then

$$f - q \in H_p^{r_1, \ldots, r_n}(M_1^*, \ldots, M_n^*; \mathbf{R}^n),$$

where

$$M_i^* = c_i \sum_{j=1}^n M_j, \tag{3}$$

and the constants $c > 0$ in (2) and $c_i > 0$ in (3) do not depend on M_i, $i = 1, \ldots, n$.

If f is periodic in all variables, then a similar description of a Nikol'skiĭ space can be given by means of best approximations of the functions by trigonometric polynomials instead of entire functions of exponential type (see [1], [4]).

Nikol'skiĭ spaces can be described by means of a Bessel−Macdonald operator applied to some class of generalized functions (see **Imbedding theorems**).

For the space $H_p^{r_1,\cdots,r_n}(\Omega)$ Nikol'skiĭ has proved transitive imbedding theorems for various dimensions and metrics (see [3] and **Imbedding theorems**), which were subsequently carried over to more general classes of functions. These theorems show that Nikol'skiĭ spaces form a closed system relative to the boundary values of the functions occurring in them: The traces of functions in Nikol'skiĭ spaces on smooth manifolds can in a certain sense be completely described in terms of Nikol'skiĭ spaces.

The properties of Nikol'skiĭ spaces make it possible to obtain necessary and sufficient conditions for the solvability of the **Dirichlet problem** in appropriate Nikol'skiĭ spaces in terms of membership of the boundary function to a certain Nikol'skiĭ space: A harmonic function u belongs to the class $H_p^r(\Omega)$, $r>1/p$, where Ω is a bounded domain in $\mathbf{R}^n$ with a sufficiently smooth boundary $\partial\Omega$, if and only if the boundary values $u\,|_{\partial\Omega}$ belong to the class $H_p^{r-1/p}(\partial\Omega)$. This implies for $p=2$, in particular, that if $u\,|_{\partial\Omega}\in H_2^\rho(\partial\Omega)$, $\rho>1/2$, then the **Dirichlet integral** $D(u)$ of u over Ω is finite, therefore, the Dirichlet problem can be solved by a direct variational method. From imbedding theorems for Nikol'skiĭ spaces it follows that if the Dirichlet integral of u over Ω is finite, then $u\,|_{\partial\Omega}\in H_2^{1/2}(\partial\Omega)$ (see [6]). A generalization of Nikol'skiĭ spaces are the Besov spaces $B_{p\theta}^{r_1,\cdots,r_n}$.

References

[1] NIKOL'SKIĬ, S.M.: 'Inequalities for entire functions of finite order and their application to the theory of differentiable functions in several variables', *Trudy Mat. Inst. Steklov.* **38** (1951), 244-278 (in Russian).

[2] NIKOL'SKIĬ, S.M.: 'Properties of certain classes of functions of several variables on a differentiable manifold', *Mat. Sb.* **33**, no. 2 (1953), 261-326 (in Russian).

[3] NIKOL'SKIĬ, S.M.: 'Imbedding theorems for functions with partial derivatives, considered in differential metrics', *Dokl. Akad. Nauk SSSR* **118**, no. 1 (1958), 35-37 (in Russian).

[4] NIKOL'SKIĬ, S.M.: *Approximation of functions of several variables and imbedding theorems*, Springer, 1975 (translated from the Russian).

[5] BESOV, O.V., IL'IN, V.P. and NIKOL'SKIĬ, S.M.: *Integral representations of functions and imbedding theorems*, Wiley, 1978 (translated from the Russian).

[6] NIKOL'SKIĬ, S.M.: 'On the solution of the polyharmonic equation by a variational method', *Dokl. Akad. Nauk SSSR* **88**, no. 3 (1953), 409-411 (in Russian).

L.D. Kudryavtsev

AMS 1980 Subject Classification: 46E15

NIL ALGEBRA - A power-associative (in particular, an associative) algebra (cf. **Algebra with associative powers**) in which every element is nilpotent (cf. **Nilpotent element**). Special cases of nil algebras are nilpotent and locally nilpotent algebras (cf. **Nilpotent algebra**; **Locally nilpotent algebra**). In the associative case the construction of a nil algebra that is not locally nilpotent is a difficult problem; essentially only one example of such an algebra is known (see [5]).

The class of nil algebras is closed under taking subalgebras and homomorphic images. The extension of a nil algebra by a nil algebra is again a nil algebra. Therefore, in every algebra the sum of all nil ideals is the largest nil ideal containing every nil ideal. It is called the *upper nil radical of the algebra*, and its quotient algebra has no non-zero nil ideals. The question whether the sum of one-sided nil ideals is a one-sided nil ideal is *Köthe's problem*. It is not known (1982) whether simple nil rings exist. When conditions 'of Burnside type' hold in a nil algebra, the algebra is frequently nilpotent or locally nilpotent; a Noetherian nil algebra is nilpotent; so is an Artinian (in particular, a finite-dimensional) nil algebra; a nil algebra of bounded index (the nilpotency indices of the elements are uniformly bounded) over a field of characteristic zero is nilpotent (*Higman's theorem*); a nil algebra satisfying a polynomial identity is locally nilpotent. It is not clear (1982) whether a finitely-generated nil algebra is nilpotent.

Of special interest are conditions under which the **Jacobson radical** $\mathrm{Rad}\,A$ of an algebra A over a field k coincides with the nil radical. Here are some of them: A is an Artinian, in particular a finite-dimensional, algebra ($\mathrm{Rad}\,A$ is even nilpotent); $\mathrm{card}\,k>\dim_k A$, in particular, when A is finitely generated over k and k is uncountable; an algebraic algebra A; A is finitely generated over k with a polynomial identity, and k is infinite. The radical of a finitely-generated algebra with a polynomial identity over a field of characteristic zero is nilpotent. This is equivalent to the condition that in such an algebra a certain standard identity holds.

Some of the assertions stated above have analogues in non-associative algebras. For example, in an alternative ring with the maximum condition for right ideals and whose additive group contains no elements of orders 2 or 3, every one-sided nil ideal is nilpotent. In a **Jordan algebra** A over a field k the Jacobson radical $\mathrm{Rad}\,A$ is a nil ideal, provided that one of the following conditions holds: $\mathrm{card}\,k>2+\dim A$ or A is an **algebraic algebra**. Every alternative or special Jordan nil ring of bounded index and without elements of order 2 in the additive group is locally nilpotent (*Shirshov's theorem*). A finite-dimensional generalized standard nil algebra is nilpotent.

Every anti-commutative algebra is a nil algebra in the sense defined above; therefore, in the class of anti-commutative rings the concept of a nil algebra is meaningless. However, various analogues of this concept turn out to be useful. Thus, in the class of Lie rings an analogue of a nil algebra is an Engel algebra, that is, one in which the inner derivations of elements are nilpotent. An Engel algebra is not necessary locally nilpotent; however, if the nilpotency indices of the inner derivations are uniformly bounded and the ground field has characteristic zero, then an Engel algebra is locally nilpotent. It is not known (1982) whether under these restrictions it is nilpotent (Higgins' problem).

References

[1] JACOBSON, N.: *Structure of rings*, Amer. Math. Soc., 1956.
[2] HERSTEIN, I.: *Noncommutative rings*, Math. Assoc. Amer., 1968.
[3] KOSTRIKIN, A.I.: 'Lie rings satisfying the Engel condition', *Izv. Akad. Nauk SSSR Ser. Mat.* **21**, no. 4 (1957), 515-540 (in Russian).
[4] SHIRSHOV, A.I.: 'On certain non-associative nil rings and algebraic algebras', *Mat. Sb.* **41**, no. 3 (1957), 381-394 (in Russian).
[5] GOLOD, E.S.: 'On nil algebras and finitely approximable p-groups', *Izv. Akad. Nauk SSSR Ser. Mat.* **28**, no. 2 (1964), 273-276 (in Russian).
[6] HIGMAN, G.: 'On a conjecture of Nagata', *Proc. Cambridge Philos. Soc.* **52** (1956), 1-4.
[7] HIGGINS, P.J.: 'Lie rings satisfying the Engel condition', *Proc. Cambridge Philos. Soc.* **50** (1954), 8-15.
[8] ZHEVLAKOV, K.A.: 'The lower nil radical of alternative rings', *Algebra i Logika* **6**, no. 4 (1967), 11-17 (in Russian).
[9] MCCRIMMON, K.: 'The radical of a Jordan algebra', *Proc. Nat. Acad. Sci. USA* **612**, no. 3 (1969), 671-678.

V.N. Latyshev

Editorial comments. The Jacobson radical of a polynomial ring $A[x_1, \ldots, x_n]$ in commutative indeterminates $\{x_i\}$ is a polynomial ring $N_n[x]$ over a nil ideal $N_n(A)$ of A. Then $N_1 \supseteq N_2 \supseteq \cdots \supseteq N_\infty = \bigcap N_n$. N_∞ is characterized as the maximal ideal such that every finitely-generated submodule of nil elements is of bounded index. If A is an algebra over an uncountable field, then all N_i are equal to the upper radical, but generally no other relation is known.

The Köthe problem is equivalent to the following problem: If A is a nil algebra, then is the matrix algebra $M_2(A)$ also nil? A similar problem is the question about algebraic algebras (*Kurosh' problem*), and the answers are unknown (1989), except in the special cases of Π-algebras.

References

[A1] AMITSUR, S.A.: 'Nil radicals, historical notes', in *Coll. Math. Soc. J. Bolyai, Keszthely (Hungary)*, Vol. 6. Rings and modules and radicals, 1971, pp. 47-65.

AMS 1980 Subject Classification: 16A22

NIL FLOW - A flow on a **nil manifold** $M = G/H$ defined by the action on M of some one-parameter subgroup g_t of a nilpotent Lie group G: If M consists of the cosets gH, then under the action of the nil flow such a coset at time t goes over in $g_t gH$.

References

[1] AUSLANDER, L., GREEN, L. and HAHN, F.: *Flows on homogeneous spaces*, Princeton Univ. Press, 1963.

D.V. Anosov

Editorial comments. The first example of a compact minimal flow that is distal but not equicontinuous was a nil flow (cf. **Distal dynamical system; Equicontinuity**).

AMS 1980 Subject Classification: 58F25, 54H20

NIL GROUP - A **group** in which any two elements x and y are connected by a relation

$$[[\cdots [[x,y]y], \ldots]y] = 1,$$

where the square brackets denote the commutator

$$[a, b] = a^{-1}b^{-1}ab$$

and the number n of commutators in the definition depends, generally speaking, on the pair x, y. When n is bounded for all x, y in the group, the group is called an **Engel group**. Every **locally nilpotent group** is a nil group. The converse is not true, in general, but it is under some additional assumptions, for example, when the group is locally solvable (cf. **Locally solvable group**).

Occasionally the term 'nil group' is used in a different meaning. Namely, a *nil group* is a group in which every cyclic subgroup is subnormal, that is, occurs in some subnormal series of the group (see **Normal series of a group**).

References

[1] KUROSH, A.G.: *The theory of groups*, 1-2, Chelsea, 1955-1956 (translated from the Russian).

A.L. Shmel'kin

Editorial comments. In [A1] it has been proved that there are periodic Engel groups that are not locally nilpotent.

References

[A1] GOLOD, E.S.: 'On nil-algebras and residually finite p-groups', *Transl. Amer. Math. Soc.* **48** (1965), 103-106. (*Izv. Akad. Nauk SSSR Ser. Mat.* **28** (1964), 273-276)

AMS 1980 Subject Classification: 20F19

NIL IDEAL

Editorial comments. A subset A of a ring R is called *nil* if each element of it is nilpotent (cf. **Nilpotent element**). An ideal of R is a *nil ideal* if it is a nil subset. There is a largest nil ideal, which is called the nil radical. One has that

$$\mathrm{Jac}(R) \supseteq \mathrm{Nil\,Rad}(R) \supseteq \mathrm{Prime\,Rad}(R),$$

where $\mathrm{Jac}(R)$ denotes the **Jacobson radical** of R and $\mathrm{Prime\,Rad}(R)$ is the *prime radical* of R, i.e. the intersection of all prime ideals of R. Each of the inclusions can be proper. If R is commutative, $\mathrm{Nil\,Rad}(R) = \mathrm{Prime\,Rad}(R)$. The prime radical is also called the *lower nil radical*, and the nil radical the *upper nil radical*.

References

[A1] FAITH, C.: *Algebra* , II. Ring theory, Springer, 1976.
[A2] MCCONNELL, J.C. and ROBSON, J.C.: *Noncommutative noetherian rings*, Wiley, 1987.
[A3] ROWEN, L.: *Ring theory*, 1, Acad. Press, 1988.

AMS 1980 Subject Classification: 16A21

NIL MANIFOLD - A compact quotient space of a connected nilpotent Lie group (cf. **Lie group, nilpotent**). (However, sometimes compactness is not required.)

References

[1] MAL'TSEV, A.I.: 'On a class of homogeneous spaces', *Transl. Amer. Math. Soc. (1)* **9** (1962), 276-307. (*Izv. Akad. Nauk SSSR Ser. Mat.* **13**, no. 1 (1949), 9-32)

D.V. Anosov

Editorial comments. Cf. also **Nil flow** and the references quoted there.

An example of a nil manifold that is rather important for various applications is the following. Consider the three-dimensional *Heisenberg group N* of all matrices of the form

$$\begin{bmatrix} 1 & y & z \\ 0 & 1 & x \\ 0 & 0 & 1 \end{bmatrix}$$

and the discrete subgroup Γ of all such matrices with integer x, y, z. The corresponding quotient space $\Gamma \backslash N$ of cosets Γn, $n \in N$, is a compact nil manifold with an invariant probability measure. It plays an important role in harmonic analysis and the theory of theta-functions.

References

[A1] AUSLANDER, L.: *Lecture notes on nil-theta functions*, Amer. Math. Soc., 1977.

AMS 1980 Subject Classification: 58D15, 22E25

NIL SEMI-GROUP - A **semi-group** with zero in which some power of every element is zero. Nil semi-groups form one of the most important classes of periodic semi-groups (cf. **Periodic semi-group**): They are precisely the periodic semi-groups with a unique idempotent, namely, the zero. *Locally nilpotent semi-groups* (that is, semi-groups in which every finitely-generated sub-semi-group is nilpotent, see **Nilpotent semi-group**) form a narrow class. For every $n > 1$, there exists a semi-group with the identity $x^n = 0$ that is not locally nilpotent (see, for example, [1], Chapt. 8 Sect. 4). A finite nil semi-group is nilpotent, and the classes of locally nilpotent semi-groups and locally finite nil semi-groups coincide (see **Locally finite semi-group**). An even narrower class is formed by the *semi-groups with an ascending annihilator series*. A semi-group S has an ascending annihilator series if it has an increasing ideal series (see **Ideal series** of a semi-group) such that for any two adjacent terms $A_\alpha, A_{\alpha+1}$,

$$SA_{\alpha+1} \bigcup A_{\alpha+1}S \subseteq A_\alpha.$$

A nil semi-group has an ascending annihilator series if and only if it has an increasing series of ideals in which all factors are finite. Every semi-group with an ascending annihilator series has a unique irreducible generating set, consisting of its indecomposable elements. An arbitrary locally nilpotent semi-group may coincide with its square. Many finiteness conditions (see **Semi-group with a finiteness condition**) imposed on a semi-group imply that is finite; for example, the minimum condition for ideals, or the maximum condition for right (or left) ideals. If all nilpotent sub-semi-groups of a nil semi-group S are finite, then so is S.

References

[1] JACOBSON, N.: *Structure of rings*, Amer. Math. Soc., 1956.
[2] SHEVRIN, L.N.: 'On the general theory of semi-groups', *Mat. Sb.* **53**, no. 3 (1961), 367-386 (in Russian).
[3] SHEVRIN, L.N.: 'Nil semi-groups with certain finiteness conditions', *Mat. Sb.* **55**, no. 4 (1961), 473-480 (in Russian).

L.N. Shevrin

AMS 1980 Subject Classification: 20M10

NILPOTENT ALGEBRA - An algebra for which there is a natural number n such that any product of n elements of the algebra is zero. If there is a non-zero product of $n - 1$ elements, then n is called the *index of nilpotency* of the algebra. Examples of nilpotent algebras are: an algebra with zero multiplication; an algebra of strictly upper-triangular matrices; direct sums of nilpotent algebras the indices of nilpotency of which are uniformly bounded; and the tensor product of two algebras one of which is nilpotent.

The class of nilpotent algebras is closed under taking homomorphic images and subalgebras. In an associative algebra (cf. **Associative rings and algebras**) the sum of finitely many nilpotent ideals is a **nilpotent ideal**, and the sum of an arbitrary set of nilpotent ideals is, generally speaking, locally nilpotent. A finite-dimensional algebra over a field of characteristic zero having a basis consisting of nilpotent elements is nilpotent. If an algebra satisfies a polynomial identity of degree d, then every nilpotent subring of it of degree $[d/2]$ belongs to a sum of nilpotent ideals. The derived algebra of a finite-dimensional **Lie algebra** over a field of characteristic zero is nilpotent. Nilpotent subalgebras that coincide with their normalizer (*Cartan subalgebras*) play an essential role in the classification of simple Lie algebras of finite dimension. A nilpotent Lie algebra has an outer automorphism. A Lie algebra with a regular automorphism (that is, one having no fixed point except zero) of prime period is nilpotent.

References

[1] JACOBSON, N.: *Structure of rings*, Amer. Math. Soc., 1956.
[2] JACOBSON, N.: *Lie algebras*, Interscience, 1962.
[3] ALBERT, A.A.: *Structure of algebras*, Amer. Math. Soc., 1939.
[4] JACOBSON, N.: 'A note on automorphisms and derivations of Lie algebras', *Proc. Amer. Math. Soc.* **6**, no. 2 (1955), 281-283.
[5] HIGMAN, G.: 'Groups and rings having automorphisms without non-trivial fixed elements', *J. London Math. Soc.* **32**, no. 3 (1957), 321-334.

V.N. Latyshev

AMS 1980 Subject Classification: 16A22, 17B30

NILPOTENT ELEMENT - An element a of a ring or semi-group with zero A such that $a^n = 0$ for some natural number n. The smallest such n is called the *nilpotency index* of a. For example, in the residue ring modulo p^n (under multiplication), where p is a prime number, the residue class of p is nilpotent of index n; in the ring of (2×2)-matrices with coefficients in a field K the matrix

$$\left\| \begin{matrix} 0 & 1 \\ 0 & 0 \end{matrix} \right\|$$

is nilpotent of index 2; in the group algebra $F_p[G]$, where F_p is the field with p elements and G the cyclic group of order p generated by σ, the element $1 - \sigma$ is nilpotent of index p.

If a is a nilpotent element of index n, then

$$1 = (1-a)(1+a+ \cdots +a^{n-1}),$$

that is, $(1-a)$ is invertible in A and its inverse can be written as a polynomial in a.

In a commutative ring A an element a is nilpotent if and only if it is contained in all prime ideals of the ring. All nilpotent elements form an ideal J, the so-called *nil radical* of the ring; it coincides with the intersection of all prime ideals of A. The ring A/J has no non-zero nilpotent elements.

In the interpretation of a commutative ring A as the ring of functions on the space $\operatorname{Spec} A$ (the spectrum of A, cf. **Spectrum of a ring**), the nilpotent elements correspond to functions that vanish identically. Nevertheless, the consideration of nilpotent elements frequently turns out to be useful in algebraic geometry because it makes it possible to obtain purely algebraic analogues of a number of concepts in analysis and differential geometry (infinitesimal deformations, etc.).

References

[1] LANG, S.: *Algebra*, Addison-Wesley, 1974.
[2] HERSTEIN, I.: *Noncommutative rings*, Math. Assoc. Amer., 1968.
[3] SHAFAREVICH, I.R.: *Basic algebraic geometry*, Springer, 1977 (translated from the Russian).

L.V. Kuz'min

Editorial comments. An element a of an associative ring R is *strongly nilpotent* if every sequence $a = a_0, a_1, \ldots$, such that $a_{n+1} \in a_n R a_n$ is ultimately zero. Obviously, every strongly-nilpotent element is nilpotent. The *prime radical* of a ring R, i.e. the intersection of all prime ideals, consists of precisely the strongly-nilpotent elements. It is a nil ideal.

References

[A1] MCCONNELL, J.C. and ROBSON, J.C.: *Noncommutative Noetherian rings*, Wiley, 1987, § 0.2.

AMS 1980 Subject Classification: 16A22

NILPOTENT GROUP - A group having a **normal series**

$$G = A_1 \supseteq A_2 \supseteq \cdots \supseteq A_{k+1} = \{1\}$$

such that every quotient A_i/A_{i+1} lies in the centre of G/A_{i+1} (a so-called *central series*). The length of a shortest central series of a nilpotent group is called its *class* (or *degree of nilpotency*). In any nilpotent group the lower (and upper) central series (see **Subgroup series**) breaks off at the trivial subgroup (the group itself), and their lengths are equal to the nilpotency class of the group.

The finite nilpotent groups are exhausted by direct products of p-groups, that is, groups of orders p^k, where p is a prime number. In any nilpotent group the elements of finite order form a subgroup, the quotient group by which is torsion free. The finitely-generated torsion-free nilpotent groups are exhausted by the groups of integral triangular matrices with 1's along the main diagonal, and their subgroups. Every finitely-generated torsion-free nilpotent group can be approximated by a finite p-group for every prime p. Finitely-generated nilpotent groups are polycyclic groups (cf. **Polycyclic group**) and, moreover, have a central series with cyclic factors.

All nilpotent groups of class at most c form a variety (see **Variety of groups**), defined by the identity

$$[[\cdots [[x_1, x_2]x_3], \ldots]x_{c+1}] = 1.$$

The free groups of this variety are called *free nilpotent groups*. Concerning completions of torsion-free nilpotent groups see **Locally nilpotent group**.

References

[1] KUROSH, A.G.: *The theory of groups*, 1-2, Chelsea, 1955-1956 (translated from the Russian).
[2] KARGAPOLOV, M.I. and MERZLYAKOV, YU.I.: *Fundamentals of the theory of groups*, Springer, 1979 (translated from the Russian).

A.L. Shmel'kin

Editorial comments. Let G be a group and R some relation (or, more generally, a predicate) that can hold between elements of groups and/or between sets of elements of groups. For instance, R could be the relation of equality, the relation 'the element g belongs to the subgroup H', or the relation of conjugacy between subsets of a group. Let $\mathscr{C}$ be a class of groups. Then one says that the group G is *approximable* by the groups from $\mathscr{C}$ with respect to the relation R if whenever the relation R does not hold in G (between elements, subsets, or between an element and a subset), then there is a homomorphism from G into a group from $\mathscr{C}$ such that the relation R also does not hold between the images under this homomorphism. I.e. the homomorphisms $G \to C$, $C \in \mathscr{C}$, suffice to *detect* whether elements (subsets) are R-different.

If the relation R is equality, then one simply says that the group G is approximable by the groups from $\mathscr{C}$.

Concerning approximability cf. also **Residually-finite group**.

AMS 1980 Subject Classification: 20D15

NILPOTENT IDEAL - A one- or two-sided ideal M in a ring or semi-group with zero such that $M^n = \{0\}$ for some natural number n, that is, the product of any n elements of M vanishes. For example, in the residue class ring $\mathbf{Z}/p^n\mathbf{Z}$ modulo p^n, where p is a prime number, every ideal except the ring itself is nilpotent. In the group ring $\mathbf{F}_p[G]$ of a finite p-group G over the field with p elements the ideal generated by the elements of the form $\sigma - 1$, $\sigma \in G$, is nilpotent. In the ring of upper-triangular matrices over a field the matrices with 0's along the main diagonal form a nilpotent ideal.

Every element of a nilpotent ideal is nilpotent. Every nilpotent ideal is also a nil ideal and is contained in the **Jacobson radical** of the ring. In Artinian rings the Jacobson radical is nilpotent, and the concepts of a nilpotent ideal and a nil ideal coincide. The latter property also holds in a **Noetherian ring**. In a left (or right) Noetherian ring every left (right) nil ideal is nilpotent.

All nilpotent ideals of a commutative ring are contained in the nil radical, which, in general, need not be a nilpotent but only a nil ideal. A simple example of this situation is the direct sum of the rings $\mathbf{Z}/p^n\mathbf{Z}$ for all natural numbers n. In a commutative ring every **nilpotent element** a is contained in some nilpotent ideal, for example, in the **principal ideal** generated by a. In a non-commutative ring there may by nilpotent elements that are not contained in any nilpotent ideal (nor even in a nil ideal). For example, in the general matrix ring over a field there are nilpotent elements; in particular, the nilpotent matrices mentioned above, in which the only non-zero elements stand above the main diagonal, but since the ring is simple, it has no non-zero nilpotent ideals.

In a finite-dimensional **Lie algebra** G there is maximal nilpotent ideal, which consists of the elements $x \in G$ for which the endomorphism $y \to [x, y]$ for $y \in G$ is nilpotent.

References

[1] LANG, S.: *Algebra*, Addison-Wesley, 1974.
[2] JACOBSON, N.: *Structure of rings*, Amer. Math. Soc., 1956.
[3] FAITH, C.: *Algebra: rings, modules and categories*, 1, Springer, 1973.
[4] HERSTEIN, I.: *Noncommutative rings*, Math. Assoc. Amer., 1968.
[5] BOURBAKI, N.: *Elements of mathematics. Lie groups and Lie algebras*, Addison-Wesley, 1975 (translated from the French).

L.V. Kuz'min

AMS 1980 Subject Classification: 16A66

NILPOTENT SEMI-GROUP - A **semi-group** S with zero for which there is an n such that $S^n = 0$; this is equivalent to the identity

$$x_1 \cdots x_n = y_1 \cdots y_n$$

in S. The smallest n with this property for a given semi-group is called the *step* (sometimes *class*) *of nilpotency*. If $S^2 = 0$, then S is called a *semi-group with zero multiplication*. The following conditions on a semi-group S are equivalent: 1) S is nilpotent; 2) S has a finite annihilator series (that is, an ascending annihilator series of finite length, see **Nil semi-group**); or 3) there is a k such that every sub-semi-group of S can be imbedded as an ideal series of length $\leq k$.

A wider concept is that of a *nilpotent semi-group in the sense of Mal'tsev* [2]. This is the name for a semi-group satisfying for some n the identity

$$X_n = Y_n,$$

where the words X_n and Y_n are defined inductively as follows: $X_0 = x$, $Y_0 = y$, $X_n = X_{n-1} u_n Y_{n-1}$, $Y_n = Y_{n-1} u_n X_{n-1}$, where x, y and $u_1, \ldots, u_n$ are variables. A group is a nilpotent semi-group in the sense of Mal'tsev if and only if it is nilpotent in the usual group-theoretical sense (see **Nilpotent group**), and the identity $X_n = Y_n$ is equivalent to the fact that its class of nilpotency is $\leq n$. Every cancellation semi-group satisfying the identity $X_n = Y_n$ can be imbedded in a group satisfying the same identity.

References

[1] LYAPIN, E.S.: *Semigroups*, Amer. Math. Soc., 1974 (translated from the Russian).
[2] MAL'TSEV, A.I.: 'Nilpotent semi-groups', *Uchen. Zap. Ivanov. Gos. Ped. Inst.* **4** (1953), 107-111 (in Russian).
[3] SHEVRIN, L.N.: 'On the general theory of semi-groups', *Mat. Sb.* **53**, no. 3 (1961), 367-386 (in Russian).
[4] SHEVRIN, L.N.: 'Semi-groups all sub-semi-groups of which are accessible', *Mat. Sb.* **61**, no. 2 (1963), 253-256 (in Russian).

L.N. Shevrin

AMS 1980 Subject Classification: 20M10

NINE-POINT CIRCLE, *Euler circle* - A circle whose periphery contains the midpoints of the sides of a triangle, the bases of its altitudes, and the midpoints of the segment connecting the orthocentre of the triangle with the vertices. Its radius is equal to one-half of the radius of the circle circumscribed about the triangle. The nine-point circle of a triangle is tangent to the circle inscribed in it and to the three escribed circles. Let H be the orthocentre of a non-equilateral triangle, let T be the centre of gravity, let O be the centre of the circumscribed circle and let E be the centre of the nine-point circle. The points H, T, O, E then lie on a straight line (*Euler's line*), E being the midpoint of the segment HO, and the pair of points H, T harmonically subdivides the pair of points O, E.

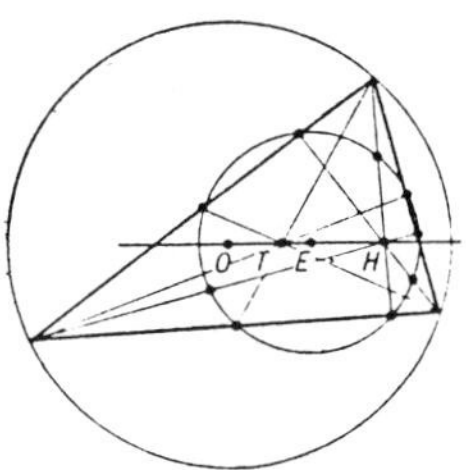

References

[1] ZETEL', S.I.: *A new geometry of triangles*, Moscow, 1962 (in Russian).
[2] PEREPELKIN, D.I.: *A course in elementary geometry*, 1, Moscow-Leningrad, 1948 (in Russian).

V.T. Bazylev

Editorial comments. Sometimes the nine-point circle is referred to as the *Feuerbach circle*. The fact that the nine-point circle is tangent to the inscribed circle and the three escribed circles is *Feuerbach's theorem*.

More generally one has the *nine-point conic* and the *eleven-point conic* determined by a *projective base* $\{a, b, c, d\}$ (giving a coordinate system) in the projective plane, cf. [A2], Sects. 16.5.5.1, 16.7.5.

References
[A1] COXETER, H.S.M.: *Introduction to geometry*, Wiley, 1961.
[A2] BERGER, M.: *Geometry* , I-II, Springer, 1987, Sects. 10.11.3, 17.5.4 (translated from the French).
[A3] VEBLEN, O. and YOUNG, J.W.: *Projective geometry*, II, Blaisdell, 1946, p. 169; 233.

AMS 1980 Subject Classification: 51M15

NODE - 1) A point of self-intersection of a curve. For a parametrically given curve a node corresponds to two or more values of the parameter. E.g. for the curve $\rho = a \sin 3\phi$ the origin of coordinates is a node.

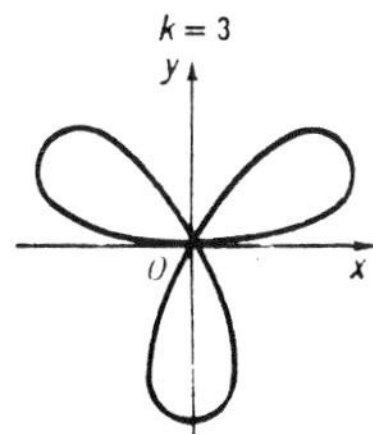

Editorial comments. A reference for 'node of a curve' is [A1].

References
[A1] COOLIDGE, J.: *Algebraic plane curves*, Dover, reprint, 1959.

2) A node is a type of arrangement of the trajectories of an **autonomous system** of second-order ordinary differential equations

$$\dot{x} = f(x), \quad x = (x_1, x_2), \quad f: G \subset \mathbf{R}^2 \to \mathbf{R}^2, \qquad (*)$$

$f \in C(G)$, G a domain of uniqueness, in a neighbourhood of a stationary point x_0. This type is characterized in the following way. There exists a neighbourhood U of x_0 such that for all trajectories of the system beginning in $U \setminus \{x_0\}$ the negative semi-trajectories leave in the course of time any compact set $V \subset U$, while the positive semi-trajectories approach x_0 while not leaving U and, moreover, being completed with x_0, touch it in well-defined directions, or vice versa. The point x_0 itself is also called a *node*, a *nodal point* or a *basis point*.

A node is either asymptotically stable in the sense of Lyapunov (cf. **Lyapunov stability**) or is totally unstable (asymptotically stable for $t \to -\infty$). The *Poincaré index* of a node is 1 (cf. **Singular point**).

For a system (*) of class C^1 ($f \in C^1(G)$) with non-zero matrix $A = f'(x_0)$, a stationary point x_0 is a node if the eigen values λ_1, λ_2 of A are real and satisfy the conditions $\lambda_1\lambda_2 > 0$, $\lambda_1 \neq \lambda_2$; it can also be a node in cases when $\lambda_1 = \lambda_2 \neq 0$, $\lambda_1 = 0 \neq \lambda_2$, $\lambda_1 = \lambda_2 = 0$. In case $\lambda_1 = \lambda_2 \neq 0$, x_0 will be a node if $f \in C^2(G)$; when this condition is not satisfied it may turn out to be **focus**. In

any of the cases listed above the trajectories of the system converging to the node x_0 touch it in well-defined directions, defined by the eigen vectors of A. If $0 \leq |\lambda_1| < |\lambda_2|$, there exist four such directions (if diametrically opposite ones are counted as distinct), and two trajectories of the system touch at x_0 in directions corresponding to the eigen value λ_1 while two trajectories touch at x_0 in directions corresponding to the eigen value λ_2 (Fig. 1). These are *ordinary nodes*. If $\lambda_2 = \lambda_1$, then the eigen directions for A at x_0 are either just two opposite directions (in this case the node is *degenerate*, cf. Fig. 2) or all directions. In this last case, under the condition $f \in C^2(G)$ every direction is tangent at x_0 to a unique trajectory of the system. Such a node is called *dicritical* (Fig. 3).

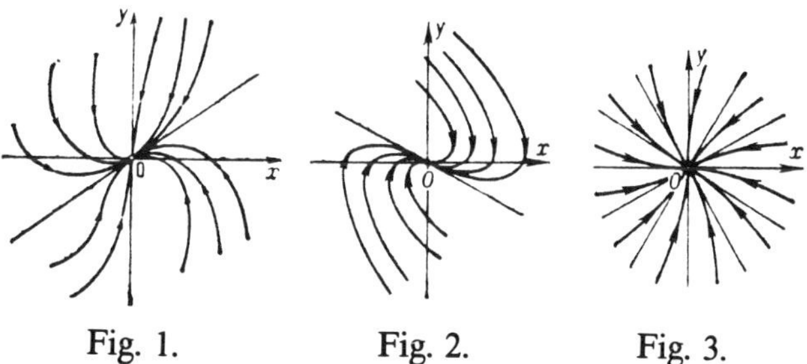

Fig. 1.   Fig. 2.   Fig. 3.

If the system (*) is linear ($f(x) = A(x - x_0)$, where A is a fixed matrix) then the point x_0 is a node only when the eigen values λ_1, λ_2 of A are real and $\lambda_1\lambda_2 > 0$. Any ray $x = x_0 + ps$ (p an eigen vector of A, $s \neq 0$ a parameter) is a trajectory for it. Ordinary, degenerate and dicritical nodes for a linear system are depicted in Figs. 4, 5 and 6.

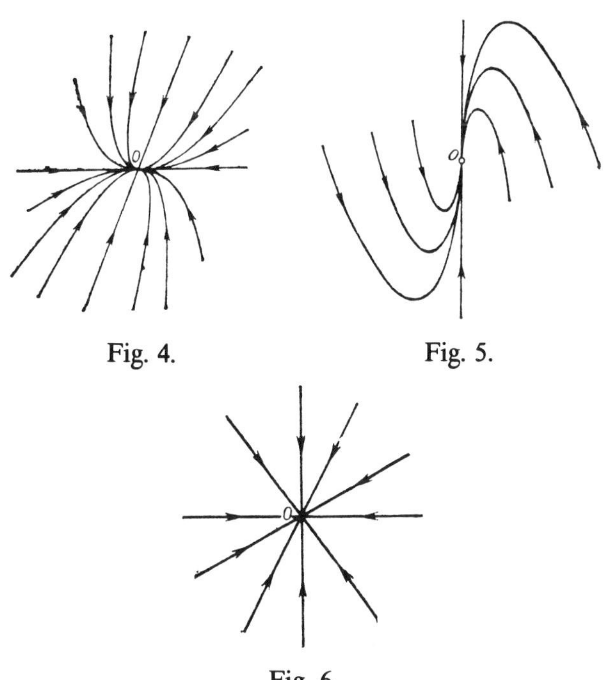

Fig. 4.   Fig. 5.

Fig. 6.

In the case of an ordinary node all curvilinear trajectories are affine images of parabolas $x_2 = c \mid x_1 \mid^{\lambda_2/\lambda_1}$, $c \in \mathbf{R} \setminus \{0\}$.

The term 'node' is also applied to a stationary point of a system of the form (*) of order $\geqslant 3$ with analogous behaviour of the trajectories in neighbourhoods of it.

For references see **Singular point** of a differential equation.

A.F. Andreev

Editorial comments.

References

[A1] LEFSHETZ, S.: *Differential equations: geometric theory*, Dover, reprint, 1977, Sect. IX.2.

[A2] BIRKHOFF, G. and ROTA, G.-C.: *Ordinary differential equations*, Ginn & Co., 1962, Sect. VI.8.

AMS 1980 Subject Classification: 53A04, 34C05, 58F99, 14H45

NOETHER − ENRIQUES THEOREM *on canonical curves* - A theorem on the projective normality of a **canonical curve** and on its definability by quadratic equations.

Let $X \subset P^{g-1}$ be a smooth canonical (non-hyperelliptic) curve of genus $g \geqslant 3$ over an algebraically closed field k and let I_X be the homogeneous ideal in the ring $k[x_0, \ldots, x_{g-1}]$ defining X in P^{g-1}. The *Noether − Enriques theorem* (sometimes called the *Noether − Enriques − Petri theorem*) asserts that:

1) X is projectively normal in P^{g-1};

2) if $g = 3$, then X is a plane curve of degree 4, and if $g \geqslant 4$, then the graded ideal I_X is generated by the components of degree 2 and 3 (which means that X is the intersection of the quadrics and cubics in P^{g-1} passing through it);

3) I_X is always generated by the components of degree 2, except when a) X is a trigonal curve, that is, has a linear series (system) g_3^1, of dimension 1 and degree 3; or b) X is of genus 6 and is isomorphic to a plane curve of degree 5;

4) in the exceptional cases a) and b) the quadrics passing through X intersect along a surface F which for a) is non-singular, rational, ruled of degree $g-2$ in P^{g-1}, $g \geqslant 5$, and the series g_3^1 cuts out on X a linear system of straight lines on F, and for $g=4$ a quadric in P^3 (possibly a cone); and for b) is the Veronese surface V_4 in P^5.

This theorem (in a slightly different algebraic formulation) was established by M. Noether in [1]; a geometric account was given by F. Enriques (on his results see [2]; a modern account is in [3], [4]; a generalization in [5]).

References

[1] NOETHER, M.: 'Ueber invariante Darstellung algebraischer Funktionen', *Math. Ann.* **17** (1880), 263-284.

[2] BABBAGE, D.W.: 'A note on the quadrics through a canonical curve', *J. London. Math. Soc.* **14**, no. 4 (1939), 310-314.

[3] SAINT-DONAT, B.: 'On Petri's analysis of the linear system of quadrics through a canonical curve', *Mat. Ann.* **206** (1973), 157-175.

[4] SHOKUROV, V.V.: 'The Noether − Enriques theorem on canonical curves', *Math. USSR Sb.* **15** (1971), 361-403. (*Math. Sb.* **86**, no. 3 (1971), 367-408)

[5] ARBARELLO, E. and SERNESI, E.: 'Petri's approach to the study of the ideal associated to a special divisor', *Invent. Math.* **49** (1978), 99-119.

V.A. Iskovskikh

Editorial comments. A smooth curve $C \subset P^{g-1}$ is called *k-normal* if the hypersurfaces of degree k cut out the complete linear system $\mid \mathcal{O}_C(k) \mid$. Instead of 1-normal, *linearly normal* is used. A curve $C \subset P^{g-1}$ is projectively normal if it is *k*-normal for every *k*. Cf. [A2], p. 140ff and 221ff for more details and results.

References

[A1] GRIFFITHS, P. and HARRIS, J.: *Principles of algebraic geometry*, Wiley, 1978.

[A2] ARBARELLO, E., CORNALBA, M., GRIFFITHS, P.A. and HARRIS, J.: *Geometry of algebraic curves*, 1, Springer, 1984.

AMS 1980 Subject Classification: 14H99

NOETHER PROBLEM - The question of the rationality of the field of invariants of a finite group acting by automorphisms on a field of rational functions. More precisely, let $K = \mathbf{Q}(x_1, \ldots, x_n)$ be the field of rational functions in n variables with coefficients in the field $\mathbf{Q}$ of rational numbers, so that K is a purely **transcendental extension** of $\mathbf{Q}$ of transcendence degree n. Also, let G be a **finite group** acting by automorphisms on K by means of permutations of the variables $x_1, \ldots, x_n$. The question is now whether the subfield K^G of K consisting of all elements fixed under G is itself a field of rational functions in n (other) variables with coefficients in $\mathbf{Q}$. This question was raised by E. Noether [1] in connection with the inverse problem of Galois theory (cf. **Galois theory, inverse problem of**). If the answer to Noether's problem were affirmative, one could construct a **Galois extension** of $\mathbf{Q}$ with a given finite group G (see [5]). The problem is also closely connected with the **Lüroth problem**.

In general, the answer to Noether's problem is negative. The first example of a non-rational field K^G was constructed in [2], and in this example G is generated by a cyclic permutation of the variables. In [3] it was established that the necessary condition for the rationality of K^G found in [2] is also sufficient. The question of rationality of K^G in the case of an Abelian group G is closely connected with the theory of algebraic tori (cf. **Algebraic torus**) (see [4]).

Frequently, Noether's problem is interpreted more generally as the problem that arises when in the original setting $\mathbf{Q}$ is replaced by an arbitrary field k. This problem has an affirmative solution, for example, when k is algebraically closed and G is Abelian.

References

[1] NOETHER, E.: 'Gleichungen mit vorgeschriebener Gruppe', *Math. Ann.* **78** (1917-1918), 221-229.

[2] SWAN, R.G.: 'Invariant rational functions and a problem of Steenrod', *Invent. Math.* **7**, no. 2 (1969), 148-158.

[3] VOSKRESENSKIĬ, V.E.: 'Rationality of certain algebraic tori', *Math. USSR. Izv.* **35**, no. 5 (1979), 1049-1056. (*Izv. Akad. Nauk SSSR Ser. Mat.* **35** (1971), 1037-1046)

[4] VOSKRESENSKIĬ, V.E.: *Algebraic tori*, Moscow, 1977 (in Russian).

[5] CHEBOTAREV, N.G.: *Grundzüge der Galois'schen Theorie*, Noordhoff, 1950, Chapt. V § 4 (translated from the Russian).

V.L. Popov

Editorial comments. For k arbitrary and G finite Abelian, there is a necessary and sufficient condition for rationality of K^G (see [A1]). For example, if $k=\mathbf{Q}$ and G is cyclic of order 8, then K^G is not rational.

For $k=\mathbf{C}$, the first examples of groups G for which K^G is not rational were constructed by D.J. Saltman [A2]. He proved that for each prime number p there exists such a group of order p^9.

References

[A1] LENTSTRA, H.W., JR.: 'Rational functions invariant under a finite abelian group', *Invent. Math.* **25** (1974), 299-325.

[A2] SALTMAN, D.J.: 'Noether's problem over an algebraically closed field', *Invent. Math.* **77** (1984), 71-84.

AMS 1980 Subject Classification: 12A55

NOETHER THEOREM - 1) *Noether's first theorem* establishes a connection between the infinitesimal symmetries of a functional of the form

$$A(u(x)) = \int L(x, u(x), u_{,j}(x))\, d^n x,$$

where $x=(x^1, \ldots, x^n)$ are independent variables, $u(x)=(u^1(x), \ldots, u^N(x))$ are functions defined in a certain domain $D \subset \mathbf{R}^n$, $u_{,j}=(\partial/\partial x^j)(u(x))$ are their partial derivatives, and L is a certain function (the Lagrangian), and the conservation laws for the corresponding system of Euler$-$Lagrange equations

$$\frac{\delta L}{\delta u^a} \equiv \frac{\partial L}{\partial u^a} - \frac{d}{dx^i}\frac{\partial L}{\partial u^a_{,i}} = 0,$$

which gives necessary conditions for an extremum of A. Namely, to an infinitesimal symmetry Z, that is, a vector field

$$Z = X^i(x)\frac{\partial}{\partial x^i} + U^a(x, u)\frac{\partial}{\partial u^a}$$

that generates a one-parameter group of transformations preserving A, corresponds the conservation law

$$v_Z = \left[LX^i + (U^a - u^a_{,j}X^j)\frac{\partial L}{\partial u^a_{,i}} \right] dx^1 \wedge \cdots \wedge \hat{dx^i} \wedge \cdots \wedge dx^n$$

(where the symbol $\hat{\ }$ indicates the omission of the corresponding factor), that is, an $(n-1)$-form depending on $u(x)$ that is closed when $u(x)$ satisfies the Euler$-$Lagrange equations.

In field theory, where $n=4$ and the coordinates x are interpreted as space-time coordinates, A is called the action and $u(x)$ the field. To fields $u(x)$ providing an extremum of the action functional correspond physically realizable fields with a given Lagrange function. If such a field $u(x)$ vanishes on the boundary of D, then by Stokes' theorem the integral of the conservation law v over a hypersurface $D \cap \{x^1=c\}$ does not depend on the choice of c. In particular, if x^1 is the time coordinate, then this integral yields a quantity that is preserved in the course of time (whence the name *conservation law*).

The invariance of the Lagrange function of distinct physical fields under parallel translations and Lorentz transformations (which is a consequence of the homogeneity and isotropy of Minkowski space-time) leads, by Noether's theorem, to the energy-momentum tensor and the angular momentum tensor of the field and to corresponding conservations laws for the energy, momentum and angular momentum of the motion. Invariance of the action functional of the electromagnetic field under gauge transformations leads to the conservation law for electric charge. Similarly, invariance of the Lagrangian of some field under gauge transformations yields conservation laws for various charges.

In classical mechanics, $n=1$ and the coordinate x^1 is interpreted as time. If the Lagrange function does not depend explicitly on x^1, then the vector field $\partial/\partial x^1$ is a symmetry, and Noether's theorem leads to the law of conservation of energy. For a mechanical system whose motion can be described as geodesic motion in some Riemannian metric, the symmetries of the corresponding action functional are Killing vector (or, more generally, Killing tensor) fields. In this case the conservation law furnished by Noether's theorem means geometrically that the magnitude of the projection of the Killing vector field in the direction of a geodesic is constant along it. The general modern formulation of Noether's theorem in the language of fibre bundles consists in the following. Let $\pi: E \to M$ be a vector bundle over an n-dimensional manifold M with a fixed volume n-form $\omega \in \Lambda^n(M)$, and let $\pi_k: J^k E \to M$ be the vector bundle of k-jets of sections of π. If x^i are local coordinates in M in which ω becomes $\omega = dx^1 \wedge \cdots \wedge dx^n$, and if x^i, u^a are local coordinates in E, then in $J^k E$ one has local coordinates x^i, u^a, u^α, where $\alpha = (\alpha_1, \ldots, \alpha_n)$ is a multi-index and $|\alpha| = \alpha_1 + \cdots + \alpha_n \leqslant k$. The value of the coordinate u^α on the k-jet $J^k_{x_0} u(x)$ of the section $u(x)$ of π is

$$u^a_{,\alpha}(x_0) = \left[\frac{\partial}{\partial x_1}\right]^{\alpha_1} \cdots \left[\frac{\partial}{\partial x^n}\right]^{\alpha_n} u^a(x_0).$$

A smooth function $L: J^k E \to \mathbf{R}$ determines an action functional A that associates with a section $s: x \to u(x)$ the number

$$A(s) = \int_M L(x, u(x), u_{,\alpha}(x))\omega.$$

An extremal $u(x)$ for this functional (in a problem with fixed ends) satisfies the Euler–Lagrange equations

$$\frac{\delta L}{\delta u^a} \equiv \sum_{\substack{\alpha=(\alpha_1,\ldots,\alpha_n) \\ |\alpha| \leq k}} (-1)^{|\alpha|} \frac{d^\alpha}{dx^\alpha} \frac{\partial L}{\partial u^a_{,\alpha}} = 0,$$

where

$$\frac{d^\alpha}{dx^\alpha} = \left[\frac{d}{dx^1}\right]^{\alpha_1} \cdots \left[\frac{d}{dx^n}\right]^{\alpha_n}$$

are the total derivatives. An infinitesimal automorphism of π, that is, a vector field Z on E of the form

$$Z = X^i(x)\frac{\partial}{\partial x^i} + U^a(x, u)\frac{\partial}{\partial u^a},$$

is called an infinitesimal symmetry of A if the Lie derivative of the Lagrange n-form $L\omega \in \Lambda^n(J^k E)$ in the direction of the vector field $Z^{(k)}$, which is generated by Z on $J^k E$, vanishes:

$$Z^{(k)}(L\omega) = 0.$$

For the Lie derivative the following fundamental *Noether formula* holds:

$$Z^{(k)}(L\omega) = \left[\overline{U}^a \frac{\delta L}{\delta u^a} + \frac{d}{dx^i} J^i\right] \omega,$$

where

$$\overline{U}^a = U^a - u^a_{,i} X^i, \quad J^i = LX^i + F^i,$$

and the F^i are the components of a certain vector field depending on $\overline{U}^a$, L and their derivatives. In particular, $F^i = \overline{U}^a(\partial L / \partial u^a_{,i})$ for $k=1$. If Z is an infinitesimal symmetry, then

$$-\overline{U}^a \frac{\delta L}{\delta u^a} = \frac{d}{dx^i} J^i,$$

that is, a certain linear combination of the variational derivatives $\delta L / \delta u^a$ of the Lagrange function L is the divergence of the vector field $J = J^i \partial / \partial x^i$. It is in this form that E. Noether stated her first theorem. The divergence of J (a so-called *Noether current*) vanishes on extremals of the action functional, and the $(n-1)$-form $v_z = J \rfloor \omega$ dual to it, which is obtained from ω by inner multiplication by J, is closed, that is, it is a conservation law.

There are important generalizations of Noether's theorem (see, for example, [5] - [7]). They are based on an extension of the concept of an infinitesimal symmetry. Instead of vector fields on E to which correspond one-parameter groups of transformations one considers vector fields on E with coefficients depending on the sections $U(x)$ and their derivatives of arbitrary order. Such fields Y no longer determine one-parameter transformation groups; however, one can

define for them by purely algebraic means the concept of a Lie derivative. A field Y is called an *algebraic infinitesimal symmetry* if the Lie derivative of the Lagrange form vanishes in the direction of this field (maybe after restricting to extremals of the action functional). The generalized Noether theorem associates a conservation law with every algebraic symmetry. When applied to various equations of mathematical physics one obtains a large number of new important conservation laws.

2) *Noether's second theorem* asserts that if the action functional admits an infinite-dimensional Lie algebra of infinitesimal symmetries whose coefficients depend linearly on p arbitrary functions $\phi^1(x), \ldots, \phi^p(x)$ and their derivatives up to order n, then the variational derivatives $\delta L / \delta u^a$ of the Lagrange function L satisfy a system of p differential equations of order m. Namely, if

$$Z = \phi^s U^a_s \frac{\partial}{\partial u^a} + \phi^s_{,\sigma}(x) U^{\sigma a}_s \frac{\partial}{\partial u^a},$$

where

$$\sigma = (\sigma_1, \ldots, \sigma_n), \quad |\sigma| = \sigma_1 + \cdots + \sigma_n \leq k,$$

is an infinitesimal symmetry for any smooth functions $\phi^s(x)$, $s = 1, \ldots, p$, then identically

$$U^a_s \frac{\delta L}{\delta u^a} + (-1)^{|\sigma|} \frac{d^\sigma}{dx^\sigma} \left[U^{\sigma a}_s \frac{\delta L}{\delta u^a}\right] = 0, \quad s = 1, \ldots, p.$$

This theorem has applications, for example, in the theory of gauge fields.

Noether proved her first and second theorem in 1918 (see [1]).

References

[1A] NOETHER, E.: 'Invarianten beliebiger Differentialausdrücke ', *Nachr. Gesellschaft. Wiss. Göttingen* (1918), 37-44; 240. Also: Gesammelte Abh., Springer, 1983, pp. 240-247.

[1B] NOETHER, E.: 'Invariante Variationsproblem', *Nachr. Gesellschaft. Wiss. Göttingen* (1918), 237-257. Also: Gesammelte Abh., Springer, 1983, pp. 248-270.

[2] BOGOLYUBOV, N.N. and SHIRKOV, D.V.: *Introduction to the theory of quantized fields*, Wiley, 1980 (translated from the Russian).

[3] GEL'FAND, I.M. and FOMIN, S.V.: *Calculus of variations*, Prentice-Hall, 1963 (translated from the Russian).

[4] ARNOL'D, V.I.: *Mathematical methods of classical mechanics*, Springer, 1978 (translated from the Russian).

[5] OVSIANNIKOV, L.V. [L.V. OVSYANNIKOV]: *Group analysis of differential equations*, Acad. Press, 1982 (translated from the Russian).

[6] MANIN, YU.I.: 'Algebraic aspects of nonlinear differential equations', *J. Soviet Math.* **11**, no. 1 (1979), 1-22. (*Itogi Nauk. i Tekhn. Sovrem. Probl. Mat.* **11** (1978), 5-152)

[7] VINOGRADOV, A.M.: 'On the algebro-geometric foundations of Lagrangian field theory', *Soviet Math. Dokl.* **18**, no. 5 (1977), 1200-1204. (*Dokl. Akad. Nauk SSSR* **236**, no. 2 (1977), 284-287)

[8] LYCHAGIN, V.V.: 'Contact geometry and non-linear second-order differential equations', *Russian Math. Surveys* **34**, no. 1 (1979), 149-180. (*Uspekhi Mat. Nauk* **34**, no. 1 (1979), 137-165)

D.V. Alekseevskiĭ

Editorial comments.

References

[A1] OLVER, P.J.: *Applications of Lie groups to differential equations*, Springer, 1986.
[A2] FUNK, P.: *Variationsrechnung und ihre Anwendung in Physik und Technik*, Springer, 1962.
[A3] LUDWIG, W. and FALTER, C.: *Symmetries in physics*, Springer, 1988.
[A4] CHENG, T.P. and LI, L.F.: *Gauge theory of elementary particle physics*, Oxford, 1984.
[A5] UHLENBECK, K.: 'Conservation laws and their application in global differential geometry', in B. Srinivasan and J. Sally (eds.): *Emmy Noether in Bryn Mawr*, Springer, 1983, pp. 103-117.

3) *Noether's normalization theorem*: In any finitely-generated commutative integral k-algebra A of transcendence degree d over a field k there are d elements $x_1, \ldots, x_d$ such that A is integral over the subalgebra B generated by them (cf. **Integral ring; Integral extension of a ring**). If A has a grading of the form $A = \oplus_{i \geqslant 0} A_i$, $A_0 = k$, then $x_1, \ldots, x_d$ can be chosen to be homogeneous.

This theorem (sometimes also called *Noether's normalization lemma*) was proved by E. Noether [1]; in the graded case it was already stated by D. Hilbert [2].

The elements $x_1, \ldots, x_d$ are algebraically independent over k, so that B is a polynomial algebra in these variables with coefficients in k. If k is infinite, then $x_1, \ldots, x_d$ can be chosen from linear combinations of generators of A over k. If k is algebraically closed, then the normalization theorem can be stated geometrically: Every irreducible affine d-dimensional algebraic variety X is a finitely-sheeted (ramified) covering of an affine d-dimensional space A^d; more accurately, it has a finite morphism onto A^d. Furthermore, if X is a closed subset of k^n, then this morphism can be realized as the restriction to X of a certain linear mapping of k^n onto a d-dimensional linear subspace.

The algebra A is finitely generated as a B-module. The subalgebra B is not unique; however, a number of properties of A as a B-module do not depend on the choice of B. For example, if A is graded, as above under the hypotheses of the theorem, and if $x_1, \ldots, x_d$ are homogeneous (so that B is also graded), then the property of A of being a free B-module does not depend on the choice of B.

References

[1] NOETHER, E.: 'Abstrakter Aufbau der Idealtheorie in algebraischen Zahl und Funktionenkörpern', *Math. Ann.* **96** (1927), 26-61.
[2] HILBERT, D.: 'Ueber die vollen Invariantensysteme', *Math. Ann.* **42** (1893), 313-373.
[3] ATIYAH, M. and MACDONALD, I.G.: *Introduction to commutative algebra*, Addison-Wesley, 1969.
[4] BOURBAKI, N.: *Elements of mathematics. Commutative algebra*, Addison-Wesley, 1972 (translated from the French).
[5] ZARISKI, O. and SAMUEL, P.: *Commutative algebra*, 1-2, v. Nostrand, 1958-1960.
[6] LANG, S.: *Algebra*, Addison-Wesley, 1974.

V.L. Popov

AMS 1980 Subject Classification: 58A15, 58E30, 58F05, 70HXX, 53-XX, 35-XX, 58F35, 58G35, 22E70, 81-XX, 73-XX, 13-XX

NOETHERIAN GROUP, *group with the maximum condition for subgroups* - A group in which every strictly ascending chain of subgroups is finite. This class is named after E. Noether, who investigated rings with the maximum condition for ideals — Noetherian rings (cf. **Noetherian ring**). Subgroups and quotient groups of a Noetherian group are Noetherian. Examples have been constructed of Noetherian groups that are not finite extensions of polycyclic groups (cf. **Polycyclic group**) [1].

References

[1] OL'SHANSKIĬ, A.YU.: 'Infinite groups with cyclic subgroups', *Soviet Math. Dokl.* **20**, no. 2 (1979), 343-346. (*Dokl. Akad. Nauk SSSR* **245**, no. 4 (1979), 785-787)

V.N. Remeslennikov

AMS 1980 Subject Classification: 20E15

NOETHERIAN INDUCTION - A reasoning principle applicable to a **partially ordered set** in which every non-empty subset contains a minimal element; for example, the set of closed subsets in some **Noetherian space**. Let M be such a set and let F be a subset of it having the property that for every $a \in F$ there is a strictly smaller element $b \in F$. Then F is empty. For example, let M be the set of all closed subsets of a Noetherian space and let F be the set of those closed subsets that cannot be represented as a finite union of irreducible components. If $Y \in F$, then Y is reducible, that is, $Y = Y_1 \cup Y_2$, where Y_1 and Y_2 are closed, both are strictly contained in Y and at least one of them belongs to F. Consequently, F is empty.

Reversal of the order makes it possible to apply Noetherian induction to partially ordered sets in which every non-empty subset contains a maximal element; for example, to the lattice of ideals in a **Noetherian ring**.

References

[1] BOURBAKI, N.: *Elements of mathematics. Commutative algebra*, Addison-Wesley, 1972 (translated from the French).

L.V. Kuz'min

Editorial comments. The term *well-founded induction* is also in use.

AMS 1980 Subject Classification: 06A10

NOETHERIAN INTEGRAL EQUATION - An **integral equation** for which Noether's theorems (see below) are valid. Let X be a Banach space, A a bounded linear operator (mapping) taking X into itself: $A : X \to X$, and let A^* be the adjoint operator of A. Let

$$Ax = y \tag{1}$$

be a linear equation, where x is an unknown and y is a given element of X. Next, let $R(A)$ be the collection of all $y \in X$ for which (1) is solvable (the *range of* A), and let $N(A)$ be the collection of all solutions of the corresponding homogeneous equation

$$Ax = 0 \qquad (2)$$

(the *null space* or *kernel* of A). The mapping A (the equation (1)) is called a **Noetherian operator** (a *Noetherian equation*) if the following conditions are satisfied.

1) The operator A defined on the whole Banach space (the equation (1)) is *normally solvable*, that is, (1) is solvable if and only if the right-hand side y is orthogonal to the solutions of the adjoint homogeneous equation

$$A^* \phi = 0, \quad \phi \in X^*, \qquad (3)$$

that is, $(\phi, y) = 0$ for every $\phi \in N(A^*)$.

2) The homogeneous equations (2) and (3) have only finitely many linearly independent solutions; the number

$$\kappa_A = k - k^*,$$

where $k = \dim N(A)$, $k^* = \dim N(A^*)$, is called the *index of the operator* (the *index of equation* (1)).

A Noetherian operator of index zero is called an (abstract) *Fredholm operator* and the corresponding equation (1) is called a *Fredholm equation*. For example, if V is a **completely-continuous operator**, $V: X \to X$, then

$$x - Vx = y \qquad (4)$$

is a Fredholm equation. It is called a *canonical Fredholm equation*. If in (4) the completely-continuous mapping on some space is an integral operator,

$$(Vx)(s) = \int K(s, t) x(t)\, dt,$$

then the equation is called a *Fredholm integral equation*. Similarly, if in the Noetherian equation (1) the linear mapping is given by means of integral operators, then it is called a *Noetherian integral equation*.

F. Noether [1] considered integral equations with a **Hilbert kernel**,

$$(A\phi)(s) = a(s)\phi(s) + \frac{b(s)}{\pi} \int_{-\pi}^{\pi} \cot \frac{t-s}{2} \phi(t)\, dt + \qquad (5)$$
$$+ \int_{-\pi}^{\pi} K(s, t)\phi(t)\, dt = f(t),$$

where the improper integral is understood in the sense of the principal value. For (5) he established the validity of three theorems, which are nowadays called *Noether's theorems*. (It is assumed that the data and the unknown function are real and Hölder continuous and that $a^2(s) + b^2(s) \neq 0$.) These are:

1) the equation is normally solvable:
2) the index of the equation is finite;
3) the index can be computed by the formula

$$\kappa_A = \frac{1}{\pi} [\arg(a - ib)]_{-\pi}^{\pi},$$

where $[\ ]_{-\pi}^{\pi}$ denotes the increment of the function

between brackets.

Theorem (3) was the first to indicate the existence of adjoint linear integral equations that have a different number of linearly independent solutions. Moreover, it follows from this theorem that the index of (5) does not depend on its completely-continuous part.

A Noetherian operator is sometimes called a *Fredholm*, a *generalized Fredholm*, a Φ-*operator*, or an *F-operator*.

References

[1] NOETHER, F.: 'Ueber eine Klasse singulärer Integralgleichungen', *Math. Ann.* **82** (1921), 42-63.
[2] NIKOL'SKIĬ, S.M.: 'Linear equations in linear normed spaces', *Izv. Akad. Nauk SSSR Ser. Mat.* **7**, no. 3 (1943), 147-166 (in Russian).
[3] ATKINSON, F.V.: 'Normal solvability of linear equations in normal spaces', *Mat. Sb.* **28**, no. 1 (1951), 3-14 (in Russian).
[4] KREIN, S.G.: *Linear equations in a Banach space*, Birkhäuser, 1982 (translated from the Russian).
[5] KRACHKOVSKIĬ, S.N. and DIKANSKIĬ, A.S.: 'Fredholm operators and their generalizations', *Progress in Math.* **10** (1971), 37-72. (*Itogi Nauk. Mat. Anal. 1968* (1969), 39-71)
[6] DANILYUK, I.I.: *Non-regular boundary value problems on the plane*, Moscow, 1975 (in Russian).
[7] PRÖSSDORF, Z.: *Einige Klassen singulärer Gleichungen*, Akad. Verlag, 1974.

B.V. Khvedelidze

Editorial comments. In modern literature the term 'completely continuous' is often replaced by 'compact'. Also the term 'Fredholm operator' is generally used for linear operators having a finite index. The class of Fredholm operators includes many important operators and there is an extensive literature on the subject. The index satisfies the logarithmic law $\kappa_{AB} = \kappa_A + \kappa_B$. For special classes of Fredholm operators, the index can be related to certain topological notions, such as the winding number of a curve (see also above). A bounded linear operator is Fredholm if and only if it invertible modulo the compact operators, i.e., if and only if it corresponds to an invertible element in the *Calkin algebra*. Normal solvability (i.e., the property of having closed range) is implied by finiteness of the index.

References

[A1] BOOSS, B.: *Topologie und Analysis, Einführung in die Atiyah − Singer-Indexformel*, Springer, 1977.
[A2] CONWAY, J.B.: *A course in functional analysis*, Springer, 1985.
[A3] GOHBERG, I.C. and KREĬN, M.G.: 'The basic propositions on defect numbers, root numbers and indices of linear operators', *Transl. Amer. Math. Soc. (2)* **13** (1960), 185-264. (*Uspekhi Mat. Nauk* **12**, no. 2 (1957), 43-118)
[A4] GOLDBERG, S.: *Unbounded linear operators*, McGraw-Hill, 1966.
[A5] KATO, T.: *Perturbation theory for linear operators*, Springer, 1976.
[A6] GOHBERG, I.C. and KRUPNIK, N.: *Einführung in die Theorie der eindimensionalen singulären Integraloperatoren*, Birkhäuser, 1979.
[A7] ZABREYKO, P.P. [P.P. ZABREĬKO], ET AL.: *Integral equations — a reference text*, Noordhoff, 1975 (translated from the Russian).
[A8] MEISTER, E.: *Randwertaufgaben der Potentialtheorie*, Teubner, 1983.

AMS 1980 Subject Classification: 45A05, 45B05, 45FXX, 45G05

NOETHERIAN MODULE - A **module** for which every submodule has a finite system of generators. Equivalent conditions are: Every strictly ascending chain of submodules breaks off after finitely many terms; every non-empty set of submodules ordered by inclusion contains a maximal element. Submodules and quotient modules of a Noetherian module are Noetherian. If, in an **exact sequence**

$$0 \to M' \to M \to M'' \to 0,$$

M' and M'' are Noetherian, then so is M. A module over a **Noetherian ring** is Noetherian if and only if it is finitely generated.

References

[1] LANG, S.: *Algebra*, Addison-Wesley, 1974.

L.V. Kuz'min

AMS 1980 Subject Classification: 13E05

NOETHERIAN OPERATOR - A linear operator (with closed range) that is simultaneously n-normal and d-normal (see **Normally-solvable operator**). In other words, a Noetherian operator A is a normally-solvable operator of finite d-characteristic ($n(A) < + \infty$, $d(A) < + \infty$). The index $\chi(A)$ (cf. **Index of an operator**) of a Noetherian operator A is also finite. The simplest example of a Noetherian operator is a linear operator acting from $\mathbf{R}^k$ to $\mathbf{R}^l$. It is named after F. Noether [1], in whose work the theory of Noetherian operators is developed parallel to the theory of singular integral equations. Linear operators generated by general boundary value problems for elliptic equations are frequently Noetherian.

In practice, as a rule one succeeds to verify the validity of the following propositions (*Noether's theorems*):

1) the equation $Ax = 0$ has either no non-trivial solutions or a finite number n of linearly independent solutions; and

2) the inhomogeneous equation $Ax = y$ is either solvable for any right-hand side y, or for its solvability it is necessary and sufficient that $<y, \psi_i> = 0$, $i = 0, \ldots, m$, where $\{\psi_i\}_0^m$ is a complete system of linearly independent solutions of the associated homogeneous equation, or it is formally adjoint to the homogeneous problem.

From 1) and 2) it follows that A is a Noetherian operator.

The property of being Noetherian is stable: If A is a Noetherian operator and B is a linear operator of sufficiently small norm or is completely continuous, then $A + B$ is also Noetherian, and $\chi(A + B) = \chi(A)$.

Suppose that $A \in L(X, Y)$, where $L(X, Y)$ is the space of linear operators from X to Y, is Noetherian. Then there is the direct decomposition

$$X = N(A) \dotplus \hat{X}, \quad Y = Z \dotplus R(A),$$

where $N(A)$ is the null space of A, $R(A)$ is the range of

A and $\dim Z = d(A)$. The general solution of the equation $Ax = y$, $y \in R(A)$, is of the form $x = \hat{A}^{-1}y + v$, where $\hat{A} \in L(\hat{X}, R(A))$, $\hat{A} = A$ on $\hat{X}$ (the restriction of A) and $v \in N(A)$ is arbitrary. If A is Noetherian with d-characteristic (n, m), then A^* is Noetherian with d-characteristic (m, n).

References

[1] NOETHER, F.: 'Ueber eine Klasse singulärer Integralgleichungen', *Math. Ann.* **82** (1921), 42-63.
[2] KREĬN, S.G.: *Linear differential equations in Banach space*, Amer. Math. Soc., 1971 (translated from the Russian).
[3] VAĬNBERG, M.M. and TRENOGIN, V.A.: *Theory of branching of solutions of non-linear equations*, Noordhoff, 1974 (translated from the Russian).

V.A. Trenogin

Editorial comments. In the Western literature a Noetherian operator is usually called a *Fredholm operator*. The index of such an operator is the number $\chi(A) = n(A) - d(A)$. The product of two Noetherian operators A and B is again a Noetherian operator, and $\chi(AB) = \chi(A) + \chi(B)$. In the first concrete applications (see Noether's paper [1]) the index was calculated as a winding number associated with a certain continuous function. The computation of the index for different classes of operators is an important problem in modern mathematics (see e.g., [A1]).

References

[A1] PALAIS, R.S.: *Seminar on the Atiyah–Singer index theorem*, Princeton Univ. Press, 1965.
[A2] GOHBERG, I.C. [I.TS. GOKHBERG] and KREĬN, M.G.: 'The basic propositions on defect numbers, root numbers and indices of linear operators', *Transl. Amer. Math. Soc. (2)* **13** (1960), 185-264. (*Uspekhi Mat. Nauk* **12** (1957), 43-118)
[A3] GOHBERG, I. [I.TS. GOKHBERG] and KRUPNIK, N.: *Einführung in die Theorie der eindimensionalen singulären Integraloperatoren*, Birkhäuser, 1979 (translated from the Russian).
[A4] GOLDBERG, S.: *Unbounded linear operators*, McGraw-Hill, 1966.
[A5] KATO, T.: 'Perturbation theory for nullity, deficiency and other quantities of linear operators', *J. d'Anal. Math.* **6** (1958), 261-322.
[A6] KREĬN, S.G.: *Linear equations in Banach spaces*, Birkhäuser, 1982 (translated from the Russian).

AMS 1980 Subject Classification: 47A53

NOETHERIAN RING, *left (right)* - A **ring** A satisfying one of the following equivalent conditions:

1) A is a left (or right) **Noetherian module** over itself;

2) every left (or right) ideal in A has a finite generating set;

3) every strictly ascending chain of left (or right) ideals in A breaks off after finitely many terms.

An example of a Noetherian ring is any principal ideal ring, i.e. a ring in which every ideal has one generator.

Noetherian rings are named after E. Noether, who made a systematic study of such rings and carried over to them a number of results known earlier only under more stringent restrictions (for example, Lasker's theory of primary decompositions).

A right Noetherian ring need not be left Noetherian

and vice versa. For example, let A be the ring of matrices of the form

$$\left\|\begin{matrix} a & \alpha \\ 0 & \beta \end{matrix}\right\|,$$

where a is a rational integer and α and β are rational numbers, with the usual addition and multiplication. Then A is right, but not left, Noetherian, since the left ideal of elements of the form

$$\left\|\begin{matrix} 0 & \alpha \\ 0 & 0 \end{matrix}\right\|$$

does not have a finite generating set.

Quotient rings and finite direct sums of Noetherian rings are again Noetherian, but a subring of a Noetherian ring need not be Noetherian. For example, a polynomial ring in infinitely many variables over a field is not Noetherian, although it is contained in its field of fractions, which is Noetherian.

If A is a left Noetherian ring, then so is the polynomial ring $A[X]$. The corresponding property holds for the ring of formal power series over a Noetherian ring. In particular, polynomial rings of the form $K[X_1, \ldots, X_n]$ or $\mathbf{Z}[X_1, \ldots, X_n]$, where K is a field and $\mathbf{Z}$ the ring of integers, and also quotient rings of them, are Noetherian. Every **Artinian ring** is Noetherian. The localization of a commutative Noetherian ring A relative to some multiplicative system S is again Noetherian. If in a commutative Noetherian ring A, $\mathfrak{m}$ is an ideal such that no element of the form $1+m$, where $m \in \mathfrak{m}$, is a divisor of zero, then $\bigcap_{k=1}^{\infty} \mathfrak{m}^k = 0$. This means that any such ideal $\mathfrak{m}$ defines on A a separable $\mathfrak{m}$-adic topology. In a commutative Noetherian ring every ideal has a representation as an incontractible intersection of finitely many primary ideals. Although such a representation is not unique, the number of ideals and the set of prime ideals associated with the given primary ideals are uniquely determined.

References

[1] WAERDEN, B.L. VAN DER: *Algebra*, 1-2, Springer, 1967-1971 (translated from the German).
[2] LANG, S.: *Algebra*, Addison-Wesley, 1974.
[3] FAITH, C.: *Algebra: rings, modules and categories*, 1, Springer, 1973.

L.V. Kuz'min

AMS 1980 Subject Classification: 13E05

NOETHERIAN SCHEME - A **scheme** admitting a finite open covering by spectra of Noetherian rings (cf. **Noetherian ring**). An affine Noetherian scheme is precisely the spectrum of a Noetherian ring. The topological space of a Noetherian scheme X is a Noetherian topological space, and the local rings $\mathcal{O}_{X,x}$ are Noetherian. If every point of a scheme has an open affine Noetherian neighbourhood, the scheme is called *locally*

Noetherian. A quasi-compact locally Noetherian scheme is a Noetherian scheme. An example of a Noetherian scheme is a scheme of finite type over a field (an algebraic variety) or over any Noetherian ring.

V.I. Danilov

Editorial comments.

References

[A1] HARTSHORNE, R.: *Algebraic geometry*, Springer, 1977.

AMS 1980 Subject Classification: 14K99

NOETHERIAN SPACE - A **topological space** X in which every strictly decreasing chain of closed subspaces breaks off. An equivalent condition is: Any non-empty family of closed subsets of X ordered by inclusion has a minimal element. Every subspace of a Noetherian space is itself Noetherian. If a space X has a finite covering by Noetherian subspaces, then X is itself Noetherian. A space X is Noetherian if and only if every open subset of X is quasi-compact. A Noetherian space X is the union of finitely many irreducible components.

Examples of Noetherian spaces are some spectra of commutative rings (cf. **Spectrum of a ring**). For a ring A the space $\mathrm{Spec}(A)$ (the spectrum of A) is Noetherian if and only if A/J is a **Noetherian ring**, where J is the nil radical of A.

References

[1] BOURBAKI, N.: *Elements of mathematics. Commutative algebra*, Addison-Wesley, 1972 (translated from the French).

L.V. Kuz'min

Editorial comments.

References

[A1] HARTSHORNE, R.: *Algebraic geometry*, Springer, 1977.

AMS 1980 Subject Classification: 14A99

NOISE IMMUNITY - A notion characterizing the ability of an information-transmitting system to resist to distortion effects of noise. The maximum attainable noise immunity for an optimal transmission method is called the *potential noise immunity*. In the theory of information transmission the noise immunity of a particular information-transmitting system is characterized by the exactness of reproducibility of information (cf. **Information, exactness of reproducibility of**) and, in particular, by the probability of erroneous decoding (cf. **Erroneous decoding, probability of**) a transmitted message.

References

[1] KHARKEVICH, A.A.: *Struggle with noise*, Moscow, 1965 (in Russian).
[2] WOZENKRAFT, J.M. and JACOBS, I.M.: *Theoretical foundations of communication techniques*, Moscow, 1969 (in Russian; translated from the English).

P.L. Dobrushin
V.V. Prelov

References
[A1] Kotelnikov, A.V.: *The theory of optimum noise immunity*, McGraw-Hill, 1959 (translated from the Russian).

AMS 1980 Subject Classification: 94A05, 94A14, 60G35

NOMOGRAPHY - A branch of mathematics in which methods of graphical representation of functional dependencies are studied. The resulting designs are called *nomograms*. Every nomogram is constructed for a definite functional dependence within specific limits of variation of the variables. In nomography computational work is replaced by the performance of the simplest geometric operations indicated in the instructions and an evaluation of the answers.

The accuracy of answers by nomography depends on the form of the nomographic presentation of the dependency, the limits of variation of the variables, the dimension of the design, and on the chosen type of nomogram. On the average, nomograms can provide answers to 2 - 3 true digits. When the accuracy of nomograms is insufficient, they can be used for provisional calculations, for finding zero-order approximations, and for the control of computations with the aim of discovering gross errors.

Nomograms can also be used to study functional dependencies, putting the nomograms at their foundation. Often such a study can be carried out by nomograms in a considerably simpler and more intuitive way than by other methods. By means of nomograms one can investigate the influence of various variables on the required variable, give an intuitive interpretation of some previously known properties of the dependency in question, and establish previously unknown peculiarities of it. Nomographic methods of investigation can be applied, for example, in problems on the selection of parameters in empirical formulas on the results of observations, in the approximation of one function by another, and for finding extremal values of functions.

The values of variables are represented in nomograms by marked points and marked lines. A set of marked points depending on a single variable is called a *scale*. The equation of scale of a variable α_1 in a rectangular coordinate system $x0y$ is written in the form

$$x = f_1, \quad y = g_1,$$

where f_1 and g_1 are functions $f_1(\alpha_1)$ and $g_1(\alpha_1)$. The scheme of the scale of α_1 is illustrated in Fig. 1.

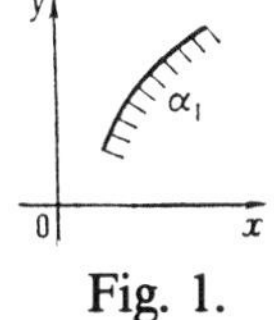

Fig. 1.

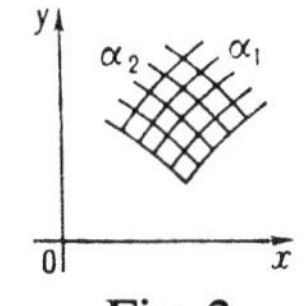

Fig. 2.

A set of marked points depending on two variables is called a *binary field*. A binary field is usually formed as a grid consisting of two families of marked lines. A point in a binary field is determined as the intersection of lines with given marks. In a rectangular coordinate system $x0y$ the binary field of two variables α_1 and α_2 is given by equations

$$x = f_{12}, \quad y = g_{12},$$

where f_{12} and g_{12} are functions $f_{12}(\alpha_1, \alpha_2)$ and $g_{12}(\alpha_1, \alpha_2)$. It is assumed that f_{12} and g_{12} are such that in the given domain of variation of the variables to every pair of values α_1 and α_2 corresponds only one pair of values x and y. The scheme of a binary field (α_1, α_2) is illustrated in Fig. 2.

Scales, families of marked lines and binary fields are formed so that it is convenient to find points and lines with given marks and to determine the marks of answer points and lines.

Elementary nomograms are those in which the answer or answers can be found as a result of performing a single geometric operation (determining a point on a scale or in a binary field; drawing a line through two points; constructing a circle with given centre and radius; dividing a segment in a given ratio; drawing a line parallel to a given one; laying off a segment of length equal to that of a given segment; or superimposing one plane to another).

To the elementary nomograms belong: the graph of a function; scale doubling alignment; nomograms from adjusted points; nomograms from equidistant points; nomograms using a compass; nomograms with parallel indices; barycentric nomograms; rhomboidal nomograms; nomograms with an oriented transparency; and nomograms with a transparency of general form.

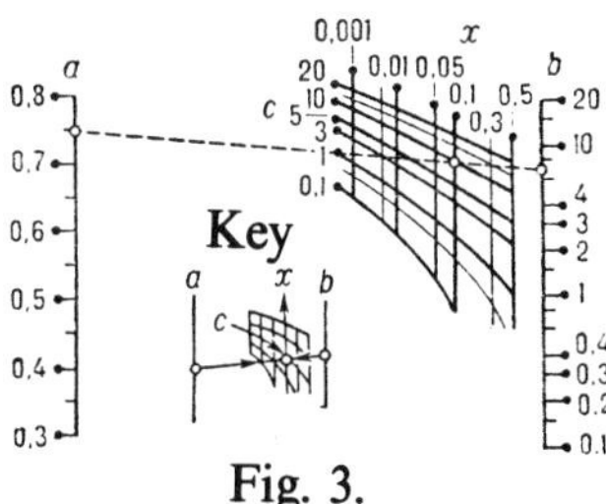

Fig. 3.

In Fig. 3 an elementary nomogram is drawn from adjusted points for the determination of the quantity x from the equation

$$x^{a-1} = b\frac{\ln(c-x)}{c-x-1},$$

which is used for thermal calculations involving ventilators. The nomogram is constructed within the limits: $0.3 \leqslant a \leqslant 0.8$; $0.1 \leqslant b \leqslant 20$; $0.1 \leqslant c \leqslant 20$; $0.001 \leqslant x < 0.5$. The variables a and b are represented on the nomogram by scales; the variables c and x by a binary field.

The nomogram shows the solution of a numerical example (data: $a=0.75$; $b=7$; $c=10$; answer: $x=0.1$).

Elementary nomograms have a simple geometric foundation: the nomographic interpretation of the condition that three points are collinear leads to a nomogram from adjusted points; the formulas for the distance between two points — to a nomogram from equidistant points and to the use of a compass; the formulas for the coordinates of a point dividing a segment in a given ratio — to a barycentric nomogram; the condition that two lines are parallel — to a nomogram with parallel indices; the formulas determining the coordinates of the fourth vertex of a parallelogram from three given vertices of it — to a rhomboidal nomogram; the formulas for the transformation of rectangular coordinates with, or without, rotation of the axes — a nomogram on two planes (with an oriented transparency or of general form).

To every elementary nomogram corresponds a canonical form of dependency that can be illustrated by the nomogram. Some canonical forms allow the construction of elementary nomograms of various types.

The most general canonical form represented by an elementary nomogram of adjusted points is

$$\frac{f_{34}-f_{12}}{g_{34}-g_{12}} = \frac{f_{56}-f_{12}}{g_{56}-g_{12}}.$$

The corresponding nomogram consists of three binary fields (α_1, α_2), (α_3, α_4) and (α_5, α_6), connected by a single alignment.

Below the frequently occurring canonical forms for individual equations and systems of equations representable by nomograms of one type or another are presented.

Canonical forms for individual equations: a) with three variables:

$$f_1 = f_{23}, \quad f_1 f_3 + f_2 g_3 + h_3 = 0, \quad f_1 + f_2 = f_3, \quad f_1 = f_2 f_3;$$

b) with four variables:

$$f_{12} = f_{34}, \quad f_1 f_{34} + f_2 g_{34} + h_{34} = 0, \quad f_{12} = f_3 + f_4;$$

c) with five variables:

$$f_{12} = f_{34} + f_{35}; \quad f_1 + f_2 + f_3 + f_4 + f_5 = 0,$$

$$f_5 = F(f_{12} + f_{34}, g_{12} + g_{34});$$

d) with six variables:

$$f_{12} + f_{13} = f_{45} + f_{46}, \quad f_1 + f_2 + f_3 + f_4 + f_5 + f_6 = 0;$$

e) with seven variables:

$$f_7 = F(f_{12} + f_{34} + f_{56}, g_{12} + g_{31} + g_{56}),$$

$$f_1 + f_2 + f_3 + f_4 + f_5 + f_6 + f_7 = 0.$$

Canonical forms for systems of equations:

$$\left.\begin{array}{l} f_3 = f_{12}, \\ g_4 = g_{12}; \end{array}\right| \quad \left.\begin{array}{l} f_{12} + f_{34} = f_{56}, \\ g_{12} + g_{34} = g_{56}; \end{array}\right| \quad \left.\begin{array}{l} f_{12} - f_{56} = f_{34} - f_{56}, \\ g_{12} - g_{56} = g_{34} - g_{78}; \end{array}\right|$$

$$\left.\begin{array}{l} f_{12} - g_7 = f_{34} - f_8 = f_{56} - f_9, \\ g_{12} - g_7 = g_{34} - g_8 = g_{56} - g_9. \end{array}\right\}$$

Composite nomograms consist of elementary nomograms of the same type or of several types. The introduction of composite nomograms considerably extends the class of dependencies that can be represented nomographically. For a summary of canonical forms that can be represented by elementary and composite nomograms see [3] and [4].

A dependency with three variables can always be represented nomographically. Dependencies with four or more variables allow the construction of nomograms only in special cases.

To extend the range of nomographically-representable dependencies one uses approximate representations. One such is based on allowing the given dependency to be nomographed with a certain admissible error.

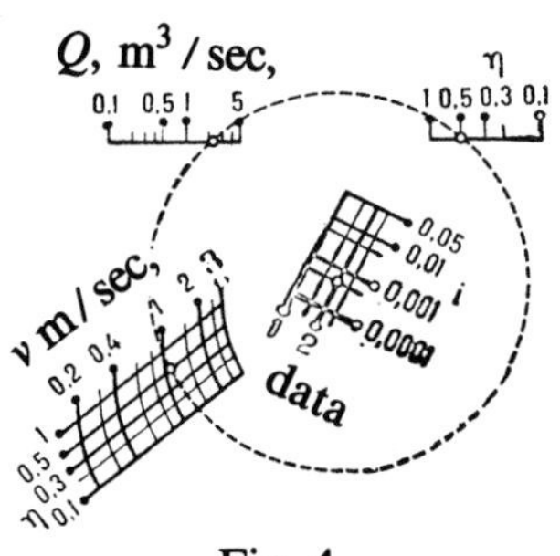

Fig. 4.

In Fig. 4 an approximate nomogram from equidistant points is drawn for the determination of the quantities η and v (or Q and v) from the system of equations

$$Q = \left[76.9 + 17.72 \log \frac{b\eta}{1+2\eta}\right] \frac{b^{2.5}\eta^{1.5} i^{0.5}}{(1+2\eta)^{0.5}},$$

$$v = \left[76.9 + 17.72 \log \frac{b\eta}{1+2\eta}\right] \frac{b^{0.5}\eta^{0.5} i^{0.5}}{(1+2\eta)^{0.5}},$$

which is used for the calculation of rectangular flows in hydraulics. To use nomography this system was replaced by the approximate system

$$Q = \frac{76 b^{2.616}\eta^{1.616} i^{0.5}}{(1+2\eta)^{0.616}}, \quad v = \frac{76 b^{0.616}\eta^{0.616} i^{0.5}}{(1+2\eta)^{0.616}},$$

with a relative error in Q and v not exceeding 2.5%. The dotted circle in the nomogram corresponds to the solution of a numerical example (data: $i=0.0005$; $b=2$ m; $Q=2.2$ m³/sec; answers: $\eta=0.5$ and $v=1.1$ m/sec).

In some cases the methods of approximate nomography make it possible to represent a table with several entries by nomograms.

To obtain the nomographic representation of a given dependency one puts it accurately or approximately into nomographic form and writes down the equations of the elements of the nomogram in a rectangular coordinate system. The transformation parameters occurring in these equations (and sometimes also arbitrary functions) are chosen so that they give to the nomogram a form convenient to use. Next one calculates tables of coordinates of the individual elements of the nomogram and then one draws the nomogram.

Machine nomography has been developed (see [4] - [6]): Systems of procedures and of standard programs have been worked out for the automatic calculation and construction of elementary nomograms by means of computers and graphical construction as well as standard programs for the automatic construction, calculation and sketching of nomograms of various types.

The main problems of theoretical nomography are representability and uniqueness. The essence of the first problem consists in clarifying whether a given equation or system of equations can be brought to some canonical form and, if possible, to determine an algorithm for this process. Solutions of this problem have been obtained for certain canonical forms. They are cumbersome and seldom used in practice. The essence of the second problem consists in finding out whether there is a unique way of bringing a given dependency to canonical form, and when it is not unique to indicate all possible ways and for each of them to establish whether the nomogram can be transformed into it.

References

[1] PENTKOVSKIĬ, M.V.: *Nomography*, Moscow-Leningrad, 1949 (in Russian).

[2] NEVESKIĬ, B.A.: *Handbook of nomography*, Moscow-Leningrad, 1951 (in Russian).

[3] KHOVANSKIĬ, G.S.: *Eléments de nomographie*, Mir, 1979 (translated from the Russian).

[4] KHOVANSKIĬ, G.S.: *Nomography and its possibilities*, Moscow, 1977 (in Russian).

[5] KHOVANSKIĬ, G.S. (ED.): *Nomography collection*, 15, Moscow, 1986 (in Russian).

[6] KHOVANSKIĬ, G.S.: *Nomography today*, Moscow, 1987 (in Russian).

G.S. Khovanskiĭ

Editorial comments.

References

[A1] REKTORYS, K. (ED.): *Survey of applicable mathematics*, Iliffe, 1969, p. Sect. 32A.

[A2] SAUER, R. and SZABÓ, I. (EDS.): *Mathematische Hilfsmittel des Ingenieurs*, III, Springer, 1968, p. Sect. G.I.3.

[A3] DAVIS, D.S.: *Nomography and empirical relations*, Wiley, 1962.

[A4] LEVENS, A.S.: *Nomography*, Wiley, 1959.

[A5] JOHNSON, L.: *Nomography and empirical equations*, Wiley, 1952.

AMS 1980 Subject Classification: 26-XX

NON-ABELIAN COHOMOLOGY - Cohomology with coefficients in a non-Abelian group, a sheaf of non-Abelian groups, etc. The best known examples are the cohomology of groups, topological spaces and the more general example of the cohomology of sites (i.e. topological categories; cf. **Topologized category**) in dimensions 0, 1. A unified approach to non-Abelian cohomology can be based on the following concept. Let C^0, C^1 be groups, let C^2 be a set with a distinguished point e, let $\text{Aff}\,C^1$ be the holomorph of C^1 (i.e. the semi-direct product of C^1 and $\text{Aut}(C^1)$; cf. also **Holomorph of a group**), and let $\text{Aut}\,C^2$ be the group of permutations of C^2 that leave e fixed. Then a *non-Abelian cochain complex* is a collection

$$C^{\bullet} = (C^0, C^1, C^2, \rho, \sigma, \delta),$$

where $\rho: C^0 \to \text{Aff}\,C^1$, $\sigma: C^0 \to \text{Aut}\,C^2$ are homomorphisms and $\delta: C^1 \to C^2$ is a mapping such that

$$\delta(e) = e \quad \text{and} \quad \delta(\rho(a)b) = \sigma(a)\delta(b), \quad a \in C^0, \ b \in C^1.$$

Define the 0-dimensional cohomology group by

$$H^0(C^{\bullet}) = \rho^{-1}(\text{Aut}\,C^1),$$

and the 1-dimensional cohomology set (with distinguished point) by

$$H^1(C^{\bullet}) = Z^1/\rho,$$

where $Z^1 = \delta^{-1}(e) \subseteq C^1$ and the factorization is modulo the action ρ of the group C^0.

Examples. 1) Let X be a topological space with a sheaf of groups $\mathscr{F}$, and let $\mathfrak{U}$ be a covering of X; one then has the Čech complex

$$C^{\bullet}(\mathfrak{U}, \mathscr{F}) = (C^0(\mathfrak{U}, \mathscr{F}), C^1(\mathfrak{U}, \mathscr{F}), C^2(\mathfrak{U}, \mathscr{F})),$$

where $C^i(\mathfrak{U}, \mathscr{F})$ are defined as in the Abelian case (see **Cohomology**),

$$(\sigma(a)(c))_{ijk} = a_i c_{ijk} a_i^{-1},$$

$$(\delta b)_{ijk} = b_{ij} b_{jk} b_{ik}^{-1},$$

$$a \in C^0, \ b \in C^1, \ c \in C^2.$$

Taking limits with respect to coverings, one obtains from the cohomology sets $H^i(C^{\bullet}(\mathfrak{U}, \mathscr{F}))$, $i = 0, 1$, the *cohomology* $H^i(X, \mathscr{F})$, $i = 0, 1$, *of the space X with coefficients in* $\mathscr{F}$. Under these conditions, $H^0(X, \mathscr{F}) = \mathscr{F}(X)$. If $\mathscr{F}$ is the sheaf of germs of continuous mappings with values in a topological group G, then $H^1(X, \mathscr{F})$ can be interpreted as the set of isomorphism classes of topological principal bundles over X with structure group G. Similarly one obtains a classification of smooth and holomorphic principal bundles. In a similar fashion one defines the non-Abelian cohomology for a site; for an interpretation see **Principal G-object**.

2) Let G be a group and let A be a (not necessarily Abelian) G-group, i.e. an operator group with group of operators G. Denote the action of an operator $g \in G$ on an element $a \in A$ by a^g. Define a complex $C^{\bullet}(G, A)$ by the formulas

$$C^k = \text{Map}(G^k, A), \quad k = 0, 1, 2,$$

$$(\rho(a)(b))(g) = ab(g)(a^g)^{-1},$$

$$(\sigma(a)(c))(g, h) = a^g c(g, h)(a^g)^{-1},$$

$$\delta(b)(g, h) = b(g)^{-1}b(gh)(b(h)^g)^{-1},$$

$$a \in C^0, \quad b \in C^1, \quad c \in C^2, \quad g \in G.$$

The group $H^0(G, A) = H^0(C^*(G, A))$ is the subgroup A^G of G-fixed points in A, while $H^1(G, A) = H^1(C^*(G, A))$ is the set of equivalence classes of crossed homomorphisms $G \to A$, interpreted as the set of isomorphism classes of principal homogeneous spaces (cf. **Principal homogeneous space**) over A. For applications and actual computations of non-Abelian cohomology groups see **Galois cohomology**. Analogous definitions yield the non-Abelian cohomology of categories and semi-groups.

3) Let X be a smooth manifold, G a Lie group and $\mathfrak{g}$ the Lie algebra of G. The *non-Abelian de Rham complex* $R_G^*(X)$ is defined as follows: $R_G^0(X)$ is the group of all smooth functions $X \to G$; $R_G^k(X)$, $k = 1, 2$, is the space of exterior k-forms on X with values in $\mathfrak{g}$;

$$\rho(f)(\alpha) = df \cdot f^{-1} + (\mathrm{Ad}\, f)\alpha;$$

$$\sigma(f)(\beta) = (\mathrm{Ad}\, f)\beta,$$

$$\delta\alpha = d\alpha - \frac{1}{2}[\alpha, \alpha],$$

$$f \in R_G^0, \quad a \in R_G^1, \quad \beta \in R_G^2.$$

The set $H^1(R_G(X))$ is the set of classes of totally-integrable equations of the form $df \cdot f^{-1} = \alpha$, $\alpha \in R_G^1$, modulo gauge transformations. An analogue of the de Rham theorem provides an interpretation of this set as a subset of the set $H^1(\pi_1(M), G)$ of conjugacy classes of homomorphisms $\pi_1(M) \to G$. In the case of a complex manifold M and a complex Lie group G, one again defines a non-Abelian holomorphic de Rham complex and a non-Abelian Dolbeault complex, which are intimately connected with the problem of classifying holomorphic bundles [3]. Non-Abelian complexes of differential forms are also an important tool in the theory of pseudo-group structures on manifolds.

For each subcomplex of a non-Abelian cochain complex there is an associated exact cohomology sequence. For example, for the complex $C^*(G, A)$ of Example 2 and its subcomplex $C^*(G, B)$, where B is a G-invariant subgroup of A, this sequence is

$$e \to H^0(G, B) \to H^0(G, A) \to (A/B)^G \to$$

$$\to H^1(G, B) \to H^1(G, A).$$

If B is a normal subgroup of A, the sequence can be continued up to the term $H^1(G, A/B)$, and if B is in the centre it can be continued to $H^2(G, B)$. This sequence is exact in the category of sets with a distinguished point. In addition, a tool is available ('twisted' cochain complexes) for describing the preimages of all — not only the distinguished — elements

(see [1], [6], [3]). One can also construct a spectral sequence related to a double non-Abelian complex, and the corresponding exact boundary sequence.

Apart from the 0- and 1-dimensional non-Abelian cohomology groups just described, there are also 2-dimensional examples. A classical example is the 2-dimensional cohomology of a group G with coefficients in a group A; the definition is as follows. Let $\mathscr{Z}^2(G, A)$ denote the set of all pairs (m, ϕ), where $m: G \times G \to A$, $\phi: G \to \mathrm{Aut}\, A$ are mappings such that

$$\phi(g_1)\phi(g_2)\phi(g_1 g_2)^{-1} = \mathrm{Int}\, m(g_1, g_2),$$

$$m(g_1, g_2)m(g_1 g_2, g_3) = \phi(g_1)(m(g_2, g_3))m(g_1, g_2 g_3);$$

here $\mathrm{Int}\, a$ is the inner automorphism generated by the element $a \in A$. Define an equivalence relation in $\mathscr{Z}^2(G, A)$ by putting $(m, \phi) \sim (m', \phi')$ if there is a mapping $h: G \to A$ such that

$$\phi'(g) = (\mathrm{Int}\, h(g))\phi(g)$$

and

$$m'(g_1, g_2) = h(g_1)(\phi(g_1)(h(g_2)))m(g_1, g_2)h(g_1, g_2)^{-1}.$$

The equivalence classes thus obtained are the elements of the cohomology set $\mathscr{H}^2(G, A)$. They are in one-to-one correspondence with the equivalence classes of extensions of A by G (see **Extension of a group**).

The correspondence $(m, \phi) \to \phi$ gives a mapping θ of the set $\mathscr{H}^2(G, A)$ into the set of all homomorphisms

$$G \to \mathrm{Out}\, A = \mathrm{Aut}\, A / \mathrm{Int}\, A;$$

let $H_\alpha^2(G, A) = \theta^{-1}(\alpha)$ for $\alpha \in \mathrm{Out}\, A$. If one fixes $\alpha \in \mathrm{Out}\, A$, the centre $Z(A)$ of A takes on the structure of a G-module and so the cohomology groups $H^k(G, Z(A))$ are defined. It turns out that $H_\alpha^2(G, A)$ is non-empty if and only if a certain class in $H^3(G, Z(A))$ is trivial. Moreover, under this condition the group $H^2(G, Z(A))$ acts simply transitively on the set $H_\alpha^2(G, A)$.

This definition of a two-dimensional cohomology can be generalized, carrying it over to sites (see [2], where the applications of this concept are also presented). A general algebraic scheme that yields a two-dimensional cohomology is outlined in [4]; just as in the special case described above, computation of two-dimensional cohomology reduces to the computation of one-dimensional non-Abelian and ordinary Abelian cohomology.

References

[1] SERRE, J.-P.: *Cohomologie Galoisienne*, Springer, 1964.

[2] GIRAUD, J.: *Cohomologie non abélienne*, Springer, 1971.

[3] ONISHCHIK, A.L.: 'Some concepts and applications of the theory of non-Abelian cohomology', *Trans. Moscow Math. Soc.* **17** (1979), 49-98. (*Trudy Moskov. Mat. Obshch.* **17** (1967), 45-88)

[4] TOLPYGO, A.K.: 'Two-dimensional cohomologies and the spectral sequence in the nonabelian theory', *Selecta Math. Sov.* **6** (1987), 177-197.

[5] DEDECKER, P.: 'Three-dimensional nonabelian cohomology for groups', in *Category theory, homology theory and their applica-*

tions (Battelle Inst. Conf.), Vol. 2, Springer, 1968, pp. 32-64.

[6] FRENKEL, J.: 'Cohomology non abélienne et espaces fibrés', *Bull. Soc. Math. France* **85**, no. 2 (1957), 135-220.

[7] GOLDSCHMIDT, H.: 'The integrability problem for Lie equations', *Bull. Amer. Math. Soc.* **84**, no. 4 (1978), 531-546.

[8] SPRINGER, T.A.: 'Nonabelian H^2 in Galois cohomology', in A. Borel and G.D. Mostow (eds.): *Algebraic groups and discontinuous subgroups*, Proc. Symp. Pure Math., Vol. 9, Amer. Math. Soc., 1966, pp. 164-182.

A.L. Onishchik
A.K. Tolpygo

AMS 1980 Subject Classification: 18G50, 12A60

NON-ABELIAN NUMBER FIELD - An algebraic number field with a non-Abelian **Galois group** over the field of rational numbers **Q**, or a field that is not normal over **Q**. Sometimes, instead of **Q**, one considers some other ground field k of algebraic numbers, and the term 'non-Abelian' is understood to refer to the Galois group over k.

L.V. Kuz'min

Editorial comments.

References

[A1] WEISS, E.: *Algebraic number theory*, McGraw-Hill, 1963.

AMS 1980 Subject Classification: 12E99

NON-ARCHIMEDEAN GEOMETRY - The totality of geometrical propositions that can be deduced from the following groups of axioms: incidence, order, congruence, and parallelism, in Hilbert's system of axioms for Euclidean geometry, and that are unrelated to the axioms of continuity (Archimedes' axiom and the axiom of completeness). In a narrower sense, non-Archimedean geometry describes the geometrical properties of a straight line on which Archimedes' axiom is not true (the *non-Archimedean line*).

To investigate geometrical relationships in non-Archimedean geometry, one introduces a calculus of segments — a non-Archimedean number system, regarded as a special number system. One defines the concepts of a segment, and the quotient, the sum and the product of two segments. In particular, one introduces a Desarguesian number system — a non-Archimedean ordered field. With the aid of these number systems one constructs a theory of similarity of figures, a theory of areas, etc. The theory of areas of polygons underlying the theory of measurements of areas in a non-Archimedean plane is based on the concept of isometry of polygons with respect to completion, which is a more general concept than that of isometry with respect dissection.

In non-Archimedean geometry there exist triangles with equal heights and bases that are isometric with respect to completion but not to dissection. Isometric polygons with respect to completion have the same measure of area, and two polygons with the same meas-ure of area are always isometric with respect to completion. Pythagoras' theorem for right-angled triangles is valid in non-Archimedean geometry.

The calculus of segments is used to introduce a system of affine (or projective) coordinates in a non-Archimedean space. For example, one selects in the plane two straight lines — coordinate axes — passing through a fixed point, and then lays off equal segments on each. In this affine coordinates system, the equation of a straight line is linear, i.e. of the form $ax + by + c = 0$, where x, y are numbers (segments) defining the coordinates of points on the line and a, b, c are fixed numbers (segments).

The construction of numerical models for non-Archimedean geometry leads to what are known as *Hilbert's transfinite (non-Archimedean) spaces*. A number space of this kind on the real line is called a *linear Veronese space*.

The numerical realization of a non-Archimedean geometry, in which the commutative law of multiplication is not necessary, also plays an important role in the construction of **Non-Desarguesian geometry**. Such a geometry must be based on axioms of incidence, order and parallelism, without congruence.

The significance of non-Archimedean geometry lies in its role in investigating the independence and consistency of Hilbert's axioms for a Euclidean space. The realization of the axioms of incidence, order, congruence, and parallelism in a numerical model proves both that they are independent of the axiom of completeness and that non-Archimedean geometry itself is consistent. On the other hand, the role of the axioms of continuity in the construction of Euclidean geometry on the basis of Hilbert's axioms is also clarified. In particular, without the axioms of continuity one cannot prove the equivalence of Euclid's axiom of parallelism to the proposition that the sum of the interior angles of any triangle is equal to two right angles.

Geometrical constructions on the non-Archimedean plane are always carried out with the aid of a straight-edge with a marked-off standard of length (marked-off segment).

References

[1] HILBERT, D.: *Grundlagen der Geometrie*, Springer, 1913.

L.A. Sidorov

Editorial comments. Cf. also Hilbert system of axioms.

AMS 1980 Subject Classification: 51B99, 51A15

NON-ASSOCIATIVE RINGS AND ALGEBRAS - Sets with two binary operations + and ·, satisfying all the axioms of **associative rings and algebras** except possibly the associativity of multiplication. The first examples of non-associative rings and algebras that are not associative appeared in the mid-nineteenth century (**Cayley**

numbers and, in general, hypercomplex numbers, cf. **Hypercomplex number**). Given an associative ring (algebra), if one replaces the ordinary multiplication by the operation $[a, b] = ab - ba$, the result is a non-associative ring (algebra) that is a Lie ring (algebra). Yet another important class of non-associative rings (algebras) is that of Jordan rings (algebras); these are obtained by defining the operation $a \cdot b = (ab + ba)/2$ in an associative algebra over a field of characteristic $\neq 2$ (or over a commutative ring of operators with a 1 and a $1/2$). The theory of non-associative rings and algebras has evolved into an independent branch of algebra, exhibiting many points of contact with other fields of mathematics and also with physics, mechanics, biology, and other sciences. The central part of the theory is the theory of what are known as *nearly-associative rings and algebras*: Lie, alternative, Jordan, Mal'tsev rings and algebras, and some of their generalizations (see **Lie algebra**; **Alternative rings and algebras**; **Jordan algebra**; **Mal'tsev algebra**).

One of the most important problems that must be solved when studying any class of non-associative algebras is the description of simple algebras, both finite dimensional and infinite dimensional. In this context, the word description is to be understood modulo some 'classical' class contained in the class being described (e.g. the description of simple algebras in the class of alternative rings is given modulo associative rings; for Mal'tsev algebras — modulo Lie algebras; for Jordan algebras — modulo special Jordan algebras; etc.). From this point of view, the various classes of non-associative algebras can be divided into those in which there are 'many' simple algebras and those in which there are 'few'. Typical classes in which there are many simple algebras are the associative algebras, the Lie algebras and the special Jordan algebras. Namely, in these classes the following imbedding theorem is valid: Any associative (Lie, special Jordan) algebra over a field can be imbedded in a simple algebra of the same type. In some classes of algebras there are many simple algebras that are far from associative — in the class of all algebras and in the class of all commutative (anti-commutative) algebras. For these classes, too, there holds an imbedding theorem analogous to that cited above.

The problem of describing the finite-dimensional simple associative (Lie, alternative or Jordan) algebras is the object of the classical part of the theory of these algebras. Subsequently, the main results about the structure of simple finite-dimensional associative (alternative, Jordan) algebras were carried over to Artinian rings of the same type — rings with the minimum condition for one-sided ideals; in Jordan rings, one-sided ideals are replaced by quadratic ideals (see **Jordan algebra**).

Classes of algebras with 'few' simple algebras are interesting. Typical examples are the classes of alternative, Mal'tsev or Jordan algebras. In the class of alternative algebras, modulo associative algebras the only simple algebras are the (eight-dimensional) Cayley–Dickson algebras over an associative-commutative centre. In the class of Mal'tsev algebras, modulo Lie algebras the only simple algebras are the (seven-dimensional) algebras (relative to the commutator operation $[a, b]$) associated with the Cayley–Dickson algebras. In the class of Jordan algebras, modulo the special Jordan algebras the simple algebras are the (twelve-dimensional) *Albert algebras* over their associative centres (algebras of the series E) (see **Jordan algebra**). In larger classes, such as those of right-alternative or binary Lie algebras, the description of simple algebras is as yet incomplete (1989). It is known that there exists no finite-dimensional simple binary Lie algebra over a field of characteristic 0 other than a Mal'tsev algebra, but it is not known whether this result is valid in the infinite-dimensional case. For right-alternative algebras it is known that, although all finite-dimensional simple algebras of this class are alternative, there exist infinite-dimensional simple right-alternative algebras that are not alternative. All simple algebras are associative for the so-called (γ, δ)-algebras (provided $(\gamma, \delta) \neq (1, 1)$); these algebras arise in a natural manner from the stipulation that the square of an ideal be an ideal. A description is known for all Jordan algebras with two generators: Any Jordan algebra with two generators is a special Jordan algebra (*Shirshov's theorem*). All Jordan division algebras have been described (modulo associative division algebras).

In the classes of alternative, Mal'tsev or Jordan algebras there is a description of all *primary rings* (i.e. algebras the groupoid of two-sided ideals of which does not contain a zero divisor), as follows. A primary alternative ring (with $1/3$ in the commutative ring of operators) is either associative or a Cayley–Dickson ring. A primary non-degenerate Jordan algebras is either special or is an Albert ring (a Jordan ring is called an *Albert ring* if its associative centre Z consists of regular elements and if the algebra $Z^{-1}A$ is a twenty-seven-dimensional Albert algebra over its centre $Z^{-1}Z$).

In a certain sense, the opposite of a simple algebra or a primary algebra is a **nil algebra**. For power-associative algebras (cf. **Algebra with associative powers**) that are not anti-commutative (such as associative, alternative, Jordan, etc., algebras), nil algebras are defined as algebras in which some power of each element equals zero; in the case of anti-commutative algebras (i.e. algebras with the identity $x^2 = 0$, such as Lie, Mal'tsev and binary Lie algebras), nil algebras are the same as *Engel algebras*, i.e. algebras satisfying a condition

$$\forall x, y \,\exists\, (\overline{(xy)\cdots y}^{\,n}) = 0.$$

In alternative (including associative) algebras, any nil algebra of bounded index (i.e. with an identity $x^n = 0$) is locally nilpotent, and if it has no m-torsion (i.e. $mx = 0 \Rightarrow x = 0$) for $m \leqslant n$, it is solvable (in the associative case — nilpotent). *Shirshov's problem* concerning the local nilpotency of Jordan nil algebras of bounded index has been solved affirmatively. It is not known (1989) whether there exists a simple associative nil ring.

In the case of Lie algebras, the problem of the local nilpotency of Engel Lie algebras is solved by *Kostrikin's theorem*: Any Lie algebra with an identity

$$[\cdots[xy]\overline{\cdots y}^{\,n}]=0$$

over a field of characteristic $p > n$ is locally nilpotent. This theorem implies a positive solution to the restricted **Burnside problem** for groups of exponent p. Recently, E.I. Zel'manov (1989) has proved the local nilpotency of Engel Lie algebras over a field of arbitrary characteristic. From this he has inferred a positive solution of the restricted Burnside problem for groups of arbitrary exponent n (using the classification of the finite simple groups). In general, all problems connected with the local nilpotency of nil algebras are known as *Burnside-type problems*. Among these is also *Kurosh problem* concerning the local finiteness of algebraic algebras (cf. **Algebraic algebra**). An alternative (in particular, associative) algebraic algebra A of bounded degree (i.e. the degrees of the polynomials satisfied by elements of A are uniformly bounded) is locally finite. In the general case, however, Burnside-type problems (such as the local nilpotency of associative nil rings, etc.) have negative solutions.

Another topic of study includes free algebras and free products of algebras in various varieties. In the variety of all non-associative algebras, any subalgebra of a free algebra is free, and any subalgebra of a free product of algebras is the free product of its intersections with the factors and some free algebra (*Kurosh theorem*). Theorems of this type are also valid in varieties of commutative (anti-commutative) algebras. These questions are most interesting for Lie algebras. Any subalgebra of a free Lie algebra is itself a free Lie algebra (the *Shirshov—Witt theorem*). However, the analogue of Kurosh theorem is no longer valid for subalgebras of a free product of Lie algebras; nevertheless, such subalgebras may be described in terms of the generators of an ideal modulo which the free product of the intersections and the free subalgebra must be factorized. Research has been done on free alternative algebras — their Zhevlakov radicals (quasi-regular radicals, cf. **Quasi-regular radical**), their centres (associative and commutative), the quotient algebras modulo the Zhevlakov radical, etc. In contrast to free associative

algebras, free alternative algebras with $n \geqslant 4$ generators contain zero divisors and, moreover, trivial ideals (non-zero ideals with zero square). There are also known instances of trivial ideals in free Mal'tsev algebras with $n \geqslant 5$ generators; while concerning free Jordan algebras with $n \geqslant 3$ generators all that is known is that they contain zero divisors, nil elements and central elements.

The theory of free algebras is closely bound up with questions of identities in various classes of algebras. In this connection one also has the problem of the basis rank of a variety (the *basis rank* is the smallest natural number n such that the variety in question is generated by a free algebra with n generators; if no such n exists, the basis rank is defined as infinity). The basis rank of the varieties of associative and Lie algebras is 2; that of alternative and Mal'tsev algebras is infinite.

The general theory of varieties and classes of non-associative algebras deals with classes of algebras on the borderline of the classical ones and with their various relationships. One characteristic result is the following. It turns out that the varieties of admissible, generalized admissible and generalized standard algebras defined at different times and by different authors actually belong to the eight-element sublattice of the lattice of all varieties of non-associative algebras, which is also made up of the varieties of Jordan, commutative, associative, associative-commutative, and alternative algebras. The variety generated by a finite associative (alternative, Lie, Mal'tsev, or Jordan) ring is finitely based, while there exists a finite non-associative ring (an algebra over a finite field) that generates an infinitely based variety. There exists a Lie algebra over an infinite field with this property. At the same time, it is still (1989) not known whether there exists a non-finitely based variety of Lie algebras over a field of characteristic zero. A.R. Kemer [18] has proved that every variety of associative algebras over a field of characteristic 0 is finitely based (a positive solution to *Specht's problem*).

The algorithmic problems in the theory of non-associative rings and algebras have been formulated under the influence of mathematical logic. It is known that the word problem in the variety of all non-associative algebras is solvable (*Zhukov's theorem*). An analogous result is valid for commutative (anti-commutative) algebras. It is known that the Lie algebras with one relation have a solvable word problem. At the same time, there exist finitely-presented Lie algebras with an unsolvable word problem. The word problem has also been investigated in the variety of solvable Lie algebras of a given solvability degree n; it is solvable for $n = 2$, unsolvable for $n \geqslant 3$. It has been proved that any recursively-defined Lie algebra (associative algebra) over a prime field can be imbedded in a finitely-presented Lie algebra (associative algebra).

References

[1] SHIRSHOV, A.I.: 'Some questions in the theory of nearly-associative rings', *Uspekhi Mat. Nauk* **13**, no. 6 (1958), 3-20 (in Russian).

[2] ZHEVLAKOV, K.A., SLIN'KO, A.M., SHESTAKOV, I.P. and SHIRSHOV, A.I.: *Rings that are nearly associative*, Acad. Press, 1982 (translated from the Russian).

[3] KOSTRIKIN, A.I.: 'The Burnside problem', *Izv. Akad. Nauk SSSR Ser. Mat.* **23**, no. 1 (1959), 3-34 (in Russian).

[4] BOKUT', L.A.: 'Imbedding theorems in the theory of algebras', *Colloq. Math.* **14** (1966), 349-353 (in Russian).

[5] BOKUT', L.A.: 'Some questions in ring theory', *Serdica* **3** (1977), 299-308 (in Russian).

[6] KUZ'MIN, E.N.: 'Mal'tsev algebras and their representations', *Algebra and Logic* **7**, no. 4 (1968), 233-244. (*Algebra i Logika* **7**, no. 4 (1968), 48-69)

[7] FILIPPOV, V.T.: 'Central simple Mal'tsev algebras', *Algebra and Logic* **15**, no. 2 (1976), 147-151. (*Algebra i Logika* **15**, no. 2 (1976), 235-242)

[8] FILLIPOV, V.T.: 'Mal'tsev algebras', *Algebra and Logic* **16**, no. 1 (1977), 70-74. (*Algebra i Logika* **16**, no. 1 (1977), 101-108)

[9] KUKIN, G.P.: 'Algorithmic problems for solvable Lie algebras', *Algebra and Logic* **17**, no. 4 (1978), 270-278. (*Algebra i Logika* **17**, no. 4 (1978), 402-415)

[10] KUKIN, G.P.: 'Subalgebras of a free Lie sum of Lie algebras with an amalgamated subalgebra', *Algebra and Logic* **11**, no. 1 (1972), 33-50. (*Algebra i Logika* **11**, no. 1 (1972), 59-86)

[11] L'VOV, I.V.: 'Varieties of associative rings', *Algebra and Logic* **12**, no. 3 (1973), 150-167. (*Algebra i Logika* **12**, no. 3 (1973), 269-297)

[12] DOROFEEV, G.V.: 'The join of varieties of algebras', *Algebra and Logic* **15**, no. 3 (1976), 165-181. (*Algebra i Logika* **15**, no. 3 (1976), 267-291)

[13] GOLOD, E.S.: 'On nil algebras and finitely-approximable *p*-groups', *Izv. Akad. Nauk SSSR Ser. Mat.* **28** (1964), 273-276 (in Russian).

[14] KUROSH, A.G.: 'Nonassociative free sums of algebras', *Mat. Sb.* **37** (1955), 251-264 (in Russian).

[15] SHIRSHOV, A.I.: 'Subalgebras of free Lie algebras', *Mat. Sb.* **33** (1953), 441-452 (in Russian).

[16] JACOBSON, N.: *Structure and representation of Jordan algebras*, Amer. Math. Soc., 1968.

[17] ZEL'MANOV, E.I.: 'Jordan nil-algebras of bounded index', *Soviet Math. Dokl.* **20**, no. 6 (1979), 1188-1192. (*Dokl. Akad. Nauk SSSR* **249**, no. 1 (1979), 30-33)

[18] KEMER, A.R.: 'Finite basis property of identities of associative algebras', *Algebra and Logic* **26**, no. 5 (1987), 362-397. (*Algebra i Logika* **26**, no. 5 (1987), 597-641)

[19] MEDVEDEV, YU.A.: 'Free Jordan algebras', *Algebra and Logic* **27**, no. 2 (1988), 110-127. (*Algebra i Logika* **27**, no. 2 (1988), 172-200)

L.A. Bokut'

AMS 1980 Subject Classification: 17-XX

NON-ATOMIC GAME - A game with the following property: If I denotes the set of all players, there is a given σ-algebra of subsets $\mathscr{C}$ on I and a **non-atomic measure** on $\mathscr{C}$ such that sets of players $C \in \mathscr{C}$ of zero measure have no effect on the outcome of the game. Non-atomic games serve as models for situations in which there is a large quantity of very 'small' individuals, like customers in an economic system, so the development of non-atomic games is closely connected with the study of economic models with large numbers of participants (see [1]). Non-atomic games are amenable to the general classification customary in game theory (cf. **Games, theory of**), and the basic game-theoretic principles of optimality (see **Core in the theory of games; Shapley value**) carry over to them in a natural way. In relation to non-atomic games, however, the non-cooperative principles of optimality are usually implemented without the usual convexity assumptions (see [2]), and the different optimality principles turn out to be more strongly interconnected. For example, for a broad range of non-atomic models of market type, the set of competitive equilibria coincides with the core, which consists of a single element — the value of the game (the analogue of the Shapley value for non-atomic games; see [1]). There are two directions of research — the theory of cooperative non-atomic games (see [1], [3], [4]) and the theory of non-cooperative non-atomic games (see [2]).

A *cooperative non-atomic game*, in analogy with an ordinary **cooperative game**, is a triple $<J, v, H>$, where $J = (I, \mathscr{C})$ is a measurable space of players; the elements $C \in \mathscr{C}$ are called *coalitions*; v is a real-valued function on $\mathscr{C}$, called the *characteristic function*; and H is a certain subset of the set FA of all finitely-additive measures of bounded variation on $\mathscr{C}$ (it is usually assumed that $\mu(I) = v(I)$ for all $\mu \in H$). In the simplest case, v is a non-atomic measure function on $\mathscr{C}$. It is assumed that the space J is standard (i.e. (I, C) is isomorphic to the unit interval, with the Borel subsets), and that the set function v is of bounded variation (i.e. is expressible as the difference of two monotone functions). Descriptions have been given [1] of various subspaces of the space BV of all functions of bounded variation on $\mathscr{C}$ for which an analogue of the Shapley value can be constructed (as a positive linear operator with values in FA).

The concepts of a balanced game, a market game, a game without side payments (see **Cooperative game**) and the related results carry over to cooperative non-atomic games (see [3], [4]).

The definition of a *non-cooperative non-atomic game* is analogous to that of a classical **non-cooperative game**. There are also analogues of the **Nash theorem (in game theory)**, as well as general results concerning the existence of equilibrium situations without convexity assumptions, in contrast with the case of games with finitely many players (see [2]).

References

[1] AUMANN, R.J. and SHAPLEY, L.S.: *Values of non-atomic games*, Princeton Univ. Press, 1974.

[2] KIRUTA, A.YA.: 'Equilibrium points in non-atomic non-cooperative games', *Mat. Met. Sots. Nauk.* **6** (1975), 18-71 (in Russian).

[3] ROSENMÜLLER, J.: *Kooperative Spiele und Märkte*, Springer, 1971.

[4] ROSENMÜLLER, J.: 'Large games without side-payments', *Operat. Res. Verfahren* **20** (1975), 107-128.

A.Ya. Kiruta
E.B. Yanovskaya

Editorial comments.

References
  [A1] HILDENBRAND, W.: *Core and equilibria of a large economy*, Princeton Univ. Press, 1974.
  [A2] OWEN, G.: *Game theory*, Acad. Press, 1982.

AMS 1980 Subject Classification: 90D99

NON-ATOMIC MEASURE - A **measure** μ on a measurable space $(X, \mathscr{S})$ for which there are no atoms of positive measure, i.e. sets $A \in \mathscr{S}$ with $\mu(A) > 0$ for which $B \subset A$ and $\mu(B) > 0$ imply $B = A$.

N.N. Vorob'ev

Editorial comments. An *atom* in a measure space $(X, \mathscr{S}, \mu)$ is a set $A \in \mathscr{S}$ for which i) $\mu(A) > 0$; and ii) $B \in \mathscr{S}$ and $B \subset A$ imply either $\mu(B) = 0$ or $\mu(B) = \mu(A)$. See also **Atom**.

A measure space $(X, \mathscr{S}, \mu)$ is called *non-atomic* if no element of $\mathscr{S}$ is an atom. In probability theory measure spaces build up completely from atoms, i.e. using *atomic measures*, frequently occur, cf. **Atomic distribution**.

A probability F decomposes as a sum $F = pF_a + (1 - p)F_c$, $0 \leq p \leq 1$, where F_a is an atomic distribution and F_c a continuous distribution, i.e. a non-atomic one. This goes by the name *Jordan decomposition theorem*.

References
  [A1] FELLER, W.: *An introduction to probability theory and its applications*, 2, Wiley, 1971, p. 135.

AMS 1980 Subject Classification: 28A12

NON-CENTRAL 'CHI-SQUARED' DISTRIBUTION, *non-central χ^2-distribution* - A continuous probability distribution concentrated on the positive semi-axis $0 < x < \infty$ with density

$$\frac{e^{-(x+\lambda)/2}x^{(n-2)/2}}{2^{n/2}\Gamma(1/2)}\sum_{r=0}^{\infty}\frac{\lambda^r x^r}{(2r)!}\frac{\Gamma(r+1/2)}{\Gamma(r+n/2)},$$

where n is the number of degrees of freedom and λ the parameter of non-centrality. For $\lambda = 0$ this density is that of the ordinary (central) **'chi-squared' distribution**. The **characteristic function** of a non-central 'chi-squared' distribution is

$$\phi(t) = (1 - 2it)^{-n/2}\exp\left\{\frac{\lambda it}{1 - 2it}\right\};$$

the **mathematical expectation** and variance (cf. **Dispersion**) are $n + \lambda$ and $2(n + 2\lambda)$, respectively. A non-central 'chi-squared' distribution belongs to the class of infinitely-divisible distributions (cf. **Infinitely-divisible distribution**).

As a rule, a non-central 'chi-squared' distribution appears as the distribution of the sum of squares of independent random variables $X_1, \ldots, X_n$ having normal distributions with non-zero means m_i and unit variance; more precisely, the sum $X_1^2 + \cdots X_n^2$ has a non-central 'chi-squared' distribution with n degrees of freedom and non-centrality parameter $\lambda = \sum_{i=1}^{n} m_i^2$.

The sum of several mutually independent random variables with a non-central 'chi-squared' distribution has a distribution of the same type and its parameters are the sums of the corresponding parameters of the summands.

If n is even, then the distribution function of a non-central 'chi-squared' distribution $F_n(x; \lambda)$ is given by $F_n(x; \lambda) = 0$ for $x \leq 0$ and for $x > 0$ by

$$F_n(x; \lambda) = \sum_{m=0}^{\infty}\sum_{k=m+n/2}^{\infty}\frac{(\lambda/2)^m(x/2)^k}{m!k!}e^{-(\lambda+x)/2}.$$

This formula establishes a link between a non-central 'chi-squared' distribution and a **Poisson distribution**. Namely, if X and Y have Poisson distributions with parameters $x/2$ and $\lambda/2$, respectively, then for any positive integer $s > 0$,

$$\mathrm{P}\{X - Y \geq s\} = F_{2s}(x; \lambda).$$

A non-central 'chi-squared' distribution often arises in problems of mathematical statistics concerned with the study of the power of tests of 'chi-squared' type. Since tables of non-central 'chi-squared' distributions are fairly complete, various approximations by means of a 'chi-squared' and a normal distribution are widely used in statistical applications.

References
  [1] BOL'SHEV, L.N. and SMIRNOV, N.V.: *Tables of mathematical statistics*, Libr. of mathematical tables, 46, Nauka, Moscow, 1983 (in Russian). Processed by L.S. Bark and E.S. Kedrova.
  [2] KENDALL, M. and STUART, A.: *The advanced theory of statistics*, 2. Inference and relationship, Griffin, 1979.
  [3] PATNAIK, P.B.: 'The non-central χ^2- and F-distributions and their applications', *Biometrica* 36 (1949), 202-232.

A.V. Prokhorov

Editorial comments.

References
  [A1] JOHNSON, N.L. and KOTZ, S.: *Distributions in statistics*, 2. Continuous univariate distributions, Wiley, 1970.

AMS 1980 Subject Classification: 62E99

NON-CLASSICAL THEORY OF MODELS, *non-classical model theory* - A theory of models (cf. **Model theory**) that differs from the classical theory in the sense that either the relevant **formal language** is not the first-order language $L_{\omega\omega}$, or the logic on which it is based is not the classical (two-valued) logic. In what follows, unless otherwise stated, the logic is assumed to be two-valued.

In the theory of models of a language L the most important problems are the following.

a) The axiomatizability of the set of identically true formulas. If there is an effective enumeration of the formulas of the language L by natural numbers, the problem becomes more precise: Is the set of numbers of the identically true formulas recursively enumerable?

b) A language L is called (α, β)-*compact* if for any set Σ of propositions of L of cardinality $\leq \alpha$, realizability of every subset $\Sigma' \subseteq \Sigma$ of cardinality $<\beta$ implies realizability of Σ. The compactness problem consists in describing the pairs of cardinal numbers $<\alpha, \beta>$ for which L is (α, β)-compact.

c) If the formulas of L form a set (rather than a proper class), then there is a cardinal number α such that every set of propositions of L that has a model of cardinality $\beta \geqslant \alpha$ also has models of arbitrarily large cardinalities. The least such cardinal number is called the *Hanf number* of L. For $L_{\omega\omega}$ this is the countable cardinal ω. The problem consists in computing the Hanf number of L and in establishing conditions for the existence of models of small cardinality.

Below the best studied non-classical languages are listed and for each of them solutions of the problems a) - c) are described.

1) The *language L_2 of second-order logic*. It is obtained from $L_{\omega\omega}$ by adding variables for predicates as well as quantifiers over such predicate-variables. A proposition Φ of the language L_2 is said to be *true* in a system $<A, F_1, \ldots, F_n, \ldots>$ (where A is a model of signature Φ and the F_n, $n \geqslant 1$, are sets of n-ary predicates on A) if Φ is true in A for the restriction of the quantifiers to the n-ary predicates in the sets F_n. If here F_n, $n \geqslant 1$, are the sets of all n-ary predicates on A, then Φ is said to be true in the model A. There is a proposition of the language L_2 that characterizes the arithmetic of natural numbers up to an isomorphism. From the **Gödel incompleteness theorem** for arithmetic it follows that the set of propositions of L_2 that are true in all models is not axiomatizable. However, there is a natural generalization Σ_2 of the axioms of first-order predicate calculus for which *Henkin's completeness theorem* holds: From Σ_2 those and only those propositions of L_2 are deducible that are true in all systems $<A, F_1, \ldots, F_n, \ldots>$ satisfying the axioms of Σ_2. In this case there is an *analogue of the Löwenheim − Skolem theorem* for $L_{\omega\omega}$: If a proposition Φ of L_2 is true together with the axioms of Σ_2 in the same system, then Φ and Σ_2 are true in a system $<A, F_1, \ldots, F_n, \ldots>$, where A and the F_n, $n \geqslant 1$, are at most countable. Some questions in the theory of models of the language L_2 are connected with problems of set theory and are unsolvable in the Zermelo − Fraenkel axiomatic set theory.

2) The *language $L_{\alpha\beta}$* (where α and β are cardinal numbers). The formulas of this language are constructed from the formulas of the first-order language by means of conjunctions and disjunctions of sets of formulas of cardinality $<\alpha$ and by negation and quantification over strings of variables of length $<\beta$. The truth of a formula in a model is defined, as in the first-order language, by induction on the structure of the formula. A cardinal number α is said to be *compact*

if for any cardinal number γ the language $L_{\alpha\alpha}$ is (γ, α)-compact. Among the languages $L_{\alpha\beta}$, the best studied after $L_{\omega\omega}$ is $L_{\omega_1\omega}$. Every countable model of countable signature can be characterized by a proposition of the language $L_{\omega_1\omega}$ up to an isomorphism. The language $L_{\omega_1\omega}$ is (α, ω_1)-compact for any α. The Hanf number of $L_{\omega_1\omega}$ is $2^{\omega, \omega_1}$, where $2^{\alpha, \nu}$ is defined by induction over the ordinal numbers ν: $2^{\alpha, 0} = \alpha$, $2^{\alpha, \delta+1} = 2^{\alpha, \delta}$ and $2^{\alpha, \delta} = \sum_{\mu < \delta} 2^{\alpha, \mu}$ when δ is a limit ordinal.

3) The *language $\mathscr{L}_\alpha$ with the quantifier 'there exist at least ω_α many'*. The language $\mathscr{L}_\alpha$ is obtained from $L_{\omega\omega}$ by adding a new quantifier Q^α. The truth of a formula is determined by induction on its length. Here a formula $(Q^\alpha x)\Phi(x)$ is true in a model A if the cardinality of the set $\{a: a \in A, \Phi(a) \text{ is true in } A\}$ is at least ω_α. Let V_α denote the set of formulas of $\mathscr{L}_\alpha$ that are true in all models of cardinality $\geqslant \omega_\alpha$. The set V_0 is not axiomatizable, but V_1 is. The language $\mathscr{L}_0$ is not (ω, ω)-compact. However, a certain compactness holds in the languages $\mathscr{L}_\alpha$. Let the symbol $\omega_\beta < < \omega_\alpha$ indicate that $\lambda < \omega_\beta$ and $\gamma_i < \omega_\beta$ $(i \in \lambda)$ imply that $\prod_{i \in \lambda} \gamma_i < \omega_\alpha$. If $\omega_\beta < < \omega_\alpha$, then $\mathscr{L}_\alpha$ is (ω_β, ω)-compact. The Hanf number of $\mathscr{L}$ is $2^{\omega, \omega}$.

In the models considered so far any proposition of a language L of signature σ was either true or false. Alternatively, one can consider models A of signature σ in which n-ary predicates are regarded not as subsets of A^n but as mappings from A^n into a set X. If on X operations corresponding to the logical connectives of the language L and quantifiers (understood as infinitary operations) are defined, then one can define the truth value $\| \Phi \|^A \in X$ of any proposition Φ of the language L in the model A. Thus one obtains a model theory with X as set of truth values. The theory is most fruitful in case X is a compact Hausdorff space or a complete Boolean algebra. In these cases many methods of the classical theory of models work. When X is a complete **Boolean algebra**, then conjunction, disjunction and negation are defined as intersection, union and complementation, respectively. The value $\| (\forall x)\Phi(x) \|^A$ is defined as the intersection of all elements of the form $\| \Phi(a) \|^A$, $a \in A$. Boolean-valued models have found wide-spread application in proofs of the compatibility of various propositions of set theory with the basic axioms of axiomatic set theory.

References

[1] CHURCH, A.: *Introduction to mathematical logic*, 1, Princeton Univ. Press, 1956.
[2] CHANG, C.C. and KEISLER, H.J.: *Model theory*, North-Holland, 1973.

E.A. Palyutin
A.D. Taĭmanov

Editorial comments. Additional references on infinitary logic are [A1], [A2].

References
[A1] KEISLER, H.J.: *Model theory for infinitary logic*, North-Holland, 1971.
[A2] DICKMANN, M.A.: *Large infinitary languages*, North-Holland, 1975.

AMS 1980 Subject Classification: 03C75, 03C80, 03C85

NON-COOPERATIVE GAME - A system

$$\Gamma = \; <J, \{S_i\}_{i \in J}, \{H_i\}_{i \in J} >,$$

where J is the set of players, S_i is the set of strategies (cf. **Strategy (in game theory)**) of the i-th player and H_i is the gain function of the i-th player, defined on the Cartesian product $S = \prod_{i \in J} S_i$. A non-cooperative game is played as follows: players, who are acting individually (do not form a coalition, do not cooperate), select their strategies $s_i \in S_i$, as a result of which the situation $s = \prod_{i \in J} s_i$ appears, in which the i-th player obtains the gain $H_i(s)$. The main optimality principle in a non-cooperative game is the principle of realizability of the objective [1], which generates the Nash equilibrium solutions. A solution s^* is called an *equilibrium solution* if for all $i \in J$, $s_i \in S_i$, the inequality

$$H_i(s^*) \geqslant H_i(s^* \parallel s_i),$$

where $s^* \parallel s_i = \prod_{j \in J \setminus i} s_j^* \times s_i$, is valid. Thus, none of the players is interested in unilaterally disturbing the equilibrium solution previously agreed upon between them. It has been proved (*Nash's theorem*) that a finite non-cooperative game (the sets J and S_i are finite) possesses an equilibrium solution for mixed strategies. This theorem has been generalized to include infinite non-cooperative games with a finite number of players [3] and non-cooperative games with an infinite number of players (cf. **Non-atomic game**).

Two equilibrium solutions s and t are called interchangeable if any solution $r = \prod_{i \in J} r_i$, where $r_i = s_i$ or $r_i = t_i$, $i \in J$, is also an equilibrium solution. They are called equivalent if $H_i(s) = H_i(t)$ for all $i \in J$. Let Q be the set of all equilibrium solutions, and let $Q' \subset Q$ be the set of equilibrium solutions which are Pareto optimal (cf. **Arbitration scheme**). A game is called *Nash solvable* and Q is said to be a *Nash solution* if all $s \in Q$ are equivalent and interchangeable. A game is called *strictly solvable* if Q' is non-empty and all $s \in Q'$ are equivalent and interchangeable. Two-person zero-sum games (cf. **Two-person zero-sum game**) with optimal strategies are Nash solvable and strictly solvable; however, in the general case such a solvability is often impossible.

Other attempts at completing the principle of realizability of the objective were made. Thus, it was suggested [4] that the unique equilibrium solution or the maximum solution (in this last situation each player may ensure his/her own gain irrespective of the strategies chosen by the other players), the choice of which is based on the introduction of a new preference relation on the set of solutions, be considered as the solution of the non-cooperative game. In another approach the solution of a non-cooperative game is defined by a subjective prognosis of the behaviour of the players [5].

References
[1] VOROB'EV, N.N.: 'The present state of the theory of games', *Russian Math. Surveys* **25**, no. 2 (1970), 77-136. (*Uspekhi Mat. Nauk* **25**, no. 2 (1970), 81-140)
[2] NASH, J.: 'Noncooperative games', *Ann. of Math.* **54** (1951), 286-295.
[3] GLICKSBERG, I.L.: 'A further generalization of the Kakutani fixed point theorem, with application to Nash equilibrium points', *Proc. Amer. Math. Soc.* **3** (1952), 170-174.
[4] HARSANYI, J.C.: 'A general solution for finite noncooperative games based on risk-dominance', in L.S. Shapley, A.W. Tucker and M. Dresher (eds.): *Advances in game theory*, Princeton Univ. Press, 1964, pp. 651-679.
[5] VILKAS, E.I.: 'The axiomatic definition of equilibrium points and the value of a non-coalition n −person game', *Theory Probab. Appl.* **13**, no. 3 (1968), 523-527. (*Teor. Veroyatnost. i Primenen.* **13**, no. 3 (1968), 555-560)

E.I. Vilkas
E.B. Yanovskaya

Editorial comments.

References
[A1] VOROB'EV, N.N.: *Game theory*, Springer, 1977 (translated from the Russian).

AMS 1980 Subject Classification: 90D10

NON-DEGENERATE REPRESENTATION - A linear

representation π of a group (ring, algebra, semi-group) X in a vector space E such that if $\pi(x)\xi = 0$ for some $\xi \in E$ and all $x \in X$, then $\xi = 0$.

A.I. Shtern

AMS 1980 Subject Classification: 22A25

NON-DESARGUESIAN GEOMETRY - A plane

geometry in which the **Desargues assumption** need not be true. The plane is then called a *non-Desarguesian plane*. Desargues's theorem cannot be proved in the plane using only the projective axioms of the plane, without recourse to the axioms of congruence (the metric axioms) or to space axioms. For example, it cannot be deduced from an axiom system consisting of all Hilbert's axioms for the plane with the exception of the axiom of congruence for triangles (cf. **Hilbert system of axioms**). The plane geometry based on such a system is non-Desarguesian; it cannot be considered as part of a three-dimensional geometry in which all axioms of Hilbert's system except the above-mentioned congruence axiom are valid. A non-Desarguesian projective 2-plane cannot be imbedded in a projective space of a higher dimension (see [1], [4], [5]).

The fact that a non-Desarguesian plane geometry can actually be constructed yields independence proofs for various groups of axioms in Hilbert's system, and also highlights the role of Desargues's theorem as an independent additional axiom of plane projective geometry (see [2]).

Attention has also been given to what are known as *non-Desarguesian systems*, in which Desargues's theorem is not valid as a configurational proposition (see **Configuration**). Non-Desarguesian systems exist, in particular, on certain surfaces and in general on certain Riemannian manifolds that are straight spaces. A simple example is the paraboloid $z = xy$, on which the points and the shortest joins constitute a non-Desarguesian system. Another example is provided by the torus: There exist metrizations of the torus without conjugate points in which the geodesics of the universal covering space constitute a non-Desarguesian system (see also [5], [6]).

References

[1] HILBERT, D.: *The foundations of geometry*, Open Court, La Salle, Illinois, 1950 (translated from the German).
[2] SKORNYAKOV, L.A.: 'Projective planes', *Uspekhi Mat. Nauk* **6**, no. 6 (1951), 112-154 (in Russian).
[3] BUSEMANN, H.: *The geometry of geodesics*, Acad. Press, 1955.
[4] MOHRMANN, H. (ED.): *Festschrift D. Hilbert: zu seinem 60.ten Geburtstag*, Springer, reprint, 1982.
[5] BIEBERBACH, L.: *Einleitung in die höhere Geometrie*, Teubner, 1933.
[6] NARAYANA RAO, M.L. and KUPPUSWAMY RAO, K.: 'A class of non-desarguesian planes', *J. Comb. Theory Ser. A* **19** (1975), 247-255.

L.A. Sidorov

Editorial comments. Some finite projective planes are non-Desarguesian, see [A2].

References

[A1] COXETER, H.S.M.: *Twelve geometric essays*, Southern III. Univ. Press, 1968, p. 248.
[A2] HALL, M.: *The theory of groups*, MacMillan, 1959, pp. 394-397.
[A3] PICKERT, G.: *Projective Ebenen*, Springer, 1975.
[A4] HUGHES, D. and PIPER, F.C.: *Projective planes*, Springer, 1973.

AMS 1980 Subject Classification: 51A35

NON-DIFFERENTIABLE FUNCTION - A function that does not have a **differential**. In the case of functions of one variable it is a function that does not have a finite derivative. For example, the function $f(x) = |x|$ is not differentiable at $x = 0$, though it is differentiable at that point from the left and from the right (i.e. it has finite left and right derivatives at that point). The continuous function $f(x) = x \sin(1/x)$ if $x \neq 0$ and $f(0) = 0$ is not only non-differentiable at $x = 0$, it has neither left nor right (and neither finite nor infinite) derivatives at that point.

The first examples of functions continuous on the entire real line but having no finite derivative at any point were constructed by B. Bolzano in 1830 (published in 1930) and by K. Weierstrass in 1860 (published in 1872). Weierstrass' function is the sum of the series

$$f(x) = \sum_{n=0}^{\infty} a^n \cos(b^n \pi x),$$

where $0 < a < 1$, b is an odd natural number and $ab > 1 + 3\pi/2$. A simpler example, based on the same idea, in which $\cos \omega x$ is replaced by a simpler periodic function — a polygonal line — was constructed by B.L. van der Waerden. Let $u_0(x)$ be the function defined for real x as the absolute value of the difference between x and the nearest integer. This function is linear on every interval $[n/2, (n+1)/2]$, where n is an integer; it is continuous and periodic with period 1. Let

$$u_k(x) = \frac{u_0(4^k x)}{4^k}, \quad k = 1, 2, \ldots,$$

then van der Waerden's function is defined by

$$f(x) = \sum_{k=0}^{\infty} u_k(x).$$

This function is continuous on the entire real line but does not have a finite derivative at any point. The first three partial sums of the series are shown in the figure.

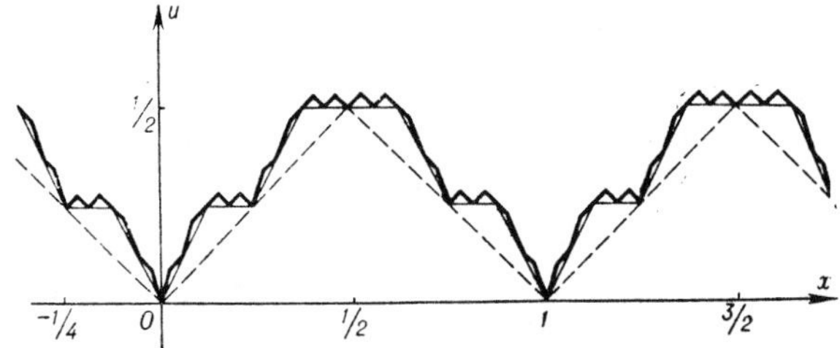

For functions of more than one variable, differentiability at a point is not equivalent to the existence of the partial derivatives at the point; there are examples of non-differentiable functions that have partial derivatives. For example, the function

$$f(x, y) = \begin{cases} \dfrac{x^2 y}{x^2 + y^2} & \text{if } x^2 + y^2 > 0, \\ 0 & \text{if } x = y = 0, \end{cases}$$

is continuous at all points of the plane and has partial derivatives everywhere but it is not differentiable at $(0, 0)$.

L.D. Kudryavtsev

Editorial comments. S. Banach proved that 'most' continuous functions are nowhere differentiable. Specifically, he showed that if C denotes the space of all continuous real-valued functions on the unit interval $[0, 1]$, equipped with the uniform metric (sup norm), then the set of members of C that have a finite right-hand derivative at some point of $[0, 1)$ is of the first Baire category (cf. **Baire classes**) in the complete metric space C. Proof of this fact and of the nowhere differentiability of Weierstrass' example cited above can be found in [A1]. A proof that van der Waerden's example has the stated properties can be found in [A2].

References

[A1] HEWITT, E. and STROMBERG, K.: *Real and abstract analysis*, Springer, 1965.
[A2] STROMBERG, K.: *Introduction to classical real analysis*, Wadsworth, 1981.

AMS 1980 Subject Classification: 26A27

NON-EUCLIDEAN GEOMETRIES - In the literal sense — all geometric systems distinct from Euclidean geometry; usually, however, the term 'non-Euclidean geometries' is reserved for geometric systems (distinct from Euclidean geometry) in which the motion of figures is defined, and this with the same degree of freedom as in Euclidean geometry. The degree of freedom of motion of figures in the Euclidean plane is characterized by the condition that every figure can be moved, without changing the distances between its points, in such a way that any selected point of the figure can be made to occupy a previously-designated position; moreover, every figure can be rotated about any of its points. In the Euclidean three-dimensional space every figure can be moved in such a way that any selected point of the figure will occupy any prescribed position; in addition, every figure can be rotated about any axis through any of its points.

The major non-Euclidean geometries are *hyperbolic geometry* or **Lobachevskiĭ geometry** and *elliptic geometry* or *Riemann geometry* — it is usually these that are meant by 'non-Euclidean geometries'. Hyperbolic geometry was the first geometric system distinct from Euclidean geometry, and the first more general theory (it includes Euclidean geometry as a limiting case). The later-discovered elliptic geometry is in some respects the opposite of hyperbolic geometry. The simultaneous investigation of the three geometries has made it possible to reveal the special features of each to a considerable degree and also to determine their relationships with other geometric systems. Below, both non-Euclidean geometries and Euclidean geometry will be compared, first as synthetic theories, then in the context of differential geometry and, lastly, in the context of group theory.

Non-Euclidean geometries as synthetic theories. Hyperbolic geometry is based on the same axioms as Euclidean geometry, with the exception of the parallelism axiom. Recall that according to the latter, through any point not lying on a given straight line a there passes exactly one straight line coplanar with a which does not intersect it; in hyperbolic geometry it is assumed that there are several such lines (and it can then be proved that there are infinitely many).

The axiom adopted in elliptic geometry is: Every straight line coplanar with another straight line intersects the latter. This axiom contradicts the system of axioms of Euclidean geometry with the parallelism axiom excluded. Thus, the axiom system underlying elliptic geometry must be different from that of Euclidean geometry not only in the replacement of one axiom — the parallelism axiom — but also in part of the other axioms. The relevant axioms of these geometries in which the differences occur are those that determine the order relations among the geometric elements. Namely, in Euclidean and hyperbolic geometries, the order of the points on a straight line is linear, i.e. similar to the order in the set of real numbers; in elliptic geometry, the points of a straight line are cyclically ordered, like the points on a circle. In addition, in Euclidean and hyperbolic geometries every straight line in a given plane divides the plane into two parts; in elliptic geometry this is not true, i.e. any two points of the plane not on a given straight line can be joined within the plane by a continuous curve that does not cut the straight line (the topological model of the elliptic plane is the projective plane).

The axioms that determine the motion of figures are the same in all three geometries.

Examples of theorems in non-Euclidean geometries.

1) In hyperbolic geometry, the sum of the interior angles of any triangle is less than two right angles; in elliptic geometry it is larger than two right angles (in Euclidean geometry it is of course equal to two right angles).

2) In hyperbolic geometry, the area of a triangle is given by the formula

$$S = R^2(\pi - \alpha - \beta - \gamma), \tag{1}$$

where α, β, γ are the interior angles of the triangle and R is a constant that depends on the specific unit chosen for the measurement of areas. In elliptic geometry, as well as in **spherical geometry**,

$$S = R^2(\alpha + \beta + \gamma - \pi) \tag{2}$$

with the same notation (in Euclidean geometry there is no fixed relationship between the area of a triangle and the sum of its angles).

3) In hyperbolic geometry there are various relations among the sides of a triangle and its angles, such as

$$\cosh \frac{a}{R} = \cosh \frac{b}{R} \cosh \frac{c}{R} - \sinh \frac{b}{R} \sinh \frac{c}{R} \cos \alpha, \tag{3}$$

where sinh, cosh denote the hyperbolic sine and cosine, a, b, c the sides of the triangle, α, β, γ the angles opposite them, respectively, and R a constant depending on the choice of scale; for a right-angled triangle (with hypotenuse c and right angle γ) one has, for example,

$$\cosh \frac{c}{R} = \cotan \alpha \cotan \beta. \tag{4}$$

Subject to a certain adjustment between the linear scale and the unit of area, the constant R in the formulas (1), (3) and (4) is the same. The number R is known as the *radius of curvature of the hyperbolic plane* (or *space*). For a given linear scale, R expresses the length of a certain segment in the hyperbolic plane (space), which is also called the radius of curvature. If the radius of curvature itself is chosen as the unit of length, then $R=1$. In elliptic geometry, as in **spherical trigonometry**, there are similar formulas:

$$\cos\frac{a}{R} = \cos\frac{b}{R}\cos\frac{c}{R} + \sin\frac{b}{R}\sin\frac{c}{R}\cos\alpha \qquad (5)$$

(for an arbitrary triangle) and

$$\cos\frac{c}{R} = \cotan\alpha\cotan\beta \qquad (6)$$

(for a right-angled triangle) with analogous notation. The number R is called the *radius of curvature of the elliptic plane* (or *space*). As is evident from formulas (4) and (6), in each of the non-Euclidean geometries the hypotenuse of a right-angled triangle is determined by its angles; moreover, the sides of any triangle are determined by its angles, implying that there exist no similar triangles that are not congruent (in Euclidean geometry there are no analogues of formulas (4) and (6), and there exist no other formulas expressing linear quantities in terms of angular ones). Replacement of R by iR converts the formulas (1), (3), (4) into (2), (5), (6), respectively; in general, replacement of R by iR transforms all metric formulas of hyperbolic geometry into the corresponding formulas of elliptic geometry, while preserving their geometric meanings. As $R\to\infty$, both systems yield formulas of Euclidean geometry (or become meaningless). Now this indefinite increase in R means that the scaling segment becomes infinitely small in comparison with the radius of curvature (as a segment). The fact that the formulas of non-Euclidean geometries tend in the limit to formulas of Euclidean geometry implies that, in non-Euclidean figures that are small in comparison with the radius of curvature, the relations between their elements differ only slightly from the Euclidean relations.

Non-Euclidean geometries in a differential-geometric context. In each of the non-Euclidean geometries, the differential properties of the plane are analogous to those of surfaces in Euclidean space. To be specific: In a non-Euclidean plane one can introduce intrinsic coordinates u, v so that the differential ds of arc length of the curve corresponding to the differentials du, dv of the coordinates is defined by

$$ds^2 = E\,du^2 + 2F\,du\,dv + G\,dv^2. \qquad (7)$$

Suppose, in particular, that the coordinate u of an arbitrary point M is defined as the length of the perpendicular dropped from M to a fixed straight line, and the coordinate v as the distance from a fixed point 0 on the line to the base of the perpendicular; the signs of u and v are chosen as in the case of ordinary Cartesian coordinates. Then formula (7) becomes, for the hyperbolic plane,

$$ds^2 = du^2 + \cosh^2\left[\frac{u}{R}\right]dv^2, \qquad (8)$$

and for the elliptic plane,

$$ds^2 = du^2 + \cos^2\left[\frac{u}{R}\right]dv^2, \qquad (9)$$

R being the same constant as in the preceding section (the radius of curvature). The terms on the right of (8) and (9) are the metric forms of surfaces of constant curvature in Euclidean space — of negative curvature $K=-1/R^2$ in the first case (e.g. a pseudo-sphere) and of positive curvature $K=1/R^2$ in the second (e.g. a sphere). For this reason, the intrinsic geometry on a sufficiently small part of the hyperbolic plane coincides with the intrinsic geometry on the corresponding part of a surface of constant negative curvature. Similarly, the intrinsic geometry of sufficiently small parts of the elliptic plane is realized on surfaces of constant positive curvature (there are no surfaces in Euclidean space that realize the geometry of the whole hyperbolic plane). Replacement of R by iR converts the metric form (8) into the metric form (9).

Since the metric form determines the intrinsic geometry of the surface, the same transformation also converts other metric relationships of hyperbolic geometry into metric relationships of elliptic geometry (as already stated previously). If $R=\infty$, each of (8) and (9) yields

$$ds^2 = du^2 + dv^2,$$

which is simply the metric form of the Euclidean plane.

In regard to their differential properties, three-dimensional non-Euclidean spaces belong to the class of Riemannian spaces, being defined in that class primarily by the fact that they have constant Riemannian curvature. In both two and three dimensions, constant curvature guarantees that the space is homogeneous, i.e. that motion of figures is locally possible (and with the same degree of freedom as in the Euclidean plane or space, respectively). A hyperbolic space is of negative curvature $-1/R^2$, and an elliptic space is of positive curvature $1/R^2$ (R is the radius of curvature). Euclidean space occupies an intermediate position: it is a space of curvature zero.

Spaces of constant Riemannian curvature can exhibit a great variety of topological structures. Among all spaces of constant negative curvature, a hyperbolic space is uniquely determined by two properties: it is complete (in the sense of completeness of a metric space) and topologically equivalent to ordinary

Euclidean space. An elliptic space is uniquely determined in the class of all spaces of constant positive curvature by one property: topological equivalence to a projective space. Analogous conditions define hyperbolic and elliptic spaces of higher dimension in the class of higher-dimensional spaces of constant Riemannian curvature.

Non-Euclidean geometries in a group-theoretic context. Consider the projective plane, with projective homogeneous coordinates (x_1, x_2, x_3), and suppose that some second-order oval curve, henceforward denoted by k, is given, say

$$x_1^2 + x_2^2 - x_3^2 = 0.$$

Every projective mapping of the projective plane onto itself that leaves the curve k invariant is called an automorphism relative to k. Every automorphism maps interior points of k to interior points. The set of all automorphisms relative to the curve k is a group. One considers only points of the projective plane lying within k; chords of k will be called 'straight lines'. Two figures will be considered equal if one of them can be transformed into the other by some automorphism. Since the automorphisms form a group, the fundamental properties of equality between figures are valid: 1) if a figure A equals a figure B, then B equals A; and 2) if A equals B and B equals C, then A equals C. In the geometric theory obtained in this way, all the axioms of Euclidean geometry are valid, except the parallelism axiom; instead, this geometry satisfies the hyperbolic axiom of parallelism (see the Figure, which demonstrates that through a point P infinitely many 'straight lines' can be drawn that do not intersect the 'straight line' a).

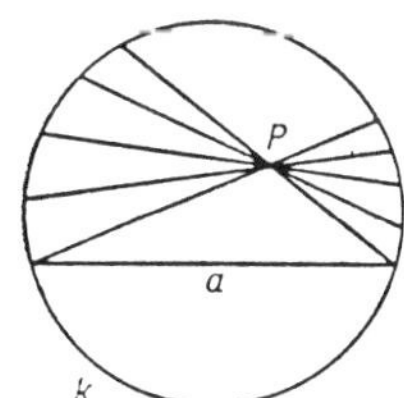

The construction just outlined yields an interpretation of (two-dimensional) hyperbolic geometry in terms of a projective plane or, as is usually said, a projective model of hyperbolic geometry; the curve k is known as the *absolute* of the model. Automorphisms relative to k play the role of motions. For this reason, hyperbolic geometry can be regarded as a theory that studies the properties of figures and the related magnitudes that remain invariant under certain automorphisms; in brief, hyperbolic geometry is essentially the theory of invariants of an automorphism group relative to an absolute oval.

Elliptic geometry (in two dimensions) admits a similar interpretation; it is the theory of invariants relative to the imaginary absolute

$$x_1^2 + x_2^2 + x_3^2 = 0. \tag{10}$$

The points and straight lines of the model are all the points and straight lines of the projective plane, the automorphisms are defined in exclusively algebraic terms as linear transformations that transform equation (10) into an equation of the same form.

Euclidean geometry may also be considered as a theory of invariants of a certain group of projective transformations: the group of automorphisms relative to the degenerate absolute

$$x_1^2 + x_2^2 = 0, \quad x_3 = 0,$$

i.e. relative to the imaginary points $(1, i, 0)$ and $(1, -i, 0)$; these are known as the *circular points at infinity*. The universe of the model comprises all points of the projective plane with the exception of the points of the straight line $x_3 = 0$, and all straight lines of the projective plane except $x_3 = 0$. In this case the automorphisms play the role of similarity transformations, but not merely motions as in non-Euclidean geometries.

The above models relate to two-dimensional geometries; the construction of projective models in a higher dimension is analogous.

References

[1] ALEKSANDROV, P.S.: *What is non-Euclidean geometry?*, Moscow, 1950 (in Russian).
[2] KLEIN, F.: *Nicht-Euklidische Geometrie: Vorlesungen gehalten während des Wintersemesters 1889-1890*, Göttingen, 1893.
[3] EFIMOV, N.V.: *Höhere Geometrie*, Deutsch. Verlag Wissenschaft., 1960 (translated from the Russian).

N.V. Efimov

Editorial comments. For models of non-Euclidean geometries see also **Beltrami interpretation**; **Klein interpretation**; **Poincaré model**.

Instead of hyperbolic geometry one also finds the appellation *Bolyai−Lobachevskiĭ geometry*, after the two codiscoverers.

References

[A1] BONOLA, R.: *Die nichteuklidische Geometrie*, Teubner, 1908 (translated from the Italian).
[A2] CARSLAW, H.S.: *The elements of non-Euclidean plane geometry and trigonometry*, Longmans, 1916.
[A3] SOMMERVILLE, D.M.Y.: *The elements of non-Euclidean geometry*, Dover, reprint, 1958.
[A4] LIEBMANN, H.: *Nichteuklidische Geometrie*, Gösschen, 1912.
[A5] COXETER, H.S.M.: *Non-Euclidean geometry*, Univ. Toronto Press, 1965.
[A6] ROSENFELD, B.A.: *A history of non-Euclidean geometry*, Springer, 1988 (translated from the Russian).
[A7] BERGER, M.: *Geometry*, 1-2, Springer, 1987 (translated from the French).
[A8] BUSEMANN, H.: *Recent synthetic geometry*, Springer, 1970.
[A9] GREENBERG, M.: *Euclidean and non-Euclidean geometry*, Freeman, 1980.
[A10] KELLY, P. and MATTHEWS, G.: *The non-Euclidean, hyperbolic plane*, Springer, 1980.

AMS 1980 Subject Classification: 51M10, 53-XX

NON-EUCLIDEAN SPACE - A space whose properties are based on a system of axioms other than the Euclidean system. The geometries of non-Euclidean spaces are the **non-Euclidean geometries**. Depending on the specific axioms from which the non-Euclidean geometries are developed in non-Euclidean spaces, the latter may be classified in accordance with various criteria. On the one hand, a non-Euclidean space may be a finite-dimensional vector space with a scalar product expressible in Cartesian coordinates as

$$(\mathbf{a}, \mathbf{b}) = \sum_{i=1}^{k} x_i y_i - \sum_{i=k+1}^{n} x_i y_i.$$

In this case one speaks of a **pseudo-Euclidean space**. On the other hand, a non-Euclidean space can be characterized as an n-dimensional manifold with a certain structure described by a non-Euclidean axiom system.

Non-Euclidean spaces may also be classified from the point of view of their differential-geometric properties as Riemannian spaces of constant curvature (this includes the case of spaces of curvature zero, which are nevertheless topologically distinct from Euclidean spaces).

L.A. Sidorov

Editorial comments.

References

[A1] GREENBERG, M.: *Euclidean and non-euclidean geometries*, Freeman, 1974.
[A2] ROSENFELD, B.: *A history of non-euclidean geometry*, Springer, 1988 (translated from the Russian).

AMS 1980 Subject Classification: 51M10

NON-FREDHOLM INTEGRAL EQUATION - An **integral equation** for which some of the **Fredholm theorems** are false. Sometimes such an integral equation is said to be a *singular integral equation*.

For example, the *Fourier integral equation*

$$\phi(x) - \sqrt{\frac{2}{\pi}} \int_0^\infty \sin(xs)\phi(s)\,ds = 0 \qquad (1)$$

has the solution

$$\phi_a(x) = \sqrt{\frac{\pi}{2}} e^{-ax} + \frac{x}{a^2 + x^2},$$

where a is an arbitrary positive constant; an infinite set of linearly independent solutions corresponds to the eigen value $\sqrt{2/\pi}$ of (1), that is, for equation (1) Fredholm's theorem that a homogeneous equation has finitely many linearly independent solutions, does not hold.

In the case of the *Lalesco−Picard integral equation*

$$\phi(x) - \lambda \int_{-\infty}^\infty e^{-|x-s|} \phi(s)\,ds = 0, \qquad (2)$$

every $\lambda \in (0, +\infty)$ is an eigen value, namely, to every positive number λ correspond the two independent solutions

$$\phi_\lambda^{(1)}(x) = e^{\sqrt{1-2\lambda}\,x}, \quad \phi_\lambda^{(2)}(x) = e^{-\sqrt{1-2\lambda}\,x}.$$

Consequently, for equation (2) Fredholm's theorem, that the set of eigen values of an equation is at most countable, does not hold.

The theory has been worked out in detail for two classes of non-Fredholm integral equations: equations in which the unknown function stands under the sign of an improper integral in the sense of the principal value (singular integral equations, cf. **Singular integral equation**); and equations in which the unknown function stands under the sign of an integral convolution transformation (cf. **Integral equation of convolution type**). For such equations, generally speaking, the Fredholm alternative is violated, as well as the fact that the numbers of linearly independent solutions of the homogeneous equation and its adjoint are equal.

References

[1] PRIVALOV, I.I.: *Integral equations*, Moscow-Leningrad, 1937 (in Russian).
[2] PETROVSKIĬ, I.G.: *Lectures on the theory of integral equations*, Graylock, 1957 (translated from the Russian).

See also the references to **Integral equation of convolution type** and **Singular integral equation**.

B.V. Khvedelidze

Editorial comments.

References

[A1] ZABREYKO, P.P. [P.P. ZABREĬKO], ET AL.: *Integral equations — a reference text*, Noordhoff, 1975 (translated from the Russian).

AMS 1980 Subject Classification: 45EXX

NON-HOLONOMIC SYSTEMS - Systems of material points that are subject to constraints among which are kinematic constraints that impose conditions on the velocities (and not only the positions) of the points of the system in its possible positions (see **Holonomic system**); these conditions are assumed to be expressible as non-integrable differential relations

$$\phi_s(x_1, \ldots, x_{3N}, \dot{x}_1, \ldots, \dot{x}_{3N}, t) = 0, \qquad (1)$$

$$s = 1, \ldots, m, \quad \phi_s(x, \dot{x}, t) \in C^1,$$

that cannot be replaced by equivalent finite relationships among the coordinates. Here, the x_ν denote the Cartesian coordinates of the points, t is the time and N is the number of points in the system. Most often one considers constraints (1) that are linear in the velocities $\dot{x}_i$, of the form

$$\sum_{i=1}^{3N} A_{s_i}\,dx_i + A_s\,dt = 0; \quad A_{s_i}(x, t), A_s(x, t) \in C^1.$$

The constraints (1) are said to be stationary if $\partial \phi / \partial t \equiv 0$. These constraints also impose conditions on the accelerations w_ν of the points:

$$\frac{\partial \phi_s}{\partial t} = \sum_{\nu=1}^{N} \operatorname{grad}_{\dot{r}_\nu} \phi_s \cdot w_\nu + \cdots = 0.$$

Following N.G. Chetaev [2], assume that the possible motions of the systems subject to the non-linear constraints (1) satisfy conditions of the type

$$\sum_{\nu=1}^{3N} \frac{\partial \phi_s}{\partial \dot{x}_\nu} \delta x_\nu = 0, \quad s = 1, \ldots, m. \qquad (2)$$

In the case of linear constraints, these conditions imply the usual relations

$$\sum_{i=1}^{3N} A_{s_i} \delta x_i = 0.$$

Unlike the situation in holonomic systems, motion between neighbouring positions at an infinitesimally-small distance from one another may be impossible in a non-holonomic system (see [1]).

In generalized Lagrange coordinates, equations (1) and (2) are written as

$$\Phi_s(q_1, \ldots, q_n, \dot{q}_1, \ldots, \dot{q}_n, t) = 0, \quad \sum_{i=1}^{n} \frac{\partial \Phi_s}{\partial \dot{q}_i} \delta q_i = 0,$$

$$s = 1, \ldots, m.$$

In a non-holonomic system, the number $n - m$ of degrees of freedom is less than the number n of independent coordinates q_i by the number m of non-integrable constraint equations.

Many and varied forms of differential equations of motion have been derived for non-holonomic systems, such as the Lagrange equation of the first kind (cf. **Lagrange equations (in mechanics)**), the **Appell equations** in Lagrange coordinates and quasi-coordinates, the Chaplygin−Voronets equations in Lagrange coordinates, the **Boltzmann equation**, the Hamel equation in quasi-coordinates, etc. (see [3]).

A characteristic feature of non-holonomic systems is that, in the general case, their differential equations of motion include the constraint equations.

References

[1] HERTZ, H.: *The principles of mechanics presented in a new form*, Dover, 1956 (translated from the German).
[2] CHETAEV, N.G.: *Izv. Fiz.-Mat. Obshch. Kazan. Univ. (3)* **6** (1932), 68-71.
[3] NEĬMARK, YU.I. and FUFAEV, N.A.: *Dynamics of nonholonomic systems*, Amer. Math. Soc., 1972 (translated from the Russian).

V.V. Rumyantsev

Editorial comments.

References

[A1] LANDAU, L.D. and LIFSHITZ, E.M.: *Course of theoretical physics*, 1. Mechanics, Pergamon, 1976.
[A2] SUDARSHAN, E.G.G. and MUKUNDA, N.: *Classical dynamics*, Wiley, 1974.

AMS 1980 Subject Classification: 70F25

NON-HOPF GROUP, *non-Hopfian group* - A group that has an endomorphism onto itself with a non-trivial kernel, that is, a group that is isomorphic to a proper quotient group of itself. (Otherwise the group is called a Hopfian group, cf. **Hopf group**.) The term stems from *Hopf's problem* (1932) whether there are such groups that are finitely generated. It turned out that there are even finitely-presented non-Hopfian groups. An example of a finitely-generated non-Hopfian group is the group with two generators x and y and the single defining relation

$$x^{-1}y^2 x = y^3.$$

Infinitely-generated non-Hopfian groups are quite easy to construct, for example, the direct product of infinitely many isomorphic groups.

References

[1] KUROSH, A.G.: *The theory of groups*, 1-2, Chelsea, 1955-1956 (translated from the Russian).
[2] MAGNUS, W., KARRASS, A. and SOLITAR, D.: *Combinatorial group theory: presentations of groups in terms of generators and relations*, Interscience, 1966.

A.L. Shmel'kin

Editorial comments. The example of the two-generated non-Hopfian group mentioned above is due to G. Baumslag and D. Solitar [A1].

References

[A1] BAUMSLAG, G. and SOLITAR, D.: 'Some two-generator one-relator non-Hopfian groups', *Bull. Amer. Math. Soc.* **68** (1962), 199-201.

AMS 1980 Subject Classification: 20E34

NON-LINEAR BOUNDARY VALUE PROBLEM - The determination in a certain domain D of variables $x = (x_1, \ldots, x_n)$ of a solution $u(x)$ of a differential equation

$$(Lu)(x) = f(x), \quad x \in D,$$

from its values on (a part of) the boundary S of this domain:

$$(Bu)(y) = \phi(y), \quad y \in S,$$

where at least one of the operators L or B is non-linear.

See also **Boundary value problem, ordinary differential equations**; **Boundary value problem, partial differential equations**.

A.P. Soldatov

AMS 1980 Subject Classification: 34-XX, 35-XX, 65NXX, 65L10

NON-LINEAR BOUNDARY VALUE PROBLEM, NUMERICAL METHODS - Methods replacing a boundary value problem by a discrete problem (see **Linear boundary value problem, numerical methods** and **Non-linear equation, numerical methods**). In many cases, especially in the discussion of boundary value problems for systems of ordinary differential equations, the description of numerical methods usually proceeds without indication of a discretization of the original problem, but such a discretization is nevertheless implicit and is realized relative to a known model.

Suppose, for example, that a two-point boundary value problem is to be discussed for a system of ordinary differential equations

$$y' = f(x, y) \quad \text{for } 0 < x < X, \qquad (1)$$

$$B(y(0)) = 0, \quad D(y(X)) = 0, \qquad (2)$$

where
$$y \equiv (y_1, \ldots, y_l)^T, \quad f \equiv (f_1, \ldots, f_l)^T,$$
$$B = (B_1, \ldots, B_{l-r})^T, \quad D \equiv (D_1, \ldots, D_r)^T,$$
$$0 \leqslant r \leqslant l.$$

Suppose that the vector $G(y) = (G_1(y), \ldots, G_r(y))$ is such that the system of equations

$$B(y) = 0, \quad G(y) = g \qquad (3)$$

uniquely determines a vector $y = \alpha(g)$ for every vector $g \equiv (g_1, \ldots, g_r)^T$; for example, it is often expedient to take
$$G_1 \equiv y_{l-r+1}, \ldots, \quad G_r \equiv y_l.$$

Then the problem (1) - (2) can be reduced to the system (operator equation)

$$\psi(g) = 0, \qquad (4)$$

where $\psi(g) = D(\omega(g))$ and $\omega(g)$ is the value at $x = X$ of the solution of (1) for the initial condition $y(0) = \alpha(g)$. To obtain the values of $\psi(g)$ one has to solve numerically (see [1], [2]) the corresponding system of differential equations, that is, to use a discretization of the original problem. To solve the system (4) one can use various iteration methods for the solution of non-linear equations. Having found a solution g^* of (4) one can determine the vector $y^* = \alpha(g^*)$ from the system (3) and, by solving (1) for the initial condition $y(0) = y^*$, obtain a solution of the problem (1) - (2). In a number of methods, apart from the values of ψ one also uses the value of its derivative, to be found either by integrating the equation

$$\eta' = \frac{\partial f}{\partial y} \eta \qquad (5)$$

(the variational equation for (1)) or by means of some formula for numerical differentiation. This way of reducing a boundary value problem to (4) is called the *shooting method*, as applied to a problem of more general form
$$y' = f(x, y, g) \quad \text{for } 0 < x < X,$$
$$y(0) = \alpha(g), \quad D(y(X(g), g)) = 0,$$

containing a certain vector parameter g explicitly, in particular, to an eigen value problem; here the function $\alpha(g)$ is assumed to be given. However, if the solution of (5) grows rapidly with x, then the error of the numerical integration leads to large errors in the values of $\psi(g)$ and its derivative and ultimately to a large error of the resulting solution.

In the method of linearization (see, for example, [1], [3]) approximations to a solution of the problem (1) - (2) are determined as solutions of a sequence of linear equations

$$y'_{n+1} = A_n(x)(y_{n+1} - y_n) + f(x, y_n) \quad \text{for } 0 < x < X \qquad (6)$$

with the linear boundary conditions

$$B_n(y_{n+1}(0) - y_n(0)) + B(y_n(0)) = 0, \quad \left.\begin{array}{l} \\ \\ \end{array}\right\} \qquad (7)$$
$$D_n(y_{n+1}(X) - y_n(X)) + D(y_n(X)) = 0,$$

where $A_n(x)$ is a square matrix of order l consisting of functions and B_n and D_n are numerical $((l-r) \times n)$- and $(r \times l)$-matrices, $n = 0, 1, \ldots$. The initial approximation $y_0(x)$ is assumed to be known. In the case of the Newton—Kantorovich method (cf. **Kantorovich process**; **Newton method**) the equations (6) and (7) take the form

$$y'_{n+1} = \frac{\partial f}{\partial y}(x, y_n)(y_{n+1} - y_n) + f(x, y_n(x)) \quad \text{for } 0 < x < X, \quad (8)$$

$$\frac{\partial B}{\partial y}(y_n(0))(y_{n+1}(0) - y_n(0)) + B(y_n(0)) = 0, \quad \left.\begin{array}{l} \\ \\ \\ \end{array}\right\} \qquad (9)$$

$$\frac{\partial D}{\partial y}(y_n(X))(y_{n+1}(X) - y_n(X)) + D(y_n(X)) = 0.$$

If $f(x, y)$, $B(y)$ and $D(y)$ are twice continuously differentiable in y, if the problems (8) - (9) are well-posed, and if $y_0(x)$ is sufficiently close to the required solution $y(x)$, then

$$\| y_n - y \|_C \leqslant K^{-1}(K \| y_0 - y \|_C)^{2^n},$$

where $K > 0$ is a constant and $\| y \|_C \equiv \max_{0 < x < X} | y(x) |$ (see [1]).

The methods mentioned generalize to the more general case of conditions of the form

$$A(y(X_0), y(X_1), \ldots, y(X_m)) = 0,$$

where

$$A \equiv (A_1, \ldots, A_l)^T, \quad 0 = X_0 < X_1 < \cdots < X_m \equiv X.$$

Many important theoretical and applied problems lead to the need of solving non-linear boundary value problems (and related problems) for equations and systems of equations of elliptic type (see, for example, [4] - [8]). For such a class of problems the basic numerical methods are projection methods (projection-grid, variational-difference, finite element) and difference methods (see [7] - [17]). Their constructions are in many respects analogous to those in the corresponding methods for linear boundary value problems (see **Laplace equation, numerical methods**; **Poisson equation, numerical methods**). However, in these methods the resulting discrete (grid) analogues to boundary value problems, being systems of non-linear equations

$$L_N(\bar{u}_N) = \bar{f}_N, \qquad (10)$$

where

$$\bar{u}_N \equiv (u_1, \ldots, u_N)^T \quad \text{and} \quad \bar{f}_N \equiv (f_1, \ldots, f_N)^T,$$

often involve considerable additional difficulties both with respect to the analysis of the systems themselves and the proximity in one sense or another of their solution to that of the original problem, and with respect to the input of numerical work for specifying the systems (10) and the search for their solutions. The specific

430

nature of the non-linearity of a boundary value problem forces one to pay special attention to the choice of Banach spaces and their grid analogues in which the problem itself and its finite-dimensional approximations lend themselves to analysis (see, for example, [5] - [12], [15]). But the most detailed study, especially of algorithms, has been made only of numerical methods for classes of non-linear problems related in a certain sense to linear problems (weakly non-linear equations, equations with restricted non-linearity), thus making it possible to use the theory of Hilbert spaces and their finite-dimensional analogues (Euclidean spaces).

Let $H_N \equiv H$ be a Euclidean space of vector-valued functions $\bar{u}_N \equiv u$ with scalar product $(u, v) \equiv \sum_{i=1}^{N} u_i v_i$, and let $l(H)$ be the set of self-adjoint positive-definite linear operators mapping H to H; alongside with H one also uses the Euclidean space H_B, $B \in l(H)$, that differs from H only in the form of scalar product:

$$(u, v)_{H_B} \equiv (u, v)_B \equiv (Bu, v), \quad \| u \|_B \equiv (Bu, u)^{1/2}.$$

Next, let
$$S_B(r) \equiv \{u: \| u \|_B \leqslant r\}, \quad r > 0.$$

If the operator L_N is continuous on $S_B(r)$ and such that for any u with $\| u \|_B = r$,

$$(L_N(u) - \bar{f}_N, u) \geqslant 0, \tag{11}$$

then the system (10) has at least one solution in $S_B(r)$. This assertion (see, for example, [9]) is one of the consequences of the classical existence theorems for solutions of non-linear operator equations based on topological principles (see, for example, [14]) and on the monotonicity of the operators (see, for example, [6], [11], [14]). In particular, if L_N is continuous everywhere and if

$$(L_N(u), u) \geqslant \delta \| u \|_B^2, \quad \delta > 0, \tag{12}$$

for any u, then the system (10) always has a solution and any one of them belongs to $S_B(r)$ with $r = \delta^{-1} \| \bar{f}_N \|_B^{-1}$. Moreover, if for any u, v, w, the inequalities

$$(L_N(u) - L_N(v), u - v) \geqslant \delta_0 \| u - v \|_B^2, \quad \delta_0 > 0, \tag{13}$$

$$| (L_N(u) - L_N(v), w) | \leqslant \delta_1 \| u - v \|_B \| w \|_B \tag{14}$$

hold, which indicate the strong monotonicity and Lipschitz continuity of L_N, then (10) and its unique solution can be found by the iteration method

$$Au^{n+1} = Au^n - \gamma(L_N(u^n) - \bar{f}_N), \tag{15}$$

where $A \in l(H)$ and the iteration parameter $\gamma > 0$ is determined by constants like δ_0 and δ_1, which are obtained when in (13) and (14) the operator B is replaced by A (see, for example, [8], [9]). Hence, for difference methods estimates of the error result as a consequence of the fact that the system (10) is well-posed and that an approximation of the original boundary value problem by a difference boundary value

problem is available. It is frequently expedient to choose the operator B so that the space H_B becomes a discrete (grid) analogue of the corresponding Sobolev space (for example, $W_2^1(\Omega)$ for second-order equations), and so it becomes possible to use either the imbedding theorems themselves (in the case of projection approximations) or their grid analogues (see, for example, [5] - [9]) for the verification of inequalities like (12) - (14). Projection methods and, in particular, projection-grid (finite-element) methods have the merit that (12) - (14) follow from the corresponding inequalities for differential operators, and the task of estimating the error of the method easily reduces to that of approximating the solutions of the original problem by elements of a selected finite-dimensional subspace (see [6] - [9], [11]). Instead of the validity of (13) - (14) in the whole space it suffices that they hold, for example, in some ball in H_B surrounding the required solution. Sometimes useful inequalities are obtained from (12) - (14) when the scalar products on the left-hand sides are replaced by scalar products in H_B and the norm $\| z \|_B$ on the right-hand sides by $\| Bz \|$. In a number of problems, on the basis of a priori estimates of their solutions, one can replace them by equivalent boundary value problems for which inequalities like (12) - (14) are satisfied (see [9]).

When (13) and (14) hold, then the iterative method (15) with $A = B$ converges for any initial approximation, and if the constants δ_0 and δ_1 are independent of N, then one can obtain a solution of (10) with accuracy ϵ in $O(|\log \epsilon|)$ iterations (usually $\epsilon \approx N^\alpha$, $\alpha < 0$, that is, $0 < c_0 \leqslant \epsilon N^{-\alpha} \leqslant c_1$). At the expense of a successful construction of a grid method one sometimes manages to choose the operator B so that the solution of the system (10) to within N^α by means of an iteration method like (15) is achieved by performing only $O(N \log^2 N)$ or even $O(N \log N)$ arithmetic operations (see **Minimization of the labour of calculation**). The use of a sequence of refining grids makes it possible to reduce these estimates to $O(N \log N)$ and $O(N)$ operations and so to obtain asymptotically-optimal numerical methods (relative to the labour involved) for the solution of certain elliptic boundary value problems. Here it is assumed that the numerical residual $L(u^n) - \bar{f}_N$ itself requires at most KN operations. Nevertheless, for certain problems (with large K) it is expedient to reduce the frequency of such computations. This can be achieved provided one can find a more accurate initial approximation by the use of some linearization of L_N and by methods of the type of continuation with respect to a parameter (see **Non-linear equation, numerical methods**). Most often **linearization methods** in conjunction with methods of continuation for non-linear operators with respect to a parameter (cf. **Continuation method (to a parametrized**

family, for non-linear operators)) are used in the most difficult situations, for example, when (13) fails to hold. Certain eigen value problems (see, for example, [9], [12], [14]) can also be considered as non-linear boundary value problems of elliptic type; the construction of the relevant numerical methods is in many respects analogous to the ones already mentioned.

Some other stationary boundary value problems can be treated numerically in the same manner as well (cf. [8], [9], [17]). The numerical methods for non-stationary non-linear boundary value problems include problems for equations and systems of parabolic type, hyperbolic type, mixed type, and others, although discrete analogues of elliptic operators similar to the ones considered above have also been used; however, for them it is even more important to have approximation methods for the individual time layers.

References

[1] BAKHVALOV, N.S.: *Numerical methods: analysis, algebra, ordinary differential equations*, Mir, 1977 (translated from the Russian).

[2] SEWELL, G.: *The numerical solution of ordinary and partial differential equations*, Acad. Press, 1988.

[3] BELLMAN, R. and KALABA, R.: *Quasilinearization and nonlinear boundary-value problems*, Amer. Elsevier, 1965.

[4] BEREZOVSKIĬ, A.A.: *Lectures on the non-linear boundary value problems of mathematical physics*, 1, Kiev, 1976 (in Russian).

[5] LADYZHENSKAYA, O.A. and URAL'TSEVA, N.N.: *Linear and quasilinear elliptic equations*, Acad. Press, 1968 (translated from the Russian).

[6] LIONS, J.-L.: *Quelques méthodes de resolution des problèmes aux limites nonlineaires*, Dunod, 1969.

[7] GLOWINSKI, R., LIONS, J.-L. and TRÉMOLIÈRES, R.: *Numerical analysis of variational inequalities*, North-Holand, 1981 (translated from the French).

[8] GLOWINSKI, R.: *Numerical methods for nonlinear variational problems*, Springer, 1984.

[9] D'YAKONOV, E.G.: *Minimization of computational work. Asymptotically optimal algorithms*, Moscow, 1989 (in Russian).

[10] KARCHEVSKIĬ, M.M. and LYASHKO, A.D.: *Difference schemes for non-linear problems of mathematical physics*, Kazan', 1976 (in Russian).

[11] VARGA, R.: *Functional analysis and approximation theory in numerical analysis*, SIAM, 1971.

[12] STRANG, G. and FIX, J.: *An analysis of the finite element method*, Prentice-Hall, 1973.

[13] ODEN, J.: *Finite elements of non-linear continua*, McGraw-Hill, 1972.

[14] KRASNOSEL'SKIĬ, M.A., ET AL.: *Approximate solution of operator equations*, Wolters-Noordhoff, 1972 (translated from the Russian).

[15] BREZZI, F., RAPPAZ, T. and RAVIART, P.: 'Finite dimensional approximation of nonlinear problems III', *Numer. Math.* **38** (1981), 1-30.

[16] AMOSOV, A.A., BAKHVALOV, N.S. and OSIPIK, YU.I.: 'Iterative processes for the problem of stationary heat exchange in a system of absolutely black bodies', *J. Comput. Math. Math. Phys.* **20**, no. 1 (1980), 110-119. (*Zh. Vychisl. Mat. Mat. Fiz* **20**, no. 1 (1980), 104-111)

[17] THOMASSET, F.: *Implementation of finite element methods for Navier−Stokes equations*, Springer, 1981.

E.G. D'yakonov

A.F. Shapkin

Editorial comments.

References

[A1] ASCHER, A.M., MATTHEIJ, D.M. and RUSSELL, R.D.: *Numerical solution of boundary value problems for ordinary differential equations*, Prentice-Hall, 1988.

AMS 1980 Subject Classification: 65L10, 65NXX

NON-LINEAR CONNECTION - A **differential-geometric structure** defined for the category of smooth fibre spaces associated with a certain principal G-bundle that determines the isomorphisms of the fibres (the parallel transfer) for the given non-linear connection along every piecewise-smooth curve in the base space of a bundle in the given category, which is compatible with the isomorphism of the corresponding fibres of the principal G-bundle. Here it is assumed that the structure in question is not identical with the classical concept of a **linear connection**, which is defined by a G-invariant **horizontal distribution** of one kind or another. A different meaning of the term non-linear connection [5] consists in the fact that the transfer for the fibres of a vector bundle defined by a horizontal distribution ceases to have a linear character, that is, is not a linear isomorphism of these fibres.

The necessity of introducing and studying non-linear connections arose from the need to study various differential-geometric structures of higher orders (such as, for example, a **Kawaguchi space**). The foundations of the general theory of non-linear connections are fairly well developed and applications of some special types (see [2] - [4]) have been investigated.

Let $\pi: X(B, G) \rightarrow B$ be a smooth principal G-bundle with structure Lie group G and canonical projection π onto the base B, and let $K(X)$ be the category of all bundles associated with X. A bundle isomorphism of $G_x = \pi^{-1}(x)$ onto $G_y = \pi^{-1}(y)$, $x, y \in B$, is defined to be a mapping $i: G_x \rightarrow G_y$ that commutes with the action of G on X. Any isomorphism i can be described by $i(\xi_0 g) = i(\xi_0)g$, $\xi_0 \in X$, $g \in G$, hence is a diffeomorphism of the fibres G_x and G_y. The set $\Gamma(X)$ of all isomorphisms between all possible fibres of a principal bundle X is a smooth bundle with structure groupoid over the base $B \times B$ (a **groupoid** is a category with inverse elements). An isomorphism $i \in \Gamma(X)$ gives rise to a corresponding isomorphism of the fibres over $x, y \in B$ of any associated bundle $Y \in K(X)$, and the groupoid $\Gamma(X)$ serves for the whole category $K(X)$.

Let $\Lambda(B)$ be the category of all piecewise-smooth curves in the base manifold B. A *connection in the category $K(X)$* of smooth bundles in the most general sense is any functor

$$\gamma: \Lambda(B) \rightarrow \Gamma(X)$$

that is the identity on the base $B \times B$. Let $\alpha \times \beta: \Gamma(X) \rightarrow B \times B$ be the canonical projection of the groupoid $\Gamma(X)$ onto its base $B \times B$, defined by the condition that if $\Gamma(X) \ni i : G_x \to G_y$, then $\alpha(i) = x$, $\beta(i) = y$. In this way B is identified with the submanifold $\mathring{B} \subset \Gamma(X)$ of all left and right units of $\Gamma(X)$. Let $\Pi(X)$ be the vector bundle over $B \equiv \tilde{B}$ formed by the fibres of the form $T_e[\alpha^{-1}(e)]$, $e \in B$, and let $T^p(B)$ be the fibre over B of p-velocities of B (the elements of $T^p(B)$ are regular p-jets of all possible smooth mappings $\mathbf{R} \to B$ with source $0 \in \mathbf{R}$). The bundles $\Pi(X)$ and $T^p(B)$ have canonical projections onto the tangent bundle $T(B)$,

$$\pi' : \Pi(X) \to T(B), \quad \pi^p : T^p(B) \to T(B).$$

A connection γ is called a *non-linear connection of order* $p = 1, 2, \ldots$, if p is the smallest number for which the functor γ determines a smooth mapping

$$\gamma^p : T^p(B) \to \Pi(X)$$

such that $\pi' \circ \gamma^p = \pi^p$. In turn, γ is determined by the γ^p corresponding to it. When $p = 1$ and the mapping $T(B) \to \Pi(X)$ is fibrewise linear, the connection degenerates to a linear one on $K(X)$. In the study of the properties of non-linear connections and in their classification a fundamental role is played by the structure equations of the mappings $\gamma^p : T^p(B) \to \Pi(X)$. These can be written in the form of Pfaffian equations connecting the differentials of the relative coordinates of the geometric objects describing the bundles $T^p(B)$ and $\Pi(X)$. In terms of the coefficients of the structure equations and by means of the operations of their differential prolongations and restrictions it has been established [2] that a non-linear connection γ^p in $X(B, G)$ gives rise to a linear connection of special structure in the smooth G-bundle $X(B, G) \otimes_B T^p(B)$ over the base $T^p(B)$ and is completely characterized by this linear connection. The forms of these linear connections have been found and also their structure equations. A non-linear analogue has been found for the theorem on the holonomy group, and its statement involves not only the curvature, but also the linear hull of the distribution of horizontal cones, which replace in the non-linear case the subspace of the horizontal distribution of a linear connection.

References

[1] VAGNER, V.V.: 'The theory of composite manifolds', *Trudy Sem. Vektor. Tenzor. Anal.* **8** (1950), 11-72 (in Russian).
[2] EVTUSHIK, L.E.: 'Non-linear connections of higher order', *Izv. Vuz. Mat.* **2** (1969), 34-44 (in Russian).
[3] EVTUSHIK, L.E.: 'Holonomy of nonlinear connections', *Sib. Math. J.* **14**, no. 3 (1973), 370-379. (*Sibirsk. Mat. Zh.* **14**, no. 3 (1973), 536-548)
[4] EVTUSHIK, L.E. and TRET'YAKOV, V.B.: 'Structures that can be defined by a system of higher order differential equations', *Trudy Geom. Sem.* **6** (1974), 243-255 (in Russian). English abstract.
[5] KAWAGUCHI, A.: 'On the theory of non-linear connections I. Introduction to the theory of general non-linear connections', *Tensor, New Ser.* **2** (1952), 123-142.

L.E. Evtushik

AMS 1980 Subject Classification: 53C05, 53B15

NON-LINEAR DIFFERENTIAL EQUATION - A differential equation (ordinary or partial) in which at least one of the derivatives of the unknown function (including the derivative of order zero: the function itself) occurs non-linearly. This term is used, as a rule, when one wishes to emphasize especially that the equation $H = 0$ in question is not linear, that is, its left-hand side H is not a **linear form** in the derivatives of the unknown function with coefficients depending only on the independent variables.

Sometimes by a non-linear differential equation one means a more general equation of a certain form. For example, a *non-linear ordinary first-order differential equation* is an equation

$$f\left(x, y, \frac{dy}{dx}\right) = 0$$

with an arbitrary function $f(x, y, u)$; here a linear ordinary first-order differential equation corresponds to the special case

$$f(x, y, u) = a(x)u + b(x)y + c(x).$$

A *non-linear partial first-order differential equation* for an unknown function z in n independent variables $x_1, \ldots, x_n$ has the form

$$F\left(x_1, \ldots, x_n, z, \frac{\partial z}{\partial x_1}, \ldots, \frac{\partial z}{\partial x_n}\right) = 0,$$

where F is an arbitrary function of its arguments; when

$$F = \sum_{i=1}^{n} A_i(x_1, \ldots, x_n, z)\frac{\partial z}{\partial x_i} + B(x_1, \ldots, x_n, z),$$

such an equation is called *quasi-linear*, and when

$$F = \sum_{i=1}^{n} A_i(x_1, \ldots, x_n)\frac{\partial z}{\partial x_i} +$$
$$+ B(x_1, \ldots, x_n)z + C(x_1, \ldots, x_n),$$

it is called *linear* (cf. also **Linear partial differential equation**; **Non-linear partial differential equation**).

N.Kh. Rozov

AMS 1980 Subject Classification: 34A34, 35F20, 35G20

NON-LINEAR EQUATION, NUMERICAL METHODS - Iteration methods for the solution of non-linear equations.

By a non-linear equation one means (see [1] - [3]) an algebraic or transcendental equation of the form

$$\phi(x) = 0, \tag{1}$$

where x is a real variable and $\phi(x)$ a non-linear function, and by a system of non-linear equations a system of the form

$$\left.\begin{array}{c} \phi_1(x_1, \ldots, x_N) = 0, \\ \cdots \cdots \cdots \\ \phi_N(x_1, \ldots, x_N) = 0, \end{array}\right\} \tag{2}$$

that is not linear; the solutions of (2) are N-dimensional vectors $x=(x_1, \ldots, x_N)$. Equation (1) and system (2) can be treated as a non-linear operator equation:

$$L(u) = f \tag{3}$$

with a non-linear operator L acting from a finite-dimensional vector subspace H_N into H_N.

Numerical methods for the solution of a non-linear equation (3) are called iteration methods if they are defined by the transition from a known approximation u^n at the n-th iteration to a new iteration u^{n+1} and allow one to find in a sufficiently large number of iterations a solution of (3) within prescribed accuracy ϵ. The most important iteration methods for the approximate solution of (3), both of a general and of a special form, characteristic for discrete (grid) methods for solving boundary value problems for equations and systems of partial differential equations of strongly-elliptic type, are the object of study of the present article. Non-linear operator equations connected with the discussion of infinite-dimensional spaces (see, for example [4] - [8]) are a very broad mathematical concept, including as special cases, for example, non-linear integral equations and non-linear boundary value problems. Numerical methods for the approximate solution of them include also methods for their approximation by finite-dimensional equations; these methods are treated separately.

One of the most important methods for solving an equation (3) is the *simple iteration method* (*successive substitution*), which assumes that one can replace (3) by an equivalent system

$$u = P(u), \tag{4}$$

where $u=(u_1, \ldots, u_N)$ is an element of a finite-dimensional normed space H_N and P, mapping H_N into H_N, is a contractive operator:

$$\exists q < 1 \, \forall u, v \, \| P(u) - P(v) \| \leqslant q \| u - v \|. \tag{5}$$

Then by a general contractive-mapping principle (see [1] - [4] and **Contracting-mapping principle**) equation (1) has a unique solution, the method of simple iteration

$$u^{n+1} = P(u^n), \quad n = 0, 1, \ldots, \tag{6}$$

converges for any initial approximation u^0, and the error z^n at the n-th iteration satisfies the estimate

$$\| z^n \| \leqslant q^n (1-q)^{-1} \| u^0 - u^1 \|.$$

Suppose that some solution $\bar{u}$ of (3) has a surrounding ball $S(\bar{u}, R) \equiv \{ v : \| v - \bar{u} \| \leqslant R \}$ such that the system (3) considered together with the additional condition

$$u \in S(\bar{u}, R) \tag{7}$$

is equivalent to (4) considered together with (7), and that (5) holds for $u = \bar{u}$ and any $v = \bar{u} + z$ with $\| z \| \leqslant R$. Then for a choice of an initial approximation

u^0 from $S(\bar{u}, R)$ in the method (6) convergence of u^n to $\bar{u}$ with an error estimate $\| z^n \| \leqslant q^n R$ is guaranteed.

For twice continuously-differentiable functions ϕ_i, if a good initial approximation to a solution of the system (2) is available, frequently an effective means of improving the accuracy is the *Newton$-$Kantorovich method*, in which an equation $\phi_1(x_1, \ldots, x_N) = 0$ from (2) determining a certain surface P_i is replaced by the equation of the tangent surface to P_i at the point x^n, where x^n is a previously obtained approximation to a solution of (2) (see [1] - [5]). Under certain additional conditions the Newton$-$Kantorovich method leads to an error estimate of the form

$$\| z^n \| \leqslant c_1 (c \| z^0 \|)^{2^n},$$

where c_i and c are certain constants. At every iteration of this method one has to solve the system of linear algebraic equations with matrix

$$A^{(n)} \equiv \left[\frac{\partial \phi_i}{\partial x_j} \right] \Bigg|_{x = x^n}.$$

Sometimes one keeps this matrix fixed for several iterations, sometimes one replaces the derivatives $\partial \phi_i / \partial x_j$ by difference approximations.

The Newton$-$Kantorovich method belongs to the group of **linearization methods** (3). Another method in this group is the **secant method**.

A large number of iteration methods (the so-called *descent methods*) (see [1] - [3], [9], [10] - [13]) are based on replacing the solving of the equations (3) by minimizing a certain functional $I(u)$ (cf. also **Descent, method of**). For example, for $I(u)$ one can take

$$I(u) \equiv \| L(u) - f \|^2. \tag{8}$$

In a number of cases when the initial non-linear equations are the Euler equations for the problem of minimizing a certain functional $I(u)$, such a variational formulation of the problem is even more natural; the operators L in similar situations are gradients of the functional $I(u)$ and are called *potential operators* (see [5] - [6]). Among the several versions of descent methods one can mention the methods of coordinate-wise descent, several gradient methods and, in particular, the method of steepest descent, the method of conjugate gradients, and others, and also their modifications (see [2], [9], [10] - [13]). A number of iteration methods for the solution of the equations (3), describing a certain stationary state, can be treated as discretizations of the corresponding non-stationary problems. Therefore, methods of this class are called stabilization methods (cf. **Adjustment method** and, for example, [2]). Examples of such non-stationary problems are problems described by a system of ordinary differential equations

$$A_1 \frac{d^2 u}{dt^2} + A_2 \frac{du}{dt} + L(u) - f = 0.$$

The introduction of an additional independent variable is characteristic for the *method of differentiation with respect to a parameter* (a *continuation method*, see [5], [13]). Its essence consists in the introduction of an auxiliary parameter $\lambda \in [0, 1]$, the choice of continuously-differentiable functions $F_i(x, \lambda)$, and replacement of (2) by the system

$$F_i(x, \lambda) = 0, \quad i = 1, \ldots, N, \quad 0 \leqslant \lambda \leqslant 1; \qquad (9)$$

for $\lambda = 0$ the system (9) must be easy to solve, and the functions $F_i(x, 1)$ must coincide with the $\phi_i(x)$, $i = 1, \ldots, N$. Generally speaking, the system (9) determines

$$x(\lambda) \equiv (x_1(\lambda), \ldots, x_N(\lambda))$$

as a function of λ, and the required solution of (2) is $x(1)$. If (9) is differentiable with respect to λ, the result is a system of ordinary differential equations

$$\sum_{i=1}^{N} \frac{\partial F_i}{\partial x_j} \frac{dx_j}{d\lambda} + \frac{\partial F_i}{\partial \lambda} = 0, \quad i = 1, \ldots, N. \qquad (10)$$

If one solves the Cauchy problem for it on $0 \leqslant \lambda \leqslant 1$ with initial conditions that are solutions of the system

$$F_i(x, 0) = 0, \quad i = 1, \ldots, N,$$

then one finds a solution of (2). Discretization of (10) with respect to λ leads to a numerical method for solving the system (2).

In the *method of continuation with respect to a parameter* (cf. **Continuation method (to a parametrized family, for non-linear operators)**) the system (9) is solved for $\lambda = \tau, \ldots, m\tau$, $m\tau = 1$, and for every one of these values of λ one applies a certain iteration method with an initial approximation that agrees with the approximation obtained by solving the system for the preceding value of λ. Both these methods are in essence iteration methods for solving (2) with a special procedure for finding a good initial approximation.

In the case of systems the problem of localizing the solution causes great difficulty. Since the majority of iteration methods converge only in the presence of a fairly good approximation to a solution, the two methods described above often make it possible to dispense with the need to localize a solution directly. For localization one also frequently uses theorems based on topological principles and on the monotonicity of operators (see [4] - [8]).

For the solution of equations (1), which are the simplest special cases of (3), the number of known iteration methods applicable in practice is very large (see, for example, [1] - [3], [12], [14]). Apart from the ones already considered one can mention, for example, the *iteration methods of higher order* (see [1], [14]), which include Newton's methods as a special case, and the many iteration methods especially oriented to finding real or complex roots of polynomials

$$p_n(z) \equiv a_n z^n + a_{n-1} z^{n-1} + \cdots + a_0,$$

where the a_i are real or complex numbers (see [1], [2]).

The problem of localizing the solutions of (1) reduces to a search for an interval at the ends of which the continuous function ϕ takes values of opposite sign. When ϕ is a polynomial, the task is less subtle, since theoretical bounds are known (see [1]) and there are methods for finding all roots with required accuracy without giving good initial approximations to them (see [12]).

Iteration methods for solving equations (3) that arise in grid analogues of non-linear boundary value problems for partial differential equations are special cases of methods for solving grid systems (see, for example, [13] , [15] - [19]). Probably one of the most intensively applied methods for solving (3) is a modified method of simple iteration, which can be can be written in the form

$$Bu^{n+1} = Bu^n - \gamma(L(u^n) - f), \qquad (11)$$

where (3) is regarded as an operator equation in the N-dimensional space $H_N = H$, and $B \in S$, where S stands for the set of symmetric positive linear operators mapping H into itself. An expedient study of such methods should proceed not in H, but in the space H_B with the new scalar product

$$(u, v)_{H_B} \equiv (Bu, v), \quad \| u \|^2 \equiv (Bu, u),$$

where (u, v) is the scalar product in H.

If the operator L satisfies the conditions of strict monotonicity and is Lipschitz continuous:

$$\forall u, v \ (L(u) - L(v), u - v) \geqslant \delta_0 \| u - v \|_B^2, \quad \delta_0 > 0, \qquad (12)$$

$$\forall u, v \ \| L(u) - L(v) \|_{B^{-1}}^2 \leqslant \delta_2 \| u - v \|_B^2, \qquad (13)$$

then (3) has a unique solution and the method (11), for a suitable choice of γ, converges for any u^0 with error estimate

$$\| z^n \|_B \leqslant q^n \| z^0 \|_B, \qquad (14)$$

where $q \equiv q(\delta_0, \delta_2, \gamma) < 1$ (see [13], [15]).

In the most general version of this theorem it suffices to require that (12) and (13) hold for a solution u and all v in a ball $S_B(u, R) \equiv \{v: \| v - u \| \leqslant R\}$ and that u^0 lies in this ball (see [13]). In this case the constants δ_0 and δ_2 may depend on R. To verify these conditions it suffices, e.g., to localize u by means of a priori estimates, obtaining $\| u \|_B \leqslant R_0$, and then to take $S_B(u, R) \equiv S_B(u, R_1 - R_0)$, provided that (12) and (13) hold for any u and v in $S_B(0, R_1)$, where $R_1 > R_0$. The constant q in (14) can be diminished if L is differentiable and for its derivatives L'_v, represented as the sum of the symmetric part $L'_{v, \text{sym}}$ and the skew-symmetric part $L'_{v, \text{skew}}$ the inequalities

$$\forall v \in S_B(u, R) \qquad \sigma_0 B \leqslant L'_{v, \text{sym}} \leqslant \sigma_1 B, \quad \sigma_0 > 0;$$

$$\forall w \ \| L'_{v, \text{skew}} w \|_{B^{-1}}^2 \leqslant \sigma_2 \| w \|_B^2$$

are known. Then $q \equiv q(\gamma, \sigma_0, \sigma_1, \sigma_2)$ (see [11], [13], [15]). Sometimes in the discussion of certain types of non-linearity it is advisable to use instead of (12) and (13) inequalities like

$$(L(u)-L(v), B(u-v)) \geqslant \bar{\delta}_0 \| B(u-v) \|^2, \quad \bar{\delta}_0 > 0;$$

$$\| L(u)-L(v) \|^2 \leqslant \bar{\delta}_2 \| B(u-v) \|^2$$

(see [13]). For the operators B in (11) one can use, for example, splitting difference operators (the method of alternating directions) or factorized difference operators (the alternating-triangle method, the incomplete matrix factorization method), and others. Most attractive from the asymptotic point of view is the use of operators B such that the constants δ_i or $\bar{\delta}_i$ do not depend on the dimension of the space H_N (see [13]) and the operators B are sufficiently simple. Along these lines one succeeds in a number of cases in constructing iteration methods that make it possible to find a solution of (3) with accuracy ϵ at the expense of altogether $O(N \log N \mid \log \epsilon \mid)$ (or even $O(N \log N)$) arithmetic operations for $\epsilon \approx N^{-\alpha}$, $\alpha > 0$ (see [13]), if the computational work in the evaluation of $L(u^n)$ for a given u^n can be estimated by $O(N)$. To verify conditions of the type (12) and (13), in many cases the grid (difference) analogues of the Sobolev imbedding theorems turn out to be very powerful (see [13]). It is important to take into account the specific nature of the non-linearity. For example, when $L(u) = \Lambda u + Pu$, where Λ is a positive linear operator and P a quadratic non-linear operator having the property of 'skew-symmetry' (that is, $(Pu, u) = 0$ for all u), one often succeeds in obtaining the constant $\delta_2(R)$ in any ball $S_B(0, R)$ and δ_0 depending only on $\| u \|_B$; then (11) converges for any u^0 (see [13]). In a number of cases one can replace the original problem on the basis of a priori estimates by an equivalent one for which the required conditions hold in the whole space.

References

[1] BEREZIN, I.S. and ZHIDKOV, N.P.: *Computing methods*, Pergamon, 1973 (translated from the Russian).
[2] BAKHVALOV, N.S.: *Numerical methods: analysis, algebra, ordinary differential equations*, Mir, 1977 (translated from the Russian).
[3] ORTEGA, J. and RHEINBOLDT, W.: *Iterative solution of nonlinear equations in several variables*, Acad. Press, 1970.
[4] KRASNOSEL'SKIĬ, M.A., ET AL.: *Approximate solution of operator equations*, Wolters-Noordhoff, 1972 (translated from the Russian).
[5] MIKHLIN, S.G.: *The numerical performance of variational methods*, Wolters-Noordhoff, 1971 (translated from the Russian).
[6] VAĬNBERG, M.M.: *Variational method and method of monotone operators in the theory of nonlinear equations*, Wiley, 1973 (translated from the Russian).
[7] GAJEWSKI, H., GRÖGER, K. and ZACHARIAS, K.: *Nichtlineare Operatorengleichungen und Operatorendifferentialgleichungen*, Akad. Verlag, 1974.
[8] LIONS, J.-L.: *Quelques méthodes de résolution des problèmes aux limites nonlineaires*, Dunod, 1969.
[9] VASIL'EV, F.P.: *Lectures on methods for solving extremal problems*, Moscow, 1974 (in Russian).
[10] PSHENICHNYĬ, B.N. and DANILIN, YU.M.: *Numerical methods in extremal problems*, Mir, 1978 (translated from the Russian).
[11] GLOWINSKI, R.: *Numerical methods for nonlinear variational problems*, Springer, 1984.
[12] VOEVODIN, V.V.: *Numerical methods of algebra. Theory and algorithms*, Moscow, 1966 (in Russian).
[13] D'YAKONOV, E.G.: *Minimization of computational work. Asymptotically-optimal algorithms*, Moscow, 1989 (in Russian).
[14] TRAUB, J.F.: *Iterative methods for the solution of equations*, Prentice-Hall, 1964.
[15] SAMARSKIĬ, A.A. and NIKOLAEV, E.S.: *Numerical methods for grid equations*, 1-2, Birkhäuser, 1989 (translated from the Russian).
[16] GLOWINSKI, R., LIONS, J.-L. and TREMOLIERES, R.: *Numerical analysis of variational inequalities*, North-Holland, 1981 (translated from the French).
[17] HACKBUSCH, W.: 'Multigrid solution of continuation problems', in R. Ansorge, Th. Meis and W. Törnig (eds.): *Iterative solution of nonlinear systems of equations*, Lecture notes in math., Vol. 953, Springer, 1982, pp. 20-44.
[18] THOMASSET, F.: *Implementation of finite element methods for Navier–Stokes equations*, Springer, 1981.
[19] HACKBUSCH, W. and TROTTENBERG, V. (EDS.): *Multigrid Methods. Proc. Köln-Porz, 1981*, Lecture notes in math., 960, Springer, 1982.

E.G. D'yakonov

Editorial comments.

References
[A1] DANIEL, J.W.: *The approximate minimization of functionals*, Prentice-Hall, 1971.

AMS 1980 Subject Classification: 65HXX

NON-LINEAR FUNCTIONAL - A special case of a **non-linear operator** defined on a real (or complex) vector space X and whose values are real (or complex) numbers. Examples of non-linear functionals are the functionals of the calculus of variations,

$$f(x) = \int_a^b F(t, x(t), x'(t)) \, dt,$$

or convex functionals, defined by the condition

$$f(\lambda y + (1-\lambda)x) \leqslant \lambda f(y) + (1-\lambda)f(x),$$

where $x, y \in X$, $0 \leqslant \lambda \leqslant 1$, and, say, $f(x) = \| x \|$ — the norm of an element in a normed space.

V.I. Sobolev

Editorial comments. See also Non-linear functional analysis.

AMS 1980 Subject Classification: 47HXX

NON-LINEAR FUNCTIONAL ANALYSIS - The branch of **functional analysis** in which one studies non-linear mappings (operators, cf. **Non-linear operator**) between infinite-dimensional vector spaces and also certain classes of non-linear spaces and their mappings. The basic divisions of non-linear functional analysis are the following.

1) Differential calculus of non-linear mappings between Banach, topological vector and certain more general spaces, including theorems on the local inver-

sion of a differentiable mapping and the implicit-function theorem.

2) The search for conditions on the action, such as continuity and compactness, of a non-linear operator acting from one specific infinite-dimensional space into another.

3) Fixed-point principles for various classes of non-linear operators (contractive, compact, compressing, monotone, and others); application of these principles to existence proofs for solutions of various non-linear equations.

4) The study of non-linear operators such as monotone, concave, convex, having a monotone minorant, and others, in spaces endowed with the structure of an ordered vector space.

5) The study of spectral properties of non-linear operators (bifurcation points, continuous branches of eigen vectors, etc.) in infinite-dimensional vector spaces.

6) The approximate solution of non-linear operator equations.

7) The study of spaces that are locally linear and of Banach manifolds — *global analysis*.

8) The investigation of extrema of non-linear functionals and variational methods for studying non-linear operators.

References

[1] VAĬNBERG, M.M.: *Variational method and method of monotone operators in the theory of nonlinear equations*, Wiley, 1973 (translated from the Russian).

[2] GAJEWSKY, H., GRÖGER, K. and ZACHARIAS, K.: *Nichtlineare Operatorgeleichungen und Operatordifferentialgleichungen*, Akad. Verlag, 1974.

[3] EELLS, J.: 'The foundations of global analysis', *Uspekhi Mat. Nauk* **24**, no. 3 (1969), 157-210 (in Russian).

[4] KRASNOSEL'SKIĬ, M.A.: *Positive solutions of operator equations*, Wolters-Noordhoff, 1964 (translated from the Russian).

[5] KRASNOSEL'SKIĬ, M.A. and ZABREĬKO, P.P.: *Geometric methods of non-linear analysis*, Springer, 1983 (translated from the Russian).

[6] LANG, S.: *Introduction to differentiable manifolds*, Interscience, 1967.

[7] LYUSTERNIK, L.A. and SOBOLEV, V.I.: *Elemente der Funktionalanalysis*, Akad. Verlag, 1968 (translated from the Russian).

[8] NIRENBERG, L.: *Topics on nonlinear functional analysis*, New York Univ., 1974.

[9] HILLE, E. and PHILLIPS, R.: *Functional analysis and semi-groups*, Amer. Math. Soc., 1957.

V.I. Sobolev

Editorial comments.

References

[A1] SCHWARTZ, J.T.: *Nonlinear functional analysis*, Gordon & Breach, 1969.

[A2] ZEIDLER, E.: *Nonlinear functional analysis and its applications*, 1-3 , Springer, 1986 (translated from the Russian).

[A3] BERGER, M.S.: *Nonlinearity and functional analysis*, Acad. Press, 1977.

AMS 1980 Subject Classification: 46-XX, 47HXX

NON-LINEAR INTEGRAL EQUATION - An **integral equation** containing the unknown function non-linearly. Below the basic classes of non-linear integral equations that occur frequently in the study of various applied problems are quoted; their theory is, to a certain extent, fairly well developed.

An important example is the **Urysohn equation**

$$\phi(x) = \lambda \int_\Omega K[x, s, \phi(s)] \, ds, \quad x \in \Omega, \tag{1}$$

where Ω is a closed bounded set in a finite-dimensional Euclidean space, $K[x, s, t]$ is a given function, the so-called kernel, which is defined for $x, s \in \Omega$, $-\infty < t < \infty$, λ is a numerical parameter, and ϕ is the unknown function.

P.S. Urysohn (see [2]) made, under certain assumptions, a complete investigation of the spectrum of the eigen values of equations (1) admitting positive eigen functions. He showed that positive eigen functions $\phi(x, \lambda)$ correspond to values λ only in a certain interval (α, β) and that $\phi(x, \lambda)$ is a monotone increasing function of λ with $\phi(x, \alpha) = 0$ and $\phi(x, \beta) = \infty$.

A special case of an Urysohn equation is the **Hammerstein equation**

$$\phi(x) = \lambda \int_\Omega K(x, s) f[s, \phi(s)] \, ds, \quad x \in \Omega, \tag{2}$$

where $K(x, s)$ and $f(s, t)$ are known functions. Existence and uniqueness theorems were first established by A. Hammerstein (see [9]). He investigated equations (2) under the assumption that the real-valued function $f(s, t)$ is jointly continuous in its arguments and that the linear integral operator generated by the kernel K is self-adjoint in $L_2(\Omega)$, positive, and acts compactly from $L_2(\Omega)$ into the space of continuous functions.

Another example of a non-linear integral equation is the **Lyapunov − Schmidt equation**

$$\sum_{\alpha, \beta} \int_\Omega \cdots \int_\Omega K_{\alpha, \beta}(x, s_1, \ldots, s_i) \times \tag{3}$$

$$\times \phi^{\alpha_0}(x) \phi^{\alpha_1}(s_1) \cdots \phi^{\alpha_i}(s_i) \times$$

$$\times v^{\beta_0}(x) v^{\beta_1}(s_1) \cdots v^{\beta_i}(s_i) \, ds_1 \cdots ds_i = 0, \quad x \in \Omega,$$

in which $K_{\alpha, \beta}$ and v are given functions, ϕ is the unknown function, i is fixed, and the summation is over all vectors $\alpha = (\alpha_0, \ldots, \alpha_i)$ and $\beta = (\beta_0, \ldots, \beta_i)$ with non-negative integer components. The left-hand side of (3) is called an *integral power series* in the two functional arguments v and ϕ.

Equations of type (3) were first considered by A.M. Lyapunov (see [1]) and later, in a more general form, by E. Schmidt (see [8]). In their research the foundations were laid of the bifurcation theory of non-linear integral equations, which aims at solving the following problem. Suppose that one searches for a solution of a non-linear problem depending on certain parameters and that for some of their values the solution may bifurcate. There arises the tasks of finding the solution itself and those parameter values for which it bifurcates (branches), the number of branches, and the representation of each branch as a function of the parameters (see [6]).

The theory of non-linear integral equations is part of the general theory of non-linear operator equations. Namely, integral equations can be regarded as specific illustrations of the corresponding operator equations. For this purpose one has to clarify general properties (continuity, compactness, etc.) of the concrete integral operators occurring in the equation.

References

[1] LYAPUNOV, A.M.: *Zap. Akad. Nauk. St. Petersburg* (1906), 1-225.

[2] URYSOHN, P.S.: *Mat. Sb.* **31** (1923), 236-255.

[3] VAĬNBERG, M.M.: *Variational methods for the study of nonlinear operators*, Holden-Day, 1964 (translated from the Russian).

[4] KRASNOSEL'SKIĬ, M.A.: *Topological metods in the theory of non-linear integral equations*, Pergamon, 1964 (translated from the Russian).

[5] KRASNOSEL'SKIĬ, M.A., ET AL.: *Integral operators in spaces of summable functions*, Noordhoff, 1976 (translated from the Russian).

[6] VAĬNBERG, M.M. and TRENOGIN, V.A.: *Theory of branching of solutions of non-linear equations*, Noordhoff, 1974 (translated from the Russian).

[7] VAĬNBERG, M.M.: *Variational methods and methods of nonlinear operators in the theory of nonlinear equations*, Wiley, 1973 (translated from the Russian).

[8] SCHMIDT, E.: 'Zur Theorie der linearen und nichtlineB ren Integralgleichungen III', *Math. Ann.* **65** (1908), 370-399.

[9] HAMMERSTEIN, A.: 'Nichtlineare Integralgleichungen nebst Anwendungen', *Acta Math.* **54** (1930), 117-176.

B.V. Khvedelidze

Editorial comments.

References

[A1] ZABREYKO, P.P. [P.P. ZABREĬKO], ET AL.: *Integral equations — a reference text*, Noordhoff, 1975 (translated from the Russian).

AMS 1980 Subject Classification: 45GXX, 47HXX

NON-LINEAR OPERATOR - A mapping A of a space (as a rule, a vector space) X into a vector space Y over a common field of scalars that does not have the property of linearity, that is, such that generally speaking

$$A(\alpha_1 x_1 + \alpha_2 x_2) \neq \alpha_1 A x_1 + \alpha_2 A x_2.$$

If Y is the set $\mathbf{R}$ of real or $\mathbf{C}$ of complex numbers, then a non-linear operator is called a **non-linear functional**. The simplest example of a non-linear operator (non-linear functional) is a real-valued function of a real argument other than a linear function. One of the important sources of the origin of non-linear operators are problems in mathematical physics. If in a local mathematical description of a process small quantities not only of the first but also of higher orders are taken into account, then there arise equations with non-linear operators. Certain problems in mathematical econom-

ics, auto-regulation, control theory, etc., also lead to non-linear operator equations.

Examples of non-linear operators.

$$1) \quad Ax = \int_a^b K(t, s, x(s))\, ds,$$

where $K(t, s, u)$, $a \leqslant t$, $s \leqslant b$, $-\infty < u < \infty$, is a function such that $g(t) = \int_a^b K(t, s, x(s))\, ds$ is continuous on $[a, b]$ for any $x(s) \in C(a, b)$ (for example, $K(t, s, u)$ is continuous on $a \leqslant t$, $s \leqslant b$, $-\infty < u < \infty$). If $K(t, s, u)$ is non-linear in u, then A is a non-linear *Urysohn operator* mapping $C[a, b]$ into itself. Under other restrictions on $K(t, s, u)$ an Urysohn operator acts on other spaces, for instance, $L_2[a, b]$ or maps one Orlicz space $L_{M_1}[a, b]$ into another $L_{M_2}[a, b]$.

$$2) \quad Bx = \int_a^b K(t, s) g(s, x(s))\, ds,$$

where $g(t, u)$ is non-linear in u and defined for $a \leqslant t \leqslant b$, $-\infty < u < \infty$. Under appropriate restrictions on $g(t, u)$ the operator B acts from one function space into another and is called a non-linear *Hammerstein operator.*

$$3) \quad F(x) = f(t, x(t))$$

is a *superposition operator*, also called a *Nemytskiĭ operator*, and, under suitable restrictions on the non-linearity in the second argument of the function, it transforms the space of measurable functions $x(t)$ into itself.

$$4) \quad D(x) = \sum_{|k| \leqslant m} D^k(a_k(t, x, Dx, \ldots, D^k x))$$

is a non-linear *differential operator of order $2m$ in divergence form* acting on the Sobolev space $W_\rho^{2m}(G)$ under suitable restrictions on the non-linear function $a_k(t, u_0, \ldots, u_m)$. Here k is the multi-index $(k_1, \ldots, k_n)$, $|k| = k_1 + \cdots + k_n$, $D^k = \partial^{|k|} / \partial t_1^{k_1} \cdots \partial t_n^{k_n}$ and G is a bounded domain in $\mathbf{R}^n$.

$$5) \quad J(x) = \int_a^b K(t, s, x(s), x'(s))\, ds$$

is non-linear *integro-differential operator* acting under appropriate restrictions on the function $K(t, s, u_0, u_1)$ in the space $C^1[a, b]$ of continuously-differentiable functions.

To non-linear operators acting from one topological vector space X into another one Y, many concepts and operations of mathematical analysis of real-valued functions of a real variable can be transferred. Thus, a non-linear operator $A: M \to Y$, $M \subset X$, is called *bounded* if $A(B \cap M)$ is a bounded set in Y for any bounded set $B \subset X$; a non-linear operator A is *continuous* at a point $x \in M$ if the inverse image $A^{-1}(U_{Ax})$ of a neighbourhood U_{Ax} of the point Ax contains $M \cap U_x$ for some neighbourhood U_x of x. As for functions, a non-linear operator that is continuous at every point of a compact set M is bounded on this set. In contrast to linear operators, if a non-linear operator A acting on a normed space is bounded on some ball, it does not follow that A is continuous on this ball. However, in certain cases continuity (boundedness) of a non-linear operator on a ball implies continuity (boundedness) of the operator in its whole domain of definition.

Among the non-linear operators acting from X to Y one can distinguish certain important classes.

1) *Semi-linear operators* $A: X \times \cdots \times X \to Y$, linear in each argument. The space $L_n(X, Y) = (I)$ of all n-linear operators is isomorphic to the space $L\{X[\cdots L(X, Y), \ldots]\} = (II)$, where $L(X, Y)$ is the space of all linear operators from X to Y. If X and Y are normed spaces, then (I) and (II) are isometric. If A is symmetric in all arguments, then $\tilde{A}(x, \ldots, x)$ is denoted by $\tilde{A}x^n$ and is called a *homogeneous operator of degree n*.

2) In spaces endowed with a partial order, *isotone operators* A and *antitone operators* $\tilde{A}$ are characterized by the conditions $x \leqslant y \Rightarrow Ax \leqslant Ay$ and $x \leqslant y \Rightarrow \tilde{A}x \geqslant \tilde{A}y$.

3) In a Hilbert space H, *monotone operators* M are defined by the condition $<Mx - My, x - y> \geqslant 0$ for any $x, y \in H$.

4) *Compact operators* transform bounded subsets in the domain of definition into pre-compact sets; among them are the *completely-continuous operators*, which are simultaneously compact and continuous.

For non-linear operators the concepts of a differential and a derivative are non-trivial and useful. An operator A acting from an open set G of a normed vector space X into a normed vector space Y is called *Fréchet differentiable* at a point $x \in G$ if there exists a continuous linear operator $A'(x): X \to Y$ such that for any $h \in X$ for which $x + h \in G$,

$$A(x + h) - A(x) = A'(x)h + \omega,$$

where $\omega / \|h\| \to 0$ as $h \to 0$. In this case the linear part $A'(x)h$ in h of the increment $A(x + h) - A(x)$ is called the *Fréchet differential* of A at x and is denoted by $dA(x, h)$, and $\omega = \omega(A, x, h)$ is called the *remainder of the increment*. The bounded linear operator $A'(x)$ is called the *Fréchet derivative* of A at x. Apart from Fréchet differentiability one also introduces Gâteaux differentiability. Namely, an operator A is called *Gâteaux differentiable* at a point x if the limit

$$\lim_{t \to 0} \frac{A(x + th) - A(x)}{t} = DA(x, h)$$

exists, which is called the *Gâteaux differential* of A at x. The Gâteaux differential is homogeneous in h, that is, $DA(x, \lambda h) = \lambda DA(x, h)$. If $DA(x, h)$ is linear in h and $DA(x, h) = A_0'(x)h$, then the linear operator $A_0'(x)$ is called the *Gâteaux derivative* of A. Fréchet differentia-

bility implies Gâteaux differentiability, and then $A_0'(x) = A'(x)$. Gâteaux differentiability does not, in general, imply Fréchet differentiability, but if $DA(x, h)$ exists in a neighbourhood of x, is continuous in h and uniformly continuous in x, then A is Fréchet differentiable at x. For non-linear functionals $f: G \to \mathbf{R}$ Fréchet and Gâteaux differentials and derivatives are defined similarly. Here the Gâteaux derivative f_0' is called the *gradient of the functional f* and is an operator from G to X^*. If $Ax = \operatorname{grad} f(x)$ for some non-linear functional f, then A is called a *potential operator*.

For operators acting on separable topological vector spaces one can in one way or another define differentiation. Let $\mathfrak{M}$ be a collection of bounded sets in a topological vector space X. A mapping $\omega: G \times X \to Y$ is called $\mathfrak{M}$-*small* if $\omega(x, th)/t \to 0$ as $t \to 0$ uniformly in $h \in \mathfrak{M}$ for any $M \in \mathfrak{M}$. A mapping $A: G \to Y$ (where $G \subset X$ is open) is called $\mathfrak{M}$-*differentiable* at $x \in G$ if

$$A(x + h) - Ax = A'(x)h + \omega(A, x, h),$$

where ω is an $\mathfrak{M}$-small mapping. Most frequently $\mathfrak{M}$ is taken to be the collection of all bounded, all compact or all finite sets of X. For non-linear operators on normed spaces the first case leads to Fréchet differentiability and the third to Gâteaux differentiability.

Higher-order derivatives $A^{(n)}(x)$ and $A_0^{(n)}(x)$ of an operator A are defined in the usual way, as derivatives of derivatives. These are symmetric multi-linear mappings. A differential of order n is then a homogeneous form $A^{(n)}(x)h^n$ of degree n. Other definitions of higher-order derivatives are possible. Suppose, for example, that X and Y are normed vector spaces, $G \subset X$ is open, and $x \in G$. If for any h for which $x + h \in G$,

$$A(x + h) - A(x) = a_0(x) + a_1(x)h + \cdots + a_n(x)h^n + \omega, \quad (*)$$

where $\omega = o(\|h\|^n)$, then the multi-linear form $k! a_k(x)$ is called the *derivative of order k*. The expression $(*)$ is then called the *bounded expansion of order n* of the difference $A(x + h) - A(x)$. Under appropriate restrictions the various definitions of higher-order derivatives are equivalent.

If a scalar countably-additive measure is given in X, then a non-linear operator can be integrated, by understanding $\int A(x)\, dx$ in the sense of the **Bochner integral**.

For a non-linear operator $A: M \to Y$, as in the case of a linear operator, the values of the parameter λ for which $(I - \lambda A)^{-1}$ exists and is continuous on $A(M)$ are naturally called *regular*, and the remaining points λ belong to the *spectrum*. In its properties the spectrum of a non-linear operator A can differ vastly from spectra of linear operators. Thus, the spectrum of a completely-continuous non-linear operator can have continuous parts; an *eigen element x_0* of an operator A, that is, an element x_0 such that $x_0 = \lambda A x_0$, can *bifur-*

cate into several eigen element branches (as λ varies), cf. **Bifurcation**.

References

[1] LYUSTERNIK, L.A. and SOBOLEV, V.I.: *Elemente der Funktionalanalysis*, Akad. Verlag, 1968 (translated from the Russian).

[2] KANTOROVICH, L.V. and AKILOV, G.P.: *Funktionalanalysis in normierten Räume*, Akad. Verlag, 1964 (translated from the Russian).

[3] VAĬNBERG, M.M.: *Variational methods for the study of nonlinear operators*, Holden-Day, 1964 (translated from the Russian).

[4] KRASNOSEL'SKIĬ, M.A. and ZABREĬKO, P.P.: *Geometric methods of non-linear analysis*, Springer, 1983 (translated from the Russian).

[5] GAJEWSKI, H., GRÖGER, K. and ZACHARIAS, K.: *Nichtlineare Operatorgleichungen und Operatordifferentialgleichungen*, Akad. Verlag, 1974.

V.I. Sobolev

AMS 1980 Subject Classification: 47HXX

NON-LINEAR OSCILLATIONS - Oscillations in physical systems described by non-linear systems of ordinary differential equations

$$\dot{x} = A(t)x + \mu X(t, x, \mu) + f(t), \qquad (1)$$

where $x \in R^n$, X contains terms of at least the second degree in the components of the vector x, f is a vector function of the time t, and $\mu > 0$ is a small parameter (or $\mu = 1$ and $X = X(t, x)$). Possible generalizations are connected with the discussion of discontinuous systems, actions with discontinuous characteristic (for example, of hysteresis type), delay and random actions, integro-differential and differential-operator equations, oscillating systems with distributed parameters described by partial differential equations, and also with the use of methods of optimal control of non-linear oscillating systems. The basic general problems of non-linear oscillations are: the search for equilibrium positions, for stationary regimes (in particular, for periodic motions, auto-oscillations), and the investigation of their stability, and problems of synchronization and stabilization of non-linear oscillations.

Strictly speaking, all physical systems are non-linear. One of the most characteristic features of non-linear oscillations is the violation of the principle of superposition of oscillations: The result of every action in the presence of a second turns out to be different from the case when the second action is absent. *Quasi-linear systems* are systems (1) with $\mu > 0$. A basic method for studying them is the method of the small parameter (cf. **Small parameter, method of the**). Above all there is the *Poincaré−Lindstedt method* for determining periodic solutions of quasi-linear systems that are analytic in the parameter for sufficiently small values of it, either in the form of power series in μ (see [1], Chapt. IX) or in the form of power series in μ with $\beta_1, \ldots, \beta_n$ added to the initial values of the components of x (see [1],

Chapt. III). For the subsequent development of this method see, for example, [2] - [4].

Another method of a small parameter is that of **averaging**. At the same time, in the study of quasi-linear systems new methods emerged: asymptotic methods (see [5], [6]), the method of V-functions (see [7]), which is based on the fundamental results of A.M. Lyapunov and N.G. Chetaev, and others.

Essentially non-linear system are those in which the small parameter prescribed in advance is absent ($\mu = 1$ and $X = X(t, x)$ in (1)). For the Lyapunov system

$$\dot{x} = Ax + X(x), \qquad (2)$$

where

$$A = \lambda J_2 + P, \quad J_2 = \left\| \begin{matrix} 0 & -1 \\ 1 & 1 \end{matrix} \right\|,$$

and where there are no multiple roots $\pm \lambda i$ among the eigen values of the $(k \times k)$-matrix P, $X(x)$ is an analytic vector function of x the expansion of which begins with terms of at least second order, and there is an analytic first integral of special form. Lyapunov (see [8], Sect. 42) has proposed a method of search for periodic solutions in the form of a series in powers of an arbitrary constant c (for which one can use the initial value of one of the two critical variables x_1 or x_2).

For systems close to Lyapunov systems,

$$\dot{x} = Ax + X(x) + \mu F(t, x, \nu),$$

where A and $X(x)$ are as in (2) and F is an analytic vector function of x and a small parameter ν, continuous and 2π-periodic in t, also a method has been proposed for determining periodic solutions (see [4], Chapt. VIII). Systems of Lyapunov type (2) in which the matrix A has l zero eigen values with simple elementary divisors, a pair of purely imaginary eigen values $\pm \lambda i$ and no multiple eigen values $\pm \lambda i$, and where $X(x)$ is as in (2), may be reduced to Lyapunov systems (see [9], Chapt. IV. 2). A study has also been made of non-linear oscillation in Lyapunov systems and in so-called *Lyapunov systems with damping*, and the general problem of energy transfer in them has been solved (see [9], Chapts. I, III, IV).

Suppose that an essentially non-linear **autonomous system** has been reduced to the Jordan form of its linear part:

$$\dot{x}_\nu = \lambda_\nu x_\nu + \delta_\nu x_{\nu+1} + x_\nu \sum_{Q \in \mathfrak{M}_\nu} f_{\nu Q} x_1^{q_1} \cdots x_n^{q_n}, \qquad (3)$$

$$\nu = 1, \ldots, n; \quad \delta_n = 0,$$

where the vector $\Lambda = (\lambda_1, \ldots, \lambda_n)$ has by assumption at least one non-zero component, $\delta_1, \ldots, \delta_{n-1}$ are zero or one according whether non-simple elementary divisors of the matrix of the linear part are absent or present, the $f_{\nu Q}$ are coefficients, and $\mathfrak{M}_\nu$ is the set of values of the vectors $Q = (q_1, \ldots, q_n)$ with integer components such that:

$$q_1, \ldots, q_{\nu-1}, q_{\nu+1}, \ldots, q_n \geq 0,$$

$$q_\nu \geq -1, \quad q_1 + \cdots + q_n \geq 1.$$

Then there exists a normalizing transformation

$$x_\nu = y_\nu + y_\nu \sum_{Q \in \mathfrak{M}_\nu} h_{\nu Q} y_1^{q_1} \cdots y_n^{q_n}, \quad \nu = 1, \ldots, n, \qquad (4)$$

reducing (3) to the **normal form** of differential equations

$$\dot{y}_\nu = \lambda_\nu y_\nu + \delta_\nu y_{\nu+1} + y_\nu \sum_{\substack{Q \in \mathfrak{M}_\nu \\ (\Lambda, Q)=0}} g_{\nu Q} y_1^{q_1} \cdots y_n^{q_n}, \qquad (5)$$

$$\nu = 1, \ldots, n; \quad \delta_n = 0,$$

such that $g_{\nu Q} = 0$ when $(\Lambda, Q) \neq 0$. Thus, the normal form (5) contains only resonance terms, that is, the coefficients $g_{\nu Q}$ in (5) can be non-zero only for those Q for which the *resonance equation*

$$(\Lambda, Q) \equiv \lambda_1 q_1 + \cdots + \lambda_n q_n = 0,$$

which plays an essential role in the theory of oscillations, holds. The convergence and divergence of the normalizing transformation (4) has been investigated (see [10], Part I, Chapts. II, III); the coefficients $h_{\nu Q}$ have been computed (by means of their symmetrization) (see [9], Sect. 5.3). In a number of problems on non-linear oscillations of essentially non-linear autonomous systems, the method of normal forms has proved effective (see [10], [9], Chapts. VI - VIII).

Of the other methods for studying essentially non-linear systems the method of point mappings (see [2], [11]), the stroboscopic method [12] and methods of functional analysis [13] have been applied.

Qualitative methods for non-linear oscillations. The starting point here is the study of the form of the integral curves of non-linear ordinary differential equations, which was undertaken by H. Poincaré (see [14]). Applications to problems of non-linear oscillations that can be described by second-order autonomous systems can be found in [2] and [15]. Questions of the existence of periodic solutions and their stability in the large for many-dimensional systems have been studied [16]; almost-periodic non-linear oscillations are treated in [13]. For applications of the theory of ordinary differential equations with a small parameter in front of some of the derivatives to problems of non-linear relaxation oscillations (cf. **Relaxation oscillation**) see [17].

For other important aspects of non-linear oscillations and further references see **Perturbation theory** and **Oscillations, theory of.**

References

[1] POINCARÉ, H.: *Oeuvres*, 1, Gauthier-Villars, 1951.
[2] ANDRONOV, A.A., WITT, A.A. and KHAĬKIN, S.E.: *Theory of oscillators*, Pergamon, 1966 (translated from the Russian).
[3] BULGAKOV, B.V.: *Oscillations*, Moscow, 1954 (in Russian).
[4] MALKIN, I.G.: *Some problems of the theory of non-linear oscilla-*

tions, Moscow, 1956 (in Russian).
[5] BOGOLYUBOV, N.N.: *Selected works*, 1, Kiev, 1969 (in Russian).
[6] BOGOLYUBOV, N.N. and MITROPOL'SKIĬ, YU.A.: *Asymptotic methods in the theory of non-linear oscillations*, Gordon & Breach, 1961 (translated from the Russian).
[7] KAMENKOV, G.B.: *Selected works*, 1-2, Moscow, 1971-1972 (in Russian).
[8] LYAPUNOV, A.M.: *Stability of motion*, Acad. Press, 1966 (translated from the Russian).
[9] STARZHINSKIĬ, V.M.: *Applied methods in the theory of nonlinear oscillations*, Mir, 1980 (translated from the Russian).
[10A] BRJUNO, A.D. [A.D. BRYUNO]: 'Analytical form of differential equations', *Trans. Moscow Math. Soc.* **25** (1972), 131-288. (*Trudy Moskov. Mat. Obshch.* **25** (1971), 119-262)
[10B] BRJUNO, A.D. [A.D. BRYUNO]: 'The analytic form of differential equations II', *Trans. Moscow Math. Soc.* **26** (1972), 199-239. (*Trudy Moskov. Mat. Obshch.* **26** (1972), 199-239)
[11] NEĬMARK, YU.I.: *The method of point mappings in the theory of non-linear oscillations*, Moscow, 1972 (in Russian).
[12] MINORSKY, N.: *Nonlinear oscillations*, v. Nostrand, 1962 (translated from the Russian).
[13] KRASNOSEL'SKIĬ, M.A., BURD, V.SH. and KOLESOV, YU.S.: *Nonlinear almost periodic oscillations*, Wiley, 1973 (translated from the Russian).
[14] POINCARÉ, H.: 'Mémoire sur les courbes définies par une équation différentielle I-IV', in *Oeuvres*, Vol. 1, Gauthier-Villars, 1951, pp. Part V; nos. 74-77.
[15] BUTENIN, N.V., NEĬMARK, YU.I. and FUFAEV, N.A.: *Introduction to the theory of non-linear oscillations*, Moscow, 1976 (in Russian).
[16] PLISS, V.A.: *Nonlocal problems of the theory of oscillations*, Acad. Press, 1966 (translated from the Russian).
[17] MISHCHENKO, E.F. and ROZOV, N.KH.: *Differential equations with small parameters and relaxation oscillations*, Plenum, 1980 (translated from the Russian).

V.M. Starszhinskiĭ

Editorial comments.

References

[A1] STOKER, J.J.: *Nonlinear vibrations in mechanical and electrical systems*, Interscience, 1950.
[A2] HALE, J.K.: *Oscillations in nonlinear systems*, McGraw-Hill, 1963.
[A3] MINORSKI, N.: *Nonlinear oscillations*, v. Nostrand, 1962.
[A4] GUCKENHEIMER, J. and HOLMES, PH.: *Nonlinear oscillations, dynamical systems and bifurcations of vector fields*, Springer, 1983.
[A5] HAYASHI, C.: *Nonlinear oscillations in physical systems*, McGraw-Hill, 1964.
[A6] NAYFEH, A.H. and MOOK, D.T.: *Nonlinear oscillations*, Wiley, 1979.

AMS 1980 Subject Classification: 34C15

NON-LINEAR PARTIAL DIFFERENTIAL EQUATION - An equation of the form

$$F(x, u, \ldots, D^\alpha u) = 0, \qquad (1)$$

where $x = (x_1, \ldots, x_n) \in \mathbf{R}^n$, $u = (u_1, \ldots, u_m) \in \mathbf{R}^m$, $F = (F_1, \ldots, F_k) \in \mathbf{R}^k$, $\alpha = (\alpha_1, \ldots, \alpha_n)$ is a multi-index of non-negative integers $\alpha_1, \ldots, \alpha_n$, and $D^\alpha = D_1^{\alpha_1} \cdots D_n^{\alpha_n}$, where $D_i = \partial/\partial x_i$, $i = 1, \ldots, n$. In the case of complex-valued functions a non-linear partial differential equation is defined similarly. If $k > 1$ one speaks, as a rule, of a vectorial non-linear partial differential equation or of a system of non-linear partial differential equations. The *order* of (1) is defined as the highest order of a derivative occurring in the equation.

One of the best known non-linear equations is the **Monge−Ampère equation**

$$\det \left| \frac{\partial^2 u}{\partial x_i \partial x_j} \right| + \qquad (2)$$

$$+ \sum_{i,j=1}^{n} A_{ij}(x, u, Du)\frac{\partial^2 u}{\partial x_i \partial x_j} + B(x, u, Du) = 0;$$

here and below $Du = (D_1 u, \ldots, D_n u)$.

If $k = m$ and F is differentiable with respect to its variables corresponding to the derivatives of highest order, then the type of (1) is defined as that of the principal linear part of F relative to these derivatives (see **Differential equation, partial**). One assigns, in general, to such a derivative (or to the resulting differential equation) a correspondingly defined weight. For example, in the non-linear heat equation

$$\frac{\partial u}{\partial x_1} = f\left[x_1, x_2, u, \frac{\partial u}{\partial x_2}, \frac{\partial^2 u}{\partial x_2^2}\right]$$

with $\partial f / \partial p_{22} > 0$, $p_{22} \Leftrightarrow \partial^2 u / \partial x_2^2$, the derivative $\partial f / \partial p_{22}$ has weight 2.

Since linearization of (1) with respect to the highest derivatives proceeds in a neighbourhood of a fixed solution, the type of (1) (in contrast to a linear equation even at a fixed point x) may depend on this fixed solution. For example, the equation

$$\frac{\partial^2 u}{\partial x_1^2} + \frac{\partial^2 u}{\partial x_2^2}\frac{\partial u}{\partial x_2} = f(x_1, x_2) \qquad (3)$$

is of elliptic type on solutions with $\partial u / \partial x_2 > 0$, but is of hyperbolic type on solutions with $\partial u / \partial x_2 < 0$.

The type of an equation determines whether boundary value (mixed) problems for this equations are well-posed and influences the method for studying them.

If the function F depends linearly on its highest derivatives, then (1) is called a *quasi-linear equation*. For example, (3) is quasi-linear. Otherwise the equation is called an *essentially non-linear equation*. For example, the Monge−Ampère equation (2) is essentially non-linear.

If the coefficients of the highest derivatives of a quasi-linear equation do not depend on the solution (or its derivatives), then the equation is called a *weakly non-linear equation*. For example, the equation

$$\Delta u = f(x, u, Du) \qquad (4)$$

is weakly non-linear.

The distinction between quasi-linear and weakly non-linear partial differential equations bears a conditional character and does not reflect an intrinsic property of the equations. Weakly non-linear equations

may have stronger non-linearity properties than quasi-linear and even essentially non-linear equations. For example, there exist weakly non-linear equations of the form (4) which have countably many distinct solutions for a given Dirichlet boundary condition in a bounded domain.

Equations of the form (1) can be considered in the whole space $\mathbf{R}^n$ or in some subdomain of it. In the first case the definition of the solution space includes conditions on the behaviour of the solutions at infinity. In the case of a domain one imposes on the boundary or on a part of it one or more boundary conditions. These boundary conditions may also involve non-linear operators.

A non-linear partial differential equation together with a boundary condition (or conditions) gives rise to a non-linear problem, which must be considered in an appropriate function space. The choice of this space of solutions is determined by the structure of both the non-linear differential operator F in the domain and that of the boundary operators. The choice of the function space for a non-linear problem is an essential feature in the investigation of the problem. For example, to the non-linear problem

$$\sum_{|\alpha|=m} (-1)^{|\alpha|} D^\alpha(|D^\alpha u|^{p-1}\operatorname{sgn} D^\alpha u) = f(x) \quad \text{with } p > 1$$

in a bounded domain $\Omega \subset \mathbf{R}^n$ and

$$D^\beta u = 0, \quad |\beta| \leq m - 1, \quad \text{on the boundary } \partial\Omega$$

corresponds the Sobolev space $\mathring{W}_p^m(\Omega)$. This problem has, for any function f from the dual space $W_q^{-m}(\Omega) = (\mathring{W}_p^m(\Omega))^*$, $q^{-1} + p^{-1} = 1$, a unique solution u in $\mathring{W}_p^m(\Omega)$. Here and below $\mathring{W}_p^m(\Omega)$ is the closure in the Sobolev space $W^m(\Omega)$ of the set of all infinitely-differentiable functions in Ω of compact support.

In the study of problems for non-linear partial differential equations one treats questions such as: the existence of a solution and the number of solutions, or the non-existence, and collapse or bifurcation (branching) of a solution. Another problem is the asymptotic behaviour of a solution when the argument tends to the boundary, in particular, to infinity in the case of unbounded domains. The theory of such equations has two aspects: a local and a global one. The local theory is relatively completely developed for general non-linear problems of elliptic, parabolic or hyperbolic type. This theory is based on the implicit-function theorem in non-linear functional analysis and on the general theory of linear problems of corresponding type.

In the case of boundary value (mixed) problems for non-linear parabolic, or hyperbolic, equations this local theory makes it possible to establish the solvability of the problem on a sufficiently small time interval or on a fixed interval under the condition that the data of the problem deviate (in the appropriate metric) by a fairly small amount from the data for a known solution (as a rule, the zero solution) of a nearby problem.

The global theory of non-linear problems is less completely developed, and then only for individual classes of equations.

Non-linear first-order partial differential equations. For a broad class of quasi-linear scalar first-order equations of the form

$$\frac{\partial u}{\partial t} + \sum_{i=1}^{n} \frac{\partial}{\partial x_i} \phi_i(t, x, u) + \psi(t, x, u) = 0 \tag{5}$$

the existence and uniqueness problem for the Cauchy problem with an initial condition for $t = 0$ has been established for all $t > 0$.

For a narrower class of equations of the form (5) the question of the asymptotic behaviour of the solutions of such problems as $t \to +\infty$ and boundary value problems have also been discussed.

The theory of systems of quasi-linear first-order partial differential equations has been developed less completely (see **Quasi-linear hyperbolic equations and systems**).

Non-linear second-order partial differential equations. *Equations of elliptic and parabolic type.* The theory of global solvability of boundary value problems for a broad class of quasi-linear scalar second-order equations of elliptic type of the form

$$\sum_{i=1}^{n} \frac{\partial}{\partial x_i} a_i(x, u, Du) + a(x, u, Du) = 0, \tag{6}$$

or

$$\sum_{i,j=1}^{n} a_{ij}(x, u, Du) \frac{\partial^2 u}{\partial x_i \partial x_j} + a(x, u, Du) = 0 \tag{7}$$

is relatively complete under the condition that an a priori estimate of $\max_x |u(x)|$ is available. Here the coefficients of the equations are subject to certain conditions.

A similar situation prevails in the theory of global solvability of boundary value (mixed) problems for the broad class of quasi-linear scalar second-order equations of parabolic type of the form

$$\frac{\partial u}{\partial t} = \sum_{i=1}^{n} \frac{\partial}{\partial x_i} a_i(x, u, Du) + a(x, u, Du), \tag{8}$$

or

$$\frac{\partial u}{\partial t} = \sum_{i,j=1}^{n} a_{ij}(x, u, Du) \frac{\partial^2 u}{\partial x_i \partial x_j} + a(x, u, Du). \tag{9}$$

This solvability theory is based on a priori estimates and the Leray–Schauder method.

A priori estimates of $\max_x |u(x)|$ for certain classes of non-linear equations (6) - (9) can be obtained on the basis of either the **maximum principle** or special integral inequalities and corresponding **imbedding theorems** for function spaces.

The global solvability theory of boundary value problems for essentially non-linear scalar equations of elliptic type has been developed for a narrow class of such equations in the case of two independent variables.

Global solvability of the boundary value (mixed) problem for essentially non-linear scalar equations of parabolic type has been established for the broad class of equations of the form

$$\frac{\partial u}{\partial t} = a\left(x, u, \frac{\partial u}{\partial x}, \frac{\partial^2 u}{\partial x^2}\right)$$

in the case of a single space variable $x \in \mathbf{R}^n$.

Questions of global solvability of problems for systems of quasi-linear equations of elliptic or parabolic type have been considered only for narrow individual classes of such systems.

One of the effective methods in the study of second-order non-linear partial differential equations of either elliptic or parabolic type is the method of super- and subsolutions. Consider, for example, the boundary value problem

$$\begin{cases} -\Delta u = f(x, u, Du) & \text{in a bounded domain } \Omega \subset \mathbf{R}^n, \\ u = 0 & \text{on the boundary } \partial\Omega \text{ of class } C^2, \end{cases} \tag{10}$$

with a continuous function f defined on $\Omega \times \mathbf{R} \times \mathbf{R}^n$ and such that Bernshteĭn's inequality

$$|f(x, u, \xi)| \leq M(|u|) \cdot (1 + |\xi|^2)$$

holds for all $(x, u, \xi) \in \bar{\Omega} \times \mathbf{R} \times \mathbf{R}^n$ with an increasing function $M: \mathbf{R}_+ \to \mathbf{R}_+ = \{t \in \mathbf{R}: t \geq 0\}$. If there exist functions $u^+, u^- \in W_p^2(\Omega)$ with $p > n$ such that $-\Delta u^+ \geq f(x, u^+, Du^+)$ almost-everywhere in Ω, $u^+ \geq 0$ on $\partial\Omega$, $-\Delta u^- \leq f(x, u^-, Du^-)$ almost-everywhere in Ω, and $u^- \leq 0$ on $\partial\Omega$ (u^+ and u^- are super- and subsolutions of the problem (10)), and $u^+(x) \geq u^-(x)$ in Ω, then (10) has a solution $u \in W_p^2(\Omega)$ and $u^-(x) \leq u(x) \leq u^+(x)$ in Ω. A good choice of super- and subsolutions of non-linear initial value problems based on solutions of model problems makes it possible to establish not only the solvability and a lower bound on the number of solutions, but also to obtain precise a priori estimates and the asymptotic behaviour of solutions of non-linear initial value problems.

The study of the global behaviour of solutions of boundary value (mixed) problems for non-linear parabolic equations is connected with that of stationary solutions of boundary value problems for corresponding non-linear elliptic equations, as it is in the case of ordinary differential equations.

Due to the fact that boundary value problems for non-linear elliptic equations do not always have a solution and that boundary value (mixed) problems for non-linear parabolic and hyperbolic equations need not have a solution for all $t > 0$, a theory of non-existence of solutions for non-linear partial differential equations has been developed.

Equations of hyperbolic type. These equations occupy a special place among the second-order non-linear differential equations. The 'loss of one derivative' in the inversion of the second-order hyperbolic operator leads to principal obstacles in the study of non-linear hyperbolic equations. Even the local theory of non-linear hyperbolic equations and systems required the development of a special theory of implicit functions in non-linear functional analysis, since the classical implicit-function theorem in functional analysis turned out to be inapplicable here.

For (essentially) quasi-linear hyperbolic second-order equations in more than two independent variables the problem of global solvability, even for the Cauchy problem, has not been studied.

In the case of two independent variables ($t \in \mathbf{R}_+$, $x \in \mathbf{R}$) the global solvability of the Cauchy problem has been established for individual equations of the form

$$u_{tt} - a^2(u_x)u_{xx} = f(t, x, u_t, u_x),$$

which reduce to special quasi-linear hyperbolic systems by the conservation principle (see **Quasi-linear hyperbolic equations and systems**).

In the case of quasi-linear hyperbolic equations of the form

$$u_{tt} - a^2 \left[\int_\Omega |Du|^2 \, dx \right] \Delta u = f(t, x) \ \text{ in } \Omega \times [0, T],$$

the global solvability (for any $T > 0$) of the mixed problem with the conditions: $u = 0$ on $\partial\Omega \times [0, T]$ and $u = \phi$, $u_1 = \psi$ at $t = 0$, $x \in \Omega$, has been established in a certain class of smooth functions in x. Here Ω is a bounded domain in $\mathbf{R}^n$ with boundary $\partial\Omega$ of class C^∞ and $|Du|^2 = \sum_{i=1}^n (\partial u / \partial x_i)^2$.

The global solvability of the Cauchy problem and also of the boundary value (mixed) problem has been established for the broad class of weakly non-linear hyperbolic equations of the form

$$u_{tt} - \Delta u = f(t, x, u, u_t, Du), \ \ t > 0, \ \ x \in \Omega \subseteq \mathbf{R}^n.$$

A special place in the theory of non-linear hyperbolic equations is occupied by the problem of existence of periodic solutions (in t) of boundary value problems for these equations. Even for weakly non-linear equations this problem has been discussed only for equations of the form

$$u_{tt} - u_{xx} = f(t, x, u)$$

in the case of two variables $t \in \mathbf{R}$ and $x \in [a, b] \subset \mathbf{R}$. The complexity of this problem stems from the fact that the kernel of the corresponding linear problem is infinite dimensional.

Weakly non-linear and quasi-linear hyperbolic equations containing dissipative terms have been studied more completely.

Non-linear partial differential equations of higher order. The solvability of boundary value (mixed) problems has been studied for the broad class of quasi-linear equations in divergence form

$$\sum_{|\alpha| \leqslant m} (-1)^{|\alpha|} D^\alpha A_\alpha(x, u, \ldots, D^\beta u) = f(x), \qquad (11)$$

$$\frac{\partial u}{\partial t} + \sum_{|\alpha| \leqslant m} (-1)^{|\alpha|} D^\alpha A_\alpha(t, x, u, \ldots, D^\beta u) = f(t, x),$$

$$|\beta| \leqslant m.$$

Concerning the functions A_α a number of conditions are assumed in this case which ensure that the non-linear operators are defined in the corresponding function spaces and satisfy certain conditions. For example, for the solvability of the boundary value problem in a bounded domain $\Omega \subset \mathbf{R}^n$ with the conditions

$$D^\omega u = 0, \ \ |\omega| \leqslant m - 1, \text{ on the boundary } \partial\Omega, \qquad (12)$$

for equation (11) the following conditions are sufficient:

1) The functions $A_\alpha(x, \xi_0, \ldots, \xi_\beta)$, $\xi_\beta \leftrightarrow D^\beta u$, are measurable in x for all $\xi_0, \ldots, \xi_\beta$, continuous in $(\xi_0, \ldots, \xi_\beta)$ for almost-all $x \in \Omega$ and satisfy the inequality

$$|A_\alpha(x, \xi_0, \ldots, \xi_\beta)| \leqslant K \left[1 + \sum_{|\gamma| < m} |\xi_\gamma|^{p-1} \right]$$

with $p > 1$ and $K > 0$, $|\alpha| \leqslant n$.

2) The following coerciveness condition holds: For any function u in the Sobolev space $\mathring{W}_p^m(\Omega)$,

$$\sum_{|\alpha| \leqslant m} \int_\Omega A_\alpha(x, u, \ldots, D^\beta u) D^\alpha u \, dx \geqslant$$

$$\geqslant a_0 \sum_{|\alpha| = m} \int_\Omega |D^\alpha u|^p \, dx - K \text{ with } a_0 > 0.$$

3) The monotonicity condition holds: For any functions u and v in $\mathring{W}_p^m(\Omega)$,

$$\sum_{|\alpha| \leqslant m} \int_\Omega (A_\alpha(x, u, \ldots, D^\beta u) +$$

$$- A_\alpha(x, v, \ldots, D^\beta v))(D^\alpha u - D^\alpha v) \, dx \geqslant 0.$$

Under the conditions 1) - 3) the boundary value problem (12) for equation (11) has a solution u in $\mathring{W}_p^m(\Omega)$ for any function f in the dual space $(\mathring{W}_p^m(\Omega))^*$.

All these conditions can be substantially relaxed. For example, the Fredholm alternative holds for differential operators of the form (11) with boundary conditions (12) that are odd and homogeneous, in principle, i.e. subject to certain conditions but without coercivity. If in this case the boundary value problem with zero boundary conditions (12) for equation (11) with $f = 0$ has only the trivial solution, then this problem is solvable for any function f in the corresponding dual space.

For a broad class of boundary value (mixed) problems for non-linear partial differential equations a theory of normal solvability has been developed, generalizing the (Hausdorff) theory of normal solvability of linear operator equations to the non-linear case. This theory gives sufficient conditions for the solvability of boundary value (mixed) problems for non-linear equations and systems of equations of parabolic type and for weakly non-linear equations and systems of hyperbolic type.

In the theory of boundary value problems for non-linear equations of elliptic type, the question of the existence of eigen functions occupies a special place. The theory of eigen functions of boundary value problems for quasi-linear equations of elliptic type has been developed for a fairly broad class of problems. In particular, the abstract Lyusternik−Shnirel'man theory on the existence of a countable set of eigen functions has been transferred to a broad class of problems.

A special chapter in the theory of non-linear equations of elliptic type of higher order and in the theory of systems of non-linear equations of elliptic type in more than two independent variables is the question of regularity of solutions of these equations and systems.

In the case of a scalar quasi-linear uniformly-elliptic second-order equation with sufficiently smooth coefficients satisfying together with their first derivatives certain growth conditions, the solution has an interior regularity exceeding the smoothness of the right-hand side by two derivatives.

In the case of quasi-linear equations of elliptic type of order higher than two, and systems of quasi-linear equations of elliptic type of order two or more, in more than two independent variables, the corresponding smoothness of the solutions does not hold everywhere in the domain, but almost everywhere. Under additional conditions one succeeds in making precise the dimensions of the sets of zero (Hausdorff) measure on which the smoothness of solutions is, in general, violated. For a special class of quasi-linear elliptic systems with bounded non-linearities it has been established that the solutions are regular everywhere in the domain.

A theory has been established of boundary value problems for a broad class of quasi-linear equations of infinite order in divergence form.

Theory of exact solutions. Among the methods for constructing exact solutions one reckons: methods based on the application of group theory in the analysis of non-linear partial differential equations; methods based on Lie−Bäcklund transformation; and methods based on the inverse problem of scattering theory.

The inverse-scattering method has made it possible to study a number of physically important equations, such as the non-linear **Korteweg−de Vries equation**

$$u_t - 6uu_x + u_{xxx} = 0;$$

the non-linear **sine-Gordon equation**

$$u_{tt} - u_{xx} + \sin u = 0;$$

the non-linear **Schrödinger equation**

$$-i\psi_t + \psi_{xx} + \kappa |\psi|^2 \psi = 0,$$

and a number of others in one space variable $x \in \mathbf{R}$. By means of this method one has been able to consider particular non-linear equations like the Korteweg−de Vries equation in two space variables.

References

[1] COURANT, R. and HILBERT, D.: *Methods of mathematical physics. Partial differential equations*, 2, Interscience, 1965 (translated from the German).

[2] LADYZHENSKAYA, O.A. and URAL'TSEVA, N.N.: *Linear and quasilinear elliptic equations*, Acad. Press, 1968 (translated from the Russian).

[3] GILBARG, D. and TRUDINGER, N.S.: *Elliptic partial differential equations of second order*, Springer, 1977.

[4] LADYZHENSKAYA, O.A., SOLONNIKOV, V.A. and URAL'TSEVA, N.N.: *Linear and quasi-linear equations of parabolic type*, Amer. Math. Soc., 1968 (translated from the Russian).

[5] FRIEDMAN, A.: *Partial differential equations of parabolic type*, Prentice-Hall, 1964.

[6] NIRENBERG, L.: *Topics in nonlinear functional analysis*, New York Univ., 1974.

[7] LIONS, J.-L.: *Quelques méthodes de resolution des problèmes aux limites nonlineaires*, Dunod, 1969.

[8] ROZHDESTVENSKIĬ, B.L. and YANENKO, N.N.: *Systems of quasilinear equations and their applications to gas dynamics*, Amer. Math. Soc., 1983 (translated from the Russian).

[9] SKRYPNIK, I.V.: *Non-linear elliptic equations of higher order*, Kiev, 1973 (in Russian).

[10] FUČIK, S., NEČAS, J. and SOUČEK, V.: *Spectral analysis of non-linear operators*, Springer, 1973.

[11] OVISANNIKOV, L.V. [L.V. OVSYANNIKOV]: *Group analysis of differential equations*, Acad. Press, 1982 (translated from the Russian).

[12] ZAKHAROV, V.E., MANAKOV, S.V., NOVIKOV, S.P. and PITAEVSKIĬ, L.P.: *Theory of solutions. The method of the inverse problem*, Moscow, 1980 (in Russian).

[13] VISHIK, M.I.: 'Quasi-linear strongly elliptic systems of differential equations in divergence form', *Trans. Moscow Math. Soc.* **12** (1963), 140-208. (*Trudy Moskov. Mat. Obshch.* **12** (1963), 125-184)

[14] GIAQUINTA, M. and MODICA, G.: 'Almost-everywhere regularity results for solutions of non-linear elliptic systems', *Manuscripta Math.* **28** (1979), 109-158.

[15] DUBINSKIĬ, YU.A.: 'Sobolev spaces of infinite order and the behaviour of solutions of some boundary value problems with unbounded increase of the order of the equation', *Math. USSR Sb.* **27**, no. 2 (1975), 143-162. (*Mat. Sb.* **98**, no. 2 (1975), 163-184)

[16] KAZDAN, J.L. and KRAMMER, R.: 'Invariant criteria for existence of solutions to second-order quasilinear elliptic equations', *Comm. Pure Appl. Math.* **31**, no. 5 (1978), 619-645.

[17] KOSHELEV, A.I.: 'Regularity of solutions of quasi-linear elliptic systems', *Russian Math. Surveys* **33**, no. 4 (1978), 1-52. (*Uspekhi Mat. Nauk* **33**, no. 4 (1978), 3-49)

[18] KRUZHKOV, S.N.: 'First order quasilinear equations in several independent variables', *Math USSR Sb.* **10**, no. 2 (1970), 217-243. (*Mat. Sb.* **81**, no. 2 (1970), 228-255)

[19] KRUZHKOV, S.N.: 'Nonlinear parabolic equations in two independent variables', *Trans. Moscow Math. Soc.* **16** (1967), 355-373. (*Trudy Moskov. Mat. Obshch.* **16** (1967), 329-346)

[20] OLEĬNIK, O.A.: 'Uniqueness and stability of the generalized solution of the Cauchy problem for a quasi-linear equation', *Uspekhi Mat. Nauk* **14**, no. 2 (1959), 165-170 (in Russian).

[21] POKHOZHAEV, S.I.: 'On the eigenfunctions of quasilinear elliptic problems', *Math. USSR Sb.* **11**, no. 2 (1970), 171-188. (*Mat. Sb.* **82**, no. 2 (1970), 192-212)

[22] POKHOZHAEV, S.I.: 'On a class of quasilinear hyperbolic equa-

tions', *Math USSR Sb.* **25**, no. 1 (1975), 145-158. (*Mat. Sb.* **96**, no. 1 (1975), 152-166)

[23] POKHOZHAEV, S.I.: 'On equations of the form $\Delta u = f(x, u, Du)$', *Math USSR Sb.* **41**, no. 2 (1982), 269-280. (*Mat. Sb.* **113**, no. 2 (1980), 324-338)

S.I. Pokhozhaev

Editorial comments. Recent developments for non-linear parabolic partial differential equations are sketched in [A2], [A3].

An important and large class of elliptic second-order non-linear equations arises in the theory of controlled diffusion processes. These are known as Bellman equations (cf. **Bellman equation**). For these equations probabilistic techniques and ideas can be used to study solvability and to obtain solutions. And indeed, many important results were first obtained in this way and were only subsequently also derived using analytic methods, cf. [A1]. It also turned out that the **Monge—Ampère equation** is a special case of the Bellman equation, [A1].

References
[A1] KRYLOV, N.V.: *Nonlinear elliptic and parabolic equations of the second order*, Reidel, 1987 (translated from the Russian).

[A2] AMANN, H.: 'Quasilinear parabolic systems', in L. Boccardo and A. Tesei (eds.): *Nonlinear parabolic equations: qualitative properties of solutions*, Longman, 1987, pp. 5-12.

[A3] CAMPANATO, S.: '$L^{2,\lambda}$ theory and nonlinear parabolic systems', in L. Boccardo and A. Tesei (eds.): *Nonlinear parabolic equations: qualitative properties of solutions*, Longman, 1987, pp. 55-62.

[A4] GARABEDIAN, P.: *Partial differential equations*, Wiley, 1964.

AMS 1980 Subject Classification: 35F20, 35G20, 35M05

NON-LINEAR POTENTIAL - A function $U_\mu(x)$ generated by a **Radon measure** μ, x being a point of the Euclidean space $\mathbf{R}^N$, $N \geqslant 2$, that depends non-linearly on the generating measure.

For example, in the study of properties of solutions of partial differential equations and of boundary properties of analytic functions, non-linear potentials of the following form turn out to be useful:

$$U_\mu(x) = U_\mu(x;p,l) = \qquad (*)$$

$$= \int \left[\int \frac{d\mu(z)}{|y-z|^{N-l}} \right]^{1/(p-1)} \frac{dy}{|x-y|^{N-l}}, \quad x \in \mathbf{R}^N,$$

where $|x-y|$ is the distance between x and y, μ is a Radon measure with compact support, and p and l are real numbers, $1 < p < \infty$, $0 < l < \infty$.

For $p = 2$ the non-linear potentials (*) turn into linear Riesz potentials (cf. **Riesz potential**), and for $p = 2$ and $l = 1$ into the classical **Newton potential**. The concepts of capacity and energy have been constructed, and analogues of certain basic theorems of **potential theory** have been proved for the non-linear potential (*) (see [1]).

References
[1] MAZ'YA, V.G. and KHAVIN, V.P.: 'Nonlinear potential theory',

Russian Math. Surveys **27**, no. 6 (1972), 71-148. (*Uspekhi Mat. Nauk* **27**, no. 6 (1972), 67-138)

E.D. Solomentsev

Editorial comments. In recent years, non-linear versions of different branches of potential theory, concrete or axiomatic, have been constructed. A sample of these developments is given by [A1] - [A6].

References
[A1] ADAMS, D.R.: 'Weighted nonlinear potential theory', *Trans. Amer. Math. Soc.* **297** (1986), 73-94.

[A2] BERTIN, E.M.J.: 'Fonctions convexes et théorie du potentiel', *Indag. Math.* **41** (1979), 385-409.

[A3] GRANDLUND, S., LINDQVIST, P. and MARTIO, O.: 'Note on the PWB-method in the non-linear case', *Pacific J. Math.* **125** (1986), 381-395.

[A4] HEDBERG, L.I. and WOLFF, TH.H.: 'Thin sets in nonlinear potential theory', *Ann. Inst. Fourier (Grenoble)* **33**, no. 4 (1983), 161-187.

[A5] LAINE, I.: 'Axiomatic non-linear potential theories', in J. Král, et al. (ed.): *Potential Theory. Survey and Problems. Prague, 1987*, Lecture notes in math., Vol. 1344, Springer, 1988, pp. 118-132.

[A6] MIZUTA, Y. and NAKAI, T.: 'Potential theoretic properties of the subdifferential of a convex function', *Hiroshima Math. J.* **7** (1977), 177-182.

AMS 1980 Subject Classification: 31C15

NON-LINEAR PROGRAMMING - The branch of **mathematical programming** concerned with the theory and methods for solving problems of optimization of non-linear functions on sets given by non-linear constraints (equalities and inequalities).

The principal difficulty in solving problems in non-linear programming is their multi-extremal nature, while the known numerical methods for solving them in the general case guarantee convergence of minimizing sequences to local extremum points only.

The best studied branch of non-linear programming is **convex programming**, the problems in which are characterized by the fact that every local minimum point is a global minimum.

References
[1] ZANGWILL, W.I.: *Nonlinear programming: a unified approach*, Prentice-Hall, 1969.

[2] KARMANOV, V.G.: *Mathematical programming*, Moscow, 1975 (in Russian).

[3] POLAK, E.: *Computational methods in optimization: a unified approach*, Acad. Press, 1971.

V.G. Karmanov

Editorial comments.

References
[A1] MINOUX, M.: *Mathematical programming: theory and algorithms*, Wiley, 1986.

AMS 1980 Subject Classification: 90C30

NON-MEASURABLE SET - A set that is not a **measurable set**. In more detail: A set X belonging to a hereditary σ-ring $H(S)$ is non-measurable if

$$\mu^*(X) > \mu_*(X);$$

here S is the σ-ring on which the measure μ is given, and μ^* and μ_* are the exterior and interior measures, respectively (see **Measure**).

For an intuitive grasp of the concept of a non-measurable set, the following 'effective constructions' are useful.

Example 1. let

$$K = \{(x,y): 0 \leqslant x \leqslant 1, \, 0 \leqslant y \leqslant 1\}$$

be the unit square and define a measure μ on the sets

$$\hat{E} = \{(x,y): x \in E, \, 0 \leqslant y \leqslant 1\},$$

where E runs through the Lebesgue-measurable sets of measure $m(E)$, by setting $\mu(\hat{E}) = m(E)$. Then the set

$$X = \{(x,y): 0 \leqslant x \leqslant 1, \, y = 1/2\}$$

is non-measurable, since $\mu^*(X) = 1$, $\mu_*(X) = 0$.

The oldest and simplest construction of a non-measurable set is due to G. Vitali (1905).

Example 2. Let $\mathbf{Q}$ be the set of all rational numbers. Then a set X (called a *Vitali set*) having in accordance with the **axiom of choice** exactly one element in common with every set of the form $\mathbf{Q} + a$, where a is any real number, is non-measurable. No Vitali set has the **Baire property**.

Example 3. Let B (respectively, C) be the set of numbers of the form $n + m\xi$, where ξ is an irrational number, m and n are integers with n even (respectively, n odd), and let X_0 be a set obtained by means of the axiom of choice from the equivalence classes of the set of real numbers under the relation:

$$x \sim y \quad \text{if} \quad x - y \in A = B \bigcup C.$$

Let $X = X_0 + B$. Then for every measurable set E:

$$\mu_*(X \bigcap E) = 0, \quad \mu^*(X \bigcap E) = \mu(E).$$

Yet another construction of a non-measurable set is based on the possibility of introducing a total order in a set having cardinality of the continuum.

Example 4. There exist a set $B \subset \mathbf{R}$ such that B and $\mathbf{R} \setminus B$ intersect every uncountable closed set. Any such set (a *Bernstein set*) is non-measurable (and does not have the Baire property). In particular, any set of positive exterior measure contains a non-measurable set.

Apart from invariance under a shift (Example 2) and topological properties (Example 3) there are also reasons of a set-theoretical character why it is impossible to define a non-trivial measure for all subsets of a given set; such is, for example, *Ulam's theorem* (see [2]) for sets of bounded cardinality.

No specific example is known of a Lebesgue non-measurable set that can be constructed without the use of the axiom of choice.

References
[1] HALMOS, P.: *Measure theory*, v. Nostrand, 1950.
[2] OXTOBY, J.: *Measure and category*, Springer, 1971.
[3] GELBAUM, B. and OLMSTED, J.: *Counterexamples in analysis*, Holden-Day, 1964.

M.I. Voĭtsekhovskiĭ

Editorial comments. See also **Measure; Gödel constructive set; Descriptive set theory.**

In certain models of ZF (not ZFC) every set of real numbers is Lebesgue measurable (a *result of Solovay*), so the axiom of choice is needed to construct a Lebesgue non-measurable set.

For Ulam's theorem see **Cardinal number.**

AMS 1980 Subject Classification: 28A05

NON-ORIENTABLE MANIFOLD - A **manifold** that does not admit an **orientation**. Such are, for example, the **Möbius strip**, the **Klein surface** and the projective spaces $\mathbf{RP}^n$ of even dimension.

M.I. Voĭtsekhovskiĭ

AMS 1980 Subject Classification: 57NXX

NON-OSCILLATION INTERVAL, *interval of disconjugacy* - A connected interval J on the real axis $\mathbf{R}$ such that any non-trivial solution $x = x(t)$ of a given ordinary linear differential equation of order n with real coefficients,

$$x^{(n)} + a_1(t)x^{(n-1)} + \cdots + a_n(t)x = 0, \qquad (*)$$

has on it more than $n - 1$ zeros, an m-fold zero counted m times. Properties of solutions of $(*)$ on a non-oscillation interval have been well studied (see, for example, [1] - [3]). There are several generalizations of the concept of a non-oscillation interval, to linear systems of differential equations, to non-linear differential equations, and also to other types of equations (difference, with deviating argument).

References
[1] HARTMAN, P.: *Ordinary differential equations*, Birkhäuser, 1982.
[2] LEVIN, A.YU.: 'Non-oscillation of solutions of the equation $x^{(n)} + p_1(t)x^{(n-1)} + \cdots + p_n(t)x = 0$', *Russian Math. Surveys* **24**, no. 2 (1969), 43-99. (*Uspekhi Mat. Nauk* **24**, no. 2 (1969), 43-96)
[3] COPPEL, W.A.: *Disconjugacy*, Springer, 1971.

Yu.V. Komlenko

AMS 1980 Subject Classification: 34C10

NON-PARAMETRIC METHODS IN STATISTICS - Methods in mathematical statistics that do not assume a knowledge of the functional form of general distributions. The name 'non-parametric method' emphasizes their contrast to the classical, parametric, methods, in which it is assumed that the general distribution is known up to finitely many parameters, and which make it possible to estimate the unknown values of these parameters from results of observations and to test hypotheses concerning their values.

Example. Let $X_1, \ldots, X_m$ and $Y_1, \ldots, Y_n$ be two independent samples derived from populations with continuous general distribution functions F and G; suppose that the hypothesis H_0 that F and G are equal is to be tested against the alternative of a shift, that is, the hypothesis

$$H_1: G(t) = F(t-\theta)$$

for all t and some $\theta \neq 0$. In the classical version it is assumed that F and G are normal distribution functions, and to test the hypothesis in question one uses the **Student test**. In the non-parametric statement of the problem no assumptions are made on the form of F and G except continuity. A typical non-parametric test for testing the hypothesis H_0 against H_1 is the **Wilcoxon test**, which is based on the sum of the ranks of the first sample in the series of joint order statistics. One rejects the hypothesis that the distributions are equal if the test statistic computed from the observations turns out to be too large or too small. The statistic of Wilcoxon's test is easy to calculate and its distribution under H_0 does not depend on F. The critical values corresponding to a given **significance level** for small values of m and n can be found in tables (see, for example, [1]); for large m and n one uses a normal approximation.

In a number of cases it is important not only to test the hypothesis of absence of a shift, but also to estimate this shift θ, which can be interpreted, for example, as the change in yield when a method of tilling the land is replaced by another, or as the additional time of sleep after taking a pill. The estimate of the parameter θ given by the quantity $\overline{Y} - \overline{X}$, which is quite satisfactory in the normal case, may be very unsuitable under departures from normality and may be even inconsistent. A non-parametric estimate of θ may enjoy vastly better properties in this respect (see [2]); for example, the median of the collection of mn numbers $Y_j - X_i$, $i=1, \ldots, m$, $j=1, \ldots, n$. This estimate is closely connected with the Wilcoxon test. One could say that it stands in the same relation to the estimate $\overline{Y} - \overline{X}$ as the Wilcoxon test to the Student test.

Notwithstanding the great variety of problems that can be solved by non-parametric methods, these problems can conventionally be divided into two large parts: problems of testing hypotheses and problems of estimating unknown distributions and parameters, which are understood as certain functionals of these distributions.

Non-parametric testing of statistical hypotheses is the most generally developed part of non-parametric methods in statistics. It is required to set up a procedure (a test) that makes it possible to accept or reject the hypothesis to be tested against a given alternative. A typical example is the **goodness-of-fit test**, and other important examples for applications are tests for symmetry, independence and randomness.

The *problem of testing goodness-of-fit* consists in the following: From a sample of a population with general distribution function G one has to test the hypothesis $G = F$, where F is a given continuous distribution function. The non-parametric nature of the problem manifests itself here in the non-parametric alternative, which can be stated, for example, in a one-sided: $F < G$ or $F > G$, or two-sided version: $F \neq G$.

The *problem of testing symmetry* consists in testing the symmetry of a general distribution function G relative to a given point x_0, that is,

$$G(x_0+t) + G(x_0-t) = 1.$$

As alternative one can take one-sided conditions

$$G(x_0+t) + G(x_0-t) \geqslant 1,$$
$$G(x_0+t) + G(x_0-t) \leqslant 1,$$

with strict inequality for at least one t, or two-sided conditions of the same type.

The *problem of testing independence* arises in cases when one has to decide whether two characteristics observed in one and the same object are independent, given independent observations on these objects.

In a similar fashion one can state the hypothesis of randomness, when it is assumed that the elements of a sample are independent identically-distributed quantities. Apart from alternatives of a general form there also occur cases when it turns out to be possible to indicate in precisely what way the distributions of the elements of a sample differ under an alternative; in this way, for example, the alternatives of trend and regression arise.

Methods of algorithmic construction of non-parametric procedures with given properties have so far only been worked out inadequately, and as a rule, intuition and heuristic arguments play a major role in the choice of a suitable procedure. In this way a large body of methods and devices for solving frequently occurring non-parametric problems has been accumulated (see [3]).

An extensive group of non-parametric tests is based on the use of empirical distribution functions. Let F_n be an empirical distribution function constructed from a sample of size n in a population with general distribution function F. By the Glivenko$-$Cantelli theorem,

$$\sup_t |F_n(t) - F(t)| \to 0 \quad \text{as } n \to \infty$$

with probability 1. Thus, the empirical and true distribution functions uniformly approach each other with probability 1 and one can base tests for goodness-of-fit with an hypothesis on the true distribution function on measures of their proximity.

The first tests of this kind were the **Kolmogorov test** and the **Cramér$-$von Mises test**, which were put forward at the beginning of the 1930's and were based, respectively, on the statistics

$$D_n = \sqrt{n}\sup_t | F_n(t) - F(t) |$$

and

$$\omega_n^2 = n \int_{-\infty}^{\infty} (F_n(t) - F(t))^2 \, dF(t).$$

It should be mentioned that both these statistics have distributions that are independent of the general distribution function F, provided only that the latter is continuous. Their limiting distributions, which were found in the middle of the 1930's by A.N. Kolmogorov and N.N. Smirnov, have been tabulated, which makes it possible to find the boundary of the critical domain corresponding to a given significance level.

Many versions of tests for goodness-of-fit based on the difference between F_n and F have been proposed and studied, for example the **Rényi test** and the tests of Anderson $-$ Darling, Watson and others (see [4]). For a successful application in the case of large samples one has to know the relevant limiting distributions in the first instance. They can be found by an approach according to which the test statistic is presented as a continuous functional of an empirical process,

$$\xi_n(t) = \sqrt{n}(G_n(t) - t), \quad 0 \leqslant t \leqslant 1,$$

where $G_n(t)$ is an empirical distribution function constructed from a sample of size n of a uniform distribution on $[0, 1]$. The process $\xi_n(t)$ converges weakly in the space $D[0, 1]$ to a Gaussian process, a so-called Brownian bridge (cf. [6]). Therefore the limiting distribution of the studied statistic coincides with the distribution of the corresponding functional on the Brownian bridge, which can be computed by standard methods.

There are modifications of the statistics D_n and ω_n^2 that are intended to test hypotheses on the distribution in the multi-dimensional case and also to test hypotheses of independence and symmetry. In these cases a number of additional difficulties arise. For example, in the multi-dimensional case all the statistics in question cease to have the property of universality (independence of the original distribution). Most important is the case of a uniform distribution on the unit cube, because a sample for a multi-dimensional distribution can in one way or another be transformed into a sample from a uniform distribution. However, neither the exact nor the limiting distributions of the Kolmogorov statistic are known (1982), not even in this simple case. Similar difficulties emerge when one has to test not a simple, but a composite hypothesis on the distribution, that is, when one assumes that the general distribution function is of the form $F(t, \theta)$, where θ is an unknown one- or multi-dimensional parameter. In

this case it is natural to estimate θ from a sample, for example, by a maximum-likelihood estimator $\hat{\theta}_n$ (cf. **Maximum-likelihood method**), and to compare $F_n(t)$ with $F(t, \hat{\theta}_n)$. The statistics D_n, ω_n^2 and their modifications can be constructed as in the case of a simple hypothesis. However, the distributions of these statistics, both exact and limiting, turn out again to depend on the form of F and in many cases also on the unknown true value of θ. The task of computing these distributions is tedious, and their exact form is not known, although for statistics like ω_n^2 tables of limiting distribution have been set up in a number of cases (see [5]). For some other statistics simulated percentage points are known.

Apart from the goodness-of-fit tests considered, their two-sample and multi-sample analogues have also been constructed, which can be used to test goodness-of-fit as well as homogeneity of certain samples (see **Smirnov test**).

A common property of tests for goodness-of-fit and uniformity based on empirical distribution functions is their consistency against arbitrary alternatives. However, the choice of one statistic or another in a practical problem is hampered because their powers are insufficiently studied. For samples of large size one can rely on a knowledge of the Pitman asymptotic relative efficiency, which has been computed for a number of the simplest statistics (see [7]).

Rank tests form another group of non-parametric tests (cf. **Rank test**). The earliest use of the rank sign test occurs in the work of J. Arbuthnott (1710), who used it to analyze statistical data on the birth-rate of boys and girls to obtain 'arguments for divine providence'. But the modern period of development of rank tests begins at the end of the 1930's. After the publication in 1945 of F. Wilcoxon's paper, in which he proposed the rank test bearing his name (cf. **Wilcoxon test**), rank methods entered into a period of intense development.

The use of rank procedures is based on the following arguments. Since the rank vector together with the vector of order statistics contains the entire information embodied in the sample, a certain part of the information is contained in the rank vector only. One can construct statistical procedures based only on the ranks without using knowledge of the sample values themselves. The advantage of such procedures is their computational simplicity, which follows from the fact that the ranks are integers. Another important feature of rank procedures is their advantage in cases when the observations are of a qualitative, not quantitative, nature, as long as they admit an ordering, which is practically important in research in sociology, psychology and medicine. Finally, the distributions of rank statistics under the null hypotheses do not depend on the underlying distribution, which makes it possible to compute these distributions once and for all.

As rank methods developed it became clear that the part of the information contained in the rank vector can prove to be significant, in which case these procedures are highly efficient. In the example studied above, connected with testing homogeneity of two samples, an extension of the domain of applicability of the test leads to a loss in power, and in the normal case Student's test (cf. **Student test**) is more powerful than any rank test. However, when there is a large number of observations, Wilcoxon's test loses little compared with Student's test. It turns out that in the normal case the asymptotic relative efficiency of Wilcoxon's test to Student's test is $3/\pi = 0.955$. But when the underlying distribution differs from the normal one, then the asymptotic relative efficiency in question can be arbitrarily large, but never drops below 0.864 (see [4]). Moreover, there is a rank test (the so-called *normal scores test*) with asymptotic relative efficiency 1 relative to Student's test in the normal case and exceeding 1 for any deviation from normality. Thus, this test turns out to be asymptotically preferable to Student's test.

Another example is connected with testing the hypothesis of symmetry. Suppose that a sample $X_1, \ldots, X_n$ is extracted from a population with general density f and that one wishes to test the hypothesis that f is symmetric with respect to zero, again with the alternative of a shift. The simplest test in this case is the **sign test**, which is based on the number of positive values among the X_i. Wilcoxon's signed rank test is based on the statistic $\sum_{X_i > 0} R_i^+$, where R_i^+ is the rank of X_i in the series of order statistics for $|X_1|, \ldots, |X_n|$. The statistic of this test uses not only the information about the signs of the observations but also about their magnitude. Therefore, one can expect that Wilcoxon's test will be more efficient than the sign test. Actually, the asymptotic relative efficiency of these tests to Student's test are $3/\pi = 0.955$ and $2/\pi = 0.637$ (in the normal case). Thus, Wilcoxon's test exceeds the sign test by a factor of $3/2$ and concedes little to Student's test.

Another example is connected with testing the hypothesis of independence. Suppose that there is a number of objects each of which has two attributes, quantitative or qualitative (the mathematical and musical talents of a student, the colour and the ripeness of berries, etc.). It is assumed that observations on the quantitative attributes can be ordered. It is required to test from n independent observations on the objects, the hypothesis that the attributes are independent against the alternative, say, that they are positively dependent. Let R_i and S_i be the ranks of the attributes corresponding to the i-th observation. A natural criterion to test independence is Spearman's coefficient of rank correlation r_s, which can be computed by the formula

$$r_s = 1 - \frac{6 \sum (R_i - S_i)^2}{n^3 - n}.$$

The hypothesis of independence is rejected for large values of r_s, i.e. close to 1.

The critical values for small n can be found in tables; for large n one uses a normal approximation. The asymptotic relative efficiency of the test based on r_s relative to that based on the sampling correlation coefficient is again fairly high, namely $9/\pi^2 = 0.912$ in the normal case (see [9]).

Since for the testing of each non-parametric hypothesis there are many rank tests, frequently proposed from heuristic arguments, the choice must be based on certain optimality arguments. As is known, uniformly most-powerful tests in the class of all possible alternatives rarely exist, even in the parametric case. Therefore, by optimal rank tests for finite sample sizes one means just locally most-powerful tests. For example, Wilcoxon's test is locally most powerful in the two-sample problem of testing homogeneity against the alternative of a shift for the logistic distribution with density $e^x(1+e^x)^{-2}$, and the normal scores test in the same problem for a normal distribution. In the asymptotic theory for the corresponding property of optimality one uses a certain concept of asymptotic efficiency, and locally most-powerful tests usually turn out to be asymptotically optimal (see [8]).

In the theory of rank tests it is assumed that the distributions of the observations are continuous, so that they can be ordered without ties and the rank statistics are uniquely determined. However, in practice observations are always rounded-off, therefore ties sometimes appear. The following two methods are most commonly used for overcoming this difficulty. The first one consists in randomly ordering tied observations. In the second method one assigns the average rank of the group to each of a group of tied observations. The merits of the two methods have not yet been sufficiently investigated.

Non-parametric estimation is a section of non-parametric statistics that deals with problems of estimating unknown distributions or functions of them such as quantiles, moments, modes, entropy, information in the sense of Fisher, etc.

The most widely used estimator for an unknown distribution function is the empirical one. The strong uniform consistency of it as an estimator of an unknown distribution function follows from the Glivenko − Cantelli theorem, and its minimax character has been established in [10]. But consistent estimation of an unknown density is a more complicated problem. For the estimation problem to be well-posed, additional a priori information is needed on the class F of densities to which the relevant density f belongs. In the classical statements the a priori family of densities is given in parametric form and is determined by a finite-dimensional vector of the unknown parameters. In the non-parametric statement the problem assumes an infinite-dimensional character, and the accuracy of the estimation of an unknown density depends essentially (cf. [11]) on a geometric characterization of the 'massiveness' of the class F.

The most extensively used estimators of an unknown density f are 'kernel estimators'

$$f_n(t) = \frac{1}{nh_n} \sum_{i=1}^{n} K\left(\frac{t - X_i}{h_n}\right),$$

where the X_i are observations, the kernel function K is absolutely integrable and satisfies the condition

$$\int_{-\infty}^{\infty} K(x)\,dx = 1,$$

and the sequence h_n is such that $nh_n \to \infty$ as $h_n \to 0$. In some cases one uses other non-parametric estimators of the density: simpler ones (the histogram, the frequency polygon) or more complicated ones, for example, Chentsov's projection estimators. The question of the accuracy of approximation by these estimators to an unknown density in relation to properties of the class F has been well studied (see [11], [12]).

An empirical distribution function and a non-parametric estimator of the density can be used to estimate functionals of unknown general distributions; for this purpose it is sufficient to replace the unknown distribution by its estimators in the expressions for the functional in question. The idea itself and the beginning of its realization go back to work of R. von Mises in the 1930's and 1940's. It has been proved that under certain restrictions on the class of functions to be estimated and on the non-parametric class of distributions there exists a minimax lower bound on the quality of non-parametric estimators (see [12]). Non-parametric estimation is closely connected with the problem of constructing robust estimates.

References

[1] BOL'SHEV, L.N. and SMIRNOV, N.V.: *Tables of mathematical statistics*, Libr. of mathematical tables, 46, Nauka, Moscow, 1983 (in Russian). Processed by L.S. Bark and E.S. Kedrova.

[2] HODGES, J. and LEHMANN, E.: 'Estimates of location based on rank tests', *Ann. Math. Stat.* 34 (1963), 598-611.

[3] WALSH, J.E.: *Handbook of nonparametric statistics*, 1-3, v. Nostrand, 1965.

[4] KENDALL, M.G. and STUART, A.: *The advanced theory of statistics*, 2. Inference and relationship, Griffin, 1979.

[5] MARTYNOV, G.V.: *The omega-squared test*, Moscow, 1978 (in Russian).

[6] BILLINGSLEY, P.: *Convergence of probability measures*, Wiley, 1968.

[7] WIEAND, H.: 'A condition under which the Pitman and Bahadur approaches to efficiency coincide', *Ann. of Stat.* 4 (1976), 1003-1011.

[8] HÁJEK, J. and ŠIDÁK, Z.: *Theory of rank tests*, Acad. Press, 1967.

[9] KENDALL, M.: *Rank correlation*, Griffin, 1968.

[10] DVORETZKY, A., KIEFER, J. and WOLFOWITZ, J.: 'Asymptotic minimax characterization of the sample distribution function and of the classical multinomial estimator', *Ann. Math. Stat.* 27 (1956), 642-669.

[11] CHENTSOV, N.N.: *Statistical decision rules and optimal inference*, Amer. Math. Soc., 1982 (translated from the Russian).

[12] IBRAGIMOV, I.A. and HAS'MINSKII, R. [R.Z. KHAS'MINSKIĬ]: *Statistical estimation: asymptotic theory*, Springer, 1981 (translated from the Russian).

[13] WAERDEN, B.L. VAN DER: *Mathematische Statistik*, Springer, 1957.

[14] LEHMANN, E.L.: *Testing statistical hypotheses*, Wiley, 1986.

[15] SCHMETTERER, L.: *Einführung in die Mathematische Statistik*, Springer, 1966.

[16] LEHMANN, E.L.: *Nonparametrics: statistical methods based on ranks*, McGraw-Hill, 1975.

Ya.Yu. Nikitin

Editorial comments. Let $X(t)$ be Brownian motion, $X(0)=0$. For fixed $t_0>0$, $x, y \in \mathbf{R}$ define the process $X(x, y; t_0)$ for $0 \leqslant t \leqslant t_0$ by

$$X(x, y; t_0)(t) = x + X(t) + \frac{t}{t_0}(y - x - X(t_0)).$$

Thus, $X(x, y; t_0)(0) = x$, $X(x, y; t_0)(t_0) = y$. This process is called *pinned Brownian motion* or the *Brownian bridge* (from x to y). Its stochastic differential equation is

$$dX(t) = dB(t) + \frac{1}{t - t_0}(y - X(t))\,dt, \quad X(0) = x.$$

Cf. [A1] for more details.

For a recent text on modern work on the direct estimation of probability densities (and regression curves) cf. [A2].

References

[A1] IKEDA, N. and WATANABE, S.: *Stochastic differential equations and diffusion process*, North-Holland, 1981, Sect. IV.8.5.

[A2] NADARAYA, E.A.: *Nonparametric estimation of probability densities and regression curves*, Kluwer, 1989 (translated from the Russian).

AMS 1980 Subject Classification: 62GXX

NON-PARAMETRIC TEST - A **statistical test** of a hypothesis H_0: $\theta \in \Theta_0 \subset \Theta$ against the alternative H_1: $\theta \in \Theta_1 = \Theta \setminus \Theta_0$ when at least one of the two parameter sets Θ_0 and Θ_1 is not topologically equivalent to a subset of a Euclidean space. Apart from this definition there is also another, wider one, according to which a statistical test is called *non-parametric* if the statistical inferences obtained using it do not depend on the particular null-hypothesis probability distribution of the observable random variables on the basis of which one wants to test H_0 against H_1. In this case, instead of the term 'non-parametric test' one speaks frequently of a 'distribution-free test'. The **Kolmogorov test** is a classic example of a non-parametric test. See also **Non-parametric methods in statistics; Kolmogorov−Smirnov test.**

References

[1] RAO, C.R.: *Linear statistical inference and its applications*, Wiley, 1965.

[2] BOL'SHEV, L.N. and SMIRNOV, N.V.: *Tables of mathematical statistics*, Libr. of mathematical tables, 46, Nauka, Moscow, 1983 (in Russian). Processed by L.S. Bark and E.S. Kedrova.

[3] IBRAGIMOV, I.A. and HAS'MINSKII, R.Z. [R.Z. KHAS'MINSKIĬ]: *Statistical estimation: asymptotic theory*, Springer, 1981 (translated from the Russian).

[4] KENDALL, M.G. and STUART, A.: *The advanced theory of statistics*, 2. Inference and relationship, Griffin, 1979.

M.S. Nikulin

AMS 1980 Subject Classification: 62GXX

NON-PASCALEAN GEOMETRY - A geometry with a non-commutative multiplication. As a consequence of the fact that in affine geometry the property of commutativity is equivalent to the **Pascal theorem**, the name non-Pascalean geometry is usually attached to a geometry in which the following theorem fails to hold: Suppose that on each of two intersecting straight lines three points A, B, C and A_1, B_1, C_1 are given, other than the point of intersection of the lines; if CB_1 is parallel to BC_1 and CA_1 is parallel to AC_1, then BA_1 is parallel to AB_1. This is sometimes called *Pappus' theorem*; it is a special case of the theorem of Pascal in the theory of conic sections (namely, when the conic degenerates to a pair of straight lines).

The possibility of constructing a non-Pascalean geometry follows from the fact that Pascal's theorem is not a consequence of the axioms of incidence, order and parallelism when the metric axioms are excluded from Hilbert's system (cf. **Hilbert system of axioms**). On the other hand, the existence of a non-Pascalean geometry is also connected with the possibility of constructing a geometry over a non-commutative skew-field, that is, a non-Pascalean geometry is at the same time a **non-Archimedean geometry**.

The significance of non-Pascalean geometry stems from the role of Pascal's theorem in research connected with establishing the independence of axiom systems and logical connections between propositions.

References

[1] HILBERT, D.: *Grundlagen der Geometrie*, Springer, 1913.

[2] BIEBERBACH, L.: *Einleitung in die höhere Geometrie*, Teubner, 1933.

[3] SKORNYAKOV, L.A.: 'Projective planes', *Uspekhi Mat. Nauk* **6**, no. 6 (1951), 112-154 (in Russian).

L.A. Sidorov

Editorial comments. The phrase '(non-) Pascalean geometry' is obsolete: it has been replaced by '(non-) Pappian geometry'.

References

[A1] COXETER, H.S.M.: *Twelve geometric essays*, Univ. Illinois Press, 1968.

AMS 1980 Subject Classification: 51A30

NON-PREDICATIVE DEFINITION - A definition that is meaningful only when the object to be defined is assumed to exist.

The formation of the set of all sets is non-predicative. So is the definition of the least upper bound of an arbitrary set of real numbers. Any non-predicative definition can be regarded as a property that selects the required object from a certain given aggregate. Since here the problem of the existence of the object in question remains open, instead of non-predicative definitions one can also speak of non-predicative properties. Once the language in which the properties are expressed is fixed, the concept of non-predicativity can be made more precise as follows. A property (more accurately, a linguistic expression for the relevant property) is called *non-predicative* if it contains a bound variable such that the object to be defined falls within its range of applicability. A property is called *predicative* if it contains no such bound variables.

The concept of non-predicativity arose in connection with the discovery of set-theoretical paradoxes, and the term itself is due to H. Poincaré (1906), who was first to raise objections against non-predicative definitions.

The antinomies (cf. **Antinomy**) discovered at the beginning of the 20-th century contain non-predicativities. For example, in Russell's paradox the set R of all sets that do not contain themselves as elements is defined by the formula:

$$\forall x\, (x \in R \Leftrightarrow \neg x \in x),$$

where x is a variable ranging over all sets. In this formula R is a possible value of the variable x. In mathematics non-predicative definitions are widely used. For example, the union S of all sets of natural numbers satisfying a condition ϕ can be defined by the formula

$$\forall n\, (n \in S \Leftrightarrow \exists M\, (\phi(M) \& n \in M)), \qquad (*)$$

where n is a variable ranging over the natural numbers and M a variable ranging over subsets of the natural numbers. In this formula S is a possible value of the bound variable. A non-predicative way of specifying an object can sometimes be replaced by a predicative one. For example, if for the property $\phi(M)$ one takes the formula $M = A \vee M = B$, where A and B are certain fixed sets, then (*) is equivalent to the predicative formula

$$\forall n\, (n \in S \Leftrightarrow n \in A \vee n \in B),$$

which expresses that S is the union of A and B.

B. Russell made an attempt of constructing mathematics on a predicative basis. In the theory of types (cf. **Types, theory of**) which he developed, sets are arranged in a hierarchy in accordance with the expressions defining them. For example, the set S in (*) must be assigned to a higher level in the hierarchy than that of M and those of the variables contained in the formula for ϕ. On the predicative basis one cannot build up analysis to its full extent. A statement is meaningful only if certain restrictions on the levels occurring in that statement are satisfied. Russell was forced to introduce an axiom of reducibility, which practically wiped out the distinction between levels. However, the predicative theory in the presence of the axioms of arithmetic makes it possible to build up analysis to an extent sufficient for many applications (see [4]).

The phenomenon of non-predicativity is based essentially on the absolute character of the understanding the word 'all' (all without any restrictions, decisively all). The division of all sets into distinct 'levels of sets' is an attempt to restrict this absolute character. A more radical revision of the method of thinking based on an absolute understanding of the concept 'all' is undertaken by the intuitionistic and constructive trends in mathematics (cf. **Intuitionism; Constructive mathematics**).

References

[1] CHURCH, A.: *Introduction to mathematical logic*, 1, Princeton Univ. Press, 1956.
[2] FRAENKEL, A.A. and BAR-HILLEL, Y.: *Foundations of set theory*, North-Holland, 1958.
[3] HILBERT, D. and ACKERMANN, W.: *Principles of mathematical logic*, Chelsea, 1950 (translated from the German).
[4] TAKEUTI, G.: *Two applications of logic to mathematics*, Shoten, 1978.

V.N. Grishin

AMS 1980 Subject Classification: 03B15, 03A05

NON-RESIDUE *of power n modulo m* - A number a for which the **congruence** $x^n \equiv a \pmod{m}$ has no solution. See also **Remainder of an integer**.

Editorial comments. Usually this term refers to the case $n=2$. It was first used by C.F. Gauss in his *Disquisitiones Arithmetica*.

AMS 1980 Subject Classification: 10A10

NON-SELF-ADJOINT OPERATOR - A linear operator in a Hilbert space the spectral analysis of which cannot be made to fit into the framework of the theory of self-adjoint operators (cf. **Self-adjoint operator**) and its simplest generalizations: the theory of unitary operators (cf. **Unitary operator**) and the theory of normal operators (cf. **Normal operator**). Non-self-adjoint operators arise in the discussion of processes that proceed without conservation of energy: in problems with friction, in the theory of open resonators, in problems of inelastic scattering, and others. Certain self-adjoint problems, in which by separation of variables an operator-valued function $\mathscr{L}(\lambda)$ appears that depends non-linearly on a spectral parameter λ, also lead to a study of non-self-adjoint operators. Many of the propositions referring to the theory of non-self-adjoint operators are valid also for operators acting in arbitrary Banach spaces, F-spaces, topological vector spaces, etc.

The most extensive method in the study of non-self-adjoint operators is that of evaluating the **resolvent**, which makes use of the theory of analytic functions, of asymptotic expansions, etc.

The first works concerned with the theory of non-self-adjoint operators were by G. Birkhoff, Ya.D. Tamarkin, V.A. Steklov, and others, in the investigation of problems for ordinary differential equations. These studies applied Cauchy's method of contour integration to the resolvent.

For non-self-adjoint partial differential operators effective methods of research were lacking for a long time. This can be explained by the complicated structure of the resolvent of such an operator as an analytic function.

In the development of a general theory of non-self-adjoint operators (in particular, partial differential operators) an important role was played by the work of M.V. Keldysh [1] (see also [2]). He studied an equation of the form

$$y = \mathscr{L}(\lambda)y, \tag{1}$$

where y is an element of a certain Hilbert space H and the operator $\mathscr{L}(\lambda)$ has the representation

$$\mathscr{L}(\lambda) = B_0 + \lambda H_0 B_1 + \cdots + \lambda^{n-1} H_0^{n-1} B_{n-1} + \lambda^n H_0^n.$$

Here H_0 is a completely-continuous invertible self-adjoint operator of finite order and the B_j, $0 \leqslant j \leqslant n-1$, are arbitrary completely-continuous operators. (A completely-continuous operator A acting on a Hilbert space is said to be an *operator of finite order* if $\sum s_k^\rho(A) < \infty$ for some ρ, $0 < \rho < \infty$; the $s_k(A)$ denote the *singular numbers* of A, that is, the eigen values of $(AA^*)^{1/2}$.) The *eigen values* of (1) are those λ for which the equation has non-trivial solutions y; these solutions are called *eigen vectors*.

Under the assumptions made above the spectrum of (1) is discrete. Since $\mathscr{L}(\lambda)$ is non-self-adjoint, apart from the eigen vectors there naturally arise (in the presence of a multiple spectrum) associated vectors. In [1] a chain is constructed of associated vectors $y_1, \ldots, y_k$ corresponding to an eigen value λ and eigen vector y according to the rule

$$y_\nu = \mathscr{L}(\lambda)y_\nu + \frac{1}{1!}\frac{\partial \mathscr{L}(\lambda)}{\partial \lambda}y_{\nu-1} + \cdots + \frac{1}{\nu!}\frac{\partial^\nu \mathscr{L}(\lambda)}{\partial \lambda^\nu}y, \tag{2}$$

$$\nu = 1, \ldots, k.$$

The system of eigen vectors and associated vectors of $\mathscr{L}(\lambda)$ is said to be *n-fold complete* if any n vectors $\phi_0, \ldots, \phi_{n-1}$ of H can be approximated in the norm of H with arbitrary accuracy by finite linear combinations of the form

$$\sum_k c_k \left[\frac{d^\nu v_k(t)}{dt^\nu} \right]_{t=0}$$

with the same coefficients c_k. Here $v_k(t)$ a is vector-valued function of the form

$$e^{\lambda t}\left[y_k + \frac{t}{1!}y_{k-1} + \cdots + \frac{t^k}{k!}y\right].$$

The definition of n-fold completeness is, naturally, connected with the solution of the Cauchy problem for the non-stationary equation corresponding to (1).

According to a theorem of Keldysh, under the assumptions made on the coefficients of $\mathscr{L}(\lambda)$ the system of all eigen vectors and associated vectors of $\mathscr{L}(\lambda)$ is n-fold complete in H. In [1] he also proved that the eigen values of $\mathscr{L}(\lambda)$ can be approximated asymptotically on rays with $\arg \lambda x = \sqrt{\pi}/n$. In the proof of completeness Keldysh developed a new method for evaluating the resolvent of an abstract completely-continuous non-self-adjoint operator of finite order. Here emerged the special role played in the completeness problem by *Volterra operators*, i.e. completely-continuous operators with a single point of the spectrum at zero. In establishing the asymptotic behaviour of the eigen values, Keldysh [3] used a new Tauberian theorem due to him.

Keldysh's research was continued by many authors. His theorem was extended in [4] to the case when the operator $\mathscr{L}(\lambda)$ depends rationally on λ.

In [5] - [7] the equation $M(\lambda)y = 0$ with $M(\lambda) = I + \lambda B + \lambda^2 C$ is considered, where C is a completely-continuous positive-definite operator and B is a bounded self-adjoint operator. A generalization of Pontryagin's theorem (see [8]) on the existence of a maximal J-non-negative invariant subspace for a J-self-adjoint operator A made it possible (see [6] and [7]) to establish the two-fold completeness of all eigen vectors and associated vectors of $M(\lambda)$ in situations important for applications, and also the one-fold completeness of the subsystem corresponding to the spectrum situated in the left (or right) half-plane. These results have been developed much further.

The summability of Fourier series in eigen vectors and associated vectors of a completely-continuous operator A of finite order ρ has been established (see [9]) if the eigen values of the quadratic form (Ax, x) lie in a sector of the complex plane with opening angle less than π/ρ. (For applications of this theorem and further generalizations of it see [10] and the references there.)

The question when the system of eigen vectors and associated vectors forms a basis in the Hilbert space has been studied in a number of papers. Most general conditions under which the system of eigen vectors and associated vectors of a dissipative completely-continuous operator forms a basis are found in [12].

In the case of singular differential operators with discrete spectrum a number of subtle results have been obtained (see [11] and [13]) on the completeness of the eigen functions and associated functions of a Sturm−Liouville operator with complex potential. Important results have been obtained for the case of an elliptic operator (see [14]). Keldysh's theorem has been generalized to the case of generalized eigen functions and associated functions of non-self-adjoint elliptic operators (see [15], [16]).

An attempt to carry over the theorem on the reduction to Jordan form of a finite-dimensional operator to the infinite-dimensional case has led to the construction of a triangular integral representation. For completely-continuous operators $B = B_R + iB_I$, where B_R and B_I are self-adjoint and B_I is of finite order, an analogue has been obtained of Schur's theorem on the unitary equivalence of B to a triangular operator (see [17]). A special place in the problem of triangular representation is occupied by Volterra operators.

A special role in this problem is played by von Neumann's theorem, stating that a completely-continuous linear operator in a Hilbert space has a non-trivial invariant subspace; an arbitrary bounded linear operator in a Banach space need not have an invariant subspace; the corresponding problem for the case of a Hilbert space is still open (1989). A Volterra operator is said to be a *unicellular operator* if for any two invariant subspaces Q_1 and Q_2 of it either $Q_1 \subset Q_2$ or $Q_2 \subset Q_1$. In [18] a necessary and sufficient condition was found for an operator B to be a unicellular operator under the assumption that B_I is nuclear and non-negative definite; this condition can be stated in terms of the growth of the resolvent of B as $\lambda \to \infty$. A simple sufficient condition for being a unicellular operator is indicated in [19].

Non-self-adjoint operators with a continuous spectrum were first studied by M.A. Naĭmark (see [20], [21]), who obtained an expansion in a Fourier integral, connected with the non-self-adjoint problem

$$l(y) = -y'' + p(x)y = \lambda y, \quad 0 \leqslant x < \infty, \tag{3}$$

$$y'(0) - \theta y(0) = 0, \tag{4}$$

where $p(x)$ is a complex-valued function subject to the condition

$$\int_0^\infty (1+x^2)\,|\,p(x)\,|\,dx < \infty \tag{5}$$

and θ is a complex number. The results in [20] imply, in particular, that in a neighbourhood of points of the real axis where $A(s) = \omega_x(0, s) - \theta \omega(0, s)$ vanishes ($s = \sqrt{x}$), the spectral projections of the operator (3) - (4) are unbounded. (Here $\omega(x, s)$ denotes the solution of (3) for which $\omega(x, s)e^{-ixs} \to 1$ as $x \to +\infty$, the *Jost solution*.) In [20], the real zeros of $A(s)$ are called *spectral singularities*. In [22] a generalization of the results of [20] to the case of the Schrödinger equation in the three-dimensional space is obtained. In a further development of [20] it has been proved (see [23]) that in general (without a restriction of type (5)) the spectral function of a differential operator must be regarded as a continuous linear functional on a certain topological space.

In [24] a system of non-self-adjoint equations is studied with singular points whose position depends on the spectral parameter. These systems occur in the theory of shells without moments (cf. **Shell theory**). For such systems asymptotic properties of the solutions have been established, and also solvability theorems of Cauchy type have been proved. Completeness theorems for the system of eigen functions and associated functions of non-self-adjoint integro-differential operators generating a non-regular problem have been established.

An important problem in the theory of non-self-adjoint operators is the expansion of the kernel of the Green operator of a biorthogonal series in eigen functions and associated functions and also the question of a basis. Tamarkin [25] has studied the expansion of a summable function in a series of eigen functions and associated functions of a regular problem, and also the question of equiconvergence with a trigonometric Fourier series. Later it was proved (see [26], [27]) that for non-regular problems equiconvergence with a trigonometric series does not hold.

For strongly-regular conditions the system of eigen functions and associated functions forms a basis in L_2. It has been proved (see [28], [29]) that this system not only forms a basis, but even a so-called Riesz basis.

In [30] an important method has been developed for studying the basis property and the uniform convergence of the expansions in eigen functions and associated functions of an ordinary non-self-adjoint operator. This method is a further development of the ideas applied in the investigation of self-adjoint problems. A new treatment for the eigen functions and associated functions has been proposed, making it possible to dispense with the specific form of the boundary conditions; general ordinary differential operators or pencils of such operators are considered and necessary and sufficient conditions have been established for the basis property of the eigen functions and associated functions of such operators, as well as a criterion for equiconvergence. The method is based only on a mean value formula for the eigen functions and associated functions (see also [31]). It also turns out that if the operator has infinitely many associated functions, then the basis property depends on the choice of the latter (see [32]).

For non-self-adjoint elliptic operators there is convergence (see [33]) of a certain sequence of means of Pois-son type of the partial sums of the biorthogonal series, that is, a summation method has been proposed.

Expansions in eigen functions and associated functions of non-regular problems were first obtained for problems of the form $y''' = \lambda y$, $y(0) = y'(0) = y(1) = 0$. It has been proved (see [26]) that expansions in uniformly convergent series of this form can hold for functions satisfying a certain analyticity condition.

Among the fundamental studies is the problem (see [34]) in which the perturbation of the spectrum of the Laplace operator under a change of domain is studied. Here for the first time the role of the capacity of the set which is varying on the spectrum of the operator became manifest. These methods were successfully applied in the study of non-self-adjoint operators.

Regularization methods, the basics of which were expounded in [35], are successfully used in the theory of non-self-adjoint operators. An example is the problem of regularized traces (cf. **Trace**) of non-self-adjoint operators. The first paper on the theory of traces was [36], where the regularized trace of a Sturm−Liouville operator was computed. More general results in the theory of operator traces were obtained in [37] and [38]. It turns out that trace formulas for ordinary non-self-adjoint differential operators depending in a complicated manner on the spectral parameter can be obtained as consequences of the formulas for regularized sums of roots of a certain class of entire functions. (The traces of singular operators and partial differential operators are considered in [38], [39].)

Among the important works in which new methods and ideas in the theory of non-self-adjoint operators were developed one counts also the survey lecture [40].

The construction of a definitive theory of non-self-adjoint operators is far from complete (1989). On the one hand, in the theory itself there are new trends of research, such as scattering theory [41], the construction of the theory of contraction operators [42], the method of the canonical Maslov operator [43], the theory of spectral operators [44], and others; on the other hand, research on applied problems, mechanics and mathematical physics suggests new ways of developing this theory.

References

[1] KELDYSH, M.V.: 'On eigenvalues and eigenfunctions of some classes of nonselfadjoint equations', *Dokl. Akad. Nauk SSSR* **77**, no. 1 (1951), 11-14 (in Russian).

[2] KELDYSH, M.V.: 'On the completeness of the eigenfunctions of some class of non-selfadjoint linear operators', *Russian Math. Surveys* **26**, no. 4 (1971), 15-44. (*Uspekhi Mat. Nauk* **26**, no. 4 (1971), 15-41)

[3] KELDYSH, M.V.: 'On a Tauberian theorem', *Trudy Mat. Inst. Steklov.* **38** (1951), 77-86 (in Russian).

[4] ALLAKHVERDIEV, D.E.: 'On completeness of a system of eigenelements and associated elements of a nonselfadjoint operator close to normal', *Dokl. Akad. Nauk SSSR* **115**, no. 2 (1957), 207-210 (in Russian).

[5] LANGER, H.K.: 'On J-Hermitian operators', *Soviet Math. Dokl.* **1**, no. 4 (1960), 1052-1055. (*Dokl. Akad. Nauk SSSR* **134**, no. 2 (1960), 263-266)

[6] KREĬN, M.G.: 'A new application of the fixed-point principle in the theory of operators on a space with indefinite metric', *Soviet Math. Dokl.* **5**, no. 1 (1964), 266-269. (*Dokl. Akad. Nauk SSSR* **154**, no. 5 (1964), 1023-1026)

[7] LANGER, H.K. and KREĬN, M.G.: *Applications of function theory in the mechanics of a continuous medium*, Moscow, 1965 (in Russian).

[8] PONTRYAGIN, L.S.: 'Hermitian operators in spaces with indefinite metric', *Izv. Akad. Nauk SSSR Ser. Mat.* **8** (1944), 243-280 (in Russian).

[9] LIDSKIĬ, V.B.: 'On completeness of a system of eigenelements and associated functions of a nonselfadjoint operator', *Dokl. Akad. Nauk. SSSR* **110**, no. 2 (1956), 172-175 (in Russian).

[10] AGRANOVICH, M.S.: 'Summability of series in root vectors of non-self-adjoint elliptic operators', *Funct. Anal. Appl.* **10**, no. 3 (1976), 1-3. (*Funktsional. Anal. i Prilozhen.* **10**, no. 3 (1976), 1-12)

[11] GOHBERG, I.C. [I.Ts. GOKHBERG] and KREĬN, M.G.: *Introduction to the theory of linear nonselfadjoint operators*, Amer. Math. Soc., 1969 (translated from the Russian).

[12] KATSNEL'SON, V.E.: 'Conditions under which systems of eigenvectors of some classes of operators form a basis', *Funct. Anal. Appl.* **1**, no. 2 (1967), 122-133. (*Funksional. Anal. i Prilozhen.* **1**, no. 2 (1967), 39-51)

[13] LIDSKIĬ, B.V.: 'Conditions for complete continuity of the resolvent of a differential operator', *Dokl. Akad. Nauk SSSR* **113**, no. 1 (1957), 28-31 (in Russian).

[14] KOSTYUCHENKO, A.G.: 'Asymptotic distribution of the eigenvalues of elliptic operators', *Soviet Math. Dokl.* **5**, no. 5 (1964), 1171-1175. (*Dokl. Akad. Nauk. SSSR* **158**, no. 1 (1964), 41-44)

[15] PONOMAREV, S.M.: 'A generalization of M.V. Keldysh's theorem concerning the completeness of eigen- and associated functions of the first boundary value problem for a nonselfadjoint elliptic operator', *Diff. Eq.* **10**, no. 12 (1974), 1777-1779. (*Differentsial. Uravnen.* **10**, no. 12 (1974), 2294-2296)

[16] KRUKOVSKIĬ, N.M.: 'The m-tuple completeness of the system of generalized eigenfunctions and associated functions of a non-self-adjoint elliptic operator', *Diff. Eq.* **12**, no. 10 (1976), 1290-1296. (*Differentsial. Uravnen.* **12**, no. 10 (1976), 1832-1851)

[17] GOHBERG, I.C. [I.Ts. GOKHBERG] and KREĬN, M.G.: *Theory and applications of Volterra operators in Hilbert space*, Amer. Math. Soc., 1970 (translated from the Russian).

[18] BRODSKIĬ, M.S. and KISILEVSKIĬ, G.E.: 'Criterion for unicellularity of dissipative Volterra operators with nuclear imaginary components', *Izv. Akad. Nauk SSSR Ser. Mat.* **30**, no. 6 (1966), 1213-1228 (in Russian).

[19] NIKOL'SKIĬ, N.K.: 'The unicellularity and nonunicellularity of weighted shift operators', *Soviet Math. Dokl.* **8**, no. 1 (1967), 91-94. (*Dokl. Akad. Nauk SSSR* **172**, no. 2 (1967), 287-290)

[20] NAĬMARK, M.A.: 'Investigation of the spectrum and the expansion in eigenfunctions of a second-order non-self-adjoint operator on a semi-axis', *Trudy Moskov. Mat. Obshch.* **3** (1954), 181-270 (in Russian).

[21] NAĬMARK, M.A.: *Linear differential operators*, Harrap, 1968 (translated from the Russian).

[22] GASYMOV, M.G.: 'Expansion in solutions of the scattering problem for a nonselfadjoint Schrödinger equation', *Dokl. Akad. Nauk AzSSR* **22**, no. 10 (1966), 9-12 (in Russian).

[23] MARCHENKO, V.A.: 'Eigenfunction expansions of non-self-adjoint singular second-order differential operators', *Mat. Sb.* **52** (1960), 739-788 (in Russian).

[24] SADOVNICHIĬ, V.A.: 'Analytic properties of solutions of system of equations with singularities that form segments', *Trudy Sem. Petrov.* **2** (1976), 211-221 (in Russian).

[25] TAMARKIN, YA.D.: *On some general problems of the theory of ordinary linear differential equations...*, Petrograd, 1917 (in Russian).

[26A] WARD, L.E.: 'An irregular boundary value and expansion problem', *Ann. of Math. (2)* **26** (1925), 21-36.

[26B] WARD, L.E.: 'A third order irregular boundary value problem and the associated series', *Trans. Amer. Math. Soc.* **34** (1932), 417-434.

[27] KHROMOV, A.P.: *Proc. Second Conf. Inst. Higher Education*, Vol. 1, Kuĭbyshev, 1962, pp. 109-113 (in Russian).

[28] MIKHAĬLOV, V.P.: 'Riesz bases in $L_2(0, 1)$', *Soviet Math. Dokl.* **3**, no. 3 (1962), 851-854. (*Dokl. Akad. Nauk SSSR* **144**, no. 5 (1962), 981-984)

[29] KESSEL'MAN, G.M.: *Izv. Vyzov. Mat.* **2** (1964), 82-93.

[30] IL'IN, V.A.: 'On the properties of a reduced system of eigenfunctions and associated functions of a Keldysh bundle of ordinary differential operators', *Soviet Math. Dokl.* **17**, no. 5 (1976), 1247-1250. (*Dokl. Akad. Nauk SSSR* **230** (1976), 30-33)

[31] MOISEV, E.N.: 'The uniform convergence of certain expansions in a closed domain', *Soviet Math. Dokl.* **18**, no. 2 (1977), 549-553. (*Dokl. Akad. Nauk SSSR* **233** (1977), 1042-1045)

[32] TIKHOMIROV, V.V.: 'On the Riesz means of expansions in eigenfunctions and associated functions of a nonselfadjoint ordinary differential operator', *Math. USSR Sb.* **31**, no. 1 (1977), 29-48. (*Mat. Sb.* **102**, no. 1 (1977), 33-55)

[33] LIDSKIĬ, V.B.: 'Fourier series expansion of principal functions of a non-self-adjoint elliptic operator', *Mat. Sb.* **57**, no. 2 (1962), 137-150 (in Russian).

[34] SAMARSKIĬ, A.A.: *Uspekhi Mat. Nauk* **5**, no. 3 (1950), 133-134.

[35] TIKHONOV, A.N.: 'Regularization of incorrectly posed problems', *Soviet Math. Dokl.* **4**, no. 6 (1963), 1624-1627. (*Dokl. Akad. Nauk SSSR* **153**, no. 1 (1963), 49-52)

[36] GEL'FAND, I.M. and LEVITAN, B.M.: 'On a simple identity for the characteristic values of a second-order differential operator', *Dokl. Akad. Nauk SSSR* **88**, no. 4 (1953), 593-596 (in Russian).

[37] LIDSKIĬ, V.B. and SADOVNICHIĬ, V.A.: 'Regularized sums of zeros for a class of entire functions', *Soviet Math. Dokl.* **8**, no. 5 (1967), 1082-1085. (*Dokl. Akad. Nauk SSSR* **176**, no. 2 (1967), 259-262)

[38] SADOVNICHIĬ, V.A.: 'The zeta-function and eigenvalues of differential operators', *Diff. Eq.* **10**, no. 7 (1974), 986-994. (*Differentsial. Uravnen.* **10**, no. 7 (1974), 1276-1285)

[39] LYUBISHKIN, V.A. and SADOVNICHIĬ, V.A.: 'Regularized traces of discrete operators', *Soviet Math. Dokl.* **24**, no. 3 (1981), 514-517. (*Dokl. Akad. Nauk SSSR* **261**, no. 2 (1981), 290-293)

[40] KELDYSH, M.V. and LIDSKIĬ, V.B.: 'On the spectral theory of nonselfadjoint operators', in *Fourth All-Union Math. Congress*, Vol. 1, Leningrad, 1963, pp. 101-120 (in Russian).

[41] LAX, P.D. and PHILIPPS, R.S.: *Scattering theory for automorphic functions*, Princeton Univ. Press, 1976.

[42] SZ.-NAGY, B. and FOIAŞ, CH.: *Harmonic analysis of operators on Hilbert space*, North-Holland, 1970.

[43] MASLOV, V.P.: *Operational methods*, Mir, 1976 (translated from the Russian).

[44] DUNFORD, N. and SCHWARTZ, J.T.: *Linear operators. Spectral operators*, 3, Interscience, 1971.

V.A. Sadovnichiĭ

Editorial comments. The main article above gives a partial picture of the theory of non-self-adjoint operators. The method of Keldysh about multiple completeness is explained in detail. A second important method, which has been developed by M.G. Kreĭn and H. Langer (see [7]), is only touched upon. The Kreĭn−Langer method uses factorization as a tool.

Consider the quadratic operator polynomial $L(\lambda) = \lambda^2 I + \lambda A_1 + A_0$, where A_0 and A_1 are bounded linear operators acting on a Hilbert space H, and let Z be a (right) operator root of $L(\lambda) = 0$, that is,

$$Z^2 + A_1 Z + A_0 = 0.$$

Then $\phi(t)=e^{tZ}x$ is a solution of the differential equation $L(d/dt)\phi=0$ for each $x\in H$. One of the problems is to find operator roots Z_1 and Z_2 of $L(\lambda)=0$ so that

$$\phi(t) = e^{tZ_1}x_1 + e^{tZ_2}x_2,$$

where x_1 and x_2 run over all vectors in H, form the set of all solutions of $L(d/dt)\phi=0$. Moreover, if such Z_1 and Z_2 have been found, one would like to know when the linear span of elementary solutions of the equations

$$\phi' = Z_1\phi, \quad \psi' = Z_2\psi$$

is dense (in a sense to be made more precise) in the space of all solutions. If Z is an operator root, then $\lambda I - Z$ is a right divisor of $L(\lambda)$, that is,

$$\lambda^2 I + \lambda A_1 + A_0 = (\lambda I - Y)(\lambda I - Z). \tag{A1}$$

The method of Kreĭn–Langer uses the theory of spaces with an indefinite metric (cf. **Space with an indefinite metric**) to obtain factorizations of the type (A1) and to analyze their properties for the important case when A_0 and A_1 are self-adjoint. Here indefinite metrics turn up naturally, because the *companion operator*

$$\begin{bmatrix} 0 & I \\ -A_0 & -A_1 \end{bmatrix}$$

is self-adjoint relative to the indefinite metric

$$\left[\begin{bmatrix} x_1 \\ y_1 \end{bmatrix}, \begin{bmatrix} x_2 \\ y_2 \end{bmatrix} \right] = <A_1x_1+x_1, x_2> + <x_1, y_2>,$$

whenever A_0 and A_1 are self-adjoint relative to the usual product $<\cdot,\cdot>$ on H. The Kreĭn–Langer method is, therefore, of a geometric nature, while Keldysh's method has an analytic character.

The Kreĭn–Langer approach has led to two directions of research. One continued to use and develop further the geometric methods of operator theory in spaces with an indefinite metric, with Langer as one of the main contributors. The other direction adapted Wiener–Hopf factorization methods for operator-valued functions and used these analytical tools to study the spectral properties of polynomials and analytic operator functions. Here the main contributors are A.S. Markus and V.I. Matsaev. The Keldysh method and the Kreĭn–Langer method are nicely presented in [A1]. Parallel, but with a stronger emphasis on the finite-dimensional case, the factorization direction was also developed in the Western literature. In the latter, connections with mathematical systems theory played an important role (see, e.g., [A2] - [A4]).

A third major method in the theory of non-self-adjoint operators, probably the oldest in this area, was initiated in the middle of the nineteen-forties by M.S. Lifshits. The main tool here is the *characteristic operator function*, which is associated with an arbitrary operator A in the following way:

$$I + 2iK^*(\lambda - A)^{-1}KJ,$$

where J is a *signature operator* (i.e. $J=J^*=J^{-1}$) and $2iKJK^*=A-A^*$. This function serves as a unitary invariant for the operator A (some trivial self-adjoint parts excluded), and in several important cases it is much easier to analyze than the original operator. The characteristic operator function has intriguing properties, for example, from its divisors

invariant subspaces of the operator A may be read off. The theory of characteristic operator functions has been extended in many different directions by M.S. Brodskiĭ, L.A. Sakhnovich, I.C. Gohberg and Kreĭn (see [A5], [A6], [11], [17]). A far-reaching development connected with the theory of characteristic operator functions concerns the study of contractive operators by B. Sz.-Nagy and C. Foiaş [42], which has led to a complete understanding of a wide class of non-self-adjoint operators [A7].

Another original method in the theory of non-self-adjoint operators is due to L. de Branges and his co-workers ([A8], [A9]). The method of de Branges is based on a deep analysis of the theory of spaces of entire vector-valued functions. Here the mapping

$$f \to z^{-1}(f(z) - f(0))$$

provides the main model. Recently this approach led to a sensational success by yielding a proof of the **Bieberbach conjecture** [A10].

References

[A1] MARKUS, A.S.: *Introduction to the spectral theory of polynomial operator pencils*, Amer. Math. Soc., 1986 (translated from the Russian).

[A2] GOHBERG, I., LANCASTER, P. and RODMAN, L.: *Matrix polynomials*, Acad. Press, 1982.

[A3] BART, H., GOHBERG, I. and KAASHOEK, M.A.: *Minimal factorization of matrix and operator functions*, Birkhäuser, 1979.

[A4] RODMAN, L.: *An introduction to operator polynomials*, Birkhäuser, 1989.

[A5] LIVŠIC, M.S. [M.S. LIFSHITS]: *Operators, oscillations, waves*, Amer. Math. Soc., 1973 (translated from the Russian).

[A6] BRODSKIĬ, M.S.: *Triangular and Jordan representations of linear operators*, Amer. Math. Soc., 1971 (translated from the Russian).

[A7] BERCOVICI, H.: *Operator theory and arithmetic in H^∞*, Amer. Math. Soc., 1988.

[A8] BRANGES, L. DE: *Hilbert spaces of entire functions*, Prentice-Hall, 1968.

[A9] BRANGES, L. DE and ROVNYAK, J.: *Square summable power series*, Rinehart & Winston, 1966.

[A10] BRANGES, L. DE: 'A proof of the Bieberbach conjecture', *Acta Math.* 154 (1985), 137-152.

[A11] BRODSKIĬ, M.S. and LIVŠIC, M.S. [M.S. LIFSHITS]: 'Spectral analysis of non-selfadjoint operators and intermediate systems', *Transl. Amer. Math. Soc. (2)* 13 (1960), 265-346.

[A12] DOWSON, H.R.: *Spectral theory of linear operators*, Acad. Press, 1978.

AMS 1980 Subject Classification: 47A45, 47A65, 47A66, 47B99

NON-SINGULAR BOUNDARY POINT, *regular boundary point* - An accessible boundary point (cf. **Attainable boundary point**) ζ of the domain of definition D of a single-valued analytic function $f(z)$ of a complex variable z such that $f(z)$ has an **analytic continuation** to ζ along any path inside D to ζ. In other words, a non-singular boundary point is accessible, but not singular. See also **Singular point** of an analytic function.

E.D. Solomentsev

Editorial comments. Note that the same point in the boundary of D may give rise to several different accessible boundary points, some of which may be singular, others regular. E.g., consider the domain $D = \mathbf{C} \setminus (-\infty, 0]$, and the function $f(z) = (h(z) - \pi i)^{-1}$, where h is the principal value of $\log z$. Then 'above' -1 there are two accessible boundary points: one singular, corresponding to approach along $z = -1 + it$, $0 \leqslant t \leqslant 1$; one regular, corresponding to approach along $z = -1 - it$, $0 \leqslant t \leqslant 1$.

References

[A1] MARKUSHEVICH, A.I.: *Theory of functions of a complex variable*, 3, Chelsea, Chapts. 2; 8 (translated from the Russian).

AMS 1980 Subject Classification: 30B40, 30D40, 30C99

NON-SINGULAR MATRIX, *non-degenerate matrix* - A square **matrix** with non-zero **determinant**. For a square matrix A over a field, non-singularity is equivalent to each of the following conditions: 1) A is invertible; 2) the rows (columns) of A are linearly independent; or 3) A can be brought by elementary row (column) transformations to the identity matrix.

O.A. Ivanova

Editorial comments.

References

[A1] KUROSH, A.G.: *Matrix theory*, Chelsea, reprint, 1960 (translated from the Russian).
[A2] MCDONALD, B.R.: *Linear algebra over commutative rings*, M. Dekker, 1984.

AMS 1980 Subject Classification: 15-XX

NON-SMOOTHABLE MANIFOLD - A piecewise-linear or topological **manifold** that does not admit a smooth structure.

A smoothing of a piecewise-linear manifold X is a piecewise-linear isomorphism $f: M \to X$, where M is a smooth manifold. Manifolds that do not admit smoothings are said to be *non-smoothable*. With certain modifications this is also applicable to topological manifolds.

Example of a non-smoothable manifold. Let W^{4k}, $k > 1$, be a $4k$-dimensional Milnor manifold (see **Dendritic manifold**). In particular, W^{4k} is parallelizable, its **signature** is 8, and its boundary $M = \partial W^{4k}$ is homotopy equivalent to the sphere S^{4k-1}. Glueing to W a cone CM over ∂W leads to the space P^{4k}. Since M is a piecewise-linear sphere (see generalized **Poincaré conjecture**), CM is a piecewise-linear disc, so that P is a piecewise-linear manifold. On the other hand, P is non-smoothable, since its signature is 8, while that of an almost-parallelizable (that is, parallelizable after removing a point) 4-dimensional manifold is a multiple of a number σ_k that grows exponentially with k. The manifold M is not diffeomorphic to the sphere S^{k-1}, that is, M is a **Milnor sphere**.

A criterion for a piecewise-linear manifold to be smoothable is as follows. Let O_n be the orthogonal group and let PL_n be the group of piecewise-linear

homeomorphisms of $\mathbf{R}^n$ preserving the origin (see **Piecewise-linear topology**). The inclusion $O_n \to \mathrm{PL}_n$ induces a fibration $BO_n \to B\mathrm{PL}_n$, where BG is the **classifying space** of a group G. As $n \to \infty$ there results a fibration $p: BO \to B\mathrm{PL}$, the fibre of which is denoted by M/O. A piecewise-linear manifold X has a linear stable normal bundle u with classifying mapping $v: X \to B\mathrm{PL}$. If X is smoothable (or smooth), then it has a stable normal bundle $\bar{v}$ with classifying mapping $\bar{v}: X \to BO$ and $p \circ \bar{v} = v$. This condition is also sufficient, that is, a closed piecewise-linear manifold X is smoothable if and only if its piecewise-linear stable normal bundle admits a vector reduction, that is, if the mapping $v: X \to B\mathrm{PL}$ can be 'lifted' to BO (there is a $\bar{v}: X \to BO$ such that $p \circ \bar{v} = v$).

Two smoothings $f: M \to X$ and $g: N \to X$ are said to be equivalent if there is a diffeomorphism $h: M \to N$ such that hf^{-1} is piecewise differentiably isotopic to g^{-1} (see **Structure** on a manifold). The sets $\mathrm{ts}(X)$ of equivalence classes of smoothings are in a natural one-to-one correspondence with the fibre-wise homotopy classes of liftings $\bar{v}: X \to BO$ of $v: X \to B\mathrm{PL}$. In other words, when X is smoothable, $\mathrm{ts}(X) = [X, \mathrm{PL}/O]$.

References

[1] KERVAIRE, M.: 'A manifold which does not admit any differentiable structure', *Comment. Math. Helv.* **34** (1960), 257-270.
[2] MILNOR, J. and STASHEFF, J.: *Characteristic classes*, Princeton Univ. Press, 1974.

Yu.I. Rudyak

Editorial comments.

References

[A1] HIRSH, M.W. and MAZUR, B.: *Smoothings of piecewise linear manifolds*, Princeton Univ. Press, 1974.
[A2] SIEBENMANN, L.C.: 'Topological manifolds', in *Actès Congrès Internat. Mathématiciens Nice, 1970*, Vol. 2, Gauthier-Villars, 1970, pp. 133-163.
[A3] SMALE, S.: 'The generalized Poincaré conjecture in higher dimensions', *Bull. Amer. Math. Soc.* **66** (1960), 373-375.

AMS 1980 Subject Classification: 57NXX

NON-STANDARD ANALYSIS - A branch of mathematical logic concerned with the application of the theory of non-standard models to investigations in traditional domains of mathematics: mathematical analysis, function theory, the theory of differential equations, probability theory, and others. The basic method of non-standard analysis can roughly be described as follows. One considers a certain mathematical structure M and constructs a first-order logico-mathematical language that reflects those aspects of this structure that are of interest to the investigator. Then one constructs by methods of **model theory** a *non-standard model* of the theory of M that is a proper extension of M. Under a suitable construction new, non-standard, elements of the model can be interpreted as limiting 'ideal' elements of the original structure. For example, if as the original structure one takes the field of real numbers, then it is natural to treat the non-standard elements of the model as 'infinitesimals', that is, as infinitely large or infinitely small, but non-zero, real numbers. Then all the usual relations between real numbers carry over to the non-standard elements, with the preservation of all their properties that can be expressed in the logico-mathematical language. Similarly, in the theory of filters on a given set the intersection of all non-empty elements of the filter determines a non-standard element; in topology this gives rise to a family of non-standard points situated 'infinitely close' to a given point. The interpretation of the non-standard elements of a model often makes it possible to give convenient criteria for ordinary concepts in terms of non-standard elements. For example, it can be proved that a standard real-valued function f is continuous at a standard point x_0 if and only if $f(x)$ is infinitely close to $f(x_0)$ for all (non-standard) points x infinitely close to x_0. The criterion thus obtained can be successfully applied to proofs of ordinary mathematical results.

Naturally, results obtained by methods of non-standard analysis can be, in principle, proved in the standard theory, but the consideration of a non-standard model has the distinct advantage that it allows one to actually introduce into the argument 'ideal' elements, making it possible to give lucid statements for many concepts connected with a limit transition from the finite to the infinite. Non-standard analysis places the ideas of G. Leibniz and his followers, about the existence of infinitely small non-zero quantities, on a strict mathematical basis, a circle of ideas which in the subsequent development of mathematical analysis was rejected in favour of the precise concept of the limit of a variable quantity.

A number of new facts have been discovered by means of non-standard analysis. Many classical proofs gain substantially in clarity when presented by means of non-standard analysis. Non-standard analysis has been used successfully in constructing a rigorous theory of certain semi-empirical methods of mechanics and physics.

References

[1] ROBINSON, A.: *Non-standard analysis*, North-Holland, 1966.
[2] DAVIS, M.: *Applied nonstandard analysis*, Wiley, 1977.

A.G. Dragalin

Editorial comments. In recent years numerous developments involving non-standard analysis, especially in stochastic analysis, the theory of dynamical systems and mathematical physics, have taken place. Some books have appeared covering such aspects, e.g., [A1] - [A5].

References

[A1] ALBEVERIO, S., FENSTAD, J.E., HØEGH-KROHN, R. and LINDSTRØM, T.: *Nonstandard methods in stochastic analysis and mathematical physics*, Acad. Press, 1986.
[A2] STROYAN, K.D. and BAYOD, J.M.: *Foundations of infinitesimal stochastic analysis*, North-Holland, 1986.
[A3] HURD, A.E. and LOEB, P.A.: *An introduction to nonstandard real analysis*, Acad. Press, 1985.
[A4] CUTLAND, N. (ED.): *Nonstandard analysis and its applications*, Cambridge Univ. Press, 1988.
[A5] NELSON, E.: *Radically elementary probability theory*, Princeton Univ. Press, 1987.
[A6] LUXEMBURG, W. and ROBINSON, A. (EDS.): *Contributions to non-standard analysis*, North-Holland, 1972.
[A7] LUXEMBURG, W. and STROYAN, K.: *Introduction to the theory of infinitesimals*, Acad. Press, 1976.

AMS 1980 Subject Classification: 03H05

NON-WANDERING POINT *of a dynamical system* - A point in the phase space of the system that is not a **wandering point**.

AMS 1980 Subject Classification: 58F-XX, 34C28

NORM - 1) A mapping $x \to \| x \|$ from a **vector space** X over the field of real or complex numbers into the real numbers, subject to the conditions:

$\| x \| \geqslant 0$, and $\| x \| = 0$ for $x = 0$ only;

$\| \lambda x \| = | \lambda | \cdot \| x \|$ for every scalar λ;

$\| x + y \| \leqslant \| x \| + \| y \|$ for all $x, y \in X$ (the *triangle axiom*). The number $\| x \|$ is called the *norm of the element x*.

A vector space X with a distinguished norm is called a **normed space**. A norm induces on X a **metric** by the formula $\mathrm{dist}(x, y) = \| x - y \|$, hence also a topology compatible with this metric. And so a normed space is endowed with the natural structure of a **topological vector space**. A normed space that is complete in this metric is called a **Banach space**. Every normed space has a Banach completion.

A topological vector space is said to be *normable* if its topology is compatible with some norm. Normability is equivalent to the existence of a convex bounded neighbourhood of zero (a *theorem of Kolmogorov*, 1934).

The norm in a normed vector space X is generated by an **inner product** (that is, X is isometrically isomorphic to a **pre-Hilbert space**) if and only if for all $x, y \in X$,

$$\| x + y \|^2 + \| x - y \|^2 = 2(\| x \|^2 + \| y \|^2).$$

Two norms $\| \cdot \|_1$ and $\| \cdot \|_2$ on one and the same vector space X are called *equivalent* if they induce the same topology. This comes to the same thing as the existence of two constants C_1 and C_2 such that

$$\| x \|_1 \leqslant C_1 \| x \|_2 \leqslant C_2 \| x \|_1 \quad \text{for all } x \in X.$$

If X is complete in both norms, then their equivalence is a consequence of compatibility. Here compatibility means that the limit relations

$$\| x_n - a \|_1 \to 0, \quad \| x_n - b \|_2 \to 0$$

imply that $a=b$.

Not every topological vector space, even if it is assumed to be locally convex, has a continuous norm. For example, there is no continuous norm on an infinite product of straight lines with the topology of coordinate-wise convergence. The absence of a continuous norm can be an obvious obstacle to the continuous imbedding of one topological vector space in another.

If Y is a closed subspace of a normed space X, then the quotient space X/Y of cosets by Y can be endowed with the norm

$$\| \tilde{x} \| = \inf\{\| x \| : x \in \tilde{x}\},$$

under which it becomes a normed space. The norm of the image of an element x under the quotient mapping $X \rightarrow X/Y$ is called the *quotient norm* of x with respect to Y.

The totality X^* of continuous linear functionals ψ on a normed space X forms a Banach space relative to the norm
$$\| \psi \| = \sup\{ | \psi(x) | : \| x \| \leqslant 1\}.$$

The norms of all functionals are attained at suitable points of the unit ball of the original space if and only if the space is reflexive (cf. **Reflexive space**).

The totality $L(X, Y)$ of continuous (bounded) linear operators A from a normed space X into a normed space Y is made into a normed space by introducing the *operator norm*:

$$\| A \| = \sup\{\| Ax \| : \| x \| \leqslant 1\}.$$

Under this norm $L(X, Y)$ is complete if Y is. When $X = Y$ is complete, the space $L(X) = L(X, X)$ with multiplication (composition) of operators becomes a **Banach algebra**, since for the operator norm

$$\| AB \| \leqslant \| A \| \cdot \| B \|, \quad \| I \| = 1,$$

where I is the identity operator (the unit element of the algebra). Other equivalent norms on $L(X)$ subject to the same condition are also interesting. Such norms are sometimes called *algebraic* or *ringed*. Algebraic norms can be obtained by renorming X equivalently and taking the corresponding operator norms; however, even for $\dim X = 2$ not all algebraic norms on $L(X)$ can be obtained in this manner.

A *pre-norm*, or *semi-norm*, on a vector space X is defined as a mapping p with the properties of a norm except non-degeneracy: $p(x) = 0$ does not preclude that $x \neq 0$. If $\dim X < \infty$, a non-zero pre-norm p on $L(X)$ subject to the condition $p(AB) \leqslant p(A)p(B)$ actually turns out to be a norm (since in this case $L(X)$ has no non-trivial two-sided ideals). But for infinite-dimensional normed spaces this is not so. If X is a Banach algebra over **C**, then the spectral radius

$$| x | = \lim_{n \to \infty} \| x^n \|^{1/n}$$

is a semi-norm if and only if it is uniformly continuous on X, and this condition is equivalent to the fact that the quotient algebra by the radical is commutative.

References
[1] KOLMOGOROV, A.N. and FOMIN, S.V.: *Elements of the theory of functions and functional analysis*, Graylock, 1957-1961 (translated from the Russian).
[2] LJUSTERNIK, L.A. [L.A. LYUSTERNIK] and SOBOLEW, W.I. [V.I. SOBOLEV]: *Elemente der Funktionalanalysis*, H. Deutsch, Frankfurt a. M., 1979.
[3] SHILOV, G.E.: *Mathematical analysis*, 1-2, M.I.T., 1974 (translated from the Russian).
[4] KANTOROVICH, L.V. and AKILOV, G.P.: *Funktionalanalysis in nomierten Räume*, Akad. Verlag, 1977 (translated from the Russian).
[5] RUDIN, W.: *Functional analysis*, McGraw-Hill, 1979.
[6] DAY, M.M.: *Normed linear spaces*, Springer, 1973.
[7] GLAZMAN, I.M. and LYUBICH, YU.I.: *Finite-dimensional linear analysis: a systematic presentation in problem form*, M.I.T., 1974 (translated from the Russian).
[8] AUPETIT, B.: *Propriétés spectrales des algèbres de Banach*, Springer, 1979.
[9] KIRILLOV, A.A. and GRISHIANI, A.D.: *Theorems and problems of functional analysis*, Springer, 1982 (translated from the Russian).

E.A. Gorin

Editorial comments. The theorem that the norms of all functionals are attained at points of the unit ball of the original space X if and only if X is reflexive is called *James' theorem*.

References
[A1] BEAUZAMY, B.: *Introduction to Banach spaces and their geometry*, North-Holland, 1982.
[A2] LINDENSTRAUSS, J. and TZAFRIRI, L.: *Classical Banach spaces*, 1-2, Springer, 1977-1979.

2) For norms in algebra see **Norm on a field** or ring (see also **Valuation**).

3) The *norm of a group* is the collection of group elements that commute with all subgroups, that is, the intersection of the normalizers of all subgroups (cf. **Normalizer of a subset**). The norm contains the centre of the group (cf. **Centre of a group**) and is contained in the second **hypercentre** Z_2. For groups with a trivial centre the norm is the trivial subgroup E.

References
[1] KUROSH, A.G.: *The theory of groups*, 1-2, Chelsea, 1955-1956 (translated from the Russian).

O.A. Ivanova

Editorial comments.

References
[A1] ROBINSON, D.J.S.: *Finiteness conditions and generalized solvable groups*, 2, Springer, 1972, p. 45.

AMS 1980 Subject Classification: 46BXX, 20D25

NORM MAP - The mapping $N_{K/k}$ of a **field** K into a field k, where K is a finite extension of k (cf. **Extension of a field**), that sends an element $\alpha \in K$ to the element $N_{K/k}(\alpha)$ that is the determinant of the matrix of the k-linear mapping $K \rightarrow K$ that takes $x \in K$ to αx. The element $N_{K/k}(\alpha)$ is called the *norm of the element* α.

One has $N_{K/k}(\alpha)=0$ if and only if $\alpha=0$. For any $\alpha, \beta\in K$,

$$N_{K/k}(\alpha\beta) = N_{K/k}(\alpha)N_{K/k}(\beta),$$

that is, $N_{K/k}$ induces a homomorphism of the multiplicative groups $K^* \to k^*$, which is also called the *norm map*. For any $\alpha\in k$,

$$N_{K/k}(\alpha) = \alpha^n, \text{ where } n=[K:k].$$

The group $N_{K/k}(K^*)$ is called the *norm subgroup* of k^*, or the *group of norms* (from K into k). If $f(x)=x^n+a_{n-1}x^{n-1}+\cdots+a_0$ is the characteristic polynomial of $\alpha\in K$ relative to k, then

$$N_{K/k}(\alpha) = (-1)^n a_0.$$

Suppose that K/k is separable (cf. **Separable extension**). Then for any $\alpha\in K$,

$$N_{K/k}(\alpha) = \prod_{i=1}^{n}\sigma_i(\alpha),$$

where the σ_i are all the isomorphisms of K into the **algebraic closure** $\bar{k}$ of k.

The norm map is transitive. If L/K and K/k are finite extensions, then

$$N_{L/k}(\alpha) = N_{K/k}(N_{L/K}(\alpha))$$

for any $\alpha\in L$.

References

[1] LANG, S.: *Algebra*, Addison-Wesley, 1984.
[2] BOREVICH, Z.I. and SHAFAREVICH, I.R.: *Number theory*, Acad. Press, 1966 (translated from the Russian).

L.V. Kuz'min

AMS 1980 Subject Classification: 12J05

NORM ON A FIELD K - A mapping ϕ from K to the set $\mathbf{R}$ of real numbers, which satisfies the following conditions:

1) $\phi(x)\geqslant 0$, and $\phi(x)=0$ if and only if $x=0$;
2) $\phi(x\cdot y) = \phi(x)\cdot\phi(y)$;
3) $\phi(x+y)\leqslant\phi(x)+\phi(y)$.

Hence $\phi(1)=\phi(-1)=1$; $\phi(x^{-1})=\phi^{-1}(x)$.

The norm of x is often denoted by $|x|$ instead of $\phi(x)$. A norm is also called an *absolute value* or a *multiplicative valuation*. Norms may (more generally) be considered on any ring with values in a linearly ordered ring [4]. See also **Valuation**.

Examples of norms. If $K=\mathbf{R}$, the field of real numbers, then $|x|=\max\{x,-x\}$, the ordinary **absolute value** or *modulus* of the number $x\in\mathbf{R}$, is a norm. Similarly, if K is the field $\mathbf{C}$ of complex numbers or the skew-field $\mathbf{H}$ of quaternions, then $|x|=\sqrt{x\cdot\bar{x}}$ is a norm. The subfields of these fields are thus also provided with an induced norm. Any field has the *trivial norm*:

$$\phi(x) = \begin{cases} 0, & x=0; \\ 1, & x\neq 0. \end{cases}$$

Finite fields and their algebraic extensions only have the trivial norm.

Examples of norms of another type are provided by logarithmic valuations of a field K: If v is a valuation on K with values in the group $\mathbf{R}$ and if a is a real number, $0<a<1$, then $\phi(x)=a^{v(x)}$ is a norm. For example, if $K=\mathbf{Q}$ and v_p is the p-adic valuation of the field $\mathbf{Q}$, then $|x|_p=(1/p)^{v_p(x)}$ is called the *p-adic absolute value* or the *p-adic norm*. These absolute values satisfy the following condition, which is stronger than 3):

4) $\phi(x+y)\leqslant\max\{\phi(x),\phi(y)\}$.

Norms satisfying condition 4) are known as *ultra-metric norms* or *non-Archimedean norms* (as distinct from *Archimedean norms* which do not satisfy this condition (but do satisfy 3)). They are distinguished by the fact that $\phi(n\cdot1)\leqslant1$ for all integers n. All norms on a field of characteristic $p>0$ are ultra-metric. All ultra-metric norms are obtained from valuations as indicated above: $\phi=a^{v(x)}$ (and conversely, $-\log\phi$ can always be taken as a valuation).

A norm ϕ defines a **metric** on K if $\phi(x-y)$ is taken as the distance between x and y, and in this way it defines a topology on K. The topology of any locally compact field is defined by some norm. Two norms ϕ_1 and ϕ_2 are said to be equivalent if they define the same topology; in a such case there exists a $\lambda>0$ such that $\phi_1(x)=\phi_2(x)^\lambda$ for all $x\in K$.

The structure of all Archimedean norms is given by *Ostrowski's theorem*: If ϕ is an Archimedean norm on a field K, then there exists an isomorphism of K into a certain everywhere-dense subfield of one of the fields $\mathbf{R}$, $\mathbf{C}$ or $\mathbf{H}$ such that ϕ is equivalent to the norm induced by that of $\mathbf{R}$, $\mathbf{C}$ or $\mathbf{H}$.

Any non-trivial norm of the field $\mathbf{Q}$ of rational numbers is equivalent either to a p-adic norm $|\cdot|_p$, where p is a prime number, or to the ordinary norm. For any rational number $r\in\mathbf{Q}$ one has

$$|r|\prod_p|r|_p = 1.$$

A similar formula is also valid for algebraic number fields [2], [3].

If ϕ is a norm on a field K, then K may be imbedded by the classical completion process in a field K_ϕ that is complete with respect to the norm that (uniquely) extends ϕ (cf. **Complete topological space**). One of the principal modern methods in the study of fields is the imbedding of a field K into the direct product $\prod_\phi K_\phi$ of all completions K_ϕ of the field K with respect to all non-trivial norms of K (see **Adèle**). If K admits non-trivial valuations, then it is dense in $\prod_\phi K_\phi$ in the adèlic topology; in fact, if $\phi_1,\ldots,\phi_n$ are non-trivial, non-equivalent norms on K, if $a_1,\ldots,a_n$ are elements of K and if $\epsilon>0$, then there exists an $a\in K$ such that $\phi_i(a-a_i)<\epsilon$ for all i (the *approximation theorem for norms*).

A norm on a field K may be extended (in general, non-uniquely) to any algebraic field extension of the field K. If K is complete with respect to the norm ϕ and if L is an extension of K of degree n, the extension of ϕ to L is unique, and is given by the formula

$$\phi'(x) = \{\phi(N_{L/K}(x))\}^{1/n}$$

for $x \in L$.

References

[1] BOURBAKI, N.: *Elements of mathematics. Commutative algebra*, Addison-Wesley, 1972 (translated from the French).
[2] BOREVICH, Z.I. and SHAFAREVICH, I.R.: *Number theory*, Acad. Press, 1966 (translated from the Russian).
[3] LANG, S.: *Algebra*, Addison-Wesley, 1984.
[4] KUROSH, A.G.: *Lectures on general algebra*, Chelsea, 1963 (translated from the Russian).
[5] CASSELS, J.W.S. and FRÖHLICH, A. (EDS.): *Algebraic number theory*, Acad. Press, 1986.

V.I. Danilov

Editorial comments. Non-Archimedean norms satisfy $\phi(n{\cdot}1) \leqslant \phi(1)$ and hence do not satisfy the **Archimedean axiom**, whence the appellation.

AMS 1980 Subject Classification: 12-XX

NORM-RESIDUE SYMBOL, *norm residue, Hilbert symbol* - A function that associates with an ordered pair of elements x, y of the multiplicative group K^* of a **local field** K an element $(x, y) \in K^*$ that is an n-th root of unity. This function can be defined as follows. Let $\zeta_n \in K$ be a primitive n-th root of unity. The maximal Abelian extension L of K with Galois group $G(L/K)$ of exponent n is obtained by adjoining to K the roots $a^{1/n}$ for all $a \in K^*$. On the other hand, there is a canonical isomorphism (the fundamental isomorphism of local **class field theory**)

$$\theta: K^*/K^{*n} \to \mathrm{Gal}(L/K).$$

The norm residue of the pair (x, y) is defined by

$$\theta(y)(x^{1/n}) = (x, y)x^{1/n}.$$

D. Hilbert introduced the concept of a norm-residue symbol in the special case of quadratic fields with $n = 2$. In [4] there is an explicit definition of the norm residue using only local class field theory.

Properties of the symbol (x, y):

1) bilinearity: $(x_1 x_2, y) = (x_1, y)(x_2, y)$, $(x, y_1 y_2) = (x, y_1)(x, y_2)$;

2) skew-symmetry: $(x, y)(y, x) = 1$;

3) non-degeneracy: $(x, y) = 1$ for all $x \in K^*$ implies $y \in K^{*n}$; $(x, y) = 1$ for all $y \in K^*$ implies $x \in K^{*n}$;

4) if $x + y = 1$, then $(x, y) = 1$;

5) if σ is an automorphism of K, then

$$(\sigma x, \sigma y) = \sigma(x, y);$$

6) let K' be a finite extension of K, $a \in K'^*$ and $b \in K^*$. Then

$$(a, b) = (N_{K'/K}(a), b),$$

where on the left-hand side the norm-residue symbol is regarded for K' and on the right-hand side that for K, and where $N_{K'/K}$ is the **norm map** from K' into K;

7) $(x, y) = 1$ implies that y is a norm in the extension $K(x^{1/n})$. (This explains the name of the symbol.)

The function (x, y) induces a non-degenerate bilinear pairing

$$K^*/K^{*n} \times K^*/K^{*n} \to \mu(n),$$

where $\mu(n)$ is the group of roots of unity generated by ζ_n. Let $\Psi: K^* \times K^* \to A$ be a mapping into some Abelian group A satisfying 1), 4) and the *condition of continuity*: For any $y \in K^*$ the set $\{x \in K^*: \Psi(x, y) = 1\}$ is closed in K^*. The norm-residue symbol has the following *universal property* [3]: If n is the number of roots of unity in K, then there exists a homomorphism $\phi: \mu(n) \to A$ such that for any $x, y \in K^*$,

$$\Psi(x, y) = \phi((x, y)).$$

This property can serve as a basic axiomatic definition of the norm-residue symbol.

If F is a **global field** and K is the completion of F relative to a place v, then by the norm-residue symbol one also means the function $(x, y)_v$ defined over $F^* \times F^*$ that is obtained by composition of the (local) norm-residue symbol (x, y) with the natural imbedding $F^* \to K^*$.

Often the norm-residue symbol is defined as an automorphism $\theta(x)$ of the maximal Abelian extension of K corresponding to an element $x \in K^*$ by local **class field theory**.

References

[1] CASSELS, J.W.S. and FRÖHLICH, A. (EDS.): *Algebraic number theory*, Acad. Press, 1986.
[2] KOCH, H.: *Galoissche Theorie der p-Erweiterungen*, Deutsch. Verlag Wissenschaft., 1970.
[3] MILNOR, J.: *Introduction to algebraic K-theory*, Princeton Univ. Press, 1971.
[4] SHAFAREVICH, I.R.: 'A general reciprocity law', *Mat. Sb.* **26**, no. 1 (1950), 113-146 (in Russian).

L.V. Kuz'min

Editorial comments.

References

[A1] IWASAWA, K.: *Local class field theory*, Oxford Univ. Press, 1986.
[A2] NEUKIRCH, J.: *Class field theory*, Springer, 1986.

AMS 1980 Subject Classification: 12B40

NORMAL *to a curve* (or *surface*) at a point of it - A straight line passing through the point and perpendicular to the **tangent** (or **tangent plane**) of the curve (or surface) at this point. A smooth plane curve has at every point a unique normal situated in the plane of the curve. If a curve in a plane is given in rectangular coordinates by an equation $y = f(x)$, then the equation of the normal to the curve at (x_0, y_0) has the form

$$(x - x_0) + (y - y_0)f'(x_0) = 0.$$

A curve in space has infinitely many normals at

every point of it. These fill a certain plane (the **normal plane**). The normal lying in the **osculating plane** is called the **principal normal**; the one perpendicular to the osculating plane is called the **binormal**.

The normal at (x_0, y_0, z_0) to a surface given by an equation $z = f(x, y)$ is defined by

$$\begin{cases} (x - x_0) + (z - z_0)\dfrac{\partial z}{\partial x} = 0, \\ (y - y_0) + (z - z_0)\dfrac{\partial z}{\partial y} = 0. \end{cases}$$

If the equation of the surface has the form $\mathbf{r} = \mathbf{r}(u, v)$, then the parametric representation of the normal is

$$\mathbf{R} = \mathbf{r} + \lambda[\mathbf{r}_u, \mathbf{r}_v].$$

BSE-3

Editorial comments. The notion of a normal obviously extends to m-dimensional submanifolds of Euclidean n-space E^n, giving an $(n - m)$-dimensional affine subspace as the normal $(n - m)$-plane to the manifold at the corresponding point. For submanifolds of (pseudo-) Riemannian manifolds, the normal planes are considered as subspaces of the tangent space of the ambient space, where orthogonality is defined by means of the (ambient) (pseudo-) Riemannian metric. See also **Normal bundle**; **Normal plane**; **Normal space (to a surface)**.

References

[A1] BERGER, M. and GOSTIAUX, B.: *Differential geometry: manifolds, curves and surfaces*, Springer, 1988 (translated from the French).
[A2] COXETER, H.S.M.: *Introduction to geometry*, Wiley, 1963.
[A3] DO CARMO, M.: *Differential geometry of curves and surfaces*, Prentice-Hall, 1976.
[A4] SPIVAK, M.: *A comprehensive introduction to differential geometry*, 1-5, Publish or Perish, 1979.
[A5] BLASCHKE, W. and LEICHTWEISS, K.: *Elementare Differentialgeometrie*, Springer, 1973.
[A6] CHEN, B.-Y.: *Geometry of submanifolds*, M. Dekker, 1973.

AMS 1980 Subject Classification: 53A04, 53A05

NORMAL ALGORITHM - The name attached to algorithms (cf. **Algorithm**) of a certain precisely characterized type. Together with recursive functions (cf. **Recursive function**) and Turing machines (cf. **Turing machine**), normal algorithms have become known as one of the most suitable refinements of the general intuitive idea of an algorithm. The concept of a normal algorithm was developed in 1947 by A.A. Markov in the course of his research on the identity problem for associative systems (see **Associative calculus**). The detailed definition and general theory of normal algorithms are set forth in [1] (Chapts. I - V); see also [2] (Chapts. I - VII).

Every normal algorithm $\mathfrak{A}$ works in a certain **alphabet** A and gives rise to a well-defined process performed on words (cf. **Word**). The specification of this alphabet occurs in the definition of $\mathfrak{A}$ as an obligatory constituent part, and in the situation in question one

says of $\mathfrak{A}$ that it is a normal algorithm in the alphabet A. Every normal algorithm in a fixed alphabet A is completely determined by indicating its *scheme* — an ordered finite list of substitution formulas in A. Every such formula is in essence an ordered pair (U, V) of words in A. Here U is called the left and V the right part of the formula. Among the formulas of a given scheme some are specially distinguished and are declared to be *conclusive*. As a rule, in the scheme of a normal algorithm a conclusive formula is written in the form $U \to \cdot V$ and an inconclusive one in the form $U \to V$.

A *normal algorithm* $\mathfrak{A}$ *in an alphabet* A is a prescription for constructing, starting from an arbitrary word P in A, a sequence of words P_i according to the following rule. The word P is taken as the initial term P_0 of the sequence and the construction process continues. Suppose that for some $i \geqslant 0$ the word P_i has been constructed and that the process of constructing the sequence in question is not yet complete. If in the scheme of $\mathfrak{A}$ there are no formulas with left parts occurring in P_i, then one puts P_{i+1} equal to P_i, and the process of constructing the sequence is considered complete. But if in the scheme of $\mathfrak{A}$ there are formulas with left parts occurring in P_i, then one takes as P_{i+1} the result of substituting the right part of the first of these formulas for the first occurrence of its left part in P_i (cf. **Imbedded word**); here the process of constructing the sequence is taken to be complete if the substitution formula to be applied at this stage is conclusive and it continues otherwise. If the process of constructing the relevant sequence terminates, then one says that the normal algorithm $\mathfrak{A}$ in question is *applicable to the word* P. The last term Q of the sequence is the *result of applying* $\mathfrak{A}$ to P and is denoted by the symbol $\mathfrak{A}(P)$. One says that $\mathfrak{A}$ *transforms* P into Q and writes $\mathfrak{A}(P) = Q$. A normal algorithm in any extension of A is said to be a *normal algorithm over this alphabet*.

There are strong reasons to suppose that the concept of a normal algorithm is an adequate precise version (formalization) of the general idea of an **algorithm in an alphabet**. More accurately, one may assume that for every algorithm $\mathfrak{A}$ in any alphabet A one can construct a normal algorithm $\mathfrak{B}$ over A that transforms an arbitrary word P in A into the same result as the original algorithm $\mathfrak{A}$. This convention is known in the theory of algorithms as the *normalization principle*. The formalization of the concept of an algorithm on the basis of that of a normal algorithm turns out to be equivalent to other known formalizations (see, for example, [3]). As a consequence, this normalization principle turns out to be equivalent to the **Church thesis**, which proposes to assume that the concept of a **partial recursive function** is an adequate formalization of the non-formal, intuitive, concept of a computable arithmetic function (cf. **Computable function**).

Originating in the first instance in connection with algebraic problems, normal algorithms have proved to be a convenient working tool in much research requiring a precise notion of an algorithm, especially when the basic objects of study are not of arithmetical nature and allow for an appropriate representation in the form of words in certain alphabets (such as situations in **constructive analysis**, for example).

References

[1] MARKOV, A.A.: *Theory of algorithms*, Israel Progr. Sci. Transl., 1961 (translated from the Russian). Also: Trudy Mat. Inst. Steklov. 42 (1954).
[2] MARKOV, A.A. and NAGORNY, N.M. [N.M. NAGORNYĬ]: *The theory of algorithms*, Kluwer, 1988 (translated from the Russian).
[3] MENDELSON, E.: *Introduction to mathematical logic*, v. Nostrand, 1964.

N.M. Nagornyĭ

Editorial comments. In 1936 - 1937 A.M. Turing formulated the thesis that his formalization of the intuitive notion of effective computation in terms of Turing numbers coincided precisely with the intuitive notion of computability. After several different formalizations turned out to be equivalent, A. Church extended Turing's thesis to the effect that all strong enough formalizations will turn out to cover precisely Turing's concept (*Church's thesis*).

AMS 1980 Subject Classification: 68C05, 03D10

NORMAL ANALYTIC SPACE - An **analytic space** the local rings of all points of which are *normal*, that is, are integrally-closed integral domains. A point x of an analytic space X is said to be *normal* (one also says that X is normal at x) if the local ring $\mathcal{O}_{X,x}$ is normal. In a neighbourhood of such a point the space has a reduced and irreducible model. Every simple (non-singular) point is normal. The simplest example of a normal analytic space is an **analytic manifold**.

In what follows the (complete non-discretely normed) ground field k is assumed to be algebraically closed. In this case the most complete results on normal analytic spaces have been obtained (see [1]) and a normalization theory has been constructed [2] that gives a natural link between arbitrary reduced analytic spaces and normal analytic spaces. Let $N(X)$ be the set of points of an analytic space X that are not normal and let $S(X)$ be the set of singular points of X (cf. **Singular point**). Then:

1) $N(X)$ and $S(X)$ are closed analytic subspaces of X, and $N(X) \subset S(X)$;

2) for $x \in X \setminus N(X)$,

$$\dim_x S(X) \leqslant \dim_x X - 2$$

(that is, a normal analytic space is smooth in codimension 1);

3) if X is a complete intersection at x and if the above inequality holds, then X is normal at that point.

A *normalization* of a reduced analytic space X is a pair $(\tilde{X}, v)$, where $\tilde{X}$ is a normal analytic space and $v : \tilde{X} \to X$ is a finite surjective analytic mapping inducing an isomorphism of the open sets

$$\tilde{X} \setminus v^{-1}(N(X)) \to X \setminus N(X).$$

The normalization is uniquely determined up to an isomorphism, that is, if $(\tilde{X}_1, v_1)$ and $(\tilde{X}_2, v_2)$ are two normalizations,

$$\tilde{X}_1 \xrightarrow{\phi} \tilde{X}_2$$
$$v_1 \searrow \swarrow v_2$$
$$X$$

then there exists a unique analytic isomorphism $\phi : \tilde{X}_1 \to \tilde{X}_2$ such that the diagram commutes. The normalization exists and has the following properties. For every point $x \in X$ the set of irreducible components of X at x is in one-to-one correspondence with $v^{-1}(x)$. The fibre at $x \in X$ of the direct image $v_*(\mathcal{O}_{\tilde{X}})$ of the structure sheaf $\mathcal{O}_{\tilde{X}}$ is naturally isomorphic to the integral closure of the ring $\mathcal{O}_{X,x}$ in its complete ring of fractions.

The concept of a normal analytic space over $\mathbf{C}$ can be introduced in terms of analytic continuation of holomorphic functions [3]. Namely, a reduced complex space is normal if and only if *Riemann's first theorem on the removal of singularities* holds for it: If $U \subset X$ is an open subset and $A \subset U$ is a closed analytic subset not containing irreducible components of U, then any function that is holomorphic on $U \setminus A$ and locally bounded on U has a unique analytic continuation to a holomorphic function on U. For normal complex spaces *Riemann's second theorem on the removal of singularities* also holds: If $\operatorname{codim}_x A \geqslant 2$ at every point $x \in A$, then the analytic continuation in question is possible without the requirement that the function is bounded. A reduced complex space X is normal if and only if for every open set $U \subset X$ the restriction mapping of holomorphic functions

$$\Gamma(U, \mathcal{O}_X) \to \Gamma(U \setminus S(X), \mathcal{O}_X)$$

is bijective. The property of being normal can also be phrased in the language of local cohomology — it is equivalent to $H^1_{S(X)} \mathcal{O}_X = 0$ (see [5]). For any reduced complex space X one can define the sheaf $\tilde{\mathcal{O}}_X$ of rings of germs of *weakly holomorphic functions*, that is, functions satisfying the conditions of Riemann's first theorem. It turns out that the ring $\tilde{\mathcal{O}}_{X,x}$ is finite as an $\mathcal{O}_{X,x}$-module and equal to the integral closure of $\mathcal{O}_{X,x}$ in its complete ring of fractions. In other words, $\tilde{\mathcal{O}}_X = v_*(\mathcal{O}_{\tilde{X}})$, where $v : \tilde{X} \to X$ is the normalization mapping.

A normal complex space can also be characterized in

the following manner: A complex space is normal if and only if every point of it has a neighbourhood that admits an analytic covering onto a domain of $\mathbf{C}^n$ (see [3], [8]).

A reduced complex space X is a **Stein space** if and only if its normalization $\tilde{X}$ has this property (see [4]). To normal complex spaces one can extend the concept of a Hodge metric (see **Kähler metric**). Kodaira's projective imbedding theorem [6] carries over to compact normal spaces with such a metric.

In algebraic geometry one examines analogues of normal analytic spaces: normal algebraic varieties (see **Normal scheme**). For algebraic varieties over a complete non-discretely normed field the two concepts are the same (see [7], [1]).

References

[1] ABHYANKAR, S.S.: *Local analytic geometry*, Acad. Press, 1964.
[2] HOUZEL, C.: 'Géometrie analytique locale I', in *Sém. H. Cartan Ann. 13 1960/61*, Vol. 2, 1963, Exp. 18-21.
[3] GRAUERT, H. and REMMERT, R.: 'Komplexe Räume', *Math. Ann.* **136** (1958), 245-318.
[4] NARASIMHAN, R.: 'A note on Stein spaces and their normalisations', *Ann. Scuola Norm. Sup. Pisa* **16** (1962), 327-333.
[5] SIU, Y.T. and TRAUTMANN, G.: *Gap sheaves and extensions of coherent analytic subsheaves*, Springer, 1971.
[6] GRAUERT, H.: 'Ueber Modifikationen und exzeptionelle analytische Mengen', *Math. Ann.* **146** (1962), 331-368.
[7] ZARISKI, O. and SAMUEL, P.: *Commutative algebra*, 2, Springer, 1960.
[8] FUKS, B.A.: *Theory of analytic functions of several complex variables*, 1, Amer. Math. Soc., 1963 (translated from the Russian).

D.N. Akhiezer

Editorial comments.

References

[A1] WHITNEY, H.: *Complex analytic varieties*, Addison-Wesley, 1972, Chapt. 8.

AMS 1980 Subject Classification: 32C20

NORMAL BUNDLE *of a submanifold* - The vector bundle consisting of tangent vectors to the ambient manifold that are normal to the submanifold. If X is a **Riemannian manifold**, Y is an (immersed) submanifold of it, T_X and T_Y are the tangent bundles over X and Y (cf. **Tangent bundle**), then the normal bundle $N_{Y/X}$ of Y is the subbundle in $T_X|_Y$ consisting of the vectors $u \in T_{X,y}$, $y \in Y$, that are orthogonal to $T_{Y,y}$.

With the help of normal bundles one constructs, for example, tubular neighbourhoods of submanifolds (cf. **Tubular neighbourhood**). The normal bundle over Y, regarded up to equivalence, does not depend on the choice of the Riemannian metric on X, since it can be defined without recourse to the metric as the quotient bundle $T_X|_Y/T_Y$ of the tangent bundle T_X restricted to Y by the vector bundle T_Y. Somewhat more general is the construction of the normal bundle of an arbitrary immersion (cf. **Immersion of a manifold**) $f : Y \to X$ of differentiable manifolds:

$$N_{Y/X} = f^* T_X / T_Y.$$

Similarly one defines the normal bundle $N_{Y/X}$ of a non-singular algebraic subvariety Y in a non-singular **algebraic variety** $\bar{X}$ or that of an analytic submanifold Y in an **analytic manifold** X; it is an algebraic (or analytic) vector bundle over Y of rank codim Y. In particular, if codim $Y = 1$, then $N_{Y/X}$ is isomorphic to the restriction to Y of the bundle over X that determines the divisor Y.

When Y is an analytic subspace of an analytic space $(X, \mathcal{O}_X)$, the normal bundle of Y is sometimes defined as the analytic family of vector spaces $N_{Y/X} \to Y$ dual to the conormal sheaf $N^*_{Y/X}$ (see **Normal sheaf**). For applications of normal bundles to the problem of contractibility of submanifolds see **Exceptional analytic set**; **Exceptional subvariety**.

References

[1] ONISHCHIK, A.L.: 'Pseudoconvexity in the theory of complex spaces', *J. Soviet Math.* **14**, no. 3 (1980), 1363-1406. (*Itogi Nauk. i Tekhn. Algebra. Topol. Geom.* **15** (1977), 93-156)
[2] MILNOR, J. and STASHEFF, J.: *Characteristic classes*, Princeton Univ. Press, 1974.
[3] ROKHLIN, V.A. and FUKS, D.B.: *Beginner's course in topology: geometric chapters*, Springer, 1984 (translated from the Russian).
[4] HIRSCH, M.: *Differential topology*, Springer, 1976.
[5] SHAFAREVICH, I.R.: *Basic algebraic geometry*, Springer, 1977 (translated from the Russian).

A.L. Onishchik

Editorial comments.

References

[A1] STEENROD, N.E.: *The topology of fibre bundles*, Princeton Univ. Press, 1951.

AMS 1980 Subject Classification: 55R10, 57RXX, 32C35

NORMAL COMPLEX *of a semi-group S* - A non-empty subset $N \subseteq S$ satisfying the following condition: For any $x, y \in S^1$ (where $S^1 = S$ when S contains a unit element and S^1 is the semi-group obtained from S by adjoining a unit element if S does not have one) and any $a, b \in N$ it follows from $xay \in N$ that $xby \in N$. A subset N is a normal complex of a semi-group S if and only if N is a class of some congruence on S (cf. **Congruence (in algebra)**).

References

[1] LYAPIN, E.S.: *Semigroups*, Amer. Math. Soc., 1974 (translated from the Russian).

L.N. Shevrin

AMS 1980 Subject Classification: 20M10

NORMAL CONVERGENCE - Convergence of a series

$$f = \sum_{k=1}^{\infty} u_k \qquad (1)$$

formed by bounded mappings $u_k : X \to Y$ from a set X into a normed space Y, such that the series with positive terms $\sum_{k=1}^{\infty} \| u_k \|$ formed by the norms of the mappings,

$$\| u_k \| = \sup\{ \| u_k(x) \| : x \in X \},$$

converges.

Normal convergence of the series (1) implies absolute and uniform convergence of the series $\sum_{k=1}^{\infty} u_k(x)$ consisting of elements of Y; the converse is not true. For example, if $u_k : \mathbf{R} \to \mathbf{R}$ is the real-valued function defined by $u_k(x) = (\sin \pi x)/k$ for $k \leq x \leq k+1$ and $u_k(x) = 0$ for $x \in \mathbf{R} \setminus [k, k+1]$, then the series $\sum_{k=1}^{\infty} u_k(x)$ converges absolutely, whereas $\sum_{k=1}^{\infty} \| u_k \| = \sum_{k=1}^{\infty} 1/k$ diverges.

Suppose, in particular, that each $u_k : \mathbf{R} \to Y$ is a piecewise-continuous function on a non-compact interval $I \subset \mathbf{R}$ and that (1) converges normally. Then one can integrate term-by-term on I:

$$\int_I f(t)\,dt = \sum_{k=1}^{\infty} \int_I u_k(t)\,dt.$$

Let $f : I \times A \to Y$, where $I \subset \mathbf{R}$ is an interval, have left and right limits at each point of I. Then the improper integral

$$\int_I f(t;\lambda)\,dt, \quad \lambda \in A, \tag{2}$$

is called *normally convergent* on A if there exists a piecewise-continuous positive function $g : \mathbf{R} \to \mathbf{R}$ such that: 1) $\| f(x;\lambda) \| \leq g(x)$ for any $x \in I$ and any $\lambda \in A$; and 2) the integral $\int_I g(t)\,dt$ converges. Normal convergence of (2) implies its absolute and uniform convergence; the converse is not true.

References

[1] BOURBAKI, N.: *Elements of mathematics. General topology*, Addison-Wesley, 1966 (translated from the French).
[2] BOURBAKI, N.: *Elements of mathematics. Functions of a real variable*, Addison-Wesley, 1976 (translated from the French).
[3] SCHWARTZ, L.: *Cours d'analyse*, 1, Hermann, 1967.

E.D. Solomentsev

AMS 1980 Subject Classification: 40A05, 40A10, 46G10

NORMAL CURVATURE *of a regular surface* - A quantity that characterizes the deviation of the surface at a point P in the direction $\mathbf{l}$ from its tangent plane and is the same in absolute value as the curvature of the corresponding **normal section**. The normal curvature in the direction $\mathbf{l}$ is

$$k_\mathbf{l} = (\mathbf{n}, \mathbf{N})k,$$

where k is the curvature of the normal section in the direction $\mathbf{l}$, $\mathbf{n}$ is the unit principal normal vector of the normal section and $\mathbf{N}$ is the unit normal vector to the surface. The normal curvature of a surface in a given direction is the same as that of the **osculating paraboloid** in this direction. The normal curvature of a surface parametrized by u and v can be expressed in terms of the values of the first and second fundamental forms

of the surface (cf. **Fundamental forms of a surface**) computed for the values (du, dv) corresponding to the direction $\mathbf{l}$ by the formula

$$k_\mathbf{l} = \frac{\mathrm{II}}{\mathrm{I}} = \frac{L\,du^2 + 2M\,du\,dv + N\,dv^2}{E\,du^2 + 2F\,du\,dv + G\,dv^2}.$$

The curvature of a regular curve lying on a surface is connected with the normal curvature of the surface in the direction of the unit tangent $\mathbf{l}$ to the curve and with the geodesic curvature k_g of the curve by the relation

$$k\mathbf{n} = k_g \mathbf{N} \times \mathbf{l} + k_\mathbf{l} \mathbf{N}$$

(see also **Meusnier theorem**). By means of the normal curvature one can construct the **Dupin indicatrix**, the **Gaussian curvature** and the **mean curvature** of the surface, as well as many other concepts of the local geometry of the surface.

D.D. Sokolov

Editorial comments.

References

[A1] BERGER, M. and GOSTIAUX, B.: *Differential geometry: manifolds, curves and surfaces*, Springer, 1988 (translated from the French).
[A2] DO CARMO, M.: *Differetial geometry of curves and surfaces*, Prentice Hall, 1976.
[A3] BLASCHKE, W. and LEICHTWEISS, K.: *Elementare Differentialgeometrie*, Springer, 1973.

AMS 1980 Subject Classification: 53A05

NORMAL DERIVATIVE - The **derivative** of a function defined in a neighbourhood of a manifold (or, respectively, on a manifold with boundary) in the direction of the **normal** to that manifold (respectively, to the boundary).

L.D. Kudryavtsev

AMS 1980 Subject Classification: 58C20

NORMAL DISTRIBUTION - One of the most important probability distributions. The term 'normal distribution' is due to K. Pearson (earlier names are **Gauss law** and **Gauss − Laplace distribution**). It is used both in relation to probability distributions of random variables (cf. **Random variable**) and in relation to the joint probability distribution (cf. **Joint distribution**) of several random variables (that is, to distributions of finite-dimensional random vectors), as well as of random elements and stochastic processes (cf. **Random element; Stochastic process**). The general definition of a normal distribution reduces to the one-dimensional case.

The probability distribution of a random variable X is called *normal* if it has probability density

$$p(x;a,\sigma) = \frac{1}{\sigma\sqrt{2\pi}} e^{-(x-a)^2/2\sigma^2}. \tag{*}$$

The family of normal distributions (*) depends, as a rule, on the two parameters a and $\sigma > 0$. Here a is the mathematical expectation of X, σ^2 is the variance of X and the characteristic function has the form

$$f(t) = \mathsf{E}e^{itX} = e^{iat - \sigma^2 t^2/2}.$$

The normal density curve $y = p(x; a, \sigma)$ is symmetric about the ordinate passing through a and has there its unique maximum $1/(\sigma\sqrt{2\pi})$. As σ decreases, the normal distribution curve becomes more and more pointed. A change in a with constant σ does not change the shape of the curve and causes only a shift along the x-axis. The area under a normal density curve is 1. When $a = 0$ and $\sigma = 1$, the corresponding distribution function is

$$\Phi(x) = \frac{1}{\sqrt{2\pi}} \int_{-\infty}^{x} e^{-u^2/2} \, du.$$

In general, the distribution function $F(x; a, \sigma)$ of (*) can be computed by the formula $F(x; a, \sigma) = \Phi(t)$, where $t = (x - a)/\sigma$. For $\Phi(t)$ (and several of its derivatives) extensive tables have been compiled (see, for example, [1], [2], and **Probability integral**). For a normal distribution the probability that $|X - a| > k\sigma$ is $1 - \Phi(k) + \Phi(-k)$ and it decreases very rapidly with increasing k (see the Table).

k	probability
1	0.31731
2	$0.45500 \cdot 10^{-1}$
3	$0.26998 \cdot 10^{-2}$
4	$0.63342 \cdot 10^{-4}$

In many practical problems, when analyzing normal distributions one can, therefore, ignore the possibility of a deviation from a in excess of 3σ — the *three-sigma rule*; the corresponding probability, as is clear from the Table, is less than 0.003. The quartile deviation for a normal distribution is 0.67449σ.

Normal distributions occur in a large number of applications. There are some noteable attempts at explaining this fact. A theoretical basis for the exceptional role of the normal distribution is given by the **limit theorems** of probability theory (see also **Laplace theorem; Lyapunov theorem**). Qualitatively, the result can be stated in the following manner: A normal distribution is a good approximation whenever the relevant random variable is the sum of a large number of independent random variables the largest of which is small in comparison with the whole sum (see **Central limit theorem**).

A normal distribution can also appear as an exact solution of certain problems (within the framework of an accepted mathematical model of the phenomenon). This is so in the theory of random processes (in one of the basic models of Brownian motion). Classic examples of a normal distribution arising as an exact one are due to C.F. Gauss (the law of distribution of errors of observation) and J. Maxwell (the law of distribution of velocities of molecules) (see also **Independence; Characterization theorems**).

The distribution of a random vector $X = (X_1, \ldots, X_n)$ in $\mathbf{R}^n$, or the joint distribution of random variables $X_1, \ldots, X_n$, is called *normal (multivariate normal)* if for any fixed $t \in \mathbf{R}^n$ the scalar product (t, X) either has a normal distribution or is constant (as one sometimes says, has a normal distribution with variance zero). For random elements with values from some vector space E this definition is retained when t is replaced by any element l of the adjoint space E^* and the scalar product (t, X) is replaced by a linear functional $l(X)$. The joint distribution of several random variables $X_1, \ldots, X_n$ has **characteristic function**

$$f(t) = \exp\left\{ i\mathsf{E}(t, X) - \frac{1}{2} Q(t) \right\},$$

$$t = (t_1, \ldots, t_n) \in \mathbf{R}^n,$$

where

$$\mathsf{E}(t, X) = t_1 \mathsf{E}X_1 + \cdots + t_n \mathsf{E}X_n$$

is a linear form,

$$Q(t) = \mathsf{E}(t, X - \mathsf{E}X)^2 = \sum_{k,l=1}^{n} \sigma_{kl} t_k t_l$$

is a non-negative definite quadratic form, and $\| \sigma_{kl} \|$ is the **covariance matrix** of X. In the positive-definite case the corresponding normal distribution has the probability density

$$p(x_1, \ldots, x_n) = C \exp\{ -Q^{-1}(x_1 - a_1, \ldots, x_n - a_n) \},$$

where Q^{-1} is the quadratic form inverse to Q, the parameters $a_1, \ldots, a_n$ are the mathematical expectations of $X_1, \ldots, X_n$, respectively, and

$$C = \frac{1}{(2\pi)^{n/2} \sqrt{\det \| \sigma_{kl} \|}}$$

is constant. The total number of parameters specifying the normal distribution is

$$\frac{(n+1)(n+2)}{2} - 1$$

and grows rapidly with n (it is 2 for $n = 1$, 20 for $n = 5$, and 65 for $n = 10$). A multivariate normal distribution is the basic model of **multi-dimensional statistical analysis**. It is also used in the theory of stochastic processes (where normal distributions in infinite-dimensional spaces are examined; see **Random element**, and also **Wiener measure; Wiener process; Gaussian process**).

Of the important properties of normal distributions the following should be mentioned. The sum X of two independent random variables X_1 and X_2 having normal distributions also has a normal distribution; conversely, if $X = X_1 + X_2$ has a normal distribution and X_1 and X_2 are independent, then the distributions of

X_1 and X_2 are normal (*Cramér's theorem*). This property has a certain 'stability': If the distribution of X is 'close' to normal, then so are the distributions of X_1 and X_2. Some other important distributions are connected with normal ones (see **Logarithmic normal distribution**; **Non-central 'chi-squared' distribution**; **Student distribution**; **Wishart distribution**; **Fisher z-distribution**; **Hotelling T^2-distribution**; **'Chi-squared' distribution**). For an approximate representation of distributions close to normal, series like **Edgeworth series** and **Gram−Charlier series** are widely used.

Concerning problems connected with estimators of parameters of normal distributions using results of observations see **Unbiased estimator**. Concerning testing the hypothesis of normality see **Non-parametric methods in statistics**. See also **Probability graph paper**.

References

[1] BOL'SHEV, L.N. and SMIRNOV, N.V.: *Tables of mathematical statistics*, Libr. of mathematical tables, 46, Nauka, Moscow, 1983 (in Russian). Processed by L.S. Bark and E.S. Kedrova.
[2] *Tables of the normal probability integral, the normal density, and its normal derivatives*, Moscow, 1960 (in Russian).
[3] GNEDENKO, B.V.: *The theory of probability*, Chelsea, 1962 (translated from the Russian).
[4] CRAMÉR, H.: *Mathematical methods of statistics*, Princeton Univ. Press, 1946.
[5] KENDALL, M.G. and STUART, A.: *The advanced theory of statistics*, 1. Distribution theory, Griffin, 1977.
[6] KENDALL, M.G. and STUART, A.: *The advanced theory of statistics*. , 2. Inference and relationship, Griffin, 1979.

Yu.V. Prokhorov

Editorial comments.

References

[A1] JOHNSON, N.L. and KOTZ, S.: *Distributions in statistics*, 2. Continuous univariate distributions, Wiley, 1970.
[A2] JOHNSON, N.L. and KOTZ, S.: *Distributions in statistics*, 3. Continuous multivariate distributions, Wiley, 1972.
[A3] PEARSON, E.S. and HARTLEY, H.O.: *Biometrika tables for statisticians*, 1, Cambridge Univ. Press, 1966.

AMS 1980 Subject Classification: 60E05, 62E15

NORMAL DYNAMICAL SYSTEM - 1) The same as a stationary **Gaussian process** (in the narrow sense), regarded as a dynamical system.

2) In the nineteen-forties and nineteen-fifties this was sometimes used for dynamical systems (measurable flows and cascades) in a **Lebesgue space** (the latter at the time was called a *normal space with a measure*).

D.V. Anosov

AMS 1980 Subject Classification: 60G15, 58FXX, 28DXX

NORMAL EPIMORPHISM - A morphism having the characteristic property of the natural mapping of a group onto a quotient group or of a ring onto a quotient ring. Let $\Re$ be a **category** with zero morphisms. A morphism $\nu: A \to V$ is called a *normal epimorphism* if

every morphism $\phi: A \to Y$ for which it always follows from $\alpha\nu = 0$, $\alpha: X \to A$, that $\alpha\phi = 0$, can be uniquely represented in the form $\phi = \nu\phi'$. The **cokernel** of any morphism is a normal epimorphism. The converse assertion is false, in general; however, when morphisms in $\Re$ have kernels, then every normal epimorphism is a cokernel. In an **Abelian category** every epimorphism is normal. The concept of a normal epimorphism is dual to that of a **normal monomorphism**.

M.Sh. Tsalenko

AMS 1980 Subject Classification: 18A20

NORMAL EQUATION, *normalized equation* - The equation of a line in a plane of the form

$$x\cos\alpha + y\sin\alpha - p = 0,$$

where x and y are rectangular Cartesian coordinates in the plane, $\cos\alpha$ and $\sin\alpha$ are the coordinates of the unit vector $\{\cos\alpha, \sin\alpha\}$ perpendicular to the line, and $p \geqslant 0$ is the distance of the coordinate origin from the line. An equation of a line of the form

$$Ax + By + C = 0$$

reduces to normal form after multiplication by the *normalizing factor* λ of absolute value $(A^2 + B^2)^{-1/2}$ and of sign opposite to that of C (if $C = 0$, the sign of λ is arbitrary).

Similarly, the equation of a plane

$$Ax + By + Cz + D = 0$$

reduces to the normal form

$$x\cos\alpha + y\cos\beta + z\cos\gamma - p = 0,$$

where $\cos\alpha$, $\cos\beta$ and $\cos\gamma$ are the direction cosines of a vector perpendicular to the plane, after multiplication by the normalizing factor λ of absolute value $(A^2 + B^2 + C^2)^{-1/2}$ and of sign opposite to that of D.

A.B. Ivanov

AMS 1980 Subject Classification: 51N20

NORMAL EXTENSION *of a field K* - An algebraic field extension (cf. **Extension of a field**) L of K satisfying one of the following equivalent conditions:

1) any imbedding of L in the **algebraic closure** $\bar{K}$ of K comes from an automorphism of L;

2) L is the splitting field of some family of polynomials with coefficients in K (cf. **Splitting field of a polynomial**);

3) any polynomial $f(x)$ with coefficients in K, irreducible over K and having a root in L, splits in L into linear factors.

For every algebraic extension F/K there is a maximal intermediate subfield L that is normal over K; this is the field $L = \bigcap_\sigma F^\sigma$, where σ ranges over all imbeddings of F in $\bar{K}$. There is also a unique minimal normal extension of K containing F. This is the composite of all fields F^σ. It is called the *normal closure of the field F relative to K*. If L_1 and L_2 are normal extensions of K, then so are the intersection $L_1 \cap L_2$ and the composite $L_1 \cdot L_2$. However, when L/K' and K'/K are normal extensions, L/K need not be normal.

For fields of characteristic zero every normal extension is a Galois extension. In general, a normal extension is a **Galois extension** if and only if it is separable (cf. **Separable extension**).

References

[1] WAERDEN, B.L. VAN DER: *Algebra*, 1-2, Springer, 1967-1971 (translated from the German).
[2] LANG, S.: *Algebra*, Addison-Wesley, 1984.
[3] POSTNIKOV, M.M.: *Galois theory*, Moscow, 1962 (in Russian).

L.V. Kuz'min

AMS 1980 Subject Classification: 12F99

NORMAL FAMILY *of analytic functions in a domain* - A family S of single-valued analytic functions $f(z)$ of complex variables $z = (z_1, \ldots, z_n)$ in a domain D in the space $\mathbf{C}^n$, $n \geq 1$, such that from any sequence of functions in S one can extract a subsequence $\{f_\nu(z)\}$ that converges uniformly on compact subsets in D to an analytic function or to infinity. Uniform convergence to infinity on compact subsets means, by definition, that for any compact set $K \subset D$ and any $M > 0$ one can find an $N = N(K, M)$ such that $|f_\nu(z)| > M$ for all $\nu > N$, $z \in K$.

A family S is called a *normal family at a point* $z^0 \in D$ if S is normal in some ball with centre at z^0. A family S is normal in D if and only if it is normal at every point $z^0 \in D$. Every compact family of holomorphic functions is normal; the converse conclusion is false (see **Compactness principle**). If a family S of holomorphic functions in a domain $D \subset \mathbf{C}^n$ has the property that all functions $f(z) \in S$ omit two fixed values, then S is normal in D (*Montel's theorem*). This criterion of normality considerably simplifies the investigation of analytic functions in a neighbourhood of an **essential singular point** (see also **Picard theorem**).

A *normal family of meromorphic functions* in a domain $D \subset \mathbf{C} = \mathbf{C}^1$ is defined similarly: A family S of meromorphic functions in D is normal if from every sequence of functions in S one can extract a subsequence $\{f_\nu(z)\}$ that converges uniformly on compact subsets in D to a meromorphic function or to infinity. By definition, $\{f_\nu(z)\}$ converges uniformly on compact subsets in D to $f(z)$ (the case $f(z) \equiv \infty$ is excluded) if for any compact set $K \subset D$ and any $\epsilon > 0$ there is an $N = N(\epsilon, K)$ and a disc $B = B(z^0, r)$ of radius $r = r(\epsilon, K)$ with centre at some point $z^0 \in K$ such that for $\nu > N$,

$$|f_\nu(z) - f(z)| < \epsilon, \quad z \in B,$$

when $f(z^0) \neq \infty$, or

$$\left| \frac{1}{f_\nu(z)} - \frac{1}{f(z)} \right| < \epsilon, \quad z \in B,$$

when $f(z^0) = \infty$. If a family S of meromorphic functions in a domain $D \subset \mathbf{C}$ has the property that all functions $f \in S$ omit three fixed values, then S is normal (*Montel's theorem*). A family S of meromorphic functions is normal in a domain $D \subset \mathbf{C}$ if and only if

$$\sup \{\rho(f(z)) : f \in S\} < \infty$$

on every compact set $K \subset D$, where

$$\rho(f(z)) = \frac{|f'(z)|}{1 + |f(z)|^2}$$

is the so-called *spherical derivative* of $f(z)$.

From the nineteen-thirties onwards great value was attached to the study of **boundary properties of analytic functions** (see also **Cluster set**, [3], [4]). A meromorphic function $f(z)$ in a simply-connected domain $D \subset \mathbf{C}$ is said to be a *normal function in the domain D* if the family $\{f(\gamma(z))\}$ is normal in D, where $\gamma(z)$ ranges over the family of all conformal automorphisms of D. A function $f(z)$ is called *normal in a multiply-connected domain D* if it is normal on the **universal covering** surface of D. If a meromorphic function $f(z)$ in D omits three values, then $f(z)$ is normal. For $f(z)$, $f(z) \neq \text{const}$, to be normal in the unit disc $G = \{z \in \mathbf{C}: |z| < 1\}$ it is necessary and sufficient that

$$\frac{|f'(z)|}{1 + |f(z)|^2} < \frac{c}{1 - |z|^2}, \quad z \in G, \quad c = c(f) = \text{const.}$$

For a normal meromorphic function $f(z)$ in the unit disc G the existence of an **asymptotic value** α at a boundary point $\zeta \in \Gamma = \{\zeta \in \mathbf{C}: |\zeta| = 1\}$ implies that α is a non-tangential boundary value (cf. **Angular boundary value**) of $f(z)$ at ζ. However, a meromorphic normal function in G need not have asymptotic values at all. On the other hand, if $f(z)$ is a holomorphic normal function in G, then non-tangential boundary values exist even on a set of points of the unit circle Γ that is dense in Γ.

References

[1] MONTEL, P.: *Leçons sur les familles normales de fonctions analytiques et leurs applications*, Gauthier-Villars, 1927.
[2] MARKUSHEVICH, A.I.: *Theory of functions of a complex variable*, 2, Chelsea, 1977 (translated from the Russian).
[3] COLLINGWOOD, E.F. and LOHWATER, A.J.: *The theory of cluster sets*, Cambridge Univ. Press, 1966.
[4] LOHWATER, A.: 'The boundary behaviour of analytic functions', *Itogi Nauk. i Tekhn. Mat. Anal.* **10** (1973), 99-259 (in Russian).

E.D. Solomentsev

Editorial comments. Let $D_1 \subseteq \mathbf{C}^m$, $D_2 \subseteq \mathbf{C}^n$ be domains. A family F of analytic mappings from D_1 to D_2 is called normal if from any sequence of mappings in F one can either extract a subsequence $\{f_\nu(z)\}$ that is uniformly convergent on compact subsets in D_1 to an analytic mapping from D_1 to D_2, or a subsequence $\{f_\nu(z)\}$ with the property that for every compact sets $K_1 \subset D_1$, $K_2 \subset D_2$ there is an N such that $f_\nu(K_1) \cap K_2 = \varnothing$ for $\nu > N$, see [A1].

References

[A1] KRANTZ, S.G.: *Function theory of several complex variables*, Wiley, 1982.

[A2] LEHTO, O. and VIRTANEN, K.I.: 'Boundary behaviour and normal meromorphic functions', *Acta Math.* **97** (1957), 47-65.

AMS 1980 Subject Classification: 30D45, 32A17

NORMAL FORM - 1) The *normal form of a matrix A* is a matrix N of a pre-assigned special form obtained from A by means of transformations of a prescribed type. One distinguishes various normal forms, depending on the type of transformations in question, on the domain K to which the coefficients of A belong, on the form of A, and, finally, on the specific nature of the problem to be solved (for example, on the desirability of extending or not extending K on transition from A to N, on the necessity of determining N from A uniquely or with a certain amount of arbitrariness). Frequently, instead of 'normal form' one uses the term '*canonical form*'. Among the classical normal forms are the following. (Henceforth $M_{m \times n}(K)$ denotes the set of all matrices of m rows and n columns with coefficients in K.)

The Smith normal form. Let K be either the ring of integers $\mathbf{Z}$ or the ring $F[\lambda]$ of polynomials in λ with coefficients in a field F. A matrix $B \in M_{m \times n}(K)$ is called *equivalent* to a matrix $A \in M_{m \times n}(K)$ if there are invertible matrices $C \in M_{m \times m}(K)$ and $D \in M_{n \times n}(K)$ such that $B = CAD$. Here B is equivalent to A if and only if B can be obtained from A by a sequence of *elementary row-and-column transformations*, that is, transformations of the following three types: a) permutation of the rows (or columns); b) addition to one row (or column) of another row (or column) multiplied by an element of K; or c) multiplication of a row (or column) by an invertible element of K. For transformations of this kind the following propositions hold: Every matrix $A \in M_{m \times n}(K)$ is equivalent to a matrix $N \in M_{m \times n}(K)$ of the form

$$N = \begin{Vmatrix} d_1 & & & & & 0 \\ & \ddots & & & & \\ & & d_r & & & \\ & & & 0 & & \\ & & & & \ddots & \\ 0 & & & & & 0 \end{Vmatrix},$$

where $d_i \neq 0$ for all i; d_i divides d_{i+1} for $i = 1, \ldots, r-1$; and if $K = \mathbf{Z}$, then all d_i are positive; if $K = F[\lambda]$, then the leading coefficients of all polynomi-

als d_i are 1. This matrix is called the *Smith normal form* of A. The d_i are called the *invariant factors* of A and the number r is called its *rank*. The Smith normal form of A is uniquely determined and can be found as follows. The rank r of A is the order of the largest non-zero **minor** of A. Suppose that $1 \leqslant j \leqslant r$; then among all minors of A of order j there is at least one non-zero. Let Δ_j, $j = 1, \ldots, r$, be the greatest common divisor of all non-zero minors of A of order j (normalized by the condition $\Delta_j > 0$ for $K = \mathbf{Z}$ and such that the leading coefficient of Δ_j is 1 for $K = F[\lambda]$), and let $\Delta_0 = 1$. Then $d_j = \Delta_j / \Delta_{j-1}$, $j = 1, \ldots, r$. The invariant factors form a full set of invariants of the classes of equivalent matrices: Two matrices in $M_{m \times n}(K)$ are equivalent if and only if their ranks and their invariant factors with equal indices are equal.

The invariant factors $d_1, \ldots, d_r$ split (in a unique manner, up to the order of the factors) into the product of powers of irreducible elements $e_1, \ldots, e_s$ of K (which are positive integers > 1 when $K = \mathbf{Z}$, and polynomials of positive degree with leading coefficient 1 when $K = F[\lambda]$):

$$d_i = e_1^{n_{i1}} \cdots e_s^{n_{is}}, \quad i = 1, \ldots, r,$$

where the n_{ij} are non-negative integers. Every factor $e_j^{n_{ij}}$ for which $n_{ij} > 0$ is called an *elementary divisor* of A (over K). Every elementary divisor of A occurs in the set $\mathscr{E}_{A,K}$ of all elementary divisors of A with multiplicity equal to the number of invariant factors having this divisor in their decompositions. In contrast to the invariant factors, the elementary divisors depend on the ring K over which A is considered: If $K = F[\lambda]$, $\tilde{F}$ is an extension of F and $\tilde{K} = \tilde{F}[\lambda]$, then, in general, a matrix $A \in M_{m \times n}(K) \subset M_{m \times n}(\tilde{K})$ has distinct elementary divisors (but the same invariant factors), depending on whether A is regarded as an element of $M_{m \times n}(K)$ or of $M_{m \times n}(\tilde{K})$. The invariant factors can be recovered from the complete collection of elementary divisors, and vice versa.

For a practical method of finding the Smith normal form see, for example, [1].

The main result on the Smith normal form was obtained for $K = \mathbf{Z}$ (see [7]) and $K = F[\lambda]$ (see [8]). With practically no changes, the theory of Smith normal forms goes over to the case when K is any principal ideal ring (see [3], [6]). The Smith normal form has important applications; for example, the structure theory of finitely-generated modules over principal ideal rings is based on it (see [3], [6]); in particular, this holds for the theory of finitely-generated Abelian groups and theory of the Jordan normal form (see below).

The natural normal form. Let K be a field. Two square

matrices $A, B \in M_{n \times n}(K)$ are called *similar* over K if there is a non-singular matrix $C \in M_{n \times n}(K)$ such that $B = C^{-1}AC$. There is a close link between similarity and equivalence: Two matrices $A, B \in M_{n \times n}(K)$ are similar if and only if the matrices $\lambda E - A$ and $\lambda E - B$, where E is the identity matrix, are equivalent. Thus, for the similarity of A and B it is necessary and sufficient that all invariant factors, or, what is the same, the collection of elementary divisors over $K[\lambda]$ of $\lambda E - A$ and $\lambda E - B$, are the same. For a practical method of finding a C for similar matrices A and B, see [1], [4].

The matrix $\lambda E - A$ is called the *characteristic matrix* of $A \in M_{n \times n}(K)$, and the invariant factors of $\lambda E - A$ are called the *similarity invariants* of A; there are n of them, say $d_1, \ldots, d_n$. The polynomial d_n is the determinant of $\lambda E - A$ and is called the *characteristic polynomial* of A. Suppose that $d_1 = \cdots = d_q = 1$ and that for $j \geqslant q+1$ the degree of d_j is greater than 1. Then A is similar over K to a block-diagonal matrix $N_1 \in M_{n \times n}(K)$ of the form

$$N_1 = \begin{Vmatrix} L(d_{q+1}) & & & 0 \\ & \cdot & & \\ & & \cdot & \\ & & & \cdot \\ 0 & & & L(d_n) \end{Vmatrix},$$

where $L(f)$ for a polynomial

$$f = \lambda^p + \alpha_1 \lambda^{p-1} + \cdots + \alpha_p$$

denotes the so-called *companion matrix*

$$L(f) = \begin{Vmatrix} 0 & 1 & 0 & \cdots & 0 & 0 \\ 0 & 0 & 1 & \cdots & 0 & 0 \\ \cdot & \cdot & \cdot & \cdots & \cdot & \cdot \\ 0 & 0 & 0 & \cdots & 0 & 1 \\ -\alpha_p & -\alpha_{p-1} & -\alpha_{p-2} & \cdots & -\alpha_2 & -\alpha_1 \end{Vmatrix}.$$

The matrix N_1 is uniquely determined from A and is called the *first natural normal form* of A (see [1], [2]).

Now let $\mathscr{E}_{A, K[\lambda]}$ be the collection of all elementary divisors of $\lambda E - A$. Then A is similar over K to a block-diagonal matrix N_2 (cf. **Block-diagonal operator**) whose blocks are the companion matrices of all elementary divisors $e_j^{n_{ij}} \in \mathscr{E}_{A, K[\lambda]}$ of $\lambda E - A$:

$$N_2 = \begin{Vmatrix} \cdot & & & 0 \\ & \cdot & & \\ & & L(e_j^{n_{ij}}) & \\ & & & \cdot \\ 0 & & & \cdot \end{Vmatrix}.$$

The matrix N_2 is determined from A only up to the order of the blocks along the main diagonal; it is called the *second natural normal form* of A (see [1], [2]), or its *Frobenius*, *rational* or *quasi-natural normal form* (see [4]). In contrast to the first, the second natural form

changes, generally speaking, on transition from K to an extension.

The Jordan normal form. Let K be a field, let $A \in M_{n \times n}(K)$, and let $\mathscr{E}_{A, K[\lambda]} = \{e_i^{n_{ij}}\}$ be the collection of all elementary divisors of $\lambda E - A$ over $K[\lambda]$. Suppose that K has the property that the characteristic polynomial d_n of A splits in $K[\lambda]$ into linear factors. (This is so, for example, if K is the field of complex numbers or, more generally, any algebraically closed field.) Then every one of the polynomials e_i has the form $\lambda - a_i$ for some $a_i \in K$, and, accordingly, $e_i^{n_{ij}}$ has the form $(\lambda - a_i)^{n_{ij}}$. The matrix $J(f)$ in $M_{s \times s}(K)$ of the form

$$J(f) = \begin{Vmatrix} a & 1 & & & 0 \\ & \cdot & & & \\ & & \cdot & & \\ & & & \cdot & 1 \\ 0 & & & & a \end{Vmatrix},$$

where $f = (\lambda - a)^s$, $a \in K$, is called the *hypercompanion matrix* of f (see [1]) or the *Jordan block of order s with eigen value a*. The following fundamental proposition holds: A matrix A is similar over K to a block-diagonal matrix $J \in M_{n \times n}(K)$ whose blocks are the hypercompanion matrices of all elementary divisors of $\lambda E - A$:

$$J = \begin{Vmatrix} \cdot & & & 0 \\ & \cdot & & \\ & & J(e_i^{n_{ij}}) & \\ & & & \cdot \\ 0 & & & \cdot \end{Vmatrix}.$$

The matrix J is determined only up to the order of the blocks along the main diagonal; it is a **Jordan matrix** and is called the *Jordan normal form* of A. If K does not have the property mentioned above, then A cannot be brought, over K, to the Jordan normal form (but it can over a finite extension of K). See [4] for information about the so-called generalized Jordan normal form, reduction to which is possible over any field K.

Apart from the various normal forms for arbitrary matrices, there are also special normal forms of special matrices. Classical examples are the normal forms of symmetric and skew-symmetric matrices. Let K be a field. Two matrices $A, B \in M_{n \times n}(K)$ are called *congruent* (see [1]) if there is a non-singular matrix $C \in M_{n \times n}(K)$ such that $B = C^T A C$. Normal forms under the congruence relation have been investigated most thoroughly for the classes of symmetric and skew-symmetric matrices. Suppose that $\operatorname{char} K \neq 2$ and that A is skew-symmetric, that is, $A^T = -A$. Then A is congruent to a uniquely determined matrix H of the form

$$H = \begin{Vmatrix} \begin{matrix} 0 & 1 \\ -1 & 0 \end{matrix} & & & & \\ & \begin{matrix} 0 & 1 \\ -1 & 0 \end{matrix} & & & \\ & & \ddots & & \\ & & & \begin{matrix} 0 & 1 \\ -1 & 0 \end{matrix} & \\ & & & & 0 \\ & & & & & \ddots \\ & & & & & & 0 \end{Vmatrix},$$

which can be regarded as the normal form of A under congruence. If A is symmetric, that is, $A^T = A$, then it is congruent to a matrix D of the form

$$D = \begin{Vmatrix} \epsilon_1 & & & & & 0 \\ & \ddots & & & & \\ & & \epsilon_r & & & \\ & & & 0 & & \\ & & & & \ddots & \\ 0 & & & & & 0 \end{Vmatrix},$$

where $\epsilon_1 \neq 0$ for all i. The number r is the rank of A and is uniquely determined. The subsequent finer choice of the ϵ_i depends on the properties of K. Thus, if K is algebraically closed, one may assume that $\epsilon_1 = \cdots = \epsilon_r = 1$; if K is the field of real numbers, one may assume that $\epsilon_1 = \cdots \epsilon_p = 1$ and $\epsilon_{p+1} = \cdots = \epsilon_r = -1$ for a certain p. D is uniquely determined by these properties and can be regarded as the normal form of A under congruence. See [6], [10] and **Quadratic form** for information about the normal forms of symmetric matrices for a number of other fields, and also about Hermitian analogues of this theory.

A common feature in the theories of normal forms considered above (and also in others) is the fact that the admissible transformations over the relevant set of matrices are determined by the action of a certain group, so that the classes of matrices that can be carried into each other by means of these transformations are the orbits (cf. **Orbit**) of this group, and the appropriate normal form is the result of selecting in each orbit a certain canonical representative. Thus, the classes of equivalent matrices are the orbits of the group $G = GL_m(K) \times GL_n(K)$ (where $GL_s(K)$ is the group of invertible square matrices of order s with coefficients in K), acting on $M_{m \times n}(K)$ by the rule $A \to C^{-1}AD$, where $(C, D) \in G$. The classes of similar matrices are the orbits of $GL_n(K)$ on $M_{n \times n}(K)$ acting by the rule $A \to C^{-1}AC$, where $C \in GL_n(K)$. The classes of congruent symmetric or skew-symmetric matrices are the orbits of the group $GL_n(K)$ on the set of all symmetric or skew-symmetric matrices of order n, acting by the rule $A \to C^T AC$, where $C \in GL_n(K)$. From this point of view every normal form is a specific example of the solution of part of the general problem of orbital decomposition for the action of a certain transformation group.

References

[1] MARKUS, M. and MINC, H.: *A survey of matrix theory and matrix inequalities*, Allyn & Bacon, 1964.
[2] LANCASTER, P.: *Theory of matrices*, Acad. Press, 1969.
[3] LANG, S.: *Algebra*, Addison-Wesley, 1974.
[4] MAL'TSEV, A.I.: *Foundations of linear algebra*, Freeman, 1963 (translated from the Russian).
[5] BOURBAKI, N.: *Elements of mathematics. Algebra: Modules. Rings. Forms*, 2, Addison-Wesley, 1975, Chapt. 4; 5; 6 (translated from the French).
[6] BOURBAKI, N.: *Elements of mathematics. Algebra: Algebraic structures. Linear algebra*, 1, Addison-Wesley, 1974 , p. Chapt. 1; 2 (translated from the French).
[7] SMITH, H.J.S.: 'On systems of linear indeterminate equations and congruences', in *Collected Math. Papers*, Vol. 1, Chelsea, reprint, 1979, pp. 367-409.
[8] FROBENIUS, G.: 'Theorie der linearen Formen mit ganzen Coeffizienten', *J. Reine Angew. Math.* **86** (1879), 146-208.
[9] GANTMAKHER, F.R.: *The theory of matrices*, Chelsea, reprint, 1977 (translated from the Russian).
[10] SERRE, J.-P.: *A course in arithmetic*, Springer, 1973 (translated from the French).

V.L. Popov

Editorial comments. The *Smith canonical form* and a canonical form related to the first natural normal form are of substantial importance in linear control and system theory [A1], [A2]. Here one studies systems of equations $x = Ax + Bu$, $x \in \mathbf{R}^n$, $u \in \mathbf{R}^m$, and the similarity relation is: $(A, B) \sim (SAS^{-1}, SB)$. A pair of matrices $A \in \mathbf{R}^{n \times n}$, $B \in \mathbf{R}^{n \times m}$ is called *completely controllable* if the rank of the block matrix

$$(B, AB, \ldots, A^nB) = R(A, B)$$

is n. Observe that $R(SAS^{-1}, SB) = SR(A, B)$, so that a canonical form can be formed by selecting n independent column vectors from $R(A, B)$. This can be done in many ways. The most common one is to test the columns of $R(A, B)$ for independence in the order in which they appear in $R(A, B)$. This yields the following so-called *Brunovskiĭ−Luenberger canonical form* or *block companion canonical form* for a completely-controllable pair (A, B):

$$\bar{A} = S^{-1}AS = \begin{Vmatrix} \bar{A}_{11} & \cdots & \bar{A}_{1m} \\ \vdots & & \vdots \\ \bar{A}_{m1} & \cdots & \bar{A}_{mm} \end{Vmatrix},$$

$$\bar{B} = S^{-1}B = (\bar{b}_1, \ldots, \bar{b}_m),$$

where $\bar{A}_{ij}$ is a matrix of size $d_i \times d_j$ for certain $d_i \in \mathbf{N} \cup \{0\}$, $\sum_{i=1}^{m} d_i = n$, of the form

$$\bar{A}_{ii} = \begin{Vmatrix} 0 & 0 & 0 & \cdots & 0 & 0 & * \\ 1 & 0 & 0 & \cdots & 0 & 0 & * \\ 0 & 1 & 0 & \cdots & 0 & 0 & * \\ & \cdot & \cdot & & & \cdot & \cdot & \cdot \\ 0 & 0 & 0 & \cdots & 0 & 1 & * \end{Vmatrix},$$

$$\bar{A}_{ij} = \begin{Vmatrix} 0 & \cdots & 0 & * \\ 0 & \cdots & 0 & * \\ \cdot & & \cdot & \cdot \\ 0 & \cdots & 0 & * \end{Vmatrix} \quad \text{for } i \neq j,$$

and $\bar{b}_j$ for $d_j \neq 0$ is the $(d_1 + \cdots + d_{j-1} + 1)$-th standard basis vector of $\mathbf{R}^n$; the $\bar{b}_j$ with $d_j = 0$ have arbitrary coefficients $*$. Here the $*$'s denote coefficients which can take any value. If d_j or d_i is zero, the block A_{ij} is empty (does not occur). Instead of $\mathbf{R}$ any field can be used. The d_j are called *controllability indices* or *Kronecker indices*. They are invariants.

Canonical forms are often used in (numerical) computations. This must be done with caution, because they may not depend continuously on the parameters [A3]. For example, the Jordan canonical form is not continuous; an example of this is:

$$\begin{Vmatrix} 1 & t \\ 0 & 1 \end{Vmatrix} \mapsto \begin{Vmatrix} 1 & 1 \\ 0 & 1 \end{Vmatrix} \quad \text{for } t \neq 0,$$

$$\begin{Vmatrix} 1 & 0 \\ 0 & 1 \end{Vmatrix} \mapsto \begin{Vmatrix} 1 & 0 \\ 0 & 1 \end{Vmatrix}.$$

The matter of continuous canonical forms has much to do with *moduli problems* (cf. **Moduli theory**). Related is the matter of canonical forms for families of objects, e.g. canonical forms for holomorphic families of matrices under similarity [A4]. For a survey of moduli-type questions in linear control theory cf. [A5].

In the case of a controllable pair (A, B) with $m = 1$, i.e. B is a vector $b \in \mathbf{R}^n$, the matrix A is cyclic, see also the section below on normal forms for operators. In this special case there is just one block $\bar{A}_{11}$ (and one vector $\bar{b}_1$). This canonical form for a cyclic matrix with a cyclic vector is also called the *Frobenius canonical form* or the *companion canonical form*.

References

[A1] WOLOVICH, W.A.: *Linear multivariable systems*, Springer, 1974.
[A2] KLAMKA, J.: *Controllability of dynamical systems*, Kluwer, 1990.
[A3] GOLUB, S.H. and WILKINSON, J.H.: 'Ill conditioned eigensystems and the computation of the Jordan canonical form', *SIAM Rev.* 18 (1976), 578-619.
[A4] ARNOL'D, V.I.: 'On matrices depending on parameters', *Russ. Math. Surv.* 26, no. 2 (1971), 29-43. (*Uspekhi Mat. Nauk* 26, no. 2 (1971), 101-114)
[A5] HAZEWINKEL, M.: '(Fine) moduli spaces for linear systems: what are they and what are they good for', in C.I. Byrnes and C.F. Martin (eds.): *Geometrical methods for the theory of linear systems*, Reidel, 1980, pp. 125-193.
[A6] TURNBALL, H.W. and AITKEN, A.C.: *An introduction to the theory of canonical matrices*, Blackie & Son, 1932.

2) A *normal form of an operator* is a representation, up to an isomorphism, of a **self-adjoint operator** A acting on a Hilbert space $\mathcal{H}$ as an orthogonal sum of multiplication operators by the independent variable.

To begin with, suppose that A is a *cyclic operator*; this means that there is an element $h_0 \in \mathcal{H}$ such that every element $h \in \mathcal{H}$ has a unique representation in the form $F(A)h_0$, where $F(\xi)$ is a function for which

$$\int_{-\infty}^{+\infty} |F(\xi)|^2 \, d(E_\xi h_0, h_0) < \infty;$$

here E_ξ, $-\infty < \xi < \infty$, is the **spectral resolution** of A. Let $\mathscr{L}_\rho^2$ be the space of square-integrable functions on $(-\infty, +\infty)$ with weight $\rho(\xi) = (E_\xi h_0, h_0)$, and let $K_\rho F = \xi F(\xi)$ be the multiplication operator by the independent variable, with domain of definition

$$D_{K_\rho} = \left\{ F(\xi) : \int_{-\infty}^{+\infty} \xi^2 |F(\xi)|^2 \, d\rho(\xi) < \infty \right\}.$$

Then the operators A and K_ρ are isomorphic, $A \simeq K_\rho$; that is, there exists an isomorphic and isometric mapping $U : \mathcal{H} \to \mathscr{L}_\rho^2$ such that $UD_A = D_{K_\rho}$ and $A = U^{-1}K_\rho U$.

Suppose, next, that A is an arbitrary self-adjoint operator. Then $\mathcal{H}$ can be split into an orthogonal sum of subspaces $\mathcal{H}_\alpha$ on each of which A induces a cyclic operator A_α, so that $H = \sum \oplus H_\alpha$, $A = \sum \oplus A_\alpha$ and $A_\alpha \simeq K_{\rho_\alpha}$. If the operator $K = \sum \oplus K_{\rho_\alpha}$ is given on $\mathscr{L}^2 = \sum \oplus \mathscr{L}_{\rho_\alpha}^2$, then $A \simeq K$.

The operator K is called the *normal form* or *canonical representation* of A. The theorem on the canonical representation extends to the case of arbitrary normal operators (cf. **Normal operator**).

References

[1] PLESNER, A.I.: *Spectral theory of linear operators*, Moscow, 1965 (in Russian).
[2] AKHIEZER, N.I. and GLAZMAN, I.M.: *Theory of linear operators in Hilbert spaces*, Pitman, 1981 (translated from the Russian).

V.I. Sobolev

3) The *normal form of an operator A* is a representation of A, acting on a **Fock space** constructed over a certain space $L_2(M, \sigma)$, where (M, σ) is a **measure space**, in the form of a sum

$$A = \sum_{m,n \geqslant 0} \int K_{n,m}(x_1, \ldots, x_n; y_1, \ldots, y_m) \times \tag{1}$$

$$\times a^*(x_1) \cdots a^*(x_n) a(y_1) \cdots a(y_m) \prod_{i=1}^{n} d\sigma(x_i) \prod_{j=1}^{m} d\sigma(y_j),$$

where $a(x), a^*(x)$ $(x \in M)$ are operator-valued generalized functions generating families of **annihilation operators** $\{a(f) : f \in L_2(M, \sigma)\}$ and **creation operators** $\{a^*(f) : f \in L_2(M, \sigma)\}$:

$$a(f) = \int_M a(x) f(x) \, d\sigma(x), \quad a^*(f) = \int_M a^*(x) \bar{f}(x) \, d\sigma(x).$$

In each term of expression (1) all factors $a(y_j)$, $j = 1, \ldots, m$, stand to the right of all factors $a^*(x_i)$, $i = 1, \ldots, n$, and the (possibly generalized) functions

$K_{n,m}(x_1, \ldots, x_n; y_1, \ldots, y_m)$ in the two sets of variables $(x_1, \ldots, x_n) \in M^n$, $(y_1, \ldots, y_m) \in M^m$, $n, m = 0, 1, \ldots$, are, in the case of a symmetric (Boson) Fock space, symmetric in the variables of each set separately, and, in the case of an anti-symmetric (Fermion) Fock space, anti-symmetric in these variables.

For any bounded operator A the normal form exists and is unique.

The representation (1) can be rewritten in a form containing the annihilation and creation operators directly:

$$A = \tag{2}$$

$$= \sum_{m,n} \sum_{\substack{\{i_1, \ldots, i_n\} \\ \{j_1, \ldots, j_m\}}} c_{i_1 \cdots i_n j_1 \cdots j_m} a^*(f_{i_1}) \cdots a^*(f_{i_n}) a(f_{j_1}) \cdots a(f_{j_m}),$$

where $\{f_i : i = 1, 2, \ldots\}$ is an orthonormal basis in $L_2(M, \sigma)$ and the summation in (2) is over all pairs of finite collections $\{f_{i_1}, \ldots, f_{i_n}\}$, $\{f_{j_1}, \ldots, f_{j_m}\}$ of elements of this basis.

In the case of an arbitrary (separable) Hilbert space H the normal form of an operator A acting on the Fock space $\Gamma(H)$ constructed over H is determined for a fixed basis $\{f_i : i = 1, 2, \ldots\}$ in H by means of the expression (2), where $a(f)$, $a^*(f)$, $f \in H$, are families of annihilation and creation operators acting on $\Gamma(H)$.

References

[1] BEREZIN, F.A.: *The method of second quantization*, Acad. Press, 1966 (translated from the Russian).

R.A. Minlos

Editorial comments.

References

[A1] BOGOLUBOV, N.N. [N.N. BOGOLYUBOV], LOGUNOV, A.A. and TODOROV, I.T.: *Introduction to axiomatic quantum field theory*, Benjamin, 1975 (translated from the Russian).
[A2] KÄLLEN, G.: *Quantum electrodynamics*, Springer, 1972.
[A3] GLIMM, J. and JAFFEE, A.: *Quantum physics*, Springer, 1981.

4) The *normal form of a recursive function* is a method for specifying an n-place **recursive function** ϕ in the form

$$\phi(x_1, \ldots, x_n) = g(\mu z(f(x_1, \ldots, x_n, z) = 0)), \tag{*}$$

where f is an $(n+1)$-place **primitive recursive function**, g is a 1-place primitive recursive function and $\mu z(f(x_1, \ldots, x_n, z) = 0)$ is the result of applying the **least-number operator** to f. *Kleene's normal form theorem* asserts that there is a primitive recursive function g such that every recursive function ϕ can be represented in the form (*) with a suitable function f depending on ϕ; that is,

$$(\exists g)(\forall \phi)(\exists f)(\forall x_1, \ldots, x_n):$$

$$[\phi(x_1, \ldots, x_n) = g(\mu z(f(x_1, \ldots, x_n, z) = 0))].$$

The normal form theorem is one of the most important results in the theory of recursive functions.

A.A. Markov [2] obtained a characterization of those functions g that can be used in the normal form theorem for the representation (*). A function g can be used as function whose existence is asserted in the normal form theorem if and only if the equation $g(x) = n$ has infinitely many solutions for each n. Such functions are called *functions of great range*.

References

[1] MAL'TSEV, A.I.: *Algorithms and recursive functions*, Wolters-Noordhoff, 1970 (translated from the Russian).
[2] MARKOV, A.A.: 'On the representation of recursive functions', *Izv. Akad. Nauk SSSR Ser. Mat.* 13, no. 5 (1949), 417-424 (in Russian).

V.E. Plisko

Editorial comments.

References

[A1] KLEENE, S.C.: *Introduction to metamathematics*, North-Holland, 1951, p. 288.

5) A *normal form of a system of differential equations*

$$\dot{x}_i = \phi_i(x_1, \ldots, x_n), \quad i = 1, \ldots, n, \tag{1}$$

near an invariant manifold M is a formal system

$$\dot{y}_i = \psi_i(x_1, \ldots, y_n), \quad i = 1, \ldots, n, \tag{2}$$

that is obtained from (1) by an invertible formal change of coordinates

$$x_i = \xi_i(y_1, \ldots, y_n), \quad i = 1, \ldots, n, \tag{3}$$

in which the Taylor–Fourier series ψ_i contain only **resonance terms**. In a particular case, normal forms occurred first in the dissertation of H. Poincaré (see [1]). By means of a normal form (2) some systems (1) can be integrated, and many can be investigated for stability and can be integrated approximately; for systems (1) a search has been made for periodic solutions and families of conditionally periodic solutions, and their **bifurcation** has been studied.

Normal forms in a neighbourhood of a fixed point. Suppose that M contains a fixed point $X \equiv (x_1, \ldots, x_n) = 0$ of the system (1) (that is, $\phi_i(0) = 0$), that the ϕ_i are analytic at it and that $\lambda_1, \ldots, \lambda_n$ are the eigen values of the matrix $\| \partial \phi_i / \partial x_j \|$ for $X = 0$. Let $\Lambda \equiv (\lambda_1, \ldots, \lambda_n) \neq 0$. Then in a full neighbourhood of $X = 0$ the system (1) has the following normal form (2): the matrix $\| \partial \psi_i / \partial y_j \|$ has for $Y \equiv (y_1, \ldots, y_n) = 0$ a normal form (for example, the Jordan normal form) and the Taylor series

$$\psi_i = y_i \sum_{Q \in N_i} g_{iQ} Y^Q, \quad i = 1, \ldots, n, \tag{4}$$

contain only resonance terms for which

$$(Q, \Lambda) \equiv q_1 \lambda_1 + \cdots + q_n \lambda_n = 0. \tag{5}$$

Here $Q \equiv (q_1, \ldots, q_n)$, $Y^Q \equiv y_1^{q_1} \cdots y_n^{q_n}$, $N_i = \{Q: \text{integers } q_j \geq 0, q_i \geq -1, q_1 + \cdots + q_n \geq 0\}$. If equation (5) has no solutions $Q \neq 0$ in $N = N_1 \cup \cdots \cup N_n$, then the normal form (2) is linear:

$$\dot{y}_i = \lambda_i y_i, \quad i = 1, \ldots, n.$$

Every system (1) with $\Lambda\neq 0$ can be reduced in a neighbourhood of a fixed point to its normal form (2) by some formal transformation (3), where the ξ_i are (possibly divergent) power series, $\xi_i(0)=0$ and $\det\|\,\partial\xi_i/\partial y_j\,\|\neq 0$ for $Y=0$.

Generally speaking, the normalizing transformation (3) and the normal form (2) (that is, the coefficients g_{iQ} in (4)) are not uniquely determined by the original system (1). A normal form (2) preserves many properties of the system (1), such as being real, symmetric, Hamiltonian, etc. (see [2], [3]). If the original system contains small parameters, one can include them among the coordinates x_j, and then $\dot x_j=0$. Such coordinates do not change under a normalizing transformation (see [3]).

If k is the number of linearly independent solutions $Q\in N$ of equation (5), then by means of a transformation

$$y_i = z_1^{\alpha_{i1}}\cdots z_n^{\alpha_{in}}, \quad i=1,\dots,n,$$

where the α_{ij} are integers and $\det\|\,\alpha_{ij}\,\|=\pm 1$, the normal form (2) is carried to a system

$$\dot z_j = z_j f(z_1,\dots,z_k), \quad i=1,\dots,n$$

(see [2], [3]). The solution of this system reduces to a solution of the subsystem of the first k equations and to $n-k$ quadratures. The subsystem has to be investigated in the neighbourhood of the multiple singular point $z_1=\cdots=z_k=0$, because the $f_1,\dots,f_k$ do not contain linear terms. This can be done by a local method (see [3]).

The following problem has been examined (see [2]): Under what conditions on the normal form (2) does the normalizing transformation of an analytic system (1) converge (be analytic)? Let

$$\omega_k = \min|\,(Q,\Lambda)\,|$$

for those $Q\in N$ for which

$$(Q,\Lambda)\neq 0, \quad q_1+\cdots+q_n < 2^k.$$

Condition ω: $\sum_{k=1}^{\infty} 2^{-k}\log\omega_k^{-1}<\infty$.
Condition $\bar\omega$: $\limsup 2^{-k}\log\omega_k^{-1}<\infty$ as $k\to\infty$.

Condition $\bar\omega$ is weaker than ω. Both are satisfied for almost-all Λ (relative to Lebesgue measure) and are very weak arithmetic restrictions on Λ.

In case $\mathrm{Re}\,\Lambda=0$ there is also *condition A* (for the general case, see in [2]): There exists a power series $a(Y)$ such that in (4), $\phi_i=\lambda_i y_i a$, $i=1,\dots,n$.

If for an analytic system (1) Λ satisfies condition ω and the normal form (2) satisfies condition A, then there exists an analytic transformation of (1) to a certain normal form. If (2) is obtained from an analytic system and fails to satisfy either condition $\bar\omega$ or condition A, then there exists an analytic system (1) that has (2) as its normal form, and every transformation to a normal form diverges (is not analytic).

Thus, the problem raised above is solved for all normal forms except those for which Λ satisfies condition ω, but not $\bar\omega$, while the remaining coefficients of the normal form satisfy condition A. The latter is a very rigid restriction on the coefficients of a normal form, and for large n it holds, generally speaking, only in degenerate cases. That is, the basic reason for divergence of a transformation to normal form is not **small denominators**, but degeneracy of the normal form.

But even in cases of divergence of the normalizing transformation (3) with respect to (2), one can study properties of the solutions of the system (1). For example, a real system (1) has a smooth transformation to the normal form (2) even when it is not analytic. The majority of results on smooth normalization have been obtained under the condition that all $\mathrm{Re}\,\lambda_j\neq 0$. Under this condition, with the help of a change $X\to V$ of finite smoothness class, a system (1) can be brought to a *truncated normal form*

$$\dot v_i = \tilde\psi_i(V), \quad i=1,\dots,n, \tag{6}$$

where the $\tilde\psi_i$ are polynomials of degree m (see [4] - [6]). If in the normalizing transformation (3) all terms of degree higher than m are discarded, the result is a transformation

$$x_i = \tilde\xi_i(U), \quad i=1,\dots,n \tag{7}$$

(the $\tilde\xi_i$ are polynomials), that takes (1) to the form

$$\dot u_i = \tilde\psi_i(U)+\tilde\phi_i(U), \quad i=1,\dots,n, \tag{8}$$

where the $\tilde\psi_i$ are polynomials containing only resonance terms and the $\tilde\phi_i$ are convergent power series containing only terms of degree higher than m. Solutions of the truncated normal form (6) are approximations for solutions of (8) and, after the transformation (7), give approximations of solutions of the original system (1). In many cases one succeeds in constructing for (6) a **Lyapunov function** (or **Chetaev function**) $f(V)$ such that

$$|f(V)|\ \leqslant c_1|\,V\,|^{\gamma} \quad\text{and}\quad \left|\sum_{j=1}^{n}\frac{\partial f}{\partial v_j}\tilde\phi_j\right| > c_2|\,V\,|^{\gamma+m},$$

where c_1 and c_2 are positive constants. Then $f(U)$ is a Lyapunov (Chetaev) function for the system (8); that is, the point $X=0$ is stable (unstable). For example, if all $\mathrm{Re}\,\lambda_i<0$, one can take $m=1$, $f=\sum_{i=1}^{n} v_i^2$ and obtain Lyapunov's theorem on stability under linear approximation (see [7]; for other examples see the survey [8]).

From the normal form (2) one can find invariant analytic sets of the system (1). In what follows it is assumed for simplicity of exposition that $\mathrm{Re}\,\Lambda=0$. From the normal form (2) one extracts the formal set

$$\mathscr{A} = \{Y:\psi_i=\lambda_i y_i a, \ i=1,\dots,n\},$$

where a is a free parameter. Condition A is satisfied on the set $\mathscr{A}$. Let X be the union of subspaces of the form

$\{Y: y_i=0, \ i=i_1, \ldots, i_l\}$ such that the corresponding eigen values λ_j, $j\neq i_1, \ldots, i_l$, $1\leqslant j\leqslant n$, are pairwise commensurable. The formal set $\tilde{\mathscr{A}}=\mathscr{A}\cap K$ is analytic in the system (1). From $\tilde{\mathscr{A}}$ one selects the subset $\mathscr{B}$ that is analytic in (1) if condition ω holds (see [3]). On the sets $\tilde{\mathscr{A}}$ and $\mathscr{B}$ lie periodic solutions and families of conditionally-periodic solutions of (1). By considering the sets $\tilde{\mathscr{A}}$ and $\mathscr{B}$ in systems with small parameters, one can study all analytic perturbations and bifurcations of such solutions (see [9]).

Generalizations. If a system (1) does not lead to a normal form (2) but to a system whose right-hand sides contain certain non-resonance terms, then the resulting simplification is less substantial, but can improve the quality of the transformation. Thus, the reduction to a 'semi-normal form' is analytic under a weakened condition A (see [2]). Another version is a transformation that normalizes a system (1) only on certain submanifolds (for example, on certain coordinate subspaces; see [2]). A combination of these approaches makes it possible to prove for (1) the existence of invariant submanifolds and of solutions of specific form (see [9]).

Suppose that a system (1) is defined and analytic in a neighbourhood of an invariant manifold M of dimension $k+l$ that is fibred into l-dimensional invariant tori. Then close to M one can introduce local coordinates

$$S = (s_1, \ldots, s_k), \ Y = (y_1, \ldots, y_l), \ Z = (z_1, \ldots, z_m),$$
$$k+l+m = n,$$

such that $Z=0$ on M, y_j is of period 2π, S ranges over a certain domain H, and (1) takes the form

$$\left.\begin{aligned} \dot{S} &= \Phi^{(1)}(S, Y, Z), \\ \dot{Y} &= \Omega(S, Y)+\Phi^{(2)}(S, Y, Z), \\ \dot{Z} &= (S, Y)Z+\Phi^{(3)}(S, Y, Z), \end{aligned}\right\} \quad (9)$$

where $\Phi^{(j)}=O(|Z|)$, $j=1,2$, $\Phi^{(3)}=O(|Z|^2)$ and A is a matrix. If $\Omega=\text{const}$ and A is triangular with constant main diagonal $\Lambda(\lambda_1, \ldots, \lambda_n)$, then (under a weak restriction on the small denominators) there is a formal transformation of the local coordinates $S, Y, Z \to U, V, W$ that takes the system (9) to the normal form

$$\left.\begin{aligned} \dot{U} &= \sum\Psi^{(1)}_{PQ}(U)W^Q\exp i(P, V), \\ \dot{V} &= \sum\Psi^{(2)}_{PQ}(U)W^Q\exp i(P, V), \\ \dot{w}_j &= w_j\sum g_{jPQ}(U)W^Q\exp(P, V), \ j=1,\ldots,m, \end{aligned}\right\} \quad (10)$$

where $P\in\mathbf{Z}^l$, $Q\in\mathbf{N}^m$, $U\in H$, and $i(P, \Omega)+(Q, \Lambda)=0$.

If among the coordinates Z there is a small parameter, (9) can be averaged by the **Krylov−Bogolyubov method of averaging** (see [10]), and the averaged system is a normal form. More generally, perturbation theory can be regarded as a special case of the theory of normal forms, when one of the coordinates is a small parameter (see [11]).

Theorems on the convergence of a normalizing change, on the existence of analytic invariant sets, etc., carry over to the systems (9) and (10). Here the best studied case is when M is a periodic solution, that is, $k=0$, $l=1$. In this case the theory of normal forms is in many respects identical with the case when M is a fixed point. Poincaré suggested that one should consider a pointwise mapping of a normal section across the periods. In this context arose a theory of normal forms of pointwise mappings, which is parallel to the corresponding theory for systems (1). For other generalizations of normal forms see [3], [6], [12] - [14].

References

[1] POINCARÉ, H.: 'Thèse, 1928', in *Oeuvres*, Vol. 1, Gauthier-Villars, 1951, pp. IL-CXXXII.

[2A] BRUNO, A.D. [A.D. BRYUNO]: 'Analytical form of differential equations', *Trans. Moscow Math. Soc.* **25** (1971), 131-288. (*Trudy Moskov. Mat. Obshch.* **25** (1971), 119-262)

[2B] BRUNO, A.D. [A.D. BRYUNO]: 'Analytical form of differential equations', *Trans. Moscow Math. Soc.* (1972), 199-239. (*Trudy Moskov. Mat. Obshch.* **26** (1972), 199-239)

[3] BRYUNO, A.D.: *Local methods in nonlinear differential equations*, 1, Springer, 1989 (translated from the Russian).

[4] HARTMAN, P.: *Ordinary differential equations*, Birkhäuser, 1982.

[5A] SAMOVOL, V.S.: 'Linearization of a system of differential equations in the neighbourhood of a singular point', *Soviet Math. Dokl.* **13** (1972), 1255-1259. (*Dokl. Akad. Nauk SSSR* **206** (1972), 545-548)

[5B] SAMOVOL, V.S.: 'Equivalence of systems of differential equations in the neighbourhood of a singular point', *Trans. Moscow Math. Soc. (2)* **44** (1982), 217-237. (*Trudy Moskov. Mat. Obshch.* **44** (1982), 213-234)

[6A] BELITSKIĬ, G.R.: 'Equivalence and normal forms of germs of smooth mappings', *Russian Math. Surveys* **33**, no. 1 (1978), 95-155. (*Uspekhi Mat. Nauk.* **33**, no. 1 (1978))

[6B] BELITSKIĬ, G.R.: 'Normal forms relative to a filtering action of a group', *Trans. Moscow Math. Soc.* **40** (1979), 3-46. (*Trudy Moskov. Mat. Obshch.* **40** (1979), 3-46)

[6C] BELITSKIĬ, G.R.: 'Smooth equivalence of germs of vector fields with a single zero eigenvalue or a pair of purely imaginary eigenvalues', *Funct. Anal. Appl.* **20**, no. 4 (1986), 253-259. (*Funkts. Anal. i Prilozen.* **20**, no. 4 (1986), 1-8)

[7] LYAPUNOV, A.M.: *Problème général de la stabilité du mouvement*, Princeton Univ. Press, 1947 (translated from the Russian).

[8] KUNITSYN, A.L. and MARKEV, A.P.: 'Stability in resonant cases', *Itogi Nauk. i Tekhn. Ser. Obsh. Mekh.* **4** (1979), 58-139 (in Russian).

[9] BIBIKOV, J.N.: *Local theory of nonlinear analytic ordinary differential equations*, Springer, 1979.

[10] BOGOLYUBOV, N.N. and MITROPOL'SKIĬ, YU.A.: *Asymptotic methods in the theory of non-linear oscillations*, Hindushtan Publ. Comp., Delhi, 1961 (translated from the Russian).

[11] BRUNO, A.D. [A.D. BRYUNO]: 'Normal form in perturbation theory', in *Proc. VIII Internat. Conf. Nonlinear Oscillations, Prague, 1978*, Vol. 1, Academia, 1979, pp. 177-182 (in Russian).

[12] KOSTIN, V.V. and LE DINH THUY: 'Some tests of the convergence of a normalizing transformation', *Dapovidi Akad. Nauk URSR Ser. A*, no. 11 (1975), 982-985 (in Russian).

[13] ZEHNDER, E.J.: 'C.L. Siegel's linearization theorem in infinite dimensions', *Manuscr. Math.* **23** (1978), 363-371.

[14] NIKOLENKO, N.V.: 'The method of Poincaré normal forms in problems of integrability of equations of evolution type', *Russian Math. Surveys* **41**, no. 5 (1986), 63-114. (*Uspekhi Mat. Nauk* **41**, no. 5 (1986), 109-152)

A.D. Bryuno

Editorial comments. For more on various *linearization theorems for ordinary differential equations* and *canonical form theorems for ordinary differential equations*, as well as generalizations to the case of non-linear representations of nilpotent Lie algebras, cf. also **Poincaré–Dulac theorem** and **Analytic theory of differential equations**, and [A1].

References

[A1] ARNOL'D, V.I.: *Geometrical methods in the theory of ordinary differential equations*, Springer, 1983 (translated from the Russian).

AMS 1980 Subject Classification: 15A21, 47B15, 03D20, 47A05, 47-XX, 81C10, 81E15, 34A25

NORMAL FUNDAMENTAL SYSTEM OF SOLUTIONS *of a linear homogeneous system of ordinary differential equations* - A **fundamental system of solutions** $x_1(t), \ldots, x_n(t)$ such that any other fundamental system $\hat{x}_1(t), \ldots, \hat{x}_n(t)$ satisfies the inequality

$$\sum_{i=1}^{n} \lambda_{\hat{x}_{i(t)}} \geqslant \sum_{i=1}^{n} \lambda_{x_{i(t)}};$$

here

$$\lambda_{y(t)} = \varlimsup_{t \to +\infty} \frac{1}{t} \log |y(t)|$$

is the **Lyapunov characteristic exponent** of a solution $y(t)$. Normal fundamental systems of solutions were introduced by A.M. Lyapunov [1], who proved that they exist for every linear system

$$\dot{x} = A(t)x,$$

where $A(\cdot)$ is a mapping

$$\mathbf{R}^+ \to \mathrm{Hom}(\mathbf{R}^n, \mathbf{R}^n) \quad (\text{or } \mathbf{R}^+ \to \mathrm{Hom}(\mathbf{C}^n, \mathbf{C}^n))$$

that is summable on every segment and satisfies the additional condition

$$\varlimsup_{t \to \infty} \frac{1}{t} \int_0^t \| A(\tau) \| \, dt < +\infty.$$

References

[1] LYAPUNOV, A.M.: *Collected works*, Moscow, 1956 (in Russian).

V.M. Millionshchikov

AMS 1980 Subject Classification: 34A99

NORMAL MATRIX - A square **matrix** A that commutes with its adjoint (that is, $AA^* = A^*A$).

Editorial comments. See also **Normal operator**.

AMS 1980 Subject Classification: 15A57

NORMAL MONOMORPHISM - A morphism having the characteristic property of an imbedding of a group (ring) into a group (ring) as a normal subgroup (ideal). Let $\mathfrak{K}$ be a **category** with zero morphisms. A morphism $\mu: U \to A$ is called a *normal monomorphism* if every morphism $\phi: X \to A$ for which it always follows from $\mu\alpha = 0$, $\alpha: A \to Y$, that $\phi\alpha = 0$, can be uniquely represented in the form $\phi = \phi'\mu$. The kernel of any morphism (cf. **Kernel of a morphism in a category**) is a normal monomorphism. The converse is not true, in general; however, if cokernels (cf. **Cokernel**) of morphisms exist in $\mathfrak{K}$, then every normal monomorphism turns out to be the kernel of its cokernel. In an **Abelian category** every monomorphism is normal. The concept of a normal monomorphism is dual to that of a **normal epimorphism**.

M.Sh. Tsalenko

Editorial comments. The above definition is not entirely standard: many authors would define a normal monomorphism to be a morphism which occurs as a kernel. (The term *normal subobject* is also in use, for an isomorphism class of normal monomorphisms.) In categories without zero morphisms, a substitute for normal monomorphisms is provided by *regular monomorphisms*, which are those morphisms which occur as equalizers (cf. **Kernel of a morphism in a category**). Every normal monomorphism is regular, but not conversely: in the category of all groups every injective homomorphism is a regular monomorphism, but a normal monomorphism $G \to H$ is an isomorphism of G onto a normal subgroup of H. However, in an additive category the concepts of normal monomorphism and regular monomorphism coincide.

References

[A1] MITCHELL, B.: *Theory of categories*, Acad. Press, 1965, Sect. I.14.

AMS 1980 Subject Classification: 18A20

NORMAL NUMBER - A real number α, $0 \leqslant \alpha \leqslant 1$, having the following property: For every natural number s, any given s-tuple $\delta = (\delta_1, \ldots, \delta_s)$ consisting of the symbols $0, \ldots, g-1$ appears with asymptotic frequency $1/g^s$ in the sequence

$$\alpha_1, \ldots, \alpha_n, \ldots, \tag{1}$$

obtained from the expansion of α in an infinite fraction in base g,

$$\alpha = \frac{\alpha_1}{g} + \cdots + \frac{\alpha_n}{g^n} + \cdots.$$

In more detail, let $g > 1$ be a natural number and let

$$(\alpha_1, \ldots, \alpha_s), (\alpha_2, \ldots, \alpha_{s+1}), (\alpha_3, \ldots, \alpha_{s+2}), \ldots, \tag{2}$$

be the infinite sequence of s-tuples corresponding to (1). Let $N(n, \delta)$ denote the number of occurrences of the tuple $\delta = (\delta_1, \ldots, \delta_s)$ among the first n tuples of (2). The number

$$\alpha = \frac{\alpha_1}{g} + \frac{\alpha_2}{g^2} + \cdots$$

is said to be *normal* if for any number s and any given s-tuple δ consisting of the symbols $0, \ldots, g-1$,

$$\lim_{n \to \infty} \frac{N(n, \delta)}{n} = \frac{1}{g^s}.$$

The concept of a normal number was introduced for $g=10$ by E. Borel (see [1], [2], p. 197). He called a real number α *weakly normal to the base g* if

$$\lim_{n\to\infty} \frac{N(n,\delta)}{n} = \frac{1}{g},$$

where $N(n,\delta)$ is the number of occurrences of δ, $0\leqslant\delta\leqslant g-1$, among the first n terms of the sequences $\alpha_1,\alpha_2,\ldots$, and *normal* if $\alpha, g\alpha, g^2\alpha, \ldots$ are weakly normal to the bases $g, g^2, \ldots$. He also showed that for a normal number

$$\lim_{n\to\infty} \frac{N(n,\delta)}{n} = \frac{1}{g^s},$$

for any s and any given s-tuple $\delta=(\delta_1,\ldots,\delta_s)$. Later it was proved (see [3], [4], and also [8]) that the last relation is equivalent to Borel's definition of a normal number.

A number α is called *absolutely normal* if it is normal with respect to every base $g>0$. The existence of normal and absolutely-normal numbers was established by Borel on the basis of measure theory. The construction of normal numbers in an explicit form was first achieved in [5]. Earlier (see [6], [7]) an effective procedure for constructing normal numbers was indicated. For other methods for constructing normal numbers and for connections between the concepts of normality and randomness see [8].

Uniform distribution of the fractional parts $\{\alpha g^x\}$, $x=1,2,\ldots$, on the interval $[0,1]$ is equivalent to α being normal.

References

[1] BOREL, E.: 'Les probabilités dénombrables et leurs applications arithmétiques', *Rend. Circ. Math. Palermo* **27** (1909), 247-271.
[2] BOREL, E.: *Leçons sur la théorie des fonctions*, Gauthier-Villars, 1928.
[3] PILLAI, S.: 'On normal numbers', *Proc. Indian Acad. Sci. Sect. A* **12** (1940), 179-184.
[4] NIVEN, I. and ZUCKERMAN, H.: 'On the definition of normal numbers', *Pacific J. Math.* **1** (1951), 103-109.
[5] CHAMPERNOWNE, D.G.: 'The construction of decimals normal in the scale of ten', *J. London Math. Soc.* **8** (1933), 254-260.
[6] SIERPIŃSKI, W.: 'Démonstration élémentaire d'un théorème de M. Borel sur les nombres absolument normaux et détermination effective d'un tel nombre', *Bull. Soc. Math. France* **45** (1917), 127-132.
[7] LEBESGUE, H.: 'Sur certaines démonstrations d'existence', *Bull. Soc. Math. France* **45** (1917), 132-144.
[8] POSTNIKOV, A.G.: 'Arithmetic modelling of random processes', *Trudy Mat. Inst. Steklov.* **57** (1960) (in Russian).

S.A. Stepanov

Editorial comments. Almost-all numbers are normal with respect to every base g (see e.g. Theorem 8.11 in [A1]). It is not known whether familiar numbers like $\sqrt{2}, e, \pi$ are normal or not. Normal numbers are potentially interesting in the context of random number generators. A normal number to a base g is necessarily irrational. The weakly-normal number (to base 10) $0.01234567890123456789\cdots$ is of course rational. The number $x=0.1234567891011121314\cdots$, obtained as $x=0.a_1a_2\cdots$ where a_i stands for the group of digits representing i to base 10, is normal to base 10 [5]. The same recipe works to obtain normal numbers to any given base.

References

[A1] NIVEN, I.: *Irrational numbers*, Math. Assoc. Amer., 1956.

AMS 1980 Subject Classification: 10A40

NORMAL OPERATOR - A closed **linear operator** A defined on a linear subspace D_A that is dense in a Hilbert space H such that $A^*A=AA^*$, where A^* is the operator adjoint to A. If A is normal, then $D_{A^*}=D_A$ and $\|A^*x\|=\|Ax\|$ for every x. Conversely, these conditions guarantee that A is normal. If A is normal, then so are A^*; $\alpha A+\beta I$ for any $\alpha,\beta\in\mathbb{C}$; A^{-1} when it exists; and if $AB=BA$, where B is a bounded linear operator, then also $A^*B=BA^*$.

A normal operator has:

1) the *multiplicative decomposition*

$$A=U\sqrt{A^*A} = \sqrt{A^*A}\,U,$$
$$A^* = U^{-1}\sqrt{A^*A} = \sqrt{A^*A}\,U^{-1},$$

where U is a unitary operator which is uniquely determined on the orthogonal complement of the null space of A and A^*;

2) the *additive decomposition*

$$A = A_1+iA_2, \quad A^* = A_1-iA_2,$$

where A_1 and A_2 are uniquely determined self-adjoint commuting operators.

The additive decomposition implies that for an ordered pair (A, A^*) there exists a unique two-dimensional **spectral function** $E(\Delta_\zeta)$, where Δ_ζ is a two-dimensional interval, $\Delta_\zeta=\Delta_\xi\times\Delta_\eta$, $\zeta=\xi+i\eta$, such that

$$A = \int_{\Delta_\infty}\zeta\,dE(\Delta_\zeta), \quad A^* = \int_{\Delta_\infty}\bar\zeta\,dE(\Delta_\zeta).$$

The same decomposition also implies that a normal operator A is a function of a certain self-adjoint operator C, $A=F(C)$. Conversely, every function of some self-adjoint operator is normal.

An important property of a normal operator A is the fact that $\|A^n\|=\|A\|^n$, which implies that the **spectral radius** of a normal operator A is its norm $\|A\|$. Eigen elements of a normal operator corresponding to distinct eigen values are orthogonal.

References

[1] PLESNER, A.I.: *Spectral theory of linear operators*, F. Ungar, 1965 (translated from the Russian).
[2] RUDIN, W.: *Functional analysis*, McGraw-Hill, 1973.

V.I. Sobolev

Editorial comments.

References

[A1] CONWAY, J.B.: *Subnormal operators*, Pitman, 1981.

AMS 1980 Subject Classification: 47B15

NORMAL p-COMPLEMENT *of a finite group G* - A normal subgroup A such that $G = AS$ and $A \cap S = 1$, where S is a Sylow p-subgroup of G (see **Sylow subgroup**). A group G has a normal p-complement if some Sylow p-subgroup S of G lies in the centre of its normalizer (cf. **Normalizer of a subset**) (*Burnside's theorem*). A necessary and sufficient condition for the existence of a normal p-complement in a group G is given by *Frobenius' theorem*: A group G has a normal p-complement if and only either for any non-trivial p-subgroup H of G the quotient group $N_G(H)/C_G(H)$ is a p-group (where $N_G(H)$ is the normalizer and $C_G(H)$ the **centralizer** of H in G) or if for every non-trivial p-subgroup H of G the subgroup $N_G(H)$ has a normal p-complement.

References

[1] GORENSTEIN, D.: *Finite groups*, Harper & Row, 1968.

N.N. Vil'yams

Editorial comments. Let G be a group of order n and let p^e be the highest power of a prime number p dividing n. A subgroup of G of index p^e (and hence of order $p^{-e}n$) is called a *p-complement* in G. A normal p-complement is a p-complement that is normal. A finite group is solvable if and only if it has a p-complement for every prime number p dividing its order. Cf. [A1], [A2] for more details; cf. also **Hall subgroup**.

References

[A1] HALL, M., JR.: *The theory of groups*, Macmillan, 1959, Sect. 9.3.
[A2] HUPPERT, B.: *Endliche Gruppen*, I, Springer, 1967, Sect. VI.1.

AMS 1980 Subject Classification: 20D20, 20E34

NORMAL PLANE *to a curve in space at a point M* - The plane passing through M and perpendicular to the **tangent** at M. The normal plane contains all normals (cf. **Normal**) to the curve passing through M. If the curve is given in rectangular coordinates by the equations

$$x = f(t), \ y = g(t), \ z = h(t),$$

then the equation of the normal plane at the point $M(x_0, y_0, z_0)$ corresponding to the value t_0 of the parameter t can be written in the form

$$(x - x_0)\frac{df(t_0)}{dt} + (y - y_0)\frac{dg(t_0)}{dt} + (z - z_0)\frac{dh(t_0)}{dt} = 0.$$

If the equation of the curve has the form $\mathbf{r} = \mathbf{r}(t)$, then the equation of the normal plane is

$$(\mathbf{R} - \mathbf{r})\frac{d\mathbf{r}}{dt} = 0.$$

BSE-3

Editorial comments.

References

[A1] DO CARMO, M.: *Differential geometry of curves and surfaces*, Prentice Hall, 1976.

AMS 1980 Subject Classification: 53A04

NORMAL RING

Editorial comments. Let R be a commutative ring with identity and S a commutative ring containing R, with the same identity element. An element $s \in S$ is *integral* over R if there are $c_i \in R$ such that $s^n + c_1 s^{n-1} + \cdots + c_n = 0$. The integral closure of R in S is the set of all $s \in S$ which are integral over R. It is a subring $\bar{R}$ of S containing R. If $\bar{R} = R$, R is said to be integrally closed in S (cf. also **Integral ring**).

A commutative ring with identity R is called normal if it is *reduced* (i.e. has no nilpotents $\neq 0$) and is integrally closed in its complete ring of fractions (cf. **Localization in a commutative algebra**). Thus, R is normal if for each prime ideal $\mathfrak{p}$ the localization $R_\mathfrak{p}$ is an **integral domain** and is closed in its field of fractions. In some of the literature a normal ring is also required to be an integral domain.

A **Noetherian ring** A is normal if and only if it satisfies the two conditions: i) for every prime ideal $\mathfrak{p}$ of height 1, $A_\mathfrak{p}$ is regular (and hence a discrete valuation ring); and ii) for every prime ideal $\mathfrak{p}$ of height $\geqslant 2$ the depth (cf. also **Depth of a module**) is also $\geqslant 2$. (Cf. [A3], p. 125.)

References

[A1] BOURBAKI, N.: *Elements of mathematics. Commutative algebra*, Addison-Wesley, 1972 (translated from the French).
[A2] NAGATA, M.: *Local rings*, Interscience, 1962.
[A3] MATSUMURA, H.: *Commutative algebra*, Benjamin, 1970.

AMS 1980 Subject Classification: 13B20, 13G05

NORMAL SCHEME - A **scheme** all local rings (cf. **Local ring**) of which are *normal* (that is, reduced and integrally closed in their ring of fractions). A normal scheme is locally irreducible; for such a scheme the concepts of a connected component and an irreducible component are the same. The set of singular points of a Noetherian normal scheme has codimension greater than 1. The following *normality criterion* holds [1]: A **Noetherian scheme** X is normal if and only if two conditions are satisfied: 1) for any point $x \in X$ of codimension $\leqslant 1$ the local ring $\mathcal{O}_{X,x}$ is regular (cf. **Regular ring (in commutative algebra)**); and 2) for any point $x \in X$ of codimension > 1 the depth of the ring (cf. **Depth of a module**) $\mathcal{O}_{X,x}$ is greater than 1. Every **reduced scheme** X has a normal scheme X^ν canonically connected with it (*normalization*). The X-scheme X^ν is integral, but not always finite over X. However, if X is excellent (see **Excellent ring**), for example, if X is a scheme of finite type over a field, then X^ν is finite over X.

References

[1] SERRE, J.P.: *Algèbre locale. Multiplicités*, Lecture notes in math., 11, Springer, 1975.

V.I. Danilov

Editorial comments. A normalization of an irreducible algebraic variety X is an irreducible normal variety X^ν together with a regular mapping $\nu: X^\nu \to X$ that is finite and a birational isomorphism.

For an affine irreducible algebraic variety, X^ν is the integral closure of the ring $A(X)$ of regular functions on X in its field of fractions. The normalization has the following universality properties. Let X be an *integral scheme* (i.e. X is both reduced and irreducible, or, equivalently, $\mathscr{O}_X(U)$ is an integral domain for all open U in X). For every normal integral scheme Z and every *dominant morphism* $f: Z \to X$ (i.e. $f(Z)$ is dense in X), f factors uniquely through the normalization $X^\nu \to X$. So also **Normal analytic space**.

Let X be a curve and x a, possibly singular, point on X. Let $X^\nu \to X$ be the normalization of X and $\bar{x}_1, \ldots, \bar{x}_n$ the inverse images of x in X^ν. These points are called the *branches* of X passing through x. The terminology derives from the fact that the $\bar{x}_i$ can be identified (in the case of varieties over $\mathbf{R}$ or $\mathbf{C}$) with the 'branches' of X passing through x. More precisely, if the U_i are sufficiently small complex or real neighbourhoods of the x_i, then some neighbourhood of x is the union of the branches $\nu(U_i)$. Let T_i be the tangent space at $\bar{x}_i$ to X^ν. Then $(d\nu)(\bar{x}_i)(T_i)$ is some linear subspace of the tangent space to X at x. It will be either a line or a point. In the first case the branch $\bar{x}_i$ is called linear. The point $(0, 0)$ on $y^2 = x^3 + x^2$ is an example of a point with two linear branches (with tangents $y = x$, $y = -x$), and the point $(0, 0)$ on $y^2 = x^3$ gives an example of a two-fold non-linear branch.

$$
\begin{array}{ccc}
X^\nu & & X^\nu \\
\downarrow\nu & & \downarrow\nu \\
X & & X
\end{array}
$$

References

[A1] HARTSHORNE, R.: *Algebraic geometry*, Springer, 1977, p. 91.
[A2] SHAFAREVICH, I.R.: *Basic algebraic geometry*, Springer, 1974, Sect. II.5 (translated from the Russian).
[A3] MATSUMURA, H.: *Commutative algebra*, Benjamin, 1970.

AMS 1980 Subject Classification: 14K99

NORMAL SECTION *of a smooth surface* Φ *at a point* P *in a direction* $\mathbf{l}$ - The section of Φ by the plane passing through the normal to the surface (cf. **Normal space (to a surface)**) at P and through the direction $\mathbf{l}$ in the **tangent plane** to Φ at P. The task of studying the local structure of a surface can be reduced to the same task for the family of curves formed by the normal sections of the surface at a given point in various directions (see **Curvature; Normal curvature**). The method of studying the local structure by means of normal sections can be generalized to surfaces of arbitrary dimension and arbitrary codimension.

D.D. Sokolov

Editorial comments.

References

[A1] KLINGENBERG, W.: *A course in differential geometry*,

Springer, 1978 (translated from the German).
[A2] CHEN, B.-Y.: *Geometry of submanifolds*, M. Dekker, 1973.

AMS 1980 Subject Classification: 53A05, 53A07

NORMAL SERIES - A series of normal subgroups

$$G = H_1 \supseteq H_2 \supseteq \cdots \supseteq H_{n+1} = \{1\}$$

of a group G (see **Subgroup series**). If each term of the series is normal not in the whole group but only in the preceding term, then the series is called *subnormal*. Apart from finite series one also considers infinite descending or ascending normal and subnormal series, the terms of which are indexed by (transfinite) ordinal numbers. One considers also more general normal and subnormal systems, the terms of which are indexed by the elements of an ordered set.

A *factor of the series* is the quotient group of some term of the series by the following (or preceding, if the series has ascending order of terms). The *length* of a series is the number of its factors other than the trivial one. A normal series that cannot be refined further is called a *chief series* (cf. **Principal series**) and a subnormal one a *composition series* (cf. **Composition sequence**). The factors of such series are called *chief* and *composition factors*. Two normal (subnormal) series are called *isomorphic* if a one-to-one correspondence can be set up between their factors such that corresponding factors are isomorphic. Any two normal (subnormal) series have isomorphic refinements (*Schreier's theorem*). In particular, any two chief (composition) series are isomorphic (the **Jordan—Hölder theorem**).

There is also another (older) terminology, in which a *normal series* is what is called above subnormal and for the concept called here a 'normal series' one uses the term 'invariant series'.

A.L. Shmel'kin

Editorial comments.

References

[A1] HALL, M., JR.: *The theory of groups*, Macmillan, 1959, Sect. 8.4.
[A2] KUROSH, A.G.: *The theory of groups*, 1, Chelsea, 1955, § 16 (translated from the Russian).

AMS 1980 Subject Classification: 20D30

NORMAL SHEAF - An analogue to a **normal bundle** in **sheaf theory**. Let

$$(f, f^\#): (Y, \mathscr{O}_Y) \to (X, \mathscr{O}_X)$$

be a morphism of ringed spaces such that the homomorphism $f^\#: f^* \mathscr{O}_X \to \mathscr{O}_Y$ is surjective, and let $\mathscr{J} = \operatorname{Ker} f^\#$. Then $\mathscr{J}/\mathscr{J}^2$ is a sheaf of ideals in $f^* \mathscr{O}_X / \mathscr{J} \cong \mathscr{O}_Y$ and is, therefore, an $\mathscr{O}_Y$-module. Here $\mathscr{N}^*_{Y/X} = (\mathscr{J}/\mathscr{J}^2)$ is called the *conormal sheaf of the morphism* and the dual $\mathscr{O}_Y$-module $\mathscr{N}_{Y/X} = \operatorname{Hom}_{\mathscr{O}_Y}(\mathscr{N}^*_{Y/X}, \mathscr{O}_Y)$ is called the *normal sheaf* *of the morphism f.* These sheaves are, as a rule, examined in the following special cases.

1) X and Y are differentiable manifolds (for example, of class C^∞), and $f: Y \to X$ is an immersion. There is an exact sequence of $\mathcal{O}_Y$-modules

$$0 \to \mathcal{N}_{Y/X}^* \xrightarrow{\delta} f^* \Omega_X^1 \to \Omega_Y^1 \to 0,$$

where Ω_X^1 and Ω_Y^1 are the sheaves of germs of smooth 1-forms on X and Y, and δ is defined as differentiation of functions. The dual exact sequence

$$0 \to \mathcal{T}_Y \to f^* \mathcal{T}_X \to \mathcal{N}_{Y/X} \to 0,$$

where $\mathcal{T}_X$ and $\mathcal{T}_Y$ are the tangent sheaves on X and Y, shows that $\mathcal{N}_{Y/X}$ is isomorphic to the sheaf of germs of smooth sections of the **normal bundle** of the immersion f. If Y is an immersed submanifold, then $\mathcal{N}_{Y/X}$ and $\mathcal{N}_{Y/X}^*$ are called the *normal* and *conormal sheaves of the submanifold Y*.

2) $(X, \mathcal{O}_X)$ is an irreducible separable scheme of finite type over an algebraically closed field k, $(Y, \mathcal{O}_Y)$ is a closed subscheme of it and $f: Y \to X$ is an imbedding. Then $\mathcal{N}_{Y/X}$ and $\mathcal{N}_{Y/X}^*$ are called the *normal* and *conormal sheaves of the subscheme Y*. There is also an exact sequence of $\mathcal{O}_Y$-modules

$$\mathcal{N}_{Y/X}^* \xrightarrow{\delta} \Omega_X \otimes \mathcal{O}_Y \to \Omega_Y \to 0, \qquad (*)$$

where Ω_X and Ω_Y are the sheaves of differentials on X and Y. The sheaves $\mathcal{N}_{Y/X}^*$ and $\mathcal{N}_{Y/X}$ are quasi-coherent, and if X is a Noetherian scheme, then they are coherent. If X is a non-singular variety over k and Y is a non-singular variety, then $\mathcal{N}_{Y/X}^*$ is locally free and the homomorphism δ in $(*)$ is injective. In this case one obtains the dual exact sequence

$$0 \to \mathcal{T}_Y \to \mathcal{T}_X \otimes \mathcal{O}_Y \to \mathcal{N}_{Y/X} \to 0,$$

so that the normal sheaf $\mathcal{N}_{Y/X}$ is locally free of rank $r = \mathrm{codim}\, Y$ corresponding to the normal bundle over Y. In particular, if $r = 1$, then $\mathcal{N}_{Y/X}$ is the invertible sheaf corresponding to the divisor Y.

In terms of normal sheaves one can express the self-intersection $Y \cdot Y$ of a non-singular subvariety $Y \subset X$. Namely, $Y \cdot Y = f_* c_r(\mathcal{N}_{Y/X})$, where c_r is the r-th **Chern class** and $f_*: A(Y) \to A(X)$ is the homomorphism of Chow rings (cf. **Chow ring**) corresponding to the imbedding $f: Y \to X$.

3) $(X, \mathcal{O}_X)$ is a complex space, $(Y, \mathcal{O}_Y)$ is a closed analytic subspace of it and f is the imbedding. Then $\mathcal{N}_{Y/X}$ and $\mathcal{N}_{Y/X}^*$ are called the *normal* and *conormal sheaves of the subspace Y*; they are coherent. If X is an analytic manifold and Y an analytic submanifold of it, then $\mathcal{N}_{Y/X}$ is the sheaf of germs of holomorphic sections of the normal bundle over Y.

References

[1] SHAFAREVICH, I.R.: *Basic algebraic geometry*, Springer, 1977 (translated from the Russian).

[2] HARTSHORNE, R.: *Algebraic geometry*, Springer, 1977.

A.L. Onishchik

Editorial comments. If X is a non-singular variety over k and Y is a subscheme of X that is locally a complete intersection, then $\mathcal{N}_{Y/X}^*$ is locally free.

AMS 1980 Subject Classification: 14A05, 14F05, 32L10

NORMAL SOLVABILITY *of an integral equation* - The property that a **linear integral equation** is solvable if and only if its right-hand side is orthogonal to all solutions of the corresponding homogeneous adjoint equation. Under appropriate conditions a **Fredholm equation**, a **singular integral equation** and an **integral equation of convolution type** are normally solvable.

B.V. Khvedelidze

Editorial comments.

References

[A1] GOLDBERG, S.: *Unbounded linear operators*, McGraw-Hill, 1966.

[A2] KATO, T.: *Perturbation theory for linear operators*, Springer, 1980.

[A3] ZABREYKO, P.P. [P.P. ZABREĬKO], ET AL.: *Integral equations — a reference text*, Noordhoff, 1975 (translated from the Russian).

AMS 1980 Subject Classification: 45-XX

NORMAL SPACE - A **topological space** satisfying the axiom T_4 (see **Separation axiom**), that is, one in which one-point sets are closed and any two disjoint closed sets can be separated by neighbourhoods (that is, are contained in disjoint open sets). Normal spaces form a special case of completely-regular spaces (Tikhonov spaces, cf. **Completely-regular space**) and are particularly important in **dimension theory**. Every closed subspace of a normal space is normal (normality is hereditary over closed sets). Spaces all subspaces of which are normal are said to be *hereditarily normal*. For hereditary normality of a space it is sufficient that all its open subspaces are normal, and it is necessary and sufficient that any two sets none of which contains an adherence point of the other are separable by neighbourhoods. A normal space is called *perfectly normal* if every closed set in it is the intersection of countably many open sets. Every perfectly-normal space is a hereditarily-normal space.

The product of two normal spaces need not be normal, and even the product of a normal space and a segment may be non-normal.

There are important classes of spaces that are more general than normal and less general than completely regular. First among such spaces close to normal were the so-called *quasi-normal*, or *π-normal*, ones [2]. These are Tikhonov spaces in which any two disjoint *π-sets* can be separated by neighbourhoods. π-sets are intersections of finitely many closed canonical sets (cf. **Canonical set**). Tikhonov spaces in which any two disjoint closed canonical sets can be separated by neighbourhoods are called κ-*normal* [3]; κ-normal spaces in which every closed canonical set is the intersection of countably many open canonical sets are called *perfectly κ-normal*. The classes of Tikhonov κ-normal, quasi-normal and perfectly κ-normal spaces are successively contained in each other and no two of them are equal.

References

[1] ALEKSANDROV, P.S.: *Einführung in die Mengenlehre und die Theorie der reelen Funktionen*, Deutsch. Verlag Wissenschaft., 1956 (translated from the Russian).

[2] ZAĬTSEV, V.I.: 'On the theory of Tikhonov spaces', *Vestnik Moskov. Univ. Mat. Mekh.*, no. 3 (1967), 48-57 (in Russian).

[3] SHCHEPIN, E.V.: 'Real functions and near-normal spaces', *Sib. Math. J.* **13**, no. 5 (1972), 820-830. (*Sibirsk. Mat. Zh.* **13**, no. 5 (1972), 1182-1196)

P.S. Aleksandrov

Editorial comments. Normal spaces are also characterized by the following two statements:

1) *Urysohn's lemma*: If $A, B \subseteq X$ are closed and disjoint, then there is a continuous function $f: X \to [0, 1]$ such that $f|_A \equiv 0$ and $f|_B \equiv 1$. In other words, any two closed sets can be separated by a continuous function.

2) The *Tietze—Urysohn extension theorem*: If $A \subseteq X$ is closed and $f: A \to [0, 1]$ is continuous, then f can be extended to a continuous $\bar{f}: X \to [0, 1]$.

A normal space X such that $X \times [0, 1]$ is not normal, a so-called *Dowker space*, was constructed by M.E. Rudin [A3].

A space X is called *collection-wise normal* if for every discrete family of subsets $\{F_\alpha: \alpha \in A\}$ there exists a discrete family $\{U_\alpha: \alpha \in A\}$ of open sets in X such that $F_\alpha \subset U_\alpha$ for all $\alpha \in A$. Here a family of subsets $\{Y_\alpha: \alpha \in A\}$ is called a *discrete family* if for every $x \in X$ there is an open neighbourhood U_x such that U_x intersects at most one Y_α.

References

[A1] ARKHANGEL'SKIĬ, A.V. and PONOMAREV, V.I.: *Fundamentals of general topology: problems and exercises*, Reidel, 1984 (translated from the Russian).

[A2] ENGELKING, R.: *General topology*, PWN, 1977.

[A3] RUDIN, M.E.: 'A normal space X for which $X \times I$ is not normal', *Fund. Math.* **73** (1971), 179-186.

[A4] ALÒ, R.A. and SHAPIRO, H.L.: *Normal topological spaces*, Cambridge Univ. Press, 1974.

AMS 1980 Subject Classification: 54D15

NORMAL SPACE (TO A SURFACE) at a point P - The orthogonal complement $N_P F$ to the tangent space $T_P F$ (see **Tangent plane**) of the surface F^m in V^n at P. The dimension of the normal space is $n - m$ (the codimension of F). Every one-dimensional subspace of it is called a **normal** to F at P. If F is a smooth hypersurface, then it has a unique normal at every of its points.

A.B. Ivanov

Editorial comments.

References

[A1] KLINGENBERG, W.: *Riemannian geometry*, de Gruyter, 1982 (translated from the German).

[A2] CHEN, B.-Y.: *Geometry of submanifolds*, M. Dekker, 1973.

AMS 1980 Subject Classification: 53A05, 53A07, 53B25

NORMAL SUB-SEMI-GROUP *of a semi-group S* - A sub-semi-group H satisfying the following condition: For any $x, y \in S^1$ (for the notation S^1 see **Normal complex**) such that $xy \in S$ and for any $h \in H$ the relations $xhy \in H$ and $xy \in H$ are equivalent. A subset of S is a normal sub-semi-group if and only if it is the complete inverse image of the unit element under some homomorphism of S onto a **semi-group** with unit element.

References

[1] LYAPIN, E.S.: *Semigroups*, Amer. Math. Soc., 1974 (translated from the Russian).

L.N. Shevrin

AMS 1980 Subject Classification: 20M10

NORMAL SUBGROUP, *normal divisor, invariant subgroup* - A subgroup H of a **group** G for which the left decomposition of G modulo H is the same as the right one; in other words, a subgroup such that for any element $a \in G$ the cosets aH and Ha are the same (as sets). In this case one also says that H is normal in G and writes $H \trianglelefteq G$; if also $H \neq G$, one writes $H \triangleleft G$. A subgroup H is normal in G if and only if it contains all G-conjugates of any of its elements (see **Conjugate elements**), that is $H^G \subseteq H$. A normal subgroup can also be defined as one that coincides with all its conjugates, as a consequence of which it is also known as a *self-conjugate subgroup*.

For any **homomorphism** $\phi: G \to G^*$ the set K of elements of G that are mapped to the unit element of G^* (the *kernel of the homomorphism* ϕ) is a normal subgroup of G, and conversely, every normal subgroup of G is the kernel of some homomorphism; in particular, K is the kernel of the canonical homomorphism onto the **quotient group** G/K.

The intersection of any set of normal subgroups is normal, and the subgroup generated by any system of normal subgroups of G is normal in G.

O.A. Ivanova

Editorial comments. A subgroup H of a group G is normal if $g^{-1} Hg = H$ for all $g \in G$, or, equivalently, if the normalizer $N_G(H) = G$, cf. **Normalizer of a subset**. A normal subgroup is also called an *invariant subgroup* because it is invariant under the *inner automorphisms* $x \mapsto x^g = g^{-1} xg$, $g \in G$, of G. A subgroup that is invariant under all automorphisms is called a *fully-invariant subgroup* or **characteristic subgroup**. A subgroup that is invariant under all endomorphisms is a **fully-characteristic subgroup**.

References

[A1] HALL, M., JR.: *The theory of groups*, Macmillan, 1958, p. 26.
[A2] CURTIS, C.W. and REINER, I.: *Representation theory of finite groups and associative algebras*, Interscience, 1962, p. 5.
[A3] KUROSH, A.G.: *The theory of groups*, 1, Chelsea, 1955, Chapt. III (translated from the Russian).

AMS 1980 Subject Classification: 20A05, 20F26

NORMAL ZERO-DIMENSIONAL CYCLE, *cycle of index zero*

A cycle $z = \sum_{i=1}^{n} a_i t_i^0$ such that $\sum_{i=1}^{n} a_i = 0$. A cycle homologous to zero is always normal; the quotient group of the group of normal cycles by that of the cycles homologous to zero is called the *reduced zero-dimensional homology group*. For a connected complex the reduced group is zero, which is convenient in any kind of definition of acyclicity. A **proper cycle** $z^0 = \{z_1^0, \ldots, z_k^0, \ldots\}$ is called *normal* if each of the cycles z_k^0 is normal.

A.A. Mal'tsev

AMS 1980 Subject Classification: 55U15

NORMALIZATION PRINCIPLE - See Normal algorithm.

AMS 1980 Subject Classification: 68C05, 03-XX

NORMALIZED SYSTEM

A system $\{x_i\}$ of elements of a **Banach space** B whose norms are all equal to one, $\| x_i \|_B = 1$. In particular, a system $\{f_i\}$ of functions in the space $L_2[a, b]$ is said to be normalized if

$$\int_a^b | f_i(x) |^2 \, dx = 1.$$

Normalization of a system $\{x_i\}$ of non-zero elements of a Banach space B means the construction of a normalized system of the form $\{\lambda_i x_i\}$, where the λ_i are non-zero numbers, the so-called *normalizing factors*. As a sequence of normalizing factors one can take $\lambda_i = 1 / \| x_i \|_B$.

References

[1] KACZMARZ, S. and STEINHAUS, H.: *Theorie der Orthogonalreihen*, Chelsea, reprint, 1951.
[2] DUNFORD, N. and SCHWARTZ, J.T.: *Linear operators. General theory*, 1, Interscience, 1958.
[3] KANTOROVICH, L.V. and AKILOV, G.P.: *Funktionalanalysis in normierten Räume*, Akad. Verlag, 1964 (translated from the Russian).

A.A. Talalyan

AMS 1980 Subject Classification: 46BXX

NORMALIZER CONDITION *for subgroups*

The condition on a **group** that every proper subgroup is strictly contained in its normalizer (cf. **Normalizer of a subset**). Every group satisfying the normalizer condition is a **locally nilpotent group**. On the other hand, all nilpotent groups, and even groups having an ascending central series (ZA-*groups*), satisfy the normalizer condition. However, there are groups with the normalizer condition and with a trivial centre. Thus, the class of groups with the normalizer condition strictly lies in between the classes of ZA-groups and locally nilpotent groups.

References

[1] KUROSH, A.G.: *The theory of groups*, 1-2, Chelsea, 1955-1956 (translated from the Russian).

A.L. Shmel'kin

Editorial comments.

References

[A1] ROBINSON, D.J.S.: *A course in the theory of groups*, Springer, 1980.

AMS 1980 Subject Classification: 20F99

NORMALIZER OF A SUBSET M of a group G in a subgroup H of G

The set

$$N_H(M) = \{h: h \in H, \, h^{-1}Mh = M\},$$

that is, the set of all elements h of H such that $h^{-1}mh$ (the conjugate of m by h) for every $m \in M$ also belongs to M. For any M and H the normalizer $N_H(M)$ is a subgroup of H. An important special case is the normalizer of a subgroup of a group G in G. A subgroup A of a group G is normal (or invariant, cf. **Invariant subgroup**) in G if and only if $N_G(A) = G$. The normalizer of a set consisting of a single element is the same as its **centralizer**. For any H and M the cardinality of the class of subsets conjugate to M by elements of H (that is, subsets of the form $h^{-1}Mh$, $h \in H$) is equal to the index $| H : N_H(M) |$.

References

[1] KARGAPOLOV, M.I. and MERZLYAKOV, YU.I.: *Fundamentals of the theory of groups*, Springer, 1979 (translated from the Russian).

N.N. Vil'yams

Editorial comments.

References

[A1] ROBINSON, D.J.S.: *A course in the theory of groups*, Springer, 1980.

AMS 1980 Subject Classification: 20F99

NORMALLY-IMBEDDED SUBSPACE

A subspace A of a space X such that for every neighbourhood U of it in X there is a set H that is the union of a countable family of sets closed in X and with $A \subset H \subset U$. If A is normally imbedded in X and X is normally imbedded in Y, then A is normally imbedded in Y. A normally-imbedded subspace of a **normal space** is itself normal in the induced topology, which explains the name. Final compactness of a space is equivalent to its being normally imbedded in some (and hence in any) **compactification** of this space. Quite generally, a normally-imbedded subspace of a finally-compact space is itself finally compact.

References

[1] SMIRNOV, YU.M.: 'On normally-imbedded sets of normal spaces', *Mat. Sb.* **29** (1951), 173-176 (in Russian).

A.V. Arkhangel'skiĭ

Editorial comments. A *finally-compact space* is the same as a Lindelöf space.

AMS 1980 Subject Classification: 54D99

NORMALLY-SOLVABLE OPERATOR - A **linear operator** with closed range. Let A be a linear operator with dense domain in a Banach space X and with range $R(A)$ in a Banach space Y. Then A is normally solvable if $\overline{R(A)} = R(A)$, that is, if $R(A)$ is a closed subspace of Y. Let A^* be the adjoint of A. For A to be normally solvable it is necessary and sufficient that $R(A) = {}^{\perp} N(A^*)$, that is, that the range of A is the orthogonal complement to the null space of A^*.

Suppose that

$$Ax = y \qquad (*)$$

is an equation with a normally-solvable operator (a *normally-solvable equation*). If $N(A^*) = \{0\}$, that is, if the homogeneous adjoint equation $A^*\psi = 0$ has only the trivial solution, then $R(A) = Y$. But if $N(A^*) \neq \{0\}$, then for (*) to be solvable it is necessary and sufficient that $<y, \psi> = 0$ for all solutions of the equation $A^*\psi = 0$.

From now on suppose that A is closed. A normally-solvable operator is called *n-normal* if its null space $N(A)$ is finite dimensional ($n(A) = \dim N(A) < +\infty$). A normally-solvable operator A is called *d-normal* if its **deficiency subspace** is finite dimensional ($d(A) = \dim {}^{\perp} R(A) < +\infty$). Operators that are either n-normal or d-normal are sometimes called *semi-Fredholm operators*. For an operator A to be n-normal it is necessary and sufficient that the pre-image of every compact set in $R(A)$ is locally compact.

Suppose that X is compactly imbedded in a Banach space X_0. For A to be n-normal it is necessary and sufficient that there is an a priori estimate

$$\| x \|_X \leqslant a \| x \|_{X_0} + b \| Ax \|_Y, \quad x \in D(A).$$

It turns out that an operator A is n-normal if and only if A^* is d-normal. Then $n(A) = d(A^*)$. Consequently, if X^* is compactly imbedded in a Banach space Z, then A is d-normal if and only if there is an a priori estimate

$$\| f \|_{Y^*} \leqslant a \| f \|_Z + b \| A^* f \|_{X^*}, \quad f \in D(A^*).$$

The pair of numbers $(n(A), d(A))$ is called the *d-characteristic* of A. If a normally-solvable operator A is n-normal or d-normal, the number

$$\chi(a) = n(A) - d(A)$$

is called the *index of the operator A*. The properties of being n-normal and d-normal are stable: If A is n-normal (or d-normal) and B is a linear operator of small norm or completely continuous, then $A + B$ is also n-normal (respectively, d-normal).

References

[1] HAUSDORFF, F.: *Grundzüge der Mengenlehre*, Leipzig, 1914. Reprinted (incomplete) English translation: Set theory, Chelsea (1978).

[2] ATKINSON, F.: 'Normal solvability of equations in Banach space', *Mat. Sb.* **28**, no. 1 (1951), 3-14 (in Russian).

[3] KREĬN, S.G.: *Linear differential equations in Banach space*, Amer. Math. Soc., 1971 (translated from the Russian).

V.A. Trenogin

Editorial comments.

References

[A1] GOHBERG, I.C. [I.TS. GOKHBERG] and KREĬN, M.G.: 'The basic propositions on defect numbers, root numbers and indices of linear operators', *Transl. Amer. Math. Soc. (2)* 13 (1960), 185-264. (*Uspekhi Mat. Nauk* 12 (1957), 43-118)

[A2] GOLDBERG, S.: *Unbounded linear operators*, McGraw-Hill, 1966.

[A3] KATO, T.: 'Perturbation theory for nullity, deficiency and other quantities of linear operators', *J. d'Anal. Math.* 6 (1958), 261-322.

[A4] KREĬN, S.G.: *Linear equations in Banach spaces*, Birkhäuser, 1982 (translated from the Russian).

AMS 1980 Subject Classification: 47A53, 47BXX

NORMED ALGEBRA - An **algebra** over the field of real or complex numbers that is at the same time a **normed space** in which multiplication satisfies some continuity condition. The simplest such condition is separate continuity. Generally speaking, this is weaker than joint continuity in the factors. For example, if one defines the algebraic operations on the set of all finite sequences $x = (\epsilon_1, \ldots, \epsilon_n, 0, \ldots)$ coordinate-wise and the norm by $\| x \| = \sum_{k=1}^{\infty} | \epsilon_k | k^{-2}$, then there arises an algebra in which the multiplication is separately, but not jointly, continuous. The joint continuity of the multiplication in a normed algebra is equivalent to the existence of a constant C such that $\| xy \| \leqslant C \| x \| \| y \|$. In this and only in this case the completion has the structure of a normed algebra that extends the original one and is a **Banach algebra**.

E.A. Gorin

Editorial comments.

References

[A1] NAĬMARK, M.A.: *Normed rings*, Noordhoff, 1964 (translated from the Russian).

AMS 1980 Subject Classification: 46H25

NORMED FIELD - A **field** on which a **valuation** is given (see also **Norm on a field**).

AMS 1980 Subject Classification: 12J05

NORMED RING - 1) The same as a normed or **Banach algebra**.

2) A ring with a **valuation** on it.

AMS 1980 Subject Classification: 46H25, 13F30

NORMED SPACE - A **vector space** over the field of

real or complex numbers with a distinguished **norm.**

E.A. Gorin

AMS 1980 Subject Classification: 46BXX

NOWHERE-DENSE SET *of a topological space* - A set A defined by the following property: Every non-empty open set $\Gamma \subset X$ contains a non-empty open set $\Gamma_0 \subset X$ such that $A \cap \Gamma_0 = \varnothing$. In other words, A is nowhere dense if it is not dense in any non-empty open set.

M.I. Voĭtsekhovskiĭ

Editorial comments. Another characterization is: The interior of the closure of a nowhere-dense set is empty. If in a topological product $X = \prod X_\alpha$ infinitely many of the spaces X_α are non-compact, then each compact subset of X is nowhere dense. A *boundary set* is the complement of a dense set, i.e. it satisfies $\overline{X \setminus A} = X$. A set whose closure is a boundary set is nowhere dense. A non-empty complete metric space is of the *second category*, i.e. in it a countable union of nowhere-dense sets is nowhere dense (the *Baire category theorem*, cf. **Baire theorem**).

References

[A1] ARKHANGEL'SKIĬ, A.V. and PONOMAREV, V.I.: *Fundamentals of general topology: problems and exercises*, Reidel, 1984 (translated from the Russian).
[A2] KELLEY, J.L.: *General topology*, v. Nostrand, 1955, p. 145.

AMS 1980 Subject Classification: 54AXX

NUCLEAR BILINEAR FORM - A bilinear form $B(f, g)$ on the Cartesian product $F \times G$ of two locally convex spaces F and G that can be represented as

$$B(f, g) = \sum_{i=1}^{\infty} \lambda_i <f, f_i'> <g, g_i'>,$$

where $\{\lambda_i\}$ is a summable sequence, $\{f_i'\}$ and $\{g_i'\}$ are equicontinuous sequences (cf. **Equicontinuity**) in the dual spaces F' and G' of F and G, respectively, and $<a, a'>$ denotes the value of the linear functional a' on the vector a. All nuclear bilinear forms are continuous. If F is a **nuclear space**, then for any locally convex space G all continuous bilinear forms on $F \times G$ are nuclear (the *kernel theorem*). This result is due to A. Grothendieck [1]; the form stated is given in [2]; for other statements see [3]. The converse holds: If a space F satisfies the kernel theorem, then it is a nuclear space.

For spaces of smooth functions of compact support, the kernel theorem was first obtained by L. Schwartz [4]. Let D be the nuclear space of all infinitely-differentiable functions with compact support on the real line, equipped with the standard locally convex topology of Schwartz, so that the dual space D' consists of all generalized functions on the line. In the special case when $F = G = D$, the kernel theorem is equivalent to the following assertion: Every continuous bilinear functional on $D \times D$ has the form

$$B(f, g) = <f(t_1)g(t_2), F> =$$
$$= \int_{-\infty}^{\infty} F(t_1, t_2)f(t_1)g(t_2) \, dt_1 \, dt_2,$$

where $f(t), g(t) \in D$ and $F = F(t_1, t_2)$ is a generalized function in two variables. There are similar statements of the kernel theorem for spaces of smooth functions in several variables with compact support, for spaces of rapidly-decreasing functions, and for other specific nuclear spaces. Similar results are valid for multilinear forms.

A continuous bilinear form $B(f, g)$ on $D \times D$ can be identified with a continuous linear operator $A : D \to D'$ by using the equality

$$B(f, g) = <g, Af>,$$

and this leads to *Schwartz' kernel theorem*: For any continuous linear mapping $A : D \to D'$ there is a unique generalized function $F(t_1, t_2)$ such that

$$A : f(t_1) \mapsto \int_{-\infty}^{\infty} F(t_1, t_2)f(t_2) \, dt_2$$

for all $f \in D$. In other words, A is an integral operator with kernel F.

References

[1] GROTHENDIECK, A.: *Produits tensoriels topologiques et espaces nucléaires*, Amer. Math. Soc., 1955.
[2] PIETSCH, A.: *Nuclear locally convex spaces*, Springer, 1972 (translated from the German).
[3] GEL'FAND, I.M. and VILENKIN, N.YA.: *Generalized functions. Applications of harmonic analysis*, 4, Acad. Press, 1964 (translated from the Russian).
[4] SCHWARTZ, L.: 'Théorie des noyaux', in *Proc. Internat. Congress Mathematicians Cambridge, 1950*, Vol. 1, Amer. Math. Soc., 1952, pp. 220-230.
[5] SCHWARTZ, L.: 'Espaces de fonctions différentielles à valeurs vectorielles', *J. d'Anal. Math.* 4 (1954-1955), 88-148.

G.L. Litvinov

Editorial comments.

References

[A1] TRÈVES, F.: *Topological vectorspaces, distributions and kernels*, Acad. Press, 1967.
[A2] SCHWARTZ, L.: *Théorie des distributions*, Hermann, 1966.

AMS 1980 Subject Classification: 47B10, 46A12, 46E10, 46E05

NUCLEAR C^*-ALGEBRA - A C^*-algebra A with the following property: For any C^*-algebra B there is on the algebraic tensor product $A \otimes B$ a unique norm such that the completion of $A \otimes B$ with respect to this norm is a C^*-algebra. Thus, relative to tensor products, nuclear C^*-algebras behave similarly to nuclear spaces (cf. **Nuclear space**) (although infinite-dimensional nuclear C^*-algebras are not nuclear spaces). The class of nuclear C^*-algebras includes all type I C^*-algebras. This class is closed with respect to the inductive limit. If I is a closed two-sided ideal in a C^*-algebra A, then A is nuclear if and only if I and A/I are. A subalgebra of a nuclear C^*-algebra need not be a nuclear C^*-algebra. The tensor product of two C^*-algebras A and B is nuclear if and only if A and B (both) are nuclear. If G is an amenable locally compact group, then the enveloping C^*-algebra of the group algebra $L_1(G)$ is nuclear (the converse is not true). Each **factor representation** of a nuclear C^*-algebra is *hyperfinite*, that is, the **von Neumann algebra** generated by this representation can be obtained from an increasing sequence of finite-dimensional factors (matrix algebras). Any factor state on a nuclear C^*-subalgebra of a C^*-algebra can be extended to a factor state on the whole algebra.

Let $L(H)$ be the C^*-algebra of all bounded linear operators on a Hilbert space H, and let A be a C^*-algebra of operators on H. If A is nuclear, then its weak closure $\bar{A}$ is an *injective von Neumann algebra*, that is, there is a projection $L(H) \to \bar{A}$ with norm one; in this case the commutant A' of A is also injective. An arbitrary C^*-algebra A is nuclear if and only if its enveloping von Neumann algebra is injective.

A C^*-algebra A is nuclear if and only if it has the *completely positive approximation property*, i.e. the identity operator in A can be approximated in the strong operator topology by linear operators of finite rank with norm not exceeding 1, and with the additional property of 'complete positivity' [1].

Every nuclear C^*-algebra has the approximation and bounded approximation properties (see **Nuclear operator**). There is, however, a non-nuclear C^*-algebra with the bounded approximation property. The C^*-algebra $L(H)$ of all bounded operators on an infinite-dimensional Hilbert space H does not have the completely positive approximation property, or even the approximation property, so that $L(H)$ is not nuclear.

References

[1] LANCE, E.C.: 'Tensor products and nuclear C^*-algebras', in R.V. Kadison (ed.): *Operator algebras and applications*, Proc. Symp. Pure Math., Vol. 38, Amer. Math. Soc., 1982, pp. 379-399.
[2] BRATTELI, O. and ROBINSON, D.W.: *Operator algebras and quantum statistical mechanics*, 1, Springer, 1979.

G.L. Litvinov

Editorial comments.

References

[A1] KADISON, R.V. and RINGROSE, J.R.: *Fundamentals of the theory of operator algebras*, 1-2, Acad. Press, 1983.
[A2] PEDERSEN, G.K.: *C^*-algebras and their automorphism groups*, Acad. Press, 1979, Sect. 8.15.15.

AMS 1980 Subject Classification: 46L05

NUCLEAR NORM, *trace norm* - A norm on the space $N(X, Y)$ of nuclear operators (cf. **Nuclear operator**) mapping a **Banach space** X into a Banach space Y.

Let X and Y be Banach spaces over the field of real or complex numbers, let $L(X, Y)$ be the space of all continuous linear operators mapping X into Y, and let $F(X, Y)$ be the linear subspace consisting of operators of finite rank (that is, with finite-dimensional range). The Banach dual of X is denoted by X', and the value of a functional $x' \in X'$ at a vector $x \in X$ by $<x, x'>$.

Every nuclear operator $A \in N(X, Y)$ can be represented in the form

$$x \mapsto Ax = \sum_{i=1}^{\infty} <x, x_i'> y_i, \tag{1}$$

where $\{x_i'\}$ and $\{y_i\}$ are sequences in X' and Y, respectively, such that

$$\sum_{i=1}^{\infty} \| x_i' \| \, \| y_i \| < \infty;$$

such representations are called *nuclear*. The quantity

$$\| A \|_1 = \inf \sum_{i=1}^{\infty} \| x_i' \| \, \| y_i \|, \tag{2}$$

where the infimum is taken over all possible nuclear representations of the form (1), is called the *nuclear norm* of A. The space $N(X, Y)$ with this norm is a Banach space that contains $F(X, Y)$ as a dense linear subspace. If $A \in N(X, Y)$, then the adjoint operator A' belongs to $N(Y', X')$, and $\| A' \|_1 \leqslant \| A \|_1$. Let $\| \cdot \|$ denote the usual operator norm in $L(X, Y)$. Then $\| A \| \leqslant \| A \|_1$ for all $A \in N(X, Y)$. If $A \in L(Y, Z)$ and $B \in N(X, Y)$, then $AB \in N(X, Z)$, and $\| AB \|_1 \leqslant \| A \| \, \| B \|_1$; if $A \in N(Y, Z)$ and $B \in L(X, Y)$, then $AB \in N(X, Z)$, and $\| AB \|_1 \leqslant \| A \|_1 \| B \|$. Any operator $F \in F(X, Y)$ can be represented in the form

$$x \mapsto Fx = \sum_{i=1}^{n} <x, x_i'> y_i. \tag{3}$$

The quantity

$$\| F \|_1^0 = \inf \sum_{i=1}^{n} \| x_i' \| \, \| y_i \|, \tag{4}$$

where the infimum is taken over all possible finite representations of the form (3), is called the *finite nuclear norm* of F. The space $F(X, Y)$ can be identified with the tensor product $X' \otimes Y$. Here, to an operator F of the form (3) there corresponds the element

$$u = \sum_{i=1}^{n} x_i' \otimes y_i \in X' \otimes Y, \tag{5}$$

and the finite nuclear norm (4) goes into the norm

$$\| u \| = \inf \sum_{i=1}^{n} \| x_i' \| \, \| y_i \|, \tag{6}$$

where the infimum is taken over all finite representations of u in the form (5). This norm is called the *tensor* (or *cross*) *product* of the norms in Y and in X'. The completion of $X' \otimes Y$ with respect to the norm (6) is denoted by $X' \hat{\otimes} Y$. The mapping $X' \otimes Y \to L(X, Y)$, under which the element (5) is mapped to the operator (3), can be extended to a continuous linear operator $\Gamma: X' \hat{\otimes} Y \to L(X, Y)$. The range of Γ is $N(X, Y)$. If Γ

establishes a one-to-one correspondence between $X' \hat{\otimes} Y$ and $N(X, Y)$, then $N(X, Y)$ coincides with the closure of $F(X, Y)$ with respect to the norm (4); in this case the restriction of the nuclear norm to $F(X, Y)$ is the same as the finite nuclear norm. But, in general, Γ may have a non-trivial kernel, so that the nuclear norm is a quotient of the norm in $X' \hat{\otimes} Y$ (see **Nuclear operator**).

Let $X = Y = H$, where H is a separable Hilbert space, let $L(H) = L(H, H)$ be the algebra of bounded operators on H, and let $L_1(H) = N(H, H)$ be the ideal of nuclear operators in $L(H)$. In this case Γ is one-to-one, for operators of finite rank the nuclear norm coincides with the finite nuclear norm, and each $A \in L_1(H)$ has a trace $\operatorname{tr} A$ (see **Nuclear operator**). The nuclear norm of an operator $A \in L_1(H)$ coincides with $\operatorname{tr}[(A^*A)^{1/2}]$, where A^* is the adjoint of A in H. The nuclear norm is connected with the **Hilbert−Schmidt norm** $\| \cdot \|_2$ by $\| A \|_2 \leqslant \| A \|_1$. The general form of a continuous linear functional on the Banach space $L_1(H)$ is given by

$$A \to \operatorname{tr} AB, \qquad (7)$$

where B is an arbitrary operator from $L(H)$, and the norm of the functional (7) coincides with $\| B \|$. Consequently, $L(H)$ is isometric to the dual of $L_1(H)$. Formula (7) also gives the general form of a linear functional on the closed subspace $L_\infty(H)$ of $L(H)$ that consists of all completely-continuous (compact) operators; here $A \in L_\infty(H)$ and B ranges over $L_1(H)$. In this case the norm of the functional (7) coincides with $\| B \|_1$, that is, the space $L_1(H)$ of nuclear operators with the nuclear norm is isometric to the dual of $L_\infty(H)$ in the usual operator norm. These results have non-trivial generalizations to the case of operators on Banach spaces.

Example. Let $X = Y = l_1$ be the space of summable sequences. An operator $A \in L(l_1, l_1)$ is contained in $N(l_1, l_1)$ if and only if there is an infinite matrix (σ_{ik}) such that A sends $\{\xi_k\} \in l_1$ to $\{\eta_i\} = \{\sum_{k=1}^\infty \sigma_{ik} \xi_k\} \in l_1$, and $\sum_{i=1}^\infty \sup_k |\sigma_{ik}| < \infty$. In this case, $\| A \|_1 = \sum_{i=1}^\infty \sup_k |\sigma_{ik}|$.

References

[1] GROTHENDIECK, A.: *Produits tensoriels topologiques et espaces nucléaires*, Amer. Math. Soc., 1955.
[2] PIETSCH, A.: *Operator ideals*, North-Holland, 1980.
[3] PIETSCH, A.: *Nuclear locally convex spaces*, Springer, 1972 (translated from the German).
[4] GOHBERG, I.C. [I.TS. GOKHBERG] and KREĬN, M.G.: *Introduction to the theory of linear nonselfadjoint operators*, Amer. Math. Soc., 1969 (translated from the Russian).
[5] GEL'FAND, I.M. and VILENKIN, N.YA.: *Generalized functions*, 4. Applications of harmonic analysis, Acad. Press, 1964 (translated from the Russian).
[6] MAURIN, K.: *Methods of Hilbert spaces*, PWN, 1967.
[7] DAY, M.M.: *Normed linear spaces*, Springer, 1958.

G.L. Litvinov

Editorial comments.

References

[A1] PIETSCH, A.: *Eigenvalues and s-numbers*, Cambridge Univ. Press, 1987.
[A2] GROTHENDIECK, A.: 'Résumé de la théorie métrique des produits tensoriels topologiques', *Bol. Soc. Mat. São Paulo* **8** (1956), 1-79.
[A3] JARCHOW, H.: *Locally convex spaces*, Teubner, 1981.

AMS 1980 Subject Classification: 47B10, 47A30

NUCLEAR OPERATOR, *nuclear mapping* - A **linear operator** mapping one **locally convex space** into another, and having a special form of approximation by *operators of finite rank* (that is, by continuous linear operators with finite-dimensional ranges). A nuclear operator has certain properties inherent in finite-dimensional operators. In particular, a nuclear operator mapping a space with a basis into itself has a finite trace (see below), which coincides with the sum of the series formed from the diagonal elements of the matrix of this operator relative to an arbitrary basis. Nuclear operators first appeared in mathematical quantum mechanics and were called 'operators with a trace' (see [1], [2]). On a Hilbert space the operators with a trace are in a one-to-one correspondence with bivalent tensors, and the trace of an operator coincides with contraction of the corresponding tensor. By using this correspondence, A.F. Ruston [3] carried over the concept of a nuclear operator to Banach spaces. Independently, in connection with the theory of nuclear spaces (cf. **Nuclear space**), A. Grothendieck carried over the concept to locally convex spaces (see [4], [5]). Let E and F be locally convex spaces over the field of real or complex numbers, let E' and F' be their duals endowed with the strong topology, let $L(E, F)$ be the vector space of all continuous linear mappings from E into F, and let $S(E, F)$ be the space of all weakly continuous mappings from E into F. Set $L(E, E) = L(E)$ and $S(E, E) = S(E)$.

A linear operator $A : E \to F$ is called *nuclear* if it can be represented in the form

$$x \mapsto Ax = \sum_{i=1}^\infty \lambda_i <x, x_i'>y_i, \qquad (1)$$

where $\{\lambda_i\}$ is a summable numerical sequence, $\{x_i'\}$ is an equicontinuous sequence in E', $\{y_i\}$ is a sequence of elements from a certain complete bounded convex circled set in F (cf. **Topological vector space**), and $<x, x'>$ denotes the value of the linear functional x' at a vector x. The representation (1) can be regarded as an expansion of the operator as a sum of operators of rank 1 (that is, with a one-dimensional range), and the corresponding series is absolutely convergent in $L(E, F)$ in the topology of uniform convergence on bounded sets. Thus, in this topology, the nuclear operator A is the limit of a sequence of operators of finite rank. If E and F are Banach spaces, then a nuclear operator A can be approximated, in the **nuclear norm**, by operators of finite rank.

The expansion (1) is called a *nuclear representation* of A. Every nuclear operator has a nuclear representation (1) such that $x_i' \to 0$, $y_i \to 0$. If E is a **barrelled space** and is complete, or at least *quasi-complete* (i.e. closed bounded sets in E are complete), then the expansion (1) is nuclear if and only if $\{x_i'\}$ and $\{y_i\}$ are bounded.

By changing the conditions on $\{\lambda_i\}$, $\{x_i'\}$, and $\{y_i\}$ one can obtain different modifications of the concept of a nuclear operator (see [4], [5], [7]). If instead of the equicontinuity of $\{x_i'\}$ one requires its elements to belong to a complete bounded convex circled set in E', then the expansion (1) defines a **Fredholm operator**; these operators form the natural domain of application of the Fredholm theory (see [4], [5]). Every nuclear operator is a Fredholm operator, and when E is endowed with the **Mackey topology**, any Fredholm operator $A : E \to F$ is nuclear. A nuclear operator A is called *strongly nuclear* (or a *nuclear operator of order* 0) if it admits a nuclear representation (1) in which $\{\lambda_i\}$ is a rapidly decreasing sequence, that is, $\sum_{i=1}^{\infty} |\lambda_i|^p < \infty$ for all $p > 0$.

Integral operators (in particular, Fredholm integral operators) provide many examples of nuclear operators and their modifications (see [4], [5], [7], [8]).

Properties of nuclear operators. Every nuclear operator $A \in L(E, F)$ is compact, that is, it maps a neighbourhood of zero in E into a set with compact closure in F. Thus, every nuclear operator is continuous, and every Fredholm operator is weakly continuous. The product (in any order) of a nuclear operator and a continuous linear operator is a nuclear operator. In particular, the set of all nuclear operators is an ideal in the algebra $L(E)$; correspondingly, the Fredholm operators form an ideal in $S(E)$. The strongly nuclear operators also form an ideal in $L(E)$. Every nuclear operator $A \in L(E, F)$ has a unique extension $\hat{A} \in L(\hat{E}, F)$, where $\hat{E}$ is the completion of E and $\hat{A}$ is nuclear. If $A \in L(E, F)$ is a Fredholm operator, then the dual mapping $F' \to E'$ is a nuclear operator. For any nuclear operator $A \in L(E, F)$ one can find Banach spaces E_1 and F_1, compact operators $K_1 \in L(E, E_1)$ and $K_2 \in L(F_1, F)$, and a nuclear operator $B \in L(E_1, F_1)$, such that $A = K_2 B K_1$. If $A \in L(E)$ is a strongly nuclear operator, then the sequence of its eigen values (in general, complex), ordered in decreasing absolute value, is rapidly decreasing.

Let E be a nuclear space and let F be a complete or quasi-complete space. Then for $A \in L(E, F)$ the following assertions are equivalent: 1) A is a nuclear operator; 2) A is a **compact operator**; 3) A is a **bounded operator**; i.e. A maps a neighbourhood of zero in E into a bounded set in F; and 4) A is a strongly nuclear operator.

Let E, F and G be Hilbert spaces, and let $K_1 \in L(E, F)$ and $K_2 \in L(F, G)$ be Hilbert$-$Schmidt operators (cf. **Hilbert$-$Schmidt operator**). Then $K_2 K_1 \in L(E, G)$ is nuclear. Conversely, every nuclear operator is the product of two operators of Hilbert$-$Schmidt type. An arbitrary completely-continuous (compact) operator $A \in L(E, F)$ is nuclear if and only if the series of eigen values of the positive-definite operator $T \in L(E)$ in the **polar decomposition** $A = UT$ converges, where U is an isometric operator mapping the range of T into F (see [9]).

Operators with a trace. Let E be an arbitrary locally convex space, and let A be a nuclear (respectively, Fredholm) operator mapping E into itself and admitting a representation of the form (1). The series $\sum_{i=1}^{\infty} \lambda_i <y_i, x_i'>$ converges absolutely; if its sum does not depend on the representation (1), then the sum is called the *trace of the nuclear (Fredholm) operator A*, and is denoted by $\mathrm{tr}\, A$. In this case the trace is *well defined* (see [4], [5]). The operator is said to be an *operator of finite trace* or a *trace class operator*. If (1) contains only a finite number of terms, then A is an operator of finite rank, and $\mathrm{tr}\, A$ is the same as the trace of the finite-dimensional operator induced in the range of A.

Let $E' \overline{\otimes} E$ be the *inductive tensor product* of E' and E, that is, the completion of the (algebraic) tensor product $E' \otimes E$ in the strongest **locally convex topology** in which the canonical bilinear mapping $E' \times E \to E' \otimes E$ $((x', x)$ goes into $x' \otimes x)$ is continuous in each variable separately. The composition of this mapping with any continuous linear form on $E' \overline{\otimes} E$ gives a bilinear form on $E' \times E$ that is continuous in each variable separately, and the correspondence between forms of this type is one-to-one. In particular, the bilinear form $(x', x) \mapsto <x, x'>$ corresponds to a continuous linear form on $E' \overline{\otimes} E$. The value of this form at a $u \in E' \overline{\otimes} E$ is denoted by $\mathrm{tr}\, u$. An element $u \in E' \overline{\otimes} E$ is called a *Fredholm kernel* if it admits an expansion of the form

$$u = \sum_{i=1}^{\infty} \lambda_i\, x_i' \otimes y_i, \qquad (2)$$

where $\{\lambda_i\}$, $\{x_i'\}$ and $\{y_i\}$ are the same as in the expansion (1) for a Fredholm operator. The Fredholm kernels form a subspace in $E' \overline{\otimes} E$, denoted by $E' \tilde{\otimes} E$.

Suppose that the algebra $S(E)$ of weakly continuous operators on E is endowed with the weak **operator topology** defined by semi-norms $A \mapsto |<Ay, x'>|$, where $A \in S(E)$, and x' and y range over E' and E, respectively. The mapping $\Gamma : E' \tilde{\otimes} E \to S(E)$ that sends an element u of the form (2) into an operator A of the form (1), is well defined, linear and continuous; also $\mathrm{tr}\, u = \mathrm{tr}\, \Gamma(u)$ if the trace of the operator $A = \Gamma(u)$ is well defined. If E and E' are complete (for example, if E is a **Fréchet space**), then Γ can be continuously extended to $E'\overline{\otimes}E$. The images of elements of $E'\overline{\otimes}E$ under this mapping are called *operators with a trace* (see [4], [5]). If E is a Banach space, then every operator with a trace is nuclear, so that in this case the classes of nuclear operators, of Fredholm operators and of operators with a trace coincide. There are operators with a trace that are not Fredholm operators (for example, in nuclear Fréchet spaces). The non-compactness of these operators makes their study difficult.

The single-valuedness problem (the 'problème de biunivocité'). If the mapping Γ is one-to-one, or at least if $\Gamma(u)=0$ implies $\operatorname{tr} u=0$, then the trace of $\Gamma(u)$ is well defined by $\operatorname{tr}\Gamma(u)=\operatorname{tr} u$.

This possibility is closely connected with the *approximation property*, which is that $L(E)$ contains a net (cf. **Net (directed set)**) of operators of finite rank that converges to the identity operator in the topology of uniform convergence on all pre-compact sets. If E is a Banach space, then the trace of any nuclear operator is well defined if and only if the approximation property holds [4]. A reflexive separable space X without the approximation property (and without a Schauder basis, thus solving a well-known problem of S. Banach) has been constructed [11]. This solves the single-valuedness problem: There is an $u\in X'\overline{\otimes}X$ such that $\Gamma(u)=0$ but $\operatorname{tr} u=1$. If a locally convex space E has the approximation property, then every nuclear operator has a well defined trace; if $\{B_\nu\}$ is a net of operators of finite rank that converges to an arbitrary operator $B\in L(E)$ uniformly on all pre-compact (or, at least, on convex balanced compact) sets, then

$$\operatorname{tr} AB = \lim_\nu \operatorname{tr} AB_\nu, \tag{3}$$

is valid for any nuclear operator A (see [12]). However, there is a locally convex space with the approximation property in which it is impossible to properly define the trace for all Fredholm operators. Any Fredholm operator on a locally convex space E has a well-defined trace if E has the *bounded approximation property*, that is, if there is a net of operators of finite rank that converges to the identity operator in the weak operator topology, and that is bounded in this topology; any space with a Schauder basis has this property. If $\{B_\nu\}$ is a bounded net that converges in $S(E)$ to an arbitrary operator B (for example, if $\{B_\nu\}$ is an arbitrary countable convergent sequence in $S(E)$), then (3) holds for any Fredholm operator A provided that the AB_ν have a well-defined trace (for example, if the B_ν are operators of finite rank, or if E has the bounded approximation property). If E has the approximation (respectively, bounded approximation) property, then for any nuclear (respectively, Fredholm) operator A and any $B\in L(E)$

(respectively, $B\in S(E)$) one has $\operatorname{tr} AB=\operatorname{tr} BA$ (see [12]).

Matrix trace. Suppose that a locally convex space E has a Schauder **basis** $\{e_i\}_{i=1}^\infty$ so that any $x\in E$ can be expanded as $x=\sum_{i=1}^\infty <x, e_i'>e_i$, where $e_i'\in E'$. Then $\sum_{i=1}^\infty <Ae_i, e_i'>$ is called the *matrix trace* of the operator A if the series is convergent. This series converges absolutely if the basis is unconditional. Any Fredholm operator on a space with a Schauder basis has a well-defined trace that coincides with its matrix trace, which in this case does not depend on the choice of the basis [13].

An arbitrary continuous operator on a Hilbert space is nuclear if and only if it has a finite matrix trace for any orthonormal basis (see [2], [8], [9]).

Nuclear trace. Let T be a compact space with a Borel measure μ, let $C(T)$ be the Banach space of continuous functions on T equipped with the topology of uniform convergence, and let $K(t, s)$ be a continuous function on $T\times T$. Then the linear integral operator

$$K:\phi(t)\mapsto \int_T K(t, s)\phi(s)\,d\mu(s)$$

on $C(T)$ (a classical Fredholm integral operator) is nuclear and has a well-defined trace; moreover,

$$\operatorname{tr} K = \int_T K(t, t)\,d\mu(t). \tag{4}$$

If K is the integral operator with kernel $K(t, s)$, acting on a space of functions on a space T with a measure μ, and if the right-hand side of (4) can be given a reasonable meaning, then this quantity is called the *nuclear trace* of K. For different classes of integral operators, conditions can be obtained that ensure the nuclearity of these operators, and enable one to give a meaning to (4) (see [4], [5], [8], [14]).

Spectral trace. Let E be a locally convex space over the field of complex numbers, and let A be a nuclear operator on E. The spectrum of A, as of any compact operator (cf. **Spectrum of an operator**), is either a finite set or is a sequence that converges to zero, and any non-zero value has finite spectral multiplicity. If the series

$$\sum_j \sigma_j(A), \tag{5}$$

formed from the non-zero eigen values of A (each eigen values appears in (5) as many times as its spectral multiplicity) converges absolutely, then its sum is called the *spectral trace* of A, and is denoted by $\operatorname{tr}_\sigma A$. Every nuclear operator on a Hilbert space has a spectral trace, which coincides with its matrix trace [15]. Let E be a *multi-Hilbert space* (a *Hilbertiable space*), that is, the topology in E can be generated by a family of semi-norms each of which is obtained from a non-negative definite Hermitian form on $E\times E$; any **nuclear**

space is an example of a multi-Hilbert space. Then any nuclear operator A on E has a well-defined trace and a spectral trace, and $\operatorname{tr}_\sigma A = \operatorname{tr} A$ (see [13]). A nuclear operator A need not have a matrix trace. A nuclear operator A on a Banach space need not have a spectral trace even when the space has a basis and $\operatorname{tr} A$ is well defined. Also, the equality $\operatorname{tr}_\sigma A = \operatorname{tr} A$ can be violated. For example, in the Banach space c_0 of sequences that converge to 0 there is a nuclear operator A such that $\operatorname{tr} A = 1$ and $A^2 = 0$, so that A does not have non-zero eigen values, and $\operatorname{tr}_\sigma A = 0$. For a nuclear operator A acting on an arbitrary Banach or locally convex space (without, perhaps, any approximation properties), it is possible to give conditions on A under which $\operatorname{tr}_\sigma A$ and $\operatorname{tr} A$ exist and are equal (see [4], [14], [16], [17]).

Example. Let E be a complex Banach space and let $L(E)$ be the algebra of continuous linear operators on E equipped with the usual operator norm. For any $A \in L(E)$ let $\alpha_r(A)$ denote the greatest lower bound of $\| A - F \|$ when F ranges over the set of all operators in $L(E)$ with *rank* (that is, dimension of the range) not exceeding $r = 0, 1, \ldots$. The set of all $A \in L(E)$ for which $\sum \alpha_r(A) < \infty$ is denoted by $l_1(E)$. Every $A \in l_1(E)$ is nuclear; if E is a Hilbert space, then $l_1(E)$ coincides with the set of all nuclear operators on E. For an arbitrary Banach space E, each operator $A \in l_1(E)$ has a trace, $\operatorname{tr} A$, and a spectral trace, and $\operatorname{tr}_\sigma A = \operatorname{tr} A$ (see [16], [17]).

References

[1] NEUMANN, J. VON: *Mathematische Grundlagen der Quantenmechanik*, Springer, 1932.
[2] SCHATTEN, R. and NEUMANN, J. VON: 'The cross-space of linear transformations II', *Ann. of Math.* **47** (1946), 608-630.
[3] RUSTON, A.F.: 'On the Fredholm theory of integral equations for operators belonging to the trace class of a general Banach space', *Proc. London Math. Soc. (2)* **53**, no. 2 (1951), 109-124.
[4] GROTHENDIECK, A.: *Produits tensoriels topologiques et espaces nucléaires*, Amer. Math. Soc., 1955.
[5] GROTHENDIECK, A.: 'La théorie de Fredholm', *Bull. Soc. Math. France* **84** (1956), 319-384.
[6] SCHAEFER, H.: *Topological vector spaces*, Springer, 1971.
[7] PIETSCH, A.: *Operator ideals*, North-Holland, 1980.
[8] GOHBERG, I.C. [I.Ts. GOKHBERG] and KREĬN, M.G.: *Introduction to the theory of linear nonselfadjoint operators*, Amer. Math. Soc., 1969 (translated from the Russian).
[9] GEL'FAND, I.M. and VILENKIN, N.YA.: *Generalized functions*, 4. Applications of harmonic analysis, Acad. Press, 1964 (translated from the Russian).
[10] PIETSCH, A.: *Nuclear locally convex spaces*, Springer, 1972 (translated from the German).
[11] ENFLO, P.: 'A counterexample to the approximation problem in Banach spaces', *Acta Math.* **130** (1973), 309-317.
[12] LITVINOV, G.L.: 'Approximation properties of locally convex spaces and the uniqueness for the trace of a linear operator', *Teor. Funkt., Funktsional. Anal. i Prilozhen.* **39** (1983), 73-87 (in Russian).
[13] LITVINOV, G.L.: 'On the traces of linear operators in locally convex spaces', *Sel. Math. Sovietia* **8**, no. 3 (1989), 203-212. (*Trudy Sem. Vekt. i Tenz. Anal.* **19** (1979), 243-272)
[14] PIETSCH, A.: 'Operator ideals with a trace', *Math. Nachr.* **100** (1981), 61-91.
[15] LIDSKIĬ, V.B.: 'Nonselfadjoint operators having a trace', *Transl. Amer. Math. Soc. (2)* **47** (1965), 43-46. (*Dokl. Akad. Nauk SSSR* **125**, no. 3 (1959), 485-487)
[16] MARKUS, A.S. and MATSAEV, V.I.: 'Analogs of Weyl inequalities and the trace theorem in Banach space', *Math. USSR Sb.* **15**, no. 2 (1971), 299-312. (*Mat. Sb.* **86**, no. 2 (1971), 299-313)
[17] KÖNIG, H.: 's-numbers, eigenvalues and the trace theorem in Banach spaces', *Studia Math.* **67**, no. 2 (1980), 157-172.

G.L. Litvinov

Editorial comments.

References

[A1] GROTHENDIECK, A.: 'Résumé de la théorie métriques des produits tensoriels topologiques', *Bol. Soc. Mat. São Paulo* **8** (1956), 1-79.
[A2] JARCHOW, H.: *Locally convex spaces*, Teubner, 1981.
[A3] SIMON, B.: *Trace ideals and their applications*, Cambridge Univ. Press, 1979.
[A4] PIETSCH, A.: *Eigenvalues and s-numbers*, Cambridge Univ. Press, 1987.

AMS 1980 Subject Classification: 47B10

NUCLEAR SPACE - A **locally convex space** for which all continuous linear mappings into an arbitrary Banach space are nuclear operators (cf. **Nuclear operator**). The concept of a nuclear space arose [1] in an investigation of the question: For what spaces are the analogues of Schwartz' kernel theorem valid (see **Nuclear bilinear form**)? The fundamental results in the theory of nuclear spaces are due to A. Grothendieck [1]. The function spaces used in analysis are, as a rule, Banach or nuclear spaces. Nuclear spaces play an important role in the spectral analysis of operators on Hilbert spaces (the construction of rigged Hilbert spaces, expansions in terms of generalized eigen vectors, etc.) (see [2]). Nuclear spaces are closely connected with measure theory on locally convex spaces (see [3]). Nuclear spaces can be characterized in terms of dimension-type invariants (approximative dimension, diametral dimension, etc.) (see [2], [4], [5]). One of these invariants is the functional dimension, which for many spaces consisting of entire analytic functions is the same as the number of variables on which these functions depend (see [2]).

In their properties, nuclear spaces are close to finite-dimensional spaces. Every bounded set in a nuclear space is pre-compact. If a nuclear space is complete (or at least quasi-complete, that is, every closed bounded set is complete), then it is *semi-reflexive* (that is, the space coincides with its second dual as a set of elements), and every closed bounded set in it is compact. If a quasi-complete nuclear space is a **barrelled space**, then it is also a **Montel space** (in particular, a **reflexive space**); any weakly-convergent countable sequence in this space converges also in the original topology. A normed space is nuclear if and only if it is finite dimensional. Every nuclear space has the *approximation property*: Any continuous linear operator in such a space can be approximated in the operator topology of pre-compact convergence by operators of finite rank (that is, continuous linear operators with finite-dimensional ranges). Nevertheless, there are nuclear Fréchet spaces (cf. **Fréchet space**) that do not have the *bounded approximation property*; in such a space the identity operator is not the limit of a countable sequence of operators of finite rank in the strong or weak operator topology [6]. Nuclear Fréchet spaces without a Schauder **basis** have been constructed, and they can have arbitrarily small diametral dimension, that is, they can be arbitrarily near (in a certain sense) to finite-dimensional spaces [7]. For nuclear spaces a counterexample to the problem of invariant subspaces has been constructed: In a certain nuclear Fréchet space one can find a continuous linear operator without non-trivial invariant closed subspaces [8].

Examples of nuclear spaces. 1) Let $\mathscr{E}(\mathbf{R}^n)$ be the space of all (real or complex) infinitely-differentiable functions on $\mathbf{R}^n$ equipped with the topology of uniform convergence of all derivatives on compact subsets of $\mathbf{R}^n$. The space $\mathscr{E}'(\mathbf{R}^n)$ dual to $\mathscr{E}(\mathbf{R}^n)$ consists of all generalized functions (cf. **Generalized function**) with compact support. Let $\mathscr{D}(\mathbf{R}^n)$ and $\mathscr{S}(\mathbf{R}^n)$ be the linear subspaces of $\mathscr{E}(\mathbf{R}^n)$ consisting, respectively, of functions with compact support and of functions that, together with all their derivatives, decrease faster than any power of $|x|^{-1}$ as $|x|\to\infty$. The duals $\mathscr{D}'(\mathbf{R}^n)$ and $\mathscr{S}'(\mathbf{R}^n)$ of $\mathscr{D}(\mathbf{R}^n)$ and $\mathscr{S}(\mathbf{R}^n)$, relative to the standard topology, consist of all generalized functions and of all generalized functions of slow growth, respectively. The spaces $\mathscr{E}$, $\mathscr{D}$, $\mathscr{S}$, $\mathscr{E}'$, $\mathscr{D}'$, and $\mathscr{S}'$, equipped with the strong topology, are complete reflexive nuclear spaces.

2) let $\{a_{np}\}$ be an infinite matrix, where $0\leqslant a_{np}<\infty$ and $a_{np}\leqslant a_{n(p+1)}$, $n,p=1,2,\ldots$. The space of sequences $\xi=\{\xi_n\}$ for which $|\xi|_p=\sum_{n=0}^{\infty}|\xi_n|a_{np}<\infty$ for all p, with the topology defined by the semi-norms $\xi\to|\xi|_p$ (cf. **Semi-norm**), is called a *Köthe space*, and is denoted by $\mathscr{K}(a_{np})$. This space is nuclear if and only if for any p one can find a q such that $\sum_{n=0}^{\infty}(a_{np}/a_{nq})<\infty$.

Heredity properties. A locally convex space is nuclear if and only if its completion is nuclear. Every subspace (separable quotient space) of a nuclear space is nuclear. The direct sum, the inductive limit of a countable family of nuclear spaces, and also the product and the projective limit of any family of nuclear spaces, is again nuclear.

Let E be an arbitrary locally convex space, and let E' denote its dual equipped with the strong topology. If E' is nuclear, then E is called *conuclear*. If E is arbitrary and F is a nuclear space, then the space $L(E,F)$ of continuous linear operators from E into F is nuclear

with respect to the strong **operator topology** (simple convergence); if E is semi-reflexive and conuclear, then $L(E,F)$ is nuclear also in the topology of bounded convergence.

Metric and dually-metric nuclear spaces. A locally convex space E is called *dually metric*, or a *space of type* $(\mathscr{D}\mathscr{F})$, if it has a countable fundamental system of bounded sets and if every (strongly) bounded countable union of equicontinuous subsets in E' is equicontinuous (cf. **Equicontinuity**). Any strong dual of a metrizable locally convex space is dually metric; the converse is not true. If E is a space of type $(\mathscr{D}\mathscr{F})$, then E' is of type $(\mathscr{F})$ (a Fréchet space, that is, complete and metrizable). Examples of nuclear spaces of type $(\mathscr{F})$ are Köthe spaces, and also $\mathscr{E}$ and $\mathscr{S}$; accordingly, $\mathscr{E}'$ and $\mathscr{S}'$ are nuclear spaces of type $(\mathscr{D}\mathscr{F})$. The spaces $\mathscr{D}$ and $\mathscr{D}'$ are neither metric nor dually metric.

Metric and dually-metric nuclear spaces are separable, and if complete, they are reflexive. The transition $E\to E'$ to the dual space establishes a one-to-one correspondence between nuclear spaces of type $(\mathscr{F})$ and complete nuclear spaces of type $(\mathscr{D}\mathscr{F})$. If E is a complete nuclear space of type $(\mathscr{D}\mathscr{F})$ and if F is a nuclear space of type $(\mathscr{F})$, then $L(E,F)$, equipped with the topology of bounded convergence, is nuclear and conuclear.

Every nuclear space of type $(\mathscr{F})$ is isomorphic to a subspace of the space $\mathscr{E}(\mathbf{R})$ of infinitely-differentiable functions on the real line, that is, $\mathscr{E}(\mathbf{R})$ is a universal space for the nuclear spaces of type $(\mathscr{F})$ (see [10]). A Fréchet space E is nuclear if and only if every unconditionally-convergent series (cf. **Unconditional convergence**) in E is absolutely convergent (that is, with respect to any continuous semi-norm). Spaces of holomorphic functions on nuclear spaces of types $(\mathscr{F})$ and $(\mathscr{D}\mathscr{F})$ have been studied intensively (see [11]).

Tensor products of nuclear spaces, and spaces of vector functions. The algebraic tensor product $E\otimes F$ of two locally convex spaces E and F can be equipped with the projective and injective topologies, and then $E\otimes F$ becomes a topological tensor product. The *projective topology* is the strongest locally convex topology in which the canonical bilinear mapping $E\times F\to E\otimes F$ is continuous. The *injective topology* (or the topology of (bi) equicontinuous convergence) is induced by the natural imbedding $E\otimes F\to L_e(E'_\tau,F)$, where E'_τ is the dual of E equipped with the **Mackey topology** $\tau(E',E)$, and $L_e(E'_\tau,F)$ is the space of continuous linear mappings $E'_\tau\to F$ equipped with the topology of **uniform convergence** on equicontinuous sets in E'. Under this imbedding $x\otimes y\in E\otimes F$ goes into the operator $x'\mapsto<x,x'>y$, where $<x,x'>$ denotes the value of the functional $x'\in E'$ at $x\in E$. The completion of $E\otimes F$

in the projective (respectively, injective) topology is denoted by $E \hat{\otimes} F$ (respectively, $E \check{\otimes} F$).

For E to be a nuclear space it is necessary and sufficient that for any locally convex space F the projective and injective topologies in $E \otimes F$ coincide, that is,

$$E \hat{\otimes} F = E \check{\otimes} F. \qquad (1)$$

Actually, it suffices to require that (1) holds for $F = l_1$, the space of summable sequences, or for F equal to a fixed space with an unconditional basis (see [12]). Nevertheless, there is a (non-nuclear) infinite-dimensional separable Banach space X such that $X \hat{\otimes} X = X \check{\otimes} X$ (see [13]). If E and F are complete spaces and F is nuclear, then the imbedding $E \otimes F \to L_e(E'_\tau, F)$ can be extended to an isomorphism between $E \hat{\otimes} F$ and $L_e(E'_\tau, F)$.

If E is a non-null nuclear space, then $E \hat{\otimes} F$ is nuclear if and only if F is nuclear. If E and F are both spaces of type $(\mathscr{F})$ (or $(\mathscr{D} \mathscr{F})$) and if E is nuclear, then $(E \hat{\otimes} F)' = E' \hat{\otimes} F'$.

Let E be a complete nuclear space consisting of scalar functions (not all) on a certain set T; let also E be the inductive limit (locally convex hull) of a countable sequence of spaces of type $(\mathscr{F})$, and let the topology on E be not weaker than the topology of pointwise convergence of functions on T. Then for any complete space F one can identify $E \hat{\otimes} F$ with the space of all mappings (vector functions) $T \to F$ for which the scalar function $t \mapsto <f(t), y'>$ belongs to E for all $y' \in F'$. In particular, $\mathscr{E}(\mathbf{R}^n) \hat{\otimes} F$ coincides with the space of all infinitely-differentiable vector functions on $\mathbf{R}^n$ with values in F, and $\mathscr{E}(\mathbf{R}^n) \hat{\otimes} \mathscr{E}(\mathbf{R}^m) = \mathscr{E}(\mathbf{R}^n \times \mathbf{R}^m) = \mathscr{E}(\mathbf{R}^{n+m})$.

The structure of nuclear spaces. Let U be a convex circled (i.e. convex balanced) neighbourhood of zero in a locally convex space E, and let p be the Minkowski functional (continuous semi-norm) corresponding to U. Let E_U be the quotient space $E/p^{-1}(0)$ with the norm induced by p, and let $\hat{E}_U$ be the completion of the normed space E_U. There is defined a continuous canonical linear mapping $E \to \hat{E}_U$; if U contains a neighbourhood V, then the continuous linear mapping $\hat{E}_V \to \hat{E}_U$ is defined canonically.

For a locally convex space E the following conditions are equivalent: 1) E is nuclear; 2) E has a basis $\mathfrak{B}$ of convex circled neighbourhoods of zero such that for any $U \in \mathfrak{B}$ the canonical mapping $E \to \hat{E}_U$ is a nuclear operator; 3) the mapping $E \to \hat{E}_U$ is nuclear for any convex circled neighbourhood U of zero in E; and 4) every convex circled neighbourhood U of zero in E contains another such neighbourhood of zero, V, such that the canonical mapping $\hat{E}_V \to \hat{E}_U$ is nuclear.

Let E be a nuclear space. For any neighbourhood U of zero in E and for any q such that $1 \leqslant q \leqslant \infty$ there is a convex circled neighbourhood $V \subset U$ for which E_V is (norm) isomorphic to a subspace of the space l_q of sequences with summable q-th powers. Thus, E coincides with a subspace of the projective limit of a family of spaces isomorphic to l_q. In particular (the case $q = 2$), in any nuclear space E there is a basis of neighbourhoods of zero $\{U_\alpha\}$ such that all the spaces $\hat{E}_{U_\alpha}$ are Hilbert spaces; thus, E is *Hilbertian*, that is, the topology in E can be generated by a family of semi-norms each of which is obtained from a certain non-negative definite Hermitian form on $E \times E$. Any complete nuclear space is isomorphic to the projective limit of a family of Hilbert spaces. A space E of type $(\mathscr{F})$ is nuclear if and only if it can be represented as the projective limit $E = \lim_{\leftarrow} g_{mn} H_n$ of a countable family of Hilbert spaces H_n, such that the g_{mn} are nuclear operators (or, at least, Hilbert $-$ Schmidt operators, cf. **Hilbert $-$ Schmidt operator**) for $m < n$.

Bases in nuclear spaces. In a nuclear space every equicontinuous basis is absolute. In a space of type $(\mathscr{F})$ any countable basis (even if weak) is an equicontinuous Schauder basis (cf. **Basis**), so that in a nuclear space of type $(\mathscr{F})$ any basis is absolute (in particular, unconditional). A similar result holds for complete nuclear spaces of type $(\mathscr{D} \mathscr{F})$, and for all nuclear spaces for which the **closed-graph theorem** holds. A quotient space of a nuclear space of type $(\mathscr{F})$ with a basis does not necessarily have a basis (see [4], [5], [6]).

Let E be a nuclear space of type $(\mathscr{F})$. A topology can be defined in E by a countable system of semi-norms $x \mapsto \| x \|_q$, $q = 1, 2, \ldots$, where $\| x \|_q \leqslant \| x \|_{q+1}$ for all $x \in E$. If E has a basis or a continuous norm, then the semi-norms $\| \cdot \|$ can be taken as norms. Let $\{e_n\}$ be a basis in E; then any $x \in E$ can be expressed as an (absolutely and unconditionally) convergent series

$$x = \sum_{n=1}^{\infty} \xi_n e_n,$$

where the coordinates ξ_n have the form $\xi_n = <x, x'_n>$, and the functionals x'_n form a bi-orthogonal basis in E'. E is isomorphic to the Köthe space $\mathscr{K}(a_{nq})$, where $a_{nq} = \| e_n \|_q$; under this isomorphism $x \in E$ goes into the sequence $\{\xi_n\}$ of its coordinates. A basis $\{f_n\}$ in E is equivalent to the basis $\{e_n\}$ (that is, it can be obtained from $\{e_n\}$ by an isomorphism) if and only if $\mathscr{K}(\| e_n \|_q)$ and $\mathscr{K}(\| f_n \|_q)$ coincide as sets [4]. A basis $\{f_n\}$ is called *regular* (or *proper*) if there is a system of norms $\| \cdot \|_q$ and a permutation σ of indices such that $\| f_{\sigma(n)} \|_q / \| f_{\sigma(n)} \|_r$ is monotone decreasing for all $r \geqslant q$. If a nuclear space E of type $(\mathscr{F})$ has a regular basis, then any two bases in E are *quasi-equivalent* (that is, they can be made equivalent by a permutation and a normalization of the elements of one of them). There are other sufficient conditions for all bases in E to be quasi-equivalent (see [4], [14]). A complete description of the class of nuclear spaces with this property is not known (1984).

Example. The Hermite functions $\phi_n(t)=e^{t^2/2}(d^n/dt^n)(e^{-t^2})$ form a basis in the complete metric nuclear space $\mathscr{S}(\mathbf{R})$ of smooth functions on the real line that are rapidly decreasing together with all their derivatives. $\mathscr{S}(\mathbf{R})$ is isomorphic to $\mathscr{K}(n^p)$.

References

[1] GROTHENDIECK, A.: *Produits tensoriels topologiques et espaces nucléaires*, Amer. Math. Soc., 1955.

[2] GEL'FAND, I.M. and VILENKIN, N.YA.: *Generalized functions*, 4. Applications of harmonic analysis, Acad. Press, 1964 (translated from the Russian).

[3] MINLOS, R.A.: 'Generalized random processes and their extension in measure', *Trudy Moskov. Mat. Obshch.* **8** (1959), 497-518 (in Russian).

[4] MITYAGIN, B.S.: 'Approximate dimension and bases in nuclear spaces', *Russian Math. Surveys* **16**, no. 4, 59-127. (*Uspekhi Mat. Nauk* **16**, no. 4 (1961), 63-132)

[5] PIETSCH, A.: *Nuclear locally convex spaces*, Springer, 1972 (translated from the German).

[6] DUBINSKY, E.: *Structure of nuclear Fréchet spaces*, Springer, 1979.

[7] ZOBIN, N.M. and MITYAGIN, B.S.: 'Examples of nuclear linear metric spaces without a basis', *Funct. Anal. Appl.* **8**, no. 4 (1974), 304-313. (*Funktsional. Anal. i Prilozhen.* **8**, no. 4 (1974), 35-47)

[8] ATZMON, A.: 'An operator without invariant subspaces on a nuclear Fréchet space', *Ann. of Math.* **117**, no. 3 (1983), 669-694.

[9] SCHAEFER, H.H.: *Topological vector spaces*, Springer, 1971.

[10] KOMURA, T. and KOMURA, Y.: 'Ueber die Einbettung der nuklearen Räume in $(s)^A$', *Math. Ann.* **162** (1965-1966), 284-288.

[11] DINEEN, S.: *Complex analysis in locally convex spaces*, North-Holland, 1981.

[12] JOHN, K. and ZIZLER, V.: 'On a tensor product characterization of nuclearity', *Math. Ann.* **244**, no. 1 (1979), 83-87.

[13] PISIER, G.: 'Contre-example à une conjecture de Grothendieck', *C.R. Acad. Sci. Paris* **293** (1981), 681-683. English abstract.

[14] DRAGILEV, M.M.: *Bases in Köthe spaces*, Rostov-on-Don, 1983 (in Russian).

G.L. Litvinov

Editorial comments. A generalized function is also called a *distribution*, and a *generalized function of slow growth* is also called a *tempered distribution*.

Let F be a topological linear space, U a neighbourhood of zero in F, A a set in F, and ϵ a (small) positive number. An ϵ-*set for A relative to a neighbourhood U of zero* is a set B such that for every $a \in A$ there is a $b \in B$ such that $a \in b + \epsilon U$. Let $N(\epsilon, A, U)$ be the smallest number of elements in ϵ-sets for A relative to U. The *functional dimension* of F is defined by

$$\mathrm{df}(F) = \sup_U \inf_V \limsup_{\epsilon \to 0} \frac{\ln \ln N(\epsilon, V, U)}{\ln \ln \epsilon^{-1}},$$

where U, V range over the neighbourhoods of zero in F. Cf. [2], Sect. I.3.8 for more details.

Let F be a locally convex space and consider two neighbourhoods of zero U, V such that U absorbs V, i.e. $V \subset \rho U$ for some positive number ρ. Let

$$\delta_r(U, V) = \inf\{\delta: \exists \text{ subspace } G \text{ of dimension } \leqslant r$$
$$\text{such that } V \subset \delta U + G\}.$$

This number is called the *r-th diameter of V with respect to U*. The *diametral dimension* of a locally convex space is the collection of all sequences $(d_r)_{r \in \mathbf{N} \bigcup \{0\}}$ of non-negative numbers with the property that for each neighbourhood of zero U there is a neighbourhood V of zero absorbed by U for which $\delta_r(U, V) \leqslant d_r$, $r \in \mathbf{N} \bigcup \{0\}$.

A locally convex space E is nuclear if and only if for some (respectively, each) positive number λ the sequence $((r+1)^{-\lambda})_{r \in \mathbf{N} \bigcup \{0\}}$ belongs to the diametral dimension of E. See [5], Chapt. 9 for more details.

Let again U, V be neighbourhoods of zero of a locally convex space F such that U absorbs V. The ϵ-*content of V with respect to U* is the supremum $M_\epsilon(U, V)$ of all natural numbers m such that there are $x_1, \ldots, x_m \in V$ with $x_1 \cdots x_k \notin \epsilon U$ for all $i \neq k$. The *approximative dimension* of a locally convex space F is the collection of all positive functions ϕ on $(0, \infty)$ such that for each neighbourhood U of zero there is a neighbourhood V of zero absorbed by U such that

$$\lim_{\epsilon \to 0} \phi(\epsilon)^{-1} M_\epsilon(U, V) = 0.$$

The number $\rho(U, V)$ is defined by the exponential rate of growth of $M_\epsilon(U, V)$ as $\epsilon \to 0$. More precisely,

$$\rho(U, V) = \limsup_{\epsilon \to 0} \frac{\ln \ln M_\epsilon(U, V)}{\ln \epsilon^{-1}}.$$

A locally convex space E is nuclear if and only if for some (respectively, each) positive number ρ the following condition is satisfied: For each neighbourhood of zero U there is a neighbourhood V of zero absorbed by U such that $\rho(U, V) \leqslant \rho$. Cf. [5], Chapt. 9 for more details.

Let U be a bounded circled neighbourhood of a topological vector space F. The *Minkowski functional* associated to U is defined by

$$q(x) = \inf_{x \in \alpha U} \alpha, \quad \alpha \geqslant 0.$$

This is well-defined for each x since U is *absorbent* (i.e. for each $x \in F$ there is an α such that $x \in \alpha U$). Cf. [A7], Sects. 15.10, 16.4.

References

[A1] GROTHENDIECK, A.: 'Résumé de la théorie métrique des produits tensoriels topologiques', *Bol. Soc. Mat. Sao-Paulo* **8** (1956), 1-79.

[A2] GROTHENDIECK, A.: *Topological vector spaces*, Gordon & Breach, 1973 (translated from the French).

[A3] JARCHOW, H.: *Locally convex spaces*, Teubner, 1981.

[A4] PISIER, G.: *Factorization of linear operators and geometry of Banach spaces*, Amer. Math. Soc., 1986.

[A5] PISIER, G.: 'Counterexamples to a conjecture of Grothendieck', *Acta. Math.* **151** (1983), 181-208.

[A6] COLOMBEAU, J.F.: *Differential calculus and holomorphy*, North-Holland, 1982.

[A7] KÖTHE, G.: *Topological vector spaces*, I, Springer, 1969.

AMS 1980 Subject Classification: 46A12

NUCLEOLUS OF A GAME - See **Core in the theory of games.**

AMS 1980 Subject Classification: 90D12

NUISANCE PARAMETER - Any unknown parameter of a **probability distribution** in a statistical problem connected with the study of other parameters of a given distribution. More precisely, for a realization of a random variable X, taking values in a sample space $(\mathfrak{X}, \mathfrak{B}, P_\theta)$, $\theta = (\theta_1, \ldots, \theta_n)$, $\theta \in \mathbf{R}^n$, suppose it is necessary to make a statistical inference about the parameters $\theta_1, \ldots, \theta_k$, $k < n$. Then $\theta_{k+1}, \ldots, \theta_n$ are nuisance parameters in the problem. For example, let $X_1, \ldots, X_n$ be independent random variables, subject to the normal law $\phi((x - \xi)/\sigma)$, with unknown parameters ξ and σ^2, and one wishes to test the hypothesis H_0: $\xi = \xi_0$, where ξ_0 is some fixed number. The unknown variance σ^2 is a nuisance parameter in the problem of testing H_0. Another important example of a problem with a nuisance parameter is the **Behrens — Fisher problem**. Naturally, for the solution of a statistical problem with nuisance parameters it is desirable to be able to make a statistical inference not depending on these parameters. In the theory of statistical hypothesis testing one often achieves this by narrowing the class of tests intended for testing a certain hypothesis H_0 in the presence of a nuisance parameter to a class of similar tests (cf. **Statistical test**).

References

[1] LINNIK, YU.V.: *Statistical problems with nuisance parameters*, Moscow, 1966 (in Russian).
M.S. Nikulin

Editorial comments.

References

[A1] LEHMANN, E.L.: *Testing statistical hypotheses*, Wiley, 1978.
[A2] LEHMANN, E.L.: *Theory of point estimation*, Wiley, 1983.

AMS 1980 Subject Classification: 62F99

NULL OBJECT OF A CATEGORY, *zero (object) of a category* - An object (usually denoted by 0) such that for every object X of the category the sets $H(X, 0)$ and $H(0, X)$ are singletons. The null object, if it exists in a given category, is uniquely determined up to isomorphism. In the category of sets with a distinguished point the null object is a singleton, in the category of groups it is the trivial group, in the category of modules it is the zero module, etc. Not every category contains a null object, but a null object can always be formally adjoined to any given category. Every category with a null object has null morphisms.

M.Sh. Tsalenko

Editorial comments. An object I of a category is called *initial* if there is just one morphism $I \to X$ for any X, and *terminal* (or *final*) if there is just one morphism $X \to I$ for any X. Thus a null object is one which is both initial and terminal. If an initial object exists in a given category, it is unique up to isomorphism, and similarly for terminal objects; but a category may have non-isomorphic initial and terminal objects. For example, in the category of sets, the empty set is an initial object and any singleton is terminal. A terminal object of a category may be regarded as a *limit* for the empty diagram in that category (cf. the editorial comments to **Limit** for the concept of a limit of a diagram in a category). Conversely, a limit of an arbitrary diagram may be defined as a terminal object in an appropriate category of cones.

AMS 1980 Subject Classification: 18A30

NUMBER - A fundamental concept in mathematics, which has taken shape in the course of a long historical development. The origin and formulation of this concept occurred simultaneously with the birth and development of mathematics. The practical activities of mankind on the one hand, and the internal demands of mathematics on the other, have determined the development of the concept of a number.

The need to count objects led to the origin of the notion of a **natural number**. All nations that have forms of writing had mastered the concept of a natural number and had developed some counting system. In its early stages, the origin and development of the concept of a number can be judged only from indirect data provided by linguistics and ethnography. Primitive man clearly had no need of counting skills to determine whether or not a given collection was complete.

Later, special names were given to definite objects or events that people often encountered. Thus, in the language of certain nations there were words for such concepts as 'three men', or 'three boats', but there was no abstract concept 'three'. In this way, probably, there arose comparatively short number series, used for the identification of individual people, individual boats, individual coconuts, etc. At these stage there was no such thing as an abstract number, and numbers were merely names.

In some primitive cultures, the necessity to communicate information about the numerical size of this or that collection has led to distinguishing certain standard collections, usually consisting of parts of the human body. In such a system of counting, each part of the body has a definite order and name. Whenever parts of the body are insufficient, bundles of sticks are used. The same purpose is served by pebbles, shells, notches on a tree or rock, lines on the ground, strings with knots, etc.

The next stage in the development of the notion of a number is connected with the transition to counting in groups; pairs, tens, dozens, etc. There arise so-called nodal numbers, and at the same time the concept of an arithmetic operation, which is reflected in the names of numbers. Definite counting methods take shape, special devices are used for counting, and numerical notations emerge. Numbers are separated from the objects being counted, and become abstract. Systems of representations of numbers begin to appear (cf. **Numbers, representations of**).

The process of formation of our modern representation system was exceptionally complicated. Only the final part of this process can be judged with definite authenticity. There are many known systems of representation. In Ancient Egypt there were several systems. In one of them there were special symbols for 1, 10, 100, 1000. Other numbers were represented by means of combinations of these symbols. The basic arithmetic operation in Ancient Egypt was addition. Well before 2000 B.C. the Babylonians used a base-60 representation system with the positional principle for writing numbers. They used only two symbols. The ancient Greeks used an alphabetic representation system, which was also used by the Slavs (cf. also **Slavic numerals**). In India, at the beginning of the new era (A.D.) there was a wide-spread oral positional decimal representation system, with several synonyms for zero (and other digits). A positional decimal representation system also arose there later. By the 8-th century A.D. this system had spread as far as the Middle East. The Europeans were introduced to it in the 12-th century.

The widening circle of objects to be counted, arising as a result of practical activities of people, and finally the inquisitiveness that is characteristic of mankind, gradually pushed back the limits of counting. The idea arose of the unbounded extension of the sequence of natural numbers, possibly attributable to the Greeks. One of Euclid's theorems states: 'There exist more than any given number of primes'. Also, Archimedes tried to convince his contemporaries that it is possible to describe a number greater than 'the number of grains of sand in the world'.

For the measurement of quantities, fractional numbers were necessary. Fractions were studied in Ancient Egypt and Babylon. Egyptian fractions were usually expressed in terms of aliquot fractions, i.e. fractions with numerator equal to 1. The Babylonians used base-60 fractions. The Chinese and the Indians were using ordinary fractions in the early centuries A.D., and were able to carry out all the arithmetic operations on them. Scholars in Central Asia, no later than the 10-th century, used a base-60 positional counting system. This system was particularly widely used in astronomical calculations and tables. Traces of it have been passed on to us in the form of the units used in the measurement of time and angles. Decimal fractions were introduced at the beginning of the 15-th century, and were widely used by the Samarkand mathematician Kashi (al'-Kashi). In Europe, decimal fractions became widespread following the publication of the book *de Thiende* (1585), written by S. Stevin. Before the intro-

duction of decimal fractions, the Europeans had used the decimal system in practice to calculate the integer part of a number, but they used base-60 fractions or ordinary fractions for the fractional part.

The further development of the concept of number proceeded mainly along with the demands of mathematics itself. Negative numbers first appeared in Ancient China. Indian mathematicians discovered negative numbers while trying to formulate an algorithm for the solution of quadratic equations in all cases. Diophantus (3-rd century) operated freely with negative numbers. They appear constantly in intermediate calculations in many of the problems in his *Aritmetika*. In the 16-th and 17-th centuries, however, many European mathematicians did not appreciate negative numbers, and if such numbers occurred in their calculations they were referred to as false, or impossible. The situation changed in the 17-th century, when a geometric interpretation of positive and negative numbers was discovered, as oppositely-directed segments.

The Babylonians had an algorithm for calculating the square root of a number to any accuracy. In the 5-th century B.C., Greek mathematicians discovered that the side and diagonal of a square have no common measure. More generally, it turned out that two arbitrary, precisely-given segments are in general not commensurable. The Greek mathematicians did not start introducing new numbers. They avoided the above difficulty by creating a theory of ratios of segments that was independent of the concept of a number.

The development of algebra, and the techniques of approximate calculation, in connection with the demands of astronomy, led Arab mathematicians to extend the concept of a number. They began to consider ratios of arbitrary quantities, whether commensurable or not, as numbers. As Nasireddin (1201 - 1278) wrote: 'Each of these ratios can be called by a number, precisely equal to the number one when one term of the ratio agrees with the other term'. European mathematics was developing in the same direction. Although G. Cardano in *Practica Arithmeticae Generalis* (1539) was still writing about irrational numbers as 'surds' (from the Latin 'surdus', 'deaf'), and as 'impossible to perceive or to imagine', Stevin in his *l'Arithmétique* (1585) stated that 'a number is that which is determined by an arbitrary quantity' and that 'no numbers are absurd, irrational, irregular, inexpressible, or surd'. And finally I. Newton in his *Arithmeticae Universalis* (1707) gave the following definition: 'By a number we understand not so much a multiple of a unit as an abstract quantity associated in a systematic way to some other quantity of the same kind that is taken as a unit. Numbers arise in three forms: integer, fraction and irrational. An integer is that which can be measured by unity; a fraction is a multiple of a portion of unity; an irrational number is incommensurable with unity'. Related to the fact that, for Newton, a quantity can be either positive or negative, the numbers in his arithmetic can also be either positive, in other words 'greater than nothing', or negative, in other words 'smaller than nothing'.

Imaginary numbers first appeared in the work of Cardano, *Ars Magma* (The Great Art, 1545). In solving the system of equations $x+y=10$, $xy=40$, he found the solutions $5+\sqrt{-15}$ and $5-\sqrt{-15}$. Cardano called these solutions 'purely negative', and later 'sophisticatedly negative'. The first to see a 'real' use of introducing imaginary numbers was R. Bombelli. In his *Algebra* (1572) he showed that the real roots of the equation $x^3=px+q$, $p>0$, $q>0$, in the case $(p/3)^3>(q/2)^2$, can be expressed in terms of radicals of imaginary numbers. Bombelli defined arithmetic operations on such quantities, and proceeded in this way to the creation of the theory of complex numbers (cf. **Complex number**). In the 17-th and 18-th centuries, many mathematicians occupied themselves in investigating the properties of imaginary numbers and their applications. Thus, L. Euler extended, e.g., the notion of a logarithm to arbitrary complex numbers (1738), and obtained a new method of integration using complex variables (1776), while earlier (1736) A. de Moivre solved the problem of extracting roots of natural degrees of an arbitrary complex number. A successful application of the theory of complex numbers was the fundamental theorem of algebra: 'Every polynomial of degree greater than zero and with real coefficients factorizes as a product of polynomials of degrees one and two with real coefficients' (Euler, J. d'Alembert, C.F. Gauss). Nevertheless, until a geometric interpretation of complex numbers as points in the plane was given (around the end of the 18-th and in the beginning of the 19-th century), many mathematicians remained distrustful of imaginary numbers.

In the early 19-th century, in connection with the great successes of mathematical analysis, many scholars realized the need for a foundation of the fundamentals of analysis — the theory of limits. Mathematicians were no longer satisfied with proofs based on intuition or on geometrical representation. There also remained the problem of constructing a unified theory of numbers. Natural numbers were often thought of as collections of unity, fractions as ratios of quantities, real numbers as lengths of line segments, and complex numbers as points in the plane. There was no complete agreement as to how arithmetic operations on numbers should be introduced. Finally, the question naturally arose of the further development of the concept of a number. In particular, was it possible to introduce new numbers, related to points in space?

The 19-th century saw intensive research in all the above directions. A general principle was formulated according to which any generalization of the concept of number should proceed — the so-called principle of permanence of formal computing laws. According to this, when constructing a new number system extending a given system, the operations should generalize in such a way that the existing laws remain in force (G. Peacock, 1834; H. Hankel, 1867). In the second half of the 19-th century, the theory of real numbers (cf. **Real number**) was constructed almost simultaneously by G. Cantor (1879), Ch. Meray (1869), R. Dedekind (1872), and K. Weierstrass (1872). Here Cantor and Meray used fundamental sequences of rational numbers, Dedekind used cuts in the field of rational numbers, and Weierstrass used infinite decimal expansions.

As a result of the work of G. Peano (1891), Weierstrass (1878) and H. Grassmann (1861), an axiomatic theory of natural numbers was constructed. W. Hamilton (1837) constructed a theory of complex numbers from pairs of real numbers, Weierstrass constructed a theory of integers from pairs of natural numbers, and J. Tannery (1894) constructed a theory of rational numbers from pairs of integers.

Attempts to find generalizations of the concept of a complex number led to the theory of hypercomplex numbers (cf. **Hypercomplex number**). Historically, the first such number system was the quaternions (cf. **Quaternion**), discovered by Hamilton. After much investigation it became clear (Weierstrass, G. Frobenius, B. Pierce) that any extension of the concept of a complex number beyond the system of complex numbers itself is possible only at the cost of some of the usual properties of numbers.

Throughout the 19-th century, and into the early 20-th century, deep changes were taking place in mathematics. Conceptions about the objects and the aims of mathematics were changing. The axiomatic method of constructing mathematics on set-theoretic foundations was gradually taking shape. In this context, every mathematical theory is the study of some algebraic system. In other words, it is the study of a set with distinguished relations, in particular algebraic operations, satisfying some predetermined conditions, or axioms.

From this point of view every number system is an algebraic system. For the definition of concrete number systems it is convenient to use the notion of an 'extension of an algebraic system'. This notion makes precise in a natural way the principle of permanence of formal computing laws, which was formulated above. An algebraic system A' is called an *extension of an algebraic system A* if the underlying set of A is a subset of that of A', if there also exists a bijection from the set of relations of the system A' onto that of A, and if for any collection of elements of the system A for which some relation of that system holds, the corresponding relation of the system A' also holds.

For example, by the system of natural numbers one usually understands the algebraic system $N = <N, +, \cdot, 1>$ with two algebraic operations: addition $(+)$ and multiplication $(\cdot)$, and a distinguished element (1) (unity), satisfying the following axioms:

1) for each element $a \in N$, $a + 1 \neq a$;

2) associativity of addition: For any elements a, b, c in N,
$$(a+b)+c = a+(b+c);$$

3) commutativity of addition: For any elements a, b in N,
$$a+b = b+a;$$

4) cancellation of addition: For any elements a, b, c in N, the equation $a+c=b+c$ entails the equation $a=b$;

5) 1 is the neutral element for the multiplication; that is, for any $a \in N$ one has $a \cdot 1 = a$;

6) associativity of multiplication: For any elements a, b, c in N,
$$(a \cdot b) \cdot c = a \cdot (b \cdot c);$$

7) distributivity of multiplication over addition: For any elements a, b, c in N,
$$(a+b) \cdot c = a \cdot c + b \cdot c;$$

8) the axiom of induction: If M is a subset of N containing 1 and the element $a+1$ whenever it contains a, then $M=N$.

From 2, 3, 4, 6, and 7 it follows that the system of natural numbers is a **semi-ring** under the operations $(+)$ and $(\cdot)$. Hence the system of natural numbers can be defined as the minimal semi-ring with a neutral element for multiplication and without a neutral element for the addition.

The system of integers $Z = <Z, +, \cdot, 0, 1>$ is defined as the minimal **ring** that is an extension of the semi-ring $<N, +, \cdot, 1>$ of natural numbers. The system of rational numbers $Q = <Q, +, \cdot, 0, 1>$ is defined as the minimal **field** that is an extension of the ring Z. The system of complex numbers $C = <C, +, \cdot, 0, 1>$ is defined as the minimal field that is an extension of the field $<R, +, \cdot, 0, 1>$ of real numbers containing an element i for which $i^2 + 1 = 0$ (cf. also **Extension of a field**).

By the system $R = <R, +, \cdot, 0, 1, >>$ of real numbers one means the **algebraic system** with two binary operations $(+)$ and $(\cdot)$, two distinguished elements (0) and (1) and binary order relation $(>)$. The axioms of R are divided into the following groups:

1) the *field axiom*: The system $<R, +, \cdot, 0, 1>$ is a field;

2) the *order axiom*: The system $<R, +, \cdot, 0, 1, >>$ is a totally and strictly ordered field (cf. **Ordered field**);

3) the *Archimedean axiom*: For any elements $a>0$,

$b>0$ in R there exists a natural number n such that
$$n \cdot a = a + \cdots + a > b \quad (n \text{ terms } a);$$

4) the *completeness axiom*: Every **fundamental sequence** $\{r_n\}_n$ of real numbers converges, i.e. if for any $\epsilon > 0$ there is a number n_ϵ such that, for any $n > n_\epsilon$ and $m > n_\epsilon$ the inequality $|r_n - r_m| < \epsilon$ holds, then the sequence $\{r_n\}_n$ converges to some element of R.

Briefly, the system of real numbers is a complete, totally, strictly-Archimedean ordered field. The system of real numbers can also be defined, in an equivalent way, as a continuous totally ordered field. In this case the Archimedean axiom and the completeness axiom are replaced by the continuity axiom:

If A and B are non-empty subsets of R such that, for any elements $a \in A$, $b \in B$, the inequality $a < b$ holds, then there exists an element $c \in R$ such that $a \leqslant c \leqslant b$ for all $a \in A$, $b \in B$.

The construction of real numbers proposed by Cantor and Meray can be used to interpret the first system of axioms for the system of real numbers, while Dedekind's construction can be used to interpret the second system. Analogously, the constructions of Hamilton, Weierstrass and Tannery are interpretations of the systems of axioms for the complex, integer and rational numbers.

As interpretations of the system of natural numbers one may use the ordinal theory of natural numbers developed by Peano, and the cardinal theory of natural numbers of Cantor.

The problem of the foundations of the concept of a number, and more broadly, the foundations of mathematics, were clearly set out in the 19-th century. This problem became a subject of mathematical logic, the intensive development of which continued into the 20-th century.

Cf. also **Arithmetic, formal**; **Constructive analysis**; **p-adic number**; **Algebraic number**; **Transcendental number**; **Cardinal number**; **Ordinal number**; **Arithmetic.**

References

[1] BEREZKINA, E.I.: *Mathematics of Ancient China*, Moscow, 1980 (in Russian).
[2] BOURBAKI, N.: *Eléments d'histoire des mathématiques*, Hermann, 1960.
[3] VAĬMAN, A.A.: *Sumero-Babylonian mathematics*, Moscow, 1961 (in Russian).
[4] WAERDEN, B.L. VAN DER: *Ontwakende wetenschap*, Noordhoff, 1957.
[5] WIELEITNER, G.: *Geschichte der Mathematik*, 2, de Gruyter, 1923.
[6] VOLODARSKIĬ, A.I.: *An outline of the history of Medieval Indian mathematics*, Moscow, 1977 (in Russian).
[7] VYGODSKIĬ, M.YA.: *Arithmetic and algebra in the Ancient world*, Moscow, 1967 (in Russian).
[8] DEPMAN, I.YA.: *The history of arithmetic*, Moscow, 1959 (in Russian).
[9] KOL'MAN, E.: *History of mathematics in Antiquity*, Moscow, 1961 (in Russian).
[10] CAJORI, F.: *A history of elementary mathematics*, Macmillan, 1896.

[11] NECHAEV, V.I.: *Number systems*, Moscow, 1975 (in Russian).
[12] RYBNIKOV, K.A.: *History of mathematics*, Moscow, 1974 (in Russian).
[13] STRUIK, D.J.: *A concise history of mathematics*, 1-2, Dover, reprint, 1948 (translated from the Dutch).
[14] FEFERMAN, S.: *The number system*, Addison-Wesley, 1964.
[15] ZEUTHEN, H.G.: *Geschichte der Mathematik in XVI und XVII Jahrhundert*, Teubner, 1903.
[16] YUSHKEVICH, A.P.: *Geschichte der Mathematik im Mittelalter*, Teubner, 1964 (translated from the Russian).
[17] *History of mathematics*, 1-3, Moscow, 1970-1972 (in Russian).
[18] *Mathematics of the 19-th century. Mathematical logic. Algebra. Number theory. Probability theory*, Moscow, 1978 (in Russian).
[19] LANDAU, E.: *Grundlagen der Analysis*, Akad. Verlagsgesellschaft, 1930.

V.I. Nechaev

Editorial comments. One version of Newton's definition of numbers in his Arithmeticae Universalis, as loosely translated above, can be found at the beginning of Caput III in the 1761 Amsterdam edition of this work.

References
[A1] GERICKE, H.: *Geschichte des Zahlbegriffs*, B.I. Mannheim, 1970.
[A2] SCRIBA, C.J.: *The concept of number, a chapter in the history of mathematics*, B.I. Mannheim, 1968.

AMS 1980 Subject Classification: 00A25, 10-XX

NUMBER FIELD - A **field** consisting of complex (e.g., real) numbers. A set of complex numbers forms a number field if and only if it contains more than one element and with any two elements α and β their difference $\alpha - \beta$ and quotient α/β ($\beta \neq 0$). Every number field contains infinitely many elements. The field of rational numbers is contained in every number field.

Examples of number fields are the fields of rational numbers, real numbers, complex numbers, or Gaussian numbers (cf. **Gauss number**). The set of all numbers of the form $H(\alpha)/F(\alpha)$, $F(\alpha) \neq 0$, forms a number field, $Q(\alpha)$, where α is a fixed complex number and $H(x)$ and $F(x)$ range over the polynomials with rational coefficients.

A.B. Shidlovskiĭ

Editorial comments. An *algebraic number field* K of degree n is an extension of degree n of the field **Q** of rational numbers. Alternatively, a number field K is an algebraic number field (of degree n) if every $\alpha \in K$ is the root of a polynomial (of degree at most n) over **Q**. A number field that is not algebraic is called *transcendental*. (Cf. also **Algebraic number theory; Extension of a field; Transcendental extension.**)

References
[A1] WEISS, E.: *Algebraic number theory*, McGraw-Hill, 1963.

AMS 1980 Subject Classification: 12AXX

NUMBER OF DIVISORS - A function of a natural argument, n, equal to the number of natural divisors of the number n. This arithmetic function is denoted by $\tau(n)$ or $d(n)$. The following formula holds:

$$\tau(n) = (\alpha_1 + 1) \cdots (\alpha_k + 1),$$

where

$$n = p_1^{\alpha_1} \cdots p_k^{\alpha_k}$$

is the canonical expansion of n into prime factors. For prime numbers p, $\tau(p) = 2$, but there exists an infinite sequence of n for which

$$\tau(n) > 2^{(1-\epsilon)\frac{\ln n}{\ln \ln n}}, \quad \epsilon > 0.$$

On the other hand, for all $\epsilon > 0$,

$$\tau(n) = O(n^{\epsilon}).$$

$\tau(n)$ is a **multiplicative arithmetic function** and is equal to the number of points with natural coordinates on the hyperbola $xy = n$. The average value of $\tau(n)$ is given by Dirichlet's asymptotic formula (cf. **Divisor problems**). The function $\tau_k(n)$, which is the number of solutions of the equation $n = x_1 \cdots x_k$ in natural numbers $x_1, \ldots, x_k$, is a generalization of the function $\tau(n)$.

References
[1] VINOGRADOV, I.M.: *Elements of number theory*, Dover, reprint, 1954 (translated from the Russian).
[2] PRACHAR, K.: *Primzahlverteilung*, Springer, 1957.

N.I. Klimov

Editorial comments.

References
[A1] HARDY, G.H. and WRIGHT, E.M.: *An introduction to the theory of numbers*, Oxford Univ. Press, 1979, Chapt. XVI.

AMS 1980 Subject Classification: 10A20

NUMBER-THEORETIC FUNCTIONS - The same as arithmetic functions (cf. **Arithmetic function**).

AMS 1980 Subject Classification: 10A20

NUMBER THEORY - The science of integers (cf. also **Integer**). Integers, together with the simplest geometrical figures, were the first and the most ancient mathematical concepts. Number theory arose from problems in **arithmetic** connected with the multiplication and division of integers.

In Ancient Greece (6-th century B.C.) divisibility of integers was studied, and particular subclasses of integers (such as prime numbers, cf. **Prime number**, composite numbers, squares) were distinguished; the structure of perfect numbers (cf. **Perfect number**) was studied; and the solution in integers of the equation $x^2 + y^2 = z^2$ was given, that is, an algorithm was described to construct right-angled triangles with sides of integer lengths. Euclid (3-rd century B.C.) in his *Elements* gave a systematic development of the theory of divisibility, based on the so-called **Euclidean algorithm** for finding the **greatest common divisor** of two integers, and proved the first theorem in the theory of prime numbers: There are infinitely many prime numbers. Somewhat later, Eratosthenes discovered a method of obtaining prime numbers, which is still called the sieve of Eratosthenes (cf. **Eratosthenes, sieve of**).

A systematization of the problems in number theory, and methods for their solution, was carried out by Diophantus (3-rd century) in his *Aritmetika*, where, in particular, he gave solutions in rational numbers of many algebraic equations of the first and second degrees in several unknowns.

In China, from the 2-nd century onwards, in connection with calendar computations, the problem arose of finding the least integer that yields given remainders on division by given numbers. This was solved by Sun-tsi (2-nd to 6-th century) and by Tsin Tsiushao (13-th century) (cf. **Chinese remainder theorem**).

In India, Bramagupta (7-th century) and Bhaskara (12-th century) gave general methods for solving in integers indefinite equations of the first degree in two unknowns, as well as of equations of the form $ax^2 + b = cy^2$ and $xy = ax + by + c$.

The golden age of number theory in Europe began with the work of P. Fermat (17-th century). He investigated the solutions of many equations in integers, and in particular conjectured that the equation $x^n + y^n = z^n$, $n > 2$, has no solution in natural numbers x, y, z (**Fermat great theorem**); he proved that a prime number of the form $4n + 1$ is a sum of two squares, and asserted one of the basic results of the theory of congruences: $a^p - a$ is divisible by p if p is a prime number (**Fermat little theorem**).

L. Euler made an exceptionally important contribution to number theory. He proved Fermat's great theorem for $n = 3$, Fermat's little theorem and a generalization of it, and a whole series of theorems about representations of numbers by binary quadratic forms (cf. **Binary quadratic form**). Euler was the first to approach solutions to problems in number theory by means of mathematical analysis, which led to the creation of **analytic number theory**. Euler's generating functions (cf. **Generating function**) appeared as the source of the **circle method** of Hardy−Littlewood−Ramanujan and the method of trigonometric sums (cf. **Trigonometric sums, method of**) of I.M. Vinogradov — the basic methods of modern additive number theory, while the function $\zeta(s)$, introduced by Euler, and its generalizations, are the basis of the contemporary analytic methods used to investigate the problem of the **distribution of prime numbers**.

In the correspondence between Ch. Goldbach and Euler, three famous problems were posed: Whether every odd $N \geqslant 5$ is a sum of three primes; whether every even $N \geqslant 4$ is a sum of two primes; and whether every odd N has the form $N = p + 2n^2$, where p is a prime number and n an integer. The first problem was

solved by Vinogradov (1937), and the other two remain unsolved (1989).

C.F. Gauss created the basic methods in and completed the construction of the theory of congruences (cf. **Congruence**), proved the **quadratic reciprocity law** (cf. also **Gauss reciprocity law**; **Reciprocity laws**), which had been formulated by Euler, laid the foundations of the theory of representations of numbers by quadratic forms $ax^2 + bxy + cy^2$ and by forms of higher degrees in several variables, introduced the so-called Gauss sums (cf. **Gauss sum**)

$$S = \sum_{n=0}^{m-1} \exp\left\{2\pi i \frac{an^2}{m}\right\},$$

which were the first trigonometric sums, and showed that they were useful in solving problems in number theory. While number theory before the time of Gauss had been merely a collection of different results and ideas, after his work it began to develop in several directions as a harmonized theory.

Great contributions to number theory have been made by many scholars in the 19-th and 20-th centuries.

One of the peculiarities and attractions of number theory is the simplicity and accessibility of the majority of its problems, and the difficulty of solving them. For example, the problem of whether there are infinitely many prime **twins** was posed by Euclid and is still unsolved (1989). Individual problems in number theory have become sources of important independent branches of mathematics. Among these are: the theory of prime numbers (cf. **Prime number**) and the related theory of the **zeta-function** and **Dirichlet series**, the theory of **Diophantine equations**; **additive number theory**, the **metric theory of numbers**, the theory of algebraic and transcendental numbers, **algebraic number theory**, the theory of **Diophantine approximations**, probabilistic number theory (cf. **Number theory, probabilistic methods in**), and the **geometry of numbers**. For example, a source of analytic number theory was the problem of the distribution of primes in series of natural numbers and the problem of representing natural numbers as sums of terms of a particular form. Solving equations in integers, in particular Fermat's great theorem, was the source of algebraic number theory. The problem of constructing a disc of unit area by means of a straightedge and a compass (cf. **Quadrature of the circle**) led to questions about the arithmetic nature of the number π, and hence to the creation of the theory of algebraic and transcendental numbers.

All the above branches of number theory are interconnected, complementing and enriching one another.

References

[1]   VINOGRADOV, I.M.: *Elements of number theory*, Dover,

1954 (translated from the Russian).

[2] VINOGRADOV, I.M.: *The method of trigonometric sums in the theory of numbers*, Interscience, 1954 (translated from the Russian).

[3] VINOGRADOV, I.M.: *Particular variants of the method of trigonometric sums*, Moscow, 1976 (in Russian).

[4] KARATSUBA, A.A.: *Fundamentals of analytic number theory*, Moscow, 1983 (in Russian).

[5] BOREVICH, Z.I. and SHAFAREVICH, I.R.: *Number theory*, Acad. Press, 1966 (translated from the Russian).

[6] DAVENPORT, H.: *Multiplicative number theory*, Springer, 1980.

[7] CHANDRASEKHARAN, K.: *Introduction to analytic number theory*, Springer, 1968.

[8] HASSE, H.: *Zahlentheorie*, Akademie-Verlag, 1963.

[9] DIRICHLET, P.G.L.: *Vorlesungen über Zahlentheorie*, Vieweg, 1894.

[10] TITCHMARSH, E.C.: *The theory of the Riemann zeta-function*, Clarendon, 1951.

[11] VENKOV, B.A.: *Elementary number theory*, Wolters-Noordhoff, 1970 (translated from the Russian).

A.A. Karatsuba

Material from the article of the same name in BSE-2

Editorial comments.

References

[A1] HARDY, G.H. and WRIGHT, E.M.: *An introduction to the theory of numbers*, Oxford Univ. Press, 1979.

[A2] LEVEQUE, W.J.: *Topics in number theory*, 1, Addison-Wesley, 1965.

[A3] SHANKS, D.: *Solved and unsolved problems in number theory*, Chelsea, reprint, 1978.

[A4] WEIL, A.: *Number theory, an approach through history*, Birkhäuser, 1984.

AMS 1980 Subject Classification: 10-XX

NUMBER THEORY, PROBABILISTIC METHODS IN - In the broad sense, that part of **number theory** in which ideas and methods from **probability theory** are used.

By probabilistic number theory in the narrow sense one means the statistical theory of the distribution of values of an **arithmetic function**.

The great majority of arithmetic functions studied in number theory are either additive or multiplicative (cf. **Additive arithmetic function**; **Multiplicative arithmetic function**). Their values are usually distributed in a very complicated way. If one traces the change of values of such functions as the argument runs through the sequence of natural numbers, one obtains a highly chaotic graph such as is usually observed when the additive and multiplicative properties of numbers are considered simultaneously. In classical investigations concerning the distribution of the values of a real arithmetic function $f(m)$, one usually studied the asymptotic behaviour of $f(m)$ itself, or of its average values. In the first case one has to find two simple functions $\psi_1(m)$, $\psi_2(m)$ such that $\psi_1(m) \leqslant f(m) \leqslant \psi_2(m)$ for all m, or at least for all sufficiently large m. For example, if $\omega(m)$ denotes the number of distinct prime divisors of a number m, then $\omega(m) \geqslant 1$ for all $m > 1$, and

$$\omega(m) \leqslant 2(\ln \ln m)^{-1}\ln m \text{ for } m \geqslant m_0;$$

$$\lim_{m \to \infty} \inf \omega(m) = 1,$$

$$\lim_{m \to \infty} \sup \omega(m)(\ln m)^{-1}\ln \ln m = 1.$$

In the second case one considers the behaviour of

$$\frac{1}{n}\sum_{m=1}^{n} f(m). \tag{1}$$

For $\omega(m)$ the average value (1) is equal to $(1 + o(1))\ln \ln n$. A solution to the first and to the second problem in the general case gives little information about the function $f(m)$, or about its oscillations. A function may deviate significantly from its average value. Here it turns out that large deviations occur rather seldom. The problem arises of finding bounds within which the values of the function $f(m)$ fluctuate for an overwhelming majority of values of the argument. If $f(m)$ is a real additive arithmetic function, and if

$$A_n = \sum_{p \leqslant n}\frac{f(p)}{p}, \quad B_n^2 = \sum_{p^\alpha \leqslant n}\frac{f^2(p^\alpha)}{p^\alpha}, \tag{2}$$

where the sums are taken over all prime numbers $p \leqslant n$ and all powers of primes $p^\alpha \leqslant n$, respectively, then

$$\frac{1}{n}\sum_{m=1}^{n}(f(m) - A_n)^2 \leqslant B_n^2\left[\frac{3}{2} + \frac{c}{\ln n}\right],$$

where c is an absolute constant. Thus, for any $t > 0$ and for all $m \leqslant n$ with the exception of $< (3/2 + c/\ln n)nt^{-2}$ numbers, the inequality

$$|f(m) - A_n| < tB_n$$

holds (an analogue of the **law of large numbers** in probability theory). For the function $\omega(m)$ this inequality can be written in the form

$$|\omega(m) - \ln \ln n| < t\sqrt{\ln \ln n}.$$

Denote by $N_n(\cdots)$ the number of natural numbers $m \leqslant n$ that satisfy a condition to be described in between the brackets in place of the dots. If one wants to characterize more exactly the distribution of the values of a real arithmetic function $f(m)$, then one is forced to consider the asymptotic behaviour of the frequencies

$$\frac{1}{n}N_n(f(m) \in E) \tag{3}$$

as $n \to \infty$, where E is an arbitrary Borel set. Among the asymptotic laws for (3), the most interesting ones are of local or integral form.

Integral laws. One studies the asymptotic behaviour of the distribution function

$$F_n(C_n + D_n x) = \frac{1}{n}N_n(f(m) < C_n + D_n x),$$

as $n \to \infty$, for given C_n, D_n.

In the case of an additive arithmetic function one searches for conditions under which $F_n(C_n + D_n x)$ converges to some distribution function $F(x)$ at all its points of continuity. Here, if $F(x)$ is not degenerate and D_n must converge to a finite (distinct from 0) or infinite limit.

In the case of a finite limit, it is sufficient to consider only $F_n(C_n+x)$. The function $F_n(C_n+x)$ has a non-degenerate limit distribution as $n\to\infty$, for some C_n, if and only if $f(m)$ has the form

$$f(m) = a\ln m + g(m),$$

where a is a constant and the function $g(m)$ satisfies the conditions

$$\sum_{|g(p)|>1}\frac{1}{p} < \infty, \quad \sum_{|g(p)|<1}\frac{g^2(p)}{p} < \infty.$$

Here C_n must be equal to

$$C_n = a\ln n + \sum_{\substack{p\leqslant n\\|g(p)|<1}}\frac{g(p)}{p}+C+o(1),$$

where C is a constant. The choice of C_n is unique up to an additive term $C+o(1)$. The limit distribution is discrete when $\sum_{f(p)\neq 0}1/p<\infty$, and continuous otherwise.

In particular, $F_n(x)$ (the case $C_n\equiv0$) has a limit distribution if and only if the series

$$\sum_{|f(p)|>1}\frac{1}{p}, \quad \sum_{|f(p)|<1}\frac{f(p)}{p}, \quad \sum_{|f(p)|<1}\frac{f^2(p)}{p}$$

converge (an analogue of the three-series theorem in probability theory).

The case $D_n\to\infty$ has not been completely investigated. Some of the simpler results are presented below, when $C_n=A_n$ and $D_n=B_n$ are defined by formula (2).

If, for any fixed $\epsilon<0$,

$$B_n^{-2}\sum_{\substack{p^\alpha=n\\|f(p^\alpha)|>\epsilon B_n}}\frac{f^2(p^\alpha)}{p^\alpha} \to 0, \tag{4}$$

as $n\to\infty$ (an analogue of the Lindeberg condition, cf. **Lindeberg–Feller theorem**), then

$$F_n(A_n+B_x x) \to \frac{1}{\sqrt{2\pi}}\int_{-\infty}^{x}e^{-u^2/2}\,du = \Phi(x) \tag{5}$$

(the normal law). If (4) is satisfied, then B_n is a slowly-varying function of $\ln n$ in the sense of Karamata. Moreover, if B_n is such a function, then (4) is a necessary condition for (5).

Let B_n be a slowly-varying function in $\ln n$. Then $F_n(A_n+B_n x)$ converges to a limit distribution with variance 1 if and only if there exists a non-decreasing function $V(u)$, $-\infty<u<\infty$, such that $V(-\infty)=0$, $V(\infty)=1$, and such that as $n\to\infty$, for all u with the possible exception of $u=0$,

$$B_n^{-2}\sum_{\substack{p^\alpha\leqslant n\\f(p^\alpha)<uB_n}}\frac{f^2(p^\alpha)}{p^\alpha} \to V(u).$$

The characteristic function $\phi(t)$ of the limit law, if it exists at all, is given by the formula

$$\phi(t) = \exp\left[\int_{-\infty}^{\infty}(e^{itu}-1-itu)u^{-2}\,dV(u)\right].$$

One has studied the rate of convergence to the limit law. For example, if $f(m)$ is a strongly-additive function and if

$$\mu_n = B_n^{-1}\max_{p\leqslant n}|f(p)| \to 0$$

as $n\to\infty$, then

$$F_n(A_n+B_n x) = \Phi(x)+O(\mu_n),$$

uniformly in x.

There are similar results for multiplicative arithmetic functions.

Local laws. Here one studies the behaviour of the frequency $N_n(f(m)=c)/n$ as $n\to\infty$ for a fixed c. In the case of a real additive arithmetic function, this frequency always has a limit, which is distinct from zero only for a countable set of values of c. Let

$$\lambda_l(f) = \lim_{n\to\infty}\frac{1}{n}N_n(f(m)=c_l), \quad l=1,2,\ldots,$$

be the collection of non-zero limits, assuming that at least one such limit exists. Then

$$\sum_l\lambda_l(f) = 1$$

and

$$\sum_l\lambda_l(f)e^{itc_l} = \prod_p\left[1-\frac{1}{p}\right]\sum_{\alpha=0}^{\infty}p^{-\alpha}e^{itf(p^\alpha)}.$$

If $f(m)$ takes only integer values and if

$$\mu_k(f) = \lim_{n\to\infty}\frac{1}{n}N_n(f(m)=k),$$

then

$$\sum_k\mu_k(f) = 1$$

if and only if

$$\sum_{f(p)\neq 0}\frac{1}{p} < \infty.$$

One has studied the rate of convergence to $\mu_k(f)$. There exists an absolute constant C such that for all integers k and all integer-valued additive arithmetic functions $f(m)$ such that $f(p)=0$ for all primes p,

$$\left|\frac{1}{n}N_n(f(m)=k)-\mu_k(f)\right| \leqslant C n^{-1/2}.$$

One has also studied the asymptotic behaviour of the frequency $N_n(f(m)=k_n)/n$ when k_n increases with n.

References

[1] KAC, M.: *Statistical independence in probability, analysis and number theory*, Math. Assoc. Amer., 1963.

[2] KUBILIUS, I.P. [I.P. KUBILYUS]: *Probabilistic methods in the theory of numbers*, Amer. Math. Soc., 1978 (translated from the Russian).

[3] KUBILYUS, I.P.: *Current problems of analytic number theory*, Minsk, 1974 (in Russian).

[4] LINNIK, YU.V.: *The dispersion method in binary additive problems*, Amer. Math. Soc., 1963 (translated from the Russian).

[5] LINNIK, YU.V.: *Ergodic properties of algebraic fields*, Springer, 1968 (translated from the Russian).

[6] POSTNIKOV, A.G.: *Ergodic problems in the theory of congruences*

and of Diophantine approximations, Amer. Math. Soc., 1967 (translated from the Russian).

[7] ELLIOT, P.D.T.A.: *Probabilistic number theory*, 1-2, Springer, 1979-1980.

I.P. Kubilyus

Editorial comments. The *Erdös–Kac theorem* gives some more precise information on $\omega(n)$. Denote by $N(x, a, b)$ the number of integers in the interval $[3, x]$ for which the inequalities

$$a \leqslant \frac{\omega(n) - \ln\ln n}{\sqrt{\ln\ln n}} \leqslant b$$

hold. Then, as $x \to \infty$,

$$N(x, a, b) = (x + o(x)) \frac{1}{\sqrt{2\pi}} \int_a^b e^{-t^2/2} \, dt.$$

See [A1].

References

[A1] NAJKIEWICZ, W.: *Number theory*, World Scientific, 1983, pp. 251-259 (translated from the Polish).

AMS 1980 Subject Classification: 10KXX

NUMBERS, REPRESENTATIONS OF - The totality of ways of representing natural numbers. In any number system, certain symbols (words or signs) denote specific numbers, called *nodal numbers*, while the other numbers are (algorithmically) obtained by certain operations from the nodal numbers. Number systems vary in their choice of nodal numbers and in their methods of forming algorithmic numbers; as written notations of numerical symbols began to appear, so number systems began to vary in the character of their numerical signs and in the principles governing the form in which they were written.

For example, the ancient Babylonians used 1, 10 and 60 as nodal numbers; the Maoris (the initial inhabitants of New Zealand) used 1, 11, 11^2, 11^3. In the Roman number system the nodal numbers are 1, 5, 10, 50, 100, 500, 1000, represented respectively by the signs I, V, X, L, C, D, M (cf. **Roman numerals**).

Number systems in which the algorithmic numbers are formed by grouping nodal numbers together are called *additive systems*. Thus, in ancient Egyptian (hieroglyphic) notation, the numbers 1, 2, 3, 4, 5, 6, 7, 8, 9, 10, 19, 40 were represented, respectively, by the symbols

The same numbers in Roman numerals are written I, II, III, IV, V, VI, VII, VIII, IX, X, XIX, XL. In this number system, the algorithmic numbers are obtained by the addition and subtraction of nodal numbers. The English method of expressing numerals is a clear example of the additive-multiplicative method of forming algorithmic numbers, for example: three hundred fifty seven.

In certain number systems, called *alphabetical systems*, numbers are represented by the same symbols as letters, plus other signs, e.g. dashes. Thus, the ancient Greeks denoted the numbers 1 to 9, as well as all the tens and hundreds, by sequences of letters of the alphabet, combined with dashes. For example, the numbers 803, 833 and 83 were written thus:

$$\overline{\omega\gamma}, \quad \overline{\omega\lambda\gamma}, \quad \overline{\pi\gamma}.$$

Alphabetical representations of numbers were used by the Slavs and many other peoples (cf. **Slavic numerals**).

Number systems are called *non-positional* if every sign used in the notation of any number has only one value. If the value of each sign depends on its position within the notation, then the system is called *positional*. The Roman number system is non-positional. Any number in the Babylonian number system could be written as a combination of two signs: a vertical wedge and a wide-angle wedge (see the example below). These signs were formed into groups from one to nine in the case of the vertical wedges, and from one to five in the case of the wide-angle wedges. The vertical wedge could represent 10 and the product of 10 and any power of the number 60. The sequential order of the digit positions was the same as it is presently. Thus,

$$= 1 \cdot 60^2 + 2 \cdot 600 + 1 \cdot 60 + 1 \cdot 10 + 6 = 4876.$$

Since the Babylonian system had no sign corresponding to no sequence (our zero), there was no guarantee that the notation of a number could be read in only one way. The exact meaning of the notation could normally be established from the context. This type of number system is therefore called a non-absolute positional system. The ancient Babylonians did subsequently introduce a special sign corresponding to our zero. The modern decimal system is positional.

All known positional number systems are additive-multiplicative systems. The positional principle of notation of numbers in these systems is explained by the following theorem of elementary **number theory**.

Let $q_0 = 1$ and let $q_1, q_2, \ldots,$ be a sequence of natural numbers greater than one. Then for any natural number a there is one and only one natural number n for which the equation

$$a_0 + a_1 q_1 + a_2 q_1 q_2 + \cdots + a_{n-1} q_1 \cdots q_{n-1} = a \quad (1)$$

has a solution in integers $a_0, \ldots, a_{n-1}$, such that

$$0 \leqslant a_0 < q_1, \ldots, 0 \leqslant a_{n-1} < q_{n-1}, \quad (2)$$

$$0 < a_{n-1} < q_n.$$

Given this, only one ordered set (tuple)

$$<a_{n-1}, \ldots, a_0> \quad (3)$$

of integers with condition (2) satisfies condition (1).

In the Babylonian number system, $q_1 = 10$, $q_2 = 6$,

$q_3 = 10$, $q_6 = 6, \ldots$, etc. In the system of the Maya Indians $q_1 = 5$, $q_2 = 4$, $q_3 = 18$, $q_4 = q_5 = \cdots = 20$.

A number system in which all terms of the sequence $q_1, \ldots, q_n$ are equal to one and the same number q and in which every number from 0 to $q - 1$ is denoted by a specific symbol is called a *q-ic number system* or a *positional number system with basis q*. In a *q*-ic system, every natural number is denoted by a sequence of the symbols shown. In order to add and multiply the numbers in a *q*-ic system, it is sufficient to have addition and multiplication tables for all numbers from 0 to $q - 1$.

References

[1] BASHMAKOVA, I.G. and YUSHKEVICH, A.P.: 'The origin of number systems', in *Encyclopaedia of elementary mathematics*, Moscow-Leningrad, 1951, pp. 11-74 (in Russian).
[2] WAERDEN, B.L. VAN DER: *Ontwakende wetenschap*, Noordhoff, 1957.
[3] YUSHKEVICH, A.P.: *Geschichte der Mathematik im Mittelalter*, Teubner, 1964 (translated from the Russian).
[4] VAĬMAN, A.A.: *Sumero-Babylonian mathematics*, Moscow, 1961 (in Russian).
[5] *The history of mathematics*, 1, Moscow, 1970 (in Russian).

V.I. Nechaev

Editorial comments.

References

[A1] BURTON, D.M.: *The history of mathematics*, Allyn & Bacon, 1985.
[A2] CAJORI, F.: *A history of mathematical notations*, 1-2, Open Court, 1952-1974.
[A3] DANTZIG, T.: *Number, the language of science*, Allyn & Unwin, 1930.
[A4] MENNINGER, K.: *Number words and number symbols*, M.I.T., 1969 (translated from the German).
[A5] CUITEL, G.: *Histoire comparée des numerations écrits*, Paris, 1975.

AMS 1980 Subject Classification: 01-XX, 00A25,

NUMERATOR *of an arithmetic fraction a / b* - The integer a that shows how many times the portion $1 / b$ is used to make up the fraction. The *numerator of an algebraic fraction A / B* is the expression A (cf. **Fraction**).

S.A. Stepanov

AMS 1980 Subject Classification: 10AXX

NUSSELT NUMBER - A dimension-less parameter characterizing the intensity of convective heat exchange between the surface of a body and the flow of a gas (fluid): $Nu = \alpha l / \lambda$, where $\alpha = Q / S \Delta T$ is the coefficient of heat exchange, Q is the amount of heat radiated (or received) by the surface of the body in unit time, $\Delta T > 0$ is the difference between the temperature of the surface of the body and that of the gas (fluid) outside the boundary layer, S is the area of the surface, l is the characteristic length, and λ is the coefficient of heat conductivity of the gas (fluid).

It is named after W. Nusselt.

BSE-3

Editorial comments.

References

[A1] CONDON, E.U.: 'Heat transfer', in E.U. Condon and H. Odishaw (eds.): *Handbook of Physics*, McGraw-Hill, 1967, § 5.5.7-5.5.8.
[A2] ECKERT, E.R.G.: *Introduction to the transfer of heat and mass*, McGraw-Hill, 1950.

AMS 1980 Subject Classification: 80A20

NYQUIST CRITERION - A necessary and sufficient condition for the stability of a linear closed-loop system formulated in terms of properties of the open-loop system.

Consider the linear single input - linear single output system with the following transfer function:

$$W(p) = \frac{M(p)}{N(p)},$$

where it is assumed that the degree of the polynomial $M(z)$ does not exceed that of the polynomial $N(z)$ (i.e. $W(p)$ is a *proper rational function*). The original Nyquist criterion gives necessary and sufficient conditions for the stability of the closed-loop system with unity feedback $u = y$. This is done in terms of the complex-valued function $z = W(i\omega)$ of the real variable $\omega \in [0, \infty)$ (the *amplitude-phase characteristic* of the open-loop system) which describes a curve in the complex z-plane, known as the *Nyquist diagram*. Suppose that the characteristic polynomial $N(z)$ of the open-loop system has k, $0 \leq k \leq n$, roots with positive real part and $n - k$ roots with negative real part. The Nyquist criterion is as follows: The closed-loop system is stable if and only if the Nyquist diagram encircles the point $z = -1$ in the counter-clockwise sense $k / 2$ times. (An equivalent formulation is: The vector drawn from -1 to the point $W(i\omega)$ describes an angle πk in the positive sense as ω goes from 0 to $+\infty$.)

This criterion was first proposed by H. Nyquist [1] for feedback amplifiers; it is one of the frequency criteria for the stability of linear systems (similar, e.g., to the *Mikhaĭlov criterion*, see [2], [3]). It is important to note that if the equations of some of the elements of the systems are unknown, the Nyquist diagram can be constructed experimentally, by feeding a harmonic signal of variable frequency to the input of the open feedback [4].

Generalizations of this criterion have since been developed for multivariable, infinite-dimensional and sampled-data systems, e.g. [5], [6], [7], [9].

References

[1] NYQUIST, H.: 'Regeneration theory', *Bell System Techn. J.* **11**, no. 1 (1932), 126-147.
[2] BULGAKOV, B.V.: *Oscillations*, Moscow, 1954 (in Russian).
[3] LAVRENT'EV, M.A. and SHABAT, B.V.: *Methoden der komplexen Funktionentheorie*, Deutsch. Verlag Wissenschaft., 1967 (translated from the Russian).
[4] ROĬTENBERG, YA.N.: *Automatic control*, Moscow, 1978 (in Russian).
[5] GNOENSKIĬ, L.S., KAAENSKIĬ, G.A. and EL'SGOL'TS, L.E.:

Mathematical foundations of the theory of control systems, Moscow, 1969 (in Russian).

N.Kh. Rozov

Editorial comments. For generalizations of the Nyquist criterion in various directions, see [A1].

References

[A1] DESOER, C.A. and VIDYASAGAR, M.: *Feedback systems: input-output properties*, Acad. Press, 1975.

[A2] DESOER, C.A.: 'A general formulation of the Nyquist stability criterion', *IEEE Trans. Circuit Theory* **CT-12** (1965), 230-234.

[A3] DESOER, C.A. and WANG, Y.T.: 'On the generalized Nyquist stability criterion', *IEEE Trans. Autom. Control* **AC-25** (1980), 187-196.

[A4] CALLIER, F.M. and DESOER, C.A.: 'On simplifying a graphical stability criterion for linear distributed feedback systems', *IEEE Trans. Automat. Contr.* **AC-21** (1976), 128-129.

[A5] VALENCA, J.M.E. and HARRIS, C.J.: 'Nyquist criterion for input-output stability of multivariable systems', *Int. J. Control* **31** (1980), 917-935.

[A6] FAURRE, P. and DEPEYROT, M.: *Elements of system theory*, North-Holland, 1977.

AMS 1980 Subject Classification: 93D10, 93E25

ω^2**-DISTRIBUTION** - See **'Omega-squared' distribution.**

AMS 1980 Subject Classification: 60E05, 62E10

*O***-DIRECT UNION** of semi-groups with zero - The **semi-group** obtained from the given family $\{S_\alpha\}$ of semi-groups with zero, pairwise intersecting at this zero, by specifying on $\bigcup_\alpha S_\alpha$ the multiplication operation that coincides with the original operation on each semi-group S_α and is such that $S_\alpha S_\beta = 0$ for different α, β. The O-direct union is also called the *orthogonal sum*. A number of types of semi-groups can be described by decomposing them in an O-direct union of known semi-groups (cf., e.g., **Maximal ideal; Minimal ideal; Regular semi-group**).

References
[1] CLIFFORD, A.H. and PRESTON, G.B.: *The algebraic theory of semigroups*, 2, Amer. Math. Soc., 1967.

L.N. Shevrin

AMS 1980 Subject Classification: 20M10

OBJECT, GEOMETRIC - See **Geometric objects, theory of.**

AMS 1980 Subject Classification: 53A55, 53C99, 53C10

OBJECT IN A CATEGORY - A term used to denote elements of an arbitrary **category**, playing the role of sets, groups, topological spaces, etc. An object in a category is an undefined concept. Every category consists of elements of two classes, the *class of objects* and the *class of morphisms*. The class of objects of a category $\mathfrak{K}$ is usually denoted by Ob $\mathfrak{K}$. With any object A of $\mathfrak{K}$ there is associated a unique identity morphism 1_A, so that different identity morphisms correspond to different objects. Hence the concept of a category can be formally defined by means of morphisms alone. However, the term 'object in a category' is a linguistic convenience which is practically always used. The division of the elements of a category into objects and mor-

phisms is only meaningful within a fixed category, since the objects of one category can be the morphisms of another. Thanks to the presence of morphisms, interrelations can be defined between the objects of a category, allowing one to single out special classes of objects (cf. **Integral object of a category; Null object of a category; Small object; Projective object of a category; Injective object**; etc.).

M.Sh. Tsalenko

Editorial comments. Cf. also Generator of a category.

AMS 1980 Subject Classification: 18A05

OBJECT LANGUAGE - A language that is an object of study. In the formalization of a meaningful theory one distinguishes two languages. One is the language of the formalized theory, or the object language, given by the construction rules of expressions in the object language and by semantic rules, denoting what its expressions stand for or how they express a reasoning. The other is the language in which one formulates the syntactic and semantic rules mentioned above. This language is called a meta-language. Usually, a meta-language is not formalized. However, it can be formalized, and it then becomes an object language for whose study one needs a new meta-language.

References
[1] KLEENE, S.C.: *Introduction to metamathematics*, North-Holland, 1951.

V.N. Grishin

AMS 1980 Subject Classification: 03BXX

OBJECTIVE FUNCTION, *target function* - The name for a to be optimized criterion function in problems of **mathematical programming.**

Editorial comments. Cf. also Operations research.

AMS 1980 Subject Classification: 90CXX

OBLIQUE DERIVATIVE, *directional derivative* - A derivative of a function f defined in a neighbourhood of the points of some surface S, with respect to a direction l different from the direction of the **conormal** of some elliptic operator at the points of S. Oblique derivatives may figure in the boundary conditions of boundary value problems for second-order elliptic equations. The problem is then called a problem with oblique derivative. See **Differential equation, partial, oblique derivatives**.

If the direction field l on S has the form $l = (l_1, \ldots, l_n)$, where l_i are functions of the points $P \in S$ such that $\sum_{i=1}^{n} (l_i)^2 = 1$, then the oblique derivative of a function f with respect to l is

$$\frac{df}{dl} = \sum_{i=1}^{n} l_i(P) \frac{\partial f}{\partial x_i}, \quad P = (x_1, \ldots, x_n),$$

where $x_1, \ldots, x_n$ are Cartesian coordinates in the Euclidean space $\mathbf{R}^n$.

References

[1] MIRANDA, C.: *Partial differential equations of elliptic type*, Springer, 1970 (translated from the Italian).

A.I. Yanushauskas

AMS 1980 Subject Classification: 26B05, 35G15

OBSTRUCTION - A concept in homotopy theory: An invariant that equals zero if a (step in a) corresponding problem is solvable and is non-zero otherwise.

Let (X, A) be a pair of cellular spaces (cf. **Cellular space**) and let Y be a simply-connected (more generally, a homotopy-simple) topological space. Can one extend a given continuous mapping $g: A \to Y$ to a continuous mapping $f: X \to Y$? The extension can be attempted recursively, over successive skeletons X^n of X. Suppose one has constructed a mapping $f: X^n \bigcup A \to Y$ such that $f|_A = g$. For any oriented $(n+1)$-dimensional cell $e^{n+1} \to Y$ the mapping $f|_{\partial e^{n+1}}$ gives a mapping $S^n \to Y$ (where S^n is the n-dimensional unit sphere) and an element $\alpha_e \in \pi_n(Y)$ (it is here that one uses that Y is homotopy simple, which allows one to ignore the base point). This defines a cochain

$$c_f^{n+1} \in C^{n+1}(X; \pi_n(Y)), \quad c_f^{n+1}(e^{n+1}) = \alpha_e.$$

Since for $e^{n+1} \subset A$ one clearly has $c_f^{n+1}(e^{n+1}) = 0$, it follows that

$$c_f^{n+1} \in C^{n+1}(X, A; \pi_n(Y)).$$

Clearly $c_f^{n+1} = 0$ if and only if f can be extended to X^{n+1}, i.e. c_f^{n+1} is an *obstruction* to extending f to X^{n+1}.

The cochain $c_f^{n+1} \in C^{n+1}(X, A; \pi_n(Y))$ is a cocycle. The fact that $c_f^{n+1} \neq 0$ does not, in general, imply that g cannot be extended to X: It is possible that f cannot be extended to X^{n+1} because of an unsuccessful choice of an extension of g to X^n. It may turn out that, e.g., the mapping $f|_{X^{n-1} \bigcup A}$ can be extended to X^{n+1}, i.e. that extension is possible by skipping back one step. It can

be shown that the cohomology class

$$[c_f^{n+1}] \in H^{n+1}(X, A; \pi_n(Y))$$

is an obstruction to this, i.e. $[c_f^{n+1}] = 0$ if and only if there is a mapping $\tilde{f}: X^{n+1} \bigcup A \to Y$ such that $\tilde{f}|_{X^{n-1} \bigcup A} = f|_{X^{n-1} \bigcup A}$ (in particular, $\tilde{f}|_A = g$). The construction of difference chains and cochains is used in the proof of this statement (cf. **Difference cochain and chain**).

Since the problem of homotopy classification of mappings $X \to Y$ can be interpreted as an extension problem, obstruction theory is applicable also to the description of the set $[X, Y]$ of homotopy classes of mappings from X into Y. Let $I = [0, 1]$ and let $A = X \times \{0, 1\}$ be a subspace of $X \times I$. Then a pair of mappings $f_0, f_1: X \to Y$ is interpreted as a mapping $G: A \to Y$, $G(x, i) = f_i(x)$, $i = 0, 1$, and the presence of a homotopy between f_0 and f_1 means the presence of a mapping $F: X \times I \to Y$ extending G. If the homotopy F has been constructed on the n-dimensional skeleton of X, then the obstruction to its extension to X is the difference cochain

$$d^n(f_0, f_1) \in C^n(X; \pi_n(Y)).$$

As an application one may consider the description of the set $[X, Y] = [X, K(\pi, n)]$, $n > 1$, where $K(\pi, n)$ is the **Eilenberg–MacLane space**: $\pi_i(K(\pi, n)) = 0$ for $i \neq n$; $\pi_n(K(\pi, n)) = \pi$. Let $f_0: X \to K(\pi, n)$ be a constant mapping and $f: X \to K(\pi, n)$ an arbitrary continuous mapping. Since $H^i(X; \pi_i(Y)) = 0$ for $i < n$, the mappings f_0 and f are homotopic on X^{n-1} and, after having chosen such a homotopy, one can define the difference cochain

$$d^n(f, f_0) \in C^n(X; \pi_n(Y)) = C^n(X; \pi).$$

The cohomology class $[d^n(f, f_0)] \in H^n(X; \pi)$ is well-defined, i.e. does not depend on the choice of a homotopy between f_0 and f (since $\pi_i(Y) = 0$ for $i < n$). Further, if two mappings $f, g: X \to Y$ are such that $[d^n(f, f_0)] = [d^n(g, f_0)]$, then $[d^n(f, g)] = 0$, and hence f and g are homotopic on X^n. The obstructions to extending this homotopy to X lie in the groups $H^i(X; \pi_i(Y)) = 0$ (since $i > n$), and hence f and g are homotopic. Thus, the homotopy class of f is completely determined by the element $[d^n(f, f_0)] \in H^n(X; \pi)$. Finally, for any $x \in H^n(X; \pi)$ there is a mapping f with $[d^n(f, f_0)] = x$, hence $[X, K(\pi, n)] = H^n(X; \pi)$. Similarly, if $\pi_i(Y) = i$ for $i < n$ and if $\dim X \leq n$, then $[X, Y] = H^n(X; \pi_n(Y))$.

In studying extension problems one has considered the possibility of extending 'by skipping back one step'. A complete solution of the problem requires the analysis of the possibility of skipping back an arbitrary number of steps. Cohomology operations (cf. **Cohomology operation**) and Postnikov systems (cf. **Postnikov**

system) are used to this end. E.g., in order to describe the set $[X, Y]$, where $\pi_i(Y)=0$ for $i<n$, $\pi_n(Y)\neq 0$, $\dim X = n+r$, it is required, in general, to study the possibility of skipping back $r+1$ steps, for which it is necessary to study the first $n+r$ levels of the Postnikov system for Y, i.e. to use cohomology operations of orders $\leq r$ (in the article **Cohomology operation** this problem is outlined for $r=1$).

The theory of obstructions is also used in the more general situation of extension of sections (cf. **Section of a mapping**). Let $p : E \to B$ be a fibration with fibre F (where $\pi_1(F)=0$ and $\pi_1(B)$ acts trivially on $\pi_i(F)$), let $A \subset B$ and let $s : A \to E$ be a section (i.e. a continuous mapping such that $ps(a)=a$). Can one extend s to B? The corresponding obstructions lie in the groups $H^{n+1}(B ; \pi_n(F))$. An extension problem is obtained from this problem if one puts $B=X$, $E=X \times Y$, $p(x, y)=x$, $s(a)=(a, g(a))$. Analogously one can also study the classification problem for sections using obstruction theory.

Finally, one can remove the restriction of homotopic simplicity of the space Y in the extension problem (as well as in the problem on sections); then one must use cohomology with local coefficients.

Obstruction theory was initiated by S. Eilenberg [2]. It was also known to L.S. Pontryagin, who did not formulate it explicitly but used it for the solution of concrete problems, see [1].

A good discussion can be found in [3] and [4].

References

[1] PONTRYAGIN, L.S.: 'Classification of continuous transformations of a complex into a sphere', *Dokl. Akad. Nauk SSSR* **19** (1938), 361-363 (in Russian).
[2] EILENBERG, S.: 'Cohomology and continuous mappings', *Ann. of Math.* **41** (1940), 231-251.
[3] HU, S.-T.: *Homotopy theory*, Acad. Press, 1959.
[4] THOMAS, E.: *Seminar on fibre spaces*, Springer, 1966.

Yu.B. Rudyak

Editorial comments. The fundamental group $\pi_1(X, x_0)$ acts on the homotopy groups $\pi_n(X, x_0)$, $n \geq 1$, cf. **Homotopy group**. The space X is called *n-simple* if this action (for this n) is trivial; X is called *simple* or *homotopy simple* if it is path connected and *n*-simple for all $n \geq 1$. Then $\pi_1(X, x_0)$ is Abelian and acts trivially on all $\pi_n(X, x_0)$. A path-connected *H*-space is simple.

References

[A1] STEENROD, N.E.: *The topology of fibre bundles*, Princeton Univ. Press, 1951.
[A2] SPANIER, E.H.: *Algebraic topology*, McGraw-Hill, 1966.
[A3] WHITEHEAD, G.: *Elements of homotopy theory*, Springer, 1978.
[A4] BAUES, H.J.: *Obstruction theory*, Springer, 1977.

AMS 1980 Subject Classification: 55S35

OCEANOLOGY, MATHEMATICAL PROBLEMS IN, *mathematical problems in oceanography* - Mathematical problems in the fields of marine physics, chemistry, geology, and biology. In marine physics, the problems mainly concern geophysical hydrodynamics (defined as the hydrodynamics of natural currents of rotating baroclinic stratified liquids). The Earth's rotation, essentially affecting large-scale currents (on global and synoptic scales), and its stratification, i.e. the change in density of the medium in the direction of the force of gravity (vertical), create a specific anisotropy of the individual hydrodynamic fields in the sea or of their statistical characteristics, which must be considered, for example, when selecting base functions to describe these fields by the **Galerkin method** through objective analysis (interpolation, extrapolation, smoothing) of empirical data on these fields and when choosing statistical models for vertical-heterogeneous random fields of turbulence and **internal waves** (cf. also **Turbulence, mathematical problems in**).

An analytic description of the sea's eigen oscillations using linearized equations of hydrodynamics is complicated by the irregular shape of its boundaries, the seabed and the shore, which makes it impossible to use solutions with separated variables. Therefore, in the theory of tides, in which the sea can have a resonant response to tidal forcing, there are analytic calculations only for model-seas of regular shape (e.g. bounded by segments of meridians and parallels). In real geometry, successive (growing) eigen frequencies must be defined as the extrema of quadratic integral functionals, related to energy, on extremals which can be selected by the Galerkin method; this approach has not yet been fully realized. In the theory of tides, both the linear and non-linear response of the sea to tidal forcing is noted, and can be described by representing the height of the tide as a functional power series with respect to the tidal forcing; the functional coefficients of this series describe the sea's properties as a resonant system.

There is a specific difference between the solutions of the equations of hydrodynamics for several classes of waves: acoustic, surface (capillary and gravitational), internal gravitational, inertial (including barotropic and baroclinic Rossby—Blinova waves, formed as a result of the change with latitude of the vertical projection of the angular speed of the Earth's rotation), and, finally, hydromagnetic waves arising from movements of an electroconductive fluid (salt sea-water) in the geomagnetic field. Separate classes of wave-solutions (and of dynamic equations for them) are constructed using asymptotic methods of non-linear mechanics, related to the van der Pol method, in a multiple time scale analysis. An example of this is the quasi-geostrophysic series, filtering fast waves from solutions of equations of hydrodynamics and separating classes of Rossby—Blinova waves.

Waves in the sea, as a rule, are non-linear. For long non-linear waves, both surface and internal, the **Korteweg — de Vries equation** holds good and its periodic (cnoidal) solutions and solitons can be used. For short waves, no general methods for finding solitons and periodic solutions have yet been constructed, and only individual examples exist (capillary Slezkin — Crapper waves, gravitational Gerstner and Stokes waves, barotropic and baroclinic Rossby solitons). Statistical theories of non-linear wave fields are also insufficiently developed, particularly those relating to the description of surface and internal gravitational waves (internal waves generating turbulent spots and blurring the spots in the layers of the vertical microstructure) and Rossby waves (with evolution of quasi-two-dimensional turbulence in a non-linear wave field).

One of the most important problems of marine hydrodynamics is the mathematical modelling of the sea's circulation (in the most general formulation, in its interaction with the atmosphere through the so-called upper mixed layer of the sea and the boundary layer of the atmosphere), while as a consequence of the broad spectrum of the scales of spatial heterogeneity (from millimetres to 10^4 kilometres) the system to be modelled here has a huge number of degrees of freedom (for millimetric elementary volumes of the order of 10^{28}), which invariably need to be taken together, for example by the method of parametrization of small-scale processes.

When approximating continuous hydrodynamic equations by difference equations, questions arise relating to the order of the approximation and the convergence and stability of the difference scheme. In contemporary so-called eddy-resolving models of the sea's circulation, spatial grids with horizontal steps of tens of kilometres are used. Schemes using spectral (including Galerkin) representations of spatial hydrodynamic fields can be competitive.

The basic problems of the mathematical processing of measurement data in marine hydrodynamics are divided into problems of sounding (functions of depth, their decomposition into modes and spectra), towing (horizontal and space-time spectra with Doppler effects) and polygon measurements (time mutual spectra, objective analysis, synchronized spatial pictures and four-dimensional analysis of wave fields).

In marine acoustics, typical problems of wave distribution in a stratified environment are examined; for an analytic description of the vertical structure of wave fields in a number of cases, the WKB approximation is used (see **WKB method**). In marine optics, multiple light scattering is a specific process for the description of which asymptotic analytical solutions and numerical solutions of the radiation transfer equation (cf. also

Radiative transfer theory), obtained by Monte-Carlo methods, are used. In marine chemistry, the major mathematical problem is the calculation of convective diffusion of non-conservative mixtures with specific sources and sinks.

In marine geology problems have arisen in connection with the development of *plate tectonics* (tectonics of lithospheric plates), concerning kinematic calculations of movements of hard plates on the surface of a sphere and their genetic explanation using a mathematical modelling of processes of density convection on the Earth's surface (arising from the transfer of heavy substances from the mantel to the core). One of the important special problems of geology is biostratigraphy, i.e. recognition of the age of strata of sedimentary rocks by means of their content of micro-palaeontological assemblies using self-teaching programs (whereas in most of the work this problem is solved approximately, without the use of a computer).

Very substantial calculating problems in the estimation of statistical characteristics of signals, noise, filtration, and form recognition arise in registering and processing data of marine multi-channel continuous deep seismo-profiling and vibro-radioscopy of the seabed, while in a number of cases both in the formulation of measurement grids (spatial distributions of emitters and receivers of signals) and in their registration, holographic methods using Fourier transforms are promising.

In marine biology the great importance of the problem of controlling the sea's biological productivity is reflected in the mathematical modelling of the structure and functioning of ecosystems and, in particular, involves population dynamics. An example of this is the problem of evolution with time t of the vertical distributions $q_i(z, t)$ of the components q_i of an ecosystem (including the concentration of a number of forms of phyto- and zoo-plankton, oxygen, carbonate, phosophoric and nitric salts, the temperature and salinity of water, and the illumination by photosynthetically active radiation), described by equations in the form

$$\dot{q}_i = A_{i\alpha}q_\alpha + B_{i\alpha\beta}q_\alpha q_\beta + \frac{\partial}{\partial z}K(z)\frac{\partial q_i}{\partial z},$$

where $A_{i\alpha}$, $B_{i\alpha\beta}$ are biological and biophysical parameters. Of interest here are numerical solutions of the Cauchy problem with specific initial data and conclusions of the qualitative theory of differential equations on the behaviour of solutions as a whole and on their dependence on the parameters contained in the equations.

References

[1] *Marine biology*, 1-2, Moscow, 1977 (in Russian).
[2] *Marine physics*, 1, Moscow, 1978 (in Russian).
[3] *Marine geophysics*, 1-2, Moscow, 1979 (in Russian).

[4] *Marine chemistry*, 1-2, Moscow, 1979 (in Russian).
[5] *Marine geology*, 1, Moscow, 1979 (in Russian).

A.S. Monin

Editorial comments. A general introduction to hydrodynamical problems in oceanography is given in [A1] - [A4]. The modelling and numerical analysis of waves in the ocean is presented in [A5] - [A7].

In the study of physical and biological processes in the ocean, some topics recently got special attention from scientists working in different fields. First, the interaction between ocean and atmosphere appears to play an important role in climatological phenomena. A well-known example is 'El Niño', a name that is used for the occurrence of excessive rainfall in certain years caused by a sea-surface temperature anomaly, see [A8]. Secondly, the storage (or buffering) of carbonicacids by the oceans is also part of this interaction. In the balance of CO_2 in the air this has to be taken into consideration. Moreover, marine organisms may deposit a part of the carbonicacids in sediments at the ocean floor.

Sediments transport (sand and mud) near coasts and in estuaries are studied in relation with human activities at these locations [A9] - [A11]. The study of pollution at the outflow of rivers is analyzed in the same context. The series of which [A8] is a volume treats many aspects of the ocean: its physics, chemistry, biology, and geology.

References

[A1] PEDLOSKY, J.: *Geophysical fluid dynamics*, Springer, 1987.
[A2] GILL, A.E.: *Atmosphere-Ocean dynamics*, Acad. Press, 1982.
[A3] MARCHUK, G.I. and SARKISYAN, A.S.: *Mathematical modelling of ocean circulation*, Springer, 1986 (translated from the Russian).
[A4] TRITTON, D.J.: *Physical fluid dynamics*, v. Nostrand Reinhold, 1988.
[A5] LE BOND, P.H. and MYSAK, L.A.: *Waves in the ocean*, Oceanography Series, Elsevier, 1978.
[A6] PUGH, D.T.: *Tides, surges and mean sea-level*, Wiley, 1987.
[A7] MADER, C.L.: *Numerical modelling of water waves*, Univ. California Press, 1988.
[A8] NIHOUL, J.C.J. (ED.): *Coupled ocean-atmosphere models*, Oceanography Series, Elsevier, 1985.
[A9] DRONKERS, J. and LEUSSEN, W. (EDS.) VAN: *Physical processes in estuaries*, Springer, 1988.
[A10] MASSEL, S.R.: *Hydrodynamics of coastal zones*, Elsevier, 1989.
[A11] NOYE, J. (ED.): *Numerical modelling: Applications to marine systems*, North-Holland, 1987.
[A12] MONIN, A.S. and OZMIDOV, R.V.: *Turbulence in the ocean*, Reidel, 1985.
[A13] CSANADY, G.T.: *Circulation in the coastal ocean*, Reidel, 1982.
[A14] MCLELLAN, H.J.: *Elements of physical oceanography*, Pergamon.

AMS 1980 Subject Classification: 86A05

OCTAHEDRAL SPACE - A space obtained from an **octahedron** by identifying its opposite triangular faces, positioned at an angle of $\pi/3$ to each other. An octahedral space is a **three-dimensional manifold** and is the orbit space of the action of a binary octahedron group on a three-dimensional sphere. It can be identified with a *cube space* obtained in an analogous way. The one-dimensional Betti group of an octahedral space is a group of order three.

M.I. Voĭtsekhovskiĭ

AMS 1980 Subject Classification: 57N10

OCTAHEDRON - A solid figure having eight triangular faces, twelve edges and six vertices, with 4 faces at each vertex. If all edges have the same length, it is one of the five *regular polyhedra* (*Platonic solids*); if the edge length is a, then the volume of the octahedron is

$$v = \frac{a^3 \sqrt{2}}{3} \approx 0.4714 a^3.$$

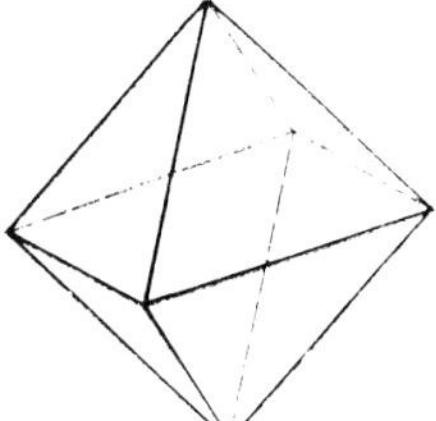

BSE-3

Editorial comments. The Schläfli symbol of an octahedron is $\{3, 4\}$. When the edges all have the same length one deals with the *regular octahedron*, reciprocal to the cube; it can be regarded either as a triangular anti-prism or as a square double-pyramid. As one of the five Platonic polyhedra (cf. **Regular polyhedra**; **Platonic solids**) it represents the ancient element *air*. It occurs in nature as a crystal of chrome alum. In terms of its circumradius as unit of measurement, its six vertices have Cartesian coordinates

$$(\pm 1, 0, 0), \quad (0, \pm 1, 0), \quad (0, 0, \pm 1);$$

thus it has edge-length $\sqrt{2}$, inradius $\sqrt{1/3}$ and volume $4/3$. Its 4 pairs of opposite faces (or the 4 diameters of the cube) are freely permuted by the *octahedral group* $\mathfrak{S}_4$ of order $4! = 24$.

References

[A1] BENNETT, G.T.: 'Deformable octahedra', *Proc. London Math. Soc. (2)* 10 (1912), 309-343.
[A2] COXETER, H.S.M.: *Regular polytopes*, Methuen, 1948, p. 5.

AMS 1980 Subject Classification: 51M20, 52A25

OCTANT - Any of the eight regions into which the 3-dimensional space is divided by three mutually-perpendicular coordinate planes.

AMS 1980 Subject Classification: 51N20

ODD FUNCTION - A function f that changes sign when the independent variable changes sign, i.e. $f(-x) = -f(x)$ for all values in the domain of f. The graph of an odd function is symmetric about the coordinate origin.

Editorial comments. See also **Even function**.

AMS 1980 Subject Classification: 26A09

ODD NUMBER - An integer that is not divisible by 2 (without remainder).

Editorial comments. An integer that is not odd is called even (cf. **Even number**).

AMS 1980 Subject Classification: 10A99

OKA THEOREMS - Theorems on the classical problems in the theory of functions of several complex variables, first proved by K. Oka between 1930 and 1950 (see [1]).

1) Oka's theorem on the **Cousin problems**: The first Cousin problem is solvable in any **domain of holomorphy** in $\mathbf{C}^n$; the second Cousin problem is solvable in any domain of holomorphy $D \subset \mathbf{C}^n$ that is homeomorphic to $D_1 \times \cdots \times D_n$, where all domains $D_\nu \subset \mathbf{C}$, except for, possibly, one, are simply connected.

2) Oka's theorem on the **Levi problem**: Any pseudo-convex Riemannian domain (cf. **Pseudo-convex and pseudo-concave**) is a domain of holomorphy.

Originally Oka proved these theorems in dimension $n = 2$; in the case of arbitrary dimension, the theorems were also proved by other mathematicians.

3) The *Oka−Weil theorem*: Let D be a domain in $\mathbf{C}^n$ and let the compact set $K \subset D$ coincide with its hull with respect to the algebra $\mathcal{O}(D)$ of all functions holomorphic in D (cf. **Holomorphic envelope**); then for any function f holomorphic in a neighbourhood of K, and for any $\epsilon > 0$, a function $F \in \mathcal{O}(D)$ can be found such that

$$\max_K |f - F| < \epsilon.$$

This fundamental theorem in the theory of holomorphic approximation is extensively used in complex and functional analysis.

4) *Oka's coherence theorem*: Let $\mathcal{O}$ be a sheaf of holomorphic functions on a complex manifold X; then for any natural number p, any locally finitely-generated subsheaf of the sheaf $\mathcal{O}^p = \mathcal{O} \times \cdots \times \mathcal{O}$ (p times) is a coherent **analytic sheaf** (cf. **Coherent sheaf**).

This is one of the basic theorems of the so-called Oka−Cartan theory, which is used essentially in proving the Cartan theorems A and B (cf. **Cartan theorem**).

References

[1] OKA, K.: *Sur les fonctions analytiques de plusieurs variables*, I. Shoten, 1961.
[2] HÖRMANDER, L.: *An introduction to complex analysis in several variables*, North-Holland, 1973.
[3] GUNNING, R. and ROSSI, H.: *Analytic functions of several complex variables*, Prentice-Hall, 1965.

E.M. Chirka

Editorial comments. Reference [A2] is an annotated English translation of Oka's fundamental papers.

References

[A1] KRANTZ, S.G.: *Function theory of several complex variables*, Wiley, 1982.
[A2] OKA, K.: *Collected papers*, Springer, 1984.
[A3] RANGE, R.M.: *Holomorphic functions and integral representations in several complex variables*, Springer, 1986.

AMS 1980 Subject Classification: 32C35, 32EXX

OMEGA-COMPLETENESS, ω-*completeness* - The property of formal systems of arithmetic in which, for any formula $A(x)$, from a deduction of $A(0), \ldots, A(\bar{n}), \ldots$, it follows that one can infer the formula $\forall x \, A(x)$, where $\bar{n}$ is a constant signifying the natural number n. If this is not true, the system is called ω-*incomplete*. K. Gödel in his incompleteness theorem (cf. **Gödel incompleteness theorem**) actually established the ω-incompleteness of formal arithmetic. If all formulas which are true in the standard model of arithmetic are taken as axioms, then an ω-complete axiom system is obtained. On the other hand, in every ω-complete extension of Peano arithmetic, every formula which is true in the standard model can be deduced.

References

[1] KLEENE, S.C.: *Introduction to metamathematics*, North-Holland, 1951.

V.N. Grishin

AMS 1980 Subject Classification: 03F25

OMEGA-CONSISTENCY, ω-*consistency* - The property of formal systems of arithmetic signifying the impossibility of obtaining ω-inconsistency. ω-*inconsistency* is a situation in which, for some formula $A(x)$, each formula of the infinite sequence $A(0), \ldots, A(\bar{n}), \ldots$, and the formula $\neg \forall x \, A(x)$ are provable, where $\bar{0}$ is a constant of the formal system signifying the number 0, while the constants $\bar{n}$ are defined recursively in terms of $(x)'$, signifying the number following directly after x: $\overline{n+1} = (\bar{n})'$.

The concept of ω-consistency appeared in conjunction with the **Godel incompleteness theorem** of arithmetic. Assuming the ω-consistency of formal arithmetic, K. Gödel proved its incompleteness. The property of ω-consistency is stronger than the property of simple **consistency**. Simple consistency occurs if a formula not involving x is taken as $A(x)$. It follows from Gödel's incompleteness theorem that there exist systems which are consistent but also ω-inconsistent.

References

[1] KLEENE, S.C.: *Introduction to metamathematics*, North-Holland, 1951.

V.N. Grishin

AMS 1980 Subject Classification: 03F25

'OMEGA-SQUARED' DISTRIBUTION, ω^2-*distribution* - The **probability distribution** of the random variable

$$\omega^2 = \int_0^1 Z^2(t) \, dt,$$

where $Z(t)$ is a conditional **Wiener process** (conditioned on $Z(1)=0$). The characteristic function of the 'omega-squared' distribution is expressed by the formula

$$\mathsf{E}e^{it\omega^2} = \prod_{k=1}^{\infty}\left[1 - \frac{2it}{\pi^2 k^2}\right]^{-1/2}.$$

In mathematical statistics, the 'omega-squared' distribution is often found in the following circumstances. Let $X_1, \ldots, X_n$ be independent random variables, uniformly distributed on $[0, 1]$, according to which an empirical distribution function $F_n(\cdot)$ is constructed. In this case, the process

$$Z_n(t) = \sqrt{n}(F_n(t)-t)$$

converges weakly to a conditional Wiener process, from which it follows that

$$\lim_{n\to\infty} \mathsf{P}\left\{\int_0^1 Z_n^2(t)\,dt <\lambda\right\} = \mathsf{P}\{\omega^2<\lambda\} =$$

$$= 1-\frac{2}{\pi}\sum_{k=1}^{\infty}(-1)^{k-1}\int_{(2k-1)\pi}^{2k\pi} \frac{e^{-t^2\lambda/2}}{\sqrt{-t\sin t}}\,dt, \quad \lambda>0.$$

See also **Cramér−von Mises test.**

References

[1] SMIRNOV, N.V.: 'On the ω^2-distribution', *Mat. Sb.* **2** (1937), 973-993 (in Russian).
[2] ANDERSON, T.W. and DARLING, D.A.: 'Asymptotic theory of certain 'goodness of fit' criteria based on stochastic processes', *Ann. Math. Stat.* **23** (1952), 193-212.

M.S. Nikulin

Editorial comments. The 'conditional Wiener process' Z is usually referred to in the Western literature as *tied-down Brownian motion, pinned Brownian motion* or as the *Brownian bridge.*

The pioneering paper is [A1].

References

[A1] DARLING, D.A.: 'The Cramér−Smirnov test in the parametric case', *Ann. Math. Stat.* **26** (1955), 1-20.
[A2] DURBIN, J.: *Distribution theory for tests based on the sample distribution function*, SIAM, 1973.

AMS 1980 Subject Classification: 60E05, 62E10

ONE-DIMENSIONAL MANIFOLD - A topological space X each point of which has a neighbourhood homeomorphic to a line (an *interior point*) or to a half-line (a *boundary point*). A connected paracompact Hausdorff one-dimensional manifold without boundary points is homeomorphic to the circle if it is compact, and to the line if it is not compact; if one or two boundary points are present, then X is homeomorphic to a half-open or closed bounded interval, respectively. Any such one-dimensional manifold can be smoothened, so that homeomorphic can be replaced by diffeomorphic in the assertions above.

A metric continuum (a connected compact metric space) K of which each point, with two exceptions, separates it, is homeomorphic to a closed interval. If every two points separate K, then K is homeomorphic to the circle. A subset $A \subset K$ *separates* K if $K \setminus A$ can be written as a union of two open disjoint subsets.

References

[1] MILNOR, J.: *Topology from the differential viewpoint*, Univ. of Virginia Press, 1965.
[2] FUKS, D.B. and ROKHLIN, V.A.: *Beginner's course in topology. Geometric chapters*, Springer, 1984 (translated from the Russian).
[3] HIRSCH, M.W.: *Differential topology*, Springer, 1976.

M.I. Voĭtsekhovskiĭ

Editorial comments. A fact related to the last paragraph above is *Wallace's theorem* (cf. [A1]): Every non-degenerate compact connected space contains at least two points that do not separate it.

References

[A1] ENGELKING, R.: *General topology*, Heldermann, 1989.
[A2] GUILLEMIN, V. and POLLACE, A.: *Differential topology*, Prentice-Hall, 1974.
[A3] GALE, D.: 'The classification of 1-manifolds: a take-home exam', *Amer. Math. Monthly* **94** (1987), 170-175.

AMS 1980 Subject Classification: 54F99, 57M99

ONE-PARAMETER SEMI-GROUP - A family of operators $T(t)$, $t>0$, acting in a Banach or topological vector space X, with the property

$$T(t+\tau)x = T(t)[T(\tau)x], \quad t, \tau>0, \quad x\in X.$$

If the operators $T(t)$ are linear, bounded and are acting in a Banach space X, then the measurability of all the functions $T(t)x$, $x\in X$, implies their continuity. The function $\|T(t)\|$ increases no faster than exponentially at infinity. The classification of one-parameter semi-groups is based on their behaviour as $t\to 0$. In the simplest case $T(t)$ is strongly convergent to the identity operator as $t\to 0$ (see **Semi-group of operators**).

An important characteristic of a one-parameter semi-group is the **generating operator of a semi-group**. The basic problem in the theory of one-parameter semi-groups is the establishment of relations between properties of semi-groups and their generating operators. One-parameter semi-groups of continuous linear operators in locally convex spaces have been studied rather completely.

One-parameter semi-groups of non-linear operators in Banach spaces have been investigated in the case when the operators $T(t)$ are contractive. There are deep connections here with the theory of dissipative operators.

References

[1] YOSIDA, K.: *Functional analysis*, Springer, 1980.
[2] KREĬN, S.G.: *Linear differential equations in Banach space*, Amer. Math. Soc., 1971 (translated from the Russian).
[3] HILLE, E. and PHILLIPS, R.: *Functional analysis and semi-groups*, Amer. Math. Soc., 1957.
[4] BUTZER, P. and BERENS, H.: *Semigroups of operators and approximation*, Springer, 1967.

[5] BARBU, V.: *Non-linear semi-groups and differential equations in Banach spaces*, Ed. Academici, 1976 (translated from the Romannian).
[6] DAVIES, E.B.: *One-parameter semigroups*, Acad. Press, 1980.
[7] GOLDSTEIN, J.A.: *Semigroups of linear operators and applications*, Oxford Univ. Press, 1985.

S.G. Kreĭn

Editorial comments.

References

[A1] PAZY, A.: *Semigroups of linear operators and applications to partial differential equations*, Springer, 1983.
[A2] CLÉMENT, PH. and HEIJMANS, H.J.A.M., ET AL.: *One-parameter semigroups*, CWI & North-Holland, 1987.
[A3] BREZIS, H.: *Operateurs maximaux monotone et semigroups de contractions dans les espaces de Hilbert*, North-Holland, 1973.
[A4] CASTEREN, J. VAN: *Generators of strongly continuous semigroups*, Pitman, 1985.
[A5] NAGEL, R. (ED.): *One-parameter semigroups of positive operators*, Springer, 1986.

AMS 1980 Subject Classification: 47D05, 47H20

ONE-PARAMETER SUBGROUP *of a Lie group G over a normed field K* - An analytic homomorphism of the additive group of the field K into G, that is, an analytic mapping $\alpha: K \to G$ such that

$$\alpha(s+t) = \alpha(s)\alpha(t), \quad s, t \in K.$$

The image of this homomorphism, which is a subgroup of G, is also called a one-parameter subgroup. If $K = \mathbf{R}$, then the continuity of the homomorphism $\alpha: K \to G$ implies that it is analytic. If $K = \mathbf{R}$ or $\mathbf{C}$, then for any tangent vector $X \in T_e G$ to G at the point e there exists a unique one-parameter subgroup $\alpha: K \to G$ having X as its tangent vector at the point $t = 0$. Here $\alpha(t) = \exp tX$, $t \in K$, where $\exp: T_e G \to G$ is the **exponential mapping**. In particular, any one-parameter subgroup of the **general linear group** $G = \mathrm{GL}(n, K)$ has the form

$$\alpha(t) = \exp tX = \sum_{n=0}^{\infty} \frac{1}{n!} t^n X^n.$$

If G is a real Lie group endowed with a two-sidedly invariant pseudo-Riemannian metric or affine connection, then the one-parameter subgroups of G are the geodesics passing through the identity e.

References

[1] PONTRYAGIN, L.S.: *Topological groups*, Princeton Univ. Press, 1958 (translated from the Russian).
[2] SERRE, J.-P.: *Lie algebras and Lie groups*, Benjamin, 1965 (translated from the French).
[3] HELGASON, S.: *Differential geometry, Lie groups, and symmetric spaces*, Acad. Press, 1978.

A.L. Onishchik

Editorial comments.

References

[A1] BOURBAKI, N.: *Elements of mathematics. Lie groups and Lie algebras*, Addison-Wesley, 1975 (translated from the French).
[A2] BOURBAKI, N.: *Groupes et algèbres de Lie*, Hermann, 1972, Chapts. 2-3.
[A3] HOCHSCHILD, G.: *Structure of Lie groups*, Holden-Day, 1965.

AMS 1980 Subject Classification: 22EXX

ONE-PARAMETER TRANSFORMATION GROUP, *flow* - The action of the additive group of real numbers $\mathbf{R}$ on a manifold M.

Thus, a one-parameter family $\{\phi_t: t \in \mathbf{R}\}$ of transformations of a manifold M is a one-parameter transformation group if the following conditions are satisfied:

$$\phi_{t+s} x = \phi_t(\phi_s x), \quad \phi_{-t} x = \phi_t^{-1} x, \quad t, s \in \mathbf{R}, \quad x \in M. \quad (*)$$

If the manifold M is smooth, then the group is usually assumed to be smooth also, that is, the corresponding mapping

$$\phi: \mathbf{R} \times M \to M, \quad (t, x) \to \phi_t x,$$

is a differentiable mapping of differentiable manifolds.

A more general concept is that of a *local one-parameter transformation group* of a manifold M. It is defined as a mapping $\phi: U \to M$ of some open submanifold $U \subset \mathbf{R} \times M$ of the form $U = \bigcup_{x \in M} (]\epsilon_-(x), \epsilon_+(x)[, x)$, where $\epsilon_+(x) > 0$, $\epsilon_-(x) < 0$ for $x \in M$, satisfying the conditions (*) for all $t, s \in \mathbf{R}$, $x \in M$ for which both sides of the equations are defined.

With each smooth local one-parameter transformation group $\{\phi_t\}$ of M one associates the vector field

$$M \ni x \to X_x = \left. \frac{d}{dt} \phi_t x \right|_{t=0},$$

called the *velocity field*, or *infinitesimal generator*, of the group $\{\phi_t\}$. Conversely, any smooth vector field X generates a local one-parameter transformation group ϕ_t having velocity field X. In local coordinates x^i on M this one-parameter transformation group is given as the solution of the system of ordinary differential equations

$$\frac{d\phi^i(t, x^j)}{dt} = X^i(\phi^j(t, x^k))$$

with the initial conditions $\phi^i(0, x^j) = x^i$, where $X = \sum_i X^i \partial / \partial x^i$.

If the local one-parameter transformation group generated by the vector field X can be extended to a global one, then the field X is called *complete*. On a compact manifold any vector field is complete, so that there is a one-to-one correspondence between one-parameter transformation groups and vector fields. This is not the case for non-compact manifolds, and the set of complete vector fields is not even closed under addition.

References

[1] ARNOL'D, V.I.: *Ordinary differential equations*, M.I.T., 1973 (translated from the Russian).
[2] PALAIS, R.: *A global formulation of the Lie theory of transformation groups*, Amer. Math. Soc., 1957.

D.V. Alekseevskiĭ

Editorial comments.

References

[A1] SELL, G.R.: *Topological dynamics and ordinary differential equations*, v. Nostrand Reinhold, 1971.

AMS 1980 Subject Classification: 57S20, 58FXX

ONE-SHEET HYPERBOLOID - See **Hyperboloid.**

AMS 1980 Subject Classification: 51N10, 51N20

ONE-SIDED AND TWO-SIDED SURFACES - Two types of surfaces, differing in the way in which they are situated in the ambient space (*one-sided position* and *two-sided position*). For example, the cylinder is a two-sided surface, while the **Möbius strip** is a one-sided surface. A characteristic distinction between these surfaces is that the boundary of the cylinder consists of two curves, while the boundary of the Möbius strip is a single curve. Among the closed surfaces the **sphere** and the **torus** are two-sided, while the **Klein surface** is one-sided. As examples of two-sided and one-sided situations one may cite imbeddings of the circle in the Möbius strip. Thus, the cycle α (see Fig.) is a one-sided curve, while the cycle β is two-sided (in general, any **desorienting path** lies in the surface one-sidedly).

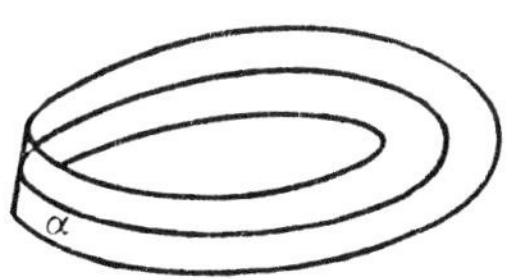
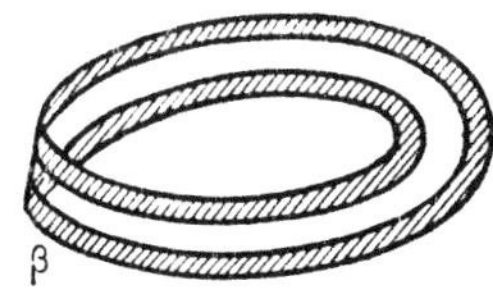

More precisely, one- and two-sided surfaces are two types of manifolds differing in the way in which they are imbedded in the ambient space (of dimension higher by one). Two-sidedness and one-sidedness are related to orientability and non-orientability (see **Orientation**), but unlike these, are not intrinsic properties of the surface and depend on the ambient space. For example, there exist orientable two-sided surfaces: $S^2 \subset S^3$, $T^2 \subset \mathbf{R}^3$; non-orientable two-sided surfaces: $RP^2 \times 0 \subset RP^2 \times S^1$; orientable one-sided surfaces: $T^2 = S^1 \times S^1 \subset RP^2 \times S^1$; non-orientable one-sided surfaces: $RP^2 \subset RP^3$ (here S^2 is the sphere, T^2 the torus, RP^2 the projective plane, RP^3 projective space, and S^1 a desorienting path on RP^2).

In an orientable space (e.g. in $\mathbf{R}^n$) a hypersurface is orientable if and only if it is two-sided.

Suppose that a normal vector is moved along a closed curve on a smooth surface immersed in some space in such a way that it remains normal. If upon return to the starting point the direction of the normal is the same as it was originally, independently of the choice of curve, then the surface is called *two-sided*; in the opposite case it is called *one-sided*. More generally, a surface Π is situated two-sidedly if and only if its **normal bundle** is trivial (there exists a non-zero section in this bundle). Conversely, the normal bundle of a one-sided surface is non-trivial: There exists a curve on Π over which the bundle is a Möbius strip.

Locally every (hyper) surface M^{n-1} in a space N^n divides N^n into two parts, that is, a point $x \in M^{n-1} \subset N^n$ has a neighbourhood $U \subset N$ such that U

consists of two components U' and U'' and $U \cap M^{n-1}$ belongs to their common boundary. On the other hand, a sufficiently small neighbourhood of M^{n-1} in N^n (if M is closed in N) has either one or two components the boundaries of which contain M. In the first case, the (hyper) surface M^{n-1} is also called *one-sided*, and in the second case, *two-sided*. Thus, although the surface locally has two sides, globally it may be one-sided, while, in contrast, a two-sided surface need not divide its neighbourhood in the space.

For a two-sided surface M^n lying in N^{n+1}, the **intersection index (in homology)** in N^{n+1} of any closed curve α in M^n satisfies the equation $(\alpha, M^n) \equiv 0 \mod 2$. But if M^n is one-sided, then $(\alpha, M^n) \neq 0$ for some curve $\alpha \in M^n$. This fact can also be taken as the definition of one- and two-sidedness (along with the movement of a normal vector and the division of a neighbourhood).

References

[1] HILBERT, D. and COHN-VOSSEN, S.E.: *Geometry amd the imagination*, Chelsea, 1952 (translated from the German).
[2] SEIFERT, H. and THRELFALL, W.: *A textbook of topology*, Acad. Press, 1980 (translated from the German).
[3] FUKS, D.B., FOMENKO, A.T. and GUTENMAKHER, V.L.: *Homotopic topology*, Moscow, 1969 (in Russian).

M.I. Voĭtsekhovskiĭ

AMS 1980 Subject Classification: 53A05, 57N35, 57R40

ONE-SIDED DERIVATIVE - A generalization of the concept of a **derivative**, in which the ordinary limit is replaced by a **one-sided limit**. If the following limit exists for a function f of a real variable x:

$$\lim_{x \to x_0 + 0} \frac{f(x) - f(x_0)}{x - x_0} \quad \left[\text{or} \lim_{x \to x_0 - 0} \frac{f(x) - f(x_0)}{x - x_0} \right],$$

then it is called the *right* (respectively, *left*) *derivative* of f at the point x_0. If the one-sided derivatives are equal, then the function has an ordinary derivative at x_0. See also **Differential calculus.**

G.P. Tolstov

AMS 1980 Subject Classification: 26A24

ONE-SIDED LIMIT - The **limit** of a function at a point from the right or left. Let f be a mapping from an ordered set X (for example, a set lying in the real line), regarded as a topological space with the topology generated by the order relation, into a topological space Y, and let $x_0 \in X$. The limit of f with respect to any interval $(a, x_0) = \{x: x \in X, a < x < x_0\}$ is called the *limit of f on the left*, and is denoted by

$$\lim_{x \to x_0 - 0} f(x)$$

(it does not depend on the choice of $a < x_0$), and the limit with respect to the interval $(x_0, b) = \{x: x \in X, x_0 < x < b\}$ is called the *limit on the right*, and is denoted by

$$\lim_{x \to x_0 + 0} f(x)$$

(it does not depend on the choice of $b > x_0$). If the point x_0 is a limit point both on the left and the right for the domain of definition of the function f, then the usual limit

$$\lim_{x \to x_0} f(x)$$

with respect to a deleted neighbourhood of x_0 (in this case it is also called a two-sided limit, in contrast to the one-sided limits) exists if and only if both of the left and right one-sided limits exist at x_0 and they are equal.

L.D. Kudryavtsev

Editorial comments. Instead of $\lim_{x \to x_0 + 0}$ (respectively, $\lim_{x \to x_0 - 0}$) one also finds the notations $\lim_{x \to x_0 +}$, $\lim_{x \downarrow x_0}$ (respectively, $\lim_{x \to x_0 -}$, $\lim_{x \uparrow x_0}$).

AMS 1980 Subject Classification: 26A15

ONE-TO-ONE CORRESPONDENCE - A correspondence between elements of two sets in which: 1) to each element of the first set there corresponds a unique element of the second set; 2) to different elements of the first set there correspond different elements of the second set; and 3) each element of the second set has been put into correspondence with an element of the first set. The relation which holds between sets if and only if there is a one-to-one correspondence between them is symmetric (the mapping inverse to a one-to-one correspondence is a one-to-one correspondence) and is transitive (the product of one-to-one correspondences is a one-to-one correspondence). If each point x of an oriented straight line is put into correspondence with its distance from a given point O (which is positive if the point is located in a positive direction from O, and is negative otherwise), the resulting correspondence is a one-to-one correspondence between the points on the straight line and the real numbers.

L.D. Kudryavtsev

Editorial comments. See also Bijection.

References
[A1] ENDERTON, H.B.: *Elements of set theory*, Acad. Press, 1977.

AMS 1980 Subject Classification: 04A05

OPEN-CLOSED SET - A subset of a **topological space** which is simultaneously open and closed in that space (cf. **Open set; Closed set**). A topological space X is disconnected if and only if it contains an open-closed set other than X and $\varnothing$. If the family of all open-closed sets of a topological space is a basis of its topology, then this space is called *inductively zero-dimensional*. Every **Boolean algebra** is isomorphic to the Boolean algebra of all open-closed sets of an appropriate inductively zero-dimensional Hausdorff compactum. The so-called *extremally-disconnected Hausdorff compacta* form

a special class of zero-dimensional compacta, and are characterized by the fact that the closure of any open set in them is also open (and closed). Every complete Boolean algebra is isomorphic to the Boolean algebra of all open-closed sets of an appropriate extremally-disconnected Hausdorff compactum.

V.I. Ponomarev

Editorial comments. An open-closed set is also called a *closed-open set* or *clopen set*.

The correspondence between Boolean algebras and inductively zero-dimensional compact Hausdorff spaces is known as *Stone duality* or *Stone topological duality*. Instead of 'inductively zero dimensional' one also finds simply '*zero dimensional*' in the literature.

A zero-dimensional compact Hausdorff space is called a *Boolean space*.

References
[A1] KOPPELBERG, S.: *General theory of Boolean algebras*, 1, North-Holland, 1989, Sect. 3.7.

AMS 1980 Subject Classification: 54A05, 54H05

OPEN MANIFOLD - A manifold without compact components, i.e. a manifold that is not a **closed manifold**.

M.I. Voĭtsekhovskiĭ

AMS 1980 Subject Classification: 57NXX, 57RXX

OPEN MAPPING - A mapping of one topological space into another under which the image of every open set is itself open.

Projections of topological products onto the factors are open mappings. Openness of a mapping can be interpreted as a form of continuity of its inverse many-valued mapping. A one-to-one continuous open mapping is a **homeomorphism**. In general topology, open mappings are used in the classification of spaces. The question of the behaviour of topological invariants under continuous open mappings is important. All spaces with the **first axiom of countability**, and only they, are images of metric spaces under continuous open mappings. A **metrizable space** which is the image of a complete metric space under a continuous open mapping is metrizable by a complete metric. If a **paracompact space** is the image of a **complete metric space** under a continuous open mapping, then it is metrizable. A countable-to-one continuous open mapping of compacta does not increase the dimensions. However, a 3-dimensional cube can be mapped by a continuous open mapping onto a cube of any larger dimension. Every compactum is the image of a certain one-dimensional compactum under a continuous open mapping with zero-dimensional fibres (i.e. inverse images of points)

Continuous open mappings under which the inverse images of all points are compact — the so-called *compact-open mappings* — are of separate interest in their own right. Spaces with a uniform base, and only they, are inverse images of metric spaces under compact-open mappings. Closed continuous open mappings are also important. All continuous open mappings of compacta into Hausdorff spaces (cf. **Hausdorff space**) fall into this category. Continuous closed open mappings preserve metrizability. Open mappings with discrete fibres play an important role in the theory of functions of one complex variable: these include all holomorphic functions in a domain. The theorem on the openness of holomorphic functions is central to proving the maximum-modulus principle, and to proving the fundamental theorem on the existence of a root of an arbitrary non-constant polynomial over the field of complex numbers.

References

[1] KURATOWSKI, K.: *Topology*, 1-2, Acad. Press, 1966-1968 (translated from the French).
[2] KELDYSH, L.V.: 'Open and monotone mappings of compacta', in *Proc. 3-rd All-Union Math. Congress*, Vol. 3, Moscow, 1958, pp. 368-372 (in Russian).
[3] STOÏLOV, S.: *The theory of functions of a complex variable*, 1-2, Moscow, 1962 (in Russian; translated from the Rumannian).

A.V. Arkhangel'skiĭ

Editorial comments.

References

[A1] ENGELKING, R.: *General topology*, Heldermann, 1989.
[A2] WHYBURN, G.T.: *Topological analysis*, Princeton Univ. Press, 1964.

AMS 1980 Subject Classification: 54C10

OPEN-MAPPING THEOREM - A continuous **linear operator** A mapping a **Banach space** X onto all of a Banach space Y is an **open mapping**, i.e. $A(G)$ is open in Y for any G which is open in X. This was proved by S. Banach. Furthermore, a continuous linear operator A giving a one-to-one transformation of a Banach space X onto a Banach space Y is a **homeomorphism**, i.e. A^{-1} is also a continuous linear operator (*Banach's homeomorphism theorem*).

The conditions of the open-mapping theorem are satisfied, for example, by every non-zero continuous linear functional defined on a real (complex) Banach space X with values in **R** (in **C**).

The open-mapping theorem can be generalized as follows: A continuous linear operator mapping a fully-complete (or B-complete) **topological vector space** X onto a **barrelled space** Y is an open mapping. The **closed-graph theorem** can be considered alongside with the open-mapping theorem.

References

[1] YOSIDA, K.: *Functional analysis*, Springer, 1980.
[2] ROBERTSON, A.P. and ROBERTSON, W.: *Topological vector spaces*, Cambridge Univ. Press, 1964.

V.I. Sobolev

Editorial comments. A recent comprehensive study of the closed-graph theorem can be found in [A1].

References

[A1] DE WILDE, M.: *Closed graph theorems and webbed spaces*, Pitman, 1978.
[A2] SCHAEFFER, H.H.: *Topological vector spaces*, Springer, 1971.
[A3] JARCHOW, H.: *Locally convex spaces*, Teubner, 1981 (translated from the German).

AMS 1980 Subject Classification: 46A30

OPEN SET *in a topological space* - An element of the topology (cf. **Topological structure (topology)**) of this space. More specifically, let the topology τ of a topological space (X, τ) be defined as a system τ of subsets of the set X such that: 1) $X \in \tau$, $\varnothing \in \tau$; 2) if $O_i \in \tau$, $i = 1, 2$, then $O_1 \cap O_2 \in \tau$; and 3) if $O_\alpha \in \tau$, $\alpha \in \mathfrak{A}$, then $\bigcup \{O_\alpha : \alpha \in \mathfrak{A}\} \in \tau$. The *open sets* in the space (X, τ) are then the elements of the topology τ and only them.

B.A. Pasynkov

Editorial comments.

References

[A1] ENGELKING, R.: *General topology*, Heldermann, 1989.

AMS 1980 Subject Classification: 54AXX

OPERAND *in programming languages* - An argument of an operation; a grammatical construction signifying an expression that gives the value of the argument of an operation; sometimes the place or position in a text where the argument of an operation is to stand is called an operand. Hence the concept of *arity of an operation*, i.e. the number of arguments of an operation.

Depending on the position of the operand relative to the sign of the operation, one distinguishes between *prefix* (e.g. $\sin x$), *infix* (e.g. $a + b$) and *postfix* (e.g. x^2) operations. Dependent on the number of operands there are *one-placed* (*unary* or *monadic*), *two-placed* (*binary* or *dyadic*) and *many-placed* (or *polyadic*) operations.

In distinguishing between the position of the operand and the operand as an actual argument, the concept of transforming or *coercing* the operand to the form required by the operation arises. For example, if a real argument is situated in the position of an integer operand, the rules of the language may imply some method of rounding the real number to an appropriate natural number. Another example of coercion is variation of the form of the representation of an object, for example, a scalar is transformed to a vector consisting of one component.

A.P. Ershov

AMS 1980 Subject Classification: 68B05

OPERATIONAL CALCULUS - One of the methods of mathematical analysis which in many cases makes it possible to reduce the study of differential operators, pseudo-differential operators and certain types of integral operators (cf. **Differential operator; Integral operator; Pseudo-differential operator**) and the solution of equations containing them, to an examination of simpler algebraic problems. The development and systematic use of operational calculus began with the work of O. Heaviside (1892), who proposed formal rules for dealing with the differentiation operator d/dt and solved a number of applied problems. However, he did not give operational calculus a mathematical basis; this was done with the aid of the **Laplace transform**; J. Mikusiński (1953) put operational calculus into algebraic form, using the concept of a function ring. The most general concept of an operational calculus is obtained using generalized functions (cf. **Generalized function**).

The simplest variant of operational calculus is as follows. Let K be the set of functions (with real or complex values) given in the domain $0 \leqslant t < \infty$ and absolutely integrable in any finite interval. The integral

$$h = f \star g = \int_0^t f(t-\tau)g(\tau)\,d\tau$$

is called the *convolution* of the functions $f, g \in K$. With the usual addition operation and the operation of convolution, K becomes a ring without zero divisors (*Titchmarsh's theorem*, 1924). Elements of the *quotient field P* of this ring are called *operators* and are written as a/b; the fact that division in K is not always possible is precisely the source of a new concept, operators, which generalizes the concept of a function. To indicate the necessary difference in operational calculus between the concepts of a function and of its value at a point, the following notation is used:

$\{f(t)\}$ for the function f of a variable t;

$f(t)$ for the value of $\{f(t)\}$ at a point t.

Examples of operators. 1) $e = \{1\}$ is the *integration operator*:

$$\{1\}\{f\} = \left\{ \int_0^t f(\tau)\,d\tau \right\}.$$

Moreover,

$$e^p = \left\{ \frac{t^{p-1}}{\Gamma(p)} \right\}$$

and, in particular,

$$e^n\{f\} = \int_0^t dt \cdots \int_0^t f(t)\,dt \ (n-\text{fold}) =$$

$$= \int_0^t \frac{(t-\tau)^{n-1}}{(n-1)!} f(\tau)\,d\tau$$

This is the Cauchy formula, a generalization of which to the case of an arbitrary (non-integer) index serves to define fractional integration.

2) $[\alpha] = \{\alpha\}/\{1\}$ (where α is a constant function) is a *numerical operator*; insofar as $[\alpha][\beta] = [\alpha\beta]$,

$[\alpha+\beta] = [\alpha]+[\beta]$, $[\alpha]\{f\} = \{\alpha f\}$, while $\{\alpha\}\{\beta\} = \{\alpha\beta t\}$, numerical operators behave as ordinary numbers. Thus the operator is a generalization not only of a function, but also of a number; [1] is the unit of the ring K.

3) $s = [1]/e$ is the *differentiation operator*, the inverse of the integration operator. So, if a function $a(t) = \{a(t)\}$ has a derivative $a'(t)$, then

$$s\{a\} = \{a'\} + [a(0)]$$

and

$$\{a^{(n)}\} = s^n\{a\} - s^{n-1}[a(0)] - \cdots - [a^{n-1}(0)].$$

Hence, for example,

$$\{e^{at}\} = \frac{1}{s-a}.$$

Of course, a non-differentiable function can be multiplied by the differentiation operator s; however, the result will be, in general, an operator.

4) $D\{f\} = \{-tf(t)\}$ is the *algebraic derivative*. It extends to arbitrary operators in the usual way. It appears that the action of this operator on a function of the differentiation operator s coincides with differentiation with respect to s.

Operational calculus provides suitable methods for the solution of linear differential equations, both ordinary and partial. For example, the solution of the equation

$$\alpha_n x^{(n)} + \cdots + \alpha_0 x = f, \quad \alpha_i = \text{const}, \ i = 0, \ldots, n,$$

satisfying the initial conditions $x(0) = \gamma_0, \ldots, x^{(n-1)}(0) = \gamma_{n-1}$ automatically reduces to an algebraic equation. It is expressed symbolically by the formula

$$x = \frac{\beta_{n-1}s^{n-1} + \cdots + \beta_0 + f}{\alpha_n s^n + \cdots + \alpha_0},$$

$$\beta_v = \alpha_{v+1}\gamma_0 + \cdots + \alpha_n\gamma_{n-v-1}.$$

The solution in its usual form is obtained by decomposition into elementary fractions with respect to the variable s, with subsequent inverse transformation by referring to appropriate function tables.

In the use of operational calculus for partial differential equations (as well as for more general pseudo-differential equations), a differential and integral calculus of *operator functions*, i.e. functions with operator values, is employed. The concepts of continuity, derivative, convergence of series, integrals, etc., must be developed for these functions.

Let $f(\lambda, t)$ be a function defined for $t \geqslant 0$ and $\lambda \in [a, b]$. A *parametric operator function* $f(\lambda)$ is defined by the formula $f(\lambda) = \{f(\lambda, t)\}$; it places operators of a certain type — functions in t — in correspondence with the values of λ being considered. An operator function is said to be continuous for $\lambda \in [a, b]$ if it can be represented as the product of an operator q and a parametric function $f_1(\lambda)=\{f_1(\lambda, t)\}$ such that $f_1(\lambda, t)$ is continuous in the ordinary sense.

Examples. 1) Using the parametric function $h(\lambda)=\{h(\lambda, t)\}$:

$$h(\lambda, t) = \begin{cases} 0 & \text{for } 0 \leqslant t < \lambda, \\ t-\lambda & \text{for } 0 \leqslant \lambda \leqslant t, \end{cases}$$

the *Heaviside function* is defined:

$$H(\lambda) = s\{h(\lambda, t)\}.$$

The values of the *hyperbolic exponential function*

$$e^{-\lambda s} \equiv sH(\lambda) = s^2\{h(\lambda, t)\}$$

are called *shift operators*, since multiplication of a given function by $e^{-\lambda s}$ requires the displacement of its graph over λ in the positive direction of the t-axis.

2) The solution of the heat equation

$$\frac{\partial x}{\partial t} = \frac{\partial^2 x}{\partial \lambda^2}$$

can be expressed using the *parabolic exponential function* (which is also a parametric operator function):

$$e^{-\lambda\sqrt{s}} = \left\{ \frac{\lambda}{2\sqrt{\pi}\, t^3} \exp\left[-\frac{\lambda^2}{4t} \right] \right\}.$$

3) A periodic function $f(t)$ with period $2\lambda_0$ has the representation:

$$\{f\} = \frac{\int_0^{2\lambda_0} e^{-\lambda s} f(\lambda)\, d\lambda}{1-e^{-2\lambda_0 s}}.$$

4) If $f(\lambda)$ has numerical values in the interval $[\lambda_1, \lambda_2]$, then

$$\int_{\lambda_1}^{\lambda_2} e^{-\lambda s} f(\lambda)\, d\lambda = \begin{cases} f(\lambda), & \lambda_1 < t < \lambda_2, \\ 0, & 0 \leqslant t < \lambda_1,\ t > \lambda_2, \end{cases}$$

i.e. multiplication of the given function $\{f\}$ by $e^{-\lambda s}$ with subsequent integration entails a *truncation* of its graph. In particular,

$$\int_0^{\infty} e^{-\lambda s} f(\lambda)\, d\lambda = \{f(t)\}.$$

Thus, with each function $f(t)$ for which the integral being considered is convergent there is a corresponding analytic function:

$$F(s) = \int_0^{\infty} e^{-st} f(t)\, dt,$$

its Laplace transform. As a result, a fairly broad class of operators is described by functions of one parameter s; moreover, this formal similarity is defined more exactly in mathematical terms by establishing a definite isomorphism.

There are various generalizations of operational calculus; for example, operational calculus of differential operators other than $s=d/dt$, for example,

$b=d/dt(t(d/dt))$, which is based on other function rings with a properly defined product.

References

[1] DITKIN, V.A. and PRUDNIKOV, A.P.: *Handbook of operational calculus*, Moscow, 1965 (in Russian).
[2] MIKUSIŃSKI, J.: *Operational calculus*, Pergamon, 1959.

M.I. Voĭtsekhovskiĭ

Editorial comments. A second edition of [2] has recently appeared, [A1], [A2]. In the examples of parametric operator functions $f(\lambda)$ above, use is made of a differential and integral calculus for operators. For more details on the truncation of an operator function $f(\lambda)$ see [A2], Part V, Chapt. 1, § 5. For $f \in K$ an operator of the form $e^{-\lambda s} f$ is identified with a Schwartz distribution with support bounded from below.

The notion of a Schwartz distribution and Mikusiński operator do not include each other, but both generalize the idea of a function and its derivatives.

The term 'operational calculus' is also used in the sense of **functional calculus**; i.e. a homomorphism of a certain algebra of functions into an algebra of operators. Finally, the phrase 'operational calculus' or '*operator calculus*' occurs in the context of the *time-ordered operator calculus* (*Feynman—Dyson time-ordered operator calculus*) developed in the 1950's for the study of quantum electrodynamics [A4], [A5], and relating to product integrals (cf. **Product integral**), [A6].

References

[A1] MIKUSIŃSKI, J.: *Operational calculus*, 1, PWN & Pergamon, 1987.
[A2] MIKUSIŃSKI, J. and BOEHME, TH.K.: *Operational calculus*, II, PWN & Pergamon, 1987.
[A3] POL, B. VAN DER and BREMMER, H.: *Operational calculus based on the two-sided Laplace transform*, Cambridge Univ. Press, 1959.
[A4] FEYNMAN, R.P.: 'An operator calculus having applications in quantum electrodynamics', *Phys. Rev.* 84 (1951), 108-128.
[A5] GILL, T.L. and ZACHARY, W.W.: 'Time-ordered operators and Feynman—Dyson algebras', *J. Math. Phys.* 28 (1987), 1459-1470.
[A6] DOLLARD, J.D. and FRIEDMAN, CH.N.: *Product integration*, Addison-Wesley, 1979.

AMS 1980 Subject Classification: 44-XX, 44A45, 44A55

OPERATIONS RESEARCH - The construction, elaboration and application of mathematical models for making optimal decisions. The theoretical side of operations research is concerned with the analysis and solution of mathematical problems of choosing from a given set X of *feasible decisions* an element satisfying some criterion of optimality, called an *optimal decision* of the problem. Sometimes 'generalized' elements of X — subsets or functions with values in X (including random variables with values in X) — are chosen. Problems of this kind are called *optimization problems*. The applied side of operations research is concerned with formulating optimization problems and with realizing their solutions.

Formulating a problem in operations research first of all includes the formal description of the set X of feasible decisions and criteria of optimality. This must reflect an informal representation of the possible and the desired under the conditions given. On the other hand, the verification of the adequacy of the informal representation itself for the objective reality is already beyond the limits of operations research. All decisions (including optimal ones) are made only on the basis of the information that is available to the subject(s) making a decision. Therefore, every problem in operations research must reflect in its formulation the knowledge of the subject(s) making a decision about the set of feasible decisions and optimality criteria. Thus, if the decision is made within a single stage of information that does not change and that is given in advance, then the problem is called *static*. In this case, the complete process of decision making can be reduced to one instantaneous act. In the alternative case, if the decision is made with several different information stages, then the decision will be arrived upon by establishing correspondences between each information stage and each decision that is feasible in it, i.e. by choosing a function expressing this correspondence. If the information stages do change one another, then the problem is called *dynamic*. In a dynamic problem it is expedient to make decision step-by-step, 'multi-step-wise', or even to model the decision making by means of a process that is continuous in time.

The information states of a decision maker may characterize its true ('physical') state in various ways. It may happen that one information state of the subject includes the set of its physical states. In this case, the decision problem is called *indefinite*. Such problems of operations research are considered in the theory of games (cf. **Games, theory of**). If an information state contains several physical states, and if the subjects also know, from their sets, the (a priori) probability of each of these physical states, the problem is called *stochastic*. Finally, if the information state consists of a unique physical state, then the problem is called *deterministic*. It is sometimes of interest to consider a family of problems, depending on a numerical or vectorial value running over some parameter set, and unifying them into a single *parametric* problem. A parametric problem differs from an indefinite problem in that solving the first consists of solving all problems corresponding to special values of the parameter, while solving the second problem consists of finding a feasible decision that is in a certain sense required, whatever the way in which the indefiniteness was actually realized. Meanwhile, solving a stochastic problem consists of finding an optimal decision that is optimal 'on the average' over the entire set of individual problems.

Problems of operations research with an arbitrary set X of feasible solutions and with arbitrary optimality criteria can be imagined theoretically. These criteria may consist of the maximization (or minimization) of the values of some numerical or vector function f on X. This function is usually called the **objective function**. In the first case one speaks of a problem of **mathematical programming** (*optimal programming*, which should not be confused with programming of a computer), while in the second case one speaks of a *problem of vector optimization*, or of a **multi-criterion problem**. Criteria expressed by a binary preference relation $\succ$ on X have also been considered. This relation need not be a total linear, or even a partial, order relation.

In mathematical programming one most often considers problems in which X is a subset of a finite-dimensional Euclidean space E^n. If X is, moreover, a convex polyhedron with a finite number of vertices and if the objective function f is linear, one has a problem of **linear programming**. If X is an arbitrary convex set and f is a convex function, one has a problem of **convex programming**. Problems of piecewise-linear programming, **quadratic programming**, etc., can be naturally defined. The set X of feasible decisions can also be a subset of a function space; hence formal variational calculus, as well as a range of problems related to the Pontryagin maximum principle, may therefore also be considered as a part of mathematical programming. In other cases X may be a finite set; such problems belong to **discrete programming**. The feasible decisions for them may be the points of an integer lattice in E^n (*integer programming*) or vectors each component of which takes only two values (*Boolean programming*). In individual cases, the elements of X are permutations on a finite number of symbols, paths in a given graph, etc. A special case of a problem of mathematical programming is that of finding a *maximin*, i.e. the maximum of a function of one variable which is obtained by minimizing a two (or more) variable function with respect to another variable (analogously, a *minimax*).

The theory of solving stochastic problems of linear programming is the subject of **stochastic programming**. Multi-criterion problems, as well as problems with preference relations, relate to the theory of games; they are classified according to their game-theoretic properties.

One of the contemporary (since the 1970's) trends in operations research is the transition from the consideration of individual problems to the study of systems, spaces, calculi of such problems, and to the study of relations between various problems or the reduction of some problems to others that have a simpler structure. The mathematical apparatus intended for and developed with the aim of solving problems of operations research is conveniently called the mathematics of operations research. The methods in question do, in principle, not differ by their nature from the mathematical methods of any other mathematical discipline with informal applications or, at least, interpretations. The stages of elaboration of mathematical methods for various problems of operations research and their classes differ. The theories of linear and convex programming are the most well developed.

Some parametric problems of operations research, distinguished by specific informal interpretations, problems and terminology, bear the name of *models of operations research*. Usually, each model has an intrinsic method for solving it. The range of models of operations research is great: from specific problems, differing only by the numerical values of their parameters (examples are the assignment and the transportation problem, some more difficult cutting problems, the allocation problem, and the theory of networks), to such specified disciplines as the theory of inventory control, allotation or reliability theory. The theory of games provides a large number of models of operations research (games of timing, Blotto-type games, poker-type games, differential pursuit games, etc.). Some advanced problems in queueing theory are also considered as models of operations research, although the majority of problems in queueing theory do not as yet have an optimization character.

The solution of such problems of operations research starts with the choice of an optimality principle. If one has a problem of mathematical programming this is a trivial matter: The optimality principle consists of the maximization (respectively, minimization) of the objective function. Thus, in this case the optimality principle of the problem formally coincides with its optimality criterion. In other cases the discovery of the optimality principle is an essential stage in the solution of the problem and it can be realized in various ways. Means for reducing a vector criterion or preference relation to numerical criteria have been used. E.g., in the case of a multi-criterion problem the optimality principle may consist in giving the individual components of the vector criterion some weights and by considering the weighted sum as objective function. Another optimality principle may consist of the maximization of the minimal component of the vector criterion (the maximin principle), etc. Optimality principles in problems with a preference relation can be extremely varied (cf., e.g., **Games, theory of**). The possibility and ways of replacing preference relations by numerical criteria constitute one of the basic problems in **utility theory**. Thus, the criterion of a problem of operations research is part of its conditions, while the optimality principle is part of its answer. In the majority of models of operations research the optimality principle is fixed.

After the choice of the optimality principle, the next stage in the solution of a problem of operations research is that of proving its realizability (i.e. the existence of a solution of the problem in the sense of this principle). The non-realizability of an optimality principle in the class of feasible decisions given by the conditions of the problem can sometimes be overcome by way of introducing decisions that are generalized in some sense or other; subsets of the set of feasible decisions or functions with values in this set.

Since the existence of optimal decisions for a problem of operations research often turns out to be non-constructive (e.g. based on some fixed-point theorem), while it is sometimes constructive but provides its indication only potentially (and not practically), the third stage of evaluating a problem of operations research is that of obtaining the optimal solution.

Problems of operations research possess a number of features, brought about by the method of formulating and solving them. First, even for the simplest parameter classes of problems it is usually not possible to express the solutions as an analytic expression in the corresponding parameters. Therefore, problems of operations research can, in the majority of cases, not be solved analytically and must be solved numerically. Secondly, the majority of problems of practical interest contain in their formulation a very large amount of numerical material, not reducible to analytic expressions; thus, the numerical solution of these problems can only be accomplished on a computer. Thirdly, the process of solving a lot of problems of operations research consists of performing simple single-type operations over large collections of numbers. Therefore, the problems of operations research impose demands on the computer involving, to a great extent, both the computer memory and its speed of operation. The actual demands of operations research have shown a great influence on the development of computers and their memory.

The range of applications of operations research is very broad. Operations research is used in the solution of technical (both constructive and technological), technico-economic and socio-economic problems, as well as in problems of control in various media and on various levels, thereby slowly solidifying the traditional 'intuitive' methods of decision making. The practical introduction of results of operations research can be found at various levels of scholarship. The first issue which was overcome relatively early is related to the construction of conceptual structures of models of real problems of decision making (or with the selection of a model of already given structure). This establishes a possibility in principle of modelling a class of real problems, including the problem under consideration, by a certain class of problems of operations research. The next issue consists in the choice of precisely that problem in this class that models the specific problem of interest. For this choice one must, in particular, measure the values of the parameters determining the problem to be solved. Now, since these parameters need not have a physical or technological character, but are often of an economic or even socio-economic character, their measurement with a required precision may pose a problem in itself. This problem in the choice of information for the construction of a specific model can be considered as the basic obstacle on the way of elaborating the optimal decision. Next, after the construction of the model, there arise the purely mathematical, including computational, difficulties of its analysis and solution. In essence, the contents of operations research consists precisely of overcoming these difficulties. Finally, after the solution has been found, the last issue arises, which is often of an organizational and psychological nature: the solution found quite often differs from traditional solutions and hence is regarded with disbelief. All this restricts the applicability of operations research. In order to overcome these difficulties more successfully, teams working in the field of operations research are often combined into inter-disciplinary groups; these are formed by, apart from mathematicians, usually also engineers, economists and specialists in the domain of the concrete discipline, and sometimes even psychologists, managers, etc.

References

[1] MORSE, P.M. and KIMBALL, G.E.: *Methods of operations research*, M.I.T., 1952.

[2] SAATY, T.L.: *Mathematical methods of operations research*, McGraw-Hill, 1959.

[3] KAUFMANN, A. and FAURE, R.: *Invitation à la récherche opérationelle*, Dunod, 1963.

[4] KAUFMANN, A.: *Methods and models of operations research*, Prentice-Hall, 1963 (translated from the French).

[5] CHURCHMEN, W., ACKOFF, R.L. and ARNOFF, E.L.: *Introduction to operations research*, Wiley, 1957.

[6] CHUEV, YU.V.: *Operations research in a military case*, Moscow, 1970 (in Russian).

[7] GERMEÏER, YU.B.: *Introduction to the theory of operations research*, Moscow, 1971 (in Russian).

[8] ACKOFF, R. and SASIENI, M.: *Fundamentals of operations research*, Wiley, 1968.

[9] VENTTSEL', E.S.: *Operations research*, Moscow, 1972 (in Russian).

[10] WAGNER, H.: *Principles of operations research*, Prentice-Hall, 1975.

[11] *Operations research. Methodological aspects*, Moscow, 1972 (in Russian).

[12] *Mathematische Standardmodelle der Operationsforschung*, Berlin, 1971.

[13] *Mathematische Standardmodelle der Operationsforschung. Mathematische Grundlagen, Methoden und Modelle*, 1-3, Berlin, 1971-1973.

N.N. Vorob'ev

Editorial comments.

References
[A1] BEALL, E.M.L.: *Introduction to optimization*, Wiley, 1987.

[A2] HILLIER, F.S. and LIEBERMAN, G.J.: *Introduction to operations research*, Holden-Day, 1967.

AMS 1980 Subject Classification: 90BXX

OPERATOR - A mapping of one set into another, each of which has a certain structure (defined by algebraic operations, a topology, or by an order relation). The general definition of an operator coincides with the definition of a **mapping** or **function**. Let X and Y be two sets. A rule or correspondence which assigns a uniquely defined element $A(x) \in Y$ to every element x of a subset $D \subset X$ is called an *operator A* from X into Y. D is called the *domain of definition of the operator A* and is denoted by $D(A)$; the set $\{A(x): x \in D\}$ is called the *domain of values of the operator A* (or its *range*) and is denoted by $R(A)$. The expression $A(x)$ is often written as Ax. The term operator is mostly used in the case where X and Y are vector spaces. If A is an operator from X into Y where $Y = X$, then A is called an *operator on X*. If $D(A) = X$, then A is called an *everywhere-defined operator*. If A_1, A_2 are operators from X_1 into Y_1 and from X_2 into Y_2 with domains of definition $D(A_1)$ and $D(A_2)$, respectively, such that $D(A_1) \subset D(A_2)$ and $A_1 x = A_2 x$ for all $x \in D(A_1)$, then if $X_1 = X_2$, $Y_1 = Y_2$, the operator A_1 is called a *compression* or *restriction* of the operator A_2, while A_2 is called an *extension* of A_1; if $X_1 \subset X_2$, A_2 is called an extension of A_1 exceeding X_1.

Many equations in function spaces or abstract spaces can be expressed in the form $Ax = y$, where $y \in Y$, $x \in X$; y is given, x is unknown and A is an operator from X into Y. The assertion of the existence of a solution to this equation for any right-hand side $y \in Y$ is equivalent to the assertion that the range of the operator A is the whole space Y; the assertion that the equation $Ax = y$ has a unique solution for any $y \in R(A)$ means that A is a one-to-one mapping from $D(A)$ onto $R(A)$.

If X and Y are vector spaces, then in the set of all operators from X into Y it is possible to single out the class of linear operators (cf. **Linear operator**); the remaining operators from X into Y are called *non-linear operators*. If X and Y are topological vector spaces, then in the set of operators from X into Y the class of continuous operators (cf. **Continuous operator**) can be naturally singled out, so are the class of *bounded linear operators A* (operators A such that the image of any bounded set in X is bounded in Y) and the class of compact linear operators (i.e. operators such that the image of any bounded set in X is pre-compact in Y, cf. **Compact operator**). If X and Y are locally convex spaces, then it is natural to examine different topologies on X and Y; an operator is said to be *semi-continuous* if it defines a continuous mapping from the space X (with the initial topology) into the space Y with the weak topology (the concept of semi-continuity is mainly used in the theory of non-linear operators); an operator is said to be *strongly continuous* if it is continuous as a mapping from X with the boundedly weak topology into the space Y; an operator is called *weakly continuous* if it defines a continuous mapping from X into Y where X and Y have the weak topology. Compact operators are often called *completely-continuous operators*. Sometimes the term 'competely-continuous operator' is used instead of 'strongly-continuous operator', or to denote an operator which maps any weakly-convergent sequence to a strongly-convergent one; if X and Y are reflexive Banach spaces, then these conditions are equivalent to the compactness of the operator. If an operator is strongly continuous, then it is weakly continuous.

The set $\Gamma(A) \subset X \times Y$ defined by the relation

$$\Gamma(A) = \{\{x, Ax\}: x \in D(A)\}$$

is called the *graph of the operator A*.

Let X and Y be topological vector spaces; an operator from X into Y is called a **closed operator** if its graph is closed. The concept of a closed operator is particularly useful in the case of linear operators with a dense domain of definition.

The concept of a graph allows one to generalize the concept of an operator: Any subset A in $X \times Y$ is called a *multi-valued operator* from X into Y; if X and Y are vector spaces, then a linear subspace in $X \times Y$ is called a multi-valued linear operator; the set

$$D(A) = \{x \in X:$$

there exists an $y \in Y$ such that $\{x, y\} \in A\}$

is called the domain of definition of the multi-valued operator.

If X is a vector space over a field k and $Y = k$, then an everywhere-defined operator from X into k is called a **functional** on X.

If X and Y are locally convex spaces, then an operator A from X into Y with a dense domain of definition in X has an **adjoint operator** A^* with a dense domain of definition in Y^* (with the weak topology) if, and only if, A is a closed operator.

Examples of operators: 1) The operator assigning the element $0 \in Y$ to any element $x \in X$ (the *zero operator*).

2) The operator mapping each element $x \in X$ to the same element $x \in X$ (the *identity operator* on X, written as id_X or 1_X).

3) Let X be a vector space of functions on a set M, and let f be a function on M; the operator on X with domain of definition

$$D(A) = \{\phi \in X: f\phi \in X\}$$

and acting according to the rule

$$A\phi = f\phi$$

if $\phi \in D(A)$, is called the *operator of multiplication by a function*; A is a linear operator.

4) Let X be a vector space of functions on a set M, and let F be a mapping from the set M into itself; the operator on X with domain of definition

$$D(A) = \{\phi \in X: \phi \circ F \in X\}$$

and acting according to the rule

$$A\phi = \phi \circ F$$

if $\phi \in D(A)$, is a linear operator.

5) Let X, Y be vector spaces of real measurable functions on two measure spaces (M, Σ_M, μ) and (N, Σ_N, ν), respectively, and let K be a function on $M \times N \times \mathbf{R}$, measurable with respect to the product measure $\mu \times \nu \times \mu_0$, where μ_0 is Lebesgue measure on $\mathbf{R}$, and continuous in $t \in \mathbf{R}$ for any fixed $m \in M$, $n \in N$. The operator from X into Y with domain of definition $D(A) = \{\phi \in X: f(x) = \int_M K(x, y, \phi(y)) \, dy\}$, which exists for almost-all $x \in N$ and $f \in Y$, and acting according to the rule $A\phi = f$ if $\phi \in D(A)$, is called an *integral operator*; if

$$K(x, y, z) = K(x, y)z, \quad x \in M, \quad y \in N, \quad z \in \mathbf{R},$$

then A is a linear operator.

6) Let X be a vector space of functions on a differentiable manifold M, let ξ be a vector field on M; the operator A on X with domain of definition

$$D(A) = \{f \in X: \text{the derivative } D_\xi f \text{ of the function } f$$

along the field ξ is everywhere defined and $D_\xi f \in X\}$

and acting according to the rule $Af = D_\xi f$ if $f \in D(A)$, is called a *differentiation operator*; A is a linear operator.

7) Let X be a vector space of functions on a set M; an everywhere-defined operator assigning to a function $\phi \in X$ the value of that function at a point $a \in M$, is a linear functional on X; it is called the δ-*function at the point a* and is written as δ_a.

8) Let G be a commutative locally compact group, let $\hat{G}$ be the group of characters of the group G, let dg, $\hat{dg}$ be the Haar measures on G and $\hat{G}$, respectively, and let

$$X = L_2(G, dg), \quad Y = L_2(\hat{G}, \hat{dg}).$$

The linear operator A from X into Y assigning to a function $f \in X$ the function $\hat{f} \in Y$ defined by the formula

$$\hat{f}(\hat{g}) = \int f(g)\hat{g}(g) \, dg$$

is everywhere defined if the convergence of the integral is taken to be mean-square convergence.

If X and Y are topological vector spaces, then the operators in examples 1) and 2) are continuous; if in example 3) the space X is $L_2(M, \Sigma_M, \mu)$, where μ is a measure on X, then the operator of multiplication by a bounded measurable function is closed and has a dense domain of definition; if in example 5) the space $X = Y$ is a Hilbert space $L_2(M, \Sigma_M, \mu)$ and $K(x, y, z) = K(x, y)z$, where $K(x, y)$ belongs to $L_2(M \times M, \Sigma_M \times \Sigma_M, \mu \times \mu)$, then A is compact; if in example 8) the spaces X and Y are regarded as Hilbert spaces, then A is continuous.

If A is an operator from X into Y such that $Ax \neq Ay$ when $x \neq y$, $x, y \in D(A)$, then the inverse operator A^{-1} to A can be defined; the question of the existence of an inverse operator and its properties is related to the theorem of the existence and uniqueness of a solution of the equation $Ax = f$; if A^{-1} exists, then $x = A^{-1}f$ when $f \in R(A)$.

For operators on a vector space it is possible to define a sum, multiplication by a number and an operator product. If A, B are operators from X into Y with domains of definition $D(A)$ and $D(B)$, respectively, then the operator, written as $A + B$, with domain of definition

$$D(A + B) = D(A) \cap D(B)$$

and acting according to the rule

$$(A + B)x = Ax + Bx$$

if $x \in D(A + B)$, is called the *sum of the operators A and B*.

The operator, written as λA, with domain of definition

$$D(\lambda A) = D(A)$$

and acting according to the rule

$$(\lambda A)x = \lambda(Ax)$$

if $x \in D(\lambda A)$, is called the *product of the operator A by the number* λ. The *operator product* is defined as composition of mappings: If A is an operator from X into Y and B is an operator from Y into Z, then the operator BA, with domain of definition

$$D(BA) = \{x \in X : x \in D(A) \text{ and } Ax \in D(B)\}$$

and acting according to the rule

$$(BA)x = B(Ax)$$

if $x \in D(BA)$, is called the product of B and A.

If P is an everywhere-defined operator on X such that $PP = P$, then P is called a *projection operator* or *projector* in X; if I is an everywhere-defined operator on X such that $I \circ I = \mathrm{id}_X$, then I is called an *involution* in X.

The theory of operators constitutes the most important part of linear and non-linear functional analysis, being in particular a basic instrument in the theory of dynamical systems, representations of groups and algebras and a most important mathematical instrument in mathematical physics and quantum mechanics.

References
[1] LIUSTERNIK, L.A. [L.A. LYUSTERNIK] and SOBOLEV, V.I.: *Elements of functional analysis*, F. Ungar, 1961 (translated from the Russian).
[2] KOLMOGOROV, A.N. and FOMIN, S.V.: *Elements of the theory of functions and functional analysis*, 1-2, Graylock, 1957-1961 (translated from the Russian).
[3] KANTOROVICH, L.V. and AKILOV, G.P.: *Functional analysis in normed spaces*, Pergamon, 1964 (translated from the Russian).
[4] DUNFORD, N. and SCHWARTZ, J.T.: *Linear operators*, 1-3, Interscience, 1958.
[5] EDWARDS, R.E.: *Functional analysis*, Holt, Rinehart & Winston, 1965.
[6] YOSHIDA, K.: *Functional analysis*, Springer, 1980.

M.A. Naĭmark
A.I. Shtern

Editorial comments.

References
[A1] KATO, T.: *Perturbation theory for linear operators*, Springer, 1976.
[A2] TAYLOR, A.E. and LAY, D.C.: *Introduction to functional analysis*, Wiley, 1980.
[A3] RIESZ, F. and SZÖKEFALVI-NAGY, B.: *Functional analysis*, F. Ungar, 1955 (translated from the French).
[A4] RUDIN, W.: *Functional analysis*, McGraw-Hill, 1973.
[A5] GOHBERG, I. and GOLDBERG, S.: *Basic operator theory*, Birkhäuser, 1981.

AMS 1980 Subject Classification: 47A05

OPERATOR ERGODIC THEOREM - A general name for theorems on the limit of means along an unboundedly lengthening 'time interval' $n = 0, \ldots, N$, or $0 \leqslant t \leqslant T$, for the powers $\{A^n\}$ of a **linear operator** A acting on a Banach space (or even on a topological vector space, see [5]) E, or for a **one-parameter semi-group** of linear operators $\{A_t\}$ acting on E (cf. also **Ergodic theorem**). In the latter case one can also examine the limit of means along an unboundedly diminishing time interval (*local ergodic theorems*, see [5], [6]; one also speaks of 'ergodicity at zero', see [1]). Means can be understood in various senses in the same way as in the theory of summation of series. The most frequently used means are the *Cesàro means*

$$\bar{A}_N = \frac{1}{N} \sum_{n=0}^{N-1} A^n$$

or

$$\bar{A}_T = \frac{1}{T} \int_0^T A_t \, dt$$

and the *Abel means*, [1],

$$\bar{A}_\theta = (1 - \theta) \sum_{n=0}^{\infty} \theta^n A^n, \quad |\theta| < 1,$$

or

$$\bar{A}_\lambda = \lambda \int_0^{\infty} e^{-\lambda t} A_t \, dt.$$

The conditions of ergodic theorems automatically ensure the convergence of these infinite series or integrals; under these conditions, although the Abel means are formed by using all A^n or A_t, the values of A^n or A_t in a finite period of time, unboundedly increasing when $\theta \to 1$ (or $\lambda \to 0$), play a major part. The limit of the means ($\lim_{N \to \infty} \bar{A}_N$, etc.) can be understood in various senses: In the strong or weak **operator topology** (*statistical ergodic theorems*, i.e. the **von Neumann ergodic theorem** — historically the first operator ergodic theorem — and its generalizations), in the uniform operator topology (*uniform ergodic theorems*, see [1], [2], [3]), while if E is a function space on a measure space, then also in the sense of almost-everywhere convergence of the means $\bar{A}_N \phi$, etc., where $\phi \in E$ (*individual ergodic theorems*, i.e. the **Birkhoff ergodic theorem** and its generalizations; see, for example, the **Ornstein−Chacon ergodic theorem**; these are not always called operator ergodic theorems, however). Some operator ergodic theorems compare the force of various of the above-mentioned variants with each other, establishing that, from the existence of limits of means in one sense, it follows that limits exist in another sense [1]. Some theorems speak not of the limit of means, but of the limit of the ratios of two means (e.g. the Ornstein−Chacon theorem).

There are also operator ergodic theorems for n-parameter and even more general semi-groups.

References

[1] HILLE, E. and PHILLIPS, R.: *Functional analysis and semi-groups*, Amer. Math. Soc., 1957.
[2] DUNFORD, N. and SCHWARTZ, J.T.: *Linear operators*, 1. General theory, Wiley, 1988.
[3] NEVEU, J.: *Mathematical foundations of the calculus of probabilities*, Holden-Day, 1965 (translated from the French).
[4] VERSHIK, A.M. and YUZVINSKIĬ, S.A.: 'Dynamical systems with invariant measure', *Progress in Math.* **8** (1970), 151-215. (*Itogi Nauk. Mat. Anal.* **967** (1969), 133-187)
[5] KATOK, A.B., SINAĬ, YA.G. and STEPIN, A.M.: 'Theory of dynamical systems and general transformation groups with invariant measure', *J. Soviet Math.* 7, no. 2 (1977), 974-1041. (*Itogi Nauk. i Tekhn. Mat. Anal.* **13** (1975), 129-262)
[6] KRENGEL, U.: 'Recent progress in ergodic theorems', *Astérisque* **50** (1977), 151-192.
[7] KRENGEL, U.: *Ergodic theorems*, de Gruyter, 1985.

D.V. Anosov

AMS 1980 Subject Classification: 47A35, 28DXX

OPERATOR GROUP - 1) A group of operators, a one-parameter group of operators (cf. **Operator**) on a **Banach space** E, i.e. a family of bounded linear operators U_t, $-\infty < t < \infty$, such that $U_0 = I$, $U_{s+t} = U_s \cdot U_t$ and U_t depends continuously on t (in the uniform, strong or weak topology). If E is a **Hilbert space** and $\| U_t \|$ is uniformly bounded, then the group $\{U_t\}$ is similar to a group of unitary operators (*Sz.-Nagy's theorem*, cf. also **Unitary operator**).

References

[1] SZÖKEVALFI-NAGY, B.: 'On uniformly bounded linear transformations in Hilbert space', *Acta Sci. Math. (Szeged)* **11** (1947), 152-157.
[2] HILLE, E. and PHILIPS, R.: *Functional analysis and semi-groups*, Amer. Math. Soc., 1948.

V.I. Lomonosov

2) A *group with operators*, a *group with domain of operators* Σ, where Σ is a set of symbols, is a **group** G such that for every element $a \in G$ and every $\sigma \in \Sigma$ there is a corresponding element $a\sigma \in G$ such that $(ab)\sigma = a\sigma \cdot b\sigma$ for any $a, b \in G$. Let G and G' be groups with the same domain of operators Σ; an isomorphic (a homomorphic) mapping ϕ of G onto G' is called an *operator isomorphism* (*operator homomorphism*) if $(a\sigma)\phi = (a\phi)\sigma$ for any $a \in G$, $\sigma \in \Sigma$. A subgroup (normal subgroup) H of the group G with domain of operators Σ is called an *admissible subgroup* (*admissible normal subgroup*) if $H\sigma \subseteq H$ for any $\sigma \in \Sigma$. The intersection of all admissible subgroups containing a given subset M of G is called the *admissible subgroup generated by the set M*. A group which does not have admissible normal subgroups apart from itself and the trivial subgroup is called a *simple group* (with respect to the given domain of operators). Every quotient group of an operator group by an admissible normal subgroup is a group with the same domain of operators.

A group G is called a *group with a semi-group of operators* Σ if G is a group with domain of operators Σ, Σ is a semi-group and $a(\sigma\tau) = (a\sigma)\tau$ for any $a \in G$, $\sigma, \tau \in \Sigma$. If Σ is a semi-group with an identity element ϵ, it is supposed that $a\epsilon = a$ for every $a \in G$. Every group with an arbitrary domain of operators Σ_0 is a group with semi-group of operators Σ, where Σ is the free semi-group generated by the set Σ_0. A group F with semi-group of operators Σ possessing an identity element is called Σ-*free* if it is generated by a system of elements X such that the elements $x\alpha$, where $x \in X$, $\alpha \in \Sigma$, constitute for F (as a group without operators) a system of free generators. Let F be a Γ-free group (Γ being a group of operators), let Δ be a subgroup of Γ, let $f \in F$, and let $A_{f,\Delta}$ be the admissible subgroup of F generated by all elements of the form $f^{-1}(f\alpha)$, where $\alpha \in \Delta$. Then every admissible subgroup of F is an operator free product of groups of type $A_{f,\Delta}$ and a Γ-free group (see [2]). If Σ is a free semi-group of operators, then, if $a \neq 1$, the admissible subgroup of the Σ-free group F generated by the element a is itself a Σ-free group with free generator a (see also [3]).

An Abelian group with an associative ring of operators K is just a K-module (cf. **Module**).

References

[1] KUROSH, A.G.: *The theory of groups*, 1-2, Chelsea, 1955-1956 (translated from the Russian).
[2] ZAVALO, S.T.: 'Γ-free operator groups', *Mat. Sb.* **33** (1953), 399-432 (in Russian).
[3A] ZAVALO, S.T.: 'S-free operator groups I', *Ukr. Mat. Zh.* **16**, no.

5 (1964), 593-602 (in Russian).

[3B] Zavalo, S.T.: 'S-free operator groups II', *Ukr. Mat. Zh.* **16**, no. 6 (1964), 730-751 (in Russian).

A.P. Mishina

AMS 1980 Subject Classification: 20F38, 47D10

OPERATOR HOMOMORPHISM - A **homomorphism** of algebraic systems (cf. **Algebraic system**) that commutes with each operator of a given set operating on these systems, i.e. a homomorphism of operator groups, operator rings, etc.

AMS 1980 Subject Classification: 08AXX

OPERATOR-IRREDUCIBLE REPRESENTATION -

A representation π of a group (algebra, ring, semigroup) X on a (topological) vector space E such that any (continuous) linear operator on E commuting with every operator $\pi(x)$, $x \in X$, is a scalar multiple of the identity operator on E. If π is a completely-irreducible representation (in particular, if π is a finite-dimensional irreducible representation), then π is an operator-irreducible representation; the converse is not always true. If π is a **unitary representation** of a group or a symmetric representation of a symmetric algebra, then π is an operator-irreducible representation if and only if π is an **irreducible representation**.

A.I. Shtern

Editorial comments.

References

[A1] Gel'fand, I.M., Graev, M.I. and Vilenkin, N.Ya.: *Generalized functions*, 5. Integral geometry and representation theory, Acad. Press., 1966, p. 149 ff (translated from the Russian).

[A2] Kirillov, A.A.: *Elements of the theory of representations*, Springer, 1976, p. 114 (translated from the Russian).

AMS 1980 Subject Classification: 16A64, 20M30, 20C99

OPERATOR RING

Editorial comments. 1) A ring consisting of linear operators on some vector space. The phrase 'operator ring' occurs occassionally in the meaning of **von Neumann algebra** or W^*-algebra.

2) A ring with operators.

AMS 1980 Subject Classification: 16A99, 46-XX, 47-XX

OPERATOR TOPOLOGY - A topology on the space $L(E, F)$ of continuous linear mappings from one **topological vector space** E into another topological vector space F, converting the space $L(E, F)$ into a topological vector space. Let F be a **locally convex space** and let $\mathfrak{S}$ be a family of bounded subsets of E such that the linear hull of the union of the sets of this family is dense in E. Let $\mathfrak{B}$ be a basis of neighbourhoods of zero in F. The family

$$M(S, V) = \{f : f \in L(E, F), f(S) \subset V\},$$

where S runs through $\mathfrak{S}$ and V through $\mathfrak{B}$, is a basis of neighbourhoods of zero for a unique topology that is invariant with respect to translation, which is an operator topology and which converts the space $L(E, F)$ into a locally convex space; this topology is called the $\mathfrak{S}$-topology on $L(E, F)$.

Examples. I) Let E, F be locally convex spaces. 1) Let $\mathfrak{S}$ be the family of all finite subsets in E; the corresponding $\mathfrak{S}$-topology (on $L(E, F)$) is called the *topology of simple* (or *pointwise*) *convergence*. 2) Let $\mathfrak{S}$ be the family of all convex balanced compact subsets of E; the corresponding topology is called the *topology of convex balanced compact convergence*. 3) Let $\mathfrak{S}$ be the family of all pre-compact subsets of E; the corresponding $\mathfrak{S}$-topology is called the *topology of pre-compact convergence*. 4) Let $\mathfrak{S}$ be the family of all bounded subsets; the corresponding topology is called the *topology of bounded convergence*.

II) If E, F are Banach spaces considered simultaneously in the weak or strong (norm) topology, then the corresponding spaces $L(E, F)$ coincide algebraically; the corresponding topologies of simple convergence are called the *weak* or *strong operator topologies* on $L(E, F)$. The strong operator topology majorizes the weak operator topology; both are compatible with the duality between $L(E, F)$ and the space of functionals on $L(E, F)$ of the form $f(A) = \sum \phi_i(A \xi_i)$, where $\xi_i \in E$, $\phi_i \in F^*$, $A \in L(E, F)$.

III) Let E, F be Hilbert spaces and let $\tilde{E}, \tilde{F}$ be countable direct sums of the Hilbert spaces E_n, F_n, respectively, where $E_n = E$, $F_n = F$ for all integer n; let ψ be the imbedding of the space $L(E, F)$ into $L(\tilde{E}, \tilde{F})$ defined by the condition that for any operator $A \in L(E, F)$ the restriction of the operator $\psi(A)$ to the subspace E_n maps E_n into F_n and coincides on E_n with the operator A. Then the complete pre-image in $L(E, F)$ of the weak (strong) operator topology on $L(\tilde{E}, \tilde{F})$ is called the *ultra-weak* (correspondingly, *ultra-strong*) *operator topology* on $L(E, F)$. The ultra-weak (ultra-strong) topology majorizes the weak (strong) operator topology. A symmetric subalgebra $\mathfrak{A}$ of the algebra $L(E)$ of all bounded linear operators on a Hilbert space E, containing the identity operator, coincides with the set of all operators from $L(E)$ that commute with each operator from $L(E)$ that commutes with all operators from $\mathfrak{A}$, if and only if $\mathfrak{A}$ is closed in the weak (or strong, or ultra-weak, or ultra-strong) operator topology, i.e. is a **von Neumann algebra**.

References

[1] Schaefer, H.H.: *Topological vector spaces*, Springer, 1971.
[2] Dunford, N. and Schwartz, J.T.: *Linear operators. General theory*, Wiley, 1988.

[3] NAĬMARK, M.A.: *Normed rings*, Reidel, 1984 (translated from the Russian).
[4] SAKAI, S.: *C*-algebras and W*-algebras*, Springer, 1971.

A.I. Shtern

AMS 1980 Subject Classification: 46H05

OPTIMAL CONTROL - A solution of a non-classical variational *problem of optimal control* (see **Optimal control, mathematical theory of**). In a typical case, an optimal control gives a solution of a problem concerning an extremum of a given functional along the trajectories of an ordinary differential equation depending on parameters (controls, inputs) (in the presence of supplementary constraints, prescribed by the formulation of the problem). Here, depending on the class of controls under consideration, an optimal control can take the form of a *function of time* (in a problem of **optimal programming control**) or a *function of time and current state* (position) of the system (in a problem of **optimal synthesis control**).

In more complicated or more specialized problems, an optimal control can take the form of a *generalized control*: A function of time with values in a set of measures, a functional of a segment of a trajectory or of a set in phase space, a boundary condition for a partial differential equation, a many-valued mapping, a sequence of extreme elements of a non-stationary problem of mathematical programming, etc.

A.B. Kurzhanskiĭ

Editorial comments. For alternative terminology and references see Optimal control, mathematical theory of.

AMS 1980 Subject Classification: 49-XX

OPTIMAL CONTROL, MATHEMATICAL THEORY OF - A part of mathematics in which a study is made of ways of formalizing and solving problems of choosing the best way, in an a priori described sense, of realizing a controlled dynamical process. This *dynamical process*, as a rule, can be described using differential, integral, functional, and finite-difference equations (or other formalized evolution relations, possibly involving stochastic aspects), depending on input functions or parameters, called *controls*, and usually subject to constraints. The sought controls, as well as the realization of the process itself, must generally be chosen according to certain constraints prescribed by the formulation of the problem.

In a more specific sense, it is accepted that the term 'mathematical theory of optimal control' be applied to a mathematical theory in which methods are studied for solving non-classical variational problems of optimal control (as a rule, with differential constraints), which permit the examination of non-smooth functionals and

arbitrary constraints on the control parameters or on other dependent variables (the constraints which are usually studied are given by non-strict inequalities). The term 'mathematical theory of optimal control' is sometimes given a broader meaning, covering the theory which studies mathematical methods of investigating problems whose solutions include any process of statistical or dynamical optimization, while the corresponding model situations permit an interpretation in terms of some applied procedure for adopting an optimal solution. With this interpretation, the mathematical theory of optimal control contains elements of **operations research; mathematical programming** and game theory (cf. **Games, theory of**).

Problems studied in the mathematical theory of optimal control have arisen from practical demands, especially in space flight dynamics and **automatic control theory** (see also **Variational calculus**). The formalization and solution of these problems have posed new questions, for example in the theory of ordinary differential equations, both in the area of generalizing the concept of a solution and generalizing conclusions from the appropriate conditions of existence, as in the study of dynamical and extremal properties of trajectories of controlled differential systems. In particular, the mathematical theory of optimal control has stimulated the study of the properties of differential inclusions (cf. **Differential inclusion**). The corresponding directions in the mathematical theory of optimal control are therefore often considered as part of the theory of ordinary differential equations. The mathematical theory of optimal control contains the mathematical basis of the theory of controlled motions, a new area in general mechanics in which the laws of creating controllable mechanical movements and related mathematical questions are investigated. In methods and in applications, the mathematical theory of optimal control is closely linked with analytical mechanics, especially with the areas relating to the **variational principles of classical mechanics**.

Although particular problems of optimal control and non-classical variational problems were encountered earlier, the foundations of the general mathematical theory of optimal control were laid in 1956 - 1961. The key point of this theory was the **Pontryagin maximum principle**, formulated by L.S. Pontryagin in 1956 (see [1]). The main stimuli in the creation of the mathematical theory of optimal control were the discovery of the method of **dynamic programming**, the explanation of the role of functional analysis in the theory of optimal systems, the discovery of connections between solutions of problems of optimal control and results of the theory of **Lyapunov stability** and the appearance of works relating to the concepts of controllability and observability of dynamical systems (see [2] - [5]). In the ensuing years the foundations have been laid of the theory of stochastic control and stochastic filtering of dynamical systems, general methods have been created for the solution of non-classical variational problems, generalizations of the basic statements of the mathematical theory of optimal control have been obtained for more complex classes of dynamical systems, and the connections with classical variational calculus have been studied (see [6] - [11]). The mathematical theory of optimal control is developing intensively, especially in the study of game problems in dynamics (see **Differential games**), problems of control with incomplete or imprecise information, systems with distributed parameters, equations on manifolds, etc.

The results of the mathematical theory of optimal control have found broad applications in the construction of control processes relating to diverse areas of modern technology, in the study of economic dynamics, and in the solution of a number of problems in the fields of biology, medicine, ecology, demography, etc.

A *problem of optimal control* can be described in general terms as follows:

1) A controllable system S is given, whose position at an instant of time t is represented by a value x (for example, by a vector of generalized coordinates and impulses of a mechanical system, or by a function in the spatial coordinates of a distributed system; by a probability distribution which characterizes the current state of a stochastic system, or by a vector of production output in a dynamic model of economy, etc.). It is assumed that a number of controls u may be applied to the system S affecting the dynamics of the system. The controls might take the form of mechanical forces, thermal or electrical potentials, investment programs, etc.

2) An equation is given which connects the variables x, u, t and describes the dynamics of the system. An instant of time is indicated at which the equation is considered. In a typical case one considers an ordinary differential equation of the form

$$\dot{x} = f(t, x, u), \quad t_0 \leqslant t \leqslant t_1, \quad x \in \mathbf{R}^n, \quad u \in \mathbf{R}^p, \qquad (1)$$

with previously stipulated properties of the function f (continuity of f in t, x, u and continuous differentiability in x are often required).

3) Information is available which can be used to construct the controls (for example, at any instant of time or at previously prescribed instances, the allowed measurable values of the phase coordinates of the system (1) or of functions in these coordinates become known). A class of functions describing the controls which can be considered is stipulated: The set of piecewise-continuous functions of the form $u = u(t)$, the set of linear functions in x of the form $u = u(t, x) = P'(t)x$ with continuous coefficients, etc.

4) Constraints are imposed on the process to be realized. At this point in particular, the conditions defining the aim of the control come into consideration (for example, for the system (1), to hit a given point or a given set of the phase space $\mathbf{R}^n$, the demand for stabilization of the solutions around a given motion, etc.). Furthermore, constraints can be imposed on the values of the controls u or on the coordinates of the position x, on functions in these variables, on functionals in their realizations, etc. In the system (1), for example, constraints on the control parameters

$$u \in U \subseteq \mathbf{R}^p \quad \text{or} \quad \phi(u) \leqslant 0, \quad \phi: \mathbf{R}^p \to \mathbf{R}^k, \qquad (2)$$

and on the coordinates

$$x \in X \subseteq \mathbf{R}^n \quad \text{or} \quad \psi(x) \leqslant 0, \quad \psi: \mathbf{R}^n \to \mathbf{R}^l, \qquad (3)$$

are possible; here, U, X are closed sets and ϕ, ψ are differentiable functions. More complex situations can also be examined, where the set U depends on t, x or when an inequality in the form $g(t, x, u) \leqslant 0$ (a case of mixed constraints) is given, etc.

5) An index (a criterion) is given of the quality of the process to be realized. It can take the form of a functional $J(x(\cdot), u(\cdot))$ in the realization of the variables x, u over the period of time under consideration. Conditions 1) - 4) are now supplemented by the requirement of optimality of the process: a minimum, maximum, minimax, etc., of the index $J(x(\cdot), u(\cdot))$.

In this way, in a given class of controls for a given system, a control u must be chosen which optimizes (i.e. minimizes or maximizes) the index $J(x(\cdot), u(\cdot))$ (under the condition that the aim of the control is achieved and that the applied constraints are fulfilled). A function (e.g. of the form $u = u(t)$ or $u = u(t, x)$, etc.) which solves the problem of optimal control is called an *optimal control* (for an example of a typical statement of an optimal control problem see **Pontryagin maximum principle**).

Among the dynamic objects embraced by the problems of the mathematical theory of optimal control, it is customary to differentiate between the finite-dimensional and the infinite-dimensional ones, dependent on the dimension of the phase space of the corresponding systems of differential equations describing them, or on the form of the constraints imposed on the phase variables.

There is a difference between problems of **optimal programming control** and of **optimal synthesis control**. In the former, the control u takes the form of a function of time; in the latter, it takes the form of a control strategy according to the feedback principle, as a function from the permissible values of the current parameters of the process. Optimal stochastic control problems are of the second type.

The mathematical theory of optimal control is concerned with questions of the existence of solutions, the derivation of necessary conditions for an extremum (optimality of a control), research into sufficient conditions, and the construction of numerical algorithms. The relations between solutions of problems of the mathematical theory of optimal control obtained in the class of programming and synthesis controls are also studied.

The different formulations of optimal control problems described above assume the existence of a correct mathematical model of the process and are calculated on the basis of complete a priori or even complete current information about the corresponding system. However, in applied formulations the accessible information on the system (for example, information about initial and final conditions, about coefficients in the corresponding equations, about the values of supplementary parameters of permissible measurable coordinates, etc.) is frequently insufficient for the direct use of the above theory. This leads to optimal control of problems formulated in other terms of information. A large section of the mathematical theory of optimal control is dedicated to problems where the description of insufficient quantities has a statistical character (the so-called theory of *stochastic optimal control*). If any statistical information on insufficient quantities is lacking, but only the areas in which they may alter are given, then the corresponding problems are examined within the framework of the theory of *optimal control under conditions of uncertainty*. Minimax and game-theoretical methods are then used to solve these problems. Problems of optimal stochastic control and of optimal control under conditions of uncertainty are especially interesting in the area of optimal synthesis control.

Although the formalized description of controllable systems can take a fairly abstract form (see [11]), the simplest classification also allows them to be divided into *continuous-time systems* (described, for example, by differential equations, either ordinary or partial, by equations with a deviating argument, by equations in a Banach space, as well as by differential inclusions, integral and integro-differential equations, etc.) and *multi-stage (discrete) systems*, described by recurrence difference equations and examined only at isolated (discrete) moments of time.

Discrete control systems, apart from being interesting in themselves, have great significance as finite-difference models of continuous systems. This is important for the construction of numerical methods for solving problems of optimal control (see [12], [13]), especially in those cases where the initial problem is subject to discretization, beginning with its own state. The basic formulations of the tasks shown above are also used for discrete systems. Although the functional-theoretic side of the research proves to be simpler here,

a transfer of basic facts of the theory of optimal control for continuous systems and their presentation in compact form entails particular difficulties and is not always possible (see [14], [15]).

The theory of optimal linear discrete control systems with constraints defined by convex functions has been dealt with in a reasonably complete way (see [15]). It goes hand in hand with methods of linear and convex programming (especially with corresponding 'dynamic' or 'non-stationary' variants, see [16]). In this theory great importance is attached to solutions which allow the optimization of a discrete dynamical system to lead to the realization of an adequate numerical discrete algorithm.

Another class of problems in the mathematical theory of optimal control is generated by questions of approximation of solutions of problems of optimal control for continuous systems by discrete ones and by questions which are closely connected with the problem of regularizing ill-stated problems (see [17]).

References

[1] PONTRYAGIN, L.S., BOLTYANSKIĬ, V.G., GAMKRELIDZE, R.V. and MISCHCHENKO, E.F.: *The mathematical theory of optimal processes*, Wiley, 1967 (translated from the Russian).

[2] BELLMAN, R.: *Dynamic programming*, Princeton Univ. Press, 1957.

[3] KRASOVSKIĬ, N.N.: *The theory of control of motion*, Moscow, 1968 (in Russian).

[4] KRASOVSKIĬ, N.N.: 'Theory of optimal control systems', in *Mechanics in the USSR during 50 years*, Vol. 1, Moscow, 1968, pp. 179-244 (in Russian).

[5] KALMAN, R.: 'On the general theory of control systems', in *Proc. first Internat. Congress Internat. Fed. Autom. Control*, Vol. 2, Moscow, 1960, pp. 521-547.

[6] FLEMING, W.H. and RISHEL, R.: *Deterministic and stochastic optimal control*, Springer, 1975.

[7] ALEKSEEV, V.M., TIKHOMIROV, V.M. and FOMIN, S.V.: *Optimal control*, Consultants Bureau, 1987 (translated from the Russian).

[8] VARGA, J.: *Optimal control of differential and functional equations*, Acad. Press, 1972.

[9] HESTENES, M.: *Calculus of variations and optimal control theory*, Wiley, 1966.

[10] YOUNG, L.: *Lectures on the calculus of variations and optimal control theory*, Saunders, 1969.

[11] KALMAN, R., FALB, P. and ARBIB, M.: *Topics in mathematical systems theory*, McGraw-Hill, 1969.

[12] MOISEEV, N.N.: *Elements of the theory of optimal systems*, Moscow, 1975 (in Russian).

[13] CHERNOUS'KO, F.L. and KOLMANOVSKIĬ, V.B.: 'Computational and approximate methods for optimal control', *J. Soviet Math.* **12**, no. 3 (1979), 310-353. (*Itogi Nauk. i Tekhn. Mat. Anal.* **14** (1977), 101-166)

[14] BOLTYANSKIĬ, V.G.: *Optimal control of discrete systems*, Wiley, 1978 (translated from the Russian).

[15] CANON, M.D., CULLUM, C.D. and POLAK, E.: *Theory of optimal control and mathematical programming*, McGraw-Hill, 1970.

[16] PROPOĬ, A.I.: *Elements of the theory of optimal discrete processes*, Moscow, 1973 (in Russian).

[17] TIKHONOV, A.N. and ARSENIN, V.YA.: *Solutions of ill-posed problems*, Winston, 1977 (translated from the Russian).

A.B. Kurzhanskiĭ

Editorial comments. In the Western literature (optimal) programming control is usually referred to as *(optimal) open-loop control*, whereas (optimal) synthesis control is usually called *(optimal) feedback* or *closed-loop control*.

Early books [A5], [A8], [A9] on the subject are based on variational methods and dynamic programming. An introduction to dynamic programming is provided in the textbooks [A3], [A7]. The measure-theoretic difficulties of this approach are treated in [A4], [A11]. The approach in which the optimal stochastic control problems with continuous state spaces are approximated by those with discrete state spaces, is explored in [A10]. Books on optimal stochastic control by variational and functional-analysis methods, written by researchers of the French school, are [A1], [A2]. A stochastic maximum principle is derived in [A6].

References

[A1] BENSOUSSAN, A.: *Stochastic control by functional analysis methods*, North-Holland, 1982.

[A2] BENSOUSSAN, A. and LIONS, J.L.: *Applications of variational inequalities in stochastic control*, North-Holland, 1982.

[A3] BERTSEKAS, D.P.: *Dynamic programming: Deterministic and stochastic models*, Prentice-Hall, 1987.

[A4] BERTSEKAS, D.P. and SHREVE, S.E.: *Stochastic optimal control: The discrete time case*, Acad. Press, 1978.

[A5] FLEMING, W.H. and RISHEL, R.W.: *Deterministic and stochastic optimal control*, Springer, 1975.

[A6] HAUSMANN, U.G.: *A stochastic maximum principle for optimal control of diffusion*, Longman, 1986.

[A7] KUMAR, P.R. and VARAIYA, P.: *Stochastic systems: Estimation, identification, and adaptive control*, Prentice-Hall, 1986.

[A8] KUSHNER, H.J.: *Stochastic stability and control*, Acad. Press, 1967.

[A9] KUSHNER, H.J.: *Introduction to stochastic control*, Holt, Rinehart & Winston, 1971.

[A10] KUSHNER, H.J.: *Probability methods for approximations in stochastic control and for elliptic equations*, Acad. Press, 1977.

[A11] STRIEBEL, C.: *Optimal control of discrete time stochastic systems*, Lecture notes in economic and mathematical systems, 110, Springer, 1975.

[A12] BRYSON, A.E. and HO, Y.-C.: *Applied optimal control*, Ginn, 1969.

[A13] LUENBERGER, D.G.: *Optimization by vector space methods*, Wiley, 1969.

[A14] BERTSEKAS, D.: *Dynamic programming and stochastic control*, Acad. Press, 1976.

[A15] DAVIS, M.H.A.: 'Martingale methods in stochastic control', in *Stochastic Control and Stochastic Differential Systems*, Lecture notes in control and inform. sci., Vol. 16, Springer, 1979, pp. 85-117.

[A16] CESARI, L.: *Optimization - theory and applications*, Springer, 1983.

[A17] NEUSTADT, L.W.: *Optimization, a theory of necessary conditions*, Princeton Univ. Press, 1976.

[A18] BARBU, V. and DA PRATO, G.: *Hamilton—Jacobi equations in Hilbert spaces*, Pitman, 1983.

[A19] LJUNG, L.: *System identification theory for the user*, Prentice-Hall, 1987.

AMS 1980 Subject Classification: 49-XX, 93-XX

OPTIMAL DECODING - A decoding which maximizes the exactness of information reproduction (cf. **Information, exactness of reproducibility of**) for given sources of information, communication channels and coding methods. In case the exactness of information reproduction is characterized by the mean probability of erroneous decoding (cf. **Erroneous decoding, probability of**), optimal decoding minimizes this probability. For example, for a transmission M of information represented by the numbers $1, \ldots, M$, the probabilities of appearance of which are $p_1, \ldots, p_M$, respectively, a discrete channel is used with a finite number of input and output signals and with transition function defined by the matrix

$$q(y, \tilde{y}) = P\{\tilde{\eta} = \tilde{y} \mid \eta = y\}, \quad y \in Y, \ \tilde{y} \in \tilde{Y},$$

where Y, $\tilde{Y}$ are the sets of values of the signals of the input η and the output $\tilde{\eta}$, respectively, while the coding is defined by the function $f(\cdot)$ for which $f(m) = y_m$, $m = 1, \ldots, M$, where $y_m \in Y$, $m = 1, \ldots, M$, is a **code**, i.e. a choice of M possible input values. Then an optimal decoding is defined by a function $g(\cdot)$ such that $g(\tilde{y}) = m'$ for any $\tilde{y} \in \tilde{Y}$, where m' satisfies the inequality

$$p_{m'} q(y_{m'}, \tilde{y}) \geqslant p_m q(y_m, \tilde{y})$$

for all $m \neq m'$. In particular, if all information has equal probability, i.e. $p_1 = \cdots = p_M = 1/M$, then the described optimal decoding is also a decoding by the method of 'maximum likelihood' (which is generally not optimal): The output signal $\tilde{y}$ has to be decoded in the information m' for which

$$q(y_{m'}, \tilde{y}) \geqslant q(y_m, \tilde{y}) \quad \text{if } m \neq m'.$$

References

[1] GALLAGHER, R.: *Information theory and reliable communication*, Wiley, 1968.

[2] WOZENCRAFT, J.M. and JACOBS, I.M.: *Principles of communication engineering*, Wiley, 1965.

P.L. Dobrushin
V.V. Prelov

Editorial comments. Cf. also **Coding and decoding**; **Information theory**.

AMS 1980 Subject Classification: 94A05, 94B35

OPTIMAL GUARANTEE STRATEGY - A strategy whose efficiency in a given situation is equal to the best guaranteed result (see **Principle of the largest sure result**). If, for example, in a situation with efficiency criterion $f(x, y)$ the undefined factor y takes values from a set Y, then the optimal guarantee strategy $\tilde{x}^*$ satisfies the equality

$$\sup_{\tilde{x}} \inf_{y \in Y} f(\tilde{x}^*, y) = \inf_{y \in Y} f(\tilde{x}^*, y).$$

If the last upper bound over $\tilde{x}$ is not attained, then the concept of an *ϵ-optimal guarantee strategy* $\tilde{x}_\epsilon^*$ arises, for which

$$\inf_{y \in Y} f(\tilde{x}_\epsilon^*, y) \geqslant \sup_{\tilde{x}} \inf_{y \in Y} f(\tilde{x}, y) - \epsilon,$$

where $\epsilon > 0$. Dependent on the set of strategies $\tilde{x} = x(y)$ and the information on the undefined factor (the conditions under which the operation is carried out), the optimal guarantee strategy is concretely defined (see [1]). So, if the set of strategies $\tilde{x}$ comprises all functions $x(y)$ and the operation contains complete information on y, then the optimal guarantee strategy $x^*(y)$ is called the *absolutely optimal strategy* and is defined by the condition

$$\sup_x f(x, y) = f(x^*(y), y) \text{ for all } y \in Y.$$

Optimal strategies corresponding to other principles of optimality are also studied (see, for example, [2] and [4]).

References

[1] GERMEĬER, YU.B.: *Introduction to the theory of operations research*, Moscow, 1971 (in Russian).
[2] GERMEĬER, YU.B.: *Games with non-conflicting interests*, Moscow, 1976 (in Russian).
[3] AUBIN, J.-P.: *L'analyse non-linéaire et ses motivations économiques*, Masson, 1984.
[4] BOROB'EV, N.N.: *Game theory. Lectures for economists and cyberneticists*, Leningrad, 1974 (in Russian).

F.I. Ereshko
V.V. Fedorov

Editorial comments. The phrase '*worst case strategy*' is also used for 'optimal guarantee strategy'. It provides a security level for the outcome of the efficiency criterion. Worst case designs naturally show up in two-person zero-sum games in which uncertainties of unknowns (the y-variable) are replaced by the worst possible. This idea is age old and universal in engineering and military analysis.

References

[A1] HO, Y.C. and OLSDER, G.J.: 'Differential games: concepts and applications', in M. Shubik (ed.): *Mathematics of Conflict*, Elsevier & North-Holland, 1983, pp. 127-186.

AMS 1980 Subject Classification: 90BXX, 90DXX, 93-XX

OPTIMAL PROGRAMMING CONTROL - A solution of a problem in the mathematical theory of optimal control (cf. **Optimal control, mathematical theory of**), in which the control $u = u(t)$ is formed as a function of time (whereby it is also supposed that, during the process, no information, apart from that which is given at the very beginning, is obtained). In this way, an optimal programming control is formed through a priori information on the system and cannot be corrected, unlike an **optimal synthesis control**.

The problem of existence of solutions to a problem of optimal programming control breaks down into two questions: explaining the possibility of realizing the aim of the control with the given constraints (the existence of an *admissible control* which realizes the aim of the control) and establishing the solvability of an extremum problem — attainability of the (as a rule, relative) extremum — in the aforementioned class of admissible controls (the existence of an *optimal control*).

It is particularly important, in relation to the first question, to study the *property of controllability of a system*. For a system

$$\frac{dx}{dt} = f(t, x, u)$$

it signifies the existence in a given class $U = \{u(\cdot)\}$ of admissible control functions $u(t)$ which transfer a phase point (see **Pontryagin maximum principle**) from any given starting position $x(t_0) = x^0 \in \mathbf{R}^n$ to any given final position $x(t_1) = x^1 \in \mathbf{R}^n$ (for a fixed or free time $T = t_1 - t_0$, dependent on the formulation of the problem). The necessary and sufficient conditions for controllability (or for *complete controllability*) are known in computable form for linear systems

$$\dot{x} = A(t)x + B(t)u, \quad x \in \mathbf{R}^n, \ u \in \mathbf{R}^p, \tag{1}$$

with analytic or periodic coefficients (these are simplest when $A \equiv \text{const}$, $B \equiv \text{const}$). For general linear systems the question of solvability of the problem of transfer from one convex set onto another (with convex constraints on u, x) has also been completely solved. In non-linear systems, only local conditions of controllability (in a small neighbourhood of the given trajectory) or conditions for particular classes of systems (see [2], [4], [5]) are known. The property of controllability has also been studied for numerous generalizations relating, in particular, to special classes U (for example, the set U of all bounded, piecewise-continuous controls $u(t)$), for controllability of some of the coordinates, or for more general classes of systems, including infinite-dimensional ones.

The question of the existence of an optimal control is in general related to a compactness property, in some topology, of minimizing sequences of controls or trajectories, and to the property of semi-continuity in the corresponding variables of the minimizing functionals. For a system

$$\dot{x} = f(t, x, u), \quad t_0 \leqslant t \leqslant t_1, \quad x \in \mathbf{R}^n, \ u \in \mathbf{R}^p, \tag{2}$$

under the constraints

$$u \in U \subseteq \mathbf{R}^p \tag{3}$$

the first of these properties is associated with convexity of the set

$$f(t, x, U) = \{f(t, x, u): u \in U\},$$

while the second (for integral functionals) is linked with convexity in the corresponding values of $J(x(\cdot), u(\cdot))$. The lack of these properties is compensated for by broadening the initial variational problems. So, the non-convexity of $f(t, x, u)$ can be compensated for by the introduction of *sliding systems*: Generalized solutions of ordinary differential equations, generated by control-measures given on U and creating an effect of 'convexification' (see [6], [7]; cf. also **Optimal sliding** regime). The absence of convexity in integral functionals $J(x(\cdot), u(\cdot))$ is compensated for by imbedding the problem in a more general one, with a new functional which is a convex minorant of the previous one, and imbedding the solution of the new problem in a broader class of controls (see [9]). In the instances shown, the existence of an optimal control often follows from the existence of an admissible control.

The theory of necessary conditions for an extremum is most developed in problems of optimal programming control. The Pontryagin maximum principle has served as a basic result in this case, as it includes necessary conditions for a strong extremum in a problem of optimal control.

General methods for obtaining necessary conditions in extremum problems were created and are used effectively for problems of optimal programming control with more complex constraints (such as phase, functional, minimax, mixed, etc.). They are based, in one way or another, on theorems of separability of convex cones (see [9], [10]). For example, let E be a vector space, let $f(x)$, $x \in E$, be a given functional, let Q_i be a set from E, let

$$Q = \bigcap_1^n Q_i,$$

let $x^0 \in Q$ be a point at which $f(x)$ reaches its minimum on Q, and let

$$Q_0 = \{x : f(x) < f(x^0)\}.$$

The essence of one wide-spread general method is that every one of the sets Q_i, $i = 0, \ldots, n$, is approximated in a neighbourhood of the point x^0 by a convex cone K_i with vertex at x^0 (the 'descent cone' for Q_0; the cone of 'admissible directions' for constraints of inequality type; the cone of 'tangent directions' for equality-type constraints, including differential links, etc.). A necessary condition for a minimum is now that x^0 is the single point common to all K_i, $i = 0, \ldots, n$, and, consequently, that the cones are 'separable' (see [8]). An analytic form is added to the latter 'geometric' condition and, where possible, is put into an appropriate form, for example, using a Hamilton function. Dependent on the initial constraints, as well as on the class of applicable variations, necessary conditions can take both a form analogous to the Pontryagin maximum principle and a form of a local (linearized) maximum principle (of a condition of a weak extremum with respect to u). The realization of this method thus depends on the possibility of describing the cones K_i analytically. Their effective description has been achieved for sets Q_i defined by smooth functions which satisfy certain supplementary regularity conditions at the examined point or by convex functions (see [9], [10]). In principle, this method also permits generalizations to the case of non-smooth constraints, including

differential ones. Here, for example, the concept of a subdifferential of a convex function and its generalization can be used, when convexity is lacking (see [11], [12]).

Conditions of the first order, analogous to the Pontryagin maximum principle, are known for solutions in the class of generalized function-measures (the so-called *integral maximum principle*), for controllable systems which can be described by differential equations with a perturbed argument, partial differential equations, evolution equations in a Banach space, differential equations on manifolds, recurrence difference equations, etc. (see [1], [6], [7], [13] - [16]).

From the necessary conditions for an extremum given in a problem of optimal control follow the well-known necessary first-order conditions of the classical variational calculus. In particular, in a two-point boundary value problem for the systems (2) and (3), where U is an open set, $J(x(\cdot), u(\cdot))$ is a standard integral functional, the Pontryagin principle implies the Weierstrass necessary condition for an extremum in the classical variational calculus.

Methods are being developed in the theory of optimal control to obtain necessary higher-order conditions (especially, of the second order) for non-classical variational problems (see [19]). Interest in higher-order conditions has largely been related to the study of degenerate problems of optimal control, leading to the so-called *special controls*, which do not have adequate analogues in the classical theory. For example, in the Pontryagin principle, the function $H(t, \psi, x, u)$ can either lead to a whole family of controls, each of which satisfies the maximum principle, or does not depend on u at all (in which case any of the admissible values of u satisfies the Pontryagin principle). This situation is quite characteristic of a whole series of applied control problems in space. In the given case, the isolation of an optimal control already requires a mastering of the first-order extremals (the so-called *Pontryagin extremals*) and the use for them of necessary second-order optimality conditions (or, in general, those of higher order). Different forms of necessary conditions have been obtained here by the use of special classes of 'non-classical' variations (for example, a 'bundle' of needle variations, etc.). The realization of special controls is also often connected with the use of sliding systems (see [17], [18]).

The theory of sufficient conditions of optimality has not been examined in much detail. Results are known which relate to conditions of local optimality and contain, among other simple requirements, conditions of non-degeneracy of the variational system and constraints on the properties of the Hessian of the right-hand sides, calculated along an admissible trajectory

530

for the corresponding ordinary differential equation. Another group of sufficient conditions is based on the method of dynamic programming and its relation to the theory of the maximum principle (see [8]). There are also formalisms which lead to sufficient conditions for an absolute minimum, based on the idea of broadening variational problems. Their domain of practical applicability embraces special classes of problems with convexity criteria and degenerate problems of optimal control (see [18]).

The complete solution of a problem of optimal programming control (necessary and sufficient conditions of optimality) is known for linear systems (1) when both the functionals and constraints on u, x are convex (in a number of cases certain extra conditions have to be fulfilled here). The concept of duality, used in convex analysis, has demonstrated a particular extremal property of the trajectories of the system, described by *conjugate variables* in the Pontryagin maximum principle. This has enabled one to reduce the boundary value problem arising from the application of necessary conditions of general type to a solution of a simpler dual extremal problem. Within the framework of this approach the theory of linear systems with *impulse controls* has been developed, which simulates objects subject to instantaneous influences (percussion, explosive, impulse), and which is formalized by the use of differential equations in generalized functions with corresponding orders of singularity. The method of *attainability domains* (see [2], [3]) has been put to effective use, especially in the theory of game systems.

In the absence of complete a priori information on the system (including the statistical description of insufficient quantities) the problem of optimal programming control is studied under *conditions of uncertainty*. In the system ($t_0 \leqslant t \leqslant t_1$)

$$\dot{x} = f(t, x, u, w), \quad x(t_0) = x^0 \in X^0, \quad w \in W, \qquad (4)$$

let the parameter $w \in \mathbf{R}^q$, realized in the form of time functions $w = w(t)$, and the vector x^0 be unknown, and let only the sets $X^0 \subseteq \mathbf{R}^n$, $W \subseteq \mathbf{R}^q$ be given. Then, assuming the existence and extendability to $[t_0, t_1]$ of the solutions

$$x(t \mid x^0, u(\cdot), w(\cdot)), \quad x(t_0 \mid x^0, u(\cdot), w(\cdot)) = x^0,$$

of equation (4) (for given $x^0, u(\tau), w(\tau), t_0 \leqslant \tau \leqslant t_1$), a bundle (ensemble) of trajectories can be formed:

$$X(t \mid u(\cdot)) =$$
$$= \bigcup \{x(t \mid x^0, u(\cdot), w(\cdot)): x^0 \in X^0, \ w(\tau) \in W,$$
$$t_0 \leqslant \tau \leqslant t\}.$$

By selecting a programming control $u(t)$ (the same one for all trajectories of the bundle), it is possible to control the position $X(t \mid u(\cdot))$ in the phase space. A typical problem of optimal programming control under conditions of uncertainty consists of the optimization of $u(t)$ along a functional Φ of maximum type:

$$\Phi(X(t_1 \mid u(\cdot))) = \max\{\phi(x): x \in X(t_1 \mid u(\cdot))\} \qquad (5)$$

(then the solution $u^0(t)$ of the problem will ensure a guaranteed result) or of an integral functional

$$\Phi(X(t_1 \mid u(\cdot))) = \int_{X(t_1 \mid u(\cdot))} f_0(x)\,dx. \qquad (6)$$

The use of the technique of inferring necessary conditions of optimality or its modifications has enabled one to formulate requirements which ensure the existence of analogues of the Pontryagin principle for the problems (5), (6) (in the first instance it takes the form of a minimax condition). For linear systems, these problems allow just as detailed a solution as for systems with complete information (see [3], [20], [21]).

References

[1] PONTRYAGIN, L.S., BOLTYANSKIĬ, V.G., GAMKRELIDZE, R.V. and MISHCHENKO, E.F.: *The mathematical theory of optimal processes*, Wiley, 1967 (translated from the Russian).

[2] KRASOVSKIĬ, N.N.: *The theory of control of motion*, Moscow, 1968 (in Russian).

[3] KRASOVSKIĬ, N.N. and SUBBOTIN, A.I.: *Game-theoretical control problems*, Springer, 1988 (translated from the Russian).

[4] KALMAN, R.: 'On the general theory of control system', in *Proc. 1st Internat. Congress Internat. Fed. on Autom. Control*, Vol. 2, Moscow, 1961, pp. 521-547 (in Russian).

[5] LEE, E.B. and MARCUS, L.: *Foundations of optimal control theory*, Wiley, 1967.

[6] GAMKRELIDZE, R.V.: *Principles of optimal control theory*, Plenum, 1978 (translated from the Russian).

[7] VARGA, J.: *Optimal control of differential and functional equations*, Acad. Press, 1972.

[8] IOFFE, A.D. and TIKHOMIROV, V.M.: 'Duality of convex functions and extremal problems', *Russian Math. Surveys* **23**, no. 6 (1968), 53-124. (*Uspekhi Mat. Nauk.* **23**, no. 6 (1968), 51-116)

[9] DUBOVITSKIĬ, A.YA and MILYUTIN, A.A.: 'Extremum problems in the presence of restrictions', *USSR Comp. Math. Math. Phys.* **5**, no. 3 (1965), 1-80. (*Zh. Vychisl. Mat. i Mat. Fiz.* **5**, no. 3 (1965), 395-453)

[10] NEUSTADT, L.W.: *Optimization: A theory of necessary conditions*, Princeton Univ. Press, 1976.

[11] PSHENICHNYĬ, B.N.: *Necessary conditions for an extremum*, Moscow, 1969 (in Russian).

[12] CLARKE, F.H.: 'Generalized gradients and applications', *Trans. Amer. Math. Soc.* **205** (1975), 247-262.

[13] SUSSMANN, H.J.: 'Existence and uniqueness of minimal realizations of nonlinear systems', *Math. Syst. Theory* **10**, no. 3 (1977), 263-284.

[14] LIONS, J.: *Optimal control of systems governed by partial differential equations*, Springer, 1971 (translated from the French).

[15] BOLTAYANSKIĬ, V.G.: *Mathematical methods of optimal control*, Holt, Rinehart & Winston, 1971 (translated from the Russian).

[16] BOLTYANSKIĬ, V.G.: *Optimal control of discrete systems*, Wiley, 1978 (translated from the Russian).

[17] GABASOV, R. and KIRILLOVA, F.M.: *Special optimal control*, Moscow, 1973 (in Russian).

[18] KROTOV, V.F., BUKREEV, V.Z. and GURMAN, V.I.: *New methods of variational calculus in flight dynamics*, Moscow, 1969 (in Russian).

[19] LEVITIN, E.S., MILYUTIN, A.A. and OSMOLOVSKIĬ, N.P.: 'Conditions of higher order for a local minimum in problems with constraints', *Russian Math. Surveys* **33**, no. 6 (1978), 97-

168. (*Uspekhi Mat. Nauk.* **33**, no. 6 (1978), 85-148)

[20] KURZHANSKIĬ, A.B.: *Control and observability under conditions of uncertainty*, Moscow, 1977 (in Russian).
[21] DEMVYANOV, V.F. and MALOZEMOV, V.N.: *Introduction to minimax*, Moscow, 1972 (in Russian).

A.B. Kurzhanskiĭ

Editorial comments. An optimal programming control is usually called an *optimal open-loop control* in the Western literature, while an optimal synthesis control is better known as an *optimal closed-loop control* or *optimal feedback control*. See also **Optimal control, mathematical theory of.**

References

[A1] FLEMING, W.H. and RISHEL, R.W.: *Deterministic and stochastic control*, Springer, 1975.
[A2] BERTSEKAS, D. and SHREVE, S.: *Stochastic optimal control, the discrete time case*, Acad. Press, 1978.
[A3] BERTSEKAS, D.: *Dynamic programming and stochastic control*, Acad. Press, 1976.
[A4] DAVIS, M.H.A.: 'Martingale methods in stochastic control', in *Stochastic Control and Stochastic Differential Systems*, Lecture notes in control and inform. sci., Vol. 16, Springer, pp. 85-117.
[A5] CESARI, L.: *Optimization - Theory and applications*, Springer, 1983.
[A6] BARBU, V. and DA PRATO, G.: *Hamilton—Jacobi equations in Hilbert spaces*, Pitman, 1983.
[A7] KUSHNER, H.: *Introduction to stochastic control*, Holt, 1971.
[A8] KUMAR, P.R. and VARAIYA, P.: *Stochastic systems: estimation, identification and adaptive control*, Prentice-Hall, 1986.
[A9] LJUNG, L.: *System identification theory for the user*, Prentice-Hall, 1987.
[A10] BRYSON, A.E. and HO, Y.-C.: *Applied optimal control*, Ginn, 1969.
[A11] KNOBLOCH, H.W.: *Higher order necessary conditions in optimal control theory*, Springer, 1981.

AMS 1980 Subject Classification: 49AXX, 93B05, 93C60

OPTIMAL QUADRATURE - A **quadrature formula** giving the best approximation to the integral

$$I(f) = \int_{\Omega} f(P)\omega(P)\,dP$$

for a class F of integrands. If

$$S_N(f) = \sum_{k=1}^{N} c_k f(P_k),$$

then

$$R_N(f) = S_N(f) - I(f)$$

is called the *quadrature error* when calculating the integral of a given function, while

$$r_N(F) = \sup_{f \in F} |R_N(f)|$$

is called the *quadrature error* in the class F. If a quadrature formula exists such that for the corresponding $r_N(F)$ the equality

$$r_N(F) = \inf_{c_k, P_k} r_N(F)$$

holds, then this formula is called the *optimal quadrature in this class*.

Optimal quadratures have only been found for cer-

tain classes of functions which, basically, depend on one variable (see [1] - [3]). Optimal quadratures are also called *best quadrature formulas* or *extremal quadrature formulas*.

References

[1] NIKOL'SKIĬ, S.M.: *Quadrature formulae*, H.M. Stationary Office, London, 1966 (translated from the Russian).
[2] BAKHVALOV, N.S.: 'On optimal convergence estimates for quadrature processes and integration methods of Monte-Carlo type on function classes', in *Numerical Methods for Solving Differential and Integral Equations and Quadrature Formulas*, Moscow, 1964, pp. 5-63 (in Russian).
[3] SOBOLEV, S.L.: *Introduction to the theory of cubature formulas*, Moscow, 1974 (in Russian).

N.S. Bakhvalov

Editorial comments.

References

[A1] ENGELS, H.: *Numerical quadrature and cubature*, Acad. Press, 1980.

AMS 1980 Subject Classification: 65D32

OPTIMAL SINGULAR REGIME, *optimal singular control* - An **optimal control** for which, at a certain period of time, the conditions

$$\frac{\partial H}{\partial u} = 0, \tag{1}$$

$$\frac{\partial^2 H}{\partial u^2} = 0 \tag{2}$$

are fulfilled simultaneously, where H is a **Hamilton function**. In the case of vectors, when the optimal singular regime occurs through k, $k > 1$, control components, condition (1) is replaced by the k conditions

$$\frac{\partial H}{\partial u_s} = 0, \quad s = 1, \ldots, k, \tag{3}$$

while instead of equality (2), the determinant

$$\left| \frac{\partial^2 H}{\partial u_s \partial u_p} \right| = 0, \quad p, s = 1, \ldots, k, \tag{4}$$

must vanish.

In an optimal singular regime, the Hamilton function H is stationary, but its second differential is not negative definite, i.e. the maximum of H as a function of u (when u is a variable within the admissible domain) is a 'mixed maximum'.

The most typical problems in which an optimal singular regime may occur are problems of optimal control in which the integrand and the right-hand sides are linearly dependent on the control.

Problems of this type are studied below, beginning with a scalar singular control.

Let the minimum of the functional

$$J = \int_0^{t_1} F(x) + u\Phi(x)\,dt \tag{5}$$

have to be determined, given the constraints

$$\dot{x}^i = f^i(x) + u\phi^i(x), \quad i = 1, \ldots, n, \qquad (6)$$

the boundary conditions

$$x(0) = x_0, \quad x(t_1) = x_1, \qquad (7)$$

and the constraints on the control

$$|u| \leq 1. \qquad (8)$$

The necessary conditions which the optimal singular regime must satisfy allow one to examine the problem (5) - (8) in advance and to isolate the manifolds of singular sections on which an optimal control lies within the admissible domain $|u| < 1$. By joining to it non-singular sections which satisfy the boundary conditions (7), with a control $|u| = 1$ defined by the **Pontryagin maximum principle**, it is possible to obtain an optimal solution of the problem (5) - (8).

In line with the maximum principle, an optimal control for any $t \in [0, t_1]$ must provide a maximum of the Hamilton function

$$H(\psi(t), x(t), u(t)) = \max_{|u| \leq 1} H(\psi(t), x(t), u), \qquad (9)$$

where

$$H(\psi, x, u) = -(F(x) + u\Phi(x)) + \sum_{i=1}^{n} \psi_i(f^i(x) + u\phi^i(x)),$$

while $\psi = (\psi_1, \ldots, \psi_n)$ is a conjugate vector function which does not vanish and which satisfies the system of equations

$$\dot{\psi}_i = -\frac{\partial H(\psi, x, u)}{\partial x^i}, \quad i = 1, \ldots, n. \qquad (10)$$

Condition (2) is fulfilled for any control, not just for an optimal singular regime.

For those periods of time at which $\partial H / \partial u \neq 0$, condition (9) defines a non-singular optimal control which takes the boundary values

$$u(t) = \mathrm{sign}\, \frac{\partial H}{\partial u} = \pm 1.$$

Thus, sections $[\tau_0, \tau_1]$ with an optimal singular control

$$|u(t)| < 1$$

can appear only when condition (1) is fulfilled:

$$\frac{\partial H(\psi(t), x(t), u(t))}{\partial u} = 0, \quad \tau_0 \leq t \leq \tau_1; \qquad (11)$$

i.e. the Hamilton function is clearly independent of the control u. Consequently, for linear control problems condition (9) does not allow a direct determination of an optimal singular control $u(t)$.

Let the function $\partial H / \partial u$ be differentiated with respect to t by virtue of the systems (6), (10) until the control u occurs in the next derivative with a non-zero coefficient. It has been proved (see [1] - [3]) that the control u with a non-zero coefficient can only occur in an even derivative, i.e.

$$\left.\begin{aligned}
\frac{\partial}{\partial u}\left[\frac{d^s}{dt^s}\left(\frac{\partial H}{\partial u}\right)\right] &= 0, \quad s = 1, \ldots, 2q-1, \\
\frac{d^{2q}}{dt^{2q}}\left(\frac{\partial H}{\partial u}\right) &= a(\psi, x) + u \cdot b(\psi, x), \\
b(\psi(t), x(t)) &\neq 0,
\end{aligned}\right\} \qquad (12)$$

and that fulfillment of the inequality

$$(-1)^q b(\psi, x) = (-1)^q \left[\frac{\partial}{\partial u}\left[\frac{d^{2q}}{dt^{2q}}\left(\frac{\partial H}{\partial u}\right)\right]\right] \leq 0 \qquad (13)$$

is a necessary condition for optimality of the singular control. If $b(\psi(t), x(t)) \neq 0$ on the whole segment, then the optimal singular control is

$$u(t) = -\frac{a(\psi(t), x(t))}{b(\psi(t), x(t))}, \quad \tau_0 \leq t \leq \tau_1.$$

In as much as conditions (12) are obtained as a result of successive differentiation of (11), then on a section of the singular regime, especially at joining points τ_0 and τ_1 of singular and non-singular sections, apart from equality (11), the following $2q - 1$ equalities are fulfilled:

$$\frac{d^s}{dt^s}\left(\frac{\partial H}{\partial u}\right) = 0, \quad s = 1, \ldots, 2q-1. \qquad (14)$$

Analysis of conditions (11), (14) shows that for cases of even and odd values of q, the character of the situation when singular and non-singular sections of the trajectory come together is different (see [4]).

When q is even, the optimal control of a non-singular section cannot be piecewise continuous. Discontinuities of the control (switching points) condense to a joining point with a singular section. Thus the optimal control is a Lebesgue-measurable function with a countable set of discontinuity points.

When q is odd, only two piecewise-smooth optimal trajectories can go to a point lying on a singular section (or come out of it). Let the dimension of the manifold of singular sections in the n-dimensional space of phase coordinates equal k. Then if q is odd, the optimal trajectories with a piecewise-continuous control fill out only a surface of dimension $k + 1$ in the phase space. Therefore, when $k \leq n - 2$, almost all remaining trajectories will have a control with an infinite number of switching points.

The hypothesis that $k = n - q$ has been formulated (see [4]). If this is true, then, when $q \geq 2$, condensation of switching points before coming out onto the singular section (or coming off it) is a typical manifestation of problems of the type (5) - (8).

An example of the joining of singular and non-singular sections of an optimal control with an infinite number of switchings is given in [5].

When q, $q \geq 2$, is even, the optimal singular control on a non-singular section bordering a singular one cannot be piecewise continuous, but has an infinite number of switching points which condense towards the point of entry τ_0, i.e. no $\epsilon > 0$ exists such that in a period $[\tau_0 - \epsilon, \tau_0]$ the optimal control is constant.

In the most commonly found optimal singular regimes, $q = 1$. In this case, singular and non-singular sections are joined by a piecewise-continuous optimal control.

In a more general case of an optimal singular regime in which k, $k > 1$, controls are examined:

$$J = \int_0^{t_1} \left[F(x) + \sum_{s=1}^{k} u_s \Phi_s(x) \right] dt, \qquad (15)$$

$$\dot{x}^i = f^i(x) + \sum_{s=1}^{k} u_s \phi_s^i(x) \qquad (16)$$

(as in the scaler case), condition (4), by virtue of linearity, is fulfilled for any control. In a section $[\tau_0, \tau_1]$, the optimal singular regime with k components with $|u_s| < 1$ must fulfill the k conditions (3):

$$M_s(\psi(t), x(t)) = \frac{\partial H(\psi(t), x(t), u(t))}{\partial u_s} = 0, \qquad (17)$$

$$\tau_0 \leq t \leq \tau_1, \quad s = 1, \ldots, k,$$

where
$$H(\psi, x) =$$
$$= - \left[F(x) + \sum_{s=1}^{k} u_s \Phi_s(x) \right] + \sum_{i=1}^{n} \psi_i \left[f^i(x) + \sum_{s=1}^{k} u_s \phi_s^i(x) \right] =$$
$$= Q(\psi, x) + \sum_{s=1}^{k} u_s M_s(\psi, x),$$

while the ψ_i are defined by (10).

Further necessary conditions of optimality for optimal singular regimes with several components differ as follows from the cases with one component examined above. On a section of an optimal singular regime, two types of necessary conditions must be fulfilled. One of them, an inequality-type condition, is the analogue of condition (13). Other necessary conditions are equality-type conditions and do not have one-component analogues (see [6]).

By differentiating (17) totally with respect to t, a system of k linear equations relative to k unknowns $u_1, \ldots, u_k$ is obtained:

$$\frac{dM_s(\psi, x)}{dt} = \sum_{p=1}^{k} u_p \left[\sum_{i=1}^{n} \left[\frac{\partial M_s}{\partial x^i} \phi_p^i - \frac{\partial M_p}{\partial x^i} \phi_s^i \right] \right] + \qquad (18)$$
$$+ \sum_{i=1}^{n} \left[\frac{\partial M_s}{\partial x^i} f^i - \frac{\partial Q}{\partial x^i} \phi_s^i \right] = 0, \quad s = 1, \ldots, k.$$

The coefficient matrix of the system (18),
$$a_{sp} = \sum_{i=1}^{n} \left[\frac{\partial M_s}{\partial x^i} \phi_p^i - \frac{\partial M_p}{\partial x^i} \phi_s^i \right], \quad s, p = 1, \ldots, k,$$

is skew-symmetric: $a_{sp} = -a_{ps}$. Hence it follows that the entries a_{ss} on the main diagonal of the matrix (a_{sp}) are equal to zero. In general, the remaining entries a_{sp}, $s \neq p$, of (18) on an arbitrary trajectory corresponding to a non-optimal control differ from zero. On an optimal singular regime with k components, the necessary conditions requiring all $k(k-1)/2$ coefficients of (18) to vanish must be fulfilled (see [6]):

$$\sum_{i=1}^{n} \left[\frac{\partial M_s}{\partial x^i} \phi_p^i - \frac{\partial M_p}{\partial x^i} \phi_s^i \right] = 0, \quad s, p = 1, \ldots, k, \quad s < p. \quad (19)$$

Apart from these conditions, the following inequality-type condition must be fulfilled (which is the analogue of condition (13) for an optimal singular regime with one component when $q = 1$):

$$\sum_{s,p=1}^{k} \frac{\partial}{\partial u_p} \left[\frac{d^2}{dt^2} \left[\frac{\partial H}{\partial u_s} \right] \right] \delta u_s \delta u_p \geq 0. \qquad (20)$$

Conditions (13), (20) can be considered as a generalization of the **Legendre condition** and the **Clebsch condition** in the case of an optimal singular regime; the inequalities shown are therefore sometimes called the *generalized Legendre — Clebsch conditions*.

The importance of optimal singular regimes in optimal control problems is explained by the following property of optimal singular regimes (see [4]): If an optimal trajectory originating at a certain point contains a section of an optimal singular regime, then all optimal trajectories originating at nearby points possess the same property.

Research has been carried out into questions relating to a definition of an order of an optimal singular regime for linear and non-linear control problems (see [8]).

All the results on optimal singular regimes mentioned above are obtained by examining the second variation of the functional. It is possible to obtain further necessary conditions for optimality of a singular regime by examining third and fourth variations of the functional (see [9]).

References

[1] KELLI, G.: *Raket. Tekhn. i Kosmonavtik.*, no. 8 (1964), 26-29.

[2] ROBBINS, G.: *Raket. Tekhn. i Kosmonavtik.*, no. 6 (1965), 139-145.

[3] KOPP, R. and MOÏER, G.: *Raket. Tekhn. i Kosmonavtik.*, no. 8 (1965), 84-90.

[4] BERSHCHANSKIĬ, YA.M.: 'Fusing of singular and nonsingular parts of optimal control', *Automat. Remote Control* **40**, no. 3 (1979), 325-330. (*Avtomatik. i Telemekh.*, no. 3 (1979), 5-11)

[5] FULLER, A.T.: 'Theory of discrete, optimal and self-adjusting systems', in *Proc. 1st Internat. Congress Internat. Fed. Autom. Control*, Moscow, 1960, pp. 584-605 (in Russian).

[6] VAPNYARSKIĬ, I.B.: 'An existence theorem for optimal control in the Boltz problem, some of its applications and the necessary conditions for the optimality of moving and singular systems', *USSR Comp. Math. Math. Phys.* **7**, no. 2 (1967), 22-54. (*Zh. Vychisl. Mat. i Mat. Fiz.* **7**, no. 2 (1967), 259-283)

[7] KRENER, A.: 'The high order maximal principle and its application to singular extremals', *Siam J. Control Optim.* **15**, no. 2 (1977), 256-293.

[8] Lewis, R.M.: 'Definition of oder and function conditions in singular optimal control problems', *Siam J. Control Optim.* **18**, no. 1 (1980), 21-32.

[9] Skorodinskiĭ, I.T.: 'The necessary condition for optimality of singular controls', *USSR Comp. Math. Math. Phys.* **19**, no. 5 (1979), 46-53. (*Zh. Vychisl. Mat. i Mat. Fiz.* **19**, no. 5 (1979), 1134-1140)

[10] Gabasov, R. and Kirillova, F.M.: *Special optimal control*, Moscow, 1973 (in Russian).

I.B. Vapnyarskiĭ

Editorial comments. In [A1] new necessary conditions for the singular control problem in the calculus of variations that generalize the classical Legendre−Clebsch condition are given. This new condition is sometimes referred to as the *Kelley condition*.

References

[A1] Kelley, H.J., Kopp, R.E. and Moyer, H.G.: 'Singular extremals', in G. Leitmann (ed.): *Topics of Optimization*, Acad. Press, 1967, Chapt. 3.

[A2] Knobloch, H.W.: *Higher order necessary conditions in optimal control theory*, Springer, 1981.

[A3] Bryson, A.E. and Ho, Y.-C.: *Applied optimal control*, Ginn, 1969.

[A4] Hermes, H. and Lasalle, J.P.: *Functional analysis and time optimal control*, Acad. Press, 1969.

[A5] Bell, D.J. and Jacobson, D.H.: *Singular optimal control problems*, Acad. Press, 1975.

[A6] Fleming, W.H. and Rishel, R.W.: *Deterministic and stochastic control*, Springer, 1975.

[A7] Bertsekas, D. and Shreve, S.: *Stochastic optimal control, the discrete time case*, Acad. Press, 1978.

[A8] Bertsekas, D.: *Dynamic programming and stochastic control*, Acad. Press, 1976.

[A9] Davis, M.H.A.: 'Martingale methods in stochastic control', in *Stochastic Control and Stochastic Differential Systems*, Lecture notes in computer and inform. sci., Vol. 16, Springer, 1979.

[A10] Cesari, L.: *Optimization - Theory and applications*, Springer, 1983.

[A11] Neustadt, L.W.: *Optimization, a theory of necessary conditions*, Princeton Univ. Press, 1976.

[A12] Barbu, V. and Da Prato, G.: *Hamilton−Jacobi equations in Hilbert spaces*, Pitman, 1983.

[A13] Kushner, H.: *Introduction to stochastic control*, Holt, 1971.

[A14] Kumar, P.R. and Varaiya, P.: *Stochastic systems: estimation, identification and adaptive control*, Prentice-Hall, 1986.

[A15] Ljung, L.: *System identification theory for the user*, Prentice-Hall, 1987.

AMS 1980 Subject Classification: 49BXX, 93C60

OPTIMAL SLIDING REGIME - A term used in the theory of **optimal control** to describe an optimal method of controlling a system when a minimizing sequence of control functions does not have a limit in the class of Lebesgue-measurable functions.

For example, suppose a minimum of the functional

$$J(x, u) = \int_0^3 (x^2 - u^2)\, dt \tag{1}$$

has to be found, given the constraints

$$\dot{x} = u, \tag{2}$$

$$x(0) = 1, \quad x(3) = 1, \tag{3}$$

$$|u| \leqslant 1. \tag{4}$$

In order to obtain a minimum of the functional (1), it is desirable for every t to have as small a value as possible of $|x(t)|$ and as large a value as possible of $|u(t)|$. The first requirement, with regard to the constraining condition (2), the boundary conditions (3) and the control constraint (4), is satisfied by the trajectory

$$x(t) = \begin{cases} 1-t, & 0 \leqslant t \leqslant 1, \\ 0, & 1 < t < 2, \\ t-2, & 2 \leqslant t \leqslant 3. \end{cases} \tag{5}$$

If the trajectory (5) could be created for a control which, for all t, has the boundary values

$$u(t) = +1 \ \text{ or } \ u(t) = -1, \tag{6}$$

the absolute minimum of the functional (1) would be obtained. However, the 'ideal' trajectory (5) cannot be created for any control function $u(t)$ which satisfies (6), since, when $1 < t < 2$, $u(t) \equiv 0$. Nevertheless, it is possible, using control functions $u_n(t)$ which, when $n \to \infty$ and $1 < t < 2$, realize ever more frequent switchings from 1 to -1 and vice versa:

$$u_n(t) =$$

$$= \begin{cases} -1, & 0 \leqslant t \leqslant 1, \\ +1, & \dfrac{k}{n} < t-1 \leqslant \dfrac{2k+1}{2n}, \ k=0, \ldots, n-1, \\ -1, & \dfrac{2k+1}{2n} < t-1 \leqslant \dfrac{k+1}{n}, \ k=0, \ldots, n-1, \\ +1, & 2 < t \leqslant 3 \end{cases} \tag{7}$$

$(n = 1, 2, \ldots)$, to create a minimizing sequence of controls $\{u_n(t)\}$ which satisfies (6) and a minimizing sequence of trajectories $\{x_n(t)\}$ converging towards the 'ideal' trajectory (5).

Each trajectory $x_n(t)$ differs from (5) only on the interval $(1, 2)$ on which, instead of being a precise path along the x-axis, it makes a 'saw-toothed' path with n identical 'teeth', positioned above the x-axis. The 'teeth of the saw' become ever finer when $n \to \infty$, such that $\lim_{n \to \infty} x_n(t) = 0$, $1 < t < 2$. In this way, the minimizing sequence of trajectories $\{x_n(t)\}$ converges towards (5), but the minimizing sequence of controls $\{u_n(t)\}$, which, when $n \to \infty$ and $1 < t < 2$, realizes ever more frequent switchings from 1 to -1 and vice versa, does not have a limit in the class of measurable (and even more so in the class of piecewise continuous) functions. This means that on the section $(1, 2)$ an optimal sliding regime occurs.

Using heuristic reasoning, it is possible to describe the optimal sliding regime obtained in the following way: An optimal control at each point of the interval $(1, 2)$ 'slides', i.e. skips from the value $+1$ to -1 and back, such that, for any interval of time, however small, the measure of the set of points t in which $u = +1$ is equal to the measure of the set of points t in which

$u = -1$, which, by virtue of equation (2), ensures a precise motion along the x-axis. The description given above of the character of change of an optimal control on part of a sliding regime is non-rigorous, for it does not satisfy the ordinary definition of a function.

It is possible to give a rigorous definition of an optimal sliding regime if, along with the initial problem (1) - (4), an auxiliary 'split' problem is introduced: To find a minimum of the functional

$$I(x, \alpha, u) = \int_0^3 (x^2 - \alpha_0 u_0^2 - \alpha_1 u_1^2)\, dt, \qquad (8)$$

given the constraints

$$\dot{x} = \alpha_0 u_0 + \alpha_1 u_1, \qquad (9)$$

$$x(0) = 1, \quad x(3) = 1, \qquad (10)$$

$$|u_0| \leqslant 1, \quad |u_1| \leqslant 1, \quad \alpha_0 + \alpha_1 = 1, \quad \alpha_0, \alpha_1 \geqslant 0. \qquad (11)$$

The split problem (8) - (11) differs from the initial one in that, instead of one control function $u(t)$, two independent control functions $u_0(t)$ and $u_1(t)$ are introduced; the integrand and the function on the right-hand side of equation (2) of the initial problem are replaced by a linear convex combination of corresponding functions, taken with different controls $u_0(t)$ and $u_1(t)$ and with coefficients $\alpha_0(t)$, $\alpha_1(t)$, which are also considered as control functions.

Thus, in problem (8) - (11) there are four controls u_0, u_1, α_0, α_1. Insofar as α_0 and α_1 are related by the equality-type condition $\alpha_0 + \alpha_1 = 1$, it is possible to drop one of the controls α_0 or α_1 by expressing it through the other. However, for the convenience of subsequent analysis, it is advisable to leave both controls in an explicit form.

Unlike the initial problem, an optimal control for the split problem (8) - (11) exists. On a section of the optimal sliding regime of the initial problem, the optimal control for the split problem takes the form

$$\alpha_0(t) = \alpha_1(t) = \frac{1}{2}, \quad u_0(t) = -1, \quad u_1(t) = +1,$$

$$1 < t < 2,$$

while on the sections of entry and exit:

$$\alpha_0(t) = 1, \quad u_0(t) = -1,$$

$$\alpha_1(t) = 0, \quad u_1(t) \text{ may be arbitrary,}$$

$$0 \leqslant t \leqslant 1,$$

$$\alpha_1(t) = 1, \quad u_1(t) = +1,$$

$$\alpha_0(t) = 0, \quad u_0(t) \text{ may be arbitrary,}$$

$$2 \leqslant t \leqslant 3.$$

On the section of an optimal sliding regime, the controls α_0 and α_1, going linearly into the right-hand side, and the integrand accept values within the admissible domain. This means that the optimal sliding regime of the initial problem (1) - (4) is an optimal singular regime, or optimal singular control, for the auxiliary split problem (8) - (11).

The same results occur for optimal sliding regimes in general problems of optimal control. Suppose that a minimum of the functional

$$J(x, u) = \int_{t_0}^{t_1} f^0(t, x, u)\, dt, \qquad (12)$$

$$f^0(t, x, u): \mathbf{R} \times \mathbf{R}^n \times \mathbf{R}^m \to \mathbf{R},$$

has to be found, given the conditions

$$\dot{x} = f(t, x, u), \quad f(t, x, u): \mathbf{R} \times \mathbf{R}^n \times \mathbf{R}^m \to \mathbf{R}^n, \qquad (13)$$

$$x(t_0) = x_0, \quad x(t_1) = x_1, \qquad (14)$$

$$u \in U. \qquad (15)$$

The optimal sliding regime is characterized by the non-uniqueness of the maximum with respect to u of the Hamilton function

$$H(t, x, \psi, u) = \sum_{i=0}^n \psi_i f^i(t, x, u),$$

where ψ_i are conjugate variables (see [2]). Under these conditions, on the section $[\tau_1, \tau_2]$ of the $(k+1)$-st 'slide' $(k > 1)$ with the maxima $u_0, \ldots, u_k$, the initial problem splits and takes the form

$$I(x, \alpha, u) = \int_{\tau_1}^{\tau_2} \sum_{s=0}^k \alpha_s f^0(t, x, u_s)\, dt, \qquad (16)$$

$$\dot{x} = \sum_{s=0}^k \alpha_s f(t, x, u_s), \qquad (17)$$

$$x(t) = x_0, \quad x(t_1) = x_1, \qquad (18)$$

$$u_s \in U, \quad \sum_{s=0}^k \alpha_s = 1, \quad \alpha_s \geqslant 0, \quad s = 0, \ldots, k. \qquad (19)$$

The Hamilton function for the problem (16) - (19),

$$H(t, x, \psi, \alpha, u) = \sum_{i=0}^n \psi_i \left[\sum_{s=0}^k \alpha_s f^i(t, x, u_s) \right],$$

after excluding α_0 and regrouping the terms, can be reduced to the form

$$H(t, x, \psi, \alpha, u) = \qquad (20)$$

$$= \sum_{s=1}^k \left[\sum_{i=0}^n \psi_i f^i(t, x, u_0) - \sum_{i=0}^n \psi_i f^i(t, x, u_s) \right] \alpha_s =$$

$$= \sum_{s=0}^k (H(t, x, \psi, u_1) - H(t, x, \psi, u_s))\alpha_s.$$

Because the $H(t, x, \psi, u_s)$, $s = 0, \ldots, k$, are equal to the maxima of H with respect to u on the set U, on the section $[\tau_1, \tau_s]$ of the optimal sliding regime with $k+1$ maxima the coefficients at the k independent linear controls $\alpha_1, \ldots, \alpha_k$ of the Hamilton function of the split problem (12) - (15) are equal to zero. An optimal sliding regime with a 'slide' through $k+1$ maxima is an optimal singular regime with k components for the split problem (16) - (19). The maximum possible value of k advisable to take when researching sliding regimes is defined by the condition of convexity of the set of values of the right-hand side vector and the convexity from below of the greatest lower bound of the set of values of the integrand of the split system obtained when the control vector (α_s, u_s), $s = 0, \ldots, k$, runs through the whole admissible domain of values. Thus, $k \leq n$ is an estimation from above for k. In the most general case, all optimal sliding regimes of the initial problem can be obtained as optimal singular controls of the split problem written out for $k = n$. In particular, in the above example, the split problem has been examined when $k = 1$, insofar as the constraints contain only one equation; investigation into the split problem (8) - (11) has proved adequate for research into the optimal sliding regime of the initial problem (1) - (4).

If $k = 1$ and controls $u_0(t), \ldots, u_k(t)$ are known which provide equal absolute maxima of the Hamilton function $H(t, x, \psi, u)$ in the admissible domain U, then the analysis of the optimal sliding regime reduces to the investigation of an optimal singular regime with k components. This investigation can be carried out by using necessary conditions of optimality of a singular control (see **Optimal singular regime**).

Investigations have been carried out into optimal sliding regimes using sufficient conditions of optimality (see [4]).

References

[1] GAMKRELIDZE, R.V.: 'Optimal sliding states', *Soviet Math. Dokl.* 3, no. 2 (1962), 559-562. (*Dokl. Akad. Nauk SSSR* **143**, no. 6 (1962), 1243-1245)

[2] PONTRYAGIN, L.S., BOLTYANSKIĬ, V.G., GAMKRELIDZE, R.V. and MISHCHENKO, E.F.: *The mathematical theory of optimal processes*, Wiley, 1967 (translated from the Russian).

[3] VAPNYARSKIĬ, I.B.: 'An existence theorem for optimal control in the Boltz problem, some of its applications and the necessary conditions for the optimality of moving and singular systems', *USSR Comp. Math. Math. Phys.* 7, no. 2 (1963), 22-54. (*Zh. Vychisl. Mat. i Mat. Fiz.* 7, no. 2 (1967), 259-283)

[4] KROTOV, V.F.: 'Methods for solving variational problems based on sufficient conditions for an absolute minimum II', *Automat. Remote Control* **24** (1963), 539-553. (*Avtomatik. i Telemekh.* **24**, no. 5 (1963), 581-598)

I.B. Vapnyarskiĭ

Editorial comments. A sliding control is also called a *chattering control*, see [A1].

For more references see also **Optimal singular regime**.

References

[A1] LEE, E.B. and MARKUS, L.: *Foundations of optimal control theory*, Wiley, 1967.

AMS 1980 Subject Classification: 49BXX, 93C60

OPTIMAL SYNTHESIS CONTROL - A solution of a problem in the mathematical theory of optimal control (cf. **Optimal control, mathematical theory of**), consisting of a *synthesis of an optimal control* (a *feedback synthesis*)

in the form of a control strategy (a feedback principle), as a function of the current state (position) of a process (see [1] - [3]). The value of the control is defined not only by the current time, but also by the admissible values of the current parameters. In this way the introduction of a positional strategy makes possible an a posteriori realization of a control u, corrected on the basis of supplementary information obtained during the process.

The simplest synthesis problem, for example, for a system

$$\dot{x} = f(t, x, u), \quad t_0 \leq t \leq t_1, \quad x \in \mathbf{R}^n, \quad u \in \mathbf{R}^p, \qquad (1)$$

with constraints

$$u \in U \subseteq \mathbf{R}^p \text{ or } \psi(u) \leq 0, \quad \psi: \mathbf{R}^p \to \mathbf{R}^k, \qquad (2)$$

and a given 'terminal' criterion

$$I(x(\cdot), u(\cdot)) = \phi(t_1, x(t_1)), \quad \phi: \mathbf{R}^{n+1} \to \mathbf{R}^1,$$

assumes that a solution u^0 is being sought to minimize the functional $I(x(\cdot), u(\cdot))$ among the functions of the form $u(t, x)$ for an arbitrary initial position $\{\tau, x\}$. The natural course is to find for every pair $\{\tau, x\}$ a solution of the corresponding problem of constructing an **optimal programming control**

$$u^0[t \mid \tau, x], \quad x = x(\tau), \quad \tau \leq t \leq t_1,$$

as a minimum of that same functional $I(x(\cdot), u(\cdot))$ and with those same constraints. It is further supposed that

$$u^0(t, x) = u^0[t \mid t, x];$$

if the function $u^0(t, x)$ is correctly defined, while the equation

$$\dot{x} = f(t, x, u^0(t, x)), \quad x(\tau) = x, \quad \tau \leq t \leq t_1, \qquad (3)$$

has a unique solution, then the synthesis problem can be solved; moreover, the optimal values of I found in the classes of programming and synthesis controls coincide (in general, conditions prevail which ensure the existence in a specific sense of the solutions of equation (3), and conditions also prevail which guarantee the optimality of all the trajectories of this equation).

The synthesized function $u^0(t, x)$, being an optimal synthesis control, leads to an optimal solution for the minimum of the functional I in the problem of optimal control for any initial position $\{\tau, x\}$. This is in contrast to an optimal programming control, which in general depends on the fixed starting point $\{t_0, x^0\}$ of the process. The solution of an optimal control problem in the form of an optimal synthesis control has many applications, especially in practical procedures for implementing the optimal control in the presence of limited information or perturbations in the dynamics. In these situations a synthesis control is preferable to a programming control.

The search for $u^0(t, x)$ in the form of a function of

the current state is immediately linked to **dynamic programming** (see [2]). The *return function* (*Bellman function, value function*) $V(\tau, x)$, being introduced as a minimum (maximum) of a quantity to be optimized (for example, the functional

$$J(x(\cdot), u(\cdot)) = \int_{\tau}^{t_1} f^0(t, x, u)\, dt + \phi(t_1, x(t_1)) \qquad (4)$$

for the system (1) if $x(\tau) = x$, $t \in [\tau, t_1]$), must satisfy the **Bellman equation** with boundary conditions depending on the aim of the control and J. For the system (1), (2) and (4), this equation takes the form

$$\frac{\partial V}{\partial t} + H\left[t, x, \frac{\partial V}{\partial x}\right] = 0, \quad V(t_1, x) = \phi(t_1, x), \qquad (5)$$

where

$$H\left[t, x, \frac{\partial V}{\partial x}\right] = \qquad (6)$$

$$= \min\left\{ \left[\frac{\partial V}{\partial x}, f(t, x, u)\right] + f^0(t, x, u) \colon u \in U \right\}$$

is the Hamilton function. This equation is connected with the equations figuring in the conditions of the **Pontryagin maximum principle**, in the same way as the Hamilton−Jacobi equation for a return function is linked in analytical mechanics to the ordinary Hamiltonian differential equations (see **Variational principles of classical mechanics**).

The derivation of equations (5) for the synthesis problem relies on the *optimality principle* asserting that a section of an optimal trajectory is also an optimal trajectory (see [2]). The viability of this approach depends on the correct definition of the informational properties of the process, particularly on the concept of position (current state, see [5]).

In the *time-optimal control problem* — regarding the minimal time $T(x)$ for a trajectory of an autonomous system (1) to hit a set M, starting from a position x — the function $V(\tau, x) = V(x)$ can be considered as a particular kind of potential $V(x) = T(x)$ with respect to M. The choice of an optimal control $u^0(t, x)$ from conditions (5), (6) now has the form

$$\min\left\{ \left[\frac{\partial V}{\partial x}, f(x, u)\right] \colon u \in U \right\} = -1,$$

$$V(x) = 0 \quad \text{when } x \in M,$$

which means that $u^0(t, x) = u^0(x)$ realizes the descent of the optimal trajectory $x^0(t)$ relative to the level surfaces of the function $V(x)$ by the fastest method permitted by the condition $u \in U$.

The use of the method of dynamic programming (as a sufficient condition of optimality) will be rigorous if the function $V(\tau, x)$ satisfies certain smoothness conditions everywhere (for example, in problems (3) - (6) the function $V(\tau, x)$ must be continuously differentiable) or

if smoothness conditions are satisfied everywhere with the exception of a 'special' set N. When certain special 'conditions of regular synthesis' are fulfilled, the method of dynamic programming is equivalent to Pontryagin's principle, which is then seen to be a necessary and sufficient condition for optimality (see [8]). Difficulties connected with the a priori verification of the applicability of the method of dynamic programming and with the need to solve the Bellman equation complicate the use of this method. The method of dynamic programming has been extended to problems of optimal synthesis control for discrete (multi-stage) systems, where the corresponding Bellman equation is a finite-difference equation (see [2], [9]).

In problems of optimal control with differential constraints, the method of dynamic programming gives an effective solution of the synthesis problem in closed form for a class of problems which embraces linear systems with quadratic performance criterion (4) (the functions f^0, ϕ are positive-definite quadratic forms in x, u and in x, respectively). This problem, related to the *analytic construction of an optimal regulator* if $\phi(t, x) \equiv 0$, $t_1 = \infty$, becomes a problem of *optimal stabilization of the system* (the property of asymptotic stability of the equilibrium position of a synthesized system follows directly from the existence of an admissible control) (see [10], [4]). The existence of a solution in the given instance is ensured by the property of stabilizability of the system (see [4]). For linear stationary and periodic systems it is equivalent to the property of controllability of the unstable models of the system (see **Optimal programming control**).

The solution of the problem of optimal stabilization has shown that the corresponding Bellman function is at the same time the 'optimal' Lyapunov function for the initial system with an obtained optimal control. Under these circumstances effective conditions of controllability have been obtained and a complete analogue of Lyapunov's theory of stability (in a first approximation and in critical cases) for problems of stabilization has been created which embraces ordinary quasilinear and periodic systems and also delay-systems. In the latter case, the role of the Bellman function is played by 'optimal' Lyapunov−Krasovskiĭ functionals, given on the sections of the trajectory that correspond to the value of the delay in the system (see [4], [5]). The theory of linear, quadratic problems of optimal control is also well-developed for partial differential equations (see [11]).

In applied problems of optimal synthesis control it is not always possible to measure all phase coordinates of the system. The following *problem of observation* therefore arises, permitting numerous generalizations: Knowing the realization $y[t] \in \mathbf{R}^m$ for an interval

$\sigma \leqslant t \leqslant \theta$ of an accessible measurement of the function $y = g(t, x)$ in the coordinates of the system (1) (if $u(t)$ is known, for example, if $u(t) \equiv 0$ and $m \leqslant n$), find the vector $x(\theta) \in \mathbf{R}^n$ at the given moment θ. Systems which, through a unique realization $y[t]$, allow one to establish $x(\theta)$, whatever its value, are called *completely observable*.

The property of complete observability, as well as the construction of the corresponding analytic operations that distinguish $x(\theta)$ and the optimization of these operations, have been well studied for linear systems. Here, a *duality principle* is known: For every problem of observation, a corresponding equivalent two-point boundary value control problem for the dual system can be established. A consequence of this is that the property of complete observability of a linear system coincides with the property of complete controllability of the dual system with control. Moreover, it turns out that the corresponding dual boundary value problems of optimal observation and optimal control can also be composed in such a way that their solutions coincide (see [3]). The properties of controllability and observability of linear systems have many generalizations to linear infinite-dimensional systems (equations in a Banach space, systems with deviating argument, partial differential equations). There is also a number of results characterizing the corresponding properties. For non-linear systems, only some local theorems on observability are known. Solutions of the problem of observation have found numerous applications in synthesis problems with incomplete information on coordinates, among them problems of optimal stabilization (see [3] - [5], [14], [15]).

The problem of synthesis control becomes especially interesting when information on the controls of the controllable process, the initial conditions and the current parameters is subject to uncertainty (perturbations). If the description of this uncertainty has a statistical character, the problems of optimal control are examined within the framework of the theory of *stochastic optimal control*. This theory, arising from the solution of stochastic problems [16], has been, to a very large degree, developed for systems of the form

$$\dot{x} = f(t, x, u) + g(t, x, u)\eta, \quad x(t_0) = x^0, \qquad (7)$$

with random perturbations $\eta(t)$ described by Gaussian diffusion processes or by more general classes of Markov processes (the initial vector is also usually taken random). In these circumstances, as a rule, it is assumed that certain probability characteristics of the variable η are given (for example, information on the moments of the corresponding distributions or on the parameters of the stochastic equations describing the evolution of the process $\eta(t)$).

Generally, the use of programming and synthesis controls gives essentially different values of the optimal performance indices J (the roles of these indices can be played, for instance, by average estimates of non-negative functionals defined on trajectories of the process). The problem of synthesis of a stochastic optimal synthesis control now has clear advantages, since the continuous measurement of coordinates of the system enables one to correct the movement with regard to the real course of the random process, not predicted earlier. The method of dynamic programming combined with the theory of generating operators for the Markov semi-groups associated with stochastic processes, has led to sufficient conditions for optimality. This has led to the solution of a number of problems of stochastic optimal control on finite or infinite time intervals, including those with complete and incomplete information on current coordinates, of stochastic problems of pursuit, etc. It is essential that, for the principle of optimality to apply, the control u exists at every moment of time t as a function of 'sufficient coordinates' z of the process which are known to have the Markov property (see [5], [6], [17], [18]).

It is in this way, in particular, that the theory of *optimal stochastic stabilization* has been developed, in conjunction with the corresponding Lyapunov theory of stability, for stochastic systems [19].

For the formulation of an optimal synthesis controller as well as for other aims of control, it is usual to evaluate the state of a stochastic system by means of measurements. The *theory of stochastic filtering* is about the solution of this question, given the condition that the measurement process is disturbed by probabilistic 'noises'. The most complete solutions known here are for linear systems with quadratic optimality criteria (the so-called *Kalman − Bucy filter*, see [13]). In applying this theory to the problem of stochastic optimal synthesis control, conditions have been developed which ensure the validity of the *separation principle*, allowing the problem of control to be solved independently of the problem of evaluating current positions on the basis of sufficient coordinates of the process (see [20]; [18] and [21] are dedicated to more general procedures of stochastic filtering, as well as to problems of a stochastic optimal control when the control itself is selected from the class of Markov diffusion processes).

A strictly formalized solution of the problem of stochastic optimal control is invariably coupled with the problem of a correct foundation for the existence questions of solutions for the corresponding stochastic differential equations. The latter circumstance generated specific difficulties in the solution of problems of stochastic optimal control when non-classical constraints are applied.

An interesting process of dynamic optimization arises in problems of optimal synthesis control under conditions of uncertainty (see **Optimal programming control**). Synthesis solutions generally permit improvement of the quality of the criteria of the process, as compared to programming solutions, which are none the less the result of statistical optimization (carried out, admittedly, in a space of dynamical systems and control functions). The concepts and methods of game theory are now used to obtain a solution of these problems.

Let there be given a system

$$\dot{x} = f(t, x, u, w), \quad t_0 \leqslant t \leqslant t_1, \tag{8}$$

with constraints

$$x^0 = x(t_0) \in X^0 \subseteq \mathbf{R}^n, \quad u \in U \subseteq \mathbf{R}^p, \quad w \in W \subseteq \mathbf{R}^q,$$

on the initial vector x^0, the control u and the disturbances w. Unlike the case of a coalition of players, represented by the initial control u subject to definition, the disturbances w are in this case treated as controls of an opponent player, and one is allowed to examine any strategies w formed from any admissible information. Moreover, the aims of control can be formulated from the point of view of each one of the players separately. If the stated aims are contradictory, then the *problem of conflicting control* arises. Research into problems of synthesis control under conditions of conflict or uncertainty is the subject of the theory of **differential games**.

The process of forming an optimal synthesis control under conditions of uncertainty can also be complicated by incomplete information on the current state. So, in the system (8), only the results of indirect measurements of the phase vector x and the realization $y[t]$ of the function

$$y(t) = g(t, x, \xi) \tag{9}$$

can be accessible; here the indefinite parameters ξ are restricted by an a priori known constraint, $\xi \in E$. The values $y[t]$, $t_0 \leqslant t \leqslant \theta$ (for a given $u(t)$), make it possible to construct a *region of information* $X(\theta, y(\cdot))$ of states of the system (8) in the phase space, along with the realization $y[t]$, equation (9) and the restriction on w, ξ. Among the elements of $X(\theta, y(\cdot)) = X(\theta, \cdot)$ there will be also an unknown true state of the system (8), which can be estimated by choosing a point $x^\star(\theta, \cdot)$ from $X(\theta, \cdot)$ (for example, the 'centre of gravity' or the 'Chebyshev centre' of $X(\theta, \cdot)$). The study of the evolution of the regions $X(\theta, \cdot)$ and the dynamics of the vectors $x^\star(\theta, \cdot)$ is the purpose of the *theory of minimax filtering*. Most complete solutions are known for linear systems and convex constraints (see [22]).

In general, the choice of a synthesis strategy of optimal control under conditions of uncertainty (for example, in the form of a functional $u = u(t, X(t, \cdot)))$ must aim at control of the evolution of the domains $X(\theta, \cdot)$ (i.e. the alternation of their configuration and their displacement in space), in accordance with prescribed criteria. For the problem shown, a number of general qualitative results is known, as well as constructive solutions in the class of special linear, convex problems (see [7], [22]). Furthermore, information containing measurements (for example, of the functions $y[t]$ in the systems (8) and (9)), permits an a posteriori re-evaluation during the process of the domain of admissible values of the indefinite parameters in the direction of their constraint. In this way, the *problem of identification* of a mathematical model of a process (for example, the parameters w in equation (8)), is solved at the same time. All that has been said enables one to treat the solutions of the problem of optimal synthesis control under conditions of uncertainty as a procedure of *adaptive optimal control*, in which a more precise definition of the properties of the model of the process gets mixed up with the choice of the controls. The questions of identification of models of dynamical processes and of the problem of adaptive optimal control are studied in detail under the assumption of existence of a probabilistic description of the indefinite parameters (see [23], [24]).

If in problems of optimal synthesis control under conditions of uncertainty the parameters w, ξ are treated as 'controls' of a fictitious opponent player, then the aims of the controls u and $\{w, \xi\}$ can be different. The latter instance leads to a non-scalar quality criterion of the process. A consequence of this is that the corresponding problems can be considered within the framework of the concepts of equilibrium situations peculiar to the multi-criterion problems of the theory of cooperative games and their generalizations.

References

[1] PONTRYAGIN, L.S., BOLTYANSKIĬ, V.G., GAMKRELIDZE, R.V. and MISHCHENKO, E.F.: *The mathematical theory of optimal processes*, Wiley, 1967 (translated from the Russian).

[2] BELLMAN, R.: *Dynamic programming*, Princeton Univ. Press, 1957.

[3] KRASOVSKIĬ, N.N.: *The theory of control of motion*, Moscow, 1968 (in Russian).

[4] KRASOVSKIĬ, N.N.: 'On the stabilization of dynamic systems by supplementary forces', *Diff. Eq.* **1**, no. 1 (1965), 1-9. (*Differentsial'nye Uravneniya* **1**, no. 1 (1963), 5-16)

[5] KRASOVSKIĬ, N.N.: 'Theory of optimal control systems', in *Mechanics in the USSR during 50 years*, Moscow, 1968, pp. 179-244 (in Russian).

[6] KRASOVSKIĬ, N.N.: 'On mean-square optimum stabilization at damped random perturbations', *J. Appl. Math. Mech.* **25** (1961), 1212-1227. (*Prikl. Mat. Mekh.* **25**, no. 5 (1961), 806-817)

[7] KRASOVSKIĬ, N.N. and SUBBOTIN, A.I.: *Game-theoretical control problems*, Springer, 1988 (translated from the Russian).

[8] BOLTYANSKIĬ, V.G.: *Mathematical methods of optimal control*, Holt, Rinehart & Winston, 1971 (translated from the Russian).

[9] BOLTYANSKIĬ, V.G.: *Optimal control of discrete systems*, Wiley, 1978 (translated from the Russian).

[10] LETOV, A.M.: *Mathematical theory of control processes*, Moscow, 1981 (in Russian).

[11] LIONS, J.L.: *Optimal control of systems governed by partial differential equations*, Springer, 1971 (translated from the French).

[12] KALMAN, R.: 'On the general theory of control systems', in *Proc. first Internat. Congress Internat. Fed. Automatic Control*, Vol. 2, Moscow, 1960, pp. 521-547.

[13] KALMAN, R. and BUCY, R.: 'New results in linear filtering and prediction theory', *Proc. Amer. Soc. Mech. Engineers Ser. 1.D* **83** (1961), 95-108.

[14] LEE, E.B. and MARCUS, L.: *Foundations of optimal control theory*, Wiley, 1967.

[15] BUTKOVSKIĬ, A.G.: *Structural theory of distributed systems*, Horwood, 1983 (translated from the Russian).

[16] KOLMOGOROV, A.N., MISHCHENKO, E.F. and PONTRYAGIN, L.S.: 'A probability problem of optimal control', *Soviet Math. Dokl.* **3**, no. 4 (1962), 1143-1145. (*Dokl. Akad. Nauk SSSR* **145**, no. 5 (1962), 993-995)

[17] LIPTSER, R.SH. and SHIRYAEV, A.N.: *Statistics of random processes*, 1-2, Springer, 1977-1978 (translated from the Russian).

[18] ÅSTRÖM, K.J.: *Introduction to stochastic control theory*, Acad. Press, 1970.

[19] KATS, I.YA. and KRASOVSKIĬ, N.N.: 'On the stability of systems with random parameters', *J. Appl. Math. Mech.* **24** (1960), 1225-1246. (*Prikl. Mat. Mekh.* **24**, no. 5 (1960), 809-823)

[20] WONHAM, W.M.: 'On the separation theorem of stochastic control', *SIAM J. Control* **6** (1968), 312-326.

[21] KRYLOV, N.V.: *Controlled diffusion processes*, Springer, 1980 (translated from the Russian).

[22] KURZHANSKIĬ, A.B.: *Control and observation under conditions of uncertainty*, Moscow, 1977 (in Russian).

[23] TSYPKIN, YA.Z.: *Foundations of the theory of learning systems*, Acad. Press, 1973 (translated from the Russian).

[24] EIKHOFF, P.: *Basics of identification of control systems*, Moscow, 1975 (in Russian; translated from the English).

A.B. Kurzhanskiĭ

Editorial comments. An optimal synthesis control is usually called an *optimal closed-loop control* or an *optimal feedback control* in the Western literature, while an **optimal programming control** is usually called an *optimal open-loop control*. See also **Optimal control, mathematical theory of.**

For a detailed discussion of when optimal open-loop controls can be used to find optimal closed-loop controls see [A11], [A12].

In the formulation of an optimal control problem one distinguishes problems with a terminal index, with an integral index or with a combination of both such, as expressed by equation (4). Instead of 'index', and depending on the particular application, one also speaks about 'cost function' (to be minimized) or 'performance index' (usually to be maximized).

In general, analytic solutions to optimal control problems do not exist. A notable exception is the case where the system is described by a linear equation:

$$x = Ax + Bu, \quad t_0 \leqslant t \leqslant t_f, \quad x(t_0) = x_0;$$

and the cost function by a quadratic equation:

$$J = \frac{1}{2} x'(t_f) Q_f x(t_f) + \frac{1}{2} \int_{t_0}^{t_f} (x' Qx + u' Ru)\, dt.$$

Here $x \in \mathbb{R}^n$, $u \in \mathbb{R}^p$, A, B, Q_f, Q, and R are matrices of appropriate sizes. Moreover, $Q_f \geqslant 0$, $Q \geqslant 0$, $R > 0$, and the transpose is denoted by $'$. The final time is supposed to be fixed here. The solution to this optimal control problem is

$$u^0(x, t) = -R^{-1}B'P(t)x,$$

where the $(n \times n)$-matrix $P(t)$ satisfies the so-called Riccati equation;

$$\dot{P} = -A'P - PA + PBR^{-1}B'P - Q, \quad P(t_f) = Q_f.$$

If the pair (A, B) is controllable and the pair (A, C) is observable, where the $(n \times n)$-matrix C is defined by $C'C = Q$, then $\lim_{t_f \to \infty} P(t_0)$ exists; it will be denoted by $\bar{P}$. The $x = 0$ solution of $\dot{x} = (A - R^{-1}B'\bar{P}B)x$ is asymptotically stable. The conditions on controllability and observability can be replaced by the weaker conditions on stabilizability and detectability, respectively; see [A13]. The notions of controllability, observability, etc. are properties of the system and as such belong to the field of mathematical system theory.

Another class of problems of which the features of the optimal control function $u^0(x, t)$ are well understood is the class of linear time-optimal control problems. The notion of reachability set helps visualizing the optimal solution; see [A14].

References

[A1] FLEMING, W.H. and RISHEL, R.W.: *Deterministic and stochastic control*, Springer, 1975.

[A2] BERTSEKAS, D. and SHREVE, S.: *Stochastic optimal control, the discrete time case*, Acad. Press, 1978.

[A3] BERTSEKAS, D.: *Dynamic programming and stochastic control*, Acad. Press, 1976.

[A4] DAVIS, M.H.A.: 'Martingale methods in stochastic control', in *Stochastic Control and Stochastic Differential Systems*, Lecture notes in control and inform. sci., Vol. 16, Springer, 1979, pp. 85-117.

[A5] CESARI, L.: *Optimization - theory and applications*, Springer, 1983.

[A6] NEUSTADT, L.W.: *Optimization, a theory of necessary conditions*, Princeton Univ. Press, 1976.

[A7] BARBU, V. and DA PRATO, G.: *Hamilton—Jacobi equations in Hilbert spaces*, Pitman, 1983.

[A8] KUSHNER, H.: *Introduction to stochastic control*, Holt, 1971.

[A9] KUMAR, P.R. and VARAIYA, P.: *Stochastic systems: estimation, identification and adaptive control*, Prentice-Hall, 1986.

[A10] LJUNG, L.: *System identification theory for the user*, Prentice-Hall, 1987.

[A11] BRUNOVSKÝ, P.: 'On the structure of optimal feedback systems', in *Proc. Internat. Congress Mathematicians Helsinki, 1978*, Acad. Sci. Fennica, 1980, pp. 841-846.

[A12] SUSSMANN, H.J.: 'Analytic stratifications and control theory', in *Proc. Internat. Congress Mathematicians Helsinki, 1978*, Acad. Sci. Fennica, 1980, pp. 865-871.

[A13] KUAKERNAEK, H. and SIVAN, R.: *Linear optimal control systems*, Wiley, 1972.

[A14] HERMES, H. and LASALLE, J.P.: *Functional analysis and time optimal control*, Acad. Press, 1969.

[A15] BRYSON, A.E. and HO, Y.-C.: *Applied optimal control*, Ginn, 1969.

AMS 1980 Subject Classification: 93E20, 93E12, 93C10, 49A10, 49A36, 49A45, 49A60, 49B10, 49B36, 49B60, 49C20, 93D15

OPTIMAL TRAJECTORY - A curve $x(t)$ in an $(n + 1)$-dimensional space of variables $t, x^1, \ldots, x^n$ along which a point $x(t) = (x^1(t), \ldots, x^n(t))$, whose motion is determined by the vector differential equation

$$\dot{x} = f(t, x, u), \quad f: \mathbf{R} \times \mathbf{R}^n \times \mathbf{R}^p \to \mathbf{R}^n, \tag{1}$$

is transferred from its original position

$$x(t_0) = x_0 \tag{2}$$

to a final position

$$x(t_1) = x_1 \tag{3}$$

under the influence of an **optimal control** $u(t)$ which minimizes a given functional

$$J = \int_{t_0}^{t_1} f^0(t, x, u)\, dt, \quad f^0: \mathbf{R} \times \mathbf{R}^n \times \mathbf{R}^p \to \mathbf{R}. \tag{4}$$

The choice of an optimal control is subject to the restriction

$$u \in U, \tag{5}$$

where U is a closed set of permissible controls, $U \subset \mathbf{R}^p$. The initial and final moments of time t_0 and t_1 are assumed to be fixed and free, respectively.

An optimal trajectory is defined in the same way for variational problems of a more general type than (1) - (5), for example, for problems with movable end-points and with constraints on the phase coordinates. For methods of tracing optimal trajectories, see **Variational calculus, numerical methods of.**

For autonomous problems, in which the functions f^0, f do not explicitly depend on the time t:

$$f^0 = f^0(x, u), \quad f = f(x, u),$$

the concept of a phase optimal trajectory proves to be more apt for the theory and its applications. A *phase optimal trajectory* is the projection of an optimal trajectory onto the n-dimensional subspace of phase variables $x^1, \ldots, x^n$. For autonomous problems, a phase trajectory does not depend on the choice of the initial moment of time t_0.

Research into the set of phase optimal trajectories which transfer the system from an arbitrary initial position to a given final position (or from a given initial position to an arbitrary final one) enables one to answer many qualitative questions arising from the variational problem being considered. The formation of the set of phase optimal trajectories is a compulsory step in the construction of a synthesis of an optimal feedback control

$$u(t) = v(x(t)),$$

which ensures a movement along an optimal trajectory at any point in the phase space.

References

[1] PONTRYAGIN, L.S., BOLTAYANSKIĬ, V.G., GAMKRELIDZE, R.V. and MISHCHENKO, E.F.: *The mathematical theory of optimal processes*, Wiley, 1962 (translated from the Russian).
[2] DERUSSO, P., ROY, R. and CLOIS, C.: *State space in control theory*, Moscow, 1970 (in Russian; translated from the English).

I.B. Vapnyarskiĭ

Editorial comments.

References

[A1] LEE, E.B. and MARKUS, L.: *Foundations of optimal control theory*, Wiley, 1967.
[A2] CESARI, L.: *Optimization - theory and applications*, Springer, 1983.

AMS 1980 Subject Classification: 49E99

OPTIMALITY PRINCIPLE - A formal description of various notions of an optimum. Optimality principles normally reflect certain characteristics of an intuitive understanding of stability, profitability and fairness. It is essential that the simultaneous realization of all (or of a sufficient large number of) such characteristics often appears to be impossible, owing to their formal incompatibility. As the theory of optimality principles becomes of an axiomatic nature, new optimality principles arise which do not always possess an intuitive transparency.

Problems of optimality principles arise, for example, when that value of a variable is sought which simultaneously extremizes a number of given functions (the so-called multi-criterion extremal problems). Problems which require non-trivial optimality principles in order to be solved actually arise in game theory (cf. **Games, theory of**). One of the simplest game-theoretic optimality principles is the **minimax principle**. Other optimality principles are realized in the form of a core or a von Neumann−Morgenstern solution (cf. **Core in the theory of games**), a **Shapley value**, etc.

For the Bellman principle of optimality see **Dynamic programming**.

N.N. Vorob'ev

Editorial comments. See also Pontryagin maximum principle; Optimal control.

AMS 1980 Subject Classification: 90D99, 49-01, 93-01

OPTIMALITY, SUFFICIENT CONDITIONS FOR - Conditions which ensure the optimality of a given solution of a problem of the variational calculus in a chosen class of comparison curves.

Sufficient conditions for optimality of a weak minimum (see [1]): For a curve $\bar{y}(x)$ to provide a weak minimum for the functional

$$J(y) = \int_{x_0}^{x_1} F(x, y, y')\, dx, \tag{1}$$

given the boundary conditions

$$y(x_0) = y_0, \quad y(x_1) = y_1,$$

it is sufficient that the following conditions are fulfilled.

1) The curve $\bar{y}(x)$ must be an extremal, i.e. it must satisfy the **Euler equation**

$$F_y - \frac{d}{dx}F_{y'} = 0.$$

2) Along the curve $\bar{y}(x)$, including its ends, the strong **Legendre condition**

$$F_{y'y'}(x, y, y') > 0$$

must be fulfilled.

3) The curve $\bar{y}(x)$ must satisfy the strong **Jacobi condition**, which requires that the solution of the Jacobi equation

$$\left[F_{yy} - \frac{d}{dx}F_{yy'}\right]\eta - \frac{d}{dx}(F_{y'y'}\eta') = 0 \qquad (2)$$

with initial conditions

$$\eta(x_0) = 0, \quad \eta'(x_0) = 1,$$

must not vanish at the points of the right-closed interval $x_0 < x \leqslant x_1$.

The coefficients of the Jacobi equation (2), which is a linear differential equation of the second order, are calculated along the extremal $\bar{y}(x)$ and represent known functions in x.

For a *strong minimum* it is sufficient that the following extra condition be fulfilled, next to those mentioned above.

4) There exists a neighbourhood of the curve $\bar{y}(x)$ such that at each point (x, y) of it, and for any y', the inequality

$$\mathscr{E}(x, y, u(x, y), y') \geqslant 0 \qquad (3)$$

hold, where

$$\mathscr{E}(x, y, u, y') = F(x, y, y') - F(x, y, u) +$$
$$-(y' - u)F_y(x, y, u)$$

is the Weierstrass function, while $u(x, y)$ is the slope of the field of extremals surrounding $\bar{y}(x)$.

At the extremal $\bar{y}(x)$, condition (3) takes the form

$$\mathscr{E}(x, \bar{y}, \bar{y}', y') = F(x, \bar{y}, y') - F(x, \bar{y}, \bar{y}') + \qquad (4)$$
$$-(y' - \bar{y}')F_{y'}(x, \bar{y}, \bar{y}') \geqslant 0.$$

Condition (4) is necessary for a strong minimum; it is called the *Weierstrass necessary condition* (cf. **Weierstrass conditions (for a variational extremum)**). Thus, unlike the sufficient conditions for a weak minimum, which require certain strong necessary conditions to be fulfilled at the points of the extremal, the sufficient conditions of a strong minimum require that the Weierstrass necessary condition be fulfilled in a neighbourhood of the extremal. In general, it is impossible to weaken the formulation of the sufficient conditions for a strong minimum once the requirement that the Weierstrass condition be fulfilled in a neighbourhood of the extremal has been replaced by the strong Weierstrass condition (condition (4) with strict inequality) at the points of the extremal (see [1]).

For non-classical variational problems, examined in the mathematical theory of optimal control (cf. **Optimal control, mathematical theory of**), several approaches to the establishment of sufficient conditions for optimality of an absolute extremum exist.

Let a problem of optimal control be posed in which the minimum of the functional

$$J = \int_0^{t_1} f^0(x, u)\,dt, \qquad (5)$$

$$f^0 : \mathbf{R}^n \times \mathbf{R}^p \to \mathbf{R},$$

has to be determined, given the conditions

$$\dot{x} = f(x, u), \quad f : \mathbf{R}^n \times \mathbf{R}^p \to \mathbf{R}^n, \qquad (6)$$

$$x(0) = x_0, \quad x(t_1) = x_1, \qquad (7)$$

$$u \in U, \qquad (8)$$

where U is a given closed set in p-dimensional space.

Using the method of **dynamic programming** [3], sufficient conditions for optimality are formulated in the following way. For a control $u(t)$ to be an optimal control in the problem (5) - (8), it is sufficient that:

a) there exists a continuous function $S(x)$ which has continuous partial derivatives for all x with the possible exception of a certain piecewise-smooth set of dimension less than n, which vanishes at the end-point x_1, $S(x_1) = 0$, and which satisfies the **Bellman equation**

$$\max_{u \in U}\left[\frac{\partial S}{\partial x}f(x, u) - f^0(x, u)\right] = 0; \qquad (9)$$

b) $u(t) = v(x(t))$ if $0 \leqslant t \leqslant t_1$, where $v(x)$ is a synthesizing function (cf. also **Optimal synthesis control**), which can be defined using the Bellman equation:

$$\frac{\partial S}{\partial x}f(x, v(x)) - f^0(x, v(x)) =$$
$$= \max_{u \in U}\left[\frac{\partial S}{\partial x}f(x, u) - f^0(x, u)\right] = 0.$$

In reality, when using the method of dynamic programming, a stronger result is obtained: Sufficient conditions for optimality for a set of different controls which transfer a phase point from an arbitrary initial state to a given final state x_1.

In the more general case of a non-autonomous system, i.e. the integrand and the vector-function of right-hand sides also depend on the time t, the function S will depend on t and a term $\partial S/\partial t$ must be added to the left-hand side of equation (9). There is a proof (see [4]) which assumes the condition of continuous differentiability of the function $S(x)$ for all x, which is very restrictive and is not fulfilled in most problems although it is usually supposed.

Sufficient conditions for optimality can be constructed on the basis of the **Pontryagin maximum prin-**

ciple. If in a region G of the phase space a regular synthesis is realized (cf. **Optimal synthesis control**), then all trajectories obtained using the maximum principle when constructing the regular synthesis are optimal in G.

The definition of a regular synthesis, although rather cumbersome, does not essentially impose any particular restrictions on the problem (5) - (8).

There is another approach to the construction of sufficient conditions for optimality (see [5]). Let $\phi(x)$ be a continuous function with continuous partial derivatives for all admissible x belonging to a given region G, and let

$$R(x, u) = \phi_x f(x, u) - f^0(x, u). \qquad (10)$$

For the pair $\overline{u}(t), \overline{x}(t)$ to establish an absolute minimum in the problem (5) - (8), it is sufficient that there exists a function $\phi(x)$ such that

$$R(\overline{x}, \overline{u}) = \max_{\substack{x \in G \\ u \in U}} R(x, u). \qquad (11)$$

Corresponding variations of the above-mentioned formulation of the sufficient conditions for optimality are permissible for the more general cases of non-autonomous systems, problems with Mayer- and Bolza-type functions (see **Bolza problem**), as well as for optimal sliding regimes (see **Optimal sliding regime** and [5]).

Research has been carried out into variational problems with functionals in the form of multiple integrals and with differential constraints in the form of partial differential equations, in which functions of several variables occur (see [6]).

References
[1] LAVRENT'EV, M.A. and LYUSTERNIK, L.A.: *A course in variational calculus*, Moscow-Leningrad, 1950 (in Russian).
[2] BLISS, G.A.: *Lectures on the calculus of variations*, Chicago Univ. Press, 1947.
[3] BELLMAN, R.: *Dynamic programming*, Princeton Univ. Press, 1957.
[4] BOLTYANSKIĬ, V.G.: *Mathematical methods of optimal control*, Holt, Rinehart & Winston, 1971 (translated from the Russian).
[5A] KROTOV, V.F.: 'Methods for solving variational problems on the basis of sufficient conditions for an absolute minimum I', *Automat. Remote Control* **23** (1963), 1473-1484. (*Avtomat. i Telemekh.* **23**, no. 12 (1962), 1571-1583)
[5B] KROTOV, V.F.: 'Methods for solving variational problems on the basis of sufficient conditions for an absolute minimum II', *Automat. Remote Control* **24** (1963), 539-553. (*Avtomat. i Telemekh.* **24**, no. 5 (1963), 581-598)
[6] BUTKOVSKIĬ, A.G.: *Theory of optimal control of systems with distributed parameters*, Moscow, 1965 (in Russian).

I.B. Vapnyarskiĭ

Editorial comments. One distinguishes *fixed end-time* and *free end-time problems*. The problem expressed by the equations (5) - (8) is a free end-time one; the end-time is determined by the moment that the state $x(t)$ equals the final state x_1.

For additional references see also **Optimal control, mathematical theory of**.

References
[A1] LEE, E.B. and MARKUS, L.: *Foundations of optimal control theory*, Wiley, 1967.

AMS 1980 Subject Classification: 49BXX

OPTIMIZATION OF A COMPUTATIONAL METHOD - Usually the same as **optimization of computational algorithms**. Sometimes, however, these concepts are treated differently. For example, it is possible to speak of the optimality of a concrete grid method for solving a boundary value problem in some class of problems, meaning that it requires the calculation of the right-hand side at a minimal number of points. When examining questions of the optimization of a computational algorithm, a number of other aspects are also studied: methods for solving the system of grid equations that arises, the program realization of these methods, etc.

N.S. Bakhvalov

AMS 1980 Subject Classification: 65KXX

OPTIMIZATION OF COMPUTATIONAL ALGORITHMS - The choice of an optimal **computational algorithm** in the solution of applied problems or in the elaboration of systems of standard programs. When solving a concrete problem, the optimal tactic can be not to optimize the solution method, but to opt for a standard program or to use the simplest method for which the composition of a program is reasonably straightforward.

The theoretical formulation of the question of optimization of computational algorithms is based on the following principles. When choosing a method for solving a problem, the researcher concentrates on certain properties, and his choice of the algorithm depends on these, while the algorithm will also be used to solve other problems possessing these properties. For this reason, in the theoretical study of algorithms one introduces a class of problems P which possess specific properties. When choosing a solution method, the researcher has a set M of solution methods at his disposal. When using a method m to solve a problem p, the solution obtained will have a certain error $\epsilon(p, m)$. The quantity

$$E(P, m) = \sup_{p \in P} | \epsilon(p, m) |$$

is called the *error of the method m in the class P*, while

$$E(P, M) = \inf_{m \in M} E(P, m)$$

is called the *optimal estimate of the error in P* for the methods from M. If a method exists such that

$$E(P, m_0) = E(P, M),$$

then this method is said to be *optimal*. A scheme of studying the problem of optimization of computational algorithms, dating back to A.N. Kolmogorov [2], considers the set of problems of computing the integral

$$I(f) = \int_0^1 f(x)\,dx$$

given the condition $|f^{(n)}| \leq A$, and where M is the set of all possible quadratures

$$I(f) \approx \sum_{j=1}^{N} C_j f(x_j).$$

Each quadrature is defined by the totality of $2N$ numbers C_j and x_j. The problem of the minimum quantity of information (see [2], [3]) needed to recover a function from a given class with required accuracy can also be included in this scheme. A more complex formulation of the problem is examined in [4], in which the amount of work involved in realizing the algorithm in a specific sense is commensurate with the amount of memory used. Optimal algorithms exist for only an insignificant number of type of problems [1]. However, for a large number of computational problems, methods have been created which are almost optimal in their asymptotic characteristics (see [5] - [8]).

Research into the characteristics of computational algorithms which are optimal in their class (see [5], [7]) comprises two parts: creating concrete solution methods with the best possible characteristics, and obtaining estimates from below of the characteristics of computational algorithms (see [2] - [4], [9]). In essence, the first part of the question is a basic problem of the theory of numerical methods and in most cases it is studied independently of the optimization problem. Obtaining estimates from below usually reduces to an estimation from below of the **ϵ-entropy** or of the **width** of the corresponding spaces; this is sometimes carried out independently, using the same techniques as are used in obtaining the estimates mentioned.

Computational algorithms may be divided into passive ones and active ones. A *passive algorithm* for solving a problem does not depend on the information obtained in solving the problem, whereas an *active algorithm* does. When computing an integral, the information used about the function is usually information on its values at N points. For a passive algorithm, the integral is computed using the formula

$$I(f) \approx \sum_{j=1}^{N} C_j f(P_j),$$

where the weights C_j and P_j from the domain Ω of definition of f are previously defined. Active algorithms for computing the integral are included in the following scheme: A point $P_1 \in \Omega$ and the functions

$$Q_q = \Phi_q(Q_1, \ldots, Q_{q-1}; y_1, \ldots, y_{q-1}), \quad q = 2, \ldots, N,$$

$$S_N(Q_1, \ldots, Q_N; y_1, \ldots, y_N)$$

are given, where y_j, S_N are numbers, $Q_j \in \Omega$. The following quantities are successively computed:

$$f(P_1), \ P_2 = \Phi_2(P_1; f(P_1)), \ f(P_2),$$
$$P_3 = \Phi_3(P_1, P_2; f(P_1), f(P_2)), \ \ldots, f(P_N),$$

and it is supposed that

$$I(f) \approx S_N(P_1, \ldots, P_N; f(P_1), \ldots, f(P_N)).$$

In convex classes of integrands which are centrally symmetric with respect to the function $f \equiv 0$, the optimal estimate in the class of passive algorithms coincides with the optimal estimate in the class of active algorithms (see [10], [11]).

In the practice of numerical integration, active algorithms of the type of integration algorithms with automatic choice of steps (see [10]) have shown their superiority over passive algorithms. This supports the generally held point of view that a formal scheme of optimization of computational algorithms does not often include specific real problems. When solving optimization problems (especially minimization problems), passive algorithms are hardly used (see [12], [13]). As an upper bound for the amount of work involved in calculating one value of a function from some class F one implicitly takes the unit of amount of work of the algorithm. There are other possible approaches to estimating the optimality of the characteristics of an algorithm. For example, the amount of work involved in computing a functional $l(f)$ from a set of functionals $L(f)$ can be taken as the unit of amount of work. In this case, a lower bound for the amount of work involved in realizing the algorithms is obtained using the theory of widths (see [14], [15]). The amount of work comprises not only the amount of work involved in obtaining information on initial data, but also the amount of work involved in processing the information. At the present time, it appears that one cannot cite examples of classes of real computational problems in which lower bounds for the amount of work involved in realizing the algorithms can be obtained which differ from the bounds on the information of the type under consideration. However, for a number of non-computational problems of this type, bounds are known (see **Algorithm, computational complexity of an** and **Algorithm, complexity of description of an**).

When research into problems of optimization of computational algorithms aims to solve problems on a computer, the problem has extra nuances related to the stability of the algorithm in relation to the error of computation, and to the restriction on the volume of the different forms of memory used (see **Computational algorithm**). The problem of optimizing computational algorithms has been treated above as a problem of optimization of computational algorithms in a class of problems. In practice, the problem of optimization of computational algorithms for a concrete problem is of vital interest (see [10], [16]). The formulation of one problem of optimization (see [16]) is as follows. A differential equation is integrated using the Runge−Kutta method with a variable step. An estimate is made of the principal term of the estimation error. This estimate is then optimized through a distribution of the points of integration (in general, for a given number of points). This approach to the problem of optimization has a crucial bearing on the development of the theory and on the practice of active algorithms of numerical integration.

References

[1] NIKOL'SKIĬ, S.M.: *Quadrature formulae*, H.M. Stationary Ofice, London, 1966 (translated from the Russian).

[2] KOLMOGOROV, A.N.: 'On certain asymptotic characteristics of totally bounded metric spaces', *Dokl. Akad. Nauk. SSSR* **108**, no. 3 (1956), 385-388 (in Russian).

[3] KOLMOGOROV, A.N. and TIKHOMIROV, V.M.: 'ϵ-entropy and ϵ-capacity of sets in function spaces', *Uspekhi Mat. Nauk.* **14**, no. 2 (1959), 3-86 (in Russian).

[4] VITUSHKIN, A.G.: *Estimation of the complexity of the tabulation problem*, Moscow, 1959 (in Russian).

[5] BABUSHKA, I. and SOBOLEV, S.L.: 'Optimization of numerical processes', *Aplikace Mat.* **10** (1965), 96-129 (in Russian).

[6] BAKHVALOV, N.S.: 'Optimal methods for solving problems', *Aplikace Mat.* **13** (1968), 27-38 (in Russian).

[7] BAKHVALOV, N.S.: 'About optimisation of numerical methods', in *Internat. Congress Mathematicians Nice, 1970*, Vol. 3, Gauthier-Villars, 1972, pp. 289-295.

[8] BAKHVALOV, N.S.: 'On optimal convergence estimates for quadrature processes and integration methods of Monte-Carlo type on function classes', in *Numerical Methods for Solving Differential and Integral Equations and Quadrature Formulas*, Moscow, 1964, pp. 5-63 (in Russian).

[9] BAKHVALOV, N.S.: 'A lower bound for the asymptotic characteristics of classes of functions with dominating mixed derivative', *Math. Notes* **12**, no. 6 (1972), 833-838. (*Mat. Zametki* **12**, no. 6 (1972), 655-664)

[10] BAKHVALOV, N.S.: *Numerical methods: analysis, algebra, ordinary differential equations*, Mir, 1977 (translated from the Russian).

[11] BAKHVALOV, N.S.: 'On the optimality of linear methods for operator approximation in convex classes of functions', *USSR Math. Math. Phys.* **11**, no. 4 (1971), 244-249. (*Zh. Vychisl. Mat. i Mat. Fiz.* **11**, no. 4 (1971), 1010-1018)

[12] WILDE, D.J.: *Optimum seeking methods*, Prentice-Hall, 1964.

[13] VASIL'EV, F.P.: *Numerical methods for solving extremal problems*, Moscow, 1980 (in Russian).

[14] OGANESYAN, L.A. and RUKHOVETS, LA.: *Variational-difference methods for solving elliptic equations*, Erevan, 1979 (in Russian).

[15] TIKHOMIROV, V.M.: *Some questions in approximation theory*, Moscow, 1976 (in Russian).

[16] TIKHONOV, A.N. and GORBUNOV, A.D.: 'Estimates of the error of a Runge−Kutta method and the choice of optimal meshes', *USSR Comp. Math. Math. Phys.* **4**, no. 2 (1964), 30-42. (*Zh. Vychisl. Mat. i Mat. Fiz.* **4**, no. 2 (1964), 232-241)

N.S. Bakhvalov

AMS 1980 Subject Classification: 65KXX

OPTIONAL RANDOM PROCESS - A stochastic process $X = (X_t(\omega), F_t)_{t \geqslant 0}$ that is measurable (as a mapping $(\omega, t) \mapsto X(\omega, t) = X_t(\omega)$) with respect to the **optional sigma-algebra** $\mathcal{O} = \mathcal{O}\,(\mathbf{F})$.

A.N. Shiryaev

Editorial comments. An optional random process is also called an *adapted random process*.

References

[A1] DELLACHERIE, C.: *Capacités et processus stochastiques*, Springer, 1972, Chapt. 3, Sect. 2.

[A2] BAUER, H.: *Probability theory and elements of measure theory*, Holt, Rinehart & Winston, 1972, Chapt. 11.

AMS 1980 Subject Classification: 60G07

OPTIONAL SIGMA-ALGEBRA, *optional σ-algebra* - The smallest σ-algebra $\mathcal{O} = \mathcal{O}\,(\mathbf{F})$ of sets (cf. **Algebra of sets**) in $\Omega \times \mathbf{R}_+ = \{(\omega, t): \omega \in \Omega, t \geqslant 0\}$ generated by all mappings $(\omega, t) \rightarrow f(\omega, t)$ of the set $\Omega \times \mathbf{R}_+$ into $\mathbf{R}$ which (for every fixed $\omega \in \Omega$) are continuous from the right (in t), have limits from the left and are adapted to a (given) non-decreasing family $\mathbf{F} = (F_t)_{t \geqslant 0}$ of sub-σ-algebras $F_t \subseteq F$, $t \geqslant 0$, where (Ω, F) is a measurable space. The optional σ-algebra coincides with the smallest σ-algebra generated by the stochastic intervals $[\![0, \tau]\!] = \{(\omega, t): 0 \leqslant t < \tau(\omega)\}$, where $\tau = \tau(\omega)$ are stopping times (relative to $\mathbf{F} = (F_t)_{t \geqslant 0}$) (cf. **Markov moment**). The inclusion $\mathscr{P}(\mathbf{F}) \subseteq \mathcal{O}\,(\mathbf{F})$ holds between the optional and predictable σ-algebras (cf. **Predictable σ-algebra**).

References

[1] DELLACHERIE, C.: *Capacités et processus stochastiques*, Springer, 1972.

A.N. Shiryaev

Editorial comments. In [A1] the optional σ-field is called the *well-measurable σ-field*.

References

[A1] DELLACHERIE, C. and MEYER, P.A.: *Probabilities and potential*, A, North-Holland, 1978 (translated from the French).

AMS 1980 Subject Classification: 60G07, 60G40

WITHDRAWN